Advances in Intelligent Systems and Computing

Volume 842

Series editor

Janusz Kacprzyk, Polish Academy of Sciences, Warsaw, Poland
e-mail: kacprzyk@ibspan.waw.pl

The series "Advances in Intelligent Systems and Computing" contains publications on theory, applications, and design methods of Intelligent Systems and Intelligent Computing. Virtually all disciplines such as engineering, natural sciences, computer and information science, ICT, economics, business, e-commerce, environment, healthcare, life science are covered. The list of topics spans all the areas of modern intelligent systems and computing such as: computational intelligence, soft computing including neural networks, fuzzy systems, evolutionary computing and the fusion of these paradigms, social intelligence, ambient intelligence, computational neuroscience, artificial life, virtual worlds and society, cognitive science and systems, Perception and Vision, DNA and immune based systems, self-organizing and adaptive systems, e-Learning and teaching, human-centered and human-centric computing, recommender systems, intelligent control, robotics and mechatronics including human-machine teaming, knowledge-based paradigms, learning paradigms, machine ethics, intelligent data analysis, knowledge management, intelligent agents, intelligent decision making and support, intelligent network security, trust management, interactive entertainment, Web intelligence and multimedia.

The publications within "Advances in Intelligent Systems and Computing" are primarily proceedings of important conferences, symposia and congresses. They cover significant recent developments in the field, both of a foundational and applicable character. An important characteristic feature of the series is the short publication time and world-wide distribution. This permits a rapid and broad dissemination of research results.

More information about this series at http://www.springer.com/series/11156

Jemal Abawajy · Kim-Kwang Raymond Choo
Rafiqul Islam · Zheng Xu
Mohammed Atiquzzaman
Editors

International Conference on Applications and Techniques in Cyber Security and Intelligence ATCI 2018

Applications and Techniques in Cyber Security and Intelligence

Set 1

Springer

Editors
Jemal Abawajy
Faculty of Science, Engineering
and Built Environment
Deakin University Geelong
Geelong, VIC, Australia

Kim-Kwang Raymond Choo
Department of Information Systems
and Cyber Security
The University of Texas at San Antonio
San Antonio, TX, USA

Rafiqul Islam
School of Computing and Mathematics
Charles Sturt University
Albury, NSW, Australia

Zheng Xu
Shanghai University
Shanghai Shi, China

Mohammed Atiquzzaman
School of Computer Science
University of Oklahoma
Norman, OK, USA

ISSN 2194-5357 ISSN 2194-5365 (electronic)
Advances in Intelligent Systems and Computing
ISBN 978-3-319-98775-0 ISBN 978-3-319-98776-7 (eBook)
https://doi.org/10.1007/978-3-319-98776-7

Library of Congress Control Number: 2018951638

This Springer imprint is published by the registered company Springer Nature Switzerland AG
The registered company address is: Gewerbestrasse 11, 6330 Cham, Switzerland

Preface

The 2018 International Conference on Applications and Techniques in Cyber Intelligence (ATCI), building on the previous successes in Ningbo, China (2017), Guangzhou, China (2016), Dallas, USA (2015), Beijing, China (2014), and Sydney, Australia (2013), is proud to be in the sixth consecutive conference year in Shanghai. Previously, the event is known as the International Workshop on Applications and Techniques in Cyber Security (ATCS 2016), held in conjunction with the International Conference on Security and Privacy in Communication Networks (SecureComm). The purpose of ATCI 2018 is to provide a forum for presentation and discussion of innovative theory, methodology, and applied ideas, cutting-edge research results, and novel techniques, methods, and applications on all aspects of cyber and electronics security and intelligence. The conference establishes an international forum and aims to bring recent advances in the ever-expanding cybersecurity area including its fundamentals, algorithmic developments, and applications.

Each paper was reviewed by at least three independent experts, and the acceptance rate was 35%. The conference would not have been a reality without the contributions of the authors. We sincerely thank all the authors for their valuable contributions. We would like to express our appreciation to all members of the Program Committee for their valuable efforts in the review process that helped us to guarantee the highest quality of the selected papers for the conference.

We would like to express our thanks to our distinguished keynote speakers, Prof. Tharam Dillon, La Trobe University, Australia; Prof. Zhou Wenbo, President of Shanghai Tunnel Engineering Co., Ltd., Prof. Lv Kangjuan, Executive Vice President, Professor and Doctoral Supervisor in SHU-UTS SILC Business School of Shanghai University, Prof. J. Christopher Westland, University of Illinois, Chicago. We would also like to acknowledge the strong support of SHU-UTS SILC Business School of Shanghai University, Shanghai, China, as well as the General Chairs, Publication Chairs, Organizing Chairs, Program Committee Members, and all volunteers.

Our special thanks due also to the editors for their assistance throughout the publication process.

Jemal Abawajy
Kim-Kwang Raymond Choo
Mohammed Atiquzzaman
Rafiqul Islam
Zheng Xu

Organization

General Chairs

Siyi Gong	Shanghai University, China
John Macintyre (Pro Vice Chancellor)	University of Sunderland

Program Chairs

Jemal Abawajy	Deakin University, Australia
Kim-Kwang Raymond Choo	University of Texas at San Antonio, USA
Mohammed Atiquzzaman	University of Oklahoma, USA
Rafiqul Islam	Charles Sturt University, Australia
Zheng Xu	TRIMPS and Shanghai University, China

Publication Chairs

Juan Du	Shanghai University, China
Mazin Yousif	T-Systems International, USA
Vijayan Sugumaran	Oakland University, USA

Publicity Chairs

Fangqi Cheng	Zhejiang Business Technology Institute, China
Kewei Sha	University of Houston, USA
Yongchun Xu	Zhejiang Business Technology Institute, China
Neil. Y. Yen	University of Aizu, Japan

Local Organizing Chairs

Min Hu	Shanghai University, China
Gang Luo	Shanghai University, China
Juan Chen	Shanghai University, China
Qigang Liu	Shanghai University, China

Program Committee Members

William Bradley Glisson	Sam Houston State University, USA
George Grispos	University of Nebraska Omaha, USA
Abdullah Azfar	KPMG Sydney, Australia
Aniello Castiglione	Università di Salerno, Italy
Florin Pop	University Politehnica of Bucharest, Romania
Ben Martini	University of South Australia, Australia
Wei Wang	University of Texas at San Antonio, USA
Neil Yen	University of Aizu, Japan
Shunxiang Zhang	Anhui University of Science and Technology, China
Guangli Zhu	Anhui University of Science and Technology, China
Tao Liao	Anhui University of Science and Technology, China
Xiaobo Yin	Anhui University of Science and Technology, China
Xiangfeng Luo	Shanghai University, China
Xiao Wei	Shanghai University, China
Huan Du	Shanghai University, China
Zhiguo Yan	Fudan University, China
Rick Church	UC Santa Barbara, USA
Tom Cova	University of Utah, USA
Susan Cutter	University of South Carolina, USA
Zhiming Ding	Beijing University of Technology, China
Yong Ge	University of North Carolina at Charlotte, USA
Danhuai Guo	Computer Network Information Center, Chinese Academy of Sciences, China
Jianping Fang	University of North Carolina at Charlotte, USA
Jianhui Li	Computer Network Information Center, Chinese Academy of Sciences, China
Yi Liu	Tsinghua University, China
Kuien Liu	Pivotal Inc., USA
Feng Lu	Institute of Geographic Science and Natural Resources Research, Chinese Academy of Sciences, China

Ricardo J. Soares Magalhaes	University of Queensland, Australia
Alan Murray	Drexel University, USA
Yasuhide Okuyama	University of Kitakyushu, Japan
Xiaogang Qiu	National University of Defense Technology, China
Xing Xie	Microsoft Research Asia, China
Wei Xu	Renmin University of China, China
Chaowei Phil Yang	George Mason University, USA
Hengshu Zhu	Baidu Inc., China
Morshed Chowdhury	Deakin University, Australia

Contents

ATCI 2018

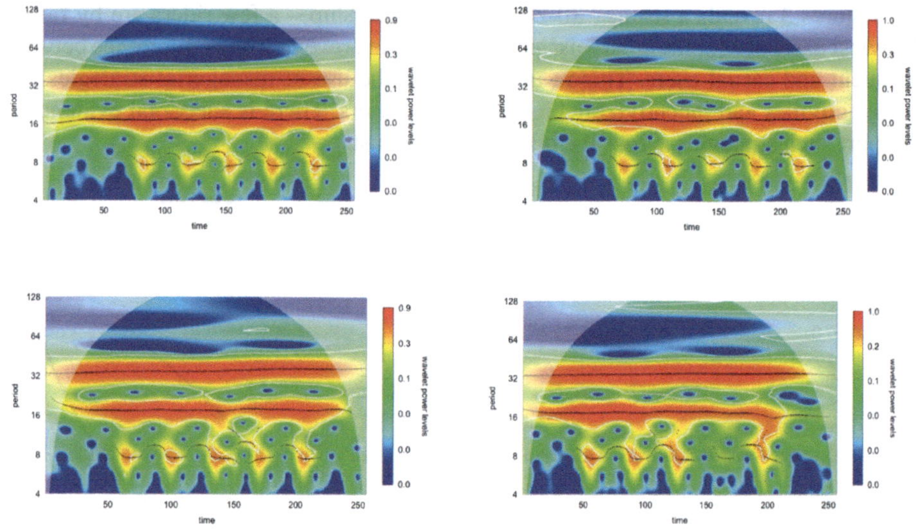

Fig. 4. Wavelet power spectrum for flattened dataset for each of four time slices: 1–256; 257–512; 513–768; 769–1024 steps (left-right, top-bottom). Each of the four represents about one-week and shows weekends and daily cycles prominently.

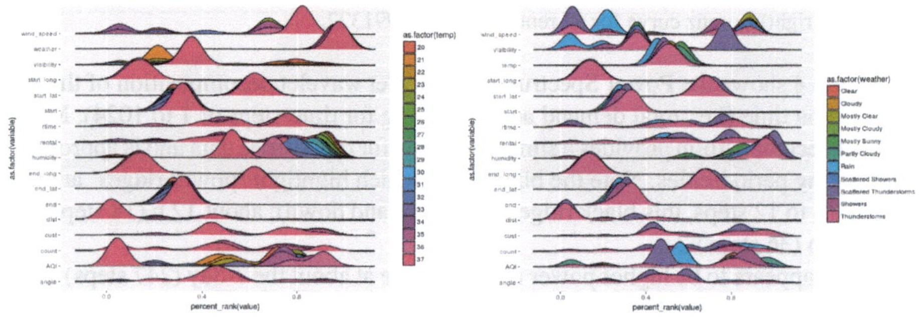

Fig. 5. Impact of selected environmental factors on bike-sharing demand: from left, temperature (degrees C); Air Quality Index.

2 Convolutional Neural Network Deep-Learning Models for Prediction of Shared Bicycle Demand

Exploratory analysis constructed a symmetric $(2^{\#\$})^{\&}$ bike-sharing space-time tensor to accommodate the exploratory wavelet analysis. Deep-learning models are significantly more flexible, and we reshaped the tensor for our main analysis to a [*time*, *latitude*, *longitude*, *environment*] = *dim*(3000, 100, 100, 10) rank 4 tensor adding 9 environmental metrics of weather, temperature, air quality, wind and so forth, and a single bike-sharing contract count (i.e., density of demand at a point in time and space). The reshaped tensor is an order of magnitude less sparse, with time resolution increased to provide a more uniform information encoding throughout the tensor cells. The model was trained on a highly parallelized Google Tensorflow platform with ~10000 CUDA hardware cores.

One-dimensional convolutional models are a standard machine learning approach to handling time-based data. Our application flattens the latitude/longitude data to a single dimension, and projects the final model onto a 9-vector of environmental variables with strides reflecting prior beliefs on any cyclical behavior, and kernel size reflecting the resolution of the data. Figure 6 summarizes the structure.

Layer	**TensorFlow Layer**	**Output Tensor**
conv1d_43	(Conv1D)	286,32
max_pooling1d_23	(MaxPooling1D)	143,32
dropout_46	(Dropout)	143,32
flatten_23	(Flatten)	4576
dense_45	(Dense)	32
dropout_47	(Dropout)	32
dense_46	(Dense)	9
Total Trainable parameters		*147081*

Fig. 6. TensorFlow optimized convolutional – fully connected model layer structure

Adding layers is expected to make the model more powerful, but at this juncture, we are interested in assuring that we have the right model, and thus have experimented with one layer, and added layers to perturb weights and flatten the results to a manageable number.

The reshaped tensor increases time resolution to 14-min time steps with Nyquist periodicities of approximately 35, 53, 105 and 723 time-steps. We use this to inform a choice of strides in the convolutional layers of the deep-learning model. Our runs showed substantially improved, and more or less comparable performance at each of these periodicities. Accuracy was in the range of 20–25% at 35, 53, 105 and 723 time-steps, whereas it was less than 5% for other periodicities.

A resolution of around 1 h in the exploratory dataset translates to about 3 h in the reshaped tensor. We use this to inform an experimental range of 3 to 10 for the kernel size in the convolutional layers of the deep-learning model. Our optimal kernel size, based on loss and accuracy, for training was 6 time-steps.

All models stabilized in validation accuracy after 100 epochs, for each of the four periodicities. The simplicity of the model (one convolutional layer, several dropouts and flattening layers) accounted for a relatively fast convergence. Validation accuracy was between 20–25% for all models, and training models approached 50% accuracy, though this was most likely an artifact of overfitting.

All models stabilized in validation accuracy after 100 epochs, for each of the four periodicities. The simplicity of the model (one convolutional layer, several dropouts and flattening layers) accounted for a relatively fast convergence. Validation accuracy was between 20–25% for all models, and training models approached 50% accuracy, though this was most likely an artifact of overfitting.

This result has implications for "optimization" and "rebalancing" policy guides for the dynamic shifting of bicycle inventory (primarily by truck transport, but possibly through contract price incentives to riders) to position bicycles in space and time in a way that is consistent with rider demand. Their performance is predicated on the accurate prediction of rider demand at a particular point in geographical space (2D) and time. Data analysis of prior contracts and other information will provide the information to facilitate this accurate prediction. (Barnes and Krizek 2005) discussed the requirements of such a predictive model but concluded that such a statistical model would be difficult to construct. Geospatial models in ecology is perhaps most advanced in providing demand models that can predict temporal-geospatial positions (Rossi et al. 1992; Cravey et al. 2001; Legendre et al. 2002; Griffith and Peres-Neto 2006). These models are intended for the analysis of consumer demand but provide models that can inform our particular approach to the problem.

Wasserstein metrics are distance functions defined between probability distributions on a given metric space, and are used in optimal transport problems such as bicycle demand rebalancing. In benchmarking our approach to demand prediction we: (a) assumed that bicycle inventories at a particular $\{time, latitude, longitude\}$ are normalized to probabilities with total weight equal to 1; (b) assumed that costs are normalized to constant 1 million chosen to scale Wasserstein distance on a [0,10] scale; and (c) used the 1st Wasserstein distance to compute relative costs of particular predictions for inventory rebalancing. Our dataset did not include a timeline of the distribution of bicycle inventory around Beijing, so for purposes of benchmarking, we established a baseline uniform distribution of bicycle inventory across the research study latitudes and longitudes. We compared actual versus convolutional neural network models at Nyquist periodicities of 35, 53, 105 and 723 time-steps (14 min time intervals). We found that moving bikes based on this model improved demand satisfaction as measured by the Wasserstein distance by 28.1% to 40.2%. As mentioned earlier, it is likely that this can be improved with models that invoke multiple convolutional layers in the model, but we feel that our data analysis proves the concept of convolutional neural networks for demand prediction where there are periodicities in consumer behavior. We surmised that 'noise' from: (1) short periods of reduction in bike demand due to poor air quality or bad weather; and (2) short periods of abnormal demand from infrequent riders were the main cause of errors introduced into our predictions of demand from the induced human cyclical demand behavior. Both could potentially be ameliorated by both more information and deeper layers in the TensorFlow model.

References

Barnes, G., Krizek, K.: Estimating bicycling demand. Transp. Res. Rec. J. Transp. Res. Board **1939**, 45–51 (2005)

Cravey, A.J., et al.: Developing socio-spatial knowledge networks: a qualitative methodology for chronic disease prevention. Soc. Sci. Med. **52**(12), 1763–1775 (2001)

El-Assi, W., et al.: Effects of built environment and weather on bike sharing demand: a station level analysis of commercial bike sharing in Toronto. Transportation **44**(3), 589–613 (2017)

Griffith, D.A., Peres-Neto, P.R.: Spatial modeling in ecology: the flexibility of eigenfunction spatial analyses. Ecology **87**(10), 2603–2613 (2006)

Legendre, P., et al.: The consequences of spatial structure for the design and analysis of ecological field surveys. Ecography **25**(5), 601–615 (2002)
Rossi, R.E., et al.: Geostatistical tools for modeling and interpreting ecological spatial dependence. Ecol. Monogr. **62**(2), 277–314 (1992)
Steinberg, J., Liza, L.: Chinese bike-sharing company Mobike valued at around $3 billion. Wall Str. J. (2017)

Analysis of Feature Selection Techniques for Correlation-Based Network Anomaly Detection

Florian Gottwalt[1(✉)], Elizabeth Chang[1], and Tharam Dillon[2]

[1] University of New South Wales, Canberra, ACT 2612, Australia
f.gottwalt@adfa.edu.au

[2] La Trobe University, Melbourne, VIC 3083, Australia

Abstract. Although the application of feature selection techniques is common practice for performance enhancements of network anomaly detection techniques, it has rarely been used for recently proposed multivariate correlation-based methods. This paper investigates qualitatively why this is the case and evaluates several feature selection techniques empirically in combination with the state-of-the-art multivariate correlation anomaly detection technique on the UNSW-NB15 network intrusion detection dataset. Our results show that there is a significant enhancement possibility using feature selection techniques. Furthermore, we believe that if the underlying nature of multivariate correlation techniques, temporal correlation changes between attributes, is considered in the selection process the performance could be further improved.

Keywords: Multivariate correlation · Network anomaly detection
Feature selection

1 Introduction

With developments like the Internet of Things and our envisioned Smart Cities, the attack surface of our IT infrastructures and with that, the risk for potential cyber-attacks is increasing dramatically [1]. In an increasingly complex and fast changing environment, generating huge amounts of data, it will be a tremendous challenge to detect intrusions [1].

To address this challenge, substantial amount of research has been performed on network anomaly detection and state-of-the-art methods employ various techniques out of the realms of statistics, machine learning and data mining [2]. Recent security incidents like [3] show, however, that there is still significant amount of research required to be able to cope with the upcoming challenge.

Research on statistical network anomaly detection techniques can generally be divided into methods utilizing first order characteristics of attributes and techniques which analyze second order characteristics. The majority of research has focused on techniques using first order characteristics, which identify anomalies by analyzing deviations from the expected mean of a feature (extreme values). In contrast, methods based on second order characteristics, also termed multivariate correlation (MC) techniques, measure changes of correlations between attributes to detect anomalies.

J. Abawajy et al. (Eds.): ATCI 2018, AISC 842, pp. 10–16, 2019.
https://doi.org/10.1007/978-3-319-98776-7_2

While first order characteristics have proven to perform well on static data. In fast-changing, dynamic environments such as that generated by the envisioned Smart Cities, these struggle to adapt to changes, resulting very often in a large amount of false alarms [4].

Addressing this issue, increasingly more work has focused on methods which utilize correlation changes between attributes to detect anomalous behavior. MC-based techniques do not require prior-distribution assumptions and are able to adapt to dynamically changing data. Furthermore, due to their linear complexity they fulfil requirements for real-time applications [4] which involve large data sets.

Regardless of the type of detection method used, an important pre-processing step for network anomaly detection techniques is to select the most relevant features employing feature selection techniques. This is crucial in the high-dimensional network traffic space to reduce complexity and increase the accuracy of detectors.

Commonly utilized feature selection techniques, such as Principal Component Analysis or approaches analyzing class-feature correlations, have resulted in significant performance enhancements for first order characteristics approaches [5].

Based on this fact, it is very surprising to see the rare application of feature selection techniques on second order characteristic approaches. To the best of our knowledge only in [6], a Principal Component Analysis technique has been used to select the most relevant features.

We believe that the reason for that lies in the underlying nature of MC-based anomaly detection techniques. Neither correlation-based feature selection techniques nor PCA are taking the changes of correlations between features over time into consideration to select features. They focus on features which describe the data best (PCA) or features that have the highest correlations with the class label [7, 8].

While these can be important characteristics in selecting the most relevant feature set we believe that the detection performance of MC-based techniques can be substantively enhanced by considering correlation changes between attributes over time as a selection metric.

To evaluate this assumption, this paper qualitatively analyses the characteristics of MC-based techniques and two commonly used feature selection methods. We furthermore empirically analyze the performance enhancement of two common feature selection techniques, PCA and Pearson Correlation on a network intrusion dataset and compare it with the results of the original correlation features.

2 Multivariate Correlation Anomaly Detection

Multivariate correlation anomaly detection techniques measure the correlations between features and based on deviations of these correlations are indicative for anomalous behavior. These correlations can be measured using either linear or non-linear correlation measures.

In early works, such as by Jin and Yeung in [9, 10], simple covariance matrices were used to classify correlation changes of normal and anomalous behavior.

The current state-of-the-art methods use transformation functions to transform features into correlation features using e.g. the Euclidean Distance [11], geometric correlation measures based on triangle areas [4, 12] or latest, addition-based functions in [13]. This allows, compared to methods based on covariance matrices, one to flag individual instances as anomalous rather than just a group of events [4].

A comprehensive survey of more non-linear entropy methods for network anomaly detection can be found in [14].

3 Feature Selection Techniques

Feature selection techniques can primarily be divided into wrapper-based, filter-based and hybrid combinations of the two [5].

Wrapper-based feature selection techniques commonly use a cross-validation approach in combination with statistical classifiers like Decision Trees or Bayesian Networks to obtain the most relevant features [5]. Filter-based methods employ statistical correlation measures, where a subset of features is selected based on their ability to describe the target class best. This can be as simple as measuring the correlation between a feature and the target class using linear correlation measures like the Pearson Correlation Coefficient [15] or non-linear measures like information gain [16]. The assumption for filter-based methods is that the larger the correlation between a feature and the class label is, the better it is at predicting it. While this can help to select the most relevant features for MC-based detection techniques, filter-based methods don't consider temporal correlations between features.

Comparing the two techniques, Wrapper-based methods usually perform better than filter-methods, however, they are computationally more expensive and hence filter-based techniques are more commonly used for large data sets such as network data [5].

Another very frequently applied filter-based selection technique is summarizing the feature set using Principal Component Analysis (PCA). To reduce redundant data and select the features that represent most of the information (high variance), PCA eliminates linear correlations between features by transforming them into higher order components. This is very convenient for machine learning techniques, which very often assume independence between features.

Looking at the application of PCA as a feature selection method for MC-based techniques it has several limitations. Firstly, only second order linear correlations are considered for the selection of features and secondly and even more critically, it does not take temporal correlation into account [17].

In summary, current filter-based feature selection techniques do not take temporal correlation changes among features into account and as this is the fundamental idea of MC-based techniques, they are only partly suitable for selecting the most relevant features.

4 Methodology

To evaluate our hypothesis, we empirically evaluated PCA as well as the Pearson Correlation feature selection techniques in combination with the state-of-the-art MC-based anomaly detection method presented in [13]. Due to space limitations we refer to [13] for implementation details of the technique.

4.1 Dataset and Pre-processing

A tremendous challenge for the development and evaluation of network intrusion detection techniques is the lack of publicly available dataset. To date, the now 19-year-old KDD99 dataset is still the most used dataset to evaluate network intrusion detection techniques even though it has been heavily criticized over the last few years for many reasons such as:

- Outdated network traffic patterns due to evolving technologies and applications [18]
- Lack of modern, sophisticated low-footprint attacks [19, 20]
- Large number of redundant records [21].

An improved version, the NSL-KDD dataset has been published by Tavalee et al. in [21], however, as it is based on the same KDD99 dataset records, it still lacks a modern representation of network traffic and state-of-the-art attacks.

To address these shortcoming, Nour and Slay have proposed a new dataset, termed UNSW-NB-15 [19, 20]. This dataset is based on a modern, simulated representation of real world traffic and furthermore includes several, modern low footprint attacks.

With that, it has addressed the biggest shortcomings of the KDD99 and the NSL-KDD datasets, yet it still has weaknesses:

- It is not based on actual real-world traffic
- It has potential simulation artefacts.

While we do acknowledge that it is very hard to create and publish a dataset based on real-world network traffic due to privacy and traffic labelling challenges, simulated datasets very often contain simulation artefacts, which affect the meaningfulness of the results. We intend to analyze this dataset for such artefacts in future work.

Regardless of that, it is the most state-of-the-art available dataset and due to that, we are using it for our evaluations.

As the dataset contains categorical and numerical features, we first had to map the categorical features to numerical values by transforming them into ordered numbers, e.g. the protocol types *TCP*, *UDP*, *ICMP* to $TCP = 1$, $UDP = 2$, $ICMP = 3$.

The dataset contains a very large number of instances and we have decided to use the first 50 k, 100 k, 150 k and 200 k instances in a 10-fold cross-validation.

For the features generated by PCA as well as by the Pearson correlation coefficient, we have selected the top 20 features based on the highest variance or the highest correlation with the class label respectively. Furthermore, as conducted in [13], we have normalized the data using a statistical normalization, where the data is normalized around the mean of a feature.

4.2 Performance Evaluation

To evaluate how the feature selection techniques would generalize on an independent dataset and to be able to use all instances for training as well as validation, we have decided to use a 10-fold cross validation.

As a comparison measure for the results of the original features and the potential enhancements with the selected features, we used the Area Under the Receiver Operating Characteristics Curve (AUC). The AUC is a measure of the whole performance, independent of a specific threshold.

In the next section the detection results are presented, compared and discussed.

5 Results and Discussion

Figure 1 shows the AUC results for the original features, the features generated by PCA and the features selected by the Pearson correlation coefficient.

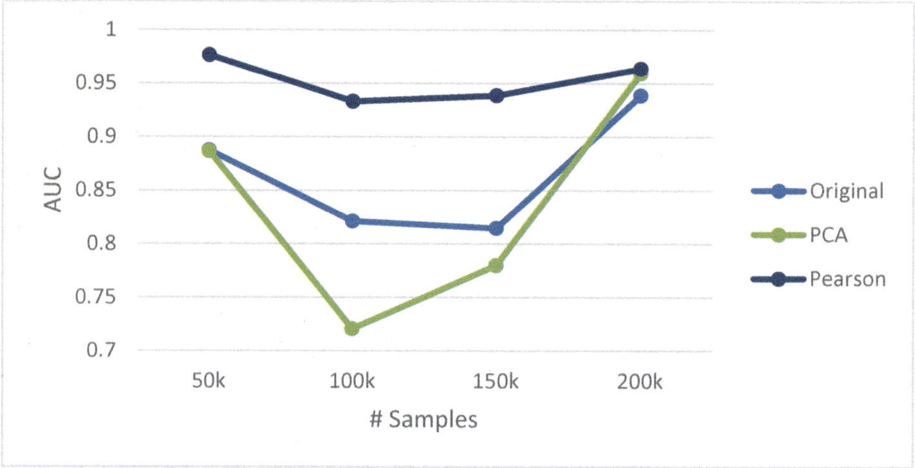

Fig. 1. Feature selection technique comparison correlated features

As can be seen, the features selected by the Pearson Correlation consistently generated the best results among all different sample sizes with an AUC as high as 0.975 for 50 k samples. On average the Pearson correlation features resulted in a ~9% performance enhancement over the original features, while the PCA components resulted in a ~3% loss of performance compared to the original features.

While the PCA results confirm the assumption that neglecting the temporal correlation changes in the selection process does not, this could also be dependent on the dataset and would need to be evaluated on different datasets.

The Pearson correlation results show that a significant performance enhancement is possible if feature selection techniques are used in combination with MC-based anomaly detection methods.

We believe that the results can be further enhanced if the temporal correlation changes are considered in the feature selection process.

6 Conclusion and Further Work

Developments like Smart Cities will result in increasingly complex and dynamic IT infrastructure, which will lead to tremendous IT-security challenges. Multivariate correlation anomaly detection techniques are one possible method to address these challenges. Despite it being common practice to use feature selection techniques for anomaly detection, they have very rarely been applied to MC-based techniques.

We have presented theoretical reasons why this is the case and have empirically evaluated several feature selection techniques on the state-of-the-art MC-based technique. We have found that there is large performance improvement potential, especially using correlation-based feature selection techniques. We believe the performance can be further enhanced if the nature of MC-based techniques, temporal correlation changes among features, is considered in the feature selection process. This is what we are going to address in future work, to develop a specific feature selection algorithm for multivariate correlation-based anomaly detection techniques.

Acknowledgement. This work was supported by the Australian Commonwealth Government through the Australian Government Research Training Program Scholarship (to Florian Gottwalt).

References

1. Gottwalt, F., Karduck, A.P.: Securing smart cities—a big data challenge. In: Information Innovation Technology in Smart Cities, pp. 183–197. Springer (2018)
2. Chandola, V., Banerjee, A., Kumar, V.: Anomaly detection: a survey. ACM Comput. Surv. (CSUR) **41**(3), 15 (2009)
3. Wong, J.C.: Uber Data Hack (2017). https://www.theguardian.com/technology/2017/nov/21/uber-data-hack-cyber-attack. Accessed Nov 2017
4. Tan, Z., et al.: A system for denial-of-service attack detection based on multivariate correlation analysis. IEEE Trans. Parallel Distrib. Syst. **25**(2), 447–456 (2014)
5. Chen, Y., et al.: Survey and taxonomy of feature selection algorithms in intrusion detection system. In: Lipmaa, H., Yung, M., Lin, D. (eds.) Information Security and Cryptology. Springer, Berlin (2006)
6. Tan, Z., et al.: Detection of denial-of-service attacks based on computer vision techniques. IEEE Trans. Comput. **64**(9), 2519–2533 (2015)
7. Wold, S., Esbensen, K., Geladi, P.: Principal component analysis. Chemometr. Intell. Lab. Syst. **2**(1), 37–52 (1987)
8. Hall, M.A.: Correlation-based feature selection for machine learning (1999)
9. Jin, S., Yeung, D.S.: A covariance analysis model for DDoS attack detection. In: 2004 IEEE International Conference on Communications. IEEE (2004)
10. Jin, S., Yeung, D.S., Wang, X.: Network intrusion detection in covariance feature space. Pattern Recogn. **40**(8), 2185–2197 (2007)

11. Tan, Z., et al.: Denial-of-service attack detection based on multivariate correlation analysis. In: Lu, B.-L., Zhang, L., Kwok, J. (eds.) Proceedings of Neural Information Processing: 18th International Conference, ICONIP 2011, Shanghai, China, November 13–17, 2011, Part III, pp. 756–765. Springer, Berlin (2011)
12. Tan, Z., et al.: Triangle-area-based multivariate correlation analysis for effective denial-of-service attack detection. In: 2012 IEEE 11th International Conference on Trust, Security and Privacy in Computing and Communications (TrustCom). IEEE (2012)
13. Li, Q., et al.: An intrusion detection system based on polynomial feature correlation analysis. In: Trustcom/BigDataSE/ICESS 2017. IEEE (2017)
14. Nychis, G., et al.: An empirical evaluation of entropy-based traffic anomaly detection. In: Proceedings of the 8th ACM SIGCOMM Conference on Internet Measurement. ACM (2008)
15. Eid, H.F., et al.: Linear correlation-based feature selection for network intrusion detection model. In: Advances in Security of Information and Communication Networks, pp. 240–248. Springer (2013)
16. Ambusaidi, M.A., et al.: Intrusion detection method based on nonlinear correlation measure. Int. J. Internet Protocol Technol. **8**(2–3), 77–86 (2014)
17. Brauckhoff, D., Salamatian, K., May, M.: Applying PCA for traffic anomaly detection: problems and solutions. In: INFOCOM 2009. IEEE (2009)
18. Gottwalt, F., Karduck, A.P.: SIM in light of big data. In: 2015 11th International Conference on Innovations in Information Technology (IIT). IEEE (2015)
19. Moustafa, N., Slay, J.: UNSW-NB15: a comprehensive data set for network intrusion detection systems (UNSW-NB15 network data set). In: Military Communications and Information Systems Conference (MilCIS). IEEE (2015)
20. Moustafa, N., Slay, J.: The evaluation of network anomaly detection systems: statistical analysis of the UNSW-NB15 data set and the comparison with the KDD99 data set. Inf. Secur. J. Glob. Perspect. **25**(1–3), 18–31 (2016)
21. Tavallaee, M., et al.: A detailed analysis of the KDD CUP 99 data set. In: Computational Intelligence for Security and Defense Applications, CISDA 2009. IEEE (2009)

An RFID Path Authentication Protocol Based on Chebyshev Polynomials

Zuming Shen and Peng Zeng(✉)

Shanghai Key Laboratory of Trustworthy Computing,
East China Normal University, Shanghai, China
zumingshenecnu@outlook.com, pzeng@sei.ecnu.edu.cn

Abstract. Radio Frequency Identification (RFID) technology has a wild range of applications in the industrial supply-chain. In particular an RFID path authentication protocol can be used to detect whether the products labeled with RFID tags have passed a complete and valid path and whether any counterfeit products have entered our supply system. In this paper, we first apply Chebyshev polynomials to RFID path authentication protocol in the supply chain. Our proposed protocol not only satisfies the three well-known security requirements (the privacy, path unlinkability and tag unlinkability), but also provides a simple identity authentication protocol between readers and tags. Further the states of tags are allowed to be exposed due to the use of Chebyshev polynomials. Compared with other related protocols, our proposed protocol reduces the calculation time and has a lower memory requirement of tags.

Keywords: RFID · Path authentication · Chebyshev polynomials
Cryptography protocol · Data security

1 Introduction

A supply chain system manages and controls the transmission of information throughout the entire logistics material and goods, from the procurement of materials to the final delivery to the user [1]. Therefore, the high efficiency and convenience of the supply chain have brought an increase in productivity. Generally the main functions of a supply chain comprise of tracking and identifying the target objects, transferring the ownerships of tags and authenticating valid paths of goods [2].

Today, with the rapid development of The Internet of Things, the industrial supply chain plays an important role in our daily-life (such as the transportation of goods, the distribution of logistics and the delivery of the product, etc. [3–8]). Nowadays, the most important technology used in the industrial supply chain is the Radio Frequency Identification (RFID), which is one of the wireless communications.

A whole RFID system consists of RFID readers, RFID tags and back-server [9]. RFID can not only identify tagged tags with specified radio signals, but also read and write the related information of tags.

RFID path authentication, an important application of RFID, has wildly used in the supply chain to ensure the complete flow of goods. In fact, in every supply chain, RFID tags attached to the products need to store the paths, the identities, and other secrets

J. Abawajy et al. (Eds.): ATCI 2018, AISC 842, pp. 17–27, 2019.
https://doi.org/10.1007/978-3-319-98776-7_3

associated with the products information. When the products reach the destination, a checkpoint reader (also named checkpoint) is required to check whether the products pass through a predetermined production process and test product quality.

In general, RFID path authentication protocols are divided into two kinds: static path, where the valid paths are written into the readers and checkpoint in advance (e.g. [10–13]), and dynamic path, where paths are dynamically generated and every reader can track tags [14, 15].

Pathchecker [16] is the first path authentication protocol in the supply chain which consists of initializer, tags, readers and checkpoint. When a tag reaches the checkpoint, the latter uses the information stored in the tag to determine whether the tag has passed a valid path. However Pathchecker needs a high time complexity of $\mathcal{O}(n)$ on tags. In 2011, Blass et al. proposed a path authentication protocol named Tracker based on multi-signature technology [17]. In Tracker, only the checkpoint can check the legitimacy of tags and detect whether tags have passed some valid paths. Based on elliptic curve Elgamal public key encryption, Wang et al. [18] made an improvement on Tracker by reducing the storage content from six group elements to five group elements. Recently, Bu et al. [12] proposed the first step-grained path authentication protocol called StepAuth by using a recursive method to merge the secrets between the successive two steps. However, StepAuth suffers to the weakness that the public key needs to be secretly saved.

In this paper, we propose a new RFID path authentication protocol based on the Chebyshev polynomials. Compared with other related schemes, the checkpoint in our proposed protocol does not need to store the private keys of the readers along with a valid path, but only the product of these private keys. Further, the state information of tags is allowed to be exposed, while in many existing path authentication protocols this kind of information need to be securely stored in tags. Our proposed protocol has a lower computational complexity compared to existing related protocols under the security guarantee.

The rest of this paper is sketched as follows. We introduce the Chebyshev polynomials, system models and system security requirements in Sect. 2. In Sect. 3, we present our RFID path authentication protocol based on Chebyshev polynomials. We give security and performance analysis in Sects. 4 and 5, respectively. Conclusions are presented in Sect. 6.

2 Preliminaries

In this section, we introduce the notion of Chebyshev polynomials, system model and system security requirements to be used in our scheme.

2.1 Chebyshev Polynomial

Definition 1 (Chebyshev polynomials [19]). Let n be an integer and p a large prime number. A Chebyshev polynomial of degree n, $T_n : \mathbb{Z}_p \to \mathbb{Z}_p$ can be defined by the following recursive relation:

$$T_n(x) = \begin{cases} 1, & n = 0; \\ x, & n = 1; \\ 2xT_{n-1}(x) - T_{n-2}(x), & n \geq 2. \end{cases}$$

It is easy to confirm that Chebyshev polynomials satisfy the following semi-group property:

$$T_r(T_s(x)) \equiv T_{sr}(x) \equiv T_s(T_r(x)) \ mod\ p$$

Definition 2 (CP-DL problem). Given $x, y \in \mathbb{Z}_p^*$, finding an integer n such that $T_n(x) = y$.

Definition 3 (CP-DH problem). Given $x \in \mathbb{Z}_p^*$ and $T_r(x), T_s(x)$, computing $T_{rs}(x)$.

The hardness of CP-DL problem and CP-DH problem has been shown in [20].

2.2 System Model

As shown in Fig. 1, an RFID path authentication protocol consists of four parties: some RFID tags $\mathcal{T}_i$, some RFID readers $\mathcal{R}_j$, a checkpoint $\mathcal{M}$, and an issuer $\mathcal{I}$.

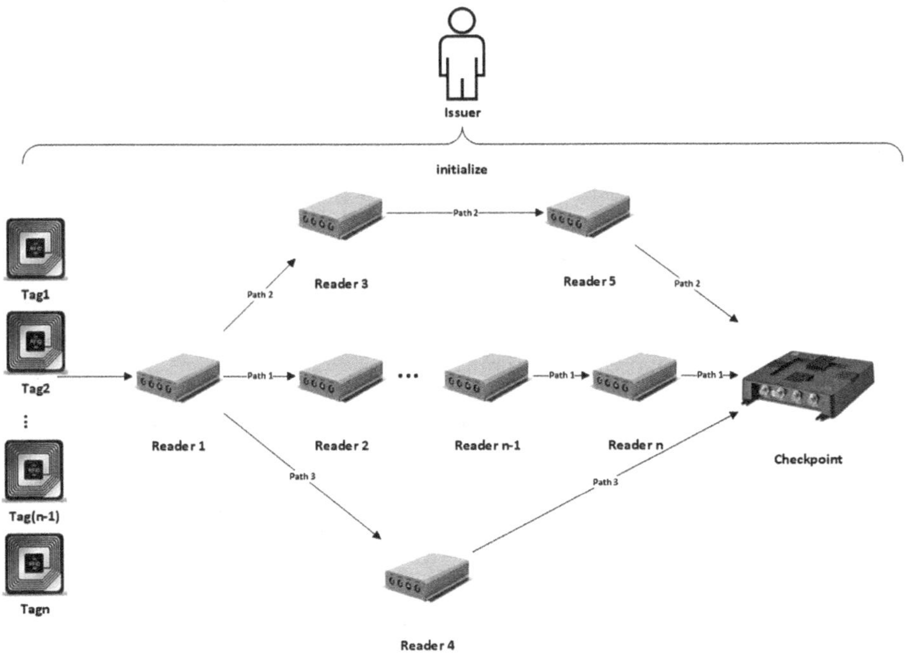

Fig. 1. System model

Tag $\mathcal{T}_i$: Each tag $\mathcal{T}_i$ is attached to a particular good. $\mathcal{T}_i$ has a re-writable memory which stores initial state (S_i) and identity (TID_i). The state can be updated during the tag authentication process with correct readers.

Reader $\mathcal{R}_j$: Each reader $\mathcal{R}_j$ is associated with some paths in the supply chain. When a tag $\mathcal{T}_i$ arrives at $\mathcal{R}_j$, the latter can read the current state of $\mathcal{R}_j$ and then $\mathcal{R}_j$ and $\mathcal{T}_i$ can authenticate each other if $\mathcal{T}_i$'s identity is stored in $\mathcal{R}_j$. If the authentication is successful, $\mathcal{R}_j$ updates $\mathcal{T}_i$ with a newly computed state.

Checkpoint $\mathcal{M}$: The checkpoint $\mathcal{M}$ is a special reader that exists at the end of each path. $\mathcal{M}$ keeps secretly a table (see Table 1) in its local database and checks whether the tags have passed predetermined paths.

Table 1. Secret information stored in the checkpoint $\mathcal{M}$

Tags	Tag's index	Tag's ID	Tag's initial state	The product of private keys involved in a valid path
$\mathcal{T}_1$	$Index_1$	TID_1	S_1	$V_1 = rsk_1 \cdot rsk_2 \cdots rsk_{j1}$
$\mathcal{T}_2$	$Index_2$	TID_2	S_2	$V_2 = rsk_4 \cdot rsk_5 \cdots rsk_{j2}$
$\mathcal{T}_3$	$Index_3$	TID_3	S_3	$V_3 = rsk_3 \cdot rsk_4 \cdots rsk_{j3}$
$\vdots$	$\vdots$	$\vdots$	$\vdots$	$\vdots$
$\mathcal{T}_{n-1}$	$Index_{n-1}$	TID_{n-1}	S_{n-1}	$V_{n-1} = rsk_6 \cdot rsk_5 \cdots rsk_{jn-1}$
$\mathcal{T}_n$	$Index_n$	TID_n	S_n	$V_n = rsk_2 \cdot rsk_4 \cdots rsk_{jn}$

Issuer $\mathcal{J}$: The issuer $\mathcal{J}$ is a trusted third party which initializes all the tags and the checkpoint $\mathcal{M}$ at the beginning of the supply chain. $\mathcal{J}$ attaches each tags $\mathcal{T}_i$ to a particular product and writes the initial states into all the tags. The predetermined paths of tags are stored in $\mathcal{M}$ at the same time. Further, in order to prevent $\mathcal{M}$ from knowing each reader's private key, $\mathcal{J}$ generates a table (see Table 1) to $\mathcal{M}$ which includes the information about tag's index, tag's ID, tag's initial state, and the product of the private keys of the readers in each tag path.

Definition 4 (RFID path authentication protocol). An RFID path authentication protocol consists of the following algorithms.

1. Initialization: Given as input a security parameter κ, the algorithm outputs the system public parameters *parmers*.
2. Setup: Given as input the system parameters *parmers* and for any tag $\mathcal{T}_i$, the algorithm assigns an identity TID_i and an index $Index_i$ to $\mathcal{T}_i$ and generates a private key tsk_i and an initial state S_i for $\mathcal{T}_i$. Similarly, for any reader $\mathcal{R}_j$ the algorithm assigns an identity RID_j to $\mathcal{R}_j$ and generates a private key rsk_j for $\mathcal{R}_j$. Finally, the algorithm outputs a table for the checkpoint.
3. Update: Given as input two random numbers r, c, a time stamp t, the current state Q_i and index $Index_i$ of tag $\mathcal{T}_i$, the algorithm outputs a new state of $\mathcal{T}_i$.
4. Verification: Given as input a challenge information S_i, two random numbers r, c, the index $Index_i$ of tag $\mathcal{T}_i$, the algorithm outputs a bit $b \in \{0, 1\}$.

2.3 Design Goal

A secure RFID path authentication protocol should achieve the following goals:

1. Authentication: Only for legal reader $\mathcal{R}$ and tag $\mathcal{T}$, they can authenticate each other. It means that any probabilistic polynomial time (PPT) adversary $\mathcal{A}$ is unable to update tag state using path information that has not been used in the supply chain.
2. Privacy: Any PPT adversary $\mathcal{A}$ is unable to know whether a tag $\mathcal{T}_i$ went through some reader in the path only based on the information stored in $\mathcal{T}_i$. Also, for some given states of two arbitrary tags $\mathcal{T}_0$ and $\mathcal{T}_1$ in a valid path, $\mathcal{A}$ is unable to distinguish $\mathcal{T}_0$ from $\mathcal{T}_1$.
3. Unlinkability: For any two tags $\mathcal{T}_0$ and $\mathcal{T}_1$, any PPT adversary $\mathcal{A}$ is unable to know whether the two tags went through the same readers in their corresponding paths.

3 Protocol

In this section, we propose a Chebyshev polynomial-based RFID path authentication protocol which consists of the following algorithms: **Initialization**, **Setup**, **Update**, and **Verification**.

Initialization: Given a security parameter κ, the issuer $\mathcal{I}$ chooses a large prime number p satisfying $|p| = \kappa$. Then, $\mathcal{I}$ sets a time threshold Δ which is based on the estimation on the interaction time between tags and readers in the system. Next, $\mathcal{I}$ chooses n random numbers $x_l \in \mathbb{Z}_p^*(1 \leq l \leq n)$ and we denote $\mathbb{X} = \{x_1, x_2, \ldots, x_n\}$. Finally, $\mathcal{I}$ chooses a secure cryptographical hash function $H : \{0,1\}^* \rightarrow \{0,1\}^\kappa$ and sets the public system parameters as

$$params = (p, \Delta, \mathbb{X}, H).$$

Setup: Given the system parameters *params*, the issuer $\mathcal{I}$ executes the following steps to generate the necessary parameters for tags and readers.

(1) When $\mathcal{I}$ receives a tag $\mathcal{T}_i$, $1 \leq i \leq n$, from factory, $\mathcal{I}$ first assigns an unique identity TID_i and an unique index $Index_i$ to $\mathcal{T}_i$.
(2) For any $1 \leq i \leq n$, $\mathcal{I}$ randomly chooses an element $tsk_i \in \mathbb{Z}_p^*$ as the private key of $\mathcal{T}_i$.
(3) For any $1 \leq i \leq n$, $\mathcal{I}$ randomly chooses an element $x_i \in \mathbb{X}$ and computes

$$S_i = Q_i \equiv T_{tsk_i}(x_i) mod\ p.$$

$\mathcal{I}$ sets the initial state of $\mathcal{T}_i$ as S_i.
(4) When $\mathcal{I}$ receives a reader $\mathcal{R}_j$, $1 \leq j \leq m$, from factory, $\mathcal{I}$ assigns an unique identity RID_j to $\mathcal{R}_j$ and randomly chooses an element $rsk_j \in \mathbb{Z}_p^*$ as the private key of $\mathcal{R}_j$.
(5) For the checkpoint $\mathcal{M}$, $\mathcal{I}$ generates an information table (see Table 1) for $\mathcal{M}$, which includes the index $Index_i$, the identity TID_i, the initial state S_i, and the product V_i of the private keys of the readers in the path that the tag $\mathcal{T}_i$, $1 \leq i \leq n$, will pass through.

Update: During the process, the tags and readers can authenticate with each other at the same time. If this authentication is successful, then the readers in the supply chain can update the states of tags. For simplicity of description, we assume that there is a tag $\mathcal{T}_i$ which is arriving at the node $\mathcal{R}_j$ of some path. The detailed interaction processes of the algorithm **Update** can be described as follows.

(1) The reader $\mathcal{R}_j$ randomly chooses a string $r \in \mathbb{Z}_p^*$ and a time stamp t_0. Then $\mathcal{R}_j$ sends r and t_0 to the tag $\mathcal{T}_i$.
(2) $\mathcal{T}_i$ checks whether $|t_1 - t_0| \leq \Delta$ after receiving r and t_0 from $\mathcal{R}_j$, where t_1 is the current timestamp.
(3) If the check is satisfied which means that the received information is fresh, then $\mathcal{T}_i$ generates a random number $c \in \mathbb{Z}_p^*$ and computes

$$M_1 = T_{r \cdot c}(Q_i) mod\ p,$$

where Q_i is the current state of $\mathcal{T}_i$.
(4) $\mathcal{T}_i$ computes

$$M_2 = c \oplus H(TID_i),$$

where TID_i is the identity of $\mathcal{T}_i$.
(5) $\mathcal{T}_i$ sends M_1, M_2, Q_i, $Index_i$, t_2 to $\mathcal{R}_j$, where $Index_i$ and t_2 are the index and the current time of $\mathcal{T}_i$, respectively.
(6) $\mathcal{R}_j$ checks the freshness of the information M_1, M_2, Q_i, $Index_i$ by comparing t_2 with the current time of $\mathcal{R}_j$. If the check is satisfied, then $\mathcal{R}_j$ can obtain the identity TID_i of $\mathcal{T}_i$ from the issuer $\mathcal{I}$ by sending the index $Index_i$ to $\mathcal{I}$.
(7) $\mathcal{R}_j$ computes

$$c' = M_2 \oplus H(TID_i)$$

and checks whether the following equation

$$T_{r \cdot c'}(Q_i) mod\ p = M_1$$

hold? If yes, $\mathcal{R}_j$ uses its private key rsk_j to calculate

$$M_3 = T_{c'}(TID_i) \oplus T_{rsk_j}(Q_i) \oplus c'$$

and

$$M_4 = T_{c'}\left(T_{rsk_j}(Q_i)\right).$$

Finally, $\mathcal{R}_j$ sends M_3 and M_4 to $\mathcal{T}_i$.
(8) Upon receiving messages M_3 and M_4 from $\mathcal{R}_j$, $\mathcal{T}_i$ computes

$$a = M_3 \oplus c \oplus T_c(TID_i) \text{ and } M_4' = T_c(a).$$

(9) If $M_4^{'} = M_4$, then $\mathcal{T}_i$ updates its state with $Q_i = a$ and stays for the interaction with the next reader in its path.

Verification: Without loss of generality, we assume that tag $\mathcal{T}_i$ has passed through a valid path $\{\mathcal{R}_1, \mathcal{R}_2, \cdots, \mathcal{R}_b\}$ of length b. According the description of the algorithm **Update**, we know that the current state Q_i of $\mathcal{T}_i$ satisfying

$$Q_i = T_{RSK}(S_i) mod\ p,$$

where $S_i = T_{tsk_i}(x_i) mod\ p$ is the initial state of $\mathcal{T}_i$ and

$$RSK = rsk_1 \cdot rsk_2 \cdots rsk_b$$

is the product of the private keys of the b readers $\mathcal{R}_j$, $1 \leq j \leq b$. When $\mathcal{T}_i$ arrives at the checkpoint $\mathcal{M}$, $\mathcal{M}$ can check whether $\mathcal{T}_i$ has passed through a valid path by the following interactions with $\mathcal{T}_i$.

(1) $\mathcal{M}$ chooses a random number $r \in \mathbb{Z}_p^*$ and sends r along with a timestamp t_0 to $\mathcal{T}_i$.
(2) $\mathcal{T}_i$ checks the freshness of r by verifying $|t_1 - t_0| \leq \Delta$, where t_1 is the current time of $\mathcal{T}_i$.
(3) $\mathcal{T}_i$ randomly chooses a number $c \in \mathbb{Z}_p^*$ and calculates

$$M_1 = T_{r \cdot c}(Q_i) mod\ p$$

and

$$M_2 = c \oplus H(TID_i).$$

Finally, $\mathcal{T}_i$ sends M_1, M_2, Q_i, $Index_i$, t_2 to $\mathcal{M}$.
(4) After receiving M_1, M_2, Q_i, $Index_i$, t_2 from $\mathcal{T}_i$, $\mathcal{M}$ can check their freshness by comparing t_2 with $\mathcal{M}$'s current time t_3. If the check is successful, $\mathcal{M}$ is able to find the initial state S_i of $\mathcal{T}_i$ and the corresponding product

$$V_i = rsk_1 \cdot rsk_2 \cdots rsk_b$$

by searching in its local table with $Index_i$ as an index.
(5) $\mathcal{M}$ checks whether we have the equation

$$M_1 = T_{r \cdot c \cdot V_i}(S_i) mod\ p.$$

If they are equal, then $\mathcal{M}$ outputs 1 which means that the tag $\mathcal{T}_i$ has passed the valid path $\{\mathcal{R}_1, \mathcal{R}_2, \cdots, \mathcal{R}_b\}$ in the system. Otherwise, $\mathcal{M}$ outputs 0 which means that the verification is failed.

4 Security Analysis

In this section, we show that our RFID path authentication protocol satisfies the three security properties: **Authentication**, **Privacy**, and **Unlinkability**.

4.1 Authentication

In our protocol, if a tag $\mathcal{T}_i$ is assigned to a valid path, then the tag can authenticate with readers $\mathcal{R}_j$ which are in this path. As shown in Sect. 3, when $\mathcal{T}_i$ wants to authenticate with $\mathcal{R}_j$, it needs to send $M_2 = t \oplus H(TID_i)$ and $Index_i$ to the reader. We assume that $\mathcal{R}_j$ only stores $\mathcal{T}_i$'s identity which can through this reader. So if an adversary $\mathcal{A}$ wants to use these forged tags or the tags which did not pass the reader, it is difficult for $\mathcal{A}$ to forge $\mathcal{T}_i$'s unique identity because the $Index_i$ and the identity of $\mathcal{T}_i$ are matched. Meanwhile, if a legal tag $\mathcal{T}_i$ does not belong to another valid path V_i and this tag wants to pass the path V_{i+1}. Then $\mathcal{R}_j$ in V_{i+1} cannot search the identity of $\mathcal{T}_i$ by $Index_i$. The state of $\mathcal{T}_i$ will not be updated when it intersects with $\mathcal{R}_j$, and $\mathcal{R}_j$ will drop this tag. Meanwhile, $\mathcal{R}_j$ also sends

$$M_3 = T_{c'}(TID_i) \oplus T_{rsk_j}(Q_i) \oplus c'$$

and

$$M_4 = T_{c'}\left(T_{rsk_j}(Q_i)\right)$$

to $\mathcal{T}_i$ The latter can use these information to authenticate $\mathcal{R}_j$.

4.2 Privacy

The privacy of tags and readers are also guaranteed in our paper. In our protocol, readers have their own private key and are independent of the checkpoint $\mathcal{M}$. We save the path information of each tag to $\mathcal{M}$ during the initialization phase, including the index, identity and the initialized state value. Most importantly, we store the V_n, the product of the private keys of all the readers involved in some valid path, in the $\mathcal{M}$'s database, which ensures that $\mathcal{M}$ is unable to get individual private key of any reader. So the privacy of readers is protected.

As shown in the Sect. 3, each tag $\mathcal{T}_i$ in the **Setup** algorithm has different initialized state value S_i that is encrypted under the tag's private key. For example, $\mathcal{T}_i$ and $\mathcal{T}_{i+1}$ have different state values $Q_i = T_{tsk_i}(x_i) mod\ p$ and $Q_{i+1} = T_{tsk_{i+1}}(x_{i+1}) mod\ p$. When the two legal tags $\mathcal{T}_i$ and $\mathcal{T}_{i+1}$ across the same reader $\mathcal{R}_j$, Q_i and Q_{i+1} will be updated to Q_i^{new} and Q_{i+1}^{new} if they are successfully authenticated with $\mathcal{R}_j$. The Chebyshev polynomials will map these two states to two different values in the finite field $\mathbb{Z}_p$. As a result, $\mathcal{A}$ cannot figure out which tag is derived from the two values. This shows that our protocol achieves the privacy of both readers and tags.

4.3 Unlinkability

Unlinkability represents that $\mathcal{A}$ cannot know whether there are two tags that have passed more than a same node. Assume that there are two tags $\mathcal{T}_i$ and $\mathcal{T}_{i+1}$ that have finished their works in the different valid path, then the states of the two tags are changed to Q_i and Q_{i+1}. In our protocol, we have $Q_i^{new} = T_{V_i}(Q_i) mod\ p$ and $Q_{i+1}^{new} = T_{V_{i+1}}(Q_{i+1}) mod\ p$. In fact, Any Chebyshev polynomial of degree n over $\mathbb{Z}_p$ can be expressed as

$$T_n(x) = a_n x^n + a_{n-1} x^{n-1} + \cdots + a_1 x + a_0 mod\ p. \quad (1)$$

It is well-known that the hardness of CP-DL problem ensures [21] that it is easy to compute the value $T_n(x)$ with the knowledge of x and n, while it is difficult to compute n with x and $T_n(x)$. So even if we publish the states of all tags, $\mathcal{A}$ is unable to get the path information or readers' private keys. As a result, $\mathcal{A}$ will be failed to judge whether there are the same paths between the two tags $\mathcal{T}_i$ and $\mathcal{T}_{i+1}$.

5 Performance Analysis

In this section, we show that our protocol is more efficient than the previous works in computation cost and memory space.

As shown in Sect. 3, a tag $\mathcal{T}_i$ in the **Update** phase communicating with a reader $\mathcal{R}_j$ needs to compute 2 Chebyshev polynomials and 3 Exclusive-OR operations, while $\mathcal{R}_j$ should execute 4 Chebyshev polynomials and 3 Exclusive-OR operations. Compared with [17] and [18], each tag needs 3 multiplication, 8 exponentiation operations and each reader needs 4 multiplication, 6 exponentiation operations, respectively. During the verification phase, the total computation cost of $\mathcal{T}_i$ and $\mathcal{M}$ includes 2 Chebyshev polynomials and 2 Exclusive-OR operations, while the corresponding cost is 2 multiplication and 2 exponentiation operations in [18] and 3 multiplication and 5 exponentiation operations in [17]. Although our protocol also needs 2 pseudo random number generator (PRNG), it is believed that PRNG operation is negligible compared with other operations. We know that the time complexity of an exponentiation operation is $\mathcal{O}(2^n)$, while the time complexity of a Chebyshev polynomials operation is $\mathcal{O}(\log n)$. Further the computation cost of Exclusive-OR operation is much less than the one of multiplication operations.

Finally we assume that p is prime number of L bits length, the hash function has an output of 128 bits length, and the identification *TID* and index of each tag has 32 and 64 bits length, respectively. Then a tag in our protocol only needs a memory space of 224+L bits, while the corresponding values are 960 bits in Tracker [17] and 800 bits in Wang et al. [18]. This shows that our scheme is more efficient because the value of L is no more than 256 bits in general. We list the detailed comparison in Table 2 in which T_M, T_E, T_C, and T_{xor} denote the times of a multiplication operation, an exponentiation operation, a Chebyshev polynomial operation, and an Exclusive-OR operation, respectively.

Table 2. Efficiency comparison

Performance	Tracker [17]	Wang et al. [18]	Our work
Update cost	$3T_M + 8T_E$	$4T_M + 6T_E$	$6T_{xor} + 6T_C$
Verification cost	$3T_M + 5T_E$	$2T_M + 2T_E$	$2T_C + 2T_{xor}$
Memory storage	960 bits	800 bits	L + 224 bits
Others	4 HMAC	None	2 PRNG

6 Conclusion

In this paper, we proposed an RFID path authentication protocol for the industrial supply chains based on the Chebyshev polynomials. We described every phase of our path authentication protocol in detail and gave the security analysis which shows that our protocol satisfies the required security properties. Different from other related schemes that need the checkpoint to store all readers' private keys, the checkpoint in our proposed protocol needs only to store the product of the private keys of the readers involved in valid paths. This not only improves the security of existing related protocols, but also results in a lower memory requirement. Further the states of tags are allowed to be exposed. These properties enable our proposed protocol to be practical and are promising to be deployed in near future for both factory and governments.

Acknowledgment. The work is supported in part by the NSFC-Zhejiang Joint Fund for the Integration of Industrialization and Informatization under Grant No. U1509219, the Shanghai Natural Science Foundation under Grant No. 17ZR1408400, the National Key R&D Program of China under Grant No. 2017YFB0802302, the National Natural Science Foundation of China under Grant Nos. 61601129, 11701179, the Shanghai Science and Technology Commission Program under Grant No. 18511105700, and the Shanghai Sailing Program under Grant No. 17YF1404300.

References

1. Mamun, M.S.I., Miyaji, A.: RFID path authentication, revisited. In: 2014 IEEE 28th International Conference on Advanced Information Networking and Applications (AINA), pp. 245–252. IEEE (2014)
2. Molnar, D., Soppera, A., Wagner, D.: A scalable, delegatable pseudonym protocol enabling ownership transfer of RFID tags. In: International Workshop on Selected Areas in Cryptography, pp. 276–290. Springer, Heidelberg (2005)
3. Deng, R.H., Li, Y., Yung, M., Zhao, Y.: A new framework for RFID privacy. In: European Symposium on Research in Computer Security, pp. 1–18. Springer, Heidelberg (2010)
4. Choo, K.-K.R., Bishop, M., Glisson, W.B., Nance, K.L.: Internet and cloud-of-things cybersecurity research challenges and advances. Comput. Secur. **74**, 275–276 (2018)
5. Lin, C., He, D., Kumar, N., Choo, K.-K.R., Vinel, A., Huang, X.: Security and privacy for the internet of drones: challenges and solutions. IEEE Commun. Mag. **56**(1), 64–69 (2018)
6. Al-Dhaqm, A.M.R., Razak, S.A., Othman, S.H., Choo, K.-K.R., Glisson, W.B., Al-Ali, A. R., Al-Rousan, M.: CDBFIP: common database forensic investigation processes for internet of things. IEEE Access **5**, 24401–24416 (2017)

7. D'Orazio, C., Choo, K.-K.R., Yang, L.T.: Data exfiltration from internet of things devices: iOS devices as case studies. IEEE Internet Things J. **4**(2), 524–535 (2017)
8. Liu, Z., Choo, K.-K.R., Groschdl, J.: Security edge devices in the post-quantum internet of things using lattice-based cryptography. IEEE Commun. Mag. **56**(2), 158–162 (2018)
9. Peris-Lopez, P., Hernandez-Castro, J.C., Estevez-Tapiador, J.M., et al.: RFID systems: a survey on security threats and proposed solutions. In: IFIP International Conference on Personal Wireless Communications, pp. 159–170. Springer, Heidelberg (2006)
10. Lee, Y., Park, Y.: A new privacy-preserving path authentication scheme using RFID for supply chain management. Adv. Electr. Comput. Eng. **13**(1), 23–26 (2013)
11. Xin, W., Sun, H., Wang, Z., Guan, Z., Chen, Z.: A RFID path-checking protocol based on ordered multisignatures. Radio Freq. Identif. Syst. Secur. **8**, 33–44 (2012)
12. Bu, K., Li, Y.: Every step you take, I'll be watching you: practical stepauth-entication of RFID paths. IEEE Trans. Inf. Forensics Secur. **13**(4), 834–849 (2018)
13. Cai, S., Deng, R.H., Li, Y., Zhao, Y.: A new framework for privacy of RFID path authentication. In: International Conference on Applied Cryptography and Network Security, pp. 473–488. Springer, Heidelberg (2012)
14. Cai, S., Li, Y., Zhao, Y.: Distributed path authentication for dynamic RFID-enabled supply chains. In: IFIP International Information Security Conference, pp. 501–512. Springer, Heidelberg (2012)
15. Wang, H., Li, Y., Zhang, Z., Cao, Z.: Two-level path authentication in EPCglobal network. In: 2012 IEEE International Conference on RFID, pp. 24–31. IEEE (2012)
16. Ouafi, K., Vaudenay, S.: Pathchecker: an RFID application for tracing products in supply-chains. In: Proceedings of RFIDSec 2009 (LASEC-CONF-2009-006) (2009)
17. Blass, E.O., Elkhiyaoui, K., Molva, R., Antipolis, E.S.: Tracker: security and privacy for RFID-based supply chains. In: 18th Annual Network and Distributed System Security Symposium, NDSS 2011, 6–9 February 2011 (2011)
18. Wang, H., Li, Y., Zhang, Z., Zhao, Y.: Efficient tag path authentication protocol with less tag memory. In: International Conference on Information Security Practice and Experience, pp. 255–270. Springer, Cham (2016)
19. Shen, Z., Zeng, P., Qian, Y., Choo, K.-K.R.: A secure and practical RFID ownership transfer protocol based on Chebyshev polynomials. IEEE Access **6**, 14560–14566 (2018)
20. Fee, G., Monagan, M.: Cryptography using Chebyshev polynomials. In: Proceedings of 2004 Maple Summer Workshop (CD-ROM) (2004). http://www.cecm.suf.ca/CAG/products2003.shtml
21. Zhang, L.: Cryptanalysis of the public key encryption based on multiple chaotic systems. Chaos, Solitons Fractals **37**(3), 669–674 (2008)

An Efficient Authentication Protocol for RFID-Based Supply Chains

Yuyin Qian and Peng Zeng(✉)

Shanghai Key Laboratory of Trustworthy Computing, East China Normal University, Shanghai, China
yuyinqianecnu@outlook.com, pzeng@sei.ecnu.edu.cn

Abstract. Radio frequency identification (RFID) is an automatic wireless data collection technology which can be used to precisely identify objects through radio waves. Nowadays, RFID has an important application in supply chain systems. When RFID tags are attached to valuable products for identification, however, they are under a potential risk of being invaded and re-written with fake tag, which will lead to economic losses and is even harmful to people's health. Therefore, path authentication has aroused great concern in modern society as it has the function of maintaining object pedigree and supply chain integrity. Existing path authentication protocols for RFID based supply chains are path grained which means that they need some readers worked as managers to preload the sets of valid paths and then to verify whether tags go through a valid path. Recently, Bu et al. proposed a novel path authentication named StepAuth in which the readers don't need to preload any valid paths before authentication. StepAuth reduces the risks of leaking valid paths. In this paper, we propose a new tag path authentication protocol for RFID-based supply chain, which further improves the efficiency of StepAuth and allows the public keys of readers to be published.

Keywords: RFID · Path authentication · Security · Privacy · Track and trace
Unlinkability · Lightweight protocol

1 Introduction

RFID technology has been widely adopted in supply chains to detect counterfeit products [1, 2]. As shown in Fig. 1, the tag first needs to register in a server/initializer to get legal path secrets and privacy information about some product. Then the registered tag is affixed to the corresponding product. According to the path secret stored in the tag, the product is assigned with a legal path. Next, legal readers along the valid path verify the product by authenticating the path secrets in the tag. Finally, legal readers update the state of the tag if the tag passes through the authentication phase.

The intention of path authentication is to provide greater protection against flooding of counterfeit products. According to the study of Counterfeiting Intelligence Bureau (CIB) of the International Chamber of Commerce (ICC), counterfeit goods make up 5% to 7% of World Trade [3]. On the one hand, such counterfeit products may be harmful to consumers' health (e.g. counterfeit medications and adulterant foods). On

J. Abawajy et al. (Eds.): ATCI 2018, AISC 842, pp. 28–37, 2019.
https://doi.org/10.1007/978-3-319-98776-7_4

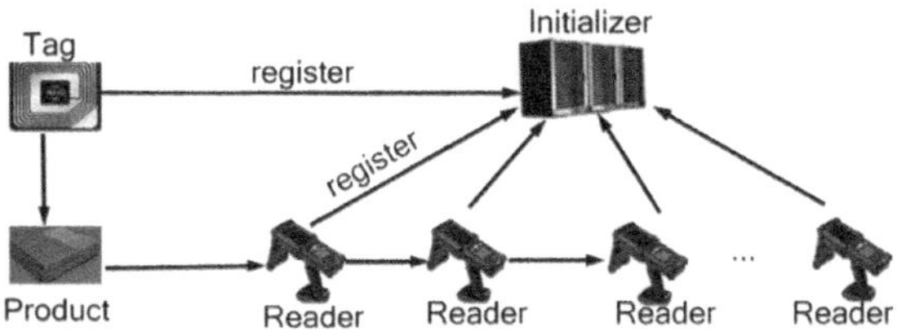

Fig. 1. RFID system workflow

the other hand, counterfeit products may cause financial losses of businesses and consumers (such as imitation cosmetics and electricity). Therefore, path authentication is significantly important in RFID-based supply chains.

Generally an RFID-based supply chain consists of the four entities: initializer, tags, readers, checkpoints (a kind of special readers) [4–8]. The initializer (worked as a trusted server) generates the public system parameters and valid paths, initializes the state of tags, and shares the verification information with the corresponding checkpoints. When a tag interacts with a reader, the latter updates the state of the tag and writes with its privacy information. The checkpoint retains the verification information and checks whether the tag has gone through a valid path. In 2009 Ouafi et al. proposed the first path authentication protocol Pathchecker for RFID-based supply chains [9]. Pathchecker needs a time complexity of $O(\log(n))$ to authenticate whether a tag has gone through a valid path by calculating the state of the tag from n valid paths. Moreover, the tags in Pathchecker are required to make some Hash calculations which maybe unaffordable since tags are assumed to be resource-restricted. In 2011 Blass et al. proposed the first real solution, named as Tracker, for RFID-based supply chains [10]. Tracker emancipates the tags from complex computation via sending a valid path set and private keys of the readers to a checkpoint. Later in 2012 Elkhiyaoui et al. proposed an on-site checking protocol Checker [11] to ameliorate Tracker by enabling every legal reader to verify whether the tags have gone through a valid subset of the valid paths. Recently, Cai et al. [12] and Wang et al. [13] further improved Tracker with stronger security requirements and lower computation cost.

As shown in [4], however, all the aforementioned path authentication protocols for RFID-based supply chains are path grained which means that they need some readers worked as checkpoints to preload the sets of valid paths and then to verify whether tags go through a valid path. This case leads to make them not well suit to enforce delivery regulation that requires each product to follow a designated path [4]. Recently, Bu et al. proposed a novel path authentication protocol named StepAuth in which the readers don't need to preload any valid paths before authentication. StepAuth is the first practically efficient protocol for distributed and stepgrained path authentication. In this paper, we propose a new tag path authentication protocol for RFID-based supply chain, which further improves the efficiency of StepAuth and allows the public keys of readers to be published.

1.1 Our Contribution

The main contribution of our proposed protocol in this paper is to further improve StepAuth [4] in the following two aspects:

- We generate the symmetric key between the initializer $\mathcal{I}$ and the corresponding readers by using the secret key sk of $\mathcal{I}$ to replace the random number r_i in ECIES, which enables the public keys of each reader to be published. On the other hand, the corresponding public keys in StepAuth are not allowed to be published.
- We use the keyed-hashing for message authentication codes (*HMAC*) instead of signatures to check the integrity of messages, which reduces significantly both the memory cost and computation cost compared with StepAuth.

1.2 Paper Organization

In the next section, we introduce the background knowledge used in our proposed protocol. We then present our path authentication protocol in Sect. 3 and analyze its security and privacy in Sect. 4. In Sect. 5, we give the performance analysis of our proposed protocol by comparing our work with StepAuth. We conclude this work in Sect. 6.

2 Adversary Model

2.1 Security

Similar to previous work [14–19], we assume that there is a probabilistic polynomial time (PPT) adversary $\mathcal{A}$ who can break the security of path authentication protocols. Further, we assume that the initializer $\mathcal{I}$ is always trusted, while the readers are possible to be disguised by $\mathcal{A}$. To define the security of path authentication protocols, we denote by $\mathcal{T}_j$ a tag and $\mathcal{P}_j = \{\mathcal{R}_1, \mathcal{R}_2, \ldots, \mathcal{R}_l\}$ is a valid path of l-length comprising of l readers $\mathcal{R}_i$, $1 \le i \le l$. Further, we denote by $S_{(all)}$ the set of all valid states in the system, $S_{(guess)}$ the set of all guess states which includes $S_{(all)}$, and $S^{(i,j)}_{(valid)}$ the set of the valid states of tag $\mathcal{T}_j$ which can pass the authentication of readers $\mathcal{R}_k$, $1 \le k \le i-1$.

We call that $\mathcal{A}$ cannot forge a valid state $s^{(i,j)}_{(adv)}$ for any reader $\mathcal{R}_i$, $1 \le i \le l$, if the following inequality holds:

$$\Pr\left[Auth\left(s^{(i,j)}_{(adv)}, K^{sec}_{\mathcal{R}_i}\right) = 1\right] \le \frac{\left|S^{(i,j)}_{(valid)}\right|}{\left|S_{(guess)}\right|} + \epsilon,$$

where $|S|$ denotes the cardinality of set S, $K^{sec}_{\mathcal{R}_i}$ denotes the private key of $\mathcal{R}_i$, and $Auth(s^{(i,j)}, K^{sec}_{\mathcal{R}_i}) = 1$ denotes that the authentication of $\mathcal{R}_i$ on the state $s^{(i,j)}$ of tag $\mathcal{T}_j$ is successful; otherwise $Auth(s^{(i,j)}, K^{sec}_{\mathcal{R}_i}) = 0$ means the corresponding authentication is failed.

2.2 Privacy

We assume there is a PPT adversary $\mathcal{A}$ that can invade the communication between the tags and the readers. $\mathcal{A}$ can launch a variety of attacks at all readers in the RFID-based supply chains. For example, $\mathcal{A}$ lurks a subordinate in advance at each step of the valid path and analyzes their acquired secrets to synthesis a comprehensive one. Although we may overestimate the real situation, it motivates us to design a more safe and privacy path authentication protocol based on RFID supply chains.

We consider privacy concerns in the two categories: privacy and unlinkability [4]. The former category is the protection of the information that we want to conceal. The latter category can be divided into two aspects: tag unlinkability and step unlinkability.

- Tag unlinkability: $\mathcal{A}$ cannot rely on the state of tags to distinguish whether they have passed through a targeted reader.
- Step unlinkability: $\mathcal{A}$ cannot know whether two random tags went through the same path or the same reader.

3 Our Protocol

3.1 Key Establishment

The initializer $\mathcal{I}$ sets a series of domain parameters in the RFID-based supply chains. $\mathcal{R}_i$ represents the i-th reader. We constructs our protocol via mainly using the generator g and the order n of g. $\mathcal{I}$ generates its private key by randomly choosing $K_{\mathcal{I}}^{sec} \in [1, n-1]$, and the pubic key of $\mathcal{I}$ is $K_{\mathcal{I}}^{pub} = K_{\mathcal{I}}^{sec} g$. Every reader $\mathcal{R}_i$ in the system generate their keys the same as initializer $\mathcal{I}$, $K_{\mathcal{R}_i}^{sec} \in [1, n-1]$, $K_{\mathcal{R}_i}^{pub} = K_{\mathcal{R}_i}^{sec} g$.

Then $\mathcal{R}_i$ sends his public key $K_{\mathcal{R}_i}^{pub}$ to the initializer $\mathcal{I}$ for generating path secret. $\mathcal{I}$ shares its public key $K_{\mathcal{I}}^{pub}$ to all readers for verifying phase.

3.2 The Initialization of Tags State

In this phase, $\mathcal{I}$ has known public keys of every valid reader in the system. Before generating the initial state of a tag named $\mathcal{T}$ for example, $\mathcal{I}$ needs to know $\mathcal{R}_i$'s symmetric key. According to the ECIES (Elliptic Curve Integrated Encryption Scheme) encryption, we uses $K_{\mathcal{I}}^{sec}$ to replace r_i, that guarantees the publicity of the public key of the reader $\mathcal{R}_i$. The initializer gets $\mathcal{R}_i$'s symmetric key as follows: $\mathcal{I}$ calculates $P = (P_x, P_y) = K_{\mathcal{I}}^{sec} K_{\mathcal{R}_i}^{pub}$ and runs $KDF(P_x, S_1^i)$ to get the symmetric key $K_{\mathcal{R}_i}^{sym}$ and K_i, S_1^i is optional shared information. We assume that there is a path $\mathcal{P} = \{\mathcal{R}_1, \mathcal{R}_2, \ldots, \mathcal{R}_{l-1}, \mathcal{R}_l\}$ of length l. $\mathcal{I}$ calculates the initial state of tag $\mathcal{T}$ as follows:

- $\mathcal{I}$ computes a ciphertext C_l as

$$C_l = Enc_{K^{sym}_{\mathcal{R}_l}}(\mathcal{R}_l,\ \mathcal{R}_l,\ \mathcal{T}).$$

- $\mathcal{I}$ computes a message digest M_l of C_l via K_l

$$M_l = HMAC(K_l, C_l)$$

and sets the state s_l as

$$s_l = \{C_l, M_l\}.$$

$\mathcal{I}$ constructs the state from $\mathcal{R}_l$ to $\mathcal{R}_1$ via following equation

$$s_i = \begin{cases} \left\{Enc_{K^{sym}_{\mathcal{R}_i}}(\mathcal{R}_i,\ \mathcal{R}_i,\ \mathcal{T}), M_l\right\}, & i = l; \\ \left\{Enc_{K^{sym}_{\mathcal{R}_i}}(\mathcal{R}_i,\ \mathcal{R}_{i+1}, s_{i+1}), M_i\right\}, & 1 \leq i \leq l-1. \end{cases}$$

- $\mathcal{I}$ writes the initial state s_1 into tag $\mathcal{T}$ and attaches the tag $\mathcal{T}$ to a product. $\mathcal{I}$ broadcasts $\mathcal{T}$ into the supply chain system.

3.3 Authentication Phase $Auth(s_i, K^{sec}_{\mathcal{R}_i})$

After reading state s_i from $\mathcal{T}$, reader $\mathcal{R}_i$ first derives $K^{sym}_{\mathcal{R}_i}$ and K_i from $KDF(P_x, S^i_1)$ via $P = (P_x, P_y) = K^{sec}_{\mathcal{R}_i} K^{pub}_{\mathcal{I}}$. $\mathcal{R}_i$ then checks whether the following equation holds:

$$HMAC(K_i,\ Enc_{K^{sym}_{\mathcal{R}_i}}(\mathcal{R}_i,\ \mathcal{R}_{i+1}, s_{i+1})) \stackrel{?}{=} M_i.$$

If it holds, $\mathcal{R}_i$ gets the result and uses the symmetric key $K^{sym}_{\mathcal{R}_i}$ to decrypt the encryption message; otherwise $\mathcal{R}_i$ rejects the tag $\mathcal{T}$ and outputs 0. Finally, $\mathcal{R}_i$ gets the plain text $(\mathcal{R}_i,\ \mathcal{R}_{i+1}, s_{i+1})$ via

$$Dec(K^{sym}_{\mathcal{R}_i},\ Enc_{K^{sym}_{\mathcal{R}_i}}(\mathcal{R}_i,\ \mathcal{R}_{i+1}, s_{i+1})).$$

If the first $|\mathcal{R}_i|$ bits of the plain text match $\mathcal{R}_i$, $\mathcal{R}_i$ accepts it and outputs 1, else he rejects the tag $\mathcal{T}$ and outputs 0. Here $|\mathcal{R}_i|$ means the length of $\mathcal{R}_i$.

3.4 Update

Upon $Auth(s_i, K^{sec}_{\mathcal{R}_i}) = 1$, $\mathcal{R}_i$ removes the first $|\mathcal{R}_i|$ bits from the plain text. $\mathcal{R}_i$ checks the first $|\mathcal{R}_i|$ bits in the remaining plain text. If $\mathcal{R}_{i+1} = \mathcal{R}_i$, it means that the tag $\mathcal{T}$ has arrived at the final targeted reader, else $\mathcal{R}_{i+1}$ means the next reader that $\mathcal{T}$ would go through. Finally,$\mathcal{R}_i$ removes the left $|\mathcal{R}_i|$ bits from the remaining plain text and gets s_{i+1}. If $\mathcal{R}_i$ is the final reader, s_{i+1} represents the information about the product which $\mathcal{T}$ attached to, else s_{i+1} means the path secret of $\mathcal{R}_{i+1}$ broadcasted by $\mathcal{R}_i$. $\mathcal{T}$ continues to go through the authentication phase and update phase with s_{i+1} and $\mathcal{R}_{i+1}$ as the same in Sects. 3.3 and 3.4.

4 Security Analysis

4.1 Security

Our protocol can satisfy the security requirement if any PPT adversary $\mathcal{A}$ cannot obtain $P = (P_x, P_y) = K_{\mathcal{I}}^{sec} K_{\mathcal{R}_i}^{pub}$, which is used to get the symmetric key $K_{\mathcal{R}_i}^{sym}$ and the *HMAC* key K_i from $KDF(P_x, S_1^i)$. The symmetric key $K_{\mathcal{R}_i}^{sym}$ and the *HMAC* key of $\mathcal{R}_i$ are used to generate path secret of $\mathcal{R}_i$ via $\mathcal{I}$. That means the probability that $\mathcal{A}$ guesses the valid path secret of path $\mathcal{P}$ is less than random guessing in Sect. 2.1. As long as there is one reader which is not compromised, $\mathcal{A}$ cannot forge the valid path secret for an counterfeit $\mathcal{T}'$.

Theorem 1. If *HMAC* is secure against adaptive chosen cipher text attacks, no PPT adversary $\mathcal{A}$ can generate a valid state for counterfeit tag $\mathcal{T}'$ which will be attached to the shoddy product with non-negligible probability.

Proof. According to the paper [22], the *HMAC* is secure against adaptive chosen cipher attacks. That means, even in such a case that adversary $\mathcal{A}$ has access to all readers in a valid path to acquire states read from tags, if $\mathcal{A}$ doesn't know $P = (P_x, P_y) = K_{\mathcal{I}}^{sec} K_{\mathcal{R}_i}^{pub}$, he can't get the symmetric key $K_{\mathcal{R}_i}^{sym}$ and the HMAC key K_i from $KDF(P_x, S_1^i)$. Thus, he cannot forge a valid state which can pass the valid path $\mathcal{P}$ with negligible probability.

4.2 Privacy

Privacy can be divided into tag unlinkability and step unlinkability.

Theorem 2. Considered that $\mathcal{A}$ can break the public key cryptography, no PPT adversary $\mathcal{A}$ with eavesdropping ability can break the tag unlinkability, in other words, $\mathcal{A}$ cannot correlate path secrets with tags.

Proof. As shown in Fig. 2. We assume that there are two path $\mathcal{P}_0 = \{\mathcal{R}_A, \mathcal{R}_B, \mathcal{R}_C\}$, $\mathcal{P}_1 = \{\mathcal{R}_A, \mathcal{R}_B, \mathcal{R}_D\}$ and they have two common readers. The product with $\mathcal{T}_0$ is assigned to go through $\mathcal{P}_0$, and another product attached with $\mathcal{T}_1$ goes through $\mathcal{P}_1$ While products with $\mathcal{T}_0$ and $\mathcal{T}_1$ go through each reader along the path, the reader can read corresponding path secrets as $\{s_{\mathcal{T}_0}^A, s_{\mathcal{T}_0}^B, s_{\mathcal{T}_0}^C\}, \{s_{\mathcal{T}_1}^A, s_{\mathcal{T}_1}^B, s_{\mathcal{T}_1}^D\}$. We assume that $\mathcal{A}$ has the ability of eavesdropping every reader in both $\mathcal{P}_0$ and $\mathcal{P}_1$. Tag unlinkability is defined as $\mathcal{A}$ cannot rely on the state of tags to distinguish whether they have passed through a targeted reader in Sect. 2.1. As the example shown in Fig. 2, the states of $\mathcal{T}_0$ and $\mathcal{T}_1$ have the same length. Without loss of generality, we discuss the tag unlinkability among reader $\mathcal{R}_A$ to reader $\mathcal{R}_B$. As mentioned before, $\mathcal{A}$ can eavesdrop $\mathcal{R}_A$ and $\mathcal{R}_B$ to grab path secret stored in $\mathcal{R}_A$ and $\mathcal{R}_B$. Then he obtains the following states:

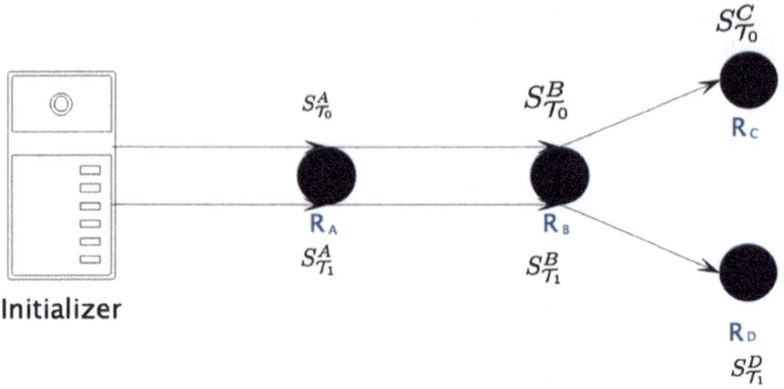

Fig. 2. Path example used in proof

$s^A_{\mathcal{T}_0} = \{Enc_{K^{sym}_{1A}}(\mathcal{R}_A, \mathcal{R}_B, s^B_{\mathcal{T}_1}), M_{1A}\}, s^A_{\mathcal{T}_1} = \{Enc_{K^{sym}_{1A}}(\mathcal{R}_A, \mathcal{R}_B, s^B_{\mathcal{T}_1}), M_{1A}\},$
$s^B_{\mathcal{T}_0} = \{Enc_{K^{sym}_{0B}}(\mathcal{R}_B, \mathcal{R}_C, s^C_{\mathcal{T}_0}), M_{0B}\}, s^B_{\mathcal{T}_1} = \{Enc_{K^{sym}_{1B}}(\mathcal{R}_B, \mathcal{R}_C, s^C_{\mathcal{T}_1}), M_{1B}\}.$

If $\mathcal{A}$ can breaks the tag unlinkability between $\mathcal{T}_0$ and $\mathcal{T}_1$, which means he can distinguish $s^A_{\mathcal{T}_0}$ and $s^A_{\mathcal{T}_1}$ and correlate $s^A_{\mathcal{T}_0}$ with $s^B_{\mathcal{T}_0}$, $s^A_{\mathcal{T}_1}$ with $s^B_{\mathcal{T}_1}$ correspondingly. To this end, $\mathcal{A}$ has to know K^{sec}_A and derives the symmetric key K^{sym}_A using $K^{pub}_{\mathcal{I}} K^{sec}_A$ to decrypt $Enc_{K^{sym}_{0A}}(\mathcal{R}_A, \mathcal{R}_B, s^B_{\mathcal{T}_0})$. However, $\mathcal{A}$ cannot acquire the private keys of $\mathcal{R}_A$ and $\mathcal{R}_B$. Thus, our protocol provides tag unlinkability.

Theorem 3. Tag unlinkability infers step unlinkablity.

Proof. Step unlinkability can be defined as $\mathcal{A}$ cannot tell whether two random tags go through the same path. We use the same assumption as in tag unlinkability. If $\mathcal{A}$ invades in $\mathcal{R}_A$ and $\mathcal{R}_B$, he will acquire $s^A_{\mathcal{T}_0}$, $s^A_{\mathcal{T}_1}$, $s^B_{\mathcal{T}_0}$, $s^B_{\mathcal{T}_1}$. If $\mathcal{A}$ can correlate $s^A_{\mathcal{T}_0}$ with $s^B_{\mathcal{T}_0}$, or $s^A_{\mathcal{T}_1}$ with $s^B_{\mathcal{T}_1}$, $\mathcal{A}$ can realize that $\mathcal{T}_0$ and $\mathcal{T}_1$ will go through the same step in the next step, in other words he can break the step unlinkability of our protocol. However, we have proved the tag unlinkability of our protocol before, thus $\mathcal{A}$ cannot correlate $s^A_{\mathcal{T}_0}$ with $s^B_{\mathcal{T}_0}$, or $s^A_{\mathcal{T}_1}$ with $s^B_{\mathcal{T}_1}$.There, our protocol provides tag unlinkability.

5 Performance Analysis

5.1 Computation Cost

Our protocol runs in the environment where the **CPU** (the Skylake Core-i5) and the frequency is 2.7e+09 Hz. As we can see, the main computation happened in initialization phase (see Sect. 3.2) and the authentication phase (see Sect. 3.3). More specifically, we first pay attention to initialization phase, which generates the initial path secret of a specific tag $\mathcal{T}$ via ECIES. The operation of one ECIES encryption costs 1.09 ms. If there is a path $\mathcal{P} = \{\mathcal{R}_1, \mathcal{R}_2, \cdots, \mathcal{R}_l\}$ for $\mathcal{T}$, it will cost $\mathcal{I}$ about $1.09 \cdot l$ ms to generate the initial path secret. At the same time, we focus on authentication phase Sect. 3.3. When $\mathcal{R}_i$ receives path secret $s_{\mathcal{T}}^{i}$, it takes $\mathcal{R}_i$ about 0.77 ms to decrypt. We can thus conclude that our protocol is efficiency (Table 1).

Table 1. Memory cost in our protocol

Memory cost	
$\mathcal{I}$	$\|secp256r1\| + 256(\|\mathcal{R}\| + 1)$
$\mathcal{R}_i$	$\|secp256r1\| + 256 + 256 + \log\|\mathcal{T}_i\| + \log\|\mathcal{R}_i\|$
$\mathcal{T}_i$	$248l$
$\|secp256r1\|$ represents the memory of domain parameters in our protocol needed	

5.2 Memory Cost

The memory cost spent in our protocol can be discussed in three aspects: $\mathcal{I}$, readers, and tags. We make choice from all Elliptic Curve domain parameters and choose *secp*256*r*1 as the domain parameters in our protocol. We can thus get the memory of domain parameters needed in ECIES is $|secp256r1| = 6 \cdot 256 = 1536$. In ECIES, the size of symmetric key is 128 bits, and we adopt HMAC-MD5 function, where the length of key $|K_i|$ is 128 bits and the size of HMAC calculation result is 128 bits.

In the initial phase, on the one hand, $\mathcal{I}$ needs to know the memory of Elliptic Curve domain parameters, which is 1536 bits. ECIES is an integrated encryption scheme, thus $\mathcal{I}$ needs to maintain all public keys of readers to calculate integrated ciphertext. On the other hand, in our protocol $\mathcal{I}$ needs id of tags $\mathcal{T}_j$ and readers $\mathcal{R}_i$ to generate path secrets. In reality, the size of reader id and tag id are 64 bits and 128 bits. In such case, we have $\log|\mathcal{R}_i| = 6$, and $\log|\mathcal{T}_i| = 7$ according to the paper [4].

In the authentication phase in Sect. 3.3 and update phase in Sect. 3.4, readers need to generate their corresponding symmetric keys and keys of *HMAC* function. Thus, readers maintain public key of $\mathcal{I}$ and domain parameters of ECIES protocol, which are 256 bits and 1536 bits. At the same time, readers need to conceive tag' id and its own id, which separately occupy $\log|\mathcal{T}_i| = 7$ bits and $\log|\mathcal{R}_i| = 6$ memories in our system. Therefore, it takes each reader about $1536 + 256 + 256 + \log|\mathcal{T}_i| + \log|\mathcal{R}_i| = 2061$ bits to support our protocol.

It is of great significance to control the tag memory, due to EPCGen2 tags, whose memory ranging from 8 KB [23] to 64 KB [24]. If the path secret generated in the initialization phase exceeds 64 bits, it will be impractical for current situation. $\mathcal{I}$ constructs the initial state from last step to beginning step. We thus calculate the size of the initial path secret from last reader, whose path secret is:$s_l = \{Enc_{K_{\mathcal{R}_l}^{sym}}(\mathcal{R}_l, \mathcal{R}_l, \mathcal{T}), M_l\}$, obviously, above path secret needs $|s_l| = 60 + 60 + 128 = 248$ bits memory in the final reader $\mathcal{R}_l$. The size of path secrets expends, when $\mathcal{T}$ comes to the next reader, whose path secret is as follows: $s_{l-1} = \{Enc_{K_{\mathcal{R}_{l-1}}^{sym}}(\mathcal{R}_l, \mathcal{R}_{l-1}, s_l), M_{l-1}\}$. Above all, we can conclude the memory of $\mathcal{R}_i$ needed is $|s_i| = 248 + (60 + 60 + 128)(l - i)$, l is the length of path $\mathcal{P}$.

Thus, we can make the conclusion that the size of the initial path secret is $|s_1| = 248 + 248(l-1) = 248l$ for a l-length path. Even for a 300-length long path, the memory of initial secret is $|s_1| = 248 \cdot 300bits = 9.08$, which is suitable for EPCGen2 on the market.

6 Conclusion

In this paper, we propose an efficient tag path authentication protocol in RFID-based supply chain. Our proposed protocol has two main improvements on StepAuth [4]. On the one hand, we reduce the memory requirement of the tag from $1024 + 896(l-1)$ to $248l$ compared with StepAuth. On the other hand, the efficiency of encryption part in our protocol has been increased to 1.09 ms and the decryption part is 0.77 ms. At the same time, the public key of the readers in our proposed protocol can be allowed to be published. In a word, our proposed protocol is efficient, less memory required, and secure under privacy and security.

Acknowledgements. The work is supported in part by the NSFC-Zhejiang Joint Fund for the Integration of Industrialization and Informatization under Grant No. U1509219, the Shanghai Natural Science Foundation under Grant No. 17ZR1408400, the National Key R&D Program of China under Grant No. 2017YFB0802302, the National Natural Science Foundation of China under Grant Nos. 61601129, 11701179, the Shanghai Science and Technology Commission Program under Grant No. 18511105700, and the Shanghai Sailing Program under Grant No. 17YF1404300.

References

1. Shen, Z., Zeng, P., Qian, Y., Choo, K.K.R.: A secure and practical RFID ownership transfer protocol based on Chebyshev polynomials. IEEE Access **6**, 14560–14566 (2018)
2. Li, X., Niu, J., Choo, K.K.R.: A robust authentication protocol with privacy protection for wireless sensor networks. In: Radio Frequency Identification and IoT Security, pp. 30–44 (2017)
3. https://en.wikipedia.org/wiki/Counterfeit
4. Bu, K., Li, Y.: Every step you take, I'll be watching you: Practical StepAuth-entication of RFID paths. IEEE Trans. Inf. Forensics Secur. (2017)
5. Burbridge, T., Soppera, A.: Supply chain control using a RFID proxy re-signature scheme. In: 2010 IEEE International Conference on RFID, pp. 29–36. IEEE (2010)
6. Mamun, M.S.I., Miyaji A.: RFID path authentication, revisited. In: 2014 IEEE 28th International Conference on Advanced Information Networking and Applications (AINA), pp. 245–252. IEEE (2014)
7. Ray, B., Chowdhury, M., Abawaiy, J.: PUF-based secure checker protocol for Networked RFID systems. In: 2014 IEEE Conference on Open Systems (ICOS), pp. 78–83. IEEE (2014)
8. Khalfaoui, M., Molva, R., Gomez, L.: Secure alert tracking in supply chain. In: 2013 International Conference on Security and Cryptography (SECRYPT), pp. 1–11. IEEE (2013)
9. Ouafi, K., Vaudenay, S.: Pathchecker: an RFID application for tracing products in supply-chains. In: Proceedings of RFIDSec 2009. (LASEC-CONF-2009-006) (2009)

10. Blass, E.O., Elkhiyaoui, K., Molva, R., et al.: Tracker: security and privacy for RFID-based supply chains. In: 18th Annual Network and Distributed System Security Symposium, NDSS 2011, 6–9 February 2011 (2011)
11. Elkhiyaoui, K., Blass, E.O., Molva, R.: CHECKER: on-site checking in RFID-based supply chains. In: Proceedings of the Fifth ACM Conference on Security and Privacy in Wireless and Mobile Networks, pp. 173–184. ACM (2012)
12. Cai, S., Deng, R.H., Li, Y., et al.: A new framework for privacy of RFID path authentication. In: International Conference on Applied Cryptography and Network Security, pp. 473–488. Springer, Heidelberg (2012)
13. Wang, H., Li, Y., Zhang, Z., et al.: Efficient tag path authentication protocol with less tag memory. In: International Conference on Information Security Practice and Experience, pp. 255–270. Springer, Cham (2016)
14. Do, Q., Martini, B., Choo, K.K.R.: A data exfiltration and remote exploitation attack on consumer 3D printers. IEEE Trans. Inf. Forensics Secur. **11**(10), 2174–2186 (2016)
15. Do, Q., Martini, B., Choo, K.K.R.: Is the data on your wearable device secure? An Android Wear smartwatch case study. Softw. Pract. Exp. **47**(3), 391–403 (2017)
16. Do, Q., Martini, B., Choo, K.K.R.: Cyber-physical systems information gathering: a smart home case study. Comput. Netw. **138**, 1–12 (2018)
17. D'Orazio, C., Choo, K.K.R.: An adversary model to evaluate DRM protection of video contents on iOS devices. Comput. Secur. **56**, 94–110 (2016)
18. D'Orazio, C.J., Choo, K.K.R.: A technique to circumvent SSL/TLS validations on iOS devices. Future Gener. Comput. Syst. **74**, 366–374 (2017)
19. D'Orazio, C.J., Choo, K.K.R., Yang, L.T.: Data exfiltration from Internet of Things devices: iOS devices as case studies. IEEE Internet Things J. **4**(2), 524–553 (2017)
20. Krawczyk, H., Canetti, R., Bellare, M.: HMAC: keyed-hashing for message authentication (1997)
21. https://en.wikipedia.org/wiki/Integrated_Encryption_Scheme
22. Naor, M., Yung, M.: Public-key cryptosystems provably secure against chosen ciphertext attacks. In: Proceedings of the Twenty-Second Annual ACM Symposium on Theory of Computing, pp. 427–437. ACM (1990)
23. A Flurry of High-Memory Tags Take Flight. http://www.Rfidjournal
24. Fujitsu Develops World's First 64 KByte High-Capacity FRAM RFID Tag for Aviation Applications. http://www.fujitsu.com/global/about/resources/news/pressreleases/2008/0109-01.html

Research on the Two Stage Full Life Cycle Management of Police Equipment

Zhiwei Tang(✉), Lingling Hu, and Jianye Li

The Third Research Institute of Ministry of Public Security, Beijing, China
99chaoyang@163.com

Abstract. "Police Equipment" is a kind of public security equipment, In this article, the "life cycle" of the "Police Equipment" is analyzed and researched comprehensively. The life cycle is divided into two phases, Approved and fitted out product stage and procurement and usage stage. The article elaborates the definition and the effect of the two phases. The two stage life cycle management system of police equipment is also elaborated in this article.

Keywords: Police equipment · Approved and fitted out product · Life cycle

1 Introduction

With the automation of the national police information network completed, the public security Organs at all levels have also built their own local area network, the traditional paper page of type, copy type, circulation-style office model is gradually being replaced by the progressive realization of the sharing of resources, authority office efficiency has been greatly enhanced. However, in Police Equipment library management, yet be able to use the Internet to achieve the implementation of automated management systems. Units complicated the management methods and models, there is no standard set of standardized management system, resulting in waste of resources, office inefficiencies have been unable to meet the future information needs of the construction and implementation of the mandate, so that the information age of the network is not yet effective play performance. In order to improve the intelligent management level of the public security equipment, through the in-depth study of the status quo of the classification of business equipment system at home and abroad, combined with public security. On the basis of modern information technology, the existing equipment and procurement and used in the actual work are based on the modern information technology, and it is put forward and developed for the actual demand of the public security. The classification method and information management system of public security equipment system to meet the requirements of management at all levels, and break through the large number of public security equipment system. According to the key technology of intelligent application, the modern intelligent application of public security equipment is realized.

J. Abawajy et al. (Eds.): ATCI 2018, AISC 842, pp. 38–43, 2019.
https://doi.org/10.1007/978-3-319-98776-7_5

2 Definition of Police Equipment

Police equipment is a kind of public security equipment, mainly including command information communication, criminal technology, detection technology, weapons, protection technology, security disposal, mobility equipment, training and other equipment mounted police. The public security organs need to set up the form and install the police equipment, and the general equipment does not need to be set up.

3 The Limits of the Two Stage Full Life Cycle of Police Equipment

For the management of police equipment, we need to manage its whole life cycle. Its life cycle is divided into two stages, namely, the stage of "approved and fitted out product" and "procurement and usage". In the Approved and fitted out product "stage, "police equipment" is referred to as "product" for short, in the "procurement and usage" stage, "police equipment" is referred to as "equipment".

3.1 "Approved and Fitted Out Product" Stage

The "Approved and fitted out product" stage mainly means that the "product" has been developed and completed, but it is not included in the sequence of police equipment. The starting point of this stage is from the "stereotype application". At this point, the product begins to enter the process of stereotype, and officially begins its stage of "layout and assembly". It is called "Approved and fitted out product" stage, because of the "product" has not been included in the series of police equipment, police equipment when it is included in the sequence, its role is changed, it is no longer a single product, but given the classification number of police equipment only. "Police equipment" "Approved and fitted out product" stage is mainly from the "Application of the type" to the beginning, to the end of the "list of police equipment".

3.2 "Procurement and Usage" Stage

The period of "procurement and usage" mainly refers to the stage of operation of "police equipment" such as purchase, outputting, warehousing, using, maintenance and scrapping. The "police equipment" starts from purchase, and it will keep this identity until it is scrapped.

4 The Function and Effect of the Full Life Cycle of the Two Stage of Police Equipment

The whole "Approved and fitted out product" stage mainly includes the application of the stereotype, the test of the pattern, the test of the team, the trial of the team, the examination of the stereotype and the review of the assembly. Setting application refers to the provincial public security organs to the Ministry of organs for setting; type testing refers to the requirements and product performance testing third party inspection agency; team by test team to test whether the products meet the requirements of tactical evaluation, proposed the establishment, equipped with side all suggestions; team refers to some products required by the trial team the trial, the trial evaluation; review is a comprehensive review of stereotypes on the front of a number of work; review to decide whether included in the column mounted police equipment series.

The stage of "procurement and usage" mainly includes procurement, warehousing, use and maintenance, discarding and loss. The equipment that is included in the police equipment sequence is released into the public security opportunity after purchase, and the police equipment is released through warehousing and outgoing. The equipment that is not out of stock cannot be used. Use and maintenance means that the police equipment can be used and maintained many times until the loss is lost.

The two stage full life cycle schematic of police equipment, such as Fig. 1.

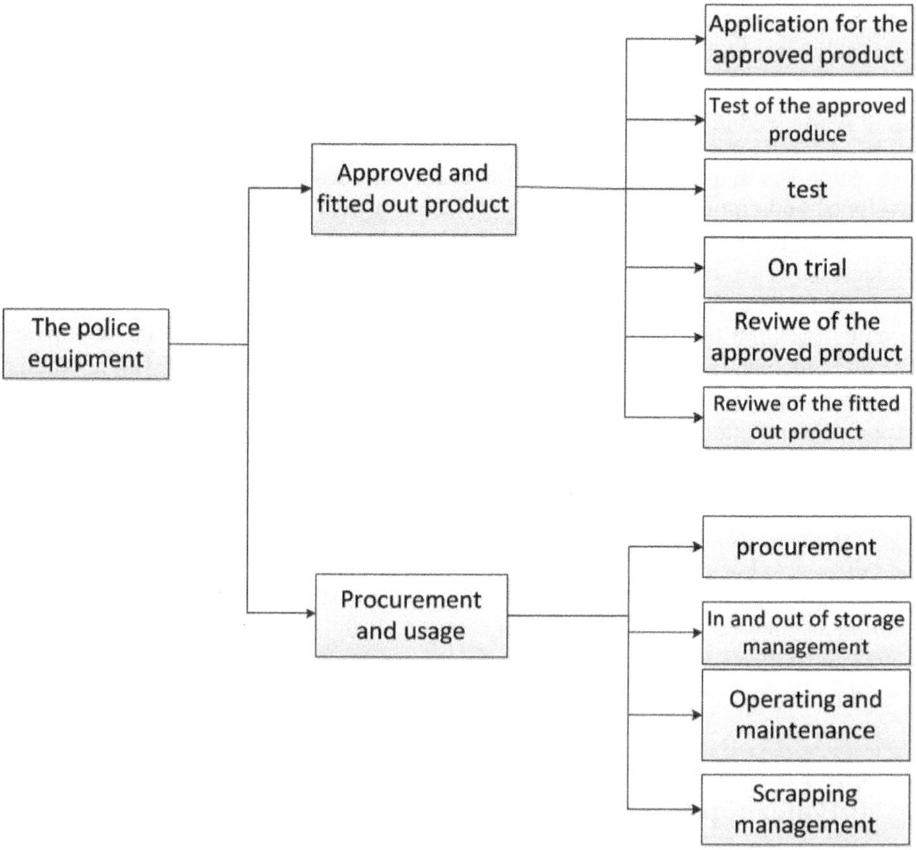

Fig. 1. Two stage full life cycle schematic of police equipment

The two stage full life cycle management flow chart of the police equipment, such as Fig. 2.

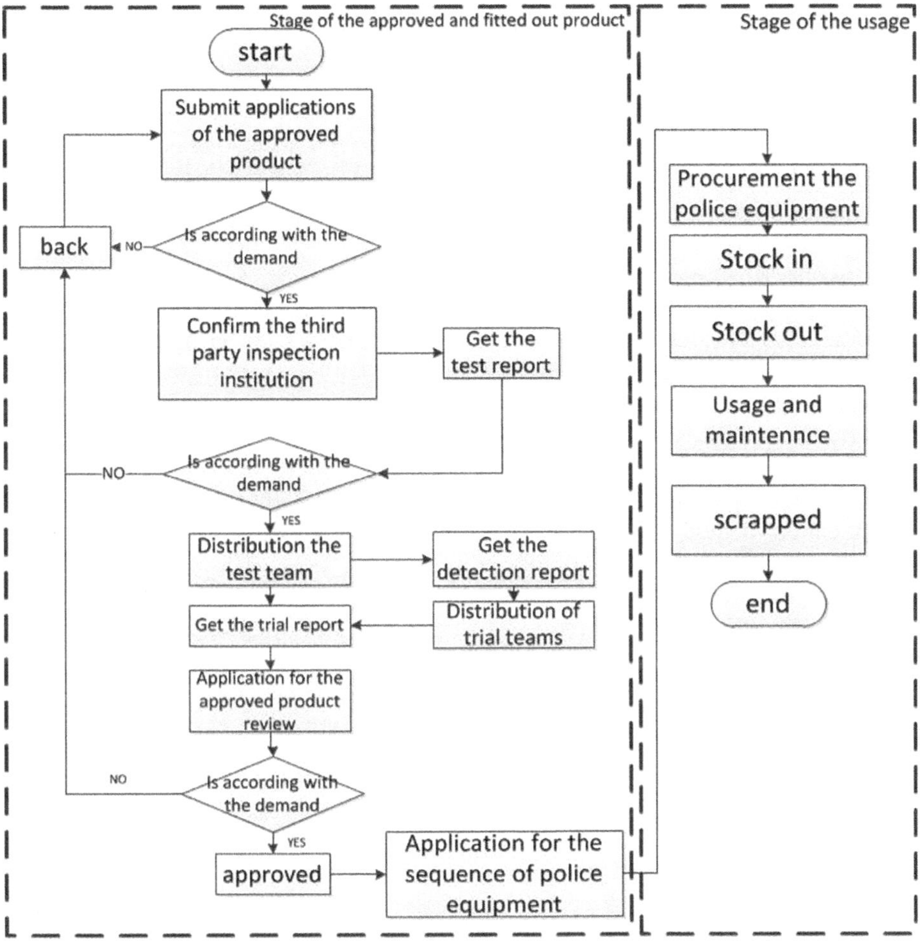

Fig. 2. Two stage full life cycle management flow chart of the police equipment

5 Two Stage Full Life Cycle Management System for Police Equipment

The two stage full life cycle management system of police equipment is the joint management of the two stages of "Approved and fitted out product stage" and "procurement and usage stage".

5.1 Equipment Cloud Platform

The system adopts the technology platform of cloud platform to uniformly store and equip all the public security information of the whole country or the whole province, so as to achieve efficient utilization of system resources and reduce the overall cost of the system operation. The use of cloud architecture has high availability and scalability, and it can deal with the problems of data scale growth, data fault tolerance, hardware failure and so on. Considering the system status of different provinces and cities, cloud platform supporting multi-level architecture can directly apply virtual applications on ministerial platforms, and also build upgraded cloud platforms, and achieve seamless integration of data.

5.2 Security Cloud Management

The equipment information system platform is run on the public security network and is separated from the Internet physics. The certificate entry mechanism is adopted in the equipment system. Users must log in with the uniform PKI issued by the police to avoid the misuse of personal user names and passwords in the management. A multi-level authorization system is built for the equipment classification system and management process to ensure the security of data.

Management platform based on CA digital certificate authentication intelligent key authentication method, through digital certificate, digital signature and other mechanisms fully guarantee the safety of the authentication process. The platform is based on the integration of resources, and through the unified authentication and access control of user identity, the service platform of single sign on in various application systems is realized.

Security cloud management architecture, such as Fig. 3.

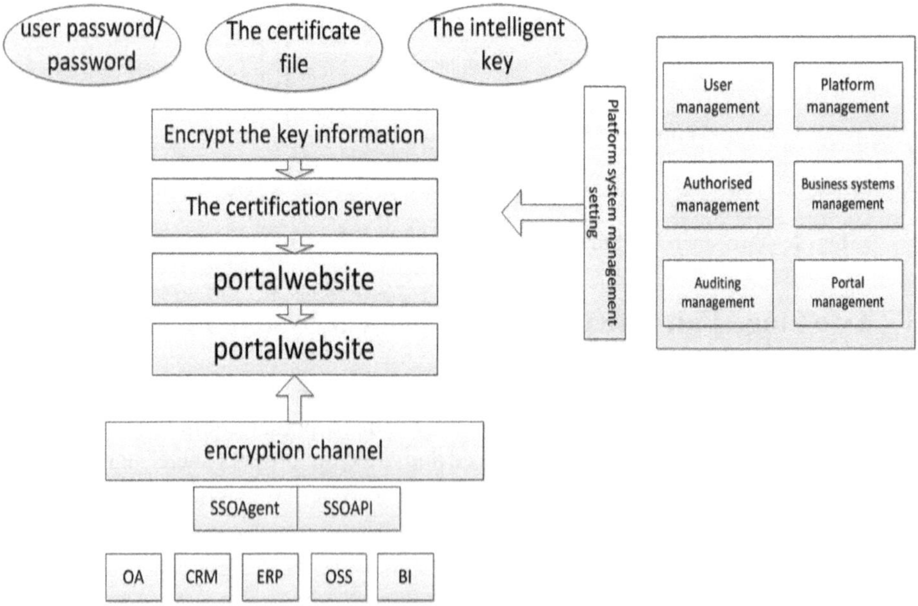

Fig. 3. Security cloud management architecture

5.3 Data Cloud Analysis

On the basis of cloud platform, we use cloud analysis technology to process massive data, that is to use hadoop and other big data processing technology to do data analysis. We emphasize that we can analyze and monitor all links of all kinds of equipment life cycle. It can not only generate the daily reports needed by the public security institutions at all levels, but also carry out the depth analysis according to the demand, and provide the data support for the leadership decision.

6 Summary

In this paper, the definition of "police equipment", the division of the two stage, the function and management system of the two stage are explained in detail. Realization of this system as a unit provides a highly efficient department, general store sales, accounting statistics and material basis for an integrated approach to information management, integrated management platform to improve the efficiency of the data sector management Police Equipment, is abroad prospect of practical application.

Acknowledgments. This work was financially supported by Funding project: the Ministry of public security technical research project project "Research on information system of public security equipment system" (2014JSYJA007).

References

1. On the application of Internet of things technology in the process of police equipment management in China. Chinese security prevention certification, (2), 22–26 (2017)
2. Police equipment market status and trend. http://www.21csp.com.cn/zhanti/2016jyzb/article/article_14635.html. 24 June 2016/11 Sept 2017
3. Police equipment stereotype is an important part of public security organs' equipment building. http://www.chinanews.com/gn/2015/12-02/7652388.shtml. 02 Dec 2015/01 Sept 2017
4. Frraiolo, D., Barkley, F., Kuhn, O.R.: A role based access control model and reference implementation within a corporate intranet. ACM Trans. Inf. Syst. Secur. **1**(2), 13–38 (1999)
5. Roman, E., Scott, A., Tyler, J.: Mastering Enterprise JavaBeans, vol. 4, pp. 13–88. Computer Publishing, New York (2002)
6. He, Y., Wang, Z.: Discussion on the safeguard mode of police equipment. Theor. Wide Angle, **1**(4), 243 (2014)

Research on Audio and Video Transmission Strategy for Android

Shaoqi Feng, Shiyang Song(✉), and Siliang Gong

Center of Internet of Things, Third Research Institute of Ministry of Public Security, Shanghai 201204, China
ssy_shu@126.com

Abstract. The Android operating system provides MediaRecoder and MediaPlayer objects for application-layer developers to implement audio and video applications. Although focused on audio and video recording and playing, these objects cannot directly be applied in real-time streaming video and audio transmission based on camera and microphone hardware. Through the deep research of Android operating system, a strategy was proposed in this paper on efficient and flexible audio and video transmission. Then a third-party stream transmission component libavstream was designed following this strategy. Finally, an App called BLife based on libavstream component was developed to test the availability and effectiveness of the proposed strategy.

Keywords: Android · Video and audio transmission · Live broadcast

1 Introduction

The application of 4G network and the popularity of smart phones help to promote the migration of traditional Internet business from PC to mobile phones. The live video broadcast based on the 3G/4G mobile network has the advantages of low cost, high flexibility, compared with the traditional satellite and microwave live broadcast. In US, live streaming media App Periscope developed by Twitter has being attracting a large number of smart phone users. People can use it to broadcast the field situation nearby and transmit the live videos to others. In China, UGC model Apps, from the earliest Qukan to the recent hottest live broadcast Douyu, have been flourished everywhere on the Internet.

Statistics show that 70%–80% of smart phone users chose to use the Android operating system. Since Android offers a wide range of application components, developers can easily use them to build their own applications in Java language. As for audio and video, Android provides an opencore multimedia framework to support all audio and video acquisition and broadcasting business. It also provides a MediaPlayer object, which supports streaming media protocols including RTSP, RTP, and HTTP, to play audio, video, and streaming media. In order to record audio and video, Android provides a MediaRecoder object for audio and video acquisition, encoding and packaging. However, MediaRecoder cannot be directly applied on streaming live business. Developers often need to repackage the packed data stream of audio and video data

J. Abawajy et al. (Eds.): ATCI 2018, AISC 842, pp. 44–55, 2019.
https://doi.org/10.1007/978-3-319-98776-7_6

before transmission, and this evidently reduces the efficiency of development process and greatly raises the development barrier [1].

2 Overall Architecture

Considering the fact that Android lacks audio and video stream transmission components, an efficient and flexible audio and video transmission strategy was designed and shown in Fig. 1. The transmission component Libavstream was also developed so as to allow developers to embed them within their Apps through simple reference implementation. The strategy is able to support the C/S broadcasting mode of RTMP, RTSP protocols and P2P mode of MPEG-TS stream format. Meanwhile, H.264 is adopted as video coding method and AAC as audio coding method. In addition, the strategy is compatible with nearly all kinds of Android devices, including smartphones, tablets, smart TV, smart glasses and smart vehicle devices [2].

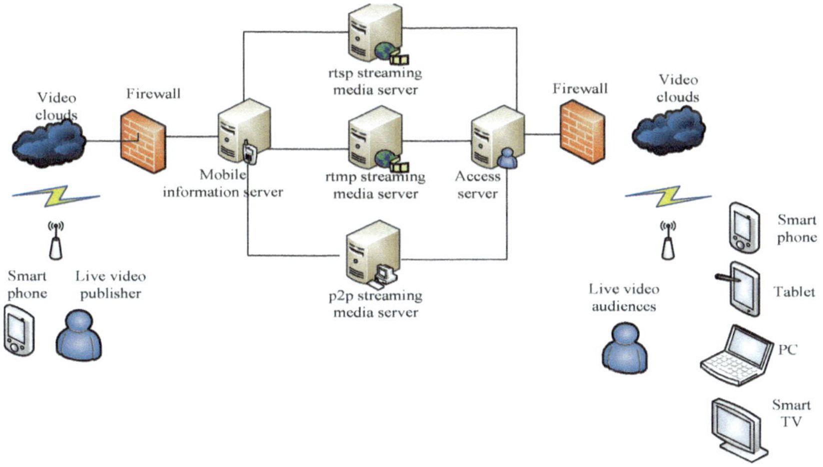

Fig. 1. Android-oriented audio and video stream transmission strategy

The implementation of the transmission strategy is divided into two sections: the Java JDK software part and the NDK C software part. The Java software part is mainly used for the configuration of stream parameters, the collection of the raw video and audio data and the opening/closing of stream coding threads. The NDK C software coding is responsible for the encoding, packaging and stream transmission of raw audio and video data and the interaction with the server.

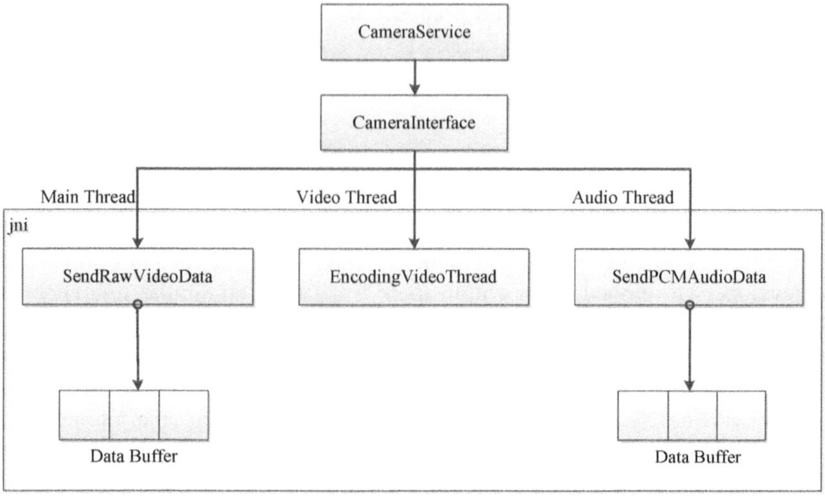

Fig. 2. Software architecture of transmission strategy on JDK layer

3 Software Architecture Based on JDK

As shown in Fig. 2, a Service component called CameraService, which is responsible for audio and video stream transmission, was designed for developers to be called in their background programs. The CameraInterface class was designed to package all the business logic functions for the CameraService to call. Hence in the CameraInterface Class, three threads are activated successively: the audio collection thread, the stream main thread, and the Service main thread. The Service main thread is responsible for setting user parameters, opening cameras and collecting raw video data. The video data are collected by the callback function onPreviewFrame in the Camera object, and the jni interface function is invoked by onPreviewFrame to transmit the collected raw video data into the pre-set video data buffer [3]. The audio collection thread and the stream main thread should be activated before video data collection by the service main thread. In the audio collection thread, the read method of AudioRecord object is adopted to collect raw audio data, and then the jni interface function SendPCMAudioData will be invoked to transmit the collected raw audio data into the pre-set audio data buffer. The stream main thread is only activated after both the audio and video data are transmitted into the buffer. The jni interface function EncodingVideoThread is then invoked by the stream main thread to open the audio and video data encoding, packaging and streaming work in the NDK layer.

In summary, three threads are opened in the transmission strategy of Java layer software architecture: the Service main thread, the raw audio data collecting thread and the stream main thread. Three jni interface functions are invoked respectively in the NDK layer, namely SendRawVideoData, SendPCMAudioData and EncodingVideoThread.

4 NDK Based Software Architecture

The transmission strategy in the NDK layer is responsible for the encoding, packaging and stream transmission of the raw video and raw audio data. In order to improve the efficiency of software development, the API functions in the FFmpeg cross-platform multimedia library are adopted to process all the audio and video data [4]. In order to implement efficient encoded transmission of audio and video data, three tasks: the encoding of raw audio data, the encoding of raw video data and the packaging and streaming work of audio and video encoded packages, are designed to work in parallel with each other in three threads. Technical details are presented respectively in the following [5] (Fig. 3).

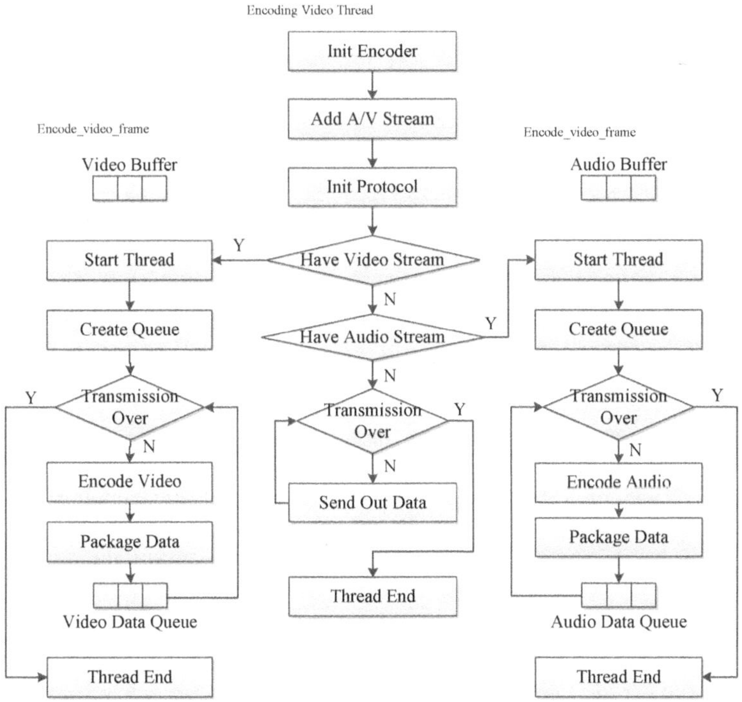

Fig. 3. NDK layer software architecture of transmission strategy

4.1 Implementation Process Based on FFmpeg

FFmpeg, which provides developers the best solution for multimedia application, is mainly used for encoding, transcoding, multiplexing/de multiplexing, streaming, filtering and playing of audio and video. FFmpeg library contains libavcodec, libavutil, libavformat, libavfilter, libavdevice, libswscale, libswresample, etc. Developers can refer to these library-related SDK in their own applications to realize audio and video processing functions [6].

Libavformat and libavcodec in FFmpeg library are used in the above transfer strategy, including libavcodec audio frequency and video coding. As for audio coding, libavcodec and FDK AAC third-party code library are adopted to package raw PCM audio data in the JDK level into AVpacket format, and then the packaged data are transmitted into audio data queue; In video coding, libavcodec and x264 third-party code library are adopted to encode video data of yuv420p format in JDK layer, and then the data are transmitted into video data queue. Another multimedia library libavformat is mainly used for packaging and streaming of encoded audio and video data. In order to support streaming protocols, three ways are provided. The first is the MPEGTS format designed for the "point to point" transmission mode, in this mode the audio and video data packaged in MPEGTS format are transmitted to the specified ports of the network destination node by UDP or TCP protocols. The second is the RTSP protocol designed for one-to-many broadcast model. The packages are pushed to the RTSP server, and the client then implements audio and video broadcast services by connecting RTSP server. The third way employs the same broadcast mode as the second one, the difference is that the client in the third way must support RTMP protocol and adobe flash media player.

It is well known that Android applications run in a virtual machine called Dalvik. Native Development Kit (NDK) allows developers to use native codes such as C/C++ to execute part of the programs, therefore in the Android operating system NDK needs to be compiled and optimized for specific mobile device processors if FFmpeg is used. For ARM processors, developers often need to open the neon multimedia instruction set to adapt the optimization of the language in FFmpeg multimedia library [7].

4.2 Sampling and Coding of Audio and Video Data

In the transfer strategy, the sampled raw audio and video data are transmitted by JDK layer through jni interface functions into the audio and video data buffer created by NDK layer. Then two threads in the NDK layer, video coding thread "encode_video_frame" and audio coding thread "encode_audio_frame", read data from the buffer as soon as possible to prevent the new data covering the old. It is noteworthy that the video frame coding process consumes a large amount of computing power and time, so the encoding of videos with high definition (1080p, 30 FPS) or with complex video coding algorithm (using h. 264 High Profile or h. 265 video encoder), are possibly slowed down in coding thread of NDK layer with some lost frames, while the speed of the video data sampling in the JDK layer remains normal. So developers should be advised to conduct reasonable assessments and settings according to the maximum resolution and frame rate of the target device.

The raw video data in yuv420p format are sampled frame by frame by the video encoding thread "encode_video_frame". After the compression in the encoder, the generated packages in AVpacket format are transmitted into video store queue in single-linked list form [8]. The transmission strategy using h. 264 encoder is responsible for the coding and compression of video frames. Here the encoding complexity of the encoder parameters is presented so that developers can freely configure the parameters in the low, normal or high quality level so as to meet the multiple needs of users.

The audio coding thread "encode_audio_frame" samples and packages the specified sampling frames each time in the audio buffer (the number of sampling frames of the AAC encoder in the transfer strategy is 1024), and the packaged frames are transmitted into the audio data storage queue in the single-linked list form. The audio data encoding process is relatively simple by using the mono transmission strategy.

4.3 Audio and Video Synchronization Based on Time Stamp

It is known that audio and video data are sampled in different frequencies in the mobile streaming media service, and that the video data occasionally lost frames, so audio and video data must be synchronized. The relatively weak CPU processing ability and insufficient memory storage of mobile devices lead to the adoption of the audio and video synchronization processing mechanism based on the time stamp in this paper. In the Android system, since audio sampling is more accurate than video sampling, it is preferable that video data be synchronized to audio data

In the FFmpeg, audio and video data are synchronized by the so-called PTS (the Presentation Time Stamp) and DTS (Decode Time Stamp) Time Stamp. In this paper, PTS is calculated to implement audio and video synchronization, and the calculation method of audio timestamp is shown in the following formula:

$$\text{audioframe\_pts} += \text{src\_nb\_samples} * \text{c\_channels} \quad (1)$$

The audioframe_pts is the timestamp for audio frame before encoding, the src_nb_samples is the number of audio data samples for audio frame, and the c_channels is the number of audio channels. Since the raw audio frames in the AVFrame format are transferred into AVpacket format, so the timestamps need to be recalculated in timestamp function av_packet_rescale_ts (&PKT, c-> time_base, st-> time_base) in the FFmpeg, where PKT is encoded data packets, c-> time_base serves as the time base value of the encoder, and st-> time_base stands for the time base value of the audio stream. The practical meaning of the function is that the encoder time base is converted into audio stream time base, so the time stamp needs to be recalculated.

Video frame timestamp calculation is roughly the same as the audio frame process; the difference is that timestamps of video frames should be calculated through more precise timing due to the lost video frames. The key codes are presented as follows [9]:

```
AVFrame *frame;
AVPacket pkt = { 0 };
int64_t frame_timebase, i_previous, i_time, left_time;
i_previous = 0;

for(/*adds stop condition here*/ ){
 i_time = x264_mdate();
 if (i_previous != 0 && (i_time - i_previous) < frame_timebase )return -1;
 if (i_previous){
   left_time = i_time - i_previous;
   frame->pts += left_time/frame_timebase;
   left_time = left_time % frame_timebase;
 }
 i_previous = i_time;
 i_previous -= left_time;
 //adds video frame encoding codes here
 av_packet_rescale_ts (&pkt, c->time_base, st->time_base);
}
```

The current time can be acquired by invoking the timing function x264_mdate () in an accuracy of microsecond. The above variable left_time is the interval between the current video sampling and the last sampling, and the incremental value of the current timestamp can be acquired by dividing left_time by the video frame time base value frame_timebase. Meanwhile left_time should be deducted from the i_previous (the current time). Finally, the real time is calculated through a timestamp of audio and video data packets, if the video time is less than the audio time, then video packets should be written into, and vice versa. The whole process ensures the accurate synchronization of audio and video data under the condition of lost video frames.

4.4 Multithreaded Synchronization Processing

As shown in Fig. 2, multi-thread parallel processing method is adopted in the NDK layer of the transmission strategy to achieve the efficient coding and stream transmission of audio and video. Audio coding thread and (or) video coding threads will be opened only after the initialization of configuration parameters from the JDK layer is finished by the stream main thread EncondingVideoThread. Stream main thread enters into the circulation state to wait for the data in the video/audio encoding data storage queue. Then once there are packaged data in either queue, stream transmission is activated based on the time stamp.

The synchronization between stream main thread and audio and video encoding thread is realized by the collaboration of the mutex pthread_mutex_t and the semaphore pthread_cond_t in threads library of pthread. In audio and video coding, function pthread_mutex_lock locks the mutex; when the encoding data are packaged into the queue, function pthread_cond_signal is activated to send a signal to notify the stream main thread to fetch packages, and then function pthread_mutex_unlock is used to unlock the mutex. In the stream main thread, function pthread_mutex_lock is firstly used to lock the mutex, then pthread_cond_wait is invoked to block threads, and pthread_mutex_unlock is used to unlock the mutex after packages are fetched out of audio and video queue.

5 Implementation and Calls

5.1 Strategy Implementation

An Android third-party library libavstream was designed on the basis of the above transmission strategy, which is perfectly fit for Android Studio development environment. The library's source code has been released on the Git Repository "code cloud", and the open source project address is: https://git.oschina.net/iavstream/BLife_libavstream.git.

The released version is built on Gradle in the jcenter warehouse, which makes it easy for developers to use. The project address is: https://bintray.com/iavstream/maven/libavstream/

Version 1.1.0 is released at present, the project contains four files: Libavstream-1.1.0-Javadoc.jar; Libavstream-1.1.0-sources.jar; Libavstream-1.1.0. aar; Libavstream-1.1.0. pom.

The first one, javadoc files in libavstream library, can be used as the API documentation. The second one works as the source file package, which contains all the source codes. The third provides aar library files which can be automatically built into the applications through the Gradle. The last file stores the descriptive information for the project file.

5.2 Developer Call

5.2.1 Dependence Settings

If libavstream is needed for App developers who use Android Studio to realize audio and video stream transmission, two files (both named build.gradle) should be configured in Gradle Scripts. One is in the Project; the other is in the Module. The function of the file build.gradle in the Project is to set the remote warehouse source as jcenter. In the other build.gradle file, the most important thing is to add "com.iavstream:libavstream:1.1.0" to its dependences. Then during construction libavstream library files can be obtained from the jcenter warehouse by Gradle to complete compile and link work in the application.

5.2.2 Permission Settings

A permission scheme is defined in Android to protect equipment resources and functions, so all applications must declare their permissions in the file AndroidManifest.xml. As developers, the libavstream library is presented in the form of Service for users to call, so declarations have to be made in AndroidManifest.xml. Add the following statement in the XML file:

```
<service
android:name="com.iavstream.libavstream.CameraService"
android:enabled="true" >
<intent-filter>
<action android:name="com.iavstream.libavstream.CameraService" />
</intent-filter>
</service>
```

After the declaration, methods provided by the CameraService in related source files can be used for live audio and video transmission. As for the relevant permissions of CameraService, which have been stated in the libavstream library, there is no need to repeat permission statements in the application. Libavstream permissions are as follows:

```
<manifest xmlns:android="http://schemas.android.com/apk/res/android"
package="com.iavstream.libavstream">
<uses-feature android:name="android.hardware.camera" />
<uses-feature android:name="android.hardware.camera.autofocus" />
<uses-permission android:name="android.permission.INTERNET" />
<uses-permission android:name="android.permission.RECORD_AUDIO" />
<uses-permission android:name="android.permission.CAMERA" />
</manifest>
```

5.2.3 API Function Call

CameraService is called to implement audio and video stream transmission services based on the camera and microphone. Three procedures are executed to implement this process. Firstly, related parameters configuration is required, especially for the assignment of MediaParameter. Secondly, the ServiceConnection object is to be defined to realize the interactions between applications and the CameraService object. Finally, the connection/disconnection function is defined to open or close the stream transmission service.

6 Implementation Results

In order to get a better evaluation and validation of the audio and video stream transmission strategy, a live audio and video App Blife was designed based on libavstream libraries. BLife allows smart phone users in mobile network to publish real-time information through audio and video broadcasts, to send messages and invite their friends to watch live broadcasts, and to share things around with friends. The application also allows users to adopt "point to point" mode to send live audio and video to the designated ports of the network node, and this provides a good solution to audio and video transmission in VPN network (Fig. 4).

Fig. 4. Blife

6.1 BLife Introduction

The BLife App integrates the functions of live broadcast and watching, so users can either invite others to watch their self-launched live broadcasts, or watch live broadcasts of friends after receiving an invitation message. BLife also supports separate audio broadcast, video broadcast, and live video and audio at the same time, which makes it easy and flexible to satisfy the requirements of users in different situations. Related parameters of the configuration can be configured in the configuration pages in the App, shown in Fig. 5. The open source project of the App lies in this address: http://git.oschina.net/iavstream/BLife_libavstream.

6.2 Performance Tests

BLife is designed on the basis of libavstream libraries to support broadcast mode and point to point transmission mode: (1) To test the broadcast model, the Darwin Streaming Server needs to be installed in advance to use SMS live broadcast, make sure

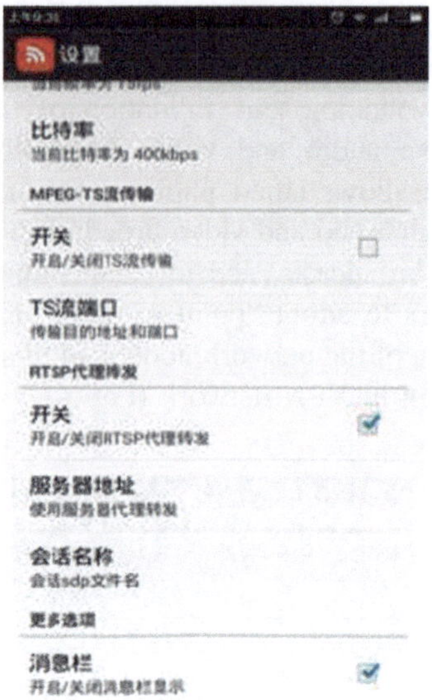

Fig. 5. Blife transfer mode settings

Fig. 6. Blife in live broadcasts

the “RTSP agent transfer” switch is selected in Parameter Settings as shown in Fig. 5, then set the address of the Server and the name of the session; (2) To test the point-to-point transmission mode, select the “MPEG - TS stream transfer” switch in Parameter Settings in Fig. 5, then set the destination address and port information of the transmission nodes.

Transmission tests were conducted in broadcast mode using RTSP model and point-to-point mode respectively under the condition of 640 * 480 @ 15 FPS video configuration and mono audio configuration. The experiment results showed that when

China Unicom's 3G network was connected in RTSP broadcast model, the output bit rate was 500 Kbps in the receiving end, the video transmission delay was 3 s, and the point-to-point mode video delay was 1 s. Results proved that the App can completely meet the needs of a variety of solutions for mobile audio and video transmission (Fig. 6).

7 Conclusions

A flexible audio and video transmission strategy was proposed for Android developers, and a standard audio and video stream transmission component libavstream was designed according to this strategy. Developers can employ the strategy in their own applications to realize fast audio-video transmission functions. Then BLife, a live audio and video application using libavstream, was designed to realize both broadcast transmission and point-to-point transmission of the audio and video. Experiment tests proved the practical value of this strategy because of its improvement in the audio and video application efficiency and its reduction of the costs of development.

Acknowledgment. The authors of this paper are members of Shanghai Engineering Research Center of Intelligent Video Surveillance. This work was supported in part the National Natural Science Foundation of China under Grant 61300202. Our research was sponsored by Program of Science and Technology Commission of Shanghai Municipality (No. 15530701300, 15XD15202000, 16511101700), in part by the technical research program of Chinese ministry of public security (2015JSYJB26).

References

1. Dong, B.: Key Technology of Real-Time Video Transmission in Mobile Terminal. Beijing University of Chemical Industry, Beijing (2015)
2. Lu, C.: Design of Real-Time Video Transmission System Based on Android. The Northern Industrial University, Beijing (2016)
3. Wang, Q.: End-to-End Real-Time Wireless Video Transmission System Based on Android. Nanjing University of Posts and Telecommunications, Nanjing (2013)
4. Wang, X.: Analysis of the passing of parameters in the NDK programming in Android. J. Fujian Comput. **12**, 156 (2014)
5. Dong, J., Xin, J., Lian, J.: Research on h. 264 video broadcast technology based on Android system. J. TV Technol. **33**(4), 11–15 (2015)
6. Shi, Y., Xuan, K., Zhang, H., Sun, Z.: JNI technology application in converged communication terminal in Android. J. Hangzhou Dianzi Univ. (Natural Science Edition) **11**, 41–44 (2015)
7. Dong, W.: JNI technology application in the network interaction. J. Electron. Des. Eng. **06**, 125–127 (2016)
8. Hua, Y.: Research on FFmpeg Coding Optimization in Android Multimedia Platform. South China University of Technology, Guangzhou (2014)
9. Hu, C., Ren, P., Li, W.: Research on FFmpeg multimedia synchronous transmission algorithm based on the Android system. Comput. Technol. Dev. **10**, 85–87 (2011)

Emojis Sentiment Analysis Based on Big Social Media Data

Dongdong Li, Xiangfeng Luo(✉), Xiao Wei, and Ruirong Xue

School of Computer Engineering and Science, Shanghai University, Shanghai, China
Dongdongli@i.shu.edu.cn, luoxf@shu.edu.cn

Abstract. As the use of emojis becomes ever-more prevalent in social media texts (e.g. twitters, microblogs), various kinds of work on emojis sentiment analysis have been carried out. However, there is little work calculating emojis sentiment quantificationally. This paper proposes an improved label propagation algorithm (LPA) called label attenuation propagation model (LAPM) to calculate emojis sentiment automatically, multi-dimensionally, and quantificationally. First, Emoji Link Network (ELN) is built to organize a huge number of emojis in social media. Second, LPA is introduced into the area of emojis sentiment analysis. Third, we measure the sentiment uncertainty of emojis and propose LAPM to calculate emojis sentiment. Experimental results show that the accuracy of LAPM is 85.2%, compared with 74.8% of LPA.

Keywords: Emojis sentiment analysis · Label propagation algorithm
PAD sentiment model

1 Introduction

Nowadays, emojis (e.g. 😂, 😭) are widely used in social media texts. Thus, various kinds of work on emojis sentiment analysis have been carried out. However, Existing approaches have the following limitations: (1) Existing methods limited emojis sentiment to negative, positive or neutral, which was hard to express human complex feelings. (2) Existing approaches rarely took into account the relations between emojis, leading to information loss between emojis. (3) Most of existing methods ignored the sentiment uncertainty of emojis. (4) There was little work calculating emojis sentiment automatically, multi-dimensionally, and quantificationally.

Aiming at the limitations mentioned above, this paper proposes LAPM to calculate emojis sentiment. (1) Six-dimensional sentiment (Love, Joy, Anger, Sad, Fear, Surprise) proposed by psychologist Parrot [9] is adopted to describe emojis. (2) ELN is built to organize a huge number of emojis in social media. (3) We measure the sentiment uncertainty of emojis based on the PAD three-dimensional sentiment model. (4) LPA is used to calculate emojis sentiment automatically, multi-dimensionally, and quantificationally. Considering the sentiment uncertainty of emojis, LAPM is proposed to improve the accuracy of emojis sentiment analysis.

J. Abawajy et al. (Eds.): ATCI 2018, AISC 842, pp. 56–63, 2019.
https://doi.org/10.1007/978-3-319-98776-7_7

2 Related Worked

With the booming of emojis, sentiment analysis based on emojis is becoming more and more popular. Novak et al. [1] have ranked emojis sentiment with negative, neutral and positive based on 1.6 million tweets labeled by 83 human annotators recruited. Miller et al. [2] asked people's interpretations of the most popular emojis to explore whether emojis' meaning is different across platforms. Dimson [3] and Eisner et al. [4] used word2vec to interpret emojis. Guibon et al. [5] aimed to build an emoji automatic recommendation system based on classification model. Xie et al. [6] recommended appropriate emojis automatically based on the contextual information. In the application, Jiang et al. [7] computed the sentiment of news events based on emojis. Hallsmar et al. [8] used emoji training heuristic for multi-class sentiment analysis.

3 Proposed Method

The process of emojis sentiment calculation by our method is shown as follows. (1) Take the relations between emojis into consideration, and then build ELN. (2) Select a few emojis as seeds, and then assign seeds with six-dimensional sentiment. (3) Update the sentiment of seed emojis involved in emojis collection with seeds. (4) LAPM uses emojis collection to calculate emojis sentiment, and then get a new emojis collection. We stop it only when emojis' sentiment is convergent, otherwise jump to step (3).

3.1 Building Emoji Link Network Based on Microblogs

Posting a microblog, users usually use several kinds of emojis to express some kind of emotion. Based on this, we propose the relation assumption, emojis that appear in the same microblog have similar sentiment.

According to relation assumption, we build Emoji Link Network (ELN) with emojis and confidence between different kinds of emojis.

$$\mathrm{ELN} = \langle \mathrm{E}, \mathrm{R} \rangle$$

where E is the set of emojis. R is the set of inner-emoji relations (such as  $\leftrightarrow$ ). If $\mathrm{E} = \{e^j | 1 \leq j \leq n\}$, the confidence that emoji e^j to e^k is denoted as $\mathrm{Conf}(e^j \rightarrow e^k)$. Like association link network (ALN) [10], ELN can be constructed automatically.

3.2 Emojis Sentiment Uncertainty Measurement

A survey using 304 participants to complete 15 emojis interpretations demonstrates that per emoji with an average of 37 interpretations [2]. So, emojis have sentiment uncertainty. Emojis with uncertain sentiment can bring the error to other emojis in propagation. In order to decrease the error, we need to measure the sentiment uncertainty of emojis. The PAD three-dimensional sentiment model proposed by psychologist Mehrabian [11] is used. PAD sentiment model is composed of P (Pleasure-displeasure), A (Arousal-nonarousal), and D (Dominance-submissiveness).

In order to get the PAD value of Love, Joy, Anger, Sad, Fear, and Surprise, we refer to the research of Zhang et al. [12] as follows:

Table 1. The PAD of five sentiment words

Sentiment words	The PAD of sentiment words		
	P	A	D
Happy	0.55	0.24	0.28
Surprise	0.34	0.34	0.04
Sad	−0.18	0.03	−0.14
Scared	−0.19	0.26	−0.13
Anger	−0.40	0.22	0.12

Joy (Happy), Surprise, Sad, Fear (Scared), and Angry are included in Table 1, whereas Love is not included. Because events of celebration usually contain the sentiment of Love and Joy, catastrophic events usually contain the sentiment of Love and Sad. We donate the P of Love as:

$$P(\text{Love}) = \left\lceil \frac{(P(\text{Joy}) + P(\text{Sad})) \times 100}{2} \right| \div 100 = 0.19 \quad (1)$$

According to the experience, we set A(Love) = 0.05, D(Love) = 0.11. In summary, we get the PAD of six-dimensional sentiment as follows (Table 2):

Table 2. The PAD of six sentiment words

Sentiment words	The PAD of sentiment words		
	P	A	D
Joy	0.55	0.24	0.28
Surprise	0.34	0.34	0.04
Love	0.19	0.05	0.11
Sad	−0.18	0.03	−0.14
Fear	−0.19	0.26	−0.13
Anger	−0.40	0.22	0.12

The six-dimensional sentiment of emoji e^j is denoted as:

$$\begin{aligned} &\text{Sent}_{ej} = \left\{ \text{Sent}_{ej}^{k} \,|\, 1 \leq k \leq 6 \right\} \\ &\text{s.t.} \quad 1 \leq j \leq n, \; e^j \in E \end{aligned} \quad (2)$$

for example $\mathrm{Sent}_{\mathrm{ej}}^{2}$ is the sentiment value of "Joy".

The sentiment uncertainty of emoji e^{j} is defined as follows:

$$\mathrm{Uc}\left(\mathrm{e}^{\mathrm{j}}\right)=\sum_{\mathrm{r}=1}^{\mathrm{k}-1}\sum_{t=\mathrm{r}+1}^{\mathrm{k}}\left|\mathrm{PA}(\mathrm{Sent}_{\mathrm{ej}}^{\mathrm{r}})-PA\left(\mathrm{Sent}_{\mathrm{ej}}^{\mathrm{t}}\right)\right|\cdot\left(\mathrm{Sent}_{\mathrm{ej}}^{\mathrm{r}}\cdot\mathrm{Sent}_{\mathrm{ej}}^{\mathrm{t}}\right) \quad (3)$$
$$\text{s.t.}\quad \mathrm{k}=6,\quad 1\leq \mathrm{j}\leq \mathrm{n},\quad \mathrm{e}^{\mathrm{j}}\in \mathrm{E}$$

where k is the dimensions of sentiment. $\mathrm{PA}\left(\mathrm{Sent}_{\mathrm{ej}}^{\mathrm{r}}\right)$ is the measure of $\mathrm{Sent}_{\mathrm{ej}}^{\mathrm{r}}$.

$$\mathrm{PA}\left(\mathrm{Sent}_{\mathrm{ej}}^{\mathrm{r}}\right)=\mathrm{P}(\mathrm{Sent}_{\mathrm{ej}}^{\mathrm{r}})\cdot\exp\left(A\left(\mathrm{Sent}_{\mathrm{ej}}^{\mathrm{r}}\right)\right) \quad (4)$$

where $\mathrm{P}(\mathrm{Sent}_{\mathrm{ej}}^{\mathrm{r}})$ is the P value of $\mathrm{Sent}_{\mathrm{ej}}^{\mathrm{r}}$, for example $\mathrm{P}(\mathrm{Sent}_{\mathrm{ej}}^{2})=\mathrm{P}(\mathrm{Joy})=0.55$. $A(\mathrm{Sent}_{\mathrm{ej}}^{\mathrm{r}})$ is the A value of $\mathrm{Sent}_{\mathrm{ej}}^{\mathrm{r}}$, for example $\mathrm{A}(\mathrm{Sent}_{\mathrm{ej}}^{2})=\mathrm{A}(\mathrm{Joy})=0.24$.

3.3 Label Attenuation Propagation Model

The main ideas of LAPM: Give emojis with certain sentiment greater permission to propagate and reduce the spread of emojis with uncertain sentiment.

We give some seeds emojis six-dimensional sentiment and donate it as matrix $\mathrm{Y_L}$. We give 0 to other emojis and donate it as matrix $\mathrm{Y_U}$. We set matrix $\mathrm{F}=[\mathrm{Y_L}:\mathrm{Y_U}]$.

Construct a $\mathrm{n}\times\mathrm{n}$ probability transfer matrix $\mathrm{P_{n\times n}}$:

$$P_{n\times n}=\begin{pmatrix} 0 & Conf(\mathrm{e}^2\rightarrow\mathrm{e}^1)\ldots & Conf(\mathrm{e}^n\rightarrow\mathrm{e}^1) \\ Conf(\mathrm{e}^1\rightarrow\mathrm{e}^2) & 0 & Conf(\mathrm{e}^n\rightarrow\mathrm{e}^2) \\ \vdots & \ddots & \vdots \\ Conf(\mathrm{e}^1\rightarrow\mathrm{e}^n) & Conf(\mathrm{e}^2\rightarrow\mathrm{e}^n)\cdots & 0 \end{pmatrix} \quad (5)$$

We calculate the sentiment uncertainty of emojis based on the matrix $\mathrm{F}=[\mathrm{Y_L}:\mathrm{Y_U}]$ and donate it as UNC:

$$\mathrm{UNC}=\mathrm{GetUc(F)}=\begin{bmatrix} Uc(\mathrm{e}^1) \\ \vdots \\ Uc(\mathrm{e}^{\mathrm{n}}) \end{bmatrix} \quad (6)$$

We donate the average of $\mathrm{Uc}\left(\mathrm{e}^{\mathrm{i}}\right)$ in UNC as AVG:

$$\mathrm{AVG}=\mathrm{GetAvg(UNC)}=\frac{\sum_{i=1}^{n}\mathrm{Uc}(\mathrm{e}^{i})}{\mathrm{n}} \quad (7)$$

We define the propagation capacity of emoji e^j as $\exp((AVG - Uc(e^j)) * \alpha)$. Then, we get a $n \times 1$ matrix PC which represents the propagation capacity of emojis.

$$PC = GetPC(UNC, AVG) = \begin{bmatrix} \exp((AVG - Uc(e^1)) * \alpha) \\ \vdots \\ \exp((AVG - Uc(e^n)) * \alpha) \end{bmatrix} \tag{8}$$

where α is a parameter which can affect the propagation capacity of emojis.

We define matrix multiplication function as follows:

$$M = Multi(A, B) = \begin{pmatrix} |A_{11} \times B_1| & \cdots & |A_{1m} \times B_1| \\ \vdots & \ddots & \vdots \\ |A_{n1} \times B_n| & \cdots & |A_{nm} \times B_n| \end{pmatrix} \tag{9}$$

We define the matrix subtraction function Sub() as follows:

$$S = Sub(A, B) = \begin{pmatrix} |A_{11} - B_{11}| & \cdots & |A_{1m} - B_{1m}| \\ \vdots & \ddots & \vdots \\ |A_{n1} - B_{n1}| & \cdots & |A_{nm} - B_{nm}| \end{pmatrix} \tag{10}$$

To make the value of emojis sentiment between 0 and 1, normalized function Nor() is defined as follows:

$$F_{n \times m} = Nor(F_{n \times m}) = \begin{pmatrix} \frac{F_{11}}{Max(F_1)} & \cdots & \frac{F_{1m}}{Max(F_1)} \\ \vdots & \ddots & \vdots \\ \frac{F_{n1}}{Max(F_n)} & \cdots & \frac{F_{nm}}{Max(F_n)} \end{pmatrix} \tag{11}$$

where $Max(F_i)$ is the maximum in the i-th row of $F_{n \times m}$.

We define function Maximum() to get the maximum from matrix:

$$Diff = Maximum(S_{n \times m}) \tag{12}$$

where Diff is the maximum in $S_{n \times m}$.

In summary, LAPM can be summarized as Algorithm 1:

Algorithm 1 Label Attenuation Propagation Model

Input: the initial six-dimensional sentiment matrix of emojis $F = [Y_L : Y_U]$. Probability transfer matrix $p_{n\times n}$. The maximum error δ.

Output: the final six-dimensional sentiment matrix of emojis $F = [Y_L : Y_U]$.

```
1.  do
2.      UNC=GetUc(F)
3.      AVG = GetAvg(UNC)
4.      PC=GetPC(UNC, AVG)
5.      M_{n×6} = Multi(F, PC)
6.      F_1 = [Y_{L1} : Y_{U1}] = P_{n×n} M_{n×6}
7.      Nor(F_1)
8.      F_1 = [Y_L : Y_{U1}]
9.      S_{n×6} = Sub(F, F_1)
10.     Diff = Maximum(S_{n×6})
11.     F = F_1
12. While( Diff >δ)
13. Return F
```

4 Experiment

4.1 Dataset

To validate the effectiveness of LAPM and LPA, we crawl six datasets including more than 70,000,000 microblogs. The number of topic and inner-emoji relations on six datasets.

4.2 Evaluation Method

In order to validate the accuracy of LAPM and LPA, we take emojis' six-dimensional sentiment acquired by Jiang et al. [7] as standard and use an important evaluation metric: the accuracy at top N.

4.3 Experiment on LPA and LAPM

LPA and LAPM need some labeled data as seeds, 20 emojis with relatively certain sentiment are selected and given six-dimensional sentiment. We take Least Time (LT) as threshold to filter out inner-emoji relations which appears infrequently in microblogs. Some experiments are conducted to pick α and we take α as 8 on our datasets.

Comparing with DATA-1 and DATA-2, the accuracy on DATA-5 and DATA-6 goes up first and then goes down because they have less inner-emoji relations. Due to many invalid inner-emoji relations are filtered out when LT goes from one to five, the accuracy on six datasets increases sharply. The best accuracy of LAPM is 13.9% higher than LPA's.

5 Conclusion

The main conclusions are summarized as followed:

1. In order to organize a huge number of emojis in social media, we take the relations between emojis into consideration and then build Emoji Link Network (ELN) which can realize sentiment propagation between emojis.
2. To decrease wrong emotions that emojis with uncertain sentiment can pass to the neighboring emojis, we put forward a method based on the PAD three-dimensional sentiment model to measure the sentiment uncertainty of emojis.
3. For the purpose of ranking emojis sentiment, LAPM and LPA are used to calculate emojis' six-dimensional sentiment automatically, multi-dimensionally, and quantificationally. The accuracy of LAPM is 85.2%, compared with 74.8% of LPA.

Acknowledgment. The research work reported in this paper was supported in part by the National Science Foundation of China under grant no. 61471232. This work was jointly supported by the Shanghai Science International Cooperation Project under Grant No. 16550720400.

References

1. Novak, P.K., Smailović, J., Sluban, B., et al.: Sentiment of emojis. PLoS ONE **10**(12), e0144296 (2015)
2. Miller, H., Thebault-Spieker, J., Chang, S., et al.: "Blissfully happy" or "ready to fight": Varying Interpretations of Emoji. In: International AAAI Conference on Web and Social Media (2016)
3. Dimson, T.: Emojineering part 1: machine learning for emoji trends. Instagram Eng. Blog **30** (2015)
4. Eisner, B., Rocktäschel, T., Augenstein, I., et al.: emoji2vec: learning emoji representations from their description, pp. 48–54 (2016)
5. Guibon, G., Ochs, M., Bellot, P.: Prédiction automatique d'emojis sentimentaux. In: CORIA, pp. 59–74 (2017)
6. Xie, R., Liu, Z., Yan, R., et al.: Neural emoji recommendation in dialogue systems (2016)
7. Jiang, D., Luo, X., Xuan, J., et al.: Sentiment computing for the news event based on the big social media data. IEEE Access **PP**(99), 1 (2017)
8. Hallsmar, F., Palm, J.: Multi-class sentiment classification on twitter using an emoji training heuristic (2016)
9. Parrott, G.W.: Emotions in Social Psychology. Psychology Press, London (2001)

10. Luo, X., Xu, Z., Yu, J., et al.: Building association link network for semantic link on web resources. IEEE Trans. Autom. Sci. Eng. **8**(3), 482–494 (2011)
11. Mehrabian, A.: Pleasure-arousal-dominance: a general framework for describing and measuring individual differences in temperament. Curr. Psychol. **14**(4), 261–292 (1996)
12. Zhang, S., Wu, Z., Meng, H.M., et al.: Facial expression synthesis using pad emotional parameters for a chinese expressive avatar. In: Paiva, A.C.R., Prada, R., Picard, R.W. (eds.) Affective Computing and Intelligent Interaction, pp. 24–35. Springer, Berlin (2007)

A Transfer Learning Approach for Credit Scoring

Wei Li[1,2], Shuai Ding[1,2(✉)], Yi Chen[1,2], and Shanlin Yang[1,2]

[1] School of Management, Hefei University of Technology, Anhui 23009, Hefei, China
dingshuai@hfut.edu.cn

[2] Key Laboratory of Process Optimization and Intelligent Decision-Making (Ministry of Education), Hefei University of Technology, Anhui 23009, Hefei, China

Abstract. Credit scoring is one of the major risks faced by banks as well as a significant part of credit risk management. To financial institutions, in most cases, samples of defaults are the minority among total loan samples, while credit client samples making repayments on time are the majority. This phenomenon is called class distribution imbalance problem that is prevailing in credit risk identification. However, existing credit scoring approaches cannot effectively solve the class distribution imbalance problem brought by the scarcity of minority samples. Thus, in this paper, a transfer learning approach is introduced, and the class distribution imbalance problem brought by the scarcity of minority samples is solved through the import of external credit information data. In this paper, a novel transfer learning model is put forward and the classification of target data is facilitated through auxiliary training data transfer so that the efficiency of external credit information using minority samples can be improved. With a new sample initial weight allocation and adjustment strategy, the ability to identify negative samples is highlighted. Through dynamic adjustments to auxiliary training sets, redundant data is duly eliminated as per the pre-set lower weight threshold, reducing the influence of the redundant data on the performance of the classifiers and enhancing the ability of transfer learning to learn imbalanced samples.

Keywords: Credit scoring · Transfer learning · Imbalanced data Credit risk

1 Introduction

Credit scoring is a significant part of daily operations of banks. A good credit scoring model can accurately identify default clients so as to enable financial institutions to eliminate credit risks and prevent loan losses. Therefore, research on credit scoring models is of great significance.

The development of credit scoring models has gone through three stages: qualitative analysis, statistical analysis, and artificial intelligence analysis [1]. Qualitative analysis was the earliest approach used for credit scoring, after which, statistical approaches were introduced into credit scoring gradually. For example, Orgler [2] is

J. Abawajy et al. (Eds.): ATCI 2018, AISC 842, pp. 64–73, 2019.
https://doi.org/10.1007/978-3-319-98776-7_8

the earliest to use regression analysis to make a scoring card to evaluate credit card clients. In recent years, with the development of data mining and artificial intelligence approaches, many of them have been introduced into credit evaluation one after another. For example, Baesens et al. put forward a self-adaptive least squares support vector machine for credit evaluation [3], Chang and Yeh adopted an artificial immune algorithm [4], Ahn et al. proposed an enterprise crisis prediction model based on rough sets and neural networks [5], and West et al. used integrated neural networks for credit evaluation [6].

Credit scoring is a matter of classification, and in other words, based on credit information, it falls into two major classes, namely, repayment on time and default. In the classification process, the distribution of the proportions of the two credit level sample classes is usually imbalanced, which is called the class distribution imbalance. That means default clients only account for a small proportion of the total samples while the vast majority of the samples are clients able to repay money on time. The class imbalance raises a serious challenge to credit scoring. For example, in a credit scoring problem where the ratio of clients repaying money on time to default clients is 9:1, general models are very likely to predict all clients as clients making repayments on time, thus obtaining an overall model accuracy of up to 90% [7]. However, such a model incorrectly predicts all the default clients that should be focused closely, and its result will impose huge credit risks to financial institutions.

To address impacts of class distribution imbalance on credit scoring, researchers have mainly come up with two types of solutions [8]. Data-level solutions belong to one of the two, which mainly use such sampling techniques as under-sampling, over-sampling and composite-sampling to balance the categorial distribution of training sets and then establish credit scoring models. Typical approaches of this type include random under-sampling (RUS) and synthetic minority over-sampling technique (SMOTE) [9]. The other type is algorithm-level solutions, which achieve a balance between the two different categories of clients by altering the inherent classification process bias of algorithms. The most typical solution is cost-sensitive learning, which corrects the categorial imbalance by assigning different misclassification costs to different categories of clients.

Whether a solution is a data-level one or an algorithm-level one, it uses internal credit information in both cases. For imbalance created by the relative scarcity, the use of internal credit information can effectively eliminate impacts of categorial imbalance since the samples of the minority category are sufficient. However, for occasions of absolute scarcity, it is usually difficult to address categorial imbalance by only using internal credit information due to the insufficiency of minority client samples. Categorial imbalance caused by absolute scarcity is common in credit scoring. On the one hand, financial institutions might only collect few client samples due to time, fund, and other resource limitations, especially client information collected in such forms as questionnaires; and on the other hand, similar situations also take place very frequently due to the limited amount of clients when a financial institution is on a small operation scale or at the initial stage.

In this paper, based on the above analysis, external credit information is imported into credit scoring, and attempts are made to integrate internal and external credit information by using transfer learning technique, so as to build a novel transfer learning model to help to solve categorial distribution imbalance problem posed by absolute scarcity to financial institutions' credit scoring of credit clients.

2 Foundations of Transfer Learning

With the concept deriving from psychology, transfer learning refers to an ability of human beings in the learning process, that is, the ability to apply the experience and skills accumulated in the learning of relevant fields to new learning tasks. For example, for a person knowing how to play table tennis, learning how to play tennis will be much easier. Some scholars introduced this concept into statistical learning field and developed a series of transfer learning modeling approaches. Ben-David and Schuller [10] gave formalized definitions to the relevance among learning tasks. Mahmud and Ray [11] used Kolmogorov to analyze the upper and lower limits of transfer learning. These theoretical explorations set theoretical foundations for transfer learning. After over a decade of development, the concept has been used to integrate external data of the system by many researchers so as to tackle data sample insufficiency problem in the target field, and it has been applied successfully in many fields such as text mining, information retrieval, and image processing [12]. However, analysis of existing papers revealed that there were very few researchers applying transfer learning to credit scoring problem, and even fewer studies using it to redress categorial imbalance of credit scoring. As for transfer learning as a modeling approach, its main idea is to utilize knowledge and information in neighbor fields to aid the modeling in the target field.

Among the numerous transfer learning approaches, instance-based transfer learning is the most common one. Instance transfer approach assumes that the data coming from relevant fields contains both useful data samples and useless ones. This approach imports useful data samples of relevant fields, processes the differences between these samples and the data samples of the target field through sampling or weighting, and integrates internal and external information of the system. Dai et al. [13] proposed a solution of taking Boosting technique as the basis and using old data in relevant fields to "boost" weak learning algorithms to strong learning algorithms through integration without compromising the classification accuracy of weak classifiers. With this technique, the effects of transfer learning can be improved effectively. However, this approach also has some problems. Firstly, TrAdaBoost algorithm is suitable for solving symmetry-based dichotomy problems, treating positive and negative data equally. But in the credit scoring problem, the sample distributions characterizing two different categories may be extremely imbalanced, and the two are quite different in significance. In addition, auxiliary data usually contains a large amount of redundant data which may be quite different from the target data set, and the existence of such data will not only affect the training speed of a model but also lead to lower classification accuracy. In this paper, based on these problems, an improved transfer learning algorithm—imbalanced ensemble transfer learning algorithm (IETLA) is proposed, which improves the contribution rate of samples in the same category (negative samples) featuring small data volume and large information volume by optimizing sample weight distribution and adjustment strategies. In the training process, "irrelevant" data is eliminated dynamically, the data with excessively low weight is eliminated according to the preset lower sample threshold, and the auxiliary training data will be more and more optimized after T rounds of iterative trainings. This makes this algorithm very suitable for solving credit scoring problem and enables it to make better predictions about client credit level than other ensemble approaches.

3 Imbalanced Ensemble Transfer Learning Model for Credit Scoring

3.1 Traditional TrAdaBoost

TrAdaBoost algorithm is an instance-based transfer learning algorithm proposed by Dai et al. [13]. Its basic idea is to use Boosting technique to screen out auxiliary data that is the least similar to the target data and find the most helpful samples by using the automatic weight adjusting mechanism. In this algorithm, AdaBoost [14] is used in the target data so as to ensure the accuracy of classification model applied to the target data.

3.2 Dynamic Elimination of Redundant Data

In the entire training process, the above TrAdaBoost algorithm is using the unchanged training set C all the time. However, in most transfer learning cases, the auxiliary training data volume used for transfer is large, in which a lot of data is quite different from the target data set. If such redundant data is reserved through the entire training process, the efficiency and accuracy of the classifier will be affected definitely, even possibly leading to "negative transfer" [15]. In this paper, a circular redundant data elimination method is used to solve this problem. After each iterative training is completed, the classifier will test the transfer auxiliary data set and alter the sample weight: correspondingly, the weight of redundant samples will decrease, and the weight of useful samples will increase, and in the meanwhile, a comparison is made between sample weight and the pre-set lower threshold, with the samples with weight lower than the threshold being eliminated. The training data set is optimized dynamically, and the efficiency and accuracy of the training are improved. The specific adjustment procedures are as follows (Algorithm 1).

3.3 Optimization of Sample Weight Distribution and Adjustments

Since it is quite common to see the imbalance between positive and negative samples in many fields, if the initial weight is assigned to all samples equally, it is more possible that a classifier is generated for the majority sample category with strong classification ability after ensemble trainings, and the weight of this sample category is increased correspondingly after adjustment. However, the sample weight is direct proportional to the probability of being chosen. Thus, it is likely to lead to the result that all the chosen ones are in the same category, further resulting in the failure of the training of the classifier. To ensure successful trainings, it is necessary to raise the contribution rate of the minority sample category (only negative samples here - namely, default samples), and thus the following improvements are made to the TrAdaBoost algorithm:

(1) Assign a relatively large initial weight to negative samples so as to ensure that such samples are more likely to be chosen. And after each round of training, an attenuation term is added to the incorrectly categorized negative sample, namely $dr \times w_i^t$ $(0 \leq dr \leq 1)$, so that the weight adjustment of the incorrectly categorized negative sample has the memory function so as to avoid it from decreasing very fast.
(2) After a subset of training samples is chosen, a judgment is made to confirm whether the subset contains two categories of samples or not. If only one category is included, some samples will be directly chosen from the absent category and put into the training subset so as to ensure that the training subset contains two categories of samples.

Algorithm 1: Redundant data dynamic elimination
Input: a mixed training data set C is extracted between target data set B and transfer auxiliary data set A pro rata according to the actual data volume. In A, the initialized weight of the sample is w_1, $w_{\min}$ is the lower weight threshold of the auxiliary training samples ($0 \leq w_{\min} \leq w_1$), $A_{\min}$ is the minimum proportion of the auxiliary data volume in the training data, and the pre-set number of iteration times is T.
Begin:
For $t = 1, 2, \ldots, T$ (1) Obtain the weak classifier through training by using weak learning algorithm P: $h_t(x): x \to \{0, 1\}$ (2) Return the transfer auxiliary data set and use the ensemble classifier h_t obtained at the last step for test, and for each sample on A, if the prediction result of h_t is identical to its real class label, its corresponding weight is increased or otherwise its weight is decreased. (3) Judge if the weight of all samples in C is smaller than $w_{\min}$, the samples with a weight lower than $w_{\min}$ will be eliminated to obtain a tailored new transfer auxiliary data set A; if the remaining samples of A after tailoring account for less than $A_{\min}$ of the total volume C, the circulation stops.
End.
Output: $C = B + A$

3.4 Imbalanced Ensemble Transfer Learning Algorithm

According to the contents mentioned above, the imbalanced ensemble transfer learning algorithm (IETLA) proposed in this paper fully considers the very common imbalance phenomenon between positive and negative samples in the real world, and the ensemble learning approach is used together with the above improvement strategies to solve the transfer learning problem of asymmetric dichotomy.

Algorithm 2: Imbalanced ensemble transfer learning

Input:

Transfer auxiliary data set A and target data set B are mixed as a training data set pro rata. In other words, the first n samples in C are data in business A, and the rest m samples are data in business B. The pre-set number of iteration times is T.

Where $x_i \in X$ is feature attribute vector of the samples, the number of dimensions is q, and $y_i \in \{0,1\}$ is the label of the sample, $i = 1,2,\ldots,N$. The size of the training subset is $N/2$.

Begin:

Initialize the sample weight:

$$w_{i_0}^1 = \begin{cases} 1/d & \textit{positive sample in A} \\ 1/n-d & \textit{negative sample in A} \\ 1/l & \textit{positive sample in B} \\ 1/m-l & \textit{negative sample in B} \end{cases}$$

where $i = 1,2,\ldots,N$. d and l are the number of positive samples in A and B.

Normalize the sample weight: $w_i^1 = \dfrac{w_{i_0}^1}{\sum_{i=1}^{N} w_{i_0}^1}$

For $t = 1,2,\ldots,T$

(1) Select the training subset D (where $D = 1/2C$)

(2) Judge if the training subset contains two categories of samples or not. If it only contains one category of samples, samples will be chosen from the other category and filled in the training subset.

(3) In the training subset, weak learning algorithm P is used to train the base classifier h_t^j, $j = 1,2,\ldots,q$, to obtain the weak classifier: $h_t = \sum_{j=1}^{q} h_t^j$

Calculate the training error rate of the weak classifier h_t on V.

$$w_t = \sum_{i=n+1}^{n+m} \frac{w_i^t \left| h_t(x_i) - y_i \right|}{\sum_{i=n+1}^{n+m} w_i^t}$$

Let $\beta = \dfrac{1}{1+\sqrt{2\ln\frac{n}{T}}}$, $\beta_t = \dfrac{w_t}{1-w_t}$, where $w_t < \dfrac{1}{2}$

(5) Adjust the weight of the samples:

if $y_i = 0$ and $h_t(x) \neq y_i$, where $1 \le i \le n$

$$w_i^{t+1} = w_i^t \beta^{\left|h_t(x_i)-c(x_i)\right|} + dr \times w_i^t \quad (0 \le dr \le 1)$$

else $w_i^{t+1} = \begin{cases} w_i^t \beta^{\left|h_t(x_i)-c(x_i)\right|}, & 1 \le i \le n \\ w_i^t \beta_t^{-\left|h_t(x_i)-c(x_i)\right|}, & n+1 \le i \le m+n \end{cases}$

(6) Apply redundant data dynamic elimination algorithm

End.

Output: final classifier

$$H(x) = \begin{cases} 1, & \sum_{t=\frac{T}{2}}^{T} \ln\left(\frac{1}{\beta_t}\right) h_t(x) \ge \frac{1}{2} \sum_{t=\frac{T}{2}}^{T} \ln\left(\frac{1}{\beta_t}\right) \\ 0, & \textit{otherwise} \end{cases}$$

The basic idea of IETLA is to apply the weak learning algorithm to feature attributes of each dimension of the data vector respectively. During initialization, positive and negative samples are assigned with different weights so as to ensure that the negative samples accounting for a small proportion of total samples but containing a large amount of information have a large initial weight. During each round of training, some samples are chosen pro rata as the training subset for training. After the training is finished, the one with the smallest error is selected as a weak classifier h_t from several simple classifiers, and the training data set is adjusted according to the redundant data dynamic elimination algorithm. In this way, a weak classifier sequence $(h_1, h_2, \cdots, h_t)$ can be obtained after t rounds of iterations, in which the classifiers with better classification effects have large weights. The final classification function $f(x)$ is generated using a voting method involving weights, which, in other words, is to boost several weak classifiers by a certain method to generate a strong classifier. The algorithm flow is shown in Fig. 1, and the algorithm is described as follows (Algorithm 2).

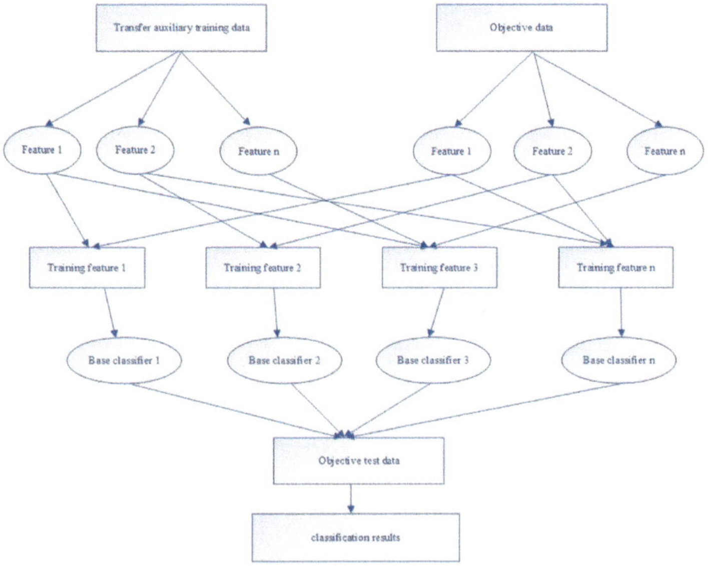

Fig. 1. Flowchart of imbalanced ensemble transfer learning

4 Experiment and Results Analysis

4.1 Data Description

Business A is a loan on credit, the feature of which is that the debtor can obtain a loan depending on his or her credit without providing any collateral, with the credit rating of the borrower as the guarantee for repayment. Business B is a cash loan, namely, a payday loan. Compared with general consumer financial products, cash loans mainly have the following five features: small amount, short period, no collateral, fast process, and high interest rate, which are also compatible with its feature of low loan threshold.

At the early stage, cash loan business got a lot of profits by adopting the high interest rate coverage risk mode of "lending money without asking details". However, this mode cannot develop sustainably, and risk control will become the key at the middle and late stages of the cash loan business. However, because clients of petty cash loans generally feature low income and education level, most of them pose potential credit risks.

Thus, with the maturity of the cash loan market, it is the top priority at present to establish a practical and effective risk control model. However, due to the insufficient product accumulation time of cash loans currently, there is not enough time to collect enough target variables (flag) to set up a model to judge the credit risk of the clients.

4.2 Experiment Results and Analysis

Business A data set contains 40,000 samples, 491-dimension feature space (including one dimension containing the flag representing whether the client defaults or not). Business B contain 4,000 samples, 491-dimension feature space (including one dimension containing the flag representing whether the client defaults or not). In the experiment, the data of business A is set as the auxiliary training data set. 75% of the data of business B is set as the target data set for training while the remaining 25% of the data is used to test the performance of the proposed model.

As for the feature selection method, in this paper, a feature sequencing approach based on learning model is used. Since the credit data set has many features, there are 490 dimensions (the last one dimension is the flag). For the prediction model, there are redundant features requiring removal.

Due to the imbalance between positive and negative samples in training samples, the algorithm features a certain degree of randomness. In the meanwhile, several experiments have been conducted, with the input being set as the test data set and the number of iteration times being 50. TrAdaBoost algorithm failed several times in the experiment. Table 2 shows a comparison table of 10 successful trainings of TrAdaBoost algorithm and the test results of IETLA algorithm. As shown in the table, TrAdaBoost algorithm only generates a low error rate twice, while IETLA algorithm reaches a low error rate in almost all cases. In addition, the test result after 10% cross-covariance on the training data set reveals that IETLA algorithm has greater error tolerance and stability than traditional TrAdaBoost algorithm.

Figures 2 and 3 show the experiment results of TrAdaBoost algorithm and IETLA algorithm with the number of iteration times being fixed as 40 and the data volume in the training subset being increased from 400 to 4,000. As shown in Fig. 2, the classification accuracy of TrAdaBoost algorithm does not change monotonically. Because redundant data exists in the transfer training set during iteration all the time, there are weak classifiers with low classification accuracy reducing the accuracy of the ensemble classifier, and with the increase of the training samples, the performance of the final classifier cannot be improved eventually. As shown in Fig. 3, the classification error rate curve of IETLA algorithm decreases monotonously, indicating that the accuracy of the final classifier can be improved by optimizing the auxiliary training sample set. The monotonicity of the training data set after 10% cross-covariance is also good, the performance of which is obviously superior to the TrAdaBoost algorithm.

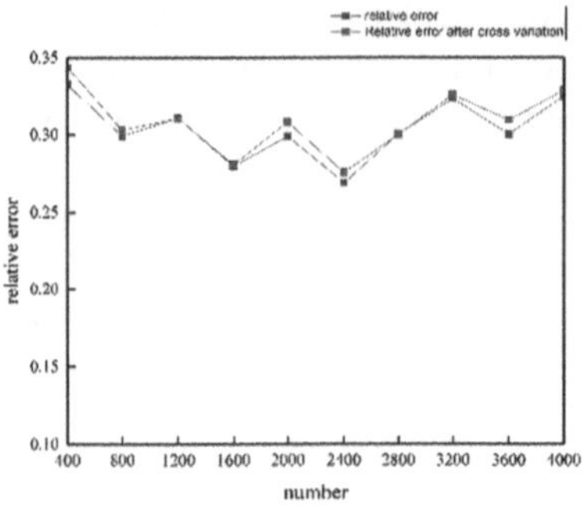

Fig. 2. Relative error of TrAdaBoost

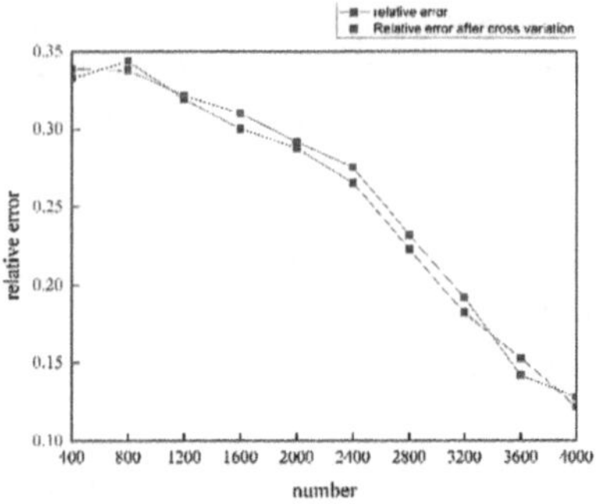

Fig. 3. Relative error of IETLA

On the one hand, IETLA algorithm distinguishes the initialization weight and adjustment strategies of positive and negative samples, highlights the ability of classifiers to identify negative samples, ensures that the subset of each sampling training round contains two categories of samples, and avoids training failures caused by imbalance between positive and negative samples. On the other hand, IETLA algorithm eliminates samples with small weight according to the preset lower weight threshold, reduces the impact of the redundant data in the transfer data set on the classification model. As the training data increases, the classification ability of the final classifier is enhanced. According to the experiment results, IETLA algorithm has some advantages over TrAdaBoost algorithm in imbalanced sample classification.

5 Conclusion

In the existing credit scoring, solutions to the categorial imbalance problem mainly include sampling, cost-sensitive learning, and so on which only use internal credit information data, unable to effectively address categorial imbalance caused by absolute scarcity. In this paper, a new solution idea is proposed. This solution introduces external credit information data, integrates internal and external credit information by using transfer learning technique, builds a balanced transfer learning model based on ensemble learning, and carries out empirical studies based on credit data set of a certain credit reporting agency. According to the study results, it is effective to introduce external credit information data through transfer learning technology to address categorial distribution imbalance caused by absolute scarcity in credit risk scoring system. Compared with scoring approaches only using internal credit information data and other transfer learning approaches, the imbalanced ensemble transfer learning model obtains greater classification accuracy. Particularly, it largely improves the accuracy of predicting minority default clients, conducive to the credit risk avoidance of financial institutions. In the practices of customer relation management, many problems are also plagued by classification imbalance, such as customer loss prediction, which can be solved using the balanced transfer learning model in the future studies. In the future, with the help of new models built through transfer learning, banks can develop their new unsecured businesses better and bank risk-control personnel can start new businesses even without enough time to accumulate data. In addition, credit reporting agencies hope they can build risk-control models for different loan scenarios using small data samples through transfer learning.

Acknowledgment. The work is partially supported by the National Natural Science Foundation of China (Nos. 71571058 and 71690235).

References

1. Crook, J.N., Edelman, D.B., Thomas, L.C.: Recent developments in consumer credit risk assessment. Eur. J. Oper. Res. **183**(3), 1447–1465 (2007)
2. Orgler, Y.E.: A credit scoring model for commercial loans. J. Money Credit Bank. **2**(4), 435–445 (1970)
3. Baesens, B., Van Gestel, T., Viaene, S., et al.: Benchmarking state-of-the-art classification algorithms for credit scoring. J. Oper. Res. Soc. **54**(6), 627–635 (2003)
4. Chang, S.Y., Yeh, T.Y.: An artificial immune classifier for credit scoring analysis. Appl. Soft Comput. **12**(2), 611–618 (2012)
5. Ahn, B.S., Cho, S.S., Kim, C.Y.: The integrated methodology of rough set theory and artificial neural network for business failure prediction. Expert Syst. Appl. **18**(2), 65–74 (2000)
6. West, D., Dellana, S., Qian, J.: Neural network ensemble strategies for financial decision applications. Comput. Oper. Res. **32**, 2543–2559 (2005)
7. Haibo, H., Garcia, E.: Learning from imbalanced data. IEEE Trans. Knowl. Data Eng. **21**(9), 1263–1284 (2009)
8. Sun, Y., Wong, A., Kamel, M.: Classification of imbalanced data: a review. Int. J. Pattern Recognit. Artif. Intell. **23**(4), 687–719 (2009)
9. Chawla, N.V., Bowyer, K.W., Hall, L.O., Kegelmeyer, W.P.: SMOTE: synthetic minority over-sampling technique. J. Artif. Intell. Res. **16**, 321–357 (2002)
10. Ben-David, S., Schuller, R.: Exploiting task relatedness for multiple task learning. In: Proceeding on 16th Annual Conference on Computational Learning Theory, Washington, DC, USA (2003)
11. Mahmud, M., Ray, S.R.: Transfer learning using Kolmogorov complexity: basic theory and empirical evaluations. In: Proceeding of the 20th Annual Conference on Neural Information Processing Systems, pp. 985–992. MIT Press, Cambridge (2008)
12. Pan, S.J., Yang, Q.: A survey on transfer learning. IEEE Trans. Knowl. Data Eng. **22**(10), 1345–1359 (2010)
13. Dai, W.Y., Yang, Q., Xue, G.R., Yu, R.: Boosting for transfer learning. In: Proceeding of the 24th International Conference on Machine Learning, pp. 193–200. ACM (2007)
14. Baig, M.M., Awais, M.M., El-Alfy, E.S.M.: AdaBoost-based artificial neural network learning. Neurocomputing **248**, 120–126 (2017)
15. Tommasi, T., Orabona, F., Caputo, B.: Safety in numbers: learning categories from few examples with multi model knowledge transfer. In: 2010 IEEE Conference on Computer Vision and Pattern Recognition (CVPR), pp. 3081–3088. IEEE (2010)

A Parallel Programming Model Research Based on Heterogeneous Multi-core Embedded Processor

YuXin Cai(✉) and ZhiWei Tang

The Third Research Institute of Ministry of Public Security,
BiSheng Road Number 339, Pudong, Shanghai, China
caiyuxin_good@163.com

Abstract. Today is the era of information explosion, efficient methods for massive data processing are becoming a hotspot. MapReduce is a popular parallel programming model and widely used in large-scale data parallel computing. This paper proposes a parallel programming model based on heterogeneous multi-core embedded processor. On the base of ordinary computer completing computing, we make optimization and improvement, and use the method of combining embedded dual-core ARM processors with multiple high performance FPGAs to complete complex processing. Xilinx Zynq device is used as the hardware platform. By selecting Sobel image processing, histogram, matrix multiply and illumination enhancement cases, we test the parallel programming model. The results show that the speedup is 84.62x versus CPU-based implementation. In addition, it is proven that this parallel programming model is suitable for the CPU-FPGA heterogeneous system which can be used as cluster cloud computing node. Meanwhile, we make a comparison of our model with other references and verify the efficient of the programming model in this paper.

Keywords: ARM · FPGA · Heterogeneous multi-core · Parallel programming

1 Introduction

With the development of multi-core processors, one of the critical challenges is the parallel programming model. MapReduce [1] is a parallel programming model proposed by Google for massive data processing, and it has been successfully applied to many applications. MapReduce is derived from the functional language, and it deals with large data sets through "Map" and "Reduce" steps. Computational tasks are divided into lots of input <key, value> pairs, the Map tasks process these <key, value> pairs and generate a series of intermediate results. When the Map tasks are completed, the intermediate <key, value> pairs which have the same key are gathered and processed by the reduce tasks. The system returns the final results when all reduce task are finished. The advantage of this programming model is that users only need to design the Map and Reduce functions, and the MapReduce runtime framework dynamically assigns the Map and Reduce functions to different computing nodes to achieve parallel processing.

FPGA is ideal for data-intensive computing, and its processing speed is very fast because of the parallelism ability, and more and more processing tasks are required to

J. Abawajy et al. (Eds.): ATCI 2018, AISC 842, pp. 74–82, 2019.
https://doi.org/10.1007/978-3-319-98776-7_9

be implemented by FPGA. Therefore, the way of programmable dedicated processor (FPGA [4] and GPU [5] etc.) combining with general processor (CPU) has become an effective method for achieving high-speed data processing.

This paper proposes a new parallel programming model for Soc heterogeneous multi-core processor. In order to enhance the system performance to maximize, the dual-core ARM is mainly used to running Linux operating system and executing control tasks, while the complex time-consuming processing tasks are migrated to FPGA. Comparing with traditional multi-board or multi-chip solutions, it is the first time to apply the MapReduce framework on a single chip consisting of CPU and FPGA. With this model, the structure of ARM plus FPGA can be used as a cloud computing node under the premise of compliance with the MapReduce standard programming model.

The rest of the paper is organized as follows. Section 2 describes architecture of the heterogeneous system. Section 3 discusses the implement of parallel model in detail. Section 4 contains experimental results and performance comparison. Finally Sect. 5 summarizes the full text.

2 Parallel Programming Model Overview

In this paper, we apply lightweight MapReduce programming model on the heterogeneous system. To address the problem of low computing performance of embedded general processor and speed up processing, we use PL to complete the most time-consuming Map tasks, and PS is responsible for the Reduce tasks, system I/O and other control functions. The system architecture is shown in Fig. 3. Meanwhile, in order to achieve parallelization, the multiple tasks can be extended to multiple FPGA platforms by means of adding much more FPGA resources, just as adding the FPGA Mapper module in Fig. 1.

As shown in Fig. 1, the system architecture is partitioned into four parts: Virtual Map Tasks, FPGA Mapper, FPGA Driver and Reducer. The Virtula Map Tasks process the initial input <key, value> pairs and call FPGA Driver. The FPGA Driver acts as the bridge between ARM and FPGA. The FPGA Mapper performs actual data processing tasks and returns intermediate results to ARM processors. The Reducer merges the intermediate <key, value> pairs which have the same keys and generate final results. In addition, the ARM processors in PS are responsible for scheduling the mappers and reducers. The detailed work flow is shown as follows.

1. CPUs accept the input files and divide them into multiple parts by the split process. For each of the split file, a Map thread is created.
2. CPUs write configuration parameters to registers on FPGA through the FPGA driver.
3. CPU scheduler assigns the map tasks to idle FPGA Mappers.
4. FPGA Mappers receive the transmission data from CPUs and generate the intermediate <key, values> pairs.
5. After the FPGA processing is completed, the results are returned to the CPU.
6. When some intermediate value <key, value> pairs are generated, CPU scheduler will assign the reduce tasks to idle reducers.

7. When all map and reduce tasks are done, CPUs merge the output data and produce the final results.

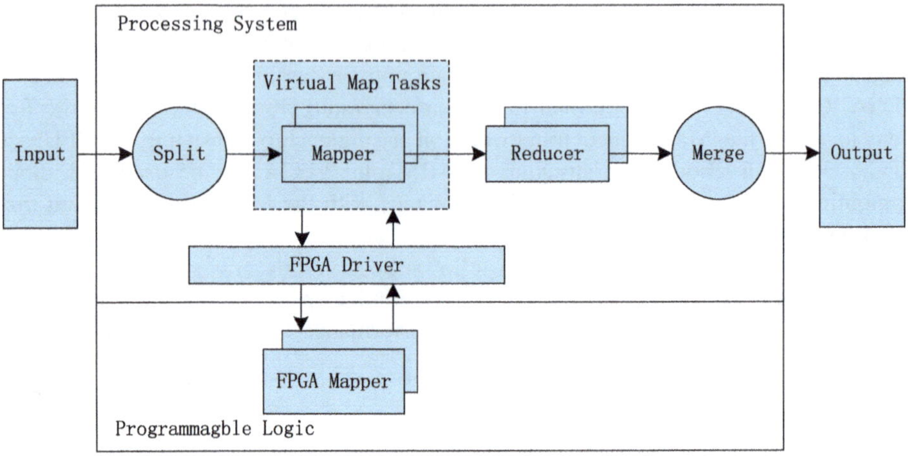

Fig. 1. Heterogeneous system architecture design with mapreduce

3 The Implement of Parallel Programming Model

In this paper, in order to maximize the system performance, the Map function is not directly realized by general-purpose processor, instead, we use CPUs to call Linux drivers of FPGA accelerators. The divided data to be processed is passed to FPGA accelerators by AXI_HP high-speed communication interfaces, and the most time-consuming processing tasks are executed by FPGA hardware layer. The Map function of Phoenix application code is achieved by calling FPGA. The interface of Map function is shown in Table 1, in which Dest_data is used for storing processed data, and the reduce_key is used to indicate from which divided whole fragments are. After FPGA accelerations executing algorithms, CPUs execute the tasks of simplify and mergence and output the final results. Additionally, CPUs can complete the initialization of FPGA accelerators and send control commands by the driver function etc.

Table 1. Part code which Map function calls FPGA function

```
void Sobel_Map(map_args_t *args)
{
    assert(args);
    call_FPGA((char*)args->data,    Dest_data,    0,    FRAME_WIDTH,
FRAME_HEIGHT/2);
    emit_intermediate((void*)&reduce_key, (void *)& args->length, 4);
}
```

In order to more clearly illustrate the relationship between CPUs and FPGA scheduling, we list part of code which CPUs call FPGA to execute algorithms. The interface of call_FPGA is shown in Table 2.

Table 2. The interface of call_FPGA

```
   void call_FPGA(void* data_in, int length_in, int cpu_id, void* data_out, int
 length_out)
   {
      if(cpu_id == 0){ //call filter engine 0
                   configureVDMA(FILTER_VDMA0_BADDR, IN_BASEADDR0,
 FILTER_OUT_BASEADDR0);
                   copy_mem2dev(((unsigned char*)data_in), IN_BASEADDR0);
                   startEngine(FILTER0_BASEADDR);
                   copy_dev2mem(((unsigned
 char*)data_out),FILTER_OUT_BASEADDR0);
                   stopVDMA(FILTER_VDMA0_BADDR);
         }
         else if(cpu_id == 1){ //call filter engine 1
            configureVDMA(FILTER_VDMA1_BADDR,          IN_BASEADDR1,
 FILTER_OUT_BASEADDR1);
                   copy_mem2dev(((unsigned char*)data_in), IN_BASEADDR1);
                   startEngine (FILTER1_BASEADDR);
                   copy_dev2mem(((unsigned
 char*)data_out),FILTER_OUT_BASEADDR1);
                   stopVDMA(FILTER_VDMA1_BADDR);
          }
      }
```

In Table 2, when software program executes, different algorithms FPGA accelerators will be called depending on CPU ID numbers. The process must firstly configure the DMA, and notify the DMA about the source address of original image (IMG_BASEADDR) and the destination address of processed data (FILTER_-OUT_BASEADDR). After that, the data which will be processed is copied into specified memory (IMG_BASEADDR), and the DMA will remove the data from memory to IP core. Then accelerators begin to process data, and the data will be sent back to specified output memory address (FILTER_OUT_BASEADDR) by DMA after processing. Finally, the DMA transmission will be stopped.

4 Experiments and Conclusions

To verify the feasibility and efficiency of proposed parallel programming model, we use zedboard evaluation board as hardware platform. The representative Sobel image processing, histogram statistics, matrix multiplication and low-light enhancement case are used to test our model. The results prove that the hardware platform which processor combined logic can be used as cluster cloud computing node under the premise of being comply with MapReduce standard programming model. Meanwhile, we compare relative performance with other reference, and verify the efficiency of our model.

In our experiment, the component which embedded platform uses is Zynq, and its model number is xc7z020clg484-1. Operating system of the platform is Linux, the processor is dual-core ARM-A9, the FPGA is Artix-7 [10].

4.1 Experimental Result with Parallel Model

To explore whether the time consumption of massive tasks will reduce or not when the parallel model is used, this section does some experiments for the four test cases by the means of dual-core ARMs combining with multiple FPGAs. The results are shown in Fig. 2.

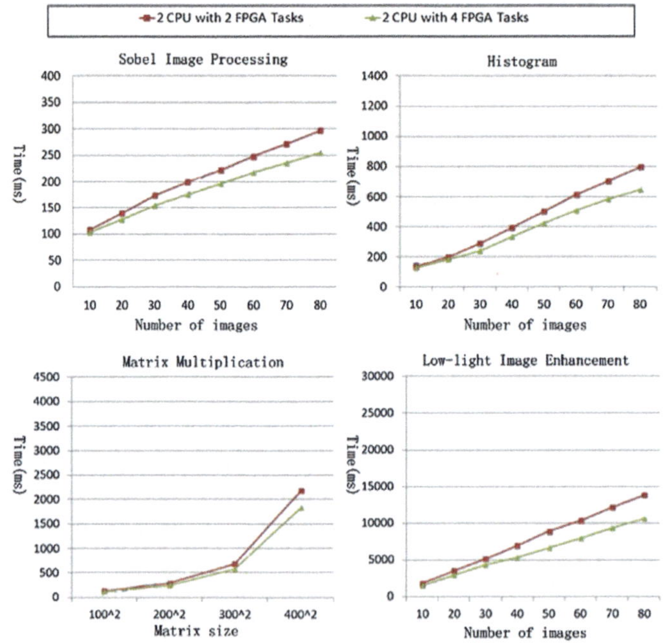

Fig. 2. The experimental result with parallel model

From Fig. 2 we can conduct an important point, namely, compared with the single FPGA, the algorithm execution time of multiple FPGA parallelism can be significantly shortened by using parallel programming model. This demonstrates that our parallel model can greatly increase the processing speed and has the feature of efficiency. In addition, a simple analysis of experimental results is made by us: Firstly, for those three image processing algorithms, the execution time of the algorithm is basically linear growth with the rise of amount of data, and the performance will not decrease with increase of the input data, which reflect the higher performance of our parallel model. Secondly, for the Sobel image processing algorithm, when input data size is small, the execution time of parallel processing is longer than the single FPGA processing. Only when the input data size is further increased, will the acceleration of parallel programming model appear. This is caused by other overhead of parallel programming model. Because the Phoenix system is based on MapReduce model, and only the Map tasks are done on FPGA, while the Reduce tasks and other procedures are still done on CPU. Meanwhile there is almost no additional software overhead besides of IP core driver when single FPGA is used for algorithm acceleration. Therefore, the parallel programming model will have even more processing time than the single FPGA when we deal with small amount of data.

4.2 Results Analysis

According to the experimental result data in Sect. 4.2 we list the speedups of different algorithm acceleration on heterogeneous system, and the result is shown in Fig. 3. All of the speedups are obtained by comparison with single CPU processing time.

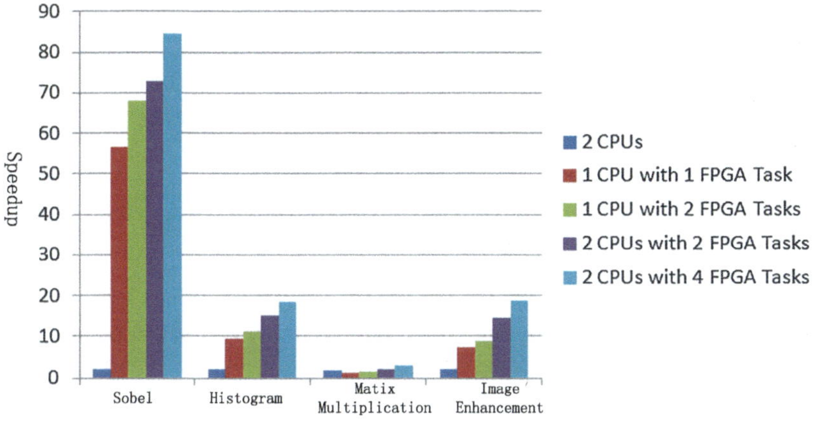

Fig. 3. The speedup of different algorithm acceleration

From Fig. 3 we can see that there is mainly a comparison of without parallel model with using model. When parallel programming model is not used, test cases include 2 CPUs, 1 CPU with single FPGA acceleration and multiple FPGA accelerations. When using the parallel programming model, we combine dual-core CPU with two FPGAs and four FPGAs. The experiment results show that the model of dual-core CPU combining with 4 FPGAs has the highest speedup, and relatively it will take most of the FPGA hardware resources. For Sobel edge detection algorithm, single FPGA acceleration ratio is 56.71 times while the highest reached 84.62 times by using parallel model. The improvement of performance is very considerable.

Table 3. Comparison of the methods in some references and in this paper

Reference	Hardware platform	Data size	Processing time	Data rate
Zhang [16]	Altera Cyclone II	64 KB	1.3 ms	48 MB/s
Sanjay [17]	Xilinx ML510	PAL@50fps	0.27 s	292.97 MB/s
Yasri [18]	Altera Cyclone II	0.5 MB	2 ms	250 MB/s
Our method	Xilinx xc7z020	135 MB	255 ms	529.41 MB/s

To further verify the efficient model in this paper, we make a comparison of this paper with other references aiming at Sobel edge detection algorithm. The result is shown in Table 3.

From Table 3 we can see that this model proposed by this paper have higher performance than other references, and the data transfer rate can reach to 529.41 MB/s. The higher speedup is not just due to Zynq device with a faster bus, it is also because that other references just used a single hardware accelerator and did not implement multiple hardware accelerators parallelization on the existing platform, while this paper takes full advantage of the feature of dual-core CPU calling multiple IP cores, designs and implements multiple hardware accelerators parallelization, and achieves the effect of reducing the processing time by increasing hardware resources. For example, a picture is divided into two same parts by CPU, the amount of data of each part is half of the original, and the processing time will be halved. Meanwhile both parts of data are processed by dual-core CPUs calling FPGA hardware accelerators in parallel, rather than CPU to perform, and the time is greatly reduced, so the whole measured speedup of parallel model will be higher.

5 Conclusion and Future Work

This paper presents a new parallel programming model based on CPU-FPGA heterogeneous multi-core embedded processor, and provides the software and hardware implementation. The communication between heterogeneous processors is realized by using high bandwidth AXI bus interfaces, and DMA is used for controlling data transmission. Based on the lightweight MapReduce, we do some improvement and transfer the most time-consuming Map tasks to FPGA. The programming model provides simple function interfaces to implement the MapReduce model. Users only need to implement the hardware accelerator on FPGA, and the parallel control and task scheduling are done by the framework. A device driver is also designed for the configuration and management of FPGA accelerator.

Acknowledgements. The authors of this paper are members of Shanghai Engineering Research Center of Intelligent Video Surveillance. In part by nation key R&D Program of China 2017YFC0803700 and Program of Science and Technology Commission of Shanghai Municipality (No. 15530701300).

References

1. Dean, J., Ghemawat, S.: MapReduce: simplified data processing on large clusters. In: Sixth Symposium on Operating System Design and Implementation (OSDI), San Francisco, CA (2004)
2. Condie, T., Conway, N., Alvaro, P., et al.: Mapreduce online [R/OL]. 2009-10. http://www.eecs.berkeley.edu/Pubs/
3. Chu, C.T., Kim, S.K., Lin, Y.A., et al.: Mapreduce for machine learning on multicore. In: Advances in Neural Information Processing Systems (2007)
4. Li, S., Yin, D., Mu, D.: Scalable mapreduce framework design and realization base on FPGA. Comput. Technol. Dev. **22**(2), 28 (2012)
5. He, B., Fang, W., Luo, Q., et al.: Mars: a mapreduce framework on graphics processors. In: 17th International Conference on Parallel Architectures and Compilation Techniques (2008)
6. Shan, Y., Wang, B., Yan, J., et al.: FPMR: mapreduce framework on FPGA. In: Proceedings of the 18th Annual ACM/SIGDA International Symposium on Field Programmable Gate Arrays (2010)
7. Yeung, J.H.C., Tsang, C.C., Tsoi, K.H., et al.: Map-reduce as a programming module for custom computing machines. In: IEEE Symposium on Field Programmable Custom Computing Machines (FCCM) (2008)
8. Tsoi, K.H., Luk, W.: Axel: a heterogeneous cluster with FPGAs and GPUs. In: Proceedings of the 18th Annual ACM/SIGDA International Symposium on Field Programmable Gate Arrays (2010)
9. High-performance-video-AXI-interconnect (2013)
10. DeHon, A., et al.: Design patterns for reconfigurable computing. In: Proceedings of 12th Annual IEEE Symposium on Field-Programmable Custom Computing Machines. IEEE CS Press, pp. 13–23 (2004)
11. Xilinx Inc. 7 series FPGAs overview (ver 1,14) (2013)
12. Zynq-7000 EPP Technical Reference Manual (2012)
13. Ranger, C., Raghuraman, R., Penmetsa, A., Bradski, G., Kozyrakis, C.: Evaluating mapreduce for multi-core and multiprocessor systems. In: Proceedings of the 13th International Symposium on High-Performance Computer Architecture (HPCA), Phoenix, AZ, February 2007
14. http://www.phoenix.com/
15. Zynq-7000 Extensible Processing Platform Overview (2012)
16. Zhang, X., Shen, P., Luo, L.: Enhancement and noise reduction of very low light level images. In: 21st International Conference on Pattern Recognition (ICPR 2012). 11–15 November 2012
17. He, K., Sun, J., Tang, X.: Signal image haze removal using dark channel prior. IEEE Trans. Pattern Anal. Mach. Intell. **33**(12), 2341–2353 (2011)
18. Zhang, L., Ling, C.: Application of Sobel edge detection based on FPGA. Electron. Appl. Technol. **37**(8), 26–29 (2011)
19. Xilinx Inc. Vivado high-level synthesis user guide (2012)
20. Singh, S., Saini, A.K., Saini, R., et al. Area optimized FPGA-based implementation of the Sobel compass edge detector. ISRN Machine Vision (2013)
21. Yasri, I., Hamid, N.H., Yap, V.V.: Performance analysis of FPGA based Sobel edge detection operator. In: 2008 International Conference on Electronic Design, 1–3 December 2008
22. Donthi, S., Haggard, R.L.: A survey of dynamically reconfigurable FPGA devices. IEEE Transaction, pp. 422–426, 8 March 2003

23. Xilinx Inc. Accelerating OpenCV application with Zynq using Vivado HLS video Libraries. 2013
24. Apache Hadoop, http://hadoop.apache.org/
25. Chu, C., Kim, S., Lin, Y., Yu, Y., Bradski, G., Ng, A.Y., Olukotun, K.: Map-reduce for machine learning on multicore. In: NIPS'07: Proceedings of Twenty-First Annual Conference on Neural Information Processing Systems. Neural Information Processing Systems Foundation (2007)
26. Adam Pisoni. Skynet, Apr. 2008. http://skynet.rubyforge.org
27. Ericson, M., Johansson, M.: A general purpose embedded system utilizing Rumime reconfiguration. In: The International Conference on Reconfigurable Computing and FPGAs ReConFig, Colima, Mexico, pp. 483–490 (2004)

An Improved Apriori-Pro Algorithm

Huaping Zhou(✉) and Daoyi Zhang

School of Computer Science and Engineering, Anhui University of Science and Technology, Huainan, 232001, China
hpzhou@aust.edu.cn 842351906@qq.com 337654199@qq.com

Abstract. An improved Apriori-Pro algorithm is proposed to solve the disadvantages of a large number of invalid candidate sets when mining association rules. The new algorithm adds the transaction label column in the item set list to calculate the support degree, and makes use of the difference of the transaction label column to determine whether the link operation is conducted, effectively avoiding the generation of the invalid candidate set. In addition, when the algorithm scans the database for the first time, the dataset is put into the list without having to scan the database multiple times. Experimental results show that the new algorithm is greatly improved compared with the original algorithm.

Keywords: Association rule mining · Apriori-pro algorithm
Invalid candidate set · Transaction label

1 Introduction

Apriori algorithm is a classical algorithm for mining association rules, which was first proposed by Agrawal in 1994 [1]. The idea of Apriori algorithm is to select the candidate set by the repeated scanning transaction database, and then the frequent item set is obtained by the prior property [2].

Although the Apriori algorithm can excavate the association rules in the mass data, the algorithm will take up a large amount of space resources and huge time consumption as the data volume increases. At the same time, because each calculation support needs to be scanned in a database, it brings a lot of I/O time overhead [3].

In order to solve the problems existing in the Apriori algorithm, this article for each of the frequent items added a TID columns, by comparing the TID column to choose whether to link, greatly reduced the incidence of candidate itemsets, reduces the overhead space; At the same time, a list is used to store data to avoid the time overhead associated with multiple scan databases.

2 Classic Apriori Algorithm

2.1 Basic Idea of Apriori Algorithm

Apriori algorithm is the simplest and most classical algorithm in mining frequent patterns. It mines frequent item sets by limiting the generation of candidate sets. The

J. Abawajy et al. (Eds.): ATCI 2018, AISC 842, pp. 83–90, 2019.
https://doi.org/10.1007/978-3-319-98776-7_10

algorithm is based on the fact that it uses prior knowledge of the nature of frequent itemsets, iterative methods using layer-by-layer search, and k-item sets to explore (k + 1)-item sets. The basic steps are as follows:

```
L1=find_frequent_1_itemsets(D);
For(k=2;Lk-1≠∅;k++){
    Ck=apriori_gen(Lk-1);
  For each transaction t∈D   {// Scan D for counting transactions
       Ck=subset(Ck,t); // get a subset of t, they're candidates
       For each candidates c∈Ct
          c.count++;}
  Lk={c(Ck|c.count>=minsupport)}}
Return L=∪kLk;
```

2.2 Apriori Algorithm Analysis

The classic Apriori algorithm addresses how to find the association rules among large Numbers of data. However, the algorithm also has disadvantages, which are embodied in:

(1) the algorithm needs to scan the database multiple times when calculating the support degree, which is too expensive. After each candidate set is generated, a global scan of the database is required to calculate the support of each candidate item, resulting in a large amount of I/O overhead.
(2) When frequent item sets are self-linked, a large candidate set is generated and space overhead is large. When linking operations, the resulting set of candidate entries is not always frequent, and as the data increases, a large number of useless candidate sets will be generated.
(3) When the candidate set and transaction are matched, the time is expensive. In support calculation, each candidate set should be matched with items in the database to consume a lot of time.

For multiple scan database problems, we can store the data in the database in a list form. Because the number of candidate sets generated is not always frequent, this article uses the method of comparing item sets to restrict the generation of candidate sets.

3 Improved Apriori-Pro Algorithm

Apriori-Pro algorithm, mainly aiming at the classic Apriori algorithm, improves the problem of producing a large number of candidate items. The algorithm is as follows.

3.1 Apriori-Pro Algorithm Thought

Based on the a priori nature (all nonempty subsets of frequent itemsets must also be frequent), we can draw an important conclusion that if two candidate k-item sets appear in p(p >= minsupport) transactions at the same time, the linked candidate The (k + 1)-item set must be frequent, in which case the two candidate k-item sets are linked; or, the link is not.

Proof: If the candidate k-item sets Ck1 and Ck2 appear simultaneously in p (p >= minimum support) transactions, it is clear that Ck1 and Ck2 are frequent. Then the linked ⟨Ck1, Ck2⟩ also appears in p transactions. Obviously ⟨Ck1, Ck2⟩ is also frequent and performs link operations. Similarly, if the number of simultaneous occurrences between Ck1 and Ck2 is lower than p, then the number of links ⟨Ck1, Ck2⟩ appearing in the transaction is also lower than p, then it is not frequent and no link operation is performed.

As shown in Table 1, a transaction database is listed. After scanning the database, a candidate 1-item set will be generated, as shown in Table 2 (where Items represents the candidate set, Sup represents the support level, and TID represents the transaction sequence). If the minimum support degree is 2, the comparison between I1 and I2 appears in transaction 1, 4, and its p value is equal to minimum support. In Table 1, it can be found that item set also appears in transaction 1, 4, knowing that is frequent, so I1, I2 link can be used. In the same way, I1 and I3 are observed, and only in transaction 5, p value is less than minimum support; In Table 1, it can be seen that is only in transaction 5, and its support degree is 1, less than the minimum support degree; Therefore, according to the prior nature, there is no need to link I1 and I3.

Table 1. Sample transaction database

TID	ID
1	I1
2	I2
3	I2, I3
4	I1, I2
5	I1, I3

Table 2. Candidate 1-set

Items	Sup	TID
⟨I1⟩	3	1, 4, 5
⟨I2⟩	4	1, 2, 3, 4
⟨I3⟩	2	3, 5

According to this conclusion, the candidate (k + 1)-item sets generated from the candidate k-item sets must be frequent, avoiding the generation of invalid candidate set, and reducing the space-time overhead.

3.2 Apriori-Pro Algorithm Details

The source code of this article is written in python language, aiming at the drawbacks of the original algorithm scanning the database too much. The improved Apriori-Pro algorithm only scans the database once. When the algorithm scans the database for the first time, the data set is stored in a set, avoiding multiple scans of the database, which greatly reduces the time complexity of the algorithm. This article uses the data in Table 3 as an example. Set the minimum support minsupport = 2.

Step 1: Traverse the new transaction database and save it in the collection

Table 3. ALLElectronics affairs database

TID	ID
1	I1, I2, I5
2	I2, I4
3	I2, I3
4	I1, I2, I4
5	I1, I3
6	I2, I3
7	I1, I3
8	I1, I2, I3, I5
9	I1, I2, I3

The unique identifier in the transaction database is TID, and the traversed data is stored in a list according to the TID. Different items under the same TID are separated by "," and different TIDs are separated by "[]," to get the list data_set = [['I1', 'I2', 'I5'], ['I2', 'I4 '], ['I2', 'I3'], ['I1', 'I2', 'I4'], ['I1', 'I3'], ['I2', 'I3'], ['I1 ', 'I3'], ['I1', 'I2', 'I3', 'I5'], ['I1', 'I2', 'I3']]. Thus, when generating frequent itemsets later, simply traverse the list without having to scan the database multiple times.

Step 2: Generate a candidate 1-item set, calculate support at the same time, delete the entry whose support is less than minsupport, and obtain a frequent 1-item set L1, as shown in Table 4.

Table 4. Frequent 1-item set L1

Items	Sup	TID
⟨I1⟩	6	1, 4, 5, 7, 8, 9
⟨I2⟩	7	1, 2, 3, 4, 6, 8, 9
⟨I3⟩	6	3, 5, 6, 7, 8, 9
⟨I4⟩	2	2, 4
⟨I5⟩	2	1, 8

This article forms the item set at the same time, add the TID column for each item set, TID column for the transaction code; At the same time, calculate the support degree according to the transaction code number, added to the Sup paired for each item set.

Step 3: Generate a frequent 2-item set from a frequent 1-item set.

From the conclusion of 3.1, we only need to cross-compare TID columns. Comparing the TID column of ⟨I1⟩ with the TID column of ⟨I2⟩, the repeated part has the transaction p = (1, 4, 8, 9), and the length of p is greater than minsupport = 2, so linking ⟨I1⟩ with ⟨I2⟩ Repeat part p as the TID column of ⟨I1, I2⟩, and the support degree 4 can be calculated at the same time. By analogy, the TID column of ⟨I1⟩ is compared with the TID columns of ⟨I3⟩, ⟨I4⟩, and ⟨I5⟩, and the TID columns of ⟨I2⟩, ⟨I3⟩, and ⟨I4⟩ are respectively started and left. The TID column of the next item is compared to determine whether the link condition is satisfied. If the link condition is satisfied, the link is made and the support degree is calculated; if the condition is not satisfied, no link is made. The frequent 2-items set resulting from the frequent 1-item set is shown in Table 5.

Step 4: Generate Frequent 3-item Sets

Table 5. Frequent 2-item set L2

Items	Sup	TID
⟨I1, I2⟩	4	1, 4, 8, 9
⟨I1, I3⟩	4	5, 7, 8, 9
⟨I1, I5⟩	2	1, 8
⟨I2, I3⟩	4	3, 6, 8, 9
⟨I2, I4⟩	2	2, 4
⟨I2, I5⟩	2	1, 8

A frequent 3-item set is generated from a frequent 2-item set, unlike the classic Apriori algorithm, which does not need to generate a candidate set. The TID columns of each item in Table 5 can be directly cross-compared. Using the aforementioned conclusions, we can draw a frequent 3-item set as shown in Table 6. Since Table 6 can no longer generate new frequent itemsets, the algorithm ends.

Table 6. Frequent 3-item set L3

Items	Sup	TID
⟨I1, I2⟩	4	1, 4, 8, 9
⟨I1, I3⟩	4	5, 7, 8, 9
⟨I1, I5⟩	2	1, 8
⟨I2, I3⟩	4	3, 6, 8, 9
⟨I2, I4⟩	2	2, 4
⟨I2, I5⟩	2	1, 8

4 Experimental Results and Conclusion Analysis

The algorithm code part is written in Python language and runs on Python 3.6.4 Shell. It uses MySQL 5.1 as the database development platform of the program. The mining of the association rules was carried out by using part of the transaction data sold by the Groceries grocery store for one month, including 1050 transactions. The maximum number of items contained in a transaction was 20 items. The running platform of the algorithm is shown in Table 7.

Table 7. Testing platform

Processor	Intel(R)Core(TM)i5-8250U CPU @ 1.80 GHz
Memory (RAM)	8.00 GB
Hard disk	256G SSD

The improved Apriori algorithm and the traditional Apriori algorithm are used to find the time of all frequent itemsets and association rules under different support degrees of frequent k item sets (k = 3), as shown in Fig. 1.

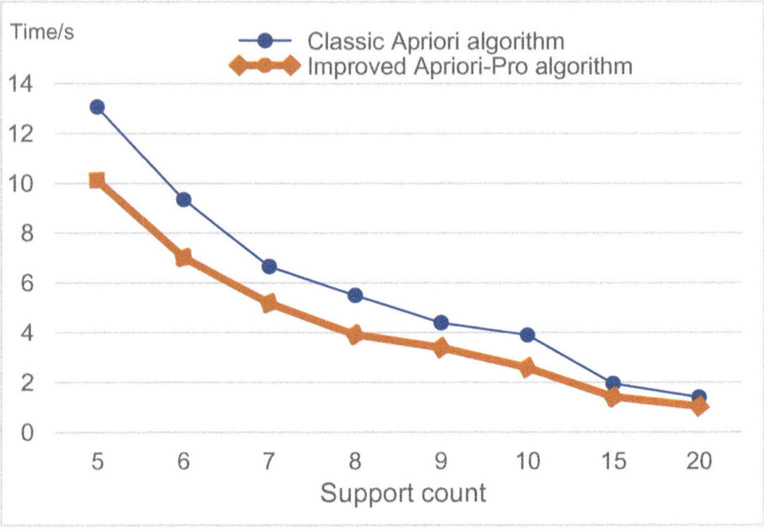

Fig. 1. Two algorithms to find frequent itemsets contrast

The minimum support count is 10, traverting the transaction data, and the comparison between the traditional algorithm and the improved Apriori-Pro algorithm in producing different frequent item sets is shown in Fig. 2.

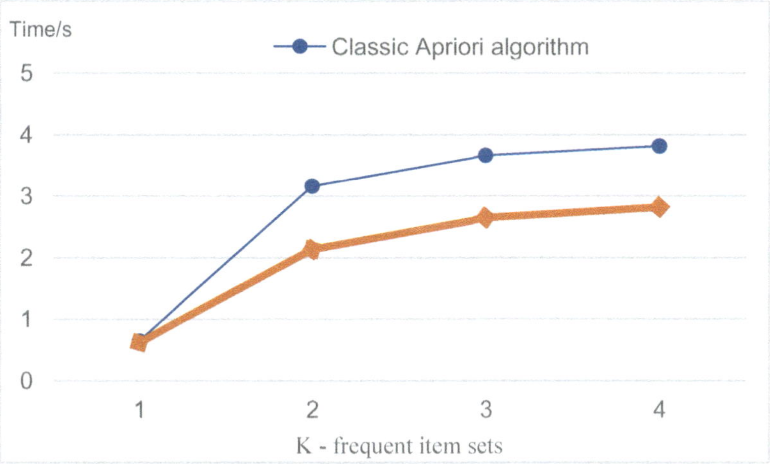

Fig. 2. Produces k-frequent itemset comparisons

The experimental results show that the improved Apriori-Pro has significantly improved compared with the traditional algorithm, and can obtain strong association rules more efficiently. Figure 1 shows that when the improved algorithm has different support degrees, the speed of mining frequent itemsets has been greatly improved compared with the original algorithm. Figure 2 shows that, with the increase of k value, the new algorithm is generated more rapidly than the original algorithm when different frequent item sets are produced. This characteristic of the new algorithm can better reflect its superiority with the increase of data.

5 Conclusions

An improved Apriori-Pro algorithm is proposed, and column of transaction items is added in the algorithm. By comparing transaction item columns, the degree of support for frequent item sets can be obtained more quickly. At the same time, whether or not the frequent (k-1) itemsets are determined by comparing the transaction item columns to determine whether the generated item sets are frequent is determined. Linking produces frequent k-item sets, avoiding the generation of invalid candidate sets. However, this algorithm still has its shortcomings, such as failing to solve the problems brought by big data mining. With the increase of data, the algorithm's time consumption will be very large. In the future, new algorithms will be merged with big data to improve the applicability of the algorithm in big data environments.

Acknowledgement. This Research work was supported by the National Science Foundation of China under (Grant No. 61703005).

References

1. Agrawal, R., Srikan, R.: Fast algorithms for mining association rules in lager databases. In: Proceedings of the Twentieth International Conference on very Large Databases, Santiago, pp. 487–499 (1994)
2. Liu, W., Chen, J., Qu, S., et al.: Improved Apriori algorithm. Comput. Eng. Appl. **47**(11), 149–151 (2011)
3. Qin, J., Hao, T., Dong, Q.: Parallel improvement of Apriori algorithm based on MapReduce. Comput. Technol. Dev. **27**(4), 64–68 (2017)
4. Wei, L., Wei, Y., Gao, C.: Improved Apriori algorithm based on bigtable and MapReduce. Comput. Sci. **42**(10) (2015)
5. Mi, Y., Jiang, L., Mi, C.: Rough association rules algorithm with negation under MapReduce. Comput. Integr. Manuf. Syst. **20**(11), 2893–2903 (2014)
6. Zhang, W.: An improved AprioriTid algotithm. J. Shenyang Univ. Technol. **38**(3), 314–318 (2016)
7. Liu, L., Yu, S., Wei, X., et al.: Animproved Apriori-based algorithm for friends recommendation in microblog. Int. J. Commun. Syst. **32**(2), 3453–3462 (2017)
8. Feng, D., Zhu, L., Zhang, L.: Research on improved Apriori algorithm based on MapReduce and HBase. In: Proceedings of 2016 IEEE Advanced Information Management, Communicates, Electronic and Automation Control Conference, pp. 887–891 (2016)

An Improved Fuzzy Support Vector Machine Algorithm Based on Refactoring Class Center

Huaping Zhou(✉), Huangli Qin, and Huali Shen

School of Computer Science and Engineering, Anhui University of Science and Technology, Huainan 232001, China
hpzhou@aust.edu.cn, qin_huangli@163.com,
759896846@qq.com

Abstract. The fuzzy support vector machine with membership function designed based on class center can effectively solve the problems of traditional support vector machine (SVM in sensitivity to noises and outliners. However it reduces the classification effect by endow support vectors with smaller membership values. Therefore, the paper proposed an improved membership function design method, which reduces the dependence on geometric distribution of the sample data while considering the essential characteristic of SVM as well. To ensure new class center closer to the surface of classification. A new class center is calculated according to the average value of the distance between the hyperplane and the data which have major impact on classification accuracy. Thus improved the classification accuracy by samples which are closer to the classification of surface of classification obtained the higher degree of membership and others get lower degree of membership.

Keywords: Support vector machine (SVM) · Membership function
Class center

1 Introduction

Support vector machine (SVM) can improve the ability of learning machine meanwhile, support vector machine using kernel function can transform nonlinear problem into linear problem. Based on the above features, support vector machine (SVM) has become an excellent algorithm for machine learning. However, some types of data in real lift are not certain, to solve this problem we need to construct the fuzzy support vector machine (FSVM) which implementation by introducing membership function in SVM. The fuzzy support vector machine gives the membership degree to all data according each sample play different role in the training process, FSVM give a small weight to noise or field point to eliminate the influence of noise and field point on training accuracy.

The design of membership function is the most critical step. There are many ways to construct membership functions. Literature [2] proposed the membership function based on the distance between samples and its class center, and the degree of membership was inversely proportional to the distance. Literature [7] proposed that the sample membership degree were not simple linear relations, thus the standard s-type

J. Abawajy et al. (Eds.): ATCI 2018, AISC 842, pp. 91–99, 2019.
https://doi.org/10.1007/978-3-319-98776-7_11

function was proposed. The algorithm proposed by literature [2] weakens the efficiency of the sample that away from class center also weakens the efficiency of vector machine. At present, most structural membership function methods are based on the distance between sample and class center, however, this method give the support vector a smaller membership which should have a larger membership in the actual classification. Based on this, literature [10] proposed to replace the class center with the hyperplane of the class, the membership function is designed as a function of distance between hyperplane of the class and samples which to enhance the dependence of the Fuzzy support vector machine on the spherical distribution of the samples.

This paper proposed a new membership function design method in reference of literature [10] and literature [2]. This method first calculate the class center of the sample, then recalculate a new class center according the sample distribution characteristics, and takes the newly acquired class center as the final class center of the sample. In this ways, the distance between the support vector and the class center is decreased while the radius of the minimum hypersphere is increased, thus, it can not only increase the membership of support vector, but also reduce the membership of noise. In this paper, the algorithm get a new class center according to the distributions of samples, in this way the algorithm can reduce the dependence of spherical of the samples, thus improving the generalization ability of the algorithm.

2 Fuzzy Support Vector Machine

Fuzzy support vector is on the basis of traditional support vector machine (SVM), according to the importance of each samples gives different membership value to reduce the impact of the outlier and noise on the classification accuracy and generalization ability of support vector machines. When using support vector machines to deal with classification problems, first, we select a membership value function for samples, use membership function to blur the sample, and get the membership value s_i of each sample x_i. Then the traditional training set becomes a fuzzy training set $S = \{(x_1, y_1, s_1), (x_2, y_2, s_2), \ldots, (x_l, y_l, s_l)\}$, $x_i \in R^n, y_i \in \{-1, 1\}$, $0 \le s_i \le 1$.

The optimization problem of constructing optimal classification surface is transformed into the following formula:

$$\begin{cases} \min & 0.5\|\omega\|^2 + c\sum_{i=1}^{l} s_i\xi_i \\ \text{s.t.} & y_i(\omega \cdot x_i + b) \ge 1 - \xi_i \\ & \xi_i \ge 0 \quad i = 1, \ldots, l \end{cases}$$

The optimal solution is the saddle point of the Lagrange function:

$$\mathrm{L}(\omega, b, \xi) = \frac{1}{2}\|\omega\|^2 + c\sum_{i=1}^{l} s_i\xi_i - \sum_{i=1}^{l} \alpha_i[y_i(\omega \cdot x_i + b) - 1 + \xi_i] - \sum_{i=1}^{l} \beta_i\xi_i$$

Because ω, b and ξ meet the following conditions at the saddle point:

$$\frac{\partial L}{\partial \omega} = \omega - \sum_{i=1}^{l} \alpha_i y_i x_i = 0$$

$$\frac{\partial L}{\partial b} = - \sum_{i=1}^{l} \alpha_i y_i = 0$$

$$\frac{\partial L}{\partial \xi} = cs_i - \alpha_i - \beta_i$$

Thus, the problem of constructing optimal hyperplane is transformed into a dual quadratic programming problem:

$$\begin{cases} \max \sum_{i=1}^{l} \alpha_i - \frac{1}{2} \sum_{i=1}^{l} \sum_{j=1}^{l} \alpha_i \alpha_j y_i y_j (x_i x_j) \\ \text{s.t. } \sum_{i=1}^{l} \alpha_i y_i = 0 \\ 0 \leq \alpha_i \leq cs_i \quad i = 1, 2, \ldots, l \end{cases}$$

At the same time, the optimal solution should satisfy the KKT conditions:

$$\begin{cases} \alpha_i [y_i(\omega \cdot x_i + b - 1 + \varepsilon_i)] = 0 \\ \beta_i \xi_i = (cs_i - \alpha_i) \xi_i = 0 \end{cases}$$

Then get the decision function:

$$f(x) = sgn(\omega \cdot x + b)$$

It can be seen from the fuzzy support vector machine model that the sample corresponding to the value of $\alpha_i = 0$ is the point of correct classification, namely the non-support vector; the samples corresponding to the value of $0 < \alpha_i < cs_i$ are the support vectors in the general sense; the samples points corresponding to $\alpha_i = cs_i$ are boundary support vectors, and boundary support vectors are sample points that are easy to be misclassified. For the membership of the fixed parameter c the membership of samples reflects the importance attached to the sample point. The greater the degree of membership is, the less likely it is to be misclassified. Therefore, in the construction of the fuzzy support vector machine, we should making noise or field point got a less membership and maximizing the membership of boundary support vectors.

3 The Design of the New Membership Function

3.1 The Design Method of the New Membership Function

The fuzzy support vector machine algorithm designed by sample and class center distance considered those sample that away from class center as isolated point and gives them a smaller membership. The further the distance from the center of the class is, the smaller the membership value is, which is consistent with the distribution characteristics of the sample. However, support vectors that is relatively far away from class center according to the thought is also seen as isolated point and gives small membership value, which greatly weakened the influence that support vector brought in the classification performance. The existing fuzzy support vector machine algorithm based on class center design can solve the above problems to some extent by adjusting the distance function.

However, this algorithm can not solve the problem of spherical dependence of the sample based on the fuzzy support vector machine algorithm based on the class center (Figs. 1 and 2).

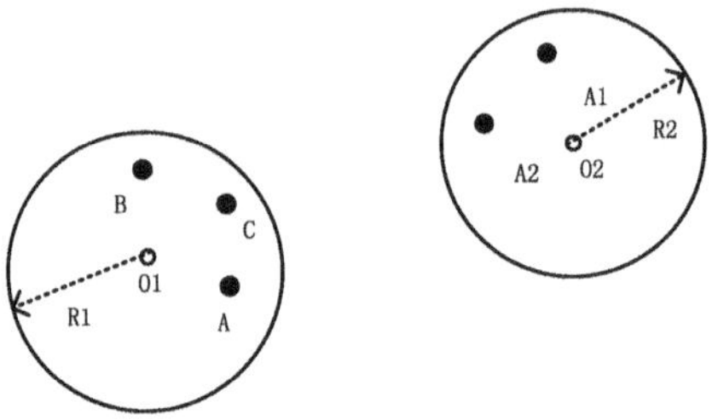

Fig. 1. Schematic diagram of membership function based on sample and class center design.

This paper designs the membership function as shown in the figure below:

As shown in Fig. 3 by introducing the offset factor, the original class center O1 is migrated to the O1, the class center O2 is shifted to position O2, the migration factor is determined by calculating the average value of the sample in two hyperplanes to the corresponding hyperplane distance. Offset factor calculation is calculated by the distance between sample that have greater influence on the structure of classification and hyperplane. The above method can overcome the dependence of the fuzzy support vector machine based on the distance between sample and the center of class. This article use the method of reconstructing class center to make the class center closer to the classification surface, in this way, the radius of the smallest hypersphere can be increased to a certain extent, and, in this way not only reduced the distance between sample that near class surface and class center to allow sample that near support vectors get greater membership value, at the same time correspondingly increase the distance between sample that away from class center and class center to make noises and outliners obtain smaller membership value.

The training samples were divided into two categories: positive and negative:

$$C^{+} = \{x_i | x_i \in C \text{且} y_i = 1\}$$
$$C^{-} = \{x_j | x_j \in C \text{且} y_j = -1\}$$

$C = C^{+} + C^{-}$ The center of the smallest super ball with two types of samples is respectively o_{+}, o_{-}, For nonlinear, we introducing mapping $\emptyset(x)$ there are:

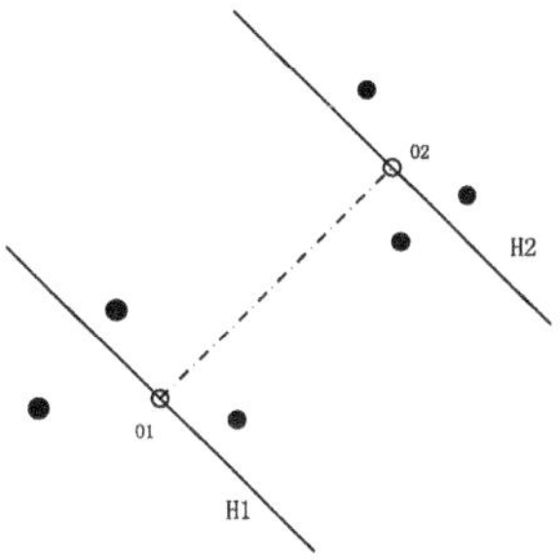

Fig. 2. A schematic diagram of membership function based on sample and in-class hyperplane design.

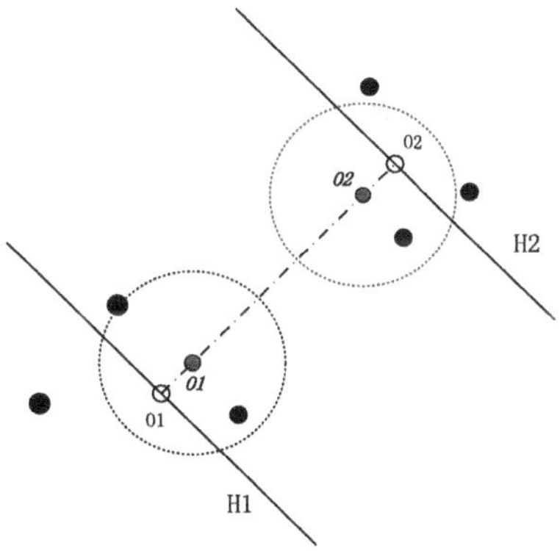

Fig. 3. Membership function design sketch.

(1) The class center of two types of samples in the feature space:

$$\varnothing(o_+) = \frac{1}{m}\sum_{i=1}^{i=m} \varnothing(x_i)$$

$$\varnothing(o_-) = \frac{1}{n}\sum_{j=1}^{j=n} \varnothing(x_j)$$

m is the number of positive samples, n is the number of negative samples.

(2) Vector $\tilde{\omega}$ consisted by two class centers:

$$\varnothing(\tilde{\omega}) = \varnothing(x_+) - \varnothing(x_-)$$

(3) The distance of two class centers D satisfied with in the feature space:

$$D^2 = \|\varnothing(x_+) - \varnothing(x_-)\|^2$$
$$= \frac{1}{n_+^2}\sum_{i=1}^{n_+}\sum_{i=1}^{n_+} K(x_i, x_j) + \frac{1}{n_-^2}\sum_{i=1}^{n_-}\sum_{i=1}^{n_-} K(x_i, x_j) + \frac{2}{n_+ \times n_-}\sum_{i=1}^{n_+}\sum_{i=1}^{n_-} K(x_i, x_j)$$
$$\mathrm{D}_{T_+} = (\varnothing(x_+) - \varnothing(x_-)) \cdot (\varnothing(x_+) - \varnothing(T_+))$$
$$= \frac{1}{n_+^2}\sum_{i=1}^{n_+}\sum_{j=1}^{n_+} K(x_i, x_j) + \frac{1}{n_-}\sum_{i=1}^{n_-} K(T_+, x_-) - \frac{1}{n_+}\sum_{i=1}^{n_+} K(T_+, x_+) - \frac{1}{n_+ \times n_-}\sum_{i=1}^{n_+}\sum_{j=1}^{n_-} K(x_i, x_j) > 0$$

(4) Samples T_+, T_- between two hyperplanes meets the following conditions:

$$\mathrm{D}_{T_+} = (\varnothing(x_+) - \varnothing(x_-)) \cdot (\varnothing(x_+) - \varnothing(T_+))$$
$$= \frac{1}{n_+^2}\sum_{i=1}^{n_+}\sum_{j=1}^{n_+} K(x_i, x_j) + \frac{1}{n_-}\sum_{i=1}^{n_-} K(T_+, x_-) - \frac{1}{n_+}\sum_{i=1}^{n_+} K(T_+, x_+) - \frac{1}{n_+ \times n_-}\sum_{i=1}^{n_+}\sum_{j=1}^{n_-} K(x_i, x_j) > 0$$

$$\mathrm{D}_{T_-} = (\varnothing(x_+) - \varnothing(x_-)) \cdot (\varnothing(T_-) - \varnothing(x_-))$$
$$= \frac{1}{n_+^2}\sum_{i=1}^{n_+}\sum_{j=1}^{n_+} K(x_i, x_j) + \frac{1}{n_-}\sum_{i=1}^{n_-} K(T_+, x_-) - \frac{1}{n_+}\sum_{i=1}^{n_+} K(T_+, x_+) - \frac{1}{n_+ \times n_-}\sum_{i=1}^{n_+}\sum_{j=1}^{n_-} K(x_i, x_j) > 0$$

(5) The migrated class center $\varnothing(x_+)^{'}\varnothing(x_-)^{'}$:

$$\varnothing(x_+)' = \frac{(\varnothing(x_-) - \varnothing(x_+)) \times d_{T+}}{D} + \varnothing(x_+)$$
$$\varnothing(x_-)' = \frac{(\varnothing(x_+) - \varnothing(x_-)) \times d_{T_-}}{D} + \varnothing(x_-)$$

(6) The samples T_+, T_- that located between two hyperplanes have mean value $d_{T_+} d_{T_-}$ of distance to its corresponding hyperplane:

$$d_{T+} = \frac{1}{n_{T+}}\sum_{i=1}^{n_{T_+}} |D_{T_i}|$$
$$d_{T_-} = \frac{1}{n_{T-}}\sum_{i=1}^{n_{T_-}} |D_{T_i}|$$

(7) The distance d_{i_+}, d_{i_-} of samples to its corresponding class center $\varnothing(x_+)^{'}\varnothing(x_-)'$:

$$d_{i_+}^2 = K\left(\varnothing(x_{i_+}), \frac{(\varnothing(x_-) - \varnothing(x_+)) \times d_{T+}}{D} + \varnothing(x_+)\right)$$
$$= K\left((\varnothing(x_{i_+}), \frac{\varnothing(x_-) \times d_{T+}}{D} + \varnothing(x_+) \times \left(1 - \frac{d_{T+}}{D}\right)\right)$$
$$= \frac{d_{T+}}{D} \times \frac{1}{n_-}\sum_{j=1}^{n_-} K(x_{i_+}, x_j) + \left(1 - \frac{d_{T+}}{D}\right) \times \frac{1}{n_+}\sum_{j=1}^{n_+} K(x_{i_+}, x_j)$$

$$\begin{aligned} d_{i_-}^2 &= K\left(\varnothing(x_{i_-}), \frac{(\varnothing(x_+) - \varnothing(x_-)) \times d_{T-}}{D} + \varnothing(x_-)\right) \\ &= K\left((\varnothing(x_{i_-}), \frac{\varnothing(x_+) \times d_{T-}}{D} + \varnothing(x_-) \times \left(1 - \frac{d_{T-}}{D}\right)\right) \\ &= \frac{d_{T-}}{D} \times \frac{1}{n_+}\sum_{j=1}^{n_+} K(x_{i_+}, x_j) + \left(1 - \frac{d_{T-}}{D}\right) \times \frac{1}{n_-}\sum_{j=1}^{n_-} K(x_{i_+}, x_j) \end{aligned}$$

According to the above ideas, membership function of this paper is defined as:

$$s(i_+) = \begin{cases} 1 - \frac{d_{i_+}}{\max(d_{i_+}) + \delta} & d_{i_+} \le \left(d_{T_+} + \frac{1}{2}\max(d_{i_+})\right) \\ \varepsilon & d_{i_+} > \left(d_{T_+} + \frac{1}{2}\max(d_{i_+})\right) \end{cases}$$

$$s(i_-) = \begin{cases} 1 - \frac{d_{i_-}}{\max(d_{i_-}) + \delta} & d_{i_-} \le \left(d_{T_+} + \frac{1}{2}\max(d_{i_+})\right) \\ \varepsilon & d_{i_-} > \left(d_{T_+} + \frac{1}{2}\max(d_{i_+})\right) \end{cases}$$

$s(i_+)$ is the membership value of the positive sample, $s(i_-)$ is the membership value of the negative samples; ε and δ as a very small positive number.

It can be seen from the definition of the membership function: by defining the migration factor to make class center moves along the support vector direction to increase the gap of value between support vector and noises or outliers, which shows that samples that have great influence in the process of structure classification surface have bigger values of membership and other have little influence in the process of structure classification surface have smaller values of membership; by introducing migration factor can solve the dependence of traditional fuzzy support vector machine on the spherical distribution, this is because the migration factor is relative small for those date set with relatively concentrated sample distribution and bigger migration factor for those data set with scattered sample distribution; by introducing ε improved classification effect by giving a smaller membership value to samples that near the classification surface, because those samples cannot be a support vector.

4 Experiments

(a) Evaluation Method

In order to verify the validity of the proposed algorithm in this paper, SVM algorithm and traditional fuzzy vector machine algorithm (OFSVM) and the algorithm in this paper are used respectively compared. The experimental data set was tested on the UCI standard data set in the machine learning reposition, and, the experimental programs are all written in MATLAB.

The experiment was carried out on a PC (CPU: 3.42 GHZ, 4 core processor, main memory: 8 GB), the operating system is Windows 7, and the experimental tool is MATLAB R2016B.

In this paper, SVM algorithm and traditional fuzzy support vector machine algorithm (OFSVM) and the algorithm in this paper is used to compare in three UCI datasets. The paper select the linear kernel function to map the sample to high dimensional space.

The penalty coefficient C was sampled in a continuous range for a certain amount of C, the classification accuracy was obtained by using different penalty coefficient C in three UCI data sets. The following figures presents our experiment results.

(b) Experimental Analysis

As shown in Fig. 4, the algorithm presented in this paper obtains the best classification effect. As shown in Fig. 6, the classification results obtained by using traditional fuzzy support vector machines are not as good as those obtained by SVM; however, the algorithm presented in the paper still got the best classification results. As shown in Fig. 5, the classification results obtained by the proposed algorithm are not lower than those obtained by SVM.

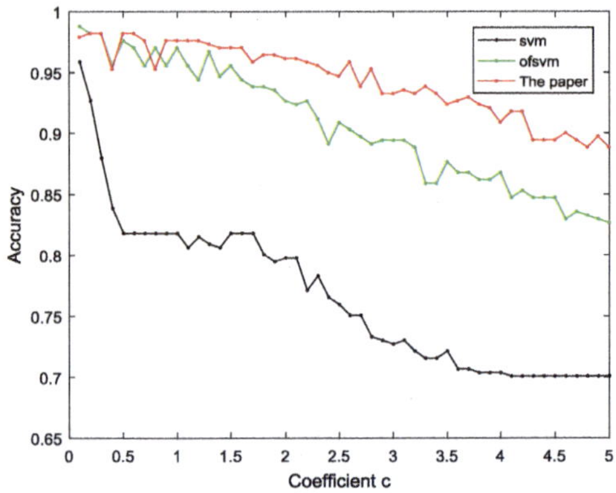

Fig. 4. The classification accuracy of different algorithm in data set Caner.

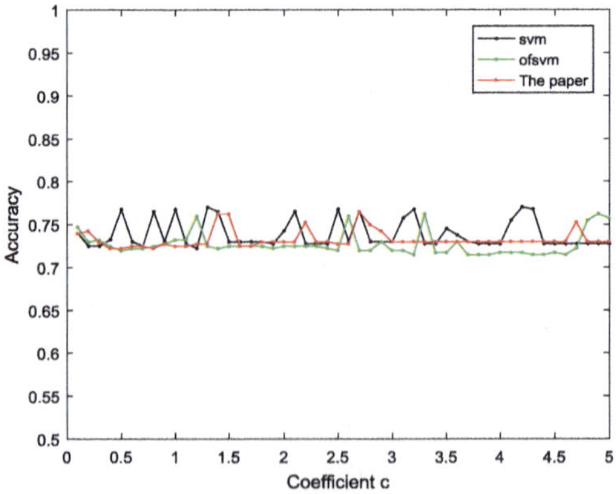

Fig. 5. The classification accuracy of different algorithm in data set German

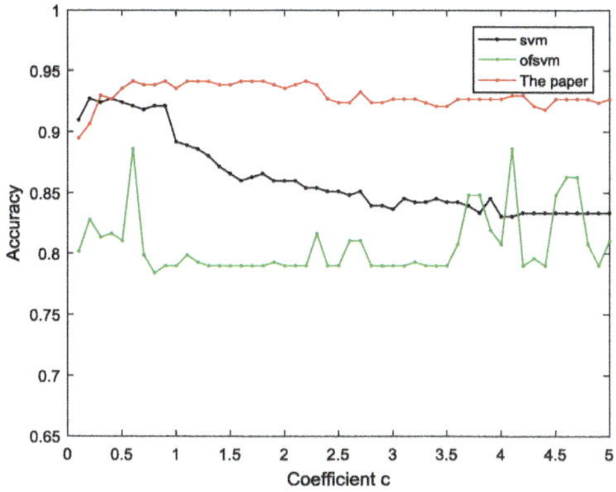

Fig. 6. The classification accuracy of different algorithm in data set Vote.

5 Conclusions

In this paper, an improved fuzzy support vector machine membership function design method based on distance between sample and class center is proposed. This method not only overcomes the dependence of traditional fuzzy support vector machine on the spherical distribution of data set, but also improves the classification accuracy. Although the algorithm proposed in this paper achieves better classification results on most data sets, there nearly the same classification effect obtained in some data set when using different algorithm. In the future we need further study about the relationship between the algorithm and the sample dataset so as to achieve better classification results.

Acknowledgements. This Research work was supported by National natural science foundation Project (Grant No. 61703005).

References

1. Vapnike, V.: The Nature of Statistical Learning Theory. Springer, Berlin (1995)
2. Jiang, X.F., Yi, Z., Lu, J.C.: Fuzzy SVM with a new fuzzy membership function. Neural Comput. Appl. **15**, 268–276 (2006)
3. Lin, C.F., Wang, S.D.: Fuzzy support vector machine. IEEE Trans. Neural Netw. **13**, 464–471 (2006)
4. Xu, C.Y., Ye, N.: An improved fuzzy support vector machine based on intra-class hyperplane. Comput. Appl. Softw. **115**, 106 (2013)
5. Li, M.M., Xiang, F.H., Liu, X.W.: A novel membership function for fuzzy support vector machines. Comput. Eng. Sci. **21**, 1263 (2009)
6. Liu, K.M., Wu, X.J.: A fuzzy support vector machine based on new membership function. Comput. Eng. **12**, 357 (2016)
7. Wu, Q., Liu, S.Y., Du, J.: Fuzzy support vector machine method based on border vector extraction. Pattern Recogn. Artif. Intell. **40**, 1–6 (2008)
8. Zhang, X., Xiao X.L., Xu G.Y.: Determination and Analysis of Fuzzy Membership for SVM (2006)
9. Kan, J.M., Li, K.Y.: Background modeling method based on improved multi-Gaussian distribution. ICCASM **42**, 214–218 (2010)

An Ensemble Data Stream Mining Algorithm for Class-Imbalanced Applications

Qin Sun(✉), Yufei Wu(✉), Huixian Duan, Jun Wang, and Lin Mei

The Third Research Institute of Ministry of Public Security, Shanghai 201204, China
sunqin14@nudt.edu.cn, wu.yufei@outlook.com

Abstract. For the category-imbalanced applications, traditional ensemble data stream mining algorithms will result in low accuracy for the small classes and fail to meet the needs of applications. This paper provides a novel class-imbalanced data learning method based on MAE named CIMAE to solve the above problem. Instead of directly using each incoming data, it acquires data blocks for online training each time by setting up a sample library and a sliding window. Compared with traditional data stream mining algorithms, the results showed that CIMAE achieves the state-the-of-art performance for class-imbalanced application.

Keywords: Ensemble data stream mining · Class-imbalanced application Online learning

1 Introduction

Data stream mining is an online mining algorithm, which means with the training data continues to come, the learning system constantly learns new training data on basis of the original learning results. At the same time, it needs of the application in terms of real-time performance and predictability. A good data stream mining system not only needs to process the incoming data in real time, but also adapts to the changing concepts [1].

Adaptive classifiers-ensemble learning method is one of the most important algorithms in the current field of data stream mining. This method divides the incoming data stream into data blocks and learns a basic classifier for each block. The adaptive classifiers-ensemble system saves a certain number of base classifiers to predict on new samples. In 2000, the SEA algorithm (Streaming Ensemble Algorithm) was proposed [2], which is the earliest ensemble data stream mining algorithm. It sets a classifier library with a certain capacity and divides the arriving data stream in order into multiple data blocks with the same size, then learns a classifier for each block and puts it into the library until the capacity reaches the upper limit. With the formation of a new classifier, heuristic method [3] is used to evaluate all classifiers in the library and individuals with poor performance are eliminated to keep the capacity of the classifier library remain stable. Finally, the category of a new sample is obtained by majority voting rule [4]. Since all base classifiers are involved in the prediction and their importance is the same, the algorithm is less adaptable to the conceptual drift of the mutation. In 2003, AWE

J. Abawajy et al. (Eds.): ATCI 2018, AISC 842, pp. 100–108, 2019.
https://doi.org/10.1007/978-3-319-98776-7_12

(Accuracy-Weighted Ensembles) algorithm [5] based on SEA was proposed. It used the weighting method to replace the average output of the basic classifiers. This method assigns a weight to each base classifier, making the classifiers with smaller classification errors occupy a higher proportion in the voting process. When the total number of classifiers is greater than the capacity of the classifier library and the proportion of the newly learned classifier is greater than some of the basic classifiers in the classifier library, the basic classifier with the smallest proportion of votes in the classifier bank are eliminated. In the prediction stage, use the weighted average value of all classifiers in library is used as forecast result. Compared to SEA, this method of weight setting improves the adaptability to concept changes. In 2005, ACE (Adaptive Classifier Ensemble) [5] is proposed. It adds concept-drifting monitor on the basis of AWE to improve the ability of adapting to sudden changes of concepts. If no concept change has been monitored, a weighted vote is used for ensemble forecasting like AWE. Otherwise, waiting until the warning window is full and re-learning a new classifier for prediction. MAE (Memorizing based Adaptive Ensemble) [6] introduces the memory and forgetting mechanism in the human learning process into ensemble data stream mining, which improves the adaptability to complex concept-drifting.

There is a serious class-imbalance problem in many applications in real life. That is, there may be some classes of samples in each arriving data block are too few or even none, and some categories have a high proportion of samples. For example, intrusion detection and failure prediction all fall into this category of applications. Traditional ensemble data stream mining algorithms don't take into account the phenomenon of class-imbalance distribution. Such problem leads to low classification accuracy and even the inability to learn. Although the memory and forgetting mechanism in the MAE algorithm can mitigate the effects of class-imbalance, there are still difficulties in learning data blocks and low prediction accuracy for serious class-imbalance problems.

Aiming at the deficiencies of the MAE algorithm in dealing with the class-imbalance problem, in this paper, we propose a data stream mining algorithm named CIMAE (class-imbalanced data learning method based on MAE) which considers the class-imbalance problem. Based on the MAE algorithm, the CIMAE algorithm mainly improves the acquisition method of the data set used for training the base classifier each time. That is, a sliding window [7] is set for each category. It's not used a newly arrived data block for online learning directly. Instead, the samples in the data block are respectively entered into the corresponding sliding window according to their own categories, and finally, a training data block for online learning is constructed based on the samples in the sliding windows of each category. The sliding windows are updated slowly for categories with few samples and updated quickly for categories with a large number of samples. Different update speeds make the number of samples per class basically equal when participating in the base classifier, Therefore, the problem of class-imbalance is transformed into a normal problem, which improves the ability of the algorithm to learn from the problem of class-imbalance.

2 Disadvantages of Traditional Ensemble Data Stream Mining

Traditional ensemble data stream mining algorithms such as SEA, AWE, ACE, etc. divide data streams arriving sequentially into data blocks of equal size. The data streams DS (Data Stream) can be seen as arriving in a data block manner. Every time a data block DB (Data Block) is reached, a new basic classifier c is learned by it, put c in a basic classifier library named ES, and then use current DB to evaluate the basic classifier in ES. If the number of basic classifiers in ES reaches the specified upper threshold k, the basic classifier with the lowest evaluation value is deleted.

Figure 1 shows the generation of basic classifiers training set in traditional ensemble data stream mining. With the continuous arrival of data stream, once the number of received samples reaches size of data block, these data constitute a data block, and then use it as a training sample set to obtain a new basic classifier. This method has the following problems: If the application is a class imbalance problem or if the sample suddenly changes within a certain period of time, resulting in a significant difference in the number of samples corresponding to each category in the data block DB, even if some classes do not have a sample, the basic classifier c obtained by using this data block as a training sample set has poor prediction performance for the application. At the same time, this data block is used to evaluate the basic classifier in the ES, the evaluation value doesn't truly reflect the actual classification effect of the basic classifiers, and it's possible to delete the good basic classifier, resulting in the effect of integrated classification is not ideal.

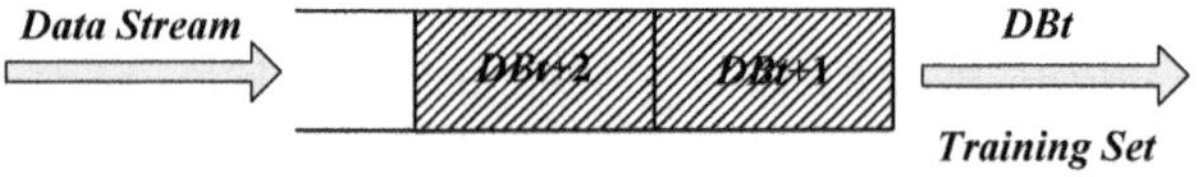

Fig. 1. Training set generation model of traditional ensemble data stream

The MAE algorithm introduces the "memory and forgetting" mechanism [8] into the learning process of the basic classifiers. The algorithm model presupposes two classifier libraries: a memory bank *MS* for storing all current useful basic classifiers; a recalling bank *ES* for storing currently recalled classifiers, and *ES* is a subset of *MS*. MAE compares every classifiers obtained by learning to the knowledge gained by the learning system. When a new data block *DB* is formed, a new classifier is learned using *DB* and placed in *MS*. At the same time, the *d* basic classifiers with the strongest correlation with the current data block in *MS* are copied into *ES*, where *d* represents the maximum capacity of the *ES*, which is a "memory" mechanism in MAE. After the "memory" is completed, the basic classifiers stored in *MS* are re-evaluated according to the results obtained in the process, and the memory weights of the basic classifiers in *MS* are updated. For the current data block, if a basic classifier is recalled, the memory intensity of this classifier is enhanced this time. If it's not, the memory intensity will be weakened. When classifying newly generated samples in the data stream, all basic classifiers in *ES* are used for classification prediction directly. Figure 2 shows training set generation based on sample library and category sliding windows.

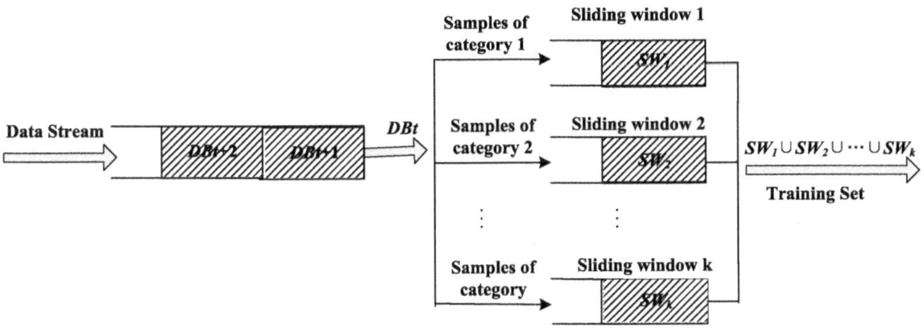

Fig. 2. Training set generation based on sample library and category sliding windows

The memory and forgetting mechanism of MAE can make the historically useful basic classifiers more stable in the "memory library" on the one hand, so as to avoid a large change in the concept of one data block causes the useful base classifier is deleted accidentally. On the other hand, selecting the basic classifiers that are most effective for predicting the current data block from the MS through the "memory" mechanism and using them for ensemble prediction, this makes full use of the time-locality effect of the data stream to improve the prediction accuracy. MAE can obtain better and more stable prediction performance than other traditional ensemble learning methods, but still cannot solve the problem of unbalanced data in the application and the missing sample of a class in the data block.

3 CIMAE Algorithm

Aiming at the defects of MAE algorithm in dealing with the imbalanced class problem, this paper presents a novel class-imbalanced ensemble data stream mining algorithm CIMAE based on MAE algorithm, which can remarkably improves the way of acquiring the data set for training the basic classifier each time.

Instead of obtaining a basic classifier by learning the latest data block *DB* directly, CIMAE sets a sample library in advance. This size of the sample library is the same as the size of *DB* and using $|DB|$ represents. The sample library contains the same number of sliding windows as the total number of categories in the data stream and using k represents the number of sliding windows. So the size of each sliding window is $|DB|/k$. When current *DB* is formed, the samples in *DB* enter the corresponding sliding window according to their own categories. While a sliding window is full, every time a new sample associated with this window arrives. According to the time series, the earliest incoming sample in this window is eliminated to update the sliding window and the sample database. Finally, a new basic classifier is learned by using the current sample database. In the Samples of each category in the sample library are always balanced in the CIMAE model. This method converts the category-imbalanced data classification into category-balanced data classification, which improves the learning ability of the algorithm and the prediction effect of the classifier. Figure 3 shows the CIMAE algorithm model. In addition to the training set acquisition method, the rest of CIMAE model is the same as MAE.

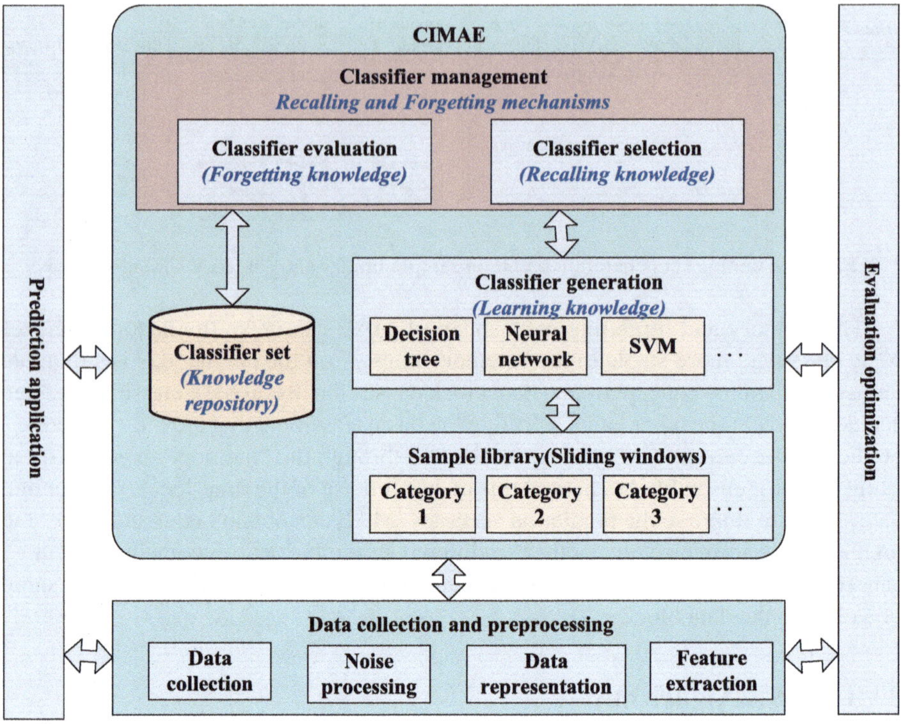

Fig. 3. The framework of CIMAE algorithm

CIMAE uses the same memory and forgetting mechanism as MAE and still uses the selective ensemble algorithm to select the recalled basic classifiers from library *MS*, and calculates the forgetting factor of each basic classifier *c* by function (1):

$$\mu_c = \frac{\beta}{\nu_c + 1} \tag{1}$$

where β is the initial value of the forgetting factor, ν_c represents the total number of times classifier c is recalled. The memory strength of c is calculated as follows:

$$w_c = e^{-\mu_c \cdot (t - \tau_c)} \tag{2}$$

where τ_c is the time when classifier *c* was last selected and *t* is the current time. The pseudo-code of CIMAE is shown in Algorithm 1 .

Algorithm 1: Class-imbalanced data learning based on MAE (CIMAE)

INPUT: *Str* :samples of data stream
  b: the capacity of memory library MS
  $|DB|$: size of data block DB
  $|DB|/k$: the capacity of each sliding window SW_i
    d: the capacity of recalling library ES, $d \leq b$
    k: Number of categories

1: Initialization: $MS \leftarrow \emptyset$; $\alpha \leftarrow 1$; t=0;
2: For all data blocks $DB_t \in DS$ do
  2.1: For all samples $e \in DB_t$ do
    2.1.1 y=Class(e);
    2.1.2 SW_y=update(SW_y, e);
  2.2: $SW = SW_1 \cup SW_2 \cup \cdots \cup SW_k$
  2.3: $c \leftarrow$ learn (SW) ;
  2.4: $w_c=1$; $\nu_c=0$; $\tau_c=t$; $\mu_c=\beta$;
  2.5: $MS \leftarrow MS \cup \{c\}$;
  2.6: ES=ensemble-prune (MS, DB_t, d);
  2.7: for all classifiers $c \in ES$
    2.8.1: $\tau_c=t$;
    2.8.2: $\nu_c = \lambda_c + 1$;
    2.8.3: compute the forgetting factor of c based on equation (1);
  2.8: end for
  2.9: for all classifiers $c \in MS$
    2.9.1: update the memory retention of c by equation (2);
  2.11: end for
  2.12: if $|MS| > b$ then delete the classifier with the lowest memory retention from MS;
  2.13: t=t+1;
3: end for

When a new prediction task arrives, it's predicted by classifiers in the ES and majority voting method is used to obtain the category of the new sample.

4 Experiment Setup and Results

There are serious class imbalance issues in the running state datasets of Tianhe-1 supercomputer' nodes [9]. The nodes operate normally most of time and the collected data is also normal. But parts of data close to the fault are failure data. The experiment uses the node operational data set of Tianhe-1. It includes 263, 582 sample data, with 10.79% of failures and 89.21% of normal. The ensemble data stream mining algorithms involved in the comparison include SEA, AWE, ACE, MAE, and CIMAE. When each data block *DB* arrives, each algorithm first predicts *DB*, obtains the prediction performance of it, and then performs online learning on it.

C4.5 decision tree [10] is used as the basic classifier of all ensemble data stream mining algorithms in the experiment, the size of each data block is set to 500 samples, and the classifiers participating in ensemble prediction are 10, that is $d = 10$. Both of the capacities of MS in MAE and CIMAE are set to $5d$. The capacity of the sample library in CIMAE is 500 samples, and the capacity of each sliding window is set to $|DB|/k = 250$ samples. In addition, the selective ensemble algorithm used in the "memory" mechanism of MAE and CIMAE in the experiment is the MDSQ algorithm [11]. The experiment uses the accuracy rate, recall rate and F-value to evaluate the performance of fault prediction comprehensively, and the final result of each index is the average test value. Experimental environment: IntelCorei5-64002.70GHzCPU, 8G memory, Windows operating system, Visual C++ 2005. Figure 4 shows the evaluation indicators of each ensemble data stream mining algorithm. Tables 1 and 2 show the average training time and average forecasting time of the corresponding algorithm on each data block.

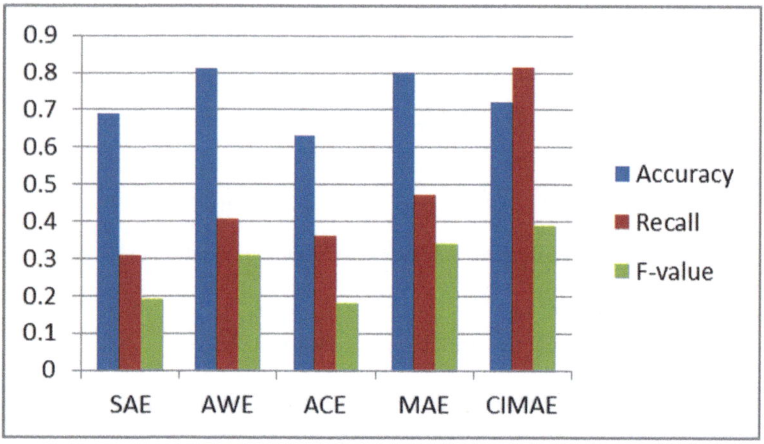

Fig. 4. Comparison of performance of multiple ensemble data stream mining algorithms

Table 1. Average training time [10^{-3} s]

Name of algorithm	SEA	AWE	ACE	MAE	CIMAE
Running time	16.13	12.86	24.78	20.52	22.93

Table 2. Average forecasting time [10^{-6} s]

Name of algorithm	SEA	AWE	ACE	MAE	CIMAE
Running time	18.39	18.71	29.35	19.66	19.97

Compared with other algorithms, although the overall accuracy of CIMAE has not been improved greatly, but the recall rate and F-value have been increased significantly. The greater the recall rate, the more failures are predicted. The higher the F-value, the better the balance effect of predictive accuracy and recall. The higher recall rate and F-values indicate that CIMAE predicts faults more failures correctly during failure prediction. For failure prediction with class-imbalanced samples, CIMAE is more practical than other ensemble data stream mining algorithms significantly.

From Tables 1 and 2, it can be seen that the average training time and average forecasting time of CIMAE are not optimal on node status data set of Tianhe-1, however, the time interval for data collection through the FPDC framework is 10 s, the time required to form *DB* is 5000 s, and the average time to train a classifier by CIMAE is 22.93×10^{-3} s. Current Basic classifiers have been trained before the next data block is formed. There is plenty of time to prepare for new block predictions.

The average time for prediction of each data block by CIMAE is 19.97×10^{-6} s. When using the CIMAE to predict the current data block, the node does not stop data acquisition, when the next *DB* is ready, the prediction result of the current *DB* has been given. If a failure is predicted, there is ample time for process migration of applications on the current node.

From the point of view of time, although the training time and prediction time of CIMAE are longer than most of the ensemble data stream classification algorithms in the experiment, it is sufficient to meet the real-time requirements of failures online learning and prediction.

5 Conclusion

Considering the application of unbalanced distribution of categories, based on MAE, an online class-imbalanced data stream learning algorithm named CIMAE is proposed. It sets a sample sliding window for each category. Each time a new data block is reached, CIMAE enters the samples in the data block into the corresponding sliding window according to its own category. The online learning of data stream is performed using the samples in all sliding windows. Compared with typical ensemble data stream mining algorithms, the recall rate and F-value of CIMAE algorithm are higher, which indicates that for categories which have small number of samples, their prediction accuracy can be greatly improved. In addition, the training and forecasting time CIMAE can meet the real-time requirements of online learning. Therefore, compared with the existing data stream online learning method, the CIMAE algorithm has better utility for the unbalanced application such as failure prediction.

Acknowledgement. This work is sponsored by the National Key Research and Development Program of China (2017YFC0806502, 2017YFC0803700, 2017YFC0821600) and by the Shanghai Rising-Star Program (17QB1401000) and by the Application Innovation Plan of Ministry of Public Security (2017YYCXSXST030) and by the program of Science and Technology Commission of Shanghai municipality (Nos. 15530701300, 1759800900).

References

1. Thamaraiselvi, G., Kaliammal, A.: Data mining: concepts and techniques. Ap Professional (2004)
2. Street, W.N.: A streaming ensemble algorithm (SEA) for large-scale classification. In: ACM SIGKDD International Conference on Knowledge Discovery and Data Mining, pp. 377–382 (2001)
3. Taillard, E.: Some efficient heuristic methods for the flow shop sequencing problem. Eur. J. Oper. Res. **47**(1), 65–74 (1990)
4. Lam, L., Suen, C.Y.: Application of majority voting to pattern recognition: an analysis of its behavior and performance. IEEE Trans. Syst. Man Cybern. A Syst. Hum. **27**(5), 553–568 (1997)
5. Wang, H., Fan, W., Yu, P.S., et al.: Mining concept-drifting data streams using ensemble classifiers. In: ACM SIGKDD International Conference on Knowledge Discovery and Data Mining, pp. 226–235. ACM (2003)
6. Jiang, Y., Zhao, Q., Lu, Y.: Adaptive ensemble with human memorizing characteristics for data stream mining. Math. Probl. Eng. **2015**, 1–10 (2015)
7. Datar, M., Gionis, A., Indyk, P., et al.: Maintaining stream statistics over sliding windows. SIAM J. Comput. **31**(6), 1794–1813 (2002)
8. Coon, D.: Introduction to psychology: exploration and application **28**(2), 89–92 (1998)
9. Hu, W., Jiang, Y., Liu, G., et al.: DDC: distributed data collection framework for failure prediction in Tianhe supercomputers. In: International Workshop on Advanced Parallel Processing Technologies, pp. 18–32. Springer International Publishing (2015)
10. Goldberg, D.E.: Genetic algorithms in search, optimization and machine learning. **xiii**(7), 2104–2116 (2006)
11. Martínez-Muñoz, G., Hernández-Lobato, D., Suárez, A.: An analysis of ensemble pruning techniques based on ordered aggregation. IEEE Trans. Pattern Anal. Mach. Intell. **31**(2), 245 (2009)

A Gaussian-Based Trust Model for Nodes of Sensor Networks

Ming Yang(✉)

The Third Research Institute of the Ministry of Public Security, Shanghai, China
ympione@163.com

Abstract. At present, the security mechanism of the sensor network is almost based on node authentication and data encryption, which can address security issues for Internet connection and information transformation. However, it lacks effective information security monitoring mechanism for node betrayal and malicious attacks occurring inside sensor networks. By analyzing factors including node performance, behavior and environment impact, etc., this thesis proposed a Gaussian-Based Trust Model (GBTM) for nodes of sensor networks against the insufficient adaptability of the current trust model. By GBTM, network monitoring cell can be utilized to build multiple Gaussian functions corresponding to monitored nodes to raise the conditional adaptability for computing the credibility of monitored nodes. And the online behavior of nodes can be evaluated by setting trust value in long or short term so that automatic credibility state and the credibility monitoring across the Internet can be realized. From the simulation analysis, it can be judged that GBTM has higher accuracy and dynamic adaptability compared with other known trust models.

Keywords: Sensor networks · Network security monitoring · Trust model Gaussian distribution

1 Introduction

Due to the intrinsic features, including no reliable center, limited resource and self-organizing, etc. of wireless sensor networks, traditional security methods such as encryption management and authentication mechanism, etc. cannot completely defend against attacks from the network, such as private nodes, Wormhole [1], Sybil attacks, etc. Regarding this question, researchers came up with a series of Soft Security [2] method. The basic concept of the trust mechanism is a quantitative evaluation by means of observing specific behavior of a node and recommendation trust to get the trust value of a node so as to identify malicious nodes in the network.

As the study of the trust mechanism continues, researchers discovered that the malicious behavior of a node was not the only reason that affects the credibility of a node. Due to the intrinsic instability of the sensor network, the node performance and network can also impact its credibility. For example, the performance of the node powered by solar energy varies between the sunshine is sufficient and the continuous cloudy condition, so its credibility is also different. Different from the credibility change of malicious nodes, the node performance and credibility change caused by the

J. Abawajy et al. (Eds.): ATCI 2018, AISC 842, pp. 109–117, 2019.
https://doi.org/10.1007/978-3-319-98776-7_13

environment are periodical and repeated. Therefore, the node performance and environment that can impact its credibility are called different 'State'. At present, most reputation models don't take consideration of the 'State' change feature in the sensor network, so the reputation values after computing are coarse and inaccurate.

As Direct Trust is the foundation of Recommendation Trust and Integrated Trust, this thesis proposed a Gaussian Based Trust Model Networks (GBTM) for Nodes of Sensor in view of the 'State' change feature of the sensor network.

2 Gaussian-Based Trust Model

Before the detailed introduction to this model, two concepts need to be introduced.

Definition 1. $C_{uv,}$ the trust value for Node v from Node u refers to the direct evaluation value for the credibility of Node v from Node u. Trust value is a subjective quantity rather than an objective quantity. Therefore, even if it is in the same 'State', the trust value for Node v from Node u might be different each time for computing.

Definition 2. $T_{v,}$ the credibility of Node v refers to the intrinsic objective value of the credibility for Node v in the network. Credibility is an objective quantity or universal view. In the same 'State', the trust value keeps the same rather than any change of trust value from one to another.

The trust value can be considered as a judge or a sample for the credibility. In the same 'State', there are lots of factors that can impact the sampling result, including its own and environmental factors, etc. and these factors are random and similar in impact intensity. Therefore, in the same 'State' of the wireless sensor network, the trust value of Node v from Node u is in accordance with the Gaussian distribution with the mean value. To correspond to multiple 'States' that may exist, M Gaussian models can be built for the evaluated Node v to the utmost. Among them, the probability density function of No. K Gaussian model is shown as below:

$$g_k(c_{uv}) = \frac{1}{\sqrt{2\pi}\sigma_k} e^{-\frac{(c_{uv}-T_k)^2}{2\sigma_k^2}}, \quad 1 \le k \le M \tag{1}$$

Each Gaussian model is corresponding to a kind of 'State'. $g_1(c_{uv})$, ranking at first is called the current model.

2.1 Model Initialization

To take K($0 \le K \le M$) as the practical Gaussian model quantity of Node u *to* Node v. When Node u never takes samples to Node v for the credibility, there is no Gaussian model of Node u to Node v. Therefore, K is initialized as 0, namely, no direct trust is judged. Assuming that each sampling process is not impacted by the previous sampling, each sampling value is independent from one another. When Node u completes sampling for Node v for the first time (X_1 as the sampling value), $K = 1$ and

$$g_1(c_{uv}) = \frac{1}{\sqrt{2\pi}\sigma_1} e^{-\frac{(c_{uv}-T_1)^2}{2\sigma_1^2}} \quad (2)$$

As sampling is proceeded only for one time, $T_1 = X_1$ and $\sigma_1 = \Psi$. The initial Mean Squared Error refers to the utmost tolerance degree of the trust value deviation for a node. The Ψ value is grater, the greater deviation for a node can tolerate. The specific value for Ψ can be specified according to the real condition.

2.2 Model Update and Dynamic Adjustment

When a new sampling event occurs, the new sampling value is recorded as X. X can be matched against the Gaussian models in line with the sequence of the Gaussian models. The matching term is $|X' - T_i| < \lambda\sigma_{i,}$ I refers to the sequence number of the corresponding Gaussian model. The parameter λ reflects the requirement for matching accuracy, so it is called the matching accuracy factor. In this thesis, the value of λ is defined as 2. According to the α [3]. Upper Quartile theory of the standard Gaussian distribution, when the λ value is defined as 2, the sampling value in line with the $g_i(c_{uv})$ distribution and matching probability can reach up to over 95%. If the matching fails, it will match against the following Gaussian model until to the end of the Gaussian model list. After matching succeeds, the mean value and mean squared error matching with the Gaussian models can be updated. The update algorithm is shown as below:

The current mean value is recorded as T_i and the mean squared error is recorded as σ_i and the mean value is recorded as T_i' and the mean squared error is recorded as σ_i' after updating. Assuming that the credibility is a steady state value in the same 'State', the relationship of T_i', σ_i', T and σ_i is shown as below:

$$\begin{cases} T_i' = \dfrac{k_i T_i}{k_i+1} + \dfrac{X'}{k_i+1} \\ \delta^2 = \dfrac{k_i}{k_i+1}[\sigma_i^2 + T_i^2 - T_i'^2] + \dfrac{(X'-T_i')^2}{k_i+1} \\ \sigma_i'^2 = \begin{cases} \delta^2, & \delta^2 \le \Psi^2 \\ \Psi^2, & \text{"else"} \end{cases} \end{cases} \quad (3)$$

k_i is the statistic quantity of the credibility value in the sampling history. But in the real condition, even if in the same 'State', the trust value T_i is not a constant quantity completely but a slowly varying variable. That is, the up-to-date sampling value has more reference significance than old sample values. Therefore, the adjustment factor is introduced as below:

ω ($\omega < k/(k + 1)$).

$$\begin{cases} T_i' = \omega + (1-\omega)X' \\ \delta^2 = \omega[\sigma_i^2 + T_i^2 - T_i'^2] + (1-\omega)(X' - T_i')^2 \\ \sigma_i'^2 = \begin{cases} \delta^2, & \delta^2 \le \Psi^2 \\ \Psi^2, & \text{“else”} \end{cases} \end{cases} \quad (4)$$

If the matching model is not the current one, the matching model ranking at first will be the current model and the models before that will be sequenced to the next accordingly. If the matching model cannot be found even if to the end of the Gaussian model list and *K*, the quantity of the current Gaussian models doesn't reach *M*, the maximum value, a Gaussian model will be created according to the model initialization procedure and it will be made as the current model. If *K*, the current Gaussian model quantity has already reached *M*, the maximum value, the model ranking in the end of the list will be deleted.

Used for the judgement and forecast of short-term behaviors for nodes, T_1, the mean value of the current model is called as the current trust value.

2.3 Comprehensive Trust Value

When forecast and judgement is implemented for the short-term credibility of the evaluated node, the current trust value can be used. However, one of the important functions of creating trust models is to provide the basis for the long-term collaboration between nodes, such as path selection and cluster-head selection, etc. Therefore, it is not enough to use the current model only. It is a must to create a comprehensive trust model that can assess long-term behaviors of nodes.

Theorem 1. *$x_1, x_2, \ldots, x_N$ are N random variables in normal distribution and they meet the followings: $x_i \sim N(u_i, \sigma_i^{2)}$, $(1 \le i \le N)$, then the variable, $x = \sum_{i=1}^{N} w_i x_i$ will meet the followings*:

$$x \sim N\left(\sum_{i=1}^{N} w_i u_i, \sum_{i=1}^{N} (w_i \sigma_i)^2\right).$$

The proof process is omitted.

The comprehensive trust value need to use the weighted sum of all sample values. Obviously, the weighted value corresponding to the sample value in the same distribution needs to be the same. The weighted value can be recorded as ξ_i (i = 1, 2, …, *K*) corresponding to the distribution $g_i(c_{uv})$. According to Theorem 1, the following conclusion can be derived as below:

$S_{uv,}$ the comprehensive trust value of the evaluated Node *v* from Node *u* meet the normal distribution. And $T_{uv} = \sum_{i=1}^{N} \xi_i T_i$, $\sigma_{uv}^2 = \sum_{i=1}^{N} (\xi_i \sigma_i)^2$.

With this Gaussian-based comprehensive trust model, the derived T_{uv} can be used to represent the comprehensive trust value and the value of σ_{uv} representing the mean squared error can be used to reflect the deviation degree of the credibility for an evaluated node from another node. The comprehensive trust value refers to the total samples of credibility and it reflects the comprehensive and long-term credibility level for an evaluated node.

There are three methods for computing ξ_i, the weighted value as below:

1. In terms of the number of samples in line with the Gaussian models. $N_{i,}$, the number of samples corresponding to $g_i(c_{uv})$ is greater, ξ_i is greater. The typical formula is shown as below: $\xi_i = \frac{N_i}{N_{sum}}$. The parameter N_{sum} is the total number of samples.
2. In terms of the time range for a corresponding Gaussian model to be the current Gaussian model. The longer the time for $g_i(c_{uv})$ to be the current Gaussian model, ξ_i is greater. The typical formula is $\xi_i = \frac{\text{Time}_i}{\text{Time}_{sum}}$. Time_{sum} is the total time length.
3. In terms of taking consideration of the parameters all above and introducing time difference as below: t_{ij} $\xi_i = \sum_{j=1} f(t_{ij}, \text{Time}_{ij}, N_{ij})$. The parameter t_{ij} is the time difference between the jth time to be the current Gaussian model and the current time. The parameter Time_{ij} refers to the time when being the current Gaussian model for the jth time. The parameter N_{ij} refers to the total sample number to be the current Gaussian model for the *j*th time.

3 Simulation Analysis

By simulation experiments, GBTM can be observed in the adaptability to the environment and the node 'State' change. The GRFSN Model [4] is an up-to-date trust model for the sensor network. The documentations [5] introduced the advantages of the GRFSN model compared with Model β, Model Beth and other classical trust models. The comparison between GBTM and GRFS in MATLAB, the simulation parameters are shown as Table 1:

Table 1. Simulation parameters

	Parameter	Description	Value
Universal settings	N	Number of nodes	2
	Bound	Trust value range	(0, 1)
	t	Simulation time range	100/200/400
GRFSN parameters	X	Initial distribution	$N(0.5, 0.5^2)$
GBTM parameters	Ψ	Initial mean squared error	0.15
	λ	Matching accuracy factor	2

3.1 Accuracy

In this experiment, in the condition that the credibility maintains the same for the evaluated Node *v*, the trust values from Node *u* to Node *v* was got in the test. Sampling was processed periodically. The trust values of the same group were tested in line with GBTM and GRFSN respectively and the trust value was drawn each 5 time lengths. The simulation result is shown as Chart 1, although the advantage of GBTM is not very obvious compared with GRFSN, GBTM can be closer to the real credibility in a faster speed at the initial period of sampling.

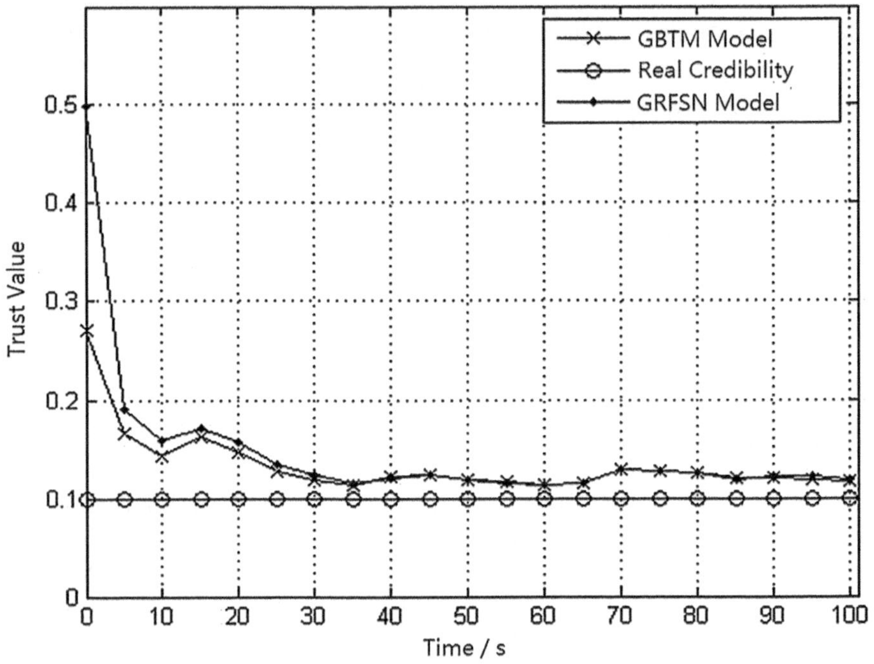

Chart. 1. Comparison of model accuracy.

3.2 Dynamic Adaptability

In MATLAB, the constant change of the 'State' in the sensor network was simulated to evaluate the dynamic adaptability of the models. In the experiment, the time range was prolonged to 200 and comparison was made by drawing the trust value each 10 time range. As shown in Chart 2, if the credibility of the evaluated node changes constantly, that is, the 'State' changes greatly, GBTM has a better dynamic adaptability to the credibility change compared with GRFSN, which has relatively slow change. Moreover, the comprehensive trust value between GBTM and GRFSN is very close to each other, it is proven that the comprehensive trust value can be used to evaluate a long-term behavior of a node and the current trust value can be used to evaluate the short-term behavior of a node.

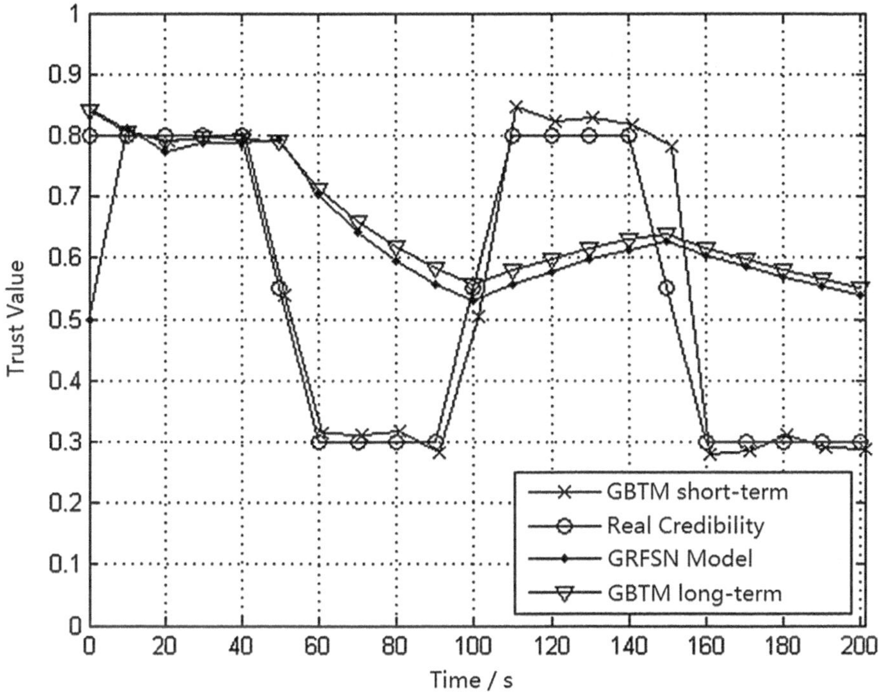

Chart. 2. Comparison of model dynamic adaptability.

3.3 Studying and Memorizing Capabilities

As shown in Chart 3, if the simulation time range is prolonged to 400, the credibility assessment for the value 0.8 was implemented. When the credibility occurred for the first time in the time range from 0 to 50, the trust value varied relatively greatly around the credibility. However, when this credibility repeated, the trust value kept a trend close to the real value in a gradually stable way. But for the tie range from 200 to 250, the credibility 0.55 occurred for the first time, the trust value varied greatly because the value never appeared before. It is shown that GBTM has stronger studying and memorizing capabilities. For the node 'State' that occurs repeatedly, it can be closer to the real credibility in a faster and more accurate way.

4 Conclusion

GBTM In view of the existing trust models which cannot accurately describe the credibility of a node in the condition that the sensor network 'State' changes, this thesis proposed a Gaussian-Based Trust Model (GBTM) for Sensor Nodes. Using this model, multiple Gaussian distributions can be created dynamically and each Gaussian distribution is corresponding to a kind of 'State' in the sensor network. In this way, the dynamic adaptability of the new model can be boosted greatly and it can vary together

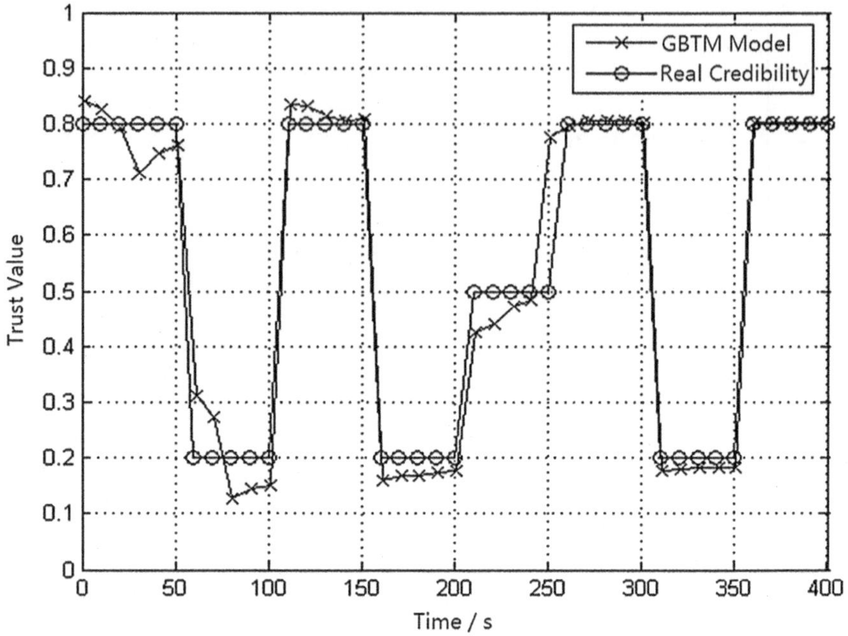

Chart. 3. Comparison of model studying and memorizing capabilities.

with the constant change of the node credibility so as to accurately and rapidly forecast the short-term credibility of an evaluated node. Meanwhile, the GBTM model introduced the concept of the comprehensive credibility used for the comprehensive and long-term credibility evaluation of nodes.

The next-step task is to expand this model further, in which the recommendation trust, time factor can be added up so as to enhance the identification capability against various attacks.

Acknowledgements. Our research was sponsored by Program of Science and Technology Commission of Shanghai Municipality (Nos. 16511101700, 17595800900, and 15530701300).

References

1. Harbin, J., Mitchell, P., Pearce, D.: Wireless sensor network wormhole avoidance using reputation based routing. In: International Symposium on Wireless Communication Systems, pp. 521–525 (2010)
2. Zorana, B., David, F., José, M., et al.: Detecting and confining sybil attack in wireless sensor networks based on reputation systems coupled with self-organizing maps. IFIP Adv. Inf. Commun. Technol. **1**, 311–318 (2010)
3. Lars, R., Sverker, J.: Simulated social control for secure internet commerce. In: Proceedings of the New Security Paradigms Workshop, pp. 18–25. ACM, New York (1996)

4. Xiao, D., Feng, J., Zhou, Q., et al.: Gaussian-based reputation framework for sensor networks. J. Commun. **29**(3), 47–53 (2008)
5. Zhou, Sheng, Shiqian, Xie, Chengyi, Pan: Probability and Statistics, pp. 56–62. Higher Education Press, Beijing (2001)

A Study of Face Recognition System's Testing and Evaluation Methods

Ming Yang(✉), Zhizong Wu, and Zhiguo Yan

The Third Research Institute of the Ministry of Public Security, Shanghai, China
ympione@163.com

Abstract. The paper made a summary of the technological development and the application of face recognition system and analyzed the deficiency of testing technique and method existing in the current national standards as well as industrial standards. The testing and assessing method aiming at face recognition system and algorithm were proposed, and the corresponding requirement and technical indexes were explained in the paper. At last, the practicability and application prospect of the technology were discussed.

Keywords: Face recognition · Algorithm · Image quality · False alarm rate Miss rate

1 Introduction

Face recognition is one of the hottest research fields. The essential technical part of achieving face recognition is the algorithm. Monitoring system and program related to face recognition mostly rely on the algorithm since the algorithm facilitates applications such as database duplication checking, passport and driving license query and verification, monitoring alarm of focused areas for the key person, criminal investigation and evidence taking. Therefore, face recognition system represents main support for information system construction of the Ministry of Public Security. An increasing number of public security authorities invest on face recognition system. However, the practical performance of the system is not satisfying as described in the indexes, which may result in the dilemma of these public security authorities. The significant cause behind this is as follows: at present, domestic technical index requirement and testing and assessing system are not complete; relevant standard specification should be more specified; the related testing standards are lacked. Therefore, it has become an urgent need to summarize the system testing techniques and methods that are scientifically suitable for public security application. This paper would analyze the current situation of the technology and the deficiency of the national standards to describe a feasible testing method to evaluate face recognition products.

J. Abawajy et al. (Eds.): ATCI 2018, AISC 842, pp. 118–123, 2019.
https://doi.org/10.1007/978-3-319-98776-7_14

2 Related Work

As the machine learning technology thrives, face recognition product enjoys a rapid development. The product mainly is mainly applied to two aspects: 1:1 facial verification and 1: n retrieval identification. The former one is used for the entry-exit person inspection (low difficulty) while the latter one is applied in fields like safeguarding, deploying surveillance, criminal investigation and searching information (high difficulty). The paper mainly discusses the testing assessment for 1: n face recognition system. A typical face recognition system's technical content includes taking a snapshot, modeling, building face sample database and developing recognition algorithm. Among these technical content, a recognition algorithm is the core content while the others are responsible for acquiring a valid facial image and establishing comparative image data basis.

Currently, The public security industry standard of China GA/T 922.2-2011 "Face Recognition Application For Security and Protection Systems-part 2: Face Image Data" has given the fundamental requirements for face sample database and test images [1]. Besides, the quality and angle of test images are clearly defined, as shown in Table 1.

Table 1. The rules of sample images and test images of GA/T922.2-2011

Index system	Picture format	Space between eyes (pixel)	Gesture and angle	Facial area	Accessories.
Image in sample database	BMP, JPEG, JPEG2000	≥ 60, ≥ 90 (recommended)	Horizontal angle, pitch angle, inclination $\leq \pm 10°$	Complete face, clear, chiseled face contour, undistorted length-width ratio, uniform illuminance and no shadow	No mask and no tinted spectacles
Images to be identified	BMP, JPEG, JPEG2000	≥ 30, ≥ 60 (recommended)	Horizontal angle: $\leq \pm 30°$ ($\leq \pm 15°$ recommended) pitch angle: $\leq \pm 20°$ ($\leq \pm 10°$ recommended) Inclination: $\leq \pm 45°$ ($\leq \pm 15°$ recommended)	No	Accessories cannot occlude main facial area

Most of domestic face recognition system products for public security have taken requirements of this standard to establish a sample face database (generally the required indexes are stricter than identification photo). However, the images to be identified vary due to the different conditions of collecting images at the community level: Some pictures are of good quality (the indexes are satisfied, and some even exceed the requirement) while some are of poor quality (with low pixels and large angle and even shadows). These factors have a significant impact on face recognition system's performance.

As for the assessment index for face recognition system's performance, the national standard GB/T 31488-2015 "Technical requirements for face identification of video surveillance in security systems" mainly proposes two indexes, namely, the miss rate of the surveillance list and the false alarm rate of non-surveillance lists. Moreover, the test method is standardized. In spite of this, some specifications about in the sample database are lacked; especially testing videos or image sets specifications. For these fundamental data that can truly measure the difference between systems or algorithms, the standard GA/T 922.2-2011 is not enough. The testing set should include multiple sorts of parameters that can measure image quality (like illumination condition, posture change scale, background and the age bracket) for assessing the system's identification capability when the image changes dynamically with multiple factors. It is indispensable to establish a set of testing specifications that can reflect changes of a number of indexes as a third-party testing basis.

The report "Face Recognition Provider's Testing Algorithms Performance" from NIST reports came up with issues like system sample database capacity, age span in testing picture sets, and error variance caused by changed image quality [3]. The report introduced several excellent face recognition algorithms from different providers. Their systems have distinguished performance responding to surging sample capacity, age distribution and images of various qualities. For those stable algorithms, their false alarm rate/miss rate logarithmic curve shows a steady linear state while those for particular purpose show a rapid change, which may result in a negative influence on the system if the sample database capacity increases, the age span enlarges or image quality declines.

With the analysis above, it can be seen that the current national standard can hardly assess the comprehensive capability of a face recognition system regarding its function and performance. We are in urgent need of a set of efficient, systematic testing method to promote face recognition system.

3 Testing and Evaluation Methods

On the basis of current technological development and application and the shortcoming of the standards, the improvement can be made from the following aspects:

In the first place, the increasing scale of testing sample database may lead to the worsening recognition function of the face recognition system. Thus, a fair and well-rounded assessment from a third-party authority is required to obtain comprehensive evaluation data, which plays an important role in the assessment of algorithms combined with machine learning. The test is classified according to the dynamic application and static application. The dynamic application (video surveillance) is based on the sample database capacity specified in national standard GB/T 31488-2015 (that is, one million scale to the maximum) [2]. Static applications (document photos or screenshots) give a full consideration to the common million-based capacity, mostly 10 million, and some even reach the billion-level sample database applications. In the new testing method, the assessment is conducted according to the scale, and the face test database is divided into three levels: 1,000,000 level, 10,000,000 level, and 50,000,000 level. This method of division enables the system to get testing assessment data for each level, demonstrating the applicability of the algorithm and the system.

In the second place, the testing method should not only standardize the sample's input in the face database, that is, meet the criteria of GA/T 922.2-2011, but also classify the quality of the image to be identified. The details are as follows: the quality of images to be identified that meet the standards is classified. The quality classification is based on specific quality indexes and is roughly divided into three levels: ① Sub-Standard: Space between eyes (pixel) 30–60; horizontal rotational angle ±30°–±20°; pitch angle ±25°–±15°; inclination ±45°–±15°; face partly occluded; ununiformed illuminance (with shadow or counter light); signal to noise ratio (SNR) 40–50 dB; distorted length-width ratio; complicated background. ② Medium-Standard: Space between eyes (pixel) 60–90; horizontal rotational angle ±20°–±5°; pitch angle ±15°–±5°; inclination ±15°–±5°; No Black Rimmed Glasses or hats; uniformed illuminance; signal to noise ratio (SNR) 50–60 dB; slightly distorted length-width ratio; simple background. ③ High-standard: Space between eyes (pixel) ≥ 90; horizontal rotational angle $\leq \pm 5°$; pitch angle $\leq \pm 5°$; inclination $\leq \pm 5°$; No Glasses; uniformed illuminance; signal to noise ratio (SNR) ≥ 60; no distorted length-width ratio; single-color background. It is advised to conduct a test for the system using images with a different level of standards to assess fault tolerance ability and false reporting rate. Only in this way can provide a similar and complex image source that the public security authorities at community level receive. Thus, a comprehensive assessment of the system dealing with images of different qualities can be given.

Besides, the testing method needs to determine the specific indexes for the system, including false alarm rate (which means the percentage of the facial image of the person to be tested from the non-sample database is falsely identified as a member of the sample database) and miss rate (which means the percentage of the facial image that belongs to sample database is not identified). Specifically, in the dynamic application test, the system's false report and miss rates are determined by "Primary Recognition" method. In the static application test, the "Top 10 and Top 50 matching" method is used to determine the system's false report and miss rates.

For a large-scale system with over 1 million samples, the specific procedure is as follows: ① The sample database whose scale is no more than 1 million is chosen from the sample database to be identified for dynamic application test; The sample database with more than 10 million samples is applied to static application test; ② 200–500 image "volunteers" with a 20-year age span are selected outside the total sample database of face recognition system to acquire video and static images of different quality levels for false report test in dynamic applications and static applications respectively; ③ The static high-standard image of the volunteer (second-generation identity photo) is added to the selected sample database (1 million for dynamic test, 10 million+ for static test) for missing report test; ④ On the basis of the determined optimization "threshold" in the system to be tested, the "Primary Recognition" method is used to obtain false report and miss rate in the dynamic application; The "Top 10 and

Top 50 matching" method is used to determine the system's false report and miss rate in the static application test. ⑤ When assessing the system, the false report and miss rate determined by "Primary Recognition" method should be closer to 0% in the dynamic test. The indexes determined by "Top 10 and Top 50 matching" method should be closer to 0% in the static test. Table 2 has listed the testing assessment based on different indexes as a reference.

Table 2. Testing index assessment reference

System/index	Sample database scale	Quality of image to be identified	Age span difference of images to be identified	Quantity of volunteers	False report rate (for reference)		Missing report rate (for reference)	
Dynamic system	100,000 – 1,000,000	Sub-standard; Medium-standard; High-standard.	0–20	Over 200	≤5%		≤5%–≤70% Classified into 4 levels	
Static system					Top10 (%)	Top50 (%)	Top10 (%)	Top50 (%)
	1,000,000	High-standard	0–20	Over 200	≤1%	≤1%	≤3%	≤2%
		Medium-standard			≤2%	≤1%	≤10%	≤7%
		Sub-standard			≤5%	≤3%	≤35%	≤32%
	10,000,000	High-standard	0–20	Over 200	≤2%	≤1%	≤7%	≤5%
		Medium-standard			≤5%	≤3%	≤15%	≤13%
		Sub-standard			≤10%	≤8%	≤3%	≤35%
	50,000,000	High-standard	0–20	Over 200	≤5%	≤3%	≤10%	≤7%
		Medium-standard			≤10%	≤8%	≤20%	≤15%
		Sub-standard			≤20%	≤15%	≤50%	≤45%

4 Conclusions

In the suggested methods above, the original index identification rate is replaced by another two indexes: false report rate and missing report rates. In this way, the face recognition algorithm and the system can be evaluated from two mutually restrained indexes, thereby avoiding introducing systematic errors in the pursuit of the high recognition rate. In addition, testing with multiple factors of a wide range such as image quality, age span, and sample database can greatly evaluate the tolerance of the face recognition system under various conditions. Hence, public security authorities can master the different performance through indexes by different systems and algorithms under different conditions, which is beneficial to choose suitable face recognition system for public security authorities at the community level. The paper mainly focuses factors including image database, testing database, and multiple image index that are concerned with face recognition system testing. Therefore some testing indexes like responding speed, storage and computing consumption are not discussed in the paper.

An advanced face recognition algorithm combined with machine learning technology greatly enhances the function and performance of the recognition system and is increasingly applied in public authorities at the community level. Assessing pros and cons or applicability of the algorithms and systems has become an urgent issue to be solved for these authorities. The standard content analysis and testing method described in this paper is a feasible solution under the existing conditions. The specific testing technology will be updated with the development of face recognition technology as well. The overall development trend of testing and assessing technology is automation, so it is important to develop and standardize face recognition test sets in a scientific and rational way. A good third-party testing sets and testing methods can give full play to assessment agency's technical advantage and provide support for the application of face recognition system.

Acknowledgement. Our research was sponsored by National Key R&D Program of China (No. 2017YFC0803700) and Program of Science and Technology Commission of Shanghai Municipality (No. 16511101700). Our research was also sponsored by CCF-Venustech Open Research Fund (Grant No. CCF-VenustechRP 2017006).

References

1. Tian, Q., et al.: GA/T 922.2-2011: face recognition application for security and protection systems-part 2: face image data, vol. 05, pp. 1–2 (2011)
2. Tian, Q., et al.: GB/T 31488-2015: technical requirements for face identification of video surveillance in security systems, vol. 12, pp. 4–8 (2015)
3. Grother, P., Ngan, M.: Face recognition vendor test (FRVT) performance of face identification algorithms. NIST Interagency Report 8009, pp. 16–137, 26 May 2014

Network Security Problems and Solutions of Video Surveillance System

Ming Yang(✉), Zhizong Wu, and Jinbo Wu

The Third Research Institute of the Ministry of Public Security, Shanghai, China
ympione@163.com

Abstract. In this paper, the application status and security problems of video surveillance system were analyzed. Solutions and design conceptions targeting the threats to terminal security were proposed, effects of solutions were described, and finally implementation and management methods of the solutions were discussed.

Keywords: Video surveillance · Terminal security · Weak password
Password-managed gateway

1 Introduction

With the rapid development of network and video surveillance technology, a new-generation video surveillance terminal featuring high definition and intelligence has been widely applied especially in public security and traffic management field where many cases that couldn't be investigated and tracked in the past are solved, which enables public security information construction to take effect to the utmost. With the increasing expansion of the construction scale, however, the number of surveillance camera grows day by day. Now, those cameras are found everywhere and confronted with information security threats from the network while they serve the public security, highlighting the security problems of terminals and data [1]. This paper describes the threats to video surveillance network, concludes video surveillance terminals' demands about security, and provides a technical proposal for enhancing the security of current terminals and systems based on the demands.

2 Related Work

2.1 Preface

With the rapid development of network and video surveillance technology, a new-generation video surveillance terminal featuring high definition and intelligence has been widely applied especially in public security and traffic management field where many cases that couldn't be investigated and tracked in the past are solved, which enables public security information construction to take effect to the utmost. With the increasing expansion of the construction scale, however, the number of surveillance camera grows day by day. Now, those cameras are found everywhere and confronted

J. Abawajy et al. (Eds.): ATCI 2018, AISC 842, pp. 124–130, 2019.
https://doi.org/10.1007/978-3-319-98776-7_15

with information security threats from the network while they serve the public security, highlighting the security problems of terminals and data [1]. This paper describes the threats to video surveillance network, concludes video surveillance terminals' demands about security, and provides a technical proposal for enhancing the security of current terminals and systems based on the demands.

2.2 Security Threats and Demands

A typical network video surveillance system consists of video surveillance terminals (camera and encoder with intelligent functions), a surveillance center (a server with the function of decoding/transcoding, storage, and video application), and a client (monitoring station) [2].

In recent years, the number of security incident related to video surveillance applications shows an upward trend, including the following known ones: a. 360 Shuidi live event (the application provides users with free cloud services and users may be live broadcast with bad intention or photographed unknowingly); b. the leakage event of urban monitoring screens (a website named Shodan posts the pictures shot by tens of thousands of web cameras, of which 49 belong to Chinese Mainland cities including Guangzhou, Beijing, Hefei, Nanjing, Chongqing, etc.); c. security risks were found in several batches of cameras (On 18 June 2017, Administration of Quality Supervision, Inspection and Quarantine checked the quality and security risks of 40 batches of smart cameras. As a result, 32 batches were found to have potential quality and security hazards with a proportion of up to 80%.); d. a new-type IoT botnet (the botnet used ELF_PERSIRAI.A, a malware to transmit infections from computer to computer; more than 1000 web cameras of several types produced by several manufacturers have been infected and manipulated by the malware).

Currently, video surveillance system is mainly confronted with the following four issues: a. terminal: weak password of video terminals (camera), lack in terminal security reinforcement (visit control and management bugs), defective codes (with trojans/virus and buffer overflow anomaly), excessive network ports, and lack of secure communication mechanism makes the terminal extremely easy to invade and therefore become a network attack target [3–5]; b. network: complicated network structure, insufficient security precautions, the connection between open and closed network, and network interconnection of different security domains may easily cause denial-of-service attacks, routing attacks, and malicious invasion of protected networks; c. data transmission: plaintext transmission of video data and the lack of an efficient and secure communication mechanism may easily lead to unexpected visits to surveillance resources and the leakage of confidential data; d. business management: the long-term absence of enabled password management, slack user account management, and the lack of audit may easily cause visits in excess of authority and that no audit is performed on security incidents.

Despite the fact that consideration has been given to the security isolation and security access to network in the video surveillance system applied to social public security, that the construction is large-sized and the mutual requirements between social resources and the system will lead to a complicated network structure without a doubt. On the other hand, the countless weak terminals and network equipment (hard to be largely updated and replaced), and the latest intranet challenge virus will inevitably bring hazards and threats to video surveillance network and system.

In reference to the above events and problems and targeting the current video surveillance systems intended for public security, requirements in terms of information security protection are concluded as follows in Table 1.

Table 1. Parallel table of security issues and requirements of video surveillance system

System element	Security issue	Security requirements
Surveillance terminal	Weak password	Strong password
	Visit control bugs	System security and reinforcement
	Defective codes	Protection against virus/trojans and bug fixes
	Port and backdoor bugs	Limited use of network ports
	Data communication threats	Support for equipment certification and data encryption
System network	Denial of service attacks	Safe communication mechanism
	Routing attacks	Protection of routers and network equipment
	Network threats at high security level	Invasion recognition, virus/trojans protection, and network isolation
Data transmission	Data leakage	Data encryption
Service management	Visits in excess of authority	password management and user privilege management
	Security incidents are unavailable to be traced	Security audit

3 Solutions

3.1 A Solution of Video Surveillance System in Which Terminals Are Connected to Private Network

The video surveillance system with local loop-locked network is the most-commonly used system now. In this environment, strong protection can be guaranteed for video data transmission, and the system design can focus on terminal password management, terminal invasion prevention, user privilege management, information system network protection at a high security level, and business management. Flowchart of the system is shown in Fig. 1. The part circled by a dashed line is the information system network at high security level. The white left-right double arrow represents the transmission of signaling and the black one-way arrow represents the transmission of video streaming data.

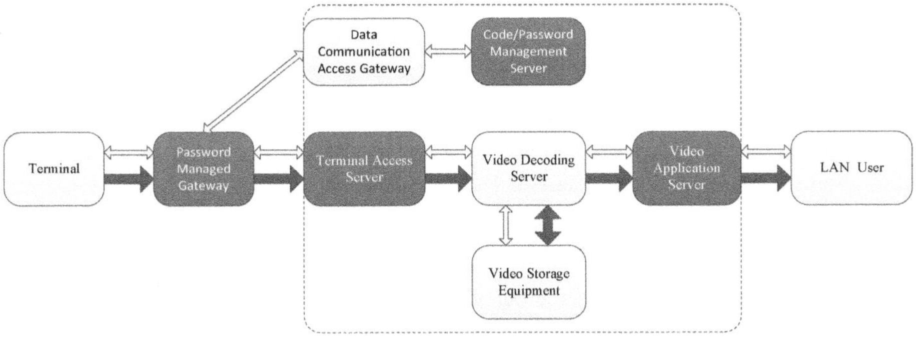

Fig. 1. Flowchart of local area network and video surveillance system

This Solution is based on the current system with an upgrade in safety which is reflected in a. use command/password to configure security gateway and manage the server; b. connect terminals to the server; c. video application server.

The main function of using command/password to configure gateway is to automatically replace the simple commands for visiting terminals by complicated ones that can be updated in accordance to safety strategies, automatically solving the command change issue for terminals once and for all (the time largely wasted in terminal labor can be saved), and this change is compatible with the current system (the modification is only performed by users and the system remain unchanged). The gateway realizes the generation, update, and configuration management of random keys for multiple terminals, multi-terminal access control agent, multi-path video streaming forwarding service agent, self-protection and visits control, as well as the support for a safe access to the dashed-line-circled part in the chart. Generally, the gateways configured by command/password can manage multi-path (over eight paths) HD video surveillance terminals by commands and forwarder video business. By configuring and managing gateways with commands/passwords, command/password-managed server can realize the management of all the mapping relations between commands and terminals, the release of safety strategies such as command update, and the management of gateway status. The system made up of the above modules can well relieve the issues about the weak passwords in terminals and indirectly prevents terminal invasion.

Connecting terminals to server realizes the functions of terminal videos' connecting to the gateway and firewall. Its main functions are: a. simple certification or binding for terminals to get access to the network (e.g., unique equipment identification certification and mac-address binding); b. protocol filtering of video streaming, namely resolving video streaming protocol and abandoning the data not in conformity with the agreed protocol; c. protecting the information system network at high safety level from virus/trojans.

Video application server is intended to reinforce the control management of visits to video resources based on the existed applications, including user authentication, command management and access permission management.

The advantage of this solution is that it fits most existed systems and reinforces the systems' information safety protection ability with low cost targeting the most widespread problems.

3.2 Solutions of Video Monitoring System or Video Encrypted System with Terminals Connected to Public Network

The biggest difference between the video monitoring system with terminals connected with public network and the system adopting loop-locked network is that the part connected to public network or the part needs encrypted video streaming transmission is placed in an open network environment and there're great threats that may cause surveillance video to be leaked into the internet or hijack surveillance terminal resources. For this reason, it's necessary to take relatively strong measures to ensure information safety. The system's structure is shown in Fig. 2.

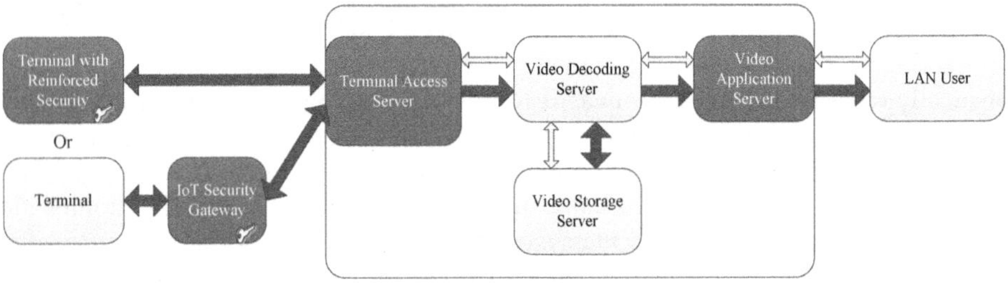

Fig. 2. Flowchart of video monitoring system or video encrypted system with terminals connected to public network system

Depending on the specific circumstance of video surveillance system, the solution has two variants: Solution A fits the systems connected with brand-new video surveillance terminals as is shown in the upper left part in Fig. 2; Solution B fits the current system, as is shown in the left lower part in Fig. 2.

Solution A is a technical solution featuring high-security level surveillance terminal, which is mainly reflected by: a. safety of the terminals is reinforced; b. the server through which terminals are connected is strengthened.

Safety of the terminals is reinforced: most video surveillance terminals in the market adopt a simplified embedded system which focuses on video capturing and application and overlooks information safety protection. To reinforce the information safety protection of terminals, safety reinforcement design must be considered. The design is expected to realize the following functions: a. operating system reinforcement; b. support the access certification and encryption of high safety level equipment; c. virus/trojans prevention; d. communication and service port limitation; e. visit control and safety log. The hardware of the design shall: a. adopts the processer with safety functions (safe algorithms and memory protection) as the master control chip or adopts security chip as the coprocessor to execute safe algorithms and store key information; b. keeps as few communication ports as possible in terminals, and physically protects program debugging ports in a closed circuit. As for software: a. an embedded system (generally the embedded Linux) with complete safety functions shall be customized; b. an equipment access certification and information channel encryption client program shall be customized and installed; c. a virus/trojans prevention program targeting IoT and video system shall be customized; d. communication and service ports shall be limited by means of firmware; e. log recording and auditing function for security incidents shall be included.

Reinforcing the server through which terminals are connected to the system not only has the protocol filtering and virus/trojans prevention function of the video surveillance system solution with terminals connected to a wired local area network but also strengthens the terminal two-way certification and VPN information channel encryption function based on public key infrastructure. These two functions can avoid the safety problems caused by weak commands on one hand, and on the other hand, guarantee the privacy of transmitted videos.

Solution B is a compatible system with the same security level as Solution A, which is realized by connecting the IoT security gateway with video surveillance terminals. The IoT security gateway needs to not only have the function of command/password configuration of video surveillance system solution with terminals connected to a wired local area network but also have the same hardware basis as Solution A and support terminal two-way certification and VPN information channel encryption function based on public key infrastructure.

In summary, the video surveillance system solution with terminals connected to the public network has higher requirements and therefore costs more in modifying and investing in the existed system.

4 Conclusions

Currently, the designs of video monitoring system are mainly focused on network information and confidential information protection. This paper put forward three proposals, to provide not only sound technical support to video surveillance network construction which serves social public security but also references for other video surveillance systems with specific requirements. The users can make their choice according to the operating status of the systems that are completely constructed or being constructed and their security requirements. It's believed that with the expansion of public security video surveillance network construction and the increasing security requirements, plans for video surveillance system will be further perfected, technical research will go deeper and deeper, and the situation awareness requirement for overall security based on local security will be gradually satisfied.

Acknowledgement. Our research was sponsored by National Key R&D Program of China (No. 2017YFC0803700) and Program of Science and Technology Commission of Shanghai Municipality (No. 16511101700).

References

1. He, X.H.: The development trend prediction of 2015 network surveillance market. China Public Secur. (Integr. Version) **2**(23), 93–99 (2014)
2. Zhang, Y.X., Li, H., Jun, C., Yang, X.G.: Analysis of information security mechanism of network video surveillance system. Telecommun. Sci. **29**(8), 69–73 (2013)
3. https://krebsonsecurity.com/2016/09/krebsonsecurity-hit-with-record-ddos/
4. Zhong, G.W., Yu, Z.C.: Analysis and measures for video surveillance security situation. China Railw. **4**, 103–107 (2015)
5. Yu, S.H.: A brief study of information security in video surveillance. A&S: Secur. Prot. Proj. Contract. **2**, 62–64 (2014)

Image Encryption Algorithms Based on Chaos Through Dual Scrambling of Dynamic Josephus Ring and Bit

Mengran Zhou[1] and M. A. Xin[2(✉)]

[1] School of Electrical and Information Engineering, Anhui University of Science and Technology, Huainan 232001, China
Mrzhou8521@163.com

[2] School of Computer Science and Engineering, Anhui University of Science and Technology, Huainan 232001, China
1341416375@qq.com

Abstract. In this paper, in order to solve the problem that the scrambling method of most chaotic image encryption algorithms is not strong in non-linear structure and the correlation of adjacent pixels is not so destructive. A new image encryption algorithms was proposed based on chaos through dual scrambling of dynamic Josephus ring and bit. The algorithm uses logistic chaotic mapping as the key generation source, and makes the encryption process closely related to plaintext image information. Firstly, the algorithm establishes the advantage of dynamic Josephus ring disarrangement by contrast and then globally scramble the original image; Then the scramble sequence is expanded by 8 times, and the bits are scrambled by logistic chaotic sequence, and changed the degree of association of adjacent pixels from a single pixel; finally, the pixel value of a single pixel is replaced diffused to generate a ciphertext image. The experimental results show that this algorithm cannot only resist the choice of Ming (Dense) attacks, but also achieve better resist the differential attacks and statistical analysis. It has the advantages of simple encryption algorithm, large key space, good encryption effect and good application prospect.

Keywords: Image encryption · Chaos sequence · Dynamic Joseph ring
Bit scrambling

1 Introduction

Since the concept of chaos proposed by Matthews in the 1990s [1], many image encryption algorithms based on chaotic systems have rapidly developed [2–6]. Because some encryption algorithms have low security of encryption, domestic and foreign scholars have made improvements based on the original algorithm and derived a series of improved algorithms. For example, in resources [7], a fractional domain image encryption algorithm based on SVD and Arnold transform is proposed. Resources [8] proposed a new chaotic image encryption algorithm, which involved a block image scrambling scheme and a new dynamic exponential diffusion scheme, and then the ciphertext image

J. Abawajy et al. (Eds.): ATCI 2018, AISC 842, pp. 131–140, 2019.
https://doi.org/10.1007/978-3-319-98776-7_16

is dynamically indexed, the results show that the algorithm has good security performance. Resources [9] has designed a new image encryption scheme by combining Logistic mapping, sine mapping and DNA coding. Resources [10] proposed a digital image encryption algorithm combining hyper-chaotic sequence and shift cipher. The image encryption algorithm proposed in this paper can also be applied to social media [11] and Weibo applications [12].

The above algorithm is basically improved from two parts of the key source and the encryption algorithm, but after improving the security of the key source, it often increases the complexity of the key system and reduces the key generation efficiency; the improved encryption algorithm often uses the similar quasi-mapping scrambling method and the similar pixel replacement method, the scrambling method is not strong in the nonlinear structure, and the pixel substitution has little damage to the adjacent pixels. Based on this, this paper proposes a new image encryption algorithm. The algorithm analyzes the classical Joseph ring problem and the structure of traditional algorithms, and establishes an image encryption method using dynamic Joseph rings and bit-pairs of double-reset chaos. The algorithm is mainly divided into two steps. The first step is to globally scramble the original image sequence by setting the parameters of the dynamic Joseph ring to get a global scrambling sequence; the second step is different from the traditional method of pixel level scrambling and substitution, adding the bit position setting. That is, global scrambling is performed on the global scrambling sequence obtained in the first step, and pixel-level pixel values are substituted for the scrambled sequence to obtain a cipher-text image, and the security of the algorithm has been strengthened, and the nonlinear structure between plain-text has been more complicated. Finally, through the simulation analysis of the experiment, it is proved that the algorithm has high security.

2 Principle of Algorithm

2.1 The Joseph Ring

The Josephus problem is a mathematical application problem and is described as follows: known n individuals (numbered 1, 2, ..., n), sitting around a big round table, starting from 1 with the number S, reporting at an interval of g, and the number of people who reported to the end number automatically Then his next one starts counting again from 1 and the interval between the numbers is g, and the number of people who reported to the finish line goes out; so the cycle continues until all the people at the round table are out. For ease of description, define the Joseph function as

$$\mathrm{Josephus}(a[], L, S, g, No) \tag{1}$$

Among them, a[] represents the original sequence, L represents the length of the sequence, S represents the position of the starting report ($1 \leq S \leq L$), g represents the interval of the report number, and No represents the report value of the dequeued position during traversal ($1 \leq No \leq L$). In this paper, we verify the static and dynamic Joseph ring scrambling.

2.1.1 Static Joseph Scrambling

Figure 1 shows the image contrast effect of Joseph scrambling and row and column swap scrambling on Lena grayscale images of size 256 ∗ 256. The results show that the effect of Joseph scrambling is better than the effect of common row and column scrambling. However, it can still be seen that there is a certain linear transformation feature, and the scrambled sequence retains part of the linear information features.

a)Joseph scrambled image b)Rank interchange scrambled image

Fig. 1. Two kinds of scrambled image contrast.

This paper further analyzes the correlation of adjacent pixels in the image. There is a correlation among the pixels of the image matrix, and the smaller the correlation of each pixel is, the better the encryption effect will be, and the correlation calculation formula is as follows.

$$\bar{x} = \frac{1}{N}\sum_{i=1}^{N} x_i \tag{2}$$

$$D(x) = \frac{1}{N}\sum_{i=1}^{N}\left(x_i - \bar{x}\right) \tag{3}$$

$$\operatorname{cov}(x, y) = \frac{1}{N}\sum_{i=1}^{N}\left(x_i - \bar{x}\right)\left(y_i - \bar{y}\right) \tag{4}$$

$$\rho_{xy} = \frac{\operatorname{cov}(x, y)}{\sqrt{D(x)}\sqrt{D(y)}} \tag{5}$$

Among them, x, y represent the gray value of two adjacent pixels in the image, and N is the correlation coefficient of two adjacent pixels. The neighboring pixel correlations were calculated for the original image, Joseph ring scrambling image, and row-and-field scrambling image respectively, and the results are shown in the following table. From the data in the table, it can also be seen that the static Joseph ring scrambling correlation is smaller than the row and column swap scrambling (Table 1).

Table 1. Scrambled image pixel correlation coefficient

Path	Plaintext image	Ranks interchangeable	Static Joseph	Dynamic Joseph
Horizontal	0.9649	0.2172	−0.0022	−0.0043
Vertical	0.9337	0.0161	0.0060	−0.0018
Diagonal	0.9094	−0.0023	−0.0118	0.0010

2.1.2 Dynamic Joseph Scrambled

From Sect. 2.1.1, the method used in Sect. 2.1.1 is to set a fixed starting position s and a fixed reporting interval m. This article continues to improve this, using a dynamic Joseph ring, Each re-reported number randomly generates the starting position s and the reporting interval m. Each time the reporting number starts from the starting position and ends at the last position in the sequence, the image after scrambling is shown in Fig. 2.

Fig. 2. Dynamic Joseph scrambled image

It can be seen from Fig. 2 that the scrambling effect of the dynamic Joseph ring is better than the two methods in Fig. 1. The scrambled image shows a more disordered state in each pixel. The adjacent pixel correlation test is performed on the image of the dynamic Joseph ring scrambling. The results are as follows: The correlations of adjacent pixels in the horizontal, vertical, and diagonal directions are −0.0043, −0.0018, and 0.0010, The data comparison is also better than the two methods in Sect. 2.1.1, which shows that the dynamic Joseph ring scrambling is more destructive to the adjacent pixels of the image and is more conducive to encryption.

2.2 Logistic Chaotic System

The logistic chaotic map is shown in Eq. (6).

$$x(k+1) = u * x(k) * [1 - x(k)], \quad (k = 0, 1, \ldots, n) \tag{6}$$

Among them, u is the control parameter, after research proves, when $3.5699456\ldots < u \leq 4$, the system is in chaos state. The iterative value $0 < x(0) < 1$, this system control parameter takes 4.

In order to make the key sequence have better uniformity, we use Eq. (7) to do homogenization of the Logistic chaotic system.

$$y = \frac{2}{\pi} \arcsin \sqrt{x} \tag{7}$$

The initial value x(0) is set to generate a chaotic sequence of length 1000. The histogram statistics of the sequence before and after the homogenization process are performed. The statistical graph is shown in Fig. 3. It can be seen from the figure that the homogeneity of the homogenized sequence is obviously better than the pre-processed key sequence and is more suitable as a key source.

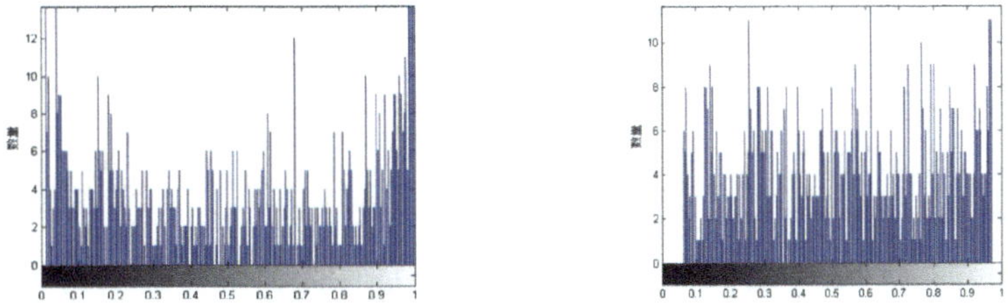

a)original sequence histogram b)homogenized histogram

Fig. 3. Comparison of the original sequence and the homogenized histogram.

3 Encryption Process

The algorithm of this paper firstly homogenizes the key sequence from the key source to make it more conform to the encryption requirements. Then it is different from the commonly used Arnold scrambling method. It uses the dynamic Joseph ring to scramble and destroys the correlation between the pixel locations of the original image through complex nonlinear operations. And for different images, the parameters and initial values of the chaotic system are different. The scrambling sequence generated by the Joseph ring scrambling method is also different, which can effectively avoid the selection of the plaintext attack; Finally, the scrambling step of bits is added between the classic pixel scrambling and replacement steps, which enhances the security of the algorithm.

3.1 Image Pixel Position Scrambling

According to the Joseph ring described in Sect. 2, the initial sequence of the Joseph ring can be set to $a[] = \{1, 2, \ldots, m \times n\}$. Next, the plaintext image P is scrambled.

Step 1. Let the original image be P, M and N respectively represent the rows and columns of the image matrix, and its image pixel matrix can be expressed as

$$P = \begin{pmatrix} p_1 & \cdots & p_N \\ \vdots & \ddots & \vdots \\ p_{N(M-1)+1} & \cdots & p_{MN} \end{pmatrix}.$$

Step 2. Converts the plaintext image pixel matrix $P_{M \times N}$ into a-long one-dimensional sequence $P = \{p_1, p_2, \ldots, p_{m \times n}\}$ of lengthen $M \times N$ in a row-first order, with a range of [0, 255].

Step 3. Set the S and No values to generate a scrambled sequence $R = \{r_1, r_2, \ldots, r_{m \times n}\}$. The scrambling sequence R is used to scramble the original sequence P.

Step 4. Takes the position of each element in the L sequence as the index of the original image sequence, and extracts the original image elements one by one according to the index of the index into a new sequence P′, P′ is the pixel matrix after the scrambling. The following example illustrates the pixel scrambling process.

3.2 Pixel Scrambling

In this subsection, the pixel value of the scrambling image pixel matrix P′ is replaced with encryption. The algorithm scrambles the bits inside the pixel value and uses the formula (12) to perform the correlation operation on the scrambled pixel matrix, Then the ciphertext image is obtained, thus avoiding the plaintext attack and improving the security of the algorithm. At the same time, the method in [13] was used to replace the encryption of pixel values, and the effect was better.

Set the algorithm's original key set to $\{x_0, y_0, H\}$, where x_0, y_0 represent the two initial state values of the Logistic chaotic system, and L represents the number of pre-iterations of the Logistic chaotic system.

The average pixel value avg of the plaintext image P is obtained by Expression (8).

$$avg = \frac{sum}{m \times n} \tag{8}$$

Where sum is the sum of all the pixel values. Then modify the initial value x(0) of the Logistic system according to the value of avg, $x(0) = avg \times x(0)$. Correlate x(0) with the content of the image so that it can counteract the chosen plaintext attack.

The number of times H of the pre-iteration of the chaotic system is obtained by Eq. (9).

$$H = 1000 + \mathrm{mod}(sum, 1000) \tag{9}$$

Step 1. Set the control parameter μ of the system and use the Logistic chaotic system to iterate L times before iteratively, and then continue iterating $8 \times m \times n$ times to generate a chaotic sequence $Q = \{q_1, q_2, \ldots, q_{8 \times m \times n}\}$ with a length of $8 \times m \times n$.

Step 2. The chaotic sequence Q is transformed according to the following formula (10), so that the sequence values of the chaotic sequence range from 0 to 255, and the chaotic image sequence $Q' = \{q'_1, q'_2, \ldots, q'_{8 \times m \times n}\}$ is obtained.

$$Q' = \mathrm{mod}(q_i \times 10^8, 256), 1 \le i \le 8 \times m \times n \tag{10}$$

Step 3. Sorts the chaotic image sequence Q′ from small to large, thereby generating a sequence $K = \{k_1, k_2, \ldots, k_{8 \times m \times n}\}$ for recording the position of each element in the sorted sequence in the original sequence Q′.
Step 4. Converts the pixel values in the original sequence P′ into a binary format and arranges them using the position sequence K to get a new binary sequence.
Step 5. Convert the new binary sequence obtained in Step 4 to decimal and obtain the intermediate ciphertext A.

3.3 Pixel Value Replacement

The initial value y_0 is set, a sequence of length L is generated, the sequence L is converted into an integer of [0, 255], and then the intermediate ciphertext image sequence A is re-encrypted according to formula (12) to obtain the ciphertext image sequence B.

$$Q_i = \mathrm{mod}(q_i \times 2^{48}, 256) \tag{11}$$

$$C_i = \mathrm{mod}(Q_i + B_i, 256) \oplus C_{i-1} \tag{12}$$

Among them, i = 1, 2, …, L, Q_i are the pixel values of the chaotic sequence, B_i is the pixel value of the middle ciphertext sequence, C_i is the pixel value of the ciphertext image after the pixel replacement of the current plaintext pixel, and C_{i-1} is the previous pixel value of the ciphertext pixel. When i = 1, a pixel value C_0 needs to be used. The value of C_0 is arbitrarily selected from 0 to 255 and can be used as a key.

4 Experimental Simulation Analysis

4.1 Simulation Results

In this paper, the classical Lena grayscale image with a size of 256 * 256 is used for simulation. The initial key parameters of the encryption system are $x_0 = 0.2234567654$, $y_0 = 0.123456789$ and $B_0 = 20$, respectively. The original image and the encrypted image obtained by using this algorithm are shown in Fig. 4.

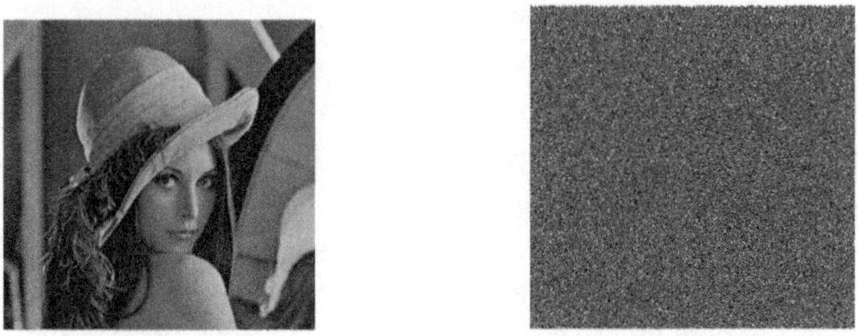

a)The original image b)Encrypted image

Fig. 4. Original image and encrypted image.

4.2 Analysis of Information Entropy

Information entropy is the measurement of the purity of information. The greater the uncertainty, the greater the information entropy. The calculation formula of information entropy H(x) is as follows.

$$H(x) = -\sum_{i=1}^{n} p(X_i) \log_2 p(X_i) \tag{14}$$

Where $p(X_i)$ represents the probability that the pixel value X_i appears in the image. A truly random source of information, including 2N symbols, with information entropy of N. If we use a 256 grayscale image as the information source, there will be 28 possible values, so the information entropy of the 256 grayscale image is 8, so if the 256 grayscale image information entropy is close to 8, explain that ciphertext images are close to random distribution. The Lena image chosen in this paper is a 256-level grayscale image, so when H(x) = 8, the image can be considered as an ideal completely random image. Table 2 lists the results of the comparison between the information entropy using this algorithm and the information entropy using the literature algorithm.

Table 2. Information entropy comparison

Plaintext	Ciphertext	References [14]	References [15]	References [16]
6.6549	7.9970	7.9951	7.9896	7.9001

4.3 Correlation Analysis

The correlation coefficient formula has been introduced in Sect. 2.1.1. Figure 5 randomly extracts 3000 adjacent pixels in the horizontal direction from the plaintext image and the ciphertext image, respectively, as a pixel value relationship diagram. It can be seen from Fig. 5 that the plaintext image (a) appears as a concentrated linear distribution, indicating that the neighboring pixels are highly correlated. The ciphertext image (b) shows a discrete distribution state, indicating that the adjacent pixels of the encrypted image are less correlated, and thus, the security of the encrypted image is improved.

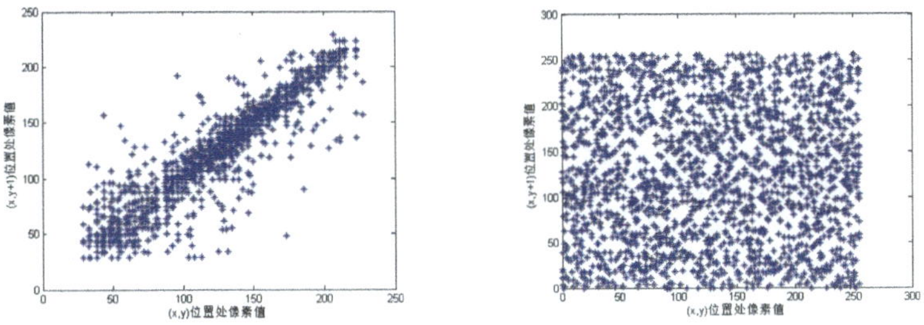

a)horizontal pixel value in plaintext b)ciphertext horizontal pixel value

Fig. 5. Adjacent pixel correlation.

Table 3 shows the correlation coefficients of all neighboring pixel values in three directions (horizontal, vertical, and diagonal) of a ciphertext image, and also gives a comparison of pixel correlation coefficients in other literatures with the correlation coefficients of pixels in this paper. The results show that the correlation coefficient is smaller and closer to zero.

Table 3. Adjacent pixel correlation

Direction	Plaintext	Ciphertext	Reference [17]	Reference [18]	Reference [19]
Horizontal	0.9649	– 0.0005	– 0.0034	– 0.0048	0.0041
vertical	0.9337	– 0.0003	0.0044	– 0.0112	– 0.0033
diagonal	0.9094	0.0002	0.0013	– 0.0045	0.0020

5 Conclusions

This article mainly introduces chaos image encryption algorithm based on dynamic Joseph ring and bit double replacement. First, the classic Joseph ring problem is introduced. Second, the detailed steps of the algorithm are introduced in two steps. In the first step, the pixels are globally scrambled using the dynamic Joseph ring, and the second step is bit reorganization of the global scrambling sequence to obtain the intermediate ciphertext. Finally, the algorithm performs experiments through matlab simulation. This paper analyzes the algorithm experimental results based on histograms, key space, information entropy, neighbor pixel correlation, and ciphertext sensitivity. Compared with the common chaotic image encryption algorithm, this algorithm combines dynamic Joseph ring traversal and bit reorganization to improve the security of the encrypted image.

Acknowledgements. This Research work was supported in part by National Natural Science Foundation of China under (Grant No. 51174258), and by the National Safe Production Major Incident Prevention Key Technology Project under (Grant No. anhui-0001-2016AQ).

References

1. Matthews, R.A.J.: On the derivation of a chaotic encryption algorithm. Cryptologia **4**(1), 29–42 (1989)
2. Xu, Y., Wang, H., Li, Y., et al.: Image encryption based on synchronization of fractional chaotic systems. Commun. Nonlinear Sci. Numer. Simul. **19**(10), 3735–3744 (2014)
3. Daneshgar, A., Khadem, B.: A Self-synchronized Chaotic Image Encryption Scheme. Elsevier, Amsterdam (2015)
4. Xie, G.B., Wang, T.: A new bit-scrambling hyperchaotic image encryption algorithm. Microelectron. Comput. **33**(07), 28–32 + 38 (2016)
5. Han, F.Y., Zhu, C.X.: Novel permutation and alternative structure image chaos encryption algorithm. J. Wuhan Univ. Sci. Ed. **60**(05), 447–452 (2014)
6. Xie, G.B., Wang, T.: Chaotic image encryption algorithm based on pixel scrambling and bit replacement. Microelectron. Comput. **33**(03), 80–85 (2016)
7. Chen, L., Zhao, D., Ge, F.: Image encryption based on singular value decomposition and Arnold transform in fractional domain. Opt. Commun. **291**(291), 98–103 (2013)
8. Xu, L., Gou, X., Li, Z., et al.: A novel chaotic image encryption algorithm using block scrambling and dynamic index based diffusion. Opt. Lasers Eng. **91**, 41–52 (2017)
9. Awudong, B., Li, G.: Research on image encryption technology based on multi chaotic mapping. In: International Conference on Smart City and Systems Engineering. IEEE Computer Society, pp. 127–131 (2017)
10. Liao, Q.N., Lu, S.D., Sun, X.B.: Digital image encryption algorithm combining hyperchaotic sequence and shift cipher. Small Microcomput. Syst. **36**(2), 332–337 (2015)
11. Xu, Z., Mei, L., Raymond Choo, K.-K., Lv, Z., Hu, C., Luo, X., Liu, Y.: Mobile crowd sensing of human-like intelligence using social sensors: a survey. Neurocomputing **279**, 3–10 (2018)
12. Zhang, S., Wei, Z., Wang, Y., Liao, T.: Sentiment analysis of Chinese micro-blog text based on extended sentiment dictionary. Fut. Gener. Comput. Syst. **81**, 395–403 (2018)
13. Deng, X.H., Liao, C.L., Zhu, C.X., et al.: Image encryption algorithms based on chaos through dual scrambling of pixel position and bit. J. Commun. **35**, 216–223 (2014)
14. Wang, X., Liu, L., Zhang, Y.: A novel chaotic block image encryption algorithm based on dynamic random growth technique. Opt. Lasers Eng. **66**(66), 10–18 (2015)
15. Liu, Y., Wang, J., Fan, J., et al.: Image encryption algorithm based on chaotic system and dynamic S-boxes composed of DNA sequences. Multimed. Tools Appl. **75**(8), 4363–4382 (2016)
16. Chai, X., Gan, Z., Yuan, K., et al.: A novel image encryption scheme based on DNA sequence operations and chaotic systems. Neural Comput. Appl. **9**, 1–19 (2017)
17. Zhang, H., Liu, S.M., Gao, M., et al.: Chaotic image encryption algorithm research based on contourlet transformation. In: International Computer Conference on Wavelet Active Media Technology and Information Processing, pp. 303–306. IEEE (2016)
18. Belazi, A., El-Latif, A.A.A., Belghith, S.: A novel image encryption scheme based on substitution-permutation network and chaos. Sig. Process. **128**, 155–170 (2016)
19. Hua, Z., Zhou, Y., Pun, C.M., et al.: 2D sine logistic modulation map for image encryption. Inf. Sci. Int. J. **297**(C), 80–94 (2015)

Statistical Analysis of Traffic-Related Social Media Data of Multiple Cities in China

Zhihao Zheng and Pu Wang(✉)

School of Traffic and Transportation Engineering, Central South University, Changsha 410075, Hunan, People's Republic of China
{vincentzheng, wangpu}@csu.edu.cn

Abstract. Social media platform has become a new data source for transportation research and practices. In this study, we analyzed the temporal-spatial features of traffic-related social media data of six big cities in China. In addition, we studied the occurrence of social media data on different levels of roads in Shenzhen, one of the six cities that we obtained road configuration data. Finally, similarities and differences of various types of traffic-related social media data were analyzed in the studied six big cities.

Keywords: Transportation data analysis · Social media data
Social transportation

1 Introduction

Social media platforms (e.g., Twitter in the U.S. and Weibo in China [1]) continuously generate huge volumes of social media messages regarding every aspect of our lives. Urban transportation is one of the topics that frequently appear in social media messages. This kind of traffic-related social media messages usually incorporate descriptions of traffic accidents, possible reasons for traffic anomalies, and suggestions on transportation planning and management. Because that social media messages are generated by human languages which inherently incorporate human intelligence, this kind of new emerging data can provide useful supplementary traffic information to traditional methods that collect traffic information using physical sensors (e.g., loop detectors, taxi GPS devices).

Many researchers have analyzed the traffic information in social media data. Mo *et al.* [2] analyzed traffic information from traffic police micro-blogs using the theories of linguistic dynamic systems. Chen *et al.* [3] found that social media tweets can sense traffic congestion. Kuflik *et al.* [4] proposed a framework for mining traffic information from social media data. Georgakis *et al.* [5] developed a framework for real-time detection of traffic events through classifying and geolocating social media data. Wang *et al.* [6] developed a traffic alert system by employing latent Dirichlet allocation method to identify traffic-related social media data. Gu *et al.* [7] extracted traffic incident information from social media data. Fu *et al.* [8] collected tweets that encapsulate real-time traffic information and generated concise and comprehensible textual information by filtering out redundant texts. D'Andrea *et al.* [9] used a support

J. Abawajy et al. (Eds.): ATCI 2018, AISC 842, pp. 141–150, 2019.
https://doi.org/10.1007/978-3-319-98776-7_17

vector machine (SVM) algorithm to identify traffic-related tweets. The authors then developed a real-time traffic monitoring system to detect traffic events. Tejaswin *et al.* [10] proposed a named entity extraction (NER) method to identify locations recorded in social media data and built an automated social media analysis system to assist traffic management. Zhang *et al.* [11] proposed a deep learning approach to detect traffic accidents in social media data. Cui *et al.* [12] collected traffic information from social media data and developed an Android-based app to distribute the traffic information. Lécué *et al.* [13–15] employed semantic web technologies to analyze, diagnose, explore, and predict traffic states. Zheng *et al.* [16] reviewed the data sources, analytical approaches, and application systems of social media based transportation research.

In the present study, we analyzed four-month Weibo data of six big cities in China. We aim to discover the temporal-spatial patterns of traffic-related social media data, and understand how the temporal-spatial patterns vary with different cities and different types of Weibo messages. Our results and findings can help researchers and transportation departments make better use of social media data in transportation research and practices.

2 Data

2.1 Traffic-Related Social Media Data

The social media data were collected from Sina Weibo, the largest micro-blog platform in China. The social media platform is used by 230 million active users, which represent one-sixth of the Chinese population. We used two sets of social media data in this study. The first dataset collects the Weibo messages post during the workdays (19 days) of October 2014 in Shenzhen. The second dataset was generated during the first four months (January – April) of 2016 by Weibo users in six big cities: Beijing, Guangzhou, Shenzhen, Zhengzhou, Nanjing, and Chengdu. For the first dataset, traffic-related Weibo data are manually identified. For the second dataset which are much larger, Weibo messages with road names and transportation keywords are selected as traffic-related Weibo data. The selected Weibo data were further categorized into three types according to the transportation keywords (Table 1). Totally 688 Weibo messages in the first dataset and 43,230 Weibo messages in the second dataset were used in this study. The number of traffic-related Weibo messages in each day of January 2016 is shown in Fig. 1. The first dataset was used to study the temporal-spatial patterns of traffic-related Weibo messages in different cities. The second dataset was used to study the occurring patterns of traffic-related Weibo on different levels of roads.

2.2 Geographical Information System (GIS) Data

The GIS data of the road networks of the studied cities (excluding Shenzhen) were obtained from an open-source online Github repository (https://github.com/ShuaiShao93/MapMatching-on-JStorm). The GIS data of each city includes road network information, rail network information, administrative district, etc. We used the road network information of each city to identify whether a Weibo message contains the name of a road. The GIS data of the Shenzhen road network were obtained from Shenzhen Transportation Authority. The data also record detailed information of road attributes (road level, speed limit, etc.).

Table 1. Transportation keywords for three types of Weibo.

Weibo type	Keywords
Traffic accident	'Accident', 'collision', 'crash', 'disabled car'
Traffic congestion	'Traffic jam', 'traffic congestion', 'slow traffic', 'large traffic volume'
Traffic control	'Road construction', 'traffic control'

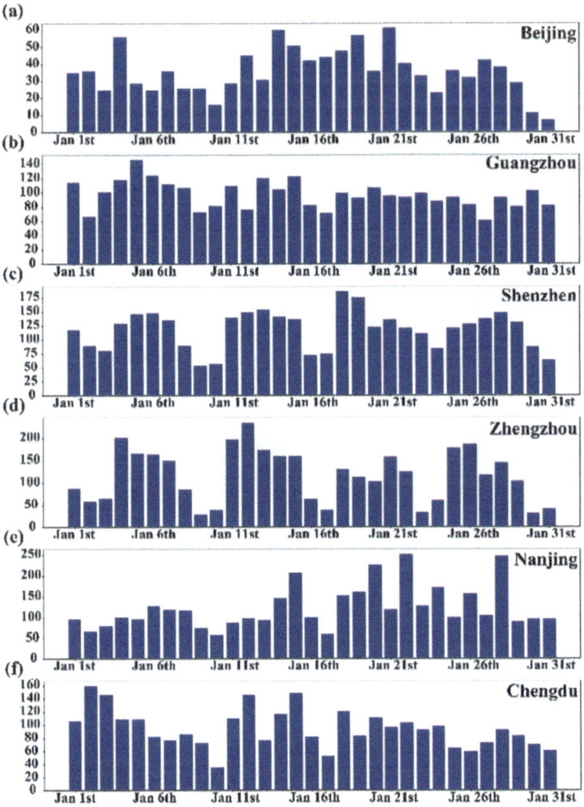

Fig. 1. The number of traffic-related Weibo in each day of January, 2016.

3 Method

We developed a web crawler to automatically collect Weibo messages from the social media platform (Fig. 2). First, the web crawler visits the Weibo stream website using the keywords shown in Table 1 at a fixed frequency. Next, the web crawler downloads each page of searching results and identifies Weibo contents using regular expression method. After decoding the contents, a standard format of Weibo message is obtained (Table 2). The post-time, message text, and location tag were recorded. Finally, the noise texts (i.e., the @ symbol, emoji, URL) that do not present actual information are eliminated.

Weibo messages containing transportation keywords are not necessarily traffic-related Weibo. Manually identifying the traffic-related Weibo is a labor-intensive work. For the much larger 2016 dataset, we used GIS data of each city to identify whether a road of the city appears in a Weibo message of the city. If a Weibo message contains a road name of this city, this Weibo message is considered as a traffic-related Weibo. We categorized Weibo into three types: (1) Traffic accident; (2) Traffic congestion; (3) Traffic control.

Table 2. Example of Weibo data with standard format.

Post time	Text (translated into English)	Location tag
17:56 Jan. 7, 2016	A bus accident slowed the traffic on Dongfeng road	*

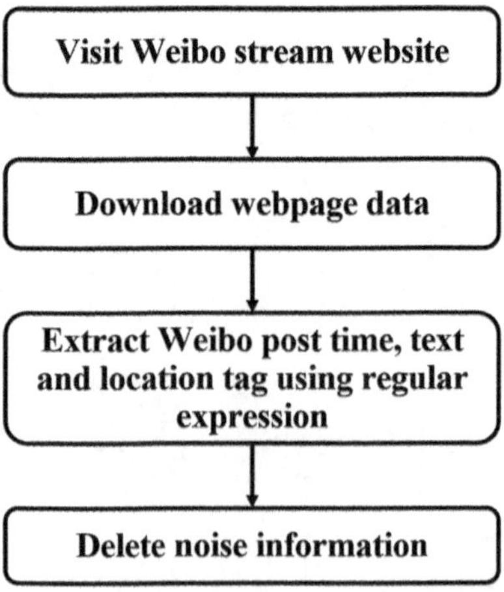

Fig. 2. Flow chart of the Weibo data collection.

4 Results

4.1 Temporal-Spatial Patterns of Weibo Messages in Different Cities

We first analyzed the daily average number of Weibo messages during holidays, weekends and workdays (Fig. 3). Interestingly, the dominant type of Weibo messages in each type of observation day varies with cities. This can be caused by the different mobility patterns and traffic conditions of the different cities. For example, Weibo reporting traffic control represent a large fraction of Weibo messages in Beijing, which may be caused by roads being blocked during important events. We find that Weibo reporting traffic accident are more on weekends for Beijing, more on weekdays for Guangzhou, Shenzhen, and Zhengzhou, and more on holidays in Nanjing and Chengdu. This kind of information is very useful for identifying the weak point of traffic safety for each city. Different temporal distributions of traffic congestion Weibo are also found for different cities. For example, we find that Weibo reporting traffic congestion are more in holidays for Beijing and Chengdu, more on weekdays for Shenzhen, Nanjing, and Zhengzhou. Traffic condition seems to be very stable during weekdays, weekends and holiday in Guangzhou. Traffic condition is much better on weekends for Zhengzhou, while traffic conditions are better on holidays for Nanjing. Different temporal distributions of traffic conditions imply different mobility patterns in the studied cities. The numbers of Weibo reporting traffic control are similar during weekdays, weekends, and holidays for all cities, albeit that slightly more traffic control Weibo are found on weekdays.

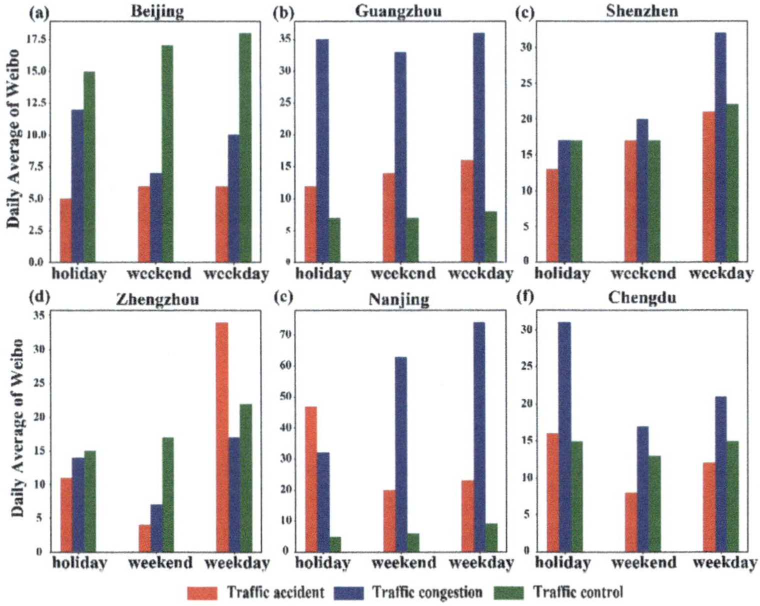

Fig. 3. Daily averages of three types of Weibo messages in holidays, weekends and workdays. (a) Beijing; (b) Guangzhou; (c) Shenzhen; (d) Zhengzhou; (e) Nanjing; (f) Chengdu.

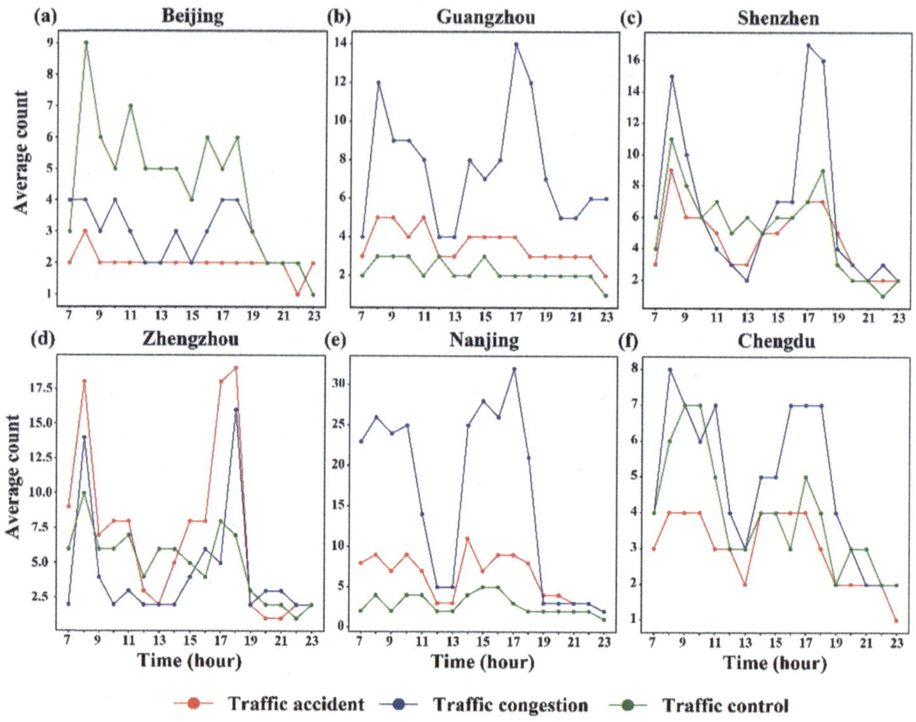

Fig. 4. Hourly averages of three types of Weibo messages in holidays, weekends and workdays. (a) Beijing; (b) Guangzhou; (c) Shenzhen; (d) Zhengzhou; (e) Nanjing; (f) Chengdu.

Next, we analyzed the temporal distributions of Weibo during each hourly time window. As shown in Fig. 4, the average number of Weibo messages tends to be higher during morning rush hours and evening rush hours for all cities. However, different cities encounter different major problems during the rush hours. More Weibo reporting traffic control during rush hours in Beijing than other cities (Fig. 4(a)). For most other cities, there are more Weibo reporting traffic congestion during the rush hours (Figs. 4(b–c, e–f)). For Zhengzhou city, however, more traffic accidents were reported by Weibo messages during the rush hours (Fig. 4(d)). The results imply that different cities may have different major traffic issues to deal with during morning and evening rush hours.

Spatial patterns of Weibo messages are also analyzed (Fig. 5). Each road is marked with different colors according to the number of Weibo messages mentioned it. For most roads, less than five Weibo are found during four months while some roads are frequently mentioned in Weibo messages. We find that Weibo messages are more likely to be collected on main roads and freeways that cross the city. This observation indicates that people tend to post traffic information if traffic events occurred in important locations. We next analyzed the number of days of each road mentioned by different types of Weibo messages (Fig. 6). For all three types of Weibo, traffic events occurred on most roads are non-recurring events which are only observed for several

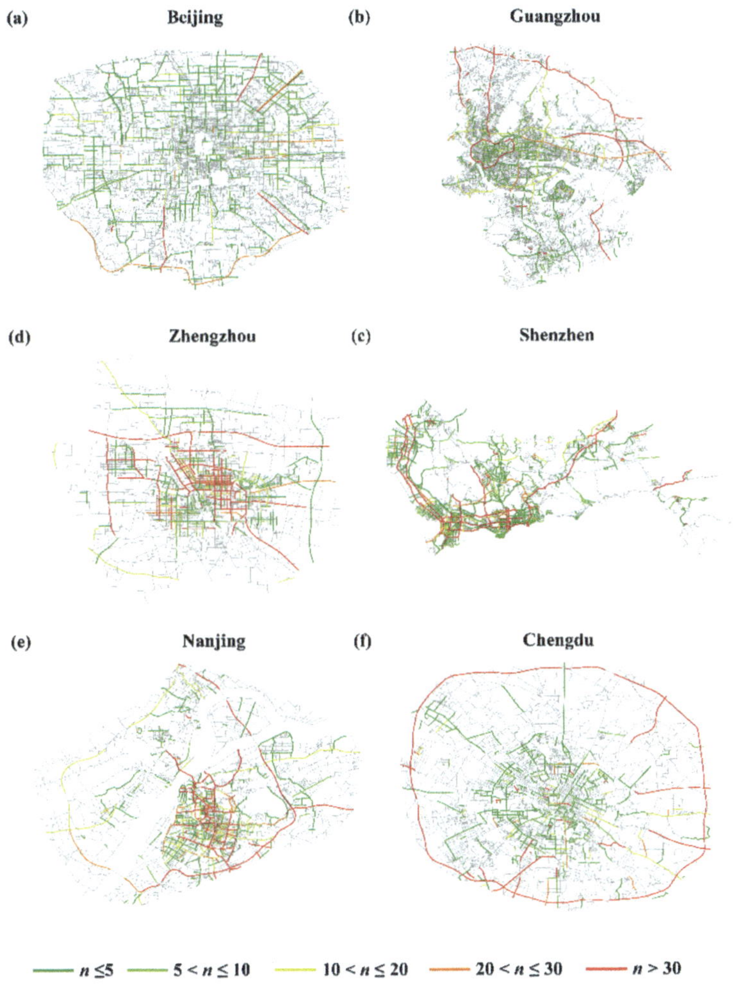

Fig. 5. Spatial distributions of Weibo messages. Parameter n denotes the number of Weibo messages mentioned each road in (a) Beijing; (b) Guangzhou; (c) Shenzhen; (d) Zhengzhou; (e) Nanjing; (f) Chengdu.

times (days). While some roads were mentioned more than 30 days during the four months, which can be regarded as recurring events. The type of seriously recurring traffic events on roads varies with different cities. For example, traffic accident and traffic control on some roads in Beijing are repeatedly reported by Weibo (Fig. 6(a)), while traffic accident and traffic congestion on some roads in Shenzhen are repeatedly reported by Weibo (Fig. 6(c)). The results present in this section show that the features of traffic related Weibo for different cities and traffic events are heterogeneous in time and space.

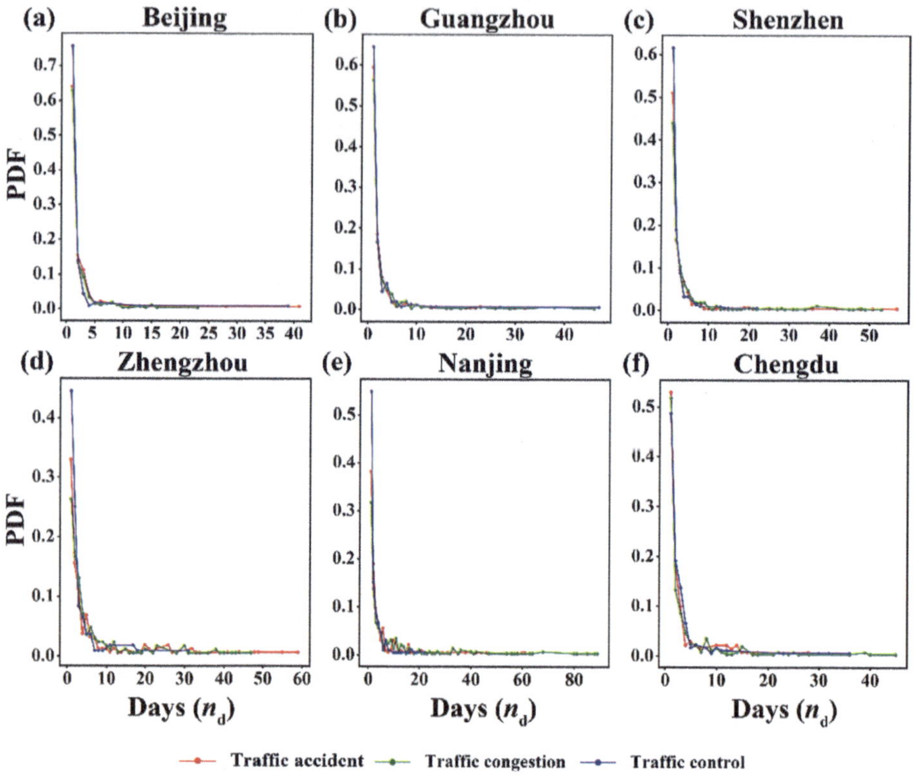

Fig. 6. Probability distributions of the number of days that each road is mentioned. Parameter nd denotes the number of days of each road mentioned by different types of Weibo messages in (a) Beijing; (b) Guangzhou; (c) Shenzhen; (d) Zhengzhou; (e) Nanjing; (f) Chengdu.

4.2 Traffic-Related Weibo Occurrence on Different Levels of Roads in Shenzhen

To further study the occurrence of various types of Weibo messages on different levels of roads, we combined the detailed road configuration information of Shenzhen with the basic GIS data. Road levels in Shenzhen are categorized into five levels: national freeways, urban freeways, main roads, secondary roads and branch roads. We calculated the percentage of three types of Weibo messages on five levels of roads. As shown in Fig. 7, the majority of Weibo messages is distributed on freeways and main roads, which is consistent with our previous investigation. Traffic accidents are more likely to be observed on urban freeways while traffic congestions are more likely to be observed on main roads.

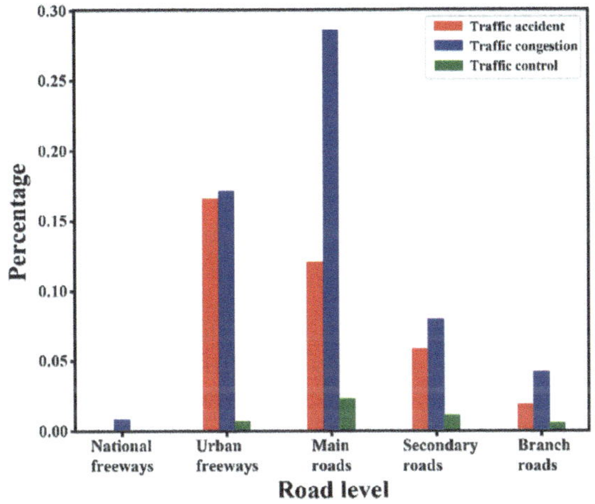

Fig. 7. Occurrence of three types of Weibo on different levels of roads.

5 Conclusion

In this study, we developed a web-crawler to collect traffic-related Weibo messages in six big cities. Tremendous amounts of social media data were used to analyze the temporal-spatial patterns of Weibo reporting different types of traffic incidents. Similarities and differences of the temporal-spatial patterns of traffic-related Weibo messages are analyzed for different cities. The results and findings may help transportation departments in theses cites identify the weak point in urban transportation management. We also explored the occurrence of traffic-related Weibo on different levels of roads in Shenzhen. The results show that traffic-related Weibo are widely distributed on major traffic passages, such as freeways and main roads. We believe that our findings can help both researchers and transportation departments better understand the spatial-temporal patterns of the new emerging transportation data source and make better use of them.

References

1. Lv, Y., Chen, Y., Zhang, X., Duan, Y., Li, N.L.: Social media based transportation research: the state of the work and the networking. IEEE/CAA J. Autom. Sin. **4**, 19–26 (2017). https://doi.org/10.1109/JAS.2017.7510316
2. Mo, H., Hao, X., Zheng, H., Liu, Z., Wen, D.: Linguistic dynamic analysis of traffic flow based on social media—a case study. IEEE Trans. Intell. Transp. Syst. **17**, 2668–2676 (2016). https://doi.org/10.1109/TITS.2016.2530698
3. Chen, P.-T., Chen, F., Qian, Z.: Road traffic congestion monitoring in social media with hinge-loss markov random fields. In: IEEE International Conference on 2014 Data Mining, pp. 80–89 (2014)
4. Kuflik, T., Minkov, E., Nocera, S., Grant-Muller, S., Gal-Tzur, A., Shoor, I.: Automating a framework to extract and analyse transport related social media content: the potential and the challenges. Transp. Res. Part C Emerg. Technol. **77**, 275–291 (2017). https://doi.org/10.1016/j.trc.2017.02.003

5. Salas, A., Georgakis, P., Nwagboso, C., Ammari, A., Petalas, I.: Traffic event detection framework using social media. In: IEEE International Conference on Smart Grid and Smart Cities, p. 5 (2017)
6. Wang, D., Al-Rubaie, A., Clarke, S.S., Davies, J.: Real-time traffic event detection from social media. ACM Trans. Internet Technol. **18**, 1–23 (2017). https://doi.org/10.1145/3122982
7. Gu, Y., Qian, Z., Chen, F.: From Twitter to detector: real-time traffic incident detection using social media data. Transp. Res. Part C Emerg. Technol. **67**, 321–342 (2016). https://doi.org/10.1016/j.trc.2016.02.011
8. Fu, K., Lu, C.T., Nune, R., Tao, J.X.: Steds: social media based transportation event detection with text summarization. In: Proceedings IEEE Conference on Intelligent Transportation Systems, ITSC, pp. 1952–1957 (2015)
9. D'Andrea, E., Ducange, P., Lazzerini, B., Marcelloni, F.: Real-time detection of traffic from Twitter stream analysis. IEEE Trans. Intell. Transp. Syst. **16**, 2269–2283 (2015). https://doi.org/10.1109/TITS.2015.2404431
10. Tejaswin, P.: Tweeting traffic: analyzing Twitter for generating real-time city traffic insights and predictions (2015). https://doi.org/10.1145/2778865.2778874
11. Zhang, Z., He, Q., Gao, J., Ni, M.: A deep learning approach for detecting traffic accidents from social media data. Transp. Res. Part C Emerg. Technol. **86**, 580–596 (2018). https://doi.org/10.1016/j.trc.2017.11.027
12. Cui, J., Fu, R., Dong, C., Zhang, Z.: Extraction of traffic information from social media interactions: methods and experiments. In: 17th IEEE International Conference on 2014 Intelligent Transportation Systems, ITSC 2014, pp. 1549–1554 (2014)
13. Lécué, F., Tallevi-Diotallevi, S.: Star-city: semantic traffic analytics and reasoning for city. In: Proceedings 19th 318201, pp. 1–8 (2014). https://doi.org/10.1145/2557500.2557537
14. Lécué, F., Tallevi-Diotallevi, S., Hayes, J., Tucker, R., Bicer, V., Sbodio, M., Tommasi, P.: Smart traffic analytics in the semantic web with star-city: scenarios, system and lessons learned in Dublin city. J. Web Semant. **27**, 26–33 (2014)
15. Lécué, F., Tucker, R., Tallevi-Diotallevi, S., Nair, R., Gkoufas, Y., Liguori, G., Borioni, M., Rademaker, A., Barbosa, L.: Semantic traffic diagnosis with star-city: architecture and lessons learned from deployment in Dublin, Bologna. Miami and Rio. Iswc **2014**(318201), 292–307 (2014). https://doi.org/10.1007/978-3-319-11915-1_19
16. Zheng, X., Chen, W., Wang, P., Shen, D., Chen, S., Wang, X., Zhang, Q., Yang, L.: Big data for social transportation. IEEE Trans. Intell. Transp. Syst. **17**, 620–630 (2015). https://doi.org/10.1109/TITS.2015.2480157

Exploring Influential Factors in Recommending Algorithms for All Items and Niche Items

Zhengye Xiao, Yi Zhang, and Yuchun Fang(✉)

School of Computer Engineering and Science, Shanghai University,
Shanghai, China
ycfang@shu.edu.cn

Abstract. The importance of recommender system is largely increasing in many E-commerce platforms. When we try to create an algorithm to complete the recommendation tasks, we should not only consider their performance on recommending all items, but also their performance on recommending niche items because of long tail effect. The successful cases of long tail effect exist in Internet. Amazon and eBay all have very success applications in this aspect. In this paper, we specifically concentrate on the factors which can influence the performance of algorithms on recommending all items and niche items. By strengthening the factors, we can enhance the ability of recommender algorithms to recommend all items or niche items to meet different needs. At the same time, these factors can provide a new view to explain the algorithms. The algorithms which we selected include collaborative filtering algorithms, the graph-based model and the content-based model. We compare the algorithms commented before on Top-N recommendation tasks and evaluate their performance on recommending long-tail items. Experimental results support the effect of factor analysis for various tasks.

Keywords: Recommender systems · Recommender algorithms
Long tail effect

1 Introduction

With the development of E-commerce platform, consumers have more choices and ways to buy products. This leads to fierce competition between platforms. Although it is not trivial to recommend consumers the most appropriate products, it is a good way to make the consumers satisfy with the platform. Hence, it is essential to develop recommender system that can adapt to various requirements of users.

There are a lot of recommender algorithms than can be classified as two ways. One way is to try to build a hybrid system like [1, 2]. They integrate the dataset and the prior knowledge which human provide to improve the performance of recommendation. The other way is to choose appropriate algorithms according to different scenes. Given that the existence of long tail effect, both of ways have to consider the performance of algorithms on recommending all items and niche items. Long tail effect means that if

J. Abawajy et al. (Eds.): ATCI 2018, AISC 842, pp. 151–160, 2019.
https://doi.org/10.1007/978-3-319-98776-7_18

you can store enough products and have enough channels of circulation, the share of the market which niche products accounted for can compare with those popular products accounted for [3]. We suppose that there are some important factors which influence the performance of algorithms on recommending all items and niche items. We select some classic but effective algorithms and some special algorithms to find them.

Collaborative Filtering (CF) algorithms are the most successful algorithms in engineering, which employ user's past behaviors and ratings to predict unknown ratings. There are two main algorithms in CF: the neighborhood model and latent factor model. The neighborhood model put emphasis on the similarity between items [4]. We predict the preference of a user to an item through the user's ratings on the similar items. The neighborhood model is good at finding the relation between items, but it usually ignores the whole ratings which the users provided.

The latent factor model, such as Singular Value Decomposition (SVD), converts the users-items rating matrix to two latent factor space [5]. One of the latent factor space expresses users' preference to different factors. If the recommender system is about movie. These factors may be the preference to action, science fiction, actors and even some factors which have not been defined. Another one expresses the latent properties of items. We can use two latent factor spaces to make an inner product to predict the unknown ratings. The latent factor model can predict the ratings from a global point of view, but it's poor at finding the relation between items and not easy to explain.

The graph-based model is similar to PageRank [6]. It is centered on the relation between users and items. The user-item matrix can be a reference to their relation. The characteristic of this model is that it does not rely on the ratings but the behavior of user rating on items. We compute the relation between users and unrated items, and if there is a strong relation between a user and an item, we consider that the user may like this item.

The content-based model relies on the prior knowledge of items' features. First, we need to use some methods to get the items' features, then we can compute the users' favorite features. We can recommend an item to a user if this item's features are similar to users' favorite features.

Except CF algorithms, we can hardly use RMSE to evaluate other algorithms. In order to use a unified method to compare algorithm, we compare the algorithms commented before on Top-N recommendation tasks which was proposed in [9] and evaluate their performance on recommending long-tail items.

After comparing with experiment results, we have several outcomes listed below:

- We find that browsing history which we may think helpful to recommend niche products makes less contribution. But it reflects the relation between users and popular trend. It can help us improve the recall of recommending on all items.
- The key factor which a personalized algorithm should have is the explicit preference of users.
- These factors can provide us a new view to explain algorithms.

The structure of the rest of the paper is as follows. We start with preliminaries in Sect. 2. Then we concretely describe the algorithms commented before in Sect. 3. Section 4 shows the experiment results. We propose our suggestion in Sect. 5. Our conclusion is put in Sect. 6.

2 Preliminaries

In order to maintain the equation of unity, we use special index to distinguish users and items. Users are donated by u, v and items are donated by i, j. The set of items is donated by $I = \{i_1, i_2, \ldots, i_k\}$ and the set of users is donated by $U = \{u_1, u_2, \ldots, u_k\}$. The known users' ratings on items are denoted by $R_{u,i}$. The predicted users' ratings on items are denoted by $r_{u,i}$. The pairs (u, i) for which $R_{u,i}$ is known are stored in $C = \{(u, i) | R_{u,i}\, is\, known\}$.

2.1 Flowchart

In this paper, we evaluate the performance of different algorithms on recommending all items and niche items. Our work in this paper can be expressed by a flowchart (Fig. 1):

Fig. 1. Flowchart of the proposed work

We summarize the algorithms which are evaluated in this paper first. Before experiments, we take measure to perform data pretreatment and set the evaluation criterion. We exhibit our experimental results for an intuitive comparison among algorithms. Detailed analysis is performed quantitatively. We illustrate our findings according to the experiments and some other information such as running time and iteration numbers. In the last step, we propose our suggestion on the important factors which influence the performance of algorithms on recommending all items and niche items.

2.2 Long-Tail

The Long-Tail means that the majority of ratings fix on a small part of all items, and the ratings on other items form a long tail [3]. When you choose products on the Internet, you may have the experience that the recommended products are fixed on some kinds of products. That is because the items on short-head are easily recommended and the items on long-tail not. However, we may want to use niche products to show our personalities. In Movielens data set [7], the 5.1% of items (200 items) have 30% ratings. In this paper, we use different algorithms to recommend other 94.9% of items to evaluate their ability to recommend niche products.

3 Methods

3.1 Neighborhood Model

In this work we use item-based neighborhood model [4]. Before we predict the unknown ratings, we should compute the similarity of i with other items. We use Pearson correlation for this propose. Pearson correlation is defined as:

$$sim(i,j) = \frac{\sum_{u \in U}\left(R_{u,i}-\bar{R}_i\right)\left(R_{u,j}-\bar{R}_j\right)}{\sqrt{\sum_{u \in U}\left(R_{u,i}-\bar{R}_i\right)^2}\sqrt{\sum_{u \in U}\left(R_{u,j}-\bar{R}_j\right)^2}}, \tag{1}$$

where $\bar{R}_i$ is the mean rating given to item i. The set of items rated by user u is donated by $S(u;i)$. The predict value $r_{u,i}$ is computed as a weight average of the ratings of the neighboring items, as:

$$r_{u,i} = \frac{\sum_{j \in S(u;i)} sim(i,j) \times R_{u,j}}{\sum_{j \in S(u;i)} |sim(i,j)|}. \tag{2}$$

3.2 Latent Factor Model

Latent Factor Model uses SVD, which can factorize the user-item matrix to two lower factor space [5, 8]. So the available rating can be computed by the inner product of two vectors. One of the vector called 'user-factor' and the other one called 'item-factor'. They can be donated by p_u and q_i. So the predicted value $r_{u,i}$ can be expressed as:

$$r_{u,i} = p_u^T q_i. \tag{3}$$

Sometimes the data may have large user and item effects. Some users tend to give higher ratings than other users and some items get higher ratings than other items. We can use b_u and b_i to express the bias of users and items. Because of the bias is relative to average value. So we can rewrite (3) as:

$$r_{u,i} = \mu + b_u + b_i + p_u^T q_i. \tag{4}$$

Here, μ is the average value of the whole matrix, and $p_u^T q_i$ becomes a deviation of average value. Because of high portion of missing ratings, we cannot directly apply SVD to get p_u and q_i. However, we can get the parameters by solving the least squares problem and avoid overfitting by adding regularization terms:

$$\min_{p*,q*,b*} \sum_{(u,i) \in C} \left(R_{u,i} - r_{u,i}\right)^2 + \lambda\left(\sum_u \|p_u\|^2 + \sum_i \|q_i\|^2 + \sum_u b_u^2 + \sum_i b_i^2\right). \tag{5}$$

where λ is a hyperparameter, and it can be determined by cross validation. We can use gradient descent to solve the least square problem, yielding:

$$\begin{aligned} b_u &\leftarrow b_u + \gamma\left(2\sum\nolimits_i \left(r_{ui} - \mu - b_u - b_i - p_u^T q_i\right) - 2\lambda b_u\right), \\ b_i &\leftarrow b_i + \gamma\left(2\sum\nolimits_u \left(r_{ui} - \mu - b_u - b_i - p_u^T q_i\right) - 2\lambda b_i\right), \\ p_u &\leftarrow p_u + \gamma\left(2\sum\nolimits_i q_i\left(r_{ui} - \mu - b_u - b_i - p_u^T q_i\right) - 2\lambda p_u\right), \\ q_i &\leftarrow q_i + \gamma\left(2\sum\nolimits_u p_u\left(r_{ui} - \mu - b_u - b_i - p_u^T q_i\right) - 2\lambda q_i\right). \end{aligned} \tag{6}$$

3.3 Graph-Based Model

It is easy to use bipartite graph to express the user-item matrix. The set U and I can be considered as two vertex sets. If a user rates or browses an item, we can match the user and the item with a line. Our goal is to compute the correlation between the user and unmatched items and give a rank of these items. There is no question that PageRank algorithms is one of the best way to compute the correlation. The PageRank algorithm can be expressed as:

$$PR(m) = \tfrac{1-d}{N} + d\sum\nolimits_{n \in in(m)} \tfrac{PR(n)}{|out(n)|}. \tag{7}$$

Here, $in(m)$ is a set of vertices which point to vertex m and $out(n)$ is a set of vertices which vertex n points to. And d is a hyperparameter. It expresses the probability of continue traversal. In PageRank, the probability of visiting each page is equal. But if we want to recommend items to user, we should set the probability of visiting the user one and the probability of visiting other items and users zero, and it can be considered as a topic-sensitive PageRank [9]. The formulation can be expressed as:

$$\begin{aligned} PR(m) &= (1-d)r_m + d\sum\nolimits_{n \in in(m)} \frac{PR(n)}{|out(n)|}, \\ r_m &= \begin{cases} 1, m = u \\ 0, m \neq u \end{cases}. \end{aligned} \tag{8}$$

Here, If the probability of visiting an item is high, the correlation between the user and the item is also high and this means that the user may like this item.

3.4 Content-Based Model

For many kinds of items, people have done pre-classification for these items. If these items are movies, they can be classified by action, adventure, animation, etc. Having a set of items I and a set of features $\{f_1, f_2, \ldots, f_k\}$, F is an M-by-K matrix. For each entry $F_{i,j}$, if item i contains feature f_j then $F_{i,j} = 1$, else $F_{i,j} = 0$. Here, there are two methods to recommend items.

The first method does not rely on ratings and is called Rating Independence Content-Based Model (RICBM). If a user rated an item, we can consider that the user has interests in the item. We can count the times of the occurrence of each feature. Then we can extract four highest features to form the user-factor. We can use Pearson correlation to compute the similarity between user-factor and item-factor and recommend the most similar N items to users. This method relies on the browsing history.

The second method is similar to neighborhood model. We can use the Pearson correlation to compute the similarity between item-feature vectors. Now that we have similarity matrix, we can also use it to predict ratings like neighborhood model, and this method obviously relies on users' ratings. We name it as Rating Dependence Content-Based Model (RDCBM). In this paper, we compare both RICBM and RDCBM on Top-N recommendation tasks.

4 Experiments

In this part, we evaluate the performance of algorithms on Top-N recommendation tasks. Given that long tail effect mentioned in Sect. 2, we also evaluate their ability to recommend niche products. Algorithms used for comparison are listed below:

MovieAvg: This algorithm is non-personalized. No matter who users are, the predict ratings of movies are their average value.
CorNgbr: This algorithm is a neighborhood model which is mentioned in Sect. 3.1.
CSVD: This algorithm is a latent factor model which is mentioned in Sect. 3.2. In our experiment, we use gradient descent to solve the least square problem. The initial value of b_u, b_i, p_u, q_i can be formulated as:

$$b_u = 0, \quad b_i = 0, \quad p_u = \frac{rand(-1,1)}{\sqrt{k}}, \quad q_i = \frac{rand(-1,1)}{\sqrt{k}}. \tag{9}$$

Here, k is the dimension of factor vector. In our experiment k is 50 and λ is 0.001. The learning rate(step size) γ is 0.00024. According to Fig. 2, this algorithm on Top-N recommendation tasks is very easy to overfit. The iterate number in our experiments is set as 9 because it has the best performance on recommending niche products.

Graph-Based: This algorithm is mentioned in Sect. 3.3. It does not rely on rating matrix. In order to make a Top-N problem converge, we adjust the iteration times to achieve it. The hyperparameter d in our experiment is set 0.6 and the iteration number is 13.
RICBM: This algorithm is a kind of content-based model which is mentioned in Sect. 3.4. It does not rely on rating matrix.
RDCBM: This algorithm is a kind of content-based model which is mentioned in Sect. 3.4. It relies on rating matrix, and the method of computing prediction ratings is similar to CorNgbr.

4.1 Data Set

The Data Set we use in this paper is Movielens data set [7]. It is provided by GroupLens. It contains 1,000,209 ratings for 6040 users to 3900 movies. The ratings are made on a 5-star scale. Each user has at least 20 ratings. These data at least satisfy the needs for CF. At the same time, the movies in the data set have been divided into different genres. So these data are helpful for content-based model to train its model.

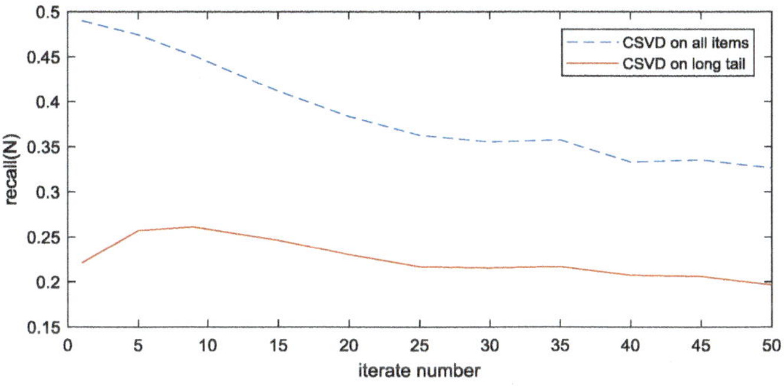

Fig. 2. Iterate number vs recall-at-N

4.2 Top-N Recommendation Tasks

In order to compare the algorithms mentioned in Sect. 3, we test their performance of Top-N recommendation tasks. The method which we use is similar to [10]. For the dataset, we do some pretreatment. Firstly, we sub-sample 10% of the ratings from the matrix to create a probe set. This probe set can be used by latent factor model to determine the regularization constants. Then, we can get a test set R which have the five-stars ratings in the probe set. If we use an item in the R and other unrated items like 400 items to be recommended, we can get a set which contains N recommended items. If the item in the R is also in the set, it is correctly recommended. To evaluate the performance of Top-N recommendation tasks, we can compute precision and recall, which can be expressed as:

$$recall = \frac{N_r}{|R|}, \tag{10}$$

$$precision = \frac{N_r}{N \cdot |R|}. \tag{11}$$

where N_r is the count of the items in the R recommended correctly.

4.3 Performance on Recommending All Items

The result of the performance of recommending all items on Top-N recommendation tasks is showed in Fig. 3. Figure 3(a) shows that the performance of both CF algorithms is in line with expectation. They have acceptable rate of recall. The performance of graph-based model is superior out of our exception because they do not rely on available ratings and they only focus on the behavior of whether user rated to items. We find that although browsing history does not reflect the preference of users, it provides us relation between users and general popular trend. The graph-based model catches this, and performs well on recommending all items. The MovieAvg also performs well. It catches the general popular trend. RICBM and RDCBM performs worse than other methods. They put emphasis on the personalized difference between users but omitting the relation among users and the general trends.

Figure 3(b) shows the precision of algorithms at a given recall. Their performance is similar to Fig. 3(a). The graph-based model also outperforms other algorithms.

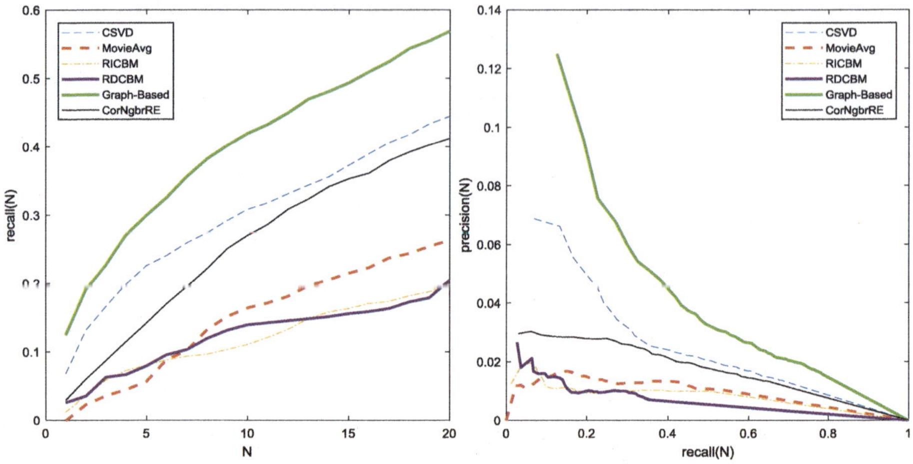

Fig. 3. (a) Recall-at-N and (b) recall vs precision on all items

We also analyze the time consumption to train the models. Although the graph-based algorithm gets the best recall, it costs too much training time. If you want to get the relation between a user and items, it needs to cost 1 min. This algorithm is easy to be accelerated by parallel computing framework like Hadoop and Spark. CorNgbr costs 10 min and CSVD costs 1 min to train the whole model. Given their performance, the cost time is still acceptable. RICBM and RDCBM cost 4 min in training.

4.4 Performance on Recommending Niche Items

The result of the performance of recommending all 94.9% items on Top-N recommendation tasks is showed in Fig. 4. Figure 4(a) reports the recall-at-N. The performance of CorNgbr and CSVD is well and they are likely to be good at recommending niche items. The graph-based model which get the best recall on items performs not well on long-tail items. Its performance is similar to MovieAvg. What the graph-based model omit is the preference of users and it leads to judging false when recommend the niche items. RDCBM which not does well on recommending all items has a good performance on recommending long-tail items because the ratings clearly express the preference of users. Given that the performance of RDCBM, we try to simply filter the ratings below three stars in *C*. We use the new training set to train graph-based model. The recall on 94.9% items increases 0.02. RICBM is still terrible. Although it tries to find the preference of users from the browsing history but the features maybe not appropriate.

Figure 4(b) reports that when recall is low, the precision of CSVD and RDCBM is better than CorNgbr. However, with the increasing of recall, CorNgbr is soon outperforming other algorithms.

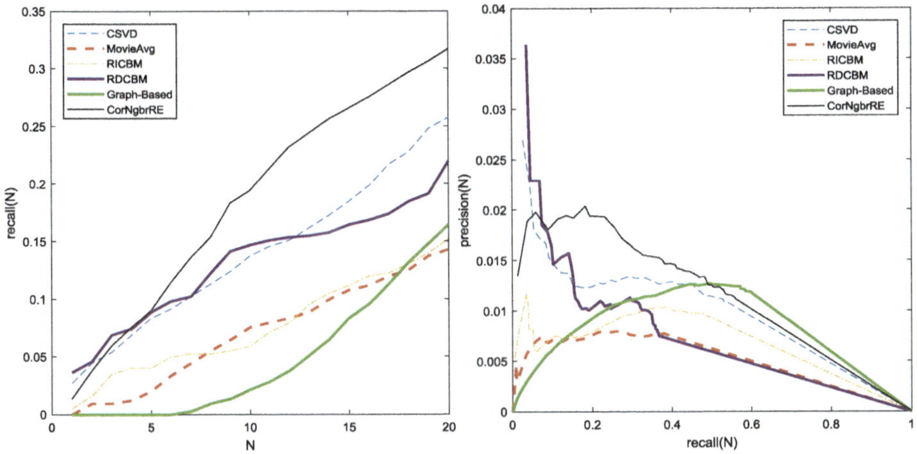

Fig. 4. (a) Recall-at-N and (b) recall vs precision on 94.9% items

5 Discussion

According to the analysis of the experimental results, we draw major conclusions as follows:

- Because of lacking information of preference in the history of users' browsing, the algorithm which only relies on browsing history can hardly performs well on recommending niche products. However, the information hiding behind browsing history such as relation between users and general popular trend is helpful to recommend all items.
- Based on the face that the performance of RDCBM and the graph-based model are easily filtered, it is notable that the detailed information of preference of users is very helpful to recommend niche items. However, fuzzy information only brings limited improvement.
- The above two factors also provide us a new view to explain algorithms. Actually, when a user rates an item we get a browsing history. We can take full advantage of ratings. For CorNgbr, the step of computing similarity is a way to get the relation between users and the step of predicting the ratings put emphasis on the preference of users. However, RDCBM relies on the pre-classified features, so it performs terrible. Comparing CSVD with SVD++ [11], the latter adds implicit feedback and it enlarges the function of browsing history in our opinion.
- For hybrid recommender system we should enlarge both of the two factors. For those people who provide few ratings, we can focus on the first factor to make conservative recommendation and for those people who provide a lot of ratings, we can focus on the second factor to satisfy personalized needs.

6 Conclusions

In this paper, we endeavor to explore the factors which can influence the performance of recommending all items and niche items. We select several algorithms, including CorNgbr, CSVD, Graph-Based model, RICBM and RDCBM, to evaluate their performance on Top-N recommendation tasks. By comparing with the experiment results, we find that the algorithms, such as the graph-based model, which simply relies on browsing history can perform well on recommending all items, but perform terrible on niche items. However, although those algorithms which can take full advantage of ratings performs not as well as the graph-based model on recommending all items, they perform well on niche items. We conclude that the relation between users and general popular trend is an important factor to recommend all items. Moreover, the detailed information of preference of users is the key factor to recommend niche items. These conclusions provide us a new view to explain recommending algorithms.

Acknowledgements. The work is funded by the Shanghai Undergraduate Student Innovation Project, the National Natural Science Foundation of China (No. 61170155) and the Shanghai Innovation Action Plan Project (No. 16511101200).

References

1. Aslanian, E., Radmanesh, M., Jalili, M.: Hybrid recommender systems based on content feature relationship. IEEE Trans. Ind. Inf. **PP**(99), 1 (2016)
2. Forbes, P., Zhu, M.: Content-boosted matrix factorization for recommender systems: experiments with recipe recommendation. In: Conference on Recommender Systems, pp. 261–264 (2011)
3. Anderson, C.: The long tail: why the future of business is selling less of more. Hyperion, July 2006
4. Sarwar, B.M., Karypis, G., Konstan, J.A., et al.: Item-based collaborative filtering recommendation algorithms. In: International World Wide Web Conferences, pp. 285–295 (2001)
5. Koren, Y.: Factorization meets the neighborhood: a multifaceted collaborative filtering model. In: Knowledge Discovery and Data Mining, pp. 426–434 (2008)
6. The Google Search Engine: Commercial search engine founded by the originators of PageRank. http://www.google.com/
7. GroupLens: MovieLens 1 M Dataset. https://grouplens.org/datasets/movielens/1m/
8. Ma, C.C.: A guide to singular value decomposition for collaborative filtering (2008)
9. Haveliwala, T.H.: Topic-sensitive PageRank. In: International World Wide Web Conferences, pp. 517–526 (2002)
10. Cremonesi, P., Koren, Y., Turrin, R., et al.: Performance of recommender algorithms on top-N recommendation tasks. In: Conference on Recommender Systems, pp. 39–46 (2010)
11. Bell, R., Koren, Y., Volinsky, C.: The BellKor solution to the Netflix Prize (2007). http://www.research.att.com/~volinsky/netflix/ProgressPrize2007BellKorSolution.pdf

A Distributed Approach of Accompany Vehicle Discovery

Yu Liang Zhao[1,2](✉), Qi Tong Wang[2], Peng Wang[2], Wei Wang[2], and Zhi Guo Yan[1]

[1] The Third Research Institute of the Ministry of Public Security, Shanghai, China
yltrimps@163.com

[2] School of Computer Science, Fudan University, Shanghai, China

Abstract. Accompany vehicle discovery is a hot topic in criminal investigation department with regard to massive vehicle data retrieval. In this paper, we study accompany vehicle discovery from traffic monitor plate recognition data. Existing methods are inefficient to query accompany vehicle. To address the problem, we propose a distributed algorithm to find accompany vehicle pair using sliding window algorithm with new prune strategies. We first load traffic monitor plate recognition data into memory by Spark, get candidate vehicle's upper bound of statistical data to prune invalidate candidates of vehicle and perform verification on these candidates to get final results. We devise some new prune strategies to find high quality candidates. Experiments with real Traffic Monitor Plate Recognition (TMPR) data in a distributed environment verify that our distributed approach can discover accompany vehicles efficiency. We also analyze the performance affecting factors from the experiments.

Keywords: Accompany vehicle discovery · Traffic monitor plate recognition Distributed approach · Prune strategy

1 Introduction

The applications of Locality Based Service are emerging nowadays with the explosive growing of position data. They are widely used in map services, such as restaurants recommendation and fastest route recommendation. They also perform a important role in other public services, such as crime investigations.

Frequent accompany vehicle finding is one of the key challenges in criminal investigations. Some serious crimes, such as a bank robbery, are usually committed by a gang of criminals. This means that they tend to have different vehicles driving along with each other. So finding the vehicles which are accompanying each other in the suspected areas is helpful for the investigators to narrow the search area. This is a typical application of frequent accompany vehicle finding, and it plays an important role in raising efficiency of crime investigations.

With more and more traffic monitoring cameras installed in large cities [1], Traffic Monitoring Plate Recognition (TMPR) data are being generated continuously. A TMPR record is extracted from the picture taken by traffic monitors. This is a reliable and easily accessed data for crime investigators. Using TMPR data, the frequent

J. Abawajy et al. (Eds.): ATCI 2018, AISC 842, pp. 161–173, 2019.
https://doi.org/10.1007/978-3-319-98776-7_19

accompany vehicle discovery is to find all the accompany vehicle pairs going with each other deliberately. However, most of the current researches are designed to work on GPS data and propose several different definitions (such as flock [2], convoy [3] and swarm [4]). As GPS positions of each vehicle are hard to track for investigator, the existing methods are not suitable for our problem.

The key challenges in frequent accompany vehicle finding lie in the following two aspects. The first challenge is that the total number of TMPR data set is very large in quantity. For example, the number of records in TMPR dataset is around 9 million per day within a single city and the average number of vehicle for 70 per 10 min within a single traffic monitor spot. So the data for a month contains around 300 million records. And if two vehicles passing the same monitor within 10 min are treated as accompanying, the total number of accompany vehicle pairs is around 630 million per day. So this leads to the second challenge that is the explosion of intermediate results. If all the intermediate pairs are generated, the size of records will span 70 times larger. This will put a high stress on both storing and computing. And as we use distributed computing frameworks, this also increase the cost of network transmission between different nodes.

To solve the two challenges, we first propose a novel and practical definition of the frequent accompany vehicle discovery problem and a distributed sliding window algorithm as our basic method to solve it. To boost the accompany vehicle pairs finding process, a pruning method based on grouped digest is devised to reduce the size of intermediate result sets. We split the occurrence information into several groups and calculate the corresponding digests first. Before generating a accompany pair, we check if it belongs to the final result set and decide whether to prune it or not. We also present and compare several different grouping strategies and pruning strategies.

In summary, we make the following contributions:

- We introduce the definition of accompany vehicle.
- We devise a basic method to solve accompany vehicle discovery problem utilizing sliding windows algorithm.
- We propose an adaptive approach and several pruning strategies to prune most of the candidates for a query to speed up the accompany vehicle discovery process.
- We conduct comprehensive experiments on real traffic data set to demonstrate the efficiency and quality of our proposed approach.

The rest of this paper is organized as follows. In Sect. 2, we review the related work. We introduce the preliminaries and formalize our problem statements in Sect. 3. In Sect. 4, we elaborate the basic accompany vehicle pair discovery algorithm and propose an adaptive approach which adopts different prune strategies to improve efficiency in Sect. 4.4. We present our experimental results in Sect. 5. Finally, we conclude the paper in Sect. 6.

2 Related Work

The increasing pervasiveness of spatio-temporal technologies has enabled collection of huge amount of records or trajectories for almost any kind of moving objects. Discovering useful patterns from their movement behaviours has a great variety of critical applications. So movement pattern discovery is a hot topic of many research studies [2, 5] in recent years. The proposed techniques have wide applications in various areas, such as transportation management, military surveillance and scientific researches [5]. Since the process of discovering patterns over large-scale data set can be quite costly, a lot of well designed techniques have been designed to improve the performance. These techniques including fast movement pattern discovery, clustering moving object gathering pattern, effective indexing structures and incremental algorithm for handling new records or trajectory arrivals. Besides, as the size of the data set may be very large and exceed the computing ability of one computer, parallel computing frameworks for big data, such as Apache Spark, are rapidly popularizing.

Fast movement pattern discovery. The problem of movement pattern discovery has been variously referred to as the searching for flocks [2], moving clusters [6], spatial-temporal joins [7], spatial co-locations [8], convoys [3], moving groups [9], swarms [4], travelling companions group [10] and so on. In the meanwhile, many researches propose many computing approaches such as Grid-based, Euclidean distance and Dynamic Time Warping. But in the original study they lack of considering the property of time for trajectories. Feng et al. propose the approach of accompany vehicle discovery applied to find accompany vehicle pairs from TMPR [10], which neglects the factor of efficiency from the point view of parallelization.

Clustering moving object gathering pattern. In [20], Zheng et al. is the first to develop a set of novel techniques to tackle the challenge of efficient discovery of gathering patterns on archived trajectory dataset. In [21], Zhang et al. proposes a moving object gathering pattern retrieving method that aims to support the retrieving of gathering patterns by using spatiotemporal graph. Yu et al. propose a density-based clustering algorithm for trajectory data stream.

Spark Overview. As the volumes of data increase, more parallel computing frameworks are developed to speed up computing. But some computing framework (such as Hadoop [11, 12], Dryad [13] and Storm [14]) aren't suitable for iterative algorithms. Apache Spark [15–17] is a general-purpose, widely-used distributed computing engine for big data processing, with APIs in Scala, Java and Python. A rich ecosystem based on Spark for in-memory big data analytics has been built, including libraries for streaming, graph processing, and machine learning Spark provides an efficient abstraction for in-memory cluster computing called Resilient Distributed Datasets (RDDs). Partitions are the basic physical units of data carried by RDD. RDDs are fault-tolerant as Spark can recover lost data using lineage graphs by rerunning operations to rebuild missing partitions. Each computer in the Spark cluster is called a worker, and there are several executors in one worker. Every executor will perform a task which is manipulating a partition, and partitions will be computed parallelly.

3 Preliminaries

In the various applications of accompany vehicles discovery, there are some common TMPR data record shared in different scenarios. We illustrate the characteristics of accompany vehicle discovery by the following example and employ the concepts of accompany vehicle discovery in this study.

Definition 1 (TMPR Data Record). A TMPR Data Record r is a trimple denoted as r = (v, s, t), where v is a vehicle plate number, s is a traffic monitor spot and t is a timestamp. Let D be the set of TMPR records denoted as D = {r1, r2, r3, …, rn}, V be the set of vehicle plate numbers denoted as V = {v1, v2, v3, …, vk}, S be the set of traffic monitor spots denoted as S = {s1, s2, s3, …, sm}, T be the set of timestamp denoted as T = {t1, t2, t3, …, tp}.

Example 1. Four vehicles' TMPR data records are recorded by traffic monitoring system. Table 1 shows simplify TMPR Data Record.

Table 1. TMPR data record

V	S	T	
*v*1	*s*1	2016-01-01	08:10:10
*v*2	*s*1	2016-01-01	08:10:15
*v*1	*s*2	2016-01-01	08:18:05
*v*2	*s*2	2016-01-01	08:18:10
*v*3	*s*2	2016-01-01	08:18:13
*v*1	*s*3	2016-01-01	09:10:10
*v*3	*s*3	2016-01-01	09:10:18
*v*4	*s*4	2016-01-01	09:12:30

Table 1 illustrates an example of TMPR Data Record. The v1, v2, v3 and v4 of V passes through traffic monitor spotID (such as s1, s2, s3 and s4 of S) at timestamp t1, t2, t3, t4, t5, t6, t7 and t8 of T.

$$D_{ij} = \{(r, r^{'})\Big|(r, r^{'}) \in D \wedge r_i \cdot v = v_i \wedge r_j^{'} \cdot \bar{v} = v_j \wedge i \neq j \wedge \Big|r_i \cdot \bar{t} - r_j^{'} \cdot t\Big| \leq \delta \wedge r_i \cdot s = r_j^{'} \cdot s\}, \ where\ r_i \in D,\ r_j^{'} \in D,\ r_i \cdot v \in V,\ r_j^{'} \cdot v \in V,\ r_i \cdot s \in S,\ r_j^{'} \cdot s \in S.$$

Definition 2 (Accompany Pair). Given a TMPR Data Record r = (v, s, t), Let Δt be the span time threshold, Accompany Pair for the TMPR data record set D, the Vehicle ID set V, and the timestamp set T can be defined as:

Example 2. As illustrated in Table 1, let Δt = 10 s, {(r1, r2), (r3, r4), (r3, r5), (r4, r5), (r6, r7)} is an accompany pair.

Definition 3 (Frequent Accompany Vehicle Pair). Let D(vi) = {r | ri.v = vi ∧ r ∈ D}, D(si) = {r | ri.s = si ∧ r ∈ D}, minS be the number threshold, Δt be the span time threshold and Dij be the set of accompany pairs, thus Frequent Accompany Vehicle Pair can be defined as:

Example 3. As in Table 1, let Δt = 10 s, minS = 2, {(v1, v2), (v1, v3)} is a frequent accompany vehicle pair. Next we define the problem of accompany vehicle pair discovery formally as follows.

Definition 4: Accompany Vehicle Pair Discovery. Given D and minS, our goal is to find all the frequent accompany vehicle pair (vi, vj)($1 \leq i \leq j \leq N$) so that $|Dij| \geq$ minS.

Example 4. As illustrated in Table 1, V = {v1, v2, v3, v4}, S = {s1, s2, s3, s4}, T = {2016-01-01 ~ 08:10:10, 2016-01-01 ~ 08:10:15, 2016-01-01 ~ 08:18:05, 2016-01-01 ~ 08:18:10, 2016-01-01 ~ 08:18:13, 2016-01-01 ~ 09:10:10, 2016-01-01 ~ 09:10:18, 2016-01-01 ~ 09:12:30}. Let m-count = 2 and t-span = 60 s, {(v1, v2)} is the only accompany vehicle.

4 Accompany Vehicle Pair Discovery Methods

In this section, we present a basic method for accompany vehicle pair discovery utilizing the idea of sliding window. Then we analyze the basic method's drawbacks and propose a general pruning strategy to boost the accompany vehicle pair discovery process.

4.1 Basic Accompany Vehicle Pair Discovery Algorithm

Let Vi be vehicle plate number set passing a monitor spot si during the span time Δt, Ti be the timestamp set corresponding to Vi, we can solve the accompany vehicle pair discovery problem by a sliding window approach. We consider accompany vehicle pair discovery per traffic monitor spot. In one spot, a sliding window of length Δt is used to generate the accompany pairs.

For example, given accompany span time Δt = 10 s for Table 1, (v1, v2) is accompany vehicle pair for traffic monitor spot s1, (v1, v2), (v1, v3), (v2, v3) are accompany vehicle pair for traffic monitor spot s2, (v1, v3) is also an accompany vehicle pair for traffic monitor spot s3.

We show an example of the accompany vehicle pair discovery process in Fig. 1. The accompany vehicle pair discovery result of the sliding window w1 is {(r1, r2), (r1, r3), (r2, r3), (r1, r4), (r2, r4), (r3, r4)} where ri.v $\neq$ rj.v($1 \leq i < j \leq 4$)$, and {(r3, r4), (r3, r5), (r4, r5), (r3, r6), (r4, r6), (r5, r6)} for the result of the sliding window w3 while ri.v $\neq$ rj.v($4 \leq i < j \leq 6$).

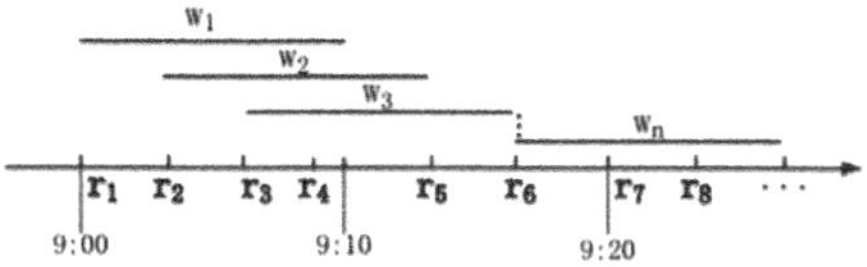

Fig. 1. A sliding window algorithm example for accompany vehicle pair discovery

If the data set D is large, its corresponding generated candidate set C may be larger than worker's memory, and then we need to split them into smaller sets which can fit into memory at once. After splitting there are some overlapping records preserved to avoid losing some candidates, but we need to remove the duplicated pairs generated by two overlapping records.

The pseudo-code for the basic accompany vehicle discovery algorithm is shown in Algorithm 1. It first filter TMPR dataset by time condition (line 1), and then group by traffic monitor spot for filtering data set (line 2), if vehicle set V and corresponding to time T is very larger, we need to split them to several segments and merge old segments into new segments for subset of vehicle set V and corresponding to time T (line 4), and finally using sliding window to get candidate accompany vehicle (line 6). If the total number of accompany vehicle no less than m, we add it to the result.

Let Nw denote the average number of records in a span of time threshold Δt, the number of candidate accompany-vehicle pairs generated holds the order of O(|D| × Nw). This is also the time and space complexity of the sliding window algorithm.

Algorithm 1 *AccompanyVehicleSlidingWindowSearch*

Input: D: TMPR data set
tr: a range of time for searching
δ: interval for accompany time
$minS$: a total number of accompany
Output: $A = \{\exists i, j \wedge i \neq j | (V_i, V_j)\}$
1: $VPFilterRDD \leftarrow$ FilterInitDataSet($D, tr, minS$)
2: $SpotsRDD \leftarrow$ VPFilterRDD.GroupByMonitor()
3: **for all** $spotRDD \in SpotsRDD$ **do**
4: $SegmentsRDD \leftarrow$ spotRDD.SplitAndMakeupNewSegment(δ)
5: **for all** $segmentRDD \in SegmentsRDD$ **do**
6: $Candidates \leftarrow$ segmentRDD.SlidingWindowSearch(δ)
7: **end for**
8: **end for**
9: $Frequents \leftarrow$ ReduceFilterCandidates($Candidates, minS$)
10: **return** $Frequents$

4.2 Matrix Pruning Framework

The sliding window algorithm is not efficient, as if the accompany time threshold Δt becomes bigger or the dataset becomes bigger, the complexity will become unacceptable. Especially the space complexity may becomes bigger the memory size. So the key challenge in accompany-vehicle pairs finding is the over-growing candidates problem.

Over-growing candidates problem means that the number of candidate accompany vehicle pairs is expanding rapidly along with the growing of the number of records or the accompany time threshold Δ. As we have mentioned, the number of records in TMPR dataset is around 9 million per day within a single city and 70 per 10 min within a single spot, making the number of candidate accompany-vehicle pairs around 630 million. This is inefficient and put a high stress on storing the candidates in memory and shuffling the candidates, as the space usage will span 70 times larger. We formalize the over-growing candidates problem in accompany vehicles finding as follows.

Definition 5 (Over-growing candidates problem). Given D, Δt, minS and Nw. The over-growing candidate problem is that it's unable to process the generated candidates with the order of O(|D| × Nw), as |D| is large in our dataset.

To solve the over-growing candidate problem, we propose a novel and general framework named Matrix Pruning. First we denote the candidates which are in the frequent accompany vehicle set are positive candidates, and those not in the set are negative candidates. Matrix pruning framework uses an intuitional method to prevent some negative candidates from generating, it utilizes the property of nonuniform distribution of the occurrence numbers of each vehicle, such as the occurrence distribution over spots or time periods to design two functions: a grouping function G and a pruning function P. The grouping function G: D → M is used to split and group the occurrences of each vehicle vi into more than one groups to get a k dimensional vector Mi. The value in one dimension stands for the total occurrence numbers in one group. The pruning function P: (Mi, Mj) → Boolean uses the occurrences vectors to identify whether a candidate pair (vi, vj) is positive or not. The framework is integrated into the sliding window algorithm by only added the positive filtered by the pruning function into the candidate set. G and P are well designed if they use a relatively small computational cost to prune most of the candidates, thus shrinking the size of the candidate set.

If we treat the vehicle plate number as another dimension besides groups, we can get a matrix whose rows correspond to vehicles and columns correspond to groups. This is the reason why it's named Matrix Pruning. We formalize the Matrix Pruning framework as follows:

Definition 6 (Matrix Pruning framework). Given D, Δt and minS, design a grouping function S and a pruning function P, the Matrix Pruning framework works as the following two steps:

Firstly, use G to calculate the grouping vector Mi for each vehicle vi.

Secondly, before generating the candidate pair for any two vehicles vi and vj, check if P(Mi, Mj) = True. If it is true, then the candidate Cij is calculated from vi and vj into the candidate set, otherwise it's dropped. Repeat this process until all the candidates is generated.

Now we will utilize the Matrix Pruning framework to solve the over-growing candidate problems existing in accompany vehicle pair discovery. Recall that there are three steps in the basic sliding window algorithm. In the first step we prune the vehicles whose occurrence is below *minS*. We can calculate the grouping vector for each vehicle at the same time. In the second step, the sliding window algorithm just put all the pairs into candidate sets. Under the Matrix Pruning framework, a pair is checked whether it's pruned before being added into the candidate set. The third step remains the same.

Integrating Matrix Pruning framework into the sliding window algorithm, we construct the grouping function G and pruning function P in the following sections.

4.3 Grouping Strategies

First we formalize the grouping function as follows.

$$G(v_i) = M_i = (m_i^1, m_i^2, \ldots, m_i^k) \tag{1}$$

$$M_i^j = \sum_{p=0}^{|S_p t_j|} Cnt_{ij}^p \tag{2}$$

Mi is a k dimension vector for vehicle vi, Gptj is the jth group of traffic spots, Cntij is the number of records of vehicle i appears in each spots in the group j. The formula is easy to understand, a grouping strategy is chosed to group the occurrences into several groups and count the numbers in each group for the corresponding dimension. We present four different grouping strategies here.

Single Spot Strategy: The trial method is to treat each spot as a group, thus Mi holds the dimension of |G|. The idea is intuitional that the most possible candidates generated by a vehicle vi in one spot with any other vehicle can't be bigger than its occurrence number in this spot, and vice versa. So the number of candidates that two vehicles generate cannot be bigger than the smaller occurrence number of them. So sum the smaller occurrence number over all the spots, we can get an upper bound for this pair. If this upper bound is smaller than the threshold minS, the pair is dropped.

Uniform Grouping Strategy: The number of spots may be large to increase computational cost. To reduce the computational cost, another strategy is to group them together into a smaller set of groups. The trial way is randomly distributing spots into equally sized groups, use these groups to group and count the occurrence numbers of each vehicle. The groups equal in size because if some groups are very large, the two occurrence numbers of the comparing vehicles are high. This will make it unable to get a lower bound, which means the grouping is not well designed.

Connection-Based Strategy: Further we can raise an optimization based on the uniform group strategy. Usually one specific vehicle only appears in a subsets of spots according the driver's need instead of random distribution, such as the route between home and office. Routes for different vehicles are different as different drivers have different driving routes or habits. The randomly distributing method in uniform group strategy just lose this routes information. To extract these into groups, we observe that if the route connecting two spots is closer, a vehicle appears in one spot will be more likely to appear in another. We group these spots into one group, so a vehicle is more likely to appear in some certain groups and different vehicles are less likely to appear in the same groups. Thus we can use this property to cluster over the spots based on their connections into equal size groups for grouping.

Spatial and Temporal Fusion Strategy: We can also expand the grouping criteria from one dimension space (spots) to the two dimensional space (time and spots). As the drivers have different preferred routes, they may also have different preferred time periods. For example the a full-time house wife may go to the market before lunch and pick up children around three pm while an office guy usually appears at around eight in the morning and five in the evening. If we split a day into three groups, before 9 am, between 9 am and 4 pm and after 4 pm. Thus the wife's occurrence will fall into the second group and the office guy's will be in the first and third groups. This means there co-occurrence number will drop to zero. So we can further split the records in a single spot group into different groups identified by time period to make the distribution over groups more leaning.

4.4 Pruning Strategies

After getting the grouped occurrence statistics into vectors, then we can use the vector to check whether a pair is a positive candidate or not. We present two different pruning strategies here. The first method is calculating the result of a function. The second method uses the idea of early stopping to provide higher efficiency.

Functional Strategy: The formalized statement of the pruning function P is as follows.

$$P(M_i, M_j) = \begin{cases} True\ if\ B < minS, \\ False\ if\ B \geq minS \end{cases} \tag{3}$$

$$B = \sum_{p=0}^{k} min(m_i^p, m_j^p) \tag{4}$$

This function is easy to understand. For two vehicle vi and vj, the number of candidates generated by them in each group can't be higher than the smaller number of the occurrences of the two vehicles in this group. So the sum of all the largest possible number of each group is a higher bound of the accompanying numbers, denoted as B. If the bound is less than the accompany threshold minS, the pair is recognized as a negative candidate and can be dropped.

Early Stopping Strategy: As the number of groups may be large, the computational cost of each pair may be relatively high. As there is a large number of pairs generated, the overall computational cost can't be neglected. We present an Early Stop Strategy to get the bound at a smaller cost.

$$P_g(M_i, M_j) = \begin{cases} True\ if\ B_g < minS, \\ P_{g+1}(M_i, M_j)\ if\ B_g \geq minS\ and\ g < k, \\ False\ if\ B_g \geq minS\ and\ g = k \end{cases} \tag{5}$$

$$B_g = B_1^g + B_g^k \tag{6}$$

$$B_1^g = \sum_{m=1}^{g} min(M_i^m, M_j^m) \tag{7}$$

The bounds extracted from each group remain the same as those in the functional strategy, and we get them one by one. We observe that the most possible co-occurrence number with any other vehicle for vi is its own occurrence number. After get the bound for the gth group, we can get the partial over-all bound B1g for groups already checked. And the bound for the remaining k-g groups, denoted as Bgk, is the sum of the occurrence numbers of vi in the rest groups. As Bgk is the biggest co-occurrence number for any vehicle with vi, the over-all bound B is small than B1g + Bgk. So if B1g + Bgk is smaller than accompany threshold, the pair is a negative candidate and dropped, otherwise we just check the next group and update the bound.

5 Experiments

In the section, we present our experimental study and demonstrate the efficiency and quality of our proposed methods.

5.1 Experiment Setup

All experiments were conducted on a cluster consisting of 5 nodes with the following configurations:

5 machines with a 4-core Intel 2.5 GHz processor and 64 GB RAM. Each node is connected to a Gigabit Ethernet switch and runs Ubuntu 14.04 with Hadoop 2.7.2 and Spark 2.0. The spark cluster is deployed in standalone mode. Our cluster configuration reserved up to 32 GB of main memory for spark.

We utilized four real data sets to evaluate our methods, which are:

- One day data set, which contains 9,357,636 valid TMPR data records, and the disk capacity is 194.3 MByte.
- Three days data set, which contains 26,242,239 valid TMPR data records, and the disk capacity is 549.0 MByte.
- Seven days data set, which contains 46,745,377 valid TMPR data records and the disk capacity is 981.5 MByte.
- Fifteen days data set, which contains 141,348,272 valid TMPR data records and the disk capacity is 2.9 GByte.

Notice that the above data set is simply for valid TMPR data records by Hashing vehicle plate and traffic monitor spot respectively to accelerately find accompany vehicle pair.

5.2 Time Costs

We first compare the performance of the basic method and the improved method, and the results are shown in Fig. 2(a). For each data set, we set the accompany span time 3 min. The cost time for accompany vehicle discovery on four data sets are shown in Fig. 2(a).

We can see that the distributed approach of accompany vehicle discovery methods are always better than basic approach. However, the cost time for four distributed algorithms had a large difference. Overall, the Fusion algorithm is the best fit for accompany vehicle discovery for other distributed algorithm.

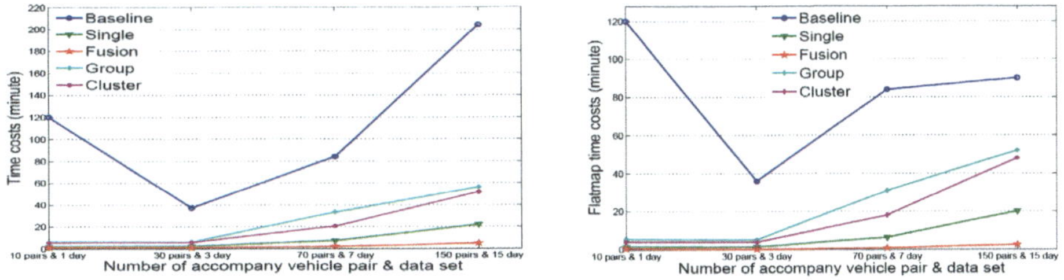

(a) Time costs (b)Flatmap time costs

Fig. 2. Accompany vehicle discovery of time costs for five algorithm

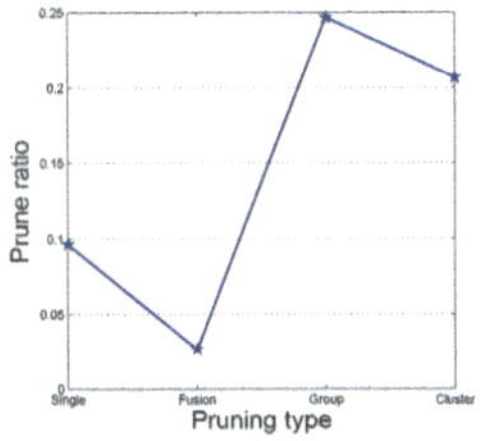

Fig. 3. Pruning ratio for four algorithms

We then compared the cost time for flatMap of the four algorithm, and the results are shown in Fig. 2(b). FlatMap in our experiments is the candidate set generation step. We can see that the improved four method performed much better than the basic method, and as the same time the Fusion algorithm remains the best. The result is consistent with the above Fig. 2(a).

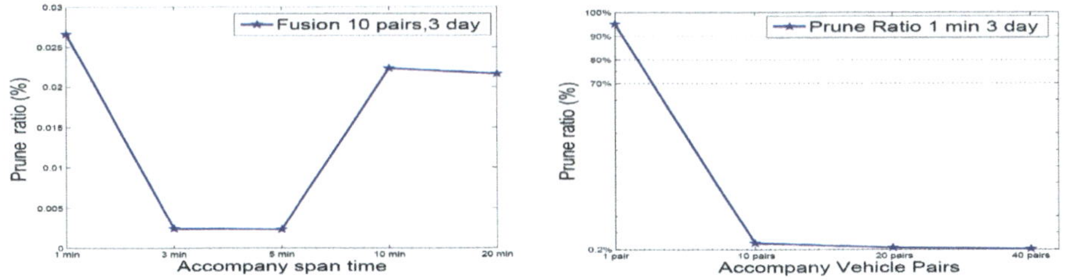

(a)Prunning on span time (b)Prunning on accompany pair

Fig. 4. Changing of pruning ratio for the fusion strategy

5.3 Pruning Quality

In this subsection, we focus on the changing of pruning quality with accompany vehicle span time increasing in the Fusing algorithm. We selected the data set of seven days and set accompany vehicle pairs threshold minS at 10. The accompany span time varied from 1 min to 20 min. The pruning ratio is defined as the number of the accompany vehicle pairs added into the candidate set divided by total accompany vehicle pairs.

Firstly we compare the prune ratio of four distributed approaches, which are shown in Fig. 2. Fusing algorithm outperformed other 3 distributed algorithm.

The pruning ratio of the Fusion approaches are shown in Fig. 3. The quality becomes better with the increasing of accompany span time. All the quality was over 0.03%.

We then compared the prune quality for increasing accompany vehicle pairs threshold from 5 to 40 pairs. The prune ratio of the Fusing approaches are shown in Fig. 4(b). With the increase of accompany vehicle pairs threshold, the quality increased. And when accompany vehicle pairs to 20 times, all the prune ratio was below 2%.

6 Conclusions

In this paper, we introduced the novel and practical definition of accompany vehicle discovery and devise a basic method to solve the problem utilizing sliding windows algorithm. Based on the sliding window algorithm, we proposed Matrix Pruning Framework to reduce the size of candidates and several corresponding grouping and pruning strategies. The comprehensive experiments on real traffic data set have demonstrated the efficiency and quality of our proposed distributed algorithm.

Acknowledgements. This work is supported by the Project of Shanghai science and technology commission project under Grant 17595800900.

References

1. Zhu, M., Liu, C., Wang, J., Wang, X., Han, Y.: Instant discovery of moment companion vehicles from big streaming traffic data. In: International Conference on Cloud Computing and Big Data, pp. 73–80. ACM (2015)
2. Gudmundsson, J., Kreveld, M.V.: Computing longest duration flocks in trajectory data. In: Proceedings of the 14th Annual ACM International Symposium on Advances in Geographic Information Systems, pp. 35–42. ACM (2006)
3. Jeung, H., Yiu, L.M., Zhou, X., Jensen, C.S., Shen, H.T.: Discovery of convoys in trajectory databases. In: VLDB, pp. 1068–1080 (2008)
4. Li, Z., Ding, B., Han, J., Kays, R.: Swarm: mining relaxed temporal moving object clusters. In: VLDB, pp. 723–724 (2010)
5. Tang, L., et al.: A Frame-work of traveling companion discovery on trajectory data streams. In: ACM Transactions on Intelligent Systems and Technology, p. 3. ACM (2013)
6. Kalnis, P., Mamoulis, N., Bakiras, S.: On discovering moving clusters in spatial-temporal data. In: SSTD. ACM (2015)
7. Benkert, M., Gudmundsson, J., Hubner, F., Wolle, T.: Reporting flock patterns. Comput. Geom. Theory Appl. **41**(3), 111–125 (2008)

8. Yoo, J.S., Shekhar, S., Smith, J., Kumquat, J.P.: A partial join approach for mining colocation patterns. In: ACM GIS. ACM (2004)
9. Aung, B.: Discovering Moving Groups of Tagged Objects. Nation University of Singapore, Singapore (2008)
10. Fang, A., Li, X., Man, S., Yue, P.: A discovery algorithm of travelling companions based on association rule mining. Comput. Appl. Softw. **29**(2), 94–96 (2012)
11. Ghemawat, S., Gobioff, H., Leung, S.T.: The Google file system. In: 19th Symposium on Operating Systems Principles, pp. 29–43 (2003)
12. Bu, Y., Howe, B., Balazinska, M., Ernst, M.D.: HaLoop: efficient iterative data processing on large clusters. In: VLDB, pp. 285–296 (2010)
13. Isard, M., Budiu, M., Yu, Y., Birrell, A., Fetterly, D.: Dryad: distributed data parallel programs from sequential building blocks. In: EuroSys (2007)
14. Storm. https://github.com/nathanmarz/storm/wiki
15. Zaharia, M., et al.: Resilient distributed datasets: a fault-tolerant abstraction for in-memory cluster computing. In: NSDI (2012)
16. Zeng, K., Agarwal, S., Dave, A., Armbrust, M., Stoica, I.: G-OLA: generalized online aggregation for interactive analysis on big data. In: SIGMOD (2015)
17. Armbrust, M., et al.: Spark SQL: Relational data processing in spark. In: SIGMOD (2015)
18. Dean, J., Ghemawat, S.: MapReduce: simplified data processing on large clusters. In: Proceedings of the 6th OSDI, pp. 137–150 (2004)
19. Yoo, J.S., Boulware, D., Kimmey, D.: A parallel spatial co-location mining algorithm based on MapReduce. In: IEEE International Congress on Big Data, pp. 25–31 (2014)
20. Zheng, K., Zheng, Y., Yuan, N.J., Shang, S.: On discovery of gathering patterns from trajectories. In: ICDE, pp. 242–253 (2013)
21. Zhang, J., Li, J., Wang, S., Liu, Z., Yuan, Q., Yang, F.: On retrieving moving objects gathering patterns from trajectory data via spatio-temporal graph. In: IEEE International Congress on Big Data, pp. 390–397 (2014)

Spatial Context for Correlation Filter Tracking

Fuhui Tang[1], Xiankai Lu[2(✉)], Xiaoyu Zhang[3], Xueqian Song[1], Jiyuan Cai[1], Hong Zhang[1], and Jiaru Song[1]

[1] School of Aeronautics and Astronautics, Shanghai Jiao Tong University, Shanghai 200240, China

[2] School of Electronic Information and Electrical Engineering, Shanghai Jiao Tong University, Shanghai 200240, China
carrierlxk@sjtu.edu.cn

[3] School of Electrical Engineering, Shanghai Dianji University, Shanghai 201306, China

Abstract. Correlation filters (CFs) have shown excellent performance in visual tracking. They exploit the circular shift of the foreground target to learn the training samples. However, the boundary effect caused by circulant assumption on a small search region limits the ability of such trackers. To overcome this problem, we propose a novel correlation filter tracking framework that takes a larger spatial context into account. In order to make the filter model generate a strong response to the foreground target but a low response to background region, we design an effective cosine window. Extensive experimental results in both quantitative and qualitative measures demonstrate the effectiveness of the proposed algorithm compared to several state-of-the-art trackers.

Keywords: Correlation filters · Visual tracking · Boundary effect
Spatial context · Cosine window

1 Introduction

Visual tracking enjoys a wide popularity due to the various applications, such as human–computer interaction, video surveillance, medical imaging and so on. The main difficulty is how to design a robust tracker to deal with a variety of challenging including illumination variation, background clutter, fast motion, and rotation, to name a few.

In recent years, correlation filter (CF) [1–3] has been introduced into visual tracking successfully. The CF tracker learns a filter model in the first frame and updates it in the subsequent image frames. By finding the maximum correlation response from candidate region, the target location can be estimated. The major advantages of such tracking method can be summarized as the following two aspects. First, the high-speed computing power in Fourier domain makes CF tracker meet the real-time requirement. Second, different from the classifier based trackers [17, 18, 20] which need draw a large number of positive and negative training samples around the target location, the CF trackers only perform a dense sampling by circularly shifting the foreground target, thereby reducing the sampling complexity. However, two issues ensue with this circulant assumption. On the one hand, the limited target search region only contains a small local neighborhood, which causes the boundary effect, i.e., the filter would have no response to the target if it

J. Abawajy et al. (Eds.): ATCI 2018, AISC 842, pp. 174–181, 2019.
https://doi.org/10.1007/978-3-319-98776-7_20

exceeds the bounding of the filter. This usually leads to tracking failure in the presence of fast motion. On the other hand, existing CF trackers always adopt a classic hann window [1] to suppress the boundary effect without considering the size of the search area. As a result, the background clutter around the target cannot be effectively restrained.

To solve these problems, we propose a novel CF tracking method based on spatial context. In our approach, for reducing the risk of tracking drift caused by fast motion, we extend the search region to include more background information. Moreover, in order to alleviate the interference from background clutter or similar objects, we design a reasonable cosine window to highlight the filter response to target foreground while suppressing the response to the background around the target.

The contributions of our work can be summarized as follows:

- We propose to learn spatial context information for correlation filter, which is able to increase the search scope of correlation filter.
- We build an effective cosine window to enhance the discriminability of the filter model.
- We validate our method by conducting numerous experiments on the benchmark dataset. We show that the proposed tracker performs favorably against several state-of-the-art trackers.

2 Related Work

2.1 CF Trackers

Since the MOSSE work of Bolme et al. [1], correlation filter has been studied as an efficient method for visual tracking. In [4], CSK method is proposed based on intensity features and further extended to KCF [3] approach with HOG features. Danelljan et al. [2] solve the problem of scale estimation by training a scale filter on pyramid. Combined with the translation filter, this method can estimate the target position and scale simultaneously. The structural correlation filter [5] is proposed to model target appearance for robust tracking by fusing part-based tracking strategy into the tracker. Liu et al. [6] introduce a part-based method which track the object parts independently by using multiple correlation filters. In [7], a reliable patch tracker is proposed to identify and exploit the reliable patches through the whole tracking process. For short-term tracking, a CSR-DCF [8] method is introduced to adapt the filter support to the part of the object. Ma et al. [9] add an online random fern classifier to correlation filter framework for long-term tracking. Choi et al. [10] propose an attentional mechanism to select the best fitting subset of all tracking modules, which increases the number of overall filters while keeping the computational efficiency. To minimize the localization errors, a landmark detection model [11] is designed for the task of object alignment.

2.2 Context Trackers

In visual tracking, a local context contains a target object and its surrounding background, there exists some relevance between them. Therefore, making good use of this relevance can improve the robustness of the tracker. In [12], the context-based distracters and supporters are designed to share similar objects and key-points respectively. Wen et al. [13] explore spatio-temporal structural context model to build a supporting field which provides much more information than the target appearance. Zhang et al. [14] present two local context models (spatial context and spatio-temporal context) to deal with different tracking challenges. In [15], the context information is incorporated into the filter training stage to improve the tracking performance. The background aware correlation filters is proposed [16] by learning real negative training examples from background instead of the circular shifted patches. We note that the aforementioned trackers only add more background information. In contrast, our approach focuses on allowing the learned filter to reasonably suppress negative examples while introducing context.

3 Proposed Algorithm

3.1 Correlation Filter

Before the discussion of our method, we first revisit the details of CF tracking. The key idea of CF is to regress the circularly shifted versions of image features to a Gaussian label, which generates dense response scores over all searching positions. The target location can thus be predicted by searching the maximum value in all response scores. Denote x as the input sample of size $M \times N \times D$, where M, N, D are the width, height, channel, respectively. The correlation filter w can be learned by minimizing the following problem:

$$w^* = \arg\min_{w} \left\| \sum_{d=1}^{D} w^d \bullet x^d - y \right\|_2^2 + \lambda \|w^d\|_2^2 \quad (1)$$

where y is the Gaussian function label, the symbol • denotes the inner product, $\lambda \geq 0$ controls the impact of the regularization term. The optimization problem in Eq. (1) can be solved in each individual feature channel with the use of fast Fourier transformation:

$$W^d = \frac{\bar{X}^d \odot Y}{\sum_{i=1}^{D} \bar{X}^d \odot X^d + \lambda} \quad (2)$$

where X, Y, W are the Fourier transformed signals of x, y, w, respectively. The operator $\odot$ is the Hadamard product and the bar means complex conjugation. Given a new image patch in the next frame, the target position can be found by maximizing the response score S:

$$S = F^{-1}\left(\sum_{d=1}^{D} W^d \odot \bar{Z}^d\right) \quad (3)$$

Here, the operator F^{-1} indicates the inverse Fourier transformation. Z is the Fourier form of the image features z. Since the filter update needs to solve a $D \times D$ linear system of equations per location, which is computationally expensive. To obtain a robust approximation, the numerator Ad and denominator Bd of the filter Wd can be updated separately:

$$A_t^d = (1 - \eta)A_{t-1}^d + \eta \bar{Y} \odot X_t^d \tag{4}$$

$$B_t^d = (1 - \eta)B_{t-1}^d + \eta \sum_{i=1}^{D} \bar{X}_t^i \odot X_t^i \tag{5}$$

$$W_t^d = \frac{A_t^d}{B_t^d + \lambda} \tag{6}$$

where t is the frame index and η is the learning rate.

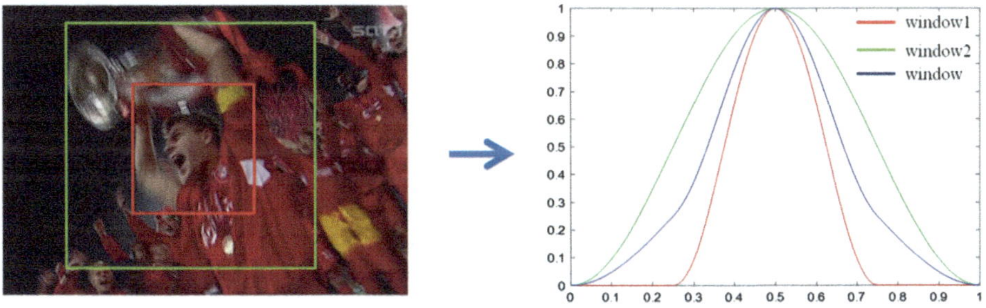

(a) Input image (b) Cosine window

Fig. 1. Visualization of cosine window for input image. (a) Input image. The red bounding box indicates the search region with 2 times of the target size, and the green bounding box is the search region with 4 times of the target size. (b) Cosine window. The window1 generates a hann window on the region inside the red bounding box while having no response on the region between red and green bounding boxes. The window2 produces a hann window on the region inside the green bounding box. The window is the combination of window1 and window2.

3.2 Spatial Context

The conventional CF based trackers [2, 3] usually adopt a small neighborhood surroundings, i.e., about 2 times the size of the target bounding box. Actually, the surroundings have a significant impact on tracking performance, especially when background clutters or similar objects occur. In addition, the image features are always multiplied by a hann window [1–3] to reduce the frequency effects of the edge of the image:

$$hann\_window = 0.5 - 0.5\cos(2\pi m) \tag{7}$$

where m denotes each position of the image. However, if the context is large, the hann window is unable to effectively suppress the boundary effects. To overcome these problem, we design a suitable cosine window for our spatial context based tracker. In our approach, we first crop an image patch with 4 times of the target size, which contains enough background information to deal with fast motion and background clutter. Increasing the search region beyond 4 only achieves a minor improvement but affects runtime. We then formulate the cosine window as follows:

$$window1 = \begin{cases} 0.5 + 0.5\cos(4\pi m - 2\pi) & \frac{1}{4} \leq m \leq \frac{3}{4} \\ 0 & else \end{cases} \tag{8}$$

$$window2 = 0.5 - 0.5\cos(2\pi m) \tag{9}$$

$$window = (window1 + window2)/2 \tag{10}$$

As shown in Fig. 1, the desired window is able to keep an effective response to the target and its small neighborhood, while suppressing the response to other regions.

4 Experiments

4.1 Experimental Setup and Methodology

We implement our method in MATLAB on a PC with Intel Core I5 3.3 GHZ and 8 G RAM. We use HOG features for image representation. The regularization parameter is set to $\lambda = 0.01$, the search region and filter size are set to 4 times of the initial target size. The learning rate is set to $\eta = 0.025$. All parameter values are kept consistent for all the image sequences.

We compare our tracker with 4 state-of-the-art trackers on 10 challenging videos. The trackers includes DSST [2], KCF [3], Struck [17], and SCM [18]. The videos from OTB-2013 benchmark dataset [19] are divided to 2 groups. Fast motion: Couple, *Tiger2, Ironman, Jumping, Matrix*. Background clutters: *Deer, Football, Skating1, Soccer, Subway*. Specifically, *Couple, Ironman, Matrix, Deer, and Soccer* sequences contain fast motion and background clutters simultaneously.

We present the results using overlap precision (OP) which is defined as the percentage of frames where the bounding box overlap surpasses a given threshold (0.5).

4.2 Quantitative Evaluation

Table 1 shows the overlap precision evaluation on 10 representative videos. The tracking results show that our tracker is consistently better than other trackers on most videos, which implies that the proposed method is able to track the target object robustly in fast motion and background clutters. In addition, our tracker achieves the best overall performance with average OP of 69.1%. This can be attributed to the use of rich spatial context information and effective cosine window.

Table 1. Overlap precision in percent on 10 videos. The best results are shown in red.

	Struck	SCM	KCF	DSST	Ours
Couple	54.3	10.7	24.3	10.7	65.7
Deer	100	2.8	81.7	78.9	100
Tiger2	64.9	11.2	36.4	29.6	83
Ironman	4.8	13.2	15	13.2	13.3
Jumping	79.9	12.1	28.1	6	91.7
Matrix	12	30	13	18	19
Football	66	58.6	70.2	69.9	76
Skating1	37	42.3	36.3	52.3	83.8
Soccer	15.6	23.7	39.3	38.8	58.7
Subway	90.9	99.4	100	22.3	100
Average	52.5	30.4	44.4	34	69.1

4.3 Qualitative Evaluation

To further demonstate the robustness of our tracker, we present several tracking results in Fig. 2. The Struck [17] can deal with fast motion (Deer, Jumping) but usually fails when background clutters occur (Football, Skating1, Soccer, Subway). The SCM [18] can track the target in the presence of slight background clutters (Subway) but drifts in the case of severe background clutters (Soccer, Skating1). Moreover, it is sensitive to fast motion (Couple, Deer, Tiger2, Jumping). The KCF [3] and DSST [2] learn CF

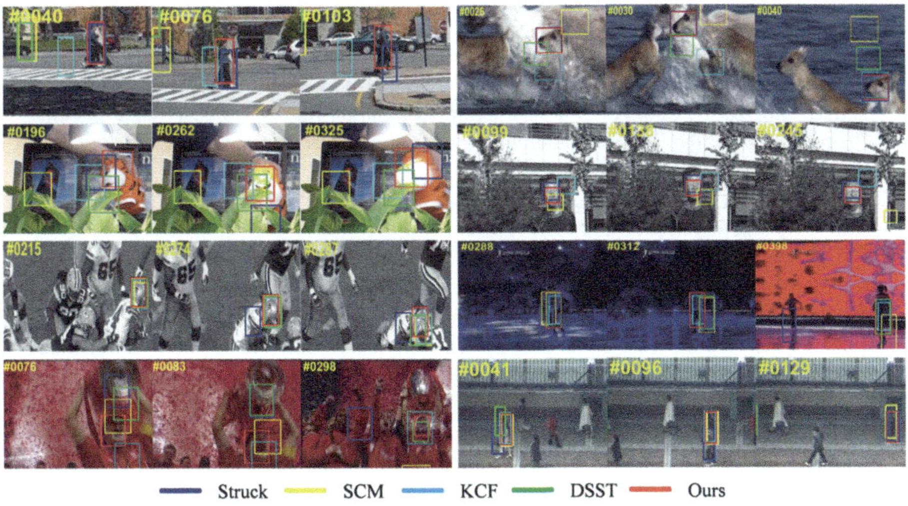

Fig. 2. Tracking results of Struck, SCM, KCF, DSST, and our tracker on eight challenging videos (from left to right and top to down are *couple, deer, tiger2, jumping, football, skating1, soccer,* and *subway,* respectively).

online to track the target. However, their limited context information leads to poor performance in fast motion (*Couple, Tiger2*) and background clutters (*Deer, Soccer*). In comparison, our tracker can provide more accurate target location in these sequences. Therefore, we draw a conclusion that our method introducing spatial context to CF tracking framewok improves the capabiliy to deal with fast motion and background clutters.

5 Conclusion

In this paper, we propose a novel CF tracking framework via spatial context and cosine window to improve the performance. The context information is used to account for fast motion, and the reasonable cosine window is able to alleviate the interference from background clutters. Experimental results on challenging videos demonstate that the proposed tracker can show good performance in the presence of fast motion and background clutters and perform favorably against several state-of-the-art trackers.

References

1. Bolme, D.S., Beveridge, J.R., Draper, B.A., et al.: Visual object tracking using adaptive correlation filters. In: 2010 IEEE Conference on Computer Vision and Pattern Recognition (CVPR), pp. 2544–2550. IEEE (2010)
2. Danelljan, M., Häger, G., Khan, F., et al.: Accurate scale estimation for robust visual tracking. In: British Machine Vision Conference, Nottingham, 1–5 September 2014. BMVA Press (2014)
3. Henriques, J.F., Caseiro, R., Martins, P., et al.: High-speed tracking with kernelized correlation filters. IEEE Trans. Pattern Anal. Mach. Intell. **37**(3), 583–596 (2015)
4. Henriques, J.F., Caseiro, R., Martins, P., et al.: Exploiting the circulant structure of tracking-by-detection with kernels. In: European Conference on Computer Vision, pp. 702–715. Springer, Berlin (2012)
5. Liu, S., Zhang, T., Cao, X., et al.: Structural correlation filter for robust visual tracking. In: Proceedings of the IEEE Conference on Computer Vision and Pattern Recognition, pp. 4312–4320 (2016)
6. Liu, T., Wang, G., Yang, Q.: Real-time part-based visual tracking via adaptive correlation filters. In: Intelligence, p. 2345390 (2015)
7. Li, Y., Zhu, J., Hoi, S.C.H.: Reliable patch trackers: robust visual tracking by exploiting reliable patches. In: 2015 IEEE Conference on Computer Vision and Pattern Recognition (CVPR), pp. 353–361. IEEE (2015)
8. Lukezic, A., Vojír, T., Zajc, L.C., et al.: Discriminative correlation filter with channel and spatial reliability. In: IEEE Conference on Computer Vision and Pattern Recognition, pp. 4847–4856 (2017)
9. Ma, C., Yang, X., Zhang, C., et al.: Long-term correlation tracking. In: 2015 IEEE Conference on Computer Vision and Pattern Recognition (CVPR), pp. 5388–5396. IEEE (2015)
10. Choi, J., Chang, H.J., Yun, S., et al.: Attentional correlation filter network for adaptive visual tracking. In: Proceedings of the IEEE Conference on Computer Vision and Pattern Recognition, p. 2 (2017)

11. Boddeti, V.N., Kanade, T., Kumar, B.V.K.V.: Correlation filters for object alignment. In: 2013 IEEE Conference on Computer Vision and Pattern Recognition (CVPR), pp. 2291–2298. IEEE (2013)
12. Dinh, T.B., Vo, N., Medioni, G.: Context tracker: exploring supporters and distracters in unconstrained environments. In: 2011 IEEE Conference on Computer Vision and Pattern Recognition (CVPR), pp. 1177–1184. IEEE (2011)
13. Wen, L., Cai, Z., Lei, Z., et al.: Online spatio-temporal structural context learning for visual tracking. In: European Conference on Computer Vision, pp. 716–729. Springer, Berlin (2012)
14. Zhang, K., Zhang, L., Liu, Q., et al.: Fast visual tracking via dense spatio-temporal context learning. In: European Conference on Computer Vision, pp. 127–141. Springer, Cham (2014)
15. Mueller, M., Smith, N., Ghanem, B.: Context-aware correlation filter tracking. In: Proceedings of the IEEE Conference on Computer Vision and Pattern Recognition (CVPR), pp. 1396–1404 (2017)
16. Galoogahi, H.K., Fagg, A., Lucey, S.: Learning background-aware correlation filters for visual tracking. In: Proceedings of the 2017 IEEE Conference on Computer Vision and Pattern Recognition (CVPR), pp. 21–26, Honolulu (2017)
17. Hare, S., Golodetz, S., Saffari, A., et al.: Struck: structured output tracking with kernels. IEEE Trans. Pattern Anal. Mach. Intell. **38**(10), 2096–2109 (2016)
18. Zhong, W., Lu, H., Yang, M.H.: Robust object tracking via sparsity-based collaborative model. In: 2012 IEEE Conference on Computer Vision and Pattern Recognition (CVPR), pp. 1838–1845. IEEE (2012)
19. Wu, Y., Lim, J., Yang, M.H.: Online object tracking: a benchmark. In: 2013 IEEE Conference on Computer Vision and Pattern Recognition (CVPR), pp. 2411–2418. IEEE (2013)
20. Babenko, B., Yang, M.H., Belongie, S.: Robust object tracking with online multiple instance learning. IEEE Trans. Pattern Anal. Mach. Intell. **33**(8), 1619–1632 (2011)

The Automobile Recommendation Method Based on Sentiment Analysis of Online Comments

Zhuangzhuang Pan and Guangli Zhu(✉)

School of Computer Science and Engineering,
Anhui University of Science and Technology, Huainan 232001, China
glzhu@aust.edu.cn

Abstract. It's difficult for consumers to select right cars from the network with the large volume of data, various types and low density of value. To solve this problem, this paper proposes a method of automobile recommendation based on online comment sentiment analysis to help consumers choose the suitable car more quickly and accurately. First, related lexicons, such as degree adverbs, negative words, automobile terminology, are constructed to extend the basic sentiment lexicon based on the analysis of the online comment corpus. Then, the multi-parameter combination of automobile attributes is carried out to calculate the sentiment value of every automobile comment according to the requirements of the consumers. Finally, the feature matrix is defined to calculate the cosine similarity by combining the automobile attributes with sentiment values of the comments, and then the top-k types of automobiles which are similar to the consumer's requirement are selected to the consumer according to the cosine similarity. Experimental results show that our method can effectively improve the accuracy of the recommended results.

Keywords: Sentiment analysis · Online comments · Feature matrix
Cosine similarity · Automobile recommendation method

1 Introduction

Nowadays, when consumers who do not understand the performance of the car want to buy a car. They are usually inclined to acquire the important attributes (e.g., automobile power, fuel consumption) by consulting vendors or network. However, these methods have two serious problems. (1) The basic parameters obtained from the vendors alone may not be in accordance with the actual situation, which will mislead the consumption choice. (2) It is difficult to get enough useful information from the major automobile websites, where vast amounts of data with large volume are stored. These data can be acquired from various ways and their update speed is fast, credibility is low [1].

To solve the above problem, this paper proposes an automobile recommendation method based on the sentiment analysis of the online comments of the Car Home (www.autohome.com.cn). The Car Home is a Chinese automobile website that provides consumers with the best and high credibility information on word-of-mouth,

J. Abawajy et al. (Eds.): ATCI 2018, AISC 842, pp. 182–193, 2019.
https://doi.org/10.1007/978-3-319-98776-7_21

evaluation, guide content and online comments. The process of the automobile recommendation method is divided into the following two steps.

(1) **The calculation of sentiment value of online comments based on sentiment lexicons.** First, comment texts from the Car Home are preprocessed and the lexicons of automobile related terms [2], degree adverb, negative words are constructed based on the comment texts. Then, the sentiment value of comment text are calculated by the analysis of positive and negative meaning of the comments and the sentiment value calculation formula proposed in this paper.
(2) **The recommendation of automobile based on the feature matrix of the automobile and the cosine similarity method.** First, the feature vector of eight indexes of automobile, the automobile model and the feature matrix of eight indexes of automobile model are defined [3]. Then, the similarity between each automobile index and consumer preference in the feature matrix is calculated by using the method of cosine similarity [4]. Finally, top-k types of automobile are recommended to consumers, which are selected from the automobiles with high similarity [5].

The motivation of the automobile recommendation method is to help consumers to select the right automobile quickly and accurately. The framework of the automobile recommendation method is shown as Fig. 1.

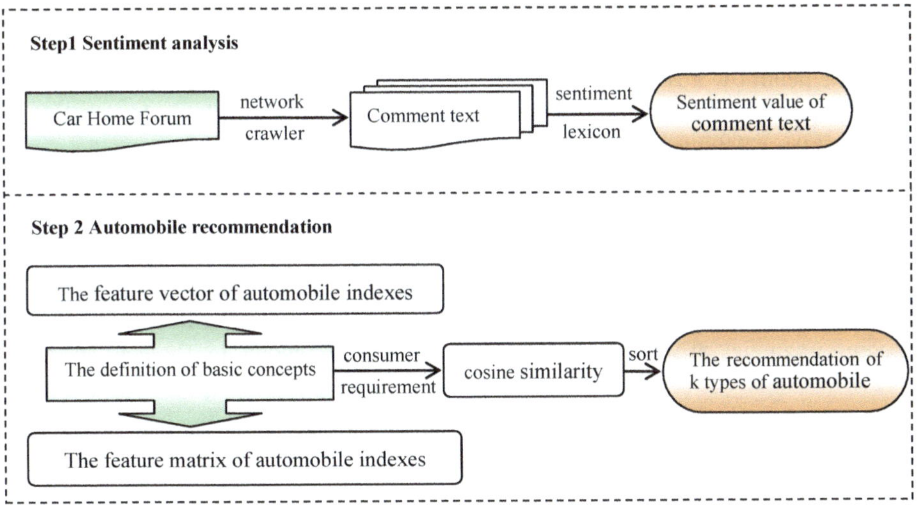

Fig. 1. The working framework of the automobile recommendation method

The rest of this paper is organized as follows: Sect. 2 constructs four sentiment lexicons. Section 3 proposes the method of calculating the comments sentiment value according to the sentiment lexicons. Section 4 proposes the method of the automobile recommendation. Section 5 is experimental evaluation and analysis of the proposed recommendation method. Conclusions are given in Sect. 6.

2 The Construction of Sentiment Lexicon

In this section, three kinds of existing sentiment lexicon are combined and negative lexicon, degree adverb lexicon, automobile terminology lexicon are constructed based on it to analyze the attributes of automobile.

2.1 Integration of Basic Sentiment Lexicon

Sentiment lexicon is the basis of analyzing words' sentiment and they are different in various direction. HowNet lexicon published by CNKI, NTUSD lexicon published by Taiwan University and Chinese sentiment vocabulary ontology library published by Dalian University of Technology are basic sentiment lexicon, which are used most frequently. HowNet, NTUSD and Chinese sentiment vocabulary ontology library are combined to get a new basic sentiment lexicon Dict. The rules of combination are:

(1) If the three lexicons have different lexical affective tendencies, HowNet is the first criterion. If the HowNet lexicon does not contain the word, NTUSD is chosen as the final basis.
(2) The order of sentiment intensity (E-intensity) is divided into five grades: 1, 3, 5, 7, 9. The last one owns the highest intensity and the first one owns the lowest intensity. The sentiment intensity refers to the Chinese sentiment vocabulary ontology library published by Dalian University of Technology. If the sentiment intensity can't be found in Chinese sentiment vocabulary ontology library, the default sentiment intensity of this word will be 3 levels.
(3) The neutral words are described by 0, the derogatory words are described by −1, the positive words are described by +1 and the words with both derogatory and positive meaning are described by ±1 in E-polarity.
(4) The sentiment polarity should be judged according to the sentiment of the whole sentence when analyze the concrete comment statement, and it is not allowed to mark the sentiment polarity with ±1.

There are 12306 positive words, 11323 derogatory words, 5475 neutral words and 86 words with both positive and derogatory meanings in the new basic sentiment lexicon. Examples of some basic sentiment lexicons are shown in Table 1.

Table 1. Examples of some basic sentiment lexicons

Words	Part of speech category	Intensity	Polarity
Like	Adj	5	1
Comfort	Adj	3	1
Innovation	Verb	5	1
Vanity	Noun	7	−1
Crowded	Adj	7	−1

2.2 The Construction of Degree Adverbs Lexicon

Degree adverbs are used to define and modify the degree of an adjective or adverb, and usually located before an adjective or adverb. As for the mood and sentiment tendency of the whole sentence is determined by the following words, and the degree adverb is only to explain its degree.

In this paper, the degree adverbs are selected from HowNet lexicon and the number of them is 219. The intensity of degree adverbs is divided into 6 grades: "ultra-6", "extremely-5", "very-4", "relatively-3", "slightly-2" and "under-1". The first one owns the highest intensity and the last one owns the lowest intensity. Examples of some degree adverbs lexicons are given in Table 2.

2.3 The Construction of Negative Words Lexicon

The negative words lexicon includes two parts: negative words and rhetorical words. If a sentence contains negative words such as "no" or "not" or a rhetorical question, it is likely to cause the sentiment tendency of the sentence to be inconsistent with the meaning of the judgment in computer language. The single negation in the sentence represents the reversal of the sentiment polarity of the sentence, while the double negation represents the positive meaning. If the polarity of each word in the negative words lexicon is 1, it will include negative words and rhetorical words. This lexicon is collected from 41 words artificially integrated by the author from Xinhua Lexicon. Some examples of negative words lexicons are shown in Table 3.

Table 2. Examples of some degree adverbs lexicons

Words	Intensity	Number
Too much...	6	30
100%, extremely...	5	69
Special, real, especially...	4	42
More, more than...	3	37
Slightly, a little...	2	29
Half, not very, relative...	1	12

Table 3. Examples of some degree adverbs lexicons

Words	Polarity	Number
No, not, have not, doesn't it mean...	−1	41

2.4 The Construction of the Lexicon on Automobile Comment Terms

The lexicon on automobile comment terms mainly covers the words that consumers use to comment on the performance of the automobile frequently. In order to construct the lexicon, the first 500 comments in the Car Home evaluation section are obtained by using crawler technology. Then, the words describing automobile attributes are manually selected and marked with polar box intensity. Finally, the words are divided into 3 and 5 levels of intensity and if the words in the lexicon on automobile comment terms are repeated with the words in the basic sentiment lexicon, delete the same words in the basic sentiment lexicon. The lexicon on automobile comment terms consists of 931 words. Examples of the lexicon on automobile comment terms are given in Table 4.

Table 4. Examples of the lexicon on automobile comment terms

Words	Intensity	Polarity
Elegant, luxurious, secure…	3	1
Spacious, soft plastic, one-button start…	5	1
Delay, frustration, fetal noise…	3	−1
Virtual position, oil tiger, broken shaft…	5	−1

3 The Sentiment Analysis of Automobile Comment Text

The sentiment analysis of comment text refers to the mining and analysis on the text's subjectivity, objectivity, viewpoint, sentiment and polarity. First, the text is preprocessed, including word segmentation and elimination of stop words. Then, the word's part of speech and sentiment polarity are judged to select the key words of the sentence and integrate the weighted text features. Finally, the sentiment tendency of the text is classified and judged, and divided into two categories, one of which is positive or negative, or the other is more detailed.

3.1 The Sentiment Values of Single Clause

In this paper, four kinds of sentiment lexicons have been constructed and sentiment words have been marked with polarity and intensity. In order to calculate the sentiment value of single clause, the sentiment score algorithm of single clause is proposed. First, the comments are divided into many words by the NLPIR Chinese word segmentation tool of the Chinese Academy of Sciences and then the part of speech of the words are marked on them. Then, the syntax that may be contained in the clause is analyzed. The words' sentiment value of the sentences with only basic sentiment words is added and the rules of multiplication are used to the sentences with modifiers and sentiment words. In the sentiment score algorithm of single clause, the sentiment value of clause is represented by S, the weight of vocabulary in the basic sentiment lexicons is represented by W, the intensity of the vocabulary in the lexicons of degree adverbs is represented by D, the number of negative words is represented by N and the weight of the words in the lexicons of automobile commentary terms, which is the product of polarity and intensity, is represented by C. The sentiment score algorithm of single clause is shown as Algorithm 1.

Algorithm 1:The sentiment score algorithm of single clause

Input: Single clause comment text of automobile attributes(V)

Output: The sentiment value of the comment text(S)

```
1.  Divide text V into several words by NLPIR tool and remove the stop word
2.  For(i=1;i++;i<=4)
3.      S=Si=0;
4.      If(only contain the words in basic sentiment lexicon)
5.          S1=W1+W2+...+Wk;
6.      If(degree adverbs appear before the sentiment words)
7.          S2=W×D;
8.      If(negative words or rhetorical words appear before an sentiment words in clauses)
9.          S3=(-1)^N×W;
10.     If(sentiment words appear before and after the automobile comments)
11.         S4=W×C;
12.         S=S+Si;
13. End for;
14. Output the sentiment value of the clause S;
```

(1) Step 1–14 are used to calculate the sentiment value of the clause, that is S.
(2) Step 2–13 is a circular statement used to calculate the sentiment value of single clause.
(3) Step 4–11 are to judge the composition of the words that may be contained in the sentence, and to give the corresponding formulas for the four cases.

The time complexity of this algorithm is O(n).

3.2 The Sentiment Scores of Paragraph Comments

The sentiment score for comment on some attribute paragraphs is indicated by $E(V)$ and the sentiment scores of each clause are S_1, S_2,... S_n. The formula for calculating is shown as formula (1).

$$E(V) = \sum_{i=1}^{n} S_i \tag{1}$$

The final results of E(V) are divided into three situations. E(V) > 0 represents that this consumer is satisfied with the performance of the automobile. E(V) < 0 represents that this consumer is not satisfied with the performance of the automobile. E(V) = 0 represents that this consumer is neutral on the performance of the automobile. An algorithm for the sentiment value of online comment text is shown as Algorithm 2.

Algorithm 2:The sentiment score algorithm of single clause

Input: The content of the comment text for an attribute of the automobile

Output: The sentiment value of the comment text *E(V)*

```
1.  For (i=1;i++;i<=n)
2.      E(V)=E(V)+S_i;
3.  End for;
4.  If(E(V)>0)
5.      Output that the consumer is satisfied ;
6.  Else if (E(V)<0)
7.      Output that the consumer is not satisfied;
8.  Else
9.      Output that this consumer is neutral;
10. End if;
11. End if;
12. Output the comment sentiment value and the consumer' s satisfaction;
13. End
```

(1) Step 1–12 are used to calculate the sentiment value of the comments and determine the satisfaction of the consumer with the automobile.
(2) Step 1–3 is a circular statement, which calculates the number of clauses in the comments by adding up the sentiment score of each clause.
(3) Step 4–11 is a judgment sentence. By judging whether the E(V) is greater than 0 or equal to 0 to judge the satisfaction of the consumer for the performance of the automobile.

The time complexity of this algorithm is O(n).

4 Recommendation Method

The main task of this section is to put forward the automotive recommendation algorithm to provide the consumers who want to buy cars with the better choices. The algorithm includes three aspects. First, the feature vectors are defined according to the evaluation from different consumers on the same automobile based on the eight indicators in the comment system of the Car Home and the average value of each performance index of the automobile is calculated respectively. Then, the feature matrix of eight indicators of different models is defined. Finally, calculating the top-k types of brand and model of the automobile which can meet the requirements of consumers to the consumers based on the consumers' preferences for each performance index and the method of cosine similarity.

4.1 Basic Concepts

Definition 1: Automobile Index Feature Vector
Assuming that in the data set, comments on some cars published by consumers who have purchased cars can be represented by feature vectors, these vectors are described in this paper as the automobile metric feature vector (AIFV). AIFV is described in the formula (2).

$$AIFV = \{I_1, I_2, I_3, I_4, I_5, I_6, I_7, I_8\} \tag{2}$$

Where, I denotes the sentiment score on the comments of car attribute from the consumes. I_1 denotes the performance-price ratio, I_2 denotes the fuel consumption, I_3 denotes the comfort, I_4 denotes the power, I_5 denotes the appearance, I_6 denotes the space, I_7 denotes the manipulation, I_8 denotes the interior decoration. These eight indicators are the eight kinds of factors in the comment system of the Car Home. Compared with the attention of consumers, it is reasonable to use these eight indicators to construct the feature vector.

To define the feature vector of the automobile can more intuitively know the consumer's satisfaction with each attribute of the automobile and more conveniently calculate the average value of the automobile attribute comments to construct the automobile index feature matrix.

Definition 2: Automobile Index Feature Matrix
Assuming that in the data set, C denotes the models of every automobile and I denotes the set of eight indicators. Considering that there is no comparability of cars with too much price difference, this paper proposes to divide the automobile set into several sets and each set takes 50,000 yuan as an interval. In each set, the value of the I index of the automobile C is derived from the average value of the indicator in the AIFV vector of the automobile. Defining the coordinate matrix AIFM, with the models C_i of every automobile taken as row coordinates and the eight indicators I_i taken as column coordinates. AIFM is described in the formula (3).

$$AIFM = \begin{bmatrix} I_{11} & I_{12} & \cdots & I_{18} \\ I_{21} & I_{22} & \cdots & I_{28} \\ \cdots & \cdots & \cdots & \cdots \\ I_{n1} & I_{n2} & \cdots & I_{n8} \end{bmatrix} \tag{3}$$

To define the feature vector of automobile index can more intuitively know the evaluation value of the eight indexes of each type of automobile and more conveniently calculate the similarity between consumers' requirements and evaluation values on different automobiles to recommend k types of model with high similarity to the consumers.

4.2 The Normalization of Each Column of the AIFM Feature Matrix

The sentiment value data of each automobile is sparse and there is no upper and lower limit, which lead to the difficulty of comparing the similarities and differences between each car. In order to solve the problem, the column of AIFM feature matrix is normalized [6], the sentiment score of the automobile is converted to the value between 0 and 1 by formula (4) and the numerical value is rounded to two decimal place. The formula of numerical normalization is described in the formula (4).

$$Y = [(X - Min) \div (Max - Min)] \tag{4}$$

In Formula (2), X represents the value before conversion and Y represents the value after conversion. Max and Min represent the max value and min value of sample data.

4.3 Automobile Recommendation Method Based on the Sentiment Analysis of Online Comment

The automobile recommendation method based on the sentiment analysis of online comment know the personalized preference of the consumer to the automobile through the way of interaction, calculate the sentiment score of the consumer's requirements for each attribute of the automobile, use the formula (4) to normalize the sentiment value of each attribute and construct the AIFV based on the consumer's requirements. Then, the similarity of each car is calculated by the AIFM feature matrix. Finally, the top-k types of automobile are recommended based on the similarity ranking. The formula of cosine similarity is described in the formula (5). The automobile recommendation algorithm based on the sentiment analysis of online comment is shown as Algorithm 3.

$$T(x, y) = \frac{x \times y}{\|x\|^2 \times \|y\|^2} = \frac{\sum x_i y_i}{\sqrt{\sum x_i^2} \times \sqrt{\sum y_i^2}} \tag{5}$$

Algorithm 3: The automobile recommendation algorithm

Input: The requirement of consumers about automobile (price R_{mon}, consumers' personalized preference for automobile *Pre*)

Output: top-k types of recommended automobile models

Search for cars at this price according to Rmon;
For (C_i(1≤i≤n))   //C_i denotes the cars that meet the price requirement
Extracting the relevant evaluation of the consumer and splicing it into a text;
The vector C_i-*AIFV* is calculated;
End for
Combining C_i-*AIFV* into matrix *AIFM* and normalizing according to column;
For (U_j(1≤j≤m))   //U_j denotes the consumers waiting for consumption
The vector *Pre-AIFV* was calculated.
End for
The row of Vector *Pre-AIFV* is normalized;
For each row (Row_i) of the matrix AIFM
The *T(Rowi, Pre-AIFV)* are calculated.
End for
Descending order according to the *T(Rowi,Pre-AIFV)*;
Output top-k types of car with high similarity to consumers

(1) Step 1–15 is used to calculate and analyze consumers' personalized preference and automobile's similarity, and then recommend top-k types of automobile brands and models with high similarity.
(2) Step 2–5 is a circular statement used to construct the feature vector C-AIFV for every automobile.
(3) Step 7–9 is a circular statement used to construct the feature vector of automobile index Pre-AIFV, which meet the consumers' personalized preference.
(4) Step 11–13 is a circular statement used to calculate the T(Rowi, Pre-AIFV) between each line of the matrix AIFM and the consumer Pre-AIFV;

The time complexity of the algorithm is determined by the sorting algorithm. The time complexity of this algorithm is $O(n^2)$.

5 Experiments

In this section, we will propose the experimental evaluation index and experimental method for this recommendation method to verify the effect of the automobile recommendation method based on online comments.

5.1 Evaluation Parameters

In this paper, the Precision and Recall are used as indicators to evaluate the recommended results, and we use the F1-Measure indicator to consider P indexes with R indexes together because of the contradiction between P and R indexes [7, 8]. The formula can be found in formulas (6), (7) and (8).

$$\mathrm{P} = \frac{V}{U \times k} \times 100\% \tag{6}$$

$$R = \frac{V}{U} \times 100\% \tag{7}$$

$$F1 = \frac{2 \times P \times R}{P + R} \times 100\% \tag{8}$$

Where, U denotes the total number of users. V denotes the total number of recommended cars to be adopted.

5.2 The Experimental Method

We solicit the intention of 1, 000 consumers to buy cars from the Car Home, and then use our recommended algorithm to recommend top-k cars, where consumers can choose any number of model not exceeding 10 to meet their requirements for the car. In order to consider the effect of this recommendation algorithm from many aspects, we assign 1, 2, 3, 4 to k separately and select two kinds of consumer groups to carry on the experimental analysis.

(1) Ordinary household: The consumers who pay 10 to 15 million yuan for the car with the main requirements of low fuel consumption, good comfort, beautiful appearance. The experimental results are shown as Fig. 2.

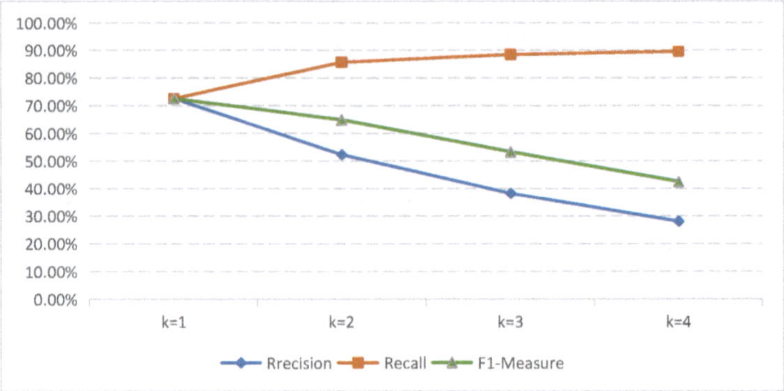

Fig. 2. The acceptability of the general household group

(2) Delivery by express delivery: The consumers who pay 10 to 15 million yuan for the car with the main requirements of low fuel consumption, strong power, large space, high price performance. The experimental results are shown as Fig. 3.

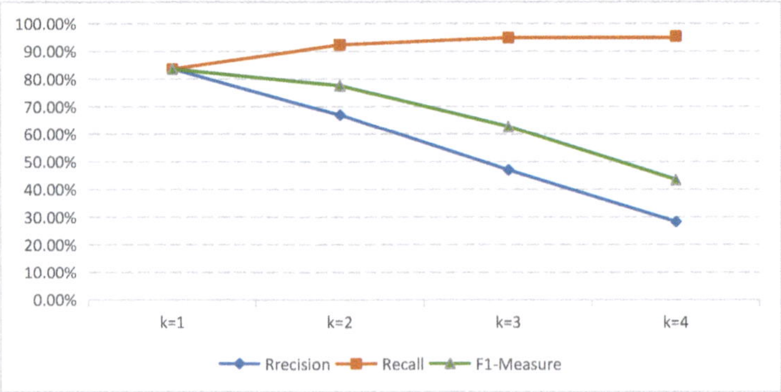

Fig. 3. The acceptance of delivery crowd by express delivery

Two conclusions can be drawn from Figs. 2 and 3.

(1) The recommended algorithm proposed in this paper has a high recall (>70%), and with the increase of K, the recall increases continuously, which ensures that most users can query the satisfied car. At the same time, the accuracy and F1 decrease significantly, but from the degree of decline, we can analyze the existence of consumers who do not choose the highest similarity rather than choose other recommended cars.

(2) From the comparison results of the two consumer groups, we can conclude that the more detailed the individualized preference to the automobile, the higher the degree of adoption of the recommended results.

6 Conclusions

We propose a automobile recommendation method for Car Home website based on the analysis of the sentiment tendency of online comments and users' preference. First of all, the implied emotion of the comment text need to be transferred into specific data. Then, the optimal recommendation result is calculated combining the individual preference of consumers. To achieve these goals, construct four sentiment lexicons are constructed, and effectively used the lexicons to calculate the evaluation score of the car consumers who have purchased the car. Then, the feature matrix is constructed according to the eight elements of automobile evaluation and the method of the similarity of cosine is used to calculate the right models and then recommend the cars of these models to consumers.

Considering the consumer's acceptance degree of the recommendation algorithm, the feasibility of the proposed algorithm is verified in the large scale data, and the adoption degree of the recommended car is also guaranteed.

References

1. Zhang, S.X., Wang, Y., Zhang, S.Y., Zhu, G.L., et al.: Building associated semantic representation model for the ultra-short microblog text jumping in big data. Cluster Comput. **19**, 1399–1410 (2016)
2. Zhao, J.Q., Gui, X.L., et al.: Comparison research on text pre-processing methods on twitter sentiment analysis. IEEE Access. **5**, 2870–2879 (2017)
3. Xu, Z., Mei, L., Raymond, K.K., Lv, C., Hu, C.P., Luo, X.F., Liu, Y.H.: Mobile crowd sensing of human-like intelligence using social sensors: a survey. Neurocomputing **279**, 3–10 (2018)
4. Gkorgkas, O., Vlachou, A., Doulkeridis, C., Nørvåg, K., et al.: Exploratory product search using top-k join queries. Inf. Syst. **64**, 75–92 (2017)
5. Hernandez, A.F.R., Garcia, N.Y.G., et al.: Distributed processing using cosine similarity for mapping big data in hadoop. IEEE Latin Am. Trans. **14**(6), 2857–2861 (2016)
6. Mingers, J., Meyer, M., et al.: Normalizing Google Scholar data for use in research evaluation. Scientometrics **112**(2), 1111–1121 (2017)
7. Wang, Y., Li, J.H., et al.: Credible intervals for precision and recall based on a K-fold cross-validated beta distribution. Neural Comput. **28**(8), 1694–1722 (2016)
8. Zhang, S.X., Wei, Z.L., Wang, Y., Liao, T.: Sentiment analysis of Chinese micro-blog text based on extended sentiment dictionary. Future Gener. Comput. Syst. **81**, 395–403 (2018)

An Adaptive Tracking Algorithm for Micro-blog Topic Based on Association Semantic Network

Shunxiang Zhang(✉) and Hongze Zhu

School of Computer Science and Engineering,
Anhui University of Science and Technology, Huainan 232001, China
sxzhang@aust.edu.cn, 928397941@qq.com

Abstract. The follow-up trend of micro-blog topic and the topic drift cannot be described detailly by the existing topic tracking algorithm. To solve this problem, an adaptive tracking algorithm for micro-blog topic based on association semantic network is proposed in this paper. First, the micro-blog text is divided into equal-time text through the time window. The training set, the first m equal-time text, is used to construct the query association semantic network (QASN). Second, the temporal association semantic network (TASN) is constructed through the a subsequent equal-time text. And the similarity of QASN and TASN is compared. Then, the training set is updated according to the similarity degree, and QASN is rebuilt. The analytical process of the subsequent equal-time text is operated repeatedly and iterative, until the last equal-time text has been analyzed. The experimental results show that the algorithm can track more precise and detailed the hot topics of micro-blog.

Keywords: Adaptive tracking algorithm · QASN · The similarity degree TASN

1 Introduction

With the rapid rise of social media, more and more users are more willing to release reviews on real-time events on micro-blog platform because of the timeliness and randomness of micro-blog. Micro-blog users will initiate micro-blog topic discussion around a certain event. When discussion volume of the topic, the context and the future trend of the topic is clear gradually. However, the fragmentation of micro-blog reviews can't make it possible for users to quickly get the development of the current topic. Therefore, as micro-blog users continue to discuss topics, adaptive tracking algorithm proposed in this paper can help users to get the subsequent trend of topics quickly and accurately.

As an important technology of text mining, topic tracking can extract subsequent development related to topic form micro-blog discussion. Therefore, how to extract the trend of the micro-blog topic more quickly and how to detect the topic drift have always been the research direction of most scholars. However, the traditional topic tracking algorithm only identifies topic related micro-blog texts through several topic

J. Abawajy et al. (Eds.): ATCI 2018, AISC 842, pp. 194–206, 2019.
https://doi.org/10.1007/978-3-319-98776-7_22

words, and roughly represents the evolution of topics as the change of several topic words. In addition, the topic drift cannot be detected or simply determined. Therefore, in this paper, the topic tracking algorithm needs to solve the following three problems:

(1) How to identify micro-blog text related to topic more accurately?
(2) How to describe the evolution of the micro-blog topic in more detail?
(3) How to better detect and respond to the micro-blog topic in the process of topic tracking?

To solve the above problems, this paper proposes an adaptive tracking algorithm for micro-blog topic based on association semantic network. The time window is introduced to divide the micro-blog text into several equal-time texts because of the fragmentation and timing of information. First, QASN is constructed according to the topic hotspots and association rules between them extracted from the training set that is the first equal-time texts. The test set, first subsequent equal-time text, is used to construct TASN. Then, the similarity threshold and the deviation threshold are introduced to divide the similarity between QASN and TASN into different ranks. The training set is updated according to the test set and the similarity rank, thus QASN is updated. Finally, the time window is moved to a unit as a test set. The analytical process of the test text is operated repeatedly and iterative, until the micro-blog text has been scanned.

To solve the above problems, this paper proposes an adaptive tracking algorithm for micro-blog topic based on association semantic network. The time window is introduced to divide the micro-blog text into several equal-time texts because of the fragmentation and timing of information. First, QASN is constructed according to the topic hotspots and association rules between them extracted from the training set that is the first equal-time texts. The test set, first subsequent equal-time text, is used to construct TASN. Then, the similarity threshold and the deviation threshold are introduced to divide the similarity between QASN and TASN into different ranks. The training set is updated according to the test set and the similarity rank, thus QASN is updated. Finally, the time window is moved to a unit as a test set. The analytical process of the test text is operated repeatedly and iterative, until the micro-blog text has been scanned (Fig. 1).

This paper is organized as follows: Sect. 2 introduces the related work of an adaptive tracking algorithm. Section 3 proposes the construction of ASN of micro-blog topic based on TF-IDF. Section 4 proposes the construction of ASN (association semantic network). Section 5 carries on the experiment for this algorithm. Section 6 proposes summary in the paper.

2 Related Work

For the work of an adaptive tracking algorithm for micro-blog topic, most scholars at home and abroad have conducted in-depth research in this regard. Micro-blog topics are tracked by building ASN in time sequence. Therefore, this section introduces the related work of extracting key words and topic tracking.

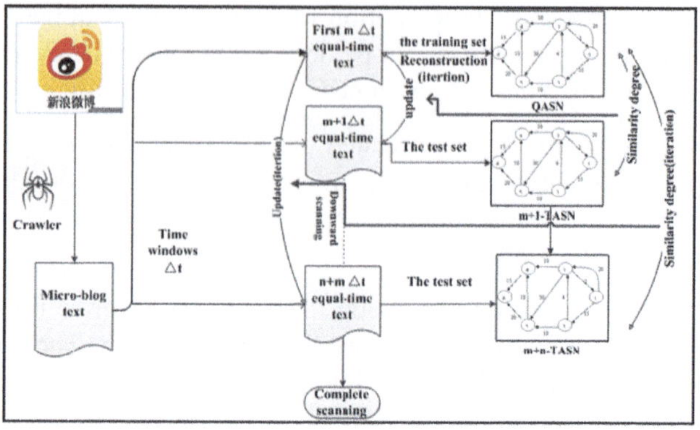

Fig. 1. Process diagram of adaptive tracking algorithm.

Nowadays, the extraction of keywords can be mainly divided into unsupervised and supervised methods. Among them, the unsupervised method mainly uses the statistical information of the words themselves to rank the importance of the words so as to extract keywords. So far, most scholars mainly use the TF-IDF value [1, 2] of words as their statistical information. TF-IDF can assess the importance of words in document collections [3] and effectively reduce the interference of high-frequency words with very small amounts of information [4]. The supervised method mainly improves the extraction effect by mining features that can characterize keywords [5]. However, there are different methods for supervising different data sets.

Topic detection and tracking (TDT) is an information processing technology designed to help people cope with the increasingly serious problem of Internet information explosion, automatic identification of new topics in news media information flow, and continuous tracking of known topics. Therefore, the topic tracking is one task of the TDT. Dai et al. [6] employs the improved single pass clustering algorithm to accomplish topic tracking and shows the topics to the users in time series order. Emerging Topic Tracking System (ETTS) [7] is used to detect the changes in the information area of our interest and generate a summary of changes. Liu et al. [8] design the techniques to extract events from topics and to infer the evolving relationship among the events.

3 Construction of ASN of Micro-blog Topic Based on TF-IDF

3.1 The Proposal of the Problem

Because the micro-blog topic is tracked only by tracking several keywords and ignoring the weight of the word in the traditional topic tracking technology, it is impossible to detect the subtle topic drift. When the keywords are constant in the current topic, the changes of the weight of each word are too large, which is called the subtle topic drift. In order to solve this problem, both the keywords and corresponding weights are tracked in this section.

Besides, the traditional topic tracking technology cannot detect the change of the correlation degree between keywords, so that it cannot predict the trend of the topic accurately. When the relevance between one keyword and other keywords is weakened, it is predicted that the keywords will be replaced in this topic, so as to predict the development direction of topic discussion. To solve this problem, the association rules between keywords are extracted and tracked to observe the development trend by Apriori algorithm in this section.

3.2 The Keyword Extraction from Micro-blog Topics Based on TF-IDF

The keywords are extracted inefficiently because of data sparsity of micro-blog text, therefore, all micro-blog text under a certain micro-blog topic within a period of time is extracted through the crawler from the micro-blog platform in this article. Each micro-blog text is joined together as a separate document to form a document set for analyzing more easily statistical information of different words. Then, to highlight the process of updating the micro-blog topic, the time window is introduced to divide the separate document into some equal-time texts.

As a traditional statistical method, TF-IDF is widely used to evaluate the importance of a word in the corpus. Each word is sorted according to its importance evaluated by TF-IDF to get the top-ranking words, that is, keywords. TF-IDF is universally used in all text, but it has no pertinence. So, the keywords are extracted more effectively by combining the word item information (length, location, etc.) of words.

Definition 1: The New Login Word Weight
The new login word is not included in the Chinese Dictionary, but it needs to be extracted to make Chinese word segmentation more accurate. As an important part of word list of Chinese word segmentation, the new login word includes proper nouns, abbreviations, new word and so on. As mentioned in document [9], the new login word can display the subject of the document better than the general vocabulary. The new login words and general words are assigned respectively different weights to highlight the superiority of new login words in document description. Therefore, the definition of the new login word weight is shown in formula (1).

$$weight_{new}(word) = \begin{cases} \alpha_{new} & word \in new\ login\ word \\ \alpha_{non-new} & word \notin new\ login\ word \end{cases} \quad (1)$$

In formula (1), $weight_{new}(word)$ is the new login word weight. α_{new} and $\alpha_{non-new}$ is the new login word weight of the new login words and general words, and α_{new} is slightly greater than $\alpha_{non-new}$. The location of the words (the beginning and end of a sentence) does not have much impact on the importance of words because of short micro-blog text. Therefore, the location of word is not taken into account in this article.

Definition 2: The Title Weight for Micro-Blog Topic
The title word can display the subject of the document better than the general vocabulary. Therefore, micro-blog topic title weight is defined as the formula (2) by analogy with the formula (1).

$$weight_{title}(word) = \begin{cases} \beta_{title} & word \in title \\ \beta_{non-title} & word \notin itle \end{cases} \tag{2}$$

In formula (2), $weight_{title}(word)$ is micro-blog topic title weight. β_{title} and $\beta_{non-title}$ is micro-blog topic title weight of the title words and the general words, and β_{title} is slightly greater than $\beta_{non-title}$.

Definition 3: The Micro-blog Topic's Single-Word Score
The micro-blog topic's single-word score represents the importance of a single word for the entire document set. As shown in formula (3), The TF-IDF is improved by combining the new login word weight, micro-blog topic title weight, and the length weight of words for micro-blog topic.

$$\begin{aligned} score(word) = {} & TF(word) \times IDF(word) \\ & \times (weight_{new}(word) + weight_{title}(word)) \end{aligned} \tag{3}$$

In formula (3), $score(word)$ represents the score of $word$. $TF(word) \times IDF(word)$ represents the TF-IDF of $word$. When the micro-blog topic's single-word score of a word is greater than the threshold, this word can be determined as keywords. Therefore, the definition of keywords is shown in formula (4).

$$keyword = \{word | word \in V, score(word) \geq k\} \tag{4}$$

In formula (4), $keyword$ is the keyword set in micro-blog text. k is threshold of the micro-blog topic's single-word score.

3.3 Construction of ASN (Association Semantic Network)

The keywords and the association rules between keywords are visualized by constructing of ASN. It is helpful for users to understand more clearly, comprehensively and quickly the main content of micro-blog's text. ASN is represented as a directed weighted diagram in this paper. And the directed weighted diagram is represented by a two-spot group G_d. The expression of the two-spot group is shown in formula (5).

$$G_d = (V_d, E_d) \tag{5}$$

In formula (5), V_d represents the vertex set of the diagram, consisting of keywords from the micro-blog document, that is, the keywords set. E_d represents the edge sets of diagrams and the weights of each edge. The frequent patterns between keywords are extracted, and then the association rules between keywords are extracted through the Apriori algorithm. The weights of ASN is dissimilarity weights, that is, the greater the weight, the greater the path between the two keywords.

$$E_d = \begin{pmatrix} conf(w_1, w_1) & \cdots & \cdots & conf(w_1, w_m) \\ \cdots & \ddots & conf(w_j, w_i) & \vdots \\ \cdots & conf(w_i, w_j) & \ddots & \vdots \\ conf(w_m, w_1) & \cdots & \cdots & conf(w_m, w_m) \end{pmatrix} \quad (6)$$

In formula (6), $conf(w_i, w_j)$ represents the confidence of the keyword w_i and the keyword $w_j(w_i \Rightarrow w_j)$. If the keyword c_i is associated with the other one w_j (Support $support(w_i, w_j)$ is less than threshold), the default value of the confidence setting is 0 ($conf(w_i, w_j) = 0$), that is, there is no corresponding edge in ASN.

4 An Adaptive Tracking Algorithm for Topic Based on ASN

4.1 Calculation of Similarity of ASN

The essence of the adaptive tracking algorithm proposed in this paper is to express the development and future trends of micro-blog topic in the form of ASN. Therefore, the QASN and TASN (the essence is ASN) is constructed separately by the first m and the following first equal-time texts. To determine whether the subsequent micro-blog text affects the trend of the topic, this section analyses the similarity between QASN and TASN.

The essence of ASN is the carrier of keywords and association rules between keywords. Therefore, the essence of similarity analysis is to analyze separately the similarity degree of the semantic set (keywords set) and the edge set (association rules between keywords).

Definition 4: Similarity of the Vertex Set
The similarity of the vertex set is the analysis of the change for the keywords and corresponding weights. The words can be compared with the semantic similarity, and the topic hot words in QASN as the semantic unit (formula 7). Therefore, the calculation of the similarity of the vertex set is converted to the similarity of two one-dimensional vectors after the keyword is quantified. From the above analysis, the specific definition of the similarity of the vertex set is as shown in the formula (8).

$$\begin{cases} V_Q = (w_{Q1} \cdot v_{Q1}, w_{Q2} \cdot v_{Q2}, \ldots w_{Qm} \cdot v_{Qm}) = (v_{Q1} \cdot 1, v_{Q2} \cdot 1, \ldots v_{Qm} \cdot 1) \\ V_T = (w_{T1} \cdot v_{T1}, w_{T2} \cdot v_{T2}, \ldots w_{Tm} \cdot v_{Tm}) \\ = (v_{T1} \cdot sim(w_{Q1}, w_{T1}), v_{T2} \cdot sim(w_{Q2}, w_{T2}), \ldots v_{Tm} \cdot sim(w_{Qm}, w_{Tm})) \end{cases} \quad (7)$$

$$sim(V_Q, V_T) = \frac{\sum_{i=1}^{m} v_{Qi} v_{Ti} \cdot sim(w_{Qi}, w_{Ti})}{\sqrt{\sum_{i=1}^{m} v_{Qi}^2} \times \sqrt{\sum_{i=1}^{m} [v_{Ti} \cdot sim(w_{Qi}, w_{Ti})]^2}} \quad (8)$$

In formula (7), $sim(w_{Qm}, w_{Tm})$ is the semantic similarity between the keyword w_{Qn} and w_{Tm}. v_{Tm} is corresponding weight of w_{Qm}. In formula (8), $sim(V_Q, V_T)$ refers to the similarity between the vertex set V_Q and the vertex set V_T of the two ASN G_Q and G_T. If the number of keywords in the vertex set V_Q is not equal to the vertex set V_T, the threshold of the micro-blog topic's single-word score should be changed so that the vertex set V_Q is equal to the number of hot topic words in the vertex set V_T. The keywords in the vertex set V_Q should correspond to the vertex set V_T as far as possible. For the rest of the key words, the two words with large semantic similarity are corresponding.

Definition 5: Similarity of the Edge Set
The similarity of the edge set is the analysis of the change for the confidence between every two keywords. Because the edge set is a $m \times m$ matrix, e_{xn} is used to express each row of the edge set. Then the calculation of the similarity between the two m-dimensional matrices can be decomposed into the m-times calculation of the similarity between the two one-dimensional vectors. The similarity of two one-dimensional vectors is cosine similarity. From the above analysis, the specific definition of the edge set similarity is shown as shown in the formula (9).

$$\begin{cases} E_x = (e_{x1}, e_{x2} \cdots, e_{xm})^T \\ sim(e_{x1}, e_{y1}) = \frac{e_{x1} \cdot e_{y1}}{|e_{x1}||e_{y1}|} \\ sim(E_x, E_y) = \sum_{i=1}^{m} sim(e_{xi}, e_{yi}) \Big/ m \end{cases} \quad (9)$$

In formula (9), E_x is the edge set of ASN, and $e_{xi}(1, 2, 3 \cdots m)$ represents each row of the edges set. $sim(e_{xi}, e_{yi})$ is cosine similarity of two one-dimensional vector e_{xi} and e_{yi}. $sim(E_x, E_y)$ is the similarity between the edge sets E_x and the edge sets E_y.

Definition 6: Similarity of ASN
The similarity of ASN is the analysis of the change for QASN and TASN. According to the similarity of the vertex set and similarity of the edge set, the specific definition of similarity of ASN is shown as shown in the formula (10). In formula (10), $sim(G_Q, G_T)$ is the similarity of ASN between G_Q and G_T.

$$sim(G_Q, G_T) = sim(V_Q, V_T) \times sim(E_Q, E_T) \quad (10)$$

4.2 Updating QASN Based on Similarity Feedback

In the course of topic tracking, the similarity threshold and the deviation threshold are introduced to divide the similarity into three grades in this section. And QASN is corresponding automatic adjusted according to different grades (Fig. 2).

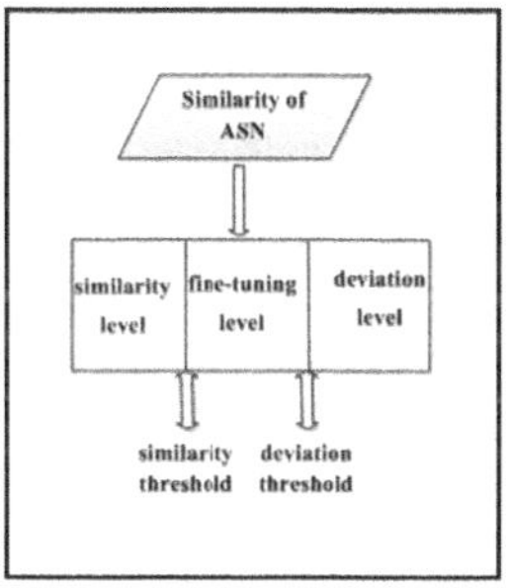

Fig. 2. Similarity grades distribution.

(1) **Similarity level:** When the similarity of ASN is lower than the similar threshold, that is, the following discussions are similar to previous discussions. So, there is no need for this equal-time text to add into training document sets, so as to avoid the reconstruction of QASN.
(2) **Fine-tuning level:** When the similarity of ASN is higher than the similar threshold and lower than the deviation threshold, this equal-time text is added into the training set, and QASN is reconstructed.
(3) **Deviation level:** When the similarity of ASN is higher than the deviation threshold, that is, the following discussion deviates from the previous discussion. Then this equal-time text is temporarily shelved and does not add into the training document set.
   (a) If among the consecutive equal-time texts are deviation level, the micro-blog topic is judged as topic drift. Then the training set is replaced by the consecutive equal-time texts.
   (b) If the consecutive equal-time texts are not all deviation level, the equal-time text is determined to temporarily deviate from the topic. So, the text is rejected by taking the information randomness of micro-blog text into account.

4.3 Algorithm Flow

Algorithm 1: a topic tracking algorithm based on micro-blog ASN

Input: micro-blog text (V), similar threshold θ_{sim}, deviation threshold θ_{dif}
Output: the change process of QASN

V is divided into some equal-time texts $V_x (x = 1,2,\cdots n)$
$V_1 \sim V_m$ is regarded as the training set S_t
for each different word w_i $score(word)$ is calculated
  if $score(word) > k$, add this word w_i to the keywords set $keyword$
  end if
end for
calculate the association rules between $keyword$
construct QASN G_Q
for (i=m+1; i<=n; i++)
$keyword$ and association rules of V_i is extracted
construct TASN G_T
Calculate $sim(G_Q,G_T)$
if ($sim(G_Q,G_T) < \theta_{sim}$) break;
else if ($\theta_{sim} \le sim(G_Q,G_T) < \theta_{dif}$) add V_i into the training set
  else V_i is considered as V_i^{dif} and add it into the deviation set S_d
    if (The previous V_x^{dif} and V_i^{dif} is discontinuous) delete V_x^{dif}
    end if
    if (the number of text is m in S_d) determine topic drift
      S_t is replaced by S_d
    end if
  end if
  Rebuilt QASN
end if
end for

From the analysis above, the specific steps of an adaptive tracking algorithm for micro-blog topic based on association semantic network are as follows. First, from the micro-blog topic, the first equal-time texts are extracted as the training set, and then QASN is constructed by the extracted keywords and association rules between keyword. In the same way, TASN is constructed by a subsequent equal-time text. Second, similarity of the vertex set and similarity of the edge set is calculated according to QASN and TASN. And then the similarity of ASN is calculated. Finally, the similarity grade of ASN is got by comparing the similarity of ASN with the similarity threshold and deviation threshold. And then QASN is corresponding updated according to the similarity grade. The analytical process is operated repeatedly and iterative, until the micro-blog text has been scanned. The specific algorithm is shown in algorithm 1.

Algorithm 1 mainly describes the construction of ASN (i.e., QASN, ATSN) and the updating of QASN by the similarity feedback. The time complexity analysis of algorithm 1 is as follows: the algorithm 1 essentially constructs or reconstructs the ASN for each equal-time text. The time complexity of constructing the ASN is, m is the number of words in each equal-time text (repetition). So, the time complexity of algorithm 1 is, n is the number of equal-time texts.

5 Experiments

(a) Experimental Method

In order to verify the feasibility of tracking the micro-blog topic in the form of ASN, first, all texts from Sina micro-blog for a period of time is obtained. Then, QASN and TASN are constructed according to the algorithm 1. And QASN is constantly updated through the constant change of topic attention.

(b) Experimental Analysis

2 groups of experiments are carried out according to the 5.1 experimental methods in this section.

(1) The first experiment: First, all discussion texts about a topic "finance" from Sina micro-blog are got. Then, the micro-blog text is divided into some equal-time text according to time window ΔT. And first two equal-time texts ($1\Delta T \sim 2\Delta T$, m = 2) are considered as a training set, thus QASN is constructed. Finally, n-TASQ (n represents which section in micro-blog texts) is constructed by each subsequent equal-time text. The ΔT = 5 min, the whole text (30 min) is divided into 6 equal-time texts. The specific QASN and n-TASN are as follows.

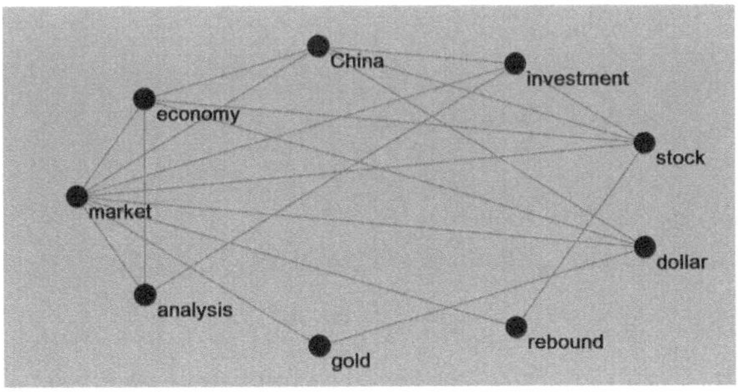

Fig. 3. QASN

As shown in Figs. 3 and 4, the keywords are effectively extracted through TF-IDF and different weights to build effectively ASN. QASN and n-TASN are only different appellations of ASN in different time periods, and the essence is not very different. Micro-blog users not only understand the event's hot words but also understand the inner links between the hot words through ASN.

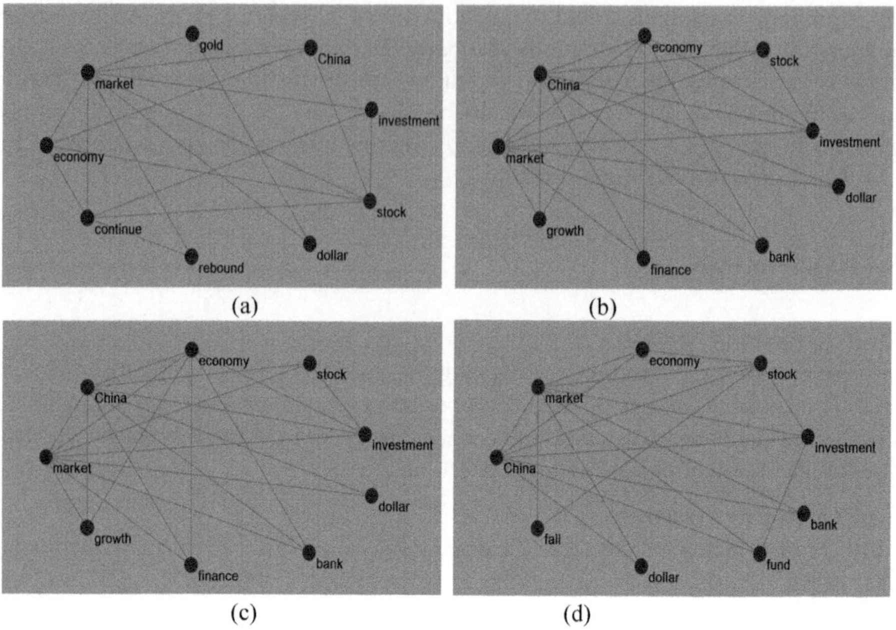

Fig. 4. n-TASN (a: 3-TASN, b: 4-TASN, c: 5-TASN, d: 6-TASN)

(2) the second experiment: the similarity of ASN is carried out through the built QASN and n-TASN to update QASN. QASN is not Fig. 4 here, but a network after a latest update from n-TASN. For example, the Fig. 4b is compared with QASN that is a network after updating with Fig. 4a. The similarity threshold is set as 0.9, and the deviation threshold is set as 0.7. The analysis of the specific similarity is shown in Table 1.

Table 1. Similarity table for QASN and TASN

QASN	3-QASN	4-QASN	5-QASN	6-QASN
Similarity of ASN	0.941	0.736	0.566	0.625
Similarity grades	Similarity	Fine-tuning	Deviation	Deviation

In Table 1, it is known that the similarity and the corresponding level of QASN and n-TASN are effectively analyzed.

(1) 3-TASN is similar with QASN, that shows the main content in the third equal-time text is same as one in the training text.
(2) The similarity grade of QASN and 4-TASN is fine-tuning level. It means the 4th equal-time text should be added into the training text and then QASN should be rebuilt.

(3) The similarity grade of QASN and 5-TASN is deviation level. It means the 5th equal-time text should be temporarily shelved.
(4) When the similarity grade of QASN and 6-TASN is deviation level, and it is judged as topic drift. So, the training text is replaced as 5th–6th equal-time text and then QASN should be rebuilt. The updating process of QASN is shown in Fig. 5.

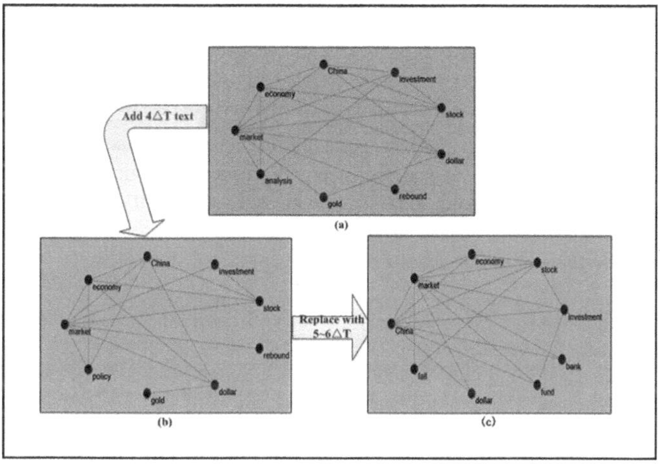

Fig. 5. Updating process of QASN (i.e., a: 1-QASN, b: 2-QASN, c: 3-QASN)

As shown in Fig. 5, QASN is effectively updated to track micro-blog topic according to the similarity of ASN. When the topic tracking to 4th text, the 4th equal-time text is added into the training text and then QASN should be rebuilt. (i.e., Fig. 5a → b, 1-QASN → 2-QASN). The words "rebound", "analysis" and other words are becoming more and more forgotten by the users, and there is a tendency to break away from the keywords set. When the topic tracking to 5th–6th equal-time text, the topic of concern is no longer the gold or the stock market rebound, but the stock market decline or bank funds, this is the topic drift (i.e., Fig. 5b → c, 2-QASN → 3-QASN).

6 Conclusions

An adaptive tracking algorithm for micro-blog topic based on ASN is used to track topics in form of QASN's updating, so that users can understand not only the replacement of micro-blog keywords, but also the change of links between keywords. The algorithm proposed in this paper applies Natural Language Processing and data mining technology to the micro-blog platform to make users clearer about the trend of the topic of micro-blog.

First, the corresponding ASN (QASN and TASN) is constructed for micro-blog text through TF-IDF and multifaceted weights. Then, the similarity between QASN and TASN is analyzed to update QASN. Similarity degree is divided into three grades (i.e., similarity, fine tuning and deviation) according to the similarity threshold and the deviation threshold. To avoid duplication calculating of similar texts, the similarity level is set. In order to avoid the mis-judgement of topic drift due to the sparsity of micro-blog text, deviation level is set. Only the consecutive m equal-time texts are deviation level, can the topical drift can be determined. The experimental results show that the algorithm can effectively track the micro-blog topic through the replacement of keywords and association rules between keywords.

Acknowledgements. This work was supported by 2018 Top talent Cultivation project of Anhui Universities (No. gxbjZD15), in part by the Natural Science Foundation of Anhui Province Universities (No. KJ2015A111), supported in part by Innovative fund for graduate students of Anhui University of Science And Technology in 2017 (No. 2017CX2110), in part by the National Science Foundation of China under Grant 61300202, and in part by the Science Foundation of Shanghai under Grant 13ZR1452900.

References

1. Jiang, D., Leung, K.W.-T., Yang, L., Ng, W.: TEII: Topic enhanced inverted index for top-k document retrieval. Knowl.-Based Syst. **89**, 346–358 (2015)
2. Hong, T.-P., Lin, C.-W., Yang, K.-T., Wang, S.-L.: Using TF–IDF to hide sensitive itemsets. Appl. Intell. **38**(4), 502–510 (2012)
3. Erra, U., Senatore, S., Minnella, F., Caggianese, G.: Approximate TF–IDF based on topic extraction from massive message stream using the GPU. Inf. Sci. **292**, 143–161 (2015)
4. Qin, P., Xu, W., Guo, J.: A novel negative sampling based on TFIDF for learning word representation. Neurocomputing **177**, 257–265 (2016)
5. Han, X., Clemmensen, L.: Regularized generalized eigen-decomposition with applications to sparse supervised feature extraction and sparse discriminant analysis. Pattern Recogn. **49**, 43–54 (2016)
6. Dai, X.Y., Chen, Q.C., Wang, X.L., Xu, J.: Online topic detection and tracking of financial news based on hierarchical clustering, IEEE, vol. 6, pp. 3341–3346 (2010)
7. Xu, Z., Mei, L., Raymond, K.K., Lv, C., Hu, C.P., Luo, X.F., Liu, Y.H.: Mobile crowd sensing of human-like intelligence using social sensors: a survey. Neurocomputing **279**, 3–10 (2018)
8. Liu, H., He, J., Gu, Y., Xiong, H., Du, X.: Detecting and tracking topics and events from web search logs. ACM Trans. Inf. Syst. **30**(4), 21 (2012)
9. Zhang, S., Zhu, H., Xu, Z.: The extraction method of new logining word/term for social media based on statistics and n-increment. J. Ambient Intell. Humaniz. Comput. **5**, 1–11 (2017)

Knowledge Extracting from Micro-Blog Topic

Shunxiang Zhang(✉) and Pengfei Ji

School of Computer Science and Engineering,
Anhui University of Science and Technology, Huainan 232001, China
sxzhang@aust.edu.cn

Abstract. Micro-blog topic knowledge extraction is the process of extracting knowledge hiding in sparse micro-blog data published by users. This paper presents a microblog topic knowledge extraction method. Firstly, the micro-blog text is preprocessed by removing the non-meaningful text (e.g. web links and animations). Secondly, part-of-speech weight and First and Last Word Weights (FLWW) are used to extract keywords candidate set which can become some entities of main knowledge (e.g., diaper and beer in diaper → beer) of a micro-blog. Finally, the non-entity keywords which are not main meanings of a micro-blog in keywords candidate set are filtered by the left and right entropy, and then the knowledge is extracted by the association rules between keywords. Experiments show that the algorithm is superior to the traditional algorithm.

Keywords: Micro-blog · Knowledge extraction · TF-IDF algorithm
Association rule algorithm

1 Introduction

The knowledge of the Micro-blog topic knowledge extraction strategy comes from users. Micro-blog is the platform for users to express their knowledge. The phenomenon of knowledge hidden in micro-blog data reflects the sparseness of big data. As the fastest growing social media platform, micro-blog has the characteristics of large amount of information, rapid dissemination, and wide participation groups. Therefore, the topic knowledge of micro-blog is sparse, scattered, and messy, which is not conducive to the study of the direction of the user's topic. Data mining, data structure and other knowledge are studied to solve how to extract micro-blog topic knowledge. As shown in Fig. 1, micro-blog information (i.e., comment, support, forward, and post content) can be obtained by crawler technology. Then Chinese word segmentation is used to filter, then TF-IDF algorithm is used to weight. And finally association rule algorithm is used to construct knowledge base. The specific steps are as follows:

(1) Data preprocessing: fetching micro-blogging data sources, deleting meaningless texts, removing web links, animations, videos, pictures, traditional Chinese characters, and English to Chinese, extracting words, and storing them in databases.
(2) Data extraction: The Chinese word segmentation system of the Chinese lexical analysis system ICTCLAS (the Institute of Computing Technology of the Chinese Academy of Sciences) is used to segment the text, the word vector was obtained.

J. Abawajy et al. (Eds.): ATCI 2018, AISC 842, pp. 207–219, 2019.
https://doi.org/10.1007/978-3-319-98776-7_23

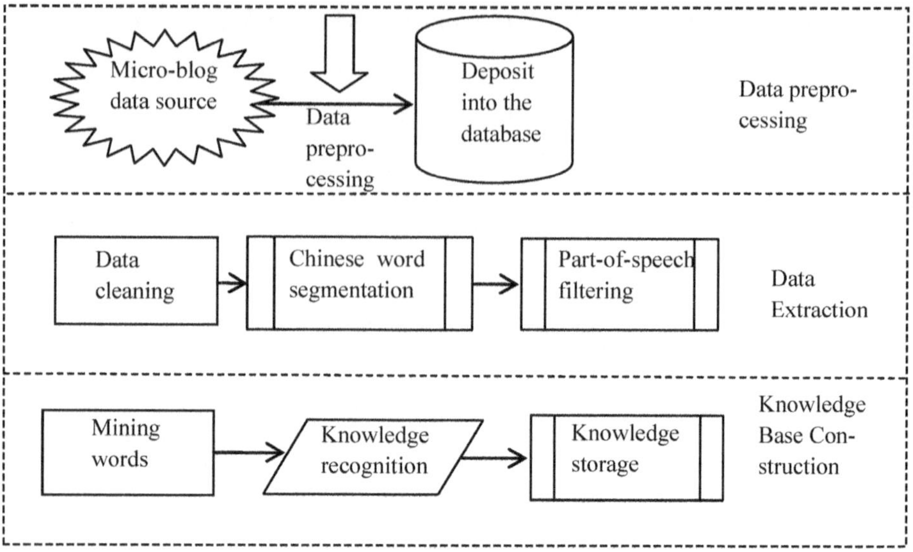

Fig. 1. Micro-blog topic knowledge extraction.

The TF-IDF algorithm is improved by word-like weighting and sentence-end-weighting to extract keywords.

(3) The knowledge base construction layer: Based on the data extraction, the key words are filtered by the left and right entropy difference, and the knowledge is extracted by using the association rules between words to store the knowledge and construct the knowledge base.

This paper is organized as follows: Sect. 3 describes the data extraction using ICTCLAS Chinese word segmentation system to get word vector deduplication, retain the actual words, through word-goal weighting, the sentence sentence at the end of the weight of the TF-IDF algorithm to improve, in order to extract keywords. Section 4 filters keywords by using left and right entropy as the difference, and uses the association rules between words to extract knowledge so that knowledge is stored and the knowledge base is constructed.

2 Related Work

This section simply responds to the three basics theoretical work on ultra-short micro-blog text representation. The first is to introduce some preprocessing of the text, store the data in the database, and then introduce the processing of the word segmentation of the text, extract the keywords of the micro-blog data, and then combine these micro-blog data features with the TF-IDF algorithm to weight the keywords. Use the association rule algorithm to get knowledge. Aldhaln Zeki [1] and others improved knowledge extraction of Hadith classification using decision tree algorithms. Joachims [2] proposed a probabilistic version of the Rocchio classifier and provided an explanation of the TF-IDF lexical weighted heuristics, the probabilistic variants of which were compared with the standard naive Bayes classifier in three text categorization tasks. Huang and Wu [3] proposed that a new algorithm DI-TFIDF be proposed by adding a new weight value, namely Distributive Degree DI (Distribution Information), to increase the weight of key feature terms. Garcia [4] Data Mining (DM) for the identification of relationships between various elements of the collected information in order to discover useful knowledge and support strategic decision making systems and sustainable development in the animal, food and health industries. Zhang [5] To efficiently mine new logining word/terms from micro-blog, this paper presents an extraction method of new logining word/term for social media based on statistics and N-increment. Hong et al. [6] propose an algorithm called SIF-IDF for modifying original databases in order to hide sensitive itemsets. It is a greedy approach based on the concept borrowed from the Term Frequency and Inverse Document Frequency (TF-IDF) in text mining. Telikani and Shahbahrami [7] A new hybrid algorithm, called Decrease the Confidence of Rule (DCR), proposed to improve a border-based solution, namely MaxMin, using two heuristics to hide the association rules for data sanitization process. Rezgui [8] A construction industry standard taxonomy was used to provide the seeds of the ontology, enriched and expanded with additional concepts extracted from large discipline-oriented document bases using information retrieval (IR) techniques. Fang et al. [9] used LDA to model each class, extract topics, and calculate the weights of each topic from three aspects: word frequency, word span, and word length. The word with the largest weight is used as the topic. Tags, and calculate the probability of document distribution for each topic.

3 Data Source Processing

3.1 Micro-Blogging Text Processing

Micro-blogging text is different from ordinary web text and has many features that ordinary web text does not have. For the convenience of research, the micro-blog text needs to be preprocessed. Web links, animations, videos, pictures are removed, and "@ +username", "#+topic+#" are removed. Micro-blogging text converts Traditional to Simplified, and English to Chinese.

3.2 Data Extraction

After preprocessing the data, the word vector is obtained by the ICTCLAS Chinese word segmentation system. Then the actual words are preserved, and words are obtained. At this point, there are still many defaming words in the words and using the TF-IDF algorithm to obtain the necessary keywords.

Term frequency (TF) refers to the frequency with which text terms appear in the file. details as follows:

$$\mathrm{tf}_{i,j} = \frac{n_{i,j}}{\sum_k n_{k,j}} \tag{1}$$

The formula (1) $n_{i,j}$ is the sum of the number of occurrences of all the words in the document.

The IDF Inverse Document Frequency is a measure of the general importance of a word. The IDF of a particular term can be obtained by dividing the total number of documents by the number of documents containing the term and obtaining the quotient logarithm. details as follows:

$$\mathrm{idf_i} = \log \frac{|\mathrm{D}|}{\left|\{\mathrm{j} : \mathrm{t_i} \in \mathrm{d_j}\}\right|} \tag{2}$$

Where $|D|$: the total number of files in the corpus $\left|\{\mathrm{j} : t_i \in d_j\}\right|$: the number of files containing the word $\mathrm{t_i}$ (the number of files in $\mathrm{n}_{i,j} \neq 0$) If the word is not in the corpus, it will result in a dividend of zero, so in general use $1 + \left|\{\mathrm{j} : t_i \in d_j\}\right|$ then

$$tfidf_{ij} = tf_{i,j} \times idf_i \tag{3}$$

In formula (3), it refers to the frequency of high words in a particular file, and the low file frequency of the word in the entire file set, which can produce a high-weight TF-IDF. Therefore, TF-IDF tends to filter out common words and retain important words. But it is not enough to fully embody key important words, so it is weighted by part-of-speech and first-end words.

Definition 1: Micro-Blog Knowledge Base

Micro-blog knowledge base is a database composed of knowledge hidden in sparse data of Micro-blog. When two or more keywords are closely related, the knowledge of these keywords constitutes. Knowledge is the relationship between entities and entities, knowledge is not entities, it has no attributes. Therefore, the knowledge base is defined by the support threshold and the confidence threshold. The specific definition is as follows.

$$\text{Knowledge} = \left\{(\mathrm{w_1} \cdots \mathrm{w_n}) | \mathrm{confident(wi, wj)} \geq \theta_c \text{ and } \mathrm{upport}(\mathrm{w_i}, \mathrm{w_j}) \geq \theta_s\right\}$$

Assume that we take $\mathrm{w_1}$ to $\mathrm{w_n}$, n entities, entities may or may not exist. If there is knowledge between two entities and using the following formula:

$$K = \{\mathrm{w_i}, R, \mathrm{w_j}\}. \tag{4}$$

The knowledge base composed of all knowledge from $\mathrm{w_1}$ to $\mathrm{w_n}$ is expressed by the following formula:

$$\mathrm{K_z} = \{\mathrm{K_1}, \mathrm{K_2}, \ldots \mathrm{K_n}\} \tag{5}$$

W is an entity, R is a knowledge, K is a knowledge of two entities and entities, and $\mathrm{K_z}$ is a knowledge base.

Examples: {diaper → beer, air → fat, hair loss → longevity, millet → price war,}

Explanation example: Many people also buy beer when Wal-Mart buys diapers. Air pollution can lead to body fatness. Hair loss people live longer than normal people. Xiaomi phones sell in the market for price wars and sell at low prices for hungry marketing.

The conditional keyword w_i and the keyword w_j satisfy both the degree of confidence greater than one entropy and the degree of support greater than one entropy, and thus the keyword w_i and the keyword w_j constitute a knowledge base. There are explanations below for the confidence and support in the knowledge base, keywords, and knowledge.

Definition 2: Micro-Blog's Part-of-Speech Weight

Micro-blog's part-of-speech weight is the degree of importance of different word-of-speech words in the topic of micro-blog. In Micro-blog, the discussion of hot topics, the number of nouns per sentence is significantly more than verbs, in order to compensate for the lack of verbs in probability, so for the majority of verbs weighted, for the nouns with low probability weighted. For example, since the outbreak of the international financial crisis, the euro zone economy was even on the brink of a "secondary recession." Most weights are given to "outbreaks" and "on the brink," and the international financial crisis is weighted with a small probability. The degree of importance of a word is extracted by assigning different weights to different parts of speech. The specific definition is as follows:

$$Wp = \begin{cases} \partial & word \in n \\ \beta & word \in v \\ \gamma & word \in adj \end{cases} \tag{6}$$

In formula (6), the weight of Wp part of speech is obtained. The weights of the words are denoted by Wp, where ∂ denotes the weight of the noun, β denotes the weight of the verb, and γ denotes the weight of the adjective. The following is an example of Micro-blog's part-of-speech weights (Table 1).

Table 1. Examples of Micro-blog's part of speech weight

Vocabulary	Entity	Weights
Noun	Xiaomi, meizu, campaign, domestic mobile phones…	0.8
Verb	Counterattack, obtain, release, application…	1.2
Adjective	United, efficient, beautiful, perseverance…	0.6

Definition 3: Micro-Blog First and Last Word Weights (FLWW)

Micro-blog First and Last Word Weights (FLWW) are weighted by the first and last words in sentences in the micro-blog topic. In the Micro-blog topic, since the first noun and the last noun of each sentence are relatively important parts, weighted emphasis is placed on the internal words of the text. In order to compensate for the deficiencies in the probability, so the weighted of verbs are improving, and the nouns are weighted with a small probability. For example, the US Pacific economic strategy will cause China's geopolitical economic dilemma, and the first and last words will emphasize the weighting of the US and the dilemma. The goal is to better extract keywords. The weights of the first and last words are denoted by W_Q, and with λ for the weight of the first word and σ for the weight of the last word. The specific definition is as follows:

$$W_Q = \begin{cases} \lambda & word \in first \\ \sigma & word \in last \end{cases} \tag{7}$$

In formula (7), the weighted weights of the first and last words of W_Q are obtained. The following are weighted examples of the three Micro-blog ending words.

(1) In this era of rapid development of **the Internet**, there are no traditional enterprises, only traditional people; in this era of speed, crisis and change, there is no homogeneous product, only homogenized **thinking**!
(2) **Xiaomi** specializes in Internet and new retail, and does a good job in brand building and internet marketing as well as Xiaomi's ecological chain and internet of things, and then replicates and **globalizes the model**.
(3) Since the development of the Internet in **China** for 20 years, it has been renamed Ma. Whether it is a network car, a shared bicycle, or a new consumer retail field such as take-out or e-commerce, Ma Yun or Ma Huateng can only be selected. Even the war can go online. Under the line, almost all **areas** of China have been affected (Table 2).

Table 2. Micro-blog word weight examples

First and last word position	Entity	Weights
First words	互联网 (the internet), 小米 (Xiaomi), 中国(China)	1.2
Sentence end words	思维 (thinking), 全球化 (globalizes the model), 领域 (areas)	1.2

Definition 4: Micro-Blog Word Weights
Micro-blog word weights are the weighting of words in the topic of micro-blog, giving the final value of the weight.

$$\mathrm{W}_{\mathrm{word}} = tf_{i,j} \times idf_i \times W_p \times W_Q \tag{8}$$

In formula (8), W_{word} is obtained. Therefore, $TF - IDF$ filters common words and retains important keyword words.

The above is the process of weighting keywords based on word weights. Thus, the keyword words are obtained through calculation, and the next step is to establish the keyword word set, establish the word set C, and sort the keyword words by the weight value. Although the keyword has been obtained, the keyword is still not very accurate, so the keyword left and right entropy is introduced to analyze the amount of keyword information. The keyword left and right entropy gives a threshold, and then filters to get more accurate keyword words.

Definition 5: Micro-Blog Keywords Left and Right Entropy
Micro-blog keyword left and right entropy is the amount of information extracted from the left and right side of the keyword in the micro-blog topic. The amount of keyword information is determined by the amount of information. Definition $E_{R\mathrm{i}}$ is the right entropy of the keyword, $E_{L\mathrm{i}}$ is the left entropy of the keyword, i is the Nth keyword, $P(\mathrm{i})$ is the probability of occurrence of the i-th word, and the probability of occurrence of the left and right words is $P(L_\mathrm{i})$, $P(R_\mathrm{i})$, respectively. The specific definition is as follows:

$$E_{\mathrm{Ri}} = -\sum_{i=1}^{n} P(R_\mathrm{i}) * \log P(R_\mathrm{i}) \tag{9}$$

$$E_{L\mathrm{i}} = -\sum_{\mathrm{i}=1}^{\mathrm{n}} P(L_\mathrm{i}) * \log P(L_\mathrm{i}) \tag{10}$$

Given the threshold θ, the left threshold is θ_L and the right threshold θ_R. When $E_{R\mathrm{i}}$ and $E_{L\mathrm{i}}$ are both greater than the threshold, the keyword is extracted, otherwise it is deleted.

Calculating the left and right entropy between a pair of words, the greater the entropy, the more a keyword. Because entropy represents uncertainty, the greater the entropy, the greater the uncertainty. This keyword is rich in terms of left and right. The more important this keyword is, the more important the extracted keywords are.

3.3 Knowledge Base Construction of Data

The key words are extracted by the keyword right and left entropy. The left and right entropy of the keyword is judged that the left and right entropy is greater than the given threshold value θ. The keyword words that are larger than the threshold value are extracted, otherwise they are deleted. This method extracts keyword words more precisely. The keyword words obtained have great information value and strong importance, but they do not constitute knowledge links between words. Therefore, the construction of the knowledge base of the keywords by the association rule algorithm makes the knowledge obtained from the keywords.

Definition 6: Micro-Blog Interword Association Semantics
The micro-blog semantic relevance between words is to extract the relation semantics between processed keywords in the micro-blog topic. The association rule algorithm is used to find the connection before the word, and then sort the keywords. The correlation semantics between keywords mainly depends on the values of support and confidence. t_i and t_i denote feature terms, i-th word, and j-th word. k indicates the number of keywords. The degree of support is expressed as a percentage of a text containing both the keywords t_i, t_j, the probability that the keyword t_i, t_j appears

simultaneously in all texts; the degree of confidence is expressed as the percentage of the text containing t_i and the text containing t_i and t_j, where $i \neq j$. If the degree of support is less than or equal to the threshold value θ_s, then there is no need to calculate the confidence level. The confidence threshold is θ_c. Support and Confidence Calculation Formula:

$$P(t_i, t_j) = \frac{Count(t_i, t_j)}{N_x} \tag{11}$$

$$\text{support}(t_i, t_j) = P(t_i, t_j) \tag{12}$$

$$\text{confident}(t_i, t_j) = \frac{P(t_i, t_j)}{P(t_i)} \tag{13}$$

$P(t_i, t_j)$ represents the probability that the feature keyword t_i, t_j appears in all micro-blog texts at the same time, which is equal to the support degree $\text{support}(t_i, t_j)$. $confident(t_i, t_j)$ represents the confidence degree that the word t_i points to the word $t_j(t_i = > t_j)$, and the constraint condition of the formula (13) is $1 \leq i \leq k, 1 \leq j \leq k$, and$i \neq j$. The degree of support is greater than the threshold and the confidence is greater than the threshold, knowledge is extracted, and the knowledge base is constructed.

3.4 Micro-Blog Topic Knowledge Extraction Association Rule Algorithm

This paper studies the strategy of extracting micro-blog topic knowledge, first preprocessing the micro-blog data, rough crawling the micro-blog data source, and ignoring the text (web link, animation, video, picture, removing "@+user name". "#+Topic+#", Traditional Chinese to Simplified, English to Chinese) Delete filter, extract words, and store them in the database. Then using the ICTCLAS provided by the Institute of Computer Science of the Chinese Academy of Sciences to segment the micro-blog text to get word vectors. The TF-IDF algorithm is improved by the part-of-speech weighting W_p and the weighted W_Q at the end of the sentence to obtain a comprehensive TF-IDF. Furthermore, the keyword is extracted by sorting the keyword's comprehensive TF-IDF. On the basis of data extraction, the keywords are filtered by the left and right entropy thresholds, and the inter-word association rules are solved by the confidence and support between words, and the support threshold θ_s and the confidence threshold θ_c are introduced to determine which keywords are used. Knowledge, then knowledge storage, build knowledge base. The specific algorithm is as follows:

Algorithm 1: Micro-blog topic knowledge extraction association statement algorithm

Enter: micro-blog topic text, giving the specified threshold, θ_{Ri}, θ_{Li}, θ_s, θ_c.
Output: Knowledge Base (G_x)
1: The micro-blog text is filtered and segmented;
2: For each word word_i in the text
3: Calculate word frequency tf-idf micro-blog part of speech weight W_p micro-blog first end word weight W_Q;
4: Calculate micro-blog word weight W_{word};
5: End for;
6: Sort the word's micro-blog word weight W_{word};
7: Extract the first n words into the keyword set C;
8: For each keyword in C
9: Calculate right entropy E_{Ri} and left entropy E_{Li};
10: If ($E_{Ri} > \theta_R$)and ($E_{Li} > \theta_L$) extract keywords
11: Else delete;
12: End if;
13: if(support$(t_i, t_j) \le \theta_s$ and confidence$(t_i, t_j) \le \theta_c$) delete keywords;
14: end if;
15: build knowledge base G_x;

Algorithm design description:

(1) The first judgment, step 1–step 5: calculating the word frequency tf of the text word, the frequency of the reverse document idf, the weight of the micro-blog part of speech, W_p, the weight of the first word of the micro-blog W_Q, and then calculating the weight of the micro-blog word W_{word} to find the weight After the words.
(2) The second judgment, that is, step 6–step 12: Sort the weighted words to generate a word set C, and then perform right and left entropy filtering on each keyword, and extract the value greater than the threshold θ, otherwise delete.
(3) The third judgment, that is, step 13–step 14: is used to correlate the feature keywords of the data mining, which is to judge whether the confidence degree confidence(t_i, t_j) is greater than the set threshold θ_c. If yes, extract keywords and delete.

4 Experiments

(a) Evaluation method

In order to verify the micro-blog topic knowledge extraction strategy, this article obtains all micro-blogs related to a topic from Sina micro-blog, and extracts keywords and association rules for a short text using Algorithm 1.

(b) **Evaluation analysis**

(1) The first experiment

The experiment is conducted by crawling hot topic topic IT on micro-blog, and then processing in Chinese, calculating TF-IDF, part-of-speech weight, first-to-last word weight, word weight, and extracting keywords. The noun ∂ is 0.8. The verb β is 1.2. The adjective γ is 0.6. The first and last words λ and σ are both 1.2. The experimental results are shown in the following table. As shown in Table 3:

Table 3. Keywords and data

Key words	TF-IDF	Part-of-speech weight	First and last word weight	Words weight
反击 (counterattack)	0.0401	1.2	1	0.04848
国内手机 (domestic mobile phone)	0.0283	0.8	1	0.02264
价格 (price)	0.0218	0.8	1.2	0.02093
魅族 (meizu)	0.0242	0.8	1	0.01936
小米 (xiaomi)	0.0199	0.8	1.2	0.01910
全面屏 (comprehensive screen)	0.0182	0.8	1	0.01456
三星 (Sansung)	0.0172	0.8	1.2	0.01651
苹果手机 (Apple)	0.0161	0.8	1	0.01288

There are 20 key words extracted from the text, of which 8 are displayed and the remaining 12 are not displayed. such as "电脑" (computer), "软件" (software), "一个" (one), "电视" (television), "互联网" (interconnected), "" (networking), "发布" (release), "乔布斯" (Job), "我们" (we), "应用" (application), "战役 (Campaign)", "价值" (value), "趋势" (trend). Word weights and then rank the word weights are analyzed. First, the Chinese word segmentation is used to extract words. And then TF and IDF of words is calculated, TF-IDF, part-of-speech weights, first-to-last words weights and word weights are also calculated, and then word weights ranking.

(2) The second experiment

Based on the keywords obtained above, the right and left entropy calculations of the keywords and the correlation between words are calculated. Filter and calculate the value of the relationship between words. Given the left and right entropy thresholds, both the left θ_L and right θ_R entropy values are 3.1. The experimental results are shown in the following table. As shown in Table 4:

20 keywords are filtered through the right and left entropy filters, and finally the above 8 keywords are obtained. The above 8 keywords are left and right entropy judgments greater than the given threshold left entropy θ_L and right entropy θ_R, and the result is extraction, otherwise it is deleted.

Table 4. Right and left entropy of keywords

Key words	Left and right entropy
反击	3.8123
国内手机	3.8213
价格	3.6982
魅族	3.5235
小米	3.5214
全面屏	3.4123
三星	3.1231
苹果	3.1012

(3) The third experiment

Extract the association rules for the keywords extracted in experiment 2 and calculate the support and confidence between the two keywords. If the degree of support and confidence is less than the corresponding threshold, it is determined that there is no close relationship between the two keywords, that is, the degree of support and confidence of the keyword are set to zero. The support and confidence thresholds are set at 0.85. Therefore, the keyword association diagram is shown in the figure below. As shown in Fig. 2:

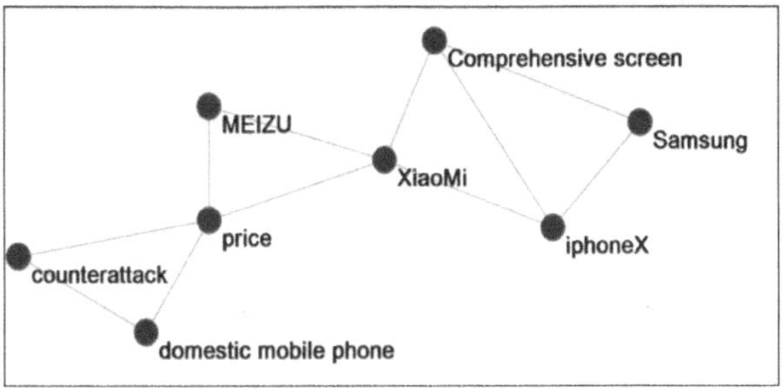

Fig. 2. Key words association diagram.

As can be seen from Fig. 2, the knowledge base is effectively constructed by the association rules between keywords. There are three groups of closely related keywords that describe three pieces of knowledge. For example, domestic mobile phones counterattack with price advantage; domestic millet and Meizu are also playing a price war; Xiaomi, Apple and Samsung also compete for a full-screen market.

(4) The fourth experiment

Traditional algorithms is Compared with the algorithm in this paper according to Accuracy, recall and F value. As shown in Table 5:

Table 5. Accuracy, recall and F value

Algorithm	Accuracy	Recall rate	F value
Algorithm in this paper	0.86	0.88	0.87
Traditional algorithms	0.75	0.66	0.71

As shown in Table 5, The accuracy and recall rate of this algorithm are obviously greater than the accuracy and recall rate of traditional algorithms. This algorithm F value = 0.8699 traditional algorithm F = 0.7051 This algorithm is significantly better than the traditional algorithm. The comparison results are as shown in Fig. 3:

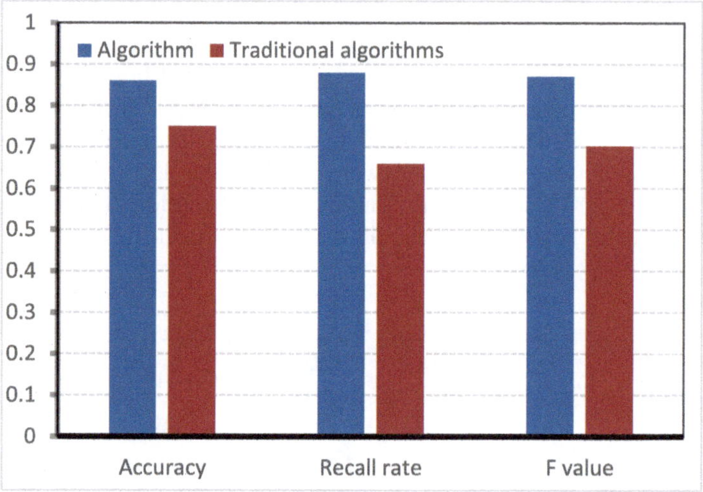

Fig. 3. Comparison between the algorithm of this paper and the traditional algorithm

Analysis: Through experiments, this algorithm is obviously superior to the traditional algorithm. Compared with the knowledge extraction strategy proposed in this paper through the accuracy, recall rate and F value, this paper has obvious advantages. The accuracy rate increased from 0.75 to 0.86, the recall rate increased from 0.66 to 0.88, and the F value increased from 0.71 to 0.87.

5 Summary of This Article

This paper conducts a research on micro-blog, which has been widely used in recent years, and proposes a micro-blog hot topic discovery method based on the characteristics of micro-blog data. This method gives a micro-blog topic knowledge extraction process, and each micro-blog topic is extracted with some keywords to describe. These keywords allow the user to have a comprehensive understanding of the topic. The main tasks of this article are:

(1) In this paper, data mining is used to obtain knowledge through TF-IDF improved algorithm weighting and association rules algorithm. The topical weighting and the weighting of the first and last sentences are better to extract the micro-blog topic knowledge. This method is better than the traditional TF-IDF algorithm. Because of the limitation of the traditional TF-IDF main dictionary, words that do not appear are masked when calculating the IDF. In the keyword calculation stage, no word is assigned to IDF = 1.0, which is obviously smaller than the IDF value of most of the words. Therefore, this text is optimized for this matter.

(2) Constructing a knowledge base is conducive to finding links between words, discovering knowledge, and obtaining value from knowledge. The knowledge of the knowledge base is hierarchical. The lower level is "factual knowledge," and the upper level is used to control "facts." Therefore, constructing a knowledge base is conducive to understanding knowledge. Therefore, this paper studies the topic extraction of microblog topics.

Acknowledgements. This work was supported by 2018 Top talent Cultivation project of Anhui Universities (No. gxbjZD15), in part by the Natural Science Foundation of Anhui Province Universities (No. KJ2015A111), supported in part by Innovative fund for graduate students of Anhui University of Science And Technology in 2017 (No. 2017CX2110), in part by the National Science Foundation of China under Grant 61300202, and in part by the Science Foundation of Shanghai under Grant 13ZR1452900.

References

1. Xu, Z., Mei, L., Raymond, K.K., Lv, C., Hu, C.P., Luo, X.F., Liu, Y.H.: Mobile crowd sensing of human-like intelligence using social sensors: a survey. Neurocomputing **279**, 3–10 (2018)
2. Joachims, T.: A probabilistic analysis of the Rocchio algorithm with TFIDF for text categorization. In: Fourteenth International Conference on Machine Learning, pp. 143–151. Morgan Kaufmann Publishers Inc. (1996)
3. Huang, L., Wu, Y., Zhu, Q.: Research and improvement of automatic keyword extraction method. Comput. Sci. **41**(6), 204–207 (2014)
4. Garcia, A.B.: The use of data mining techniques to discover knowledge from animal and food data: examples related to the cattle industry. Trends Food Sci. Technol. **29**(2), 151–157 (2013)
5. Zhang, S., Zhu, H., Xu, Z.: The extraction method of new logining word/term for social media based on statistics and n-increment. J. Ambient Intell. Humaniz. Comput. **5**, 1–11 (2017)
6. Hong, T.P., Lin, C.W., Yang, K.T., Wang, S.L.: Using TF-IDF to hide sensitive itemsets. Appl. Intell. **38**(4), 502–510 (2013)
7. Telikani, A., Shahbahrami, A.: Optimizing association rule hiding using combination of border and heuristic approaches. Appl. Intell. **47**(2), 544–557 (2017)
8. Rezgui, Y.: Text-based domain ontology building using TF-IDF and metric clusters techniques. Knowl. Eng. Rev. **22**(4), 379–403 (2007)
9. Fang, X., Huang, X., Wang, R.: Identification of hot topics in mobile complaint text based on LDA model. Data Anal. Knowl. Discov. **1**(2), 19–27 (2017)

Personal Relations Extraction from Microblog Text Based on Relations Trigger Words

Qian Chen and Shunxiang Zhang(✉)

School of Computer Science and Engineering,
Anhui University of Science and Technology, Huainan 232001, China
417515111@qq.com, sxzhang@aust.edu.cn

Abstract. Microblog data contains a lot of personal relations (e.g., colleagues, friendships). It is difficult to extract personal relations from the great micro-blog data. In order to solve this problem, this paper proposes a Personal Relations Extraction Model (PREM) to help users effectively extract personal relationships of microblog texts. First, trigger word lexicon is constructed manually. Combining five kinds of personal relations, each trigger word is represented as a trigger word vector. Second, we use Semantic Dependency Parsing to distinguish the feature words between the primary words and the non-primary words in the sentence. The feature vector of a sentence is composed of these feature words. Finally, the personal relation is extracted through set intersections and relational weights. The experimental results show the proposed method has high accuracy.

Keywords: Microblog text · Personal relations · Semantic Dependency Parsing

1 Introduction

More and more people are participating in microblog community. Microblog has a lot of active users making data grow rapidly every minute [1]. How to quickly and accurately extract the personal relations from the great microblog data has become the current research hotspot.

This paper proposes a Personal Relations Extraction Model (PREM), which helps the user to extract the personal relations quickly. This model is divided into two steps. The working framework of the PREM is shown as Fig. 1.

Step1. Construction of a Trigger word lexicon. We manually constructed a relations trigger lexicon. Each trigger word is given two attributes, one is relational classification and the other is relational weighting. The trigger word vector is constructed to represent these attributes.

Step2. Extraction of personal relations. This sentence is first preprocessed to remove meaningless symbols. The syntax tree of the sentence is drawn through the semantic dependence parsing (SDP). Using the syntax tree of the sentence, the words that best express the meaning of the sentence are extracted, and these words are called primary words. The remaining words of the sentence are considered to supplement the expression of the sentence, and these words are called non-primary

J. Abawajy et al. (Eds.): ATCI 2018, AISC 842, pp. 220–229, 2019.
https://doi.org/10.1007/978-3-319-98776-7_24

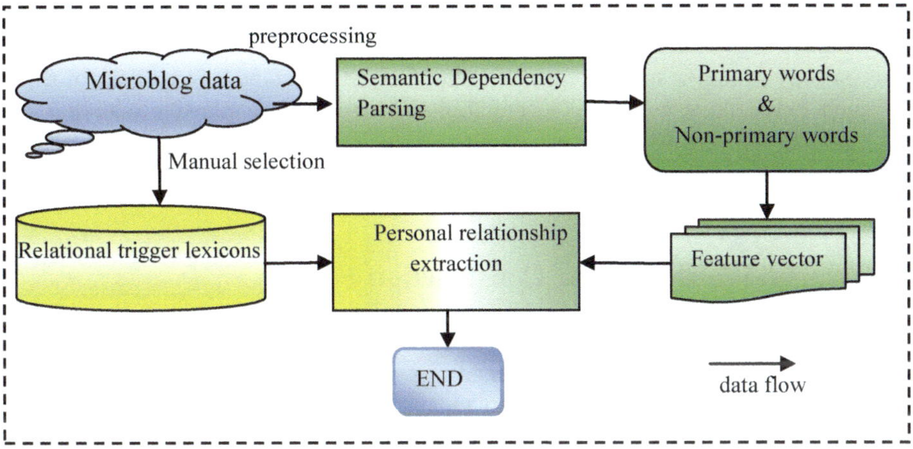

Fig. 1. Working framework of PREM

words. The primary and non-primary words constitute the feature vector of the sentence. Personal relationships are calculated by relational classification calculations and relationship weight calculations.

The rest of the paper is organized as follows: Sect. 2 introduces the related works of entity relationship extraction. Section 3 presents the construction of trigger lexicon. Section 4 presents the extraction of personal relations. Section 5 is the experiment and analysis. Section 6 summarizes this paper.

2 Related Works

Relations extraction refers to the relations between two entities extracted from related corpus after entities are extracted. Entity relations recognition is also called relation extraction. There are mainly three methods for relation extraction.

(1) Artificial construction rules. This method is based on the template that the expert has defined to analyze the statements that need to be processed. Chinatsu Aone uses a lexicon-driven relational extraction method. The method uses corpus data to identify names and noun phrases. The relations recognition module uses a lexicon-driven model to achieve the purpose of relation extraction.
(2) Machine learning methods. This method essentially transforms relational extraction into a classification problem. A classifier is constructed by training a small amount of corpora and then using machine learning. Machine learning methods are divided into three categories: supervised learning, semi-supervised learning, and unsupervised learning. Bootstrap is a kind of semi-supervised method [2]. The Bootstrap method uses a small number of instances as a set of initial seeds, learning with a pattern. New instances are extracted from the unstructured data. The new pattern continuously learns from the new instance and expands the pattern set.

(3) Open information extraction. This method does not require manual tagging of training sets as training corpus, nor is it limited to specific fields. This method is parsed through a small corpus and automatically extracts some relational triples and labels them according to credibility. These triplets train the classifier as a training set. Using a trained classifier to perform a large number of relations extractions, the credibility of each triple is finally calculated [3].

3 Trigger Word Lexicon Construction

This paper uses trigger word matching for personal relations extraction. The trigger word refers to certain words that can clearly represent the personal relations. For example: tom and Jerry weddings held today, the 'wedding' in this sentence is the trigger word for family relations. The trigger word lexicon is a set of trigger words. This paper selects five representative relations from microblogs: family relations, colleague relations, business relations, classmate relations, hostile relations, and artificially screened 8159 words to construct the trigger word lexicon.

3.1 Trigger Word Attributes

Trigger words include two kinds of attributes, one is relational classification, and the other is relational weighting. Relations classification refers to whether the trigger word belongs to this personal relation. Trigger words can only belong to one type of relations, but they can also belong to multiple types of relations.

The relation weight of the trigger word refers to the extent to which this trigger word distinguishes the relations [4]. The higher the degree of distinction, the higher the relations weighting of the word. A trigger word has five relations weights (Tables 1 and 2).

Table 1. Example of classification of trigger words

Personal relation	Trigger words
Family	Weddings, chat, party, sons
Colleague	Work, party, meeting, salary, chat
Business	Managers, investment, chat, meeting, negotiations
Classmates	School, student, teacher, graduation
Hostility	Angry, beat, negotiated

Table 2. Example of relation weights of trigger words

Trigger words	Relation weights
Manager	Family: 0; Colleague: 1; Business: 3; Classmate: 0; Hostile: 0
Chat	Family: 1; Colleague: 1; Business: 1; Classmate: 1; Hostile: 0
Company	Family: 0; Colleague: 2; Business: 2; Classmate: 0; Hostile: 0
Wedding	Family: 1; Colleague: 2; Business: 0; Classmate: 0; Hostile: 0
Retirement	Family: 0; Colleague: 2; Business: 1; Classmate: 0; Hostile: 0

3.2 Construction of Trigger Word Vector

We can use vector to represent the trigger word, which is recorded as the trigger word vector TGV. Each TGV corresponds to one trigger word, and five components in TGV represent five relations. Each component has an integer in the range 0–4. The size of the value indicates the size of the relations weight. The components take zero and non-zero values indicate whether they belong to personal relations of the components. See the formula for a description of TGV. TGV description see formula (1)

$$TGV = \{R_1, R_2, R_3, R_4, R_5\} \quad (1)$$

Where, R indicates the relations weight, R_1 indicates family relations, R_2 indicates colleague relations, R_3 indicates commercial relations, R_4 indicates student relations, and R_5 indicates hostile relations.

4 Extraction of Personal Relations

Microblog data is real-time, with no more than 140 words. It may contain special characters. Microblog users will forward a lot of text. For these features, we need to preprocess the microblog text. The sentence obtained primary words and non-primary words through SDP. The primary words and non-primary words constitute the sentence feature vectors. Relation extraction algorithm to extract the personal relations.

4.1 Pre-treatment

A microblog text has characters and sentences that do not help the extraction of the personal relations. In order to filter these contents, the microblog text needs to be preprocessed [5]. Preprocessing has three steps.

Step1: The sentence with two names in the microblog text is selected.
Step2: Filter the useless information in the sentence. Such as "@ user" format message, "# topic name #" part, emoticons. These have no substantive effect on the extraction of entities.
Step3: Remove web links, animations, videos, pictures. Although this part contains a lot of information, it has not yet been processed by a better method and is therefore removed.

4.2 Construction of a Sentence Feature Vector

A sentence is composed of many words. Words can be divided into nouns, verbs, adjectives, etc. according to part of speech. Words are divided into the subject and the predicate according to the grammar. This paper uses the LTP Chinese natural language processing system of Harbin Institute of Technology to semantically analyze sentences and obtain a syntactic tree. We can extract the primary and non-primary words of sentences from the syntax tree. These personal names, primary words and non-primary words constitute a vector called sentence feature vector (SFV) [6].

Semantic Dependency Parsing (SDP) analyzes the semantic relations between the various linguistic units of a sentence and presents the semantic relations as a dependency structure. It can obtain long-range collocation information that is independent of the physical location of words. We can get a dependency tree through semantic dependency analysis [7]. SDP analysis of sentences is shown in Fig. 2.

The construction of SFV is divided into three steps:

Step 1. Tag the sentence part of speech [8]. The word is marked with the part of speech in the sentence. The person's name is extracted as the first two components.
Step 2. In order, the first word pointed by the root arrow is extracted, then the arrow from the word is extracted from the word on the right side of the sentence, and so on, until the end of the sentence pointed to by the arrow is extracted. This part of the word is called the primary word. According to part of speech, the nouns and verbs of the primary words are preserved. These reserved primary words are placed in SFV in order of extraction. The first word is recorded as the third component of SFV, and the last one is recorded as the K-th component of SFV.
Step 3. From the beginning to the end of the sentence, the remaining nouns and verbs are extracted in sequence. These words are non-primary words. These non-primary words are placed in SFV in the order of extraction. SFV description see formula (2)

$$SFV = \{E_1, E_2, P_1 \cdots P_k, W_1 \cdots W_i\} \tag{2}$$

where, E_1 and E_2 indicate the names of person's name in a sentence. P_1 to P_k presents the primary word of the sentence. W_1 to W_i represent non-primary word of the sentence.

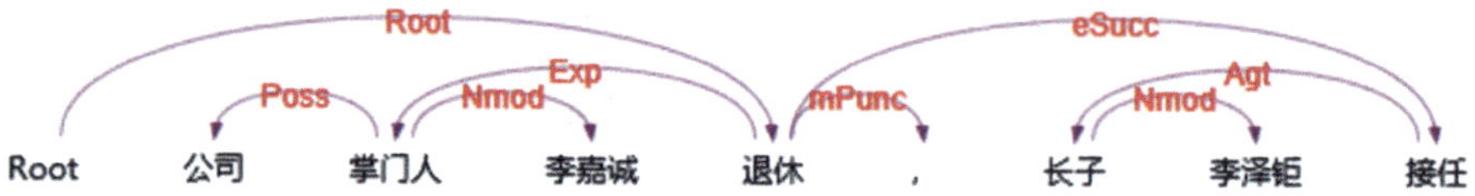

Fig. 2. The Semantic Dependency Parsing result of the sentence

4.3 Personal Relations Extraction Algorithm

This paper constructs the trigger word lexicon and the sentence feature vector. The sentence feature vector is divided into three parts. The first part is the name of the person, the second part is the primary word, and the third part is the non-primary word. The primary word is more representative of the subject of the sentence than the non-primary word. In the algorithm, the primary words are preferentially calculated for their intersection. If there is only one personal relation in the intersection, it is extracted. If there is more than one result in the intersection, the relation weights of the non-primary words are calculated, and the unique relations are extracted in the plurality of relations in the intersection.

Based on the above detailed discussion, an algorithm can be designed to extract the personal relations of a microblog statement. The relations extraction algorithm is as follows:

Algorithm: Personal Relations Extraction Algorithm

```
Input: A sentence containing two personal relations, Trigger word lexicon
Output: Name, Personal relations
1: SFV is constructed through SDP
2: Extract E1, E2 as Name 1, Name 2
3 : for (the number of primary word P)
4 :        for(the number of TGV)
5 :            if (P==TGV)
6:                 These TGVs form a set called PV
7:             End if
8:         End for
9:  End for
10: for(the number of TGVs in PV)
11:      A new vector which is the transpose of the previous vector multiplied by a subsequent vector called RS
12  End for
13 : if(Number of non-zero components in RS ==1)
14 :       Output non-zero component personal relations
15 : End if
16 : if(Number of non-zero components in RS!=1)
17 :       Get these non-zero component relations CR
18 : End if
19 : for (Number of P and W in SFV)
20 :       for(the number of TGV)
21 :           if (P and W==TGV)
22 :                Sum of relations weights for these TGV
23 :           End if
24 :       End for
25 : End for
26 : Output the personal relations corresponding to the component with the largest relation weight in the CR
```

Algorithm description:

(1) Step1–step2: The sentence obtained the name of the person, the primary word, and the non-primary word through the SDP. These words constitute the SFV.
(2) Step3–step9: A double FOR loop. The primary words in the SFV match the vectors of the trigger word dictionary. The IF statement is that if the subject is a trigger, the trigger is put in a set named PV.
(3) Step10–step12: This is a for loop. Calculating the intersection of TGV relationship classification in PV.
(4) Step13–step15: This is an IF statement. If the non-zero component in the RS is 1, the relationship is classified as 1. The person relations corresponding to this component are output.
(5) Step16–step18: This is an IF statement. If there are multiple character relationships in the RS, put these character relationships into the CR.
(6) Step19–step25: This is a double FOR loop. There is an IF statement in the loop. These statements determine whether the primary words and non-primary words in SFV are trigger words. The sum of the relation weights of the trigger words is calculated.
(7) Step26: The relation weights of those personal relationships existing in the CR are compared. Personal relation with the greatest relationship weight is output.

5 Experiments

5.1 Experimental Methods

The experimental data is 9867 microblog sentences crawled from Sina Weibo. This paper will verify the accuracy of the model from two perspectives. We take the judgment of the experimenter as the correct standard.

(a) Ordinary people's relationship extraction

1,200 microblog sentences of ordinary people were randomly selected. The PREM and experimenters judged these statements separately.

(b) Public figures' relationship extraction

900 microblog sentences of public figures were randomly selected. The PREM and experimenters judged these statements separately.

The environment for this experiment is win10, and Python2.7.

5.2 Experimental Procedure

The evaluation index adopts accuracy rate P, recall rate R, and F value. The evaluation index adopts accuracy rate, recall rate R and F value. Accuracy and recall are two metrics that are widely used in information retrieval and statistical classification to evaluate the quality of results. The accuracy rate is the ratio of the number of relevant documents retrieved to the total number of retrieved documents, and the precision of

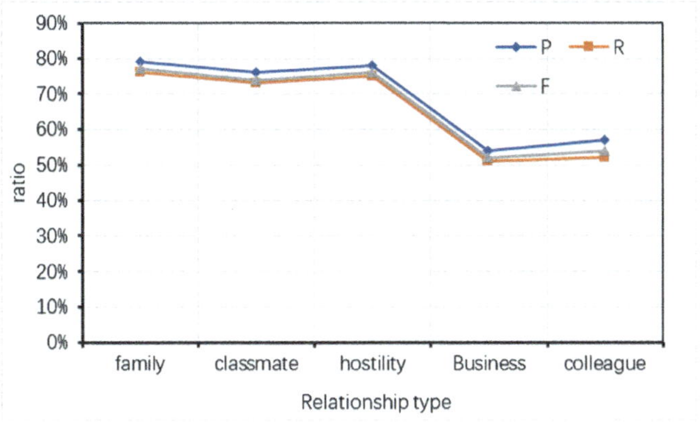

Fig. 3. Extraction of ordinary people relations

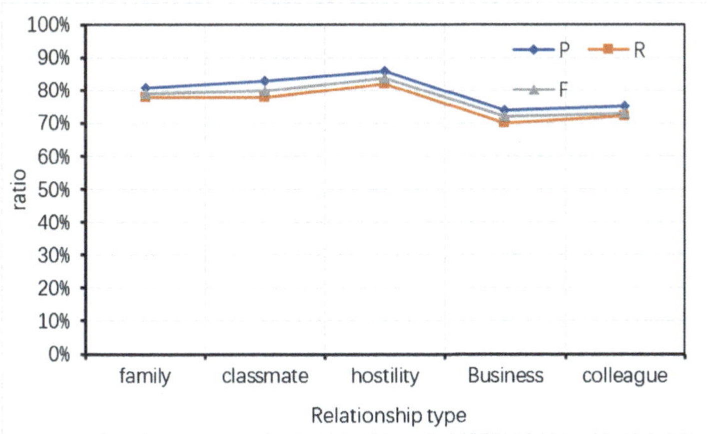

Fig. 4. Extraction of public figures relations

the retrieval system is measured. The recall rate refers to the ratio of the number of relevant documents retrieved and the number of related documents in the document library. The recall rate of the retrieval system is measured. The F-value is the weighted harmonic average of the recall rate and accuracy. See formulas 3, 4 and 5.

$$\mathrm{P} = \frac{C}{T} \times 100\% \tag{3}$$

$$\mathrm{R} = \frac{C}{S} \times 100\% \tag{4}$$

$$\mathrm{F} = \frac{P \times R \times 2}{P + R} \times 100\% \tag{5}$$

Where C is the number of instances of a class that were correctly extracted; T is the number of instances actually extracted; S is the number of related instances in the experimental data set.

(1) The first set of experiments: we select 1,200 microblog sentences about Ordinary people for processing. Finally, the results of the experimenter were compared with the results of PREM. It can be seen from the figure that this method has a higher correct rate for family relations, classmates relations and hostile relations. The result is shown in Fig. 3.
(2) The second set of experiments: we select 1200 microblog sentences about public figures for processing. Finally, the results of the experimenter were compared with the results of PREM. It can be seen from the figure that this method has a better effect on the extraction of public figures. The result is shown in Fig. 4.

5.3 Experiment Analysis

Two sets of experiments show that PREM of this paper works well, and it is only relatively poor for business relations and colleague relations. After analysis, this is because the selection of trigger words that describe business relations and colleague relations is difficult, and the selection of relational trigger words are very important for the effectiveness of relation extraction. The second set of experiments proves that the method of this paper has a very significant effect on the extraction of public figures. After analysis, the microblog texts of public figures are standardized and the words are relatively fixed.

6 Conclusion

Microblog text has features such as short text and non-standard grammar. This paper proposes a personal relations extraction model. The experiment can prove that the method of PREM effectively extract the personal relations. The effect of the extraction of public figures is very significant especially. Future work, Because the trigger words lexicon a greater impact on the results, we will improve the method of constructing the trigger word lexicon.

Acknowledgements. This work was supported by 2018 Top talent Cultivation project of Anhui Universities (No. gxbjZD15), in part by the Natural Science Foundation of Anhui Province Universities (No. KJ2015A111), supported in part by Innovative fund for graduate students of Anhui University of Science And Technology in 2017 (No. 2017CX2108), in part by the National Science Foundation of China under Grant 61300202, and in part by the Science Foundation of Shanghai under Grant 13ZR1452900.

References

1. Zhang, S.X., Wei, Z.L., Wang, Y., Liao, T.: Sentiment analysis of Chinese micro-blog text based on extended sentiment dictionary. Future Gener. Comput. Syst. **81**, 395–403 (2018)
2. Wang, F., Guo, Q., Lei, J., Zhang, J.: Convolutional recurrent neural networks with hidden Markov model bootstrap for scene text recognition. Comput. Vis. **11**(6), 497–504 (2017)
3. Mausam, M.S., et al.: Open language learning for information extraction. In: Proceedings of Conference on Empirical Methods in Natural Language Processing and Computational Natural Language Learning (EMNLP-CONLL), pp. 523–534. ACM, New York (2012)
4. Kim, H.K., Kim, M.: Model-induced term-weighting schemes for text classification. Appl. Intell. **45**, 30–43 (2016)
5. Uysal, A.K., Gunal, S.: The impact of preprocessing on text classification. Inf. Process. Manage. **50**, 104–112 (2014)
6. Zhang, H., Zhong, G.Q.: Improving short text classification by learning vector representations of both words and hidden topics. Knowl.-Based Syst. **102**, 76–86 (2016)
7. Hao, T.Y., Xie, W.X., Wu, Q.Y., et al.: Leveraging question target word features through semantic relation expansion for answer type classification. Knowl.-Based Syst. **133**, 43–52 (2017)
8. Sun, W.W., Wan, X.J.: Towards accurate and efficient Chinese part-of-speech tagging. Comput. Linguist. **42**(3), 391–419 (2016)

Recommendation Algorithm Based on Latent Semantic of Tags

Kelei Sun(✉), Huali Shen, Jiaming Yu, and Shunxiang Zhang

School of Computer Science and Engineering,
Anhui University of Science and Technology, Huainan 200031, China
{klsun, sxzhang}@aust.edu.cn, 759896846@qq.com,
996205763@qq.com

Abstract. Aiming at the problem that traditional based on user ratings collaborative filtering recommendation algorithm only considering the user's history rating information, does not Preprocessing user-item rating matrix, leads to the sparse serious of the rating matrix. To solve this problem, puts forward an based on latent semantic tags recommendation algorithm. Firstly, the label attributes of user's directly or indirectly are used to calculate the similarity between users. Secondly, use the user's first M nearest neighbors to predicted ratings of items which were not rated and user's predicted ratings is filled into the user-item rating matrix. Finally, combined with collaborative filtering algorithm to make the top-N interested items are recommended to the target user. The experimental results on MovieLens dataset show that compared with the traditional recommendation algorithm based on user ratings, our algorithm not only solves the sparse problem of the rating matrix effectively, but also improves the precision of recommendation.

Keywords: Data sparse · Latent semantic tags · Collaborative filtering
Label attributes · Precision

1 Introduction

Because of the rapid development of information technology online networks application software such as Google, Tencent, Baidu and Facebook have closely related to our everyday life. These applications produce massive information every day lead to explosive growth of data in the network, which will make users abundant waste of time to search own self required information. In order to solve the problem of information overload effectively, construct a reasonable recommendation algorithm model has attracted much attention. At present, there are a lot of recommended algorithm models: based on association rules [1–3], content-based [4, 5], collaborative filtering (CF) hybrid recommendation and social network-based recommender systems [6–8].

Collaborative filtering techniques are used most commonly in several common technologies. The CF mainly consists of two parts, data based collaborative filtering and model based collaborative filtering. At present, there are many ways to improve have been proved its superiority by many literature. Literature [1] use a clustering method for choosing neighbors based on the item attribute. This clustering method is

J. Abawajy et al. (Eds.): ATCI 2018, AISC 842, pp. 230–238, 2019.
https://doi.org/10.1007/978-3-319-98776-7_25

used to fill the rating matrix, which can effectively reduce the sparse of the input matrix. Literature [2] the corresponding algorithm is used to calculate the relationship value between the user and the movie, according to the relationship between the two values, generated Top-N recommender. Literature [3] k-means algorithm is used to cluster the item. Then the user pre-rating is predicted on the basis of similar projects and According to the have been filled user-item rating matrix to calculate the similarity between any two users. Literature [4] found out the tag set between users through the historical behavior of users, and use the semantic distance of tag set to measure the similarity between them to achieve recommendation. Literature [5] Ba et al. proposed a new method combining the clustering algorithm with SVD algorithm. And users are gathered into clusters according to attributes, e.g., gender, age, and occupation. Then After getting the populated user-item rating matrix and recombined into a new rating matrix to calculate the similarity between any two users. They considered using item attributes or user attributes, resulting in a similarity measure better than that proposed in other methods. Due to the filling of the input matrix, the pre-rating is more precision.

The methods proposed a recommendation method by calculating the similarity of the user characteristic or item characteristic above. The methods proposed a recommendation method by calculating the similarity of the user characteristic or item characteristic above. Nevertheless, the user's clustering center or item's clustering center is not easy to determine. In this paper, we proposed a personalized recommendation algorithm based on Latent Semantic of Tags. Similar users are determined by comparing the similarity of each attribute between users. Because the smaller the user's property distance is, the more similar the user's interests and hobbies are. Then we proposed a matrix filling means. We fill in the rating matrix with user's predicted ratings by user's age, sex, occupation and Association attributes between users and items. Associated attribute include the movie type that the user likes and the number of movie that the user has evaluated. We use the filled matrix and combined with collaborative filtering algorithm to make the top-N interested items are recommended to the target user.

2 Recommendation Algorithm Based on Latent Semantic of Tags

According to the analysis in the previous section, there are mainly two problems in collaborative filtering algorithm based on user ratings: Scoring matrix sparseness and low precision problems. To solve this problem, we proposed a Recommendation Algorithm Based on Latent Semantic of Tags. The algorithm is mainly divided into two parts: The first part is to construct user-item rating matrix, user similarity matrix and filled user-item rating matrix by using user's historical ratings information and user's tag information. The second part based on the filled rating matrix, and combined with the User-Based CF algorithm to produce Top-N recommendation for the target users.

2.1 Filled Sparse Rating Matrix

Based on the user association information, the commonality between users can be tapped, so as to determine the similar neighboring users. Because some of the user's own inherent attributes are permanent, such as geographical location, occupation, age, sex, etc. These attribute features have no special meaning to themselves, but they can be used to reflect a group's interest preferences if they are associated with some scenarios. In addition to the user's own inherent attributes can produce similarity between users, there are some user-based non-inherent properties can also dig out the association between users. For example, this paper uses the movie type that the user likes and the number of movie that the user has evaluated as the factors in calculating the similarity between users. Next, use the correlation information between users to calculate the similarity between users, construct user similarity matrix. Then get the user's pre-scoring. Finally fill pre-scoring into the sparse rating matrix.

In this paper, the calculated similarity between users is divided into three parts: calculating the similarity between user's inherent attributes, calculating the similarity between user's non-inherent attributes and calculating the total similarity of users.

2.1.1 Calculating the Similarity Between User's Inherent Attributes

The types of values between the inherent attributes ($\boldsymbol{A_k}$) of the user may be different, which can be divided into numerical attributes, two element attributes and classified attributes.

(1) when $\boldsymbol{A_k}$ is a two element attributes

$$u\_sim(i,j) = \begin{cases} 0, & A_{ik} \neq A_{jk} \\ 1, & A_{ik} = A_{jk} \end{cases} \tag{1}$$

where $\boldsymbol{A_k}$ represents the value of the ***k*-th** attribute of user ***i***, $\boldsymbol{A_k}$ represents the value of the ***k*-th** attribute of user ***j***.

(2) when $\boldsymbol{A_k}$ is numerical attributes

$$u\_sim(i,j) = \begin{cases} 1, & |A_{ik} - A_{jk}| \leq 5 \\ \frac{5}{A_{ik} - A_{jk}}, & |A_{ik} - A_{jk}| > 5 \end{cases} \tag{2}$$

After many experiments, the user's age difference within the age of 5 defines a similarity of 1 is more appropriate.

(3) when $\boldsymbol{A_k}$ is classified attributes

In order to obtain the degree of relevance between the classification attributes, this paper builds the classification attribute hierarchy tree according to the classification dictionary, the structure of the tree is as follows (Fig. 1):

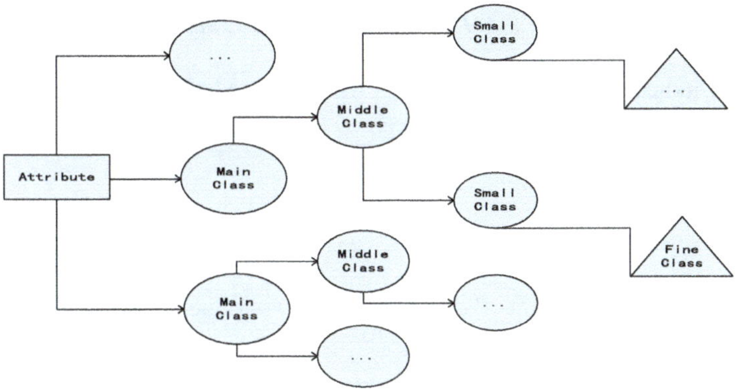

Fig. 1. Classification attribute hierarchy tree

The similarity between user classification attributes is calculated as follows:

$$u\_sim(i,j) = \frac{depth(father(A_{ik}, A_{jk}))}{depth(T)} \tag{3}$$

where father($\boldsymbol{A_{ik}}$, $\boldsymbol{A_{jk}}$) represents two users $\boldsymbol{A_{ik}}$, $\boldsymbol{A_{jk}}$ attribute corresponds to the common parent node of the leaf node in the tree, depth (father($\boldsymbol{A_{ik}}$, $\boldsymbol{A_{jk}}$)) represents the depth of the node, depth($\boldsymbol{T}$) represents the depth of the tree.

2.1.2 Calculating the Similarity Between User's Non-inherent Attributes

According to the data set selected in this paper, we choose the movie type that the user likes and the number of movie that the user has evaluated is the user's non-inherent attributes, and there are 19 kinds of movie types. The user's non-inherent attributes above can be divided into two types: collection attributes and value attributes. The similarity of each attribute is calculated as follows:

(1) when A_k is collection attributes

$$Definition: \boldsymbol{A}_{ik} = \{\boldsymbol{MT}_{i1}, \boldsymbol{MT}_{i2}, \ldots, \boldsymbol{MT}_{iw}, \ldots, \boldsymbol{MT}_{i19}\}$$
$$\boldsymbol{A}_{jk} = \{\boldsymbol{MT}_{j1}, \boldsymbol{MT}_{j2}, \ldots, \boldsymbol{MT}_{jw}, \ldots, \boldsymbol{MT}_{j19}\}$$

where $\boldsymbol{MT_{iw}}$ represents whether user $\boldsymbol{i}$ likes the $\boldsymbol{w}$**-th** movie type, $\boldsymbol{MT_{jw}}$ represents whether user $\boldsymbol{j}$ likes the $\boldsymbol{w}$**-th** movie type, $\boldsymbol{MT_{iw}}$ and $\boldsymbol{MT_{jw}}$ the value is 0 or 1, the value 1 means like, and the value 0 means dislike. The similarity is calculated as follows:

$$u\_sim(i,j) = \frac{sum(F(A_{ik} \cap A_{jk}))}{sum(F(A_{ik} \cup A_{jk}))} \tag{4}$$

$$F(A_{ik} \cap A_{jk}) = (MT_{i1} \cap MT_{j1}, MT_{i2} \cap MT_{j2}, \ldots, MT_{iw} \cap MT_{jw}, \ldots, MT_{i19} \cap MT_{j19}) \tag{5}$$

$$F(A_{ik} \cup A_{jk}) = (MT_{i1} \cup MT_{j1}, MT_{i2} \cup MT_{j2}, \ldots, MT_{iw} \cup MT_{jw}, \ldots, MT_{i19} \cup MT_{j19}) \quad (6)$$

sum **(*F*)** indicates that the elements in the set are added.

(2) when $\boldsymbol{A_k}$ is numerical attributes

$$u\_sim(i,j) = \frac{\min[A_{ik}, A_{jk}]}{\max[A_{ik}, A_{jk}]} \quad (7)$$

where min $[A_{ik}, A_{jk}]$ represents the minimum of two values, max $[A_{ik}, A_{jk}]$ represents the maximum of the two values.

2.1.3 Calculating the Total Similarity of Users

Before calculating the total similarity of users, it is necessary to determine the weight $\boldsymbol{w}$ $(\boldsymbol{A_k})$ of each attribute of the user in the similarity calculation. After determining the weight of each attribute, the similarity between users is calculated as follows [6]:

$$s(i,j) = \sum_{k}^{n} w(A_k) \times u\_sim(A_{ik}, A_{jk})$$

$$\sum_{k}^{n} w(A_k) = 1 \quad (8)$$

where $\boldsymbol{n}$ represents the number of user attributes, $\boldsymbol{a\_sim(A_{ik}, A_{jk})}$ represents the similarity of user $\boldsymbol{i}$ and user $\boldsymbol{j}$ in the $\boldsymbol{k}$**-th** attribute.

According to the above formula (8), the similarity between any two users can be calculated, so that the similarity matrix $\boldsymbol{S_{m,m}}$ of the user can be constructed, then according to $\boldsymbol{S_{m,m}}$, the unobserved items of the target users are pre-ratingd, and the user pre-scoring is filled into the sparse scoring matrix $\boldsymbol{R_{m,n}}$. User pre-scoring method is as follows:

$$p_{ai} = \overline{r_j} + \frac{\sum_{j \in NBS\_a \cap U_a} s(a,j) * (r_{ji} - \overline{r_j})}{\sum_{j \in NBS\_a \cap U_a} |s(a,j)|} \quad (9)$$

where $\overline{r_j}$ represents the average value of user $\boldsymbol{j}$, $\boldsymbol{s(a,j)}$ represents the similarity between user $\boldsymbol{a}$ and user $\boldsymbol{j}$, $\boldsymbol{r_{ji}}$ represents user $\boldsymbol{j}$ to rating item $\boldsymbol{i}$, $\boldsymbol{NBS\_a \cap Ua}$ is a collection of users who the first $\boldsymbol{k}$ similar users and user $\boldsymbol{a}$ have evaluated the common item.

2.2 Generate Top-N Recommendation List

From Sect. 2.1, the filled rating matrix($\boldsymbol{Fill\_R_{m,n}}$) can be obtained. On the basis, combined with the User-Based CF algorithm to make the top-N interested items are recommended to the target user. The implementation process can be divided into three steps: constructing a user similarity matrix, calculating the final pre-rating of the target user, and generating a Top-N recommendation list.

(1) Constructing a user similarity matrix

According to $Fill\_R_{m,n}$, the similarity between users can be obtained through the similarity calculation formula so as to construct the user similarity matrix ($S_{m,m}$). The similarity between users is calculated as follows [4]:

$$sim(i,j) = \frac{\sum_{u \in U_{ij}} (r_{iu} - \overline{r_i}) * (r_{ju} - \overline{r_j})}{\sqrt{\sum_{u \in U_{ij}} (r_{iu} - \overline{r_i})^2} * \sqrt{\sum_{u \in U_{ij}} (r_{ju} - \overline{r_j})^2}} \quad (10)$$

where r_{iu} represents user i for rating on item u, r_{ju} represents user j for rating on item u, $\overline{r_i}$ is the verage rating of user i, $\overline{r_{ij}}$ is the average rating of user j, U_{ij} is the set of items ratingd by user i and j together.

(2) Calculating the final pre-rating of the target user

Using the historical rating data of the k nearest neighbors of the target user to predict the pre-rating of the non-graded item. User pre-scoring formula is as follows [7]:

$$P_{ai} = \overline{r_j} + \frac{\sum_{j \in NBS\_a \cap U_a} \mathrm{sim}(a,j) * (r_{ji} - \overline{r_j})}{\sum_{j \in NBS\_a \cap U_a} |\mathrm{sim}(a,j)|} \quad (11)$$

where $\overline{r_j}$ is the average rating of user j, $sim(a,j)$ represents the similarity between user a and user j, r_{ji} represents user j for rating on item i, NBS_a represents the first k user sets similar to user a, U_a represents the item evaluated by user a, $NBS\_a \cap Ua$ is a collection of users who the first k similar users and user a have evaluated the common item.

(3) Generating a Top-N recommendation list

Based on the pre-rating of the target user, the first N items with the highest pre-rating are recommended.

3 Experimental Results and Analysis

3.1 Experimental Data Sets

This experiment uses MovieLens data set in this paper. The dataset includes item information table, user item rating information table, user information table, movie type information table, user occupation information table and the like.

In order to ensure experimental rationality, a total of 100000 user item scoring data were selected, and as a test set by a user who selected not less than 10 items of common items and each item of no less than 10 items. Eighty-two divides 100000 data, 80% is used as training data, 20% is used as test data, the number of user comments in training data is 80,000, and the number of user comments in test data is 20,000.

In order to verify the effectiveness of the improved algorithm, this paper choose the mean absolute error (MAE) as the measure of the result. MAE refers to the error value of the user's pre-scoring data calculated by algorithm and the user's actual scoring data obtained The smaller the error value, the closer the algorithm pre-scoring is to the real rating of the user, which in turn makes the recommendation more in line with the user's interest and higher the precision rate. MAE calculation formula is as follows [8]

$$MAE = \frac{\sum_{t}^{n} |P_t - R_t|}{n} \tag{12}$$

where $\boldsymbol{n}$ is the number of items evaluated by the target user in the test set, $\boldsymbol{P_t}$ is the pre-rating of the ***t*-th** item by the target user, $\boldsymbol{R_t}$ is the real rating of the ***t*-th** item by the target user.

3.2 Experimental Results and Analysis

This paper chooses the user-based collaborative filtering algorithm (User-Based CF), Collaborative Filtering Recommendation Algorithm Incorporating Social Network Information (ISNI) as a comparison experiment with the algorithm proposed in this paper (LST). There are two key parameters of the improved algorithm proposed in this paper: the nearest neighbor number $\boldsymbol{M}$ when first compute user pre-rating and user each attribute weight $\boldsymbol{w_i}$. The experimental results show that the value of $\boldsymbol{M}$ and the weight value of the attribute label have a great impact on the precision of the proposed algorithm. Therefore, before carrying out experiments need to determine the value of the corresponding parameters in the experiment. After repeated experiments, when the neighbor number $\boldsymbol{M} = 30$ and the corresponding age and sex, occupation, users love movie type and evaluate the number of films of the five kinds of attributes weights for $\boldsymbol{w1} = 0.1$, $\boldsymbol{w2} = 0.2$, $\boldsymbol{w3} = 0.3$, $\boldsymbol{w4} = 0.3$, $\boldsymbol{w5} = 0.1$, the whole system performance is better.

In the experiment, the prediction user's precision is different when the number of the nearest neighbor number $\boldsymbol{M}$ is calculated for the first time in the user's pre-rating. When making Top-N recommendation, make the nearest neighbor number k of the

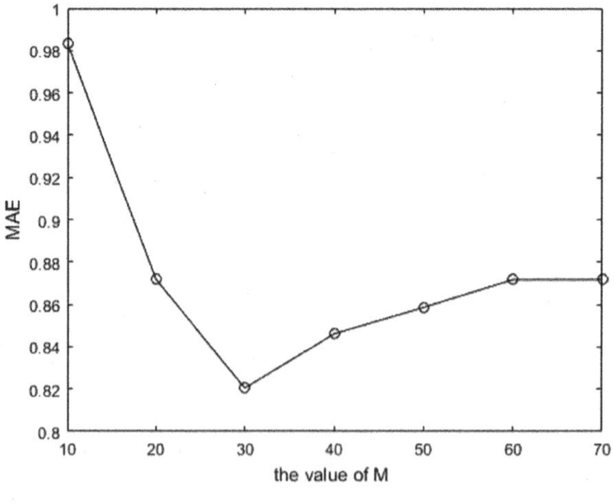

Fig. 2. MAE value with M take different values change

target user in User Based CF algorithm fixed, and the user's each attribute weights are $w1 = 0.1$, $w2 = 0.2$, $w3 = 0.3$, $w4 = 0.3$, $w5 = 0.1$, the changed process of the value of MAE with the value of M is shown in Fig. 2.

From Fig. 2, MAE decreases continuously as the value of M increases from the beginning, indicating that the algorithm pre-scoring of the target user gradually approaches the true rating of the target user. When $M = 30$, MAE obtains the minimum, and the algorithm pre-grading is the most accurate. As the value of M further increases, the value of MAE gradually increases, indicating that the pre-rating calculated by the algorithm is deviating from the actual rating of the user, and the performance of the algorithm begins to deteriorate.

When $M = 30$, the weight of user's label attribute is $w1 = 0.1$, $w2 = 0.2$, $w3 = 0.3$, $w4 = 0.3$, $w5 = 0.1$, use average absolute error of the recommended result is taken as the evaluation standard. The experimental results of the algorithm in 3 are shown in Fig. 3 below.

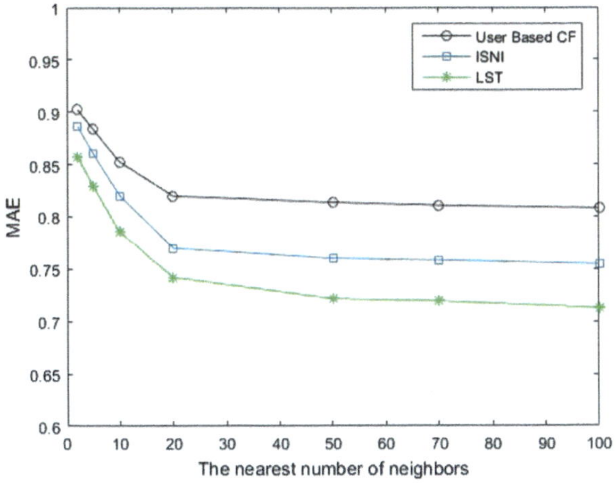

Fig. 3. Comparison of the MAE values of 3 algorithms

From Fig. 3, can see that with the gradual increase of the nearest neighbor number k of User-Based CF algorithm, the MAE value of algorithm in 3 is gradually decreasing, and the magnitude of reduction is smaller. By comparison, the improved algorithm (LST) proposed in this paper is lower than the User-Based CF and ISNI algorithm under the same k value.

4 Conclusions

In this paper, we have presented a Recommendation Algorithm Based on Latent Semantic of Tags (LST). Since recommender systems often depend on users history ratings to calculate their pre-rating. There is no reasonable use of the user's related tags information. We believe that these tags information related to the user is of key importance in the Recommendation Algorithm field. This paper algorithm (LST) can effectively use the semantic of label implicitness to calculate the similarity between two users. Pre-scoring is calculated based on the top M nearest neighbors of the target user, and the pre-ratings are filled in the sparse scoring matrix. On this basis, the final pre-scoring is obtained through the collaborative filtering algorithm, and the Top-N recommendation is generated according to the user's pre-scoring. The user rating in the range of 1–5 points. This paper validates the algorithm in the MovieLens data set, and the experimental results show that the algorithm (LST) can effectively reduce the MAE value.

Acknowledgements. This Research work was supported in part by the National Science Foundation of China under (Grant No. 51504010), in part by the Natural Science Foundation of Anhui Province under (Grant No. 1408085QE94) and in part by the Major Teaching Reform Research Project of Anhui University of Science and Technology under (Name. Research on Reform of Examinations Database of Computer in High School).

References

1. Xu, Z., Mei, L., Raymond, K.K., Lv, C., Hu, C.P., Luo, X.F., Liu, Y.H.: Mobile crowd sensing of human-like intelligence using social sensors: a survey. Neurocomputing **279**, 3–10 (2018)
2. Zhang, S.X., Zhang, S.Y., Yen, N.Y., Zhu, G.L.: The recommendation system of micro-blog topic based on user clustering. MONET **22**(2), 228–239 (2017)
3. Sen, S., Vig, J., Riedl, J.: Tagommenders: Connecting Users to Items Through Tags, pp. 671–680. ACM Press, New York (2009)
4. Zhao, S., Du, N., Nauerz, A., et al.: Improved Recommendation Based on Collaborative Tagging Behaviors, pp. 413–416. ACM Press, New York (2008)
5. Ba, Q., Li, X., Bai, Z.: Clustering collaborative filtering recommendation system based on SVD algorithm. In: IEEE International Conference Software Engineering and Service Science, pp. 963–967 (2013)
6. Meng, X.W., Liu, S.D., Zhang, Y.J., et al.: Research on social recommender systems. J. Softw. **26**(6), 1356–1372 (2015)
7. Guo, L.J., Liang, J.Y., Zhao, X.W.: Collaborative filtering recommendation algorithm incorporating social network information. Pattern Recog. Artif. Intell. **29**(03), 281–288 (2016)
8. Sarwar, B., Karypis, G., Konstan, J., et al.: Item-Based Collaborative Filtering Recommendation Algorithms, pp. 285–295. ACM Press, New York (2001)

Ontology-Supported Case-Based Reasoning Approach for Double Scenario Model Construction in International Disaster Medical Relief Action

Min Zhu[1,2], Ruxue Chen[1], Shi Chen[1], Shaobo Zhong[2], Tianye Lin[2], and Quanyi Huang[2(✉)]

[1] Navy General Hospital, PLA, Beijing, China
zhuml4@tsinghua.mail.edu.cn
[2] Institute of Public Safety Research (IPSR), Tsinghua University, Beijing, China
qyhuang@tsinghua.edu.cn

Abstract. Man-made and natural disasters have affected people worldwide. This paper presents an ontology-supported case-based reasoning (OS-CBR) approach, to support a double scenario model recommending possible medical needs for international medical relief action. Thus, the action's decision makers could effectively respond to disasters concerning not only the disaster scenario but also the regional scenario which related to the local health and medical service condition. The advantages of the OS-CBR approach is that it builds a case retrieving process, which provides a more convenient system for decision support based on knowledge from, and solutions provided for past events. The OS-CBR approach reasoning includes a set of algorithms that have been successfully implemented in four components: retrieval; adaption; revision and preservation; and knowledge presentation. A case studies validated the OS-CBR approach and demonstrate its usage.

Keywords: Disaster relief · Ontology · Case-based reasoning
Double scenario model

1 Introduction

Over the past years, man-made and natural disasters have affected large numbers of people worldwide. In the disaster, local health services may be disrupted and require international help, especially in some low and middle income countries [1].

But in recent years, foreign medical rescue teams deployed to sudden-onset disasters (SOD) have been criticized for over focusing on trauma care and for being ill-adapted to dominating health needs of the affected region [1]. In some experience

R. Chen and S. Chen—Co-authors

J. Abawajy et al. (Eds.): ATCI 2018, AISC 842, pp. 239–250, 2019.
https://doi.org/10.1007/978-3-319-98776-7_26

summary articles, the influences of the local health service coverage, and the local people health condition are instantly mentioned [2, 3].

The basis of international action is making the medical preparedness plan, which is based on determining the likely medical needs during the action [1, 4]. So we promoted a conceptual double scenario model for predicting medical service needs in the international disaster relief action and tried to make it to be a part of decision support system (DSS) to support international medical relief decision makers to effectively respond to disasters especially the SOD.

2 Previous Study Basis and the Main Purpose of This Paper

In previous studies, we promoted the conception of the model (Fig. 1) and constructed an ontology-based knowledge base to structure, store and inquire related international medical relief cases (Fig. 2).

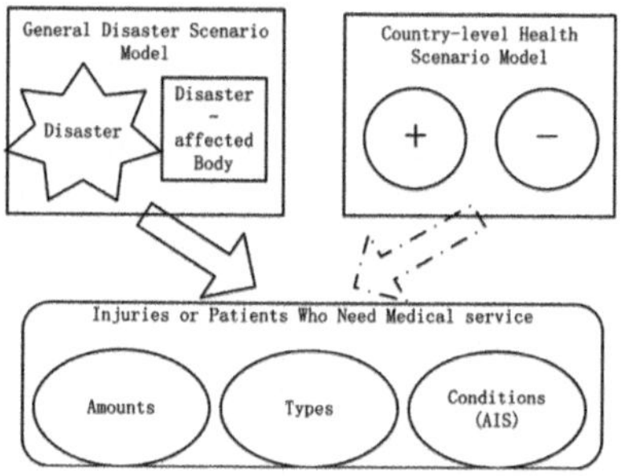

Fig. 1. Abridged general view of the double-level scenario model

The conception model has two different levels of ambition: a disaster scenario, in which mortality, casualty figure, composition of injuries are constrained by the disaster features, and a country scenario, in which the medical needs are related to the local health and service conditions.

We used an ontology-based approach to construct the model knowledge base. As a vocabulary, ontology enables us to define case structure. As a terminology, ontology allows us to define the query vocabulary.

This study aims to enable the model find similar cases and recommend proper medical needs. We use Case Based Research (CBR) as main method.

3 The OS-CBR Approach and Related Algorithms

The OS-CBR approach has 5 steps with the following components:

(1) an information semantic extraction algorithm for knowledge acquisition,
(2) a set of ontologies,

(3) a knowledge base with related rules and cases, and
(4) a CBR component that includes an ontology case retrieval algorithm (Algorithm 1), case adaptation algorithm (Algorithm 2), case revision and retain, and
(5) knowledge presentation.

We have constructed a knowledge base using an ontology-based approach in previous study. This paper will focus on step 4 and 5. (If you want to know more about details about step 1–3, please refer to our previous paper "An Ontology-based Approach to Double Scenario Model Knowledge Base Development for International Disaster Relief (BDKE)")

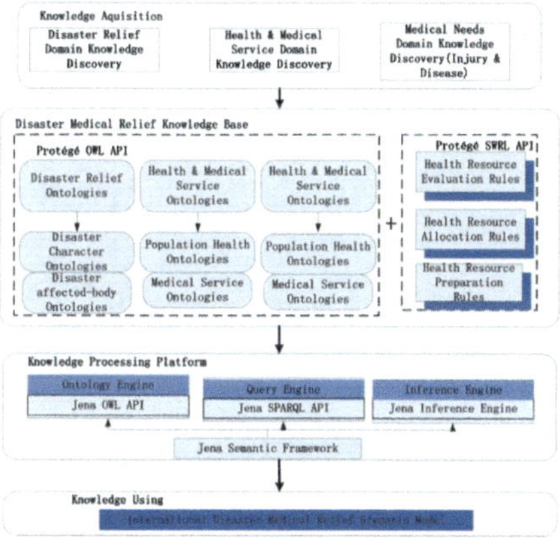

Fig. 2. General view of the construction of ontology-based scenario model knowledge base

3.1 Outline of OS-CBR Component

Figure 3 shows the outline of CBR component. The component mainly consists of case retrieval, case adaptation, and case preservation. We do case retrieval and adaption on first level model and second level model respectively. This stage consists of a cycle of revise and re-uses stages that repeat until a suitable solution is generated by the system.

3.2 Case Retrieval

The case retrieval is the most important process in CBR design. First, we do case retrieval on general disaster level to obtain rough medical needs, for example, the total number of injuries, the composition of injuries and so on. Then, we do case retrieval on local level to adjust the medical needs in 3 aspects, which could give the international medical relief team a more proper recommendation according to the local health and medical condition.

The value of similarity is between 0 (not similar) and 1 (most similar). The total similarity value is calculated by [5]:

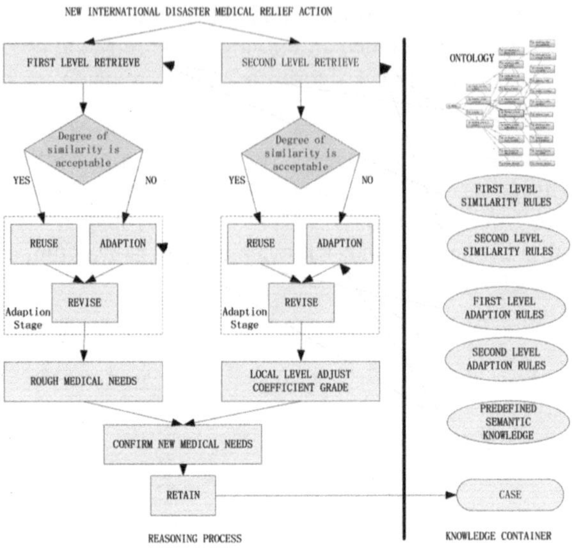

Fig. 3. Outline of reasoning

$$S = \frac{\sum_{i=1}^{n}\sum_{j=1}^{m}\left(S_{ij}\right)}{N} \tag{1}$$

where S_{ij} is similarity of parameter i of a new case, compared to feature j of an old case. n is the number of the new case feature, m is the number of the old case feature, and N is the total number of features. The similarity calculation of S_{ij} depends on the level of model and data type.

3.2.1 First Level Model Case Retrieval

The first level model is the general disaster relief scenario model which is mainly composed of the situational elements related to the features of the disaster itself, as well as the feature elements of the regional disaster bearing body affected by the disaster. This level model has impacts in the first time of disaster on the main structure of medical rescue demand. Because of the multidisciplinary and complexity of disaster relief, the possible data types in the cases of first level model could be numerical, minimum and maximum numerical, text string and linguistic.

For the numerical data type, the similarity value is calculated by:

$$S_{ij} = \left(1 - \sqrt{\left(\frac{C_{ij} - X_i}{\max(C_i) - \min(C_i)}\right)^2}\right) \times \frac{W_i}{10} \tag{2}$$

where S_{ij} is the similarity parameter i of a new case, compared to parameter j of an old case. C_{ij} is the value of the parameter i of the case j, X_i is the input value of the parameter i, W_i is the weight factor of the parameter i, and $\max(C_i)-\min(C_i)$ is the largest scope of the C_i of the case j.

The interval value [a, b] is calculated by the following rules:if data - type is 'INTERVAL' and $a \leq C_{ij} \leq b$ then

$$S_{ij} = \frac{W_i}{10} \text{ otherwise } S_{ij} = 0 \tag{3}$$

The minimum and maximal numerical value is calculated by the following rules:if data- type is 'MIN' and $X_i \leq C_{ij}$ then

$$S_{ij} = \frac{W_i}{10} \text{ otherwise } S_{ij} = 0 \tag{4}$$

if data- type is 'MAX' and $X_i \geq C_{ij}$ then

$$S_{ij} = \frac{W_i}{10} \text{ otherwise } S_{ij} = 0 \tag{5}$$

The textual string type parameter similarity is calculated in a straightforward way, so that, if the input value X_i is included in the textual string value of C_{ij}, the similarity $S_{i,j} = 1$ (Eq. (6)) [6].

if data- type is 'STRING' and $C'(\propto_i) = C(\alpha)$ then

$$S_{ij} = \frac{W_i}{10} \text{ otherwise } S_{ij} = 0 \tag{6}$$

If X_i and $C_{i,j}$ are totally different to each other, the similarity $S_{i,j} = 0$.

The linguistic value similarity (high, medium and low) is calculated in the following way (see Eqs. (7) and (8)):

$$\text{if } X_i = C_{ij} \text{ then } S_{ij} = \frac{W_i}{10} \tag{7}$$

$$\text{if } X_i \neq C_{ij} \text{ and } X_i \text{ or } C_{ij} = \text{medium then } S_{ij} = \frac{0.5 \times W_i}{10} \text{ otherwise}$$
$$S_{ij} = \frac{0.1 \times W_i}{10} \tag{8}$$

The threshold type parameter similarity is calculated in the following way, so that, if the input value X_i is less than the value of C_{ij}, the similarity $S_{ij} = 1$ [5].

if data- type is 'THRESHOLD' and $C'(\propto_j) > X_i$ then

$$S_{ij} = \frac{W_i}{10} \text{otherwise } S_{ij} = 0 \tag{9}$$

The case retrieval algorithm works as follows: We assume that the case base has stored the historical data related to a given disaster scenario. When a new disaster scenario is input into the system, the case base is searched to retrieve all cases with a similar profile. The similarity of the new problem to the stored cases is determined by calculating the distance between case features. Then, once the most similar cases have been obtained from the case base, they will be used in the next adaptation phase to generate a rough medical needs.

3.2.2 Second Level Model Case Retrieval

With the extension of the rescue time, the impact of second level model gradually increased [13]. The features of cases in second level model are selected according to the Sustainable Development Goals (SDG) of 2030 of World Health Organization (WHO), which made monitoring and statistical analysis of health services and health status in 194 different countries about 1000 related health indicators.

Unlike doing only once case retrieval in the first level model comparing each parameter's similarity and add them all, we do the case retrieval 3 times in second level model. Certain feature's parameters of the second level model case is described as shown in Table 1.

Most of these feature's data type are numerical but in different dimensions. In order to deal with the different dimensions, we choose Standardized Euclidean Distance (SED) to calculate the similarity. We obtain SED by the following rule:

$$d_{kj} = \sqrt{\sum_{i=1}^{n} \left(\frac{X_i - C_{ij}}{s_i}\right)^2} \quad (k = 1, 2, 3) \tag{10}$$

where d_{kj} is the SED of a set of parameters in type k in a new case, compared to an old case. n is the total number of parameters in type k. X_i is the input value of the parameter i, C_{ij} is the value of the parameter i of the case j.

The value of SED is between 0 (most similar) and 1 (not similar). When comparing parameters of type 1 in new case, C_1 is more similar to the target query Q than C_2, if the d_{11} is shorter than d_{12}.

The case retrieval algorithm works as follows: We assume that the case base has stored the historical data. When a new region is input into the system, the case base is searched to retrieve all cases with a similar profile type by type. The similarity of the new region to the stored cases is determined by calculating the SED between case features. Then, once the most similar cases of 3 types have been obtained from the case base, they will be used in the next adaptation phase to generate recommend medical needs adjustments.

Table 1. Certain feature parameters of the second level model case description

Case retrieve	Type name	Abbreviation	Index	Data Type presentation	Indicator	Effectiveness
Type 1	General Health Condition	HALE	Health Adjusted Life Expectancy	Age	life expectancy weighted or adjusted for the level of health-related quality of life	Lower age and higher percentage indicate poorer health condition which would aggravate the severity of injury and patients as well as the number of patients
		HTN	Prevalence of raised blood pressure	Percent	Percent of defined population with raised blood pressure (SBP $\geq$ 140 OR DBP $\geq$ 90)	
		DM	Prevalence of raised blood glucose	Percent	Prevalence of raised fasting blood glucose	
	Health Coverage	HC1	Basic Hospital Access	Ratio	Hospital beds (per 10 000 population)	The higher percentage or ratio could help the country or area to reduce the injury number, release severe wounded and reduce patient number after disaster
		HC2	Health-worker Density	Ratio	Density of physicians (per 1 000 population)	
		HC3	Access to Essential Medicines	Statistic	Median availability of selected generic medicines (%)	
		HC4	Compliance with the International Health Regulations	Percent	International Health Regulations (IHR) core capacity index	
		Other	Mobilization of resources;	Percent	General government expenditure on health as a percentage of total government expenditure	
Type 2	Special Population Health	Motality	Maternal mortality; Under-five mortality	Ratio	Maternal mortality ratio (by 100,000 live birth); Probability of dying by 5 per 1000 live birth	Higher ratio or percentage indicate poorer health condition among children and women which would aggravate the severity of injury and increase special patients
		Malnutrition	Stunting among children under 5 years of age	Percent	Prevalence of stunting among children under 5 years of age	

(*continued*)

Table 1. (*continued*)

Case retrieve	Type name	Abbreviation	Index	Data Type presentation	Indicator	Effectiveness
	Professional Service to Special Population	PSSP1	Family planning	Percent	Married or in-union women of reproductive age who have their need for family planning satisfied with modern methods	The higher percentage could help reduce child or women injury, release severe child wounded and reduce child or women patient number
		PSSP2	Antenatal delivery and care	Percent	Antenatal care coverage - at least four visits (%)	
		PSSP3	Health-seeking behavior for child illness	Percent	Children with suspected pneumonia taken to an appropriate health provider (%)	
Type 3	Infectious Disease Distribution	HIV	HIV/AIDS	Number	Number of people (all ages) living with HIV Estimates	Higher number indicate higher probability of occuring new infectious disease patients after the disaster
		TB	Tuberculosis	Number	Number of new Tuberculosis cases (per 100 000 population)	
		Malaria	Malaria	Number	Number of new malaria cases (per 1000 population at risk)	
	Infectious Disease Control	HIV	HIV Therapy	Percent	Estimated antiretroviral therapy coverage among people living with HIV (%)	The higher percentage indicates the country or area could prevent new infectious disease outbreaking at a higher level after disaster
		TB	Tuberculosis Therapy	Percent	Tuberculosis treatment coverage for all forms of tuberculosis	
		SH	Sanitation and hygiene	Percent	Population using at least basic sanitation services (%)	
		IMMU	Immunization Coverage	Percent	Diphtheria tetanus toxoid and pertussis (DTP3) immunization coverage among 1-year-olds(%)	

3.3 Case Adaption

Case adaptation is the process of transforming the most similar cases retrieved from the case base into a medical needs appropriate for the current scenario. Our research uses the substitution adaptation strategy. Substitution is used when parts of the old medical needs are in conflict with, or contradict the new scenarios. Substitution is performed by using predefined semantic knowledge.

3.4 Case Revision and Preservation

When a new disaster occurs, the most similar significant disasters and the respective adjustments are retrieved. The medical needs of retrieved cases are then adapted to meet the new case. The suggested medical needs is revised for its suitability by experts. The revised solutions are then retained temporarily in database. When the suggested medical needs are actually applied, the medical readiness plan for the current situation can then be evaluated. Finally, the new case is retained in the case base to help find medical needs for future disasters.

3.5 Case Knowledge Presentation

Most cases would record the number of injuries or patients as well as epidemic of infectious disease. But not all cases would record the medical relief readiness plan. What is more, along with the development of medical technology, even the medical needs are similar, the readiness plan would be improved. Appropriate medical needs which reflect the actual local situation is the basis of international medical relief team's preparedness plan. So we choose medical needs to be the outcome of the system. According to the double scenario model, two kinds of case knowledge are presented through the system. One is the rough emergent medical needs after disaster out breaking, such as the total number of injuries, the number of severe injuries and so on. The other is the adjustment of injury treatments, general medical requirements and disease control requirements, such as medical needs from children, pregnant women, elders suffering chronic disease and so on.

3.5.1 Identification of Need for Knowledge

It is essential to specify the requirements of the needed knowledge for a current case to facilitate an accurate search for knowledge, or to create new knowledge.

In the first model, one knowledge required is the rough total number of injuries, the other is the possible composition of injuries which reflect the possible medical needs for severe injuries treatments.

In the second model, the knowledge requirements are mainly three types. The first type reflect the adjustment for treatment of injuries, such as Injury Amount Adjust (IA-A) and Composition of Injury Condition Adjust (CIC-A). The second type related to the disease treatment needs, such as New Patient Amount (NPA), Disease Composition (DC) and so on. The third type reflect the disease control condition after disaster, such as Infectious Disease Epidemic (IDE).

- Definition 3. Knowledge Requirements (KR).

$$\mathrm{KR} = \{\mathrm{T}(\mathrm{S_i}) | \mathrm{S_i} = (\mathrm{S_i^1}, \mathrm{S_i^2}, \cdots, \mathrm{S_i^n})\}$$

Where KR is knowledge requirement need, $\mathrm{T(S_i)}$ is the requirement type, $\mathrm{S_i^n}$ is requirement steps, and i is the ith knowledge requirement need.

3.5.2 Knowledge Retrieval

The knowledge retrieval (KR) model is based on the information retrieval (IR) system. An IR system retrieves documents based on a user's query as input and the documents are sorted by their relevance to the query [4]. The IR system is based on keyword indexing systems and Boolean logic queries that are sometimes supplied with statistical methods [22]. The basic idea of the IR system is that the user asks the system and obtains desired information. The obtained information is then used to solve a given problem or situation.

4 Application Results and Discussion

We apply the OS-CBR approach in the international medical relief action to armed conflicts happening in Syrian Arab Republic (Syrian).

The first level model input attributes of this new case is shown in Table 2.

Table 2. Input attributes of armed conflicts happening in Syrian

Nation	Disaster type	Population	Conflict source	Start time	Hit area	Degree
Yemen	Armed conflicts	25,235,000	Arab spring	2011	Whole country	Severe

The calculate results of similarity in first level model using OS-CBR approach is shown in Table 3. (Assume the w_i of each attributes are equal.)

Table 3. Similarity results from first level model using OS-CBR appraoch

Attribute	Similarity of Case 1	Similarity of Case 2	Similarity of Case 3	Similarity of Case 4	Similarity of Case 5	Similarity of Case 6	Similarity of Case 7
Disaster type	1.00	1.00	1.00	1.00	1.00	1.00	1.00
Source of disaster	1.00	0.87	0.73	0.37	0.52	0.34	0.28
Degree	1.00	1.00	1.00	0.50	0.75	0.50	0.50
Hit population	0.89	0.92	0.48	0.73	0.63	0.28	0.49
Hit area	1.00	1.00	1.00	0.50	0.50	0.50	0.50
Start time	1.00	0.45	0.83	0.83	0.72	1.00	0.63
Location	0.92	0.89	0.27	0.72	0.48	0.36	0.50
Total ($S_i * w_i$)	0.97	0.88	0.76	0.66	0.66	0.57	0.56

As shown in Table 3, the most similar case in the case base is case 1, the armed conflicts happened in Yemen. Yemen's armed conflicts has caused 49963 injuries.

Then we calculate the SED in second level model using OS-CBR approach. The results are shown in Table 4.

Table 4. SED results from second level model using OS-CBR appraoch

Case retrieve	Name	Abbreviation	Index	Syrian	Yemen	SED (Yemen)	Outcome Case	SED (Outcome)	Globle Average	SED (Globle)
Type 1	Health Coverage	HC1	Basic Hospital Access	15	7	0.7836	17	0.2385 (Iran)	19	0.4192
		HC2	Health-worker Density	1.546	0.311		1.49		2.9	
		HC4	Compliance with the International Health Regulations	63	46		66		60	
		Other	Mobilization of resources	4.8	3.93		9.4		9.1	
	General Health Condition	HALE	Health Adjusted Life Expectancy	63	54		61		59	
		HTN	Prevalence of raised blood pressure	24.5	30.7		25		27.5	
		DM	Prevalence of raised blood glucose	14.6	11.3		12.1		9.5	
Type 2	Professional Service to Special Population	PSSP1	Family planning	87.5	69.6	0.8134	85	0.3157 (Jamaica)	76	0.6935
		PSSP2	Antenatal delivery and care	91.1	49.1		87		75	
		PSSP3	Health-seeking behavior for child illness	76.8	34		75		60	
	Special Population Health	Motality	Maternal mortality; Under-five mortality	68;17.4	385;47		93;23		216; 40.8	
Type 3	Infectious Disease Control	HIV	HIV Therapy	18	No data	0.8392	27	0.4213 (Egypt)	50	0.5192
		TB	Tuberculosis Therapy	80	59		80		75	
		IMMU	Immunization Coverage	DTP370	DTP3:84		DTP3:79		DTP3:85	
		S&H	Sanitation and hygiene	93	60		93		88	
	Infectious Disease Distribution Malaria	HIV	HIV/AIDS	9,900	No data		1913		190,155	
		TB	Tuberculosis	21	48		15		142	
		Malaria	Malaria	0	22.2		0		94	

As shown in Table 4, although the armed conflicts in Yemen are similar to Syrian, the 3 types of features' SED in the second model of Yemen are longer than globle average to Syrian. As the system's result, Syrian is more similar to Iran, Jamaica and Egypt in different feature type. Comparing these three country's situation after disaster with Yemen, medical needs in Syrian should be less than Yemen.

According to the WHO reports, the number of people in need of humanitarian assistance is 20.7 million in Yemen while the number in Syria is 13.1 million. The number of people in acute need of humanitarian assistance is 9.8 million in Yemen while the number in Syria is 5.6 million. In Yemen, 1.8 million children acutely malnourished. In Syria, the situation is better, 19,000 children acutely malnourished. Cholera out broke in Yemen and the cumulative total from 27 April 2017 to 14 Jan 2018 is 1,035,676 suspected cholera cases.

From these facts, it can be concluded that more international actions are needed in Yemen. From the WHO's new response plan for these two countries, we could see that the main points in Yemen's plan are more emergent medical resources delivery and life-saving medical services supports and trainings. Syria's plan focuses mainly on helping the local health service system to recover and supporting polio vaccination.

The outcomes of the OS-CBR for double-level scenario model fit the reality.

5 Conclusion and Further Study

In order to establish a more fit-to-reality medical readiness plan, this research aims to establish a OS-CBR approach for the double scenario model.

The approach can be consummate by some improvement, such as: features of first level model should be more clear and calculable. Each feature might have different weight in the model and the weight might change with time and different purpose. The sensitivity and precision analysis of the approach is also needed. More case studies are needed to check the effectiveness of the approach as well. Further research plans are making and more research measures are taking.

Acknowledgements. This study is Project in the National Key Research and Development Program of China (No. 2017YFC1405300); Army logistics key projects (#AWS14L012; #BHJ14L010); Army standard program (#BHJ17B042).

References

1. Reinhardt, J.D., Li, J., Gosney, J., Rathore, F.A., Haig, A.J., Marx, M., DeLisa, J.A.: Disability and health-related rehabilitation in international disaster relief. Glob. Health Action **4**, 7191–7197 (2011)
2. Zeid, S.: Women's, children's, and adolescents' health in humanitarian and other crises. BMJ **351**(Suppl. 1), 56–60 (2015)
3. Yonekura, T., Ueno, S., Iwanaka, T.: Care of children in a natural disaster: lessons learned from the Great East Japan earthquake and tsunami. Pediatr. Surg. Int. **29**, 1047–1051 (2013)
4. Guha-Sapir, D.: Rapid assessment of health needs in mass emergencies: review of current concepts and methods. World Health Stat. **44**, 171–181 (1991)
5. Virkki-Hatakka, T., Reniers, G.L.L.: A case-based reasoning safety decision-support tool: next case/safety. Expert Syst. Appl. **36**(7), 10374–10380 (2009)
6. Wang, X., Zhang, H., Xu, Z.: Public sentiments analysis based on fuzzy logic for text. Int. J. Softw. Eng. Knowl. Eng. **26.09**(10), 1341–1360 (2016)
7. Wang, X., Luo, X., Liu, H.: Measuring the veracity of web event via uncertainty. J. Syst. Softw. **102**, 226–236 (2015)

Web Sanitization from Malicious Code Injection Attacks

Hussein Alnabulsi(✉) and Rafiqul Islam

School of Computing and Mathematics, Charles Sturt University, Albury 2640, Australia
{halnabulsi,mislam}@csu.edu.au

Abstract. We propose a new methodology to sanitize web pages to prevent code injection attacks. One of a common programming error that usually happens in the web application is using of an improper encoding method to sanitize the source code of the web page. Our methodology provides a proper encoding method to the webpages which have an improper encoding of untrusted data, so it can stop and prevent code injection attacks caused by improper encoding of untrusted data from occurring. Our framework is an automatic encoding method to sanitize web browser contains multiple interpreters, such as: JavaScript, CSS, HTML, and URI. In this methodology we also need to detect zero- day attack (XSS vulnerabilities) which may not be detected by detection tools. Our methodology can prevent a many types of code injection vulnerabilities, such as: XSS injection vulnerabilities.

There is a study that sponsored by Google showed that thirty percent usage of encoding method is incorrect. This incorrect encoding leads to code injection vulnerabilities in the webpages. In some encoding cases we should utilize more than one encoding method in the context, such as: URI and JavaScript encoding methods.

Keywords: Code injection attack · XSS attack · Encoding · Sanitization

1 Introduction

Some of research selected improper validation of a user input as one of the most popular vulnerabilities that occurring in the webpages. The top ten of vulnerabilities in webpages of Open Web Application Security Project (OWASP) research puts improper validation of a user input as the first point of webpages vulnerabilities.

Validation of a user input is an input checking method, in this method the website apply a validation of users' input parameters to ensure that these parameters are not a malicious input and valid to enter the webserver or presented in a web browser. Here is an example of input validation, e.g., a website checks how many items that a user want to buy, so it should be an integer numeric value not a float or a nonnumeric string. Also the most important thing is to be sure that a user input does not contain any code injection attack such as SQL injection or XSS injection attacks [6].

Cross Site Scripting (XSS) is vulnerability in webpages happens by inserting malicious syntaxes in a webpage that does not supply sanitization of user input. When a

J. Abawajy et al. (Eds.): ATCI 2018, AISC 842, pp. 251–261, 2019.
https://doi.org/10.1007/978-3-319-98776-7_27

user sends malicious syntaxes to a webserver through a vulnerable webpage, if there is no input sanitizing in a webserver, then the malicious syntaxes could be executed in a webserver and affected a web browser. XSS attack could redirect a webpage to a malicious website, steal a user's cookie and a session user id to access user's data [5].

We use the term "URI" (Uniform Resource Identifier) to refer to what are commonly called "URL", but the "URI" is more general term for the same concept.

Most of sanitization programs sanitize webpages just for HTML contexts. In the following example for context insensitive sanitization, Django's auto-escape did not provide protection for user input which is located in the place of a Href link.

Example 1:

```
<i Href="UntrustedLink"> user link
</i>.
```

The data hole in this example is the template variable "UntrustedLink", which is a variable that refers to a user input. The "UntrustedLink" may lead to XSS injection attack (JavaScript: URI), but Django's auto-escape just sanitizes HTML Syntax. Django did not provide any sanitization to untrusted content just for HTML Syntax sanitization [7].

Consider the following example of simple Ctemplate template for context sensitive sanitization.

Example 2:

```
<Html>
  <Body>
    <Script>
    Function N1()
    {
     Document.getElementById("NameSpan").Textcontent = "name: " + "name";
    }
    </Script>
    <Span ID="NameSpan" OnClick="N1()"> Show N1
    </Span> <Br/>
    <i Href="URI"> PageName
    </i>
  </Body>
</Html>
```

In this example, there are 3 data holes; the first data hole is ("name") that defined as JavaScript syntax. The auto-escape sanitizer applied :Javascript_escape to the first data and it should be "name :JavaScript_escape" instead of "name". The second data hole is ("URI"), the auto-escape sanitizer applied :URI_Escape_with_Arg = HTML to the data hole because it is a URI context. The third data hole is (PageName), which is a HTML context and the auto-escape sanitizer applied HTML_Escape to encode the third data hole [7].

Here is an example 3 from the GRAMPS application authored on Django.

```
The example 3:
{
 If Header.Sortable
 X Href= "Header.URI|Escape"
 EndIf
}
```

In the above example, the programmer applied manually escaping instead of auto-sanitization on "Header.URI|Escape", but the escape in the example is an improper sanitization in a XSS vulnerability. The right sanitization should be by using JavaScript: URI vector [1].

The next example is about a fragment of JSP program [8]:

```
The example 4:
{
  string f = (String)Request.getParameter("f");
  string c = (String) Request.getParameter("loc");
  <h Href="JavaScript:Void(0)" OnClick="Action(EscapeHTML(f) ');" > page1
  </h>
  <P>
  <EscapeHTML(c)>
}.
```

In example 4, a user enters 2 inputs through: f, loc variables at lines 1, 2. Variable f used for rendering an Html link as shown in line 3. The loc variable is for displaying in Html body as shown in line 4. A malicious JavaScript input for loc may look as: <Script> attack1() </Script>. When the encoding function EscapeHTML(c) is run, it replaces <character to < and <character to > so it can stop the malicious JavaScript attack from occurring.

The encoding function EscapeHTML(f) does not work for a malicious input f which may be this JavaScript attack: '); attack2(); //. The JavaScript attack will be executed without any changes from the Html encoding "EscapeHTML(f)". For preventing this sort of attack JavaScript attack, the JavaScript encoding should replace (') with (\') to stop the attack.

1.1 Determining if the HTML Output Includes Any Input Parameters

In this subsection, we need to analyse the website's design and the source code of the webpage to determine if the output contains any input parameters it may available in the source code of the webpage from many places. Here is a list of input parameter resources [10]:

1. Form Fields, such as:
   A. Response.write (UserName.text)
   B. Response.write (Request.form ("UserName"))

2. Query Strings, such as:
   A. Response.write (Request.querystring ("name1"))
3. Data Access Method and Database such as:
   A. Sqldatareader read1 = cmd.Executereader()
   B. Response.write (read1.getstring())
4. Cookie collection, such as:
   A. Response.write()
   B. Request.cookies ("User1").values ("ID"))
5. Session and application variables, such as:
   A. Response.write (session ("Name1"))
   B. Response.write (application ("Name1")).

1.2 Problem Statement

One of big challenges in the web industry is cyber security, OWASP lists over 80 vectors that can be targeted using cross site scripting attacks. Cyber security developers still have too much work to do for protecting the web applications against cyber security especially code injection attacks. One of the Google's sponsored study showed that around thirty percent usage of encoding method is incorrect. This incorrect encoding leads to code injection vulnerabilities in the webpages. Also we need to protect the websites from zero-day attacks by finding XSS vulnerabilities in the web applications and protecting them from XSS attack before it starts which is a problem that still occurs in the web applications.

2 Related Works

In comparison between commercial web sanitizer tools. Weinberger et al. [1] developed a framework model to sanitize XSS vulnerability in the web application. The authors evaluated XSS injection attacks in 14 commercial web sanitizer tools. The result shows that many instances of the commercial web sanitizer tools do not solve several troubles of the XSS vulnerability. When the commercial tool sanitizes for many context, including: URI attributes, CSS keywords, and JavaScript string contexts is more precise than the commercial tool sanitizes only for HTML tag context. The authors showed the analysis of 8 webpages to get the requirements of XSS sanitization from the webpages. One of the result shows that there is a gap between the abstractions provided by commercial tools and the requirements of the webpages.

The four commercial tools, ClearSilver, Smarty, Ctemplate, and Django sanitized emitting untrusted data into a JavaScript string. The Ctemplate tool sanitized the data emitted into JavaScript outside of the string literal context. There is no tool allowed untrusted JavaScript syntax to be emitted into JavaScript contexts.

For sanitizing user input variables in webpages. Sadeghian et al. [2] proposed a model for detecting and escaping a malicious syntax. The model can sanitize variables that a user enters as inputs in the webpage. The implementation shows that the authors' model succeeded in sanitization of all sorts of HTTP header requests, including: Get, Post, and Request. So it can defend a user from SQL injection attacks with limited changes in the original code of the webpage. Also the model is implemented as a library that can be inserted in the vulnerable webpage by calling one line of code.

Medeiros et al. [3] proposed a mechanism for correcting vulnerabilities in webpages and a framework to apply the mechanism for PHP applications and to validate the input vulnerabilities in the webpage. The framework searches for vulnerabilities by combining two techniques: the first technique called the static source code analysis and data mining. Data mining is used for identifying the false positive by using a machine learning classifier. The combination of detection techniques could not obtain a complete correct result. The problem of the static analysis is undecidable. The evaluation supposes that the framework could detect and correct the vulnerabilities of the classes. The authors present the hybrid of methods to detect vulnerabilities with less false positives rate. The hybrid of methods obtains more precise in detection, and automatically correction the code of the vulnerabilities by applying modifies in the HTML code of the web application. The experimental result used a dataset of open source PHP web applications, and the framework executed in the WAP tool.

The research [4] focuses on defense mechanism for Cross Site Scripting (XSS) attack, which is the highest threat on web application nowadays. The mechanism depends on input filtering and validation that could prevent XSS attack on the server side. Bisht & Venkatakrishnan (2008) discussed many real world XSS attacks, then they analyse the reasons of failure in input filtering mechanism. The conclusion showed that the filtering mechanism is benefit just in the first level of defense against XSS attack, but it is improper in preventing many sorts of attacks, such as: when a user input includes a rich content of HTML. So the authors proposed XSS-Guard, which is a new tool designed as a prevention mechanism against XSS attacks. It works by automatically learning a set of scripts that a webpage wants to create for HTML requests. XSS-Guard has a robust mechanism to identify scripts at the server side and removes any scripts in the output.

3 Proposed Model

In this section we present the model of our framework as it shows in Fig. 1.

Figure 1 shows how our PHP framework works; it consists of a form box to enter the HTML data source of the webpage, and the process button to start the sanitization process. We used six different WYSIWYG editors [11] as a dataset to classify our framework.

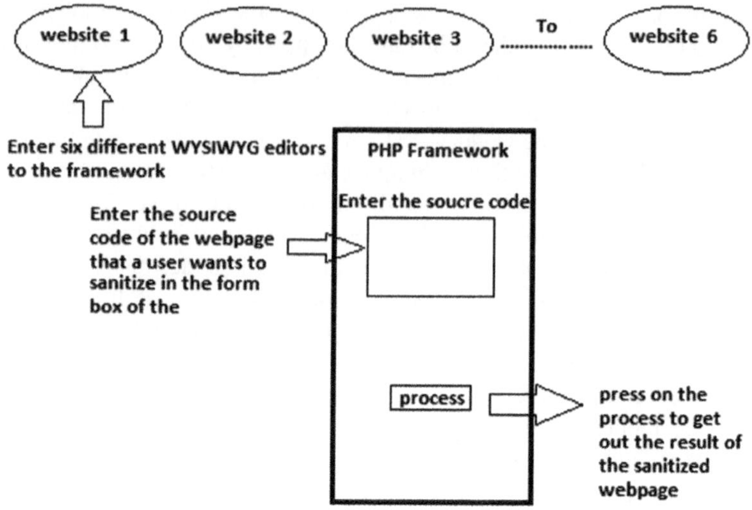

Fig. 1. The proposed model of our framework

3.1 Proposed Algorithm

In this subsection, we preview the algorithm of our PHP program, as shown below.

Algorithm 1. Web page vulnerability checking algorithm

Step 1: Define the tag black list:
var $tagBlacklist = array('applet', 'body', 'bgsound', 'base', 'basefont', 'embed', 'frame', 'frameset', 'head', 'html', 'id', 'iframe', 'ilayer', 'layer', 'link', 'meta', 'name', 'object', 'script', 'style', 'title', 'img', 'xml');
Step 2: Define the attributes black list:
var $attrBlacklist = array ('action', 'background', 'codebase', 'dynsrc', 'lowsrc', 'href', 'src', 'onclick', 'lowsrc');
Step 3: For cleaning a submitted HTML form, insert a setup of the filter and call.
$_POST = $myFilter->process($_POST);
$_GET["name"] = $myFilter->process($_GET["name"]);
Step 4: we have control over which tags and attributes to auto-escape or to auto-strip. Auto-escape tags whose tag-names contain non-alphabetic characters (such as <?php), attributes containing JavaScript.
Auto-strip any blacklisted that the user can select and declare from tags and attributes.
Step 5: Select which sanitizer function should be used such as:
Name: JavaScript_escape
Href="URI": encodeURIComponent
Onclick, for strings : HTML_Escape.
Step 6: Enter the source code of the webpage that a user want to sanitize with the blacklisted for any tags or attributes in the platform of our framework, and press on Process Query button to apply the sanitization process on the HTML source code of the webpage, then get the result of sanitized webpage against code injection attacks.

4 Experimental Setup

In our experimental, we have built our framework using the PHP programming language, and we have checked each data line of the webpage source code automatically by using our framework. We have downloaded XAMPP Control Panel v3.2.2 with an Apache server. The dataset that have been used in our framework classification is from popular WYSIWYG website editors [11]. We used the Zed Attack Proxy (ZAP), which is a white box security tool used by developer for security test to detect web applications vulnerabilities.

4.1 Algorithm Evaluation

In this subsection, we preview the results of six different WYSIWYG editors [11] that classified by our framework as it shows in Table 1. It represents six different websites, the highest value of the accuracy result is website 4, which is 0.9523.

Table 1. The result table of six different websites that classified by our framework

Data set A	TP	TN	FP	FN	TNR	PR	RR	Accuracy
Website 1	28	0	2	0	0	0.9333	1	0.9333
Website 2	39	0	3	2	0	0.9512	0.9512	0.8863
Website 3	26	0	2	0	0	0.9285	1	0.9285
Website 4	20	0	1	0	0	0.9523	1	0.9523
Website 5	17	0	1	0	0	0.9444	1	0.9444
Website 6	34	0	3	1	0	0.9189	0.9714	0.8947
Average	27	0	2	1	0	0.9381	0.9871	0.9232

5 Methodology

To prevent cross site scripting attacks, the sanitizer's framework must sure that all variables output in a web page are encoded before sending it to end users. Encoding variables output substitutes HTML markup with alternate representations called entities. When entities are encoded the web application displays them but does not run them, such as <script> gets converted to < script>;. Table 2 shows the entity name that we can use for HTML encoding for some HTML characters.

Table 2. Entity names for HTML characters

Result	Description	Entity name	Entity number
&	Ampersand	&	&#38
>	Greater than	>	>
<	Less than	<	<
	Non Breaking Space		

In our auto sanitization framework, we need to show which variables need to sanitize. We can know them from potentially dangerous HTML tags and attributes:

The following HTML tags and attributes may allow an attacker to inject XSS attacks: HTML tags:<applet>, <body>, <bgsound>, <base>, <basefont>, <embed>, <frame>, <frameset>, <head>, <html>, <id>, <iframe>, <ilayer>, <layer>, <link>, <meta>, <name>, <object>, <script>, <style>, <title>, <xml>, <img>.

HTML attributes: <action>, <background>, <codebase>, <dynsrc>, <lowsrc>, <href>, <src>.

Then we need to know which encoding or escaping method we can apply on HTML tags and attributes to sanitize a web browser. An attacker can use HTML attributes such as href, style, and script for preceding tags to inject cross site scripting. Such as, the <src> attribute of the <img> tag can be a source of injection, as shown below:

<Img Src="Java
Script:Alert('xss attack');">
Also, a hacker may use the <style> tag to inject a script, such as:
<Style Type="Text/JavaScript"> Alert('code injection attack');
</Style>

5.1 Encoding and Escaping Rules

In this subsection we describe the encoding and escaping method to sanitize a web browser contains multiple interpreters, such as: JavaScript, CSS, HTML, and URI [12].

A. Escape

Encoding is transforming data from one format into another format, while as the escaping is a subset of encoding, where not all characters need to be encoded. Only some characters are encoded (by using an escape character). Escape does not encode [+], because it may make problems if the decoder treats [+] as a space (when [+] comes from HTML form). Escape does not encode [/], so it may occur troubles if the escape operation encodes syntax with [/] in it.

B. JavaScript Encoding Rules

The [Escape] function has been provided by JavaScript and produced a URI-Encoded version, but the URI-Encoded does not sanitize properly because it does not escape some characters that should be escaped for sanitization, and does not handle Non-ASCII characters in proper way. The newer version of JavaScript version provides another 2 escaping functions: [encodeURI] and [encodeURIComponent]. The difference between these functions is the characters that they do escape. All of these 3 functions do not escape any characters of ASCII numeric or alphabetic characters.

The [escape] function does not escape: / @ _ + - . *
The [encodeURI] does not escape: / : ; = ? @ _ & ' ( ) * + , - . ~! # $
The [encodeURIComponent] does not escape: - . _ ~( ) *! '

C. URI Encoding Rules

The URI syntax declares a set of reserved characters, such as: [:], [/], [?], [#], [&], [=]. The reserved characters are used as delimiters in the URI syntax, such as: the [&] separates between key and value pairs in the query string, and the question mark [?] indicates the beginning of a query string. When you insert a string into the URI, it is important to escape characters and avoid confusing them with delimiters. E.g., if a user wants to sanitize a page called [M&N], and the URL link as: [?Page = M% 26 N&Action = View]. The [%26] is a URL encoded for the [&] character. The URL encoded for the space characters [] is [%20]. The EncodeURI does not encode any URI delimiters, e.g. [:], [/], [?], [#], [&], [=], cause of that the sanitization using EncodeURI would not be right (EncodeURI does not escape [&]). Another example of sanitization using EncodeURI is [k = encodeURI ("?Page = " + PageName + "&Action = View"]. If the PageName called [My Role], then the EncodeURI would encode the URI in the right way to [?Page = My%20R%C3%B4le&Action = View].

D. EncodeURIComponent

To solve the problem that we mentioned above, it is better to use encodeURIComponent, because encodeURIComponent can escape more characters than EncodeURI. In the previous example, if we used encodeURIComponent, it would be like: [k = "?Page = " + encodeURIComponent(PageName) + "&Action = View]. So the above example will encoded in the right way whatever characters are inserted in the PageName. The [&] character in the PageName will encoded to [%26], and the [&] in the [&Action = View] will stay as a [&] delimiter. The encodeURIComponent does not escape some characters such as: [!], ['], [(], [)], [*], these characters consider as reserved characters, but these characters do not harm like others do.

6 Comparison with Other Approaches

We used the Zed Attack Proxy (ZAP) which is testing tool for finding vulnerabilities in web applications. It is an integrated penetration testing tool for finding vulnerabilities in web applications and is maintained by hundreds of security developers. ZAP provides automated scanners as well as a set of tools that allow you to find security vulnerabilities manually [9]. We applied our dataset on ZAP and compared the results with our framework results, as it shows in Table 3. The dataset is from of three different WYSIWYG editors.

Our framework presents a better result comparing with ZAP security tool in classifying zero-day vulnerabilities. The accuracy of ZAP tool is 45.71%, while the accuracy of our frame work is 93.33%. Most of false positives are found from using

Table 3. The result table of three different websites that classified by ZAP and compared with our framework

	TP	TN	FP	FN	PR	RR	Accuracy
ZAP							
Website 1	17	0	11	11	0.6071	0.4358	0.4358
Website 2	23	0	16	16	0.5897	0.4281	0.4281
Website 3	16	0	9	10	0.64	0.4444	0.4571
Our framework							
Website 1	28	0	2	0	0.9333	1	0.9333
Website 2	39	0	3	2	0.9512	0.9512	0.8863
Website 3	26	0	2	0	0.9285	1	0.9285

HTML encoding in JavaScript context, such as: <i OnClick = "Function('< % = EscapeHTML(n) %>')" >page1</i>.

7 Conclusion

We previewed the results of six different WYSIWYG editors that classified by our framework. The results of the evaluation with six different websites are as follows: the highest value of the accuracy result was 95.23%, the precision rate was 95.23% and the recall rate was 100%. We compared our framework with ZAP tool, which is a security white box tool that can provide some sort of security to users, and our framework has obtained a precise accuracy result was 93.33%, while the accuracy result of the ZAP tool was 45.71%, the precision rate was 64% and the recall rate was 44.44%.

References

1. Weinberger, J., Saxena, P., Akhawe, D., Finifter, M., Shin, R., Song, D.: A systematic analysis of XSS sanitization in web application frameworks. In: 16th European Conference on RESEARCH in Computer Security (ESORICS 2011), pp. 150–171. Springer, Heidelberg (2011)
2. Sadeghian, A., Zamani M., Manaf, A.: SQL injection vulnerability general patch using header sanitization. In: IEEE International Conference on Computer, Communication, and Control Technology, (I4CT2014), pp. 239–242. IEEE (2014)
3. Medeiros, I., Neves, N., Correia, M.: Automatic detection and correction of web application vulnerabilities using data mining to predict false positives. In: WWW 2014 Proceedings of the 23rd International Conference on World Wide Web, pp. 63–74. IEEE (2014)
4. Bisht, P., Venkatakrishnan, V.N.: XSS-GUARD: precise dynamic prevention of cross-site scripting attacks. In: Detection of Intrusions and Malware, and Vulnerability Assessment, vol. 5137, pp. 23–43. Springer, Berlin (2008)
5. An Introduction to Cross Site Scripting, April 2018. https://www.cybrary.it/0p3n/an-introduction-to-cross-site-scripting

6. Balzarotti, D., Cova, M., Felmetsger, V., Jovanovic, N., Kirda, E., Kruegel, C., Vigna, G.: Saner: composing static and dynamic analysis to validate sanitization in web applications. In: IEEE Symposium on Security and Privacy (sp 2008), pp. 387–401. IEEE (2008)
7. Weinberger, J., Saxena, P., Akhawe, D., Finifter, M., Shin, R., Song, D.: An Empirical Analysis of XSS Sanitization in Web Application Frameworks (2011)
8. Mohammadi, M., Chu, B., Lipford, H.: Detecting cross-site scripting vulnerabilities through automated unit testing. In: IEEE International Conference on Software Quality, Reliability and Security, pp. 364–373. IEEE (2017)
9. Zed Attack Proxy (Zap), April 2018. https://www.utest.com/tools/zed-attack-proxy-zap
10. Meier, J.D., Mackman, A., Wastell, B., Bansode, P., Wigley, A.: How To: Prevent Cross-Site Scripting in ASP.NET, April 2018. https://msdn.microsoft.com/en-us/library/ff649310.aspx
11. Javed, A.: Revisiting XSS sanitization. In: ISACA Ireland Conference 2014 (2014)
12. Matt Giuca, URI Encoding Done Right, April 2018. https://unspecified.wordpress.com/2008/05/24/uri-encoding/

Event Element Recognition Based on Improved K-means Algorithm

Tao Liao[1(✉)], Wenhua Yang[1], Shunxiang Zhang[1], and Zongtian Liu[2]

[1] School of Computer Science and Engineering, Anhui University of Science and Technology, Huainan 232001, China
tliao@aust.edu.cn

[2] School of Computer Engineering and Science, Shanghai University, Shanghai 200072, China

Abstract. Event element recognition is a difficult problem in the extraction of event information, and its important content includes the identification of time and location elements. At present, the identification of time and location elements mainly uses the method of machine learning, but the method based on machine learning is susceptible to sparseness of corpus. An event element recognition method based on improved K-means algorithm is proposed. Firstly, the method is preprocessed to get the annotation corpus, and then the K-means algorithm is improved. The K value of the cluster is obtained by using the Canopy algorithm, and then the clustering analysis is carried out according to the improved algorithm. Finally, Feature filtering unrelated data, and re-use the clustering algorithm to cluster the data, to achieve the event time and location elements of the identification. The experimental results show that the method has achieved good results, and the F value of time and location recognition is 73.82% and 72.38%.

Keywords: Event element recognition · Canopy algorithm · Clustering K-means algorithm · Part of speech feature

1 Introduction

With the rapid development of Internet, people are increasingly aware of the events they care about from the Internet. It is with this demand that the Natural Language Processing (NLP) research on events is quietly rising. Information Extraction (IE) is a hot area in the field of natural language processing. Event extraction is an important research direction in the field of information extraction. At present, it is attracting wide attention.

Event extraction is one of the tasks of the ACE (Automatic Content Extraction, ACE) [1] evaluation conference. The ACE Evaluation Conference, sponsored by the National Institute of Standards and Technology (NIST), is designed to advance and develop technologies for information extraction. In 2005, ACE included the task of event extraction in the evaluation content.

Including two aspects of event extraction, that is, what happened (event recognition) and the information is closely related to the event (event element information such as time and place and people). Event Recognition [2–4] to determine whether a sentence

J. Abawajy et al. (Eds.): ATCI 2018, AISC 842, pp. 262–270, 2019.
https://doi.org/10.1007/978-3-319-98776-7_28

containing an event trigger word is an event occurring in the real world, event element recognition mainly identifies elements such as time, place, people, object in the event.

This paper focuses on the identification of events' time and location. The temporal relationship of events [5] is an important research content in the field of natural language processing. Its main purpose is to identify the specific relationship between events and their temporal information and to reason between the temporal expressions of events. Event location information is also an important part of research, it is an integral part of the incident, which reflects the true location of the incident. The location of the event includes geographical entities such as provinces, prefectures, and counties, as well as institutional entities such as hospitals, schools, and companies. Therefore, the event location recognition mainly includes two kinds of work of location entity name recognition and institutional name recognition. It can be seen that the elements of time and location are important elements in the incident and they play an important role in the fields of information retrieval system, question answering system and machine translation [6–8].

Example 1. A great earthquake occurred in [Sichuan Province] in [2008], houses collapsed, many people died, and the economy suffered serious losses.

From Example 1, we can see that the time element of the event is "2008" and the location element is "Sichuan Province".

At present, the identification of event elements mainly adopts the method of machine learning. This kind of learning requires a large-scale artificial annotation of mature corpus as a training set to obtain relevant knowledge of event elements. The learning effect depends on the quality and scale of the corpus [9]. If the corpus is not enough, it often makes the recognition effect is not ideal. Therefore, an event element recognition method based on improved K-means algorithm is proposed. The method incorporates part-of-speech features and uses the part-of-speech of each term to help identify the time and location features of the event.

2 Related Work

Event element recognition is to recognize the words, such as the time, place, and people in the sentences that indicate the events. At abroad, in 2006, Ahn [10] put forward the identification of event elements as a multi-class classification problem, and based on classification learning, event recognition and event element recognition were performed on ACE English corpus. Ahn's method has the problem of unbalanced positive and negative cases, and the fact that each type of event element constructs a multivariate classifier alone has the problem of sparse data when the corpus size is small. Tan et al. [11] firstly identified the event using local feature selection and fusion of positive and negative features, and then used multi-layer pattern matching to identify the event element on the ACE Chinese corpus. Due to the limited rules adopted, the identification of event elements is not very effective. In the study of event time, Lin et al. [12] used dynamic methods to deal with implicit time expressions, used new scoring models to determine the webpage's attention time, and designed a time-text retrieval ranking based on time and text relevancy method.

At home, Zhao et al. [13] proposed an improved method on the basis of Ahn's research, this method put forward to expand the trigger words, and the same event elements in different categories are merged to construct a multivariate classification model to identify the event elements. Experiments on ACE Chinese corpus demonstrate that the event element recognition effect is effectively improved. Ding et al. [14] used a filtering method based on keywords and trigger words to identify event types, and then event element recognition by using maximum entropy classification method. However, the dependence of this method on learning corpus Strong. Fu [15] adopted feature weighting method to identify event elements. Firstly, ReliefF feature selection algorithm in classification algorithm was improved and applied to clustering algorithm, and then assign different weights according to the different contributions of cluster based on each feature, and finally clustering event elements using clustering algorithm.

According to the above research status at home and abroad, it is found that the identification of event elements is realized by machine learning method, but some accuracy may not be high in recognition process. Aiming at this problem, this paper adds a part-of-speech rules optimization program based on the improved K-means algorithm.

3 Improved K-means Algorithm

The traditional K-means algorithm is a simple and fast clustering algorithm with scalability and high efficiency for processing large data sets, but it also has many disadvantages [16]. For example: the sensitivity to outliers, easily lead to the center point offset; the complexity of the algorithm should not be controlled, the number of iterations may be more; can not determine the number of K and so on. In view of the problem that the number of K can not be determined, the paper introduces Canopy algorithm, which can effectively solve the problem of K value.

Canopy clustering can effectively reduce the complexity of calculating the distance between points in the K-means algorithm. Algorithm steps shown in Fig. 1.

Input: The original array data, determine two thresholds T1, T2
Output: get the number of clusters K

1) The original state of the data is not yet classified, so remove a little from the set P, P as the first class, we also class called canopy.

2) Continue to take points from the set, such as P, calculate the distance d from P to any canopy that has been generated, add P to the canopy if the distance to a canopy is less than T1, and P if the distance from P to all canopy centers Is greater than T1, then P as a new canopy.

3) If P to the canopy distance is less than T2, P and the canopy are close enough, then P will be removed from the collection to avoid adding repeatedly to other canopy.

4) Perform the above operation on the points in the collection until the collection is empty. The algorithm ends, the clustering is completed, and the K value is output.

Fig. 1. The steps of canopy algorithm

It can be seen from the algorithm steps that this stage mainly carries out the "rough" clustering on the data. That is, the main data is placed in a subset, which is the canopy subset. Canopy is overlap, there will be no one object does not belong to which canopy. Therefore, the number of canopy based on "rough" clustering is the K value of the following K-means clustering.

4 Event Time and Location Elements Recognition

4.1 The Overall Process of Event Time and Location Elements Recognition

In this paper, an improved K-means algorithm is used to identify event time and location elements, which is divided into five steps. The specific process is shown in Fig. 2.

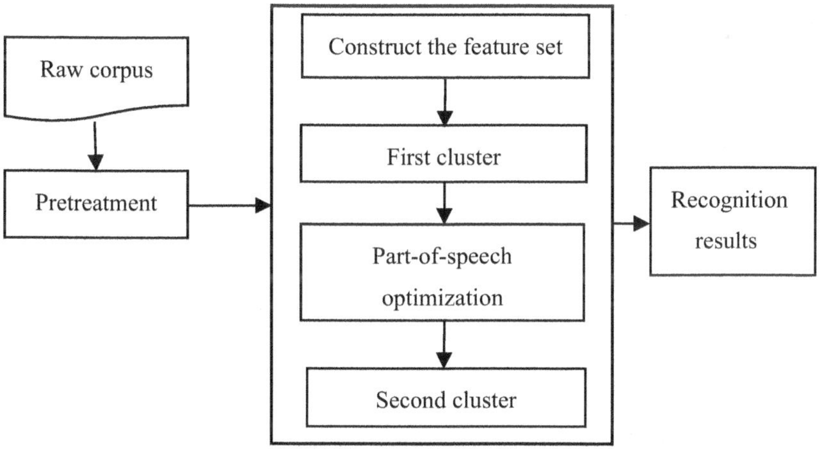

Fig. 2. The overall process of event time and location elements recognition

Step 1: First of all, LTP preprocessing of the obtained raw corpus to generate annotating corpus. Each term in the annotation corpus corresponds to a variety of features.
Step 2: Select the annotating corpus part-of-speech features, dependent syntax features and semantic dependency analysis and other features to construct the feature set.
Step 3: According to the index table to quantify the feature set to get the experimental data set, and a cluster is carried out.
Step 4: Analyze the clustering results, and use the part-of-speech features for screening optimization.
Step 5: Perform secondary clustering to identify the elements of time and location.

4.2 Preprocessing and Construction of Feature Set

The paper uses the CEC (Chinese Emergency Corpus) corpus [17], which is composed of 200 news articles on five types of emergencies such as earthquake, fire, traffic accident, terrorist attack and food poisoning, and removes its annotations as raw corpus in our experiment. The CEC corpus mainly contains the six most important data structures: Event, Denoter, Time, Location, Participant, and Object, which describe the event and its time elements, respectively. Although the size of the CEC corpus is small, the annotation of the events and event elements is comprehensive. In the process of preprocessing, we use the Language Technology Platform (LTP) developed by social computing and information retrieval research center of Harbin Institute of Technology to process the raw corpus to get the experimental data.

The main selections of the feature set are the part-of-speech features, the dependency parsing features and the semantic dependency parsing, and they all have their own different functions. The part-of-speech features are mainly part-of-speech tagging of sentences, and a part-of-speech task for each word. The part-of-speech categories may be nouns, verbs, adjectives, or others. Dependency parsing mainly analyzes the dependencies between words and words in sentences and reveals the features of the syntactic structure. Semantic dependency parsing analyzes the semantic association between each word in a sentence and the semantic association are presented in a dependency structure. Using semantic dependency to depict sentence semantics, we do not need to abstract the word itself. Instead, we describe the word through the semantic framework [18]. Based on the role of these three features, we get the feature set. This paper does not adopt the role of other features in the sentence because other features have little to do with the identification of the event elements in this paper.

4.3 First Cluster

By the quantitative processing of the feature set, we obtained the experimental data set. According to the result of Canopy algorithm, we can add the obtained K value to K-means algorithm. Therefore, in the process of implementing the algorithm we do not need to give K value in advance.

In clustering analysis, it is often confused with the classification algorithm. Obviously, the clustering algorithm is not a classification algorithm. The classification algorithm is to give a data, and then determine it belongs to the class which has already been sorted. The clustering algorithm is to give a bunch of raw data and then gather the similar features in a class.

Therefore, the basic steps of the improved K-means clustering algorithm are as follows:

Input: The original data (u1, u2, u3, …, uk) are all experimental data sets quantified by part-of-speech features, dependency parsing features and semantic dependency parsing. Random data (x_1, x_2, x_3, …, x_n) are initialized according to the K value obtained by Canopy algorithm. u_k and x_n are vectors.

Output: get clustering results (some classes)

(1) Calculate the distance between each object and these center objects according to the average (center object) of each clustering object, and repartition the corresponding objects according to the minimum distance, and the formula is:

$$c_i = argminx_i - u_j^2 \tag{1}$$

c_i show that the data closest to each initial data, x_i show that the ith the original data, u_j show that the jth distance average.

(2) Recalculate the mean (center object) of each (changing) class, with the formula:

$$u_j = \frac{\sum_{j=1}^{m} \{c_i = u_j\} x_i}{\sum_{j=1}^{m} \{c_i = j\}} \tag{2}$$

c_i show that the data closest to each initial data, x_i show that the ith the original data, u_j show that the jth distance average.

(3) Loop (1) through (2) until each class no longer changes.

After repeated calculation by the algorithm, K classes are obtained. By detecting each class, we found that the time and location elements are basically concentrated in one class. Considering that there are a large number of "impurity data" in this class, which affects the identification of time and place, so we are going to process these "impurity data" further.

4.4 Part-of-Speech Optimization and Second Cluster

In natural language processing, a text is usually represented by a vector of words, while word weights are usually calculated statistically. However, the linguistic features of words themselves are also crucial in measuring their weights and characterizing texts [19, 20]. Taking the part-of-speech as an example, words of different parts of speech have different contributions in text representation. Functional words or imaginary words such as interjections, pronouns, conjunctions, etc. have little effect and can be removed as stop words.

Example 2. China new network April 20: According to the information of the offical website of China Earthquake Administration, Beijing time at 1:22 on April 20, 2016, in Dingjie County, Shigatse, Tibet (28.13° north latitude, 88.16 east longitude), the 3.3 magnitude earthquake occurred, the depth of the earthquake was 7 km.

The word segmentation after LTP preprocessing is as follows:

China new network/n April 20/nt According to the information/n of the offical website/n of China/ns Earthquake Administration/n, Beijing/ns time/n at 1:22/nt on April 20, 2016/nt, in Dingjie County/ns, Shigatse/ns, Tibet/ns (28.13°/m north latitude/b, 88.16°/m east longitude/b), the 3.3 magnitude/m earthquake/n occurred/v, the depth/n of the earthquake/n was 7 km/m.

After the improved K-means clustering algorithm, due to the more impurity data in the class of time and location, which affects the precision, we use the part-of-speech feature to remove the impurity data in the class. In the example, nouns (n), other noun modifiers (b), quantifiers (q), and numerals (m) are useless for extracting time and location, but the nouns for annotated event are reserved.

The new participle is annotated as follows:

April 20/nt China/ns Beijing/ns at 1:22/nt on April 20, 2016/nt, in Dingjie County/ns, Shigatse/ns, Tibet/ns, the earthquake/n occurred/v.

After part-of-speech feature optimization, the obtained data set is again clustered by the improved K-means algorithm to separate the elements of time and location. In the second cluster, the data set is mainly composed of part-of-speech features and semantic dependency parsing. The data set is composed of the class whose time and location are concentrated after the optimization of the part-of-speech features, and then the K-means clustering algorithm is applied to the new data. Data sets are clustered to achieve separation of event time and location elements.

5 Experimental Results and Analysis

The experimental data in this paper uses a CEC corpus that removes annotation, and the HIT LTP module is used for word segmentation and parsing analysis. In order to achieve the experimental results in this paper, we used Matlab software platform for experiments. The precision (P), recall (Recall) and F value were used as the criteria to evaluate the event elements recognition results.

Because the number of Canopy can be used as K value, the blindness of choosing K is reduced to a certain extent. But if you just calculate the value of K, T1 has no effect, as long as specified, half of the average distance of all points is used as T2. When the distance is less than the threshold T2, the data already belong to a class, and T1 just look for the data involved in the next cluster, for the calculation of K value is not directly related, so do not need to be considered. The impact of different thresholds on the experimental results is shown in Table 1.

Table 1. The effect of different thresholds on experimental results

Thresholds	K value	Experimental results					
		Precision		Recall		F value	
		Time	Location	Time	Location	Time	Location
51.97	3	72.49	70.21	75.21	74.69	73.82	72.38
44.60	4	63.45	62.42	68.43	66.73	65.85	64.50
21.57	5	65.03	62.86	56.91	59.77	60.70	61.28

As can be seen from Table 1, the different thresholds result in different values of K, which have different effects on the precision, recall and F value in the experimental results. When the threshold T2 is greater than or equal to 51.97, the canopy algorithm can be grouped into three classes, that is, the K value is 3. When the canopy algorithm is clustered into four classes from 44.60 to 51.97, that is, the K value is 4. When the value is smaller than 44.60 and greater than 21.57, Can be grouped into five classes, namely K is 5. From observation, we found that the result of clustering into 3 classes is better.

In addition, this paper compare the experimental results with some other event element recognition results at home and abroad, and our comparison results are listed in Table 2.

Table 2. Comparison between different algorithms

Compare objects	Precision (%)	Recall (%)	F value
Reference [10]	68.80	47.70	56.34
	50.00	48.20	49.08
Reference [13]	65.28	63.49	64.37
Reference [15]	67.28	/	/
Our method	72.49	75.21	73.82
	70.21	74.69	72.38

It can be seen from Table 2 that the precision and recall in the reference [10] are not too high, mainly due to the small size of the corpus and result in the sparsity of data. The recall of the reference [13] is obviously higher than that in the reference [10], but the precision is not too high. The main reason is that the feature extraction is not comprehensive enough and triggered word expansion is not sufficient. The precision of the reference [15] is higher than that of the previous two. It proposed the method of feature weighting, which uses the clustering algorithm to identify the features, assigns different weights according to different features, and finally performs clustering on the results. but it did not give the recall. As the use of different experimental corpus, this paper inconvenience direct compares our experimental results to domestic and foreign event element recognition methods, but also achieves good results.

6 Conclusion

In this paper, we analyzed the current status of the study of event elements and proposed an event element recognition algorithm based on improved K-means algorithm. We improved the traditional K-means algorithm, added the Canopy algorithm, obtained the K value by the Canopy algorithm, which solved the blindness of the K value determined in advance, and then used the K-means algorithm to cluster and used the Part-of-speech characteristics to filter the clustering results. Finally, the clustering algorithm is used again to identify the event time and location elements. However, it can be seen from the final results of the experiment that there is still much room for improvement in the effect of current event factor identification.

Acknowledgments. The authors would like to thank the editors and anonymous reviewers for their valuable comments. This paper is supported by the Natural Science Foundation of China (No. 61273328), the Anhui Province College Natural Science Foundation (No. KJ2016A202), and the Anhui Province College Excellent Young Talents Support Program (gxyq2017007).

References

1. Doddington, G., Mitchell, A., Przybocki, M., et al.: The automatic content extraction (ACE) program-tasks, data, and evaluation. In: Proceedings of the LREC 2004, pp. 837–840 (2004)
2. Fu, J., Liu, Z., Fu, X., et al.: Identification of events based on dependency analysis. Comput. Sci. **11**, 217–219 (2009)

3. He, Z., Liu, L., Xin, X.: Chinese event recognition based on semantic role. Comput. Eng. Sci. **04**, 181–185 (2013)
4. Meng, H.: In the Field of Emergency Incident Extraction Technology Research. Shanghai University, Shanghai (2015)
5. Li, L., Zhang, Y., Wang, H., et al.: Application of time ML in text temporal relation analysis. J. Sci. Technol. China **5**, 95–103 (2014)
6. Yang, H., Chua, T.S., Wang, S., et al.: Structured use of external knowledge for event-based open domain question answering, pp. 33–40. ACM Press, New York (2003)
7. Zhang, S., Zhang, S., Yen, N.Y., Zhu, G.: The recommendation system of micro-blog topic based on user clustering. MONET **22**(2), 228–239 (2017)
8. Xu, Z., Xuan, J., Liu, Y., Choo, K.-K.R., Mei, L., Hu, C.: Building spatial temporal relation graph of concepts pair using web repository. Inf. Syst. Front. **19**(5), 1029–1038 (2017)
9. Liu, W., Liu, F., Wang, D., et al.: A method for extracting text event elements based on event ontology. J. Chin. Inf. Act **30**(04), 167–175 (2016)
10. Ahn, D.: The stages of event extraction. In: The COLING-ACL 2006 Workshop on Annotating and Reasoning About Time and Events, pp. 1–8 (2006)
11. Tan, H., Zhao, T., Zheng, J.: Identification of Chinese event and their argument roles. In: 2008 CIT Workshops 2008 IEEE 8th International Conference on Computer and Information Technology Workshops, pp. 14–19 (2008)
12. Gurev, V., Pathmanathan, P., Fatyebert, J.L.: A high-resolution computational model of the deforming human heart. Biomech. Model. Mechanobiol. **14**(4), 829–849 (2015)
13. Zhao, Y., Qin, B., Che, W., et al.: Study on Chinese event extraction. J. Chin. Inf. Act **22**(1), 3–8 (2008)
14. Ding, X., Song, F., Qin, B., et al.: Study on the extraction of typical events in music. J. Chin. Inf. Process. **25**(2), 15–20 (2011)
15. Fu, J., Liu, Z., Liu, W., Shan, J.: Event feature recognition based on feature weighting. Comput. Sci. **03**, 239–241 (2010)
16. Shi, K., Liu, H., Song, W.: Text clustering based on part-of-speech and center-point improvement. J. Pattern Recogn. Artif. Intell. **06**, 996–1001 (2012)
17. Liao, T.: Text-oriented representation and its application. Shanghai University, Shanghai (2014)
18. LTP. http://www.ltp-cloud.com/intro/[OL]
19. Li, J.: Chinese text in the time, event and attribute identification. Shanxi University, Shanghai (2013)
20. Huang, Y.: Research on Chinese event extraction. University of Suzhou, Suzhou, China (2014)

A Comparative Method of Legal Documents Based on LDA

Zhanxing Hao, Xiao Wei(✉), and Hong Hu

School of Computer Science and Information Engineering, Shanghai Institute of Technology, Shanghai 201418, China
shawnwei@outlook.com

Abstract. In order to further strengthen the consistency and coordination of legal documents for crimes between public security bureau and prosecutors office and court, expand the fairness of cases and legal supervision channels, we will see whether there are any flaws in legal documents by comparing the basic facts of legal documents. In view of whether the description of the basic facts of the legal documents is consistent, a comparative method of legal documents based on the Latent Dirichlet Allocation (LDA) model. First, the legal text pretreatment, and then use LDA to build the MCMC corpus, by Gibbs sampling methods for reasoning, indirect calculation of model parameters, the text representation is fixed on the set of feature words implied probability distribution, the use of JS (Jensen-Shannon) distance to calculate the similarity between the texts. Through the analysis and comparison of relevant experiments, the results show that the comparison of legal documents based on the LDA model can achieve better accuracy and the clustering effect of the text.

Keywords: Comparison of legal documents · LDA model · Gibbs sampling JS distance · K-means algorithm

1 Introduction

In recent years, the Internet as an open and distributed information platform has been developed rapidly, the number and types of information are also exponentially explosive growth, the increase of the amount of information makes the information processing and the importance of the use of prominent in many fields. In the era of big data and artificial intelligence, how to deal with text is faced with so many texts information. It is a hot topic in the field of information processing in the era of big data and artificial intelligence. In text processing, text similarity is widely applied to other related fields, such as text mining, information retrieval, classification, clustering, machine translation, speech recognition and other related fields. The quality of text similarity results directly affects the result of text comparison.

The rapid development of information technology has provided strong technical support for the construction of judicial institutions. In the current society, the number of cases continues to grow, new types of cases complex has gradually increased, constantly enhance the legal awareness of the public in the new period and solve the

J. Abawajy et al. (Eds.): ATCI 2018, AISC 842, pp. 271–280, 2019.
https://doi.org/10.1007/978-3-319-98776-7_29

disputes, the public channels are also diversified, but often because there is no comparison and do not understand, which did not form a unified standard for identification of similar facts in a legal case the judicial reform, as a core issue of concern, the different sentence is one of the main reasons affecting the judicial public trust. The comparison of legal documents as a kind of application of text analysis provides a reference standard for unified referencing scale and promoting the standardization of similar cases and sentencing.

In the field of text mining, researchers both at home and abroad have done a great deal of research. The topic model originates from Latent Semantic Indexing (LSI) [1]. When using the LSI model to excavate the text, it takes into account the semantic relations among the words, and has a good dimensionality reduction effect, but LSI is not a probability model. Based on the theme idea of LSI, Hofmann put forward the Probabilistic Latent Semantic Indexing (pLSI) in 1999 [2], and pLSI is considered as a real topic model. Blei put forward the LDA topic model in 2003, and used LDA to model the text, and then used the Support Vector Machine (SVM) to classify the modeled text, which improved the accuracy of text classification [3]. Wang *et al.* [4] proposed one kind based on the LDA (Latent Dirichlet Allocation) method to calculate the similarity of text topic model, using LDA as the corpus modeling, inference using Gibbs sampling in MCMC, introduction of model parameters, exploring the relationship between the different main inscription hidden in the text in the text can be the subject distribution, then to calculate the similarity between the texts. Potential semantic analysis of text by using LDA model by Liu *et al.* [5], the semantic is divided into low, medium and high semantic, web free text detection based on the semantics of low frequency semantic region, and the high frequency region as text features for text clustering. The clustering results are corrected using text classification and semantic interoperability the mechanism, obtained good clustering effect. Li [6] proposed a LDA model for additional labels (Labeled-LDA), into the category of text information through the traditional LDA model, improve the classification ability of the model, it can calculate the implied theme distribution in each of the categories on the amount, in order to overcome the defects of the traditional LDA model for classification of compulsory distribution implied theme the effective improvement of the performance of text classification. Wilson [7] have proved that a reasonable weighting strategy based on LDA model can make high-frequency vocabularies naturally distribute, but not be allocated or dispersed randomly to all subjects. In the standard LDA model, each characteristic word is considered as equally important. Obviously, this assumption is not applicable in the legal texts. In legal texts, the most characteristic words are the time, place, person and event attribute. Therefore, it is necessary to combine weighting methods to adjust the weight of different feature words to improve the result of the analysis of text similarity.

Comprehensive analysis of above text features and text similarity literature method, the Latent Dirichlet Allocation model weighted to model text, using the statistical characteristics of the text, can effectively reduce the dimension of text representation, and can solve the problem of synonyms and polysemy, calculation method and similarity without using external dictionary. And the text is represented by the multi granularity feature, and the comparison between text is carried out by calculating the similarity of the important word of the text.

2 LDA Model

LDA (Latent Dirichlet Allocation) is a document topic generation model proposed by Beli for discrete data sets, also known as a three level Bayesian probability model, which contains three layers of words, themes and documents [3]. LDA is an unsupervised learning technology, which can be used to identify the latent topic information in large scale document sets or corpus. It uses the way of bag of words to transform text information into digital information that is easy to model. The LDA model is similar to the word clustering approach will be similar to the word clustering as a theme, with the semantic meaning between the theme and the theme, in the same theme in the lexical entry with synonym characteristics, and the same word in different theme is polysemy characteristic, thereby eliminating the calculation the similarity between the lexical entry in the process of the calculation of text similarity, distribution and utilization of the theme of the text can calculate the text similarity. Documents, topics, and words can be expressed in a directed probability graph.

In Fig. 1, a square frame represents a loop, the hollow node represents an implicit variable, and the arrow indicates a dependency. The parameter α represents the relative strong and weak relations between the different hidden subjects in the text centralization, and the parameter β represents the probability distribution of the subject itself. K is the number of topics, M is the total number of documents, and Z represents the hidden theme, and W represents each word of the word list. θ represents the text-subject probability distribution, and φ represents the probability distribution of the subject-word.

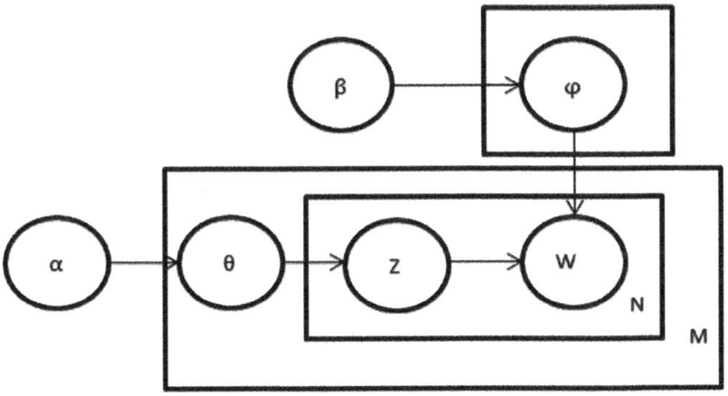

Fig. 1. LDA model directed probability graph

In the construction of the LDA model, the parameter estimation of the model needs to be done, the commonly used methods include expectation propagation algorithm, variational Bayesian inference and Gibbs sampling, parameter inference method based on Gibbs sampling can be effectively extracted from the text theme, Gibbs extraction algorithm is the most popular [8, 9] extraction algorithm in LDA model. In the LDA model, the most important thing is that the two sets of parameters are the probability distribution of word items and the probability distribution of the subject matter of each document. According to the graph model, the probability value of a text can be obtained as follows:

$$p(w|\alpha,\beta) = \int p(\theta|\alpha)\left(\prod_{n=1}^{N}\sum_{Z_n} p(z_n|\theta)p(w_n|z_n,\beta)\right)d\theta \tag{1}$$

The Gibbs method is the first to sample the subject of each word. As long as the subject of the word is determined, then the parameters can be calculated after the statistical frequency. Therefore, the parameter estimation problem becomes the conditional probability of calculating the subject sequence under the word sequence. The formula is as follows:

$$p(z_i = k|\overrightarrow{z_{-i}},\vec{w}) = \frac{p(\vec{w},\vec{z})}{p(\vec{w},\overrightarrow{z_{-i}})} \propto \frac{n_{k,-i}^{t}+\beta_t}{\sum_{t=1}^{V} n_{k,-i}^{t}+\beta_t}\left(n_{m,-i}^{k}+\alpha_k\right) \tag{2}$$

In the formula, z_i represents a topic variable corresponding to a word i, $-i$ indicates the remove of the item i, n_t^k represents the number of word items t in the topic K; β_t is the Dirichlet prior of the word term t; n_m^k represents the number of occurrences of topic k in text m; α_k is Dirichlet prior of subject k. The theme of each word label is obtained, the calculation formulas of parameters can be obtained by the following formula.

$$\varphi_{k,t} = \frac{n_k^t+\beta_t}{\sum_{t=1}^{V} n_{k,-i}^{t}+\beta_t} \tag{3}$$

$$\vartheta_{m,k} = \frac{n_m^k+\alpha_k}{\sum_{t=1}^{K} n_m^k+\alpha_t} \tag{4}$$

In (3) and (4), $\varphi_{k,t}$ represents the probability of the term t in topic k; $\vartheta_{m,k}$ represents the probability of subject k in text m. Knowing the subject mark of each word, you can estimate the parameters by simple count.

3 Text Contrastive Analysis

3.1 Text Preprocessing

Data preprocessing is an important early process of data mining. Through preprocessing, the original data is structured and standardized, and it is converted into an inputtable form of LDA model. Before studying the application of the comparison of legal instruments, we must first preprocess the text. In this process, it is first necessary to segment the text, and the segmentation uses the language technology platform [10] (LTP-Cloud) to segment the text. By inputting legal texts, the Chinese character sequence is cut into word sequences using a word segmentation system. After word segmentation filters out words that appear in the text with high frequency but no legal text event information, Include stop words, prepositions, conjunctions and other non-characteristic words. It can not only reduce the dimension of the text, but also improve the accuracy of the subsequent processing of the text.

3.2 Feature Extraction

In the above LDA model, the weight of each feature word and the multi-granularity feature of the text are not considered, which will affect the accuracy of the topic model for the legal text processing. In the legal text, the words which can better express their characteristics are mainly the time, place, character and event attribute in the legal text. The degree of importance of different feature in the text is different, so the addition of appropriate weights applied to the process based on LDA model to generate feature words, then different feature with different weights, thus obtained the feature words more representative of the significance of legal instruments, which can improve the contrast effect of the text.

In the text of the method to make some adjustment and improvement on the standard LDA model, reduce the probability distribution in the generation process for those words that appear frequently but that are insufficiently expressed for textual ability. At the same time, improve the probability distribution in the generation process for words that have a low frequency but have a significant meaning of textual features. The generation of characteristic words is shown as follows.

$$p(w|\theta, \varphi) = \frac{weight(t, d) \cdot \varphi_{k,t}}{\sum_{t=d} weight(t, d)} \tag{5}$$

The characteristic word generation process in the standard LDA model is introduced into the process of Gibbs Sampling, and the adjusted formula is shown as follows:

$$p(z_i = k|z_{-i}, w, \alpha, \beta) \propto p(z_i = k, w_i = t|z_{-i}, w_{-i}, \alpha, \beta) \propto \frac{weight(t, d)n^t_{k,-i} + \beta_t}{\sum_{t=1}^{V}(weight(t, d)n^t_{k,-i} + \beta_t} \cdot \frac{weight(t, d)n^k_{m,-i} + \alpha_k}{\sum_{k=1}^{K}(weight(t, d)n^k_{m,-i} + \alpha_k} \tag{6}$$

Combining the mutual information, the point mutual information calculation of the feature word w_i and the text d is shown by the formula:

$$PMI(w_i, d) = log_2 \frac{p(w_i, d)}{p(w_i)P(d)} \tag{7}$$

Adjust the weight of each feature word and extend it to the LDA model. The adjusted feature word w_i and the point mutual information of the text d are calculated as follows:

$$weight(t = w_i, d) = \sum_{j=1}^{t} \lambda \cdot \mu_j \cdot PMI(w_i) \tag{8}$$

Where μ_j represents the weight, and $\mu_1 > \mu_2 > \cdots > \mu_n$, $\sum_{j=1}^{t} \mu_j = 1$, λ is an adjustment factor.

For text d_i, generate the feature words through the above improved LDA model, and the feature vectors are obtained $d_i = (t_{i1}, t_{i2}, \ldots, t_{im})$. Considering that the feature words obtained by using only the LDA model are more generalized, the ability to represent text details is insufficient. Therefore, this article introduces some feature words that can effectively describe the details of the text. First, we calculate the TF-IDF value of each feature word in the text [11]. Sort them and use words above a certain threshold δ to represent the text, and get the $d_i = (w_{i1}, w_{i2}, w_{i3}, \ldots, w_{in})$. Combining the features of the above two types, the feature vectors of the text can be finally obtained $d_i = (t_{i1}, t_{i2}, t_{i3}, \ldots, t_{im}, w_{i1}, w_{i2}, w_{i3}, \ldots, w_{in})$.

3.3 Similarity Calculation

After the Gibbs sampling algorithm, the word vector space of the text can be mapped to the feature word vector space. Next, the result can be used as a text similarity calculation. In the subject representation of the text, the similarity of two texts can be calculated by calculating the probability distribution of the theme corresponding to it. Since the topic is a mixed distribution of word vectors, the JS distance is used to represent the similarity between two text vectors p and q. The formula of distance calculation for JS is as follows:

$$D_{js}(p, q) = \frac{1}{2} \times \left(D_{kl}\left(p, \frac{p+q}{2}\right) + D_{kl}\left(\frac{p+q}{2}\right)\right) \tag{9}$$

Where D_{kl} is KL distance and KL distance is from the perspective of information entropy, also called relative entropy, to measure the difference between two probability distributions in the same event space. The calculation formula is as follows.

$$D_{kl}(p, q) = \sum\nolimits_{j=1}^{T} p_j \ln \frac{p_j}{q_j} \tag{10}$$

As for all j, when $p_j = q_j$, $D_{kl}(p, q) = 0$. But the KL distance is not symmetric, and $D_{kl}(p, q) \neq D_{kl}(q, p)$, so often use the symmetric version:

$$D_{kl}(p, q) = \lambda D_{kl}(p, \lambda p + (1-\lambda)q) + (1-\lambda) D_{kl}(p, \lambda p + (1-\lambda)q) \tag{11}$$

When $\lambda = \frac{1}{2}$时, $D_{kl}(p, q) = D_{kl}(q, p)$, distance range of JS is [0,1]. In this paper, we use the JS distance formula as a standard to measure the similarity between texts.

3.4 Evaluation Method

Using a balanced indicator commonly used in information retrieval, measure the similarity of text by the F-metric in the K-means algorithm. The F measure consists of the precision *P(i, j)* and the recall rate *R(i, j)*, and the precision and recall rate are shown as follows.

$$P(i,j) = \frac{n_{ij}}{n_j},\ R(i,j) = \frac{n_{ij}}{n_i} \tag{12}$$

n_i represents the number of texts that belong to category J; n_i represents the number of texts that actually belong to category i; n_{ij} represents the number of texts that belong to j and also belongs to i. The F metric is defined as follows.

$$F(i,j) = \frac{2 \times p(i,j) \times R(i,j)}{p(i,j) + R(i,j)} \tag{13}$$

The F metric values of text clustering are defined as follows.

$$F = \sum\nolimits_i \frac{n_i}{n} max(F(i,j)) \tag{14}$$

The F value quantitatively analyzes the accuracy and comprehensiveness of the extracted topics from the macro perspective, and the ensuing topical information entropy quantitatively analyzes the degree of confusion in the topic extraction from a microscopic point of view [12]. The greater the value of the topic information entropy, the less clear the semantic information of the extracted topic and the higher the degree of confusion, the worse the effect of the topic extraction. The expression of subject information entropy is as follows.

$$H(T) = -\sum\nolimits_{n=1}^{N} p(w_n|T) log_2 p(w_n|T) \tag{15}$$

In the above formula $T = p(w_n|z_n, \beta)$, it represents the probability distribution of the characteristic word ω_n under the specified topic z_n and parameter β.

4 Experiments

4.1 Corpus Selection

In this paper, we collected and sorted the corpus of legal texts. The corpus was divided into four categories: traffic accidents, drug trafficking, theft and fraud. There were 1680 articles in each category.

4.2 Experimental Setup

In the modeling process of LDA model, the parameters of LDA mainly include the number of topics K, hyperparameters α and β, the empirical values of α and β are used, which are 50/K and 0.1 respectively, the experimental results are better [12], and the value of K is 20. In order to ensure the accuracy of the experimental results, Gibbs sampled iterations need to reach more than 1000 times. This experiment is carried out under the Windows7 operating system, and the computer CPU is Intel (R) Core (TM)

2 CPU E8400@ 3.00 GHz and memory 12G. The software for extracting feature words and modeling LDA model is Python27.

4.3 Analysis of Experimental Results

In the calculation of text similarity, we first need to know the word similarity threshold μ can affect the specific performance and distribution of clustering results in selected text feature weighted LDA model, through the classical K-means clustering algorithm to cluster the experimental results, the experimental results are as follows (Fig. 2).

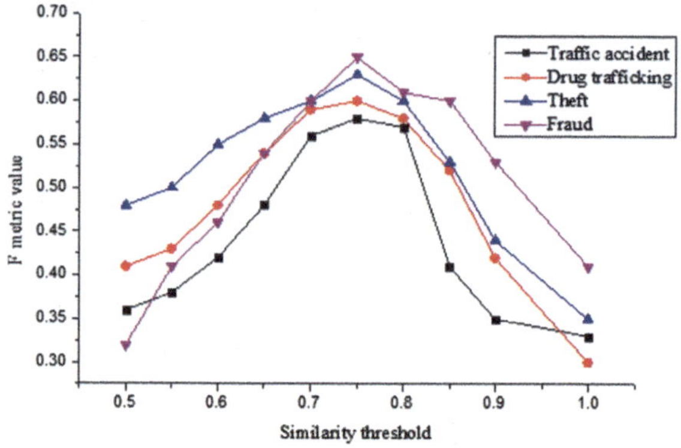

Fig. 2. Similarity threshold value and F measurement value

From the graph, we can see that the clustering result is related to the similarity threshold value of the characteristic words in the same cluster. According to the experimental results, the threshold value of word similarity is 0.75.

Before calculating the precision, recall and F value, we need to determine the correct theme number, effective theme number and the correct number of subjects. The accuracy, recall, and F value of the four types of text data topic extraction are shown in Table 1.

Table 1. Precision, recall, and F value of text data topic extraction

Text type	Precision rate	Recall rate	F value
Traffic accident	76.1	75.21	75.65
Drug trafficking	72.46	70.78	71.60
Theft	73.51	72.76	73.13
Fraud	80.67	76.54	78.50

In Table 1, the precision, recall and F value of the type of fraudulent text are higher, and the type of traffic accident text is the second. Among these four types of text, the precision, recall and F value are not only related to the type of text, but also to the length of text. The limitation of the number of short text may not be able to fully express the logical relation of text.

The above is the comparison of different types of text data, and the effect of theme extraction is compared. Next, we compare the theme effect from the perspective of thematic information entropy. We use the (15) formula to calculate the information entropy extracted from LDA for different texts, as shown in Table 2.

Table 2. Information entropy of text data extraction topics

Text type	Thematic information entropy
Traffic accident	6.45
Drug trafficking	5.89
Theft	5.86
Fraud	6.01

The subject information entropy of theft is the lowest in Table 2. The extraction effect of the theme is the best. The extraction effect of traffic accident topic extraction effect relative theft is poor.

5 Conclusions

LDA model is an important method of data mining, this paper proposes a method of comparative legal text based on the LDA model, not only can the dimension be reduced effectively in the model, but also multi-granularity features can be used to express texts, which effectively avoids the sparsity caused by too fine granularity and improves the computational efficiency of text similarity. In the process of text modeling and computing similarity, no external dictionaries are used. The effect of similarity computation is entirely based on the characteristics of different text sets, so the computation method in text is relatively good. Experiments show the effectiveness of this method in comparing legal texts. The next work will continue to study on the basis of the LDA model, optimize the text modeling method, and improve the accuracy of the text comparison.

Acknowledgments. Research work reported in this paper was supported by the Science Foundation of Shanghai under grant no. 16ZR1435500.

References

1. Deerwester, S., Dumais, T., Landauer, G., et al.: Indexing by latent semantic analysis. J. Am. Soc. Inf. Sci. **41**(6), 391–470 (1990)
2. Hofmann, T.: Probabilistic latent semantic indexing. In: Proceeding of the 22th Annual International SIGIR Conference on Research and Development in Information Retrieval, pp. 50–57. Association for Computing Machinery, Univca (1990)
3. Blei, D.M., Ng, A.Y., Jordan, M.I.: Latent Dirichlet allocation. J. Mach. Learn. Res. **3**(3), 933–1002 (2003)
4. Wang, Z.Z., He, M., Du, Y.P.: Computing of text similarity based on LDA theme model. Comput. Sci. **40**(12), 229–231 (2013). (in Chinese)

5. Liu, Z.L., Wang, D., Feng, S., et al.: A LDA based potential partition and Web document clustering algorithm. Chin. J. Inf. 25(1), 60–67 (2011). (in Chinese)
6. Li, W.B., Sun, L., Huang, R.H., et al.: A new algorithm for text classification based on Labeled-LDA model. J. Comput. **31**(10), 1780–1788 (2008). (in Chinese)
7. Li, W.B., Sun, L., Zhang, D.K.: Text classification based on labeled-LDA model. Chin. J. Comput. **31**(4), 620–627 (2008)
8. Wilson, A.T., Chew, P.A.: Term weighting schemes for latent Dirichlet allocation. In: Human Language Technologies: The 2010 Annual Conference of the North American Chapter of Association for Computational Linguistics, pp. 465–473. Association for Computational Linguistics (2010)
9. Doucet, A., Godsill, S., Andrieu, C.: On sequential Monte Carlo sampling methods for Bayesian filtering. Stat. Comput. **10**(3), 197–208 (2000)
10. Ma, H.Y.: Based on Gibbs sampling test cast generation count study. Autom. Instrum. (3), 11–12 (2011). (in Chinese)
11. Liu, T., Chen, W.X., Li, Z.H.: Language technology platform. Chin. J. Inf. **16**(11), 53–61 (2011). (in Chinese)
12. Huang, C.H., Hou, F.: A combination of lexical entry semantic information and TF-IDF method of text similarity measure method. Chin. J. Comput. **34**(5), 856–863 (2011)
13. Ruthven, I., Lalmas, M.: A survey on the use of relevance feedback for infor mation access systems. Knowl. Eng. Rev. **18**(2), 95–145 (2013)

Feature Selection Approach for Twitter Sentiment Analysis and Text Classification Based on Chi-Square and Naïve Bayes

S. Paudel[1(✉)], P. W. C. Prasad[1], Abeer Alsadoon[1], MD. Rafiqul Islam[1], and Amr Elchouemi[2]

[1] School of Computing and Mathematics, Charles Sturt University, Sydney, Australia
cwithana@studygroup.com
[2] Walden University, Minneapolis, USA

Abstract. With the rapid growth of web and mobile technology, Social networking services like Twitter are widely used, resulting in large amounts of data being generated daily in social networking sites. Efficient Sentiment analysis of such data is very important for a range of applications and improvement of accuracy in detecting sentiment is the main aim of this research. This report examines the combination of a Chi-Squared feature selection algorithm, k-mean clustering and TF-IDF for attribute weighting based on Naïve Bayes, for classification of text and sentiment in communications generated on Twitter. This approach is compared with other approaches based on Naïve Bayes to give an account of their relative strengths and weaknesses. When running experiments on multi-domain twitter datasets, results indicate that the proposed method shows superior performance across a range of. The main aim of this research is to enhance the performance of the Naïve Bayes classifier using a feature selection technique.

Keywords: Chi-squared · Feature selection · Naïve Bayes · TF-IDF
Twitter sentiment analysis

1 Introduction

The increasing use of social networking sites like Facebook, Twitter, google and similar platforms, where users can post their thoughts and opinions about any topic, has changed the web from a static repository to a dynamic forum, where information is changing continuously. In this type of medium, sentiment analysis is significant for the validation of people's opinion and to facilitate modifications to user interfaces (UI) [1]. In terms of expressed opinions, Twitter is one of the biggest social network platforms with 326 million active users worldwide and 500 million tweets per day. Twitter Dataset is widely accepted as a sentiment analysis dataset because of its simplicity and easy accessibility [2]. Using different machine learning techniques, sentiment analysis classifies emotions contained in tweet messages as positive, negative and neutral.

J. Abawajy et al. (Eds.): ATCI 2018, AISC 842, pp. 281–298, 2019.
https://doi.org/10.1007/978-3-319-98776-7_30

Machine learning is a technique which allows computers to learn and predict results based on a topic [3]. Among different machine learning algorithms, Naïve Bayes is considered one of the most effective algorithms and it is widely used for classification [4]. Chi square is an algorithm designed for feature selection, which combined with Naïve Bayes improves accuracy and performance of classification to overcome the limitations of the previous method and provide an optimal solution for twitter sentiment analysis.

A review of different approaches to sentiment analysis revealed that, in general, the machine learning based methods yield more precise results than the lexicon-based methods [5]. However, machine learning based approaches require more time and effort, because a large set of manually classified data needs to be prepared to train the classifier [6]. This report aims to enhance the classification approach of twitter sentiment analysis based on the Naïve Bayes algorithm with the aim of increase accuracy.

Twitter Sentiment analysis is the measurement of feelings; attitude and emotions on tweets, and people express their feelings in diverse ways, text, images and emoticons [7]. In most of the approach emoticons and acronyms used in tweets are removed as noise but with only 140 characters are allowed on each tweet on twitter people are using a different kind of slang and emoticons on twitter. Therefore, the sentiment should be classified by measuring these emoticons and acronyms as well.

Basically, there are two approaches for sentiment analysis: one is a lexicon-based approach and other is machine learning approach. Lexicon based approach can be further divided into two sections, mainly corpus-based approach, and Dictionary based approach. Corpus based approach can be subdivided into two methods; one is a semantic method, and another is statistic method. Support Vector machine (SVM) and neural network is type of linear classifiers. Naïve Bayes, Bayesian Network and Maximum Entropy are the types of probabilistic classifier [8].

The purpose of this study is to increase accuracy of twitter sentiment classification as a positive or negative. To enhance the classification approach, we apply clustering using attribute weighting and feature selection approach. In addition, the secondary aim of this study is to use emoticons and acronyms on tweets for twitter sentiment classification. The implementation of feature selection algorithm in machine learning classifier is expected to increase the accuracy of the twitter sentiment classification based on Naïve Bayes as well as the addition of emoticons and acronym on preprocessing is expected to helpful for better performance of twitter sentiment analysis.

Deep learning in twitter sentiment analysis requires complex hardware and computational resources and not all companies are willing or able to invest heavily to obtain sentiment analysis. This paper targets twitter sentiment classification using limited resources using a combination of Naïve Bayes and feature selection algorithms to enhance performance and accuracy. Obstacles to accurate sentiment analysis come from the detection of sarcasm detection, identification of negation, implicit product attributes handling, and the complexity of sentences and documents [9].

The remainder of the report is structured into five main sections as follows: Background and literature Review in Sect. 2. The proposed work is given in Sect. 3. Section 4 presents experiments and results. And finally, the conclusion and future research are in Sect. 5.

2 Related Work

2.1 Twitter Sentiment Analysis

There are three approaches to sentiment analysis, lexicon based, machine learning based and a hybrid of the former. For the purpose of this research, the review of literature is confined to the machine learning technique.

Using different machine learning techniques, sentiment analysis classifies a tweet message and emotion as positive, negative or neutral. Figure 1 shows the basic structure of machine learning sentiment analysis demonstrated on an example from twitter [10].

Sentiment analysis is the study of people's opinions and emotions on a particular topic, identifying the sentiment expressed in the text and classifying it as positive, negative or neutral. Sentiment analysis is a process of classification that takes place at document, sentence and aspect level [11]. The analysis process includes data acquisition, data pre-processing, sentiment extractions and sentiment analysis.

Sentiment analysis can be carried out using a range of techniques. However, the main methods currently used are based on machine learning. One machine learning technique is using a binary shuffled frog algorithm for twitter sentiment classification [12]. The work is carried out by using Information Gain (IG), Genetic Algorithm (GA) and Binary Shuffled Frog (BFA) algorithm. The classifiers used for the extracted features are Naïve Bayes, K-Nearest Neighbour (KNN), RBF network and Logistic Model Tree (LMT). The researchers observed that RBF performed better than other algorithms. When the classifier is applied in conjunction with feature selection algorithms GA and IG, accuracy is increased by 1.57%. Experiments revealed that this approach enhances both performance and accuracy. Furthermore, GA combined with the IG algorithm as feature selection improves the performance.

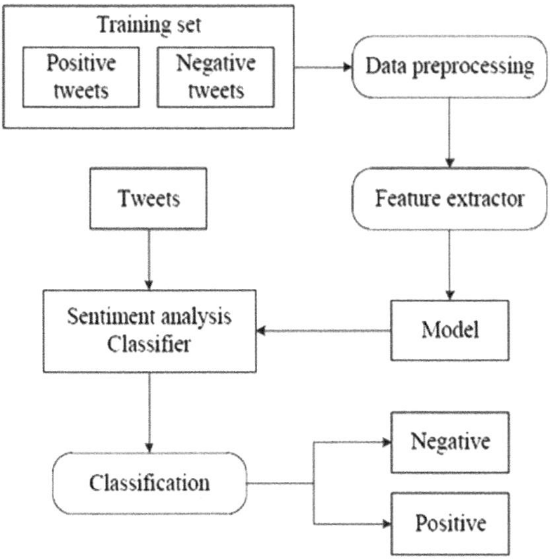

Fig. 1. Structure of twitter sentiment analysis

Devika [3] carried out a comparative study on different approaches to sentiment analysis and the level of analysis. Machine learning techniques like Naïve Bayes, support vector machine, and maximum entropy methods were studied and compared. In this study, researchers reported that all approaches meet accuracies of more than 80%.

Hai et al. [13] worked on ratings and reviews and aimed to identify semantic aspects as well as sentiments at the aspect level. They developed a new model, the Supervised Joint Aspect and Sentiment Model (SJASM), which represents documents as opinion pairs and models aspect terms and opinion words within the document for unseen aspects and detection of sentiments, while the online reviews and ratings are time stamped. This model does not provide any approach for time and is therefore, of no further use.

In a later study, Pletneva et al. [14] presented a new text sentiment analysis method based upon the weight of a sentiment word. For search space reduction of optimal weights, the co-clustering is used for words and documents. Dual valued sentiment (positive or negative) is selected as a base, yet the outcomes can be general on N-valued scale. This method requires a sentiment lexicon with a high degree of emotional words as well as corpus that contains test and training documents. Text collection from the Information Retrieval Seminar (ROMIP) which is publicly available is used in the experiment which confirms the effectiveness of this approach in comparison with other classifiers.

3 Current Best Solutions

The operation flow of the current best scheme developed by Song [10] is presented in this section. Figure 2 illustrates the operation flow of a novel attribute weighting and feature selection approach based on Naïve Bayes for Twitter sentiment analysis. The two methods are adopted in this scheme one is for attribute weighting, and another is for feature selection. The feature selection method is completely dependent on the attribute weighting method. The performance and results of the attribute weighting has an effect on feature selection which ultimately affects the classification of tweets. The data pre-processing [10] has removed emoticons and acronym used in tweets as noise, but they are important properties because people use emoticons and different acronyms in their tweet to express their feelings and sentiment due to the 140-character limitation in one tweet.

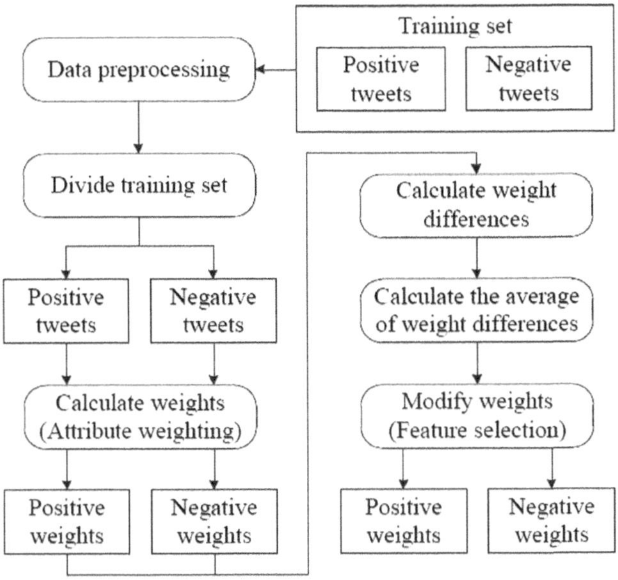

Fig. 2. Operation flow of current best solution

4 Proposed Scheme

The Proposed scheme is presented in this section, which consists of Term frequency and inverse document frequency (TF-IDF); attribute weighting and Chi-square feature selection using Naïve Bayes for twitter sentiment analysis. The Fig. 3 illustrates the structure and the operation flow of the proposed approach.

4.1 Data Pre-processing

Data from twitter contain different kinds of words and sentences, which are ex- pressed by different users in many different ways.

So, elimination of unnecessary and redundant data is an important process for sentiment analysis, and datasets need to be labelled as positive or negative to train the machine learning algorithms [15]. Pre-processing of data also contains removal of non-English tweets, numbers, URLs and email addresses.

4.1.1 Tokenisation

The first task of pre-processing twitter data sets is to tokenize the tweets. This is done by using Tweet NLP, which is a twitter specific tagger; it tags the twitter specific entries like Hashtag, emoticons and mentions (@) too.

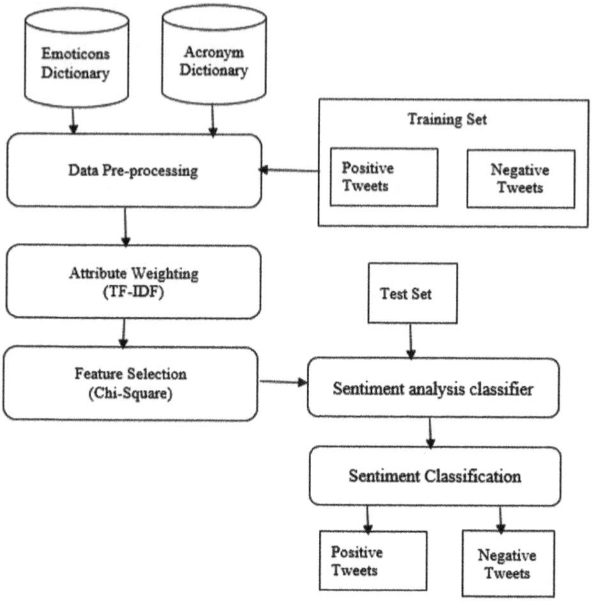

Fig. 3. Operation flow of proposed approach

4.1.2 Non-English Tweet Removal

Twitter supports more than sixty languages but this research only focuses on the English language requiring the removal of non-English tweets.

4.1.3 Emoticon Replacement

As previously established, in the process of determining sentiment, emoticons play a very important role. Hence, they are replaced with sentiment polarity obtained from the Emoticons Dictionary.

4.1.4 Acronym Replacement

Because of character limitation in twitter, acronym is used in most of the tweets, requiring the replacement of such acronyms with a term from by the Acronym Dictionary. For example, "ASAP" is replaced by "as soon as possible".

4.1.5 URL Removal

URLs in tweets are shortened because of character limitations, and, in any case, do not carry any information regarding sentiment - they need to be removed.

4.1.6 Target Removal

'@' is used to mention the handle of a twitter account which is also not important regarding sentiment analysis. Hence, they are removed.

4.1.7 Hashtags

Hashtags are basically used for summarisation and are very important. However, the tag itself (#) is a special character and needs to be removed together with other punctuation symbols.

4.1.8 Remove Numbers

Numbers are not important in terms of the measurement of sentiment, so numbers are removed.

4.1.9 Remove Nouns and Prepositions

If most commonly used form of a word is a noun, those words are removed because nouns do not carry sentiment and are of no use. The same applies for preposition.

4.1.10 Remove Stop-Word

In sentiment classification, 'stop' words occur in both positive and negative training sets and do not carry any sentiment information. Stop word are he, she, at, a, the, etc. and are removed.

4.2 Attribute Weighting

Bag of words (BoW) is one method to represent a document, where words most frequently occurring in a document represent each document. Term frequency and inverse document frequency (TF-IDF) weighting scheme balance the number of occurrences of a word in a document [16].

$$w_{(t,d)} \; \log(1 + tf_{(t,d)}) \times \log_{10}(N/df_t) \tag{1}$$

The Eq. (1) ascribes a value of the feature to that document and a weight for each word. The formula for the proposed method is modified as follows:

$$\arg\max_{c \in C} P(c_j) \prod_{t \in A} P(t|c) \tag{2}$$

4.3 Feature Selection

The main problem for text classification is the high dimensionality of feature space. Sets of unique terms or words that occur in all documents are the feature set of a text document. Using feature selection has some advantages. It provides better visualization and understandability, reduction over fitting and generalization of the model, and it provides efficiency in terms of time and space complexity [17].

The feature selection approach can be defined as: select m features from the original n features ($m \leq n$). To represent the contents of the text, the m feature can be more concise and efficient. The initial step in sentiment classification is feature selection, to extract and select text features [18]. Some of the features are: Term presence and Frequency. These features are individual word n-grams and the frequency count of these words. It gives the words 1 or 0 (binary weighting). Parts of Speech (POS): Adjectives are important indicators of sentiment. Opinion words and phrases: Commonly used words for sentiment expression like good or bad [19]. Different approaches for feature selection can be classified as filters and wrappers. Using the filter method, features or variables are selected independently in relation to the classification approach used. The wrapper method depends on the classification methods used [20].

(a) Chi-Square (X2)

In feature selection, occurrences of the class and occurrence of the feature are two events. If the two events are independent occurrences of the feature can be used for occurrence of the class. Our aim is to select features of highly dependent class. In case of two independent events, if the difference between observed count and expected count is small, or they are close, a small chi square score and a high value of X2 signifies a greater likelihood of a correlation between feature and class [21].

(b) Chi-square test for feature selection.

Suppose a training instance which belongs to negative and positive class. To calculate the chi square (x2) score of feature X is shown in following Table 1 [22].

Table 1. Term document matrix

	Positive class	Negative class	Total
Feature X occurs	A	B	A + B = M
Feature X not occurs	C	D	C + D = N − M
Total	A + C=P	B + D = N − P	N

It is basically used to measure the absence of independence between 't' and 'c' Where, t = term and c = class and compared to the x2 distribution with one degree of freedom. x2 static is expressed as Eq. 3.

$$X^2_{(t,c)} = \frac{D \times (PN - MQ)^2}{(P + M) \times (Q + N) \times (P + Q) \times (M + N)} \quad (3)$$

Where, D = total no. of documents
P = no. of documents of class 'c' containing term 't'
Q = no. of documents containing 't' occurring without 'c'
M = no. of document class 'c' occurring without 't'
N = no. of documents of other classes without 't'

4.4 K-Mean Clustering

K-mean clustering is one of the simplest and most popular algorithms. The process of partitioning a group of data into a small number of clusters is clustering. K-mean stores k centroids that it uses to define clusters. If a point is closer to that clusters centroid than any other centroid, it is considered to be a cluster [23]. K-mean algorithm works as:

K is the one input that k-mean algorithm expects, which is the number of the cluster. Initialize Centre of the cluster

$\mu i = \text{value}, i = 1, \ldots, k$

Find attribute which is closest cluster to data point.

$$c_i = \{j{:}d(X_j, \mu_l), l \neq i, j = 1, \ldots, n\} \quad (4)$$

Set position to mean of every data points of that cluster.

$$\mu_i = \frac{1}{|c_i|} \sum\nolimits_{j \in ci} x_j, \forall i \quad (5)$$

Repeat step b – c until convergence

Where, $|c|$ is number of elements in c.

4.5 Proposed Algorithm

Description of algorithms: The following algorithms accept two parameters:

- The input document matrix corresponding to the tweets: TM
- A float input for threshold: thresh (For this case thresh is taken as '0')

Step 1. Apply feature selection technique based on X2 (chi-square) on the TM (D*W) to compute (CH) value for each word.
Step 2. Select words whose value is greater than thresh.
Step 3. Form a document matrix (M) with words selected from step 2.
Step 4. Transpose (M). Each word is represented by a row. Transposed (M) is represented by 'N'.
Step 5. Create 'nc' clusters on transposed matrix N.
Step 6. Select the most demonstrative words which are closest to the centre.
Step 7. Add them to F such n(F) = nc.
Step 8. Calculate the Euclidian norm for every point in a cluster between the point and centre. Select nearest one.
Step 9. Return Features set F.

5 Experiments, Results and Analysis

5.1 Methodology

For the experiment, the steps below are followed

a. Text documents are stripped of spaces and punctuation.

b. Stop words, numbers, hash tags, targets, email addresses and URL are removed from twitter data.
c. Emoticon and Acronym replacement is carried out using respective dictionaries.
d. Term document matrix is prepared.
e. Stemming is applied.
f. TF-IDF weighting method is used.
g. .Term document matrix is used in the experiment.
h. Comparison of results with other classifiers with proposed feature selection and without feature selection.

5.2 Hardware and Software Used

Intel Core – i5 2430 M 2.4 GHz
Memory: 4 GB DDR3
Operating System: Windows 10 Enterprise 64 bit
Rapid Miner Studio 7.2 with extensions of Weka Packages

5.3 Framework and Tools

The platform of choice is Rapid Miner which is a software platform for data preparation, machine learning, and predictive model deployment. The list of packages installed are Rapid Miner, WEKA, Python scripting, and Text processing.

5.4 Dataset Information

To calculate accuracy and performance of the purposed algorithms, different subsets of 1,600,000 data sets supplied by sentiment 140 are used. A 5:5 ratio of positive and negative classes are used in the test sets as well as the training sets. Table 2 shows the number of words before processing and after processing in our dataset.

Accuracy Calculation of twitter sentiment analysis is presented as an Eq. (8), explained in Table 2 below:

Table 2. Dataset information

Number of documents		Before processing	After processing
5,000	Total	12,412	7,619
	Positive	7,421	4,827
	Negative	4,991	2,792
10,000	Total	25,242	14,438
	Positive	13,532	8,357
	Negative	11,710	6,081
15,000	Total	35,256	18664
	Positive	19,322	9,254
	Negative	15,934	9,410

$$\text{Accuracy} = \frac{\text{TP} + \text{TN}}{\text{TP} + \text{TN} + \text{FP} + \text{FN}} \times 100\% \quad (6)$$

TP = Number of True Positive documents
TN = Number of True Negative documents
FP = Number of False Positive documents
FN = Number of False Negative documents

TP: True Positive, FP: False positive, FN: False Negative, TN: True Negative.

The experiment was conducted on four public data sets, which are summarized in Table 3; Including Hotel, computers, Books, DB World downloaded from UCI Machine Learning Repository.

Table 3. Confusion matrix

Actual\predicted	Positive	Negative
Positive	TP	FN
Negative	FP	TN

5.5 Results and Analysis

Figure 4 shows the accuracy of twitter sentiment analysis with the proposed algorithm using 1,000 test sets and different training sets. The accuracy with the Multi Nominal Naïve Bayes (MNB), and the Bernoulli Naïve Bayes (BNB) schemes are also compared. From Fig. 5, it is evident that the accuracy of the proposed scheme is higher than the other schemes. As the size of the training set increases, accuracy of all the schemes also increases as expected.

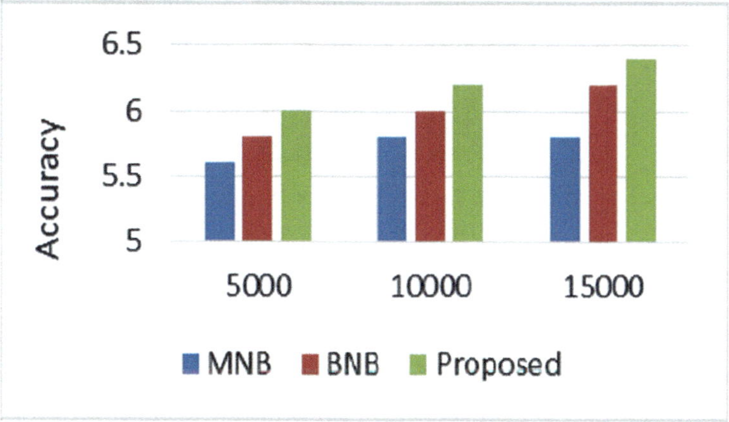

Fig. 4. The accuracies of twitter sentiment analysis for 1000 test set

Figure 5 shows accuracy with 2,000 test data sets. The results achieved and the figures below show that an equivalent result as for the 1,000 test set is achieved.

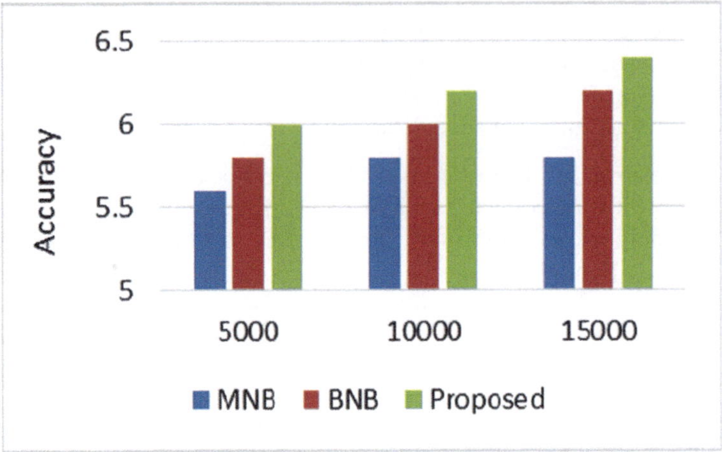

Fig. 5. The accuracies of twitter sentiment analysis for 2000 test set

As mentioned earlier, feature selection and attribute weighting are important steps in sentiment analysis, as a new algorithm for improving accuracy was proposed. Figure 6 shows the relative effectiveness with 1000 test sets.

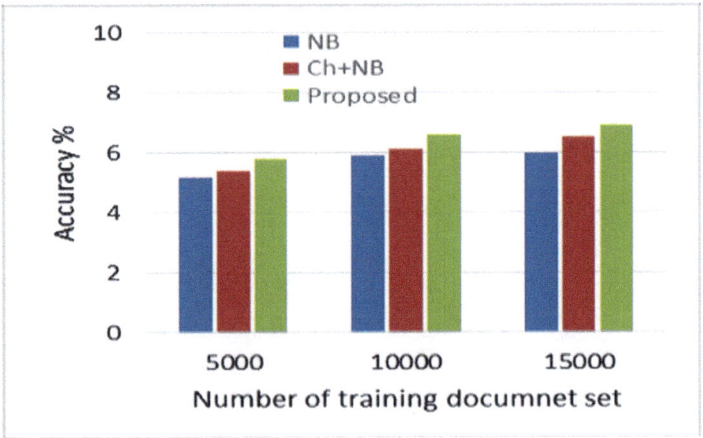

Fig. 6. The relative effectiveness with 1000 test sets

Here the attribute weighting is for the case of sentiment analysis based on the proposed approach without feature selection. It can be noticed that feature selection is always more influential in terms of accuracy than attribute weighting regardless of the number of training documents. Also, the proposed approach is always more effective than the existing approach as identified from the comparison.

Classification Accuracy on the twitter dataset using (a) Naïve Bayes (b) chi-square with Naïve Bayes (c) proposed algorithm with Naïve Bayes. Comparison of the result of proposed algorithm with the Naïve Bayes and other standard classifiers for 2000 test sets given in Fig. 7. With the proposed algorithm, feature space can be reduced. Using (a) chi-squared and (b) proposed algorithm, the total number of features and the reduced number of features are shown in Fig. 8.

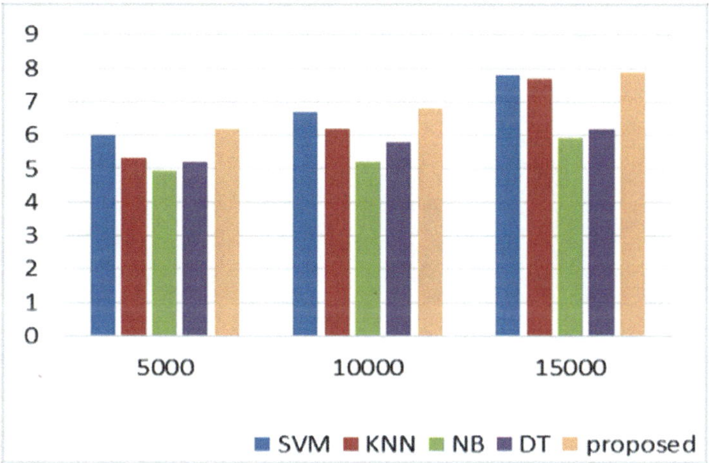

Fig. 7. Classification accuracy rate of proposed algorithm with standard classifier for 2000 test sets

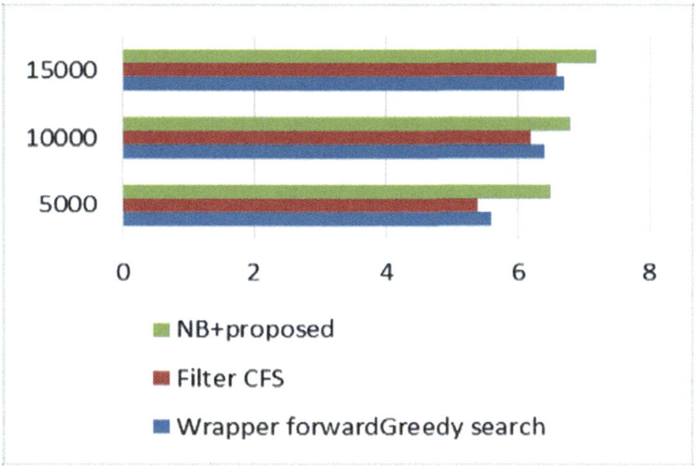

Fig. 8. Comparison of greedy search with greedy search and filter CFS

(a) Results
Classification accuracy was used, which is a measure of how well a document is classified into its appropriate class. It is simply the percentage of Correctly Classified Documents from the Total Documents. All the classification accuracies have been computed on testing datasets. We present the following evaluation and comparison. Classification accuracy on the test dataset using (a) Naïve Bayes, (b) chi-squared with naïve Bayes, and (c) proposed algorithm with naïve Bayes is computed. The result is summarized in Fig. 9 below:

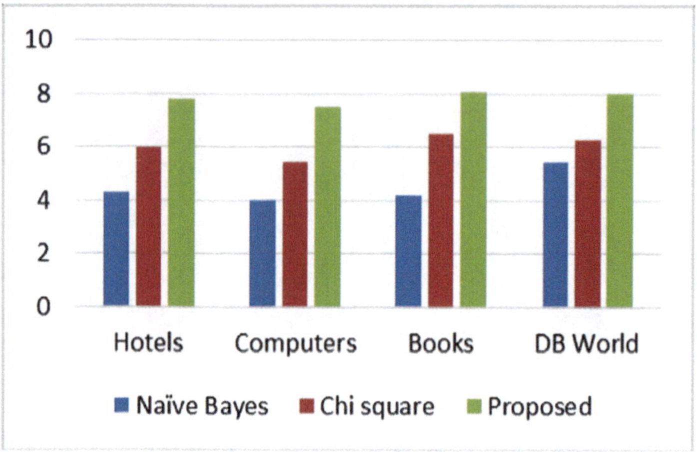

Fig. 9. Text classification accuracy

With the proposed algorithm, feature space is significantly reduced. The total number of features and reduced number of features using (a) chi-squared and (b) proposed approaches are displayed in Table 4 below:

Table 4. Dataset information

Datasets	Number of documents	Number of terms	Number of categories
Hotel	50	3360	2
Computers	50	3358	2
Books	50	3300	2
DB world	64	3723	2

The results of proposed approach and naïve Bayes with other classifiers like kNN and SVM and decision tree (DT) are compared, which makes naïve Bayes (NB) comparable with other classifiers; the results are summarized in Fig. 10 (Table 5).

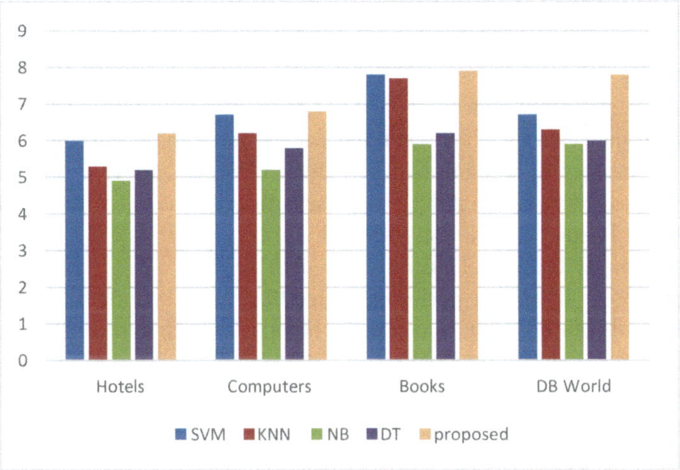

Fig. 10. Comparison with other standard classifiers

Table 5. Feature space comparison

Datasets	Total features	Feature using (X2)	Proposed
Hotels	3360	6	3
Computers	3358	8	4
Books	3300	4	3
DB world	3723	5	2

(b) Analysis
Both, the experiment twitter sentiment analysis and text classification results are analysed in four different bases. The first is in terms of accuracy, demonstrating how the proposed algorithm increases accuracy of Nave Bayes; the second is a reduction in feature space and shows how much reduction of feature space is obtained using the proposed approach. The third is a comparison with other classifiers, demonstrating how well it performs compared to other standard classifiers and finally a comparison with other feature selection algorithms. In the data pre-processing steps, two dictionaries have been added, to convert Emoticons and Acronyms which were helpful in increasing accuracy of sentiment classification on twitter.

(c) Increase in accuracy
From Fig. 4 it can be observed that there is higher accuracy for the twitter sentiment analysis with the proposed algorithm using 1,000 test sets and different training sets. Accuracy in comparison with the Multi Nominal Naïve Bayes (MNB) and the Bernoulli Naïve Bayes (BNB) schemes was also established. From Fig. 5, it can be observed that the accuracy of the proposed scheme is higher than the other schemes. As the size as the training sets increases, accuracy of all

the schemes also increases as expected. The experiment on text classification using the proposed algorithm achieved significant improvements over Naïve Bayes (Fig. 9).

(d) Reduction in feature space
Figure 6 and Table 4 show that the use of only the chi square algorithm has reduced features by a significant amount. When, in addition, the proposed algorithm is used, it shows that the feature selection is done effectively and better than with other feature selection algorithms.

(e) Comparison with other classifiers
The proposed algorithm performs better than other standard classifiers (Fig. 7; 10). Classification results have been compared with other standard classifiers, Support Vector Machine, KNN, Naïve Bayes and Decision Tree and the proposed approach outperformed other standard classifier in both twitter sentiment experiments as well as text classification.

(f) Comparison with other feature selection approaches
Accuracy with the greedy forward search-based wrapper method and the CFS based multivariate filter method was evaluated, which employs the breadth first approach in Fig. 8, and the proposed approach achieved significantly better results on accuracy.

6 Conclusions

For text classification, Naïve Bayes has been dismissed as an inferior classifier. Nevertheless, we have proposed a feature selection algorithm based on chi-square, TF-ID and k-mean clustering which can be used with Naïve Bayes classifier for better performance. Our proposed algorithm has been evaluated with twitter datasets as well as compared to other classifiers and feature selection methods. The proposed Algorithm aids Naïve Bayes performance improvement. Combined with Naïve Bayes, the proposed algorithm gives better performance in comparison with other classifiers. The proposed algorithm combined with Naïve Bayes also gives better accuracy.

This clearly indicates that the proposed algorithm for feature selection improves Naïve Bayes performance for twitter sentiment analysis as well as for text classification.

Future work will include enhancement of the proposed algorithm by applying other advanced clustering techniques and by increasing the number of training datasets for twitter sentiment analysis.

Acknowledgement. We are grateful to Mrs. Angelika Maag for proof reading and making corrections to this article. Without her support, it would have not been possible to submit this in the current form.

References

1. Kumar, A., Sebastian, T.M.: Sentiment analysis on Twitter. Int. J. Comput. Sci. Issues (IJCSI) **9**(4), 372–378 (2012)
2. Bigonha, C., Cardoso, T.N.C., Moro, M.M.: Sentiment-based influence detection on Twitter. J. Braz. Comput. Soc. (2012). https://doi.org/10.1007/s13173-011-0051-5
3. Devika, M. D., Sunitha, C., Amal, G.: Sentiment analysis: a comparative study on different approaches. Procedia Comput. Sci. 44–49 (2016)
4. Kotsiantis, S.B.: Supervised machine learning: a review of classification techniques. Informatica, 249–268 (2007)
5. Giachanou, A., Crestani, F.: Like it or not: a survey of twitter sentiment analysis methods. ACM Comput. Surv **49**, 28 (2016). http://doi.org/10.1145/2938640
6. Zainuddin, N., Selamat, A., Ibrahim, R.: Hybrid sentiment classification on twitter aspect-based sentiment analysis. Appl. Intell. 1–9 (2007)
7. Wilson, T., Wiebe, J., Hoffmann, P.: Recognizing contextual polarity: an exploration of features for phrase-level sentiment analysis. Comput. Linguist. **35**(3), 399–433 (2009)
8. Sonawane, V.A.: Sentiment analysis of Twitter data: a survey of techniques. Int. J. Comput. Appl. **139**(11), 5–15 (2016)
9. Ebrahimi, M., Amir, H.Y., Sheth, A.: Challenges of sentiment analysis for dynamic events. IEEE Intell. Syst. **32**(5), 70–75 (2017)
10. Song, J., Tae, K.K., Lee, B., Kim, S., Yong Youn, H.: A novel classification approach based on Naïve Bayes for Twitter sentiment analysis. KSII Trans. Internet Inf. Syst. **11**(6), 2996–3011 (2017)
11. Rana, T.A., Cheah, Y.: Aspect extraction in sentiment analysis: comparative analysis and survey. Artif. Intell. Rev. **46**(4), 459–483 (2016)
12. Yuvaraj, N., Sabari, A.: Twitter sentiment classification using binary shuffled frog algorithm. Intell. Autom. Soft Comput. **23**, 373–381 (2016)
13. Hai, Z., Cong, G., Chang, K.: Analyzing sentiment in one go: supervised joint aspect and sentiment model. Futur. Comput. Inform. J. **29**, 1172–1185 (2016)
14. Pletneva, E.V.: Text sentiment classification based on a genetic algorithm. J. Comput. Syst. Sci. Int. **55**, 106–114 (2016)
15. Sebastiani, F.: Machine learning in automated text categorization. ACM Comput. Surv. **34**, 1–47 (2002)
16. Lunghan Zhang, L.J.: Two feature weighting approaches for Naive Bayes text classifiers. Knowl. Based Syst. **100**(15), 137–144 (2016)
17. Subhajit Dey Sarkar, S. G.: A novel feature selection technique for text classification using Naïve Bayes. Int. Sch. Res. Not. (2014). Article no. 717092
18. Ang Yang, J.Z.: Enhanced Twitter sentiment analysis by using feature selection and combination. In: Proceedings of International Symposium on Security and Privacy, pp 52–57 (2015)
19. Schuller, B., Knaup, T.: Learning and knowledge-based sentiment analysis in movie review key excerpts. In: Esposito, A., Esposito, A.M., Martone, R., Müller, V.C., Scarpetta, G. (eds.) Toward Autonomous, Adaptive, and Context-Aware Multimodal Interfaces. Theoretical and Practical Issues. LNCS, vol. 6456. Springer, Berlin (2011)
20. Yelena Mejova, P.S.: Exploring feature definition and selection for sentiment classifiers. In: Proceedings of the Fifth International AAAI Conference on Weblogs and Social Media (2011)

21. Rachburee, N., Punlumjeak, W.: A comparison of feature selection approach between greedy, IG-ratio, chi-square, and mRMR in educational mining. In: IEEE Conference Proceedings, pp. 420–424. The Institute of Electrical and Electronics Engineers, Inc. (2015)
22. Qiu, Y.F., Wang, W., Liu, D.Y.: Research on an improved CHI feature selection method. Appl. Mech. Mater. 241–244 (2012)
23. Saroj, T.: Kavita, R: Review: study on simple k mean and modified K mean clustering technique. Int. J. Sci. Eng. Comput. Technol. **6**(7), 279–281 (2016)

Prospect of Application of Hyperspectral Imaging Technology in Public Security

Cheng Cheng(✉) and Bo Zhao

The Third Research Institute of the Ministry of Public Security,
339 Bisheng Road, Shanghai, China
13916201644@139.com

Abstract. Hyperspectral technology has both the characteristics of imaging and spectroscopy. It can detect the external shape of the object and the internal information of the object. Therefore, it has been widely studied in recent years, especially in remote sensing monitoring, crop detection, biological characteristics research, The field of medical research and other fields have been initially applied. This paper analyzes the principles, architecture, data acquisition methods and data processing of hyperspectral technology, and combines the characteristics of hyperspectral technology to predict and forecast the development and application of this technology in the field of social public safety.

Keywords: Hyperspectral imaging · Image processing · Public security

1 Introduction

Like other spectral imaging, Hyperspectral imaging is a technique for spatially scanning an object to be measured while acquiring the characteristic spectrum information of the object over the entire electromagnetic spectrum. Hyperspectral imaging technology can obtain the spatial information and spectral information of the target at the same time, and the obtained spectrum has fine wavelength resolution and large wavelength range. In particular, some measured objects have unique "fingerprints" on the electromagnetic spectrum (Spectral spectrum), which has obvious advantages in target material identification, abnormal target detection, camouflage target recognition, complex background suppression, etc., and gradually becomes a new generation of photoelectric detection technology [1–4]. Hyperspectral imaging equipment constructed using hyperspectral sensors and data processing modules has been initially applied in astronomy, agriculture, biomedicine, earth science, physics, etc. Currently, practical hyperspectral imaging spectrometers and related detection Performance and detection methods are the research hotspots in this field.

2 Hyperspectral Imaging System Structure

A typical hyperspectral imaging system consists of five parts: an area array camera, a spectroscopic device, a light source, a transmission mechanism, and computer hardware and software. The light source provides illumination for the entire imaging

J. Abawajy et al. (Eds.): ATCI 2018, AISC 842, pp. 299–304, 2019.
https://doi.org/10.1007/978-3-319-98776-7_31

system; the light splitting device disperses the mixed light into monochromatic light of different frequencies through the optical component, and projects it onto the area array camera; the light generated by the light source acts on the object to be detected and becomes a carrier of physical or chemical information. And then projected to the area array camera through the beam splitting component; computer software and hardware are used to control the hyperspectral imaging system to collect data, to process and analyze image and spectral data for specific applications, and to provide storage space for hyperspectral images.

3 Hyperspectral Imaging Data Acquisition

Hyperspectral imaging data can be described as a data cube of two-dimensional space {x, y} and one-dimensional spectrum {λ}, which can be divided into four types according to different acquisition methods (Fig. 1):

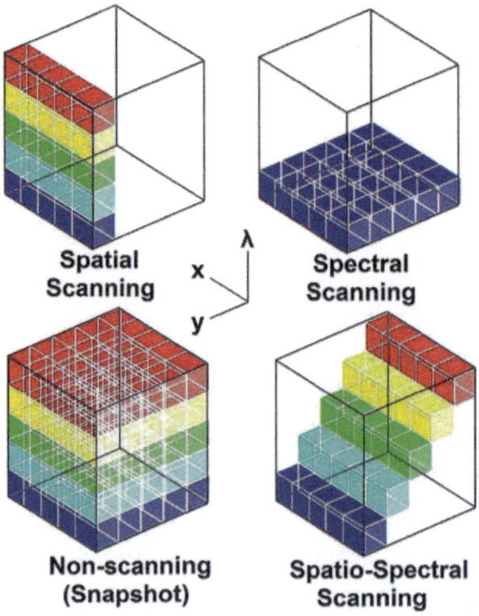

Fig. 1. Data acquisition methods of hyperspectral imaging

(1) Spatial scanning: Also known as line scanning, that is, {x} in one direction of the target, scanning with incident light {λ} of different wavelengths; moving one point in the y direction, scanning again, successively get the entire target data. This method is often used in remote sensing or real-time monitoring on conveyor belts.
(2) Spectral scanning: Scanning the target surface {x, y} with a single wavelength λ; successively using different wavelength scanning to complete the target data acquisition. This approach is typically used in hyperspectral imaging systems where the number of images of the desired wavelength is small. In order to obtain hyperspectral images, it is necessary to frequently move the spectral camera or the

detection object, which is not conducive to rapid detection and is often used for the detection of microscopic objects.

(3) Non-scanning/snapshot: A complete data cube {x, y, λ} can be obtained in one measurement. The most prominent advantages of this approach are higher luminous flux and shorter acquisition times.

(4) Spatio-spectral scanning: The sensor output of this mode is a spatially scanned map {(x, y)|(λ)} based on the wavelength encoding of the target. This scanning method formed an experimental device in 2014, which has some advantages of spatial scanning and spectral scanning and avoids the deficiencies of the two methods.

Figure 2 shows the output data of four hyperspectral imaging methods. From left to right, the first scan spectrum, the monochromatic spatial spectrum, the one-time imaging snapshot, and the wavelength-encoded spatial scan.

Fig. 2. Four data outputs of hyperspectral imaging

4 Hyperspectral Data Processing

Hyperspectral imagery is a three-dimensional data cube with "maps in one", so the processing and analysis of hyperspectral images can be image processed and analyzed in the spatial domain at a specified wavelength, and in the spectral domain at the specified pixel coordinate position. Spectral processing and analysis can also be processed and analyzed in both the spatial and spectral domains.

Generally, the hyperspectral image processing flow includes three levels of hyperspectral image data acquisition and correction layer, hyperspectral image processing and analysis layer, and application layer. Hyperspectral image data acquisition and correction layer includes acquisition of sample hyperspectral image and reference image, hyperspectral image correction; hyperspectral image processing and analysis layer including spectral processing and analysis and image processing and analysis; application layer including internal quality and safety Detection and external defect identification and extraction.

5 Application Prospects in Public Security Business

Hyperspectral imaging technology is a fusion technique of traditional image and spectrum, which can simultaneously acquire the spatial and spectral information of the research object. Because the image data can reflect the external characteristics of agricultural products, and the spectral data can analyze the internal physical structure and chemical composition of the object, it has become a research hotspot in the fields of agriculture, food processing, biomedicine, remote sensing, and aerospace. For example, in the agricultural field, it is possible to remotely monitor crop growth, growth defects and pests and diseases; to detect protein in animal compound feed to avoid mad cow disease; to detect pesticide residues in fruits and vegetables, and to perform non-destructive testing on agricultural products such as cucumber, apple and potato. In the food processing industry, full-spectrum imaging combined with intelligent software can be used for online food sorting, removal of debris, and removal of defective products. In biomedical applications, neurological studies, retinopathy examinations, etc. can be used. In the field of remote sensing, metal deposits exploration, geological health environment genetic information detection, vegetation geochemical information and physical and chemical information extraction can be carried out.

However, there are few reports on the application of hyperspectral imaging in the field of social public safety. The author believes that this technology will be widely studied and applied in the following fields.

5.1 Monitoring

Hyperspectral monitoring is a typical application of hyperspectral imaging technology and can play an important role especially in the military field. For example, military equipment anti-reconnaissance usually relies on visual camouflage, infrared camouflage, etc., while hyperspectral imaging scans in multiple frequency bands, and it is easy to find the characteristic spectral information of these facilities and identify the location.

As shown in Fig. 3, the ambient temperature is -15°C under outdoor conditions, and is photographed by Specim's LWIR hyperspectral imager. The relative radiation spectrum of different targets is as indicated by the blue arrow. It can be seen that the infrared spectra of different targets have distinct characteristics. The contrast shows the temperature of the target.

Fig. 3. Hyperspectral infrared radiation measurement chart

Since conventional thermal infrared hyperspectral imaging systems require liquid nitrogen or liquid helium cooling, it is difficult to find applications in most applications. Specim has recently developed a thermal infrared hyperspectral camera that can be used outdoors or in drone applications without the need for an external light source. In the public security business, such equipment can be used for security in important places, monitoring of crowded areas, and so on.

5.2 Remote Chemical Gas Imaging

For chemicals in environments such as toxic gases, it is difficult to observe with the naked eye; the newly developed Telops HyperCam hyperspectral imaging device can detect harmful gases of the order of several PPM within 5 km. Figure 4 shows a remote chemical imaging of SF_6 and NH_3 at 1.5 km using a Telops HyperCam imaging spectrometer.

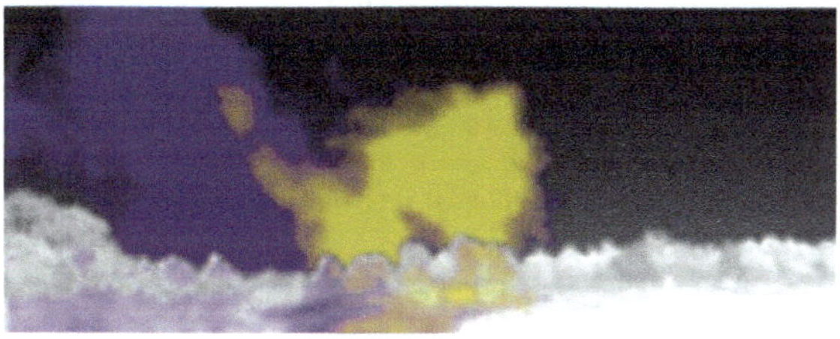

Fig. 4. Remote chemical imaging of SF_6 and NH_3 at 1.5 km

In the field of public safety, this technical equipment can be used to detect the leakage of dangerous chemical gas transport vehicles, remotely monitor industrial hazardous gas emissions such as chemical plants, and real-time warning of dangerous gases in public places. Especially when searching for a drug-making factory, it is possible to remotely monitor the gas emissions of the toxic intermediates over a large area to locate the hidden manufacturing dens.

5.3 Identification of Physical Evidence

In the field of physical evidence identification, a new detection method is provided.

As shown in Fig. 5, three hair samples are analyzed by microscopic hyperspectral imaging system. The results and spectral curves can be determined that two hair samples are from the same person. In addition, hyperspectral imaging technology can also be used for document identification, handwriting identification, gun and ammunition residue analysis, material analysis, and flammable liquid residue analysis.

5.4 Other Applications

Drug Investigation: Since most drugs are difficult to be discovered through careful camouflage and hiding, or mixed with drugs, food, daily necessities, etc., can be identified by hyperspectral imaging technology, and the Ministry of Public Security's physical evidence identification center is carrying out relevant research.

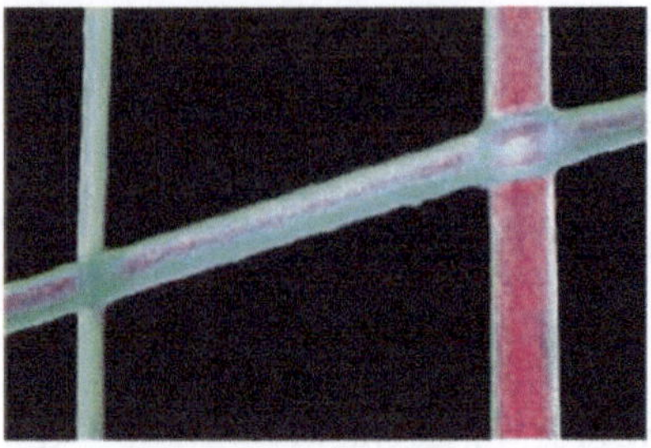

Fig. 5. Hair samples analyzed by microscopic hyperspectral imaging system

Fire Accident Investigation: Scanning the fire residue in the accident scene with hyperspectral imaging, detecting the burning characteristics in the spectral image of the residue in the scene, looking for the fire source, and providing a technical means for the fire accident investigation.

Acknowledgement. This work is supported by National Key R&D Program of China (No. 2018YFC0809500).

References

1. Chang, C.-I.: Hyperspectral Imaging: Techniques for Spectral Detection and Classification. Springer, Heidelberg (2003). ISBN 978-0-306-47483-5
2. Grahn, H., Geladi, P.: Techniques and Applications of Hyperspectral Image Analysis. Wiley, Hoboken (2007). ISBN 978-0-470-01087-7
3. Holma, H.: Thermische Hyperspektralbildgebung im langwelligen Infrarot, Photonik, May 2011
4. Chamberland, M., et al.: High-performance field-portable imaging radiometric spectrometer technology for hyperspectral imaging applications. In: Proceedings of SPIE, vol. 5994, p. 59940N, September 2005

Design and Development of Automatic Packaging Equipment for Boxed Drugs

Zhongyuan Zhang[1(✉)], Yu Xie[1], and Junwei Zhang[2]

[1] Wenzhou Vocational and Technical College, Wenzhou 325035, Zhejiang, China
jwk208@163.com

[2] Ruiming Industrial Co., Ltd., Wenzhou 325200, Zhejiang, China

Abstract. In order to make the packaged drugs neat and beautiful, with a superior packaging quality, the market is in urgent need for new efficient, economical and flexible packaging equipment. This paper designed the automatic packaging equipment for boxed medicine, adopting PTP three-degree-freedom pneumatic manipulator to carry out feeding for the packing box and the tablet board. Besides, the boxing used four-degree-freedom manipulator. The pneumatic mechanism was cooperated with mechanical structure to complete boxing craft in the workbench movement process, simplifying the mechanical structure and improving the production efficiency.

Keywords: Drug packaging · PLC · Touch screen · Pneumatic manipulator

1 Introduction

Modern society has increasingly high requirements on the packaging of goods, with no exception of Drugs. In order to make the packaged drugs neat and beautiful, with good packaging quality, the packaging equipment is required to have precise action, high positioning accuracy, high productivity and a certain degree of flexibility. Therefore, requirements are getting higher for the mechanical structure and control of packaging machine equipment. The market is in urgent need for new efficient, economical and flexible packaging equipment [1]. This paper designed the automatic packaging equipment for boxed drugs, mainly to solve the problems including cumbersome packaging process and high artificial packaging costs of flake material in pharmaceutical, chemical, food and other industries.

2 Design

2.1 Equipment Functions and Technical Indicators

The device uses non-standard design, mainly for sheet material packaging in pharmaceutical, food, chemical and other industries, realizing automatic feeding, sealing box, coding and boxing for box and pill board. It adopts PLC and touch screen to conduct

J. Abawajy et al. (Eds.): ATCI 2018, AISC 842, pp. 305–311, 2019.
https://doi.org/10.1007/978-3-319-98776-7_32

integrated control, with three kinds of operation modes, namely manual debugging, resetting back to zero and automatic running.

The equipment can effectively guarantee the packaging quality. Specifically, according to the requirements of packaged items and the demanded form and size of enterprises, it guarantees consistent specifications for the packaging, and the manual packaging cannot achieve. In addition, it is conducive to labor protection by reducing the labor intensity and the improvement of the working conditions. For some products with serious impact on health, such as dusty, toxic, irritating and radioactive products, handmade packaging is harmful to health. However, the mechanical packaging can avoid the harm, effectively protecting the environment from contamination. At the same time, the device has a good man-machine interface, simple operation panel, manual and automatic operation, screen display, parameter settings, fault shutdown, etc.

2.2 Overall Program

On the basis of full investigation and study, feasibility analysis and demonstration can determine the application scope of packaging equipment, packaging technology, the primary structure of the various components and the overall layout. As for formulating the overall program, the design is carried out for the packaging machine of a certain brand of drugs in order to make the identified program both advanced and cost-effective [2]. Thus, the packaged materials have large production quantities and changeless specifications.

2.2.1 Main Content of the Investigation

The investigation is mainly to understand the type, nature, size, shape, packaging methods and forms, packaging requirements, specifications, characteristics and supply condition of the packaging materials. The survey also includes the understanding of the operators' technical level for the same type of packaging, the working environment of the new packaging machine, the temperature and humidity of the packaging environment.

2.2.2 Referring to Related Technical Document

The accessed document mainly includes the existing drawings, technical data and information on the use condition for the same type of packaging equipment. At the same time, a variety of related domestic and foreign scientific literature is also involved for understanding the status and development trends of the same type of domestic and foreign machinery.

2.2.3 Initial Design

The application scope and productivity are determined for the packaging machine, with the analysis of packaging process analysis. The overall layout is completed, with the main technical parameters set. Figure 1 shows the packaging process.

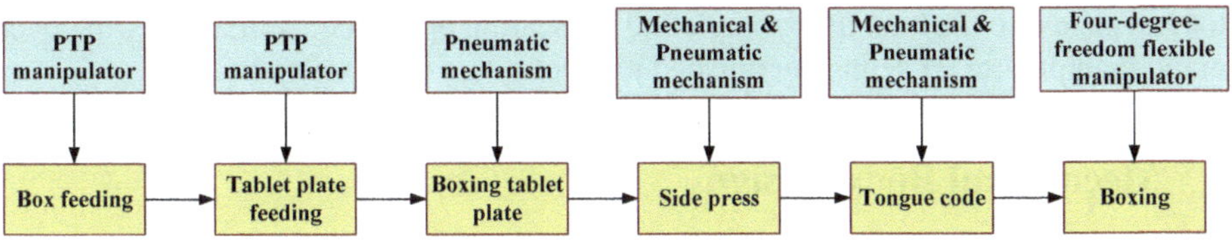

Fig. 1. Boxed medicine automatic packaging equipment process flow chart

Figure 2 shows the overall program for boxed drug automatic packaging equipment. Its working principle is as follows: At Station 1, Pneumatic Manipulator 4 conducts packing box feeding. At Station 2, the Pneumatic Manipulator 10 feeds the tablet plate onto the feeding chute 12, then the feeding cylinder 13 put the tablet plate into the box. At Station 3, the cylinder 15 drives the pendulum 14 to swing at 30 to 90° in the horizontal direction, thus completing pressing the left side of the package box. At this time, the workbench moves to the right. Then, the guide rod 16 finishes pressing the right side of the box, keeping the left side page in the pressed state. Further, the workbench continues to move to the right, with two elastic metal sheets 19 pressing the tongue page and the top of the box. At Station 4, the tongue cylinder 17 drives the tongue tab 18 to put the

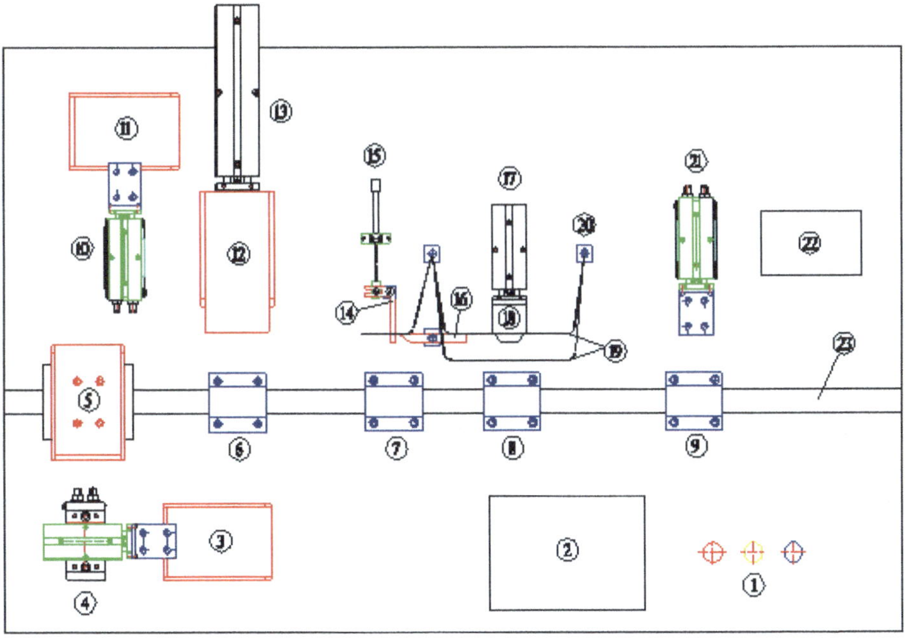

Fig. 2. Overall program for boxed drug automatic packaging equipment. 1-Button group; 2-Touch screen; 3-Box trough; 4, 10, 21-Pneumatic manipulator; 5-Station; 6-Station 2; 7-Station 3; 8-Station 4; 9-Station 5; 11-Pill sheet trough; 12-Feeding chute; 13-Feeding cylinder; 14-Pendulum; 15-Pendulum cylinder; 16-Guide bar; 17- Tongue cylinder; 18-Tongue sheet; 19-Elastic metal plate; 20-Pillar; 22-Finished product discharge port; 23-Linear guide

tongue page into the box, thus completing the sealing operation. At Station 5, the pneumatic manipulator 21 sends the finished product to the discharge port.

3 Mechanical Body Design

3.1 Mechanical Design

The structure design was conducted for pneumatic manipulator, workbench, feeding mechanism and sealing mechanism, etc., with the system working pressure, clamping force, running speed and stroke of the moving parts calculated. The parameters such as system pressure loss and structural strength were checked. Moreover, the type selection was made for the manipulator cylinder, turntable, sucker, reversing valve, valve seat and pneumatic auxiliary element [3].

3.2 Pneumatic System Design

Figure 3 shows the pneumatic system schematic diagram. The system pressure is 0.45 MPa [4]. A, F and M are the robot swing cylinder; B, E and J the robot lift cylinder; C, G and K the robot telescopic cylinder; D, H and L the vacuum suction; N is the feeding cylinder; P the pendulum cylinder; S the tongue cylinder.

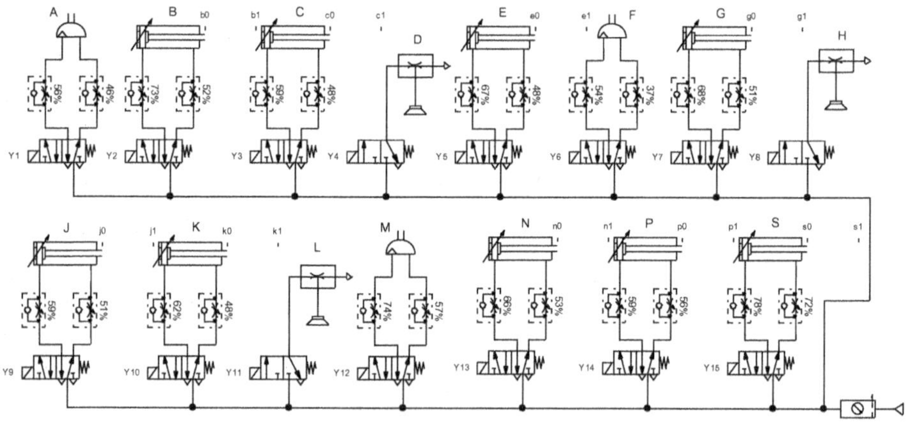

Fig. 3. Pneumatic system schematic diagram

4 Equipment Control System Design

The core components of the packaging equipment control system were the Mitsubishi FX3U series PLC, with Weiluntong MT6070IH as the system man-machine dialogue interface. According to the packaging equipment control requirements, functions needed to be achieved including manual debugging, resetting to zero and automatic operation [5].

Manual debugging mode achieves the left/right movement and fixed-point movement of the workbench, separate control lifting up and down of each manipulator and the implementation parts, stepping forward and backward, sucker suction and deflation, positioning clamping cylinder clamping and releasing, loading cylinder lifting up and down, etc. Automatic operation mode includes 2 kinds of work conditions, namely single cycle running and continuous operation, which can be selected by pressing the button in the host automatic operation interface. The automatic operation mode achieves each packaging process according to the workflow, then showing the number of packaged products in the production count window. In the packaging process, the procedure can be stopped by pressing the emergency button in case of malfunction. With troubleshooting, the process can be restarted at the origin point through pressing the reset button.

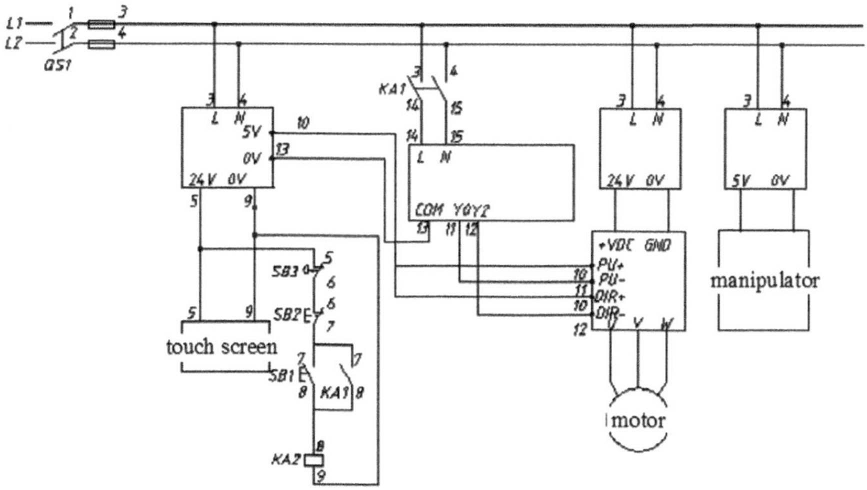

a) ectrical schematic diagram

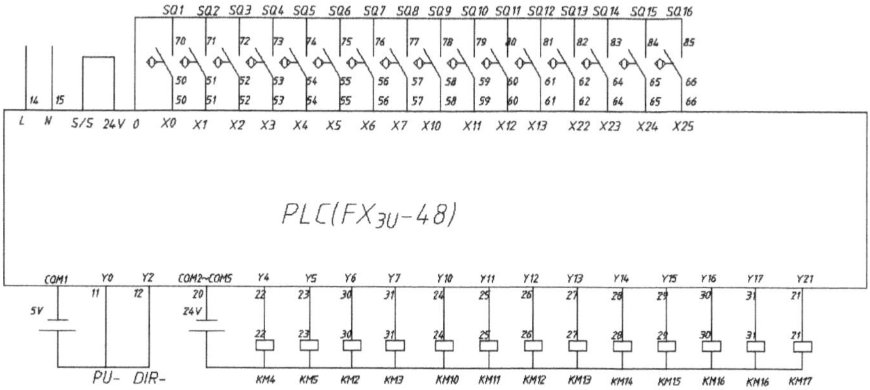

b) PLC external wiring diagram

Fig. 4. Boxed medicine automatic packaging equipment electrical schematic diagram and PLC wiring chart

Figure 4 shows boxed medicine automatic packaging equipment electrical schematic diagram and PLC wiring chart.

5 Prototype Production and Debugging

General parts and standard parts are purchased, with non-standard parts designed and processed. The assembly and commissioning are completed for the machine body and the control system. In the debugging process, failures are resolved including the cylinder positioning inaccuracy, the polarity incompatibility between the two sensors and the IST instruction switching operation exception, etc. Therefore, the local structure is optimized, with the desired function achieved. Figure 5 shows the prototype model of the boxed medicine automatic packaging equipment.

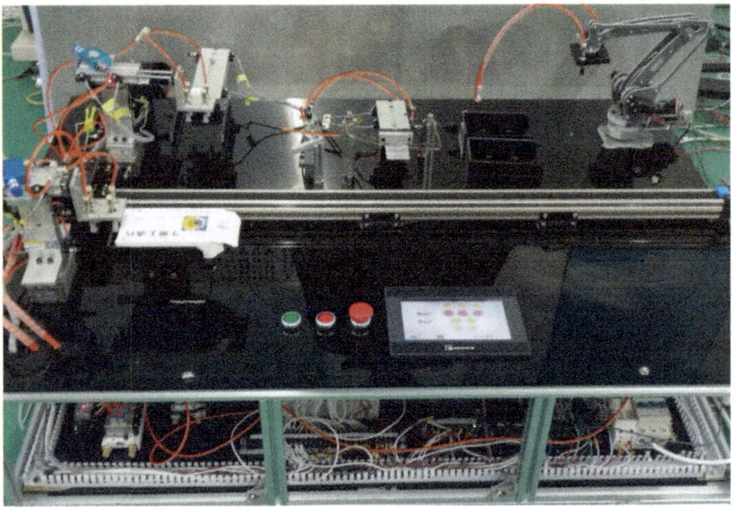

Fig. 5. Boxed medicine automatic packaging equipment prototype model

6 Conclusions

The boxed drug automatic packaging equipment uses air pressure, electrical and mechanical transmission with PLC and pneumatic control. It integrally achieves automatic feeding, sealing and boxing, etc., adopting PLC and touch screen for integrated control. It has complete functions and stable operation, with three operating modes, namely the manual debugging, resetting back to zero and automatic operation. The equipment is characterized by compact structure, safety, reliability, small area and simple operation. Although there are many domestic produced pharmaceutical packaging equipment, their price is relatively high. Developing pill packaging equipment has practical value and application prospects in line with the requirements above.

References

1. Sun, F., Ma, X.: Introduction to Packaging Machinery. Printing Industry Press, Beijing (2015)
2. Cheng, D.: Mechanical Design Manual. Chemical Industry Press, Beijing (2010)
3. Zhang, Z.: Pneumatic Drive and Control System Design, p. 8. Wenzhou Vocational & Technical College, Wenzhou (2016)
4. SMC Company: Modern Practical Pneumatic Technology. Mechanical Industry Press, Beijing (2003)
5. Wang, C.: Gas, Liquid, Electric Control Technology. Shanghai Science and Technology Press, Shanghai (2010)

Development of Automatic Material Sorting Machine Based on PLC

Yu Xie(✉) and Zhongyuan Zhang

Wenzhou Vocational and Technical College, Wenzhou 325035, Zhejiang, China
189153298@qq.com

Abstract. Automatic material sorting is one of important links in assembly line work of manufacturing industry and logistics industry. The automatic sorting machine designed and developed in this work was comprehensively controlled by air pressure, mechanical drive, PLC and touch screen. Aiming at solving problems of low sorting efficiency, high error rate, etc. of manual sorting, this machine could improve manufacturing effectiveness and reduce product cost for enterprises. The machine was able to sort metal and plastic articles in different sizes. Transporting materials with conveyor belt, this machine could identify different material textures and sizes with photovoltaic sensor and magnetic sensor. Then, sorted materials could be placed in corresponding positions after being grabbed with pneumatic robotic manipulator with three degrees of freedom.

Keywords: Material sorting · PLC · Touch screen
Pneumatic robotic manipulator

1 Introduction

In manufacturing industry, the most effective and cost-saving method is assembly line work where automatic material sorting is one of important links [1]. With the flourishing development of e-commerce in China, material sorting is also a complex link in which mistakes may easily occur. This work aims at designing and manufacturing automatic sorting machine based on manufacturing and industry and logistics to solve problems of low sorting effectiveness, high error rate, etc. Therefore, enterprises can improve production efficiency and reduce production cost.

2 Design Plan

2.1 Machine Functions and Technical Indexes

With non-standard design, the machine is mainly used in sorting metal articles and plastic articles in size of $\Phi 20 \times 30 \sim \Phi 30 \times 60$. The machine has three operating models, manual debugging, resetting to zero and automatic operation. By manual debugging, functions, including independent control of conveyer belt, lifting and grabbing of feeding (unloading) robotic manipulator and rotation of material storage, can be realized.

J. Abawajy et al. (Eds.): ATCI 2018, AISC 842, pp. 312–317, 2019.
https://doi.org/10.1007/978-3-319-98776-7_33

By resetting to zero, the original point can be set for the system to reset the whole system to zero, stopping the conveyor belt in automatic material sorting machine and turning all manipulators and material storage to initial state. By automatic operation, procedures, including automatic feeding, transportation, material identification, height identification and automatic selection of shipping position, will operate successively.

2.2 Overall Plan

Figure 1 shows the overall plan of machine. The working principle is as follows: Conveyor belt will transport materials. After the shape sensor detects the material, conveyor belt will stop and detect the material shape. Meanwhile, Manipulator 1 operates for shape sorting to transport materials to Storage Unit 1. After the material sensor in Storage Unit 1 detects the materials in different textures in position, Manipulator 2 operates for material sorting to transport the materials in Storage Unit 1 to Unit 2. When the sensor in storage unit detects the materials in position, Manipulator 3 operates to transport sorted materials to conveyor belt.

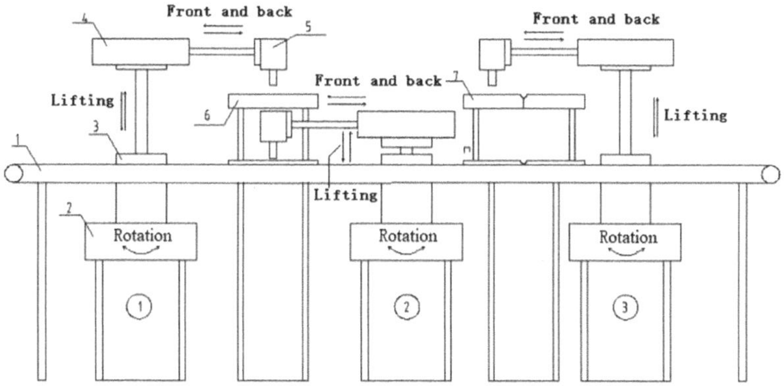

1-Conveyor belt; 2-Oscillating cylinder; 3-Lifting cylinder; 4-Telescopic cylinder; 5-Pneumatic gripper; 6, 7-Storage units

Fig. 1. Overall plan of automatic material machine

3 Design of Mechanical Body

3.1 Design of Mechanical Structure

Structure was designed for pneumatic robotic manipulator, storage unit, conveyor belt, etc.; working pressure and clamp force of system, operating speed and journey of all movement parts were calculated for analysis; pressure loss, structural strength and other parameters were verified; types were selected for cylinder of robotic manipulator, rotary table, pneumatic gripper, reversing valve, valve seat, pneumatic auxiliary components; three-dimensional modeling was designed for relevant components and the whole machine with SOLIDWORKS software [2, 3]. Figures 2 and 3 show the structures of manipulator and storage unit.

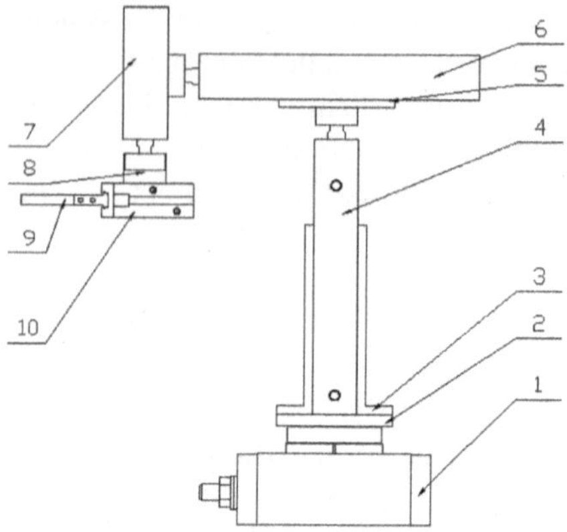

Fig. 2. Structural diagram of pneumatic robotic manipulator. 1-Oscillating cylinder; 4, 7-Lifting cylinder; 6-Telescopic cylinder; 2, 3, 5 and 8-Junction plate; 9, 10- Pneumatic gripper

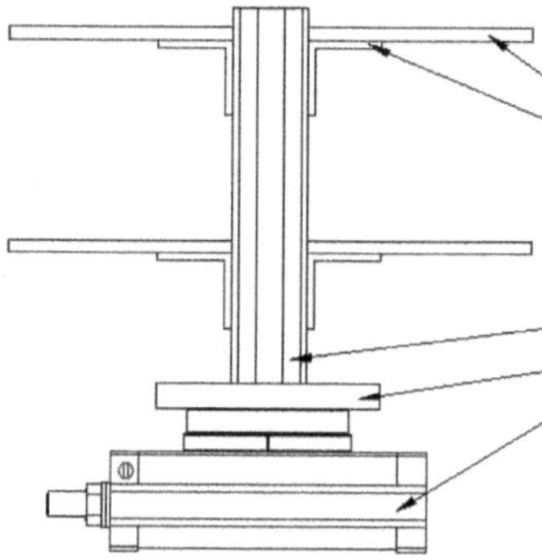

Fig. 3. Structural diagram of storage unit. 1-Oscillating cylinder; 2, 4-Junction plate; 3-Stand; 5-Storage unit

3.2 Design of Pneumatic System

Figure 4 is the diagram of pneumatic system where B, E and J are the lifting cylinders of manipulator; C, G and K the telescopic cylinders of manipulator; A, F and I the oscillating cylinders of manipulator; D, H and L the manual grippers; M is the oscillating cylinder of storage unit; N, O and P are the lifting cylinders of pneumatic gripper. The system pressure is 0.3–0.6 MPa [4].

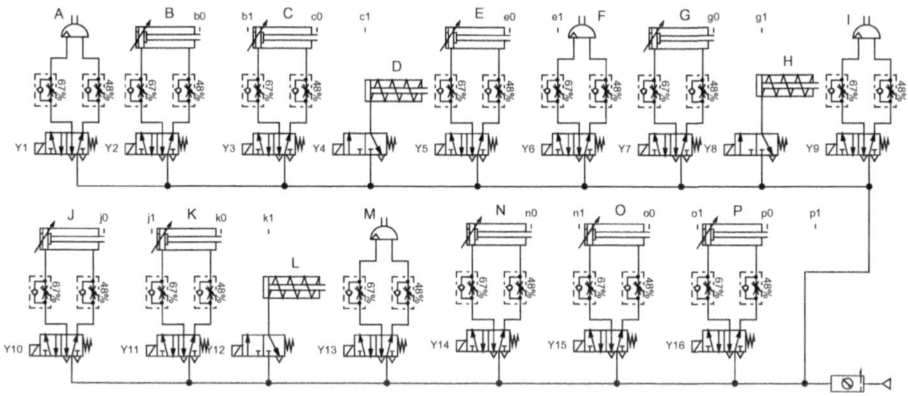

Fig. 4. Diagram of pneumatic system

4 Control System Design of PLC and Touch Screen

The core part of control system in automatic material sorting machine is Mitsubishi FX3U PLC. Moreover, Weinview TK6070IP is used as the interface of man-machine interaction.

4.1 Design of Hardware Circuit

After analyzing three control plans, the control by PLC and touch screen was selected when designing the diagrams of gas circuit, electrical circuit, etc. Then, PLC input and output ports were set according to the analysis on control signals of pneumatic and electrical elements. Table 1 shows the selected electrical components. Figure 5 shows the electrical diagram and PLC wiring diagram of automatic material sorting machine.

Table 1. Electrical elements

No.	Element name	Type	Quantity
1	PLC	FX3U-48MT	1
2	Alternating current motor	YZ-57BLS120	1
3	Low-voltage circuit breaker	DZ47-60 3P 10A	1
4	Fuse protector	RT18-32X 2P	1
5	24 V voltage output switching power supply	S-100-24	1
6	Two-position five-way magnetic valve	4V110-06	16
7	Magnetic switch	LJ18A3-8-Z/BY	2
8	Photoelectric sensor switch	E3F-DS10P1	6
9	Cylinder-induced magnetic switch	CS1-J	18
10	Touch screen	TK6070IP	1
11	Intermediate relay	JTX-3C	2

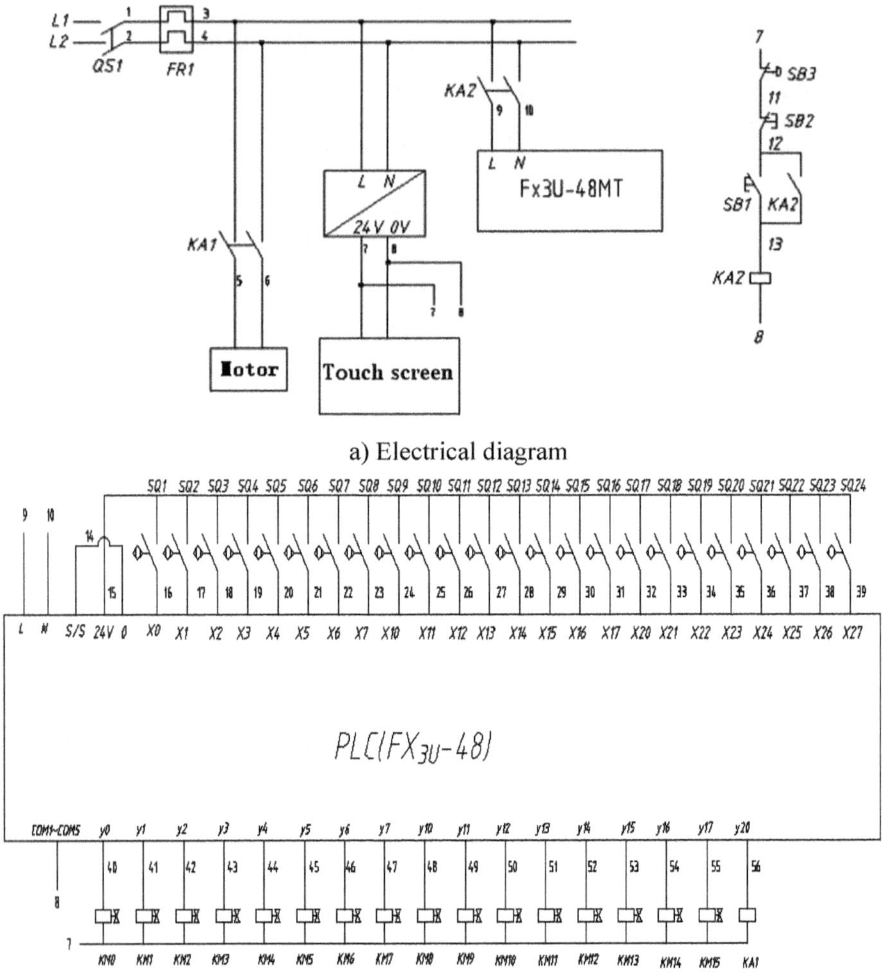

a) Electrical diagram

b) PLC external wiring diagram

Fig. 5. Electrical diagram and PLC wiring diagram of automatic material sorting machine

4.2 Design of System Software

Internal I/O port is set according to functional requirements of automatic material sorting machine. Figure 6 shows the interface design of startup, manual debugging, single station operation and automatic operation.

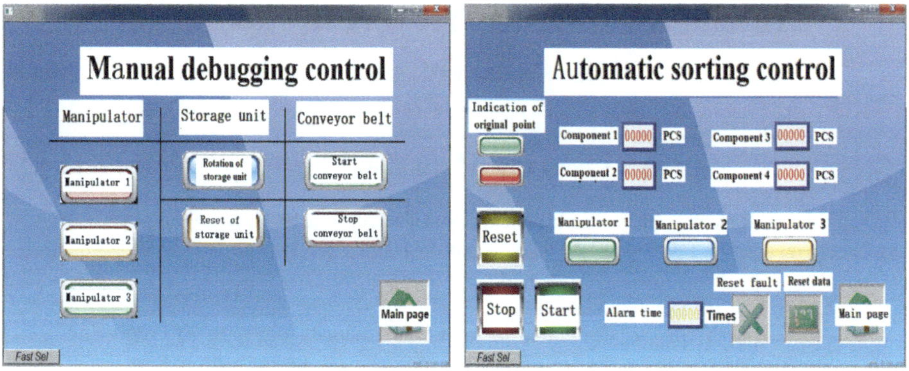

Fig. 6. Interface of touch screen

5 Conclusions

Controlled by pneumatic and mechanical drive, as well as PLC and touch screen, automatic material sorting machine was able to sort metal and plastic articles in different sizes. This machine that transported materials with conveyor belt could identify different material textures and sizes with photovoltaic sensor and magnetic sensor. Then, sorted materials could be placed in corresponding positions after being grabbed with pneumatic robotic manipulator with three degrees of freedom. The practical operation of sample machine shows that this machine with compact structure and reliable security can satisfy regulated functions.

References

1. Wang, C. (ed.): Gas, Fluid and Electric Control Technology. Shanghai Scientific & Technical Press, Shanghai (2010)
2. Cheng, D. (ed.): Mechanical Design Manual. Chemical Industry Press, Beijing (2010)
3. Zhang, Z. (ed.): Pneumatic Transmission and Control System Design. Wenzhou Vocational & Technical College, Wenzhou (2015)
4. SMC Company: Modern Practical Pneumatics. China Machine Press, Beijing (2002)

Intelligent Counter-Terrorism Equipment Status and Development

Cheng Cheng(✉) and Bo Zhao

Third Research Institute of the Ministry of Public Security, Shanghai, China
13916201644@139.com

Abstract. With the strengthening of public security agencies in combating terrorist crimes, the activities of the three forces of domestic national separatist forces, violent and terrorist forces, and religious extremist forces have been greatly suppressed, and the maintenance of the situation has generally improved. Especially in the Xinjiang region, the situation of social security has improved markedly, and violent and tragic events have been declining year by year. The number of tourists visiting Xinjiang continues to grow, presenting a scene of national harmony and social stability. However, ethnic conflicts have not been completely resolved, and the three forces in the country have not yet been eliminated. The horrors and cloud have not yet been dissipated. Therefore, we still need to further strengthen anti-terrorism efforts, increase the research and development of counter-terrorism science and technology, weapons and equipment, and advance counter-terrorism efforts from passive defense to active precaution, from post-disposal to advance warning and from one by one search to precision strike. Through scientific and technological means and intelligent means, efforts will be made to contain violence and terror in the initial stages.

Keywords: Counter-terrorism equipment · Intelligent equipment
Video structured description

1 Intelligence Counter-Terrorism Equipment Construction

After years of efforts, public security agencies have achieved some achievements in the application and management of smart equipment and have played an important role in the fight against terrorism.

1.1 The Application of Internet of Things Technology Has Achieved Initial Success

The Internet of Things (IoT) technology connects counter-terrorism equipment with the network, and carries out information exchange and communication, realizing data intercommunication and remote control. Such as smart pistol management system, the combination of IoT technology and firearms, can identify user identities, leaving the user automatically lock, to achieve the gun to prevent stealing and rob; can also record the use of guns to achieve full-time gun monitoring. Another example is the intelligent

J. Abawajy et al. (Eds.): ATCI 2018, AISC 842, pp. 318–324, 2019.
https://doi.org/10.1007/978-3-319-98776-7_34

public security IoT prevention and control system, which is a prevention and control alarm system for multi-service areas such as public security prevention, information collection, emergency management and control. It can realize active alarm, integrated command and dispatch, and the interaction between police and people. Can effectively complete the city grid management, improve the ability to prevent crime, and enhance the people's sense of security and well-being.

1.2 Biometrics Identification Technology is Widely Used

The application of biometric identification technology in the field of anti-terrorism is most prominent, and the technology system is constantly enriched, such as fingerprints, human faces, DNA, iris, sound, finger veins, and gait recognition. For example, in an important location, the dynamic face recognition system adopts deep learning image detection, recognition and tracking technologies to dynamically identify and automatically track the target person. The remote iris recognition device utilizes the characteristics of uniqueness of the iris recognition, high reliability, and strong anti-counterfeiting characteristics to quickly and accurately determine the identity of the target person. In addition, there are biometric fusion recognition technologies that make full use of the independence and complementarity between different biometrics to accurately identify the identities of people involved in terrorism and accurately grasp the trends of key personnel and key groups involved in terrorism.

1.3 The Security Equipment Has Been Continuously Upgraded

There have been metro terrorist attacks in Russia, France, and the United Kingdom, with disastrous consequences. Therefore, security inspections are necessary measures to discover and eliminate hidden dangers and prevent accidents. The domestic security inspection technology and equipment have roughly gone through three stages of development: the first is the technology integration period, such as X-ray inspection technology from single-source to multi-source and multi-perspective, spiral scanning and other methods; the second is the industry integration period, such as X-ray inspection instruments and gas checker, liquid checker, metal detector door, and terahertz human body security instrument; third is system integration period, such as integrated smart security system, based on security equipment and radio frequency identification, face recognition, radiography, video and image analysis and deep learning and other technologies. The new security inspection system has the advantages of automatic acquisition of personnel information, simultaneous inspection of characters, automatic image analysis, etc., and carries out intelligent inspections from the individual process of entering and leaving a station, greatly improving the security inspection work. The efficiency and pertinence have ensured the safety and quick travel of the people.

2 The Development of Intelligent Anti-terrorism Equipment

We continue to increase our investment in anti-terrorism technology and equipment and advance the development of counter-terrorism equipment. Recently, we carefully analyzed the new types of anti-terrorism equipment and found that the level of equipment intelligence in areas such as intelligence warning, synthetic warfare, and single-police comprehensive equipment is high.

2.1 The Field of Intelligence Warning

The counter-terrorism intelligence work has enriched the sources of information through means such as communication and networking; at present, it is still implementing early warning, prevention, and preemption through intelligence-aware devices and wisdom-based systems. Here are two technical equipments: First, integrated reconnaissance system. The integrated reconnaissance system is based on the real-time and accurate requirements of anti-terrorism intelligence, and establishes a three-dimensional reconnaissance network, space reconnaissance low-altitude reconnaissance, ground reconnaissance, and water reconnaissance to achieve all-terrain and full-time intelligence acquisition and early warning of terrorist activities. Space reconnaissance mainly uses high resolution satellites, high bandwidth satellites, and Beidou navigation satellites to obtain intelligence on terror. Its characteristics are high orbits, rapid discovery targets, and a wide range of reconnaissance, which can guarantee large-span anti-terrorism operations and all-round information coordination. Low-altitude reconnaissance uses unmanned aerial vehicles, aerostats and other mounted observation equipment to conduct reconnaissance and surveillance on the activities and positions of terrorists involved in atmospheric space, which is flexible, mobile, and accurate. The ground reconnaissance work can use the patrol robot portable observation and analysis equipment to replace the combat personnel to perform reconnaissance missions in the dangerous area. Water reconnaissance is the use of unmanned boats, submersibles and other equipment to detect various dangerous targets on the water surface and underwater. The integrated reconnaissance system eliminates the technical barriers between proprietary intelligence systems. Through information acquisition, convergence, and collaboration, it improves the timeliness and accuracy of target discovery and effectively suppresses the occurrence of terrorism-related activities. The second is Video structured description [1]. This technology converts video data into textual information that can be understood by humans and computers according to the semantic relationship of video content, and can quickly retrieve targets such as people, vehicles, and objects involved in terror from massive video data. For example, video enrichment technology can remove video content that is not related to content and retain video content related to the target. Target image search technology can quickly retrieve content in a video that is highly similar to objects involved in terrorism. Text retrieval technology can search in video to find clues related to terrorism using text described by witness statements. This series of technologies has shortened the time for research and determination in actual combat, and has increased the efficiency of the discovery of criminal clues.

2.2 Synthetic Combat Areas

The anti-terrorism synthesis combat system is usually composed of individual soldier communication equipment, link communication equipment, front-line command equipment, and an integrated command and control platform. It has the ability to connect multiple system information in the face of emergencies, forms a visualized and flattened system, and improves cross-regional, cross-police, cross-departmental, cross-level counter-terrorism synthesis combat capabilities. In response to the actual needs of sudden incidents, environmental complexity, and command accuracy at the current emergency department, the anti-terrorism emergency department's command and control equipment can provide faster, more accurate, and more reliable emergency response and complete process command and control functions. Complete high-precision map of three-dimensional battlefield environments in high-mountain mountains, desert Gobi and other harsh environments can be reconstructed rapidly, and real-time video, positioning, and voice information can be integrated to realize real-time situational awareness of multiple dynamic targets, wireless communication positioning, and command control assistance. The system has assisted anti-terrorist combat units in handling hundreds of terrorist cases.

2.3 Single-Police Comprehensive Combat System

At present, the handling of vicious and violent terrorist incidents is inseparable from the anti-terrorist combat team. The combat equipment of these team members is developing from the combat unit based on people, weapons, and protective clothing to the integrated combat system supported by the battlefield integrated information system. The single-police integrated combat system is usually composed of subsystems such as helmets, glasses, weapons, and protective equipment. It can realize "panorama" perception of the battlefield, "real-time" information sharing, "integration" of combat command, and "diversified attack" under conditions of information, and can enhance the combat capabilities of anti-terrorist members. For example, the smart helmet is lightweight and bullet-proof. It has the functions of voice image video capture, data transmission, positioning, night vision, etc. The smart glasses can dynamically compare targets, quickly lock suspects or targets, and share information with team members, unmanned equipment and command centers. Smart combat garments have built-in bio-micro sensors that can monitor the physiological status of players in real time to facilitate post-injury medical care and rescue. Exoskeletal systems as a wearable device that can enhance human capabilities to run faster, carry more equipment, and can improve the survivability on the battlefield.

3 Trends in Counter-Terrorism Technology and Equipment Development

The counter-terrorism situation in China is constantly changing and anti-terrorism equipment has also been upgraded. Combining anti-terrorism task requirements, anti-terrorism equipment presents the following three trends in the direction of intelligence.

3.1 Big Data Helps Intelligence Early Warning

Over the years, breakthroughs from "post-mortem verification" to "pre-warning" have been an urgent need for public security counter-terrorism work. The in-depth digging function of big data technology can help public security agencies to scientifically, accurately, and quickly discover the information related to terrorism from video and information resources [2]. In the era of big data, video resources can be used to carry out emotional and behavioral analysis of high-risk groups. In key locations, analyze each person's emotional state (such as aggressiveness, stress and stress), activity status (such as abnormal stay, aggregation, and wait-and-see), and identify individuals with potentially dangerous emotions and violent tendencies. And for its early warning and follow-up, once there is excessive behavior, it will immediately be intervened and quickly stopped [4]. At the same time, information resources can be used to conduct in-depth fusion analysis of terrorism-related information. Any terrorist-related activities can be traced, while big data technology can assist anti-terrorist scouts to find these clues in the information space, gather different dimensions of text, voice, images and other terrorist-related information, collide, merge and judge the nature and value of information, and warn of the possible harms and propose countermeasures to contain violent terrorist activities in the bud.

3.2 Artificial Intelligence Helps Combat Terrorism

The haze of international terrorism has infiltrated from time to time, and the internal and external collusion of national separatism has attempted to create troubles, which has brought difficulties to border control in uninhabited areas. With the rapid development of sensors, artificial intelligence and other technologies, intelligent unmanned machines have emerged. This type of unmanned machines can autonomously plan paths, autonomously search, identify, track, and dispose of targets, and is suitable for carrying out daily counter-terrorism tasks in unmanned land. For example, the Long voyage UAV system has the functions of battlefield monitoring, communication relay, and over-the-horizon target tracking. It can perform long-range unmanned area border patrols. The full-featured intelligent anti-terrorism robot can comprehensive monitor the border environment, timely detect of problems and alarm disposal from the perspective of bionic vision, hearing, and smell. The future battlefield against terrorism is likely to consist of commanders, technicians, and peripheral unmanned machines. More and more unmanned machines will replace humans and perform missions such as global reconnaissance and precision strikes in unmanned areas and dangerous areas, and will reduce casualties of anti-terrorist combat personnel.

3.3 Wisdom Judgment Helps Countering Cyber Terrorism

After terrorism entered the Internet era, it continued to expand and spread, using information technology to disrupt and destroy, and gradually formed cyber terrorism [5]. Terrorists have used hidden and untraceable technologies such as deep web to open up new financing channels; they even attempted to use the Internet to destroy target communications, hydropower, and other infrastructure, and brought major security risks to the country. Therefore, cyber anti-terrorism has caused the attention of the governments of all countries in the world. To strengthen cyber anti-terrorism, the first is to strengthen international and domestic cooperation. Gather information on the activities of communication, socialization, and life related to terrorist targets at home and abroad, closely track terrorist trends, accurately intercept information on the activities of terrorists, and curb terrorist plots of extremists. The second is to strengthen wisdom research. Establish key regional intelligence investigation and identification mechanisms, make full use of data collisions, knowledge maps, business understanding and other technologies to strengthen intelligence analysis, and master key personnel and organizational structures so that they can be easily implemented. The third is to strengthen online and offline collaboration. Discover keenly in the network, strike severely in the reality, and cooperate with each other in the network and reality to combat terrorism activities and maintain social security.

4 Conclusions

The counter-terrorism work has a long way to go, and intelligent equipment is in the ascendant. In the current and future period, anti-terrorism work is still a key task in safeguarding public security. It is our unshirkable responsibility to strengthen anti-terrorism smart equipment construction and help combat violence and terrorist activities. The development of counter-terrorism equipment needs to strengthen the integration of government, scientific research, and industry, and take the route of comprehensive technological equipment development. The in-depth application of new technologies such as big data, and artificial intelligence, and smart anti-terrorist tactics research should be regarded as the growth point of counter-terrorism combat capabilities. Anti-terrorism equipment should be promoted in the direction of intelligence.

Acknowledgment. This article is supported by The National Key Research and Development Program of China. (Project No. 2016YFC0801034).

References

1. Xu, Z., Mei, L., Hu, C., Liu, Y.: The big data analytics and applications of the surveillance system using video structured description technology. Cluster Comput. **19**, 1283–1292 (2016)
2. Pelzer, R.: Policing of terrorism using data from social media. Eur. J. Secur. Res (2018)
3. Ball, L.: Automating social network analysis: a power tool for counter-terrorism. Secur. J. **29**, 147–168 (2016)
4. Kebbell, M.R., Porter, L.: An intelligence assessment framework for identifying individuals at risk of committing acts of violent extremism against the West. Secur. J. **25**, 212–228 (2012)
5. Stohl, M.: Networks, terrorists and criminals: the implications for community policing. Crime Law Soc. Change **50**, 59 (2008)

A Monitoring System of Ground-Air Interaction Multi-sensory Fusion Plant Growth Condition

Li Jingzhao[1(✉)] and Guo Mingming[2]

[1] School of Electrical and Information Engineering, Anhui University of Science and Technology, Huainan 232001, China
254662583@qq.com

[2] School of Computer Science and Engineering, Anhui University of Science and Technology, Huainan 232001, China
1014215792@qq.com

Abstract. In view of the shortage of resources and environment in China, the development of intelligent agriculture is relatively backward. The traditional wireless sensor networks (WSNs) use fixed sink node to receive data and communications use multi-hop transmission, which easily creates the problem of "energy hole". In this paper, a monitoring system of plant growth condition based on ground-air interaction multi-sensory fusion is designed, which embodies modern agriculture. The system uses the unmanned aerial vehicle (UAV) loading mobile sink node to work together with the ground fixed cluster head nodes to achieve single-hop transmission of node communication, and effectively solves the imbalance of energy consumption of each node. In the first two sections, the overall design of the monitoring system is introduced. In sections three and four, the communication methods of the system for mobile sink node are given. In section five, the trajectory planning of UAV mobile sink node is proposed. The sixth section designs the test scheme, collects the related data and analyzes it. The seventh section gives a summary. It is proved that on the basis of the monitoring function of plant growth, the system improves the performance of the traditional static sensor network and achieves the perfect combination of the internet of things and agriculture.

Keywords: Agricultural monitoring system · Wireless sensor networks
Mobile sink · Unmanned aerial vehicle

1 Introduction

The problem of agricultural monitoring in areas where there is a large area of agricultural management, far away from urban areas and backward communication facilities and other infrastructure. Ordinary observations cannot fully understand the growth cycle of crops and cannot provide effective measures, which in turn affect farmland harvest.

In traditional static sensor networks, sensor nodes and sink node are statically distributed in the network, and data information is transmitted from source nodes to

J. Abawajy et al. (Eds.): ATCI 2018, AISC 842, pp. 325–332, 2019.
https://doi.org/10.1007/978-3-319-98776-7_35

sink node by multi-hop. This causes the sensor nodes which nearby the sink node to become the hotspot of data transmission. These hotspots need to process their own data, and data information needs to be forwarded for other nodes. Therefore, its energy consumption rate is much higher than that of sensor nodes elsewhere in the network. When the energy consumption of these hotspots is exhausted, the connectivity of the network is destroyed, the so-called "energy hole" problem occurs [1]. At the same time the network is paralyzed. The problem of "energy hole" has seriously affected the life cycle of sensor networks and restricted the further development of sensor networks.

In order to solve the problem of "energy hole", the researchers proposed to introduce mobile sink node into sensor networks. In sensor networks based on mobile sink node, the sink node is no longer statically distributed in a certain position of the network, but is constantly moving in the network. And the sensor nodes around the sink node are constantly changing. Therefore, the problem of "energy hole" can be effectively solved or alleviated. In addition, mobile sink node can enhance the efficiency of network data transmission and balance the rate of energy consumption in different regions of the network. In general, the introduction of sink node mobility has greatly improved the performance of various aspects of sensor networks, and new vitality has been injected into the development of sensor networks.

2 Design and Implementation of Monitoring System

The ground-air interaction multi-sensory fusion plant growth monitoring system is composed of wireless sensor data acquisition network and intelligent monitoring processing center. The function of the wireless sensor data acquisition network is to realize the establishment of the network and the collection of the sensing data [2]. The sensor nodes build the Mesh network based on the hardware based on the ZigBee wireless communication protocol. The sensor nodes are further divided into ground sensor nodes, fixed cluster head nodes and mobile sink node. The sensor terminal nodes of the farmland are arranged to collect environmental parameters such as temperature, humidity, light intensity, and CO2 in real time. When the UAV of the reloaded mobile sink node is near, the fixed cluster head nodes and the mobile sink node complete network connection, and the data is sent to the UAV to be saved. After the UAV completes the data collection [3], it returns to the control room and can transfer the data to the upper computer management system through the U disk.

The intelligent monitoring and processing center consists of a database server and a central computer, which realizes the design of the upper computer. First, the intelligent monitoring and processing center are responsible for the user management of the system, and the addition, deletion, and query operations for the users of the system are realized. Secondly, storing and displaying the received data. The data are plotted as dynamic curves for analysis, and then threshold values are set for the environmental parameters. When the collected data are not within the threshold range, the system sends an alarm prompt, so that the real-time monitoring of the crop growth environment is realized. Finally, publishing data information and transplanting the web protocol into it. The system can be connected to the Internet to realize the structure of "the underlying (sensor)-Internet network-remote monitoring". The information of the whole system can be published to the Internet, and the data can be read on the made APP and the Internet access can be realized.

By analyzing the related data of plant monitoring on APP. Farmers can monitor the crops and some key measures can also be implemented in real time according to the analysis report. Which makes the healthy growth of crops can be guaranteed. The mobile phone APP used by farmers has complete functions, strong practicability and certain personality, which meets the needs of modern people. The entire APP includes modules such as cloud records, plant status, functional measures, and growth forecasts (Fig. 1).

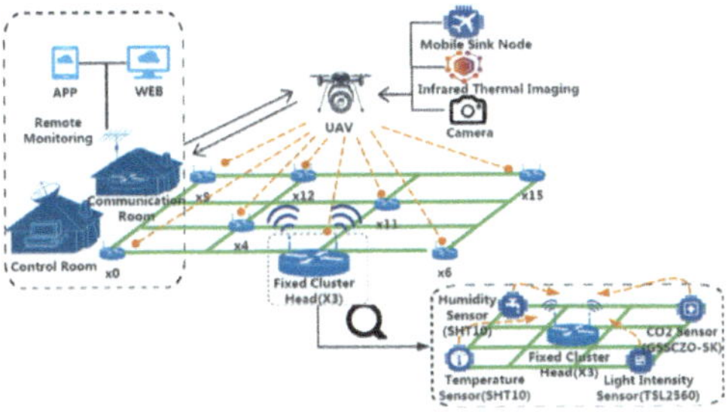

Fig. 1. The UAV monitoring system.

3 Communication among Ground Fixed Nodes

3.1 Establishment of Ground Fixed Cluster Structure

In the monitoring system of this paper, the ground sensor network is divided into multiple logical clusters. The nodes in each cluster can be divided into cluster heads, cluster members and spare nodes. In this protocol, each node is marked [4] with the node flag bit r. As shown in Fig. 2, The formation process of cluster structure is analyzed. The a node that serves as a network coordinator is a cluster head of the area, and its flag bit r is denoted as 0. The fixed cluster head node a connects self-organized with other nodes in this area to form a logical cluster, and each node is broadcasted a notification to determine type of node a. After receiving the cluster head broadcast, other cluster members determine their own node type and set their own flag. The node b with the most remaining energy is determined as a spare node, and the flag bit r is set to 1. The remaining nodes are normal cluster members and the flag bit r is set to 2. In this way, each regional node can form separately its own logical cluster.

3.2 Updating Cluster Head and Intra-cluster Communication

The energy consumption of the cluster heads is relatively large. When the remaining energy of the cluster head node is lower than the set threshold, the spare node is selected as the new cluster head node, and its flag bit r is changed from 1 to 0. The node with the most remaining energy will be select as a new standby node according to the new cluster head node in the logic cluster. Conversely, the original cluster head node

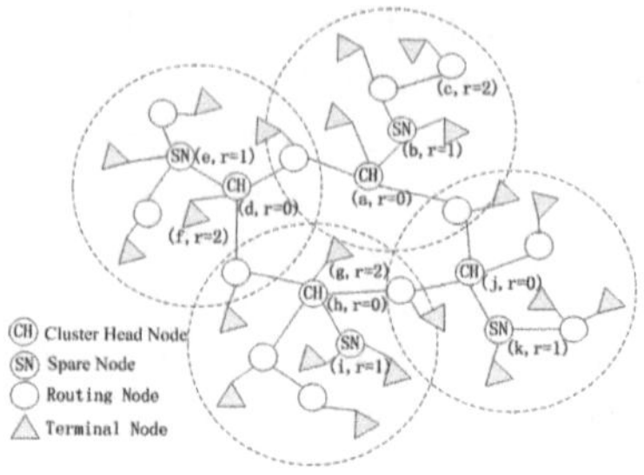

Fig. 2. Formation of ground sensor clusters.

becomes a common cluster member, and the flag bit r changes from 0 to 2. The intra-cluster communication can be set as the data transmission from the child node to the parent node until the cluster head completes the data acquisition.

4 Ground-Air Communications Between Static Node and Dynamic Node

4.1 The Initialization of UAV

The mobile sink node loaded on UAV moves along a fixed trajectory and broadcasts periodically beacon signals to all sensor nodes. All nodes near the trajectory of UAV receive the beacon message. At the same time, the information in each cluster (i.e., size of each regional cluster [5] and the location of cluster head nodes et al.) is analyzed the mobile sink node. The fixed cluster head nodes are identified as communication nodes in each area of the ground. It is only needs to communicate with cluster head nodes [6] for collecting the plant growth data in the whole area.

4.2 Communication with Air UAV for Ground Cluster Head

The final stage of the communication protocol involves a problem about communicating between ground cluster head and UAV. a key requirement that needs to be solved urgently is to determine when the connection between the cluster head and the mobile sink node is available. So we use a confirmation-based protocol [7] between the cluster head and the mobile sink node. During the work traversal of the UAV, the UAV broadcasts periodically along the moving trajectory to indicate that it is in a connected active state. The cluster head can determine whether the mobile sink node is in the connectivity range based on this broadcast signal. When connectivity is confirmed, the cluster head will start sending packets to the mobile sink node. After receiving the data packet, the mobile sink node will send a receipt signal to the cluster head nodes so that the cluster head can determine that the data has been reliably transmitted [8]. Finally, the cluster head can clear the already transmitted packet cache.

5 Experiments

The ground-air interaction multi-sensory fusion plant growth condition monitoring system uses mobile sink node and ground fixed sensor nodes to work together. In terms of performance, compared to the traditional static network, it is greatly improved. In this section, we design a test plan which collects data from various aspects and analyze the superiority of the monitoring system.

(a) Evaluation Method

In a farm field, as shown in Fig. 3, the entire wireless sensor network is designed like a grid layout structure. Each temperature, humidity, carbon dioxide, and illumination sensor nodes are deployed in each fixed grid. Each grid is equivalent to a cluster which has fixed cluster head nodes (x0, x1……x15). When we analyze the data, the fixed cluster head nodes can represent other sensor nodes in its grid [9]. The UAV loads the sink node to realize nodes move. The movement trajectory is A→B→C→D. In order to control variables and compare and analyze data easily, the fixed sink node is placed at the projection point X16 of the initial position A of the UAV. Then we prepare the UAV which has already loaded the sink node, check whether the communication equipment is in good condition, record the relevant experimental data and analyze it after the operation.

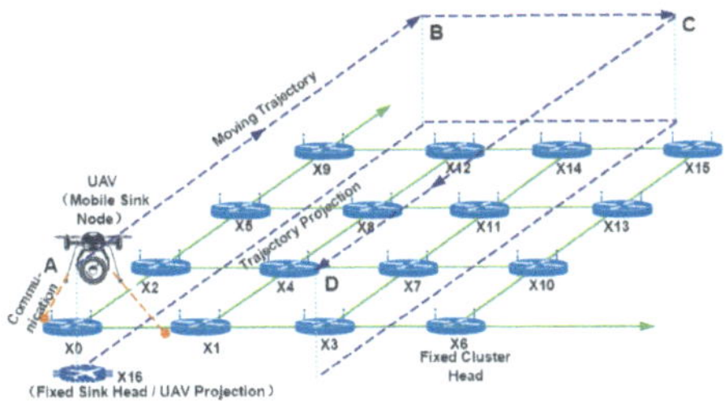

Fig. 3. The mobile roadmap of UAV test program.

In order to operate the experiment and collect data statistics easily, this paper will arrange the grid into a square area. We perform multiple rounds of experiments under the conditions of fixed sink node and mobile sink node. The temperature, humidity, energy consumption and time delay parameters of each node are collected respectively for comparison.

(b) Experimental Analysis

Figure 4 shows the average temperature and humidity curves of each node after several rounds of data collection. Due to the different collection time, the collected data is slightly different. However, from the average temperature and humidity, it can be seen that the test data curves are very close, proving that the collected data under the conditions of the mobile sink node has the same effect as it under the conditions of the fixed sink node.

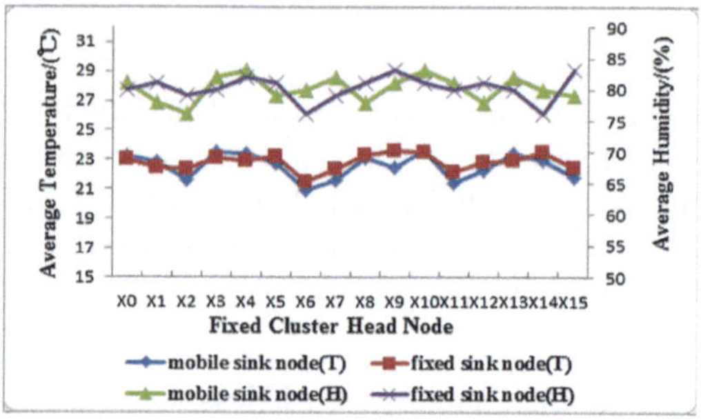

Fig. 4. Curve of average temperature and humidity of cluster head nodes.

It can be seen from Fig. 5 that under the condition of fixed sink node, the nodes (such as nodes X1, X3, X4, X7, X8) which are closer to the sink node and closer to the middle of the experimental scene have higher energy consumption. The further away from the scene, the lower the energy consumption of the nodes (such as nodes X9, X12, and X15). The cause why these nodes consume much higher energy than other nodes is because they not only need to send their own data, but also need to forward the data sent by other nodes. Analyzing nodes X1, X3, and X4, it can be seen that their high energy consumption caused "energy hole". These nodes consuming high energy will directly affect the lifecycle of entire network. It can be seen that under the condition of mobile sink node the energy consumption of each cluster head node does not change much. This is because the UAV flies to the nearby cluster head nodes to receive data and communicates by single hop. This means makes each fixed cluster nodes' consumption relatively balanced and improves "energy hole".

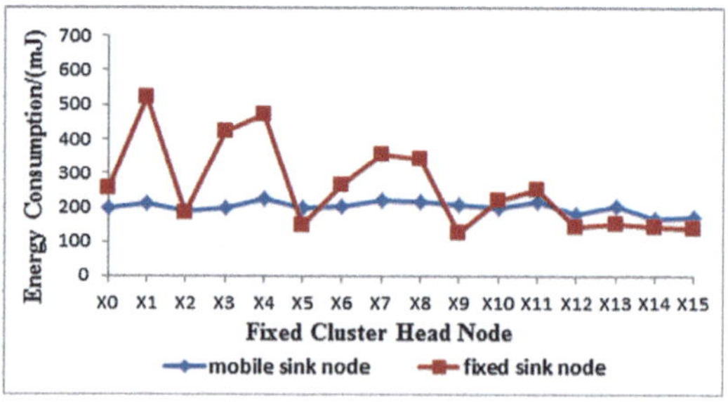

Fig. 5. Curve of each cluster head node's energy consumption.

From the two curves showing in Fig. 6, we can also see that the use of mobile sink nodes does not decrease delay. The mobility of sink nodes makes nodes relatively unstable. Therefore, all the problems affect the real-time performance of UAV receiving data. We do not require high real-time performance when monitoring the status of plant growth, so it has little impact on the performance of monitoring system.

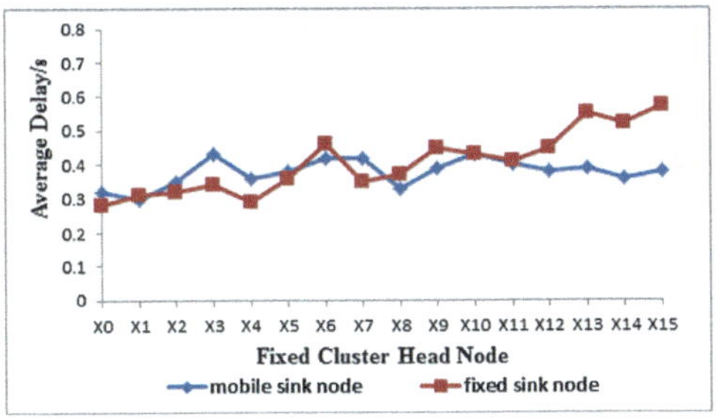

Fig. 6. Curve of average delay of cluster head nodes

In summary, this article can draw two results and the specific analysis is as follows.

1. By analyzing test data, we can deploy relevant sensors more rationally, plan out the movement trajectory and flying height of UAV reasonably in this test program. From Fig. 4 where shows the comparison of temperature and humidity, it can be seen that both schemes can complete the basic data transmission and collection tasks. Figure 5 shows the performance in energy consumption, and we can get that the system using mobile sink node can improve the performance.
2. It can be seen from Fig. 6 that the system using mobile sink node makes average delay curve of each cluster head node slightly stable, but it has not improved a lot, and still has shortcomings. For the inadequacies of the system, planning the trajectory, flight altitude and speed [10] of UAV, as well as UAV's stay time [11] in the collection area rationally can further improve the system' performance. We do not require high real-time performance when monitoring the status of plant growth, so it has little impact on the performance of monitoring system.

6 Conclusions

In this paper, we introduce a ground-air interaction multi-sensor fusion plant growth condition monitoring system in which UAV with infrared thermal imaging and camera auxiliary functions loads sink node to achieve movement and combines with ground fixed cluster sensor nodes to complete data acquisition. We also achieve a comprehensive intelligent monitoring of plants through remote monitoring using APP. Our system saves a large number of manpower and material resources, makes the collected data more accurate, and avoids the errors caused by human fatigue. The system can analyze the collected data in time and take response measures as soon as possible to reduce farmland economic losses. According to the result of experiments, the designed monitoring system with high reliability and applicability integrates technology of the Internet of Things with the agriculture perfectly and reflects the intelligence of modern agriculture.

Acknowledgements. This work is jointly supported by the Major Innovation Platform and Innovative Talent Team of Universities. Research and Development Team Foundation of Mine Internet of Things (Grant No. 2017A053). Anhui university of Science and Technology Post-graduate Innovation Fund (Grant No.2017CX2111).

References

1. Liu, H., Cheng, L.: Design of environmental monitoring system with mobile sink. Comput. Eng. Appl. **46**(19), 7–9 (2010)
2. Cheng, Z., Liu, Y.: Efficient data collection method for wireless sensor networks involved in mobile sink nodes. J. Beijing Jiaotong Univ. **35**(2), 48–54 (2011)
3. Somasundara, A.A., Kansal, A., Jea, D.D., et al.: Controllably mobile infrastructure for low energy embedded networks. IEEE Trans. Mob. Comput. **5**(8), 958–973 (2006)
4. Qian, Z., Zhu, S., Wang, X.: An cluster-based ZigBee routing algorithm for network energy optimization. Chin. J. Comput. **36**(3), 485–493 (2013)
5. Chen, G., Li, C., Ye, M., et al.: An unequal cluster-based routing protocol in wireless sensor networks. Wirel. Netw. **15**(2), 193–207 (2009)
6. Nayak, A., Stojmenovic, I.: Sink Mobility in Wireless Sensor Networks. Wireless Sensor and Actuator Networks: Algorithms and Protocols for Scalable Coordination and Data Communication, pp. 153–184. Wiley, Hoboken (2010)
7. Ma, M., Yang, Y.: SenCar: an energy efficient data gathering mechanism for large scale multihop sensor networks. IEEE Trans. Parallel Distrib. Syst. **18**(10), 1476–1488 (2007)
8. Stojmenovic, I., Nayak, A., Kuruvila, J.: Design guidelines for routing protocols in ad hoc and sensor networks with a realistic physical layer. IEEE Commun. Mag. **43**(3), 101–106 (2005)
9. Mamalis, B., Gavalas, D., Konstantopoulos, C., Pantziou, G.: Clustering in wireless sensor networks. In: Zhang, Y., Yang, L.T., Chen, J. (eds.) RFID and Sensor Networks: Architectures, Protocols, Security and Integrations, pp. 324–353. CRC Press, Boca Raton (2009)
10. Li, Z., Hong, T., Wang, N., et al.: Data transmission performance for 2.4 GHz in-field wireless sensor network. In: International Conference on Computer Engineering and Technology. IEEE, V1-465-V1-469 (2010)
11. Ma, Y., Chen, J., Hao, Y.: Study of energy balancing routing protocol based on sink mobility. Comput. Eng. Appl. **49**(14), 77–80 (2013)

RSCHE: A Virtual Grid Structure Based Routing Scheme with Cell-Header Group Exchanged for Wireless Sensor Network

Jin Wang[1,2](✉), Junming Niu[1], Xiang Yin[1], and Kai Wang[1]

[1] School of Information Engineering, Yangzhou University, Yangzhou, China
{jinwang,yinxiang}@yzu.edu.cn, njm_2016@163.com, kennwong99@163.com

[2] School of Computer and Communication Engineering, Changsha University of Science and Technology, Changsha, China

Abstract. In wireless sensor network (WSN), sensor nodes equipped with batteries are randomly deployed to collect massive data. Since nodes' energy source is usually not rechargeable or replaceable, energy efficiency has always been a chanlleging task for WSN in terms of network lifetime. Among existing works, introducing mobile sinks to WSNs has been considered as an energy efficient way for data collection. Under these circumstance, the network topology becomes dynamic due to sink's mobility. Consequently, it incurres more communication cost to maintain the routing path and thus decreases the network life time. In this paper, we present a routing scheme with cell-header group exchanged (RSCHE) based on virtual grid structure for WSN. Our method partitions the sensor area into uniform-sized grid cells, in which cell-header is elected to transmit information. Those cell-headers are divided into two groups: Indirect communication cell-header (ICH) group and Direct communication cell-header (DCH) group. Despite their mission of collecting data from cell members, cell-headers in DCH group are also responsible for gathering information from cell-headers which belong to ICH group. Furthermore, we set an exchange time (Exch_t) value for ICH and DCH to optimize the routing path. Simulation results show that our proposed protocol performs better compared with existing works in terms network lifetime and energy consumption.

Keywords: Wireless sensor networks (WSNs) · Mobile sink · Cell-header Energy efficiency

1 Introduction

Over recent years, wireless sensor networks (WSN) [1] have been proved a promising technology in many applications. Generally speaking, WSN is a distributed network consisting of tiny sensor nodes and base station. In WSN, sensor nodes are deployed in sensor field for monitoring environment and transmitting data to mobile sink (MS) or base station. Thus a wireless sensor network could be used for smart homes, environment monitoring [2], surveillance and many more applications [3]. For example, in wireless

J. Abawajy et al. (Eds.): ATCI 2018, AISC 842, pp. 333–343, 2019.
https://doi.org/10.1007/978-3-319-98776-7_36

image sensor networks, a usage of WSN where sensor nodes are equipped with tiny cameras to capture images has been a good choice in applications like intrusion detection, weather monitoring, and disaster management etc. Based on WSNs system, we can detect particular environment parameters such as pressure, humidity, temperature, and gas in dangerous areas.

Usually in WSN, tiny sensor nodes are embedded devices with sensing, storage and transmitting capability for collecting information. However, they are always equipped with limited battery power in most applications and battery recharge seems impossible since its deployment environment is dangerous or unreachable. Therefore, energy efficiency has become the key point to extend the network lifetime. Recently, in order to find appropriate method, lots of works have been made to balance network energy consumption. Among those existing works [4, 5] introducing sink's mobility to WSNs has been considered as the most efficient way to prolong network lifetime. Typical routing protocols, like cluster-based data gathering algorithm [6–8], divides all sensor nodes into groups named cluster and each cluster consists of Cluster Head (CH) and Cluster Members (CM). Sensor information is collected by cluster members, aggregated in cluster heads and finally uploaded to MS.

In recent years, typical protocols [9] like chain-based, grid-based routing algorithm [10] have been proposed for WSN. In particularly, grid-based routing schemes overlaying a virtual infrastructure over the sensor field has been considered as a good strategy for WSNs. For those schemes like two-tier data dissemination (TTDD), dynamic directed backbone (DDB) [11], only a few nodes selected as cell-headers are responsible for maintaining the routes toward MS. In this context, cell-headers are used to gather data from source nodes and report them to the mobile sink. Besides, in order to make sure finish the data gathering mission, cell-headers are always selected with higher residual energy. Consequently, the network's energy consumption could be balanced and its lifetime becomes longer. For more relative algorithms, we will describe them in the next section.

The rest of the paper is organized as follows. In Sect. 2, related work is presented. Section 3 shows network model and details our proposed algorithm. Section 4 gives the simulation results. Finally, we conclude this paper with future works in Sect. 5.

2 Related Work

In [12], Two-Tier Data Dissemination (TTDD) routing schemes is proposed for WSN, in which once source node detect a new event, the virtual grid structure will be built for data dissemination. Firstly, the source node is supposed to be located at cross line position in the grid. And then it broadcasts a notification to its all adjacent crossing points of the grid. Nodes received notifications further forward the data information to its own adjacent nodes. This process continues until the entire grid network is completely built. A query from mobile sink could be flooded along the grid cell to the source node. During the process of query dissemination, a reverse route is also selected for reporting data. Then, sensor data could be transmitted toward the mobile sink. However, the construction of virtual grid weaken the network lifetime.

In [13], a virtual infrastructure named rail (Railroad) is built in the middle of the network area. Consequently, nodes locate inside the rail are named rail nodes. Information can be gathered by source nodes and then transmitted to the nearest rail nodes. When a sink needs to collect data, it sends a query to the rail and message will be transmitted among rail nodes until finally uploaded to sink. In [14], Line-based Data Dissemination (LBDD) splits the sensor field into two equal parts by defining a vertical line. Meanwhile, nodes inside the boundary of this wide line are called line-nodes. Sensor data is gathered by source nodes and stored among those line-nodes. Similarly, sink could collect information by sending a query towards inline nodes. Another algorithm called Geographic Cellular-like Architecture (GCA) [15] constructs hexagonal virtual grids for mobile sink collecting data. In GCA, nodes close to the center of the grid cell is selected as cell-header. For data collection, mobile sink disseminates its query packet to the closest cell-header and then propagates it to other cell-headers. When the sink moves into another cell, the routing path could be reconstructed by sending change information to the older cell-header. In spite of GCA's advantage, since it does not optimize the routing path, the reconstructed routes cost is increased which consequently will decrease the network lifetime. In 2015, a Virtual Grid-Based Dynamic Routes Adjustment Scheme (VGDRA) [16] was proposed for wireless sensor networks. It partitions the sensor area into virtual grids and constructs a virtual backbone structure consisting of cell-headers. In each grid, nodes closest to the center of a grid cell are appointed as cell-headers. For data collection, mobile sink movies around the sensor field and interacts with the closest cell-header. With a set of communication rules are made, only a few cell-headers are required for reconstructing process. Thus it could reduce the overall communication cost and prolong the network lifetime.

In this work, our proposed algorithm RSCHE divides the sensor area into several virtual grid of equal sized cells and constructs the backbone network using a few grid cell-headers. For nodes in each virtual grid, taking residual energy, nodes centrality and distance to MS into consideration, cell-header candidates with highest priority will be elected as grid cell-header. For those cell-headers, we split them into two groups: Indirect communication cell-header (ICH) and Direct communication cell-header (DCH) group. For both of them, cell-headers are used to collect data from nodes in their own grid cell, while cell-headers in DCH group are also responsible for gathering information from cell-headers which belong to ICH group. Furthermore, we set ICH and DCH an exchange time (Exch_t) value to optimize routing path toward the mobile sink.

3 Proposed Protocol

3.1 Energy Model

Energy consumption model here is adopted just as the traditional LEACH. The transmitting energy consumption for sensor node is given in Eq. 1, the receiving energy consumption is given in Eq. 2, and d_0 is calculated in Eq. 3.

$$E_{Tx}(k,d) = \begin{cases} E_{elec} * k + \varepsilon_{fs} * k * d^2, if\ d < d_0 \\ E_{elec} * k + \varepsilon_{mp} * k * d^4, if\ d \geq d_0 \end{cases} \tag{1}$$

$$E_{Rx}(k) = E_{elec} * k \tag{2}$$

$$d_0 = \sqrt{\varepsilon_{fs}/\varepsilon_{mp}} \tag{3}$$

For sensor node, E_{elec} represents the energy consumption while transmitting or receiving per bit information. Here k is the number of bits, ε_{fs} is the transmit amplifier energy consumption in free space channel model, while ε_{mp} is used in multipath fading channel model. Besides, the distance for transmitter to receiver is d.

3.2 Network Model

Traditional routing algorithms like LEACH, VGDRA, consider the optimal number of cluster head is 5% of the total sensor nodes, Eq. 4 shows the relationship between the squared number K and total sensor nodes' number N. which means K is the number of cell-headers while N is the total number of sensor nodes.

$$\mathrm{K} = \begin{cases} 4 & N \times 0.05 \leq 6 \\ 6 & 6 < N \times 0.05 \leq 12 \\ 9 & 12 < N \times 0.05 \leq 20 \end{cases} \tag{4}$$

In this work, three hundred nodes are randomly scattered in $200 \times 200\ \mathrm{m}^2$ fields to collect information. Thus we divide the sensor field into 16 parts. Figure 1 shows the grid construction of our work. We set the cell-headers into two groups, for one of them, number in its grid cells are all odd number while the other are all even number. This two

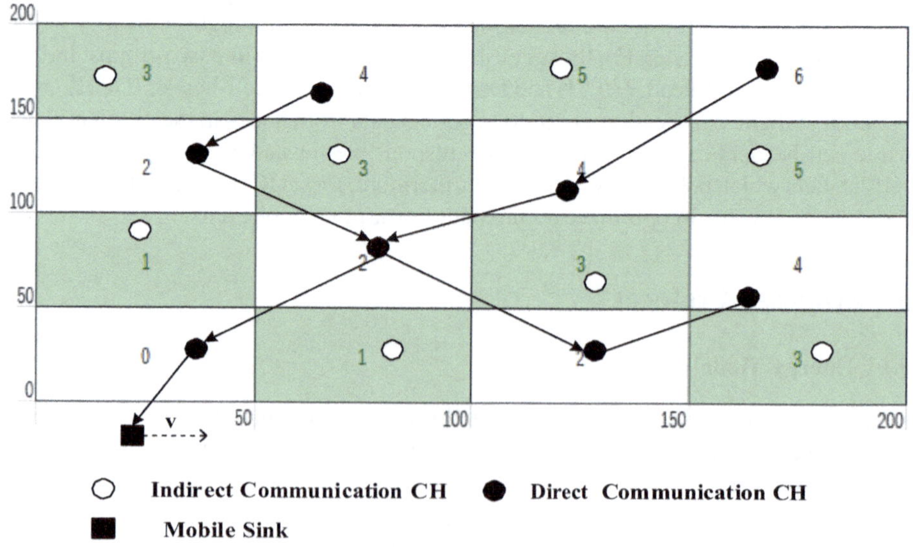

Fig. 1. Network model

groups are named indirect communication group and direct communication group respectively, thus result in electing Indirect communication Cell-Header (ICH) and Direct communication Cell-Header (DCH). Therefore the routing path could be constructed by cell-headers only in DCH group. Here we give an example to explain how we calculate the number in each grid cells. Knowing the node's geographic coordinates (X, Y) in sensor field, and when 50 m is the side length of grid cells. We could get each grid's number by summing $\lfloor X/50 \rfloor$ and $\lfloor Y/50 \rfloor$.

In this network model, data is transmitted by cell-headers scattered in each grid to MS. With mobile sink's movement, only a few nodes are required to adjust their data delivery path when routing path needs reconstruction.

3.3 Routing Scheme

We take three parameters into consideration to calculate the priority opportunity of becoming a cell-header. Consequently, node with highest value of priority becomes the cell-header in each grid cells. These cell-headers in each grid is elected according following rules:

(1) Remaining Energy (RE): since cell-header must have enough energy to transmit data, nodes with acceptable energy level should be considerable in selecting cell-header candidates.
(2) Nodes Centrality (NC): every node has its own geographic coordinates, thus we could evaluate each node's centrality among its neighbors within grid cells. The lower value of node's centrality means less energy consumption required to collect data from other nodes.
(3) Distance to MS: undoubtedly, there always exists one node among those cell-headers to upload the whole network's information to MS. The lower value of distance between node and MS, the more chance to be cell-header candidates.

Since the network becomes dynamic caused by MS, we set exchange time value for ICH and DCH to exchange each other. Appropriate time value will make sure the cell-header, which is responsible for uploading the whole network information to MS, always be the closest one among all cell-headers. In other words, cell-header communicating with mobile sink in grid cells must belong to DCH group when MS is nearby. Figure 2 shows our scheme which select only a few cell-headers maintain the routing path to the closest location of MS. The routing scheme is shown as below.

Step 1: If time value T is a multiple of Exch_t, ICH and DCH group switch to each other. Otherwise do nothing.
Step 2: In ICH group, once finishing data collecting mission, cell-headers are supposed to transmit information to its closest cell-header neighbor which belongs to DCH group.
Step 3: In DCH group, cell-headers are used for constructing routing path.
Step 4: Data is transmitted through routing path and finally uploaded to MS.
Step 5: If there exists surviving sensor nodes, go to step 1. otherwise end the process.

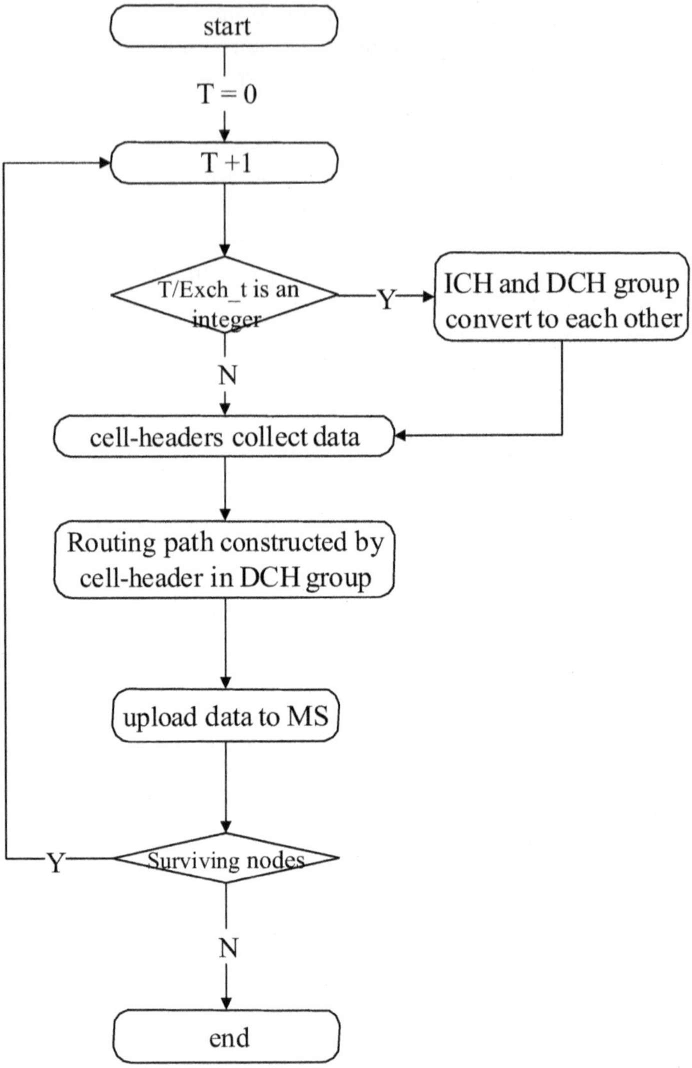

Fig. 2. Flow chart of our routing scheme

Besides, we should notice that the exchange time value differs according mobile sink's speed. With MS moving around the sensor field collecting data via grid cell-header, distance from this cell-header to MS should be the shortest among all cell-headers. In fact, only cell-header belongs to DCH group can uploading data to MS, thus appropriate exchange time is needed to make sure that happens.

4 Simulation Results and Analysis

In our experiment, we firstly varied the mobile sink's speed and CH group's exchange time value to find efficient parameter, which will prolong the network lifetime. Then we made a comparison with other algorithms to prove our protocol could indeed balance the network energy consumption and consequently extend the network lifetime. Table 1 list the simulation parameters which will be used in our simulating work.

Table 1. Parameters for Simulation

Network size	200×200 m^2
Number of sensor nodes	300
Number of rounds	1500
Initial energy	1 J
Sink speed	10, 15, 20 m/s
E_{elec}	50 nJ/bit
ε_{fs}	10 pJ/bit/m^2
ε_{mp}	0.0013 pJ/bit/m^4

4.1 Experiment of Network Lifetime

As we know, the network lifetime is relayed on the duration of time when the first sensor node dies. Thus the round number of first dead node appears is considered to evaluate the network's performance in different Exch_t values. we presented its changes in Fig. 3.

It is obviously that the network lifetime is much longer when Exch_t is 5. Furthermore, we list the first, half and final dead round number in Table 2. Noticed that the final dead round number is largest than others when Exch_t is 40, we analyzed the network topology changes at the end phases of simulation. Then we make a conclusion that at the end part of simulation, the most alive nodes always located at the center of sensor field. Shorter Exch_t will increase distance and change dramatically between MS and sensor nodes, which consequently deplete node's energy quickly. As presented in Table 2, we learned that when Exch_t is 5 and 10, there is little difference the number of node's dead round as well as Exch_t is 20 and 40. So we only depicted two of them to show the alive number's changes in Fig. 4, from which difference between first/final dead round number is obviously depicted.

Table 2. Round number of dead node

Exch_t value	5	10	20	40
First dead round	345	327	293	257
Half dead round	803	775	804	819
Final dead round	1322	1263	1428	1435

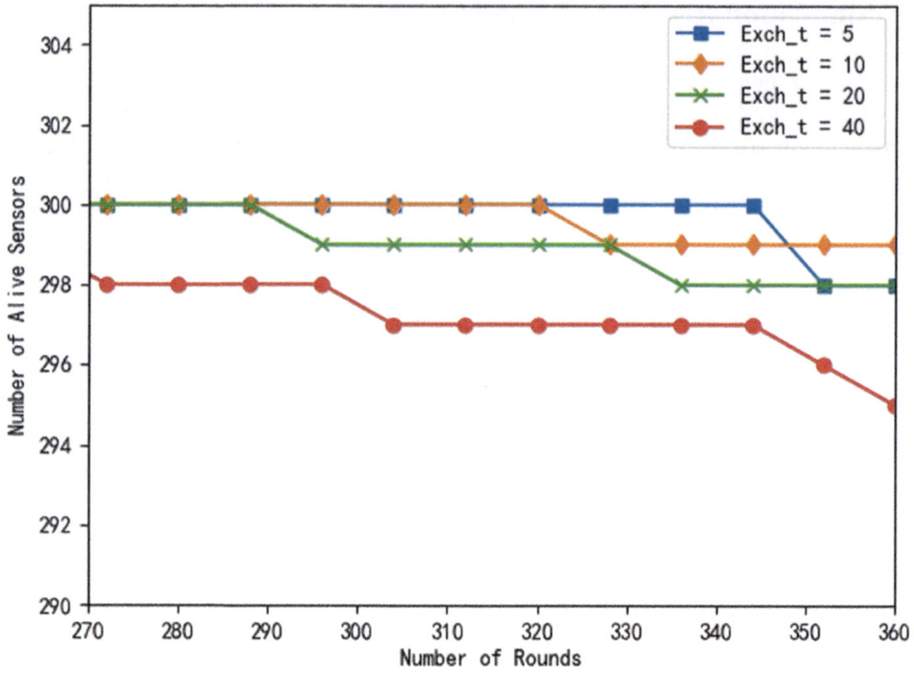

Fig. 3. Comparison of network lifetime in varied Exch_t when MS's velocity is 10 m/s

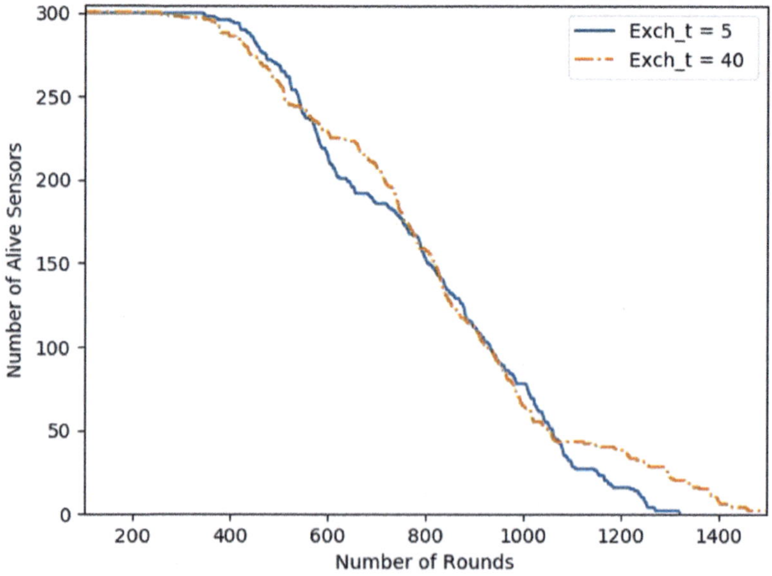

Fig. 4. Number of alive nodes each round

Moreover, we also tested the network performance in the same way when MS's speed is 15 and 20 m/s, and the results are listed in Table 3. In a conclusion, appropriate Exch_t values for extending network lifetime differ when mobile sink's velocity varies.

Table 3. Round number of dead node

MS's velocity	15	15	20	20	20
Exch_t value	7	10	10	20	40
First dead round	**376**	323	307	245	**341**
Half dead round	819	798	793	811	814
Final dead round	1305	1336	1356	1529	1327

4.2 Simulation of Varied Methods

Existing works like TTDD, GCA, VGDRA etc. are also using a virtual infrastructure for wireless sensor networks, so we compared our proposed protocol with some of them to test our algorithm's efficiency. We randomly deployed 300 sensor nodes in a sensor area of 200 × 200 dimension. Mobile sink moves around the sensor field counterclockwise with speed varied from 10 to 20 m/s. And CH group's exchange time values were adopted when mobile sink moves pass through the first grid, 1/4, 1/2 and the whole length of sensing area's border. For example, when velocity of MS is 10 m/s, we may

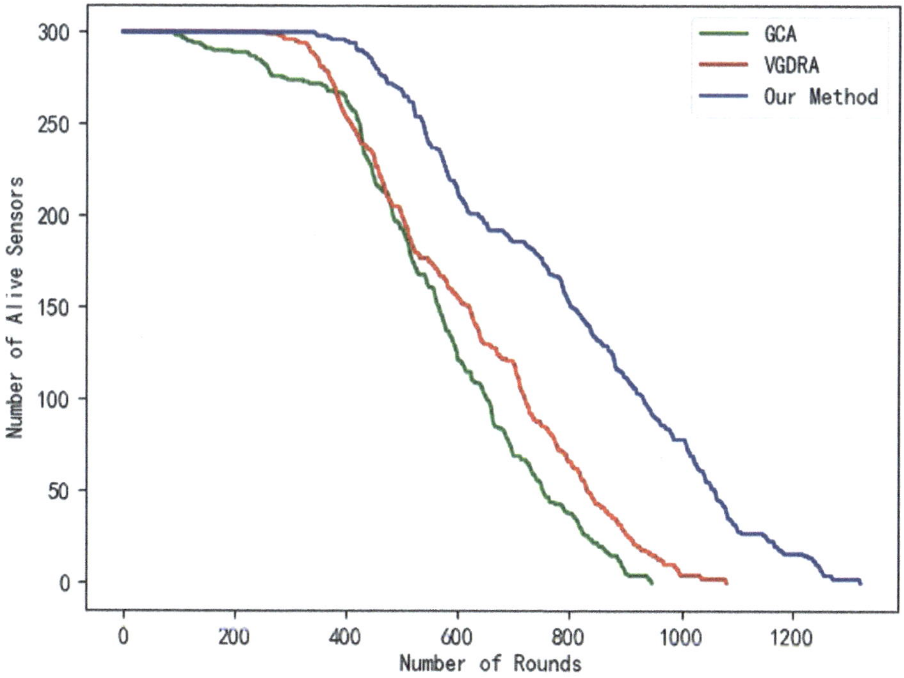

Fig. 5. Number of alive nodes each round for different algorithms

set the CH group's exchange time value as 5, 10, 20 and 40. After the simulation of different parameters, we may seek the best one to extend the network life time longer. And the simulation results are shown in Figs. 5 and 6.

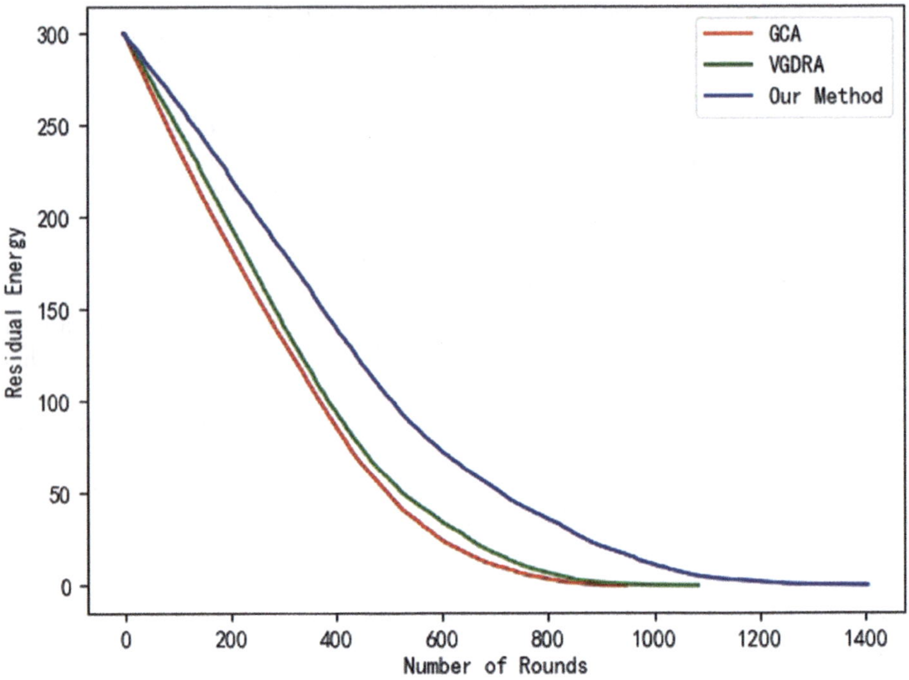

Fig. 6. Total residual energy each round

As Fig. 4 shows, in our proposed routing protocol the number of first dead node round number is 347, which is much larger than VGDRA and GCA. Besides, it has much more nodes alive as compared to VGDRA and GCA in each round. Last picture Fig. 6 describes the network's energy consumption in each round, from which we may find out our algorithm is superior in terms of network lifetime and energy consumption.

5 Conclusion

In this paper, we proposed a routing scheme with cell-header group exchanged (RSCHE) based on virtual grid structure for WSNs. It is capable to maintain the optimal routes to MS with limited communication cost. Our RSCHE method divides the sensor area into several virtual grid cells and constructs the backbone network using a few grid cell-headers. We make rules to select cell-header with higher residual energy. And split them into two groups: Indirect communication cell-header (ICH) and Direct communication cell-header (DCH) group. Furthermore, we set ICH and DCH an exchange time (Exch_t) value to optimize routing path towards the mobile sink. Simulation results have also proved that our protocol performs better compared with other routing scheme in terms of energy consumption and network lifetime. In our future work, the mobile sink is supposed to move randomly within the sensor field for collecting data.

Acknowledgments. This work is supported by the National Natural Science Foundation of China (61772454, 6171101570). It is also supported by the Natural Science Foundation of Jiangsu Province (BK20150460). Professor Jin Wang is the corresponding author.

References

1. Demigha, O., Hidouci, W.K., Ahmed, T.: On energy efficiency in collaborative target tracking in wireless sensor network: a review. IEEE Commun. Surv. Tutor. **15**(3), 1210–1222 (2013)
2. Lara-Cueva, R.A., Gordillo, R., Valencia, L., et al.: Determining the main CSMA parameters for adequate performance of WSN for real-time volcano monitoring system applications. IEEE Sens. J. **17**(5), 1493–1502 (2017)
3. Wang, H., Dong, L., Wei, W., et al.: The WSN monitoring system for large outdoor advertising boards based on ZigBee and MEMS sensor. IEEE Sens. J. **PP**(99), 1 (2017)
4. Pantazis, N.A., Nikolidakis, S.A., Vergados, D.D.: Energy-efficient routing protocols in wireless sensor networks: a survey. IEEE Commun. Surv. Tutor. **15**(2), 551–591 (2013)
5. Yetgin, H., Cheung, K.T.K., El-Hajjar, M., et al.: A survey of network lifetime maximization techniques in wireless sensor networks. IEEE Commun. Surv. Tutor. **19**(2), 828–854 (2017)
6. Singh, S., Kumar, P., Singh, J.: A survey on successors of LEACH protocol. IEEE Access **PP**(99), 1 (2017)
7. Nayak, P., Devulapalli, A.: A fuzzy logic-based clustering algorithm for WSN to extend the network lifetime. IEEE Sens. J. **16**(1), 137–144 (2015)
8. Sasirekha, S., Swamynathan, S.: Cluster-chain mobile agent routing algorithm for efficient data aggregation in wireless sensor network. J. Commun. Netw. **19**(4), 392–401 (2017)
9. Liu, X.: Atypical hierarchical routing protocols for wireless sensor networks: a review. IEEE Sens. J. **15**(10), 5372–5383 (2015)
10. Hamida, E.B., Chelius, G.: Strategies for data dissemination to mobile sinks in wireless sensor networks. Wirel. Commun. IEEE **15**(6), 31–37 (2008)
11. Bhatti, R., Kaur, G.: Virtual grid based energy efficient mobile sink routing algorithm for WSN. In: International Conference on Intelligent Systems and Control, pp. 30–33. IEEE (2017)
12. Luo, H., Ye, F., Cheng, J., et al.: TTDD: two-tier data dissemination in large-scale wireless sensor networks. Wirel. Netw. **11**(1–2), 161–175 (2005)
13. Shin, J.H., Kim, J., Park, K., et al.: Railroad: virtual infrastructure for data dissemination in wireless sensor networks. In: ACM International Workshop on Performance Evaluation of Wireless Ad Hoc, Sensor, and Ubiquitous Networks, pp. 168–174. ACM (2005)
14. Hamida, E.B., Chelius, G.: A line-based data dissemination protocol for wireless sensor networks with mobile sink. In: IEEE International Conference on Communications, pp. 2201–2205. IEEE (2013)
15. Chen, X., Xu, M.: A geographical cellular-like architecture for wireless sensor networks. In: International Conference on Mobile Ad-Hoc and Sensor Networks, pp. 249-258. Springer, Heidelberg (2005)
16. Khan, A.W., Abdullah, A.H., Razzaque, M.A., et al.: VGDRA: a virtual grid-based dynamic routes adjustment scheme for mobile sink-based wireless sensor networks. Sens. J. IEEE **15**(1), 526–534 (2015)

An Improved Person Reidentification Method by Selective Convolutional Descriptor Aggregation

Shuren Zhou[1,2(✉)], Maolin Ke[1,2], and Peng Luo[1,2]

[1] Hunan Provincial Key Laboratory of Intelligent Processing of Big Data on Transportation, Changsha University of Science and Technology, Changsha 410114, China
zsr@csust.edu.cn

[2] School of Computer and Communication Engineering, Changsha University of Science and Technology, Changsha 410114, China

Abstract. Person re-identification (person re-ID) is refers to pedestrians matching under a multi camera network in a non-overlapping field of view, that is, whether the pedestrian targets taken by different cameras are the same. Therefore, person re-ID is also a sub problem of image retrieval. In this paper, fine grained image retrieval method SCDA (Selective Convolutional Descriptor Aggregation) is used in person re-ID, and the framework of identification-verification is improved. The method can select useful deep descriptors, at the same time, remove background by localizing the main object.

Keywords: Person re-identification · Computer vision · Image retrieval Deep learning

1 Introduction

Person re-ID is a computer vision technology used to judge whether there is a specific pedestrian in the image or video sequence. Given a monitored pedestrian image, the pedestrian image is retrieved from multiple monitoring devices in a region. Therefore, person re-ID is also a sub problem of image retrieval. Compared with face recognition, person re-ID is more difficult, which leads to the difficulty of applying mature face recognition technology to person re-ID. First, the pixel of the camera may be poor, especially the face in the remote camera is very small, and there is a wide range of back and side face in the pedestrian image. It is difficult to use face recognition. In addition, the possible posture of the same pedestrians on different cameras is different, and the clothes between different pedestrians may be the same. Therefore, it is difficult to distinguish the pedestrians from the color of the clothes and the pose of the pedestrians. In general, the research of person re-ID faces many challenges, such as low resolution of image, change of view, change of pose, change of light, and occlusion. In addition, the scene under the actual video surveillance is very complex. Many monitoring scenes have large flow of people, complex scenes, and the screen is easily occluded. All of these have brought great challenges to the research of person re-ID.

J. Abawajy et al. (Eds.): ATCI 2018, AISC 842, pp. 344–350, 2019.
https://doi.org/10.1007/978-3-319-98776-7_37

Compared with the artificial design mentioned, the deep convolution neural network can obtain the intrinsic features of the image more, and has shown good performance in target detection, image classification and image segmentation [1].

With the development of deep convolution neural network (ResNet [2], GoogleNet [3], etc.), the design method based on loss function is commonly used in person re-ID. Since CNN can automatically extract Representation features from the task requirements automatically from the original image data, some researchers regard the person re-ID problem as an Identification or Verification problem ([4–7]): the Identification problem is to use the ID or attribute as a training label to train the model; the verification problem refers to the input of a pair of pedestrian pictures, allowing the network to learn whether these two pictures belong to the same pedestrian.

In this paper, we use identification-verification [8] network and use ResNet-50 [9] as pre-trained CNN models, and combine the advantages of the two models for data training. In the last convolution layer of ResNet-50, the feature descriptors are selected by SCDA (Selective Convolutional Descriptor Aggregation) [10] method, and the feature descriptors with strong response are retained, and the feature descriptors are aggregated with avgPool and maxPool in order to improve the precision of recognition. The following are some of the work of this article: first introduced the pre-trained CNN models, then introduced the identification-verification framework, and introduced its advantages, and finally introduced the SCDA method and the application and improvement in this experiment. In the experimental part, we introduced the relevant dataset. In this experiment, we used the maket1501 [11] dataset and compared the experimental results of other algorithms with ours.

2 Related Work

2.1 ResNet-50

In experiment, we used ResNet-50 as pre-trained CNN models. The deep residuals network is a deep convolution network proposed in 2015. Once it was born, it captured three titles of image classification, detection and location in ImageNet. The idea of Deep Residual Learning is as follows:

If there is a good network A that can work, then a network B is deeper than it is, just to make the previous part of the B exactly the same as the A, and the latter part only implements an identity mapping, so that B can at least get the same performance as A, not even worse. The idea of deep residual learning is also generated. Since the following part of B completes the identity mapping, why not add this prior to the training of the network (in the process of network training, it is very important to add prior information guidance and reasonable priori can get very good results), so the network is constructed. In the shortcut connection, the output of each layer is not the mapping of the input in the traditional neural network, but the superposition of the input mapping and input. As shown in the Residual net in Fig. 1: (1) identity: for identity mapping, this path is always there. (2) F(x): residual function for learning: $H(x) - x = F(x)$.

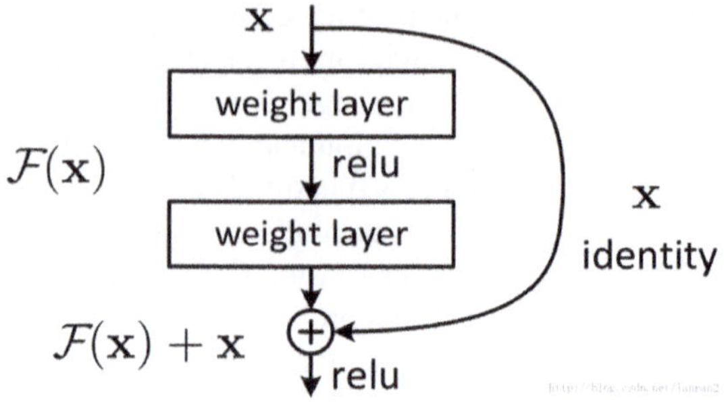

Fig. 1. Shortcut connection

2.2 Identification-Verification Model

In our work, we use the Identification-verification model. As shown in Fig. 2, identification loss and verification loss are used to train the network, and the network input is a number of pedestrians' pictures, including the classified subnetwork (Classification Subnet) and the verification subnetwork (Verification Subnet). The classification sub network performs ID prediction and calculates the classification error loss according to the predicted ID. The verification sub network converged the characteristics of the two pictures to determine whether the two pictures belong to the same pedestrian. The sub network is essentially equivalent to a two classification network. The network can predict ID and calculate similarity scores simultaneously. The loss that needs to be optimized is three target loss, that is, two identification loss and a verifying loss, and the author transforms identification loss and verification loss into a cross entropy loss [12] to monitor the update of model parameters in the real training model. After training enough data, once again input a test picture, the network will automatically extract a feature, which is used for person re-ID task.

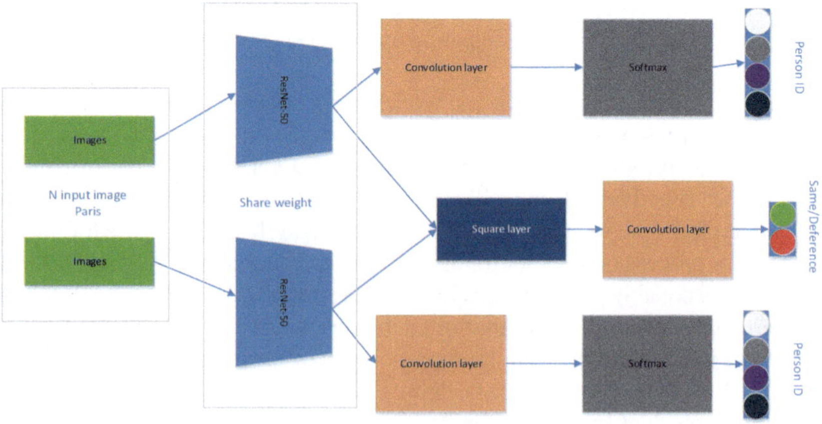

Fig. 2. The identification-verification network

As shown in Table 1, it is the results on market1501 which obtain by identification and verification loss and jointly. In contrast, the experimental results obtained using the identification-verification model are the best.

Table 1. The results for identification and verification loss

Method	rank-1	mAP
CaffeNet [13] (I)	50.89	26.79
CaffeNet (V)	41.24	22.47
CaffeNet (V+I)	62.14	39.61
VGG16 [14] (I)	65.02	38.27
VGG16 (V)	42.99	24.29
VGG16 (V+I)	70.16	47.45
ResNet-50 (I)	73.69	51.48
ResNet-50 (V)	64.58	44.94
ResNet-50 (V+I)	79.51	59.87

2.3 Selective Convolutional Descriptor Aggregation

Selective Convolutional Descriptor Aggregation (SCDA) is the first fine grained image retrieval method based on deep learning recently proposed. As in the image retrieval work under other deep learning framework, in SCDA, the fine-grained image as input into the pre-trained CNN model gets convolution feature. Different from the traditional image retrieval, a deep learning method is used to select the convolution descriptor for the problem of fine grained image retrieval because the convolution features are all over the all connected layer features. In image retrieval, there is no fine part annotation, even image level markup cannot be obtained. SCDA algorithm can still locate objects under unsupervised conditions, and then select convolution feature descriptors based on localization results. The reserved depth features are processed by averaging and maximum pooling respectively, and then cascaded to form the final image representation.

We add up the obtained last convolution feature map through the pre-train model ResNet-50, and then the convolution feature (H × W × D) is added in the depth direction, so that we can get an aggregation map (H × W × 1). In this two-dimensional graph, the mean value of all H × W elements can be calculated, and the mean value $\bar{a}$ is the key to the location of the object: the convolution feature of the element location larger than the M in the aggregation map needs to be retained; the smaller one is discarded.

$$M_{i,j} = \begin{cases} 1 \ if \ a_{i,j} > \bar{a} \\ 0 \ otherwise \end{cases} \tag{1}$$

The location of a fine-grained object is responsive to most of the channels of the convolution feature tensor, and the convolution features are added in the depth direction to accumulate the response of the position of these objects. The mean, as a "ruler", marks the response of "not qualified" as noise, and marks the location of "reaching the standard" as the object. The location preserved is corresponding to the location of the convolution feature descriptor. The reserved depth features are processed by averaging and maximum pooling respectively, and then cascaded to form the final image representation.

$$P_{avg} = \frac{1}{N}\sum\nolimits_{i,j} x_{(i,j)} \tag{2}$$

$$P_{max} = \max\nolimits_{i,j} x_{(i,j)} \tag{3}$$

3 Experiment

The common dataset of pedestrian data is market1501, CUHK03 [15] and DukeMTMC4ReID [16]. We used the market1501 data set in our experiment, it contains 751 identities and 12936 images from six dis-joint cameras. This dataset also includes 2793 false alarms detected by DPM, which as distractors for the real scenario.

In our experiment, we change all the training images to 256 × 256. Then randomly change images size to 224 × 224 for ResNet-50. The batch size is set to 48. The learning rate is initialized as 0.001, then set to 0.0001 for the final. In testing, we extract features from the final convolution layer, we add up obtained feature map and then the convolution feature (H × W × D) is added in the depth direction, so that we can get an aggregation map (H × W × 1) and select convolution feature descriptors based on localization results. The reserved depth features are processed by averaging and maximum pooling, and then cascaded to form the final image feature. Final, between query and gallery features, we use the cosine distance to obtain the ranking result. As shown in Table 2, the rank-1 and mAP precision are shown. Compared with other methods, our experiment result is better.

Table 2. Person re-ID methods by market-1501

Method	rank-1	mAP
DADM [20]	39.4	19.6
BoW + KISSME [17]	44.42	20.76
Multiregion CNN [19]	45.58	26.11
CAN [21]	48.24	24.43
Fisher network [23]	48.15	29.94
SL [18]	51.9	26.35
DNS [22]	55.43	29.87
Gate reid [24]	65.88	39.55
ResNet-50 (I+V)	79.51	59.87
Ours	81.41	61.54

4 Conclusion

Person re-ID is widely considered as a sub problem of image retrieval, so it can also be solved by traditional fine-grained image analysis. In this paper, fine grained image retrieval method SCDA (Selective Convolutional Descriptor Aggregation) is used in person re-ID, and the framework of identification-verification is improved. For person re-ID, the future work is to find more novel applications for the proposed method.

Acknowledgments. This work was supported by the Scientific Research Fund of Hunan Provincial Education Department of China (Project No. 17A007); and the Teaching Reform and Research Project of Hunan Province of China (Project No. JG1615).

References

1. Zheng, L., Yang, Y., Hauptmann, A.G.: Person re-identification: past, present and future (2016). arXiv preprint arXiv:1610.02984
2. He, K., Zhang, X., Ren, S., Sun, J.: Deep residual learning for image recognition. In: Proceedings of the IEEE Conference on Computer Vision and Pattern Recognition, pp. 770–778 (2016)
3. Szegedy, C., Liu, W., Jia, Y., Sermanet, P., Reed, S., Anguelov, D., Rabinovich, A.: Going deeper with convolutions. In: Proceedings of the IEEE Conference on Computer Vision and Pattern Recognition, pp. 1–9 (2015)
4. Li, W., Zhao, R., Xiao, T., Wang, X.: Deepreid: deep filter pairing neural network for person re-identification. In: Proceedings of the IEEE Conference on Computer Vision and Pattern Recognition, pp. 152–159 (2014)
5. Yi, D., Lei, Z., Liao, S., Li, S.Z.: Deep metric learning for person re-identification. In: 2014 22nd International Conference on Pattern Recognition (ICPR), pp. 34–39. IEEE
6. Varior, R.R., Haloi, M., Wang, G.: Gated siamese convolutional neural network architecture for human re-identification. In: European Conference on Computer Vision, pp. 791–808. Springer, Cham (2016)
7. Ahmed, E., Jones, M., Marks, T.K.: An improved deep learning architecture for person re-identification. In: Proceedings of the IEEE Conference on Computer Vision and Pattern Recognition, pp. 3908–3916 (2015)
8. Zheng, Z., Zheng, L., Yang, Y.: A discriminatively learned CNN embedding for person reidentification. ACM Trans. Multimed. Comput. Commun. Appl. (TOMM) **14**(1), 13 (2017)
9. He, K., Zhang, X., Ren, S., Sun, J.: Deep residual learning for image recognition. In: Proceedings of the IEEE Conference on Computer Vision and Pattern Recognition, pp. 770–778 (2016)
10. Wei, X.S., Luo, J.H., Wu, J., Zhou, Z.H.: Selective convolutional descriptor aggregation for fine-grained image retrieval. IEEE Trans. Image Process. **26**(6), 2868–2881 (2017)
11. Zheng, L., Shen, L., Tian, L., Wang, S., Wang, J., Tian, Q.: Scalable person re-identification: a benchmark. In: Proceedings of the IEEE International Conference on Computer Vision, pp. 1116–1124 (2015)
12. Rubinstein, R.Y., Kroese, D.P.: The cross-entropy method. Technometrics **50**(1), 92 (2008)
13. Krizhevsky, A., Sutskever, I., Hinton, G.E.: Imagenet classification with deep convolutional neural networks. In: Advances in Neural Information Processing Systems, pp. 1097–1105 (2012)

14. Simonyan, K., Zisserman, A.: Very deep convolutional networks for large-scale image recognition (2014). arXiv preprint arXiv:1409.1556
15. Li, W., Zhao, R., Xiao, T., Wang, X.: Deepreid: deep filter pairing neural network for person re-identification. In: Proceedings of the IEEE Conference on Computer Vision and Pattern Recognition, pp. 152–159 (2014)
16. Gou, M., Karanam, S., Liu, W., Camps, O., Radke, R.J.: Dukemtmc4reid: a large-scale multi-camera person re-identification dataset. In: IEEE Conference on Computer Vision and Pattern Recognition Workshops (2017)
17. Zheng, L., Shen, L., Tian, L., Wang, S., Wang, J., Tian, Q.: Scalable person re-identification: a benchmark. In: Proceedings of the IEEE International Conference on Computer Vision, pp. 1116–1124 (2015)
18. Chen, D., Yuan, Z., Chen, B., Zheng, N.: Similarity learning with spatial constraints for person re-identification. In: Proceedings of the IEEE Conference on Computer Vision and Pattern Recognition, pp. 1268–1277 (2016)
19. Ustinova, E., Ganin, Y., Lempitsky, V.: Multi-region bilinear convolutional neural networks for person re-identification. In: 2017 14th IEEE International Conference on Advanced Video and Signal Based Surveillance (AVSS), pp. 1–6. IEEE (2017)
20. Su, C., Zhang, S., Xing, J., Gao, W., Tian, Q.: Deep attributes driven multi-camera person re-identification. In: European Conference on Computer Vision, pp. 475–491. Springer, Cham (2016)
21. Liu, H., Feng, J., Qi, M., Jiang, J., Yan, S.: End-to-end comparative attention networks for person re-identification. IEEE Trans. Image Process. **26**(7), 3492–3506 (2017)
22. Zhang, L., Xiang, T., Gong, S.: Learning a discriminative null space for person re-identification. In: Proceedings of the IEEE Conference on Computer Vision and Pattern Recognition, pp. 1239–1248 (2016)
23. Wu, L., Shen, C., van den Hengel, A.: Deep linear discriminant analysis on fisher networks: a hybrid architecture for person re-identification. Pattern Recognit. **65**, 238–250 (2017)
24. Varior, R.R., Haloi, M., Wang, G.: Gated siamese convolutional neural network architecture for human re-identification. In: European Conference on Computer Vision, pp. 791–808. Springer, Cham (2016)

A Survey of Multi-object Video Tracking Algorithms

Shuren Zhou[1,2], Maolin Ke[1,2], Jia Qiu[1,2], and Jin Wang[1,2(✉)]

[1] Hunan Provincial Key Laboratory of Intelligent Processing of Big Data on Transportation, Changsha University of Science and Technology, Changsha 410114, China
jinwang@csust.edu.cn

[2] School of Computer and Communication Engineering, Changsha University of Science and Technology, Changsha 410114, China

Abstract. Video multi-object tracking is one of the important research topics in the field of computer vision, which is widely used in military and civil areas. At present, the research of single object tracking algorithm is quite mature, however the research of multi-object tracking is still ongoing. This paper focuses on four important stages in the multi-object tracking process: feature extraction, detector, data association and the tracker. The feature extraction part introduces the current methods of feature extraction, as well as the merits and demerits of each method; In the stage of detection, the tracking effect of the object appearance model in specific applications is described, and then the paper analyze the multi-object tracking algorithm based on detection and tracking as well as the multi-object tracking algorithm based on deep learning; In the tracking stage, the establishment of object motion model and multi-object tracking with different tracker hybrid algorithm are introduced; During the stage of data association, the paper introduce the multi-object tracking based on energy minimization and commonly used data association algorithm, respectively. Then the current mainstream datasets and evaluation methods are introduced. Finally, the future development of the multi-object tracking is discussed and forecasted.

Keywords: Video analysis · Computer vision · Multi-object tracking Deep learning

1 Introduction

Target tracking [1] as a branch of computer vision, has been a hotspot for many years. Especially in the complex background, the target tracking is more challenging when the target environment changes or the target is blocked. According to the number of tracking targets, target tracking can be divided into two categories: single target and multi-target. The research on single target tracking is quite mature at home and abroad, but the research on multi-target tracking is still in development. Video target tracking is widely used in the fields of man-machine intelligent interaction, video surveillance and military weapon application [2]. Because of the wide application, target tracking has always attracted much attention.

J. Abawajy et al. (Eds.): ATCI 2018, AISC 842, pp. 351–369, 2019.
https://doi.org/10.1007/978-3-319-98776-7_38

Up to now, there are two main research frameworks for multi-target tracking algorithm: one is the framework based on detection and tracking (tracking-by-detection) [3], the other is the deep learning framework [4] which is popular in recent years. In the framework based on detection and tracking, two models are often used, the appearance model and the motion model. The appearance model is mainly the overall appearance features modeling, which separate the target and background as much as possible; the motion model is mainly the movement characteristics modeling, which predicting the position of target, mining related information between frames, and then gets the track of the target by event analysis.

With the arrival of deep learning and big data, a number of new research methods have emerged in the field of multi-target tracking. Target detection and recognition are the precondition of target tracking. In recent years, deep learning has achieved great success in target detection and target recognition, and greatly promoted the development of deep learning in target tracking. It can be said that deep learning has brought a new revolution to the field of computer vision.

2 Difficulty of Research

At present, multi-target tracking methods have achieved good results in some simple scenarios, but there are still many problems. The complexity of background, the diversity of targets, the similarity between background and target, and the interaction between multiple objectives will cause difficulties in the research. The main difficulties of current multiple target tracking are as follows:

Occlusion problem: occlusion is a common phenomenon in the process of target tracking. The occlusion can be divided into self-occlusion, the static object occlusion in the background and the occlusion by other targets. The process of target occlusion can be divided into two stages: In the first stage, the target is slowly blocked. In this process, the information that the target can detect is less and less. Therefore, the phenomenon of target loss often occurs. In the second stage, also known as target reproduction, the target is followed again after it is blocked. The information of the target is gradually restored during the process of recurrence. It is also a difficult problem to ensure the accurate tracking again after the reappearance of target. Occlusion can lead to the loss of effective information of the target, especially when the target is blocked for a long time. At present, most of the systems cannot solve the problem of long time occlusion. Finding an effective solution will be the purpose of continuous research.

Background complexity: the complexity and stability of the target background will determine the difficulty and effect of the tracking. The interference in the background includes the change of brightness, the change of the object in the background, the transformation of the scene and the shadow. The color of the background is easy to change with the change of illumination condition, and the target is moving in real time, so it is necessary to update in time. Because there is more than one target in the background, other objects similar to the target can cause misdiagnosis, which will greatly increase the difficulty of target tracking. In addition, the shadow will lead to differences in background color, which cause difficulties to the target detection.

Data association: data association is also the difficulty of multiple target tracking. The efficiency and complexity of association algorithm will directly affect the efficiency and accuracy of tracking. Data association is a process of matching the tracking target with the trajectory of prediction to get the correct track of the target. A single target does not need to establish data association. When there are multiple targets and they are very close, data association is particularly important, and the process of association is more complex. At present, although there are many data association algorithms, how to establish a more efficient data association algorithm is still one of the difficulties.

3 Video Multi-target Tracking Framework

At present, there are many frameworks for multiple target tracking, and many improved frameworks have emerged as a result of continuous development. The general multiple target tracking system framework is shown in Fig. 1.

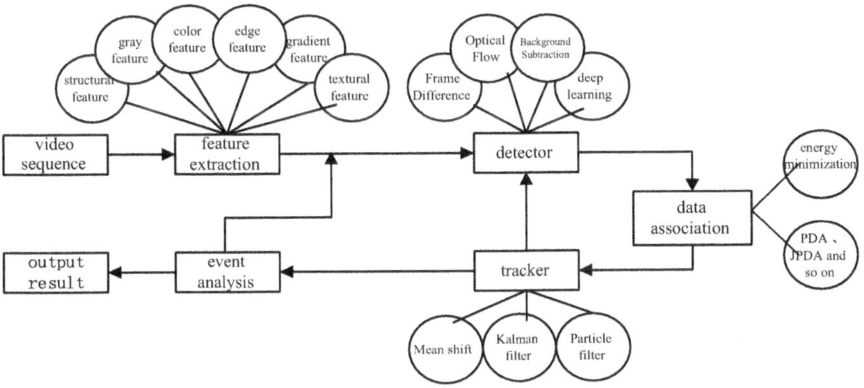

Fig. 1. Multi-object tracking system framework

The detector includes processing of the appearance model, and the tracker contains processing of the motion model. The appearance model includes three parts: feature extraction, classification matching and update. Then mining association information between frames by data association. And by the searching and sampling spatial information of potential targets in motion model, provide a reliable sample for later classification matching, then calculate and compare the reliability scores to predict the candidate samples. The highest score sample is the prediction results. In addition, part of the target location, which predicted by the motion model, will be imported into the detector to deal with the target reappearance after occlusion. Finally, the noise interference is eliminated by the event analysis, and the continuous motion trajectory is obtained.

4 Feature Extraction

Feature extraction is the first step of target detection. a suitable feature directly affects the effect of the following detection and tracking [5]. With the change of target shape in detection and tracking, the ideal feature need to be highly distinguishable, and has strong robustness for the change of targets. The features selection often depends on the representation of the target. At present, there are mainly 6 ways to represent objects: grayscale, color, edge, texture, gradient and structure. But in the actual study, due to the complexity of the scene, it is often necessary to mix several methods together.

Gray feature is the simplest and intuitionistic feature expression, and the most representative is Haar-like feature proposed by Viola [6] and so on. Gary feature has three forms, which are original gray scale feature, gray histogram and Haar-like feature. The first two feature gray the video image firstly, and then use the pixel gray information. The representative algorithm is CSK algorithm [7] and LSHT algorithm [8]. Haar-like feature series is a typical scalar feature, which has the fast computation speed. It ignites the hope for the practicability of pedestrian detection. The representative algorithm is CT algorithm [9]. However, the scalar features are easily influenced by other factors such as illumination. Later, some scholars improved the characteristics of Haar-like. The literature [10] proposed a compression tracking method based on Random Channel Haar-like Feature (RCCT), which effectively solved the phenomenon of target drift and improved the accuracy of target tracking.

Color feature [11] is the most basic feature representation, which marks the appearance of the target by different colors. Color histogram is a common form of color feature, it calculates the proportion of each color to replace the calculation of each color's position in the image space. The advantage is that it has little dependence on the attributes of the image itself and has good robustness. The disadvantage is that it is greatly influenced by the change of illumination and is easily interfered by other background colors. In recent years, a new color feature method, Color Name feature [12], has been developed. In order to deal with the label noise in image, the literature [13] presented a self-supervised convolutional neural network framework (Self-supervised CNN, SS-CNN) for feature extraction of Color Name. A lot of noise samples are filtered out by neural network, and good feature information is obtained.

Edge feature [14] is generated by violent changes of objects and background edges. In tracking, edge detection can be used to detect changes in edges. Compared with the color feature, the generation of edge feature depends on the target and the background, and does not depend on the subjective perception of the color. In order to solve the problem of blurring of image edge, the method proposed in literature [15] has extracted the details and edge information from low frequency domain, which effectively prevented the destruction of image soft threshold, and ensured the clarity of image edge feature information.

Texture features regard the image as a plane. By judging the change of plane density, it quantifies the smoothness and regularity of the image, so as to describe the spatial color and intensity distribution. Texture features are not obtained directly, they are generated by image preprocessing. Texture features can be divided into four types: statistical texture feature, model texture feature, signal processing texture feature and structural texture feature. LBP [16] (Local Binary Patterns, local two value mode) is commonly used to describe texture features. It is widely applied for its good rotation invariance and grayscale invariance.

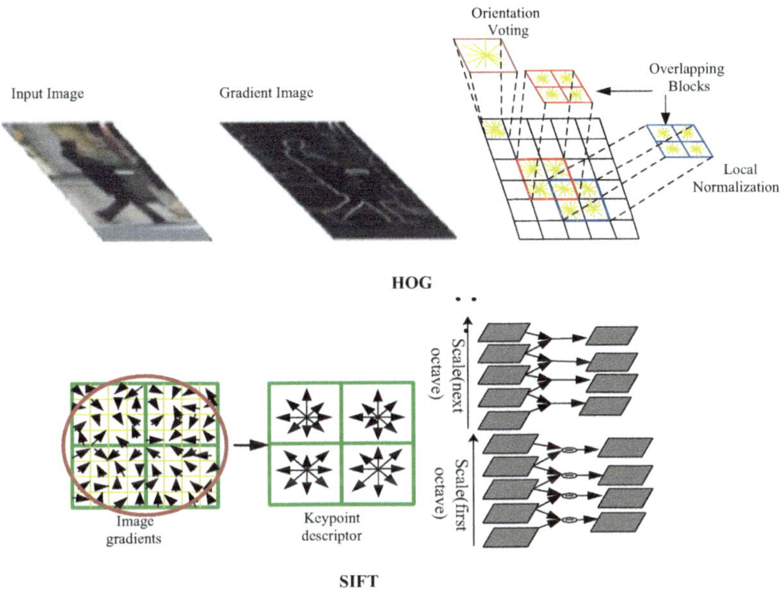

Fig. 2. SIFT and HOG gradient features

The gradient feature consists of global feature by summarizing the gradient direction of the local region of the image. The commonly used gradient features are SIFT feature [17] and HOG feature [18], as shown in Fig. 2. Compared to the SIFT and HOG features, the same point is using local features, but the latter is better in real time. SIFT features have scale invariance, which can also maintain good detection results even if the images are rotated, scale scaling or luminance changes. The feature of HOG is to use local statistical information to represent the whole image information by using the gradient information of local block units. The effects of the gradient on the environment changes are relatively small and have stable performance, but it can not directly reflect the information of the size and attitude.

The structural features [19] are trained by neural networks, the feature from the bottom to the top is more and more complex. High level features are represented by a layer by layer combination of the underlying features. As the number of layers increases, more available feature information and reference information can be obtained, and the accuracy will be improved. The structural features obtained by the neural network are very distinguishable, and high precision classification results can be achieved with some simple classifiers. However, there are also shortcomings in this method. Because of the increased number of layers, the computation becomes more complex and the search space increases. The training time will also be longer, so it is not the more training layers, the better.

5 Detector

5.1 Appearance Model

After we finish the feature extraction, we need to detect the target, and use the appearance model in the process of detection. Appearance model is an important step in video object tracking [20]. A good appearance model can improve the performance of tracking. The literature [21] proposed that the exact appearance model has better effect than the motion model, and has obtained good experimental results which has recognized by many scholars. Currently, there are many detection algorithms in the field of target detection, such as Optical Flow, Background Subtraction and Frame Difference. In recent years, with the application of deep learning in target detection field, the appearance model of detector has been greatly developed. There are many detection algorithms based on deep learning, such as RCNN algorithm [22], Fast-RCNN algorithm [23] and Faster-RCNN algorithm [24].

According to the motion relationship between the target and the camera, the target detection can be divided into two types, that is, the target detection based on the static background and the target detection based on the dynamic background. In the static background, the target detection camera is immobile in the whole monitoring process, only the monitored object is moving; in dynamic background, target detection cameras and surveillance targets are moving, resulting in complex relative motion.

Static Background

1. Optical Flow

Optical Flow is the first method to be applied to target detection. It calculates the motion information between adjacent frames of moving objects based on the characteristics of image pixels varying with time. Optical Flow can be used both in dynamic scenes and in static scenes, without knowing any information in the scene in advance. But it is complex to calculate, easy to be disturbed by the environment changes, very poor in real time, and easy to produce noise. Literature [25] proposes a detection algorithm combining Optical Flow and three frame difference method to solve the complex problem. The algorithm can simplify the computation by selecting the representative Harris corner points and calculating the Optical Flow information only for these points. Some scholars have applied the neural network to the Optical Flow. In literature [26], the convolution neural network framework is used to transform the Optical Flow estimation problem into a supervised learning task, which has achieved good results.

2. Background Subtraction

Background Subtraction can distinguish target background and environment background in video sequence. The algorithm realizes detection by differentiating the pixel value between the current frame and the background image. With the movement of the target, the environment is changing all the time, so the key to the Background Subtraction is the modeling and updating of the background. In view of the above reasons, the Background Subtraction method is vulnerable to environmental interference. When the environment changes, such as illumination and shadow, it will affect the detection effect of the algorithm. But the Background Subtraction has very good real-time and small computational complexity, so it is widely applied. Literature [27] combines HOG feature and Background Subtraction to propose a new pedestrian detection method. The new method meets the real-time and accuracy requirements at the same time. In literature [28], the Gauss hybrid model was used to track the moving characters by Background Subtraction.

3. Frame Difference

The Frame Difference method obtains the moving object contour by the difference operation of two adjacent frames or multiple frames in the image sequence, and it has good detection effect for the scene with multiple moving targets. The Frame Difference has a strong ability to adapt to the environment, and is less affected by the illumination change interference, and its complexity of the algorithm is small. However, due to the problem of the algorithm itself, there will be a discontinuous hole, which leads to incomplete target contour in the detection. The three frames difference method and Optical Flow are used in literature [29], which not only solves the problem of poor real-time and large computational complexity, but also reduces the complexity of the algorithm. The maximum between-cluster variance (OTSU) is used to deal with the three frames difference image, which make the detection area more complete in the incomplete region.

4. Neural Network

Based on deep learning detection method, it uses neural network to sample the information of the first frame of the video, and then trains a deep model that can track the target motion. However, most of the objects will be deformed during the movement, which leads to the difficulty of target recognition and tracking, and then leads to the loss of the target. In order to solve this problem, people use the idea of migration learning and expand the training samples by using Data Augmentation [30]. By pre training network for non-tracking target, the model can learn the general features of different objects in different environments, such as illumination change, background transformation and other factors. In tracking, the first few frame information is used to adjust the network to adapt the network to target tracking in different scenes.

In Table 1, the above four methods are summarized. Each method has its strong points. When applied in practice, different inspection methods should be adopted according to the specific environment and conditions. At the same time, in order to achieve the best experimental results, several joint methods will be used.

Dynamics Background. As all know, detecting objects in dynamics background is more complex than in static background. In general, the motion of the camera has two types: one type is the stent of camera fixed, the camera can move from left-to-right or up-to-down. Another is setting the camera on an object which move regularly. So, the foreground and the background with a global movement. A global movement estimated

Table 1. Comparison of four detection methods

	Optical Flow	Background Subtraction	Frame Difference	Neural network
Detection principle	Time-varying Optical Flow	Current frame and background model for difference	Two consecutive frames or multiple frames for difference	Training features to identify target objects
Detection effect	A more complete foreground target area	A more complete foreground target area	There are discontinuous holes in the target	The effect is better, there is a complete foreground area
Environmental robustness	Easy to be disturbed by the environment	Easy to be disturbed by the environment	Strong adaptability to environment	Strong adaptability to environment
Complexity	Complex	General	Simpleness	Complex
Real-time	Poor	Good	Good	Good

with the image is necessary. It can be solved by a same model and parameters Even with different move vector from different area of the image. The matter of global movement estimated simplify to the matter of global movement model and parameters estimated. Two main method to get it: one is Block matching method [31], which contain three basic elements:

- Match rule: Maximum correlation and minimum error.
- The search method: Three step search or Cross search.
- Block size to select: Classification and adaptation.

One is Optical Flow estimation with a flow equals to get the moving pixels' speed, which is not robust under the complex background.

5.2 Multi-target Tracking Algorithm Based on Detection Tracking

As the Complex of the ground. it's difficult to express the object using single feature. So [32] proposed a method of fusing multi-features and adaptive templates. In detail, combining the color, Hog and the edge feature to distinguish different coverage. multi-combine methods achieve the image order features' match, rebuild the track of the covered object. [33] proposed combining the Kalman filter algorithm and Mean-shift algorithm. Dividing tracker into different levels for different detection templates to track the object. For extended coverage problems. [34] using particle filter to track with linear estimation particle to get mobile route. [35] based on [34], proposed multi-features combination HPF framework. For HPF, get the different feature information not only for the extended coverage, but also in complex environment can get a good performance.

With developing in multi-object tracking. More and more methods are proposed. The research orientation tends to relate the specific application to innovate. Of course, object detection would be the first step for the object tracking, which can increase the accuracy.

5.3 Multi-target Tracking Algorithm in Deep Learning

Deep learning has made great breakthroughs in speech recognition, image recognition and natural language processing. However, in the field of target tracking, the achievements of deep learning are not outstanding, and the main reasons are as follows.

- Only the first frame of video is annotated. Training data is small.
- Deeping learning is not easy to satisfy the custom of the real-time and accuracy.

Even though, researchers have attempted to optimize the structure of network to achieve the requirements.

TLD [36] (Tracking-Learning-Detection) proposed a method add a learning training stage into the process of original detection and tracking, so the algorithm can learn the target characteristics quickly and with little prior knowledge to track. However, the algorithm has its own disadvantages. First, the algorithm can only track the single target, and it requires off-line training. But with continuous improvement, some new TLD algorithms are derived. Literature [37, 38] proposed the ATLD (Accelerated TLD) algorithm through improved TLD algorithm, and the new algorithm not only improved the running speed, but also realized the tracking of multiple targets. Literature [39] the TLP algorithm And AOG (And-Or Graph), combining with AOG net build tree. Or, And, Inverter, the three kinds of doors are used to extract the different information. Compared to the previous TLD algorithm, using TLP algorithm for dynamic analysis of tracking target has more features, more flexible. The method of learning tracking gradually becomes the focus of multi-target tracking research.

As CNN (Convolutional Neural Networks) get great breakthrough in the field of target detection, also indirectly promote the development of the study of target tracking. The basic structure of convolution neural network is shown in Fig. 3.

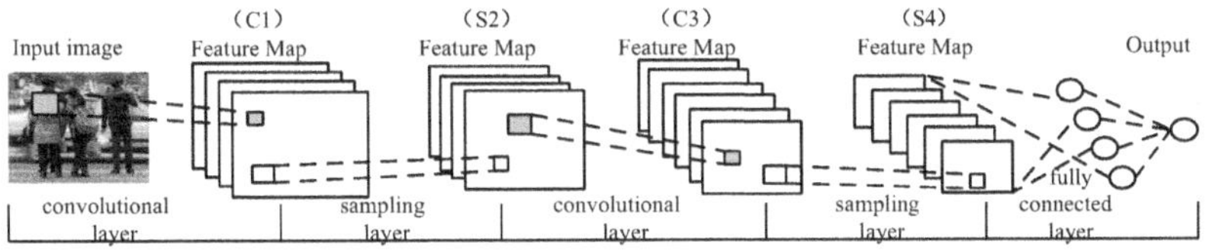

Fig. 3. Convolution neural network structure.

The basic structure includes: input layer, convolution layer, sampling layer, full connection layer and output layer. The feature extraction is the core, which is alternately carried out by the convolution layer and the sampling layer. The feature vector is obtained from multiple feature map, and finally access to the full connection layer and output layer.

Literature [40] introduced the data correlation method based on CNN. Training Siamese convolutional neural network to learn image features. Then matching the feature image output by CNN and the image processed by gradient enhancement classifier; at last, the location of the target is determined by the probability size. Some think a high performance detection can get a good tracking performance. Paper [41] with CNN as the detector, combining the POI (Person of Interest) and Kalman filter. [42] Hedge algorithm, as shown in Fig. 4, this paper proposed combining the KCF trackers and CNN features is a powerful tracker to track targets. Literature [43] also propose a CNN-based framework for online Multi-object Tracking. This framework utilizes the merits of single object trackers in adapting appearance models and searching for target in the next frame. The framework achieves computational efficiency by sharing features and using ROI-Pooling to obtain individual features for each target. And the framework also introduces spatial-temporal attention mechanism (STAM) to handle the drift caused by occlusion and interaction among targets. [44] with RNN (Recurrent Neural Networks) model, compared with CNN training method, RNN which forming a chain sequence, and then use the LSTM [45] (Long Short-term Memory) for data association, finally achieved the end-to-end tracking. Literature [46] designs a CNN network to obtain appearance cue tailored towards person Re-ID, and an LSTM network for motion cue to predict target position, respectively. Both cues are combined with a triplet loss function, which performs end-to-end learning of the fused features in a desired embedding space.

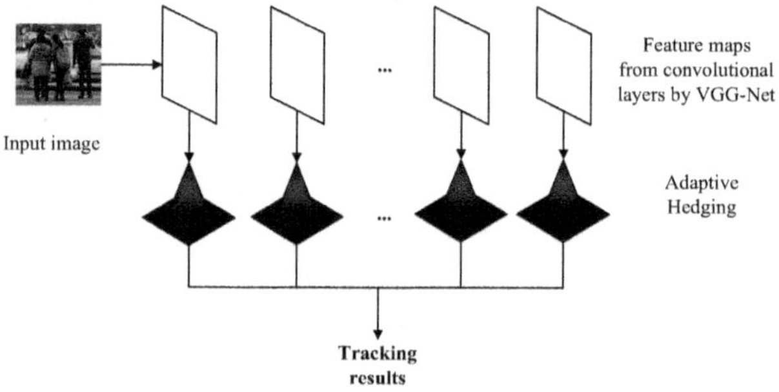

Fig. 4. Procedure for Hedge deep tracker

6 Tracker

6.1 Movement Model

The model generates effective template information, and the motion model predicts the position of the next frame when the target is tracked. The prediction position generated is multiple choice, which requires the search for an optimal method to allocate. The classical allocation Algorithm has Hungarian Algorithm [47]. When selecting the optimal location, it is necessary to correlate the information between the previous frame and even several frames, and then selecting the efficient correlation algorithm will improve the tracking efficiency. With the information of inter-frame, the efficiency of the tracking will be improved by selecting the efficient correlation algorithm. The markov motion model is often used in motion models, and we assume that the state of a target is expressed as

$$Z_t = (x_t,\ y_t,\ u_t,\ v_t) \tag{1}$$

$$(x_t,\ y_t) = (x_{t-1},\ y_{t-1}) + (u_s,\ v_s) \cdot V_t + \varepsilon(x,\ y) \tag{2}$$

$$(u_t,\ v_t) = (u_s,\ v_s) + \varepsilon(u,\ v) \tag{3}$$

$(x_t,\ y_t)$ is the location of the object, $(u_t,\ v_t)$ represents the horizontal and vertical components of the velocity. $(u_s,\ v_s)$ is the instantaneous velocity of the target. $\varepsilon(x, y)$, $\varepsilon(u, v)$ represent the target position and the velocity of the Gaussian noise. Δt as the frame of per second. With the markov model, the target can be easily captured even when the target is rapidly changing. At present, the motion model is divided into four types as follow:

- Mean-shift

Mean-shift is a non-parametric method based on the increase of density gradient. Iterate along the gradient descent direction of the probability density. Basically is to optimise Bhattacharyya coefficient function.

- Slide window

Sliding window, like Struck [48], KCF [49], is the most basic motion search model, the general search box will be set to square or circle, scanning the whole image through multi-Scales. It is obvious that the target area is likely to be searched, so compute global image may cost much.

- Kalman filter

Kalman filter [50] is a classical algorithm, which can predict the position and speed of the target tracking, and it is very efficient. But this prediction with biased and noisy. Recently, researchers proposed Nonlinear-Kalman filter like Extended Kalman Filter (EKF) [51], Unscented Kalman Filter(UKF) [52] and so on. [53] introduce the merit and demerit of EKF and UKF in detail.

- Particle filter

Particle filter [54] from noisy data to estimate motion state, in the state space by using the weighted and random spread a large number of discrete variables, to approximate probability distribution and recursive. At Nonlinear and non-Gaussian condition, it performs well.

6.2 Combined-Tracking Algorithm

Different tracking algorithms have their own characteristics. So, combined-tracking algorithm is popular. [55] proposed a method based on ACF (Aggregate Channel Features) and Particle filter. Two specific ACF detectors were used to detect the target, then, training Ada-boost classifier for classification, Particle filter applied to track. [56] image segmentation technologies also apply into multiple target tracking as the pre-treatment of the image, [57] gave a multi-objective solution, by adding the identity information, to avoid the error caused by the similarity, and automatically determine potential tracking error, further to improve the tracking accuracy.

Many multi-target tracking methods rely on independently detecting objects in each frame, grouping them into short, reliable trajectories and then grouping them further into the full trajectory. This grouping usually relies on applying local smoothing constraints, but almost never executes more global constraints on the trajectory. Literature [58] proposed a non-markov method to guide the tracking algorithm by using behavioral patterns to enhance global consistency.

7 Data Association

The purpose of data association is to calculate the correlation probability between each observation value detected in the current frame from the detector and the possible tracking targets in the previous frame tracker, and to connect the information between the frames before and after the matching degree of the probability, thus forming a continuous track. There are two kinds of data association based on energy minimization and probability based.

7.1 Energy Minimization

The multi-target tracking problem is transformed to the energy minimization problem. Energy minimization: in the solution space, each solution corresponds to a loss, and the loss function is constructed to find a suitable method to find the optimal solution. In the process of solving the problem, all the feedback information and all the possible trajectories of the feedback information are often known, and each combination has a loss. The result of the solution is to seek the optimal combination.

[59] is a typical based on energy minimization for multiple target tracking, the paper clearly expounds the composition and minimize the loss function, by constructing a continuous loss function, jump out of local optimum to find the global optimal, the optimal solution obtained. [60] proposed the discrete continuous energy function, the function contains a single target data association and track the estimate information, tags loss describes the physical properties of each track, at last, based on the gradient of the continuous energy function to update the shape of the trajectory. [61] uses alpha-expansion frame to do graph partitioning to minimize energy.

A typical data association method includes finding a graph matching or network flow to minimize the sum of pairs of associated costs. Literature [63] shows that the characteristics of data association based on network flow can be learned by back-propagation, and the smooth network flow problem is regarded as a differentiable function of the cost of a pair.

7.2 Data Association Method Based on Probability

Data association method based on probability like Multiple Hypothesis Tracking (MHT) [64], Near Neighbor Data Association (NNDA) [62], Probability Data Association (PDA) [65], Joint Probabilistic Data Association (JPDA) [66], NNJPDA [69] and so on. In [67], JPDA algorithm was improved to reduce the processing time of the associated information and the effect was significant in the case of chaotic application scenarios. [68] decrease the complexity of the traditional JPDA algorithm and solve the occlusion problem caused by the size of the target.

In future, researchers tend to combine multi-feature technology, multi-motion model, multi-sensor and other comprehensive methods for tracking. Following the development of the deep learning, Neural network structure with data association would be popular.

Document [70] proposed Quadruplet Convolutional Neural Networks (Quad-CNN) for multiple target tracking, using quadruplet losses to associate cross frame target detection. The proposed network considers the appearance of targets and their time adjacency for data association. Different from the traditional sort loss, the quadruplet losses implement additional constraints, making the time proximity detection closer than the time proximity detection with a larger time gap. The whole network is trained at end to end. In order to track, target association is performed by using the minimax label propagation of the metric learned.

8 Datasets

8.1 Data

An authoritative dataset is critical to the credibility of the results. At classification and object detection, we can know like as ImageNet [71], Pascal VOC [72], COCO [73] image dataset for training DNN. In tracking, VTB dataset [67], VOT dataset [74], MOT Challenge [75] etc. MOT contains many open video sequences like PETS 2009, Town-Centre, PETS 2016 Challenge. In PETS 2009 and Town-Centre, pedestrians are frequently blocked by other targets or obstacles (e.g. traffic signs). In PETS 2016 Challenge, Changes in posture and scale of pedestrians vary greatly. For various complex environments, multi-target video tracking datasets update every year, and the specific data content reference in [76].

8.2 Evaluation Measures

A good algorithm need meet high accuracy, high robustness and high efficiency. The CLEAR MOT standard [77] is currently accepted degree evaluation methods, respectively from the following two aspects to evaluate the merits of the algorithm: Multiple Object Tracking Precision (MOTP), the indicator is evaluating the accuracy of the algorithm in target position: Multiple Object Tracking Accuracy (MOTA) is the Accuracy of the evaluation algorithm in determining the number of targets and other relevant attributes of the target. In addition, recall rate comparison (Recall), track the success rate (Mostly Tracked, MT), mostly lost(ML), miss rate (Missed Detections,

MS), error detection rate (False Positives, FP), the number of Frames Per Second (Frames Per Second, FPS) as well as the total number of ID used in the switch (ID Switches, IDS), etc.

9 Development Trend

Video multi-target tracking technology has better performance in simple scenes, but it is still not mature in the complex environment. In the era of big data, deep learning gives new direction to multi-target tracking. The future research and development trend as follow:

1. Focusing on the situation when the target is blocked after a long time and re-tracking.
2. A full range of observation objectives, different perspectives can help to deal with the problem of occlusion of the target.
3. In addition, currently, tracking methods are 2D modeling. How to apply 3D modeling to target tracking and achieve target holographic projection tracking is also the focus of later research.
4. The eye is basic to observe and track target. People can ignore complex scenes through the processing of human brain and brain nerves. In the future, the observation and tracking ability of human eyes can be simulated. By bionic technology, the machine can have human visual thinking. How to implement this technology is also worth studying.
5. Multiple camera [78]. Multiple cameras have two configurations. The first is that multiple cameras record the same scene, that is, multiple views. However, a key issue in this setup is how to integrate information from multiple cameras. The second is that each camera records different scenes, that is, non-overlapping multiple camera networks. In this case, data association between multiple cameras becomes a reidentification problem.
6. Other computer vision tasks. Although multiple target tracking is in service to other advanced computer visual tasks, there is a tendency to use some complementary computer vision tasks to solve this problem. Possible combinations include target segmentation, pedestrian re identification, pedestrian pose estimation and motion recognition.

10 Finally

This paper summarizes the research methods of multi-target video tracking in recent years, and then provides readers with rich multi-target tracking knowledge. Describing the whole process of multi-target tracking, several important stages which contain feature extraction, detector, data association and tracker filters.

This paper elaborates the appearance and motion model, enumerates some popular multi-target tracking algorithms, introduces the algorithm based on deep learning framework and the algorithms of various data associations. In addition, introduces the datasets and evaluation methods related to multi-target tracking. At last, the paper analyzes the new opportunities come from deep learning for multi-target tracking and the future development trend of multi-target tracking under the new environment. We believe that the research of multi-target tracking will have a broader prospect.

Acknowledgments. This work was supported by the Scientific Research Fund of Hunan Provincial Education Department of China (Project No. 17A007); and the Teaching Reform and Research Project of Hunan Province of China (Project No. JG1615).

References

1. Wu, Y., Lim, J., Yang, M.H.: Online object tracking: a benchmark. In: 2013 IEEE Conference on Computer Vision and Pattern Recognition (CVPR), pp. 2411–2418. IEEE (2013)
2. Huang, K.Q., Chen, X.T., Kang, Y.F., Tan, T.N.: Intelligent visual surveillance: a review. Chin. J. Comput. **38**, 1 (2015)
3. Zeng, Q., Wen, G., Li, D.: Multi-target tracking by detection. In: 2016 International Conference on Audio, Language and Image Processing (ICALIP), pp. 370–374. IEEE (2016)
4. Nam, H., Han, B.: Learning multi-domain convolutional neural networks for visual tracking. In: 2016 IEEE Conference on Computer Vision and Pattern Recognition (CVPR), pp. 4293–4302. IEEE (2016)
5. Huang, K.Q., Ren, W.Q., Tan, T.N.: A review on image object classification and detection. Chin. J. Comput. **37**(6), 1225–1240 (2014)
6. Viola, P., Jones, M. J., Snow, D.: Detecting pedestrians using patterns of motion and appearance. In: Null, p. 734. IEEE (2003)
7. Henriques, J.F., Caseiro, R., Martins, P., Batista, J.: Exploiting the circulant structure of tracking-by-detection with kernels. In: European Conference on Computer Vision, pp. 702–715. Springer, Berlin (2012)
8. He, S., Yang, Q., Lau, R.W., Wang, J., Yang, M.H.: Visual tracking via locality sensitive histograms. In: 2013 IEEE Conference on Computer Vision and Pattern Recognition (CVPR), pp. 2427–2434. IEEE (2013)
9. Zhang, K., Zhang, L., Yang, M.H.: Real-time compressive tracking. In: European Conference on Computer Vision, pp. 864–877. Springer, Berlin (2012)
10. Chen, J., Liu, Y., Li, N., Guo, Z.: Compressive tracking based on random channel haar-like feature. In: 2015 Sixth International Conference on Intelligent Control and Information Processing (ICICIP), pp. 151–154. IEEE (2015)
11. Possegger, H., Mauthner, T., Bischof, H.: In defense of color-based model-free tracking. In: Proceedings of the IEEE Conference on Computer Vision and Pattern Recognition, pp. 2113–2120 (2015)
12. Danelljan, M., Shahbaz Khan, F., Felsberg, M., Van de Weijer, J.: Adaptive color attributes for real-time visual tracking. In: IEEE Conference on Computer Vision and Pattern Recognition (CVPR), Columbus, Ohio, USA, 24–27 June 2014, pp. 1090–1097. IEEE Computer Society (2014)
13. Wang, Y., Liu, J., Wang, J., Li, Y., Lu, H.: Color names learning using convolutional neural networks. In: 2015 IEEE International Conference on Image Processing (ICIP), pp. 217–221. IEEE (2015)

14. Wang, Y., Wang, Q., Chen, D.: Research on target recognition based on edge features. In: 2012 5th International Congress on Image and Signal Processing (CISP), pp. 1312–1315. IEEE (2012)
15. Wu, J.Z., Huang, Y.D.: Image denoising algorithm based on edge feature extraction in curvelet domain. In: 2012 International Conference on Wavelet Analysis and Pattern Recognition (ICWAPR), pp. 7–10. IEEE (2012)
16. Ojala, T., Pietikainen, M., Harwood, D.: Performance evaluation of texture measures with classification based on Kullback discrimination of distributions. In: Proceedings of the 12th IAPR International Conference on Pattern Recognition, 1994. Vol. 1-Conference A: Computer Vision and Image Processing, vol. 1, pp. 582–585. IEEE (1994)
17. Lowe, D.G.: Distinctive image features from scale-invariant keypoints. Int. J. Comput. Vision **60**(2), 91–110 (2004)
18. Dalal, N., Triggs, B.: Histograms of oriented gradients for human detection. In: IEEE Computer Society Conference on Computer Vision and Pattern Recognition, 2005. CVPR 2005, vol. 1, pp. 886–893. IEEE (2005)
19. Hayakawa, Y., Oonuma, T., Kobayashi, H., Takahashi, A., Chiba, S., Fujiki, N.M.: Feature extraction of video using deep neural network. In: 2016 IEEE 15th International Conference on Cognitive Informatics and Cognitive Computing (ICCI* CC), pp. 465–470. IEEE (2016)
20. Zhang, H.L., Hu, S., Yang, G.: Video object tracking based on appearance models learning. J. Comput. Res. Dev. **52**(1), 177–190 (2015)
21. Supancic, J.S., Ramanan, D.: Self-paced learning for long-term tracking. In: Proceedings of the IEEE Conference on Computer Vision and Pattern Recognition, pp. 2379–2386 (2013)
22. Girshick, R., Donahue, J., Darrell, T., Malik, J.: Rich feature hierarchies for accurate object detection and semantic segmentation. In: Proceedings of the IEEE Conference on Computer Vision and Pattern Recognition, pp. 580–587 (2014)
23. Girshick, R.: Fast R-CNN. In: IEEE International Conference on Computer Vision, pp. 1440–1448. IEEE Computer Society (2015)
24. Ren, S., He, K., Girshick, R., Sun, J.: Faster R-CNN: towards real-time object detection with region proposal networks. IEEE Trans. Pattern Anal. Mach. Intell. **39**(6), 1137–1149 (2017)
25. Yuan, G.W., Chen, Z.Q., Gong, J., Xu, D., Liao, R.J., He, J.Y.: A moving object detection algorithm based on a combination of optical flow and three-frame difference. J. Chin. Comput. Syst. **34**(3), 668–671 (2013)
26. Dosovitskiy, A., Fischer, P., Ilg, E., Hausser, P., Hazirbas, C., Golkov, V., van der Smagt, P., Cremer, D., Brox, T.: Flownet: learning optical flow with convolutional networks. In: Proceedings of the IEEE International Conference on Computer Vision, pp. 2758–2766 (2015)
27. Bing-Bing, W., Zhi-Xin, C., Jia, W., Liquan, Z.: Pedestrian detection based on the combination of HOG and background subtraction method. In: 2011 International Conference on Transportation, Mechanical, and Electrical Engineering (TMEE), pp. 527–531. IEEE (2011)
28. Suresh, S., Deepak, P., Chitra, K.: An efficient low cost background subtraction method to extract foreground object during human tracking. In: 2014 International Conference on Circuit, Power and Computing Technologies (ICCPCT), pp. 1432–1436. IEEE (2014)
29. Han, X., Gao, Y., Lu, Z., Zhang, Z., Niu, D.: Research on moving object detection algorithm based on improved three frame difference method and optical flow. In: 2015 Fifth International Conference on Instrumentation and Measurement, Computer, Communication and Control (IMCCC), pp. 580–584. IEEE (2015)
30. Liu, W., Anguelov, D., Erhan, D., Szegedy, C., Reed, S., Fu, C.Y., Berg, A.C.: SSD: single shot multibox detector. In: European conference on computer vision, pp. 21–37. Springer, Cham (2016)

31. Fard, M.K., Yazdi, M., MasnadiShirazi, M.: A block matching based method for moving object detection in active camera. In: 2013 5th Conference on Information and Knowledge Technology (IKT), pp. 443–446. IEEE (2013)
32. Li, W., Yao, J., Dong, T., Li, H.: Object tracking based on fragment template and multi-feature adaptive fusion. In: 2015 8th International Symposium on Computational Intelligence and Design (ISCID), vol. 2, pp. 481–484. IEEE (2015)
33. Zhang, J., Presti, L.L., Sclaroff, S.: Online multi-person tracking by tracker hierarchy. In: 2012 IEEE 9th International Conference on Advanced Video and Signal-Based Surveillance (AVSS), pp. 379–385. IEEE (2012)
34. Guan, Y., Chen, X., Yang, D., Wu, Y.: Multi-person tracking-by-detection with local particle filtering and global occlusion handling. In: 2014 IEEE International Conference on Multimedia and Expo (ICME), pp. 1–6. IEEE (2014)
35. Gan, M., Cheng, Y., Wang, Y., Chen, J.: Hierarchical particle filter tracking algorithm based on multi-feature fusion. J. Syst. Eng. Electron. **27**(1), 51–62 (2016)
36. Wang, N., Yeung, D.Y.: Learning a deep compact image representation for visual tracking. In: Advances in Neural Information Processing Systems, pp. 809–817 (2013)
37. Jin, Z., Liu, C.: Accelerated TLD algorithm and its application in multiple target tracking. Comput. Syst. Appl. **25**(6), 196–201 (2016)
38. Sharma, S., Khachane, A., Motwani, D.: Real time multi-object tracking using TLD framework. In: International Conference on Inventive Computation Technologies (ICICT), vol. 2, pp. 1–6. IEEE (2016)
39. Lu, Y., Wu, T., Chun Zhu, S.: Online object tracking, learning and parsing with and-or graphs. In: Proceedings of the IEEE Conference on Computer Vision and Pattern Recognition, pp. 3462–3469 (2014)
40. Leal-Taixé, L., Canton-Ferrer, C., Schindler, K.: Learning by tracking: Siamese CNN for robust target association. In: Proceedings of the IEEE Conference on Computer Vision and Pattern Recognition Workshops, pp. 33–40 (2016)
41. Yu, F., Li, W., Li, Q., Liu, Y., Shi, X., Yan, J.: POI: multiple object tracking with high performance detection and appearance feature. In: European Conference on Computer Vision, pp. 36–42. Springer, Cham (2016)
42. Qi, Y., Zhang, S., Qin, L., Yao, H., Huang, Q., Lim, J., Yang, M.H.: Hedged deep tracking. In: Proceedings of the IEEE Conference on Computer Vision and Pattern Recognition, pp. 4303–4311 (2016)
43. Chu, Q., Ouyang, W., Li, H., Wang, X., Liu, B., Yu, N.: Online multi-object tracking using CNN-based single object tracker with spatial-temporal attention mechanism. arXiv preprint arXiv:1708.02843 (2017)
44. Milan, A., Rezatofighi, S.H., Dick, A.R., Reid, I.D., Schindler, K.: Online multi-target tracking using recurrent neural networks. In: AAAI, pp. 4225–4232 (2017)
45. Hochreiter, S., Schmidhuber, J.: Long short-term memory. Neural Comput. **9**(8), 1735–1780 (1997)
46. Xiang, J., Zhang, G., Hou, J., Sang, N., Huang, R.: Multiple target tracking by learning feature representation and distance metric jointly. arXiv preprint arXiv:1802.03252 (2018)
47. Munkres, J.: Algorithms for the assignment and transportation problems. J. Soc. Ind. Appl. Math. **5**(1), 32–38 (1957)
48. Hare, S., Golodetz, S., Saffari, A., Vineet, V., Cheng, M.M., Hicks, S.L., Torr, P.H.: Struck: structured output tracking with kernels. IEEE Trans. Pattern Anal. Mach. Intell. **38**(10), 2096–2109 (2016)
49. Henriques, J.F., Caseiro, R., Martins, P., Batista, J.: High-speed tracking with kernelized correlation filters. IEEE Trans. Pattern Anal. Mach. Intell. **37**(3), 583–596 (2015)

50. Kalman, R.E., Bucy, R.S.: New results in linear filtering and prediction theory. J. Basic Eng. **83**(1), 95–108 (1961)
51. Julier, S.J., Uhlmann, J.K.: New extension of the Kalman filter to nonlinear systems. In: Signal processing, sensor fusion, and target recognition VI, vol. 3068, pp. 182–194. International Society for Optics and Photonics (1997)
52. Wan, E.A., Van Der Merwe, R.: The unscented Kalman filter for nonlinear estimation. In: Adaptive Systems for Signal Processing, Communications, and Control Symposium 2000. AS-SPCC. The IEEE 2000, pp. 153–158. IEEE (2000)
53. Yang, S., Li, H.: Application of EKF and UKF in target tracking problem. In: 2016 8th International Conference on Intelligent Human–Machine Systems and Cybernetics (IHMSC), vol. 1, pp. 116–120. IEEE (2016)
54. Gordon, N., Ristic, B., Arulampalam, S.: Beyond the Kalman Filter: Particle Filters for Tracking Applications, vol. 3, pp. 1077–2626. Artech House, London (2004)
55. Kokul, T., Ramanan, A., Pinidiyaarachchi, U.A.J.: Online multi-person tracking-by-detection method using ACF and particle filter. In: 2015 IEEE Seventh International Conference on Intelligent Computing and Information Systems (ICICIS), pp. 529–536. IEEE (2015)
56. Milan, A., Leal-Taixé, L., Schindler, K., Reid, I.: Joint tracking and segmentation of multiple targets. In: 2015 IEEE Conference on Computer Vision and Pattern Recognition (CVPR), pp. 5397–5406. IEEE (2015)
57. Yu, S.I., Meng, D., Zuo, W., Hauptmann, A.: The solution path algorithm for identity-aware multi-object tracking. In: Proceedings of the IEEE Conference on Computer Vision and Pattern Recognition, pp. 3871–3879 (2016)
58. Maksai, A., Wang, X., Fleuret, F., Fua, P.: Non-Markovian globally consistent multi-object tracking. In: 2017 IEEE International Conference on Computer Vision (ICCV), pp. 2563–2573. IEEE (2017)
59. Milan, A., Roth, S., Schindler, K.: Continuous energy minimization for multitarget tracking. IEEE Trans. Pattern Anal. Mach. Intell. **36**(1), 58–72 (2014)
60. Milan, A., Schindler, K., Roth, S.: Multi-target tracking by discrete-continuous energy minimization. IEEE Trans. Pattern Anal. Mach. Intell. **38**(10), 2054–2068 (2016)
61. Wei, L., Xingwei, L.: Multiple object tracking based on energy minimization. In: 2016 4th International Symposium on Computational and Business Intelligence (ISCBI), pp. 206–211. IEEE (2016)
62. Singer, R.A., Stein, J.J.: An optimal tracking filter for processing sensor data of imprecisely determined origin in surveillance systems. In: 1971 IEEE Conference on Decision and Control, vol. 10, pp. 171–175. IEEE (1971)
63. Schulter, S., Vernaza, P., Choi, W., Chandraker, M.: Deep network flow for multi-object tracking. IEEE Conference on Computer Vision and Pattern Recognition, pp. 2730–2739. IEEE Computer Society (2017)
64. Kim, C., Li, F., Ciptadi, A., Rehg, J.M.: Multiple hypothesis tracking revisited. In: Proceedings of the IEEE International Conference on Computer Vision, pp. 4696–4704 (2015)
65. Bar-Shalom, Y., Tse, E.: Tracking in a cluttered environment with probabilistic data association. Automatica **11**(5), 451–460 (1975)
66. Cox, I.J.: A review of statistical data association techniques for motion correspondence. Int. J. Comput. Vision **10**(1), 53–66 (1993)
67. Hamid Rezatofighi, S., Milan, A., Zhang, Z., Shi, Q., Dick, A., Reid, I.: Joint probabilistic data association revisited. In: Proceedings of the IEEE International Conference on Computer Vision, pp. 3047–3055 (2015)

68. Shi, X., Song, Y. Q., Yang, Z., Chen, J.: Multiple target tracking under occlusions using modified Joint Probabilistic Data Association. In: 2015 IEEE International Conference on Communications (ICC), pp. 6615–6620. IEEE (2015)
69. Chen, S.L., Xu, Y.B., Zhu, M.: K nearest neighbor joint possibility data association algorithm. In: 2010 2nd International Conference on Information Engineering and Computer Science (ICIECS), pp. 1–4. IEEE (2010)
70. Son, J., Baek, M., Cho, M., Han, B.: Multi-object tracking with quadruplet convolutional neural networks. In: Proceedings of the IEEE Conference on Computer Vision and Pattern Recognition, pp. 5620–5629 (2017)
71. Deng, J., Dong, W., Socher, R., Li, L.J., Li, K., Fei-Fei, L.: Imagenet: a large-scale hierarchical image database. In: IEEE Conference on Computer Vision and Pattern Recognition, 2009. CVPR 2009. pp. 248–255. IEEE (2009)
72. Everingham, M., Van Gool, L., Williams, C.K., Winn, J., Zisserman, A.: The pascal visual object classes (voc) challenge. Int. J. Comput. Vis. **88**(2), 303–338 (2010)
73. Lin, T.Y., Maire, M., Belongie, S., Hays, J., Perona, P., Ramanan, D., Dollár, D., Zitnick, C. L.: Microsoft coco: common objects in context. In: European Conference on Computer Vision, pp. 740–755. Springer, Cham (2014)
74. Kristan, M., Pflugfelder, R., Leonardis, A., Matas, J., Porikli, F., Cehovin, L., Nebehay, G., Fernandez, G., Vojir, T., Gatt, A., Khajenezhad, A.: The visual object tracking vot2013 challenge results. In: Proceedings of the IEEE International Conference on Computer Vision Workshops, pp. 98–111 (2013)
75. Leal-Taixé, L., Milan, A., Reid, I., Roth, S., Schindler, K.: Motchallenge 2015: towards a benchmark for multi-target tracking. arXiv preprint arXiv:1504.01942 (2015)
76. Milan, A., Leal-Taixé, L., Reid, I., Roth, S., Schindler, K.: MOT16: a benchmark for multi-object tracking. arXiv preprint arXiv:1603.00831 (2016)
77. Bernardin, K., Stiefelhagen, R.: Evaluating multiple object tracking performance: the CLEAR MOT metrics. J. Image Video Process. **2008**, 1 (2008)
78. Zhang, Z., Wu, J., Zhang, X., Zhang, C.: Multi-target, multi-camera tracking by hierarchical clustering: recent progress on DukeMTMC Project. arXiv preprint arXiv:1712.09531 (2017)

An Overview of Image-Based Person Re-identification

Shuren Zhou[1,2(✉)], Maolin Ke[1,2], and Peng Luo[1,2]

[1] Hunan Provincial Key Laboratory of Intelligent Processing of Big Data on Transportation, Changsha University of Science and Technology, Changsha 410114, China
zsr@csust.edu.cn

[2] School of Computer and Communication Engineering, Changsha University of Science and Technology, Changsha 410114, China

Abstract. Person re-identification (person re-ID) is refers to pedestrians matching under a multi camera network in a non-overlapping field of view, that is, whether the pedestrian targets taken by different cameras are the same. In this paper, the image based traditional person re-ID is introduced from feature description and metric learning. In addition, with the extensive application of the deep learning algorithm in recent years, it also brings about the change of person re-ID algorithm. From the three aspects of loss function design, local feature and data augmentation, this paper introduces some work of person re-ID algorithm based on deep learning. Finally, it summarizes the development of pedestrian datasets in recent years, and looks forward to the future development trend of person re-ID.

Keywords: Person re-identification · Computer vision
Generative Adversarial Networks · Deep learning

1 Introduction

Person re-ID is a computer vision technology used to judge whether there is a specific pedestrian in the image or video sequence. Given a monitored pedestrian image, the pedestrian image is retrieved from multiple monitoring devices in a region. Therefore, person re-ID is also a sub problem of image retrieval. Person re-ID is designed to make up for the visual limitations of the current fixed camera and can be combined with pedestrian detection and pedestrian tracking technology, and is widely used in intelligent video surveillance, intelligent security and other fields. This technology has gained more and more attention in the field of computer vision [1–5].

Compared with face recognition, person re-ID is more difficult, which leads to the difficulty of applying mature face recognition technology to person re-ID. First, the pixel of the camera may be poor, especially the face in the remote camera is very small, and there is a wide range of back and side face in the pedestrian image. It is difficult to use face recognition. In addition, the possible posture of the same pedestrians on different cameras is different, and the clothes between different pedestrians may be the same. Therefore, it is difficult to distinguish the pedestrians from the color of the

J. Abawajy et al. (Eds.): ATCI 2018, AISC 842, pp. 370–378, 2019.
https://doi.org/10.1007/978-3-319-98776-7_39

clothes and the pose of the pedestrians. In general, the research of person re-ID faces many challenges, such as low resolution of image, change of view, change of pose, change of light, and occlusion. In addition, the scene under the actual video surveillance is very complex. Many monitoring scenes have large flow of people, complex scenes, and the screen is easily occluded. All of these have brought great challenges to the research of person re-ID.

Compared with pedestrian detection, the research of person re-ID is not yet mature. In 2006, the concept of person re-ID was put forward for the first time on CVPR [6], and related research was emerging. In 2007, Gray proposed a database VIPeR [11] that was significant for the research of person re-ID. Since then, more and more scholars began to pay attention to the research of person re-ID. After 2014, deep learning was applied to the field of person re-ID; since 2016, person re-ID has ushered in the growth of blowout, and dozens of related papers have appeared in the conferences of major computer vision; the related papers on CVPR 2018 have reached more than 30. At the same time, related datasets, such as market-1501 [7], cuhk03 [8], MARS [9] and so on, are also constantly expanding, and the results on each dataset have been greatly improved. At present, the problem of person re-ID has become a hot issue in computer vision.

2 Traditional Person Re-id Algorithm

The traditional person re-ID algorithm can be roughly divided into two categories based on the feature description method and the metric learning method. The method based on feature description is concerned with finding a good model to describe the features of pedestrian appearance. Based on the method of metric learning, it is concerned about finding the effective measure criterion of pedestrian feature similarity.

The focus of the method based on feature representation is to design a robust and reliable pedestrian image feature representation model, that is, to distinguish different pedestrians from the influence of light and visual angle changes. The most commonly used is the color histogram. In most cases, the color structure of the pedestrian is simple, so the color representation is an effective feature, which is usually expressed in RGB, HSV histogram. The image of RGB space is transformed into HSL and YCbCr color space, and the distribution of the target pixel value in the logarithmic color space is observed. The color feature has a certain invariance in the discomfort environment of pedestrian recognition, such as different illumination or angle. Shape features such as Histogram of Oriented Gradients (HOG) [10], and local features such as local invariant feature Scale-Invariant Feature Transform (SIFT) [11], SURF [4] and Covariance descriptors [12]. The feature selection technique improves the performance of recognition rates for pedestrians re recognition, such as Fisher vector [13] coding, extracting color or texture histogram, predefining block or fringe shape image regions, or coding region feature descriptors to establish advanced visual feature [14].

Because of the change of camera angle, scale, light, clothing and pose, different resolution and occlusion, different cameras may lose continuous position and motion information. The measurement of the similarity between pedestrian apparent features, such as Euclidean distance and barren distance, cannot get good recognition. Therefore, researchers propose ways of metric learning. This method obtains a new distance metric space, so that the distance between the same pedestrians and different images is less than the distance from different people. Distance metric learning is usually based on Mahalanobis distance.

3 Study on Person Re-id Based on Deep Learning

Compared with the characteristics of the artificial design mentioned above, the deep convolution neural network can obtain the intrinsic features of the image more, and has shown good performance in target detection, image classification and image segmentation. The person re-ID method based on deep learning is mainly divided into the following ideas.

3.1 Design of Loss Function

With the development of deep convolution neural network (ResNet [15], GoogleNet [16], etc.), the design method based on loss function is commonly used in person re-ID. Since CNN can automatically extract Representation features from the task requirements automatically from the original image data, some researchers regard the person re-ID problem as a Identification or Verification problem [8, 17–19]: the Identification problem is to use the ID or attribute as a training label to train the model; the verification problem refers to the input of a pair of pedestrian pictures, allowing the network to learn whether these two pictures belong to the same pedestrian. The former is multi class modeling, and the latter is two classification modeling or modeling of similarity regression tasks.

The former is trained by the classification loss, the disadvantage is that the input is independent and does not consider the fine-grained difference, such as the similarity of the image pair. But there is a clear relationship when a feature is embedded through the cross-entropy loss learning. When the test is tested, the last layer of the classification layer needs to be removed, and the feature is embedded in the similarity calculation with the characteristics of the learning. One advantage of matching is that when testing, gallery can extract all the features of all pedestrian blocks at once, and the computation efficiency is high.

The latter is also a classification loss in essence, but its essence is also a process of similarity measure or feature selection, building a robust computing model for visual angle changes and comparing loss. When the dataset is not large enough, the model of training for a specific fixed angle of view can also be very good. The disadvantage is that it is not easy to train, because the input is the image pair, often the positive pair is less, the positive and negative sample pair is extremely uneven, the training model is biased toward the negative sample learning, but it can train the model by some regularization strategies or data enhancement or partial training by proportion.

Literature [20] uses Identification loss and verification loss to train the network, and the network input is a number of pedestrians' pictures, including the classified subnetwork (Classification Subnet) and the verification subnetwork (Verification Subnet). The classification sub network performs ID prediction and calculates the classification error loss according to the predicted ID. The verification sub network converged the characteristics of the two pictures to determine whether the two pictures belong to the same pedestrian. The sub network is essentially equivalent to a two classification network. The network can predict ID and calculate similarity scores simultaneously. The loss that needs to be optimized is three target loss, that is, two identification loss and a verifying loss, and the author transforms identification loss and verification loss into a cross entropy loss to monitor the update of model parameters in the real training model. After training enough data, once again input a test picture, the network will automatically extract a feature, which is used for person re-ID task.

The commonly used methods of loss function are Triplet loss [21–23], Triplet hard loss with batch hard mining (TriHard loss) [26], Quadruplet loss [24] and Margin sample mining loss (MSML) [25]. Triplet loss needs three input pictures. Unlike the contrast loss, an input Triplet consists of a pair of positive sample pairs and a pair of negative sample pairs. Triplet can close the distance between the positive sample pairs and open the distance between the negative sample pairs. Finally, the pedestrian images of the same ID can be clustered in the feature space to achieve the purpose of person re-ID. Quadruplet loss is another improved version of Triplet loss. As the name implies, Quadruplet needs four input pictures. Unlike Triplet, it adds a negative sample picture. Compared to the relative distance between positive and negative samples, Triplet loss considers the absolute distance between positive and negative samples at the same time, so that the model can make the model learn better. MSML is a metric learning method that considers both relative distance and absolute distance and introduces the idea of difficult sample sampling.

3.2 Person Re-ID Method Based on Local Feature

The main concern of early Person re-id research is in global feature, which is to get a feature vector by integer graph for image retrieval. But they gradually found that the global feature encountered bottlenecks and began to gradually study local feature.

Image dicing is a common way to extract local feature. [8] proposes a filter paring neural network (FPNN), which divides a pedestrian picture horizontally into a M (=20) horizontal stripe, each horizontal stripe has W1 (=9) blocks and performs block matching in the same stripes. On this basis, [19] introduced the cross-input neighborhood difference layer, matching the neighborhood, using a square neighborhood. The local correlation is calculated according to the convolution feature graph on the image. Then sum feature is used to add the neighborhood of the output feature map. Finally, the correlation of the distant pixels is calculated. However, this disadvantage is that the requirements for image alignment are high. If two images are not aligned up and down, the contrast between the head and the upper body is likely to lead to the error of the model.

The existing big datasets begin to adopt automatic detection methods, such as DPM to detect pedestrians and to cut pedestrians from the background. However, automatic detection will contain errors, which affect the learning and recognition of the next stage. [27] let the deep learning method correct the input itself, learn a two-dimensional transformation, and align the pedestrians, and then identify the next step.

3.3 Data Augmentation

Re-ID has a very big problem, that is, data acquisition is difficult. DukeMTMC-reID is currently the largest person re-ID dataset, with a total of 702 training data, with an average of 23.5 training data per class (each person). Since the Generative Adversarial Networks (GAN) image is generated from the original training data, and the original dataset is more similar on the bias (such as light, contrast), it is easier to improve the final result by approaching the actual test dataset.

The principle of GAN is shown in Fig. 1, assuming that we have two networks, G (Generator) and D (Discriminator). G is a network that generates pictures, which receives a random noise Z, generates pictures through the noise, is recorded as G (z), and D is a discriminant network to distinguish whether a picture is "true". Its input parameter is x, which represents a picture, and the output D (x) represents the probability that x is a real picture. If it is 1, it represents a real picture, and the output is 0, which means that it is impossible to be a real picture.

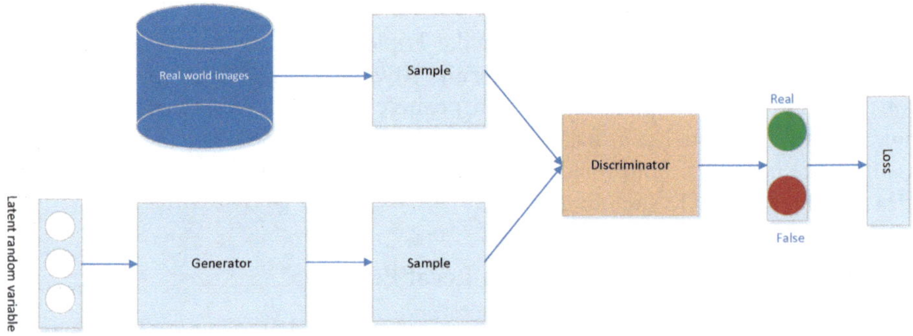

Fig. 1. Generative adversarial networks

In training process, the goal of generating network G is to generate real pictures as far as possible and deception to discriminate network D. The goal of D is to separate the G generated images from the real ones as far as possible. In this way, G and D constitute a dynamic "game process". In the most ideal state, G can generate G (z) that is "false". For D, it is difficult to determine whether the image generated by G is real, so D (G (z)) = 0.5. In this way, we get a generative model G, which can be used to generate pictures.

Literature [28] first proposes a method of GAN generating images to expand training datasets. This method generates more training data for training with only original training data and uses a label smoothing regularization method to assign unifying labels to unmarked images. Because the method of generating pictures is random, it cannot solve the problem of pose of the same pedestrian. Literature [29] uses GAN to create a series of standard pose images. A total of 8 pose were extracted, and the 8 pose basically covered all angles. Each picture generates 8 pose of such standard, so pose different problems will be solved. Finally, the feature of these pictures is used to carry out a average pooling to get the final feature, which integrates the information of each pose. For the problem of pedestrian picture occlusion, [30] algorithm uses the antagonistic network (GAN) to generate the occlusion samples. As the generated samples can shield some of the original feature, the network can study another part of the feature, thus the potential Feature can be excavated.

Because of the difference between different datasets (dataset bias), when the model trained on a set of data is applied directly to another dataset, the performance of the re-ID will decline substantially. To overcome this bias, the paper [31] uses GAN to migrate pedestrians of one dataset to another dataset. To achieve this migration, the loss of GAN is slightly designed, one is the absolute error loss of the foreground, and the other is the normal discriminator loss. The discriminator loss is used to determine which domain the generated graph belongs to, and the loss of the foreground is to ensure that the pedestrian foreground is as realistic and unchanged as possible. In view of the transfer learning of re-ID, [32] constructs the basic framework of Learning via Translation from the angle of image style conversion and puts forward Similarity Preserving cycle-consistent Generative Adversarial (SPGAN), which further improves the performance of the framework. SPGAN constructs unsupervised Self-Similarity and Domain-Dissimilarity relationships to constrain learning of Source-target Translation.

4 Datasets

Table 1 is some common pedestrian datasets.

Table 1. Person re-identification datasets

Dataset	Release time	Identities	Cameras	Images	Label method	Crop size
VIPeR	2007	632	2	1264	Hand	128 × 48
QMUL iLIDS	2009	119	2	476	Hand	Vary
GRID	2009	1025	8	1275	Hand	Vary
CAVIAR4ReID	2011	72	2	1220	Hand	Vary
PRID2011	2011	934	2	24541	Hand	128 × 64
WARD	2012	70	3	4786	Hand	128 × 48
CUHK01	2012	971	2	3884	Hand	160 × 60
CUHK02	2013	1816	10 (5 pairs)	7264	Hand	160 × 60
CUHK03	2014	1467	10 (5 pairs)	13164	Hand/DPM	Vary
RAiD	2014	43	4	6920	Hand	128 × 64
Market1501	2015	1501	6	32217	Hand/DPM	128 × 64
PRW	2016	932	6	34304	Hand	Vary
DukeMTMC-reID	2017	1812	8	36441	Hand	Vary
DukeMTMC4ReID	2017	1852	8	46261	Doppia	Vary
Airport	2017	9651	6	39902	ACF	128 × 64

Market1501 [33] contains 751 identities and 12936 images from six dis-joint cameras. This dataset also includes 2793 false alarms detected by DPM, which as distractors for the real scenario. The PRW [34] (Person Re-identification in the Wild) dataset is an extension of Maretk1501 dataset. Instead of only provide bounding boxes, the author released the full frames with annotations. The DukeMTMC [28]) dataset is a large scale multi target and multi camera tracking dataset with labels. A total of over 2700 people are marked as unique identities in 8 cameras. With access to all information (full frame, frame level ground live, calibration information, etc.), this dataset has great potential.

The changes in the dataset in recent years are mainly in a number of aspects: first, the size of the dataset is expanding, the early datasets are relatively small, but the recent datasets are relatively large; and secondly, bounding boxes is beginning to be given by pedestrian detectors, such as DPM, ACF, rather than manual; Finally, more cameras were used when collecting datasets.

5 Finally

This paper mainly introduces the person re-ID algorithm based on traditional method and deep learning method, and then introduces the development of pedestrian database in recent years. With the continuous development of deep learning, we think that person re-ID has several main development directions:

1. Natural language retrieval. The pedestrians' language description is used to find the designated pedestrians, such as the dress color, whether they are carrying a bag, long hair or short hair.
2. Use generated data to assist in training. Using generated GAN to generate more data, supplement the original dataset, and strengthen the training of deep learning network models.
3. Transfer learning. Due to the difference between datasets, the trained model on dataset A performs poorly on dataset B. At present, there is still much room for research in transfer learning.
4. Combined with pedestrian detection and tracking. The quality of pedestrian detection and tracking will directly affect the accuracy of re-ID, which is rarely considered in re-ID. Re-ID is currently regarded as an independent research task, and the combination of pedestrian detection and tracking will be the main development direction in the future.

Acknowledgments. This work was supported by the Scientific Research Fund of Hunan Provincial Education Department of China (Project No. 17A007); and the Teaching Reform and Research Project of Hunan Province of China (Project No. JG1615).

References

1. Gray, D., Tao, H.: Viewpoint invariant pedestrian recognition with an ensemble of localized features. In: European Conference on Computer Vision, pp. 262–275. Springer, Berlin (2008)
2. Prosser, B.J., Zheng, W.S., Gong, S., Xiang, T., Mary, Q.: Person re-identification by support vector ranking. In: BMVC, vol. 2, no. 5, p. 6 (2010)
3. Mignon, A., Jurie, F.: Pcca: a new approach for distance learning from sparse pairwise constraints. In: IEEE Conference on Computer Vision and Pattern Recognition (CVPR), 2012, pp. 2666–2672. IEEE (2012)
4. Zhao, R., Ouyang, W., Wang, X.: Unsupervised salience learning for person re-identification. In: IEEE Conference on Computer Vision and Pattern Recognition (CVPR), 2013, pp. 3586–3593. IEEE (2013)
5. Zheng, W.S., Li, X., Xiang, T., Liao, S., Lai, J., Gong, S.: Partial person re-identification. In: Proceedings of the IEEE International Conference on Computer Vision, pp. 4678–4686 (2015)
6. Gheissari, N., Sebastian, T.B., Hartley, R.: Person re-identification using spatiotemporal appearance. In: IEEE Computer Society Conference on Computer Vision and Pattern Recognition, 2006, vol. 2, pp. 1528–1535. IEEE (2006)
7. Roth, P.M., Hirzer, M., Köstinger, M., Beleznai, C., Bischof, H.: Mahalanobis distance learning for person re-identification. In: Person Re-identification, pp. 247–267. Springer, London (2014)
8. Li, W., Zhao, R., Xiao, T., Wang, X.: Deepreid: deep filter pairing neural network for person re-identification. In: Proceedings of the IEEE Conference on Computer Vision and Pattern Recognition, pp. 152–159 (2014)
9. Zheng, L., Bie, Z., Sun, Y., Wang, J., Su, C., Wang, S., Tian, Q.: Mars: a video benchmark for large-scale person re-identification. In: European Conference on Computer Vision, pp. 868–884. Springer, Cham (2016)
10. Oreifej, O., Mehran, R., Shah, M.: Human identity recognition in aerial images. In: IEEE Conference on Computer Vision and Pattern Recognition (CVPR), 2010, pp. 709–716. IEEE (2010)
11. Jüngling, K., Bodensteiner, C., Arens, M.: Person re-identification in multi-camera networks. In: IEEE Computer Society Conference on Computer Vision and Pattern Recognition Workshops (CVPRW), 2011, pp. 55–61. IEEE (2011)
12. Tuzel, O., Porikli, F., Meer, P.: Pedestrian detection via classification on Riemannian manifolds. IEEE Trans. Pattern Anal. Mach. Intell. **30**(10), 1713–1727 (2008)
13. Ma, B., Su, Y., Jurie, F.: Local descriptors encoded by fisher vectors for person re-identification. In: European Conference on Computer Vision, pp. 413–422. Springer, Heidelberg (2012)
14. Chen, D., Yuan, Z., Hua, G., Zheng, N., Wang, J.: Similarity learning on an explicit polynomial kernel feature map for person re-identification. In: Proceedings of the IEEE Conference on Computer Vision and Pattern Recognition, pp. 1565–1573 (2015)
15. He, K., Zhang, X., Ren, S., Sun, J.: Deep residual learning for image recognition. In: Proceedings of the IEEE Conference on Computer Vision and Pattern Recognition, pp. 770–778 (2016)
16. Szegedy, C., Liu, W., Jia, Y., Sermanet, P., Reed, S., Anguelov, D., Rabinovich, A.: Going deeper with convolutions. In: Proceedings of the IEEE Conference on Computer Vision and Pattern Recognition, pp. 1–9 (2015)

17. Yi, D., Lei, Z., Liao, S., Li, S.Z.: Deep metric learning for person re-identification. In: 22nd International Conference on Pattern Recognition (ICPR), 2014, pp. 34–39. IEEE (2014)
18. Varior, R.R., Haloi, M., Wang, G.: Gated siamese convolutional neural network architecture for human re-identification. In: European Conference on Computer Vision, pp. 791–808. Springer, Cham (2016)
19. Ahmed, E., Jones, M., Marks, T.K.: An improved deep learning architecture for person re-identification. In: Proceedings of the IEEE Conference on Computer Vision and Pattern Recognition, pp. 3908–3916 (2015)
20. Zheng, Z., Zheng, L., Yang, Y.: A discriminatively learned CNN embedding for person reidentification. ACM Trans. Multimed. Comput. Commun. Appl. (TOMM) **14**(1), 13 (2017)
21. Schroff, F., Kalenichenko, D., Philbin, J.: FaceNet: a unified embedding for face recognition and clustering. In: Proceedings of the IEEE Conference on Computer Vision and Pattern Recognition, pp. 815–823 (2015)
22. Liu, H., Feng, J., Qi, M., Jiang, J., Yan, S.: End-to-end comparative attention networks for person re-identification. IEEE Trans. Image Process. **26**(7), 3492–3506 (2017)
23. Cheng, D., Gong, Y., Zhou, S., Wang, J., Zheng, N.: Person re-identification by multi-channel parts-based cnn with improved triplet loss function. In: Proceedings of the IEEE Conference on Computer Vision and Pattern Recognition, pp. 1335–1344 (2016)
24. Chen, W., Chen, X., Zhang, J., Huang, K.: Beyond triplet loss: a deep quadruplet network for person re-identification. In: Proceedings of CVPR, vol. 2 (2017)
25. Xiao, Q., Luo, H., Zhang, C.: Margin Sample Mining Loss: A Deep Learning Based Method for Person Re-identification. arXiv preprint arXiv:1710.00478 (2017)
26. Hermans, A., Beyer, L., Leibe, B.: In defense of the triplet loss for person re-identification. arXiv preprint arXiv:1703.07737 (2017)
27. Zheng, Z., Zheng, L., Yang, Y.: Pedestrian alignment network for large-scale person re-identification. arXiv preprint arXiv:1707.00408 (2017)
28. Zheng, Z., Zheng, L., Yang, Y.: Unlabeled samples generated by gan improve the person re-identification baseline in vitro. arXiv preprint arXiv:1701.07717 (2017)
29. Qian, X., Fu, Y., Wang, W., Xiang, T., Wu, Y., Jiang, Y.G., Xue, X.: Pose-Normalized Image Generation for Person Re-identification. arXiv preprint arXiv:1712.02225 (2017)
30. Huang, H.J., Li, D.W., Zhang, Z., Huang, K.Q.: Adversarially occluded samples for improving generalization of person re-identification models. In: CVPR (2018). (in press)
31. Wei, L., Zhang, S., Gao, W., Tian, Q.: Person Transfer GAN to Bridge Domain Gap for Person Re-identification. arXiv preprint arXiv:1711.08565 (2017)
32. Deng, W., Zheng, L., Kang, G., Yang, Y., Ye, Q., Jiao, J.: Image-Image Domain Adaptation with Preserved Self-similarity and Domain-Dissimilarity for Person Re-identification. arXiv preprint arXiv:1711.07027 (2017)
33. Zheng, L., Shen, L., Tian, L., Wang, S., Wang, J., Tian, Q.: Scalable person re-identification: a benchmark. In: Proceedings of the IEEE International Conference on Computer Vision, pp. 1116–1124 (2015)
34. Zheng, L., Zhang, H., Sun, S., Chandraker, M., Tian, Q.: Person re-identification in the wild. arXiv preprint arXiv:1604.02531 (2017)

Electrical Data Recovery with Weather Information via Collective Matrix Factorization

Xushan Han[1], Qian Dang[2(✉)], Huafeng Zhang[1], Yanwen He[1], and Lin Gao[3]

[1] State Grid Gansu Electric Power Corporation, Lanzhou 730000, China
[2] Information Communication Corporation,
State Grid Gansu Electric Power Corporation, Lanzhou 730050, China
shuiqiao9999@163.com
[3] Jinchang Power Company, State Grid Gansu Electric Power Corporation,
Lanzhou 737100, China

Abstract. It is an important application in power systems to monitor load and calculate the electricity cost using data collected from smart meters. As the development of smart grid and energy internet, this leads to a significant increase in the amount of data transmitted in real time. Due to the mismatch with communication networks that were not designed to carry high-speed and real time data, data losses and data quality degradation may happen constantly. The electrical data is incomplete. Therefore, the electrical data recovery is an important issue. The electrical data is produced by the action and feelings of the human being which has strong season feature and depends on the weather. We take the weather information as aid to recover electrical data via collective matrix factorization. We perform extensive simulations based on the real power system and weather data sets and the simulation results validate the efficiency and efficacy of the proposed scheme compared without weather information.

Keywords: Collective matrix factorization · Electrical data recovery
Weather information

1 Introduction

It is an important application in power systems to monitor load and calculate the electricity cost using data collected from smart meters. Currently, for the resident user, only the daily average data is required to be collected, that is, the daily data in zero clock of smart meters is automatically frozen and then is uploaded in batches.

However, as the development of smart grid and energy internet [1], opening up the sales-side market in No. 9 document of the new power reform, the electrical data collection of users has new requirement to collect multiple electrical data in real time daily. This leads to a significant increase in the amount of data transmitted in real time. Due to the mismatch with communication networks that were not designed to carry high-speed and real time data, data losses and data quality degradation may happen constantly. There is no guarantee that collection will be successful. The electrical data is incomplete and some data is unknown which is unable to satisfy the need of the collection and

J. Abawajy et al. (Eds.): ATCI 2018, AISC 842, pp. 379–388, 2019.
https://doi.org/10.1007/978-3-319-98776-7_40

analysis of power consumption data under the new market and technological development prospects. Therefore, how to recover the unknown data according to the collected data is an important problem in power systems.

Matrix factorization (MF) [2] decomposes a (sparse) matrix X into the production of two (low-rank) matrices, which denote the latent variables of users and items respectively. The production of the two matrices can approximate matrix X, therefore helping fill the missing values in X. MF has been widely used in recommendation and traffic estimation [3–6].

Advanced methods use collective matrix factorization (or called context-aware matrix factorization) [7, 8] to accommodate different data sets with different matrices, which share a common dimension between one another. By decomposing these matrices collaboratively, we can transfer the similarity between different objects learned from a data set to another one, therefore complementing the missing values more accurately.

Therefore, we want to recover the electrical data by collective matrix factorization. The electrical data is produced by the action and feelings of the human being which has strong season feature and depends on the weather. For example, the air condition works more in winter and summer, which leads a higher value of smart meter. The weather data share the time dimension with the electrical data. Above all, we take the weather data as the aid context to recover electrical data accurately.

The main contributions of this work can be summarized as follows:

- We first model the time and users' value of smart meter that we take as input. We analyze the rank of electrical data matrix and the electrical data matrix is low rank which is the basis of using MF.
- We model the time and weather attributes which consist of maximum temperature, minimum temperature, weather conditions, wind direction, wind power. The weather attributes are coded to be number for MF.
- We apply a collective matrix factorization method to mine the relationship among user, weather and time, and use them to fill the missing value of electrical data matrix.
- We perform extensive simulations based on the real power system and weather data sets and the simulation results validate the efficiency and efficacy of the proposed scheme.

The remaining of this paper is organized as follows. The Preliminaries and related Work is reviewed in Sect. 2. Section 3 introduces our problem formulation with the power system data and weather data. Section 4 provides the algorithm of the solution. Section 5 provides simulation results and analysis. In the end, we conclude this work in Sect. 6.

2 Preliminaries and Related Work

2.1 Matrix Factorization

MF techniques approximate a low rank matrix **X** as a product of two much smaller matrices:

$$\mathbf{X} \approx \mathbf{U}\mathbf{V}^{\mathrm{T}} \tag{1}$$

where **U** is an $I \times K$ and **V** is a $J \times K$ matrix. **X** has many unknown elements which cannot be treated as zero. The subset of its entries x_{ij}, $(i,j) \in \Omega$ are known. The subset Ω can be formed with randomly selected entries of the matrix, and the sampling operator $P_\Omega:\mathbb{R}^{I\times J} \to \mathbb{R}^{I\times J}$ is defined by

$$[P_\Omega(\mathbf{X})]_{ij} = \begin{cases} x_{ij}, & (i,\ j) \in \Omega \\ 0, & otherwise \end{cases} \tag{2}$$

For this case, the approximation task can be defined as follows. Let $\mathbf{U} \in \mathbb{R}^{I\times K}$ and $\mathbf{V} \in \mathbb{R}^{J\times K}$. Let u_{ik} denote the elements of **U**, and v_{jk} the elements of **V**. Let $\mathbf{U}_{i*}$ denote a row of **U**, and $\mathbf{V}_{j*}$ a row of **V**. We can calculate the dot product of the two vectors corresponding to $\mathbf{U}_{i*}$ and $\mathbf{V}_{j*}$ as Eq. 2 to get the estimation of x_{ij}.

$$\hat{x}_{ij} = \mathbf{U}_{i*}\mathbf{V}_{j*}^{\mathrm{T}} = \sum_{k=1}^{K} u_{ik}v_{jk} \tag{3}$$

X can be recovered by solving the optimization problem.

$$\begin{aligned} &\min_{\mathbf{U},\mathbf{V}} \frac{1}{2}\left\|P_\Omega\left(\mathbf{X} - \mathbf{U}\mathbf{V}^{\mathrm{T}}\right)\right\|_{\mathrm{F}}^2 \\ &= \min_{\mathbf{U},\mathbf{V}} \frac{1}{2} \sum_{(i,j)\in\Omega} e_{ij}^2 = \min_{\mathbf{U},\mathbf{V}} \frac{1}{2} \sum_{(i,j)\in\Omega} (x_{ij} - \hat{x}_{ij})^2 \end{aligned} \tag{4}$$

where e_{ij} denotes the training error on the (i, j)-th element. Problem (5) states that the optimal **U** and **V** minimizes the sum of squared errors only on the known elements of **X**. We can use a simple incremental gradient descent method to find a local minimum, where one gradient step intend to decrease the square of prediction error. We compute the gradient of $\frac{1}{2}e_{ij}^2$:

$$\frac{\partial}{\partial u_{ik}}\frac{1}{2}e_{ij}^2 = -e_{ij}\cdot v_{jk}, \quad \frac{\partial}{\partial v_{jk}}\frac{1}{2}e_{ij}^2 = -e_{ij}\cdot u_{ik} \tag{5}$$

Having obtained the gradient, we can now formulate the update rules for u_{ik} and v_{kj} as follows:

$$u'_{ik} = u_{ik} + \eta\cdot e_{ij}\cdot v_{jk}, \quad v'_{jk} = v_{jk} + \eta\cdot e_{ij}\cdot u_{ik} \tag{6}$$

where η is a small value that determines the rate of approaching the minimum. To avoid over fitting, a regularized MF by penalizing the square of the Euclidean norm of weights is introduced [9–11].

$$\min_{\mathbf{U},\mathbf{V}} \frac{1}{2}\left\|P_\Omega\left(\mathbf{X} - \mathbf{U}\mathbf{V}^{\mathrm{T}}\right)\right\|_{\mathrm{F}}^2 + \frac{\lambda}{2}\left(\|\mathbf{U}\|_{\mathrm{F}}^2 + \|\mathbf{V}\|_{\mathrm{F}}^2\right) \tag{7}$$

2.2 Collective Matrix Factorization

Collective matrix factorization accommodates different data sets with different matrices, which share a common dimension between one another. For example, there is a known matrix $\mathbf{L} \in \mathbb{R}^{l \times W}$ which shares the column with **X**. The main idea is to propagate the information among **X** and **L** by requiring them to share low-rank matrices **U** in a collective matrix factorization model. As matrix **L** are known, we can obtain an accurate decomposition of them, i.e. matrices **U** and **V**. Thus, Matrix **X** can be complemented more accurately.

Collective Matrix Factorization has been widely used to improve the accurate of recovery or recommendation. Zheng et al. [12] propose a collective matrix factorization method to enable location-activity recommendation. A location-activity matrix is built based on many users' location histories. If this location-activity matrix is completely filled, we can recommend a set of locations for a particular activity by retrieving the top k locations with a relatively high frequency from the column that corresponds to that activity. The information from another two matrices are incorporated into the matrix factorization. One is a location-feature matrix; the other is an activity-activity matrix. Such kind of additional matrices are usually called contexts, which can be learned from other data sets.

Shang and Zheng et al. [13] propose a collective matrix factorization method to instantly estimate the travel speed on each road segment throughout an entire city, based on the GPS trajectory of a sample of vehicles (such as taxicabs). To address this issue, four context matrices are built. First is the historical traffic patterns on road segments in which an entry denotes the average travel speed derived from historical data over a long period. Second is the physical features of a road segment, such as the shape of a road, number of lanes, speed constraint, and the distribution of surrounding POIs. To capture high-level traffic conditions, a city is divided into uniform grids. By projecting the recently received GPS trajectories into these grids, the third matrix is built which indicates the historical high-level traffic patterns. The fourth is the real-time high-level traffic conditions in a city.

How to select the context and obtain the context is the key problem in collective matrix factorization.

3 Problem Formulation

3.1 Electrical Data and Low Rank Analysis

For N users of power systems in a given area. The difference between the values of the smart meter at the two time points is the electricity value, which indicates how much electricity is used during this period of time. We define a matrix $\mathbf{X} \in \mathbb{R}^{H*N}$ to hold the electrical data, which contains the data within a H times measurement. In the electrical matrix, a row corresponds to a time slot and a column corresponds to a user. An entry represents the electrical data on a particular time slot and user, as shown in Fig. 1.

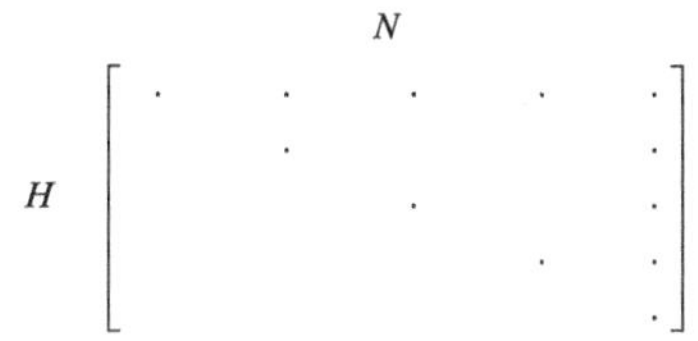

Fig. 1. The matrix of electrical data. The row is time. The column is user.

Because of data losses and data quality degradation, there is no guarantee that collection will be successful. The electrical matrix is incomplete and has many unknown elements. The subset of its entries x_{ij}, $(i, j) \in \Omega$ are successfully collected and known. We also use the sampling operator $P_\Omega : \mathbb{R}^{H \times N} \rightarrow \mathbb{R}^{H \times N}$ similarly to Eq. (2).

We want to fill the missing values of **X**. Due to the persistence of user behavior, electricity consumption remains relatively stable for some time, and electricity consumption is relatively stable. The electrical matrix can be viewed as a low-rank matrix.

We first apply the singular value decomposition (SVD) [14, 15] to examine whether the matrix has a good low-rank structure. Consider the distribution of singular value of a electrical matrix composed of 160 users in 388 days, as shown in Fig. 2. The X-axis is the id of the singular value. The Y-axis is the value of the singular value. The number of non-zero singular values represents the rank of the matrix. In Fig. 2, the first five feature values account for 95% of all feature values, and the electrical matrix can be considered as a low-rank matrix approximately in all the scenarios under investigation. The low-rank feature is the prerequisite for using matrix factorization.

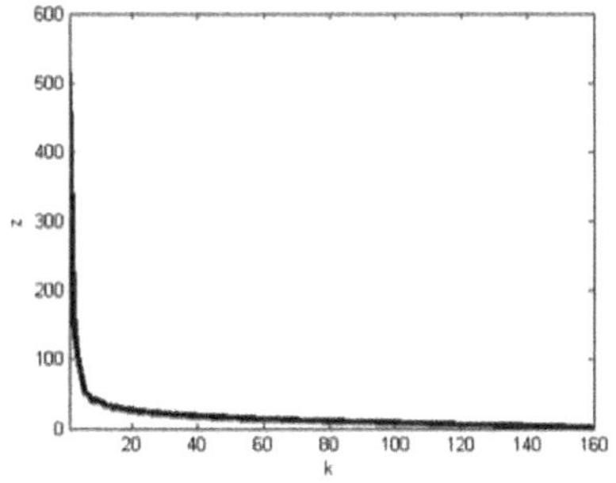

Fig. 2. The rank analysis of electrical data matrix. The X-axis is the id of the singular value. The Y-axis is the value of the singular value.

3.2 Weather Data

There is a lot of contact between the user's electrical behavior and the season and the weather. The weather information is publicly accessible. Therefore, the weather information can be used as an auxiliary to recover electrical data. The weather information is the weather of the user's area in the electrical matrix, including five attributes:

maximum temperature, minimum temperature, weather conditions, wind direction, and wind power.

The maximum and minimum temperatures are between −60 and 50 °C. The weather conditions include 44 kinds of situations, such as sunny, cloudy, overcast, showers, thunderstorms, thunderstorms accompanied by hail, light rain, moderate rain, heavy rain, heavy rain, heavy rainstorms, heavy rainstorms, small to moderate rain, moderate to heavy rain, heavy rain, heavy rain to heavy rain, Heavy rainstorms to heavy rainstorms, sleet, snow showers, light snow, medium snow, heavy snow, snowstorms, small to medium snow, medium to heavy snow, large to heavy snow, fog, dense fog, strong fog, frost, freezing rain, sandstorm, dust, blown sand, sandstorm, strong sandstorm, extremely strong sandstorm, slight haze, mild haze, moderate haze, heavy haze, extreme strong haze, haze, haze line. Wind direction values include 8 wind directions: east wind, south wind, west wind, north wind, southeast wind, northeast wind, northwest wind, and southwest wind. The degree of wind power includes a total of 18 levels and is on a scale of 0 to 17.

In order to build a matrix, the values of the attributes should be represented by a non-negative number according to encoding. The maximum and minimum temperatures are encoded by removing the unit and adding 60 to the original value. The weather conditions are encoded in the order of 1 for sunny, 2 for cloudy, and 44 for squatting. The direction of the wind direction is the same as the weather conditions. The east wind is 1, the south wind is 2, and the southwest wind is 8. The wind power code directly adopts the wind level and removes the unit level.

The weather information is encoded to form a weather matrix $\mathbf{L} \in \mathbb{R}^{H*W}$ in which the number of rows is equal to the number of rows in the H-times of the electrical matrix, and the number of columns W represents the five attributes: the maximum temperature, the minimum temperature, the weather, the wind direction, and the wind power. The element value of the i-th row and j-th column of the matrix L represents the coded value of the j-th attribute of the i-th time.

4 Collective Matrix Factorization for Electrical Data Recovery with Weather Information

According to the concept of MF, the low-rank electrical data matrix $\mathbf{X}$ can be decomposed into two smaller matrix products.

$$\mathbf{X} \approx \mathbf{U}\mathbf{V}^{\mathrm{T}} \tag{8}$$

Where $\mathbf{U} \in \mathbb{R}^{H \times K}$ is the feature matrix of the time, $\mathbf{V} \in \mathbb{R}^{K \times N}$ is the user's feature matrix, and $K \ll \min(H, N)$.

Weather information is openly shared, easily accessible, and known. It is considered to take the weather matrix as the context information to help with electrical matrix recovery. Similarly, the weather matrix $\mathbf{L}$ can also be factorized into the product of two smaller matrices $\mathbf{U} \in \mathbb{R}^{H*K}, \mathrm{T} \in \mathbb{R}^{K*W}$. Since the weather matrix and the electrical data matrix share time information, the time feature matrix $\mathbf{U}$ can also be shared after factorization, as shown in Fig. 3.

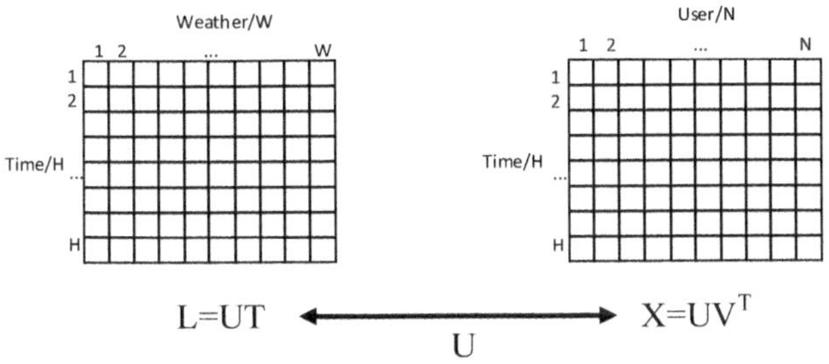

Fig. 3. Collective matrix factorization for electrical data and weather data

$$\mathbf{L} \approx \mathbf{UT} \tag{9}$$

The object of the collective matrix factorization with weather matrix is as follows.

$$\min_{(\mathbf{U},\mathbf{V},\mathbf{T})} L(\mathbf{U},\mathbf{V},\mathbf{T}) \tag{10}$$

$$L(\mathbf{U},\mathbf{V},\mathbf{T}) = \frac{1}{2}\left\|\mathrm{P}_{\Omega}\left(\mathbf{X}-\mathbf{UV}^{\mathrm{T}}\right)\right\|_{\mathrm{F}}^{2} + \frac{\lambda_1}{2}\|\mathbf{L}-\mathbf{UT}\|_{\mathrm{F}}^{2} + \frac{\lambda_2}{2}\left(\|\mathbf{U}\|_{\mathrm{F}}^{2} + \|\mathbf{V}\|_{\mathrm{F}}^{2} + \|\mathbf{T}\|_{\mathrm{F}}^{2}\right) \tag{11}$$

Where $\|\cdot\|_{\mathrm{F}}$ represents Frobenius norm. The first two terms in the objective function are used to control the error in the matrix factorization process. The last item is the Euclidean paradigm of the factorized sub-matrix. The regularization penalty term prevents the matrix item from appearing negative values.

The objective function is not conjointly convex for all variables **U**, **V**, **T**. We solve it by gradient descent. The partial derivative of the variable is used as a gradient. × represents matrix multiplication.

$$\begin{aligned}
\frac{\partial L(\mathbf{U},\mathbf{V},\mathbf{T})}{\partial \mathbf{U}_{i*}} &= (\hat{x}_{ij} - x_{ij})\mathbf{V}_{j*}^{\mathrm{T}} + \lambda_1(\mathbf{U}_{i*} \times \mathbf{T} - \mathbf{L}_{i*}) \times \mathbf{T} + \lambda_2 \mathbf{U}_{i*} \\
\frac{\partial L(\mathbf{U},\mathbf{V},\mathbf{T})}{\partial \mathbf{V}_{j*}} &= (\hat{x}_{ij} - x_{ij})\mathbf{U}_{i*} + \lambda_2 \mathbf{V}_{j*} \\
\frac{\partial L(\mathbf{U},\mathbf{V},\mathbf{T})}{\partial \mathbf{T}} &= \lambda_1(\mathbf{U}_{i*} \times \mathbf{T} - \mathbf{L}_{i*}) \times \mathbf{U}_{i*} + \lambda_2 \mathbf{T}
\end{aligned} \tag{12}$$

Having obtained the gradient, we can now formulate the update rules as follows:

$$\begin{aligned}
\mathbf{U}_{i*} = \mathbf{U}_{i*} - \eta\frac{\partial L(\mathbf{U},\mathbf{V},\mathbf{T})}{\partial \mathbf{U}_{i*}} &= \mathbf{U}_{i*} - \eta\left((\hat{x}_{ij} - x_{ij})\mathbf{V}_{j*}^{\mathrm{T}} + \lambda_1(\mathbf{U}_{i*} \times \mathbf{T} - \mathbf{L}_{i*}) \times \mathbf{T} + \lambda_2 \mathbf{U}_{i*}\right) \\
\mathbf{V}_{j*} = \mathbf{V}_{j*} - \eta\frac{\partial L(\mathbf{U},\mathbf{V},\mathbf{T})}{\partial \mathbf{V}_{j*}} &= \mathbf{V}_{j*} - \eta\left((\hat{x}_{ij} - x_{ij})\mathbf{U}_{i*} + \lambda_2 \mathbf{V}_{j*}\right) \\
\mathbf{T} = \mathbf{T} - \eta\frac{\partial L(\mathbf{U},\mathbf{V},\mathbf{T})}{\partial \mathbf{T}} &= \mathbf{T} - \eta\left(\lambda_1(\mathbf{U}_{i*} \times \mathbf{T} - \mathbf{L}_{i*}) \times \mathbf{U}_{i*} + \lambda_2 \mathbf{T}\right)
\end{aligned} \tag{13}$$

where η is a small value that determines the rate of approaching the minimum. The algorithm is described in Algorithm 1.

Algorithm 1 Algorithm description for our model

Input: **X**, **L**, Error threshold ε

Output: **U**, **V**, **T**

1.Random initialization $\mathbf{U} \in \mathbb{R}^{N*K}, \mathbf{V} \in \mathbb{R}^{K*H}, \mathbf{T} \in \mathbb{R}^{K*W}$

2. η is the step, t is the number of iteration which is set to 1

3.While (t<M and L_t-L_{t+1}> ε)

4. for all $\omega_{ij} \neq 0$

$$\hat{x}_{ij} = \mathbf{U}_{i*} \times \mathbf{V}_{j*}^{\mathrm{T}}$$

$$\mathbf{U}_{i*} = \mathbf{U}_{i*} - \eta\lambda_2\mathbf{U}_{i*} - \eta(\hat{x}_{ij} - x_{ij})\mathbf{V}_{j*}^{\mathrm{T}} - \eta\lambda_1(\mathbf{U}_{i*} \times \mathbf{T} - \mathbf{L}_{i*}) \times \mathbf{T};$$

$$\mathbf{V}_{j*} = \mathbf{V}_{j*} - \eta\lambda_2\mathbf{V}_{j*} - \eta(\hat{x}_{ij} - x_{ij})\mathbf{U}_{i*};$$

$$\mathbf{T} = \mathbf{T} - \eta\lambda_2\mathbf{T} - \eta\lambda_1(\mathbf{U}_{i*} \times \mathbf{T} - \mathbf{L}_{i*}) \times \mathbf{U}_{i*};$$

$$L_{t+1} = \sum_{\omega_{ij} \neq 0} \| \hat{x}_{ij} - x_{ij} \|^2$$

$$t = t + 1$$

5.Return **U**, **V**, **T**

5 Simulation

We get the real electrical data from Lanzhou power system company and select the electrical data of 163 users and 31 times which are all known. The users are located in Jiuquan of the Lanzhou province. The times are from August 1, 2016 to August 31, 2016. We obtain the weather information in August 2016 from the weather online website and encode the weather information according to the code rule in Sect. 3.2.

In this paper we adopt the evaluation measure, the root mean squared error (RMSE), which is defined as:

$$\mathrm{RMSE} = \sqrt{\frac{\sum_{(i,j) \notin \Omega} (x_{ij} - \hat{x}_{ij})^2}{S}} \tag{14}$$

where $(i,j) \notin \Omega$ contains time-user pairs on which the values are predicted, S is the number of unknown elements. Smaller RMSE indicates higher accuracy.

We evaluate our algorithm (CMF) performance by comparing with the non-negative matrix factorization (NMF) by the Eq. (7). The parameters of NMF are set as follows: λ=0.002, η= $1/\sqrt{t + 10^8}$. The main parameters of our algorithm are set as follows: M = 800, λ_1=λ_2=0.002, η= $1/\sqrt{t + 10^8}$, ε= 1^{-4}.

The collection ratio is rate between the number of known elements and total elements. According to the collection ratio, we random set the elements to unknown to simulate the unsuccessful collection and form a collected electrical data matrix **X**. Based on this electrical data matrix **X**, NMF and our algorithm (CMF) are used to recover it successively.

The relationship between the collection ratio and the recovery accuracy is shown in Fig. 4. Compared with NMF, the CMF has a higher recovery accuracy by the help of the weather matrix.

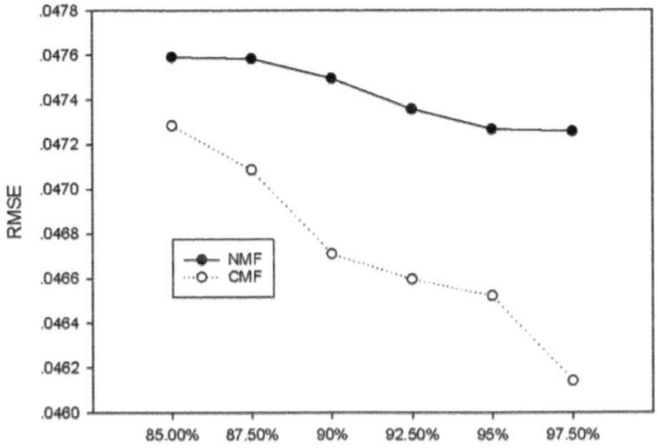

Fig. 4. Recovery accuracy with different collection rates

6 Conclusion

We use the weather information to help for the recovery of electrical data by the collective matrix factorization. It is the first work for electrical data recovery with the help of weather information. We take our algorithm's simulation on the real power system and weather data sets and the simulation results validate the efficiency and efficacy of the proposed scheme. Future work is to use more real data set to test the algorithm and take fine-grained time into account.

References

1. Tsoukalas, L.H., Gao, R.: From smart grids to an energy internet: assumptions, architectures and requirements. In: 3rd International Conference on Electric Utility Deregulation and Restructuring and Power Technologies, DRPT 2008, pp. 94–98. IEEE (2008)
2. Tikk, D.: Investigation of various matrix factorization methods for large recommender systems. In: IEEE International Conference on Data Mining Workshops, pp. 1–6. IEEE (2008)
3. Bao, J., Zheng, Y., Mokbel, M.F.: Location-based and preference-aware recommendation using sparse geo-social networking data. In: ACM SIGSPATIAL Conference on Advances in Geographic Information Systems (GIS 2012), pp. 199–208. ACM (2012)
4. Hsieh, H.P., Lin, S.D., Zheng, Y.: Inferring air quality for station location recommendation based on big data. In: 21st SIGKDD Conference on Knowledge Discovery and Data Mining, pp. 437–446. ACM (2015)
5. Gao, P., Wang, R., Wang, M., et al.: Low-rank matrix recovery from noisy, quantized and erroneous measurements. IEEE Trans. Signal Process. **PP**(99), 1 (2018)

6. Zheng, Y., Zhang, L., Ma, Z., et al.: Recommending friends and locations based on individual location history. ACM Trans. Web **5**(1), 5–44 (2011)
7. Nguyen, D.M., Tsiligianni, E., Deligiannis, N.: Learning discrete matrix factorization models. IEEE Signal Process. Lett. **PP**(99), 1 (2018)
8. Singh, A.P., Gordon, G.J.: Relational learning via collective matrix factorization. In: ACM SIGKDD International Conference on Knowledge Discovery and Data Mining, pp. 650–658. ACM (2008)
9. Hoyer, P.O.: Non-negative matrix factorization with sparseness constraints. J. Mach. Learn. Res. **5**, 1457–1469 (2004)
10. Lee, D.D., Seung, H.S.: Algorithms for non-negative matrix factorization. In: Advances in Neural Information Processing Systems, pp. 556–562. IEEE (2011)
11. Gligorijevic, V., Panagakis, Y., Zafeiriou, S.P.: Non-negative matrix factorizations for multiplex network analysis. IEEE Trans. Pattern Anal. Mach. Intell. **PP**(99), 1 (2018)
12. Zheng, V.W., Zheng, Y., Xie, X., et al.: Collaborative location and activity recommendations with GPS history data. In: International Conference on World Wide Web (WWW'10), pp. 1029–1038. IEEE (2010)
13. Shang, J., Zheng, Y., Tong, W., et al.: Inferring gas consumption and pollution emission of vehicles throughout a city. In: ACM SIGKDD Conference on Knowledge Discovery and Data Mining (KDD 2014), pp. 1027–1036. ACM (2014)
14. Golub, G.H., Reinsch, C.: Singular value decomposition and least squares solutions. Numer. Math. **14**(5), 403–420 (1970)
15. Klema, V., Laub, A.J.: The singular value decomposition: its computation and some applications. IEEE Trans. Autom. Control **25**(2), 64–176 (1980)

Two-Phases Detection Scheme: Detecting Android Malware in Android Markets

Xin Su[1,2], Jiuchuan Lin[3(✉)], Fuhui Shen[1,2], and Yi Zheng[4]

[1] Hunan Provincial Key Laboratory of Network Investigational Technology, Hunan Police Academy, Changsha, China
suxin@hnu.edu.cn
[2] Department of Information Technology, Hunan Police Academy, Changsha, China
[3] Key Lab of Information Network Security of Ministry of Public Security, The Third Research Institute of Ministry of Public Security, Shanghai, China
linjiuchuan@stars.org.cn
[4] Zhejiang Economic Information Center, Hangzhou, China

Abstract. Recently, Android application becomes popular and important in human's daily work, life, entertainment. However, because of open source of Android application, more and more malware aim to this platform and launch various malicious attacks to threaten Android users' security. Previous research works focus on using static behavioral analysis to detect Android malware, which cannot capture dynamic behaviors and in-efficiency to detect Android malware. In this paper, we present a Android application two-stage detection scheme that using two kinds of dynamic behavioral characteristics to detect Android malware. This framework first uses system call statistics to identify potential malicious apps. After verification, if the software is clean, the application will then be released to the relevant markets. To mitigate against false negative cases, users who run new apps can invoke our network traffic monitoring (NTM) tool which triggers network traffic capture upon detecting some suspicious behaviors e.g. detecting sensitive data being sent to output stream of an open socket. The network traffic will be analyzed to see if it matches network characteristics observed from malware apps. If suspicious network traffic is observed, the relevant Android markets will be notified to remove the application from the repository. We trained our system call and network traffic classifiers using 32 families of known Android malware families and some typical normal apps. Later, we evaluated our framework using other malware and normal apps that used in the training set. Our experimental results using 120 test apps (which consist of 50 malware and 70 normal apps) indicate that we can achieve a 94.2% and 99.2% accuracy with J.48 and Random forest classifier respectively using our framework.

Keywords: Android security · Data mining · System call monitor
Network traffic monitor

J. Abawajy et al. (Eds.): ATCI 2018, AISC 842, pp. 389–399, 2019.
https://doi.org/10.1007/978-3-319-98776-7_41

1 Introduction

Android applications (apps) have become the predominant means of accessing personalized computing services such as email, banking, etc. The Android platform is the computing base for a majority of the smartphones in usage, and it acts as a host to thousands of mobile apps created to enrich the users' mobile experience. However, this rapid deployment and extensive availability of mobile apps has made them attractive targets for various malware. Malware authors generally take advantage of the update mechanism of mobile apps to infect existing mobile apps with malicious code and thus compromise the security of the smartphone. Recent statistical data show that malware based on Android platform accounts for 97% of mobile malware [1]. The private data of the users, such as IMEI, contacts list, and other user specific data are the primary target for the attackers, which is a serious threat for network users security and privacy. Consequently, there is an urgent need to identify and cope with the malware from normal apps for the Android platform.

Recent years, many research works focus on detecting Android malware. Static behavioral analysis is an efficiency approach to detect Android malware [2–5]. These approaches reverse Android app install file, and extract several static behavioral features from configuration file and smali code files. Then, combine these features with machine learning algorithms to build classification model for purposes of Android malware detection. However, static behavioral features is one aspect of Android malware, and many Android malware execute malicious activities during runtime [6]. Therefore, the approach of static behavioral analysis cannot capture the runtime behaviors, and fail to detect the Android malware which contain dynamic malicious activities.

In this paper, we present a smartphone dual defense protection framework for detecting new malicious Android apps in the Android markets. Our solution consists of two main phases: (i) a verification service in Phase 1 which checks for potential malicious codes in new apps using system call statistics, (ii) a network monitoring tool in Phase 2 that allows users who install new apps to check for potential false negative cases generated from the verification process during Phase 1.

2 Two-Phases Detection Scheme Framework

In our framework, we assume that an application verification service (AV) is provided by either the Android Official Market and/or Alternative Android Markets. The AV service will verify if any application is clean of any potential malwares. Such a service can be provided for free or at a low fee paid by the application developers. Android Markets can refuse to let any application developer upload a new application to their repositories unless that application received a verification approval by the application verification service. Every application developer who would like to release his new application to an Android market must first submit that new application for application verification. He uploads his new application to an Application Repository, and indicates which Android Markets he would like his new application to be uploaded to (Fig. 1).

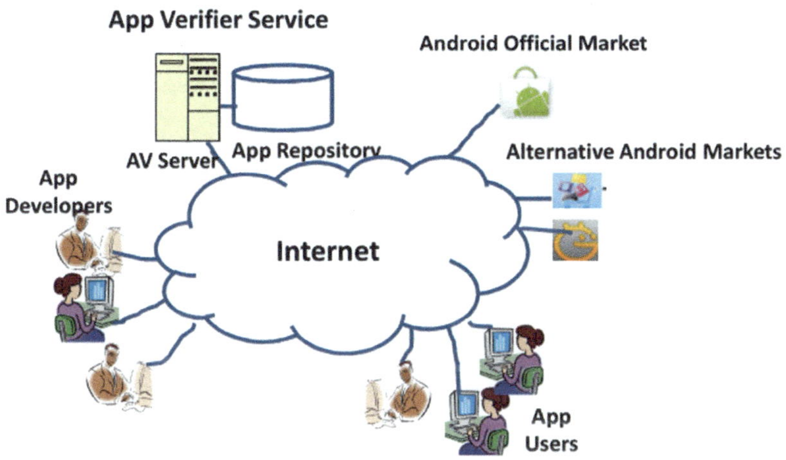

Fig. 1. Two-phases detection scheme framework

Periodically, the AV server runs those new apps which are submitted during a certain time period, and so that it can collect system call statistics using the system call monitoring tool which we develop. The system call monitoring tool generates a log of system calls generated by any new application. The AV server analyzes the log files of all new apps in that batch to extract relevant features that can be fed to a WEKA-based classification tool. Our SCM classification tool will determine if the new application that undergoes verification contains any malware characteristics.

After Phase 1, the new apps will be labeled as "normal" or "malware". The AV server will then use trigger rules for normal or malware apps to decide if network monitoring tool will be triggered to collect network and analyze network traces in Phase 2. Phase 2 gives us an opportunity to correct some of the mislabeled apps. After the final labels for the apps are generated, the AV server will release all apps that are labeled as normal to the Android markets, and send an email to the application developer if his application fails the verification process.

2.1 System Call Monitor

Figure 2 shows how our system call monitor (SCM) interacts with the application undergoing test (AUT) as well as the Android Operating System (AOS). Our SCM interacts with the AOS to obtain the process identifiers that are associated with the AUT. The AUT is run multiple times at the verification server, and relevant system call statistics for each session are recorded in a log file. A dataset of system call related statistics will be created for every application that undergoes evaluation. The more users use a particular category of apps, the more complete and accurate our system will be. The server will release the AUT to the relevant Android markets if the classification result indicates that it is very likely to be benign.

Android Operation System is based on linux kernel. In Linux, a system call is how a program requests a service from the operating system's kernel. System calls provide

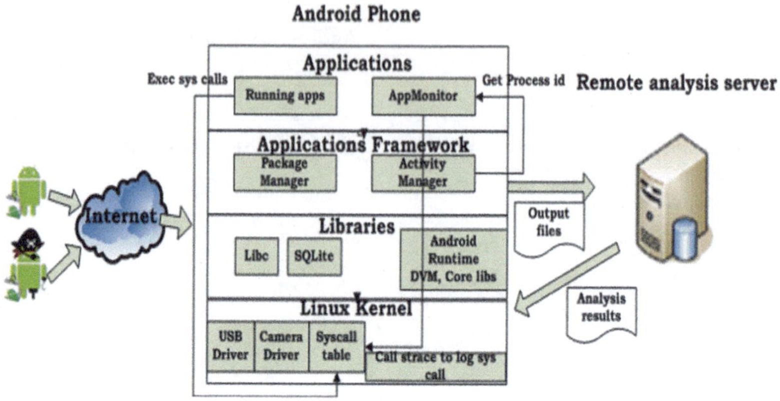

Fig. 2. System call monitor

useful functions to application programs like network communication, file management or process related operations. When an application from user space makes a request to the Operation System, the request goes through glibc library, system call interface, kernel and finally to the hardware. Functions like getpid(), open(), read(), chmod(), and socket() are some of the functions that glibc provides for apps to invoke a system call.

All requests made from upper layers pass through the kernel using the system call interface before they are executed in hardware. Thus, capturing and analyzing the system calls that pass through system call interface will provide accurate information about the behavior of an application. Our design makes use of a tool available in Linux

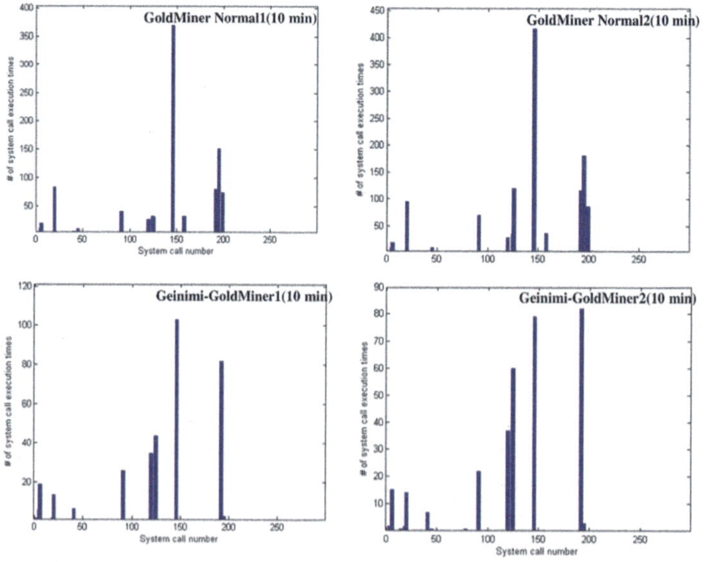

Fig. 3. System call statistics of normal and malicious GoldMiner application

called strace to collect the system calls. The aim of monitoring such system calls is to generate a log file with all events generated by that Android application under test. This file contains useful information e.g. counts of different system calls made by the application, clone processes, opened and accessed files.

To illustrate why system call statistics can be used to effectively identify potential malware apps, we collect system call statistics of two sets of apps, namely (a) Goldminer and (b) Bgserv-Chinamobile. In each set, we have a clean application from the official Android market, and another that contains malware from a 3rd party market. We plot the collected system call statistics for two runs of Goldminer and Chinamobile apps in Figs. 3 and 4 respectively. The x-axis shows the system call identifiers while the y-axis shows the number of times that particular system call is activated. One can observe that the numbers of clone (callId = 121) and mprotect (callId = 126) calls in the malware version of Goldmine are significantly higher than its normal version while the numbers of sigprocmask (callId = 127) and mmap (callId = 193) system calls in the malware version of Chinamobile application are significantly higher than its normal version.

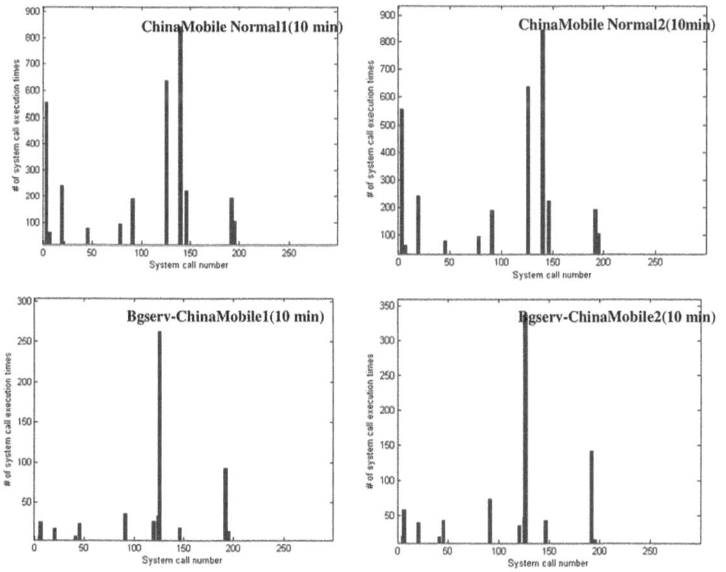

Fig. 4. System call statistics of normal and malicious ChinaMobile application

There are 288 system calls but we merely focus on fifteen that are related to process, memory and I/O management since typically the statistics of such calls are different between normal and malicious apps. The fifteen system calls used in our SCM classifier are: sys_chmod, sys_chown, sys_mount, sys_access, sys_open, sys_clone, sys_getpriority, sys_mmap, sys_read, sys_exit, sys_kill, sys_brk, sys_execve, sys_kill, sys_times, and sys_nice.

Next, we trained our system call classifier by using system call statistics collected by running 200 randomly selected normal, and 180 malware apps previously discovered (chosen from the android malware genome dataset [6]). The chosen malware families and the categories of normal apps are shown in Tables 1 and 2. We fed the collected system call statistics to the J.48 [7] and Random forest [8] classifiers in the WEKA tool and conducted a ten-fold cross validation experiment. The classification results were tabulated in Table 3. It shows that J48 classifier achieves a 91.6% accuracy rate while the Random forest achieves a 91.6% accuracy rate. The trained classifier identifies five of the system calls as important attributes that can be used to distinguish between normal and malware apps: sys_read, sys_mmap, sys_access, sys_brk, and sys_clone. For J.48 classifier, 15 malwares are classified as normal while 17 normal apps are classified as malwares. 7 of the 15 mislabeled malwares can be classified as malwares via the network monitoring tool which we will discuss in the next Sect. 3 of the malwares do not produce any network traffic.

Table 1. Top 10 of chosen malware families

Malware family	#	Malware family	#
droidkungfu	26	basebridge	21
anserverbot	18	Geinimi	14
droiddream	13	droiddreamlight	13
GoldDream	12	Pjapps	12
adrd	7	Zsone	6

Table 2. Chosen normal application categories

App category	# of apps	App category	# of apps
Book	11	Photograph	9
Game	10	Business	8
Travel	11	Shopping	10
Social	9	Finance	9
Tools	10	Lifestyle	16
Entertainments	10	Personalize	1
Communications	13	Weather	3
Education	3	News	12
Productivity	10	Sports	10
Comic	10	Medical	10
Transportation	10	Health	10

Table 3. System call classifier results

Algorithm	Correctly classified	Incorrectly classified	Confusion matrix
J48	348/91.58%	32/8.42%	m n
			165 15 m
			17 183 n
Randomforest	348/91.58%	32/8.42%	m n
			168 12 m
			20 180 n

2.2 Network Traffic Monitor

Next, we describe the second component of our dual defense protection framework, namely the network traffic monitor. Any user that uses our dual protection framework will install a network traffic monitoring tool. Our traffic monitoring tool has a set of triggering rules that look for suspicious application behaviors e.g. high read and mmap system call statistics. In current implementation, our network traffic monitoring tool was built using the idea we borrowed from [9]. When an application writes to a socket's output stream, the buffer is sent to the sendStream() method within the OSNetworkSystem core library. We modified sendStream so that if the buffer is tainted by data that should not be sent to its intended destination, we trigger the network capture tool (tcpdump [10]) available in the system.

When a user runs an application, and one of the triggering rules fires, then it will trigger a tcpdump capture action to log network traffic for a configurable period (e.g. 5 or 10 min). Once the logging completes, our program will parse the pcap files, extract TCP/IP flows, and analyze their statistics. The statistics will be recorded in an arff file. This arff file will be sent to a remote verification server which then invokes a J.48 based WEKA network classifier which has been previously trained to determine if the traffic characteristics are similar to those obtained from previously seen malwares. If the classifier output indicates that this application is a potential malware, then the verification server will remove this application and the user will be notified of the suspicious nature of this application. Our network traffic monitoring tool also maintains a whitelist such that all apps that have gone through such network testing do not undergo such testing again when they are run subsequently unless new versions of the apps are installed (Fig. 5).

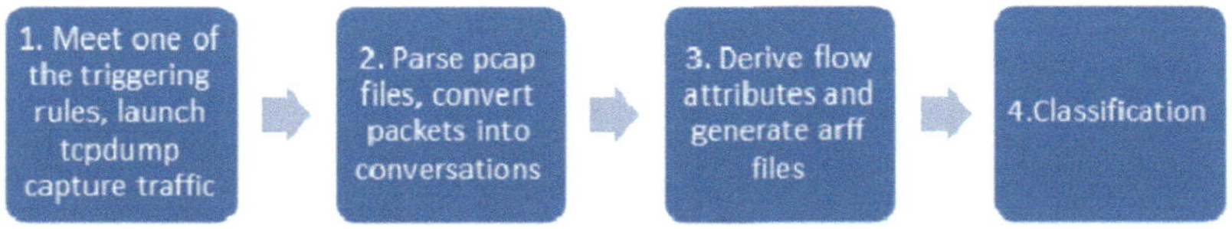

Fig. 5. Network traffic monitoring procedure

2.2.1 Training Our NTM Classifier

To train our NTM classifier, we installed 49 malware and 60 normal apps, and collected network traffic generated when each of these apps was run. The 49 malware apps were chosen from 22 malware families as shown in Table 4. Then, we extract TCP/IP conversations from these collected traces. Each of the TCP/IP conversation is referred to as an instance. We remove those conversations with known legitimate advertisement websites. We manually labelled each conversation as malware or normal instances based on whether the conversation is with a blacklisted website or carries sensitive information e.g. IMEI etc. There are 90 malware instances and 373 normal instances. Then, we extracted the following 9 statistics for each instance, namely the average and standard deviation of the number of sent/received packets, the average and the standard deviation of the number of bytes sent/received and the average TCP/IP session duration. We also randomly selected 60 normal apps e.g. Facebook, Twitter, Gmail, Tencent. qq, MSN, Google Play, Ebay, Wiki, Amazon etc, and analyzed the TCP/IP conversations (instances) generated by running these apps too. Then, we fed these statistics with the appropriate manually generated labels (malware or normal) to a J.48 and Random forest classifiers in WEKA tool.

Next, we randomly selected 50 malware apps and 70 normal apps that were not in the training set, collected their network traffic, remove the ad libraries network traffic, and extract the relevant statistics related to the TCP/IP conversations seen in these network traffic. The collected network traces consist of 571 TCP/IP conversations (instances), out of which 55 are malware instances (accessing blacklisted web sites or sending sensitive information) and 516 are normal instances. Then, we fed the statistics of each TCP/IP conversation as a test instance to our trained J.48 and Random forest network traffic classifiers.

Table 4. Malware families used in traffic classifier training

Malware family	#	Malware family	#
adrd	3	golddream	3
anserverbot	3	gingermaster	1
asroot	1	geinimi	8
basebridge	4	Gone60	1
beanbot	1	hippoSMS	1
bgserv	2	jsmshider	2
ddlight	2	MMMarketPay	1
droidkungfu	4	pjapps	2
droidkungfuapp	1	plankton	3
droidcoupon	1	zhash	1
ggtracker	1	other	3

The test results showed that our trained J.48 and Random forest network traffic classifiers produce an accuracy rate of 91.6% and 96.7% respectively. 21 (11) malware instances are labeled as normal while 27 (8) normal instances are labeled as "malware" by J48 and Random forest classifiers respectively. Since a malware application may generate both "malware" and "normal" instances, we studied the results more carefully and observed that 33 (41) of the malware apps can be identified as "malware" if we use the rule that an application is a malware as long as 25% of the traffic instances that it generates are classified as "malware" by either the J.48 or the Random forest classifier. With such a rule, 33 (41) of our malware apps will be labeled as "malwares" and 69 (70) of normal apps will be labeled as "normal" by our J.48 (Random forest) classifier. Overall, Random forest classifier produces better results.

2.2.2 Heuristic Rules for Triggering Network Monitoring Tool

Now, we discuss how the two components, namely SCM and NTM are used simultaneously in our framework. Any new application will first be tested using the SCM classifier. If the SCM classifier deems this to be a normal application, the new application will be released to the Android Market for user downloads. Any application that is labeled as malware by SCM can be further verified using the NTM tool by the verification server. If the NTM tool indicates the new application is harmless, it will still be released. When a user who installs our NTM tool first runs a new application, our NTM tool will also check the triggering rules to see if the NTM tool needs to be triggered. If so, the NTM tool collects network traffic statistics and sends them to the AV server. The NTM classifier will determine if this application is a potential malware. If it is, then the application will be removed from the Android markets.

Since we want to minimize the number of apps that will trigger the NTM tool, we define certain heuristic rules after we analyzed different categories of normal apps. Our heuristic rules focus on 4 system calls, namely, access, mmap, read and brk for we found that known malwares seem to have different statistics in these 4 types of system calls. In Table 5, we show the different categories which our 50 normal apps fall under, e.g. book, game, travel, social network, tools, entertainments etc. and the chosen conditions. Our NTM will be triggered if((access > Th1)||((mmap > Th2) && (read < Th3))||(brk > Th4) and internet permission is requested by the application. Currently, the thresholds were chosen by studying the statistics collected from limited number of normal apps. However, we find triggering using such rule not to be too effective. Thus, instead, we use the following rule (referred to as Rule 2) for apps labeled as "normal" to trigger network traffic monitoring: an application that requests for INTERNET permission and READ_PHONE_STATE and either (SEND_SMS or RECEIVE_SMS or WRITE_HD).

3 Evaluation

In this section, we describe how we evaluate our framework. We developed our prototyped system call and traffic monitoring tools. Then, we randomly select 70 normal and 50 malware apps. The malware apps are chosen from the 16 of the 49 families of malware samples provided by the authors in [6] as well as 3 new ones which appear in July 2012. In particular, we included the four malware families, i.e., BaseBridge, DroidKungFuUpdate, AnserverBot, and Plankton, that adopt update attacks [6]. The 70 normal apps are chosen from 17 categories e.g. comics, sports, news, lifestyle, medical, communication, transportation, health, game, book, education, entertainment, etc.

We run each of these 120 apps, and collect system call statistics using our system call monitoring tool. Then, we test them using our trained J.48 and Random forest WEKA-based classifiers. The results we obtained are tabulated in Table 6. We can achieve an accuracy rate of 90.8%/96.7% for J.48 and Random forest classifier respectively using statistics from the 15 system calls that we describe in Sect. 2. Eight of the malwares are labeled as normal and three of the normal apps are labeled as malwares.

Next, we collect network traffic traces from apps that are labeled as "malwares" from Phase 1 which satisfy the triggering rule Rule 1 described in Sect. 2, then we can identify these mislabeled normal apps as "normal". In addition, we collect network traffic from apps labeled as "normal" which satisfy rules in Sect. 2 of the misclassified malwares from the J.48/Random forest system call classifiers qualified, and the NTM classifier was able to infer that they are indeed malicious apps. Thus, our final accuracy rate is 94.2% (99.2%) are classified correctly after the 2 pronged approach) using J.48 and Random forest classifiers respectively.

Table 5. Results from our system call classifier

Algorithm	Correctly classified	Incorrectly classified	Confusion matrix
J48	109/90.83%	11/9.17%	m n
			42 8 m
			3 67 n
Randomforest	116/96.67%	4/3.33%	m n
			46 4 m
			0 70 n

4 Conclusion

In this paper, we have presented a smartphone dual defense protection framework that can be used to detect malware application in Android-based markets. Our framework consists of two major components, namely (a) system call monitoring tool which is used in Phase 1 to evaluate new application for potential malwares, (b) network monitoring tool which is used to identify new apps that are false negatives of the system call monitoring tool. Our preliminary results indicate that our framework indeed provide good detection accuracy. Random forest classifier produces better results than J.48 classifier. We intend to evaluate our framework using more malware samples from [6] and normal apps. We also intend to determine the appropriate rules which minimize the triggering of network traffic monitoring tool for normal apps. In addition, we are looking into whether adding Android system call behaviors of malware apps can further improve our system. Last but not least, we intend to deploy a cloud-based system that mimics a third-party Android market that provides the verification service described in our framework and advertise this in Android markets to gather more evidence about the effectiveness of our approach.

References

1. Mobile Malware. http://www.forbes.com/sites/gordonkelly-/2014/03/24/report-97-of-mobile-malware-is-on-android-this-is-the-easy-way-you-stay-safe/
2. Su, X., Zhang, D., Li, W., Zhao, K.: A deep learning approach to Android malware feature learning and detection. In: Proceedings of the 15th IEEE International Conference on Trust, Security and Privacy in Computing and Communications, Trustcom, pp. 244–251 (2016)

3. Wang, X., Wang, W., He, Y., Liu, J., Han, Z., Zhang, X.: Characterizing Android apps' behavior for effective detection of malapps at large scale. Future Gener. Comput. Syst. **75**, 30–45 (2017)
4. Zhu, H., You, Z., Zhu, Z., Shi, W., Chen, X., Cheng, L.: DroidDet: effective and robust detection of android malware using static analysis along with rotation forest model. Neurocomputing **272**, 1–9 (2017)
5. Aung, Z., Zaw, W.: Permission-based Android malware detection. Int. J. Sci. Technol. Res. **2**(3), 228–234 (2013)
6. Zhou, Y., Jiang, X.: Dissecting Android malware: characterizing and evolution. In: IEEE Symposium on Security and Privacy (2012)
7. Decision tree. https://en.wikipedia.org/wiki/Decision_tree
8. Random forest. https://en.wikipedia.org/wiki/Random_forest
9. Hornyack, P., Han, S., Jung, J., Schechter, S., Wetherall, D.: These aren't the droids you're looking for: retrofitting Android to protect data. In: 18th ACM Conference on Computer and Communications Security, CCS, pp. 639–652 (2011)
10. TCPDUMP. http://www.tcpdump.org/

A Software Quality Testing Evaluation Method and Its Power Grid Application

Keshou Wu[1], Xingyu Gao[2], Zhenyu Chen[3(✉)], and Jin Wang[4]

[1] Engineering Research Center for Software Testing and Evaluation of Fujian Province, Xiamen University of Technology, Xiamen, China
[2] Institute of Microelectronics, Chinese Academy of Sciences, Beijing, China
[3] China Electric Power Research Institute, Beijing, China
czy9907@gmail.com
[4] School of Computer and Communication Engineering, Changsha University of Science and Technology, Changsha, China

Abstract. Due to the software quality demand and features of power grid, this paper presents a methodology design for software quality testing evaluation tree of power grid dispatching automation system. This design includes three quality levels, including quality domain, quality category and quality set, it also combines different software quality testing measurement to enrich quality contents in each level. Therefore, we can see that, this hierarchical structure of quality testing evaluation tree can reach its own destination and achieve its own featured properties such as well performance, stable running ability, strong compatibility and so forth, especially for power grid dispatching automation system.

Keywords: Software quality · Methodology design
Testing evaluation · Hierarchical structure · Power grid

1 Introduction

As the smart grid enters the comprehensive construction phase, a large number of new power grid automation systems and new equipments will be put into operation one after another. It is necessary to conduct comprehensive and detailed evaluation and testing to ensure that new systems and new equipments can meet the requirements of standardization, also need they run stably and reliably at the same time. In order to better complete the software testing of such systems, it needs to systematically evaluate and detect them. The current testing methods have some drawbacks and issues in terms of high cost of detecting environmental conditions, high participation of detection personnel, and low efficiency of detection resources, especially for concurrent user testing, large data volume testing, long-term reliability testing, compatibility testing and so on [1,2]. It is difficult for the development unit of power dispatching automation system to perform the test itself, because that the requirements of difficulty of manual completion, the

J. Abawajy et al. (Eds.): ATCI 2018, AISC 842, pp. 400–404, 2019.
https://doi.org/10.1007/978-3-319-98776-7_42

complexity of detection environment, and the performance of concurrent processing are seriously high [3,4]. Therefore, it is urgent to introduce new software quality testing evaluation technologies, so as to enhance the standardized automatic detection capability of the power grid automation system, also reduce the input of manpower and material resources required for software detection, which can benefit for software quality related researches [5–12] and applications [13–19].

Some organizations have conducted a variety of researches on cloud testing and provided cloud testing services. Cloud Testing is a cloud test platform based on UI automation test. It supports multiple operating systems and multiple browser platforms. The common user employs Selenium to write automated test scripts locally and upload them to their website [20]. Keynote allows users to record scripts on the Kite browser, upload them and then run tests in different available locations at Keynote to see the results [1]. While Keynote's Kite has few nodes available, and it uses specialized automated testing tools; however this may have an impact on the results of the testing. SOASTA provides functional testing and performance testing to provide a cloud testing environment for web applications. There is not much researches on software such as C/S architecture [2]. Japanese scholar T. Hanawa studies the distributed system test based on cloud computing environment service for the inconvenience caused by the complication and scalability of distributed systems, and proposes the D-Cloud model [3]. D-Cloud not only automates system configuration, automates the execution of test cases, but also runs multiple test processes simultaneously with the flexibility to mimic hardware errors. However, the author did not implement synchronization detection for distributed systems. At the same time, IBM provides cloud-based test management solutions for some enterprises [4,21]. These researches and development of cloud testing simplify the test management process to a certain extent, improve the level of test automation and enhance the test efficiency. However, the detection of the dispatching automation system has special requirements on the environment, business process and information security. Therefore, as a key part of the software life cycle, it needs to be comprehensively and thoroughly evaluated and tested to ensure that the new system and new equipments can meet the requirements of standardization and stability, and better complete the system software testing and evaluation for the power grid of dispatching and control.

2 The Methodology Design of Software Quality Testing Evaluation

Considering abovementioned features of software quality testing evaluation in power gird, we present our methodology design structure for dispatching automation system as shown in Fig. 1. This hierarchical structure is the quality testing evaluation tree of power grid, which contains three different levels, i.e. quality domain, quality category and quality set. Take dispatching automation for an example, there are seven quality categories, including functionality, reliability, ease of use, efficiency, maintainability, compatibility. Further, for such quality categories, these sub-tree nodes are comprised of their own quality set contents.

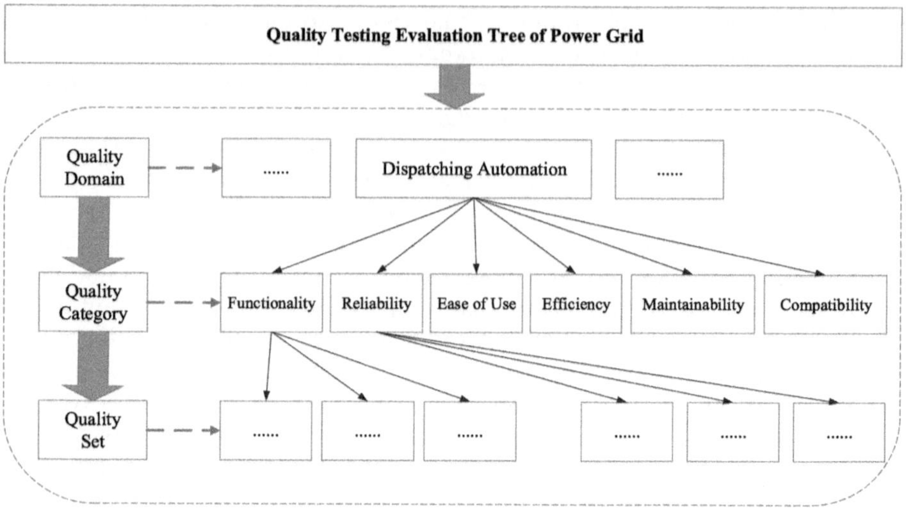

Fig. 1. The hierarchical structure for methodology design of software quality testing evaluation.

(1) Functionality: The ability of a software product to meet the requirements of explicit and implicit requirements when the software is used under specified conditions.
(2) Reliability: The ability of a software product to maintain a specified level of performance when used under specified conditions.
(3) Ease of use: The ability of a software product to be understood, learned, used, and attracted to users when used under specified conditions.
(4) Efficiency: The ability of a software product to provide adequate performance, under specified conditions, relative to the amount of resources used.
(5) Maintainability: The ability of a software product to be modified. Modifications may include improvements, or changes in the software's adaptation to the environment, requirements, and functional specifications.
(6) Compatibility: The ability of a software product to move from one environment to another.

Besides, take functionality for an example, this ability includes suitability, accuracy, interoperability, confidentiality security and functional compliance. In detail we can see that, (a) Suitability: The ability of a software product to provide a suitable set of functions with specified tasks and user goals. (b) Accuracy: The ability of a software product to have the correct or consistent results and effects of the required accuracy. (c) Interoperability: The ability of a software product to interact with one or more defined systems. (d) Confidentiality security: The ability of a software product to protect information and data so that unauthorized persons or systems cannot read or modify such information and data, but does not deny access to authorized personnel or systems. (e) Functional compliance: Refers to the ability of software products to adhere to standards, conventions or regulations and similar regulations related to functionality.

3 Conclusion

In this paper, considering features of software quality testing evaluation in power gird, a software quality evaluation method has been put forward. To meet the demands of dispatching automation system's software testing, our designed hierarchical structure of quality testing evaluation tree aims to achieve its own featured properties in quality domain of power gird, such as well performance, stable running ability, strong compatibility and so forth. Thus, it is of great significance to improve the software quality of dispatching control system, and to ensure the safe, stable and reliable operation of the entire of power gird.

Acknowledgment. This work was supported by National Nature Science Foundation of China (Nos. 61702491, 61772454, 61811530332, 61502404), Distinguished Young Scholars Foundation of Fujian Educational Committee (No. DYS201707), Xiamen Science and Technology Program (No. 3502Z20183059).

References

1. Banzai, T., Koizumi, H., Kanbayashi, R., et al.: D-cloud: design of a software testing environment for reliable distributed system using cloud computing technology. In: Proceedings of the 2nd International Symposiums on Cloud Computing in Conjunction with CCCGrid, May 2010
2. Tilley, S.: 1st International Workshop on Software Testing in the Cloud, STITC (2009)
3. Candea, G., Bucur, S., Zamfir, C.: Automated software testing as a service. ACM (2010)
4. Romano, B.L., Braga, G., de Campos, H.F., et al.: Software testing for web-applications non-functional requirements. IEEE (2009)
5. Li, D., Chen, Z., Deng, Z.: A wide area service oriented architecture design for plug and play of power grid equipment. Proc. Comput. Sci. **129**, 353–357 (2018)
6. Chen, Z., Li, D., Deng, Z., et al.: The application of power grid equipment plug and play based on wide area SOA. In: Proceedings of 2nd IEEE International Conference on Energy Internet, pp. 19–23. IEEE, Beijing (2018)
7. Chen, Z., Chen, Y., Gao, X., et al.: Unobtrusive sensing incremental social contexts using fuzzy class incremental learning. In: Proceedings of International Conference on Data Mining, pp. 71–80. IEEE, USA (2015)
8. Chen, Z., Chen, Y., Wang, S., et al.: Inferring social contextual behavior from Bluetooth traces. In: Proceedings of the 2013 ACM Conference on Pervasive and Ubiquitous Computing, pp. 267–270. ACM, USA (2013)
9. Gao, X., Chen, Z., Tang, S.: Adaptive weighted imbalance learning with application to abnormal activity recognition. Neurocomputing **173**, 1927–1935 (2016)
10. Gao, X., Hoi, S.C., Zhang, Y., et al.: SOML: sparse online metric learning with application to image retrieval. In: Proceedings of AAAI, USA, pp. 1206–1212 (2014)
11. Gao, X., Hoi, S.C., Zhang, Y., et al.: Sparse online learning of image similarity. ACM Trans. Intell. Syst. Technol. (TIST) **8**(5), Article no. 64 (2017)

12. Xiang, Z., Chen, Z., Gao, X., et al.: Solving large-scale TSP using a fast wedging insertion partitioning approach. Math. Probl. Eng. **2015**, 1–9 (2015)
13. Zhang, H., Yuan, J., Gao, X., et al.: Boosting cross-media retrieval via visual-auditory feature analysis and relevance feedback. In: Proceedings of the 22nd ACM International Conference on Multimedia, pp. 953–956. ACM (2014)
14. Chen, Z., Chen, Y., Hu, L., et al.: ContextSense: unobtrusive discovery of incremental social context using dynamic Bluetooth data. In: Proceedings of the 2014 ACM Conference on Pervasive and Ubiquitous Computing, pp. 23–26. ACM, USA (2014)
15. Wang, R., Chen, F., Chen, Z., et al.: StudentLife: assessing mental health, academic performance and behavioral trends of college students using smartphones. In: Proceedings of the 2014 ACM Conference on Pervasive and Ubiquitous Computing, pp. 3–14. ACM, USA (2014)
16. Chen, Z., Wang, S., Shen, Z., et al.: Online sequential ELM based transfer learning for transportation mode recognition. In: Proceedings of the 6th IEEE International Conference on Cybernetics and Intelligent Systems, pp. 78–83. ACM, USA (2014)
17. Chen, Z., Lin, M., Chen, F., et al.: Unobtrusive sleep monitoring using smartphones. In: Proceedings of the 7th International ICST Conference on Pervasive Computing Technologies for Healthcare, pp. 145–152. ICST, Venice (2013)
18. Chen, Z., Wang, S., Chen, Y., et al.: InferLoc: calibration free based location inference for temporal and spatial fine-granularity magnitude. In: Proceedings of the 10th IEEE International Conference on Embedded and Ubiquitous Computing, pp. 453–460. IEEE, Paphos (2012)
19. Chen, Y., Chen, Z., Liu, J., et al.: Surrounding context and episode awareness using dynamic Bluetooth data. In: Proceedings of the 2012 ACM Conference on Pervasive and Ubiquitous Computing, pp. 629–630. ACM, USA (2012)
20. SOASTA. http://www-alldaytest.com/index.do?lan=:cn
21. Leon, D., Masri, W., Podgurski, A.: An empirical evaluation of test case filtering techniques based on exercising complex information flows. In: Proceedings of the 27th International Conference on Software Engineering, pp. 412–421 (2005)

An Index Display Software Evaluation Architecture for Power Grid Application on Big Data Platform

Keshou Wu[1], Xingyu Gao[2], Zhenyu Chen[3](✉), and Jin Wang[4]

[1] Engineering Research Center for Software Testing and Evaluation of Fujian Province, Xiamen University of Technology, Xiamen, China
[2] Institute of Microelectronics, Chinese Academy of Sciences, Beijing, China
[3] China Electric Power Research Institute, Beijing, China
czy9907@gmail.com
[4] School of Computer and Communication Engineering, Changsha University of Science and Technology, Changsha, China

Abstract. For the reason that existing index display systems can not reach comprehensive and friendly reflection and measurement for current power grid operation status, based on power grid dispatching and control big data platform, this paper studies the features of multiple management systems for power grid dispatching and control, then construct an index display software evaluation architecture using the layered structure. At last, over this built structure, the architecture is able to achieve multi-dimensional dynamic visualization display of various indices, which makes it more convenient to reach good effects for analysis and presentation of the power gird's operation status, and enhance the management level of power grid dispatching and control.

Keywords: Index display · Power grid · Big data · Visual editing Architecture design

1 Introduction

With the construction of smart grids, the data sources, scale, and types of power grid for dispatching and control are rapidly expanding and enriching. Due to the inconsistency of these data structures and codes, there are a large number of interfaces among various systems, and even the emergence of "information islands", many valuable data of power grid operation and management have not been effectively integrated and utilized. It is necessary to centralize the management of power grid operation and management data, comprehensively analyze and mine potential applications in power grid operation and management data [1,2]. At present, some power dispatching and control centers of State Grid have carried out researches on key technologies for big data, proposed and developed the system architecture of power dispatching and control big data

J. Abawajy et al. (Eds.): ATCI 2018, AISC 842, pp. 405–409, 2019.
https://doi.org/10.1007/978-3-319-98776-7_43

platform [3–5], further realized the unified management of multi-source heterogeneous system data, such as Energy Management system (EMS), Operation Management System (OMS), Production Management System (PMS), etc., which provides a good data foundation for big data related researches [6–14] and applications [15–21].

Under this situation, to meet the needs of comprehensive analysis for personnel at all levels in the power grid of dispatching and control center, this paper studies index display building based on power grid big data analysis and mining, integrates existing indices [22–24] to establish the entire index software evaluation architecture across system and professional departments. Thus, this paper presents an index display software evaluation architecture based on the big data platform of power grid dispatching and control, employs the big data visualization analysis technology to realize the multi-dimensional dynamic interactive display of new indices, which can help the dispatcher and other professionals to utilize more advanced and more intuitive tools and means to achieve timely and effective monitoring and analysis of power grid.

2 Architecture Design of Index Display Software Evaluation Application

The architecture of the index display software evaluation as shown in Fig. 1. This hierarchical structure is divided into different layers, i.e. data storage layer, connection data layer, visual editing layer, scheduling management layer and index display layer, where the bottom layer is based on the storage layer and interface layer provided by the big data platform.

(1) The data storage layer contains HBase, Hive, and real-time analysis databases [25–28]. HBase and Hive mainly store the power grid dimension information, which service for various dimensions of the power grid information index display. The real-time analysis database mainly stores the calculation result data information of the indices. Above the storage layer, there is the interface layer, which provides data access interface for interaction with the database.
(2) The connection data layer contains three sub-modules: connection data, data query and storage process. The connection data module connects the Hive, the real-time database through the defined query conditions, and creates the index data source. The data query module queries the filtered data through the connected data source, then generate the final subject form filtered by the index theme. A stored procedure module is a set of SQL statements to accomplish a specific function.
(3) The visual editing layer includes interface management, chart components, filter components and other components. This layer mainly creates to form the overall layout through graphical creation, modification, and deletion of various graphical components.

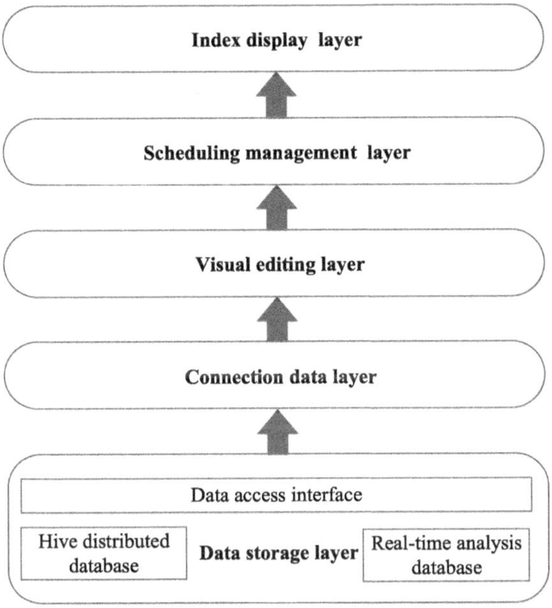

Fig. 1. The architecture of index display software evaluation application.

(4) The scheduling management layer includes job management, task management, and manual execution. This layer is mainly based on different types of tasks to achieve the automatic update of the timing for different topic indices interface, synchronous query data, incremental synchronization data, and automatic export display interface and so forth.
(5) The index display layer consists of the display release management and indicator display interface. The presentation and release interface provides the analysis report publishing management function, so that the index display interface implements multi-dimensional dynamic analysis of all the indicators that have been calculated based on subject areas, topic sets, and themes.

3 Conclusion

Based on the big data platform of power grid dispatching and control, this paper faces with the business characteristics of multiple management systems for power grid dispatching and control related data, employs the big data analysis and mining manners to establish a cross-system multi-professional index display software evaluation architecture. This architecture utilizes visualization analysis technology to achieve multi-dimensional dynamic interactive display of various indices, so as to achieve a good analysis and display effectiveness of the power grid's operation and management.

Acknowledgment. This work was supported by National Nature Science Foundation of China (Nos. 61702491, 61772454, 61811530332, 61502404), Distinguished Young Scholars Foundation of Fujian Educational Committee (No. DYS201707), Xiamen Science and Technology Program (No. 3502Z20183059).

References

1. China Electrical Engineering Society Information Specialized Committee: China Electric Power Big Data Development White Paper. China Electric Power Press, Beijing (2013)
2. Su, W., Li, L., Yang, J.: Research on data integration and management system of power grid. China Electr. Power Educ. **S3**, 369–370 (2008)
3. Zhu, C., Wang, J., Deng, C.: Research and design of power big data platform. Electr. Power Inf. Commun. Technol. **13**(6), 1–7 (2015)
4. Yue, Y., Zhang, X., Gao, Y.: Technology system of electric power big data based on Hadoop. Power Energy **1**, 16–20 (2015)
5. Wu, K., Liu, W., Li, Y., et al.: Cloud-computing based power big data analysis technology and its application. Electr. Power **48**(2) 111–116, 127 (2015)
6. Zhao, Y., Liu, H.: Research on big data technique application in electric power industry. Telecommun. Sci. **30**(1), 57–62 (2014)
7. Li, D., Chen, Z., Deng, Z., et al.: A wide area service oriented architecture design for plug and play of power grid equipment. Procedia Comput. Sci. **129**, 353–357 (2018)
8. Chen, Z., Li, D., Deng, Z., et al.: The application of power grid equipment plug and play based on wide area SOA. In: Proceedings of 2nd IEEE International Conference on Energy Internet, pp. 19–23. IEEE, Beijing (2018)
9. Chen, Z., Chen, Y., Gao, X., et al.: Unobtrusive sensing incremental social contexts using fuzzy class incremental learning. In: Proceedings of International Conference on Data Mining, pp. 71–80. IEEE, USA (2015)
10. Chen, Z., Chen, Y., Wang, S., et al.: Inferring social contextual behavior from Bluetooth traces. In: Proceedings of the 2013 ACM Conference on Pervasive and Ubiquitous Computing, pp. 267–270. ACM, USA (2013)
11. Gao, X., Chen, Z., Tang, S., et al.: Adaptive weighted imbalance learning with application to abnormal activity recognition. Neurocomputing **173**, 1927–1935 (2016)
12. Gao, X., Hoi, S.C., Zhang, Y., et al.: SOML: sparse online metric learning with application to image retrieval. In: Proceedings of AAAI, pp. 1206–1212, USA (2014)
13. Gao, X., Hoi, S.C., Zhang, Y., et al.: Sparse online learning of image similarity. ACM Trans. Intell. Syst. Technol. (TIST) **8**(5), Article no. 64 (2017)
14. Xiang, Z., Chen, Z., Gao, X., et al.: Solving large-scale TSP using a fast wedging insertion partitioning approach. Math. Probl. Eng. **2015**, 1–9 (2015)
15. Zhang, H., Yuan, J., Gao, X., et al.: Boosting cross-media retrieval via visual-auditory feature analysis and relevance feedback. In: Proceedings of the 22nd ACM International Conference on Multimedia, pp. 953–956. ACM (2014)
16. Chen, Z., Chen, Y., Hu, L., et al.: ContextSense: unobtrusive discovery of incremental social context using dynamic Bluetooth data. In: Proceedings of the 2014 ACM Conference on Pervasive and Ubiquitous Computing, pp. 23–26. ACM, USA (2014)

17. Wang, R., Chen, F., Chen, Z., et al.: StudentLife: assessing mental health, academic performance and behavioral trends of college students using smartphones. In: Proceedings of the 2014 ACM Conference on Pervasive and Ubiquitous Computing, pp. 3–14. ACM, USA (2014)
18. Chen, Z., Wang, S., Shen, Z., et al.: Online sequential ELM based transfer learning for transportation mode recognition. In: Proceedings of the 6th IEEE International Conference on Cybernetics and Intelligent Systems, pp. 78–83. ACM, USA (2014)
19. Chen, Z., Lin, M., Chen, F., et al.: Unobtrusive sleep monitoring using smartphones. In: Proceedings of the 7th International ICST Conference on Pervasive Computing Technologies for Healthcare, pp. 145–152. ICST, Venice, Italy (2013)
20. Chen, Z., Wang, S., Chen, Y., et al.: InferLoc: calibration free based location inference for temporal and spatial fine-granularity magnitude. In: Proceedings of the 10th IEEE International Conference on Embedded and Ubiquitous Computing, pp. 453–460. IEEE, Paphos, Cyprus (2012)
21. Chen, Y., Chen, Z., Liu, J., et al.: Surrounding context and episode awareness using dynamic Bluetooth data. In: Proceedings of the 2012 ACM Conference on Pervasive and Ubiquitous Computing, pp. 629–630. ACM, USA (2012)
22. Ni, J., He, G., Shen, C., et al.: A review of smart grid in America. Autom. Electr. Power Syst. **34**(8), 9–13 (2010)
23. Wang, Z., Li, H., Li, J., et al.: Assessment index system for smart grids. Power Syst. Technol. **33**(17), 14–18 (2009)
24. Tan, W., He, G., Liu, F., et al.: Elementary research on smart grid's low-carbon index system. Autom. Electr. Power Syst. **34**(17), 1–5 (2010)
25. Dean, J., Ghemawat, S.: MapReduce: simplified data processing on large clusters. In: Proceedings of the 6th Symposium on Operating System Design and Implementation, pp. 137–150 (2004)
26. Ghemawat, S., Gobioff, H., Leung, S.T.: The Google file system. In: Proceedings of the 19th ACM Symposium on Operating System Principles, pp. 29–43, New York (2003)
27. Chang, F., Dean, J., Ghemawat, S., et al.: Bigtable: a distributed storage system for structured data. In: Proceedings of the 7th Symposium on Operating System Design and Implementation, pp. 205–218 (2006)
28. Apache. Apache Hadoop coreEB/OL

An Uniform Access Method of Heterogeneous Big Data with Power Grid Application

Zhenyu Chen[1], Dapeng Li[1], Lixin Li[1], Ruili Ye[1], Xingyu Gao[2](✉), Fangchun Di[1], Lingxu Guo[3], Jie Zhang[3], and Zhengping Chen[4]

[1] Beijing Key Lab of Research and System Evaluation of Power Dispatching Automation Technology, China Electric Power Research Institute, Beijing, China
[2] Institute of Microelectronics, Chinese Academy of Sciences, Beijing, China
gxy9910@gmail.com
[3] State Grid Tianjin Electric Power Company, Tianjin, China
[4] State Grid Fujian Electric Power Company, Fuzhou, Fujian, China

Abstract. In order to solve the problems faced with uniform access to heterogeneous power grid big data in the smart dispatching control system, we study the uniform access technologies for structured and unstructured data, multiple-database transparent access technologies for differential interfaces. As for various professionals' diverse demands on the power grid database, a framework for the uniform access of massive heterogeneous data to typical businesses of the power grid has been proposed. The solution has the capability of effective support various requirements for various power grid applications to uniform access, and provides fast and accurate data information for upper layer application, analysis, and decision-making of the power grid data platform of dispatching and control.

Keywords: Power grid · Big data · Dispatching and control system Structured and unstructured Data · Uniform access

1 Introduction

With the large-scale construction of UHV AC/DC hybrid power grids, and the deepening and advancement of the "big operation" construction, the service scope of power grid dispatching and control systems has rapidly expanded, and the data source for utilization of smart detection equipment has quickly increased, then diverse data types, data scale and processing pressure have been augmented dramatically. The real-time data acquisition, transmission, storage, and access technologies for the power grid, as the key supporting technology for power grid dispatching and control, have been paid important attention on both of academic and industry communities in recent years [1], and also benefit from intelligent Computer Science researches [5–12] and applications [13–19].

J. Abawajy et al. (Eds.): ATCI 2018, AISC 842, pp. 410–416, 2019.
https://doi.org/10.1007/978-3-319-98776-7_44

For state-of-the-art researches and systems, United States Electric Power Research Institute (EPRI) has began to conduct the first-batch research on smart power grid, and proposed that the smart power grid dispatching and control system should have features such as self-healing, interaction, optimization, prediction, collaboration, integration, security and so on [2]. France RTE power grid operation technical support system has the characteristics of uniform construction, uniform operation and management. Due to the use of a completely uniform technology system, it is possible for RTE to achieve the mode of nationwide centralized construction and maintenance, then the cost advantages of development, engineering and maintenance are very obvious [3]. In terms of architecture, search engine giant Google has dominated the world for its powerful clustering technology. At present, there are tens of thousands of nodes in Google's cluster, which is very large and rare all over the world. The distinctive feature of Google's cluster system is to build a stable, high-performance, highly-available, and scalable system using low-cost mature technologies, to successfully solve a large amount of computation, large storage, and high real-time application requirements [4].

Under the environment of smart power grid, the requirements for uniform access to massive multi-source heterogeneous data have been continuously improved in the monitoring, real-time control, and management businesses of power grid units at all levels. In the context of power grid dispatching and control cloud platform, this paper focuses on the uniform access problems of massive heterogeneous data in the power grid, studies the uniform access technology for both structured and unstructured data, also in order to meet the needs of different application businesses in the power grid, we propose a framework for uniform access to massive amounts of heterogeneous data by power grid's typical businesses.

2 Power Grid Big Data

Power grid big data comes from power generation, transmission, transformation, distribution, electricity, and other production and operation segments, which can be approximately divided into three aspects: power grid operation and equipment detection, power enterprise marketing, and power enterprise management [20].

Power grid big data has four characteristics: (1) Large volume of data: PB level, conventional dispatching automation system contains 100,000 collection points, and data collection point of power distribution system will reach tens of millions; (2) Various data types: Structured, semi-structured, and unstructured data such as real-time data, historical data, time series, text data, images, videos, etc.; (3) Low value density: Most collected data are normal data. There is only a very small amount of abnormal data, while these abnormal data is the most important foundation for dispatching and control, maintenance and detection, etc. (4) Fast processing speed: A large amount of data is analyzed in a few seconds or even a fraction of a second to support decision making.

3 Power Grid Data Components and Features

3.1 Data Components

The power grid structured data is stored in the relational database as two-dimensional data, while structured data has a variety of formats and standards. The source and format of typical power grid data is shown in Table 1. With the input into operation of smart detection/monitoring equipment, the proportion of unstructured data has gradually increased, and the growth rate of unstructured data exceeds structured data and semi-structured data. From the perspective of computer processing, on the one hand, unstructured data is more difficult to understand, unify, and standardize than structured data. On the other hand, multiple related database system sets form the heterogeneous database, whose data sharing and transparent access has yet to be resolved.

Table 1. Typical data source and format of power gird system.

Data source	Data format
Plant automation system	Structured data/unstructured data
Production management system	Structured data/unstructured data
Geographic information system	Semi-structured/unstructured data
Dispatching automation system	Structured data
Electricity information collection	Structured data
Load control system	Structured data
Load monitoring system	Structured data/Semi-structured data
Marketing business application system	Structured data
Power quality monitoring system	Structured data/Semi-structured data
Power grid weather information system	Unstructured data
95598 customer service system	Unstructured data
ERP system	Structured data
Regional socio-economic data	Structured data/unstructured data

3.2 Data Feature Modeling

For structured data, generally it has the normal feature modeling as well. While for unstructured data, in order to uniformly deal with different types of data and use descriptive information to unstructured data on the cloud platform, Li Wei et al. proposed the uniform management of unstructured data combined with the tetrahedron model [21–23].

As shown in Fig. 1, the tetrahedron model uses four facets to describe the basic attributes, semantic features, underlying features, and raw data of an unstructured data, also describes the internal relationships among these facets.

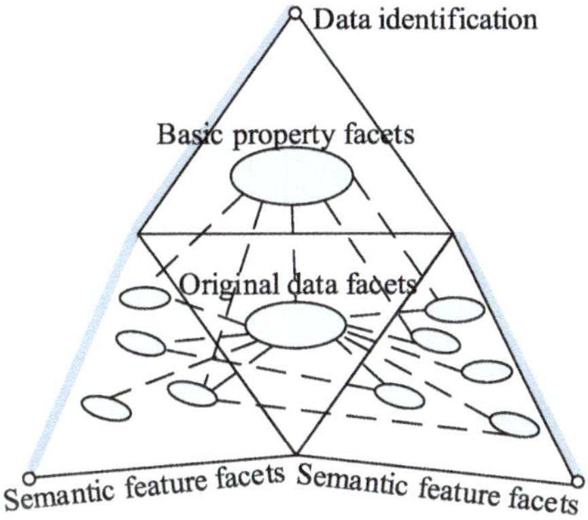

Fig. 1. The tetrahedral model diagram.

- Basic attribute: All unstructured data has their general attributes. These attributes do not involve the semantics of data, including name, type, creator, creation time, and so on.
- Semantic feature: The semantic attributes unique to unstructured data expressed in the text, including semantic elements such as the author's intentions of creation, data subject descriptions, and underlying feature meanings, etc.
- Underlying features: Unstructured data characteristics obtained through various special processing technologies (such as image, voice, video, etc.), such as color, texture, shape, etc. of the image data.
- Raw data: The original ecological file of unstructured data.

The tetrahedron model is used to model unstructured data and support the extension of semantic features, underlying features, etc., which can effectively manage data in a uniform manner and improve the precision and recall rate of uniform access.

4 The Uniform Access Architecture

The framework power grid's typical service architecture for massively heterogeneous data access is shown in Fig. 2.

Power grid big data comes from all aspects of production and operation, and serves various business needs of different units such as state-level, branch-level, province-level, region-level and county-level dispatching. For various types of business needs, with models, data, and calculations at the core, the same type of business is constructed physically or logically to form a unified business processing mode. The application of this processing mode can meet the demand for uniform access of the same type of business data, and at the same time meet the quick response requirements for large-scale power grid data access.

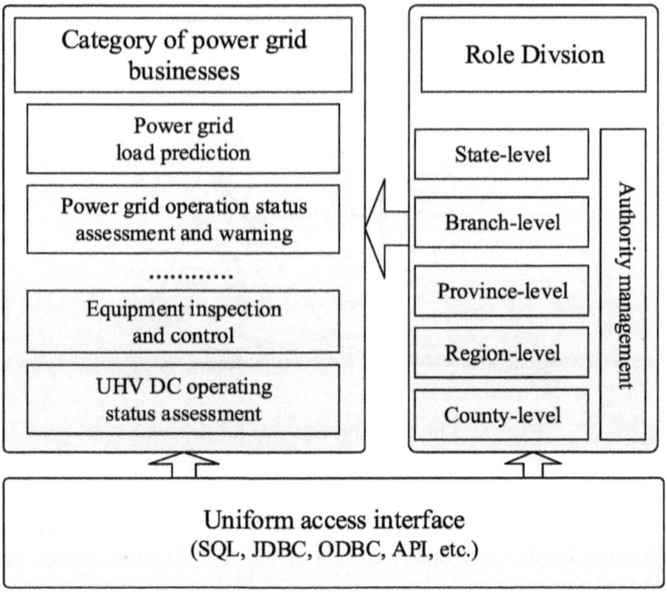

Fig. 2. The framework of power grid's typical service architecture for massively heterogeneous data access.

The typical businesses of power grid include power load forecasting (short-term, medium-term, long-term), grid operation status assessment and warning, equipment detection and control, evaluation and early warning of UHV DC system operation status.

5 Conclusion

This paper analyzes the characteristics of power grid big data in the background of smart power grids for large-scale heterogeneous data in the power grid dispatching support system, and studies uniform access technology for structured and unstructured data in power grid, then proposes typical business's uniform access architecture for massively heterogeneous data, which is able to have such solutions for efficient, uniform, and rapid access to multiple databases in the future of smart power grid dispatching and control system.

Acknowledgment. This work was supported by Science and Technology Program of State Grid Corporation of China (No. 5442DZ170019), National Nature Science Foundation of China (No. 61702491), and Science and Technology Innovation Program of China Electric Power Research Institute (No. 5242001700DZ).

References

1. Song, Y., Zhou, G., Zhu, Y.: Present status and challenges of big data processing in smart grid. Power Syst. Technol. **37**(4), 927–935 (2013)
2. Cleveland, F.: IntelliGrid architecture: power system functions and strategic vision. Utility Consulting International (2005)
3. Li, Z., Pang, B., Li, G., et al.: Development of unified european electricity market and its implications for China. Autom. Electr. Power Syst. **41**(24), 2–9 (2017)
4. Zhang, M., Xu, H., Wang, X., et al.: Google TensorFlow machine learning framework and applications. Microcomput. Appl. **36**(10), 58–60 (2017)
5. Li, D., Chen, Z., Deng, Z., et al.: A wide area service oriented architecture design for plug and play of power grid equipment. Procedia Comput. Sci. **129**, 353–357 (2018)
6. Chen, Z., Li, D., Deng, Z., et al.: The application of power grid equipment plug and play based on wide area SOA. In Proceedings of 2nd IEEE International Conference on Energy Internet, pp. 19–23. IEEE, Beijing (2018)
7. Chen, Z., Chen, Y., Gao, X., et al.: Unobtrusive sensing incremental social contexts using fuzzy class incremental learning. In: Proceedings of International Conference on Data Mining, pp. 71–80. IEEE, USA (2015)
8. Chen, Z., Chen, Y., Wang, S., et al.: Inferring social contextual behavior from bluetooth traces. In: Proceedings of the 2013 ACM Conference on Pervasive and Ubiquitous Computing, pp. 267–270. ACM, USA (2013)
9. Gao, X., Chen, Z., Tang, S., et al.: Adaptive weighted imbalance learning with application to abnormal activity recognition. Neurocomputing **173**, 1927–1935 (2016)
10. Gao, X., Hoi, S.C., Zhang, Y., et al.: SOML: sparse online metric learning with application to image retrieval. In: Proceedings of AAAI, pp. 1206–1212, USA (2014)
11. Gao, X., Hoi, S.C., Zhang, Y., et al.: Sparse online learning of image similarity. ACM Trans. Intell. Syst. Technol. (TIST), **8**(5), Article 64 (2017)
12. Xiang, Z., Chen, Z., Gao, X., et al.: Solving large-scale TSP using a fast wedging insertion partitioning approach. Math. Probl. Eng. **2015**, 1–9 (2015)
13. Zhang, H., Yuan, J., Gao, X., et al.: Boosting cross-media retrieval via visual-auditory feature analysis and relevance feedback. In: Proceedings of the 22nd ACM international conference on Multimedia, pp. 953–956. ACM (2014)
14. Chen, Z., Chen, Y., Hu, L., et al.: ContextSense: unobtrusive discovery of incremental social context using dynamic bluetooth data. In Proceedings of the 2014 ACM Conference on Pervasive and Ubiquitous Computing, pp. 23–26. ACM, USA (2014)
15. Wang, R., Chen, F., Chen, Z., et al.: StudentLife: assessing mental health, academic performance and behavioral trends of college students using smartphones. In Proceedings of the 2014 ACM Conference on Pervasive and Ubiquitous Computing, pp. 3–14. ACM, USA (2014)
16. Chen, Z., Wang, S., Shen, Z., et al.: Online sequential ELM based transfer learning for transportation mode recognition. In Proceedings of the 6th IEEE International Conference on Cybernetics and Intelligent Systems, pp. 78–83. ACM, USA (2014)
17. Chen, Z., Lin, M., Chen, F., et al.: Unobtrusive sleep monitoring using smartphones. In: Proceedings of the 7th International ICST Conference on Pervasive Computing Technologies for Healthcare, pp. 145–152. ICST, Venice, Italy (2013)

18. Chen, Z., Wang, S., Chen, Y., et al.: InferLoc: calibration free based location inference for temporal and spatial fine-granularity magnitude. In: Proceedings of the 10th IEEE International Conference on Embedded and Ubiquitous Computing, pp. 453–460. IEEE, Paphos, Cyprus (2012)
19. Chen, Y., Chen, Z., Liu, J., et al.: Surrounding context and episode awareness using dynamic bluetooth data. In Proceedings of the 2012 ACM Conference on Pervasive and Ubiquitous Computing, pp. 629–630. ACM, USA (2012)
20. Sheng, W., Wang, J., Wang, J., et al.: Design and implementation of a unified data acquisition and monitoring system for medium and low voltage distribution networks. Autom. Elect. Power Syst. **36**(18), 72–76 (2012)
21. Wang, L., Tao, J., Rajiv, R., et al.: G-Hadoop: map reduce across distributed data centers for dataintensive computing. Future Gener. Comput. Syst. **29**(3), 739–750 (2013)
22. Li, W., Lang, B.: A tetrahedron data model of unstructured database. Sci. China **40**(8), 1039–1053 (2010)
23. Cai, Y., Fu, T., Ni, S., et al.: Study on key technology of unstructured data modeling features. Power Syst. Clean Energy **33**(1), 13–17 (2017)

A Cloud Computing Based District Power Grid Dispatching and Control Integrated Standby Framework

Zhenyu Chen[1], Lixin Li[1], Dapeng Li[1], Fangchun Di[1], Xingyu Gao[2](✉), Can Cui[1], Zhengping Chen[3], Lingxu Guo[4], and Jie Zhang[4]

[1] Beijing Key Lab of Research and System Evaluation of Power Dispatching Automation Technology, China Electric Power Research Institute, Beijing, China
[2] Institute of Microelectronics, Chinese Academy of Sciences, Beijing, China
gxy9910@gmail.com
[3] State Grid Fujian Electric Power Company, Fuzhou, Fujian Province, China
[4] State Grid Tianjin Electric Power Company, Tianjin, China

Abstract. In order to adapt the situation of county power grid development and the integration of dispatching and control, complete the company's five-stage standby dispatching architecture for meeting the needs of emergency dispatching work, this paper studies the key technologies of cloud computing, and proposes the solution framework of clouding computing based district dispatching and control integrated standby system combining the construction mode of multiple masters and one reserve. The entire solution has reached the integration effect of multiple masters and one reserve for the region-county power grid dispatching and control system, reduced the construction investment of standby centers and enhanced the construction level of standby dispatching.

Keywords: Standby dispatching cloud center
Power grid dispatching and control system · Cloud computing
Multiple masters and one reserve

1 Introduction

Power grid dispatching and control is the backbone of power grid operation. It undertakes important tasks such as organizing power grid operation monitoring and conducting accident processing and recovery. It is necessary to improve the ability of the power grid to control and resist various types of unexpected accidents, natural disasters and social events to ensure its uninterrupted operation in real time, which is also a prerequisite for ensuring the security of the power grid, and an important basis for the emergency command work of the power grid. Currently, large-scale power grids in the world have generally adopted different forms of standby dispatching and have shown rapid development trends. In some countries, there has been no significant difference between primary and standby

J. Abawajy et al. (Eds.): ATCI 2018, AISC 842, pp. 417–421, 2019.
https://doi.org/10.1007/978-3-319-98776-7_45

scheduling. All backups and hot spares for dispatching have been implemented. Standby dispatching functions have also been changed from data backup and technical system backup to the deployment of dispatching service functions, to better meet the needs of large-scale power grid security [1–3] researches [4–11] and applications [12–18]. This paper proposes a cloud computing technology based framework for district dispatching and control integrated standby system, which achieves a construction mode of multiple masters and one reserve standby system, and also solves the standby problem for power grid dispatching and control.

2 Functional Standby System Needs Analysis

As shown in Fig. 1, there are logical links between the province and the region to achieve the interconnection, any node failure does not affect data transmission. Under such a standby dispatching situation, how to provide a mutual standby mode that can meet the requirements of seamless switching between mutual devices and mutual systems in local and wide areas, and how to implement load balancing in a wide area and how to implement undifferentiated application access support without geographical restrictions remains the problem to be solved [19–23]. Moreover, the standby dispatching center is responsible for completing the business continuity support for the dispatching business system in the event of a disaster, and has high requirements for the reliability of the system [24].

Therefore, the district dispatching and control integrated and mutual standby mode proposed in this paper, can meet the requirements of the state grid's

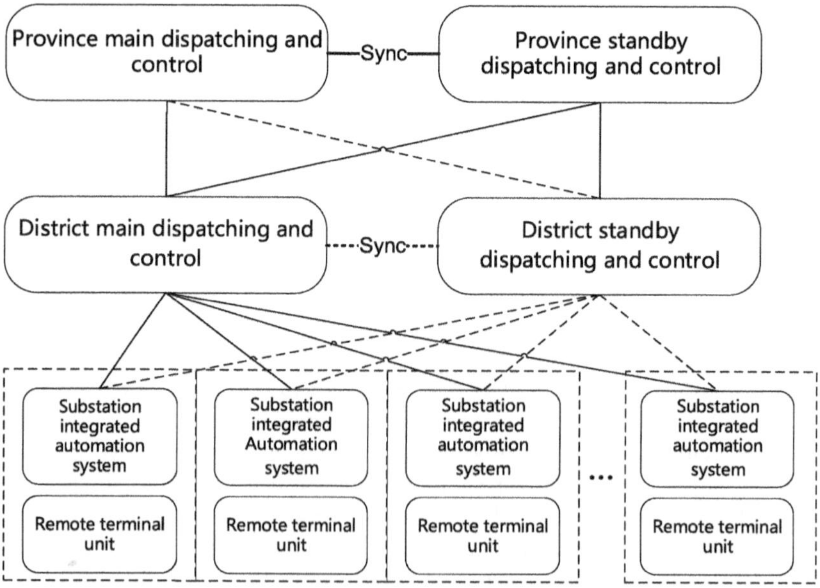

Fig. 1. Power dispatching data network by convergence point backup.

standby dispatching construction requirements, even on this basis, a unified big data platform can be built to further develop more advanced applications and create more advanced next generation power grid dispatching and control system [25].

3 Cloud Standby System Framework Design

The regional dispatching integrated and mutual standby scheme adopts the layout mode of multiple masters and one reserve, to realize the reuse of hardware and software resources, which can save construction funds compared to the traditional one-master-one-reserve mode. The virtualized standby center provides the whole province's standby services, unified data services, and unified advanced application services, to fulfill the requirements of district dispatching control integrated standby system. The virtual standby center is comprised of two sub-centers, data collection in the nearby areas is completed according to the proximity principle, high-speed network interconnection is used between sub-centers, and a virtualized full-network standby center is formed based on the integrated support platform.

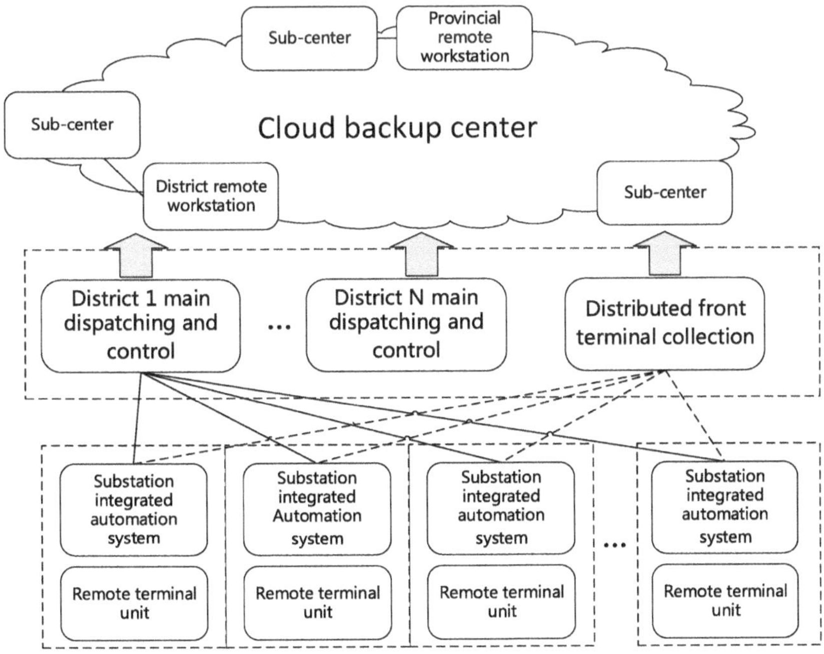

Fig. 2. Construction framework of standby dispatching and control system.

As shown in Fig. 2, this framework meets the requirements for hot standby requirements at the three levels of data, business, and technical support systems for standby dispatching in the county. First, the acquisition equipment is deployed on the second convergence point, where the regional communication network overlaps with the dispatching data network to implement dual convergence collection and distribution processing, for real-time data of 110 kV and 35 kV power grids. The data is sent to the standby center for processing, and implement the data layer to be standby. Secondly, remote stations at the provincial, regional, and county levels can access the virtualized standby center system to implement centralized standby for real-time dispatching services. Finally, the functions of layered power grid operation monitoring and substation centralized monitoring have been developed at the virtual standby center, to achieve the standby of power grid operation monitoring.

4 Conclusion

This paper presents a cloud computing based integrated standby framework for district power grid dispatching and control, and studies the key technologies of cloud computing like virtualization. Based on the cloud computing technology, this framework of the integrated standby system can build a cloud standby system to achieve pre-acquisition, unified platform, unified model maintenance, master and standby switch, master and standby synchronization, which can simultaneously support the integration standby of several master dispatching centers, and achieve the integrated standby effect for multiple masters and one reserve of power grid dispatching and control systems.

Acknowledgment. This work was supported by Science and Technology Program of State Grid Corporation of China (No. 5442DZ170012), National Nature Science Foundation of China (No. 61702491).

References

1. Ding, F., Zhu, H., Zhang, M.: The construction of SCADA disaster backup system for TCSC region power grid dispatching. Power Distrib. Utilizat. **22**(2), 41–43 (2005)
2. Zhang, J.: The solution of data synchronization and acquisition for power grid reserved dispatching system. Power Syst. Commun. **30**(8), 47–50 (2009)
3. Mai, S., Liang, S.: The solution of data synchronization for reserved dispatching EMS system. Power Syst. Commun. **31**(7), 46–49 (2010)
4. Li, D., Chen, Z., Deng, Z., et al.: A wide area service oriented architecture design for plug and play of power grid equipment. Proc. Comput. Sci. **129**, 353–357 (2018)
5. Chen, Z., Li, D., Deng, Z., et al.: The application of power grid equipment plug and play based on wide area SOA. In: Proceedings of 2nd IEEE International Conference on Energy Internet, pp. 19–23. IEEE, Beijing (2018)
6. Chen, Z., Chen, Y., Gao, X., et al.: Unobtrusive sensing incremental social contexts using fuzzy class incremental learning. In: Proceedings of International Conference on Data Mining, USA, pp. 71–80. IEEE (2015)
7. Chen, Z., Chen, Y., Wang, S., et al.: Inferring social contextual behavior from bluetooth traces. In: Proceedings of the 2013 ACM Conference on Pervasive and Ubiquitous Computing, USA, pp. 267–270. ACM (2013)

8. Gao, X., Chen, Z., Tang, S., et al.: Adaptive weighted imbalance learning with application to abnormal activity recognition. Neurocomputing **173**, 1927–1935 (2016)
9. Gao, X., Hoi, S.C., Zhang, Y., et al.: SOML: sparse online metric learning with application to image retrieval. In: Proceedings of AAAI, USA, pp. 1206–1212 (2014)
10. Gao, X., Hoi, S.C., Zhang, Y., et al.: Sparse online learning of image similarity. ACM Trans. Intell. Syst. Technol. (TIST), **8**(5) (2017). Article 64
11. Xiang, Z., Chen, Z., Gao, X., et al.: Solving large-scale TSP using a fast wedging insertion partitioning approach. Math. Probl. Eng. **2015**, 1–9 (2015)
12. Zhang, H., Yuan, J., Gao, X., et al.: Boosting cross-media retrieval via visual-auditory feature analysis and relevance feedback. In: Proceedings of the 22nd ACM International Conference on Multimedia, pp. 953–956. ACM (2014)
13. Chen, Z., Chen, Y., Hu, L., et al.: ContextSense: unobtrusive discovery of incremental social context using dynamic bluetooth data. In: Proceedings of the 2014 ACM Conference on Pervasive and Ubiquitous Computing, USA, pp. 23–26. ACM (2014)
14. Wang, R., Chen, F., Chen, Z., et al.: StudentLife: assessing mental health, academic performance and behavioral trends of college students using smartphones. In: Proceedings of the 2014 ACM Conference on Pervasive and Ubiquitous Computing, USA, pp. 3–14. ACM (2014)
15. Chen, Z., Wang, S., Shen, Z., et al.: Online sequential ELM based transfer learning for transportation mode recognition. In: Proceedings of the 6th IEEE International Conference on Cybernetics and Intelligent Systems, USA, pp. 78–83. ACM (2014)
16. Chen, Z., Lin, M., Chen, F., et al.: Unobtrusive sleep monitoring using smartphones. In: Proceedings of the 7th International ICST Conference on Pervasive Computing Technologies for Healthcare, pp. 145–152. ICST, Venice (2013)
17. Chen, Z., Wang, S., Chen, Y., et al.: InferLoc: calibration free based location inference for temporal and spatial fine-granularity magnitude. In: Proceedings of the 10th IEEE International Conference on Embedded and Ubiquitous Computing, pp. 453–460. IEEE, Paphos (2012)
18. Chen, Y., Chen, Z., Liu, J., et al.: Surrounding context and episode awareness using dynamic bluetooth data. In: Proceedings of the 2012 ACM Conference on Pervasive and Ubiquitous Computing, USA, pp. 629–630. ACM (2012)
19. Zhao, R.: The study and construction of SCADA remote disaster recovery for power grid dispatching system. Zhejiang University (2009)
20. Chen, M.: The research and implementation of key technologies for "dispatching cloud" of power system. Guangxi University (2013)
21. Zhang, W.: The study and application of converged dispatching automation backup system. Shandong University (2013)
22. Wang, Y.: The research and implementation of dispatching automation technology for the integration of region and county. Shandong University (2013)
23. Dang, J.: The research on the integration of Hohhot power grid dispatching and controlling. North China Electric Power University (2012)
24. Li, L., Xie, Q., Yuan, R., et al.: Optimization analysis and design of cloud disaster backup system for power dispatching. Autom. Electr. Power Syst. **36**(23), 82–86 (2012)
25. Yan, H., Di, F., Yuan, R.: The research on big data and its application scenario in the intelligent dispatching of power grid. Electr. Power Inf. Commun. Technol. **12**(10), 7–12 (2014)

Detection of Anomaly in Firewall Rule-Sets

Victor Clincy[1(✉)] and Hossain Shahriar[2]

[1] Department of Computer Science, Kennesaw State University, Kennesaw, USA
vclincy@kennesaw.edu

[2] Department of Information Technology, Kennesaw State University, Kennesaw, USA
hshahria@kennesaw.edu

Abstract. Firewall is generally the first layer of defense in network security, where rules are specified to allow or reject incoming and outgoing network traffic. However, firewall rules may be prone to 'erratic' operations because of human errors involved in the rule-set creation. In this paper, we provide an overview of anomaly detection in firewall rule-sets and their applications.

Keywords: Firewalls · Ruleset correctness · Anomaly detection
Literature survey

1 Introduction

The advancement in the field of networking has led to emerging platforms namely cloudlets, fog, mobile and edge computing [15]. It has been argued that while these are focused on bringing computational power near to the user, they are complex to avail the full potential. Software defined networking is currently thought to be the next generation communication [15]. Even though the evolution of communication will continue, the need for privacy and security of applications running on the network cannot be ignored. On the other hand, we are looking at the popularity of smart home devices that are often dubbed as IoT. These IoT devices being on network allows attackers to see information on user activity (e.g., actions performed by users) [14]. All attacks start from the network connectivity and flow of packets between hosts. Firewall is the first layer of defense for any network. It is a special software or hardware that operates on a set of rules to accept or drop packets.

Being the first layer, or one of the first layers, of defense in a typical network, it is important that rules in firewalls are specified, written and enforced correctly [10]. This topic is not new, as it has been researched extensively (e.g., [4–10]). For example, since 2000, several researchers have researched on firewall rule-set correctness [1, 3] that include mathematically defined rule-set correctness based on of rule-set anomalies. Methods to find and identify anomalies in firewall rule-sets have also been designed [7–9].

Despite the maturity of this topic, there is no recent literature survey focusing on firewall technology, particularly what methodologies exist for firewall rule-set generation, inspection and remediation. Hence, in this paper, we conduct a preliminary

J. Abawajy et al. (Eds.): ATCI 2018, AISC 842, pp. 422–431, 2019.
https://doi.org/10.1007/978-3-319-98776-7_46

survey of the literature, and also determine whether Free Open Source Software (FOSS) software are currently capable of considering rule-set correctness for the full range of the TCP and UDP protocol functionalities. Writing effective firewall rules can prevent malicious traffic coming from remote hosts intended to attack web applications. Despite web applications are major access points for end users, and a significant number of mitigation approaches are available (as can be seen from literature survey by Prokhorenko et al. [11]), correctly writing firewall rules will be essential to mitigate attacks.

Our work is motivated by a number of recent works. For example, Cintuglu et al. [16] conducted a survey to develop taxonomy on smart grid domains, research goals, test platforms, and communication infrastructures, followed by an evaluation of test-beds. Baktir et al. [15] performed a literature survey on edge computing techniques and concluded that while these techniques can bring computation to user devices, focus to be given on the challenges they bring on, and suggested software defined networking as an alternate communication model. Tourani et al. [17] conducted a survey on security, privacy and access control enforcement mechanisms for information centric networking, which includes internet, mobile and ad-hoc networks. They analyze the drawback of approaches, attacks (e.g., denial of service, cache pollution), privacy (anonymity, content privacy), access control (e.g., attribute-based, session-based, encryption-based). Peng et al. [12] provided a taxonomy of user profile intended to detect intrusions in the network.

The paper is organized as follows. Section 2 provides an overview of firewall. Section 3 discussed rule-set complexity. Section 4 highlights some approaches on rule-set anomaly detection. Finally, Sect. 5 concludes, discusses future research directions.

2 Background: Firewalls

The original firewall is, perhaps, first developed by engineers at Digital Equipment Corporation in 1988 [2]. Like many other technologies, firewalls have evolved throughout the years, where each generation includes capabilities from previous generation(s), as well as additional enhancements to mitigate limitations in the previous generation technologies. The packet filtering mechanics in all the firewall generations work identically, and will be introduced when we describe the first generation firewalls. The protocols considered in this work are UDP and TCP.

The source is the originating IP address and port number tuple, and the destination is also an IP address and port number tuple. The firewall compares both tuples to an ordered rule-set. The firewall will then apply the action of the first matching rule to the packet. That action is either accept or reject[1] that packet, either forwarding or processing. A firewall can also have a simple default policy, which is to either accept or decline a packet that does not match any of the rules in the list. Thus, a firewall can be configured with one to many rules. The number of rules is largely determined by the number of needs serviced by network resources.

[1] This is a simplification: possible rules include accept, drop, reject, log, amongst others.

Stateful firewall filters were developed by AT&T Bell Laboratories in 1990 [3]. In addition to packet filtering (source and destination), these firewalls can also inspect the state of a connection associated with an incoming IP packet. As not all protocols use connection states (UDP does not, for example), state filtering is not useful in every use case. However, TCP connections are all stateful, and thus state filters are useful for inspecting TCP packets.

To be more precise, let us consider a host with a web browser initiating a connection to a secure web server, and waiting for requests on a network interface on port 443 (SSL). Once a connection is established, the session does not continue on the server's listening port of 443, but is re-established on a port in the ephemeral port range[2] until its conclusion. Thus, it makes no sense to allow TCP packets flagged as part of an *existing* session to ports to be used only for *new* sessions. A second generation packet firewall can reject such packets.

The third generation firewall inspects packets and network communications from the application level. This generation is commonly referred to as Application layer firewalls[3]. They are frequently used to stop Trojan and malware, and thus serve as components in anti-malware applications [4].

This level of firewall is unique in that it inspects traffic from the operating system level of a host, rather than inspecting traffic at the interface level. Specifically, rather than listening on ports, these firewalls listen to processes by monitoring socket communications[4] between a process and a remote host. By monitoring the socket communications of a process, such firewalls can potentially detect various types of communication abuse.

By monitoring remote process communications, the application firewall may disallow hostile remote entities, in addition to disallowing communications from a local process deemed to be malicious or suspicious.

3 Firewall Rule-Set Complexity

Before discussing rule-set correctness, it may help to understand the source of error-conditions. Rule-sets are created by network engineers – the administrator(s) of a network – to facilitate the enforcement of network security policies. There may be one or more network administrators, and especially in the case of multiple administrators, firewall systems are naturally prone to inconsistent states due to human errors. There are a number of challenges associated with rule-set correctness, such as:

1. Interdependence of rules to enforce a network security policy [5];
2. Semi-continuous evolution of network environments; and

[2] It is suggested by IANA that the range to be [49152, 65535], although it is generally constrained to [1024, 65535]; please see http://www.iana.org/assignments/port-numbers (last accessed June 8th, 2018).

[3] The application layer is the seventh layer of the OSI model, which is the highest and most abstracted from the mechanics of network communications.

[4] Sockets simply represents the combination of IP address and port number that a packet is sent to.

3. Difficulty of introducing a new rule into what is already a complex firewall rule-set.

For a subjective indication of how complex a firewall rule-set is, we can use Wool's.

Equation [5] shown below as an example:

$$RC = Rules_{fw} + Objects_{network} + Interfaces * (Interfaces - 1)/2$$

where, RC is rule complexity, $Rules_{fw}$ indicates the number of firewall rules, $Objects_{network}$ implies the number of objects defined on network; and *Interfaces* indicate total firewall interfaces.

Wool's Equation has been used by Kahn et al. [6] to describe firewall rule-set complexity. The definition of "Network Object" was not defined by Wool in his article, but we may rely on Microsoft's definition [7]. This equation yields a scalar, and simply a numeric value. This value may not be useful in quantifying the complexity of a rule-set, but it serves as a comparison basis between rule-sets. We may consider that a large number of RC indicates how prone a firewall rule-set is to logical errors in the ordered sequence.

4 Firewall Rule-Set Correctness

Correctness is the lack of overlapping rules in a rule-set and incorrect rule-sets can result in conflict(s). These conflicts represent certain logical overlaps between two or more rules, which, ideally, should not exist. It is true that a network administrator may deliberately introduce rules with overlapping logic, use precedence in the rule-set to reduce the total number of rules in the set [5]. Various combinations can produce a properly operating firewall. There appears to be a lack of research focusing on the correctness in stateful connections.

4.1 Firewall Rule-Set Anomaly

Logical overlaps are termed as anomalies, which can be broadly defined into four categories [1, 6, 7, 9]. We will now describe these four anomaly classifications.

Shadowing Anomaly: Arule is shadowed when a previous rule matches all the packets that match this rule, such that the shadowed rule will never be activated. For example, rule Ry is shadowed by rule Rx, if Ry follows Rx in the order, and Ry is a subset match of Rx, and the actions of Rx and Ry are different.

Correlation Anomaly: Two rules are correlated if the first rule matches some packets that match with the second rule, and the second rule matches some packets that match with the first rule. Rules Rx and Ry have a correlation anomaly, if Rx and Ry are correlated, and the actions of Rx and Ry are different.

Generalization Anomaly: Arule is a generalization of another rule, if this general rule can match all the packets that match a specific rule that precedes it. Rule Ry is a generalization of rule Rx, if Ry follows Rx in the order, and Ry is a superset match of Rx, and the actions of Ry and Rx are different.

Redundancy Anomaly: A redundant rule performs the same action on the same packets as another rule, such that if this (redundant) rule is removed, the security policy will not be affected. Rule Ry is redundant to rule Rx, if Rx precedes Ry in the order, Ry is a subset or exact match of Rx, and the actions of Rx and Ry are similar. If Rx precedes Ry in the order, and Rx is a subset match of Ry, and the actions of Rx and Ry are similar, then Rule Rx is redundant to rule Ry provided that Rx is not involved in any generalization or correlation anomalies with other rules preceding Ry.

4.2 Anomaly Detection with State Diagram

There are two methodologies for identifying anomalies in rule-sets. One can use a *state transition diagram* to iteratively compare pairs of rules, which reveals logical overlaps. The other methodology is to use *Packet Space Segmentation* to construct *grid representations* of security policies to discover policy anomalies.

State Diagram: Astate transition diagram uses relational algebra to determine what distinct relationship both rules R_x and R_y have. Relations are defined between the attributes of R_x and R_y, where the attributes are {protocol, source, destination, action}. Each *i*th attribute, from 1 to 4, is considered between R_x and R_y. In the Fig. 1 below, a state transition diagram is shown (adapted from [7]), which can be used to determine whether two packets logically overlap each other in any way.

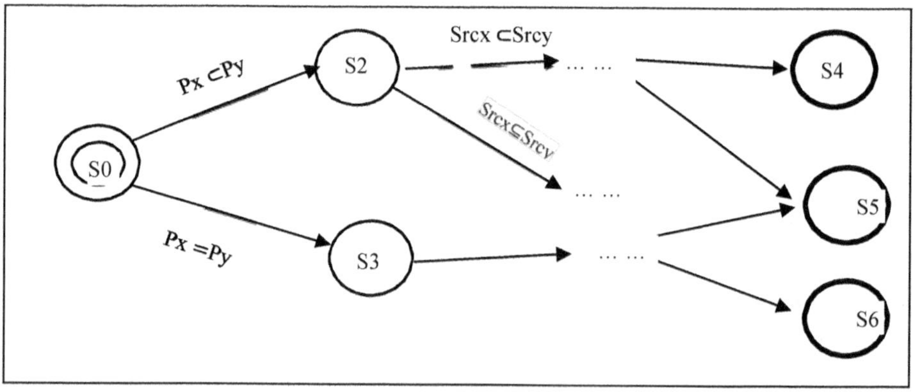

Fig. 1. State diagram of rule-set. Adapted from [7]

In a rule-set {1 … n}, each *i*th rule from (1 … n − 1) is checked against every rule from (i + 1 … n) to determine the number of anomalies present. Optimally, the relationship between Rx and Ry is none, which is state 15. Otherwise, it will be one of the anomaly classifications (states S4, S5, S6, for example).

4.3 Anomaly Detection with Packet State Segmentation

Hu et al. [4] provided an alternative method of determining rule-set correctness, and a graphical representation [5]. For example, let us consider the following rule-set in Table 1.

Table 1. Example of firewall rules

Rule	Protocol	Src IP	Src port	Dest IP	Dst port	Action
R_1	UDP	20.1.2.*	*	172.31.1.*	53	Allow
R_2	UDP	20.1.*.*	*	172.31.1.*	53	Allow
R_3	TCP	20.1.*.*	*	192.168.*.*	25	Deny
R_4	TCP	20.1.1.*	*	192.168.1.*	25	Allow
R_5	*	20.1.1.*	*	*	*	Allow

By applying the packet state segmentation algorithm, we perform a sequence of segment space transformations. This yields a more intuitive overview of anomalies in the security policy. Several types of space are identified in the packet space segments:

1. Non-overlapping: A space is associated with at most one rule; that is to say that no two rules overlap a segment space. For this, the anomaly classification result is none.
2. Conflicting Overlap: A space is associated with two or more rules with conflicting actions. For this approach, anomaly classifications include shadowing, correlation, and generalization.
3. Non-conflicting Overlap: A space is associated with two or more rules with the same action. For this, anomaly classification is redundancy.

The transformation then allows one to construct a grid-representation of the policy space, which facilitates easy identification of anomalous policy spaces (see Fig. 2).

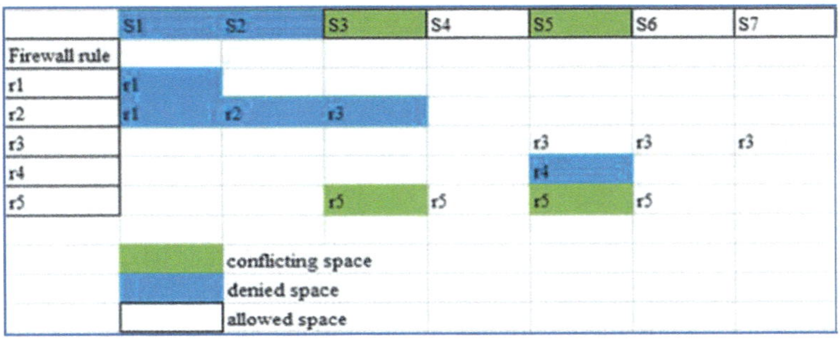

Fig. 2. Policy space representation. Adapted from [4]

There are several algorithms involved in producing these representations, and we refer interested reader to [4] for the complete listing.

Note that in our inspection of the first generation firewall correctness, we must consider at least four dimensions, namely: protocol, source, destination, action. A more elaborate system separates IP address and port number from both the source and destination; thus, deriving a total of four dimensions. This cannot be visualized in three dimensions, except for topological projections from m-dimensions to $m - n$ dimensions, where $0 < n < m$.

4.4 Anomaly Detection with Policy Tree

Firewall rule-sets can be represented graphically [9] using a *Policy tree* (nodes and edges). A policy tree includes rule-sets, and one can trace the tree down to its leaf nodes for actions that represent decision making process when a packet is received on a network interface. The algorithm to build such a tree is presented in [9].

The root of the tree represents the protocol choice between TCP and UDP. There are two branches from the root node: UDP and TDP, which are further represented as edges to the next nodes. The latter is Source IP, and such nodes have a series of edges representing individual IP addresses, or ranges of IP addresses, depending on the configured rules. These edges lead to the children nodes of the Source IP, namely: Source port.

In summary, each node serves as the entry point for a determination of protocol, source, destination, and action. Each edge from a node represents a possible result for each determination. A policy tree's leaf nodes may be labeled with anomalies. Anomaly identification within the rule-set comes *after* the policy tree is created. This is a separate process, and the leaf nodes are labeled as anomalous to each other afterwards.

4.5 Anomaly Detection with Stair Case Modeling

Bilal et al. [6] demonstrated another way to analyze the logical relationships between rules in an ordered set using what is known as the Raining 2-D Box Modeling. While the authors did not present a precise algorithmic mechanism for creating the models shown in their work, one may intuitively put it together with minimal effort.

Table 2. Example of ruleset

Order	Protocol	Src IP	Dest IP	Dst port	Action
1	TCP	160.0.0.1	192.0.0.*	22	Accept
2	TCP	142.183.40.*	165.0.0.20	80	Accept
3	TCP	170.0.0.*	175.0.0.7	21	Deny
4	TCP	170.0.0.4	175.0.0.*	21	Accept
5	TCP	170.0.0.*	175.0.0.*	21	Deny
6	TCP	Any	Any	Any	Deny

In Table 2, rule 5 clearly matches all the packets that rule 3 matches, and so one discerns that rule 5 matches all packets that rule 3 matches, plus some. Rule 3 takes precedence, but it is entirely unnecessary – and thus redundant – to rule 5. After analysis of the rule-set, example anomalies in a series of Raining Boxes are shown below:

- Correlation Anomaly between rules 3 and 4 (Fig. 3);
- Redundancy Anomaly between rules 3 and 5; and
- The combined anomalies between rules 3, 4 and 5.

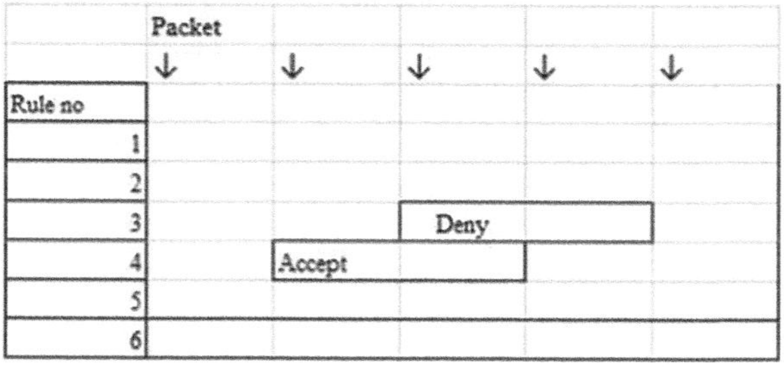

Fig. 3. Correlation anomaly (rule 3 and 4)

In Fig. 3, one may see that rule 3 overhangs rule 4. As such, some packets which could match to rule 4 will never be checked against rule 4 because rule 3 takes precedence. It is also clear that simply removing rule 3 will not resolve the anomalous situation; a more sophisticated algorithm is required to resolve the anomaly.

4.6 Intuitive Mitigation

Redundant rules may be removed in the case that their removal in no way changes the security policy of the rule-set [6]. For example, in Table 2, we clearly see that removal of rule 3 would create a change in policy because of its correlation with rule 4. However, no packet is compared against rule 5 because it is *completely* shadowed by rules 3 and 4. Thus, rule 5 is unnecessary, and may be removed. This removal does *not* affect the correlation between rules 3 and 4; the rule-set is still anomalous. However, with the removal of rule 5, the rule-set is shorter, which is considered an improvement over the previous rule-set.

4.7 Assisted Rule Insertion

In [1], an algorithm is presented to assist an administrator in correctly placing a rule in the correct order, i.e., placing the rule to introduce minimal anomalous overlap. Note that this is not automatic rule insertion, and it does not prevent the addition of anomalous overlaps. It is only a tool to

- make the best possible judgment about rule placement;
- help the administrator to be more aware of how the new rule will interact with the rules immediately preceding and proceeding it; and
- inform the administrator if insertion will result in anomalous overlaps.

4.8 Semi Automated Rule Management for Conflict Resolution

Hu et al. [4] introduce an approach for conflict resolution. To summarize the process above, the mechanism is used to re-order rules in the rule-set. An action constraint *ac* is assigned to a conflicting segment in the packet space. When a packet arrives which would be processed by the conflicted segment, the *ac* is applied, and the best rule, which matches the *ac*, is moved to precedence.

Risk levels RL for a partial conflicted segment can be generated, with upper and lower thresholds (UT and LT, respectively) determined by the administrator before the firewall is activated. RL metrics were determined for a conflicted segment using the Common Vulnerability Scoring System (*CVSS*). The authors accounted for the *exploitability* and *severity* of a vulnerability, in addition to the *asset importance value* of the vulnerable service in the conflicted packet space.

In large networks, firewall rule-sets can be thousands of rules long. In the current threat landscape, organizations are frequently looking for ways to increase their cyber and network security, which can result in ever increasing rule-sets. Inspecting thousands of rules is a tedious task, if not impractical in the real-world, to correct rule-set. Thus, an automated system which uses industry acceptable tools to make decisions and/or alert administrators will be required.

5 Conclusion and Future Research

In this study, we provided an overview of existing approaches for firewall rule-set error and anomaly detection. There appeared to be limited information about anomalies and rule-sets for second generation firewalls. Specifically, *TCP connection states* were generally not considered. One could argue that factoring an new dimension into the rule-set analysis could potentially complicate things. However, we posit that an eighth dimension could in fact simplify the decision making process. For example, let us consider this rule-set in pseudo-code:

- Accept *new* connections to ports [p1,p2, … pn] where the protocol is TCP;
- Accept packets where state is *related, established* and protocol is TCP and interface is Intx;
- Default Policy: DROP ALLPACKETS

These three rules can secure *many* firewalls (not all, of course). Whereas the rule-sets we discussed in the paper do not account for UDP, UDP is used most often for insecure web page access. The implication is that UDP does not need as many rules as TCP. Current versions of Linux iptables can account for ranges of ports, as well as discreetly specified ports. This can significantly reduce the rule-set length, when appropriately applied. The pseudo-code listed above can be directly implemented in iptables. Future research include capturing stateful consideration in rule-set analysis, and to compile a list of firewall rules that should be present in all firewalls to help mitigate against common protocol abuses.

References

1. Al-Shaer, E.S., Hamed, H.: Design and implementation of firewall policy advisor tools. Technical report CTI-techrep0801, School of Computer Science Telecommunications and Information Systems, DePaul University, August 2002
2. Ingham, K., Forrest, S.: A history and survey of network firewalls. In: Proceedings of National Conference on Recent Developments in Computing and Its applications, p. 4, August 2009. Accessed 10 June 2018
3. Software Firewalls: Made of Straw? Part 1 of 2, Symantec Connect Community (2010). Accessed 03 May 2018
4. Hu, H., Ahn, G., Kulkarni, K.: FAME: a firewall anomaly management environment. In: Proceedings of the 3rd ACM Workshop on Assurable and Usable Security Configuration (SafeConfig), pp. 17–26. ACM, New York (2010). https://doi.org/10.1145/1866898.1866902
5. Wool, A.: A quantitative study of firewall configuration errors. Computer **37**(6), 62–67 (2004)
6. Khan, B., Khan, M.K., Mahmud, M., Alghathbar, K.S.: Security analysis of firewall rule sets in computer networks. In: Proceedings of 4th International Conference on Emerging Security Information, Systems and Technologies, pp. 51–56 (2010)
7. Al-Shaer, E., Hamed, H.: Firewall policy advisor for anomaly detection and rule editing. In: The IEEE/TFIP International Symposium on Integrated Network Management Conference, USA, March 2003, pp. 17–30 (2003)
8. A systematic approach for conflict resolution in firewall policies. Technical report ASU-SCIDSE-10-2, Arizona State University, Tempe, May 2010. http://sefcom.asu.edu/confres/confres.pdf
9. Al-Shaer, E.S., Hamed, H.H.: Modeling and management of firewall policies. IEEE Trans. Netw. Serv. Manag. **1**(1), 2–10 (2005)
10. Radomskiy, S.: Security policy rules optimization and its application to the Iptables firewall. M.Sc. thesis, University of Tampere (2011). https://tampub.uta.fi/bistream/handle/10024/82342/gradu04878.pdf?sequence=1
11. Prokhorenko, V., Choo, K.-K.R., Ashman, H.: Web application protection techniques: a taxonomy. J. Netw. Comput. Appl. **60**, 95–112 (2016)
12. Peng, J., Choo, K.-K.R., Ashman, H.: User profiling in intrusion detection: a review. J. Netw. Comput. Appl. **72**, 14–27 (2016)
13. Do, Q., Martini, B., Choo, K.-K.R.: Cyber-physical systems information gathering: A smart home case study. Comput. Netw. **138**, 1–12 (2018)
14. Haddad Pajouh, H., Dehghantanha, A., Khayami, R., Choo, K.-K.R.: A deep recurrent neural network based approach for internet of things malware threat hunting. Future Gener. Comput. Syst. **85**, 88–96 (2018)
15. Baktir, A.C., Ozgovde, A., Ersoy, C.: How can edge computing benefit from software-defined networking: a survey, use cases, and future directions. IEEE Commun. Surv. Tutor. **19**(4), 2359–2391 (2017)
16. Cintuglu, M.H., Mohammed, O.A., Akkaya, K., Uluagac, A.S.: A survey on smart grid cyber-physical system testbeds. IEEE Commun. Surv. Tutor. **19**(1), 446–464 (2017)
17. Tourani, R., Misra, S., Mick, T., Panwar, G.: Security, privacy, and access control in information-centric networking: a survey. IEEE Commun. Surv. Tutor. **20**(1), 566–600 (2018)

Classification and Coding of Pipe Gallery Bim Model for Operation and Maintenance

Song-Tao Lu[1,2(✉)], Min Hu[1,3], Li Teng[1,4], and Li Zhou[1,3]

[1] Shanghai University-Shanghai Urban Construction Industrialization Research Center, Shanghai 200444, China
470577273@qq.com
[2] Department of Civil Engineering, Shanghai University, Shanghai 200444, China
[3] Sydney Business College, Shanghai University, Shanghai 201800, China
[4] Shanghai Underground Space Design and Research Institute, Shanghai 200020, China

Abstract. The pipelines are densely arranged in the pipe gallery. A large number of professions are involved in the design of pipe gallery. Different classification and coding systems are used in the design, construction, and operation & maintenance stage, forming the "information island". This information island is difficult to be unified managed. Based on the characteristics of the pipe gallery and its operation and maintenance, the classification and coding system of pipe gallery BIM model for operation and maintenance is established. It lays a foundation for the information transmission and sharing during the whole life cycle of the pipe gallery. Finally, a small plug-in for automatic coding of the facility and equipment components is developed, achieving a deep integration of coding and BIM model, and ensuring the implementation of the coding system.

Keywords: Comprehensive pipe gallery · Classification · Coding

1 Introduction

Information management is the guarantee of automatic and pipelined operation and maintenance of integrated pipe gallery [1]. A standardized and uniform application model for information and data is provided by information classification and coding, clearing the way for the communication and data transfer within the whole life cycle of the integrated pipe gallery. The earliest classification and coding of building information was fragmented and one-sided, which originated from the attempts made by western countries in 1920 in order to solve the non-standard of information management. Afterwards, based on different classification methods, countries in the world gradually established and perfected their own classification systems according to the specific classification requirements and classification objects, presenting a situation of "letting a hundred flowers blossom and a hundred schools contend". The main representatives are the *MasterFromat* and *UniFormat* of the American Building Standard Association, the *CI/SfB* (Construction Index/Standard for Buildings) of the Royal

J. Abawajy et al. (Eds.): ATCI 2018, AISC 842, pp. 432–442, 2019.
https://doi.org/10.1007/978-3-319-98776-7_47

Institute of British Architects, etc. However, the diversified classification systems are not conducive to the international exchange and sharing of building information, and therefore, the *ISO* (International Organization for Standardization) has studied and formulated a unified standard for the classification and coding of building information since 1990 [2], such as *ISO/DIS 12006-2*, *ISO12006-3*, etc., opening a new era of classification and coding system for building information. The *ISO* standard provides basic guidelines and frameworks for classification of building information in different countries and changes the situation that the classification system varies from country to country, achieving the mutual conversion of various standard systems within the *ISO* framework. Based on this, the classification system of building information has made great progress, forming mature international common standards, such as *OmniClass* and *Uniclass*.

The classification and coding of building information started relatively late in China, and a GB "Model Classification and Coding Standard of Building Engineering Design Information" was established in line with China's conditions in 2014 under the framework of *ISO12006-2*. And based on it, industrial and local classification and coding standards of building information emerge fast [3]. However, regarding the current national standard and the local classification and coding standards derived therefrom, the macro regulations on information classification and coding are more focused, while for the specific projects on the micro, there is no practical guidance. Therefore, based on the characteristics of integrated pipe gallery, the GB "Model Classification and Coding Standard of Building Engineering Design Information" is deeply refined, and a model classification and coding method of integrated pipe gallery building information for operation and maintenance is proposed in this paper, laying a foundation for the information management of the operation and maintenance of integrated pipe gallery.

2 Existing Classification System of Building Information

The earlier classification and coding standards of building information, such as *MasterFromat* and *UniFormat*, have been gradually marginalized as they cannot meet the increasingly complicated project information due to their small coverage, although they have also been updated since the uniform classification and coding standard is released by *ISO*, such as the publication of *MasterFromat1997* Outline and *MasterFormat2004* Outline. *ISO12006-2*, *OmniClass* and *Uniclass* are the current mainstream classification and coding systems, and they, especially the *ISO12006-2*, are used as reference by the "Model Classification and Coding Standard of Building Engineering Design Information". The three mainstream standards are briefly introduced next.

The early classification standard *12006-2:2001* released by *ISO* in 2001 is the basis of modern building information classification. Based on the data feature and application target involved in the whole life cycle of a building, 17 recommended classification tables in accordance with different classification principles are provided by *12006-2:2001*, and the isolation of building project's information on each stage is relieved by correlating the construction process. With the development of the times, BIM technology has been emerging, and the *IFC* standard format is developed by the neutral organization *buildingSMART* and has been widely used in order to import and convert the model data of building information in different software. Due to the lack of clarity in the relevance of classified objects, *12006-2:2001* can no longer meet the requirements of *IFC* and needs to be revised urgently. To this end, the *ISO* revises *12006-2* and releases the *12006-2:2015* version, which integrates support for BIM, defines the classification framework of building environment and details how the object category in each table relates to each other, making up for the defect of the 2001 version. However, only a framework rather than a detailed and complete classification is provided by the *12006-2:2015*. In actual application, the classification system is required to be reorganized and the classification table is required to be listed in accordance with the actual national conditions.

In order to adapt to the national conditions of European and American countries, a more detailed classification method and a more concise project coding system are developed by *OmniClass* Editorial Board based on *ISO12006-2*, *MasterFormat*, *UniFormat* and other classification systems, making up for the deficiencies of the past classification systems. And then the information collection and collation become more standard, which is conductive to the information exchange and project management of the project participants, as well as the connection with the *IFC* standard. *OmniClass* standard has better compatibility with *ISO* and stronger ability to exchange information with other classification systems as it mainly refers to the *ISO2006-2*. Currently, the *OmniClass* coding system is supported by most BIM software, so it is always used as a reference by other countries when their own classification and coding standards are compiled.

UniClass is a voluntary classification released by *CPI* in 1997 and has been widely promoted due to the government recognition. However, *Uniclass* is not uniform and its tables vary in many ways. For example, numbers below level 1 are used for coding by most of *Uniclass* tables, while alphanumeric characters are used by two tables of it. Besides, *Uniclass* does not fully comply with *ISO 12006-2* due to its premature release time. To this end, *UniClass 2* which complies with the *ISO 12006-2* standard is released by *Uniclass Editorial Board* and has become the "*OmniClass*" for the UK. *UniClass 2* enables different types of projects, such as building, garden, municipality and facility to be classified in the same unified classification system, and provides a coding system that is flexible enough to meet the future classification needs. In order to better support BIM applications, *UniClass 2015* has been developed by *the UniClass Editorial Board* through the research of 3 years based on *UniClass 2*. *UniClass 2015* supports BIM Toolkit and is able to parse and check BIM model parameters, benefiting ordinary users.

3 Classification System of BIM Model for Pipe Gallery

Components in the pipe gallery are various, and the properties of various components are also different. The database will be very disordered if the component objects are not classified by a classification system, and the property entries (database fields) will even up to thousands with the addition of only one component. In order to check and modify the model data of the pipe gallery and to facilitate the communication and data transmission between the project participants, it is necessary to study the industrial classification and coding method which is macroscopic and boring. In addition, any classification and coding system is not easily fabricated, and a more complete and targeted classification and coding system can only be established by taking full advantage of the existing ones. *ISO12006-2* is the core of the current mainstream classification and coding systems, and classification and coding under its framework is good for the China BIM technology to keep up with international standards. Therefore, based on the characteristics of the pipe gallery and the "*Model Classification and Coding Standard of Building Engineering Design Information*", which is a successful application of *ISO12006-2* in China, classification system of BIM model for pipe gallery is established.

Integrated pipe gallery is an underground structure designed to hold various pipelines. As long as there is a pipe gallery, the power, communication, gas, water and other pipelines should be allocated within it. It avoids the waste caused by the repeated excavation of pavement, ensures the smooth city traffic, prolongs the service life of the road and the pipeline, and improves the utilization efficiency of the underground space. However, once the pipe gallery is constructed, its structure cannot be changed. Therefore, in the design phase of the pipe gallery, enough space for pipeline layout must be reserved with a consideration of urban planning. In addition, due to the large number of pipelines within the pipe gallery, huge losses will be caused once an accident happens [4]. As a whole life cycle construction project, the pipe gallery construction involves a large number of professions. During the construction process, coordination between different participants is difficult; the exchange and sharing of internal information are inefficient.

Based on the above analysis on the characteristics, and considering that the pipe gallery is long vertical with a relatively small cross-sectional area, the pipe gallery is divided according to the "first vertical and then horizontal" principle.

In the vertical direction, the pipelines are densely arranged within the pipe gallery with great danger. Considering that the fire is a main hazard for the pipe gallery, it is stipulated that the entire pipe gallery model should be divided according to the fire zones from the viewpoint of fire safety management. Fire doors are doors that meet the requirements of fire-resistant stability, integrity and insulation for a certain period of time [5]. Here the fire zone refers to the fire prevention unit formed by a number of fire doors with relatively high fire endurance, and it can prevent the spread of fire to other areas of the pipe gallery in a certain period of time [6]. The fire zone can be self-set according to the height of pipe gallery, fire risk and fire-fighting capacity in line with the relevant provisions of public security fire, for example, the area between the five consecutive fire doors can be defined as a fire zone. A solid foundation for the fire safety design and management in the future can be laid if the fire zones have been considered from the model division stage.

In the horizontal direction, the pipe gallery can be divided into three parts according to the different functions of various parts in it, that is, the main project, the pipeline project and the other project. The main project is the underground infrastructures made up of reinforced concrete. It is built to protect the pipe gallery and can be divided into structures, ancillary facilities, pipeline introduction and surface facilities according to the functions. The pipeline project is the most important part as the purpose of building a pipe gallery is to store various pipelines. It can be divided into electricity, communication, gas, water and heat parts according to the type of intake pipelines. The other project refers to the auxiliary facilities used to better operate and maintain the various pipelines in the pipe gallery. It can be divided into fire protection system, power distribution system, lighting system, monitoring and alarm system, communication system, ventilation system, drainage system, water supply system and identification system according to the function.

In summary, the final classification system of BIM model for pipe gallery is shown in Table 1 (partial).

Table 1. Division of pipe gallery model (partial)

Level-1 category	Level-2 category	Level-3 category	Level-4 category
Fire zone-1	Main project	Structures	Standard section, hoisting port, substation, branch port, inlet and exhaust outlets
		Ancillary facilities	…
		Pipeline introduction and surface facilities	…
	Other project	Fire protection system	…
		Power distribution system	
		Lighting system	
		Monitoring and alarm system	
		Communication system	
		Ventilation system	
		Drainage system	
		Identification system	
	Pipeline project	Communication pipeline	
		Electricity pipeline	
		Gas pipeline	
		Heat pipeline	…
		Water pipeline	…
Fire zone-2	Main project	Structures	Standard section, hoisting port, substation, branch port, inlet and exhaust outlets
		Ancillary facilities	…
		…	…

4 Coding System of BIM Model for Pipe Gallery

The safety of pipelines in the pipe gallery is very important, and accurate monitoring and real-time alarm are required to avoid any possible accident. Therefore, it is crucial to establish a pipe gallery model using BIM technology and to build a management platform for the operation and maintenance of pipe gallery, in order to improve the monitoring efficiency and implement the real-time alarm and linked emergency measures. The uniform coding of the pipe gallery model is the prerequisite for all the above. Coding is a marking method with reference and identification functions. It marks the component model with a set of symbols that are regular, meaningful and easy to be identified and processed by computers and humans [7]. Based on a single coding, the project, location, and name of its corresponding component can be clearly determined. Operation and maintenance management platform can realize the precise positioning, retrieval and management of component models by using the "coding" as the retrieval content. Therefore, the construction of coding is very important, and its scientificity and rationality will directly affect the management efficiency of the operation and maintenance of the pipe gallery. Based on the macroscopic regulations on the coding in "*Model Classification and Coding Standard of Building Engineering Design Information*" and combined with the characteristics of the pipe gallery, a coding system of BIM model for pipe gallery is established in this paper. The proposed coding system is not only in line with international standards but also a practical guidance for engineering project.

Based on the characteristics of the pipe gallery and its operation and maintenance, the coding of the pipe gallery is divided into four parts, that is, the project code, the division code of management grid, the classification code of facility and equipment, and the sequence code, a total of 18-digit codes, as shown in Fig. 1.

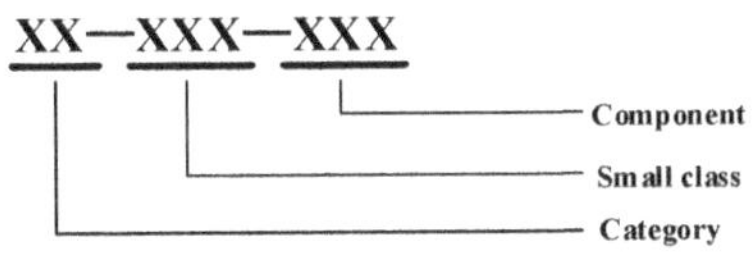

Fig. 1. Coding structure of facility and equipment

Among them, the project code is represented by three letters based on the project name, and the Chinese pinyin initials are commonly used. Use the first three letters if the number of Chinese pinyin initials is more than three, for example, represent the North Island Pipe Gallery ("Bei Dao Guan Lang" in Chinese) by BDG.

Management grid represents the current location of facility and equipment. It consists of partition code and location code, represented by a 4-digit numeric code, as shown in Fig. 2 [8].

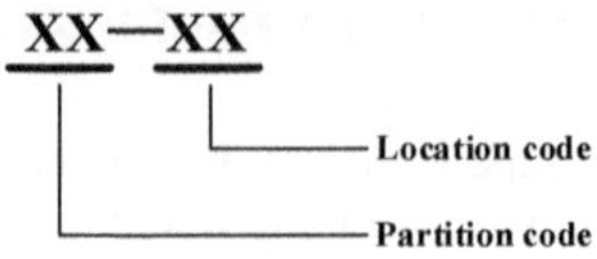

Fig. 2. Division code structure of management grid

The partition code indicates the partition where the facility and equipment are currently located, and is represented by a 2-digit numeric code. The location code indicates the specific location of the facility and equipment, such as hoisting port, substation, etc., and is represented by a 2-digit numeric code. The division and coding of management grid is shown in Table 2.

Table 2. Division and coding of management grid

Partition	Partition coding	Grid division	Grid coding	Merged name	Merged coding
Fire zone-1	01	Standard section	01	Standard section of fire zone-1	0101
		Hoisting port	02	Hoisting port of fire zone-1	0102
		Substation	03	Substation of fire zone-1	0103
		Branch port	04	Branch port of fire zone-1	0104
		Inlet and exhaust outlets	05	Inlet and exhaust outlets of fire zone-1	0105

The classification code of facility and equipment are letters in accordance with the three-level division structure, including a total of eight characters. According to the divided level of the model, the classification code is divided into category, subcategory and name, as shown in Fig. 3.

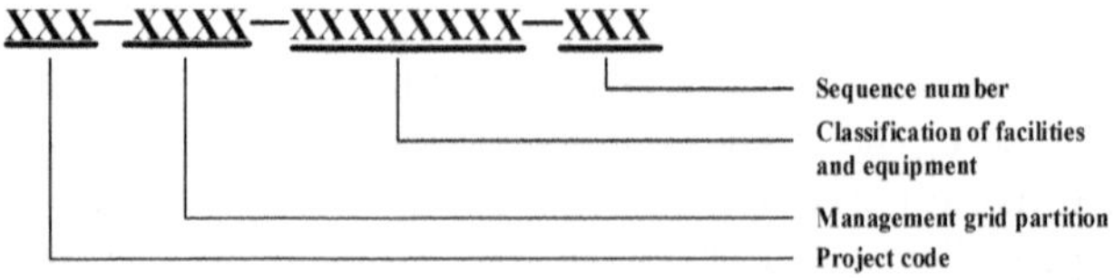

Fig. 3. Classification code of facility and equipment

The category, subcategory and name are respectively two, three, three characters represented by capital English letters, and the Chinese pinyin initials are commonly used. If a component cannot be divided any longer after it is divided to the subcategory, letters “OOO” can be used for the last three digits, as shown in Table 3.

Table 3. Classification code of facility and equipment

Category	Category coding	Subcategory	Subcategory coding	Name	Name coding	Merged coding
Structures	GZ	Standard section	BZD	Wall	OOO	GZBZDOOO
		Hoisting port	DZK	Pillar	OOO	GZDZKOOO
		Substation	FDS	Window	OOO	GZFDSOOO
		Branch port	FZK	Branch port	OOO	GZFZKOOO
		Inlet and exhaust outlets	JPF	Inlet and exhaust outlets	OOO	GZJPFOOO
Ancillary facilities	FS	Drainage facility	PSS	Open channel	FMG	FSPSSFMG
			PSS	Collecting sump	JSK	FSPSSJSK
			PSS	Collecting well	JSJ	FSPSSJSJ
		Outlet	TFK	Outlet	OOO	FSTFKOOO
		Cat ladder	FPT	Cat ladder	OOO	FSFPTOOO
		Railing	FLG	Railing	OOO	FSFLGOOO
		Feeder nose	TLK	Feeder nose	OOO	FSTLKOOO
		Fire door	FHM	Fire door	OOO	FSFHMOOO
		Bridge support	QZJ	Bridge support	OOO	FSQZJOOO

The classification can be expanded by the management units according to their own requirements, as requirements for the model depth are not the same regarding different operation and maintenance. For components with deeper operation and maintenance, such as parts, they can be expanded on the basis of three-level division. According to the linear classification method, the categories in the same level are classified according to their subordinate relations [9] and the categories of the same branch are classified according to their juxtapositions; each expanded new level consists of three characters represented by capital English letters, and the Chinese pinyin initials are also commonly used.

5 Realization of Automatic Coding for BIM Model

The classification and coding system of pipe gallery proposed in this paper has been applied to the integrated pipe gallery project of Beidao West Road, Lingang, Shanghai. The integrated pipe gallery is 951 m in length with a standard size of B * H = 5.7 * 4.85 m and an excavation depth of 7 m. The intake pipelines include water supply, electricity, communication, sewage and gas pipelines. Currently, the project is under construction with a project cost of about RMB 100 million.

Due to the large number of facilities and equipment involved in the pipe gallery, artificial sequential coding is inefficient and error-prone. To this end, a small plug-in to achieve automatic coding of the pipe gallery facilities and equipment is developed in this paper. The plug-in is a secondary development based on Revit, with Visual Studio 2015 as the development tool, C# language.net as the programming language, and WinForm interface as the framework. In order to avoid the read of shaft, marking and other unrelated factors, the component coding is required to be imported. Firstly, the excel describing the component coding is required to be converted to a csv format and then imported into the plug-in. Secondly, read the components. All elements are required to be read, and the elements existed in the component code table are required to be displayed in the window. Finally, code the components and write the coding into the flag properties. It is worth noting that sequence code is required to distinguish the multiple component units of the same type. The sequence code is three characters coding from 001, and it sorts in ascending order according to the component's coordinate with the X-axis as the first consideration and the Y-axis next. Here a branch port is taken as an example and the specific process is shown in Figs. 4 and 5. In Fig. 4, the first column is the name,the second column is the code, the third column is the component ID.

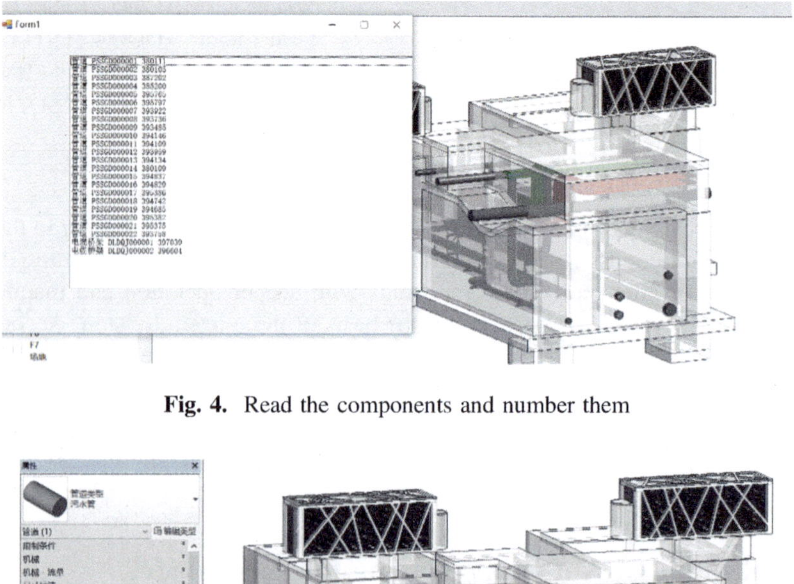

Fig. 4. Read the components and number them

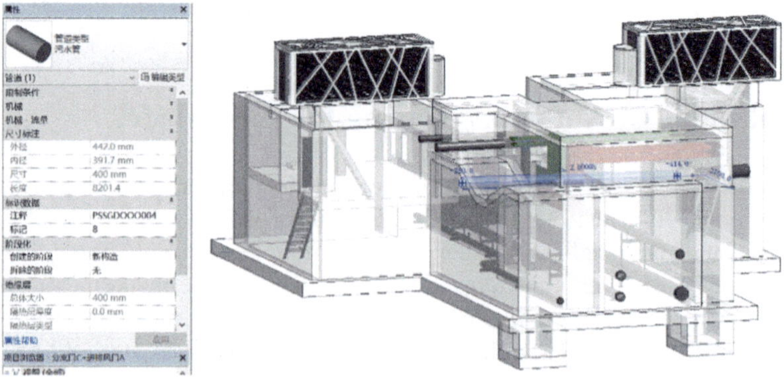

Fig. 5. Coding has been written to the component annotation

6 Conclusions

Combined with the integrated pipe gallery project of Beidao West Road, Lingang, Shanghai, the pipe gallery is unified classified and coded in the modeling process, which facilitates the information transfer and sharing between the design, construction and operation & maintenance units, avoiding the appearance of "information island" and laying a good foundation for the later operation and maintenance management of the project. Conclusions are as follows:

(1) Considering that the pipe gallery is long vertical with a relatively small cross-sectional area, the divided principle "first vertical and then horizontal" is proposed. In the vertical direction, the pipe gallery is divided according to the fire zones in order to meet the needs of fire-fighting; in the horizontal direction, the pipe gallery can be divided into three parts according to the different functions of various parts in it, that is, the main project, the pipeline project and the other project, and a more detailed division is carried out according to the specific function.
(2) Based on the characteristics of the pipe gallery and its operation and maintenance, the coding of the pipe gallery is divided into four parts, that is, the project code, the division code of management grid, the classification code of facility and equipment, and the sequence code. The rule that the management grid code consists of partition code and location code is proposed, the formation rule of the classification code of facility and equipment is emphasized, and the method for extending coding is explained in this paper.
(3) A small plug-in for automatic coding of the facility and equipment is developed in this paper, achieving a deep integration of coding and BIM model, and ensuring the implementation of the coding system. This plug-in is not only applied to the pipe gallery project, but also universal for coding of facility and equipment in rails, roads and other engineering projects.

References

1. Zhou, L.: Research on key technology of construction of safety monitoring system for urban underground comprehensive pipe corridor. Mod. Surv. Mapp. **39**(6), 39–41 (2016)
2. Zhu, C., Yang, X.: The reference role of architectural information classification and coding system for railway BIM. Railw. Technol. Innov. **5**, 35–38 (2014)
3. Chen, S.: Research on Design and Information Management of Steel Truss Arch Bridge Based on Bim Technology. Southwest Jiao Tong University, Chengdu (2016)
4. Zhang, H.: Thinking of the monitoring and alarm system of intelligent comprehensive corridor. Constr. Technol. **21**, 20–22 (2016)
5. Liao, X.: Analysis of factors affecting fire resistance of fireproof doors. Doors Windows **9**, 20–21 (2012)
6. Liu, H.: The effective application of building fire prevention and separation technology. Urban Archit. **11**, 242 (2014)

7. Li, J.: Comparative analysis of segmental coding and level code. Account. Comput. China **11**, 11–12 (2003)
8. Bi, X., Chen, H., Lai, H., et al.: Research on classification and coding of urban rail transit facilities based on building information model (BIM). Res. Urban Rail Transit **19**(1), 5–9 (2016)
9. Yang, B., Sun, L., Yu, X.: Research on the classification system of general manufacturing technology. Machinery **47**(8), 68–69 (2009)

A UAV Operational Situation Similarity Measurement Method Based on Empirical Mode Decomposition

Yao Lu(✉), DongSheng Li, and Yang Gao

National University of Defense Technology, Hefei 230037, China
lu_yao_mail@126.com

Abstract. Due to the large amount of the operational situation data of UAV, the similarity measurement efficiency is low. Therefore, this paper proposes a UAV operational situation similarity measure method based on empirical mode deconposition. Firstly, the situation elements and its representation mode are determined. Then we use the empirical mode decomposition method to extract the sequence trends of the empirical situation and the new situation, and then measure the similarity between the trend sequences of the new situation and experience situatin by vector distance. At last, the method extracts the part of the experience situation similar to the new situation as the basis of operational decision making under the new situation. Through the use of common data sets for simulation analysis, it is shown that the method can effectively measure the degree of similarity between two sequences. The measurement accuracy is high and the time consuming is short, which has certain practical value.

Keywords: Operational situation · Similarity measure · Time series
EMD

1 Introduction

With the increasing complexity of operational missions and operational environments, UAVs have become widely used in modern warfare because of their unique advantages. They have been gradually applied to early warning detection, close to reconnaissance, wide-area surveillance, electronic countermeasures, and special operations. UAV's intelligent decision-making capability is a key technology for its wide application [1]. It can use the knowledge of experience to complete the judgment of the current situation, discover the similar characteristics of the new situation and experience, and thus make "empirical" decisions.

The UAV's operational situation can be represented by the space-time information and attribute information of each operational element [2]. There is a large amount of data for historical experience, and the use of new situation data and empirical situation data for comparison metrics will consume a lot of time. This approach is obviously not applicable to the ever-changing battlefield environment. Therefore, the UAV operational situation information is represented as various time series data in this paper. The similarity measure of time series data is used to measure the similarity of UAV

J. Abawajy et al. (Eds.): ATCI 2018, AISC 842, pp. 443–451, 2019.
https://doi.org/10.1007/978-3-319-98776-7_48

operational situation. Common methods for measuring similarity of time series include Euclidean distances, dynamic time warping distances, and multidimensional morphological representations. Euclidean distance has the advantages of being intuitive and efficient, but it is very sensitive to the translation of data, and the calculation accuracy is not high [3]. The dynamic time warping distance can overcome the shortcomings of Euclidean distance that cannot reflect the overall development trend of data, but the complexity is high [4, 5].

In view of the above problems, this paper proposes a similarity measurement method for UAV operational situation based on empirical mode decomposition method (EMD). Firstly, the UAV operational situation elements are characterized, time series information is used to represent the situation data, and then Empirical Mode Decomposition (EMD) is used to extract trend information of the situational time series. Based on this, use the PLR piecewise linear representation method to perform piecewise linear representation on this basis. By comparing the trends of different granularities, we gradually reduce the empirical situation data similar to the new situation, and eventually extract the part of the experience situation that has a high degree of similarity to the new situation, and achieve the purpose of measuring the similarity.

2 UAV Operational Situation Measurement Elements and Their Characterization

The battlefield situation refers to the deployment of all forces, weapons and equipment and battlefield environment by all the participating troops and supporting support forces in the battlespace space [6]. Due to the complexity of the battlefield situation, this paper focuses only on the local situation when a specific combat entity (i.e. UAV) performs its task. According to the commonly used expression of operational situation, combat information is usually divided into space-time information and attribute information. The space-time information reflects where the enemy is, where the military is, the current time, and other basic operational situations. It directly reflects information such as target location, speed, and time and plays a greater role in decision-making. Therefore, space-time information is the information type that this article focuses on.

When UAVs perform combat missions, they often face the threat of detection by enemy radar, the threat of missiles and anti-aircraft gun damage, and their own flying height, flight speed, and azimuth of the target will also affect the outcome of the operation. Therefore, considering the situational information under confrontational conditions, the UAV's own flying height, speed, and the relative distances and relative angles between the enemy's defense equipment and mission objectives are taken as the metrics of the situation, as shown in Fig. 1. The unit of relative distance d is km, the unit of relative angle θ is, and the unit of flying speed v is m/s, and the unit of flying height h is m.

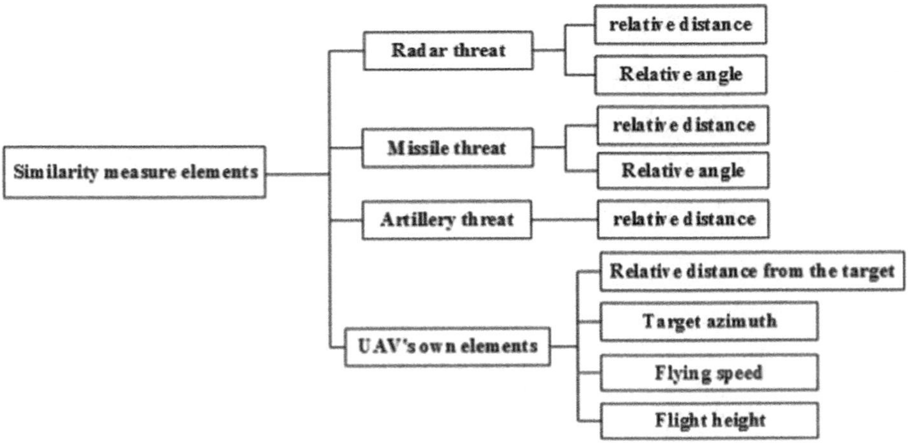

Fig. 1. Similarity measure elements

3 Similarity Measurement Method of UAV Operational Situation Based on EMD

3.1 Trend of Situation Sequence Extraction by EMD Method

The Empirical Mode Decomposition (EMD) method is a method proposed by Hilbert-Huang et al. for non-stationary time signal analysis [7]. Its core idea is to decompose the data being analyzed into a combination of multiple data sequences with different characteristics. Its purpose is to decompose the signal into a stack of multiple eigen-mode functions (IMFs). The IMF must satisfy the following two conditions [8]:

(1) In the entire time range of the function, the number of local extremum points and zero-crossings must be equal, or at most one difference;
(2) At any point in time, the envelope maximum of the local maximum and the local minimum must be zero.

EMD requires the input signal to have at least two extreme points, a maximum point and a minimum point. The characteristic time scale is defined by the time between two extreme points. If the data lacks extremum points but has deformation points, the extremum points can be obtained once or several times through data differentiation, and then the decomposition results can be obtained through integration [9]. The algorithm specific measurement steps are as follows:

Step 1. Initialization: $r_0 = x(t)$, $i = 1$.
Step 2. Get the ith IMF:
- (a) Initialization: $h_0 = r_{i-1}(t)$, $j = 1$.
- (b) Find the local extreme point of $h_{j-1}(t)$.
- (c) Three-spline interpolation of the maximum and minimum points of $h_{j-1}(t)$ to form the upper and lower envelopes.

(d) Calculate the average value $m_{j-1}(t)$ of the upper and lower envelopes.
(e) $h_j(t) = h_{j-1}(t) - m_{j-1}(t)$.
(f) If $h_j(t)$ is an IMF function, then $imf_i(t) = h_j(t)$, otherwise, $j = j+1$, go to step (b).

Step 3. $r_i(t) = r_{i-1}(t) - imf_i(t)$.

Step 4. If $r_i(t)$ has more than two extreme points, then $i = i+1$, go to step 2, otherwise, the decomposition ends and $r_i(t)$ is the residual component. The algorithm finally gets:

$$x(t) = \sum_{i=1}^{n} imf_i(t) + r_n(t) \tag{1}$$

where $x(t)$ represents the original time series signal, where $imf_i(t)$ represents the sub-signals that satisfy the IMF. $r_n(t)$ is a trend sequence.

Figure 2 shows an empirical mode decomposition process. Signal is the first line of the ChlorineConcentration time series signal in the UCR Time Series Classification Archive dataset. The process is the detail of the decomposition process, and res is the trend sequence obtained after the decomposition. Accumulating the imf signal can substantially fit the original sequence.

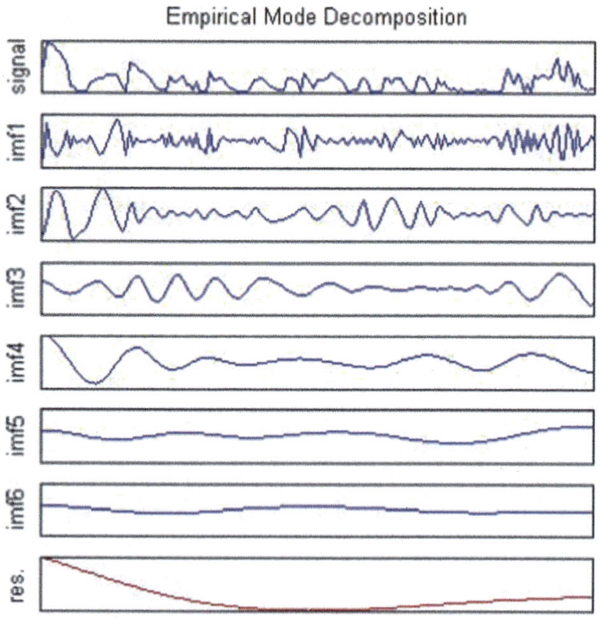

Fig. 2. Empirical mode decomposition process

The segmented representation of the original sequence is computationally intensive and susceptible to noise. The fitting sequence can filter noise on the basis of retaining basic trends and reduce the data dimension. Using the PLR segmentation method for the fitting sequence, the extracted segmentation sequence features are more accurate and can improve the accuracy of the similarity measure.

3.2 Piecewise Linear Representation Method

The PLR piecewise linear representation method has a good data compression effect [10]. The basic idea of the PLR method is to select a local extreme point or inflection point in the sequence. If the ratio of the point to the end point is greater than the parameter ε, the point is considered as a feature point. By connecting all the feature points with a straight line, a segmented time series based on the feature points can be obtained. The size of the parameter ε determines the segmentation granularity of the segmentation representation.

$S = [u_1, u_2, \ldots, u_i, \ldots, u_N]$ is the original sequence, where $i = 1, 2, \ldots, N$, N is the sequence length, and $u_i = (t_i, x_i)$ is the i-th point in the original sequence, indicating that the value of data at time t_i is x_i. The basic idea of using the PLR method is to select the feature points in the sequence to form a series of line segments. Therefore, the PLR segment sequence is represented as $S' = [u'_1, \ldots, u'_j, u'_{j+1}, \ldots, u'_m]$, where u'_j, u'_{j+1} are the start and end points of the jth segment, and m is the number of feature points. The vector sequence of S' is defined as $\mathbf{P} = [\vec{p}_1, \ldots, \vec{p}_j, \ldots, \vec{p}_{m-1}]$, where $\vec{p}_j = u'_{j+1} - u'_j$ is the two-dimensional vector corresponding to the segment j in the segment sequence, representing the direction of the segment j, reflecting the development trend of the current segment.

3.3 Similarity Measure Steps

Assuming that UAV has acquired a lot of experience situation information in a battlefield environment, the UAV will enter the environment through a new track, and the situation under the track is a new situation. The command system can measure the similarity between the experience and the new situation, extract a part with a high degree of similarity, and shift the decision strategy of this part to a new situation, so as to achieve the goal of obtaining a better decision in a short period of time under the new situation. The specific measurement steps are as follows:

Step 1. Divide the time series. A large number of situation data in the empirical situation is divided into different time series according to different tracks, which is expressed as $L = \{L_1, L_2, \ldots, L_n\}$, where n is the total number of sequences; UAV obtains data information under a new situation through a new track and is expressed as S.

Step 2. Time series representation. After completing the division of the time series, consider the above metrics for each set of time series and express any set of time series as

$$L_i = \begin{bmatrix} d_{r1} & d_{r2} & \cdots & d_{rN} \\ \theta_{r1} & \theta_{r2} & \cdots & \theta_{rN} \\ d_{m1} & d_{m2} & \cdots & d_{mN} \\ \theta_{m1} & \theta_{m2} & \cdots & \theta_{mN} \\ d_{g1} & d_{g2} & \cdots & d_{gN} \\ d_{D1} & d_{D2} & \cdots & d_{DN} \\ \phi_{D1} & \phi_{D2} & \cdots & \phi_{DN} \\ v_1 & v_2 & \cdots & v_N \\ h_1 & h_2 & \cdots & h_N \end{bmatrix} \tag{2}$$

Among them, d_r, θ_r, d_m, θ_m, d_g, d_D, ϕ_D, v and h indicate the relative distance and relative angle of our UAV and enemy radar, the relative distance and relative angle to the enemy missile, the relative distance to the enemy antiaircraft gun, and the relative distance to the target azimuth, own flight speed and flight altitude, subscript $1, 2, \ldots, N$ indicates the length of the sequence. This article adopts the method of sub-elements measurement for situation data, that is, measures the similarity between each element separately, and finally integrates the overall similarity according to the proportion of each element. The following uses the relative distance d_r from the radar as an example to measure the similarity of the element in the old and new situation. The remaining elements can be measured in the same way. So this time series can be expressed as:

$$L_i = [d_{r1}, d_{r2}, \ldots, d_{rN}] \tag{3}$$

Step 3. Extract time series trends. The EMD method was used to perform empirical mode decomposition and fit each trend sequence to obtain a fitting time trend sequence.
Step 4. The PLR piecewise linear representation method is used to segment the trend fitting sequence of each metric element to obtain each metric element segment sequence.
Step 5. Equal length processing is performed on different sequences of segmented sequences and a vector sequence of segmented sequences is calculated.
Step 6. Calculate the similarity between different vector sequences. The two equal-length vector sequences are $\mathbf{P} = [\vec{p}_1, \ldots, \vec{p}_i, \ldots, \vec{p}_{m-1}]$, $\mathbf{Q} = [\vec{q}_1, \ldots, \vec{q}_i, \ldots, \vec{q}_{m-1}]$, respectively, and the angle between each segment vector is $\varphi_i = <\vec{p}_i, \vec{q}_i>$, $i = 1, 2, \ldots, m-1$. To make the similarity proportional to the distance, define the segment vector distance as:

$$s_i = 1 - \cos\varphi_i = 1 - \frac{\vec{p}_i \cdot \vec{q}_i}{|\vec{p}_i||\vec{q}_i|} \tag{4}$$

The segmented vector distance interval is $[0, 2]$. The larger the angle between different segment vectors is, the larger the vector distance is, and the lower the similarity is; the smaller the included angle is, the smaller the vector distance is, and the higher the similarity is. In order to improve the accuracy of the calculation result, the weight $(t_{i+1} - t_i)/t_N$ is introduced in the similarity formula to represent the weight of the similarity s_i of the segment i in the overall similarity, where t_N represents the length of the entire sequence, and $t_{i+1} - t_i$ represents the length of the segment i. Then the overall similarity of the two sequences can be obtained by segmented similarity:

$$Sim(\mathbf{P}, \mathbf{Q}) = \sum_{i=1}^{N} (t_{i+1} - t_i)s_i/t_N \tag{5}$$

Using EMD to extract sequence trends before segmentation to reduce the impact of noise, can more accurately reflect the morphological changes of the sequence, so that the segmentation of the extracted trend features more accurate. By measuring the similarity between new situation and experience, we can extract experience knowledge with a high degree of similarity in the new situation and provide a good decision-making basis for UAV operations.

4 Simulation Analysis

4.1 Experimental Data Set

The experiment selects the ChronineConcentration, MALLAT, Adiac, and Phoneme time series data in the UCR Time Series Classification Archive data set as the analysis object, and compares the existing similarity measurement algorithm with the classification performance. The relevant information of the data set is shown in Table 1. The performance parameters of this experimental platform are shown in Table 2.

Table 1. Data set parameters

Name	Training set	Test set	Sequence length	Category number
ChlorineConcentration	467	3840	167	3
MALLAT	55	2345	4024	8
Adiac	390	391	176	37
Phoneme	214	1896	2310	39

Table 2. Experimental platform parameters

Structure	CPU	RAM	Operating system	Software platform
Configuration	AMD A8-5550 M 21 GHz	8.00 GB	64bit Windows7	Matlab R2014a

4.2 Classification Effect Comparison Test

In order to prove the feasibility and superiority of this algorithm, we use the similarity measure algorithm of this paper and the traditional Euclidean distance method and the dynamic time warping distances method to perform experiments. The test sequence is found most similar sequence in the training center to realize the nearest neighbor classification. Then observe and compare the classification results of each algorithm, and measure the classification effect by judging the consistency of the tags between the test sequence and the most similar sequences. The experimental results and time-consuming of each algorithm are shown in Table 3.

Table 3. Experimental results

Name	Algorithm of this paper		Euclidean		DTW	
	Error rate	Time consuming	Error rate	Time consuming	Error rate	Time consuming
ChlorineConcentration	0.113	23 s	0.217	41 s	0.115	50 s
MALLAT	0.219	35 s	0.379	53 s	0.252	63 s
Adiac	0.384	16 s	0.390	29 s	0.392	37 s
Phoneme	0.148	24 s	0.170	40 s	0.155	58 s

From the experimental error rate, we can see that for the ChlorineConcentration sequence and the MALLAT sequence, the classification error rate based on the similarity measurement algorithm of this paper is obviously lower than the traditional Euclidean distance measurement method, which is equivalent to the DTW method. The error rates of the three algorithms in the experiments of Adiac sequence and Phoneme sequence are similar, but the algorithm in this paper is still slightly lower than the traditional Euclidean distance method and the DTW method. From the time-consuming point of view of the experiment, for the four sequences, the time consuming of the algorithm in this paper is lower than that of the Euclidean distance method and the dynamic time bending distance method, and only about half of them. In summary, the similarity measurement algorithm proposed in this paper can effectively measure the degree of similarity between different time series, and the calculation result is stable and effective, and the time is short. It is suitable for the similarity analysis and related research of the sequence.

5 Summary

Aiming at some deficiencies of the existing methods, an EMD-based sequence similarity measurement method is proposed to measure the similarities between UAV operational situations and verify the effectiveness of the algorithm on the common data set. The method can extract the trend information of the sequence while reducing the influence of noise, effectively measure the morphological trend similarity of the sequence, and overcome the feature that the Euclidean distance is sensitive to the data translation, and the method is simple and short in time. This method is used to measure the degree of similarity between the new situation and the experience situation in UAV operations, and can accurately extract the similarity in the experience situation to the new situation. The use of experience in operational decision-making knowledge can make the best decision-making program for combat operations in the current situation and improve the completion probability of combat tasks.

References

1. Wang, X.G., Zhang, W.G., Chen, W.: BVR air combat decision making and simulation for UAV formation. Control Decis. **2**, 328–334 (2015)
2. Fang, J., Wang, Y., Yuan, J.: Measurement of information superiority based on set distance. Syst. Eng. Electron. **39**(1), 114–119 (2017)
3. Zhang, Y., Wang, Y.Z., Cao, Z.S.: Calculating the distance of time serier by form-fitting. J. Huazhong Univ. Sci. Technol. (Nat. Sci. Ed.) **40**(8), 72–76 (2012)
4. Berndt, D.J.: Using dynamic time warping to find patterns in time series. In: KDD Workshop, pp. 359–370 (1994)
5. Chen, Y., Hu, B., Keogh, E., et al.: DTW-D: time series semi-supervised learning from a single example. In: ACM SIGKDD International Conference on Knowledge Discovery and Data Mining, pp. 383–391. ACM (2013)
6. Guo, R., He, X.Y.: Preprocessing method for intelligent analysis of battlefield situation data. Electron. Technol. Softw. Eng. **16**, 157 (2017)
7. Bellare, M., Ristenpart, T.: Multi-property-preserving hash domain extension and the EMD transform. In: Advances in Cryptology—ASIACRYPT, pp. 299–314. Springer, Berlin (2006)
8. Huang, N.E., Shen, Z., Long, S.R., et al.: The empirical mode decomposition and the Hilbert spectrum for nonlinear and non-stationary time series analysis. Proc. R. Soc. A Math. Phys. Eng. Sci. **1998**(454), 903–995 (1971)
9. Damerval, C., Meignen, S., Perrier, V.: A fast algorithm for bidimensional EMD. IEEE Signal Process. Lett. **12**(10), 701–704 (2005)
10. Keogh, E.J., Pazzani, M.J.: An enhanced representation of time series which allows fast and accurate classification, clustering and relevance feedback. In: International Conference on Knowledge Discovery and Data Mining, pp. 239–243. AAAI Press (1998)

Research on Parking Guidance Information System Based on "Internet Plus"

Yang Li(✉)

College of Information Technology, Shanghai Jianqiao University, Shanghai 201306, China
16083@gench.edu.cn

Abstract. In view of city parking difficulties, the parking guidance information system (PGIS) based on "Internet Plus" is researched. After analyzing the shortcomings of the traditional PGIS, in order to realize the information sharing and the unified hierarchical management, the open and multilevel management mode PGIS based on "Internet Plus" is proposed, the overall structure and main functions of the PGIS are designed. Two kinds of parking information publication methods based on variable message sign (VMS) and parking app are put forward. The composition and level division of VMS are studied. Functions, software modules, parking guidance process and development strategy of the parking app are researched. The PGIS can realize parking information and resource sharing, effectively integrate existing parking resources, improve the utilization rate of parking space, and guide drivers to find and arrive the parking space efficiently, it can effectively alleviate city parking difficulties.

Keywords: Parking guidance information system (PGIS) · Viarable message sign (VMS) · Parking app · Internet plus

1 Introduction

With the rapid development of social economy, parking is a common problem in big cities, which is unfavorable to the sustainable development of urban economy and society. Therefore, it is imperative to make reasonable research on parking problems in urban traffic and put forward reasonable solutions.

Parking guidance information system (PGIS) is an effective measure to improve the utilization ratio and turnover rate of parking lots from the aspect of parking management. PGIS is an important branch of the intelligent transportation system (ITS). PGIS collects the information of location and space usage of parking lots, obtains the traffic information of the surrounding roads through ITS. After the data fusion processing, the information related to the parking is published to the drivers in real time and accurately, which leads the drivers to find the spare parking space quickly and efficiently. PGIS can consolidate the existing parking resources in the urban area, improves the utilization of parking space, and reduces the traffic induced by the vehicle seeking parking space. It can also help to alleviate traffic congestion, less car cruise time and traffic pollution, and build a good overall environment for urban traffic [1, 2]. This paper put forward PGIS based on "Internet Plus", the overall structure and the main functions of the PGIS

J. Abawajy et al. (Eds.): ATCI 2018, AISC 842, pp. 452–461, 2019.
https://doi.org/10.1007/978-3-319-98776-7_49

was designed, two kinds of parking information publication methods based on viable message sign (VMS) and app were researched.

2 Architecture Design of PGIS Base on "Internet Plus"

2.1 Disadvantages of Traditional PGIS

Considering the development of modern cities, there are two disadvantages of traditional PGIS:

- It is in the rapid development of the network age, the development speed of each industry which is also driven by the emerging network and information technology, and the transportation industry is inevitably associated with new technology such as Internet, Internet of Things (IOT) and Big Data. In this situation, if the data collected by one or more information sources is not shared and used interactively, will create the information islands and the accuracy of information will be reduced. In other words, if the parking information can be shared and the data of other information platforms can be used, traffic congestion and travel costs will be reduced.
- Now, it is mainly the construction of regional PGIS, and with the continuous acceleration of urbanization process, more and more cities have emerged the development trend of multi centers and groups. If there is a lack of relevancy among regional PGISs, it is not conducive to the management and control of city.

2.2 The Proposal of PGIS Based on "Internet Plus"

In view of the above two disadvantages of traditional PGIS, this paper proposed "Internet Plus" based PGIS with open and multi-level management mode.

The "Internet Plus" based open mode is for the integration and sharing of parking information, refers to the parking information collected by the PGIS can be shared and interactive used with not only all regional PGISs, but also other information platform such as geographic information public service platform, public transportation information control center, public security management information center and so on.

The "Internet Plus" based open mode PGIS is an effective way to solve the parking problem. The problem of parking difficulty can be due to the blocking of parking information to a large extent, and high efficiency and big data operation under the Internet can solve parking information blocking issue [3]. Through the Internet and intelligent sensing technology, a large amount of parking information can be timely transmitted to the driver's mobile phone. At the same time, "Internet Plus" mode makes use of the technology of IOT, mobile payment and so on, it simplifies the link of parking processing [4]. In addition, its real-time publishing of parking information and other add-value services, through the sharing of information on-line and off-line, can significantly improve the utilization of parking resources.

The multi-level management and control is proposed for the multi centers and groups city development mode. It refers to the establishment of a management and control center with three or four levels of multi-level structure to achieve the level by level management for all the PGISs in city wide. The three-level structure of PGIS is used for small to middle scale parking lot, includes region center, administrative district center and city center. The four-level structure of PGIS is used for large scale parking lot, includes parking lot management center, region center, administrative district center and city center.

2.3 The Overall Architecture of PGIS Based on "Internet Plus"

"Internet Plus" based PGIS collects the dynamic parking information through a variety of sensors, relying on the Internet, consolidates the information of scattered parking resources, realize the sharing of parking information, open up the information barriers between parking resources and the drivers, and also implement the information sharing between PGIS and other management systems of the city, can effectively alleviate the city parking difficulty and improve the overall city management capability [5].

"Internet Plus" based PGIS consists of four subsystems, which are parking information acquisition subsystem, parking information transmission subsystem, parking information processing subsystem and parking information publication subsystem [6]. The overall architecture is shown as Fig. 1.

Parking Information Acquisition Subsystem. It is through manual acquisition or automatic acquisition of video detector, acoustic detector and geomagnetic detector or other methods to collect the relevant information of the parking lots in the region. The information mainly includes the name and location of parking lots, parking space number and occupation status. Parking information acquisition is the basis for the implementation of PGIS. The accuracy of acquired information is a prerequisite for the accurate operation of PGIS, will affect the efficiency of dynamic traffic in city.

Parking Information Transmission Subsystem. It is the use of wired network or wireless network for information transmission and interaction. The information collected from the parking information acquisition devices is transmitted to the central management center. After processing and fusion, the information is published to the publication subsystem in order to guide the drivers parking.

Parking Information Processing System. It is a series of operations of processing, fusion, storage and interaction on the information related to parking and surrounding roads condition transmitted to the parking information management and control center. The function is the core of PGIS, it also provides favorable conditions for information sharing between PGIS and other city management systems.

Parking Information Publication Subsystem. It is the last section of PGIS, it mainly publishes the information processed by the central management center to the drivers who need to park. The information usually includes: parking lot location, spare parking space number, operation time and traffic condition of surrounding roads. The methods of information publication mainly include parking guidance signs, mobile app, Internet, broadcasting and vehicle navigation devices.

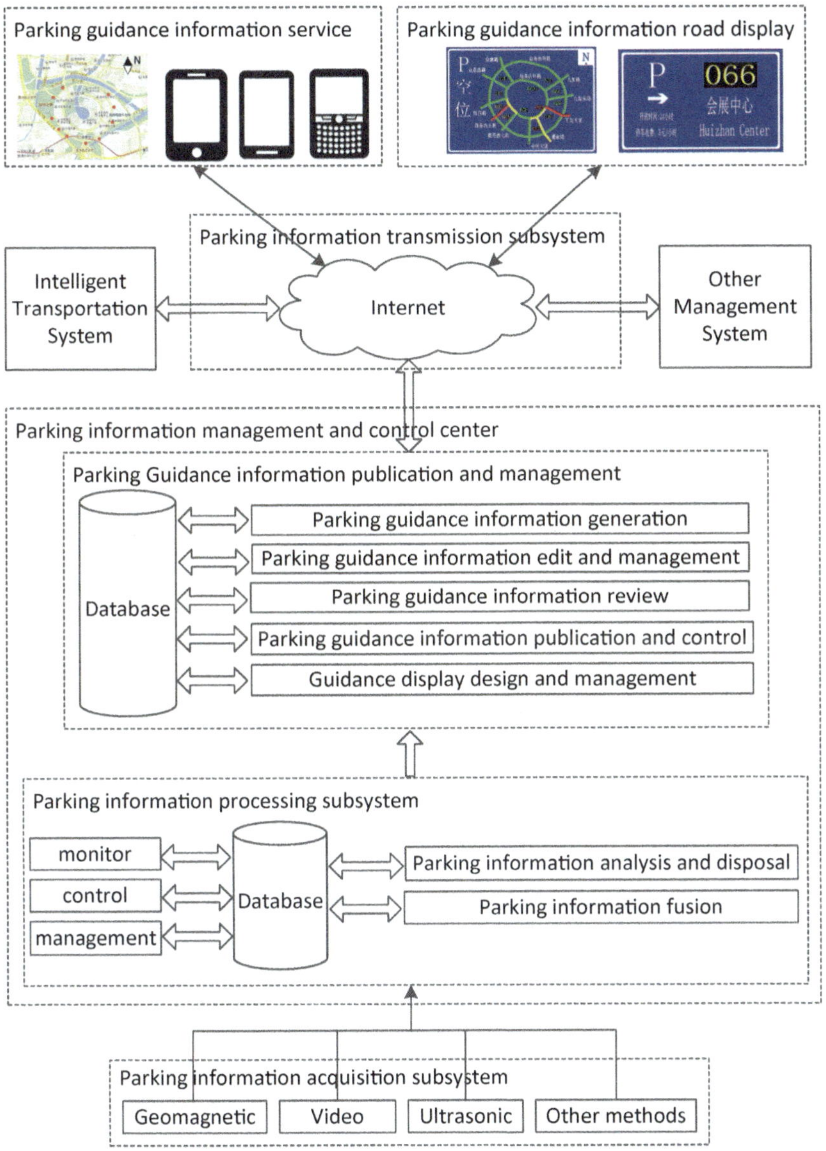

Fig. 1. Overall architecture of the PGIS

2.4 The Main Functions of PGIS Based on "Internet Plus"

The main work flow of the PGIS is the parking information acquired by acquisition devices from various parking lots, then transmitted to the regional guidance management center after consolidated by the parking lot management center, and then after the information processing, fusion, interaction and storage, the parking guidance information is published by parking information publication subsystem, and the information is also transmitted to the higher level management centers. The main function of the PGIS is to provide the real time parking information to the parking demanders, process and store parking information, and share data to other management systems of the city [7]. The direct function is to provide necessary parking information for city managers, parking managers and parking demanders. The analysis shows that the PGIS function needs to include parking information acquisition, processing, transmission and publication. The main work flow and function is shown as Fig. 2.

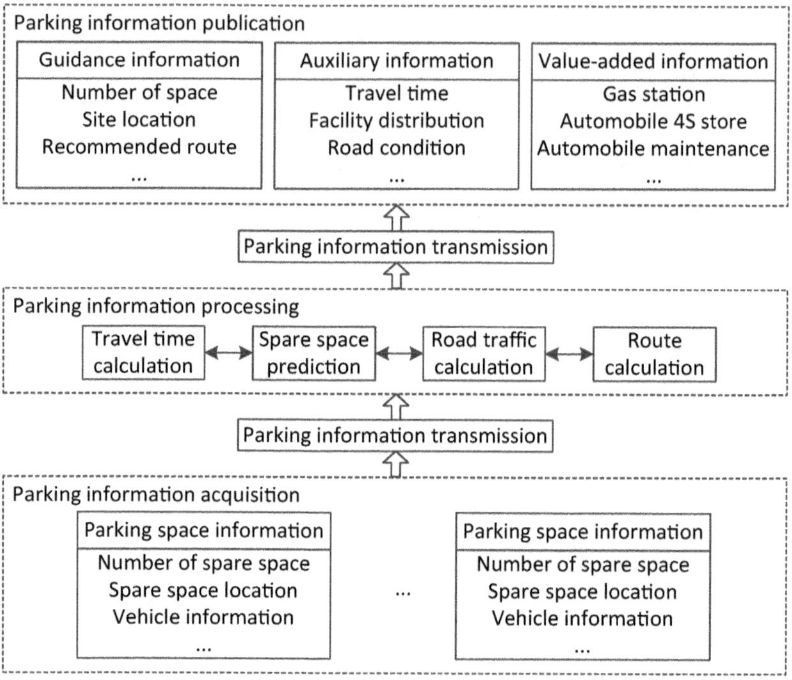

Fig. 2. The main flow and functions of the PGIS

The main purposes and functions of the PGIS include the following aspects:

- The location and the driving path of the parking lot are specified for easy search.
- The number of spare parking space in parking lots is provided in real time.
- Service of spare parking space enquiry for target parking lots is provided.
- Quickly find the target parking lot and reduce the cruising time.
- Improve utilization of parking lots, relieve parking pressure and improve safety.
- The parking information is fused and stored to achieve information sharing with other management systems of the city.
- Balance the distribution of vehicles in the road network, improves utilization of road resources.
- Through sharing the private and enterprise parking resources, create more parking spaces for the public parking demanders.
- Based on "Internet Plus" mode, provide related value-added services.

3 Parking Guidance Information Publication Based on VMS

Variable message sign (VMS) is an important device for parking guidance information publication, it is used as the basic display unit with LED light emitting device, and has the function of graphic and text display. According to the instructions from traffic, parking information and command and dispatch department, the operator can send various announcement and traffic control information to the VMS, so as to effectively induce the city traffic flow, effectively improve the utilization of the parking lot and the traffic capacity of the road network, and provide the high quality service for safe and fast parking. Because of the advantages of VMS in dynamic information publication, it has become the main force of information publication for PGIS [8].

3.1 The Composition of Information Publication System Based on VMS

In order for the VMS to display accurate, timely and reliable parking information, the VMS must include the following modules: the communication module, the LED display module, the controller and internal software, and other auxiliary devices.

The communication module is responsible for the data exchange between the information processing subsystem and the VMS, receives and demodulates the parking information and transfers it to the controller. The controller analyzes and processes the received information, control the order and process of information publication. At last, the controller sends the information that needs to be published to the LED display module, and then completes the publication of parking guidance information.

3.2 Hierarchical Division of VMS

The parking information required by drivers varies in different regions and situations. When it is in the far area, the driver pays more attention to the distribution of the parking lots. When it is closer to the parking lot, the driver begins to pay attention to whether there are spare spaces in the parking lot, and then the driver will pay attention to the entrance location of the selected parking lot. In view of the need to convey different information to drivers in different regions and locations, the VMS can be classified into four levels, and different levels have different guidance functions [9].

First Level VMS. It is used to publish the location, direction, occupation and road traffic information of parking lots, shows the overall parking space information in the area. It is usually set up on the main road of the area. The driver can choose the appropriate parking area according to the guidance information provided by the VMS. The first level VMS is also known as the regional parking guidance sign.

Second Level VMS. It is usually set at the key intersection of the parking path, or in the streets with dense parking lots, guides the driver to choose which parking lot. The difference between the second level and the first level is that the first level VMS only provides rough guidance information, and the second level VMS provides the specific information of the parking lot, includes name, number of spare parking space, direction and distance, to help the driver choose the parking lot.

Third Level VMS. It is also called the entrance guidance sign of the parking lot, which mainly provides the specific information of the parking lot, includes entrance location, number of spare parking space, opening time and parking fee. When the driver arrives at the parking lot, he can know the utilization information of the parking lot through the VMS, and decide whether to choose the parking lot.

Forth Level VMS. When the driver enters the parking lot, it is necessary to provide the specific location of the spare parking spaces inside the parking lot to the driver. This requires the guidance of the spare parking spaces by the forth level VMS inside the parking lot, so that the driver can find the spare parking space for parking quickly and reduce the driver's cruising inside the parking lot.

4 Parking Guidance Information Publication Based on Parking App

4.1 Main Functions of Parking App

With the development of new technologies, and under the background of "Internet Plus" new era, parking app on mobile devices is the product of the combination of Internet and parking, can improve the utilization and balance of existing parking resources, it is an effective way to solve the parking difficulty. Through the app of the mobile phone, drivers can query the spare parking space around the location or the destination, and carry out continuous positioning and navigation [10]. Drivers can also reserve the parking space by app before travelling, can share spare parking space through sharing private parking spaces and enterprise parking spaces, to maximize the utilization of existing parking resources. Parking fee charging and settlement through the modern electronic settlement platforms, improve the efficiency of entering and leaving the parking lot, alleviate city parking difficulties with the minimum cost.

4.2 Parking App Software Modules

Parking app on mobile client mainly includes parking information query, parking space sharing, navigation, parking space reservation and other value-added service functions. The key modules of the app software are described as below.

Parking Space Information Query Module. The module is the core part of the app, it belongs to parking information publication subsystem of PGIS. Parking app publishes parking information to the driver, the driver can query the spare parking space of the parking lots in the region on the app, and then reach the parking space smoothly through the navigation function. When shortage of parking space, the driver can make a reservation to the parking space on the app first, if the reservation is successful, the smart parking lock on the parking space will lift the landing lever to realize the possession of the parking space, and then the driver will navigate to the parking lot.

Parking Space Sharing Module. It can reallocate the parking resource, improve the utilization of spare parking resources, and alleviate the difficulty of parking to a certain extent. The function usually needs to be used with the smart parking lock. Private parking space owners can share their spare parking spaces through the parking app. Other drivers can carry out information query of the shared parking space through the app. The driver can book the spare parking space through the app if needed, and then the private parking space owner sends a temporary password of the smart parking lock to the driver, and the driver can unlock the smart parking lock through the temporary password when arrive the parking space, and finally the parking is successful.

Value-Added Service Module. The module mainly provides the driver with other service information related to the vehicle through parking app, such as gas station location, automobile 4S store information, automobile maintenance information and so on. This module can provide more convenient information for drivers, and it can also improve the economic benefits of app provider.

4.3 Information Publication and Guidance Process of Parking App

Parking information published by parking app mainly comes from the parking information acquired by the information acquisition subsystem of PGIS, and a small part of the parking information comes from the private parking space owners. The parking information is published after the processing of the app background management platform. The driver can check the parking information by the app, and can book the parking space in advance if necessary. Then the navigation function of app can guide the driver with the parking route, to improve the convenience of parking. After parking, the driver can also pay the parking fee through the app online payment function.

4.4 The Development Strategy of Parking App

The development of the parking app can be divided into three steps. Through the step by step mode to guide the driver's parking behavior, to gradually build the habit of all parking must be booked in advance; otherwise, parking can not be successful.

First Step - Parking Query. The first step launched by app is to implement the function of parking query only, and it is free of charge, so as to achieve the goal of popularizing and occupying market quickly. The necessary premise is that the app must be convenient and practical for drivers to use. Only in this way, the app can have vitality.

Second Step - Parking Space Sharing. On the basis of the first step, when the app users reach a certain volume, the parking space sharing function can be launched. Parking space sharing function is mainly for private spare parking spaces owners to rent out parking spaces through the app and obtain certain benefits. There is a small number of parking reservation functions in this process, but at this step, the app only provides free information services, only through this way to train the app users' parking habits, to lay the foundation for the third step.

Third Step - Parking Reservation. On the basis of the previous two steps, and people's habit of parking through app has been initially developed, the parking reservation function will be launched at this stage. This step is also an important step for the app investors to get return on investment.

1. Through parking reservation, the app platform can set up a cash flow, and all parking settlement must be settled through the platform. Through rechargeable and preferential activities, a huge cash pool can be quickly established and lay an economic foundation for the strategic investment later.
2. After all vehicle parking reservations are realized, the app platform can accumulate huge travel data. Through the data analysis, the commercial value of the data can be fully excavated, such as the OD survey, road planning, restaurant location selection, and so on. These useful travel data will produce considerable economic value.
3. After all parking fees are settled through the app platform, when the driver booked the parking space, the parking fee began to be calculated. After vehicle arrival, the smart parking lock was unlocked. When vehicle departure is detected by the parking lock, the billing is ended. The parking cost is calculated and deducted from the deposit amount automatically. During operation time, the gate bar of parking lot entrance keeps opened all the time, the vehicle can drive directly into and out of the parking lot, avoiding the phenomenon of queuing and waiting at the entrance of parking lot, so as to realize the unimpeded parking.
4. Through the app, the intelligent management of parking lots can be realized, and the management cost can be reduced by reducing staffing. Each parking lot only need one or two personnel to maintain order, and all other jobs can be done by parking app and smart parking locks.

5 Conclusions

This paper uses the idea of "Internet Plus" to solve the parking problems in the city, PGIS based on "Internet Plus" was proposed, the overall structure and the main functions of the PGIS was designed, two kinds of parking information publication methods based on VMS and app were researched. The PGIS can not only realize the parking information and resource sharing, provide efficient guidance to drivers for parking, effectively alleviate the difficulties of parking in the city, but also realize the information sharing between the PGIS and other management systems in the city using the "Internet Plus" mode, to improve the overall management level of the city. Not only provides parking related services, but also provides other value-added services, which is conducive to the gradual development from an intelligent parking service provider to become a comprehensive service provider, not only to provide more convenience for users, but also to upgrade the mode of profit.

References

1. Jonkers, E., Van Noort, M.: Parking guidance modelling simulation and impact assessment. In: International IEEE Conference on Intelligent Transportation Systems, pp. 858–864 (2011)
2. Wang, C.S., Yao, G.X.: Study on parking guidance system design in urban. J. Road Traffic Saf. **2009**(1), 41–44 (2009)

3. Feng, M., Wang, M.: Intelligent parking system based on internet plus. J. Anqing Norm. Univ. **23**(2), 69–71 (2017)
4. Luo, Z.M., Ma, F.H.: Design of intelligent parking guidance and payment system based on mobile phone application. J. Comput. Era **2017**(7), 17–21 (2017)
5. Wang, L.F., Chen, H., Li, Y., Shao, H.P.: Active intelligent parking guidance system. J. Comput. Appl. **31**(4), 1141–1144 (2011)
6. Hu, Y.X.: Parking guidance information integration structure model design and implement. J. Road Traffic Saf. **2010**(1), 43–47 (2010)
7. Chen, H.R., Cai, Y.G.: Research and implementation of smart city parking system. J. Ind. Control Comput. **30**(3), 55–57 (2017)
8. Sun, D.J., Ni, X.Y., Zhang, L.H.: A discriminated release strategy for parking variable message sign display problem using agent based simulation. J. IEEE Trans. Intell. Transp. Syst. **17**(1), 1–10 (2015)
9. Zhang, B.Y., Zhou, X.T.: The method of dividing area and gradation in design of metropolises PGIS. J. Traffic Transp. **2006**(02), 69–71 (2006)
10. Gu, Y.J., Hu, F.Y.: Design and implementation of parking guidance and information system based on iOS. J. Suzhou Univ. Sci. Technol. **28**(3), 77–80 (2015)

Dynamic Multi-objective Optimization for Mixed Traffic Flow Based on Partial Least Squares Prediction Model

Juan Chen(✉), Zhengxuan Xue, and Dongxiao Han

SHU-UTS SILC Business School, Shanghai University, Shanghai 201899, China
chenjuan82@shu.edu.cn

Abstract. A dynamic multi-objective genetic algorithm based on partial least squares prediction model DNSGA-II-PLS is presented in this paper to solve the mix traffic flow multi-objective timing optimization problem with time-varying traffic demand. Take motor vehicle delay, non-motor vehicle delay and urban road service level as objectives to solve the problem. Then, the algorithm is tested under the simulation environment. And making comparation with three improved dynamic multi-objective genetic algorithms based on prediction strategy, the results show that the proposed DNSGA-II-PLS algorithm can effectively reduce the delay of motor vehicles, non-motor vehicles and pedestrian waiting time.

Keywords: Single intersection · Mixed traffic flow · Least squares
Dynamic multi-objective optimization

1 Introduction

In China, the mixed traffic flow formed by motor vehicles, non-motor vehicles and pedestrians is one of the main problems leading to serious traffic jam at the intersection.

At present, among the researches on the optimization of mixed traffic flow signal timing around the world, one method is to consider a single optimization performance objective. Based on the arriving principle of motor vehicles flow between adjacent intersections, Another approach is to synthetically consider a number of performance indicators, and weighting sum method is used to form a general goal. Its essence is to transform the multi-objective optimization problems to single-objective optimization problems to solve. For example, the delay model proposed by Wu (2017) chooses three indicators as objective function, which is to maximize traffic capacity of signalized intersection, minimize average vehicle delay and minimize the number of stopping at intersections, and the stability of traffic capacity is taken as the second objective function. Finally, it is converted into a total objective function by means of linear weighting. Although it is easier to solve the problem by transforming multi-objective to single-objective, it still belongs to the single-objective optimization model.

In solving multi-objective optimization problems, the non-dominated sorting algorithm (NSGA-II) proposed by Deb et al. (2002) has been widely used due to the fact that it can reduce the complexity of non-inferior sequencing genetic algorithms and

J. Abawajy et al. (Eds.): ATCI 2018, AISC 842, pp. 462–468, 2019.
https://doi.org/10.1007/978-3-319-98776-7_50

has the advantages of fast running speed and good convergence of solution sets and other advantages. In order to improve the efficiency of traffic signal control at over-saturated intersections, Li (2013) proposes a traffic signal control optimization algorithm based on non-dominated sorting algorithm (NSGA-II), which shows good performance.

Based on the multi-objective evolutionary algorithm NSGA-II,this paper constructs the multi-objective dynamic model of single-intersection under mixed traffic environment with the goal of vehicle delay, non-motor vehicle delay and traffic capacity at intersections.

2 Signal Timing Model of Isolated Intersection

2.1 Selection of Performance Indicators

2.1.1 Vehicle Delay

The calculation of vehicle delays at intersections is currently widely using a mathematical analysis model in the Road Traffic Capacity Manual HCM2000 (Road Capacity Manual 2007), but it is not suitable to be applied to the special case of reality in China. When calculating vehicle delays at intersections in China, the average signal control delays for every vehicle in every lane group must be calculated separately. Every vehicle delay consists of the following three parts (Xiao 2014):

$$d = d_1 + d_2 + d_3 \tag{1}$$

where d is the average signal control delay per vehicle, d_1 is the signal control delay assuming vehicles arrivals uniformly, d_2 is the incremental delay to account for the effect of random arrivals and over-saturation queues, d_3 is the initial queue incremental delay which accounts for the incremental delay caused by all accumulated vehicles in previous analysis period to initial queue at start of analysis period.

2.1.2 Non-motorized Vehicle Delay

In Chinese mixed traffic conditions, if bicycle flow arriving at the intersection is constantly changing, the average delay of bike at t can be expressed as:

$$\bar{d}_b(\mathrm{t}) = \frac{q_s^b \cdot (t_R^b)^2}{2(q_s^b - q_b(t))(t_G^b + t_R^b)} \tag{2}$$

where t_R^b is the signal red light time. When the green light at the downstream intersection lights up, the vehicle will leave the intersection with the saturated flow rate q_s^b. t_G^b is the green light time. $q_b(t)$ is the traffic flow of the bicycle at time t arriving at the intersection.

2.1.3 Pedestrian Delay

This article assumes that the average pedestrian arrival time at the intersection is uniform, and the average waiting time of pedestrians is only related to the red light time of signal, it can be expressed as:

$$\bar{T}_w = \frac{1}{2} \cdot \frac{(t_R)^2}{C} \tag{3}$$

where t_R is the red light time for pedestrians crossing street.

2.2 Dynamic Multi-objective Signal Optimization Model for Mixed Traffic Flow

2.2.1 Objective Function

This paper firstly designs a dynamic multi-objective optimization model for intersection signal control. The control variables include the green signal-to-signal ratio and the period of the intersections. The objective function takes the average delay of the vehicles at the intersection, the traffic capacity of vehicles and the average delay of non-motor vehicles three indicators into account. The objective function is as follows:

$$\min \begin{cases} f_1(t) = \bar{d}(t) \\ f_2(t) = \bar{Q}(t) \\ f_3(t) = \bar{T}_w(t) \end{cases} \tag{4}$$

2.2.2 Constraint Condition

The actual signal cycle must be satisfied with formula (5).

$$C_{\min} < C < C_{\max} \tag{5}$$

where $C_{\min}$ is the minimum long cycle of intersection; $C_{\max}$ is the maximum long cycle of intersection. This paper takes 100 s and 200 s, respectively.

The signal cycle should be more than the sum of the green time of every phase.

$$\sum_{i=1}^{n} g_i < C, i = 1, 2, \ldots, n \tag{6}$$

The maximum saturation of every motor vehicle flow should be less than 1.

$$X_i = \frac{q_i}{q_{si}\lambda_i} < 1 \tag{7}$$

where q_i is the ith vehicle actual flow rate, and q_{si} is the ith saturation flow rate. λ_i is the i-phase green ratio.

3 Prediction Model

The working principle of partial least squares prediction model used in this paper is as follows: When changes in the external environment is detected, the best population $Q_t, \cdots Q_1$ recorded in historical time can provide the optimal POS information for the time of $(t+1)$, which means that there exists a functional relationship between the optimal $POS(Q_{t+1})$ of $(t+1)$ and the best population $Q_t, \cdots Q_1$ of the historical time:

$$Q_{t+1} = F(Q_1, \cdots Q_t, t) \tag{8}$$

To predict a new solution x_{t+1} at time $(t+1)$, we need to use the optimal solution $(x_t, x_{t-1}, \ldots, x_{t-K+1})$ at K moments before time $(t+1)$. For an individual x_t in the population Q_t stored at time t, the parent individual x_{t-1} in the last time $t-1$ can be defined as the most similar individuals in the population Q_{t-1}.

$$x_{t-1} = \arg \min_{y \in Q_{t-1}} \|y - x_t\|_2 \tag{9}$$

After constructing a time series for each individual in the population and for individuals most similar to historical moments, this paper uses partial least-squares regression to predict the individuals at the next moment.

Let the size of the population be N and the dimension of the decision vector be P. In order to improve the computational efficiency, the partial least squares prediction mechanism selects the $n(n \leq N)$ individual fitting prediction models according to non-inferior rankings every time. Taking Q_t as the dependent variable and Q_{t-1} as the independent variable, the normalized array of observation variables of dependent variables and independent variables can be expressed as:

$$Q_{t-1} = \begin{bmatrix} x_{t-1}^{11} & \cdots & x_{t-1}^{1p} \\ \vdots & & \vdots \\ x_{t-1}^{n1} & \cdots & x_{t-1}^{np} \end{bmatrix}, Q_t = \begin{bmatrix} x_t^{11} & \cdots & x_t^{1p} \\ \vdots & & \vdots \\ x_t^{n1} & \cdots & x_t^{np} \end{bmatrix} \tag{10}$$

Conducting Partial least squares regression on Q_{t-1} and Q_t yields the regression equation PLS: $x_t = PLS(x_{t-1})$, where $x_t = (x_t^1, \ldots, x_t^p), x_{t-1} = (x_{t-1}^1, \ldots, x_{t-1}^p)$. Then the solution at time $t+1$ can be predicted as $x_{t+1} = PLS(x_t)$ based on the above PLS model.

4 Optimization Algorithm

For dynamic evolutionary algorithms, the algorithm needs to know if there is any change in the external environment in time in order to implement corresponding response measures. The similarity test is the tool used to determine whether the environment changes. This paper selects similarity detection operator (Farina et al. 2003) based on population statistics.

4.1 Dynamic Multi-objective Evolutionary Algorithm Based on Partial Least Squares Prediction Model

This paper has improved the NSGA-II (Deb et al. 2002) algorithm, adding the above prediction model. The algorithm framework is as follows:

Step 1: Population size pop, evolutionary algebra gen = 1, environmental change parameters n_t, τ_t, threshold δ, Initial time t_{begin}, the maximum number of environmental changes $Time_{\max}$;
Step 2: Let Time = 0, current time $t = t_{begin}$. Initialize randomly to get the initial population P_{gen} of size pop.
Step 3: The crossover and mutation operation of P_{gen} was carried out to obtain the population $P_{cross,mutation}$, and the non-inferior ranking of $P_{gen} \cup P_{offspring}$ was performed, and the sub-generation population $P_{offspring}$ with the size of pop was selected. Let $gen = gen + 1$, $P_{gen} = P_{offspring}$;
Step 4: If $Time = Time_{\max}$, output the optimal Pareto optimal solution $Q_0, Q_1, \ldots, Q_{Time_{\max}}$ for each environment state, and end; otherwise, turn to Step 4.
Step 5: If gen is the integer multiple of τ_t, we randomly select $n_\varepsilon = 10$ individuals from the current Q_t for the test population, to calculate the similarity detection operator $\varepsilon(t)$ value. If $\varepsilon(t) \geq \delta$, it is considered that the environment changed significantly. We record the optimal solution Q_t in the current environment, and let $Time = Time + 1$. If $t = t_{begin} + \frac{1}{n_t}\left\lfloor \frac{gen}{\tau_t} \right\rfloor$, turn to step 5; otherwise turn to step 2.
Step 6: If $t < 2$, then turn to step 6; otherwise turn to step 7;
Step 7: We mutate some individuals of Q_{Time}, and make the modified $Q_t^{mutation}$ as the initial solution of the new moment, that is, to make $P_{gen} = Q_{Time}^{mutation}$, then turn to step 2.
Step 8: We use the partial least squares prediction mechanism to predict the best POS position $P^{Q_{Time+1}} = PLS(Q_{Time})$ in the new time, which is used as the initial solution for the new moment solution, that is, $P_{gen} = P^{Q_{Time}}$, then turn to step 2.

5 Case Study and Simulation

5.1 Basic Information of Single Intersection

In order to test the intersection control method's, which is based on partial least-squares prediction of dynamic multi-objective evolutionary algorithm, signal control effect under the condition of mixed traffic flow, using the actual case of intersection in literature (Xiao 2014), that is, taking Nanjing Taiping North Road- Zhujiang Road's junction's intersection points' intersection as the research object. This paper uses interval TOPSIS (Technique for Order Preference by Similarity to an Ideal Solution) decision-making method (Xu 2004) for decision-making.

In order to measure DNSGA-II-PLS and other existing intelligent dynamic optimization algorithms' signal optimal control performance on the same intersection, we use the DNSGA-II-PLS to solve the signal optimization control model, and through

TOPSIS decision to get the average of 10 independent operation results. Then we contrast the algorithm with dynamic multi-objective evolutionary algorithm based on vector auto-regression prediction (DNSGA-II-AR) (Zhou et al. 2007), dynamic multi-objective evolutionary algorithm DNSGA-II-PREM (Hatzakis 2006) based on simple prediction, and mutation-based DNSGA-II-MUT (Deb et al. 2007) of Deb. The parameters are set as follows: the population size pop is 100, the crossover probability $P_c = 0.9$, and the mutation probability $P_m = 0.1$. The experimental results are shown in Table 1.

It can be seen from the calculation results in Table 1 that the DNSGA-II-PLS algorithm has better performance metrics of $\bar{d}$, $\bar{d}_b$, $\bar{T}_w$ traffic performance than those of other optimization algorithms. This indicates that compared with DNSGA-II-AR, DNSGA-II-PREM, DNSGA-II-MUT three kinds of dynamic algorithms, DNSGA-II-PLS proposed in this paper can to some extent improve the service level of urban road under the dynamic traffic demand, reduce the average delay of vehicles and the average delay of non-motor vehicles, improving the signal control efficiency in the intersection effectively. However, the above only validates the feasibility of the algorithm theoretically, and the simulation method is used to test the adaptability of the algorithm under actual road conditions below.

Table 1. Control results comparsion intersection signal optimization

	Comparison algorithm	$\bar{d}$	$\bar{d}_b$	$\bar{T}_w$
Time Window 1	DNSGA-II-PLS	35.5	26.588	25.084
	DNSGA-II-AR	45.783	36.384	34.978
	DNSGA-II-PREM	48.317	35.745	34.332
	DNSGA-II-MUT	50.963	35.139	33.720
Time Window 2	DNSGA-II-PLS	38.74	30.8	29.338
	DNSGA-II-AR	42.359	31.04	29.580
	DNSGA-II-PREM	44.02	31.296	29.839
	DNSGA-II-MUT	51.997	31.548	30.093
Time Window 3	DNSGA-II-PLS	153.711	44.758	43.435
	DNSGA-II-AR	166.024	43.72	42.387
	DNSGA-II-PREM	164.619	44.269	42.941
	DNSGA-II-MUT	166.672	43.641	42.307
Time Window 4	DNSGA-II-PLS	54.211	31.318	29.861
	DNSGA-II-AR	67.021	24.455	22.929
	DNSGA-II-PREM	63.353	24.629	23.105
	DNSGA-II-MUT	82.411	24.078	22.549
Time Window 5	DNSGA-II-PLS	37.261	31.377	29.921
	DNSGA-II-AR	37.787	30.746	29.283
	DNSGA-II-PREM	37.664	30.991	29.531
	DNSGA-II-MUT	38.046	30.702	29.239

6 Conclusion

Aiming at the traffic jam problem caused by mixed traffic flow at single intersection, this paper establishes a dynamic multi-objective optimization model based on the delay of vehicles, the delay of non-motor vehicles and the service level of urban roads, and proposes the dynamic multi-objective genetic algorithm DNSGA-II-PLS to solve the objective function. According to the simulation, the proposed DNSGA-II-PLS algorithm can effectively improve the service level of urban roads at the intersection of dynamic traffic demand and reduce the delay of motor vehicles and non-motor vehicles.

References

Wu, J.S.: Optimization model of intersection of urban road signal control intersection. Highw. Automot. Appl. (6) (2017)

Deb, K., et al.: A fast and elitist multi-objective genetic algorithm: NSGA-II. IEEE Trans. Evolut. Comput. **6**(2), 182–197 (2002)

Li, Y.: NSGA-II based traffic signal control optimization algorithm for over-saturated intersection group. J. Southeast Univ. **29**, 211–216 (2013)

American Transportation Research Council: Road Capacity Manual: Metric. People's Transportation Press, Portland (2007)

Farina, M., Deb, K., Amato, P.: Dynamic multi-objective optimization problems: test cases, approximation, and applications. In: Proceedings of the Second International Conference on Evolutionary Multi-criterion Optimization, EMO 2003, Faro, Portugal, 8–11 April 2003 (2003)

Xu, Z.S.: Uncertain Multi-attribute Decision Making Method and Application. Tsinghua University Press, Beijing (2004)

Zhou, A., et al.: Prediction-based population re-initialization for evolutionary dynamic multi-objective optimization. In: International Conference on Evolutionary Multi-criterion Optimization (2007)

Hatzakis, I., Wallace, D.: Dynamic multi-objective optimization with evolutionary algorithms: a forward-looking approach (2006)

Deb, K., Bhaskara Rao, N.U., Karthik, S.: Dynamic multi-objective optimization and decision-making using modified NSGA-II: a case study on hydro-thermal power scheduling. In: International Conference on Evolutionary Multi-criterion Optimization (2007)

Multi-objective Genetic Algorithm Based Prefabricated Component Production Process Optimization Considering the Demand Fluctuations in the Construction Site

Juan Du[1,2], Peng Dong[1,2](✉), and Xin Wang[1]

[1] SHU-UTS SILC Business School, Shanghai University, Shanghai 201800, China
pippen@shu.edu.cn

[2] SHU-SUCG Research Centre for Building Industrialization, Shanghai University, Shanghai 200072, China

Abstract. Based on the prefabricated component supply chain, the production of precast components contains series of segmentation processes, and the fluctuations in demand of components of assembly phase also affect the production scheduling. This article develops the corresponding mathematical model to characterize the production and assembly phases of prefabricated component supply chain, and uses multi-objective genetic algorithms to optimize the production processes considering the demand fluctuations in the assembly phase. Through comparative analysis, the optimal production schedule under the influence of different demand fluctuations can be obtained, which is of great significance for shortening the production period and reducing the production cost.

Keywords: Precast component · Production optimization
Multi-objective genetic algorithm · Demand fluctuation

1 Introduction

Prefabricated construction project developed rapidly in China due to the advantages of fast construction, high quality and labor saving, compared with traditional casting construction project. In prefabricated construction project, a large number of wet works at the construction site are transferred to the prefabricated component factory, which can greatly reduce the labor required at the construction site and the stacking of raw materials. As a result, the production of prefabricated component is a crucial stage in the entire PC supply chain.

However, there are still some issues impact the prefabricated component production efficiency. Firstly, the repeating usage rate of casting mold is very low. As the components are not fully standardized and the property developers prefer the unique design of architecture, the shape of components differs widely and usually the factory should build new mold for different project causing high resource waste. So, how to minimize the number of molds while satisfying on-time delivery is a question worthy of study.

J. Abawajy et al. (Eds.): ATCI 2018, AISC 842, pp. 469–476, 2019.
https://doi.org/10.1007/978-3-319-98776-7_51

Secondly, the production schedule of prefabricated components is a complex issue and there is no exact method to solve it. Currently, producers mainly perform component scheduling based on their experience. However, when the scale of the problem becomes larger, it is difficult for the decision maker to obtain an optimal solution. Therefore, it is necessary to look for a scientific algorithm to implement the production scheduling of prefabricated components.

Lastly, the demand fluctuation at the assembly stage will affect the production scheduling significantly. There are some unpredictable events may occur in the assembly stage, which will lead to the delivery date of some components being advanced or delayed for several days. So, it is very important to take into account the demand fluctuation while making production scheduling.

At present, genetic algorithm has been well applied in the production scheduling problem. Some scholars used genetic algorithm to study the production scheduling of PCs and developed some mathematical models. However, most studies have only considered the production stage, ignoring that the whole supply chain interactions will affect each stage significantly.

Based on the analysis above, this paper developed a production scheduling model for prefabricated components considering the demand fluctuation. For the production stage, this paper will develop the scheduling model with the integration of the literatures [1–3]. For the transportation stage, this paper considers that the transportation resources are sufficient since the actual research results shows that transportation vehicle is not a scarce resource. For the assembling stage, this article will use stochastic event simulation to model unpredictable events, and the occurrence of these events will have influence on production stage. Besides, genetic algorithm is used to optimize the model, and the model is verified through the data in actual project.

2 Literature Review

The research to the prefabricated component production optimization already has the following published contributions.

Chan and Hu divided the operation process of PCs into interruptible and non-interruptible operations. The daily working time is divided into normal working hours and abnormal working hours [4]. However, this model assumes that there are no resource constraints and the number of molds is sufficient. In Leu and Hwang' research the number of cranes, the size of the steam curing space and the storage space of rebar cage are regarded as resource constraints [5]. However, the article did not propose a related mathematical model. Benjaoran proposed the bespoke precast flowshop scheduling model (BP-FSSM) for production planning. This model takes the reuse of molds into account and puts forward multiple optimization goals [6]. But the model does not consider the transportation of components, which actually is an important factor affecting the timely delivery. Chien-Ho Koetal integrates the size of the buffer between each workstation in to the model [7]. In another study, the author proposed a genetic algorithm-based decision support system (GA-DSS) to assist production managers in arranging production planning [8]. Yang proposed a multi-production line workflow scheduling model for prefabrication [1]. This study sets the minimization of component type change as the optimization goal and sets the number of production pallets as a constraint. Anvari first apply the multi-objective genetic algorithm to the production, transportation and assembly of prefabricated components as a whole [2]. However, this paper mainly deals with the problem of resource scheduling among production processes and assembly processes but it does not optimize the production schedule between different types of components. Wang and Hu changed the six operations in the previous study to nine operations. Mold production, PC storage, and transportation processes were integrated to modify the traditional production scheduling model [3]. Wang proposed a two-hierarchy simulation-GA hybrid model for precast production (TSGH_PP) [9]. The model considers the uncertainty of the processing time for each operation. At the same time, it optimizes the allocation of resources to reduce waste of resources. However, the impact of the assembly phase is not considered in the model.

Although the research in this field already has many works, this paper innovatively integrates the three processes of prefabricated components, including production, transportation and assembly. By setting randomly occurring events during the assembly phase, sudden changes in the delivery date of the components will result in the synergy of the three processes. This article will use multi-objective genetic algorithm to achieve optimization. Additionally, the optimization goal is the shortest completion time, minimum additional cost and maximum resource utilization. Finally, a series of near-optimal production scheduling methods are obtained.

3 Mathematical Model

In this section, the paper develops the corresponding mathematical model to characterize the production and assembly phases of prefabricated component supply chain. The proposed mathematical model will play important role for the next step in using multi-objective genetic algorithms to optimize the production processes considering the demand fluctuations in the assembly phase.

3.1 Production Process Model

The production scheduling of PCs could be viewed as a multi-objective mixed flow shop scheduling problem. Based on the literatures [1–3], this study described the model as follows: n components are processed through m work stations, each component is processed through all work stations and the sequence cannot be changed. As is shown in Fig. 1, the paper divides the production process of prefabricated components into 8 steps, (1) Mold assembling; (2) reinforcement setting; (3) concrete casting; (4) curing; (5) mold removal; (6) cleaning and repairing; (7) storing; (8) transportation. Among them, curing, storing and transportation stations can process multiple components simultaneously, while other work stations can only process one component at a time. The daily time is divided into eight normal working hours and abnormal working hours, and the overtime hours do not exceed four hours at most. The goal of the optimization is to find the production order of all the components so as to meet on-time delivery and minimize costs.

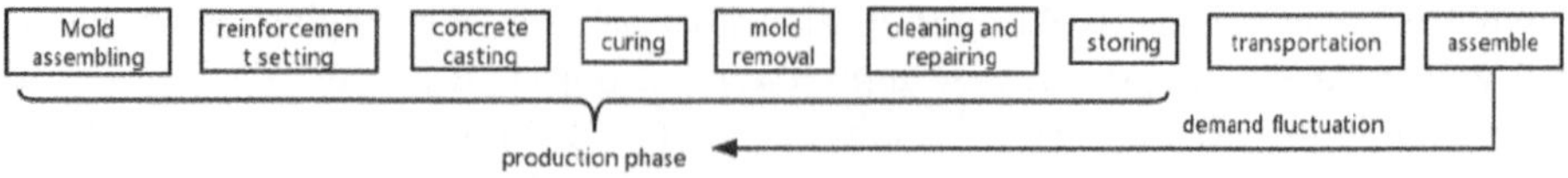

Fig. 1. Production process of prefabricated component

The completion time of component J_i at the work station k is denoted by $C(J_i, k)$. And $t_{i,k}$ represent the operation time of the component J_i at the station k, then:

$$C(J_i, k) = max\{C(J_{i-1}, k), C(J_i, k-1)\} + t_{i,k} \tag{1}$$

Considering the actual production process of prefabricated components, all operating steps are divided into interruptible and non-interruptible operations. For interruptible operations such as Mold assembling, reinforcement setting, mold removal, cleaning and repairing, storing, if workers' work exceeds normal working hours, they can stop working and continue on the next day, These will not affect the quality of the precast components. The completion time of these operations can be expressed as:

$$C(J_i, k) = \begin{cases} T & if\ T \leq 24d + H_w \\ T + H_N & if\ T > 24d + H_w \end{cases}, \quad k = 1, 2, 5, 6 \tag{2}$$

where H_w is the normal working hours per day, $H_w = 8$; H_N is the non-normal working hours per day, $H_N = 24 - H_w$; T is the cumulative completion time calculated under the condition that the normal working hours of 8 h are not considered; D is the number of days used from the start of production to this step. The calculation of T and D can be expressed as:

$$\mathrm{T} = \max\{C(J_{i-1}, k), C(J_i, k-1)\} + t_{i,k} \tag{3}$$

$$\mathrm{D} = \mathrm{integer}(T/24) \tag{4}$$

For non-interruptible operations such as concrete casting, curing, and transportation, these operations must be completed once they are started. Otherwise, the quality of the components will be affected. If these operations can be completed within the overtime hours, workers have to work overtime to complete the operation. If these operations cannot be completed within the overtime hours, the beginning of these operations should be delayed to the next day.

For the casting operation, the calculation of its completion time can be expressed as:

$$C(J_i, \mathrm{k}) = \begin{cases} T & if\ T \leq 24D + H_w + H_o \\ 24(D+1) + t_{i,k} & if\ T > 24D + H_w + H_o \end{cases}, \quad k = 3 \tag{5}$$

where H_O is the overtime hours, $H_O = 4$.

For curing and storing, these stations can handle many components simultaneously, therefore, this operation can be performed as soon as the operation of the component at the previous station is completed. Since these two operations hardly require human resources, these operations can also be performed during non-normal working hours. If the completion time is within the abnormal working hours, the calculation of the completion time should be postponed to the start of the next day. The completion time for curing and storing can be expressed as:

$$C(J_i, \mathrm{k}) = \begin{cases} 24(D+1) & if\ (24D+H_w) < T^* < 24(D+1) \\ T^* & if\ T^* \leq 24D+H_w\ or\ T^* \geq 24(d+1) \end{cases}, \quad k = 4, 7 \tag{6}$$

where $\mathrm{T}^* = \mathrm{C}(\mathrm{J_i}, \mathrm{k}-1) + \mathrm{t_{i,k}}$.

For the transportation phase, the transport vehicle can transport multiple components at a time and the transport vehicle is sufficient in the actual project, so the component can transport to the construction site as soon as the storage is completed. However, this operation requires human resources, if the completion time is within the non-normal working hours, the start time of the operation needs to be moved to the next day. The completion time of transportation can be expressed as:

$$C(J_i, \mathrm{k}) = \begin{cases} 24(D+1) + t_{i,k} & if\ T^* > 24D + H_w + H_O \\ T^* & if\ T^* \leq 24D + H_w + H_O \end{cases}, \quad k = 8 \tag{7}$$

3.2 Optimization Constraints

(1) Constraint of mold

Due to the high cost of molds and the low rate of reuse, the waste is severe. So the manufacture of molds is usually limited, which requires us to distribute the molds reasonably. A component must have a corresponding type of mold to start production, otherwise it must wait until other component removes the mold. The time that component $\mathrm{J_i}$ waits for its corresponding type ϕ mold is expressed as:

$$C(J_{i,\phi}, 0) = min\{\forall y\{C(J_{y,\phi}, 5)\}\} \tag{8}$$

where $\mathrm{J_{i,\phi}}$ is the component which mold type is ϕ and production sequence is i.

$\mathrm{C}(\mathrm{J_i}, 0)$ is the start time of component $\mathrm{J_i}$ in work station 1. $\mathrm{C}(\mathrm{J_i}, 5)$ is the completion time of mold removal of component $\mathrm{J_i}$.

(2) Constraint of storage zone between work stations

If the next station is occupied, the component should wait in the storage zone. If the storage capacity between station k and station k + 1 is $\mathrm{B_k}$, The completion time of the component $\mathrm{J_i}$ in work station k must be later than the start time of the component $\mathrm{J_{i-B_k}}$ in work station k + 1, so as to ensure that the number of components in the storage zone is less than $\mathrm{B_k}$.

$$C(J_i, k) \geq S(J_{l-B_k}, k+1) \tag{9}$$

3.3 Demand Fluctuation Modeling of Assemble

During the assembly process, if the construction progress is carried out normally, the demand of component can be satisfied according to the predetermined production plan. However, due to a variety of uncertainties, it often affects the progress of construction. Changes in the construction schedule will lead to changes in the demand for components, and the production schedule should be changed according to this effect. For example, when foundation pits were excavated, if runoff and piping occurred, it would take a long time to process these factors, causing the delivery dates of all components in the order to be postponed for several days. When the construction progress of a certain floor is particularly fast, the delivery date of the components of the next floor needs to be several days in advance; when the construction progress of a certain floor is slower than usual, the delivery date of the components of the next floor needs to be delayed several days. This formed the demand fluctuations of the assembly side, and the production side needs to dynamically perform the production scheduling of the components in accordance with such fluctuations in demand.

Because the influence of this kind of demand fluctuation is difficult to model, this paper uses the form of random events to simulate the uncertain events in the assembly process. The paper divides the demand fluctuations into two forms, global demand fluctuations and local demand fluctuations. If random events that represent local demand fluctuations occur, some components will be randomly selected and the delivery date parameters of these components will be changed. If random events that represent global demand fluctuations, the delivery date parameters of all components in the order will be advanced or delayed by several days.

$$if\left(\text{rand} < P_g\right)\forall d_i = d_i \pm \Delta d \tag{10}$$

where rand is a random number between 0-1; P_g is the probability of occurrence of global demand fluctuations; Δd is the number of days changed.

$$if(\text{rand} < P_l)d_{randn} = d_{randn} \pm \Delta d \tag{11}$$

where P_l is the probability of occurrence of local demand fluctuations; d_{randn} represents the due date of a randomly selected component.

3.4 Optimization Objective

(1) Total completion time

$$f_1(x) = C(J_n, 8) \tag{12}$$

where $\mathrm{C(J_n, 8)}$ is the completion time of the last component

(2) Machine idle time

$$f_2(x) = \sum_{k=1}^{8}\left[C(J_n, k) - \sum_{i=1}^{n} t_{i,k}\right] \tag{13}$$

(3) Extra costs
Delays in the completion of components will result in an increase in penalty, while early completion will require additional sites for stacking, this will also result in additional costs. The extra cost can be expressed as:

$$f_3(x) = \sum_{i=1}^{n}[\alpha_i \cdot max(0, C_i - d_i) + \beta_i \cdot max(0, d_i - C_i)] \tag{14}$$

where α_i is the unit storage fee due to early completion; β_i is the unit penalty due to delay; C_i is the completion time of component J_i; d_i is the delivery time of component J_i in contract.

4 Analysis and Discussion

This paper introduces a method to optimize the production schedule of prefabricated components by considering the fluctuation of demand in the assembly stage, so as to shorten the completion time and cost of the production. The future research intends to use software to conduct experiments. The parameters of the genetic algorithm will be selected in conjunction with the references [1–3]. The experiment will simulate the fluctuation of the demand of the assembly stage by introducing random events. By conducting the experiments under different scenarios and comprehensively comparing various experimental results, a more robust production scheduling scheme can be finally obtained, which can better adapt to fluctuations of the demand in the assembly stage.

Acknowledgements. This work was supported in part by the National Natural Science Foundation of China under Grant 71701121, in part by the Chinese Ministry of Education of Humanities and Social Science Project under Grant 17YJC630021.

References

1. Yang, Z., Ma, Z., Wu, S.: Optimized flowshop scheduling of multiple pro duction lines for precast production. J. Autom. Constr. **72**, 321–329 (2016)
2. Wang, Z., Hu, H.: Improved precast production–scheduling model considering the whole supply chain. J. Comput. Civ. Eng. **31**(4), 04017013 (2017)
3. Wang, Z., Hu, H., Gong, J.: Framework for modeling operational uncertainty to optimize offsite production scheduling of precast components. J. Autom. Construct. **86**, 69–80 (2018)

4. Chan, W.-T., Hu, H.: An application of genetic algorithms to precast production scheduling. J. Comput. Struct **79**, 1605–1616 (2001)
5. Leu, S.-S., Hwang, S.-T.: GA-based resource-constrained flow-shop scheduling model for mixed precast production. J. Autom. Constr. **11**, 439–452 (2002)
6. Benjaoran, V., Dawood, N., Hobbs, B.: Flowshop scheduling model for bespoke precast concrete production planning. J. Constr. Manag. Econ. **23**, 93–105 (2005)
7. Ko, C.-H., Wang, S.-F.: Precast production scheduling using multi-objective genetic algorithms. J. Expert Syst. Appl. **38**, 8293–8302 (2011)
8. Ko, C.-H., Wang, S.-F.: A-based decision support systems for precast production planning. J. Autom. Constr. **19**, 907–916 (2010)
9. Anvari, B., Angeloudis, P., Cheng, W.Y.O.: A multi-objective GA-based optimization for holistic manufacturing, transportation and assembly of precast construction. J. Autom. Constr. **71**, 226–241 (2016)

Anomaly Detection in Time Series Data Based on Unthresholded Recurrence Plots

Min Hu[1,2], Xiaowei Feng[1,2(✉)], Shengchen Zhou[1,2], and Wei Xu[1,2]

[1] SHU-UTS SILC Business School, Shanghai University, Shanghai 201800, China
fengxiaoweill@163.com
[2] SHU-SUCG Research Centre for Building Industrialization, Shanghai 200072, China

Abstract. The problem of detecting anomalies in time series has attracted much attention recently due to its numerous applications. In this paper, we proposed a novel framework based on nonlinear methods. Firstly, we use time-dependent URP (Unthreholded Recurrence Plots) to represent time series, so as to capture the nonlinear characteristic of time series. Secondly, we present ELM-AE (Extreme Learning Machine Auto-Encoder) based algorithm to learn the main features of URPs. Finally, we compute residual errors as anomaly score for each time series points. This framework can perform well on complex nonlinear systems using unlabeled datasets.

Keywords: Anomaly detection · Time series · Recurrence plots
Extreme learning machine

1 Introduction

Anomaly detection, also called outlier detection, refers to the problem of finding unusual patterns in data that do not conform to expected behavior [1]. It can be used to discover interesting events in a wide variety of application domains, such as intrusions in network data, mechanical faults in system sensor data, climate disturbances in temperature data, and disease patterns in health monitoring data. Numerous anomaly detection methods have been proposed in literatures. Majority of them can be categorized into classification based, nearest neighbor based, clustering based and statistical test-based methods.

The first work of anomaly detection in time series data was performed in 1972 [2]. Conventional methods are mainly based on training models and compute the outlier score for a sequential point or subsequence as its deviation from the predicted value (e.g. AR model, neural networks [3]) or pre-learned normal profile (e.g. immune system [4]). However, in practice, neither normal nor labeled data are available for model training which highlights the demand for unsupervised anomaly detection algorithms [5]. Another crucial is to develop a valid representation of the time series for feature extraction. Discrete Wavelet Transform (DWT) [6], Discrete Fourier Transform (DFT) [7] and Symbolic Aggregate Approximation (SAX) [8] are common methods to transform time series data into a more meaningful feature space. In the field of physics,

J. Abawajy et al. (Eds.): ATCI 2018, AISC 842, pp. 477–484, 2019.
https://doi.org/10.1007/978-3-319-98776-7_52

there is a more simple and flexible way to represent and extract features from time series, called time-dependent recurrence quantification analysis (RQA). It is based on recurrence plots (RPs), a nonlinear time series analysis tool proposed by Eckmann et al. in 1987 [9]. RQA refers to quantitative measures that describe the structures of RPs and is highly effective to detect transitions in the dynamics of systems from time series [10]. However, there exist some limitations that affect RPs and RQA's application such as the challenging process of tuning parameters like the recurrence threshold ε [15].

In this paper, we propose a novel anomaly detection framework based on unthresholded RPs (URPs) and ELM autoencoder model. URPs are believed to be more informative compared with RPs and do not require tuning parameters. ELM autoencoder is adopted as an unsupervised method to learn main features from window-sliding RPs, while the dissimilarity between reconstructed RP and the original one is computed as anomaly score. We will detailedly introduce the proposed algorithm for finding anomalies of the time series data in the next section.

2 Anomaly Detection Algorithm Based on URPs and ELM-AE

In this section, we introduce our framework for anomaly detection in time series using URPs and ELM-AE. We adopt the unthresholded recurrence plots (URPs) as representation method of time series data, and followed by the extreme learning machine as the unsupervised anomaly detection algorithm to find the final anomalies. The proposed framework is shown in Fig. 1 and details of the key techniques and algorithm are described as follows.

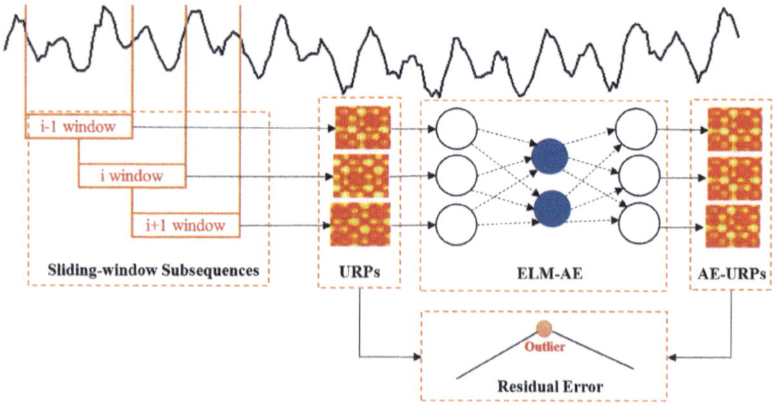

Fig. 1. The framework for time series anomaly detection based on URPs and ELM-AE

2.1 Recurrence Plots

The method of recurrence plots (RPs) was introduced to visualize the time dependent behavior of the dynamics of systems [9]. Given a d-dimensional time series dataset

$x = \{x_1, x_2, \ldots, x_i, \ldots, x_n\}$ where x_i corresponds to the i-th observed sample. The time delay embedding technique is applied to implement phase space reconstruction on the dataset x [24].

$$X_i = \{x_i, x_{i+\tau}, \ldots, x_{i+(m-1)\tau}\} \tag{1}$$

where x_i is a d-dimensional sample at timestamp i, m is the embedding dimension and τ is the delay factor. Consequently, the reconstructed trajectory X in phase space is a $(t - (m - 1)\tau) \times m$ array where every element is a vector with length d. In our experiments, we chose $m = 1$ for all datasets since recurrence plots' structure seems to remain qualitatively unchanged with changing embedding dimension [28].

Next, the recurrences of the trajectory X_i in phase space can be measured using Eq. (2).

$$R_{i,j}(\varepsilon) = \Theta(\varepsilon - \|X_i - X_j\|), i, j = 1, 2, \ldots, N \tag{2}$$

where N is the number of trajectories and equals $(t - (m - 1)\tau)$. ε is the threshold distance, $\|\cdot\|$ is the norm (e.g. Euclidean norm) and Θ is the Heaviside function (i.e. $\Theta(x) = 0$ if $x < 0$, and $\Theta(x) = 1$ otherwise).

The threshold distance ε is a vital parameter for RPs, which influences the formation of line structures. Determining an appropriate value for this parameter, however, is not intuitive. As a rule of thumb, a value of approximately 1% recurrence [29], 10% of the diameter of the reconstructed phase space [11] and 5% of the maximal phase space diameter [15] are all possible solution to threshold ε.

In order to eliminate the threshold distance parameter, we just compute the pairwise distance of each pair of trajectories, as shown in Eq. (3). This representation is known as unthresolded RPs (URPs) or distance plot. Figure 2 shows an example of a thresholded and an unthresolded recurrence plot for the same time series.

$$R_{i,j} = \|X_i - X_j\|, i, j = 1, 2, \ldots, N \tag{3}$$

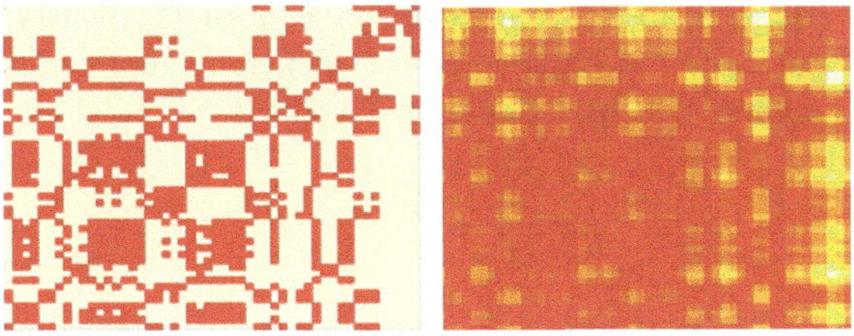

Fig. 2. Examples of a thresholded (left) and an unthresholded (right) recurrence plot generated from the same time series data

2.2 Extreme Learning Machine

Extreme Learning Machine (ELM) was originally proposed by Huang, et al. [30]. It is a machine learning algorithm based on Single Layer Feed-forward Neural Networks (SLFN). Unlike conventional neural network algorithms, the hidden nodes of ELM are independent from the input data and do not need to be tuned. The input data are mapped into L dimensional ELM random feature space and the network output is given by Eq. (4):

$$f_L(x) = \sum_{i=1}^{L} \beta_i h_i(x) = h(x)\beta \tag{4}$$

where $\beta = [\beta_1, \ldots, \beta_L]^T$ is the output weight matrix between the hidden layer of L nodes and the output node. $h(x) = [h_1(x), \ldots, h_L(x)]$ are outputs of the hidden nodes for the input x and $h_i(x)$ is the output of the i-th hidden node. In general, the activation function $h_i(x)$ can be expressed as Eq. (5):

$$h_i(x) = H(\omega_i, b_i, x), \omega_i \in \mathbb{R}^d, b_i \in \mathbb{R} \tag{5}$$

where the input weights ω_i and biases b_i are randomly assigned based on some continuous probability distribution function. $H(\cdot)$ is a nonlinear piecewise continuous function satisfying ELM universal approximation capability theorems. Given N training samples $\{(x_i, t_i)\}_{i=1}^{N}$, ELM is to resolve the following linear system optimization problem:

$$\min_{\beta} ||H\beta - T||^2 \tag{6}$$

where $T = [t_1, \ldots, t_N]^T$ are the target labels and $H = [h^T(x_1), \ldots, h^T(x_N)]^T$. The output weights β can be calculated by least-squares solution with minimum norm:

$$\hat{\beta} = H^{\dagger} T \tag{7}$$

where $H^{\dagger}$ is the Moore-Penrose generalized inverse of the matrix H. In practice, a regularization term is added to achieve a better generalization performance and make the solution more robust, which is shown in Eq. (8):

$$\hat{\beta} = (\frac{I}{C} + H^T H)^{-1} H^T T \tag{8}$$

In our experiments with ELM in this paper, the activation function is a sigmoid function: $g(x) = 1/(1 + e^{-x})$, while the probability density function is a uniform distribution function in the range from 0 to 1.

2.3 Extreme Learning Machine Auto Encoder

Extreme Learning Machine Auto-Encoder (ELM-AE) [31, 32] is similar to the basic ELM described in Sect. 2.2 except for: (1) input data are used as the output data, namely $t = x$, (2) random weights and biases of the hidden nodes are made orthogonal. Input data can be represented in three different architectures: compressed architecture, sparse architecture and equal dimension architecture. The network structure can be shown in Fig. (3).

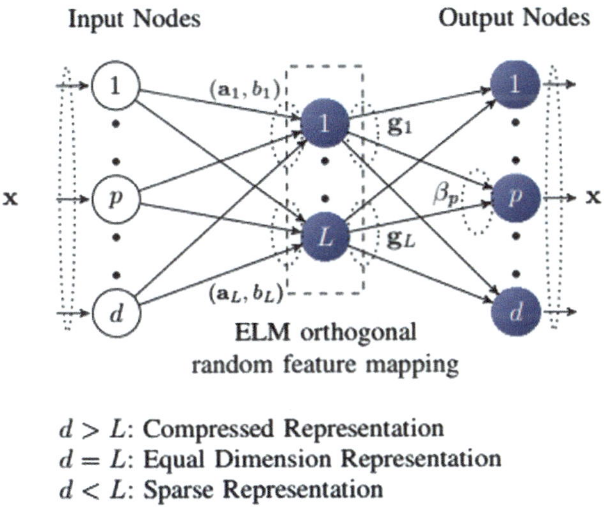

Fig. 3. Extreme learning machine auto-encoder network structure [31, 32]

With Johnson-Lindenstrauss Lemma [33], ELM-AE project the input data to a different or equal dimension space by Eq. (9):

$$h = g(ax + b), a^T a = \mathrm{I}, b^T b = 1 \tag{9}$$

where $a = [a_1, \ldots, a_L]$ are the orthogonal random weights and $b = [b_1, \ldots, b_L]$ are the orthogonal random biases between the input nodes and the hidden nodes. The output weights β are calculated by Eq. (10):

$$\hat{\beta} = \left(\frac{I}{C} + H^T H\right)^{-1} H^T X \tag{10}$$

2.4 Proposed Algorithm for Anomaly Detection

The main idea of our algorithm is to combine the strengths of URPs and ELM-AE. URPs are used to represent the time series at different periods of time with nonlinear dynamic features, and ELM-AE can learn the features automatically and identify the structure difference among a series of URPs.

The algorithm begins by normalizing the original input data according to Eq. (11):

$$z_i = \frac{y_i - \mu}{\sigma} \tag{11}$$

where μ and σ represent the mean and standard deviation of the values in the time series. The value z_i is also called the Z-score of the time series. We then perform slide a moving-window extract subsequence of length w from the normalized data and obtain a set of subsequences $\{S_i\}_{i=1}^{N-w+1}$. We then generate URPs for each subsequence, namely the distance matrix $\{R_i\}_{i=1}^{N-w+1}$ according to Sect. 2.1. During the process of phase space reconstruction, we set the constant lag $\tau = 1$ in this paper in order to get anomaly score for each point in the time series.

After the procedure of transforming the original time series into the form of URPs, we now apply the ELM-AE method to the URPs to gain the output weight matrix β. For β is responsible for the projection from feature space to output data, we can obtain the reconstructed URPs by the learn β.

Finally, we compute the residual errors between each original URP and the reconstructed one by performing $L2$-normalization. The anomaly score can also be normalized according to Eq. (11). It's worth mentioning that if we score X_t as an anomaly point, the anomaly score for $\{X_i\}_{i=t}^{t+w-1}$ all would be high for the sake of windows-overlapping effect. Thus, we should choose a small sliding-window size w in practice.

3 Conclusion

In this paper, we formulated the problem of anomaly detection in time series as the problem of comparing series of unthresholded recurrence plots (URPs). Our contributions in this paper were to extend the existing RPs-based time series analysis method, and to propose to use ELM-AE as an unsupervised learning and detecting framework. Our methods avoid the problem of parameter setting in conventional RQA analysis and are more robust to processing complex time series than other types of anomaly detection methods.

In the future, we will extend our framework to cope with multivariate time series data and missing value, and further apply the proposed methods to build a risk feature database to improve our understanding of tunnel disasters.

Acknowledgements. This work is supported by Shanghai Science and Technology Commission's Scientific and Technological Innovation Plan (No.16DZ1201704 and No.18DZ1205502).

References

1. Chandola, V., Banerjee, A., Kumar, V.: Anomaly detection: a survey. ACM Comput. Surv. (CSUR) **41**(3), 15 (2009)
2. Fox, A.J.: Outliers in time series. J. R. Stat. Soc. Ser. B (Methodol.) 350–363 (1972)

3. Hill, D.J., Minsker, B.S.: Anomaly detection in streaming environmental sensor data: a data-driven modeling approach. Environ. Model Softw. **25**(9), 1014–1022 (2010)
4. Dasgupta, D., Forrest, S.: Novelty detection in time series data using ideas from immunology. In: Proceedings of the International Conference on Intelligent Systems, pp. 82–87 (1996)
5. Eskin, E., Arnold, A., Prerau, M., Portnoy, L., Stolfo, S.: A geometric framework for unsupervised anomaly detection. In: Applications of Data Mining in Computer Security, pp. 77–101. Springer (2002)
6. Kim, S.S., Reddy, A.N., Vannucci, M.: Detecting traffic anomalies using discrete wavelet transform. In: Information Networking. Networking Technologies for Broadband and Mobile Networks, pp. 951–961. Springer, Heidelberg (2004)
7. Khalid, S.: Activity classification and anomaly detection using m-mediods based modelling of motion patterns. Pattern Recogn. **43**(10), 3636–3647 (2010)
8. Keogh, E., Lin, J., Fu, A.: Hot sax: efficiently finding the most unusual time series subsequence. In: Fifth IEEE International Conference on Data Mining, p. 8. IEEE
9. Eckmann, J.P., Kamphorst, S.O., Ruelle, D.: Recurrence plots of dynamical systems. Europhys. Lett. **4**(9), 973–977 (1987)
10. Marwan, N., Romano, M.C., Thiel, M., Kurths, J.: Recurrence plots for the analysis of complex systems. Phys. Rep. **438**(5), 237–329 (2007)
11. Marwan, N., Wessel, N., Meyerfeldt, U., Schirdewan, A., Kurths, J.: Recurrence-plot-based measures of complexity and their application to heart-rate-variability data. Phys. Rev. E **66** (2), 026702 (2002)
12. Marwan, N., Donges, J.F., Zou, Y., Donner, R.V., Kurths, J.: Complex network approach for recurrence analysis of time series. Phys. Lett. A **373**(46), 4246–4254 (2009)
13. Fabretti, A., Ausloos, M.: Recurrence plot and recurrence quantification analysis techniques for detecting a critical regime. Examples from financial market inidices. Int. J. Mod. Phys. C **16**(05), 671–706 (2005)
14. Nichols, J.M., Trickey, S.T., Seaver, M.: Damage detection using multivariate recurrence quantification analysis. Mech. Syst. Signal Process. **20**(2), 421–437 (2006)
15. Schinkel, S., Dimigen, O., Marwan, N.: Selection of recurrence threshold for signal detection. Eur. Phys. J. Spec. Top. **164**(1), 45–53 (2008)
16. Zbilut, J.P., Webber, C.L.: Recurrence Quantication Analysis. Wiley Encyclopedia of Biomedical Engineering
17. Marwan, N., Schinkel, S., Kurths, J.: Recurrence plots 25 years later—gaining confidence in dynamical transitions. EPL (Eur. Lett.) **101**(2), 20007 (2013)
18. Palmieri, F., Fiore, U.: Network anomaly detection through nonlinear analysis. Comput. Secur. **29**(7), 737–755 (2010)
19. Acharya, U.R., Sree, S.V., Chattopadhyay, S., Yu, W., Ang, P.C.A.: Application of recurrence quantification analysis for the automated identification of epileptic EEG signals. Int. J. Neural Syst. **21**(03), 199–211 (2011)
20. Liu, X., Bo, L.: Identification of resonance states of rotor-bearing system using RQA and optimal binary tree SVM. Neurocomputing **152**, 36–44 (2015)
21. Luo, W., Gallagher, M., Wiles, J.: Parameter-free search of time-series discord. J. Comput. Sci. Technol. **28**(2), 300–310 (2013)
22. Silva, D.F., De Souza, V.M., Batista, G.E.: Time series classification using compression distance of recurrence plots. In: 2013 IEEE 13th International Conference on Data Mining (ICDM), pp. 687–696. IEEE (2013)
23. Spiegel, S., Jain, J.B., Albayrak, S.: A recurrence plot-based distance measure. In: Translational Recurrences, pp. 1–15. Springer International Publishing (2014)

24. Hu, M., Zhou, S., Wei, J., Deng, Y., Qu, W.: Change-point detection in multivariate time-series data by recurrence plot. WSEAS Trans. Comput. **13**, 592–599 (2014)
25. Marchette, D.J., Solka, J.L.: Using data images for outlier detection. Comput. Stat. Data Anal. **43**(4), 541–552 (2003)
26. Wei, L., Kumar, N., Lolla, V.N., Keogh, E.J., Lonardi, S., Chotirat (Ann) Ratanamahatana. Assumption-free anomaly detection in time series. In: SSDBM, vol. 5, pp. 237–242 (2005)
27. Hu, M., Wu, F.F., Zhu, B., Lu, B., Pu, J.L.: A new hazard identification method-state transition graph. In: Applied Mechanics and Materials, vol. 48, pp. 71–78 (2011)
28. Iwanski, J.S., Bradley, E.: Recurrence plots of experimental data: to embed or not to embed? Chaos Interdiscip. J. Nonlinear Sci. **8**(4), 861–871 (1998)
29. Zbilut, J.P., Zaldivar-Comenges, J.M., Strozzi, F.: Recurrence quantification based Liapunov exponents for monitoring divergence in experimental data. Phys. Lett. A **297**(3), 173–181 (2002)
30. Huang, G.B., Zhu, Q.Y., Siew, C.K.: Extreme learning machine: theory and applications. Neurocomputing **70**(1), 489–501 (2006)
31. Kasun, L., Zhou, H., Huang, G., Vong, C.: Representational learning with extreme learning machines for big data. IEEE Intell. Syst. **28**(6), 31–34 (2013)
32. Johnson, W.B., Lindenstrauss, J.: Extensions of Lipschitz mappings into a Hilbert space. Contemp. Math. **26**(189–206), 1 (1984)
33. R Core Team, R: A Language and Environment for Statistical Computing, R Foundation for Statistical Computing, Vienna, Austria (2015). http://www.R-project.org/

A Two-Stage Temporal Anomaly Detection Algorithm Based on Danger Theory

Min Hu[1,2], Zhiwei Ji[3(✉)], Ye Guo[1,2(✉)], Ke Yan[4], and Haigen Hu[5(✉)]

[1] SHU-UTS SILC Business School, Shanghai University, Shanghai 201800, China
vicky_ye49@163.com
[2] SHU-SUCG Research Centre for Building Industrialization, Shanghai 200072, China
[3] School of Information and Electronic Engineering, Zhejiang Gongshang University, Hangzhou 310018, China
jzw18@hotmail.com
[4] College of Information Engineering, China Jiliang University, Hangzhou 310018, China
[5] Computer Science and Technology, Zhejiang University of Technology, Pingfeng Campus, Hangzhou 310023, Zhejiang, China
hghu@zjut.edu.cn

Abstract. Anomaly detection is to identify the abnormal states from time series, which is widely used for early diagnosis or warning in various areas. In recent years, there are many algorithms developed for anomaly detection. However, the time series from different areas have obvious specificities, which limits the generalization of the current anomaly detection approaches. In this study, we proposed a two-stage anomaly detection algorithm (TSAD) to identify discords in univariate time series. Based on Danger Theory, the TSAD approach determined an abnormal subsequence using a two-stage recognition strategy, recognition signal is referred as a danger signal to report the candidate abnormal subsequences, and the response signal is to determine if a candidate subsequence is an anomaly based on its statistical properties. To validate the effectiveness, we applied our developed approach on extensive time series. The simulation results indicate that TSAD is capable of identifying discords from time series. Besides, comparison with other typical algorithms also proved that it is outstanding to others.

Keywords: Anomaly detection · Danger theory · Single time series

1 Introduction

Anomaly detection refers to the problem of finding patterns from univariate/multivariate data which do not conform to expected behavior [1]. It was widely applied in many areas, such as fault detection, network security and climate monitoring [2]. Using anomaly detection technologies, we can identify abnormal or dangerous events in time to avoid risks or accidents [3].

J. Abawajy et al. (Eds.): ATCI 2018, AISC 842, pp. 485–494, 2019.
https://doi.org/10.1007/978-3-319-98776-7_53

In general, anomaly detection algorithms can be roughly categorized as three classes. The first one is statistic-based approaches [1], which was designed by using statistic models and evaluated with discordancy test. They are relatively simple and robust, but data distribution in many cases is unknown [4]. The second category is clustering-based approaches [5], which identify the outliers as abnormal patterns from all the subsequences segmented by a fixed-length sliding window. However, the detection accuracy mainly depends on the similarity metrics of pairwise subsequences, and computational complexity will be obviously increased when dealing with high-dimensional datasets [6]. The third category is intelligent computing methods, including neural networks, fuzzy theory and rough sets [7–9]. The challenge is that training and testing stages of these methods are mutually independent, so they lack continuous learning ability [10].

Due to the complexity of time series in real world, the above algorithms may not solve the special problems well. Recently, the concepts of immune theory (also called danger theory) has been increasingly applied in the fields of web mining and dynamic self-organizing SDM, which show the characteristics of self-adaptive and efficient [11, 12]. However, danger theory-based anomaly detection algorithms were rare studied.

In this study, we proposed a two-stage anomaly detection algorithm (TSAD) inspired by Danger Theory. Our TSAD approach worked on univariate time series and identified the abnormal states in two phases: (1) some potential abnormal sub-sequences were initially detected in the primary-recognition stage; (2) each potential sub-sequence was further evaluated with statistical features to confirm as an anomaly in the secondary-recognition stage. To validate the effectiveness, we tested the TSAD approach on some datasets, including *Npr44* [13], *chfdb_chf13_45590* [13] and *mit-dbx_mitdbx_108* [13]. The results show that our algorithm is capable of identifying abnormal states.

The rest of the paper is structured as follows: In Sect. 2 the proposed computational approach TSAD is presented. In Sect. 3, the testing datasets and the experiment design are described in detail. In Sect. 4, the simulation results are analyzed and discussed, while in Sect. 5 conclusions are drawn and suggestion for future work are presented.

2 A Two-Stage Temporal Anomaly Detection Algorithm (TSAD)

Based on Danger Theory, our proposed algorithm TSAD is used to efficiently identify discord from univariate time series data via two-stage recognition. At first, we introduced the concept of danger theory and how we mapped it to anomaly detection.

The immune system in host (e.g. human), can recognize various invading viruses or metastatic tumor cells after the initial activation of T cells, and try to lysis or kill these non-self molecules through the sustained clonal amplification within the microenvironment. The biological mechanism of immune response motivated researchers to develop Danger Theory for identifying abnormal patterns in computer science area [14, 15].

The danger theory was firstly proposed by Polly and colleagues in 1999 [16]. In general, the danger theory based computational models obtain the optimal solutions by simulating the whole process of immune response via two stages: primary-recognition stage and secondary-recognition stage [17]. They were widely used in many applications, including intrusion detection system [15], risk assessment [18], etc. So we developed a two-stage anomaly detection algorithm (TSAD), the flowchart of TSAD is illustrated in Fig. 1. The details of each component will be described in the following sections.

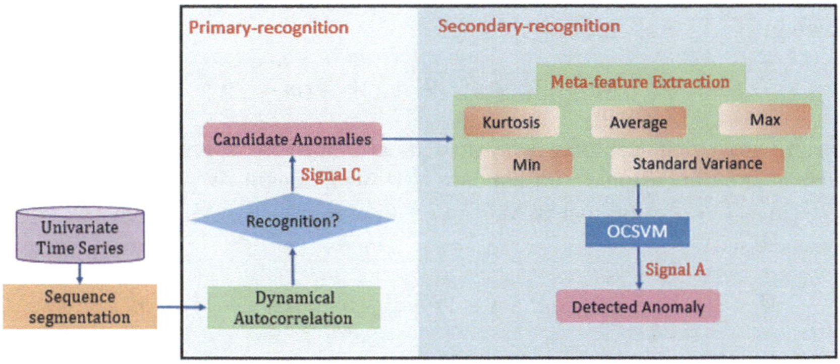

Fig. 1. The flowchart of the TSAD algorithm.

2.1 Sequence Segmentation

Before identifying the discords, data normalization was addressed with Eq. (1):

$$X'(i) = \frac{X(i) - \min(X(i))}{\max(X) - \min(X)} \tag{1}$$

Where $X(i)$ denotes the i-th data point of variable X ($1 \leq \mathrm{i} \leq \mathrm{T}$), and min(*) and max(*) are the minimal and maximal values of series X. T is the total number of observed time points. After normalization, the value of each element in X will be restricted in the range from 0 to 1. The first step of TSAD framework is to segment the original time series with a fixed sliding window. After obtained a set of equal length subsequences, the change points will be identified. The challenging is that how to determine the size of sliding window (W). In this study, for the periodic sequence, we used an autocorrelation-based strategy to determine the value of W. For the non-periodic series, we chose a fixed value for W (100, or 200) were widely used in other studies [19].

2.2 Primary-Recognition Reports Candidate Anomalies

Similar to the binding affinity between the antigen representing MHC molecule of a T cell receptor and a peptide of a target cell, Signal C shown in Fig. 1 is a recognition signal as an early abnormal indicator. According to the study of early-warning signals, an abnormal subsequence can be captured based on the local dynamical changes of temporal data [20]. In this study, autocorrelation was used to evaluate dynamical changes of temporal data as it refers to the correlation of a time series with its own past and current values [21].

Before we calculate the autocorrelation vector of the original series, state space reconstruction was implemented at first. Given a time series $\boldsymbol{X} = \{x(t) \in R^d, t = 1, 2, \ldots, N\}$, the t-th subsequence segmented by sliding window can be described as: $s(t) = \{x(t), x(t+1), \ldots, x(t+(W-1)\}$.

Using delay coordinates to reconstruct the state space trajectory [22], we will obtain a state-space matrix M for each sub-sequence s(t), which can be represented as following:

$$M = \{M_i, i = 1, 2, \ldots, W - (m-1)\tau\} \tag{2}$$

where $\tau = \alpha\Delta t$ denotes time delay and m is embedded space dimension. In other words, a one-dimensional subsequence s(t) with length W was transformed to a $[(W - (m-1) * \alpha) * m]$ matrix M. The $i-$th row in matrix M, denotes as M_i, can be represented as:

$$M_i = \{x(t+i-1), x((t+i-1)+\tau), \cdots, x((t+i-1)+(m-1)\tau)\} \tag{3}$$

where $1 \le i \le W - (m-1) * \alpha$. Therefore, for each subsequence s(t), we will calculate the autocorrelation coefficient with its corresponding state-space matrix. The Autocorrelation Function [23] is defined as shown in Eq. (4):

$$\begin{aligned} C_{s(t)}(\varepsilon) &= P(M_i - M_j < \varepsilon) \\ &= \frac{2}{(W-\alpha m)(W-\alpha(m-1))} \sum_{i=1}^{W-\alpha m} \sum_{j=i+1}^{W-\alpha(m-1)} \Theta(\varepsilon - M_i - M_j) \end{aligned} \tag{4}$$

Where $\Theta(*)$ is Heaviside Step Function [21, 24]. Actually, $C_{s(t)}(\varepsilon)$ is the probability of which M_j is in the ball of radius ε centred at M_i in reconstruction space. However, the rigid boundary of $\Theta(*)$ in Eq. (4) may lead to loss of proximity information and instability of dynamics measures in regard to the time series [24]. To overcome this weakness, $\Theta(*)$ is replaced by exp(*), so that Eq. (4) was represented as Eq. (5):

$$\begin{aligned} C_{s(t)}(\varepsilon) &= P(M_i - M_j < \varepsilon) \\ &= \frac{2}{(W-\alpha m)(W-\alpha(m-1))} \sum_{i=1}^{w-\alpha m} \sum_{j=i+1}^{W-\alpha(m-1)} \exp(\varepsilon - M_i - M_j) \end{aligned} \tag{5}$$

Although $C_{s(t)}(\varepsilon)$ has potential to distinguish potential dynamics structures, it is still not enough to be used as an important criterion for identifying the similarity involved in chaotic time series. Hence, we defined dynamical autocorrelation to instead of the traditional autocorrelation [21]:

$$S\_C(s(t)) = \lim_{\varepsilon < 0} \left| ln \frac{C_{s(t)}(\varepsilon)}{C_{s(1)}(\varepsilon)} \right| \tag{6}$$

By using Eq. (6), function S_C(*) calculated the dynamics of the whole time series, and returned a vector $\mathrm{SC} = \{\mathrm{S\_C}(\mathrm{s}1), \mathrm{S\_C}(\mathrm{s}2), \ldots\}$ of dynamical autocorrelation. Given threshold θ, we obtained a subset of SC, named as $SBC = \{c_{v1}, c_{v2}, \ldots c_{vk}\}$, where $c_{vl} > \theta$. Each value c_{vl} in can be considered as a signal C, which reports a candidate abnormal subsequence $\mathrm{s}(vl)$ to the next stage.

2.3 Secondary-Recognition Determines Anomalies

In the first stage of TSAD approach, we filtered out the candidate abnormal subsequences with higher matching "binding affinity". In the second stage, we evaluated each candidate subsequences and further excluded all the false negative. To reach this goal, we adopt five statistical features (Kurtosis, average, max, min, and standard variance) to represent all the candidate abnormal subsequences for identifying the outliers. All the candidate abnormal subsequences were then represented as data points in a 5-dimensional space so that we can use OCSVM to determine the potential anomalies.

3 Experiment Design

To prove the effectiveness, we designed simulation experiments to test TSAD approach with three datasets collected from real world. The details are given as follows.

3.1 Data Pre-processing

For evaluating the performance of our approach, we firstly chose some time series datasets, including *Npr44* [13], *chfdb_chf13_45590* [13] and *mitdbx_mitdbx_108* [13].

Npr44, belongs to patient monitoring data, showing a patient's respiration, as they wake up. This sequence includes two discords.

Chfdb_chf13_45590 is ECG datasets. An ECG is a bioelectrical signal which records the heart's electrical activity versus time. It is also widely used for testing anomaly detection algorithms by many researchers.

The time series *mitdbx_mitdbx_108*, is from the PhysioNet Web Server. It is a two features time series, here we are looking at just the ML II column. Cardiologists have annotated three discords in this time series [25].

3.2 Experiment Design

To validate the effectiveness, we tested our proposed approach TSAD on extensive data sets, which were described in above section. In addition, the time series *mitdbx_mitdbx_108* is two-dimensional, and we selected the first variable for testing. The second variable in *chf13* was selected to test our algorithm. For *Nprs44*, all subsequence in the time region was selected for algorithm testing.

To further prove the outstanding performance, we compared our approach with three well-established algorithms on all the above datasets, including, Brute Force[19, 26] and

Hot SAX [19]. The performance of anomaly detection was reflected with three statistical measurement, including Accuracy, specificity, and sensitivity (see the formula (7–9)).

$$\text{Accuracy} = \frac{TN + TP}{TN + TP + FP + FN} \tag{7}$$

$$\text{Sensitivity} = \frac{TP}{TP + FN} \tag{8}$$

$$\text{Specificity} = \frac{TN}{TN + FP} \tag{9}$$

The variables TP, FP, TN, and FN were defined as the number of normal subsequences correctly detected as normality (true positives), the number of abnormal subsequences that are detected as normality (false positives), the number of abnormal subsequences that are recognized as anomaly (true negatives), and the number of normal states that are recognized as anomaly (false negatives).

3.3 Experimental Parameters

The source codes of TSAD algorithm was debugged and implemented under the environment of Matlab R2017b and Libsvm 3.2.2 with Intel Core i7-6000 processor, 8G RAM and 120G SSD. An OCSVM classifier has been implemented to identify the anomalies from the whole data samples. In OCSVM model, Gaussian RBF kernel was employed, and the kernel parameters C and γ were optimized by grid search [27]. In the grid search, we set C = 2^a and γ = 2^b. Variable a changes from −5 to 10 with step 0.1, and variable b changes from −10 to 10 with step 0.25.

4 Results and Discussion

In our experiments, we applied our TSAD approach on three time series datasets to verify its effectiveness. In addition, three well-known methods for anomaly detection were implemented on the same datasets for comparing the performance with TSAD.

4.1 Experiment Results and Analysis of TSAD

To test our TSAD, we firstly investigated a time series (*Nprs44*) showing a patient's respiration (measured by thorax extension), as they wake up (Fig. 2). In Fig. 2A, we see two discords in this series, which was reported by other researchers [19]. The 1st discord is very obvious deep breath take as the patient opened their eyes. In contrast, the 2rd discord is much more subtle and difficult to see at this scale. The clinical doctors noticed a few shallow breaths that indicated the transition of sleeping states. In both cases, the discords straddle the change in sleeping cycle. We tested our TSAD with this dataset, and identified the 1st discord, but missed the 2rd one (Fig. 2B).

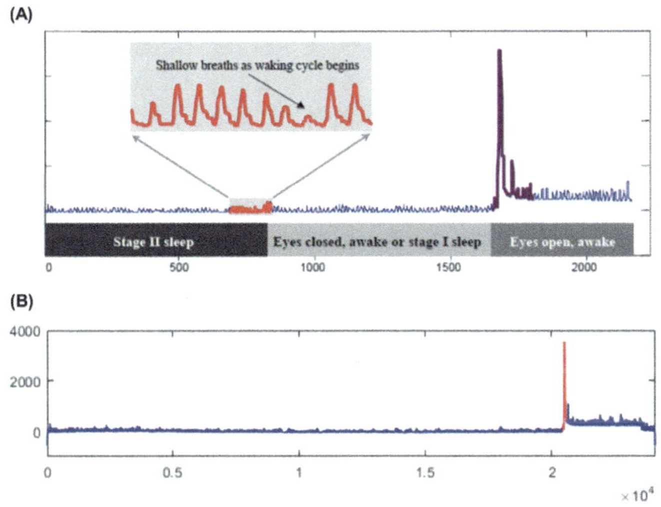

Fig. 2. (A) The time series *Nprs44* and two well-known discords reported in previous study [13]. (B) The anomly detected by our TSAD approach.

Electrocardiograms (ECGs), are arguably the important time series, and as such, there are many annotated test datasets we can consider. We then test our TSAD approach on the 2st variable in *chf13* dataset. The prediction of our TSAD shown in Fig. 3B is consistent as the known discord in Fig. 3A. In Fig. 4A, we consider an ECG dataset *mitdbx_mitdbx_108* that has three different types of anomalies. In this figure, we could perhaps spot the anomalies by eye: the second discord indicates supraventricular escape beat, and the other two denote premature ventricular contraction [25]. The TSAD can identified the first two discords but missed the 3rd one (see Fig. 4B). Overall, above experiments revealed that our proposed method TSAD is effective in anomaly detection.

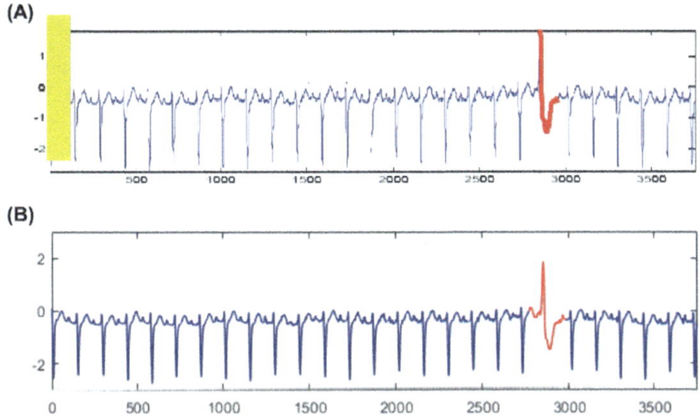

Fig. 3. (A) The time series *chf13* and one well-known discord reported in previous study [13]. (B) One anomly was detected by our TSAD approach

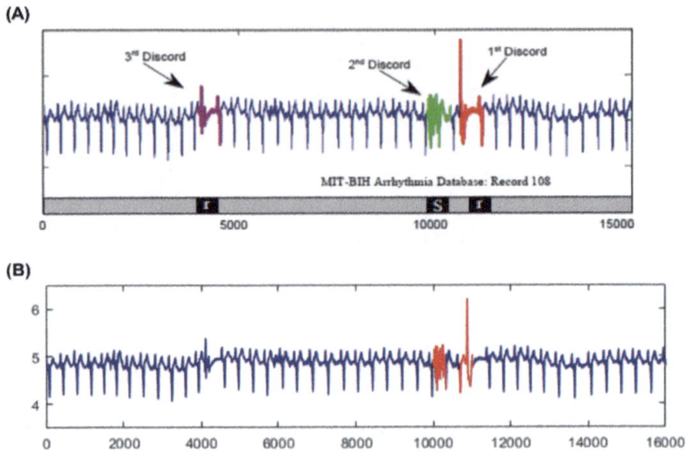

Fig. 4. (A) The time series *mitdbx_108* and three well-known discords reported in previous study [13]. (B) Our TSAD approach detected the first two discords.

4.2 Comparison with Other Algorithms

To further verify the performance of TSAD, we tested all of the above datasets on two golden methods: Brute Force [19] and Hot SAX [19]. The performance of each approach was evaluated through three statistical metrics as shown in formula (7–9). In Table 1, we can found that the performances of three methods on ECG datasets, e.g. *chf13* is very close due to these time series signals are periodic and the discord in each sequence is very obvious. For the other datasets, TSAD is better than other two methods. We concluded that TSAD represented a significant performance rather than other two approaches. The details of the comparisons was represented in Table 1.

Table 1. The comparison of TSAD with other two golden algorithms. "*Chf13*" denote the dataset *Chfdb_chf13_45590.*

Approach	Dataset	Variable	Accuracy	Sensitivity	Specificity
TSAD	Nprs44		0.9979	1.0	0.5
Hot SAX	Nprs44		0.9979	1.0	0.5
Brute Force	Nprs44		0.9959	1.0	0.5
TSAD	Chf13	Variable 2	1.0	1.0	1.0
Hot SAX	Chf13	Variable 2	1.0	1.0	1.0
Brute Force	Chf13	Variable 2	1.0	1.0	1.0
TSAD	Xmit108		0.9888	0.985	1.0
Hot SAX	Xmit108		0.9095	1.0	0.2
Brute Force	Xmit108		0.9138	0.9536	0.5

5 Discussion and Conclusion

This paper proposed a novel computational framework for identifying abnormal subsequence from univariate time series. Based on Danger Theory, the TSAD approach identified the abnormal states through two phases: (1) some candidate abnormal subsequences were initially detected in the primary-recognition stage; (2) each potential subsequence was further evaluated with a set of statistical features to confirm as an anomaly in the secondary-recognition stage. To prove the effectiveness, TSAD was tested on three popular time series datasets and compared with two well-known algorithms, Hot SAX, and Brute Force. The simulation experiments show that the performance of TSAD is significantly outstanding than other two methods.

As to other sliding window-based strategy, TSAD also suffer from this challenging. Although we selected periodic and non-periodic series in our experiments, more testing is needed to further prove the generalization of this algorithm.

Acknowledgements. This work was supported by Science and Technology Committee of Shanghai (No. 18DZ1205502).

References

1. Chandola, V., Banerjee, A., Kumar, V.: Anomaly detection: a survey. ACM Comput. Surv. **41**(3), 1–58 (2009)
2. Lazarevic, A., Ertöz, L., Kumar, V., et al:. A comparative study of anomaly detection schemes in network intrusion detection (2003)
3. Hu, M., Jiang, P., Zhou, S.C., et al.: Visual early-warning signal detection for critical transitions. J. Comput. **28**(2), 1–13 (2017)
4. Ye, N., Chen, Q.: An anomaly detection technique based on a chi-square statistic for detecting intrusions into information systems. Qual. Reliab. Eng. Int. **17**(2), 105–112 (2015)
5. Wei, J.: Research of outlier detection and data recovery based on statistical method. Nanjing Univ. Posts Telecommun. (2016)
6. Mehrotra, K.G., Mohan, C.K., Huang, H.M.: Clustering-based anomaly detection approaches (2017)
7. Ilhan, A., Mehmet, K., Erhan, A.: An approach for automated fault diagnosis based on a fuzzy decision tree and boundary analysis of a reconstructed phase space. ISA Trans. **53**(2), 220 (2014)
8. Khajavi, M.N., Keshtan, M.N.: Intelligent fault classification of rolling bearings using neural network and discrete wavelet transform. J. VibroEng. **16**(2), 761–769 (2014)
9. Yao, X.-H., Fu, J.-Z., Chen, Z.-C.: Intelligent fault diagnosis using rough set method and evidence theory for NC machine tools. Int. J. Comput. Integr. Manuf. **22**(5), 472–482 (2009)
10. Dong, L.I., Liu, S., Zhang, H.: A method of anomaly detection and fault diagnosis with online adaptive learning under small training samples. Pattern Recogn. **64**, 374–385 (2017)
11. Salmon, H.M., De Farias, C.M., Loureiro, P., et al.: Intrusion detection system for wireless sensor networks using danger theory immune-inspired techniques. Int. J. Wireless Inf. Netw. **20**(1), 39–66 (2013)
12. Secker, A., Freitas, A.A., Timmis, J.: A danger theory inspired approach to web mining. Lecture Notes in Computer Science, vol. 2787, pp. 156–167 (2003)

13. Li, J., Pedrycz, W., Jamal, I.: Multivariate time series anomaly detection: a framework of hidden markov models. Appl. Soft Comput. **60**, 229–240 (2017)
14. Azmi, R., Pishgoo, B.: SHADuDT: secure hypervisor-based anomaly detection using danger theory. Comput. Secur. **39**(4), 268–288 (2013)
15. Sommer, P.: Intrusion Detection Systems as Evidence, pp. 2477–2487. Elsevier North-Holland, Inc., Hoboken (1999)
16. Matzinger, P.: The danger model: a renewed sense of self. Science **296**(5566), 301–305 (2002)
17. Bretscher, P., Cohn, M.: A theory of self-nonself discrimination. Science **169**(3950), 1042–1049 (1970)
18. Zainal, K., Jali, M.Z.: A perception model of spam risk assessment inspired by danger theory of artificial immune systems. Procedia Comput. Sci. **59**, 152–161 (2015)
19. Keogh, E., Lin, J., Fu, A.: HOT SAX: efficiently finding the most unusual time series subsequence, pp. 226–233 (2006)
20. Brock, W.A.: Early-warning signals for critical transitions. Recent Dev. Seismol. **6**(7260), 53–59 (2009)
21. Li, C.G., Pei, L.Q.: Method for distinguishing dynamical species in chaotic time series. Acta Phys. Sin. **52**(9), 2114–2120 (2003)
22. Takens, F.: Detecting strange attractors in turbulence. Lecture Notes in Mathematics, vol. 898, pp. 366–381. Springer, Heidelberg (1981)
23. Savit, R., Green, M.: Time series and dependent variables. Phys. D **50**(1), 95–116 (1991)
24. Godfrey, L.B., Gashler, M.S.: A continuum among logarithmic, linear, and exponential functions, and its potential to improve generalization in neural networks, pp. 481–486 (2016)
25. http://www.cs.ucr.edu/~eamonn/discords/ICDM05_discords
26. Zheng, D., Li, F., Zhao, T.: Self-adaptive statistical process control for anomaly detection in time series. Expert Syst. Appl. **57**(C), 324–336 (2016)
27. Ji, Z., Meng, G., Huang, D., et al.: NMFBFS: a NMF-based feature selection method in identifying pivotal clinical symptoms of hepatocellular carcinoma. Comput. Math. Methods Med. **2015**(3), 1–12 (2015)

A Data-Driven Framework for Tunnel Infrastructure Maintenance

Yufeng Sun[1], Min Hu[1(✉)], Wenbo Zhou[2], and Wei Xu[1]

[1] SILC Business School, Shanghai University, Shanghai 201400, China
minahu@163.com
[2] Shanghai Tunnel Engineering Co., Ltd., Shanghai 200032, China

Abstract. The availability and reliability of tunnel infrastructure lies on the health of tunnel structure, therefore, scientific maintenance is needed to enhance the tunnel's structure reliability and reduce the inspection and maintenance cost through optimized maintenance schedule. This paper aims to develop a data-driven framework for the operation and maintenance of tunnels, which consists of eight stages: maintenance data profile, tunnel failure mode and effect analysis (FMEA), data processing, feature engineering, anomaly detection, failure inspection, failure prediction and maintenance schedule. These Stages are interpreted into fifteen steps in a sub-system with data fusion techniques used for the whole process of tunnel components maintenance. The proposed framework for tunnel's maintenance ensures the safety and reliability of tunnel's operation, improves the efficiency of failure detection and reduces the maintenance cost as well.

Keywords: Data-driven · Anomaly detection · Predictive maintenance

1 Introduction

The highway tunnel is an important part of the transportation infrastructure system. By the end of 2016, the total length of the highway tunnel is over 14 million meters in china [1]. With the increase of service period, the accumulation of damage of the tunnel structure threats the facilities' operation severally if these tunnels are not repaired in time. However, the cost of tunnel maintenance or replacement is very high. According to Russell, some tunnels' maintenance cost up to $1.5 million per kilometer per year, while the replacement cost is $250 million per kilometer [2]. Therefore, it is essential to detect the defects of the tunnel early and make a reasonable maintenance plan to guarantee the safety, reliability and cost-effectiveness of the tunnel operation.

The gradual deterioration of tunnel is mainly caused by environmental effects, which possibly leads to structural failure and even result in catastrophic consequences if failures are not repaired in time [3]. In order to avoid the above consequence, maintenance strategy is essential. There are three basic maintenance strategies: corrective maintenance, preventive maintenance and predictive maintenance. The evolution of maintenance strategies can be summarized from corrective maintenance and preventive maintenance to predictive maintenance [4]. The predictive maintenance predicts the trend of deterioration and identifies appropriate maintenance opportunities,

J. Abawajy et al. (Eds.): ATCI 2018, AISC 842, pp. 495–504, 2019.
https://doi.org/10.1007/978-3-319-98776-7_54

which is an effective way of ensuring tunnel structure safety and controlling the maintenance cost as well.

In recent years, structure health monitoring systems that could obtain health data of tunnel structure easily have been designed and applied in the tunnel operation, which provide monitoring data for researchers to predict the degradation of the tunnel. However, the obtained sensor data is heterogeneous and the frequency of the monitoring data is different from that of the inspection data, which brings difficulties for data fusion strategy.

It is well recognized that the efficiency of maintenance depends on sensing technology, associated diagnostic and maintenance strategies [5]. Prognostics and health management (PHM) is one type of maintenance activity, which has been gradually applied in various fields such as aerospace, mechanical, electric engineering and etc. To our best knowledge, there is not a data-driven based framework considering the whole process in the tunnel maintenance engineering. In order to bridge the gap, this paper proposes a tunnel infrastructure maintenance framework that integrates data fusion strategy within the traditional framework of PHM system.

2 The Proposed Framework

The maintenance of the tunnel infrastructure is a wide field and includes structure health management system, environment monitoring system and tunnel ventilation fan system, etc. [6] From these systems, various types of data such as online monitoring data and offline inspection data could be obtained. Thus a data-driven based framework for tunnel maintenance with data-fusion techniques is introduced. The detail architecture is depicted in Fig. 1.

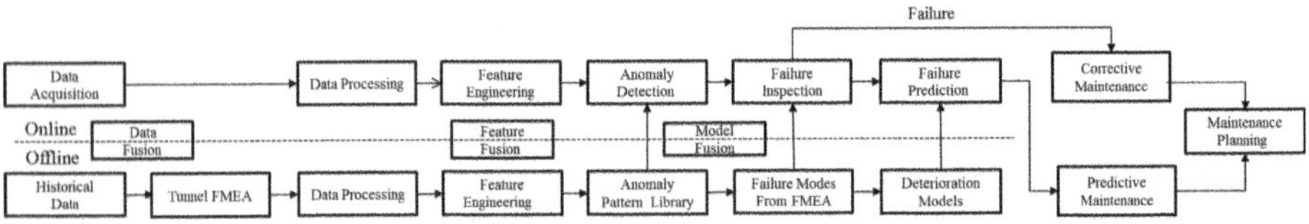

Fig. 1. Data-driven framework for tunnel component maintenance

The framework can be depicted into two main areas: the offline algorithm development and online application. In the development of offline algorithm, the main work is to identify critical failure modes, establish anomaly pattern library for failure detection and estimate failure probability based on deterioration models. The process involves data processing, feature engineering, model creation, failure inspection and perdition. In the online application stage, online monitoring data is acquired firstly and necessary feature extraction should be conducted before online anomaly detection. According to the alarm threshold based on offline development phase, the online anomaly detection application would issue an early warning signal when abnormal indicator exceeds the threshold, which is followed by a failure inspection based on identified failure modes. If the component's damage is diagnosed as serious, then corrective maintenance is put forward, otherwise a damage accumulation model would be established to assess the risk and calculate the residual life of component for planning the predictive maintenance. In this framework, offline learning models and online prediction are organically combined, where the online prediction application is based on the offline models and the offline models are updated through the online prediction application' effects conversely.

Data fusion techniques are applied in the whole framework, which mainly include three levels: the signal level, feature level and model level. The main purpose of data fusion technology used is to reduce the uncertainty of failure prediction and thus improve the accuracy maintenance for tunnel component. A detailed sub-system flowchart is developed as shown in Fig. 2.

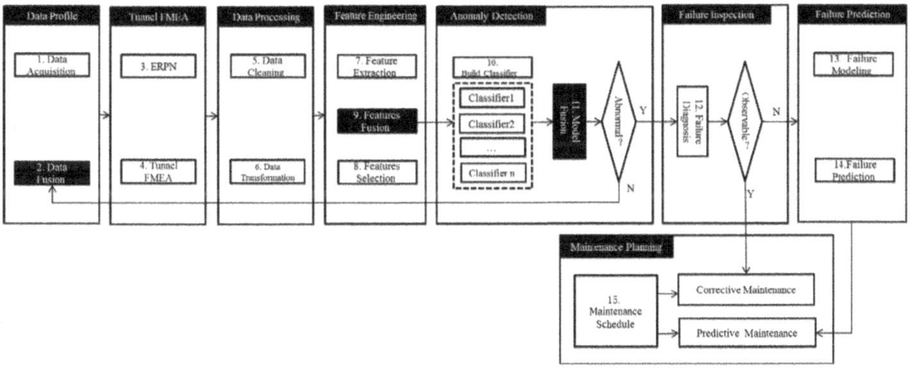

Fig. 2. Flowchart of the sub-system for tunnel component maintenance

2.1 Data Profile

Step 1: Data Acquisition. In order to evaluate a tunnel's current state, detect abnormal monitoring data, establish deterioration models and plan the maintenance schedule scientifically, tunnel related information should be acquired as much as possible. These data profile is described as in Table 1.

Step 2: Data Fusion. Selecting appropriate sensors to monitor the tunnel is essential. The accuracy, reliability, availability and cost are main factors needed to consider to a sensor system [7]. In the tunnel real-time health monitoring system, tunnel components are equipped with different sensors to obtain monitoring data and these sensors' sampling frequency is different. Multi-source same or similar sensors should be fused at first, which means signals from different sensors are combined to create a new signal with a less noise than the original sensors. Then, offline historical data such as inspection data and repair data could be used to develop accurate models for estimating the risk of tunnel components, which helps to reduce the uncertainly warning signals since online continuous monitoring raw signals are always mixed with noise [8].

Table 1. Tunnel Maintenance Data Profile

Data source	Data information	Data type
BIM model information	Mechanical and electrical model	Static data
	Civil engineering model	Static data
Construction information	Construction parameters	Static data
Operation information	Leakage data	High frequency real-time data
	Cracks data	High frequency real-time data
	Settlement data	High frequency real-time data
	Humidity data	High frequency real-time data
	Traffic flow data	High frequency real-time data
	Regular inspection data	Low frequency scheduled data
	Maintenance history data	Low frequency scheduled data

2.2 Tunnel FMEA

Failure modes and effects analysis (FMEA) was first described in US-military in 1949 [9], which is an important risk assessment method to evaluate the risk and reduce the probability of failure occurring by the potential failure modes.

Step 3: Exponential Risk Priority Number (ERPN). The Risk Priority Number (RPN) is a classical method to evaluate the risk of failure in FMEA through three parameters the severity(S), occurrence (O) and detection (D). Each parameter ranges from 1 to 10 and thus the RPN value ranges from 1 to 1000, which means a failure mode with a large RPN value is serious, likely to occur and hard to detect. A certain failure mode could be conducted by a mathematical product of the three parameters described in Eq. (1).

$$\mathrm{RPN} = \mathrm{S} * \mathrm{O} * \mathrm{D} \tag{1}$$

However, there are several shortcomings of the traditional RPN method such as RPN have many duplicated number and does not consider the weight of S, O and D. Therefore, an exponential RPN (ERPN) approach was presented to enhance the FMEA by Chang el al. [10], which provides a more accurate and reasonable risk assessment. The ERPN method is described in Eq. (2).

$$\mathrm{ERPN} = 3^{W_S * S} + 3^{W_O * O} + 3^{W_D * D} \tag{2}$$

In this study, the ERPN method is employed to prioritize failure modes for the maintenance of tunnel components, as presented in step 4.

Step 4: Tunnel FMEA. FMEA has been widely used in many important fields such as manufacturing industry, aerospace industry and engineering fields. However, there are few studies about the evaluation on the safety of tunnel system based on this theory. Yuan et al. [11] conducted a comprehensive research to analyze 31 failure modes of tunnel system such as water leakage and cracks, etc. Yuan et al. [12] then proposed a tunnel

FMEA study by employing these failure modes in the FMEA and used RPN approach to prioritize with each identified failure modes for predictive maintenance of a tunnel.

In this framework, the ERPN method is employed in the FMEA to prioritize failure modes for the maintenance of tunnel components. The relative importance eights of S, O and D are obtained by FMEA evaluation experts based on their judgment. For example: we assume $W_S = 0.4$, $W_O = 0.35$, and $W_D = 0.25$, with different (S, O, D) combination of (6, 5, 4) and (5, 4, 6) for two potential failure modes segment-cavitation and segment-spalling, respectively. The results show the two failures modes have a same RPN value of 120 but a different ERPN values with 23.81 and 18.65, which means the segment-cavitation has a higher rank than segment-spalling. This example indicates that the ERPN method is more practical in the application of tunnel FMEA.

2.3 Data Processing

Step 5: Data Imputation. Missing data is an unavoidable problem in tunnel data profile, thus some imputation approaches such as mean or median imputation, interpolation method, or the modeling based method are employed to fill the missing points. Each of the above methods has advantages and disadvantages, for example: the mean imputation technique is easy for complement of the missing data, however, when the missing data is not random or the distribution of data is not a proper distribution then the dataset may be biased. Therefore, choosing which method depends on the dataset's characteristics.

Step 6: Data Transformation. Monitoring data and inspection data are always contain various kinds of variables. Some variables such as seepage location and segment are linguistic and other variables such as settlement data and operating record are numeric. It is hard to create a reliable model for tunnel assessment based on different type of data, thus the data normalization is necessary for scaling the variables into the same range. A MinMaxScaler method is usually adopted to scale the data to a given minimum and maximum value, usually between 0 and 1, which is calculated as follows:

$$X = \frac{x - Min}{Max - Min} \tag{3}$$

2.4 Feature Engineering

Step 7: Feature Selection. Feature selection selects a feature without any transformation, which aims to select the most relevant attributes to build health indicators that can be used for anomaly detection of the tunnel health state. Feature selection algorithm may use ranking method such as correlation method to select features, and further methods may need to search for feature subsets by trail.

Step 8: Feature Extraction. With the development of data collection and storage, large amount of information has been recorded in the whole life cycle of tunnel systems. It is commonly that the feature dimension is much larger than the number of samples and the effect is not ideal if taking these data for modeling directly because the characteristics of redundancy may bring noise and affect the calculation results. Feature extraction is a feature reduction process, which aims to map the original feature space to a new feature space with lower dimensions.

Step 9: Feature Fusion. The extracted features are used for constructing indicators for anomaly detection. However, each indicator has its own advantages and disadvantages, thus fusion strategy at feature level is carried out to eliminate the impact of individual feature and the fused features are used as selected variables for anomaly detection.

2.5 Anomaly Detection

From the perspective of data-driven methods, in the tunnel health monitoring system, anomaly detection technique is employed to provide an effective way to detect abnormal condition for potential failure.

Step 10: Build Classifier. Variation in high frequency monitoring data from sensors has been widely used in anomaly detection for predictive maintenance. There are two mainstream approaches for anomaly detection: statistical and machine learning methods [13]. In this study, a correlation-based anomaly detection method is adopted as an example to analyze the abnormal condition, which is a simple and low-cost solution [14]. The process of the proposed anomaly detection is illustrated in Fig. 3. First, historical normal sensor data are extracted to calculate the correlations between each two sensors data, the correlation is measured by the Pearson correlation coefficient as shown in Eq. (4), where *X*, *Y* are data of two sensors, and r is the Pearson correlation coefficient value (CC). Secondly, the selection of correlation coefficient is necessary when with the increase of the number of sensors. The CC between two sensors that are highly correlated should be reserved. Thirdly, the CC is merged with Sensor data, and then employs a Vector Quantization clustering method to generate a normal state model. Finally, the model is applied to detect abnormal of the online sensors data, in normal situation, the CC varies in a certain range when the CC exceeds alarm threshold, it means an anomaly detected.

$$\mathrm{r} = \frac{\sum_{i=1}^{n}(X_i - \bar{X})(Y_i - \bar{Y})}{\sqrt{\sum_{i=1}^{n}(X_i - \bar{X})^2 \sum_{i=1}^{n}(Y_i - \bar{Y})^2}} \tag{4}$$

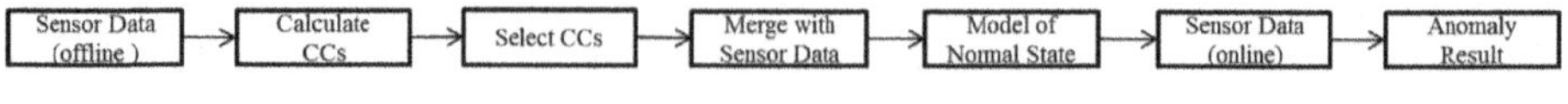

Fig. 3. Process of the anomaly detection (correlation-based)

Step 11: Model Fusion. At model fusion level, all classifiers are grouped as an abnormal indicator which is employed to estimate the tunnel component's abnormal state. In this step, the main purpose is to provide early warnings for tunnel component's failure, and the main challenge is to reduce the false positive rate and false negative rate of the detection model effectively. In this paper, the anomaly detection is considered as a binary classification, which means abnormal or normal is the output of prediction. Then Adaboost algorithm as a linear weighted method is employed to show how the model fusion works, which is a common approach for model fusion and is introduced in 1995 by Freund and Schapire [15]. $H(i,t)$ stands for the result of abnormal detection of the ith tunnel component at time t, which is calculated by Eq. (5) where α_k as the weight of the kth classifier, $h_k(i,t)$ as the abnormal value of ith tunnel component evaluated by the kth classifier at time t. The pseudocode for AdaBoost algorithm is given in Fig. 4, in which the function $h_k(x) \neq y_i$ determines the output 0 or 1, thus the classifier with low error rate is given a larger voting weight.

$$H(i,t) = \sum_{k=1}^{K} \alpha_k * h_k(i,t) \tag{5}$$

Given: $(X_1, Y_1), \ldots, (X_m, Y_m)$ where $x_i \in X,\ y_i \in Y = \{-1, +1\}$

Initialize weights. $w_i(k) = 1/\mathrm{m}, (\mathrm{i} = 1,2,\ldots,\mathrm{m}), k = 1;$

Do for $k = 1,2,\ldots,\mathrm{K}$:

  (1) Based on sample weight $w_i(t)$ and algorithm P to train to get weak classifier $h_k(x): x \to \{-1, +1\}$

  (2) Calculate the error rate of $h_t(x)$:

  $\varepsilon_k = \sum_{k=1}^{m} w_i(k)[h_k(x) \neq y_i] / \sum_{k}^{m} w_i(t);$

  (3) Choose $\beta_k = \varepsilon_k/(1-\varepsilon_k)$, $\alpha_k = \frac{1}{2}\log(1/\beta_k);$

  (4) Update weights: $w_i(k+1) = w_i(k) * \begin{cases} e^{-\alpha_k}, & if\, h_k(x) = y_i \\ e^{\alpha_k}, & if\, h_k(x) \neq y_i \end{cases}$

End for.

The finally strong classifier $H(x) = sgn(f(x))$,

$$where\ f(x) = \sum_{k=1}^{K} \alpha_k h_k(x)$$

Fig. 4. The pseudocode of AdaBoost algorithm

2.6 Failure Inspection

Step 12: Failure Diagnosis. In step 11, an anomaly warning signal would be issued if indicator value exceeds the predefined threshold. Then failure diagnosis is performed, which aims to identify failure mode and output the diagnosis results. If the result shows that damage exists in one component, then a corrective maintenance plan should be put forward in step 14. Otherwise, the failure prediction is employed to predict the possible damage of the tunnel component in the future.

2.7 Failure Prediction

Step 13 & 14: Failure Modeling and Failure Prediction. Although the failure diagnosis shows there is not an existing failure, the possible damage may exist but could not be observed. The aim of failure prediction is to evaluate the risk of tunnel component and analyze the possibility of a failure happens in the future.

There are two types of data-driven approaches used to establish risk models for analyzing the failure probability of tunnel components, which are statistical prediction methods and machine learning methods. Both methods are established models to predict the probability of future risk by making use of empirical data that includes past inspection data and related influencing factors.

From the perspective of statistics, Weibull distribution is widely used to establish failure model in engineering system, which is also suitable for identifying failure modes in tunnel maintenance. The cumulative distribution function for the exponentiated Weibull distribution is defined as follows:

$$F(t) = 1 - R(t) = 1 - \exp\left[-\left(\frac{t}{\eta}\right)^{\beta})\right] \quad t, \eta, \beta > 0 \tag{6}$$

Where $\beta > 0\, and\, \eta > 0$ are the shape parameter and scale parameter of the distribution. The function $F(t)$ and $R(t)$ are the failure probability and reliability functions, respectively.

In addition, the machine learning methods can be utilized to perform models are artificial neural network, support vector machines, Bayesian networks, etc. Based on the selected features that correlate with failure degradation process, perdition of defects is produced based on a training process in which model of weights will adjust automatically based on the relationships between the prediction results and the correlated features.

2.8 Maintenance Planning

Step 15: Maintenance Planning. Maintenance management refers to the set of maintenance action to guarantee the performance of tunnel stay in a predetermined level. In order to prevent fast deterioration in tunnel structural performance, two maintenance options (i.e. Corrective maintenance and predictive maintenance) can be selected in this framework. Corrective maintenance require repair or replacement immediately when the failure detected, however, recent studies showed that the corrective maintenance for those failures have little impact on the tunnel operation can be deferred to a suitable time to carry out a group maintenance [16], which helps to minimize the total maintenance and maximize the operation efficiency. Compared to corrective maintenance, predictive maintenance aims to prevent a failure from occurring. Thus the failure prediction determines whether it is necessary to carry out a predictive maintenance for tunnel components.

3 Conclusion

This paper proposes a data-driven based framework for improvement of tunnel maintenance performance. The objective of this paper tries to employ data fusion strategy in the traditional tunnel maintenance architecture, which bridge the gap of the maintenance in the process of the tunnel operation. There are two main phases in this framework: the offline algorithm development phase and online implementation phase. The online phase includes getting data from different sensors, data preprocessing, selecting the features that are the most useful for anomaly detection, failure diagnosis, predicting of defects and scheduling relevant maintenance. The offline phase makes use of historical data (monitoring data, failure data, deterioration data, etc.) to establish models to fit the online application. In addition, a detailed sub-system is presented to depict how the framework could provide early warning detection, identify failures and predict the degradation of the tunnel components in order to help tunnel operation staffs take suitable actions.

The advantages of the proposed framework could be summarized as follows: Firstly, this framework ensures the safety and reliability operation of tunnel infrastructure through the combination of online and offline algorithms, which reduce the uncertainty of prediction of tunnel component's health state. Secondly, applying data fusion techniques helps to provide early warning of deteriorating condition, which improves the efficiency of maintenance schedule greatly. Finally, the predictive maintenance helps to assess the current state of tunnel's operation and predict the possible failures, which could avoid unnecessary maintenance activities and thus reduce the total maintenance costs.

References

1. Transportation Industry Development Statistical Bulletin. http://zizhan.mot.gov.cn/zfxxgk/bnssj/zhghs/201704/t20170417_2191106.html
2. Russell, H.A., Gilmore, J.: Inspection policy and procedures for rail transit tunnels and underground structures. Tcrp Synth. Transit Pract. (1997)
3. Ai, Q., Yuan, Y., Mahadevan, S., et al.: Maintenance strategies optimisation of metro tunnels in soft soil. Struct. Infrastruct. Eng. **13**, 1–11 (2017)
4. Niu, G.: Data-driven technology for engineering systems health management (2017)
5. Jiang, X.: Recent development in structural damage diagnosis and prognosis. Recent Pat. Eng. **4**(2), 102–121 (2010)
6. Hu, M., Zhang, M.Z., Shi, Y.Q., et al.: Research on tunnel operation and maintenance decision support system based on big data. J. Inf. Technol. Civ. Eng. Archit. **8**(1), 48–52 (2016)
7. Cheng, S., Prognostics, M.A., Lab, H.M., et al.: Sensor system selection for prognostics and health monitoring. In: ASME 2008 International Design Engineering Technical Conferences and Computers and Information in Engineering Conference. American Society of Mechanical Engineers, pp. 1383–1389 (2008)
8. Niu, G., Yang, B.S., Pecht, M.: Development of an optimized condition-based maintenance system by data fusion and reliability-centered maintenance. Reliab. Eng. Syst. Saf. **95**(7), 786–796 (2010)

9. United States Department of Defense. MIL-P-1629—Procedures for Performing a Failure Mode Effect and Critical Analysis. Department of Defense (US). MIL-P-1629, 9 November 1949
10. Chang, K.H., Chang, Y.C., Lai, P.T.: Applying the Concept of Exponential Approach to Enhance the Assessment Capability of FMEA. Springer, New York (2014)
11. Yuan, Y., Bai, Y., Liu, J.: Assessment service state of tunnel structure. Tunn. Undergr. Space Technol. Inc. Trenchless Technol. Res. **27**(1), 72–85 (2012)
12. Yuan, Y., Jiang, X., Liu, X.: Predictive maintenance of shield tunnels. Tunn. Undergr. Space Technol. Inc. Trenchless Technol. Res. **38**(3), 69–86 (2013)
13. Hashemian, H.M., Bean, W.C.: State-of-the-art predictive maintenance techniques*. IEEE Trans. Instrum. Meas. **60**(1), 226–236 (2011)
14. Zhao, P., Kurihara, M., Tanaka, J., et al.: Advanced correlation-based anomaly detection method for predictive maintenance. In: IEEE International Conference on Prognostics and Health Management, pp. 78–83. IEEE (2017)
15. Freund, Y., Schapire, R.E.: A decision-theoretic generalization of on-line learning and an application to boosting. In: European Conference on Computational Learning Theory, pp. 23–37. Springer, Berlin (1995)
16. Dan, M.F., Liu, M.: Time-dependent bridge network reliability: novel approach. J. Struct. Eng. **131**(2), 329–337 (2005)

A Heuristic for Legacy Information System Re-modularization

Lifeng Mu, Lihui Chen, and Fangyuan Wang(✉)

SHU-UTS SILC Business School, Shanghai University, Shanghai 201899, China
mulifeng@i.shu.edu.cn, 272853113@qq.com,
chloechenn827@163.com

Abstract. During software evolution, the architecture of the system, which underwent continuous modifications, usually drifts away from its original design and often causes quality declining. Nevertheless, redeveloping systems is risky and expensive and it could take years before the legacy system is replaced by a reliable and equivalent system. Re-modularization is a feasible technique used in automatically restructuring a legacy information system to improve architecture quality and to protect the previous investment. A combinatorial optimization model is proposed to formulate the legacy system re-modularization problem. Unlike previous studies, a new metric of balancing between coupling complexity and cohesion complexity is introduced to avoid the over-cohesive issue. In addition, making use of the problem characteristics, a heuristic based on edge-contraction and vectorization is designed to solve the re-modularization model, and numerous experiments illustrate the fact that the proposed heuristic may lead to better solutions than GA (genetic algorithm) in most cases, and it needs shorter computation times than GA and the heuristic without vectorization.

Keywords: Heuristics · Decision making · Software clustering problem
Information system re-modularization · Architecture optimization

1 Introduction

After making extensive investments in IT asset over the course of many years, various commercial information systems, which are built up over generations and are based on outdated technologies, are still running at numerous enterprises. Complex internal logic relationships, high maintenance costs and heterogeneous technical solutions prevent them from unfolding their full potential in future market changes. It is irrational to discard them completely and to replace the whole legacy system with a new system. In addition, a legacy system contains critical business logics which are valuable assets of the enterprises. The business logics are very difficult to be captured and be redeveloped, since they are usually embedded in the source code, and may not be explicitly and accurately documented anywhere else. Therefore, an operable method called re-modularization based on the concept of reusable components is usually employed to deal with a legacy system reconstruction. Through this method, functional and valuable components of a legacy system are identified, adapted, encapsulated, reused and finally reorganized in a new system.

J. Abawajy et al. (Eds.): ATCI 2018, AISC 842, pp. 505–516, 2019.
https://doi.org/10.1007/978-3-319-98776-7_55

The re-modularization problem is defined as follows: given a set of identified legacy components and a set of relations among them, in order to achieve a high-quality architecture, how to allocate identified components into modules, and how many modules should be included in the architecture? (Note that putting the number of modules into the scope of decision variables differentiates our research from the traditional software clustering problems). The issue generally requires deep analysis and understanding of complex internal logic relationships among identified components in a legacy system, that is a human dependent, difficult and time-consuming task. Therefore, an automatic optimization method would be a valuable tool for software engineering practitioners. Some previous studies have made various attempts to address issues mentioned above. We survey them briefly hereafter.

1.1 The Principle of Loose Coupling and Tight Cohesion

The proper architecture design is the critical issue to the success of any information system. Two criteria, i.e., coupling and cohesion, are widely regarded as two primary quantitative measures of the architecture quality. Cohesion refers to the strength of the relationship between components within a given module, and coupling is usually defined as the strength of the relationship between components which belong to different modules. Low coupling and high cohesion are firstly introduced by Stevens et al. [1] and are widely regarded as a primary principle of architecture design. For years, numerous coupling and cohesion metrics have been introduced to measure relationships among components for different programming paradigms.

In terms of procedural programming paradigm, empirical analysis and structural analysis are used to define distinct metrics of coupling and cohesion. Adamov et al. propose metrics of coupling and cohesion to measure indirectly the structural complexity of software [2]. Schach et al. propose 8 coupling and cohesion measures based on empirical results from the analysis of data for the repeated maintenance of three software products [3]. Darcy et al. describe and compare measures in this paradigm in term of software complexity [4].

Regarding the object-oriented (OO) paradigm, Chidamber et al. firstly introduce static coupling and cohesion metric—the CBO (Coupling Between Objects) and the LCOM (Lack of Cohesion in Methods) [5]. Recently, Revelle et al. introduce new feature coupling metrics based on structural and textual source code information and extend the unified framework for coupling measurement to include these new metrics [6]. Perepletchikov et al. present a controlled experiment to prove that the SOC-specific coupling metrics can facilitate coupling related design decisions with the aim of producing more maintainable service-oriented software products [7]. Kagdi et al. present an approach that combines conceptual coupling and evolutionary coupling to support change impact analysis in source code [8]. Using statistical techniques, Al Dallal constructs models based on the selected internal attributes (cohesion and coupling) to predict class maintainability [9].

Other methods based on architectural complexity are widely accepted at software engineering community. Coppick et al. firstly attempt to extend the McCabe and Watson complexity measures [10] to complexity metrics in OO systems [11]. Subsequently, other researchers [12–16] focus mostly on how to characterize a single component according to its interactions with other components. These researchers propose that the architectural complexity can be measured by combining with different per-class criteria. For example, the average number of attributes, abstract attributes, ancestor classes and methods.

Despite such achievements, according to the study of Anquetil et al. [17], even the widely-accepted principle of tight cohesion and loose coupling has become controversial according to recent researches. The principle that good architecture should exhibit tight cohesion and loose coupling is based on the assumption that future system changes will occur in the coupling parts with high probability. However, according to hands-on experiences, people gradually realize that even experienced engineers can not accurately predict where future changes will happen in the architecture, especially when decoupling a complex monolithic system depended heavily upon expert experience rather than scientific optimization approaches. Once important maintenance behaviors are required inside of a cohesive module, it will be a disaster for enterprises which are requested to react quickly to market changes. Therefore, the principle of complexity balance will be a wiser choice for risk-sensitive enterprises. Another shortcoming of the traditional principle is that reckless pursuit of high cohesion and low coupling will lead re-modularization algorithms to an over-cohesive solution with a single module containing all components.

In our work, a metric based on the principle of complexity balance is proposed to address the over-cohesive issue. Not only can it be used to reduce the risk of high maintenance costs at the architecture level, but the metric can also be used to design a heuristic to get a high-quality solution to re-modularization.

1.2 Search-Based Re-modularization Methods

Mancoridis et al. early try to employ conventional genetic algorithms and hill-climbing to solve the modularization problem of a software system from its source code [18]. Inspired by their work, a tool called Bunch is developed to automate software module clustering [19]. Recently, search technologies have become mainstream methods to solve modularization problems. Mitchell et al. describe their search algorithms for software clustering and demonstrate how altering the clustering parameters impact on the behavior and performance of their algorithms [20]. Harman et al. design a heuristic in which the fitness functions are derived from measures of module granularity, cohesion and coupling [21]. Mahdavi et al. adopt a multiple hill climbing approach to solving the automated modularization problem, and they show that results from a set of multiple hill climbs can be combined to locate good 'building blocks' for subsequent searches, and can reduce the search space [22]. Research is not only limited to single-objective approach, Praditwong et al. introduce two novel multi-objective formulations of the software modularization problem, in which several different objectives (including cohesion and coupling) are represented separately [23]. Bavota et al. propose an automated approach to improving the quality of software modularization, and the approach analyze underlying latent topics in source code as well as structural dependencies to recommend refactoring operations aiming at moving a class to a more suitable package [24].

It is worth mentioning that some researchers also propose global optimization algorithms for the same problem. Representative researchers, Kramer et al., make an outstanding contribution in this field. they apply the Dantzig–Wolfe decomposition and Staged Column Generation to solve the software clustering problem [25]. Algorithms have been proven by 45 instances to be invaluable for addressing medium or small scale problems (the number of components is less than 100). The reason why we adopt heuristic rather than global optimization algorithms is that we attempt to address large-scale problems (the number of components is more than 100), and the problem (NP-hard [26]).

The primary contributions of this article can be summarized as follows:

(1) A metric of complexity balance is proposed to balance the risk of high maintenance costs at the architecture level and to guide search algorithms to get high-quality solutions of re-modularization.
(2) We also elaborate on that the re-modularization problem considering complexity balance and provide the formal description of the problem.
(3) We propose a heuristic based on edge-contraction and vectorization to guarantee solution quality of re-modularization problems. Comparative experiments prove that the proposed algorithm has better performance in speed than traditional heuristics and evolutionary algorithms.

The paper is organized as follows. In Sect. 2, we present a mathematical description of the problem and formulate an optimization model. A heuristic based on vectorization is introduced in Sect. 3. Experiments are performed to compare the proposed heuristic with PHWV (the proposed heuristic without vectorization) and a GA (genetic algorithm) in Sect. 4. Finally, we conclude this research and suggest future research opportunities in Sect. 5.

2 Mathematical Formulation

According to the study of Darcy et al. [4], coupling and cohesion also can be used to indicate the architectural complexity. Inspired by their study, we introduce a new metric named *FCB* (the function of complexity balance) to balance between coupling complexity and cohesion complexity at the architecture level. We define an interaction as a relationship between two legacy components and should be exposed to each other through interfaces. The interaction degree is defined as the complexity degree of an interaction, which can be quantified using the number of variables, events and functions provided by two component's interfaces to each other. More variables, events and functions mean more functionality provided, meanwhile, also mean more complex logic relations and more efforts involved during reuse, secondary development, integration and maintenance. Coupling is defined as the sum of degrees of all interactions between legacy components belonging to different modules, and cohesion is defined as the sum of degrees of all interactions inside of the same module. The *FCB* can be formulated as following:

$$FCB = C + \max(d_i), \quad i = 1, 2, \ldots, M \tag{1}$$

where C is coupling of the whole architecture, d_i is cohesion of the ith module, M is the number of modules (a decision variable).

Through repeated experiments, we notice that C is monotonically increasing with M, and max(di) is monotonically decreasing with M, and FCB is a linear combination of M and max(di). FCB thus initially decreases, but turns to increase after reaching the minimum at a certain M. Unfortunately, the observation cannot be used directly to derive analytical a closed-form solution to the problem (because the problem is a binary integer programming), it still can be used to design a heuristic to find a high-quality solution quickly.

The re-modularization problem can be described abstractly by using graph theory. All identified components in a legacy system can be regarded as a set of vertices V, all interactions among components can be described as a set of edges E connecting vertices V, and an interaction degree between two components can be denoted as a positive weight $c(e)$ associated with each edge $e \in E$. Accordingly, this undirected weighted graph $G = (V, E)$ can be used to represent an legacy information system, where $|G|$ (the order of G) is the number of vertices in G. To represent an edge between adjacent two vertices, c_{ij} for $i,j = 1, 2, \ldots |G|$ and $i,j = 1, 2, \ldots |G|$ is used to denote a positive weight associated an edge between vertex i and vertex j, otherwise, $c_{ij} = 0$. Let a partition of $V = \{V_1, V_2, \ldots V_M\}$ represent a feasible re-modularization solution, where M is an integer decision variable, and $1 \leq M \leq |G|$. $G_k = (V_k, E_k)$ induced by V_k represents the kth subgraph (the kth module) and $F = E - \{E_1, E_2, \ldots E_M\}$ indicates a subset of edges whose deletion leaves a graph G with M subgraphs $\{G_1, G_2, \ldots, G_M\}$ disconnected with each other. FCB can be formulated as $\sum_{e \in F} c(\mathrm{e}) + \max(\sum_{e \in E_k} c(\mathrm{e}))$. The re-modularization problem can be described as finding a partition of V to minimize the value of FCB. In order to formulate the problem mathematically, A binary decision variable, x_{ik} for $i = 1, 2, \ldots, N, k = 1, 2, \ldots, M$ is introduced and $x_{ik} = 1$ denotes that vertex i belongs to the kth subgraph, otherwise $x_{ik} = 0$.

With these notations, equations, and inequalities, the re-modularization problem can be presented as the following programming model:

$$\min \left(\sum_{i=1}^{|G|} \sum_{j=i+1}^{|G|} c_{ij} - \sum_{k=1}^{M} \sum_{i=1}^{|G|} \sum_{j=i+1}^{|G|} \left(c_{ij} x_{ik} x_{jk} \right) + \max_{k=1,2,\ldots,M} \left(\sum_{i=1}^{|G|} \sum_{j=i+1}^{|G|} \left(c_{ij} x_{ik} x_{jk} \right) \right) \right) \tag{2}$$

s.t.

$$\sum_{k=1}^{M} x_{ik} = 1, \quad i = 1, 2, \ldots, |G| \tag{3}$$

$$\sum_{i}^{|G|} x_{ik} \geq 1, \; k = 1, 2, \ldots, M \tag{4}$$

$$x_{ik} \in \{0, 1\}, \quad i = 1, 2, \ldots |G|, k = 1, 2, \ldots, M \tag{5}$$

$$M \in \{1, 2, \ldots, |G|\} \tag{6}$$

In the objective (2), $\sum_{i=1}^{|G|}\sum_{j=i+1}^{|G|} c_{ij}$ represents the total sum of $c(e)$, $e \in E$, and $\sum_{k=1}^{M}\sum_{i=1}^{|G|}\sum_{j=i+1}^{|G|} (c_{ij}x_{ik}x_{jk})$ denotes the total sum of $c(e)$, $e \in E_1 \cup E_2 \cup, \ldots, \cup E_M$. Therefore, $\sum_{e \in F} c(e) = \sum_{i=1}^{|G|}\sum_{j=i+1}^{|G|} c_{ij} - \sum_{k=1}^{M}\sum_{i=1}^{|G|}\sum_{j=i+1}^{|G|} (c_{ij}x_{ik}x_{jk})$. $\sum_{i=1}^{|G|}\sum_{j=i+1}^{|G|} (c_{ij}x_{ik}x_{jk})$ represents the sum of $c(e)$ where $e \in E_k$, $k = 1, 2, \ldots, M$, thus $\max(\sum_{e \in E_k} c(e)) = \max_{k=1,2,\ldots,M} \left(\sum_{i=1}^{|G|}\sum_{j=i+1}^{|G|} (c_{ij}x_{ik}x_{jk})\right)$. Accordingly, the objective function (2) is to minimize *FCB*. Constraints (3) denote each vertex only belongs to one of V_k for k = 1, 2, …, *M*. Constraints (4) represent each V_k for k = 1, 2, …, *M* contains at least one vertex. Constraints (5) denote x_{ik} is binary decision variable. Constraint (6) represents the value range of the integer decision variable *M*.

The model (2)–(6) is already proved that this kind of problem is an NP-hard combinatorial optimization [16], and two kinds of approaches mainly were used to solve the problems, i.e. approximation algorithms and search-based algorithms. One of the most important paradigms in the design and analysis of approximation algorithm is the notion of a divide-and-conquer algorithm, but according to the study of Arora et al. [16], approximation algorithms designed in this way are either too complicated to analyze or have poor quality guarantees. As for search-based algorithms, although it is also hard to guarantee the high-quality solution. But by making use of the problem characteristics found in Sect. 2, we designed a heuristic to guarantee the speed and the high-quality solution in most cases.

3 A Heuristic for Re-modularization

The basic principle is simple: At first, the graph includes all vertices, and each vertex is attached with an initial cohesion value equal to 0. The two vertices, for which the merge produces the largest reductions in *FCB* value are identified, merged into a new vertex, and the edge joining merging nodes is contracted. The newly merged vertex is attached with a new cohesion value (the sum of the contracted edge's weight and the cohesion value of two vertices before the merge). All edges from the two vertices to remain vertices are replaced by one edge weighted by the sum of all these edge weights. The graph is iteratively shrunk by repeating edge contracts and vertex merges until there is only one vertex left. The minimal *FCB* found during this shrinking process is regarded as the good objective value, and the corresponding modularization solution is taken as the final solution.

A simple example shown in Fig. 1 is given to illustrate this algorithm. The example is based on the architecture of an airline travel system specifically designed for FlighByNight, Inc. Details of UML descriptions, which include class diagrams, component diagrams, sequence diagrams, can be found on http://www.phloxblog.in/projectuml/. Because we focus on legacy assets re-modularization at the level of components rather than classes, Fig. 1 only provides a component diagram.

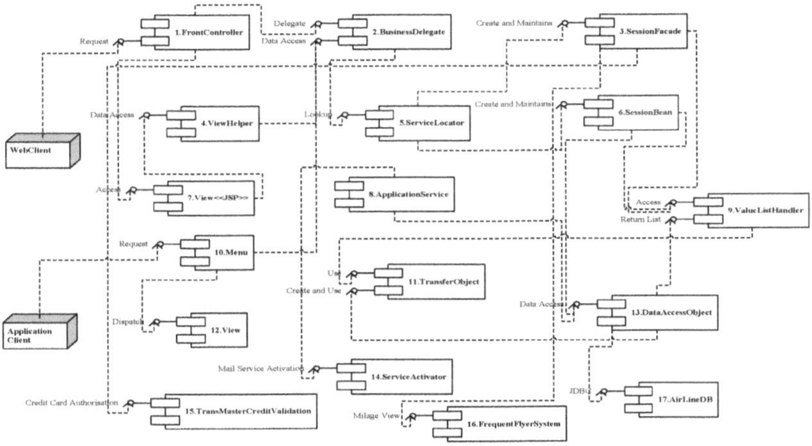

Fig. 1. The architecture of an airline travel system

The original graph $G = (V, E)$ abstracted from the architecture of the airline travel system is provided in Fig. 2, each vertex attached with a cohesion value (the value inside a box), all vertices' cohesion value are zero at first, and summation of all edge weights is cohesion of the graph G. Since the merge of vertex 8 and vertex 14 will produce the largest reduction in the *FCB* value of G, they are identified to be merged firstly. A new graph after the first merge is given in Fig. 2, there are two sub-graphs, i.e., the 1st sub-graph includes vertex 8 and vertex 14, and the 2nd sub-graph includes all other vertices. The cut F includes one edge, i.e., the edge between vertex 8 and vertex 13, and coupling $C = 2$. The cohesion of the 1st sub-graph $d_1 = 5$, the cohesion of the 2nd sub-graph $d_2 = \mathit{total\ edge\ weight} - C - d_1 = 53 - 5 - 2 = 46$. So $FCB = 2 + 46 = 48$. This shrinking process is repeated until one vertex left, and the best *FCB* and the corresponding solution is recorded. *FCB* gets the minimal value is 23 where $C = 6$ and $\max(di) = 17$. The corresponding re-modularization plan, i.e. (1, 2, 4, 7, 10, 12), (3, 5, 6, 9, 15, 16), and (8, 11, 13, 14, 17), is the solution found by the algorithm.

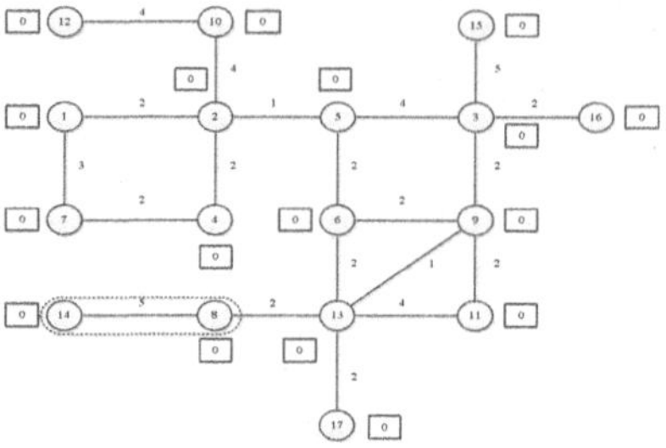

Fig. 2. C = 0, $\max(d_i) = 53$, $FCB = 53$

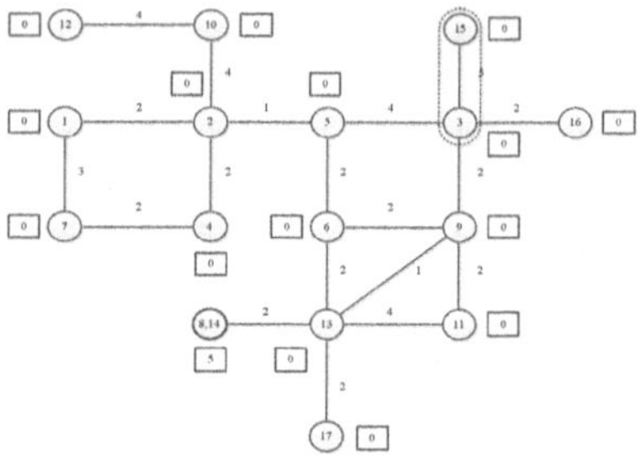

Fig. 3. $C = 2$, $\max(d_i) = 46$, $FCB = 48$

4 Comparison Experiments

All experiments were performed on a computer with a 3.2 GHz Intel Core i5-3470 CPU (4 × 256 KB L2 cache and 6 M L3 cache) and DDR3 8G memory. To evaluate objectively solution quality and efficiency of the proposed heuristic, two groups of comparison experiments are performed based on real cases.

Regarding the first group, 17 test data sets derived from actual cases are used to compare algorithms. Full descriptions and component diagrams of software systems can be found from the URLs listed in Table 1.

GA is usually employed to tackle the re-modularization problem [27–30]. In order to prove the advantage of the proposed heuristic, each specific case is solved using three different algorithms respectively, including the proposed heuristic, PHWV (the proposed heuristic without vectorization), and a GA boosted by vectorization techniques.

4.1 The Enhanced GA Algorithm for Comparison

A Chromosome is represented as a simple array mapping components (array indices) to modules (array elements used to identify modules). The number of modules is determined by the number of different integer values in a chromosome. For example, the array [2, 5, 5, 7, 7, 5, 2, 5, 7, 7] denotes a re-modularization solution where eight components are grouped into three modules, identified by the numbers 2, 5 and 7, respectively. The choice of numbers of module identifier is arbitrary, so this re-modularization solution is equivalent to [1, 3, 3, 8, 8, 3, 1, 3, 8, 8] and [4, 1, 1, 6, 6, 1, 4, 1, 6, 6]. each integer (gene) is generated by using a uniform distribution $U(1, |G|)$ for the sake of diversity.

In order to make an unbiased comparison, some state-of-the-art technologies which is proven to be efficient are adopted in GA [31, 32], for example Laplace crossover [31] and power mutation [32]. Other parameter settings are described as follows:

The size of the population is going to be $\max(\min(10 \times |G|, 200), 40)$, and the size of the elitist archive is assigned with $0.05 \times \max(\min(10 \times |G|, 200), 40)$. Laplace crossover is used 100% of the time to breed individuals, and each gene has a 1% probability of getting a mutation, independent of other genes. The number of generations is constant at 400 for observing the long-term tendency. The GA stops when the average relative change of the best fitness function value over 200 generations is equal to 0. More importantly, we further improve the efficiency of calculating the fitness function by using vectorization techniques.

4.2 Results of Comparison

Evolutionary algorithms like GA are based on the concept of natural evolution which select individuals randomly, and the GA return slightly different results after each run, so just running GA once can prove nothing. The customized GA is thus used 20 times to solve each specific problem, and the median value of 20 solutions is adopted in each comparison experiment. The proposed heuristic and PHWV also run 20 times to solve a given problem, because when more than one minimal *FCB* are identified in a *FCB* vector, one of them is selected randomly. Just like GA, the median value of solutions is adopted in comparisons.

Table 1. Software systems are used to verify the algorithms.

ID	Name	URL
1	Open Power Libs	static.invenzzia.org/docs/opl/3_0/book/en/installation.html
2	XML Schema	www.w3.org/TR/2004/REC-xmlschema-1-20041028/
3	PDM-UG	www.wikistep.org/index.php/PDM_Usage_Guide
4	org.eclipse.xsd	www.w3.org/TR/xmlschema-1/
5	Ygor	safe.nrao.edu/wiki/bin/view/Kbandfpa/ ArrayMnCSoftwareDesign#Overview
6	CTRP	wiki.nci.nih.gov/pages/viewpage.action?pageId=65732646
7	Handbook diagram	gamedev.stackexchange.com/questions/55429/do-you-actually-use-diagrams-to-model-games
8	org.eclipse.xsd Part2	www.w3.org/TR/xmlschema-2/
9	MPK	java4life.tripod.com/scea/projectscea011003.html#purpose
10	j2ee. webapplication	eclipse.org/webtools/jst/components/j2ee/api/j2ee_models_ overview.html
11	FBN	java4life.tripod.com/scea/project.html#assumptions
12	MPK Class Diagram	java4life.tripod.com/scea/projectscea011003.html#purpose
13	com.ibm.etools. xsd	www.w3.org/TR/xmlschema-1/
14	SOLERES	www.ual.es/acg/soleres/jucs/#
15	RSF	www2.caret.cam.ac.uk/rsfwiki/Wiki.jsp?page=SystemInfo
16	Microsoft Private Cloud	social.technet.microsoft.com/wiki/contents/articles/13188.system-center-2012-integration-guide.aspx
17	MDDS	kohei.us/

Table 2 provides the results of the experiment and the best *FCBs*, the fastest speeds are highlighted using the gray background in Table 2.

Table 2. Results of comparison experiments based data derived from software systems

ID	$\|G\|$	$\|E\|$	Heuristic			PHWV			GA		
			FCB	*M*	Time(s)	*FCB*	*M*	Time(s)	*FCB*	*M*	Time(s)
1	13	16	30	2	0.0093	30	2	0.0077	27	3	8.2901
2	13	22	42	2	0.0081	42	2	0.0101	42	3	6.7996
3	16	20	30	3	0.0126	30	3	0.3679	30	3	8.6787
4	17	20	58	2	0.0114	58	2	0.0133	58	4	8.3064
5	18	20	38	3	0.0139	38	3	0.0128	36	3	9.096
6	18	23	45	2	0.0144	45	2	0.0122	45	4	8.2156
7	25	24	27	5	0.0177	27	5	0.0163	27	6	9.6406
8	25	26	33	5	0.017	33	5	0.016	31	5	7.3997
9	26	26	22	5	0.0215	22	5	0.021	24	6	10.093
10	26	27	51	6	0.0199	51	6	0.0165	51	7	7.9543
11	26	27	23	6	0.0191	23	6	0.0182	25	6	9.0566
12	27	31	32	4	0.022	32	4	0.0226	35	5	8.3556
13	29	29	58	2	0.0217	58	2	0.02	58	5	10.1716
14	33	44	57	5	0.0296	57	5	0.0412	63	4	10.9372
15	38	44	49	6	0.0332	49	6	0.0408	53	7	14.1567
16	56	58	49	11	0.0428	49	11	0.0728	73	8	15.6963
17	152	158	141	9	0.2046	141	9	1.421	341	4	14.2487

According to results in Table 2. Results of comparison experiments based data derived from software systems, the GA only finds better solutions than the proposed heuristic and PHWV in three cases (i.e. Case 1, Case 5, and Case 8), three algorithms find exactly the same objective value in seven cases (i.e. Case 2, Case 3, Case 4, Case 6, Case 7, Case 10, and Case 13), and the proposed heuristic and PHWV find better solutions than the GA in seven cases (i.e. Case 9, Case 11, Case 12, Case 14, Case 15, Case 16 and Case 17). In most cases, the proposed heuristic and PHWV show the better performance against GA in solution quality. It also can be concluded that the GA takes much more time to converge to a solution even though vectorization is adopted to improve its efficiency. In contrast, the proposed heuristic and PHWV can find better solutions in less time than the GA in most cases. The results of Mann–Whitney–Wilcoxon tests also prove that the proposed heuristic outperforms the customized GA obviously.

5 Discussion and Conclusion

This paper discusses a practical management problem of software engineering, i.e., reconstruction of a legacy monolithic information system. A new metric, *FCB*, is introduced to balance complexity between coupling and cohesion at the architecture level, and it is proven that *FCB* is suitable as an objective of the re-modularization problem. A heuristic based on edge-contraction and vectorization is designed to solve the proposed problem, and experiments illustrate the fact that it can guarantee the high-quality solution and excellent efficiency in most cases.

According to our current research, the solution quality might be further improved based on solutions found by the proposed heuristic through reallocation of components among different modules. Moreover, solutions provided by the proposed heuristic can be used as seeds in an initial population, then GA could be used to find better solutions [33]. All of those will be addressed in the future work as an extension of this paper. By the way, the proposed heuristic cannot replace experienced experts to accomplish re-modularization optimization totally, it is only designed as a useful tool to provide potential high-quality re-modularization solutions for further consideration.

Acknowledgements. This work was supported in part by grants from NSFC under grant No. 71302051.

References

1. Stevens, W.P., Myers, G.J., Constantive, L.L.: Structured design. Ibm Syst. J. **38**(2.3), 231–256 (1999)
2. Adamov, R., Richter, L.: A proposal for measuring the structural complexity of programs. J. Syst. Softw. **12**(1), 55–70 (1990)
3. Schach, S.R., et al.: Determining the distribution of maintenance categories: survey versus measurement. Empir. Softw. Eng. **8**, 351–365 (2003)
4. Darcy, D.P., et al.: The structural complexity of software an experimental test. IEEE Trans. Softw. Eng. **31**(11), 982–995 (2005)
5. Chidamber, S.R., Kemerer, C.F.: Towards a metrics suite for object oriented design. In: Sigplan Notices (1991)
6. Revelle, M., Gethers, M., Poshyvanyk, D.: Using structural and textual information to capture feature coupling in object-oriented software. Empir. Softw. Eng. **16**(6), 773–811 (2011)
7. Perepletchikov, M., Ryan, C.: A controlled experiment for evaluating the impact of coupling on the maintainability of service-oriented software. IEEE Trans. Softw. Eng. **4**(37), 449–465 (2011)
8. Kagdi, H., Gethers, M., Poshyvanyk, D.: Integrating conceptual and logical couplings for change impact analysis in software. Empir. Softw. Eng. **18**(5), 933–969 (2013)
9. Dallal, J.A.: Object-oriented class maintainability prediction using internal quality attributes. Inf. Softw. Technol. **55**(11), 2028–2048 (2013)
10. McCabe, T.J., Watson, A.H.: Software complexity. J. Def. Softw. Eng. **7**(12), 5–9 (1994)
11. Coppick, J.C., Cheatham, T.J.: Software metrics for object-oriented systems. In: Conference on Scientific Computing (1992)
12. Zhou, Y., Leung, H., Xu, B.: Examining the potentially confounding effect of class size on the associations between object-oriented metrics and change-proneness. IEEE Trans. Softw. Eng. **35**(5), 607–623 (2009)
13. Simons, C.L., Parmee, I.C., Gwynllyw, R.: Interactive, evolutionary search in upstream object-oriented class design. IEEE Trans. Softw. Eng. **36**(6), 798–816 (2010)
14. Kwong, C.K., et al.: Optimization of software components selection for component-based software system development. Comput. Ind. Eng. **58**(4), 618–624 (2010)
15. Li, H., Li, B.: A pair of coupling metrics for software networks. J. Syst. Sci. Complex. **24**(1), 51–60 (2011)

16. Dixit, V.P., Vishwkarma, R.: Comparison of class-level versus object-level static and dynamic coupling and cohesion measures in object oriented programming. In: Wireless and Optical Communications Networks (2014)
17. Anquetil, N., Laval, J.: Legacy software restructuring: analyzing a concrete case. In: Conference on Software Maintenance and Reengineering (2011)
18. Mancoridis, S., et al.: Using automatic clustering to produce high-level system organizations of source code. In: International Conference on Program Comprehension (1998)
19. Mancoridis, S., et al.: Bunch: a clustering tool for the recovery and maintenance of software system structures. In: International Conference on Software Maintenance (1999)
20. Mitchell, B.S., Mancoridis, S.: Using heuristic search techniques to extract design abstractions from source code. In: Genetic and Evolutionary Computation Conference (2002)
21. Harman, M., Hierons, R.M., Proctor, M.: A new representation and crossover operator for search-based optimization of software modularization. In: Genetic and Evolutionary Computation Conference (2002)
22. Mahdavi, K., Harman, M., Hierons, R.M.: A multiple hill climbing approach to software module clustering. In: International Conference on Software Maintenance (2003)
23. Praditwong, K., Harman, M., Yao, X.: Software module clustering as a multi-objective search problem. IEEE Trans. Softw. Eng. **37**(2), 264–282 (2011)
24. Bavota, G., et al.: Improving software modularization via automated analysis of latent topics and dependencies. ACM Trans. Softw. Eng. Methodol. **23**(1), 4 (2014)
25. Kramer, H.H., et al.: Column generation approaches for the software clustering problem. Comput. Optim. Appl. **64**(3), 843–864 (2016)
26. Arora, S., Lee, J.R., Naor, A.: Euclidean distortion and the sparsest cut. J. Am. Math. Soc. **21** (1), 1–21 (2008)
27. Hadaytullah, et al.: Tool support for software architecture design with genetic algorithms. In: International Conference on Software Engineering Advances (2010)
28. Mitchell, B.S., Mancoridis, S.: On the evaluation of the Bunch search-based software modularization algorithm. Soft. Comput. **12**, 77–93 (2007)
29. Jiang, H., et al.: A foundational study on the applicability of genetic algorithm to software engineering problems. In: Congress on Evolutionary Computation (2007)
30. Tseng, T., et al.: Applying genetic algorithm for the development of the components-based embedded system. Comput. Stand. Interfaces **27**(6), 621–635 (2005)
31. Deep, K., et al.: A real coded genetic algorithm for solving integer and mixed integer optimization problems. Appl. Math. Comput. **212**(2), 505–518 (2009)
32. Deep, K., Thakur, M.: A new crossover operator for real coded genetic algorithms. Appl. Math. Comput. **188**(1), 895–911 (2007)
33. Sayyad, A.S., et al.: Scalable product line configuration: a straw to break the camel's back. In: Automated Software Engineering (2013)

E-maintenance Platform Design for Public Infrastructure Based on IFC and Semantic Web Technologies

Min Hu[1,2](✉) and Yunru Liu[1,2]

[1] Sydney Institute of Language and Commerce, Shanghai University, Shanghai 201800, China
minahu@163.com, liuyunru0814@163.com
[2] SHU-SUCG Research Centre for Building Industralization, Shanghai 200072, China

Abstract. As an important kind of infrastructure, a tunnel's life is estimated to be 100 years or more. During its service life, effective maintenance strategy plays an essential role to keep its availability and safety. Maintenance work involves a set of participants, activities and resources, bringing difficulties to implement effective maintenance. This paper proposes an e-maintenance framework combining BIM, IFC and semantic web technologies to help maintenance managers and maintenance personnel to get information they need timely and accurately, making effective maintenance decisions. A cost analysis case is provided in this paper to illustrate the implementation mechanism and the validation of this method.

Keywords: E-maintenance · BIM · IFC · Semantic web

1 Introduction

Infrastructure plays an important role in sustainable development and maintenance is of crucial importance to keep and improve system availability and safety. The actual maintenance work involves a great number of different and complex activities, associating various parties and requiring considerable resources. To support this work and reduce the potential security risks brought by the jumbled maintenance process, the concept of e-maintenance emerges. It refers to the integration of information and communication technologies to meet needs for supporting maintenance strategies and plans [1]. By dint of wireless and Internet technologies, any manager, operator or expert is capable of remotely linking to a company's equipment diagnosis, fixing, performance monitoring, and data collection and analysis. It has almost unlimited potential to reduce the complexity of traditional maintenance guidance through online guidance based on the results of decision-making and analysis of product condition [2]. The application of e-maintenance complies with the management requirements of PPP (Public–Private Partnership) projects.

Nevertheless, the establishment of e-maintenance also faces some challenges due to the large scale, big volume and high diversity and complexity of infrastructure facilities

J. Abawajy et al. (Eds.): ATCI 2018, AISC 842, pp. 517–526, 2019.
https://doi.org/10.1007/978-3-319-98776-7_56

and equipment. Firstly, an e-maintenance system involves a variety of cross-platform data, information and knowledge integration issues [1], which leads to the difficulty in information retrieval and tracking. In addition, quantity take-off and cost estimation (CE) should be efficient and accurate because of their high significance to a project's success. They require extracting information based on the knowledge of domain experts about the rules and processes throughout the projects lifecycle [3]. Heterogeneous models used by different designers also brings challenge to predict and diagnose failures and disturbances and to estimate the remaining lifetime of components [4].

To support the e-maintenance objectives, this paper proposes a semantic web services framework utilizing IFC-based industry ontology to address the e-maintenance problems. Section 2 reviews some studies relevant to e-maintenance and building information integration and analysis. Section 3 proposes an IFC-based framework of e-maintenance platform. Section 4 illustrates the role of IFC in the platform and one application case shown in Sect. 5 proves the operability of this platform.

2 Related Work

2.1 E-maintenance

The term e-maintenance can be considered as a maintenance strategy [1], a maintenance plan [5], a combination of Web service technology [6] or a distributed artificial environment [7]. Unlike the conventional maintenance, information transfer need to get through a hierarchical reporting process, an e-maintenance platform introduces an unprecedented level of transparency and efficiency into the entire industry and it can be adequate support of business process integration [8]. According to Lee, at an e-maintenance platform, goods are checked out from stores against a work order or a location and the transaction is recorded in real time. Then, the massive data bottlenecks between the plant floor and business systems can be eliminated by converting the raw machine health data, product quality data, and process capability data into information and knowledge for dynamic decision-making [9].

Despite the perceived advantages of e-maintenance, there are also some challenges for the establishment of e-maintenance waiting to be responded through future research [1]. What matters most is that a reliable, scalable and common informatics platform has to be developed in order to implement e-maintenance successfully [10]. The development of e-maintenance requires information integration among various systems [11]. To support the objectives of e-maintenance, a multi-tasking and multi-user operating environment, and a fast and easy-to-manage database for international experts are necessary to retrieve or store their aggregated knowledge and experiences [12].

2.2 BIM and IFC

Building Information Modeling (BIM) technology has been receiving an increasing attention in the AEC (Architecture, Engineering and Construction) industry. Compared with the traditional Computer Aided Design (CAD) technology, BIM is capable of restoring both geometric and rich semantic information of building models, as well as their relationships, to support lifecycle data sharing [13]. And it is expected to facilitate the collaboration and interaction among different actors involved in different stages of a building project since decisions made by each expert working on a particular part have an impact on the overall project [14]. One of the major challenges to realize coordination is interoperability problem due to data heterogeneity, modeling software diversity and lack of common modeling standard [15].

As the most widely used data exchange standard for BIM, IFC (Industry Foundation Classes) developed by building SMART, which provides the commonly shared concepts, attributes and relationships of BIM resources [13], has been widely supported by the market-leading BIM software vendors in the AEC industry [15]. The IFC schema is based on a set of concepts such as classes, attributes, relationships, property sets, and quantity definitions, to describe information of building models to be exchanged between different software applications so that BIM data can be extracted from proprietary software and be exchanged with other applications [14]. However, as a general purpose data model that facilitates the exchange among different software applications, the IFC schema cannot anticipate the multiple ways of representing information of different BIM software. In addition, its ability to define the characteristics of a model in different ways might hinder its interpretation. The use of semantic web technologies to address these problems in the AEC sector has been proposed by several authors [13, 14, 16, 17]. Zhang and Issa have proposed an ontology-based approach based on OWL DL language for capturing, interpreting, and reusing the semantics of product information [16]. Pauwels has suggested using the semantics and syntax of RDF graphs to combine models from different CAD/BIM applications [17]. They point out that the definition of mapping rules is one of the real challenges to address in the future.

2.3 Semantic Web Technologies

Researchers started to propose the use of semantic web technologies in the AEC industries in the early 2000s. The purpose of the Semantic Web is to apply these technologies to the Web of data. Pauwels et al. concludes three main added value brought by semantic web technologies, which are information exchange, data integration and complex search queries across several data sources, and construction document information management [15]. To make this possible, the data on the Web should be available in the standard formats required by the tools developed by the Semantic Web.

According to Costa and Madrazo, an ontology can be described as "a formal and explicit specification of a shared conceptualization" [14]. These specifications are defined by means of classes, attributes, values, relationships, roles and rules. RDF/OWL is the most widely used language created by the World Wide Web Consortium to describe ontologies in a formal way.

Recently, several IFC-based ontologies have been developed for particular application needs [16–18]. For instance, Zhang and Issa [16] used an IFC ontology to extract partial model from a complete IFC model. Pauwels et al. [17] utilized an IFC ontology to semantic rule checking. Nepal et al. [18] presented an approach for converting the IFC schema into the OWL format, which is a remarkable effort to lift the IFC specification onto the ontology level.

Up to now, there is little research on IFC-based ontologies applied in e-maintenance. Since both IFC and domain ontologies contain some semantic information and relationships, focusing on the information retrieval problem of e-maintenance and interoperability problem of BIM, we propose an e-maintenance platform for public infrastructure based on IFC-ontology and semantic web services.

3 The Framework of E-maintenance Platform for Public Infrastructure

3.1 Framework Overview

The framework of e-maintenance we developed is represented in Fig. 1 and includes two regions, six layers and a bidirectional data stream.

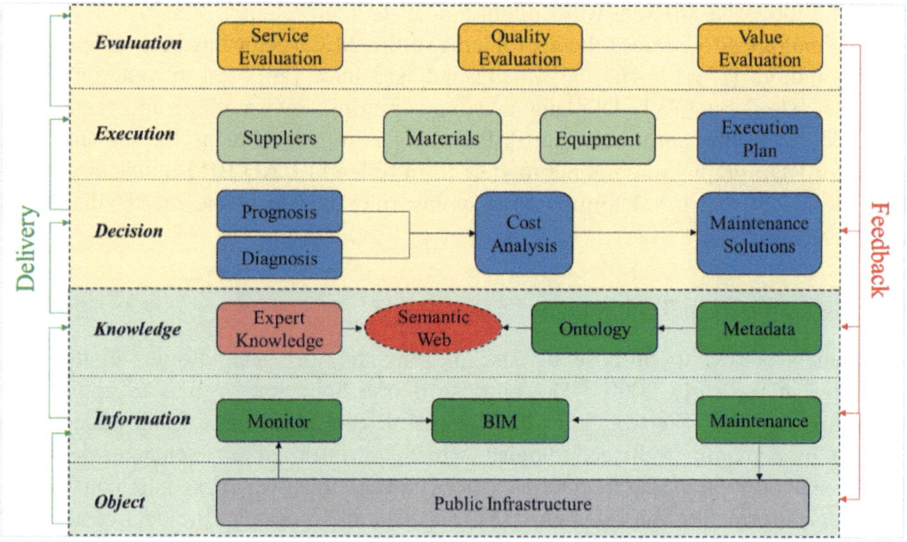

Fig. 1. The framework of E-maintenance

Two Regions: Support Region and Service Region. The support region provides underlying information and knowledge support for the whole framework while the service regions aim to realize various functions, such as decision making, execution and evaluation.

Six Layers: The framework is divided into six layers according to the roles of two regions, which consist of object, information, knowledge, decision, execution and evaluation.

(1) Object layer refers to the public infrastructure, including all series of facilities and equipment.
(2) Information layer covers monitoring data (status information), BIM data (structural information) and maintenance data (activity information).
(3) Knowledge layer includes metadata, ontology and expert knowledge, as well as a semantic web linking ontologies and expert knowledge.
(4) Decision layer includes prognosis, diagnosis, cost analysis and maintenance solutions determination process.
(5) Execution layer mainly focuses on the formulation of execution plan, which consists of service supplier selection, materials and equipment procurement.
(6) Evaluation layer evaluates the performance of maintenance, including service evaluation, quality evaluation and value evaluation.

Bidirectional data stream: Forward Transferring and Backward Feedback.

3.2 Service Mechanism

The main purpose of this framework is to retrieve infrastructure information in combination with the complexity of components' relationship, and to integrate the complex information to support maintenance decision making, improving maintenance efficiency and quality.

The information layer accurately describes three kinds of information: status information (real-time monitoring information), structural information (geometric features and spatial location) and activity information (maintenance data, traffic flow, equipment control and emergency treatment etc.). The main part of knowledge layer is to create ontologies based on these discrete and independent data from a structural or spatial perspective and link them through a semantic web, taking expert knowledge into consideration at the same time. Based on the semantic web established in knowledge layer, four engines are developed in decision layer, which are prognosis, diagnosis, cost analysis and maintenance solutions determination. The execution layer will generate an execution plan including selection of suppliers, materials and equipment procurement to support the maintenance work planned in the maintenance solution. Finally, the evaluation layer can track the execution performance and generate evaluation reports of service, quality and value automatically. At the same time, the evaluation results will get back to the support region, and the maintenance solution can be adjusted to obtain better maintenance performance.

The service mode consists of three modules: online activities, offline activities and interaction mechanism. Online activities mainly focus on the generation process of maintenance solutions. Offline activities include the whole procurement process (materials, equipment and services). The execution plan and evaluation results help realize the interaction of online and offline activities.

Online Activities: The maintenance staff can retrieve various data about facilities and equipment stored in the information layer with the help of ontology and semantic web. And they can use these data to prognoses and diagnose some potential defects. In addition, they can do a cost analysis about eliminating these defects according to some similar history records. After the cost analysis process, a final maintenance solution will be generated online.

Offline Activities The execution plan includes the labor, materials and equipment needed to perform the maintenance solution, managers can select the most suitable suppliers of materials, equipment and services combined with the market price, company credit and some other related factors.

Interaction Mechanism: The execution plan generated online provides guidance to the offline procurement process. And the evaluation results generated offline need to be uploaded online to modify the maintenance solution.

4 Information Interconnection Based on IFC

4.1 Information Expression of IFC

The IFC standard contains and manages all kinds of information through a hierarchical and modular framework, which is divided into four levels from bottom to top, namely, resource layer, core layer, interoperability layer and domain layer. Each level contains several modules consisting of a variety of entities, defining types, selecting types, enumerating types, rules, functions and property sets. The entities in the IFC standard can be divided into two categories. One is the entity that can be used independently for data exchange. Such entities are derived entities of IfcRoot, which are distributed in core layer, interoperability layer and domain layer. They all have the GlobalId attribute, which enables the globally unique identification of entities. The other is the entity distributed in resource layer, which cannot exchange data independently. They are not derived from IfcRoot so they cannot be identified in the process of data exchange. These entities usually exist in the form of properties of IfcRoot derived entities.

IfcRoot has a direct property named GlobalId, and instances of its inherited entities have a globally unique identification, which can be independently used for data exchange. The three derived abstract entities from IfcRoot are IfcObjectDefinition, IfcPropertyDefinition and IfcRelationship. These three entities and their derivative entities are distributed in core layer. IfcObjectDefinition and its derived entities are used to define objects. IfcPropertyDefinition and its derived entities are used to define properties. IfcRelationship and its derived entities realize the definition of relationships. Through the definition of objects, properties and relationships, the core framework of IFC model is formed. By dint of this core framework, users can find a formal way to describe entities and properties and define the complex relationships between entities and properties or property sets, interconnecting information from various sources.

4.2 Building Information Expression in IFC

To automatically apply IFC data, identifying useful design information and analyzing how these information is expressed by IFC are necessary. After analysis, useful design information includes building product category, material information, building product detailed classification, geometric features and so on. Their interconnected relationship expressed in IFC can be illustrated by Fig. 2.

In IFC, the expression of building products is realized through the entity IfcProduct and its derived entities. In addition, IfcBuildingElementProxy can also be used to express the undefined building products in IFC. The geometric features are expressed by IfcProductRepresentation entity. The derived entities of IfcProductRepresentation define a variety of geometric expressions, including SweepSolid, BoundingBox, Brep, etc., which can be used to describe the geometric features of a building product in various shapes. Material information is expressed by the selection of five material structures listed in the entity IfcMaterialSelect, which can be used to express the structure of the building products and material information. Then the relationship entity IfcRelAssociatesMaterial can be used to associate the building product and its material information.

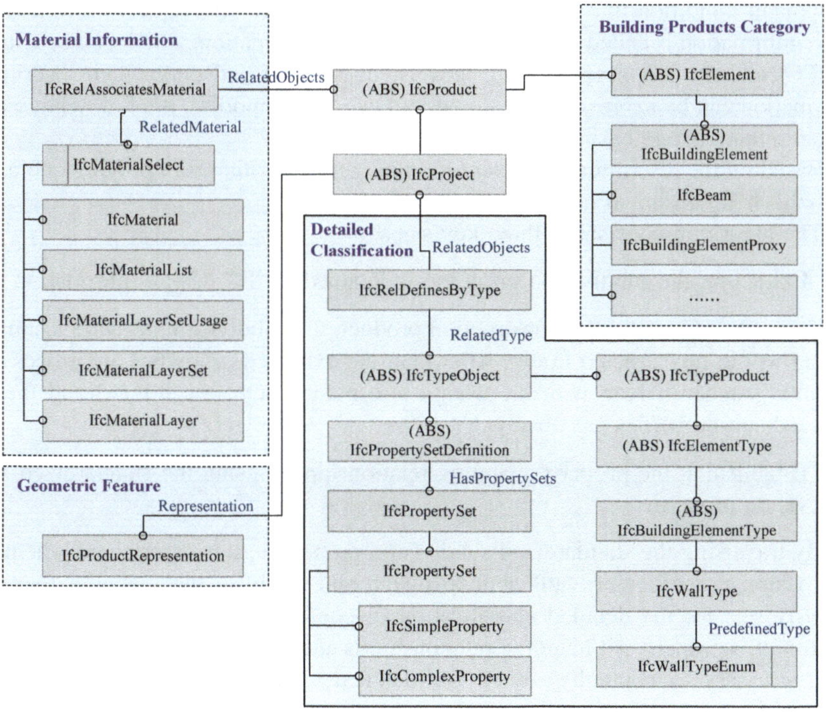

Fig. 2. Building Information Interconnection in IFC

5 System Application

Taking the division of cost items in tunnel operation and maintenance as an example, this paper briefly introduces the cost analysis process based on E-maintenance framework, and shows how it can improve the working efficiency and accuracy, supporting effective maintenance. Cost analysis based on quota is a widely used method in AEC industry. It basically includes quantity take-off and valuation. Under the traditional mode, it is necessary to remodel the design drawing to generate the bill of quantities, and the repeated input of the design information leads to low work efficiency. At the same time, the expression of design information is often inconsistent with the information expression in quota due to the designer preference, regional differences and so on, which leads to a semantic problem of information expression. In this paper, we introduce a method to automatically match the items in BIM model with cost items in quota through e-maintenance platform.

The automatic cost item generation mechanism mainly includes two parts: discrimination rules and the algorithm. The cost items in "Quota for Operation and Maintenance of Tunnel Project" are divided according to the operation and maintenance tasks, and each cost item has a unique code. According to the analysis, as long as the following five aspects of information are available, that is, the type of building product, geometric characteristics, material information, detailed classification and component condition, the corresponding quota item can be uniquely identified. And these information is called quota item discriminant information, which can be extracted by IFC, and the mapping relation between quota item and quota item discriminant information can be expressed and stored in database, supporting the following work as a discrimination rule base.

Based on the discrimination rules, an intelligent algorithm for applying quota items is established, as shown in Fig. 3.

The algorithm consists of three key steps:

(1) Extracting the subdivision properties contained in IFC design information

For each IFC entity of a construction product, the subdivision properties contained therein would be extracted firstly. After that, the extracted properties are processed by semantic standardization with the design information semantic library, and the standardized sub-properties are obtained.

(2) Establishing the property mapping relationship by using the standardized subdivision properties.

By traversing the standardized subdivision properties, it is determined whether the subdivision properties are sufficient to match and establish the mapping relation of property sets for the detailed classification of construction products. If the match is successful, a property set mapping relationship is automatically established. Otherwise, it is necessary to manually set up the detailed classification of building products beforehand and then the building property set mapping relationship can be established.

(3) Determining the quota items to be applied to the construction products by using the complete subdivision properties.

Based on the detailed classification, the detailed content of the property set of the detailed classification mapping is automatically set by using the standardized subdivision information. If the property set is set up automatically, the corresponding quota items are applied to the rule base of quota items automatically. Otherwise, it is

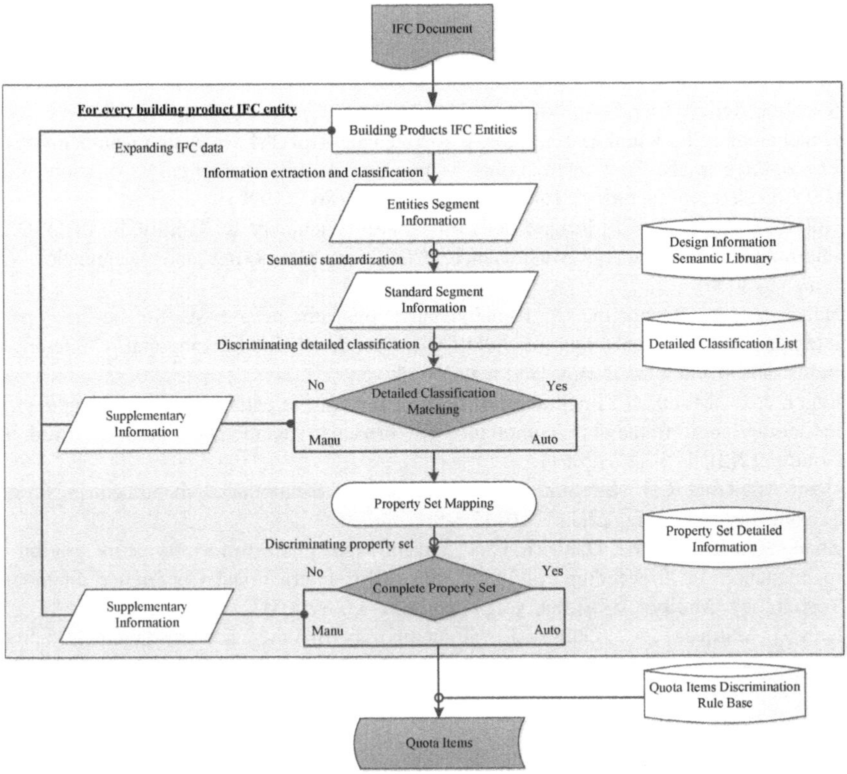

Fig. 3. Intelligent algorithm for obtaining quota items

necessary to manually complete the property sets, and then corresponding quota items can be found from the rule base of quota items.

6 Conclusions

In this paper, we have proposed an e-maintenance framework for public infrastructure based on IFC and semantic web technologies to deal with the interoperability problem in AEC industry. It is able to gather and process heterogeneous information and output specific information for particular users, supporting more effective maintenance work. Taking tunnel projects as an example, this paper implemented the method into a tunnel maintenance cost analysis process, providing more detailed information about the automatic implementation mechanism of the proposed e-maintenance framework. This idea can also be applied to other infrastructure domain to improve maintenance efficiency.

Acknowledgment. Our research was sponsored by Shanghai Municipal Science and Technology Commission (No. 16DZ1201700).

References

1. Muller, A., Marquez, A.C., Iung, B.: On the concept of e-maintenance: review and current research. Reliab. Eng. Syst. Saf. **93**, 1165–1187 (2008)
2. Goncharenko, I., Kimura, F.: Remote maintenance for IM [inverse manufacturing]. In: Proceedings of the First International Symposium on Environmentally Conscious Design and Inverse Manufacturing, Tokyo, Japan, pp. 862–867 (1999)
3. Holmberg, K., Halme, J.: Prognostics for industrial machinery availability. In: PHOTO 2005 International Seminar on Maintenance, Condition Monitoring and Diagnostics, Oulu, Finland (2005)
4. Hausladen, I., Bechheim, C.: E-maintenance platform as a basis for business process integration. In: Proceedings of INDIN04, Second IEEE International Conference on Industrial Informatics, Berlin, Germany, pp. 46–51 (2004)
5. Leger, J.-B., Morel, G.: Integration of maintenance in the enterprise: towards an enterprise modelling-based framework compliant with proactive maintenance strategy. Prod. Plan. Control **12**(2), 176–187 (2001)
6. Ucar, M., Qiu, R.G.: E-maintenance in support of E-automated manufacturing systems. J. Chin. Inst. Ind. Eng. **22**(1), 1–10 (2005)
7. Zhang, W., Halang, W., Diedrich, C.: An agent-based platform for service integration in E-maintenance. In: Proceedings of ICIT 2003, IEEE International Conference on Industrial Technology, Maribor, Slovenia, vol. 1, pp. 426–433 (2003)
8. Crespo Marquez, A., Gupta, J.: Contemporary maintenance management: process, framework and supporting pillars. Omega **34**(3), 313–326 (2006)
9. Lee, J.: E-manufacturing: fundamental, tools, and transformation. Robot. Comput. Integr. Manuf. **19**(6), 501–507 (2003)
10. Koc, M., Ni, J., Lee, J., Bandyopadhyay, P.: Introduction of e-manufacturing. In: Proceedings of the 31st North American Manufacturing Research Conference (NAMRC), Hamilton, Canada (2003)
11. Li, Y., Chun, L., Nee, A., Ching, Y.: An agent-based platform for web-enabled equipment predictive maintenance. In: Proceedings of LAT'05 IEEE/WIC/ACM International Conference on Intelligent Agent Technology, Compiegne, France (2005)
12. Wang, J., Tse, P., He, L.S., Yeung, R.: Remote sensing, diagnosis and collaborative maintenance with web-enabled virtual instruments and mini-servers. Int. J. Adv. Manuf. Technol. **24**(9–10), 764–772 (2004)
13. Gao, G., Liu, Y.S., Lin, P., et al.: BIMTag: concept-based automatic semantic annotation of online BIM product resources. Adv. Eng. Inform. **31**, 48–61 (2017)
14. Costa, G., Madrazo, L.: Connecting building component catalogues with BIM models using semantic technologies: an application for precast concrete components. Autom. Constr. **57**, 239–248 (2015)
15. Industry Foundation Classes (IFC). IFC4 Release Candidate 4. http://www.buildingsmart-tech.org/ifc/IFC2x4/rc4/html/index.htm (2014)
16. Zhang, L., Issa, R.R.A.: Ontology-based partial building information model extraction. J. Comput. Civ. Eng. **27**(6), 576–584 (2013)
17. Pauwels, P., Deursen, D.V., Verstraeten, R., et al.: A semantic rule checking environment for building performance checking. Autom. Constr. **20**(5), 506–518 (2011)
18. Nepal, M.P., Staub-French, S., Pottinger, R., et al.: Querying a building information model for construction-specific spatial information. Adv. Eng. Inform. **26**(4), 904–923 (2012)

Building and Application of Shield Construction Big Data Analysis Platform

Gang Chen(✉) and Gang Li

Shanghai Tunnel Engineering Construction Company Ltd.,
TBM Construction Branch, Shanghai 20092, People's Republic of China
cgx9000@21cn.com, lgl218@126.com

Abstract. Big data technology can high-speed, efficient, multi-dimensional analysis of real-time data of devices, predict fault equipment. The paper builds a big data analysis platform based on the Aster database, performs ETL on shield construction data, and synchronizes data from cloud server to platform. Using mode of correlation analysis and auto-grouping model, the data of Hongmei South Road shield engineering was analyzed. The result is presented visually. The abnormal detection rule of automatic motor current is implemented to provide decision support for equipment preventive maintenance.

Keywords: Teradata Aster · Big data analysis · Fault diagnosis
Anomaly detection

1 Introduction

The full name of shield machine is shield tunnel boring machine. It is a kind of engineering machinery for tunnel driving. The fault diagnosis of shield is still in semi-automatic state. Troubleshooting still mainly relies on manual overhaul. It is easily affected by technical level, experience and subjective factors. Diagnosis efficiency is low and it is easy to cause misjudgment [1]. When skilled technicians remove fault, the time to determine cause and location of fault may account for more than 70% of total time. This makes shield equipment is not fully used, and indirectly increases project cost [1]. In recent years, Shanghai Tunnel Engineering Co., Ltd. has continuously expanded the production scale, fault of shield machine equipment has become more and more. The company obviously feels that it is ineffective in equipment fault diagnosis.

Applying data analysis technology to equipment fault diagnosis, intelligent diagnosis technology based on knowledge has become main developing direction of equipment fault diagnosis. Currently, analysis and research of shield data mainly focuses on performance evaluation of shield and establishment of applicability models. S. Okubo, K. Fukui, and W. Chen have developed expert systems to evaluate the reliability of shield machines in Japan [2]. Peter E. Sutherland have proposed a voltage comparison method to evaluate performance of shield machine [3]. M Alvarez Grima have used fuzzy logic and neural networks to evaluate performance of TBM shield machine [4]. With rapid development of information technology, big data platform technology has become an important technology in the field of modeling and simulation [5]. Big data technology can high-speed, effective, multi-dimensional analysis of

J. Abawajy et al. (Eds.): ATCI 2018, AISC 842, pp. 527–535, 2019.
https://doi.org/10.1007/978-3-319-98776-7_57

real-time operating data of equipment, predict of equipment fault. Analysis results can for production and management service.

The paper builds a big data analysis platform based on Aster database, performs ETL on data of shield tunneling, and synchronizes data from cloud server to platform. The model of correlation analysis and auto-grouping is used to analysis data. The result is presented visually. Abnormal detection rule of motor current is implemented to provide decision support for preventive maintenance of equipment.

2 Shield Machine

The basic working principle of shield machine is cylindrical steel component, it advances along axis of tunnel and excavates soil. The cylindrical assembly shell is shield, plays temporary support role in tunnel driving that have not yet been lining, withstands pressure of surrounding earth. Sometimes it also withstands pressure groundwater, and keeps out groundwater [6]. Excavation, dumping, lining and other operations of shield are carried out. Shield machine structure is shown in Fig. 1, the working steps of it are as follows [7]:

1. Shield excavation

The hydraulic motor drives cutter head to rotate and open propelling cylinder of shield, push shield machine forward. As propelling cylinder advances, cutter head rotate continuously, and soil that has been cut is filled with soil chamber. Screw conveyor discharge residue to belt conveyor, which is transported by the belt conveyor to earth box of dirt truck, and transported to ground through shaft.

2. Controlling soil discharge and dumping speed in driving

Pressure of soil chamber is balanced with pressure of excavation face and groundwater. Excavation face is stable and don't correspond to subsidence or uplift of ground surface. Keeping amount of earth removed from screw conveyors and soil chamber is balanced with amount of soil that has been cut down, excavation work can be carried out smoothly.

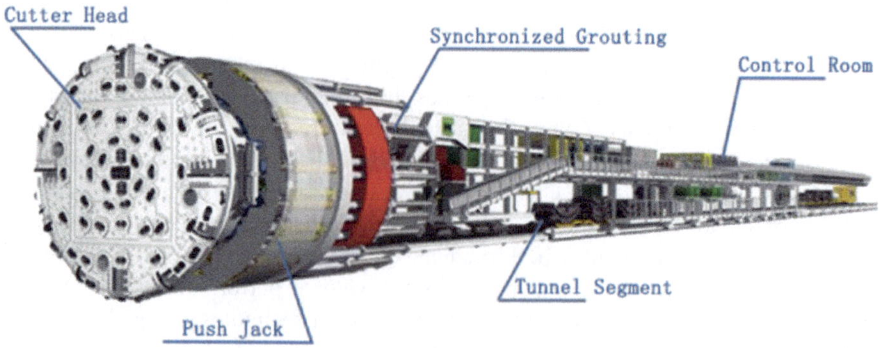

Fig. 1. Shield machine

3. Segment erection

After shield machine has excavated a certain distance, assembly machine is operated to segment erection to form tunnel.

3 Big Data Analysis Platform

The paper builds big data analysis platform which is based on Teradata Aster. The Aster database embeds big data analysis functions of Map Reduce for multi-structured data sources and data types, can provide massively parallel computing performance and scalability analysis capabilities. Aster database uses MPP (Massively Parallel Processor) architecture, which internally supports row storage and column storage. It can flexible design based on front-end application, typically using row storage. In certain scenarios, column storage can improve I/O efficiency, reduce storage requirements, and increase system efficiency [8]. Aster database takes advantage of parallel nature of Map Reduce process. It combines MPP and Map Reduce to ensure maximum parallel processing and improve overall processing performance.

For independent deployment, shield construction database is installed on cloud server. It uses a variety of access methods to store and integrate real-time data including of structured data and unstructured data. Using ETL (extract, transform, load) tools to synchronize data to Aster database, and provide data retrieval, data analysis, and data mining services for operation, maintenance, and decision-making of big data analysis platform. The platform network structure is shown in Fig. 2. The details are as follows:

Cloud server: It stores real-time data of shields which is collected and transmitted from industrial IPC. The entire application layer functions are deployed in Ali Cloud Server.

ETL server: It periodically extracts, loads, cleans, converts, and exports data from cloud database. ETL scheduling system is performed by ETL server.

Queen node: It is Aster's master node, used to manage and control worker nodes, and serves as the only external storage for clusters. It provides service of SQL query

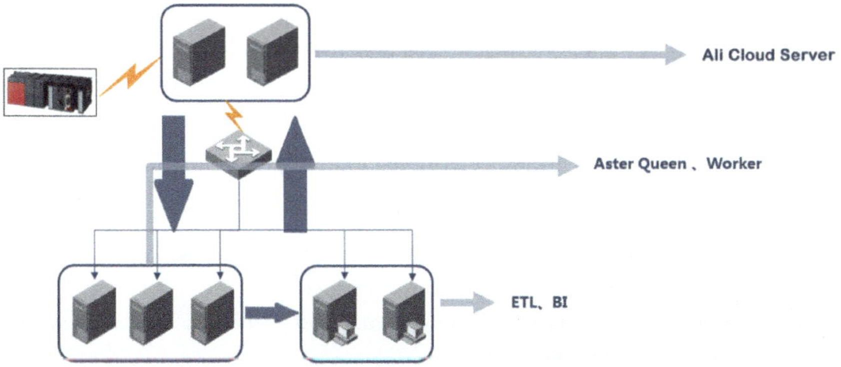

Fig. 2. Platform network structure

and data loading. The Queen node decomposes external SQL request into multiple tasks that are given to various work nodes.

Work node: It is Aster's work node, used to store data and perform various complicated calculations.

BI Server: Data visualization is based on BI server. It can visually display relevant data. It needs to integrate analysis results into application systems, provide concurrent services. It has richer display capability, and faster visualization performance.

4 Application of Big Data Technology

Shield machine generates a large amount of real-time data during digging. Through OPC communication, data acquisition system accesses PLC address to collect data once per second. Shield machine of Hongmei South Road Project is equipped with a large number of sensors and PLC controllers. The collected data has more than 760 parameters. There are about 4 million per 100 rings. The project includes two tunnels: west line tunnel and east line tunnel. West line tunnel is the first excavated tunnel, of which had been collected data from No. 157 ring to No. 1695 ring. Data of east line tunnel had been collected from No. 11 ring to No. 1209 ring.

4.1 Data Sieving

Sieving and cleaning out valid data from a large number of measuring points in the entire project is important for data analysis. As cutter heads are working in shield tunneling stage, related data was selected for analyzing data. All data of west line tunnel are total of 760 parameters. Data is divided into two groups of tunneling phase and assembling phase. As equipment fault occurs in phase of shield advancing, the data of assembling phase is removed out. All of data can't be directly used to distinguish phases of shield. As communicating with equipment expert, motor current of cutter head is used to indirectly determine for shield phase. The records of all motor current of cutter head that are not all at 0 conditions are initially identified as data for tunneling phase. Then parameters for tunneling phase are selected, all 0 parameters and Boolean parameters are deleted. Differential calculation of parameters, delete absolute value of result less than 10^{-3}. Finally, artificial selecting is performed to screen out apparently irrelevant parameters, such as control signals and switching signals. After data sieving, effective amount of parameters was reduced from 760 to 60.

1. Parameters that reflect working conditions: total of propelling oil pressure, partition pressure (A–F), total thrust, pressure of excavating chamber(No. 1–No. 5), air pressure of working pressure cabin.
2. Parameters that reflect overall operating result: shield advance speed, shield advance speed(plc), stroke of propelling cylinder.
3. Parameters that directly relate to cutter head, and there is a high degree of correlation between current and torque.

4.2 Model Foundation

The working characteristic and structure of shield machine have brought challenges to data analysis. It mainly reflects in two aspects: diversity of working conditions and immeasurability of important indicators.

1. Diversity of working conditions

Shield machine works in underground. It is directly affected by soil structure, soil pressure, angle of shield. Soil layer structure is qualitative result of coarse grain sampling. During shield machine advancing, soil type, depth and angle are different. Shield may produce data under multiple combinations of working conditions. Different Shields have different data in different soil conditions. It is difficult for data mining (Fig. 3).

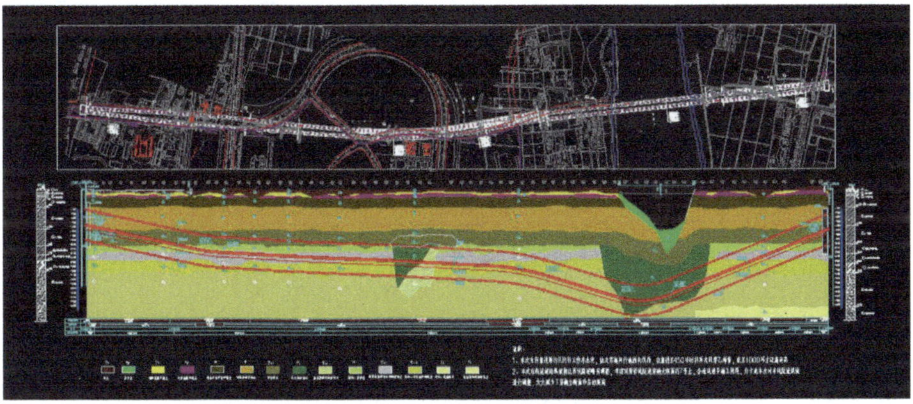

Fig. 3. Soil layer of Hongmei South Road Project

Before shield tunneling, the piling method is used to collect information of soil. It can be seen in soil data of Hongmei South Road project, there are more than 10 different soil types, sampling once per 10 rings. Then it obtain relationship between ring number and soil. According to principal component analysis, soil is marked ten types.

Soil types are qualitative data. It is impossible to quantify effects of soil quality on cutter head. In addition to soil, the angle of shield machine is also an important factor. It doesn't have valid value in collected data.

2. Immeasurability of important indicators

The parameters that can be used are associated with multiple subsystems, such as motors, gearboxes, linkage devices. Many parameters have high correlation among themselves, including seal grease, small motors, cutters, gears, and so on. Oil grease sensor did not return valid data. Drive sensor collect parameters, there are high design relevance. Main bearing of shield is oil seal. It is difficult to directly deploy sensors for monitoring. It can only be indirectly inferred from other peripheral measurable indicators.

4.3 Abnormality Detection Model of Cutter Head Motor

According to relationship of driving parts of shield, there are associated relationship between cutter head motor and main bearing. Abnormality of main bearing can cause abnormality of cutter head motor. Abnormality of individual cutter head motor may also cause abnormality of main bearing. During shield tunneling, it continuous operate in one ring. And working conditions of 15 cutter heads (actually there are 19 cutter-heads, No. 5, No. 6, No. 15 and No. 16 cutter head have not been activated) are consistent. Same to soil layer, earth pressure and angle. There should be correlation between motors under normal conditions. Relatively low correlation reflects that motor is affected, abnormality may occurs.

1. Correlation Analysis

   Current and torque of cutter head are taken as objects to calculate correlation coefficient. Correlation coefficient is to calculate similarity between two equal-length vectors. Similarity coefficient is calculated from each dimension of vector. Common methods is as follows:

(1) Inner product method: $Inner(x, y) = \sum_i x_i y_i = \langle x, y \rangle$
(2) Cosine method: $\mathrm{Cos}Sim(x, y) = \frac{\sum_i x_i y_i}{\sqrt{\sum_i x_i^2}\sqrt{\sum_i y_i^2}} = \frac{\langle x, y \rangle}{\|x\|\|y\|}$
(3) Person method: $P_{x,y} = E[(x - \bar{x})(y - \bar{y})]/\sigma_x \sigma_y$; σ is standard deviation of vector x.

   Cosine method is used here. In one ring, equivalent length vector of 15 parameters is formed by current of motor. Then Cos-Similarity value between these 15 vectors is calculated. A total of $N \times C_{15}^2$ correlation coefficients are available.

2. Abnormal detection

   $N \times C_{15}^2$ correlation coefficients are subjected to BIN statistics (bin_size = 5) and histogram is obtained. It can be seen that most of correlation coefficients are above 0.8 or even 0.9, as shown in Fig. 4. In No. 614 and No. 615 ring, correlation coefficients of No. 8 cutter head motor is lower than 0.9.

   From data of west line tunnel between No. 614 ring and No. 615 ring, it can be seen that current of No. 8 motor is 0, as shown in Fig. 5. Manual records can be verified that No. 8 motor has been repaired.

3. Automated process implementation

   After analyzed data of eastern line tunnel, parameters of anomaly detection were adjusted to follow conditions:

(1) Calculate cos-Similarity value of [ring number + cutter head];
(2) Finding the number of ring which threshold is 0.8 based on histogram;
(3) The ring data of half of motor cos-Similarity values under 2 standard deviations are selected. If most motors have low cos-similarity, it is overall inconsistent, non-motor problems.

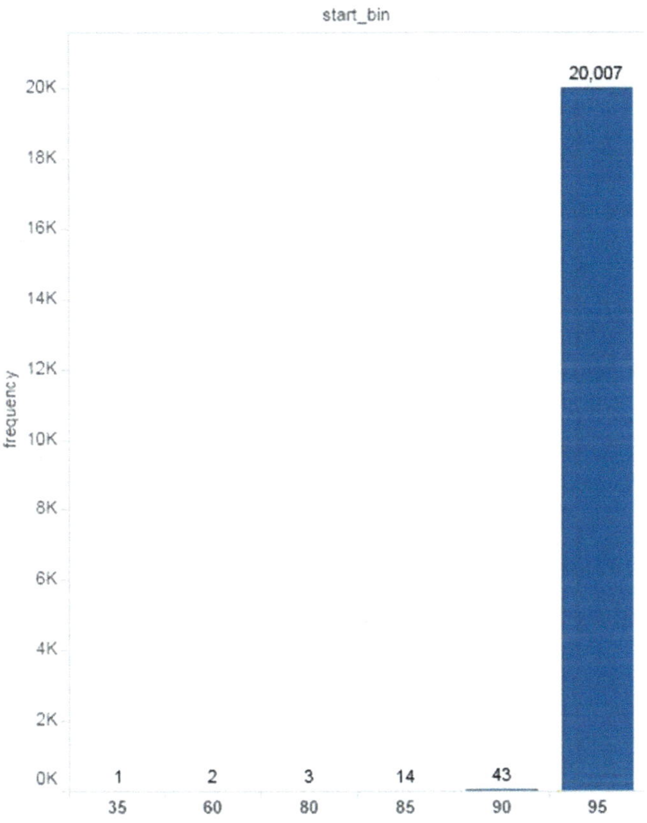

Fig. 4. Histograms of correlation coefficients grouping

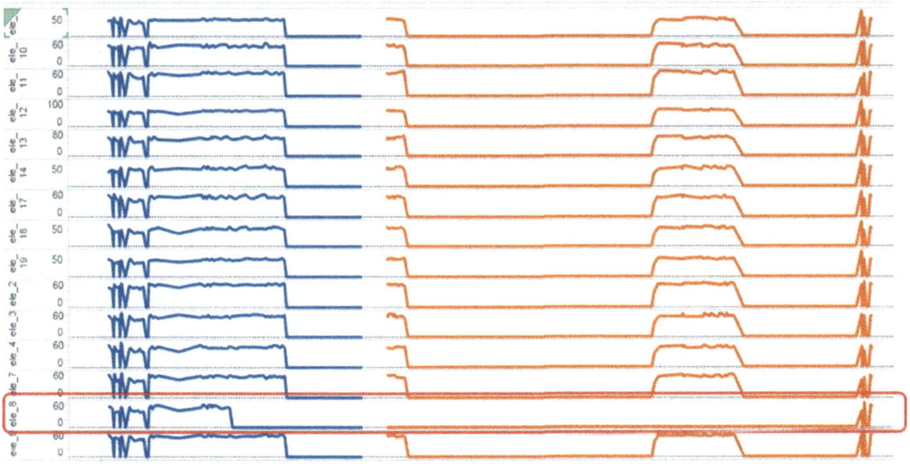

Fig. 5. Abnormal Characterization of Ring Motor Current in No. 614 and No. 615 Ring

Based on above conditions, seven motors which had problem have been identified from data of east line tunnel and west line tunnel. There are No. 4, No. 8, No. 12, No. 14, No. 17, No. 18 and No. 19 motor respectively. After finishing project, No. 4, No. 8, No. 17, and No. 18 were confirmed that there were problems with equipment during equipment inspection. No. 19 was not mentioned in inspection, but there was a serious abnormality in eight rings of data in motor of east line tunnel. It is also worth noting that there are various abnormality motors during shield driving of No. 947 ring on east line tunnel. In nearly an hour, almost half of the motors did not work and relied on rest of motor. Whether it is problem for main bearing, it is worth paying attention.

By analyzing data, the fault motors which have mentioned in manual report are found, and accuracy rate is 100%. It also found that the report didn't mention fault motor. By monitoring real-time data of motor current, it is confirmed that these detected motors have abnormal characterizations:

(1) No. 4, No. 8, No. 14, No. 17, No. 18 and No. 19 motor have been found (As shown in Fig. 6, the red is fault motor which has been mentioned in manual report);
(2) No. 8, No. 18 and No. 19 motors have problems in more than 5 rings data, and they occur on both east and west lines tunnel.

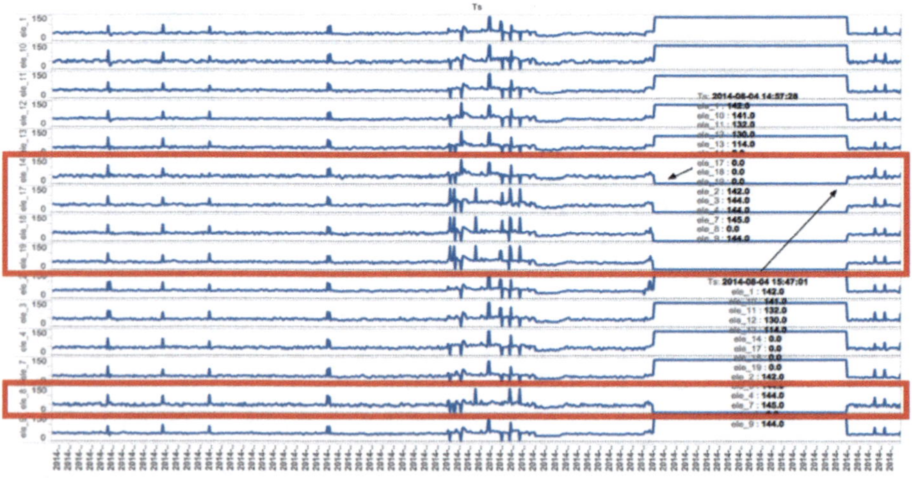

Fig. 6. East Line 947 Ring Motor Current Abnormality

5 Conclusions

In era of big data, the big data analysis platform is essential infrastructure. It can break through the bottleneck of shield construction technology. Big data analysis platform of Teradata-based shield construction has completed. The system is operating normally and stable. By using association rules and automatic grouping model, the data of Hongmei South Road is analyzed. Then model of abnormal motor current is established for automatic detection. The model can find all fault motor. By monitoring real-time data of motor current, it is confirmed that these detected motors have abnormal characterizations. The model can be applied to any other project.

Acknowledgment. This work was supported in part by Enterprise Technology Innovation and Energy Level Improvement Project of Shanghai SASAC contracts 2017014.

References

1. Huang, K., Zhao, J., Zhou, Q.-C., Xiong, X.-L.: Fault diagnosis on shield machines based on multivariable statistical process monitoring. Chin. J. Constr. Mach. **10**(02), 222–227 (2012)
2. Okubo, S., Fukui, K., Chen, W.: Expert system for applicability of tunnel boring machines in Japan. Rock Mech. Rock Eng. **36**, 305–322 (2003)
3. Sutherland, P.E.: Performance calculation for tunnel boring machine motors. In: Conference Record of the 2000 IEEE Industry Applications Conference, vol. 4, pp. 2668–2673 (2000)
4. Alvarez Grima, M., Brumes, P.A., Verhoef, P.N.W.: Modeling tunnel boring machine performance by neuro-fuzzy methods. Tunn. Undergr. Space Technol. **15**(11), 25–269 (2000)
5. Gong, X.-Y., Li, B.-H., Chai, X.-D., Gu, M.: Survey on big data platform technology. J. Syst. Simul. **26**(03), 489–496 (2014)
6. Zhou, W.-B.: Shield Tunnelling Technology. China Architecture and Building Press, Beijing (2004)
7. Liu, J.-H., Zhang, Y.-Q., Pan, Z.-L.: Tunnel Engineering. Shanghai Scientific and Technical Publishers, Shanghai (1999)
8. Zhang, Q.-H., Li, M.: Design and implementation of ETL strategy in tele-communication based on teradata data warehouse. Manuf. Inf. Eng. China. **41**(13), 10–13 (2012)

A Professional Competence Measurement Approach Based on a Modified PageRank Algorithm

Shanshan Lyu, Zhao Wu, and Nan Jing(✉)

SHU-SILC Business School, Shanghai University, Shanghai 201800, China
minahu@163.com

Abstract. Recruiting talents requires companies to accurately measure applicants' professional competence. Traditionally hiring companies rely on applicants' documentation and third-party recommendations, of which the validity, trustworthiness and effectiveness are often questionable. Nowadays, the social networking web sites which focus on professional profile can help hiring companies to find applicants' information online which are under the scrutiny of the public audience, thus are more trustworthy and effective. This paper presents an approach to measure the professional competence of the job applicants by applying a modified PageRank algorithm on the user data mined from the social networking web sites. It discusses the technical details of this measurement approach and introduces a prototype system which implements the approach. The application of this measurement approach in real life shows its superiority in the measurement of applicants' professional competence, compared with traditional hiring procedures.

Keywords: Social networks · Data mining · PageRank algorithm
Professional competence measurement

1 Introduction

Recruiting talents is one of the most challenging tasks for organizations, as having talents means higher productivity, better performance, and more revenue. To recruit real talents, organizations need to accurately understand the skills and experiences of the job applicants in order to make an effective and informed hiring decision. Wrong decision in talents recruiting will incur significant loss of value and high turnover of human resources [14].

Traditionally, organizations need to review applicant's documentations (e.g. resume, CV, cover letter) and request third parties for their recommendations to measure applicants' professional competence [8]. However, these documentations and recommendations vary in the degrees of trust and validity. In other words, not all information is fully honest and trustworthy, as it may be affected by the relationship between applicants and third-parties, or the conducting code of applicants or third parties themselves [5, 12]. In order to be hired, for example, the applicant can hide or modify some of her information (e.g. educational background, work history), or collude

J. Abawajy et al. (Eds.): ATCI 2018, AISC 842, pp. 536–545, 2019.
https://doi.org/10.1007/978-3-319-98776-7_58

with those who make the recommendations in order to get hired. These drawbacks are especially severe since most of the applicant's documentations or recommendations are provided only to the hiring companies, instead of being under the scrutiny of the public audience who may be able to point out the dishonest information or behavior.

In order to reduce these drawbacks in the traditional recruiting process, we consider the social networking revolution as an opportunity to help hiring companies to effectively and accurately measure the applicants' professional competence, i.e. their knowledge, expertise, experience and competencies. The social networking web sites, especially the ones which focus on professional users (such as LinkedIn) can help the hiring companies obtain valid and trustworthy information, including educational background, work history and recommendation. Since all this information is visible online and under the scrutiny of the public audience, it is considered more creditable than the information provided by the applicants and the third parties directly to the hiring companies. Following this direction, the authors present an approach to measure the professional competence of the job applicants based on the information on the social networking web sites. Section 2 reviews the literatures that are related to this work. Section 3 presents the technical details of this approach and Sect. 4 discusses a research prototype that is developed to evaluate this approach based on the information gathered by crawling users' profiles on LinkedIn. Section 5 points out the future work and concludes this paper.

2 Literature Review

2.1 Traditional Recruiting Procedure

The major application of this research is to help hiring companies to improve their recruiting procedure by measuring applicant's professional competence based on her information on the social networking sites. So first we will review the traditional recruiting procedure to put this research into context.

Generally speaking, the traditional recruiting process consists of the following three steps:

a. Reviewing applicant's qualifications

Hiring companies review candidates' application materials and evaluate candidates' qualifications against the requirements listed in the job description. Such qualifications normally include degrees, certificates, and job history.

b. Gathering references:

Hiring companies request and review recommendations from third parties to qualitatively learn the impressions that others received from the applicant.

c. Conducting: On-site interview:

In the final stage, the hiring company will organize a panel of people to evaluate the applicant in a formal interview.

This three-step procedure has several shortcomings: First, the step of reviewing applicant's qualification heavily depends on the knowledge and experience of the employee who is doing the review, and is prone to human errors, due to the lack of knowledge in applicant's institution and experience in applicant's specialty field; Second, the credibility of the gathered recommendations often suffers from the leniency bias [13], especially when the reference is made only visible to the employer, not to the public; Third, the on-site interview is always very costly to the employer, considering the time spent by the entire interview panel, but without an effective scrutiny due to the last two drawbacks, the cost of the on-site interview cannot be well compensated.

In this article, we will discuss how to use the information on the social networking web sites to improve the recruiting process and reduce the aforementioned drawbacks. In particular, we define a set of factors that the hiring companies can observe online, and calculation methods that they can use to measure the professional competence of the job seekers.

2.2 Social Network

A social network in our research work refers to the computer network that connects people or organizations [17]. It can be seen as a group of people and connections between them [7]. More specifically it is defined as "a category of Web sites with profiles, semi-persistent public commentary on the profile, and a traversable publicly articulated social network displayed in relation to the profile" [4]. Social networks have gained the attention of the research community that tries to understand users profile and their interconnection [16, 19] as well as interactions among users [15].

Within various kinds of social networking web sites, the sites that aim at professionals, such as LinkedIn, let individuals share their educational background, career paths and skill-sets. Also on LinkedIn users can make recommendations that are similar to traditional references, stating the relation between the recommender and the applicant and the applicant's position at the time they worked together. Compared with the traditional CV which may only mention the most recent positions, a professional social networking site can list an individual's entire work history. And since the online site is under the scrutiny of the public audience, it adds credibility and trustworthiness to the information provided to the hiring companies.

2.3 PageRank Alogorithm

Google's PageRank algorithm determines the importance of all the pages based on regression relationship, i.e. if a page is linked by many high-quality pages, that page must also be a high-quality page [18]. In other words, a hyperlink to a page counts as a vote to that page, and the PageRank approach determine the ranking of the web pages based on the number of votes. Also if a page is already marked as high-quality, its vote bears a higher value than the votes of other pages, and the pages that this high-quality page links to will get higher value, too. For a page V_i, if there are pages $V_j (j = 1, 2, 3..., n)$ pointing to the page A, PageRank value of V_i and V_j are named PR(V_i) and PR(V_j) respectively. Then the formula of calculating PR(V_i) is listed as follows:

$$PR(V_i) = \frac{(1-d)}{n} + d \times \sum\nolimits_{V_j \in In(V_i)} \frac{PR(V_j)}{|Out(V_j)|} \tag{1}$$

In formula (1), n is the total number of pages. d is the damping factor which means the probability of clicking the hyperlink on the current page, and it is equal to 0.85. $1-d$ is the probability of moving to a new page. $Out(V_j)$ is the number of out-link of page V_j, where $j = 1, 2, \ldots, n$.

In summary, PageRank ranks the value of a page based on the value of other linked pages. Since the calculation of PageRank depends on the association among items on a graph, it is often used in graph-based data structure like social network.

3 Professional Competence Measurement Approach

The goal of our research is to measure an individual's professional competence based on her information on social networking web sites. This idea of using PageRank algorithm to calculate a user's influence in her social network lays the ground of our work in measuring an individual's professional competence based on professional recommendations. If user u has k users who trust her, which means u has k in-links, the influence of user u is calculated as follows:

$$PA(u_i) = \frac{1-d}{|S|} + d \sum\nolimits_{i=1}^{n} \frac{PA(u_j)}{|Out(u_j)|} \tag{2}$$

$|S|$ is the number of users in the user's social network. d is the damping factor which means the probability of recognizing the effect of friends' professional ability on the current user, and it is equal to 0.85. $1-d$ is the probability of recognizing the effect of other people's professional ability on the current user. $PA(u_i)$ and $PA(u_j)$ represent the professional ability of user u_i and u_j respectively. $Out(V_j)$ is the number of friends of u_j, where $j = 1, 2, \ldots, n$.

To better illustrate our approach, we start our work with the public information on LinkedIn. In more details, for measuring an individual's professional competence, we refer to her academic background, work history, recommendation and endorsement on the LinkedIn page. Based on this information, similar as how PageRank approach works, our approach calculates a score that can indicate the degree of the user's professional competence and help reviewer make a more informed decision.

By definition, competence is the ability of an individual to do a job properly. An individual's ability is to achieve certain outcomes through a set of measurable features, functions, processes, or services. In our data domain, i.e. the social networking site, we use one's academic background to measure the "features" and "functions", use work history to measure the "processes" and "service", and then use recommendation and endorsement to measure the person's achievement. This measurement approach can be represented by the following formula:

$$PA(u) = ES + WS + RS \tag{3}$$

Where PA(u) is user u's professional ability, ES is her educational background score, WS is her work history score, RS is her recommendation score. Below we discuss in details the formulas to calculate ES, WS, and RS.

a. Educational Background Score

$$ES = \frac{\sum_{i=1}^{N_e} Deg(i)Ins(i)}{MAX\{AS\}} \tag{4}$$

Where $MAX\{ES\}$ is the maximum value of all users' educational score; N_e is the total number of degrees or certificates that user u have earned; $Deg(i)$ indicates the level of the degree i (see Table 1), where for each degree we use the number of years that earning the specific degree generally takes and then normalize each value.

Table 1. The level of the degree

Degree	Score
Ph.D.	0.5 (5 years)
Master's degree	0.2 (2 years)
Bachelor's degree	0.3 (3 years)

The value of $Ins(i)$ depends on the ranking of the institution i. There are many institutions ranking available, and for illustration purpose we use the 2017 QS World University Rankings in this work. Since this ranking list 100 institutions, for simplicity, the top 1 gets 1, top 2 gets 0.99, and so on. The top 100 gets 0.01. If the user's institution is not listed in this ranking, we give it 0.01 for now.

b. Work History Score

$$WS = \frac{\sum_{i=1}^{N_w} Dur(i)}{MAX\{WS\}} \tag{5}$$

Where N_w is the total number of jobs in user u's work history; $Dur(i)$ is the duration of u's i^{th} job, using the number of months; $MAX\{AS\}$ is the maximum value of all users' work history score.

c. Recommendation Score

$$RS(u) = EndScore(u) + RecScore(u) \tag{6}$$

Where *RecScore* is calculated based on the recommendations that the user u gets on LinkedIn and *EndScore* is based on her endorsements. In details, they are calculated as follows:

$$EndScore(u) = \sum_{i=1}^{N_s} \left[\left(\sum_{j=1}^{N_{Erec(i)}} \frac{EndScore(ij)}{N_{Egiv(ij)}} \right) \right] \tag{7}$$

where N_s is the total number of the user u's skills; $N_{Erec(i)}$ is the total number of endorsement that user u gets for her i^{th} skill; $EndScore(ij)$ is the endorsement score of the user ij, i.e., the jth user who endorses the ith skill of user u; $N_{Egiv(ij)}$ is the total number of endorsements that user ij gives to other users.

Note that we include $EndScore(ij)$ and d $N_{Egiv(ij)}$ in our equation because of two reasons: first, one skill of one user can be endorsed by many users, and we want to differentiate the value of different endorsements based on their owners; second, the more endorsements that a user makes, the less value her endorsement bears. Both reasons refer to the similarly original ideas in the PageRank algorithm [6].

Similarly, the recommendation score is calculated as follows:

$$RecScore(u) = \sum_{i=1}^{N_{Rrec}} \frac{RecScore(i)}{N_{Rgiv(i)}} \tag{8}$$

Where N_{Rrec} is the total number of recommendations that user u gets; $RecScore(i)$ is the recommendation score that i^{th} user who makes a recommendation for user u; $N_{Rgiv(i)}$ is the total number of recommendation that this i^{th} user makes for all other users.

With all components ready, the mathematical equation to calculate the professional competence is as follows:

$$\begin{aligned} PA(u) &= ES + WS + RS \\ &= \frac{\sum_{i=1}^{N_e} Deg(i)Ins(i)}{MAX\{AS\}} + \frac{\sum_{i=1}^{N_w} Dur(i)}{MAX\{WS\}} + \sum_{i=1}^{N_s} \left[\left(\sum_{j=1}^{N_{Erec(i)}} \frac{Edm(ij)}{N_{Egiv(ij)}} \right) \right] + \sum_{i=1}^{N_{Rrec}} \frac{Rcm(i)}{N_{Rgiv(i)}} \end{aligned} \tag{9}$$

4 Research Prototype and Experiment Results

In order to exercise this approach, the authors have developed a Web-based software prototype which implements the proposed system using computer technologies. The uniqueness of this prototype is in two aspects: first, because social networking web sites often have a large amount of data, the technical design of this prototype has taken big data-centered technologies into the consideration for better feasibility and scalability; second, the prototype design put each technical module in different layers, so that within each layer the modules have high cohesion as they all process same data or same kind of data, and within all layers we can keep low coupling as only data is communicated. Figure 1 shows the technical design of this prototype.

Our prototype uses a Java development framework JeeSite as the major implementation platform. Jeesite is an open-source foundation for building Chinese-based enterprise-level information system. It uses the Twitter Bootstrap as the front-end user

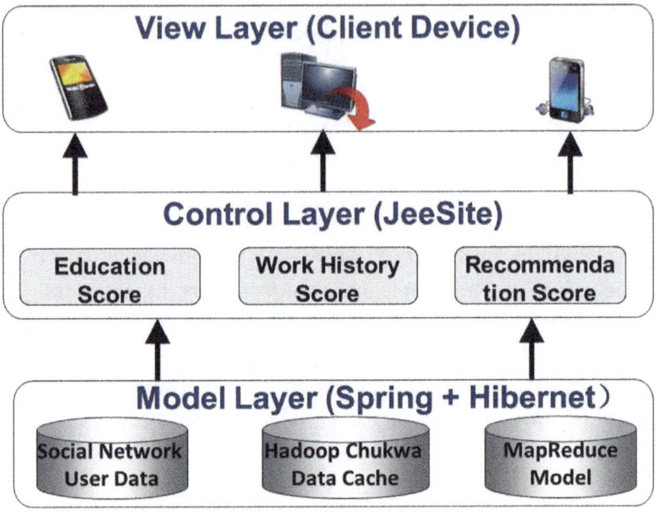

Fig. 1. Technical architecture of system prototype

interface, the Java Spring framework [10] as its programming platform, and the Java Hibernate framework [9] as its database management system. All frameworks are used in various layers of our prototype.

This prototype design has three layers in accordance with the well adopted Model-View-Controller standard [1]. The view layer is the user interface component that runs on the client devices and shows the result information to users on PC, smart phone, or other applicable device. The controller layer implements the business logic of calculating competence score based on user's educational background, work history and recommendations. The model layer interacts with the social networking web site to access social data, then uses the Hadoop Chukwa data caching services [2] and the MapReduce programming model to cache the relevant data. Hadoop is an open source software framework that processes large-scale datasets on clustering storage [3]. MapReduce is a programming model that provides parallel and distributed algorithms to analyze the data sets on a cluster [11]. Both Hadoop and MapReduce are well-used technologies for Big Data. They will provide technological foundations for our approach to operate on large-scale datasets.

The major functionality of this prototype includes loading the profile from each specified URL (i.e. a user's LinkedIn page), parsing the web page into specific information, calculating the competence score based on this information, and display the calculation results that ranks the users list according to their competence scores. Figures 2 and 3 show samples of these calculation results.

Then, we applied this prototype into a hiring company to validate its effectiveness in real life. Twenty applicants competed for five positions and four of the twenty applicants lied on the resume, which is designed purposely and admitted by the CEO of the hiring company. We extracted the professional information of applicants in the LinkedIn, got their competence scores through the proposed professional competence

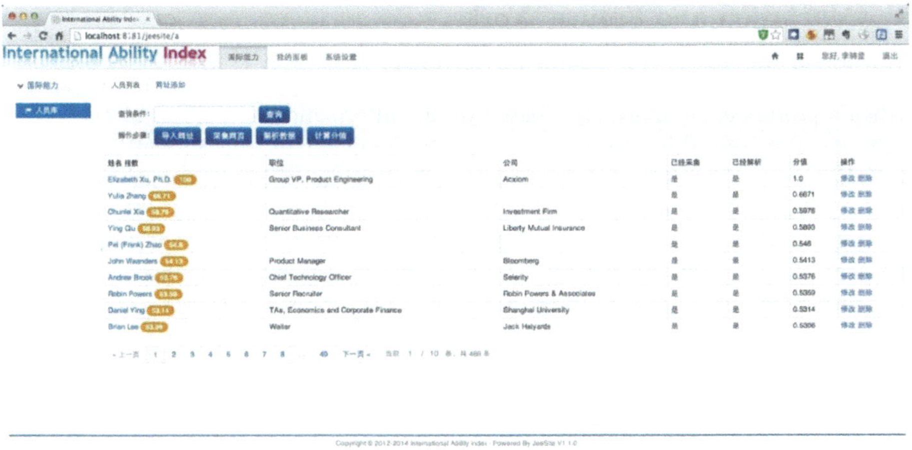

Fig. 2. Users list sorted by competence score (1/2)

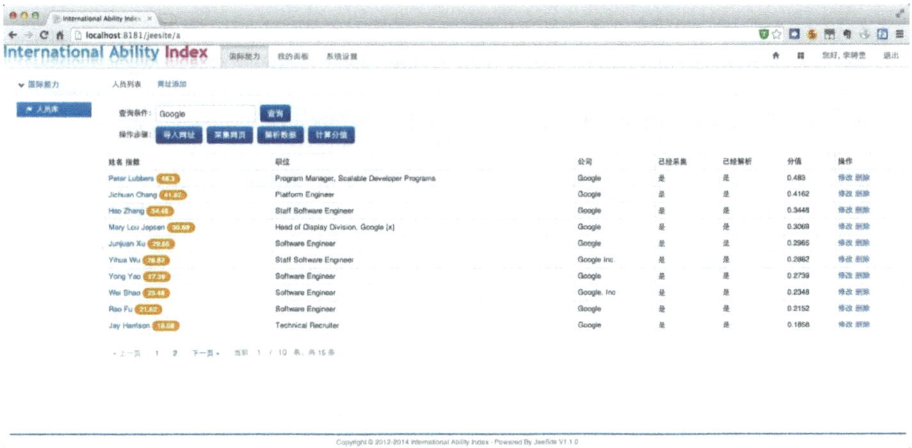

Fig. 3. Users list sorted by competence score (2/2)

measurement approach; at the same time, as a comparison experiment, we measured applicant's professional competence through the traditional recruiting procedure. The results show that the rank of applicants through the proposed approach and the rank of applicants through the traditional recruiting procedure have 76% similarity. However, the proposed approach only takes 1/60 of time of the traditional procedure. In addition, none of the four lied applicants can get a position in the rank provided by the proposed approach while one can get a position in the rank provided by traditional procedure. These results demonstrate that the professional competence measurement approach is much more effective and trustworthy than the traditional procedure.

5 Conclusion and Future Work

In this paper, we have proposed a modified PageRank approach to measure an individual's professional competence based on the information on the social networking web site. Precisely this approach accesses the individual's educational background, work history, and third-party recommendation to calculate the professional competence score. Since all the information is public online, it brings more valid and trustworthy information to the hiring companies. This paper presents the mathematical methods for the competence score calculation, introduces a research prototype that is developed based on these methods, and validate the effectiveness of this approach in real life. It is our next step to further evaluate the approach and research prototype. We will have to consider how to reasonably and effectively include more available information on the social networking web sites into our mathematical model, such as comments, publications and news, in order to yield a more accurate competence measurement approach.

References

1. Alur, D., Crupi, J., Malks, D.: Core J2EE Patterns: Best Practices and Design Strategies, pp. 1745–1769. Prentice Hall Press, Upper Saddle River (2003)
2. Jose, A.S., Binu, A.: Automatic detection and rectification of DNS reflection amplification attacks with Hadoop MapReduce and Chukwa. In: International Conference on Advances in Computing & Communications, pp. 195–198. IEEE (2014)
3. Shvachko, K., Kuang, H., Radia, S., et al.: The Hadoop distributed file system. In: IEEE Symposium on MASS Storage Systems and Technologies, pp. 1–10. IEEE Computer Society (2010)
4. Boyd, D.: Friends, Friendsters, and Top 8: writing community into being on social network sites. First Monday **11**(12) (2006)
5. Brass, D.J., Butterfield, K.D., Skaggs, B.C.: Relationships and unethical behavior: a social network perspective. Acad. Manag. Rev. **23**(1), 14–31 (1998)
6. Brin, S., Page, L.: Reprint of: the anatomy of a large-scale hypertextual web search engine. Comput. Netw. ISDN Syst. **56**(18), 3825–3833 (2012)
7. Jeremy, R., Cooke, E.: Link prediction and link detection in sequences of large social networks using temporal and local metrics. Online Submiss. **4**(4), 266–267 (2006)
8. Fazel-Zarandi, M., Fox, M.S.: Constructing expert profiles over time for skills management and expert finding. In: International Conference on Knowledge Management and Knowledge Technologies, pp. 1–6. DBLP (2011)
9. Ottinger, J.B., Minter, D., Linwood, J.: Beginning Hibernate. Apress, New York (2014)
10. Mak, G., Rubio, D., Long, J.: Spring Recipes: A Problem-Solution Approach, 2nd edn. Apress, New York (2010)
11. Mmel, R.: Google's MapReduce programming model—Revisited. Sci. Comput. Program. **70**(1), 1–30 (2008)
12. Yang, M., Zhang, Z., Li, X., et al.: An empirical study of free-riding behavior in the Maze P2P file-sharing system. In: International Conference on Distributed Computing Systems, pp. 56–56. IEEE (2005)
13. Mamykina, L., Manoim, B., Mittal, M., et al.: Design lessons from the fastest Q&A site in the west. In: Sigchi Conference on Human Factors in Computing Systems, pp. 2857–2866. ACM (2011)

14. Naveh, Y., Richter, Y., Altshuler, Y., et al.: Workforce optimization: Identification and assignment of professional workers using constraint programming. IBM J. Res. Dev. **51** (3.4), 263–279 (2007)
15. Rafailidis, D., Kefalas, P., Manolopoulos, Y.: Preference dynamics with multimodal user-item interactions in social media recommendation. Expert Syst. Appl. **74**, 11–18 (2017)
16. Tang, J., Musolesi, M., Mascolo, C., et al.: Temporal distance metrics for social network analysis. In: ACM Workshop on Online Social Networks, pp. 31–36. ACM (2009)
17. Vienna, S.: Social network analysis and mining. ACM Trans. Intell. Syst. Technol. **2**(3), 22 (2018)
18. Wang, J., Liu, Z., Zhao, H.: Group recommendation based on the PageRank. J. Netw. **7**(12) 2012
19. Xie, H., Li, X., Wang, T., et al.: Incorporating sentiment into tag-based user profiles and resource profiles for personalized search in Folksonomy. Inf. Process. Manag. **52**(1), 61–72 (2016)

Particle-Filter-Based Prediction for Anomaly Detection

Xinwen Gao, Guoyao Xu(✉), and Yufan Wu

Shanghai University, Shanghai, China
xu78guo6@outlook.sg

Abstract. Anomaly detection in video aims to automatically detect the unusual occurred event and model was trained from some normal samples. Previously, reconstruction-cost-based methods were the most commonly used way to solve this problem. However, because of the variety of unusual event in real world and the capacity of deep learning methods in fitting, the high cost may not always be guaranteed well on abnormal event videos. In this paper, we propose to tackle the anomaly events within a particle-filter-based method. It is based on an assumption that normal event is easy to predict, and abnormal event is hard to. We use particle filter to give the prediction's eigenvalues and compare it with the next frame's eigenvalues, if the prediction doesn't conform the real data well then, we can discriminate the abnormal events from the normal data. Experiments on UMN dataset validate the effectiveness of our method.

Keywords: Anomaly detection · Particle filter · Optical flow

1 Introduction

Anomaly detection problem in video tend to automatically tackle the abnormal events, which can be widely used in real monitor-based security system. In the recent several years, this problem attracts many researchers' attention. However, it's still an extremely challenging task because in real world, the abnormal events are so various but sparse, especially compared to the normal ones.

In the previous works, the idea of reconstruction cost comparison between normal training data and abnormal one, whose features were extracted from normal training data, is a commonly used strategy [1–4]. Furtherly, in terms of features that extracted from videos before discriminator works, almost all existing approaches can be roughly separated into two parts: (i) hand-crafted-features-based approaches [1, 2, 5]. These methods use some specific and interpretable ways to extract feature as a representation of motion and appearance information, such as HOF, HOG and trajectory. Based on these stable features, then the dictionary-learnt step or some other algorithms are applied to them to identify the judgement baseline of the discriminator. The more and better the features contain the difference information between normal and abnormal behavior, the better the discriminator will work. (ii) Deep-learning-based approaches. These methods give an end-to-end way to solve this problem, which have created the most progressive achievement. Instead of hand-crafted features, deep learning methods use convolutional neural network as the most commonly used way to extract features

J. Abawajy et al. (Eds.): ATCI 2018, AISC 842, pp. 546–553, 2019.
https://doi.org/10.1007/978-3-319-98776-7_59

that contain the target information such as motion and appearance, which would be sent to an auto-encoder and enforced to reconstruct the normal behavior with a small error later. However, because of the strong capacity of deep neural network, the high error may not be guaranteed well on abnormal event videos, which is the major loophole of these methods.

In this article, we summarize our contributions as follows: (i) We propose a particle-filter-based prediction frame for anomaly detection, which tries to identify the abnormal events by comparing their eigenvalues to their expectation. Our solution agrees with the concept of anomaly detection that normal events are predictable while abnormal ones are unpredictable. Inspired by some related works [6, 7, 17], we use a different way to achieve the prediction which was generated by the former information. With the sequence data generated from the features, we enforce the predicted sequence's value to be as close as possible to the ground truth's sequence value, which will mostly cause the larger deviation when abnormal events occur. (ii) Other than generating the whole frame as the prediction [17], we design a way to get the eigenvalues of hand-crafted features, which build up the eigenvalues' sequence for particle filter to process. Such a frame is proved working swiftly on our PC in experiments.

2 Related Work

In [17], the authors used the whole frame generator (U-net and GAN) as the predictor, which utilize the previous sequence information. Their method both considered the appearance and motion information and achieved a new baseline for anomaly detection.

In [6], the authors proposed a hybrid agent approach to detect anomalous behavior in crowded scenarios, and they divided the behavior into individual and group models, proposing a hybrid agent system with static and dynamic agents so that Effectively distinguish between individual and group behaviors. They propose using group behavior as a Bag-of-words model to determine the abnormal behaviors of groups by integrating static and dynamic proxy information.

In [7], the authors use the convolutional network to generate prediction frames for a given input sequence. In order to deal with the intrinsic fuzzy prediction obtained from the standard mean square error (MSE) loss function, they proposed three characteristic learning strategies: multi-scale structure, confrontation training method and image gradient difference loss function.

In [8], the authors studied the problem of activity recognition and abnormal behavior detection in patients with Alzheimer's disease, and studied three variants of re-current neural network (RNN): Vanilla RNN (VRNN), long-term short-term RNN (LSTM) and gated recursive unit RNN (GRU). Activity identification is treated as a sequence tag problem, and abnormal behavior is marked by a deviation from the normal mode. In addition, they also proposed a method to expand the sample set size which can reflect certain behaviors of dementia patients.

3 Particle-Filter-Based Prediction for Anomaly Detection Method

Our method can be roughly separated into two parts, generation of sequence and prediction. The following flow chart shows how it works (Fig. 1).

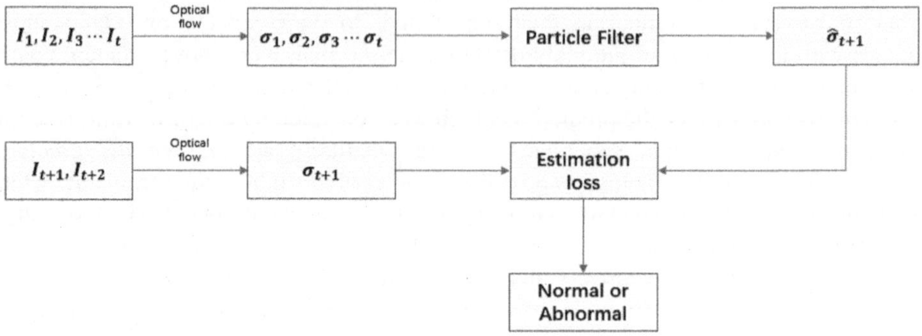

Fig. 1. The pipeline of our eigenvalues prediction network, $I_1, I_2, I_3 \cdots I_t$ refer to the image sequence

Our goal in this paper is to mostly test if the particle filter-based prediction is useful for video prediction. So, in our method, to reduce the computational overhead and validate our ideas as soon as possible, we simplify each whole frame's motion feature as a specific eigenvalue, which may damage effectiveness and cause the inability of localize the event in a specific image. We will perfect our method in the later research.

3.1 Build the Eigenvalues' Sequence

For the image-based representation of motion, histograms of optical flow directions weighted with their norm values are usually used [9]. Here, we compute global optical flow OF using Farneback's method as described in [10]. The OF between two frames can be represented in the following equation.

$$\mathrm{OF}(I_{t+1}(\mathrm{x}, \mathrm{y})) = I_t(x + u(t+1), y + v(t+1)) \tag{1}$$

The motion vector $(u(\delta t), v(\delta t))$ represents the motion between two frames. Instead of the commonly used HOF feature, we use L2-norm to give an eigenvalue as the simple representation of adjacent frames' motion information.

$$\sigma(t) = \sum_{i=1}^{width} \sum_{j=1}^{height} \sqrt{u_{ij}^2(t) + v_{ij}^2(t)} \tag{2}$$

By Eq. (2), we convert the image sequence into a numerical sequence, such an implementation is proved working for abnormal events fill the whole screen but doesn't work for localizing the small events.

3.2 Use Standard Particle Filter to Estimate the Sequence's Value

Particle filter, also known as Sequential Monte Carlo methods, is a commonly used method in signal processing and target tracking area [11–13]. Compared with some classical estimator such as Kalman filter, this method is easier to implement and allows complex nonlinear and non-Gaussian estimation problems to be solved efficiently in an on-line manner, which can handle the normal events well. On the other hand, the particle filter's capacity is much weaker than deep neural network in fitting the data, so that it may always cause a large error in estimating the untrained mode, especially for anomaly detection.

In this paper, considering our target is to predict the normal samples' eigenvalues well and the prediction on abnormal events' eigenvalues would be inaccurate, we use standard particle filter as described in [14], without any optimization steps.

The standard particle filter is consisting of two main steps: (i) Sequential Importance Sampling. (ii) Resampling

In the first step, in order to compute an expectation $\mu_f = E[f(x)]$.

$$u_f = \int f(x)p(x)\mathrm{d}x \tag{3}$$

we use a series of particles' mean to indicate it. Specifically, we assume that these particles respect a target distribution $q(x) \sim N(u, \sigma^2)$, So, the computation of Eq. (3) can be converted into the following equation:

$$u_f = E[w(x)f(x)], w(x) = \frac{p(x)}{q(x)} \tag{4}$$

Therefore, a sample of independent draws $x^{(1)}, \ldots, x^{(m)}$ from $q(x)$ can be used to estimate u_f by the following equation:

$$\hat{u}_f = \frac{1}{m}\sum_{i=1}^{m} f\left(x^{(i)}\right)w\left(x^{(i)}\right) \tag{5}$$

In the first step, we use Eq. (5) to give the estimation of our numerical sequence. However, we still need a method to measure the accuracy of each $\hat{u}_f$, that's step 2—resampling actually doing.

Learn from [15], we use the following equation to measure how close $\hat{u}_f$ fitting the u_f. The particles with large variance will be deleted and with small variance will be copied. The variance of $\hat{u}_f$ can be represent in this way:

$$V\left(\hat{u}_f\right) = \sqrt{\frac{1}{m}\sum_{i=1}^{m}\left(\frac{w_k^j}{\bar{w}_k} - 1\right)^2}, \bar{w}_k = \frac{1}{m}\sum_{i=1}^{m} w_k^j \tag{6}$$

w_k^j refer to each particle's weight.

For IS to be accurate (with a limited number m of draws) this variance must be small, which requires $q(x)$ be approximately proportional to $p(x)$ for most x [15]. So, in our experiments, we use some normal frames' eigenvalues to adjust the distribution $q(x)$ where particles come from and enforce it to be close to the distribution of normal frames' eigenvalues.

4 Experiments and Discussion

In this part, we evaluate our proposed method on UMN dataset [16] and UCF-crime [18], our target is to test that if our method can separate the global abnormal events from normal samples.

UMN dataset consists of three different scenes of crowded escape events and the total frame number is 7739 (1450, 4415 and 2145 for scenes 1–3, respectively) with a 320 * 240 resolution. Its normal events are pedestrians walking randomly filling the whole screen, and the abnormal events are pedestrians swiftly running almost at the same time. There are total of 11 abnormal events in the whole dataset (Figs. 2, 3 and 4).

UCF-crime consists of long untrimmed surveillance videos which cover 13 real world anomalies, including Abuse, Arrest, Arson, Assault, Accident, Burglary, Explosion, Fighting, Robbery, Shooting, Stealing, Shoplifting, and Vandalism. These anomalies are selected because they have a significant impact on public safety[1].

Fig. 2. UMN Normal frame

Fig. 3. UMN Abnormal frame

[1] http://crcv.ucf.edu/projects/real-world/.

Fig. 4. Examples of UCF-crime dataset, including both normal and abnormal samples

We use the frame-level detection as our evaluation. Once the estimation error is larger than all errors ever occurred in the training step, then the frame would be seen as abnormal events' start.

On the other hand, to detect the end of abnormal events is a problem for our method. It is because that the strong tracking capacity of particle filter. Once an unexperienced mode occurs, the particles cannot estimate it very well and the large error will appear. However, with the resampling step, the estimator will swiftly track the later abnormal values if they respect the same distribution as the first unusual value. So, considering the motion mode always changes in most abnormal events, here we use a trick to identify the end of abnormal events. If the large errors don't occur and the small error mode can be maintained to over 0.5 s, we deem that the abnormal events have ended, and the start of this 0.5 s is seen as the end time.

In practice, we initialize the particles from the first several frames of each scene, enforcing particles fitting the data well, and leave the others for testing. After getting the optimal estimation of target distribution $q(x) \sim N(u, \sigma^2)$ in Eq. (4) on normal events, when test data's eigenvalues are input into particle filter, we use the following normalizing step to preprocess these eigenvalues σ_{test}:

$$\sigma'_{\text{test}} = \frac{\sigma_{\text{test}} - u}{\sigma^2} \tag{7}$$

The following figure shows detection result of first scene in UMN.

Based on our discriminator's rule, the first value beyond the largest error occurs will be seen as the abnormal events happen. As shown in Fig. 5, the first large error occurs on 479 with the ground truth is 526 behind, which means our method can totally detect the abnormal events in this dataset. As shown in Fig. 6, one of the normal video in UCF-crime, based on the first 500 training samples, all errors are smaller than those in first 500 training values, which means there are no anomalies in the video.

In terms of running time, our framework is implemented with MATLAB 2016b on CPU i5-7300HQ. The average running time is about 10fps, we have analyzed that in our framework, the computation of global optical flow cost the most time.

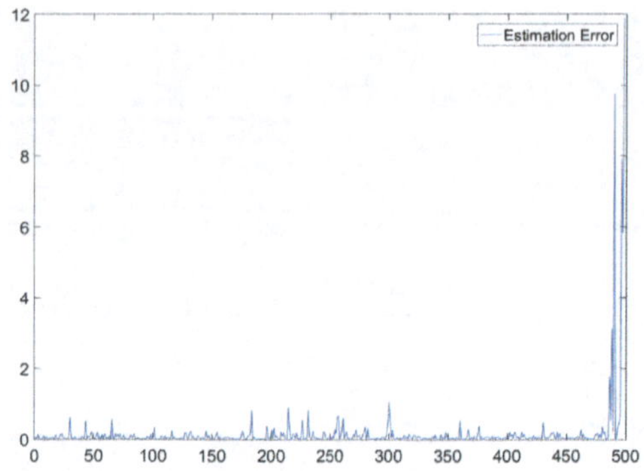

Fig. 5. The error of particle filter's estimation, the front 300 values are the training normal data

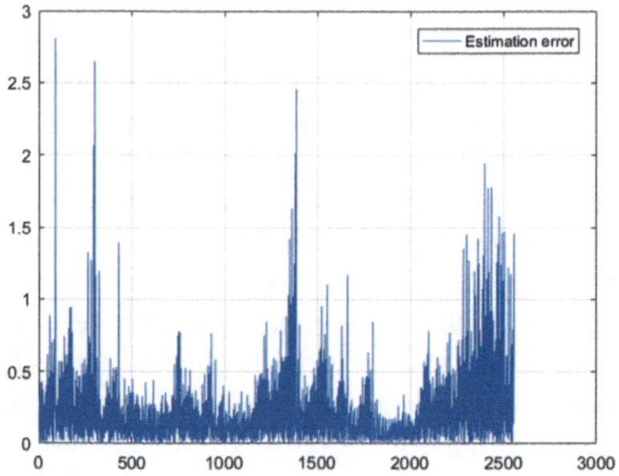

Fig. 6. The result of Normal_Videos_050_x264.mp4 in UCF-crime, the front 500 values are the training normal data, all values are smaller than those in front 500 values, which means no anomalies.

Considering that the newest flownet [17] has strongly speed up the running time of optical flow. Our further work will concentrate on a combination with deep learning method and try to solve the localization problem in our framework.

5 Conclusion

In this paper, we propose a particle filter-based prediction frame for anomaly detection. In anomaly detection area, prediction-based method is a new direction for such problem's solving. Compared with the reconstruction-based method, this kind of methods have a much more reasonable assumption: normal events are easier to predict than the abnormal ones. Inspired by such a new idea, we migrate the particle estimator to this problem. As the predictor that take advantage of both the history information and the newest observation, we guess it may work for prediction task in predict the normal events' eigenvalues and may not handle the suddenly occurred abnormal events' eigenvalues well. We also consider that because of the strong capacity of particle filter in tracking robustness, even though the new unexperienced mode occurs, the estimator may still work after a short confusion. So, we utilize such a short confusion as the judging criteria for detecting the abnormal events.

References

1. Cong, Y., Yuan, J., Liu, J.: Abnormal event detection in crowded scenes using sparse representation. Pattern Recognit. **46**, 1851–1864 (2013)
2. Lu, C., Shi, J., Jia, J.: Abnormal event detection at 150 FPS in MATLAB. In: 2013 IEEE International Conference on Computer Vision, pp. 2720–2727 (2013)
3. Zhao, B., Fei-Fei, L., Xing, E.P.: Online detection of unusual events in videos via dynamic sparse coding. In: CVPR 2011, pp. 3313–3320 (2011)
4. Hasan, M., Choi, J., Neumann, J., Roy-Chowdhury, A.K., Davis, L.S.: Learning temporal regularity in video sequences. In: 2016 IEEE Conference on Computer Vision and Pattern Recognition (CVPR), pp. 733–742 (2016)
5. Tung, F., Zelek, J.S., Clausi, D.A.: Goal-based trajectory analysis for unusual behaviour detection in intelligent surveillance. Image Vis. Comput. **29**, 230–240 (2011)
6. Cho, S., Kang, H.: Abnormal behavior detection using hybrid agents in crowded scenes. Pattern Recognit. Lett. **44**, 64–70 (2014)
7. Mathieu, M., Couprie, C., LeCun, Y.: Deep multi-scale video prediction beyond mean square error. CoRR, abs/1511.05440 (2015)
8. Arifoglu, D., Bouchachia, A.: Activity Recognition and abnormal behaviour detection with recurrent neural networks. In: FNC/MobiSPC (2017)
9. Baccouche, M., Mamalet, F., Wolf, C., Garcia, C., Baskurt, A.: Sequential deep learning for human action recognition. In: HBU (2011)
10. Gehrig, D., Kuehne, H., Wörner, A., Schultz, T.: HMM-based human motion recognition with optical flow data. In: 2009 9th IEEE-RAS International Conference on Humanoid Robots, pp. 425–430 (2009)
11. Farnebäck, G.: Two-frame motion estimation based on polynomial expansion. In: SCIA (2003)
12. Zhang, T., Liu, S., Xu, C., Liu, B., Yang, M.: Correlation particle filter for visual tracking. IEEE Trans. Image Process. **27**, 2676–2687 (2018)
13. Wang, F., Lin, B., Li, X.: An ant particle filter for visual tracking. In: 2017 IEEE/ACIS 16th International Conference on Computer and Information Science (ICIS), pp. 417–422 (2017)
14. Vermaak, J., Andrieu, C., Doucet, A., Godsill, S.J.: Particle methods for Bayesian modeling and enhancement of speech signals. IEEE Trans. Speech Audio Process. **10**, 173–185 (2002)
15. Wang, F., Lu, M., Zhao, Q., Yuan, Z.: Particle filtering algorithm. Chin. J. Comput. **8**(37), 1679–1694 (2014)
16. Tokdar, S.T., Kass, R.E.: Importance sampling: a review (2009)
17. Liu, W., Luo, W., Lian, D., Gao, S.: Future frame prediction for anomaly detection—a new baseline. CoRR, abs/1712.09867 (2017)
18. Sultani, W., Chen, C., Shah, M.: Real-world anomaly detection in surveillance videos. CoRR, abs/1801.04264 (2018)

Analysis on Building a Green Eco-smart City Based on Block Chain Technology

Pengfei Zhang(✉)

Shanghai Expo Development (Group) Co., Ltd., Shanghai 200125, China
Bureauzhangpengfei39@126.com

Abstract. This article gives a brief introduction to what a block chain is, its technical features and Bitcoin. Block chains allow shared economic rewards to soar. What is a green eco-smart city; The application and prospect of block chain technology in urban areas, how to create green ecology and wisdom, block chain landing mode and project introduction are looked forward and discussed.

Keywords: Internet · Block chains · Bitcoin · A shared economy
Green ecology, smart city · (a) mode of operation · Landing mode

1 Introduction

We're in an age of technology outbursts, with smartphones, smart drones, drones, robots, VR, QR codes, and so on impacting our lifestyle by bringing a smartphone out to dinner, shopping, watching movies, and shopping. Or you can pay at will at roadside stalls. No one is driving….

The Internet is also the aggregate of global information resources. It is a collection of information resources and resource sharing. The Internet has more than 60,000 networks and is officially connected to 86 countries. E-mailboxes can reach more than 150 countries. More than 4.8 million hosts are connected through it. There are more than 25 million users, and the daily information flow reaches trillions of bits. or more, each month, more than 1 billion electronic letters are sent (see Fig. 1). The Internet + the Internet of things + big data every day is impacting the economic trajectory of our traditional industries, sharing the economy to subvert the traditional economic model, and change is spreading rapidly… Everyone feels every day that the Internet is changing our work and life styles. However, the security of the Internet is increasingly worrying. Credit problems, payment security make everyone uneasy. At the same time, the Internet solves the problem of efficient transmission of information. At the same time, because the cost of credit collection has soared, changes have no doubt taken place on top of changes.

What is a block chain, saving our digital information files (data or source code) to keep records permanently in chronological order, generating what we call a block BLOCK. The blocks BLOCK is formed one by one in time. The blocks record all valuable digital information creation and exchange of activity information. The chain of these blocks is connected to form a chain of blocks. The size of each block in the block chain contains the compressed value of the transaction information of the

J. Abawajy et al. (Eds.): ATCI 2018, AISC 842, pp. 554–564, 2019.
https://doi.org/10.1007/978-3-319-98776-7_60

Fig. 1. Global internet

Fig. 2. Time lock

previous block, thus forming a long chain that contains the creation block (the first block) to the block generated at this moment – that is, the block chain. This is to be a centralized network link, continue to record all activity information in chronological order, can be traced back to the source through time stamps (Fig. 2 Time Lock), and the information written cannot be altered by forgery. And based on the value exchange protocol, distributed information sharing within the block chain is realized.

From Buddhism, Taoism, Islam, and Christianity, the teachings of honesty and credit are all covered. Ancient and modern Chinese and foreign dynasties, from saints, bachelors, civilians, and rulers, all dream of building an ideal society of integrity system. The struggle between humanity and morality has accompanied thousands of years of documented human history... Block chains have solved credit problems in some areas. Fei currency from Yap Island; From shells to copper coins to paper money to electronic money; Block chain digital currency is a system composed of credit records and liquidation. The block chain's consensus protocol rule(low) is deeply embedded in the transaction. Through open source code and data, it can be permanently queried, cannot be tampered with, and transactions can be mutually authenticated. What a great invention a block chain is.

Said that the block chain application has to say BITCION (Fig. 3), which is the first real application case of the block chain. It is an electronic currency (digital currency) generated through open source P2P software, and it is also a virtual asset of the network. Based on a set of password codes, its first 256 instructions, 75 were retained; Set multiple signatures, set at least 3 keys; The Bitcoin with the initial amount of solidification is generated by complex algorithm; Realize value exchange and trading functions. During the flow process, transactions cannot be forged or tampered with, so that the user owner can truly implement security authentication, trading, and payment. Although Bitcoin has many negative reports in public media, online media, and mainstream media, it cannot stop its development and application expansion, and it cannot ignore the increasingly powerful influence.

Fig. 3. Bitcoin

Block chains can be constructed as an infinite expansion of the Internet of Things and build a block chain ecology. Large data generated through the Internet of Things can provide a great deal of space in the block chain. The zero-marginal cost society brought about by the shared economy, the creation of a shared economy ecosystem, and the realization of new asset and service sharing become possible. The reputation of crowdfunding in the block chain reduces people's trust costs, low rates (initiation, management), easy circulation, transparent rules and auditing, block chain technology allows the crowdfunding platform to take off, and the crowdfunding model promotes the huge use of block chains. Space area. Block chains have been partially used in finance, banking, insurance, government services, military, real estate, energy conservation, environmental protection, education, and copyright protection. Its honest, efficient and safe trading system has the potential to subvert the traditional financial currency, banking and insurance industry, lawyer industry, accounting industry, etc.

2 Relevant Application of Green Ecological Wisdom City in Block Chain

The Department of National Housing and Construction and other departments of the characteristic cities issued a notice in 2016, proposing to cultivate 1,000 characteristic cities by 2020. At present, there are many cases of failure in the country, causing damage to the ecological environment and cultural heritage. What is a green eco-smart city? Our interpretation of four aspects: First, the inheritance of cultural characteristics (architecture + local heritage and customs + diet + language + music); Second, green environmental protection and ecology(environmental protection + soil and water conservation + ecological maintenance + zero-row goals + recycling); Third, the introduction of characteristic industries(smart and intelligent technology industry + local characteristic industry + cultural and entertainment industry import); Fourth, solve the issue of agriculture, rural areas and farmers. Block chain technology opens a new world for green eco-cities (Fig. 4). How to use the block chain to build, the core is still the industry. What kind of industry is it really about to bring in?

Fig. 4. Block chain keys

2.1 What is a Green Ecological City

To build a block chain technology to create a green ecological city is a feature of the characteristics. First, solve the problems of agriculture, rural areas and farmers, through the smart digital agricultural industry, to achieve the industrialization of agricultural cultivation, poverty eradication and wealth to change the face of the countryside, and simultaneously develop tourism and tourism agriculture. Secondly, it is implanted in the energy-saving and environmental protection industry. Based on the economic model of block chain sharing, high efficiency and distributed characteristics, solar energy and wind energy can be promoted in large quantities; Emission charges and emission reduction have the driving force to achieve the zero emission targets of cities and scenic spots; Sharing environmentally friendly cars can reduce the number of vehicles in the area, reduce land resources, save energy and reduce emissions, and promote a circular economy (Fig. 5). This is truly a green ecological city.

Fig. 5. Circular economy

2.2 What is a Block Chain of Smart Cities

Projects in smart cities across the country have been fully implemented, but the exchange of information, collaborative management and governance between government departments to achieve trustworthy intelligence and efficiency is very limited. The problem of silos of large data across sectors is causing an alarming waste of resources (Fig. 6).

The creation of a smart city in a chain of blocks cannot be achieved without the full

Fig. 6. Smart city

support of the government. We do not believe that block chains can create anarchic self-running societies. The characteristics of de-centralized, efficient, cheap, personalized block chains, It can rovide big data and intelligent supervision for the government in such areas as smart dynamic population management, medical and social security management, tax administration, smart municipal services, smart fire protection, smart food safety, energy conservation, ecological and environmental protection, and real estate management. We will truly achieve openness, transparency in trust and seamless integration of government affairs.

The main body of the green eco-smart city is human. We build a smart management system based on population management as an entry point. In the past, the population management Department of the city was a police station, and population management was data on households and some floating populations. In urban areas, according to the township and village party committees, people with property rights (permanent population) and renters (floating population) are registered. The registered population and the floating population correspond to houses, and the dynamic population data are updated from time to time.

With regard to safety supervision, the establishment of an intelligent fire control and intelligent food safety prevention system can set up files for units in the green eco-smart urban areas: fire control key units, food and beverage stores, accommodation and hotels. Industry and commerce, security inspection, fire protection, Anjian, and urban management departments share information, update management supervision from time to time, and adopt a multi-pronged coordinated and cooperative supervision to avoid the past long management or the death or oversight of supervision. Due to the low cost, high efficiency, transparency, and non-tamper-proof characteristics of the block chain, the information recorded in time stamps provides efficient government administration guarantees for the city government. For energy-saving ecological and environmental protection, for enterprises or units in areas such as production enterprises, restaurants, hotel and hotel public toilets, according to the block chain technology, the corresponding address registration information, reduction and operation information, pollution and emission information are established. The operating mechanism of who discharges who pays and who manages who charges can fully reflect the advantages in the block chain, reducing operations and supervision.

Block chains provide a technical support platform for intelligent and intelligent government affairs. The image of a local government that rebuilds the trust of the general public (fairness and honesty) is far-reaching.

2.3 The Application of Block Chain Technology in the Green Ecological Intelligent City

2.3.1 The Use of Block Chains in the Industrial Planning, Planning, Design, and investment Construction of the Township Government

To build a green eco-smart city, we first have a deliberate long-term plan and design. Planning does not create a castle in the air. It needs to be based on industrial planning and import. Therefore, what kind of IP is the focus we have introduced, and first of all, the core industry. Industry leading industries and block chain enterprises that can represent future development trends; The second is the tourism industry, which can introduce people's education camps and experience sports. The second is the basic supporting aspects of accommodation, scenic spots and urban transportation, solar and wind energy, shared cars and so on. With a clear industrial direction, we can make short-term and long-term plans. Block chains can be applied to our planning management and implementation, policies can be open, transparent, planning has continuous implementation, and it is more convenient for local investment promotion.

Investment construction, whether PPP or EPC, has been made more enforceable, especially for long-term contracts. It is also possible to use a shared economy crowdfunding platform to solve more profitable individual projects such as hotels, accommodation, transportation, open-air camps, entertainment facilities and small roads and bridges, and to reduce the investment of some of the heavy assets of local governments.

2.3.2 The Application of Block Chains in the Operation and Management of Green Eco-smart Cities

Block A Chain Green Platform

Through block chain technology, Internet of Things sensors are implanted in various places in the city, such as visitor centers, factory areas, water treatment plants, garbage collection stations, public toilets, etc. The use of block chains can enable regulatory units, sewage units, and the public to participate in ecological and environmental protection platforms. They always understand the discharge of waste water, exhaust gas, and waste residue. The block chains cannot be modified and all information can be accurately recorded in accordance with the time. Supervision, prevention and treatment of the air environment, water environment, soil environment, major pollution emissions, ecological restoration, and rural environment should be carried out in a more detailed and dynamic manner. We will achieve the goal of zero emissions and recycling and maintain the integrity and continuity of local ecosystems. Many parties actively participate in the sharing of wind, solar, geothermal and biomass energy (Fig. 7).

Block B Chain Intelligent Population Management:

Fig. 7. Green energy

In the smart cities of the block chain, the large-scale data government obtained by the town, township, and village system (described in more detail) through population management provides fast public information services and information on convenient people's initiatives and makes the process and results more transparent and traceable. Provide accurate social security, education, old-age, employment, medical services and other services. The use of intravenous palmprint technology to enter personal information, its only non-replicable can replace any bank card, access card, social security card.

Block C Chain Smart Security Supervision Platform:

The Internet of Things establishes an enterprise APP, and equipment and facilities set up a two-dimensional code. Through the block chain technology, it can comprehensively supervise fire safety, food safety, public security, and production safety. It can enter information anytime, anywhere, and cannot be changed. The supervision information process and processing results cannot be changed. Food and beverage, hotel, theater, entertainment, factory warehouse no dead ends.

D Block Chain Wisdom Education Platform

Consistent with the concept of fairness, accuracy, and quality, it is supported by block chains and the Internet of Things, and mobile Internet. It uses educational data, home-school interaction, intelligent robots, and other products as tools to effectively promote interregional education balance and improve quality. It is possible to issue game interactive coins in conjunction with 10 years of education courses, combine learning games, increase interest, and learn online and offline independent interactive learning in class (Fig. 8).

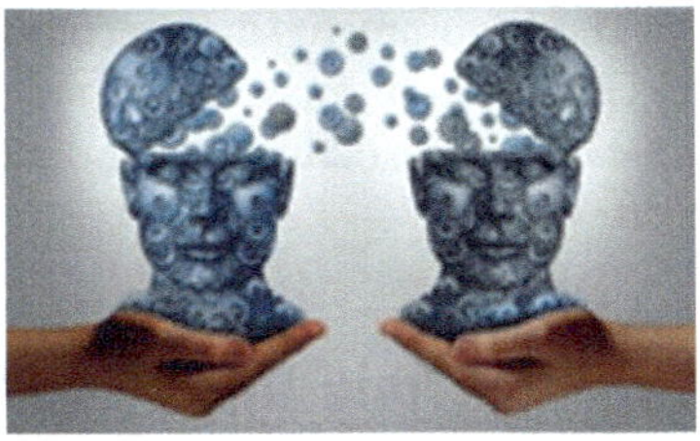

Fig. 8. Self-study

3 Analysis of Investment Benefits and Application Patterns

3.1 Operating Methods

The overall use of investment + operations, that is, social capital investment + government purchase of some services model, reduce local financial pressure. For block chain projects involving public services, investment platforms can be established through funds and social capital financing, basic algorithms, technical support, and software service platforms can be provided to help blocks and units and enterprises establish blocks and establish urban block chain ecology. For the introduction of urban IP projects, an urban industry guidance fund can be established to support both residential and residential, smart transportation, smart digital agriculture, or education industries, which can allow enterprises to be deeply planted in the local area to achieve corporate profits and development. Create a crowdfunding service platform to achieve the sharing of resources, manpower, funds, and channels (Fig. 9).

Fig. 9. Is fully shared

3.2 Investment Benefits

According to the standard 30,000 cities, they invest in green ecological platforms, smart population management platforms, smart security platforms, and smart education platforms. According to traditional smart city big data and cloud computing centers, they need 200 to 300 million investment, and the government purchases services every year. 5–10 million service fees. According to the block chain model, the government does not need input, and only the original scattered unit data, that is, the block, is connected together. The input only needs 1/3 of the original, and the block chain operates on its own (Fig. 10). Block chain runs on its own), The cost of services has also been significantly reduced, with only a small number of personnel to maintain, approximately 2 to 5 million. The government saves resources and improves efficiency; The people have no cost of credit, enjoy more convenient government services, and have low-cost access to educational resources; Environmental ecology is perfectly preserved. The benefits are obvious.

4 Business Model of Block Chain Business Project

4.1 Landing Mode: Establish a "Block Chain Incubator" Industrial Guidance Mechanism

Block chains are very cutting-edge technologies that represent the future. The green eco-smart city will not be reached overnight, and the business model must be step-by-step. The first is to land on the "block chain creation center" that is, the incubator. Its task is to provide basic algorithm services for block chain companies, provide block chain green eco-smart city government planning framework and industrial planning, build block chain ecology for urban government investment, and provide administrative, financial, and legal services for block chain companies entering cities. Business services and other business services Provide financial services for block chain companies entering the city, train and introduce block chain talents for the city. The second is to establish an industrial fund; Establish a block chain digital currency or integral trading platform and establish a sky account book. The second is the core enterprise to stay. Fourth, infrastructure investment and construction.

4.2 Project Self-running (Fig. 10)

Green eco-smart cities must be built with block chain technology to create block chain industry ecology. There are many industrial projects. This article only takes solar energy and wind energy as an example, each unit or farmer installed solar energy and wind energy (Fig. 11). It can be invested in or paid for to solve equipment inputs. In urban areas, through block chain technology, all services can be exchanged and automatically settled through sky accounts(books) through installation, operation fees, maintenance and overhaul. The account credit is free of cost, permanent check, and cannot be tampered with. After the scale, it can issue "energy coins" to increase liquidity, online circulation and value exchange. To truly create a green eco-smart city.

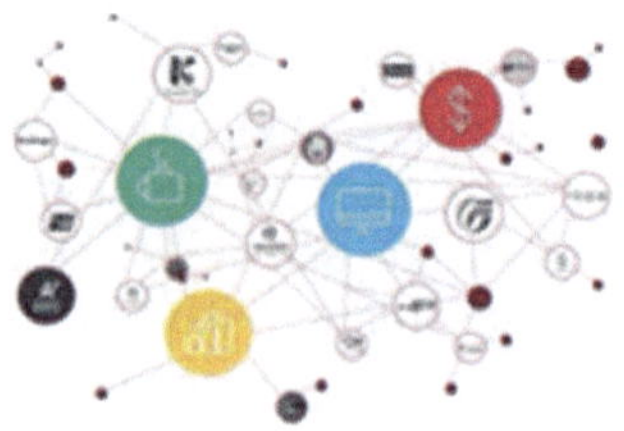

Fig. 10. Block chain self-running

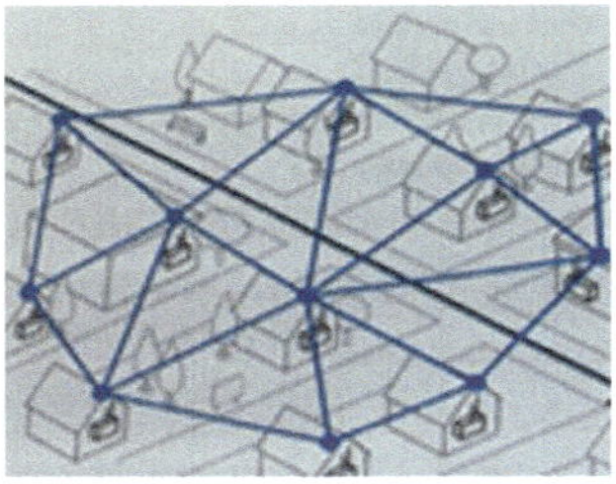

Fig. 11. Application motogram

4.3 On the Basis of Building a City Based on the Block Chain Technology

It is planned to build a practical case in an island in Guangxi-a block chain green ecological wisdom city. At present, it is in the construction stage. There are 16,000 islanders, 24 km, and there are abundant resources for humanities and natural landscapes. This will be a historic step in China's block chain technology, with far-reaching implications (Fig. 12).

Fig. 12. Proposed to build a block chain of green eco-smart islands

Fig. 13. Continuation of new history

5 Conclusion and Suggestions

Block chain is a great invention with far-reaching impact. She has led the future changes in the green and environmentally friendly lifestyle and work style. The quantitative changes will inevitably take place. She will continue to write a new page in the history of the new generation of the Internet (Fig. 13). It is recommended that cooperation on the basis of better policies and an ecological environment by all levels of government can be simpler, more efficient and less costly, leading to the effective integration of resources and the creation of more economic and social values. What is more important is that the commercial green eco-smart city is the result of the joint efforts of each market body and is also a participant in the block chain. Each participant can find his own position in the block chain and contribute a skill. And get the corresponding remuneration to realize its own value.

Green smart city design and block chain application is an interdisciplinary, cross-stage comprehensive design process, and the BIM model just responds to this demand and realizes the coordinated design and data set of various types of jobs on a single data platform. It is recommended to add 5D information to BIM modeling software so that cross-stage management, design, and block chain applications are fully involved in the information model. The implementation of BIM technology at the headquarters of the World Expo Central Enterprise has significantly advanced the analysis of physical information of various buildings from the late design stage, which helps architects and owners to make decisions related to green buildings in the program and even in the conceptual design stage. Therefore, using BIM to run through the application of green smart cities and block chains, integrating technology informatization and management informatization to achieve more effective project life management and enterprise resource planning, is the inevitable trend of the future development of China's construction field. 5.

References

1. Zhu, L., Yang, H., Cai, Z.: Internet + Model Construction-Deep Interpretation of Internet+'s 8 Core Technologies. Machinery Industry Press
2. Alu.sadan, Translated by Zhouxun: Sharing the outbreak of the economy. Wenhui Publishing House, India
3. Zhu, W.M., Hua, L., et al. (trans. by): Business Block Chain-Opening a New Age of Encryption Economy. CITIC Publishing Group Publishing
4. Wangtian, J., Wuyuanwen, Zhaoguodong, Liuwenxian (eds.): Block Chain and Big Data-Building an Intelligent Economy. People's Post and Telecommunications Press
5. Zhangpengfei: The Life-Cycle Management of Large Engineering Based on BIM. Published by Tongji University Press (2016)

3D Indoor Network Model for Emergency Response of Multi-utility Tunnel

Gang Yu[1,2], Zeyu Mao[1,2(✉)], Min Hu[1,2], and Li Zhou[1,2]

[1] SHU-UTS SILC Business School, Shanghai University, Shanghai 201800, China
1056866878@qq.com
[2] SHU-SUCG Research Centre for Building Industrialization, Shanghai 200072, China

Abstract. With the expansion of the construction scale of underground multi utility tunnel and the increase of internal complexity, the complexity of emergency response decision-making has been caused. In this regard, some studies have accelerated the emergency response decision process by integrating indoor space and semantic information into an indoor network model with spatial analysis capabilities. However, the current proposed indoor network model and its research in emergency response focus on the field of housing construction. Few studies have proposed a method for using the indoor network model for multi utility tunnel emergency response. Although there have been very few studies that have attempted to apply the indoor network model to the emergency response of multi utility tunnel, their models all show some limitations. Therefore, this paper proposes an indoor network model for emergency response of integrated pipeline corridors, which presents its construction method and application framework to prove the applicability of the model.

Keywords: BIM · Multi utility tunnel · Emergency response
Indoor navigation model

1 Introduction

The underground multi utility tunnel, which integrates various engineering pipelines such as electricity, communications, gas, heat supply, and water supply and drainage, is an important infrastructure and lifeline for ensuring the operation of the city [1]. As of the end of 2017, the total size of the city's underground integrated pipe gallery exceeded 2000 km [2]. With the expansion of the scale of underground multi utility tunnel, the hidden dangers of underground space security have increased, resulting in increased difficulty in responding to emergencies such as rescue and disaster relief and accident disposal. However, emergency response, as an important component of the underground space maintenance and operation phases such as multi utility tunnel, generally depends on the actual experience of operators [3]. This makes it necessary for emergency repair personnel to plan for a repair path after arriving at the scene, which significantly delays the emergency response process [4].

J. Abawajy et al. (Eds.): ATCI 2018, AISC 842, pp. 565–571, 2019.
https://doi.org/10.1007/978-3-319-98776-7_61

In order to make the emergency response quick and accurate, many studies have analyzed the internal monitoring data of the building [5], combined with the internal structure of the building to calculate the optimal path [6] from the entrance to the target and then accelerate the emergency response process. Thus, the concept of indoor network models was proposed. Unlike traditional models, the indoor network model aims to extract key building information (rooms, doors, stairs, etc.) inside the building and ultimately serve the emergency response.

However, the existing method of using the indoor network model for emergency response is mainly concentrated in the field of house construction. Few studies have proposed a method for using the indoor network model for emergency response of a multi utility tunnel. Although there are very few engineering examples that have attempted to apply the indoor network model to the emergency response of multi utility tunnel, their models have shown some limitations [7]. On the one hand, the indoor network model of the doors, rooms, and stairs in the interior space that are collectively expressed as nodes cannot express the complexity of laying pipes inside the multi utility tunnel [8]. This makes the existing emergency response method based on the indoor network model unable to assist managers to make decisions in the face of issues such as blasting and closing valves. On the other hand, the two-dimensional computer-aided design (CAD)-based indoor network model created by the traditional method is limited by the non-object-oriented geometric representation, so that the indoor network model constructed based on the method is difficult to accurately express indoor entities in an actual environment [9].

In order to solve the problems above, this paper proposes a 3D indoor network model for emergency response of multi utility tunnel. Based on the traditional indoor network model, this model defines new classes for the important model elements in the multi utility tunnel so that it can be applied to the comprehensive pipeline emergency response. At the same time, in order to eliminate the limitations of the existing two-dimensional expressions of existing models, this article introduces BIM technology and uses the BIM model of multi utility tunnel as an information source. Revit API is used to parse the corresponding data from the BIM model and complete the UTDM construction.

The next section will detail the proposed model, the construction method, and use UTDM for a comprehensive framework of comprehensive corridor emergency response to demonstrate model applicability.

2 The Indoor Network Model for Emergency Response of Utility Tunnel

This section aims to propose an indoor network model UTDM for integrated pipeline emergency response, as shown in Fig. 1. First, the model is constructed by extracting information from BIM and finally converted into database data of the operation and maintenance platform. Through the real-time data interaction with the operation and maintenance personnel in the emergency response phase, emergency response repair path planning is completed.

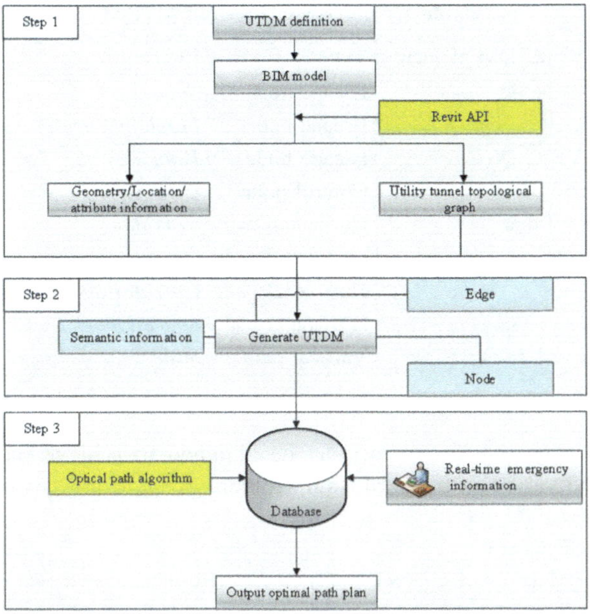

Fig. 1. Flow chart of emergency response method based on UTDM model

2.1 UTDM Definition

The proposed UTDM is an indoor geometric network model. This model determines the adjacency relationship of indoor space by integrating the interior topological diagram of the multi utility tunnel, which combines the geometric information to construct. At the same time, UTDM further improves the geometric and attribute information of traditional GNM on the basis of its specific definition as follows:

$$\text{UTDM} = (V, E, I) \tag{1}$$

Where $V = \{v_1, v_2, \cdots v_i\}$ is a set of UTDM nodes, and $E = \{e_1, e_2, \cdots, e_j\}$ is a UTDM edge set. The specific structure is shown in Table 1. Among them, the node element that needs to be operated in the emergency response of valves, fire hydrants, etc. is defined as "Control", and the different spaces connected to the upper and lower ends of the stairs are marked as special parts of the internal space (including vertical connections). The name is LandingSpace.

I is a set of UTDM semantic information. The content is further subdivided into I_v and I_e, representing the semantics of points and edges, respectively. The UML diagram of the UTDM semantic model is shown in Fig. 2. It is worth mentioning that the defined semantic model contains the topological relationship between the pipeline and the corresponding control. When the node type is Control, the semantic information I_{v_i} contains the "Pipe", "Control", and the "Pipe" shown in Fig. 2. "Connection" class. By accessing the pipelines, the corresponding control points (valves, fire hydrants) and

Table 1. Nodes and edges of MTDM

MTDM elements	Type	Element name
Node	Escape node	*Space*
	Escape node	*LandingSpace*
	Escape node	*Door*
	Control node	*Control*
Edge	Horizontal path	*Corridor*
	Horizontal path	*Corridor-control*
	Horizontal path	*Corridor-door*
	Horizontal path	*Space-Door*
	Vertical path	*Stair*

connectors (two-way and three-way) can be obtained to achieve rapid retrieval and positioning of the valves in the explosion problem of multi utility tunnel.

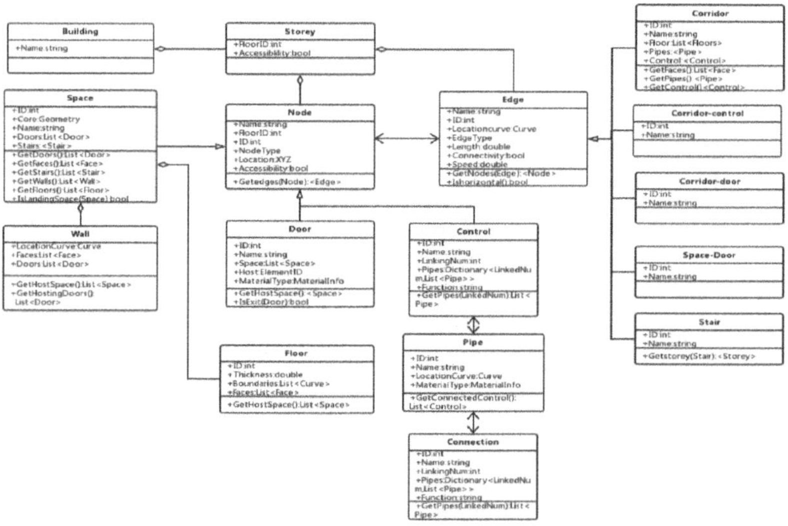

Fig. 2. UTDM semantic model

2.2 UTDM Model Construction

To obtain the information necessary to construct UTDM, this article uses the Revit API to extract information from the BIM model, which is based on the Element class in the Revit API. The "Element class" contains most of the objects (walls, families, pipelines, axes, etc.) that are visible in the multi utility tunnel. It is the base class for most classes in the BIM model of integrated pipe corridors. Its UML diagram is shown in Fig. 3 [10].

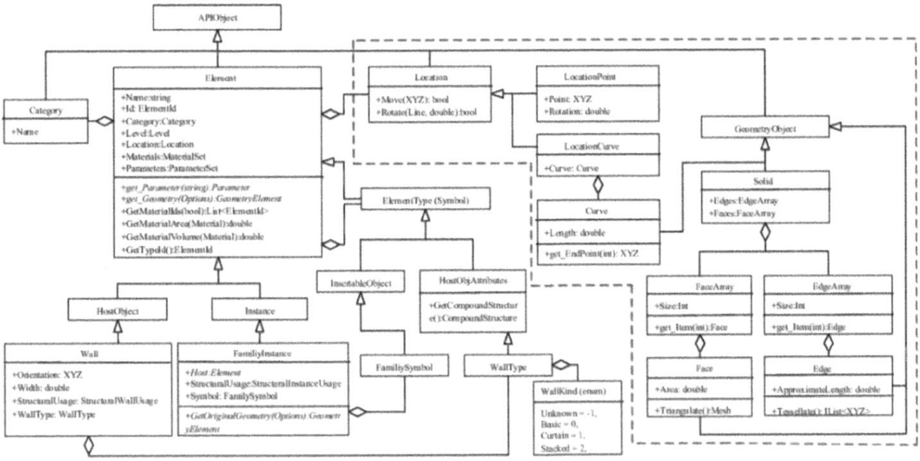

Fig. 3. UML class diagram of multi utility tunnel elements

2.2.1 Construct UTDM Nodes

Space: The spatial node is established by the "Wall" and "Floor" elements that enclose the space. We can get the geometry information by calling the "GeometryElement" of the corresponding wall/floor element by calling the "get_Geometry()" method in the "Element" class to calculate the space geometry coordinates.

Door: The property and location information of the internal door of the multi utility tunnel is stored in the "FamilyInstance" class of the BIM model. Some doors are used as the entrances and exits connecting the internal and external corridors of the multi utility tunnel to determine the starting point of the planning path. Therefore, it is necessary to determine whether each gate element is currently integrated. The entrance and exit of the corridor can call the WallType.get_Parameter(BuiltInParameter.FUNCTION_PARAM) method in "WallType" to determine whether the component is an exterior wall.

Control: In the BIM model, "Control" is embodied as "FamilyInstance" and its corresponding coordinates. The attributes can be extracted from the corresponding family instances. In the multi utility tunnel BIM model, the topological relationship between the pipeline and the connection can be obtained by calling the "Domain" attribute of the pipe connection (Fig. 4b) to further determine the topology diagram of the pipeline, valve, and connection.

2.2.2 Construct UTDM Edges

Corridor: The horizontal route of the corridor is composed of the central axis. Therefore, the route of the UTDM model needs to be automatically generated from the boundary information of the corridor. The MAT algorithm was first proposed in the literature in 1967 [11]. By acquiring the corridor geometry, the corresponding central axis can be calculated. Based on the Revit API, the corridor profile can be obtained by calling the "get_Geometry()" method in the "Element" class and entering the profile into the mid-axis transformation (MAT) algorithm to obtain the corridor path.

Door/Control to Corridor: According to the acquired corridor path $M(G_{corridor})$, the edge of the door/control to the corridor can be calculated. As mentioned earlier, the coordinates of the door/control are obtained from "FamilyInstance" in BIM. So the edge of the door/control to the corridor can be generated by projecting the door perpendicular to the corridor. If the vertical point does not exist, connect the door/control to the nearest point in the corridor.

Space to Door: The topological relationship between the door and the space can be obtained by calling the "HostObject" subclass corresponding to the door element to obtain its subordinate wall. By judging the space to which the wall belongs, the topological relationship between the door and the corresponding space can be obtained. In the integrated pipeline BIM model, the space is represented by the "Space" class. The "GerBoundarySegments()" method of the "Space" class can be used to obtain the wall elements corresponding to the space (Fig. 4a). If two points are linked, the path is determined based on the connecting node.

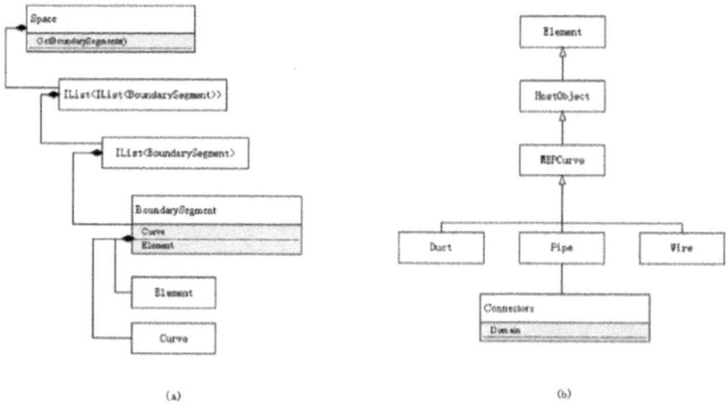

Fig. 4. Spatial boundary related class diagram/MEPCurve class diagram

Stair: As a vertical edge, the stairs connect the different levels of the multi utility tunnel. By searching the geometrical profile of the floor corresponding to the room, the vertical connection relationship of the corresponding room can be obtained. Connecting the geometric coordinates of the corresponding room can form a complete vertical route within the multi utility tunnel.

3 Conclusion

This paper proposes an indoor network model based on BIM, UTDM, in combination with the internal structure of an multi utility tunnel. This model can be used for various indoor route planning applications (such as emergency response). Indoor route planning requires the use of a large number of interconnected building model information to calculate the navigation path. The BIM model serves as the most detailed resource for semantic information and can be used as an ideal data source for generating UTDM models. In this paper, the transition process from BIM model to UTDM is given to verify the feasibility and effectiveness of this scheme, and further expand the application scenario of BIM model. Compared with the existing indoor network model, the UTDM model not only improves the efficiency in terms of extraction, but also improves its details (such as adding pipe attributes to controls) to make it more applicable to integrated corridors. In the application phase, this paper proposes that UTDM can be converted into a corresponding database in the operation and maintenance platform, and further expand the model analysis function by establishing the connection between outdoor and indoor network functions.

Acknowledgements. This work was supported in part by the National Natural Science Foundation of China under Grant 71701121, in part by the Shanghai Municipal Science and Technology Commission Project under Grant 16DZ1201704.

References

1. Chang, J.R., Lin, H.S.: Preliminary study on application of building information modeling to underground pipeline management. In: Geo-Hubei 2014 International Conference on Sustainable Civil Infrastructure, pp. 69–76 (2014)
2. Hailong, B.: Research on the development trend of urban integrated pipeline. China Munic. Eng. **6**, 78–81 (2015)
3. Eskesen, S.D., Tengborg, P., Kampmann, J., Veicherts, T.H.: Guidelines for tunnelling risk management: international tunnelling association, working group no. 2 ☆. Tunn. Undergr. Space Technol. **19**(3), 217–237 (2004)
4. Occupational Safety and Health Service: Guidelines for the Management of Work in Extreme Temperatures. Department of Labour, New Zealand (1997)
5. Chen, Y., Sun, L., Wang, F., Zhou, X.: Congestion-aware indoor emergency navigation algorithm for wireless sensor networks. In: Global Telecommunications Conference, vol. 122, pp. 1–5. IEEE (2012)
6. Inoue, Y., Sashima, A., Ikeda, T., Kurumatani, K.: Indoor emergency evacuation service on autonomous navigation system using mobile phone. In: Second International Symposium on Universal Communication, pp. 79–85. IEEE Computer Society (2008)
7. Qiang, T., Jian, W., Lining, Z., Jianbo, T.: Research on intelligent operation management technology of urban underground comprehensive pipeline. Technol. Mark. **22**(12), 27–28 (2015)
8. Lee, J.: A spatial access-oriented implementation of a 3-D GIS topological data model for urban entities. Geoinformatica **8**, 237–264 (2004)
9. Teo, T.-A., Cho, K.-H.: BIM-oriented indoor network model for indoor and outdoor combined route planning. Adv. Eng. Inform. **30**(3), 268–282 (2016)
10. AutodeskAsiaPteLtd: Autodesk revit Secondary Development Basic Tutorial. Tongji University Press, Shanghai (2015)
11. Blum, H.: A Transformation for Extracting New Descriptors of Shape, Models for the Perception of Speech and Visual Form, pp. 326–380. The MIT Press, Cambridge (1967)

Computer Data Integration Analysis of the Personal Credit Reporting System

Sunping Guan(✉)

Aba Teachers University, Aba County 624000, Sichuan, China
iweiming@foxmail.com

Abstract. The Internet finance is growing rapidly in China. The third-party payment, the P2P loan and the network financial management and so on are constantly changing people's living habits. At the same time, the problems of the credit default exposed in the process of the development of the Internet finance also hindered its further development. The traditional credit evaluation mode is based on a person's loan history and repayment performance, and uses the logic regression to judge the person's credit. The data sources of the big data credit are very extensive, including the electronic commerce, the social networking and the searching behaviors, which generated a lot of data.

Keywords: Personal credit · Credit reporting system · Computer Data integration · Big data analysis

For a long time, China's credit market is a single pattern dominated by the Central Bank. The Central Bank's credit reporting system mainly collects the business and the personal information based on the bank credit information. Therefore, in the coverage of the users and the diversity of the data, the Central Bank's credit reporting system has great limitations. How to meet the financial needs of more than 500 million people without credit records is the market opportunity that the big data credit covets.

1 New Patterns of the Credit Market

On January 5th of this year, the Central Bank issued the "Notice on preparations for personal credit reporting business", and the personal credit business was officially "opened". Tencent Credit, Sesame Credit, Shenzhen Qianhai Credit, Pengyuan Credit, China Chengxin Credit, China Intellicredit, Kaola Credit, and Beijing Sinoway Credit are allowed to carry out the preparations of the personal credit business. Among them, the Sesame Credit relying on the Ant Financial and the Tencent Credit under the banner of the Tencent are most competitive because of their wide coverage, the large scale of the data, and the strong technical strengths. In addition, Qianhai Credit is also regarded as a strong competitor because of his backing to the comprehensive financial group – China Ping'an.

In addition, Kaola Credit collects the personal data of the credit card repayment, transfer and public payment on the La Carla platform. The data of the Sinoway Credit is mainly from the two shareholders: a large number of the gas data of the residents By

J. Abawajy et al. (Eds.): ATCI 2018, AISC 842, pp. 572–577, 2019.
https://doi.org/10.1007/978-3-319-98776-7_62

the ENN Capital, and Emay Softcom under the banner of INF Tech which is the senior strategic partner of the three largest Telecom operators in China. Among the other three institutions, the major shareholders of China Chengxin Credit and Pengyuan Credit are the old-brand enterprise credit rating companies. China Intellicredit is a private third-party credit reporting company, mainly based on the "anti-fraud" businesses.

In fact, in addition to the first batch of the eight institutions approved to carry out the personal credit rating businesses, a lot of the companies with the massive data and the technological bases are in the dark layout of the big data collection market. According to the media reports, more than 30 companies including Jingdong Finance, Baidu Finance, Xiaomi, and CreditEase and so on are all interested in applying for the second batch of the personal credit rating license. It is understood that the two sides have announced the establishment of a joint venture called JD-Zest Finance Gaia, which aims to build the Jingdong's own credit evaluation system using the technologies in the big data mining and processing. Another strong competitor in the big data collection market, ZC-credit under the banner of CreditEase, also released a risk-management product for the P2P agencies, the microfinance institutions and the bank credit departments in June this year, including the credit score, the personal loan data, and the personal risk list data.

In addition, the P2P lending institutions, including the PPDAI and the Jimubox, also use their data accumulated over the past few years to establish their own credit evaluation system. Taking PPDAI as an example, it launched a risk management model based on the big data at the beginning of the year – the magic mirror system. In addition to those companies holding the data and resources, the third-party technology and service institutions around the field of the big data credit also began to emerge, such as "Baifendian" focusing on the big data mining, and "Icekcredit" and "Wecash" that accomplished the data analysis and the credit evaluation in the upper layer of data source.

2 The Present Situations of the Development of the Credit Collection System in China

The construction of the credit collection system in China is started by the credit investigation. At present, it has basically formed a diversified pattern, which is based on the Central Bank's public credit and information collection system, and the market-based credit agencies are supplemented. The Central Bank's credit information system basically establishes credit files for every company and individual with the credit activities, and provides the basic credit information services to the outside world. The market-oriented credit institutions are based on the segmentation market, give full play to their advantages of the subdivision data and provide the diversified credit and information services to the society.

The Central Bank's credit information system includes the basic database of the enterprise credit information and the basic database of the personal credit information. In July 2006, it realized the nationwide networking query. The data of the Central Bank's credit information system mainly comes from the traditional credit institutions, such as the banking and the financial institutions, and it also includes the social security, the provident fund and other public information. After obtaining the credit data from these institutions, the credit reporting, the credit rating and other credit products are formed through the rating of models. The demand sides of the credit system of the Central Bank include banks, government departments, industrial and commercial enterprises, and individual natural persons and so on, and most of them occur in the personal housing purchase and car purchase, the personal microfinance, the business credit, the bond trading and other scenes.

The credit is an important infrastructure of the modern financial system. The credit is a kind of activity where the professional third-party institutions collect, sort out, record and process the credit information of the natural people, the legal persons and other organizations according to the laws, and provide the credit reports and the credit information consulting services according to the laws. It provides a platform of the credit information sharing for the professional credit institutions, thus helping the customers to judge and control the credit risks and carry on the credit management activities. The credit evaluation activities carry on the credit evaluation to each subject in the financial activities, greatly alleviating the adverse selection caused by the information asymmetry in the credit activities and reducing the credit risks and transaction costs of the financial system, to ensure the healthy and efficient operation of the financial systems. According to the nature of the ownership, the credit agencies can be divided into the public credit institutions and the private credit institutions. The public credit institutions refer to the public credit information registration institutions established by the governments (usually the Central Bank). The private credit agencies are generally independent from the governments. They are the credit institutions with the private property rights and the marketable operations, including the credit agencies operated by the private enterprises and the industrial associations.

According to the different subjects of the credit information, the credit agencies can be divided into the corporate credit institutions and the personal credit agencies. The enterprise credit agency refers to the credit agency that collects and processes the credit information of the enterprise legal persons, and provides the enterprise credit evaluation services for the credit institutions with compensation. The personal credit institutions refer to the credit institutions that collect and process the natural person's credit information and provide the personal credit evaluation services according to the requirements of the users.

The profit model of the credit industry and the application of the data in the industrial chain are as follows. The natural people, the legal persons and other organizations, in the daily life and management, have produced a large number of the credit-related data and information in the financial institutions, the government departments and other enterprises and institutions. Through the cooperation with the data collection agencies, the credit collection agencies aggregate and integrate the credit information scattered in various departments, and use their own models and analytical methods to process the information, to generate the credit scores, the credit reports and other products. The credit agencies earn profits by selling the credit products and providing the value-added services to the users of the credit information.

According to the calculation, the future personal credit market of our country is not less than 100 billion, and the individual credit business of our country is still in the beginning stage. The private credit information sharing platforms are basically in the state of their respective battalion and free from the social credit systems. Because of the different genes of each organization, the data collection and the scoring systems are different, with the lack of the sharing channels and mechanisms among each other. We must establish the public credit service platform covering all individuals and enterprises relating to the Internet finance, and allowing any internet financial institutions to access.

Although a number of the market-oriented credit agencies have emerged in China, because of the government's policies and scale advantages in the credit market, some large foreign credit agencies have also entered the domestic credit market to participate in the competition. The credit agencies with the market management have not been able to obtain more opportunities for their development, and are generally small in scale and difficult to collect data, and low in their professional levels and poor in their quality of the credit products. And the imperfect credit information sharing mechanism makes it difficult for them to fully enhance their market competitiveness. Therefore, as long as the People's Bank Credit Center and the main e-commerce platforms, the P2P and the public funding sites establish a cooperative relationship, establish the data interface, and unify the data collection standards, they can effectively obtain most of the Internet financial credit information data.

The development of the foreign credit acquisition industries has a certain reference to the rapid leap of China's credit acquisition industries. With the arrival of the era of the big data, the future credit development in China should be constantly reformed and innovated from several aspects, such as the systems, technologies, supervision and protection. At present, our country has not yet issued a formal privacy protection law, and the current laws and regulations also lack the clear regulations in the scope of the credit information collection and the scope of its application. What model of the credit market in China and what role the central bank credit center should play is still to be resolved.

3 Computer Data Integration and Analysis of the Personal Credit Information System

In the traditional credit collection activities, the collection and processing of the personal information covers the personal loans, credit cards, guarantee and other credit information, as well as 8 kinds of public information of the deposit information of the personal housing provident funds, the social insurance payment and distribution information, the vehicle transaction and mortgage information, the court decision and execution information, the tax information, the telecommunications information, the personal subsistence assistance information, the practice qualification and reward and punishment information altogether, involving more than 80 items. However, under the background of the big data, the personal information in the credit reporting activities has a new extension. The personal information formed by the mining and analysis of the big data is widely used in the personal credit reporting activities, and we will be the "data portrait" through the mining and analysis of the big data.

The researches of the data mining are most common in the field of the information technology, and the special research on the data mining is scarce in law, and more researches are based on the whole concepts of the big data collection. The data mining is a computational process of exploring the big data sets involving the machine learning, the statistics and the database systems. This is a branch of the interdisciplinary computer science. The general goal is to extract information from a set of the data and transform it into a comprehensible model, so that we can further use it. In addition to the original analysis steps, it involves the database and data management, the data preprocessing, the analysis of the model and reasoning, the measurement of the degrees of interests, the analysis of the complexity, and the post processing of the discovery structure, the visualization and the online updating. The data mining is the process of finding the knowledge in the database. The data mining is not the technology produced in the era of the big data. The data mining has already had a set of technical rules before the big data era.

The data mining is a process to extract the information and knowledge that is hidden in it but is potentially useful from the large, incomplete, noisy, fuzzy, and random practical application data, so as to classify, analyze and visualize the worthless data, and form the valuable personal information. In the end, the purpose of predicting the individual behaviors is achieved. Before the big data age, the traditional data mining can not only deal with the smaller databases, but can only deal with the structured data of the characters, numbers, symbols and other unstructured data, such as hypertexts, graphics and images.

In the context of the big data, the distributed data mining, as the new data mining technology, should reasonably divide the data into several small modules and carry out the parallel processing. The characteristic of this heterogeneity can make up for the shortage of the traditional data mining. Compared with the traditional data mining technology, the big data mining technology can deal with more data at a lower cost in a shorter time, and can keep the processed data for a long time. It realizes the deep mining of the extensive information, and can better reflect the credit situations of the information subjects, and can provide the richer credit products and services that conform to the scenes.

The characteristics of the high efficiency and the concealment of the data mining make the traditional means of the supervision invalid. On the one hand, the mining of the big data can analyze and process the individual credit information in a few hours, and even within a few minutes, and the regular inspection and filing and other methods are not able to respond to the flexible process of the information processing in time. On the other hand, the mining of the big data is to integrate and process the scattered and fuzzy data, and the credit extension individuals are unable to know what data is used and how to deal with it. We cannot all prohibit the use of this convenient and concealed technology in the use of the information collection activities, thus hindering the development of the financial science and technology. Therefore, we should clarify the fair boundaries through the legislation, and effectively allocate the legal resources between the personal information protection and the financial innovation, which is a proper way to deal with the orderly conduct of the information collection activities under the big data.

4 Conclusion

For the financial institutions, the credit can help them to fully understand the credit status of the customers, improve the approval and supervision efficiency before, during and after the loan, and reduce the operation costs and the credit risks. For the loan customers, the credit can help the clients with the good credit to shorten their examination and approval time, reduce the costs of borrowing, and enjoy the high-quality financial services. At the same time, the credit activities can also produce the invisible supervision and supervision, help the financial customers to actively perform their contracts and ensure the healthy development of the financial system.

References

1. Zheng, Z.: Applications of the data warehouse technology in the distance education. Sci. Technol. Innov. Herald (03), 256–261 (2014)
2. Xie, X.: Design and implementation of the enterprise energy utilization information management system based on the data warehouse. Resour. Ind. (09), 141–144 (2014)
3. Yan, L.: Research on the role of the financial infrastructure in promoting the economic development – taking China's credit information system as an example. Sci. Technol. Inf. (03), 89–93 (2015)
4. Xu, C.: Social credit system: the mode of China's establishment of the efficient credit information system. Inner Mongolia Sci. Technol. Econ. (17), 58–62 (2015)
5. Lu, Y.: The application of the frame structure of the nuclear emergency decision support system and the fuzzy decision-making method in the evaluation subsystems. Comput. Telecommun. (01), 98–102 (2016)
6. Zhou, Q., Luo, J.: The risk management using limit theory of statistics on extremes on the big data era. J. Comput. Theor. Nanosci. **12**, 6237–6243 (2015). https://doi.org/10.1166/jctn.2015.4661
7. Zhou, Q.: Multi-layer affective computing model based on emotional psychology. Electron. Commer. Res. **18**(1), 109–124 (2018)

Digital Image Encryption Algorithm Based on Affine Cipher

Qi Ke(✉), Qi-Nan Liao, Ai-Quan Li, and Rong Gao

School of Information and Statistics,
Guangxi University of Finance and Economics, Nanning 530003, China
keqikeruzhen@126.com

Abstract. In order to improve the security of cipher code in encryption system and obtain better diffusion effect of the encrypted image, an image encryption algorithm based on affine cipher is presented. The algorithm first scrambles pixels position by using Lorenz chaos sequence, and then introduces the affine cipher of classical cryptograph to diffuse and confuse pixel values. Theoretical analysis and experimental results show that the algorithm has high sensitivity keys, better security of encryption, and excellent diffusion and confusion properties. The algorithm has a better diffusion effect compared with the XOR encryption performance.

Keywords: Lorenz chaos sequence · Affine cipher · Image encryption Classical cryptograph · Pixels diffusion

1 Introduction

At present, the method of digital image encryption is mainly about two aspects: the pixel position displacement and the pixel value transformation. Pixel position displacement method can achieve very good scrambling effect by using chaotic system [1] which produces better random sequences. Many digital image encryption algorithm are based on pixel position displacement method, whereas pixel value transformation method uses bitwise XOR or XNOR operations in many researches [3–5]. But XOR operation encryption has less performance on diffusion and known or chosen plaintext attack [6]. In order to achieve global diffusion which means any change of plaintext pixel can influence the whole encryption pixles. We proposed a novel digital image encryption algorithm based on affine cipher by using Lorenz chaos sequence and affine cipher.

2 Lorenz Chaos Sequence and Affine Cipher

2.1 Lorenz Chaotic System

Lorenz chaotic system [2] is a nonlinear dynamic system, the autonomous dynamic differential equations three variables described below:

J. Abawajy et al. (Eds.): ATCI 2018, AISC 842, pp. 578–585, 2019.
https://doi.org/10.1007/978-3-319-98776-7_63

$$\begin{cases} dx/dt = \sigma(y - x) \\ dy/dt = \gamma x - zy - y \\ dz/dt = xy - bz \end{cases}. \quad (1)$$

When system into chaos state, where σ = 10, r = 28, b = 8/3. For a given initial condition, the Pseudo random chaotic sequence {(Xi, Yi, Zi)|i = 1, 2, 3…} which is very sensitive to initial condition can be produced.

2.2 Affine Cipher

Affine cipher is a kind of classical substitution ciphers in cryptography, originally designed for number of 26 English letters encryption. Its encryption function described as:

$$\mathrm{e}(x) = (ax + b)\mathrm{mod}26 \quad a, b \in Z_{26}. \quad (2)$$

If and only if gcd (a, 26) = 1, ax + b = y mod26 congruence equation has a unique solution for each y.

3 Proposed Encryption Algorithm

In this paper, we used the combination of pixel position displacement and value transformation methods for encryption. First, Lorenz chaotic system is used to generate chaotic sequence to carry out the pixel position displacement and provide the parameters that need to be encrypted. Then using affine cipher to produce pixel value diffusion and confusion, so as to get a good encrypted image.

The pixel size is expressed as: m × n, (i, j) stands for each pixel's location, i, j is the pixel coordinates, i ∈ [1, m], j ∈ [1, n]. Using the formula (1) to produce three groups (X, Y, Z) of Lorenz chaotic sequences which length of m × n: {(Xi, Yi, Zi)| i = 1, 2, 3…m × n}.

3.1 Pixel Position Replacement

The Lorenz chaotic system generating pseudo random sequence to change the plaintext pixels of the original position (i, j) to a new position (inew, jnew), which disrupts the connection between the pixels of the plaintext image and obtains good scrambling effect. Where inew ∈ [1, m], jnew ∈ [1, n].

Choosing any one of three groups (X, Y, Z) of Lorenz chaotic sequences list_permute, then ascending or descending the list_permut sequence to get a new ordered sequence listsorted. where list_permute ∈ {X, Y, Z},{listsorted (i) |i = 1, 2, 3…m × n}. Calculating every list_permute(i) index in the sequence of listsorted which expresses as: {idx(i)|i = 1, 2, 3…m × n}. Scrambling pixel position of encrypted image can be obtained, according to the new location shifted between the idx sequence and plain text pixel's original location (i, j).

The formula for calculating the position of the pixel from the original position (i, j) to the new location (inew, jnew) as follow:

$$\begin{cases} inew = \left\lceil \frac{idx((i-1)n+j)}{n} \right\rceil \\ jnew = idx((i-1)n+j) \bmod n \end{cases}. \tag{3}$$

Where i ∈ [1, m], j ∈ [1, n], when jnew = 0, inew = inew-1, jnew = n.

3.2 Affine Pixel Encryption

This paper used affine cipher for further confusion and diffusion pixel value of scrambling image, changing the distribution of pixel values and weakening the statistical characteristics of the plain image.

The affine function of formula (2) can be used to encrypt the 256 gray levels image and RGB color image by changing the parameter from 26 to 256.

Affine function can be changed to:

$$fc(x) = (ax + b)\text{mod}256 \tag{4}$$

Where a, b ∈ Z256, and gcd(a, 256) = 1.

Affine decryptions function as follow:

$$fd(y) = a^{-1}(y - b)\text{mod}256. \tag{5}$$

Where a − 1 is the inverse of a.

Pixel Value Diffusion Encryption

Using founction_gcd() method to find all the parameters of a to meet gcd(a, 256) = 1 of affine function formula (4). We can get the total solution of a is 128, expressed as A = {aj |j = 1, 2…128}.

list_B is one of three groups (X, Y, Z) of Lorenz chaotic sequences, where list_B ∈ {X, Y, Z}. Integer function and modular operation 256 of list_B:

$$b_j = (\text{round}(list\_B(j) \times 10^k))\text{mod}256. \tag{6}$$

Where k is an integer which stands for shifting k decimal points on the right of the new chaotic sequences, and k ≥ 4. We can get the new chaotic sequences B = {bj | j = 1, 2, …m × n}.

The encryption affine algorithm for the diffusion of the pixel value is expressed as:

$$\begin{cases} cfs(0) = b_j \\ cfs(i) = (a_j \times p(i) + cfs(i-1))\text{mod}256, 1 \leq \text{i} \leq m \times n \end{cases}. \tag{7}$$

Where aj ∈ A, bj ∈ B, aj is the scaling factor of affine function, bj is the initial cipher text of the diffusion encryption. The cfs(0) is the initial cipher text which can be cipherkey; p(i) is the current pixel value by spread; cfs(i) is the encrypted pixel value; cfs(i − 1) is the excipher pixel value.

Confusion Encryption of Pixel Value

The list_confusion is one of three groups (X, Y, Z) of Lorenz chaotic sequences. Where list_confusion ∈ {X, Y, Z}. Assumed that S(i) is the new chaotic sequences coming from integer function and modular operation 256 of list_confusion. Where i = 1, 2... m × n, k ≥ 4.

$$S(i) = (\mathrm{round}(list\_confusion(i) \times 10^{k}))\mathrm{mod}256. \tag{8}$$

Pixel value confusion encryption of affine cipher can be defined as:

$$C(i) = (a_l \times cfs(i) + S(i))\mathrm{mod}256. \tag{9}$$

Where al ∈ A, l = {1, 2...128}.

The encrypted image pixels are stored in C(i), where i = 1, 2 ...m × n, which is the encrypted image in the paper. The decryption algorithm is the inverse of the encryption algorithm. Formula (5) can decrypt original plaintext image by using right cipher key parameters of a and b.

4 Simulation Results and Analysis

4.1 Experimental Results

The simulation experiment environment of this paper is Matlab 7.11 platform. Lorenz system parameter: σ = 10, r = 28, b = 8/3, x0 = −1.25, y0 = 0, z0 = 1.13, h = 0.001, iteration number is 106, x0, y0, z0 are the initial value, h is the integration step; In pixel position replacement step, parameters for pixel value diffusion encryption as: aj = 53, bj = 111, k = 5; Chaotic sequence Z = list_confusion of pixel value confusion encryption step, where al = 99, k = 8. This paper tested many digital images to verify the validity of the algorithm. Figure 1 shows the contrast between original image and encrypted image by the proposed algorithm. From Fig. 1(b) and (d), we can see that he

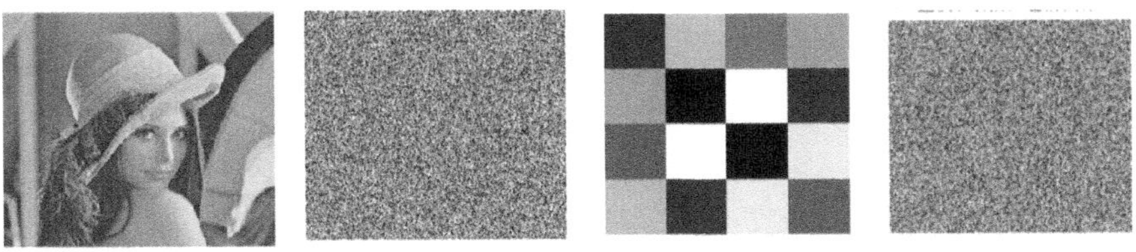

(a) Original Lena (b) Encrypted Lena (c) Original Lena (d) Encrypted Gird

Fig. 1. Encrypted effect by the proposed algorithm

proposed algorithm can achieve ideal encrypted effect and can completely disturb the original image.

4.2 Algorithm Performance Analysis

Analysis of Gray Histogram

In Fig. 2(a) and (c) are the gray histogram of original image of Lena and gird. Figure 2 (b) and (d) are the gray histogram which encrypted by the proposed algorithm. We can see that the proposed algorithm can completely change the histogram of the original image, getting uniform gray value distribution, and the statistical value of the encrypted pixel is consistent, which shows that the encryption algorithm can effectively resist the statistical attack of the pixel values.

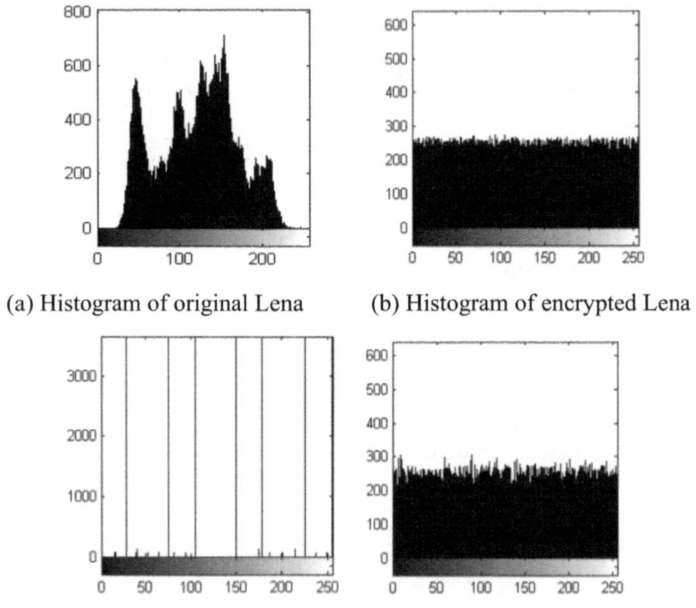

(a) Histogram of original Lena (b) Histogram of encrypted Lena

(c) Histogram of original Gird (d) Histogram of encrypted Gird

Fig. 2. Comparison of gray histograms by the encryption algorithm

Analysis of Pixel Correlation

In this paper, formula [7] is used to analyze the correlation between pixels:

$$\begin{cases} Rxy = Conv(x,y)/\sqrt{D(x)D(y)} \\ Conv(x,y) = \frac{1}{N} \times \sum_{i=1}^{N} (x_i - E(x))(y_i - E(y)) \\ E(x) = \frac{1}{N} \times \sum_{i=1}^{N} x_i \\ D(x) = \frac{1}{N} \times \sum_{i=1}^{N} (x_i - E(x))^2 \end{cases} . \quad (10)$$

Where xi and yi represent the gray value of two adjacent pixels. N is the total number of pixels of the image. We take all the pixels of level, vertical and diagonal both the original and encrypted images to test the correlation between pixels.

Table 1. Correlation between adjacent pixels before and after comparison algorithm encryption

	Lena image	Lena encrypted image	Gird image	Gird encrypted image
Horizontal	0.9260	0.0017	0.9534	0.0024
Vertical	0.9492	0.0022	0.9568	0.0031
Diagonal	0.8996	0.0019	0.9046	0.0028

As we can see from Table 1, the correlation between pixels of the two images is very small, almost zero which means algorithm effecty destroy correlation of the image and the pixel of the image has very good decentrality. Therefore, the algorithm has better resistance of known cipher attacks.

Information Entropy

Information entropy is a measure of the probability distribution of the gray value in the image. When the probability of each gray value is 1/256, the information entropy can achieve the ideal maximum value 8. The formula [7] as follows:

$$H(x) = -\sum_{i=1}^{n} p(x_i)\log_2 p(x_i). \quad (11)$$

Where p (xi) is the probability of gray value xi.

From Table 2, we can see that the information entropy of the image Lena and gird is 7.9993 and 7.9982 respectively, which is very close to the ideal value of gray image information entropy. Experiments show that the encryption algorithm can make the gray value of the encrypted image become uniform distribution.

Table 2. Information entropy of the image before and after encryption

	Lena encrypted image	Gird encrypted image
Information Entropy	7.9993	7.9982

Diffusion Comparison of Affine Encryption and XOR Encryption

From Fig. 3(a) and (e), we can see that the affine encryption algorithm can achieve better diffusion effects. The pixel distribution is uniform and the image information is

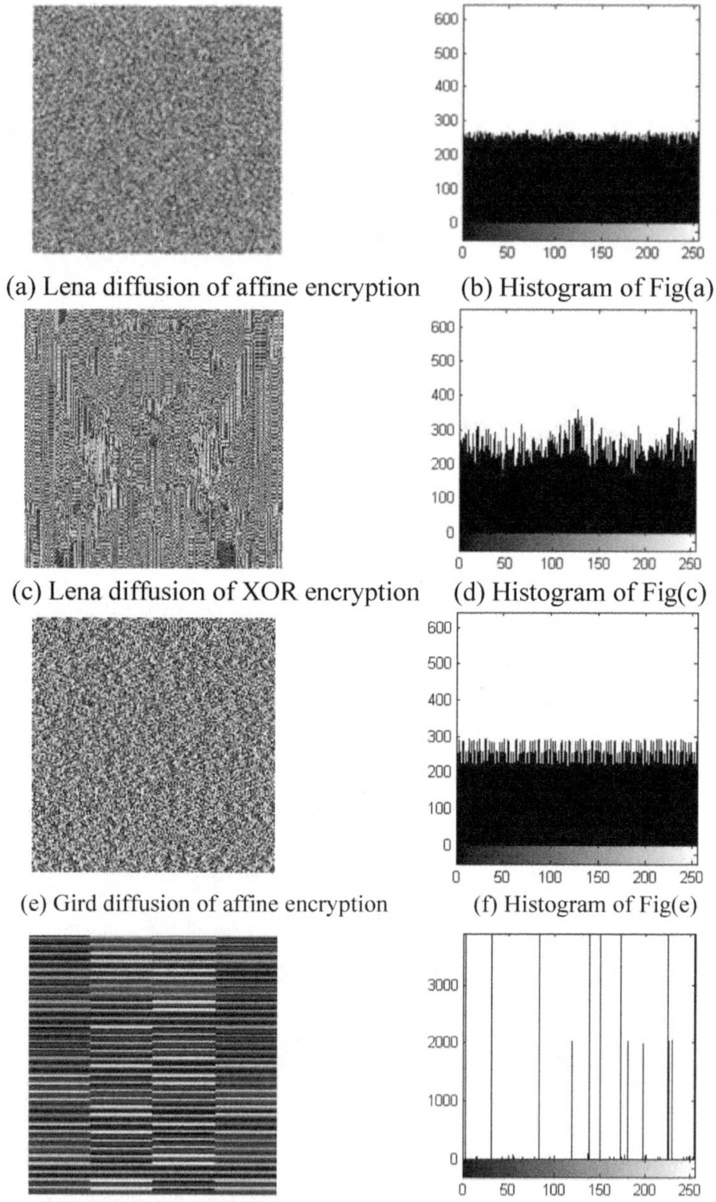

(a) Lena diffusion of affine encryption (b) Histogram of Fig(a)

(c) Lena diffusion of XOR encryption (d) Histogram of Fig(c)

(e) Gird diffusion of affine encryption (f) Histogram of Fig(e)

(g) Gird diffusion of XOR encryption (h) Histogram of Fig(g)

Fig. 3. Diffusion comparison of affine encryption and XOR encryption

completely disturbed without any information of the original image. From Fig. 3 (c) and (g), we can see part of the original image information and contour, the pixel value more simple XOR encryption diffusion effect is more poor. Comparing Fig. 3(b) and (d), (f) and (h) are the comparing histogram after diffusion by affine encryption and XOR encryption respectively. Obviously, statistic characteristics of affine encryption after diffusion is much weakened than XOR encryption. The more singleness of gray value of original image, the more difference of statistic characteristics will be achieved. The proposed algorithm has good performance of diffusion, and the statistical structure of the image is completely spread to the cipher text.

5 Summary

In this paper, we proposed a digital image encryption algorithm based on affine cryptography which uses the Lorenz chaotic sequence and affine cryptography. Using the non-periodic non-convergence characteristics of Lorenz chaotic system to get a better random sequence to conduct pixel position replacement and receive good scrambling effect. Classical affine cryptography is used to encrypt the image can be very good to disturb the initial value of pixels, effectively weaken the statistical characteristics. The diffusion of pixel values using an affine cipher can make plain image statistical structure fully diffused to the cipher text, diffusion effect is far better than the XOR encryption.

References

1. Li, D., Lu, J., Wu, X., et al.: Estimating the ultimate bound and positively invariant set for the lorenz system and a unified chaotic system. J. Math. Anal. Appl. **323**(2), 844–853 (2006)
2. Lian, Q.: Digital image encryption algorithm based on chaos sequence and CBC mode. J. Microelectron. Comput. **3**, 165–169 (2011)
3. Wang, J.: Image encryption algorithm based on improved ergodic matrix and pixel value diffusion. J. Comput. Appl. **32**(16), 1646–1649 (2012)
4. Yan, S., Qian, H., Li, Z.: Symmetric image encryption scheme based on chaotic system. J. Comput. Eng. **34**(14), 155–157 (2008)
5. Zhong, H., Li, J., Lin, Z., Wang, Z.: Image encryption scheme based on hyper-chaotic sequence. J. Appl. Res. Comput. **30**(10), 3110–3113 (2013)
6. Tong, X.-J., Cui, M.-G.: Image encryption with compound chaotic sequence cipher shifting dynamically. J. Image Vis. Comput. **26**(6), 843–850 (2008)
7. Liao, Q.: New color image encryption and decryption algorithm. J. Comput. Eng. **36**(17), 144–146 (2010)
8. Zhou, Q., Liu, R.: Strategy optimization of resource scheduling based on cluster rendering. Clust. Comput. **19**(4), 2109–2117 (2016). https://doi.org/10.1007/s10586-016-0655-9

Research on the Evaluation Methods of the Guidance and Test on the Student Employment Based on the Computer Model

Wenzhao Zhang(✉)

Zhengzhou University of Science and Technology,
Zhengzhou 450064, Henan, China
isermi@yeah.net

Abstract. The development of our society is inseparable from the talents, and the personnel are the important driving force for the economic development, whose distribution mode will affect the growth rate of our social economy. As the most important human resources in our country, the university students can not only rely on the regulation of the market, because the market has its own limitations. We also need the joint efforts of the governments and the relevant colleges and universities, to guide them together. The greatest significance of the employment guidance is to give full scope to the talents, and everybody was made to display his talents fully, so as to achieve the purpose of the rational allocation of the personnel resources, and finally form the harmonious development of the talents and our society.

Keywords: Computer model · Student employment · Guidance and test
Evaluation method · Entrepreneurial guidance

The computer simulation model refers to a powerful conceptual model for observing and predicting the situation of the objective system by using the computer's ability to deal with the information in a large amount and at high speed, setting certain environment in the computers and implementing some laws and rules in the objective system and running at high speed in a computer program.

1 Connotation and Significances of the Employment Guidance and Employment Management for University Students

1.1 The Meaning of the College Students' Employment Guidance

College students' employment guidance belongs to the category of the school education. In general, the meaning of the college students' employment guidance includes the following levels. First, help the students to plan their future rationally according to their own circumstances. Second, the employment guidance for college students is a process. It is a practical activity and requires the extensive participation of the university students. Therefore, the definition of the connotation of the college students' employment guidance should be summarized as follows. The college students'

J. Abawajy et al. (Eds.): ATCI 2018, AISC 842, pp. 586–592, 2019.
https://doi.org/10.1007/978-3-319-98776-7_64

employment guidance is an educational guidance practical activity to help the students to plan their future career effectively and promote the rational allocation of the social personnel resources based on the characteristics of the students and combined with the relevant social employment situation, so as to realize the harmonious development of the humans and the society. The employment management of the college students refers to the systematic induction and generalization of the information and data of the employment of the college students, so as to guide their occupational planning and occupational choice of the college students.

1.2 The Significances of the College Students' Employment Guidance and Their Employment Management

In the current society, the employment guidance for college students is of great significance. It has an important guiding role for the choice of the college students' career path, and the significance of the students' employment guidance is mainly reflected in the following aspects.

First, it helps the college students to set up a correct view of employment. Through the employment guidance of the college students, the college students can understand the social employment situations at the present stage, and choose their own profession with their own professional experience and educational backgrounds. Especially for college students who have never left their campus, they lack knowledge of the overall situation of the employment, and are confused about their occupational planning. Therefore, for the college students, through the employment guidance, through the employment guidance information provided by the teachers, and according to the actual situations of the development, we believe that they will have a clear goal for the choice of their future career, which is of great practical reference values to them.

Second, it is conducive to promoting the reform of the education in colleges and universities and establishing the prestige brands in colleges and universities. The employment guidance of the college students is not only of great significance for the college students themselves, but also of great significance for the development of colleges and universities. In the development of the present social colleges and universities, the criteria for judging the achievements of colleges and universities are generally based on the employment situation of the students. By conducting the employment guidance for the college students, it will be conducive to enhancing the competitiveness of the university students in the process of their job hunting. According to the employment situations of the university students, the schools will rationalize the professional courses to meet the needs of the modern society.

2 Existing Problems in the Employment Guidance in Colleges and Universities

First of all, the employment guidance is restricted by the personal qualities. After the expansion of the enrollment in colleges and universities, the scale of students is enlarged, and the quality of students is affected, and the quality of the graduates has been significantly differentiated, but the employers also pay more attention to the comprehensive qualities of the graduates. Wang Weili believes that the graduates' employability is insufficient. 1. There is a lack of the correct analysis and understanding of the employment situation and they do not have the clear target in their job seeking. 2. Some of the graduates do not attach importance to their self recommendation. 3. They lack confidence, and the initiatives. 4. They will fail because of their self defeats. 5. They have the poor ability to choose the jobs. Gao Baoyan believes that 1. The graduates are timidity, with the lack of the self-confidence. 2. There is a lack of the self-awareness. 3. They have the individualism, without the team consciousness and collective ideas. 4. They are difficult in the interpersonal communication. 5. They are poor in the ability to strain. Therefore, the effects and ways of the employment guidance for the graduates in the process of their employment are closely related to their comprehensive qualities.

Secondly, there are problems in the ways, methods and contents of the employment guidance. In the ways of the guidance, Sun Bin pointed out that, although the employment guidance in colleges and universities in our country is constantly reforming, it is only limited to the shallow services, such as the information release, the policy preaching, the skill training, and the agreement handling and so on. In the functions and contents of the employment guidance, it is disconnected with the vocational orientation of the students, with the lack of the scientific education of the vocational concepts. In the methods and means of the employment guidance, the scope is narrow, and the means are old, and there are few ways. In the courses for the employment guidance, the courses for the employment guidance are monotonous, and the process education is neglected in the specific operation of the employment guidance. 1. Schools do not pay enough attention. 2. The teaching materials are too old, with few actually living examples. 3. In the teaching process, the theory is less connected with the practice, and the teaching method is too monotonous. 4. There is a lack of the professional teachers.

Finally, there are problems of the employment guidance staff. For a long time, our college graduates take up an occupation in the designated units strictly in accordance with the personnel distribution system. After the two-way selection of the employment of college students, the demand for the employment guidance appears. Although the employment guidance work has been put on the agenda, as a result of various reasons, such as funds, the capacity of teachers and the teaching levels and so on, the employment guidance in colleges and universities has attracted more and more attention from leaders and the society at all levels, but the specialization and professionalization of the employment guidance teams are mostly oral.

3 Applications and the Main Practice of the Computer Application Technologies in College Students' Employment Guidance Education

First, applying the computer technologies to guide the college students' employment education is conducive to fully adapting to the development of the times and sticking to the hot topics in the employment of the university students. The employment guidance education has very strong practicality. The important principle to follow is to fully adapt to the development of the times and sticking to the hot issues that are closely related to the social employment. Therefore, in the teaching activities, we should not stay only in the narration of the theories and the preaching of the boring sermon, but should train and guide the students to use their knowledge to analyze and solve the difficult problems that may be encountered in the employment, and to apply what they have learnt to practice. Therefore, in the teaching of the employment guidance, a large number of the case analysis links should be introduced, to increase the teaching information about the employment. The development and applications of the computer multimedia technology provide an effective means to realize this requirement. Using the multimedia teaching means, it is convenient and effective to integrate the hot issues of the employment, the major events and the typical cases of the employment into the teaching activities. In this way, it not only makes the teaching content closely related to the development of the times and reflects the pulse of the times, but also enriches the teaching contents, so that students can understand and grasp the situations, opportunities and challenges facing the employment. At the same time, all kinds of the successful and failed cases in the society can also provide reference and helpful help for college students' employment.

Second, the use of the computer teaching means is conducive to creating a realistic teaching situation, to facilitate the employment guidance education and stimulate the students' learning interests, and facilitate the students to understand and grasp the problems that need to be paid attention to in the employment. The computer multimedia teaching means can synthesize the animations, images, audios and video information and rich variety of the text instructions, and can be good at creating the realistic, vivid and interesting teaching situations. It can make the learners fully mobilize various senses in the process of their learning and accept different forms of the information stimulation in all directions and eliminate the monotonous feeling caused by the mere text contents in the traditional teaching. It can stimulate the interests in learning and arouse the enthusiasms of the learners, so as to improve their learning efficiency. Therefore, the modern computer multimedia teaching methods are of great significance for optimizing the classroom teaching, implementing the employment education and imparting the knowledge of entrepreneurship to students in many ways and multiple channels. In the teaching of the employment guidance, the real and vivid situational teaching can not only stimulate the students' interests in learning, but can also arouse the thinking of the college students more easily, so that the students can really participate in the active learning state. It can be said that the teaching of the computer multimedia technology in the education of the employment guidance of the college students is an important trend. It is the main direction to study, develop and utilize the multimedia technology to make it better to serve the teaching activities.

Third, the use of the computer technologies is conducive to the construction of a wide range of the employment guidance and service platform for the college students, and providing a large number of the employment information and opportunities. At present, our society is already an information society. The employment information from all sides is very important for the college students' employment. The practice of many employment processes shows that one of the difficult problems that puzzle college students' employment is the lack of the effective employment information and the problem of the information asymmetry. To some extent, whether the graduates can get enough employment depends to a large extent on the full mastery of the information on the employment demand. Therefore, in the guidance and education of the college students' employment, it is necessary to make full use of the computer technologies. On the one hand, guide the students to collect and collate the employment information effectively, and on the other hand, guide the students to solve various problems with the computer technologies in different stages, such as the resume, the written examination, the interview and other stages. In particular, as a college counselor, it is necessary to make full use of the computer application technology to do a good job in the information services for the students, guide the students to design a characteristic electronic resume, and give full play to the communication function of the information network.

4 Application and Improvement Measures of the Employment Guidance for College Students

4.1 Strengthen the Methods of the Employment Guidance and Conform to the Development of the Times

In order to improve the employment guidance of the students, it is necessary to conduct a comprehensive and systematic guidance and planning for the employment guidance strategy and adaptation skills of colleges and universities, to adapt the national economic structures to the curriculums for the development of the times, to cater for the trend of the development of the globalization economy. We should create the appropriate opportunities to guide the students to participate in the training and planning of the social employment planning, guide the students' employment and practice closely, guide the students to keep pace with the times in the face of the employment, and promote the development of the reform, the social stability and the improvement of the people's living standards.

4.2 Widen the Employment Guidance Methods and Make Use of the Employment Resources

The departments of universities and social organizations should actively implement the website construction of the employment centers for university students, and create various forms of the recruitment activities and large job-fairs. We should give full play to the publicity of the network, make use of the platform of the network, establish the employment information network, and use the modern network media in order to provide a broad employment platform for the students. The colleges and universities should establish the employment guidance departments, to keep in close contact with the employers in all walks of life, and use the media publicity and education, to keep the dynamic changes of the employment information and expand the resources of the employment.

4.3 Colleges Should Promote the Science and Specialty of the Employment Guidance Team

The employment guidance is designed to the work in a number of fields and multiple levels, and we should carry out the specific management of the specific items in the employment guidance. We should offer the professional and scientific employment guidance, and create the professional employment guidance quality, to guarantee the employment guidance in line with the scientific outlook on development. In terms of the personnel system, the laws and regulations for the employment guidance, the employment training, the professional ethics and other aspects, we must have the scientific professional abilities and qualities. Therefore, we should improve the professional levels and the scientific outlook on development of the employment guidance team, enhance the awareness of the professional services for the students, carry forward the education and the employment as the service tenet, make the full and comprehensive guidance for the students' employment, and make the professional employment guidance smoothly conform to the needs of the market.

4.4 Improve the Teaching Practicality of the Employment Guidance in Colleges and Universities

The employment guidance in colleges and universities should be based on the practice, the practicality and the students' active participation. This requires that the teaching of the employment guidance should not only have the profound theoretical knowledge, but should also have the rich and practical applications, apply the methods of the employment guidance to the practical teaching, and guide the students to actively invest in the society. Through the social practice, we can improve the students' management abilities, communication abilities and abilities to cope with the unexpected accidents.

5 Conclusion

Under the new situation of the development of our higher education and the prominent social employment problems, the importance of the employment guidance in colleges and universities has attracted more and more attention. The reasonable and perfect employment guidance can effectively improve the employability of the college students, arouse the enthusiasms of the college students and enhance their confidence in employment of the college students. In order to realize these functions, it is necessary to carry out the employment guidance work as a systematic project, which cannot be limited to the level of the graduates, but it is necessary to carry out a long-term and systematic guidance to the students of all grades in colleges and universities, so that college students can truly understand their careers, genuinely accept the employment and actively consider their job selection.

References

1. Cheng, X.: On the roles and importance of the occupational guidance in promoting the higher-quality employment. Soc. Sci. J. Univ. Shanxi (11), 117–119 (2012)
2. Shang, L.: On the importance of the employment guidance in promoting the higher-quality employment. Mod. Ind. Econ. Inf. (18), 109–112 (2014)
3. Li, C.: On the roles and importance of the employment guidance in promoting the realization of the higher-quality employment – innovation and applications of the professional guidance in the management of the modern technical education. Mod. Manag. Sci. (05), 232–236 (2015)
4. Ma, X.: On the roles and importance of the employment guidance in promoting the graduates to achieve the higher-quality employment. Soc. Sci. Rev. (New Theor. Version), (19), 126–133 (2015)
5. Liu, X.: On the important roles of the vocational guidance in promoting the higher-quality employment. Market Weekly, (11), 81–83 (2016)

Research on the Ideological and Political Assessment System Based on the Data Visualization

Sunping Guan(✉)

Aba Teachers University, Aba County 624000, Sichuan, China
iweiming@foxmail.com

Abstract. The data visualization is a scientific and technological research on the forms of the visual expression of the data. Among them, the form of the visual representation of the data is defined as a kind of the information extracted in a certain outline form, including various attributes and variables of the corresponding information unit. It is a constantly evolving concept and its boundaries are constantly expanding. In this paper, starting with the understandings of the theoretical connotation of the data visualization, combining with the understanding of the specific background of the construction of the current ideological and political assessment system, the author explores the ideological and political assessment system based on the data visualization.

Keywords: Data visualization · Ideology and politics · Assessment system
Big data evaluation

The system of the ideological and political assessment is to provide a good tool to manage and standardize the ideological and political work for the scientific standard formulation of the ideological and political work, the concrete methods and the reasonable quantification of the results, making the ideological and political work as the entity of the evaluation and assessment, and making the ideological and political workers of the Party and government organs and the enterprises and institutions become the targets of the evaluation and assessment. Through the formulation of the scientific and reasonable working standards and working methods, we can find some ideological problems in the organizations or personnel in time, so as to determine the main objectives of the current ideological work and achieve a definite target, to greatly improve the efficiency of the work and the effects of the work.

1 The Connotation Cognition of the Theories of the Data Visualization

The data visualization mainly refers to the more advanced technical methods, which allow the use of the graphics, the image processing, the computer vision and the user interface to visualize the data by the expression and modeling and displaying the solid, surfaces, attributes and animations. Compared with the special technologies such as the solid modeling, the technical methods of the data visualization are much more extensive.

J. Abawajy et al. (Eds.): ATCI 2018, AISC 842, pp. 593–598, 2019.
https://doi.org/10.1007/978-3-319-98776-7_65

The scientific computing visualization is an important research direction in the current computer science. It is the theory, method and technology of using the computer graphics and the image processing technology to transform the data and calculation results from the scientific computing into graphics or images displayed on the screen to carry out the interactive processing. As a new discipline, the scientific computing visualization has been developing rapidly since its establishment. Many famous universities, research institutes, supercomputer centers and large companies in the United States and Western Europe and Japan have carried out researches on the visualization theories and methods of scientific computing. Many commercialized visual software and hardware products have been introduced, and the visual application market has been expanded year by year. The scientific computing visualization has become a hot topic in the recent international academic conferences.

Since 1990, the international and the regional academic conferences related to visualization have been very frequent. Especially since 1991, the American Institute of Electrical and Electronics has held an annual academic conference on visualization of scientific computing and published the collection of papers. In 1995, a new publication was added to the American IEEE transactions. At the same time, the supercomputer centers, research institutes and WP = 65 companies in the U. S. and German have begun to develop the software systems for the scientific computing visualization and gradually formed the goods, which were introduced to the market, such as the AVS system introduced by Stardent Computer Company in the United States, the apE system developed by the Ohio Supercomputer Center in the United States, and the VIS-A-VIS system developed by the SV Darmstadt FHG-AGD Research Center in Germany and so on. Over the past 10 years, the researches on the theories of visualization in scientific computing have been booming in the world, and have been applied. We also attach great importance to the visualization researches in China. The National Natural Science Foundation has listed the scientific computing as the "85" key funding project, and the National Science and Technology Commission has also listed it as a special project for the basic researches. After entering the 90s, some of the major graphics research centers in China have carried out the research work in succession, and some achievements have been achieved in the visualization application. Since the CAD/CG National Key Laboratory of the Zhejiang University carried out the visualization in 1991, the research of the algorithm is focused on the body illumination model and the volume rendering algorithm, and the combined scattering model and the corresponding volume rendering algorithm are put forward, and are applied to the medical images and the limited original data fields and so on. In the system researches, a general visualization system platform based on the data flow mechanism for the introduction of the control node has been developed, and the matching graphics and image processing library, and the advanced level of the current international similar software has been reached.

The visualization technology is mainly composed of three technologies, namely, the establishment of an empirical model, the mapping of an abstract visual object and the realization of visual objects. According to the classification method of Brodlie, the visualization technologies can be divided into the point data visualization, the scalar field visualization, the vector field visualization and the tensor field visualization technology. The visualization technology is mainly used in the fields of the medicine, the weather forecast, the computational fluid dynamics and the finite element analysis. With the development of the computer technologies, the application of the visualization technologies will be more and more extensive.

2 The Concrete Background Understanding of the Construction of the Ideological and Political Examination System at Present

The evaluation of the ideological and political education is the starting point and the foundation of the ideological and political education in colleges and universities in the new situation. It is an important means to test the effectiveness of the ideological and political education. The evaluation of the ideological and political education includes the evaluation of the main subjects of our education, the evaluation of the educational environment (policies and systems and so on), and the evaluation of the educational media, methods and the ideological status of the educated. Through the mining technology of the big data and the artificial intelligence algorithm, the big data evaluation system of the ideological and political education is established, to monitor and objectively evaluate the actual effect of the ideological and political education, and promote the combination of the ideological and political education with the information technology, to escort the all-round development of people, the cultivation of the builders and successors of the socialist causes and the socialism with Chinese characteristics.

Evaluate the inevitable trend of the times of the subjects. At present, the student groups in colleges and universities are the "post-95s" network "aboriginal people" with the rapid development of the Internet, and the subjects of the ideological and political education has changed greatly. First, the "multiple subjects" coexist. The educational evaluation is no longer the "one corner opera" of the educators in the traditional concept. The students appraise the teaching, the management, and the services, and the teachers appraise the students. They influence each other, restrict each other and exist alternately. Second, the consciousness of the subjects awakens. Whether it is the evaluator or the evaluator, the concept of "I" and "personality" is magnified unprecedentedly in the virtual world of the Internet. The user experience and the user feeling are the key to determining whether the evaluation can be carried out smoothly and can reflect the problems effectively, and the contents, methods and means of the evaluation have been challenged. The third is the demands of the values of the subjects. The value orientation of the traditional evaluation is gradually deconstructed, and a large number of the new pluralistic values have penetrated into all aspects of the subject's behaviors. It is necessary to establish an evaluation system which conforms to the characteristics of the main age, so that the evaluation "feels the ground" and is easy to be operated, rather than becoming a mere formality.

It is the inevitable requirement of the institutional orientation of the evaluation standard. This clear policy orientation must always run through the evaluation system of the ideological and political education. It is the inevitable way of the innovative development of the evaluation methods. Since the 1980s, researchers in the ideological and political education have devoted themselves to the evaluation of their effects, and have formed a comprehensive evaluation system. Since the reform and opening up for more than 30 years, the forms of the educational elements have changed greatly, but the evaluation system of the ideological and political education has not been improved and optimized in time. It lacks the relatively scientific and simple methods, easy for the operation and keeping pace with the times.

In terms of the ways of the evaluation, the form is more than the substance, most of which draw conclusions through questionnaire, statement, field discussion and investigation and so on. The evaluation method is not scientific, and the subjective experience of the educators is still the mainstream. The empirical analysis is over quantified, and the evaluation process lacks the dynamic updating. In the whole evaluation process, the blind "scientific rationality" and the conservative "empiricism" interweave with each other, resulting in the evaluation losing its due functions. The "network survival" has become one of the most important ways of the survival of the teachers and the students in the contemporary colleges and universities. A large number of the behavioral data left by the Internet and the information system terminals have laid an objective foundation for exploring the characteristics of the behaviors and mastering the laws of our education. On the basis of the analysis of the mass data and combining with the educational standards and requirements, we can construct the evaluation system which meets the requirements and characteristics of the educational era and respects the characteristics of the subjective behaviors, to make the evaluation more accurate, objective, dynamic and long-term. It is an inevitable way for the innovation of the theories and ways of the educational evaluation brought by the big data.

3 The Ideological and Political Assessment System Based on the Data Visualization

The data visualization is mainly based on the graphical means to communicate and convey the information clearly and effectively. However, this does not mean that the data visualization must be boring because of its functional use, or that it is extremely complex to look colorful. In order to communicate the concepts and thoughts effectively, the aesthetic forms and functions need to go hand in hand, through the intuitively conveying the key aspects and features, so as to realize the deep insight of the rather sparse but complex data sets. However, designers often do not have a good grasp of the balance between the design and the function, thus creating a nonrealistic form of the data visualization that cannot achieve its main purposes, that is, to communicate and convey the information.

The data visualization is closely related to the information graphics, the information visualization, the scientific visualization and the statistical graphics. At present, the data visualization is a very active and critical aspect in the fields of researches, teaching and development. The term "data visualization" realizes the unification of the mature scientific visualization and the younger information visualization.

Adhere to the establishment of the high moral values and the cultivation of people, and clarify the evaluation criterion. Construct the evaluation system with the educational subjects, the educational environment (policies and systems and so on), the education media and the educational methods, which should be tested by the individual and the whole ideological states of the college students. We should pay attention to the contents in the following three aspects. The first is whether the evaluation indexes of the ideological and political education are reasonable and can play their roles. Secondly, under the impact of the multiple values, under the influence of various social thoughts and surrounded by the massive information, the evaluation of the ideological and political education should always adhere to the standard of establishing the high moral values and cultivating the people. Thirdly, the big data evaluation of the ideological and political education should follow the laws of our education, respect the behavioral characteristics and laws of the main objects and subjects of our education, and follow the laws of the human development. And the guidance of the setting of the evaluation indexes should "be centered on the students, take care of the students and serve the students", and lay the scientific ideological basis and the moral standards for the growth and success of the students.

Set up the big data model and optimize the evaluation indexes. With the help of the social resources of the big data and the data and information system for the "smart campus", through the analysis and excavation of the big data technology, we can establish the evaluation index model for the big data ideological and political education with the individual and the whole ideological situations of the college students as the dependent variable and the various factors in the ideological and political educational process as the independent variables. Once the model is built, on the one hand, the reliability and the validity of the evaluation indexes of the current education can be detected effectively. On the other hand, we can further excavate the potential and undeveloped related factors that have a real impact on the ideological and political education, and further optimize the current monitoring and evaluation index system.

Because of the long-term, objective and multi-angle observation and monitoring of the standard, this kind of the index can reflect the effects of the educational factors on the effect of our education. We can carry out the scientific qualitative and quantitative comprehensive study of the scope of the indexes, the degree of their influence, and the weight and the ratio between the indexes, which will be independent of the subjective will of people. The scientific evaluation index system can reflect the essential and epochal characteristics of the evaluation subjects. We should adhere to the evaluation orientation of establishing the high moral values and cultivating the people, and do not stay in the stage of the social stereotype, so that the evaluation will keep pace with the times.

The evaluation system is constructed by using the data mining technology. The construction of the big data evaluation system of the ideological and political education is the complex and dynamic system engineering. It needs the theoretical support of many disciplines, such as psychology, sociology, behavioral science, and ideological and political education and so on. It needs the professional technical support of the big data application, and the extensive participation of the evaluation subjects and objects. The big data evaluation system of the ideological and political education is not a closed system once and for all, but an open and constantly improving dynamic cycling system. The construction of its algorithms and models will also conform to the development of the science and technology in the times and will be constantly revised and promoted. And the results of the scientific calculation are fed back to the evaluators, and the effective evaluation and guidance of the educational process will be made in real time, so as to enhance the pertinence and effectiveness of the ideological and political education. With the popularization and application of the big data technology in colleges and universities, we can establish the data network connecting countries, our society and inter-schools in the future, and improve and guarantee the scientific development of the big data evaluation system of the ideological and political education from the aspects of the data interaction and the privacy protection and so on.

4 Conclusion

The data visualization has always been an evolving concept, and its boundaries are constantly expanding. Therefore, it is best to define it broadly. The data visualization refers to the more advanced technical methods that allow the use of the graphics, the image processing, the computer vision, and the user interfaces to visualize the data through the expression and modeling, and displaying the solid, the surface, the attributes and the animation. Compared with the special technologies, such as the solid modeling, the technical methods of the data visualization are much more extensive.

References

1. Zhang, L.: Influence of the multiculturalism on the contemporary college students and the countermeasures. Educ. Forum (16), 99–103 (2014)
2. Meng, X.: Applications of the positive psychology in improving college students' ideological and political educational methods. J. Nanjing Univ. Aeronaut. Astronaut. (11), 212–213 (2015)
3. Sun, M.: The influence of the youth subcultures on the college students' ideological and political education and its countermeasures. J. Shenyang Norm. Univ. (02), 51–53 (2016)
4. Wang, L.: Research on the operational mode of the mobile advertising alliance in the cloud computing environment. News Res. Guide (17), 71–74 (2015)
5. Hou, S.: The development mode and strategies of the digital advertising industry in the era of the big data. J. Cap. Univ. Econ. Bus. (02), 261–263 (2016)
6. Zhou, Q., Luo, J.: The risk management using limit theory of statistics on extremes on the big data era. J. Comput. Theor. Nanosci. **12**, 6237–6243 (2015). https://doi.org/10.1166/jctn.2015.4661
7. Xie, J., Luo, J., Zhou, Q.: Data mining based quality analysis on informants involved applied research. Clust. Comput. **19**(4), 1885–1893 (2016). https://doi.org/10.1007/s10586-016-0657-7

Research on the Interaction Mechanism Between the Ideological and Political Education and the Computer Network Under the New Media Environment

Jiasheng Ma(✉)

Zhengzhou University of Science and Technology,
Zhengzhou 450064, Henan, China
zqybws@126.com

Abstract. Educators can take the lead in establishing the special forum on the ideological and political education, or establishing the corresponding special groups through the micro-blog and We-Chat and other new ways of communication, to promote the equal communication between the students and the workers. The introduction of the new media teaching methods in the computer teaching is also to adapt to the needs of the cultivation of the urban and applied high-quality talents. In this paper, starting with the analysis of the connotation and characteristics of the new media environment and combining the cognition of the problems of the ideological and political educational activities under the new media environment, the author analyzes the interaction mechanism of the ideological and political educational activities and the computer network under the new media environment.

Keywords: New media environment · Ideological and political education
Computer network · Interaction mechanism

The use of the new media needs certain computer basis, so there are many older educators in the college teaching who do not have the knowledge of this aspect. Although some educators know the use of the modern new media, they do not know how to effectively combine the ideological and political education with the new media. These have affected the implementation of the ideological and political education under the background of the new media in colleges and universities.

1 The Analysis of the Connotation and Characteristics of the New Media Environment

The attention of the European and American scholars to the new media is mainly focused on the two aspects of the historical evolution of the new media and the related applications of the new media, of which the research on the new media is especially widespread. Roger·D·Wimmer and Joseph·R·Dominick believe that the media research has four stages of development. The first phase focuses on the themes of the media

J. Abawajy et al. (Eds.): ATCI 2018, AISC 842, pp. 599–605, 2019.
https://doi.org/10.1007/978-3-319-98776-7_66

itself, and the second phase deals with the use of the media, such as the user characteristics, the user perception, the user attitudes, and the usage patterns. The third phase involves the media effects, such as its influence on cultures, the society, our daily life, the political systems and the public [1], and the last stage is the study of the media improvement, including the future orientation, the updated applications, the technological development, and the establishment of theories and models and so on.

The European and American educational fields also attach great importance to the application of the new media in their education. Bush, the former president of the United States, also referred to the informatization of education to the height of "determining the status and the competitiveness of the United States in the world in the future". The government of the United States has always believed that the information technology should be a powerful tool for schools to make efforts to improve the students' academic achievements, and they are calling on all the teachers and the students to use the modern multimedia computers in the classrooms, and every classroom is directly connected to the information highway [2, 3].

With the advanced development of the modernization of the educational technologies, with the rapid development of the new media technologies with the computers as the core, the applications of the computer aided teaching tools and the auxiliary teaching media have become more and more extensive and mature. And the teaching with the combination of various new media can stimulate a variety of our organs, such as the human vision and hearing, which is the best way to improve the teaching efficiency [4, 5]. In particular, in the course of our teaching, the use of various interactive teaching websites, computer network resources and computer network experimental equipment can further increase the students' interests in their learning, so as to achieve a good teaching effect.

2 Cognition of the Problems in the Ideological and Political Educational Activities in the New Media Environment

Under the background of the new media, the ideological and political education in colleges and universities has achieved remarkable results. The ideological and ideological ideas of the educators have been changed, and the ways of our education and the contents of our education have increased. However, at present, there are still some challenges and problems.

2.1 Backward Ideas and a Lack of Understandings of the New Media

The use of the new media in the ideological and political education is short, and most people lack understandings of the new media. The application of the new media in the ideological and political education in colleges and universities is regarded as the teaching with the multimedia tools, and even the use of the multimedia in the ideological and political education in colleges and universities is also considered as a kind of arrangement. This backward ideology leads to the lack of emphasis on the application of the new media by the ideological and political education workers in colleges and universities, which, to some extent, hinders the development of the new media in the ideological and political education in colleges and universities.

2.2 Unsound Infrastructure

China has a vast territory and a complex environment. At present, China has not completed the comprehensive construction of the national information highway, and the distribution of resources between the urban and the rural areas is not equal. For example, there is a certain gap between the central and the western regions in terms of the economy and education and so on. The construction of the telecommunications and the network facilities in these areas is not in place, which, to a certain extent, hinders the application of the new media in the political education in colleges and universities. In addition, the network has been popularized in our universities, but there are still some problems in the application of the multimedia in universities. It is mainly as follows. The overall infrastructure of some colleges and universities is not in place as a whole and the future planning is not in place, and the later input cannot keep up and the maintenance is not in place, which leads to the backwardness of the multimedia equipment in colleges and universities. The multimedia equipment cannot be used at all. This has hindered the applications of the multimedia in the ideological and political education of colleges and universities to a certain extent.

2.3 The Ideological and Political Educators Are Backward in Their Concepts, with Improper Operations

The traditional concepts and ideas cannot be changed overnight. At present, the traditional concepts of the ideological and political educators in colleges and universities are deep-rooted. Many ideological and political educational workers have become accustomed to the traditional teaching mode of the cramming education. They excludes the new mode of the ideological and political education under the new media, such as the computer network, the mobile network and the television network and so on, which affects the smooth development of the ideological and political education in the background of the new media. For example, colleges and universities have basically been equipped with the multimedia assisted instruction, but some teachers use this new media only as a photocopier or a projector. This new mode of the traditional teaching method has improved the enthusiasms of the college students to a certain extent, but cannot fundamentally play the role of the new media.

2.4 The Difficulty in Controlling the New Media

There are many advantages in the use of the new media, but inevitably there are also some shortcomings. For example, the new media has the features of openness, sharing, randomness and anonymity, which easily makes the information of the new media disorderly and facilitates the infiltration of the negative thoughts, and has caused great difficulty to the ideological and political education of college students. In today's era, the speed of the economic globalization is accelerating, and the world is closely linked together. Some foreign negative thoughts and wrong values have gradually gone deep into our country. The process of the ideological and political education is mainly the internalization, externalization and feedback, while the new media has the features of the two-way, the diversity, and the difficulty in controlling, which, to a certain extent, has increased the difficulty in the feedback in the process of the ideological and political education, and is not conducive to the workers' timely grasp of the students' ideological trends and the smooth implementation of the ideological and political education in colleges and universities.

3 Interaction Mechanisms Between the Ideological and Political Educational Activities and the Computer Networks Under the New Media Environment

How to innovate the moral education of the college students in the modern situation of the informatization and the globalization with the new media has become an important topic. Using the advantages of the traditional media, we should interpret the essence of the phenomena, analyze the intension of the events deeply, present the more profound and more regular information to the students, and effectively improve the pertinence and effectiveness of the ideological and political education work, to realize the whole media pattern of the ideological and political education.

3.1 Inter-Kingdom Protoplast Fusion and Collaborative Innovation

To achieve the integration of the new media and the cross-border integration and collaborative innovation between the new media and the traditional media is the only way to improve the effectiveness of the ideological and political education in colleges and universities. Colleges and universities should establish the new media operation centers, and integrate the official micro-blog, We-Chat, micro vision, campus network, forum and the micro-blog and We-Chat of all departments to a platform, to facilitate the students to obtain the information and convenient services. First, circle the groups of the teachers and the students. At present, the new media platforms in colleges and universities are separately managed by different departments, which are scattered in the layout and poor in the influence, causing difficulties for teachers and students. Through the integration of the new media, teachers and students are firmly attracted to the new media platform, and then the topics are released through the topic setting function. Through the messages and discussions of the teachers and the students, the related issues will be clearer and clearer, so that the teachers and the students can receive the information and ideas in time and quickly to be transmitted by the schools. The website managers will set top the messages and posts that can reflect the mainstream values are clear and lively in the language expression, and recommend them as the wonderful speeches, to improve the sense of belongings of the teachers and the students, to meet the desire of the teachers and the students, and to consolidate their loyalty to the new media in the campus.

3.2 The New Media and the Traditional Media Will Integrate with Each Other, and Act in Cooperation with Each Other, and Complement Each Other, to Realize the All Media Structure of the Ideological and Political Education

The new media has many advantages, such as the fast dissemination, the large audience, and the strong ability of the topic setting and the strong influence of the public opinions. In the short term, we can focus on a topic and quickly create the social hot spots. But the new media is not universal. The new media information and discussion are often only the subjective impression of the individuals, with the lack of the rational thinking, and the lack of depth, which is not professional and easy to misread the information. Therefore, while fully paying attention to the new media, we must not ignore the power of the traditional media. The traditional media, such as the school newspaper, the publicity board and the electronic display screens, should be spread out in the traditional media after combing the heated and highly concerned questions of the new media in time, and make professional and authoritative answers to the questions, puzzles and mistakes in the new media in time, to clarify the misreading and guide the public opinions.

3.3 Build the New Type of the Team of the Ideological and Political Educational Work

Master the new media technology and establish the new media thinking. The team engaged in the ideological and political education in colleges and universities is basically graduated from the liberal arts and other related majors, with the lack of the theoretical and technological knowledge of the new media. We should take various measures, such as training, learning and introducing talents, to improve the media literacy of the team of the ideological and political educational work and the levels of using the new media technology. Build a team, deepen the student's new media world, gain the position of the opinion leaders, and persuade and guide the students to accept the mainstream values. The new media is an open platform, and anyone can express any opinion or idea. Some views seem correct, but they are very confusing and provocative, but they are very wrong. College students are limited by their theoretical levels and life experiences. They are not always able to distinguish the truth from the false, understand the right and the wrong, and are easy to be used by others. Therefore, on the new media platform, we must have the participation of the ideological and political education workers with a high political awareness and a solid theoretical level. They discuss with the students on an equal basis, discover the deviations in time, use their profound knowledge, clear logic and correct views to criticize and expose the errors, remove the false and save the truth, and actively guide the college students to set up the correct values.

3.4 Enrich the Resources of the Ideological and Political Education Activities Based on the Computer Network Interaction Mechanism

We should integrate the resources of the ideological and political education. Strengthen the teaching of the course of the "Situations and Policies", start the meeting every semester before the class, and organize the teachers to carry out the research activities of the rural society, the special topic discussion and the collective lesson preparation, so as to effectively guide the hot spots and respond to the concerns of the students. Build a large platform for our education, play the roles of the large and open network courses in the ideological and political education, and promote the integration of the students' interactive community with the new media, such as the micro-blog, We-Chat, Renren network, the mobile newspaper and other new media. We should strengthen the construction of the professional teams. Set up the professional team of the teachers of the ideological and political theories, the school Party and government cadres, the cadres of the Communist Youth League, the propaganda cadres and the student cadres as the main subjects, so that the ideological and political working system is formed with the combination inside and outside the schools and the online with the offline. The leaders of the mobile Internet operation companies should be invited as the honorary tutors and the school counselors to participate in the ideological and political work. We should strengthen the training of the professional skills of the new media team, set up the courses of the new media literacy education, cultivate a keen awareness of the media information and the network communication technologies, and improve the ability to guide the public opinions on the Internet.

3.5 Interaction Mechanisms Between the Ideological and Political Education Activities and the Computer Networks Under the New Media Environment

With the advent of the era of the educational informatization, the ideological and political education workers in colleges and universities should carry forward the main melody of the times through the network, eulogize the true, the good and the beauty of the times, lead and spread the positive energy of the network cultures, and make the educated people enlightened and infected. The correct guidance of the public opinions is the first point to carry out the ideological and political education in the network. The education of the network political thoughts in colleges and universities should take the Marx's theories as the guiding principle, strengthen the depth and the breadth of the positive propaganda, strengthen the education of the patriotism, strengthen the national self-confidence and pride, and cultivate the sense of the national mission and the sense of the national responsibility of college students. Vigorously carry forward the national spirits and the advanced cultures, and give full play to the positive roles of the Internet in promoting the correct world outlook, outlook on life and values. Convincing the scientific connotation of the theories of the Marx's doctrine and strengthening the infectious force of the ideological and political education cannot be separated from the vigorous propaganda of the network. Only when the socialist core values are proclaimed in a clear-cut stand, and the core value system of socialism is fundamental and the network position is occupied with a strong positive voice can we ensure correct direction of our the network ideological and political education, form a strong public opinions on the Internet, and firmly grasp the correct directions of the public opinions.

4 Conclusion

To set up the new media thinking, it is required that the ideological and political educators should lay down their teachers' identity, participate in the students' new media experience on an equal position, respond quickly to the hot issues of the students' attention, and give feedback and guidance in time. The tedious and monotonous expression should be transformed into the lively and short and vigorous expression, and we should learn to use the interactive means to serve the students pertinently. We should fully respect the students' individuality. Regardless of the language expression or the topic selection, they should be close to the students' needs. According to the new media rules, maintain the official platform. We should strengthen the maintenance of the school official micro-blog and We-Chat when the students focus on the Internet in the evening and at noon, and use the rich forms of the information and the contents to attract the students.

References

1. Zhao, Z.: Research on the innovation of the university students' moral education in the new media perspective. J. Xinjiang Norm. Univ. **34**(2), 85–89 (2013)
2. Guo, D.: Challenges and countermeasures of the new media information on college students' ideological and political education. News Front **20**(16), 78–83 (2013)
3. Lai, G.: Research on the ideological and political education in colleges and universities under the background of the new media. China Educ. Netw. **23**(8), 153–157 (2014)
4. Chen, W.: Research on the innovation of the ideological and political education in colleges and universities under the new media environment. Mod. Commun. (J. Commun. Univ. China) **35**(9), 122–128 (2015)
5. Zhang, J.: Research on the ideological and political work to deal with the civic participation of the micro-blog. Inf. Secur. Res. **12**(1), 68–72 (2016)

Research on the Key Technology of the Computer Graphics and Image Processing

Weina Zhang(✉)

College of Computer and Information Engineering, Chifeng University, Chifeng 024000, Inner Mongolia, China
iccc2013@yeah.net

Abstract. With the continuous development of science and technologies, the computer technology has also made full progress. The graphics and image processing technology is an important part of the computer technology, and it is also widely used in the modern applied science. The computer graphics and image processing technology has brought many conveniences to people's daily life, work and study, and it also has a strong impact on people's vision. This paper describes the computer graphics and the computer graphics and image processing technology, and analyzes the differences and contacts between the computer graphics and image techniques, and discusses the practice and application of the key technologies of the computer graphics and image processing, to provide some ideas for how to promote the researches on the orderly and healthy development of the computer graphics and image processing technology.

Keywords: Computer graphics · Image processing · Key technology
Technology intelligentization

Since the twenty-first century, computers have been developing rapidly. The technology of the computer graphics and image development has gone through 30 years of development, and has become one of the most active technologies. The computer graphics and image technology is also widely used in all aspects of our life, from the packaging design and the picture processing, to the medical drawings and the engineering design. These applications not only make our life more colorful, but also make our vision opener.

1 Overview of the Computer Graphics

With the rapid development and progress of the computer technology and the graphics and image processing technology, we also gradually break through the limitation of the static images, and gradually developed the animation technologies. In addition, in the field of the modern commercial art, designers can also use the computer graphics and image processing technology to complete the related art design, such as the applications of 3DMAX and ALINS and other software. Among them, the computer graphics and image processing technology has some functions that the ordinary artists can't reach.

J. Abawajy et al. (Eds.): ATCI 2018, AISC 842, pp. 606–611, 2019.
https://doi.org/10.1007/978-3-319-98776-7_67

1.1 The Main Contents of the Computer Graphics

The research contents of the computer graphics include many aspects, including the graphic hardware, the graphic interaction technology, the curved surface modeling, the virtual implementation and the physical modeling. This is to use the mathematical algorithm to transform the corresponding two-dimensional and three-dimensional graphics into the computer. The main purpose to set up the subject of the computer graphics is to make the images that the computers convert out more real, and to make the graphics that the computers transform with the stronger sense of reality, it is necessary to establish the geometric representation of the scene of the graphic description, to calculate the effect of the virtual light sources, textures and material properties. Therefore, the connection between the computer graphics and the geometric design is very close. In the computer graphics, the main research contents include the curve and surface modeling technology in the geometric scenes and the solid modeling technology. The graphics transformed by the computers usually need to process the graphics once again, so the computer graphics and the corresponding graphics and image processing technologies need to be closely linked, so that we can produce the better sense of reality of the graphics.

1.2 Basic Concepts of the Graphics and Image Processing Technology

The graphics and image processing technology mainly carries out the modification, storage and consummation of the geometric data and the geometric models in the mathematical description with the use of the computer technologies. Its main processing techniques include the geometric transformation, the image digitalization, the modeling design, the elimination of the hidden lines and surfaces, and various color designs. In the graphics and image processing technology, the computer software usually needed to use includes the general design-aided software CAD, the manufacturing software CAM, and the education assisted software CAI and so on, and also we need the applications of the knowledge and technical fields of the computer art design, the corresponding simulation, the animation and the virtual realization. In the designing and rendering process of a variety of the graphics, the graphic system must have the capabilities of input, output, storage, dialogue and computing. And therefore, the functions of the graphics system mainly include: the function to input various commands and the geometric parameters, the output function to make the graphics maintain the displaying status and be preserved permanently, the storage function to store all the geometry data and to carry out the maintenance and detection of the data, the dialogue function to carry out the man-machine communication, and a variety of the calculation and analysis functions in the design process.

1.3 Systems and Functions of the Computer Graphics

The computer graphics system is mainly composed of the corresponding hardware devices and the graphics and image software. The hardware devices in the computer graphics system mainly include the graphic input, the output devices and the corresponding graphic processing devices. The graphics processing equipment is the core of the hardware equipment, and the main function of the graphics processing equipment is to store the processed graphics, and also to complete the complex calculation of the functions in the process of the graphics processing. Therefore, the graphics processing devices can also reduce the burden of the CPU of the system, and eventually will be able to improve the displaying quality and speed of the graphic system. The hardware input devices in the graphical system are mainly the keyboard and the mouse, and all kinds of graphics are plotted by executing the corresponding commands through the mouse and the keyboard. For example, in the design aided software CAD, the two hardware devices are mainly dependent. With the continuous development of the computer technologies, now the hardware equipment also includes the space ball, the data glove, the light pen and the touch screen and so on. The graphics output equipment in the hardware equipment of the graphics system mainly refers to the displaying system and the drawing system in the system.

2 The Applications of the Computer Graphics and Image Design in Our Life

Because the computer graphics and image processing is convenient, flexible, effective and accurate, there are many applications of the computer graphics and image design in our life, such as the web design, the graphic design, the packaging design, the interior design, and the software design and so on. For example, the PS software, which is common in life, can be used to deal with photos, including adding texts, cutting, or local processing of the photos.

There are applications of the computer graphics in the engineering drawing, which can be realized through the CAD software. It can make things specific, and can express the graphics with the stereo forms, and gradually expand them from points to lines, from lines to surfaces and from surfaces to bodies, to achieve the overall structures of the experimental model. The computer graphic design can be used in the medicine, and the computer graphic design can transform the image information in the medicine into the 3D views, to make the diagnosis of the doctors greatly improved.

The computer graphics and image processing technology can be applied in the packaging design. The computer graphic technology can be used to carry out the layout, size, and color adjustment and so on, to make the package closer to people's life. The computer graphic design can be applied in the animation design. Animations are stitched up by pictures, and the computer graphic technology can make the pictures into animations. So the dynamic pictures become more realistic, and also can reduce the workload, so that the animations can show more real rules and can be easier for people to accept. Nowadays, the computer graphics and image processing has been distributed in all aspects of our production and life. This technology has played a great role, giving people a new perspective and multi-dimensional space to see the world, and making our life more colorful.

3 Operating Environment

3.1 Hardware and Software Configuration

The hardware configuration can be divided into two major categories, the workstations and the microcomputers. In terms of the software configuration, the workstation software includes the TDI and the Alias, which are the specialized software for the production of the graphics in the workstation, with different functions. The microcomputer software includes various versions of the 3DS, and the 3DS is the production software of the graphics used in the micro computers, which has been widely used in our country. PS is the professional software that specializes in image processing. Win morph is the two-dimensional image processing software that turns one image into another.

3.2 Functional Comparison

The workstations, in the aspects of the CPU speed, the memory capacity and the graphic display and so on, have the superiority incomparable of the microcomputers. The image texture that the workstation generated (refers to the use of different means of expression, to show the characteristics of different objects) is better than that of the micro machines, which is because the calculation methods of the image processing and the number of the calculation parameters of the two systems are very different. TDI refers to the interface layer between the protocols of various transport layers (such as SPX, TCP, etc.) in the Windows NT operating system and the receiving software (or the redirected software interface). For every pixel of the images, TDI uses the method of subdividing every pixel into many small blocks to calculate the colors separately. Therefore, the computation of a 720 * 576 standard image is far more than that of the microcomputer. The overall texture of the image is better than the microcomputer.

From the aspect of the modeling, 3DS has no function of the curved surface modeling, which is not very convenient for some complex curves and surfaces. And TDI provides a powerful function of the curve and surface modeling. From the user interface, the producers who used 3DS will feel that the material on the object is difficult. We have to mix up the material in a model, but the effect of the material on the object is not seen. Only after the image is generated can we see the real effect. The IPR module in TDI provides the method of regulating the materials of objects and light sources directly in the final image, so it has the fastest visual feedback, and the result is the same as the effect of the final generation. Compared with the simulation effect of TDI on the objects in the natural world, 3DS lacks a sense of authenticity. For example, 3DS is just a simulation of the real things. In TDI, AMAP absorbed the research results of the botanists on the plants in nature, which is shown through the computer, so the topology of the plants generated by it is exactly the same as that of the plants in nature, so TDI simulation results are quite good and vivid.

4 The Development Prospect of the Computer Graphics and Image Technology

The hardware and the software developed according to the computer graphics and image technology have a great development prospect in our daily production and life. This requires that we should establish a more complete and scientific management system and a standardized system, to enable the computer graphics and image processing technology to benefit the mankind and make people's lives colorful.

4.1 Applications in Industries

The industrial automation is the core of the industrial development, and computer technology is the technical support of the industrial automation, and the computer image processing technology is one of the inevitable means that will be used. In the industrial production, the computer image processing technology aims mainly at the identification of the parts and the components and the production line. The control system will receive the structures of the industrial production and the transmission of the order of the control, and realize its automatic production. In the workshop of the auto parts, there are systems for the automatic identification and detection of the parts. They mainly use the computer image processing technology, which can save the time and the efficiency and can improve the production quality.

4.2 Applications in Agriculture

In agriculture, the computer image processing technology is mainly used in the process of picking the agricultural products and the production of the agricultural products. The machine operates automatically and widely. The main thing is that we can reduce the human resources and the pressure of the farmers. The most important thing is that the processing efficiency of the agricultural products has been improved. For example, while picking the tree-ear, we can carry out the picking through the automatic system. We can carry out the picking with the use of the robot's visual system, which can not only improve the picking efficiency, but can also reduce the damage degree of the tree-ear and improve the quality of the tree-ear. At present, the foreign robots have already been used in the agricultural production, such as the fruit picking and the vegetable extraction and so on. The development of them China in this respect is still relatively slow compared with the developed countries. It is also necessary to increase the researches on the computer image processing technology in the agriculture to improve the rapid development of our agriculture in China.

4.3 Applications in Traffic

There are many applications in traffic, such as the cameras, which can be seen everywhere in traffic roads, with the main use of the computer graphics processing technology. Using the video surveillance system, we can record and detect the illegal driving behaviors, which is helpful for the traffic policemen to investigate and punish the drivers who violate the traffic regulations. The record of the real-time traffic is beneficial to create a good traffic environment and to ensure the safe and orderly operation of the traffic. The public security departments mainly use the surveillance videos to supervise the traffic behaviors and the driving behaviors, and urge people to obey the traffic rules to ensure the driving safety.

4.4 Applications in the Textile Process

In the textile industry, it is mainly reflected in the inspection of the textile materials, which is also the application of the computer image processing technology. It is mainly to evaluate the finished textiles, which has obvious advantages compared with the previous manual inspection. The inspection of the textiles is more objective, rigorous and fair, which can promote the development of the textile industry.

5 Conclusion

With the development of the computer graphics and the graphics and image processing technology, they have been applied to our daily life, and we need to apply this advanced computer technology in many fields. At the same time, the computer graphics and image processing technology has also played a great role in the development of various fields. This technology can create a new visual effect and give full play to people's design abilities and creativity. Therefore, it will be very important to integrate the computer graphics and the image processing technology into all fields, and promote the development of the computer graphics and the graphics and image processing technology.

References

1. Li, J.: Computer graphics and image processing technology. J. Beijing Univ. Aeronaut. Astronaut. (13), 28–31 (2013)
2. Zhou, Q.: The application of the computer technology in graphic and image processing and researches on the key technologies. Comput. Knowl. Technol. (15), 87–90 (2013)
3. Wang, S.: Research and practice of the vector-based temporal graphics and image processing technology. Electron. Test (06), 77–79 (2014)
4. Liu, M.: Researches on the applications of the computer graphics and image technology in the spiral bevel gear detection. Heilongjiang Sci. Technol. Inf. (13), 79–82 (2015)
5. Zhou, Q.: Research on heterogeneous data integration model of group enterprise based on cluster computing. Clust. Comput. **19**(3), 1275–1282 (2016). https://doi.org/10.1007/s10586-016-0580-y
6. Xia, H., Liu, G.: Research on the image stitching technology based on the point features. Shandong Ind. Technol. (01), 36–40 (2016)

Research on the Security Management Information System Based on the Data Mining

Xiaodong Meng(✉)

School of Computing, Hulunbuir College, Hulunbuir 021000, Inner Mongolia, China
yzulingchen@yeah.net

Abstract. With the continuous advancement and development of the data mining technology, the level of the information management has been kept in sync with the pace of the times. The data mining is a network, informatized and intelligent technology with the highly integrated information technology and the depth integration of the information applications. It is the independent information management system with the intelligent operations ultimately formed based on the temporal and special information, relying on the Internet of things, the cloud computing, the sensing devices and other information technologies, through their mutual optimization and combination. Based on this, this paper analyzes the application of the data mining technology in the security management information system.

Keywords: Data mining technology · Information system
Security management · Application

In recent years, along with the extensive applications of our national safety management information system in the entire urban construction, we apply the data mining technology to the construction of the information system, and accelerate the development of the social and economic industries to perfect the important ways of the construction of the information system. In the promotion the development of the data mining technology in our country, not only need we focus on the basis of the construction of the information system, but more importantly we should widely express the steps and prospects of our city development through the development of the information system, in order to perfect the new directions of the construction and development of our country and society. Therefore, the innovation and development of the data mining technology will be the most important development link in the construction of the future safety management information system, and also an effective guarantee for our country to promote the social progress.

1 A Brief Introduction to the Data Mining Technology

The data mining is an important method of the data analysis in the field of the computer information, and it is a reliable way to analyze the data in the system. The method is widely used in the modeling of the observation data that can meet the distribution. In the samples that are observed in reality, integrate the prior information of the sample parameters, and the prior knowledge is often the valuable experience obtained through

J. Abawajy et al. (Eds.): ATCI 2018, AISC 842, pp. 612–617, 2019.
https://doi.org/10.1007/978-3-319-98776-7_68

the long-term practice and accumulation of the learning summaries. And then using the data mining theories, deduct the maximum posteriori distribution of the parameters, and finally obtain the maximum expected values of the maximum of the parameter posterior distribution through the inferring method. The expected value is the best parameter of the information system, so as to obtain the best information expression system. The data mining technology can be recognized and used widely in the field of the computer information. Its main advantage is that the probability can be directly used to determine and analyze the problem of uncertainty, while the disadvantage is that the traditional data mining information system relies more on the assumption of the parameters of the information system. That is to say, the expression ability of the data mining information system mainly depends on the priori assumptions of the parameters in the information system. In the information system of the data mining of the parameters, the prior information of the parameters is usually selected based on the experience of people. However, the prior assumptions cannot accurately describe the features of the unknown, and it is difficult to select the most reasonable prior distribution in the face of the unknown.

The data mining is the product of combining the advantages of the non-parametric method and the data statistical learning method. The non-parameter means that the parameters of the information system are flexible and variable, and the number of the parameters is not limited, and the parameters will vary with the size of the data. When the information system is added to the priori information without the parameter constraints, the expression ability of the information system will be greatly improved, which can explain any signal. Through the data mining learning method, we can acquire the number of the features of the information system directly from the observation data, enabling the information system to better explain and describe the real data. With many advantages, the method of the non-parametric information system has received the extensive attention and researches in recent years. The data mining can solve the problem of the priori parameter setting when meeting the complex information systems or the large amounts of data. However, the traditional information system of the parameter data mining is prone to over-fitting or under-fitting.

The essence of the modeling method of the data mining is to use the non-parametric modeling method based on the data mining framework, and then change the spatial dimensions f the parameters according to the sample data, to make the information system have more flexibility and scalability in the signal expression. Combined with the real observation data, we can adaptively learn and select the optimal parameters of the information system. The data mining information system is the probability information system without the prior fixed parameter hypothesis. By using the wide distribution hypothesis, replace the fixed parameter hypothesis, and learn about the probability distribution of the parameters of the information system directly from the sample data, and then determine the specific parameters of the prior wide distribution, so that the information system can be more flexible and can adapt to all kinds of the uncertain problems. The common modeling methods of the non-parametric data mining include the Beta Process, the Dirichlet Process and the mixed information system derived from the data.

2 Applications of the Data Mining in the Security Management Information System

As far as the current development trend of the computer field is concerned, it is the common understanding of the Chinese people in the information age to speed up the construction of the industrial informatization. The pace of the construction of the industrial informatization of the information system is accelerating, and in the near future, the information technology will be an indispensable tool for the security management of the information systems. The application of the data mining technology not only provides convenience for the status of the information system industry, but also provides convenience for the life and work of people. Meanwhile, it also reveals the industrial characteristics and the social applicability of the services of the information system, and has won the opportunity to promote the development of the information system well and quickly.

2.1 The Data Mining Can Promote the Intelligent Development of the Information Security Management System

In the construction of the safety management information system, we need to effectively coordinate all aspects of the departmental work and improve the quality and levels of every aspect in the operation of the system, such as the lifestyle of the residents and all kinds of the unexpected events and so on in the system and the public services that need to be carried out by the government departments. The data mining plays an important role in the improvement of the coordination abilities of these aspects, which is an indispensable and important carrier. For example, we can specify the traffic safety in the system. For the effective application of the data mining, we can detect and control various dangerous sources in the traffic, carry out the effectively query for the traffic information, optimize the design the optimal road shortcut, and carry out the real-time monitoring of the traffic flow of each section. At the same time, in the application of the data mining effectively, we can make the best route in the transportation, reduce the transportation costs, and improve the efficiency of the safety management. Through the acquisition of the validity of all kinds of the information and data and the organic integration of all kinds of the data and information, we can promote the intelligent development of many fields in the system.

2.2 The Data Mining Promotes the Construction of the Internet of Things of the Security Management System

The construction of the perfect system of the Internet of things plays an important role of the catalyst in the safety management information system. It not only can ensure the orderly implementation of the work of the safety management information system at the technical level, but at the same time it can also play an important role in promoting the security management information system in the energy saving, the economic development and other aspects. However, the construction of the scientific Internet of things often requires the abundant data mining as a basis, which indirectly determines the support of the data mining for the safety management information system. Specifically, the data mining can promote the construction of the Internet of things of the system mainly in two aspects. First, the data mining system can provide a full range of the surface material information for the construction of the Internet of things of the system, including the spatial attribute, marking and positions and so on, so that the designers and managers of the system can understand and master the development state of various substances with the help of the application of the network monitoring technologies and devices. Second, through the overall analysis of the data mining, we can summarize the overall actual coverage of the sensor networks, so as to make more rational layout of their sensing devices of the Internet of things.

2.3 The Data Mining is an Important Tool for Expanding the Intelligent Services of the Security Management Information System

The construction of the safety management information system includes many aspects, including the traffic, the power grid, and the economy and many other links. The safety management information system is put forward following the concept of the sustainable development. In the construction, we need to increase the application of the new technologies. In order to adapt to the development of the times, we will establish a standard geographic information platform. When building, if we need data to consult, we can query on this platform. In the future development, we need to publicize the information platform and apply this platform as widely as possible, so as to provide convenience for the construction of the safety management information system. When the safety management information system is built, it is necessary to increase the application of the environmental protection technologies and save our resources. In the actual application of the data mining, it is necessary to increase the adaptability of the intelligence, actively find the existing problems and improve the quality of the services.

2.4 The Data Mining is the Core Software Support of the Security Management Information System

In the safety management information system, using the data mining methods, we can carry out the multi-scale, multi-temporal and multi-type data description. On the basis of the Internet of things, we can realize the dynamic, real-time and continuous full coverage of the acquisition of the information system. With the application of the mathematical methods, combined with the related theories of intelligence science and technologies, construct the decision-making model of the system construction and the development space. Understand and solve the related non-structural problems in the process of the system development, carry out the in-depth study of the structural features of the system, predict the trend of the changes, realize the monitoring, simulation, deduction, spatial analysis and evaluation of the construction and development of the system, and improve the scientificalness and efficiency of the decision making in the construction of the system. The data mining enables the safety management information system not only to have the Omni-directional ability of the information collection, but also to have the stronger capabilities of the information processing, analysis, sharing and collaboration. In the more and more practice of the safety management information systems, the spatial information and the spatial information sharing platform based on it, and the derived ecosystem of the spatial information services have become the core support of the safety management information system.

2.5 The Data Mining Accelerates the Transformation of the Security Management Information System

The application of the data mining in the security management information system can effectively speed up its transformation. At the present stage, the main body of the system construction is still the digital system, which plays a very important role in improving the degree of humanization and the functions of the system. However, in the process of the continuous improvement of the system construction, for example, we will further deepen the transformation of the digital system, and the direction of the transformation is to build the safety management information system. To achieve this goal, we must know this is not an easy process, and we need to recognize the complexity of the construction of the safety management information system. Under the premise of the full attention to the safety management information system, we should actively carry out the creative work, in order to facilitate the smooth development of the safety management information system. In the process, the application of the data mining system can combine the intelligent technologies and the Internet technologies effectively, making it possible to transform the digital systems into the safety management information system.

2.6 The Data Mining Can Promote the Intelligent Decision Making of the Safety Management Information System

It is an important role of the data mining to carry out the intelligent decision making to realize the intelligence of the management of the safety management information system. For example, the effective application of the data mining makes the construction and management of the safety management information system work have a unified standard for the decision-making, management, and services and so on, which can improve the accuracy of the application of the information resources and ensure the effectiveness of the information circulation. If we can implement the intelligent environmental protection, we can collect the environmental information through various sensors and conduct the comprehensive monitoring and analysis, so as to make the real-time and digitalized environmental decision-making and management. The development trend of the knowledge, intelligence, speech and intellectualization of the data mining promotes the data mining technology to become the cornerstone of building the safety management information system. Through the use of the satellite positioning, the remote sensing images, it can conveniently and automatically respond to the human and material factors of the system, scientifically and accurately deal with the massive information technologies with the comprehensive automation and intelligence, analyze the relationship between various elements and the situations of their changes, to provide services for our decision making. Without the data mining, the security management information system cannot be promoted.

3 Conclusion

In a word, in the future development of the security management information system, the application of the data mining technology has become an inevitable trend. In order to effectively build the safety management information system, we need to constantly build a better information basis, to ensure that the application of various information systems can meet the needs of the masses. In the field of the information security management in the future, we will seriously implement the "three services", create a "five-type" information system, construct the harmonious information system, innovate the management ideas, mine the existing resources, continue to deepen and perfect the informatization level of the information system industry, and build a professional team who is expert in the management and has a deep understanding of the information technologies. We should often have a thought of sufferings, and often bear the consciousness of self-alertness in heart. We should continue out learning, but should not slack, and we should also forge ahead and advance bravely, so that the information system security management levels and the information construction step to a new level.

References

1. Zhang, T., Zhang, L.: Research on the application of the data mining technology in the coal mine monitoring. Fujian Comput. **2** (2013)
2. Wang, Y., Liu, Z.: Strategies for the construction of the safety management information system in groups of the coal enterprises. Coal Mach. **2** (2013)
3. Liao, M.: The safety monitoring systems in coal mines based on the spatial data mining. Manuf. Autom. **24** (2012)
4. Liu, X., Tian, Y.: Research on the safety early-warning management system against the gas explosion in coal mines based on the data mining. Technol. Dev. Enterp. **15** (2011)
5. Zhou, Q., Luo, J.: Artificial neural network based grid computing of E-government scheduling for emergency management. Comput. Syst. Sci. Eng. **30**(5), 327–335 (2015)
6. Zhou, Q., Luo, J.: The service quality evaluation of ecologic economy systems using simulation computing. Comput. Syst. Sci. Eng. **31**(6), 453–460 (2016)
7. Zhou, Q.: Research on heterogeneous data integration model of group enterprise based on cluster computing. Clust. Comput. **19**(3), 1275–1282 (2016). https://doi.org/10.1007/s10586-016-0580-y
8. Zhou, Q.: Multi-layer affective computing model based on emotional psychology. Electron. Commer. Res. **18**(1), 109–124 (2018)
9. Xie, J., Luo, J., Zhou, Q.: Data mining based quality analysis on informants involved applied research. Clust. Comput. **19**(4), 1885–1893 (2016). https://doi.org/10.1007/s10586-016-0657-7

Advances in Intelligent Systems and Computing

Volume 842

Series editor

Janusz Kacprzyk, Polish Academy of Sciences, Warsaw, Poland
e-mail: kacprzyk@ibspan.waw.pl

The series "Advances in Intelligent Systems and Computing" contains publications on theory, applications, and design methods of Intelligent Systems and Intelligent Computing. Virtually all disciplines such as engineering, natural sciences, computer and information science, ICT, economics, business, e-commerce, environment, healthcare, life science are covered. The list of topics spans all the areas of modern intelligent systems and computing such as: computational intelligence, soft computing including neural networks, fuzzy systems, evolutionary computing and the fusion of these paradigms, social intelligence, ambient intelligence, computational neuroscience, artificial life, virtual worlds and society, cognitive science and systems, Perception and Vision, DNA and immune based systems, self-organizing and adaptive systems, e-Learning and teaching, human-centered and human-centric computing, recommender systems, intelligent control, robotics and mechatronics including human-machine teaming, knowledge-based paradigms, learning paradigms, machine ethics, intelligent data analysis, knowledge management, intelligent agents, intelligent decision making and support, intelligent network security, trust management, interactive entertainment, Web intelligence and multimedia.

The publications within "Advances in Intelligent Systems and Computing" are primarily proceedings of important conferences, symposia and congresses. They cover significant recent developments in the field, both of a foundational and applicable character. An important characteristic feature of the series is the short publication time and world-wide distribution. This permits a rapid and broad dissemination of research results.

More information about this series at http://www.springer.com/series/11156

Jemal Abawajy · Kim-Kwang Raymond Choo
Rafiqul Islam · Zheng Xu
Mohammed Atiquzzaman
Editors

International Conference on Applications and Techniques in Cyber Security and Intelligence ATCI 2018

Applications and Techniques in Cyber Security and Intelligence

Set 2

Springer

Editors
Jemal Abawajy
Faculty of Science, Engineering and Built Environment
Deakin University Geelong
Geelong, VIC, Australia

Kim-Kwang Raymond Choo
Department of Information Systems and Cyber Security
The University of Texas at San Antonio
San Antonio, TX, USA

Rafiqul Islam
School of Computing and Mathematics
Charles Sturt University
Albury, NSW, Australia

Zheng Xu
Shanghai University
Shanghai Shi, China

Mohammed Atiquzzaman
School of Computer Science
University of Oklahoma
Norman, OK, USA

ISSN 2194-5357 ISSN 2194-5365 (electronic)
Advances in Intelligent Systems and Computing
ISBN 978-3-319-98775-0 ISBN 978-3-319-98776-7 (eBook)
https://doi.org/10.1007/978-3-319-98776-7

Library of Congress Control Number: 2018951638

This Springer imprint is published by the registered company Springer Nature Switzerland AG
The registered company address is: Gewerbestrasse 11, 6330 Cham, Switzerland

Preface

The 2018 International Conference on Applications and Techniques in Cyber Intelligence (ATCI), building on the previous successes in Ningbo, China (2017), Guangzhou, China (2016), Dallas, USA (2015), Beijing, China (2014), and Sydney, Australia (2013), is proud to be in the sixth consecutive conference year in Shanghai. Previously, the event is known as the International Workshop on Applications and Techniques in Cyber Security (ATCS 2016), held in conjunction with the International Conference on Security and Privacy in Communication Networks (SecureComm). The purpose of ATCI 2018 is to provide a forum for presentation and discussion of innovative theory, methodology, and applied ideas, cutting-edge research results, and novel techniques, methods, and applications on all aspects of cyber and electronics security and intelligence. The conference establishes an international forum and aims to bring recent advances in the ever-expanding cybersecurity area including its fundamentals, algorithmic developments, and applications.

Each paper was reviewed by at least three independent experts, and the acceptance rate was 35%. The conference would not have been a reality without the contributions of the authors. We sincerely thank all the authors for their valuable contributions. We would like to express our appreciation to all members of the Program Committee for their valuable efforts in the review process that helped us to guarantee the highest quality of the selected papers for the conference.

We would like to express our thanks to our distinguished keynote speakers, Prof. Tharam Dillon, La Trobe University, Australia; Prof. Zhou Wenbo, President of Shanghai Tunnel Engineering Co., Ltd., Prof. Lv Kangjuan, Executive Vice President, Professor and Doctoral Supervisor in SHU-UTS SILC Business School of Shanghai University, Prof. J. Christopher Westland, University of Illinois, Chicago. We would also like to acknowledge the strong support of SHU-UTS SILC Business School of Shanghai University, Shanghai, China, as well as the General Chairs, Publication Chairs, Organizing Chairs, Program Committee Members, and all volunteers.

Our special thanks due also to the editors for their assistance throughout the publication process.

Jemal Abawajy
Kim-Kwang Raymond Choo
Mohammed Atiquzzaman
Rafiqul Islam
Zheng Xu

Organization

General Chairs

Siyi Gong	Shanghai University, China
John Macintyre (Pro Vice Chancellor)	University of Sunderland

Program Chairs

Jemal Abawajy	Deakin University, Australia
Kim-Kwang Raymond Choo	University of Texas at San Antonio, USA
Mohammed Atiquzzaman	University of Oklahoma, USA
Rafiqul Islam	Charles Sturt University, Australia
Zheng Xu	TRIMPS and Shanghai University, China

Publication Chairs

Juan Du	Shanghai University, China
Mazin Yousif	T-Systems International, USA
Vijayan Sugumaran	Oakland University, USA

Publicity Chairs

Fangqi Cheng	Zhejiang Business Technology Institute, China
Kewei Sha	University of Houston, USA
Yongchun Xu	Zhejiang Business Technology Institute, China
Neil. Y. Yen	University of Aizu, Japan

Local Organizing Chairs

Min Hu	Shanghai University, China
Gang Luo	Shanghai University, China
Juan Chen	Shanghai University, China
Qigang Liu	Shanghai University, China

Program Committee Members

William Bradley Glisson	Sam Houston State University, USA
George Grispos	University of Nebraska Omaha, USA
Abdullah Azfar	KPMG Sydney, Australia
Aniello Castiglione	Università di Salerno, Italy
Florin Pop	University Politehnica of Bucharest, Romania
Ben Martini	University of South Australia, Australia
Wei Wang	University of Texas at San Antonio, USA
Neil Yen	University of Aizu, Japan
Shunxiang Zhang	Anhui University of Science and Technology, China
Guangli Zhu	Anhui University of Science and Technology, China
Tao Liao	Anhui University of Science and Technology, China
Xiaobo Yin	Anhui University of Science and Technology, China
Xiangfeng Luo	Shanghai University, China
Xiao Wei	Shanghai University, China
Huan Du	Shanghai University, China
Zhiguo Yan	Fudan University, China
Rick Church	UC Santa Barbara, USA
Tom Cova	University of Utah, USA
Susan Cutter	University of South Carolina, USA
Zhiming Ding	Beijing University of Technology, China
Yong Ge	University of North Carolina at Charlotte, USA
Danhuai Guo	Computer Network Information Center, Chinese Academy of Sciences, China
Jianping Fang	University of North Carolina at Charlotte, USA
Jianhui Li	Computer Network Information Center, Chinese Academy of Sciences, China
Yi Liu	Tsinghua University, China
Kuien Liu	Pivotal Inc., USA
Feng Lu	Institute of Geographic Science and Natural Resources Research, Chinese Academy of Sciences, China

Ricardo J. Soares Magalhaes	University of Queensland, Australia
Alan Murray	Drexel University, USA
Yasuhide Okuyama	University of Kitakyushu, Japan
Xiaogang Qiu	National University of Defense Technology, China
Xing Xie	Microsoft Research Asia, China
Wei Xu	Renmin University of China, China
Chaowei Phil Yang	George Mason University, USA
Hengshu Zhu	Baidu Inc., China
Morshed Chowdhury	Deakin University, Australia

Contents

ATCI 2018

ATCI 2018 Companion

ATCI 2018 Companion

The Development and Application of Data Management System Based on B/S Architecture

Jingjie Yang(✉)

Liaoning Jianzhu Vocational College, Liaoyang 111000, Liaoning Province, People's Republic of China
331857956@qq.com

Abstract. With the era of big data, data management needs to be informationized and strict. Sales management needs to be standardized and process oriented, which is an urgent technology need for the data manager. In this paper, the XX company's development of invoicing data management system as an example, to describe how to solve the confusion caused by manual records, and the problems of sales order management, purchasing plan and the data of warehouse management, and how to realize the standardization and clear processing of the data management. The main development methods and key technology of the data management system are as follow in detail.

Keywords: Invoicing · Information · Data management

1 Development Background

In order to achieve strict management and standardization of the goods information and clear processing and standardization of sales, The XX company authorized me to develop a set of integrated data management system for procurement, sales and stock. Its aim is to achieve integrated management of clear, transparency warehouse management, solving the manual record of chaos, and sales, procurement plan, and warehouse management.

Engaged in teaching software technology for many years, in order to improve the students' practice level, I with three students developed the system. With the object-oriented method, we first set up a logical model of the system, and then a corresponding physical model [1]. System logic model is roughly divided into three steps. The first step is to describe the requirements. The second step is to establish the static structure of the system according to the demand, and the third step is to describe the behavior of the system. Among the established models, the first step and the second step are static. They can be described with case diagram, class diagram and object diagram. But the third step is dynamic, it can be described with four figures, that is, state diagram, activity diagram, sequence diagram and collaboration diagram, which is belong to the behavior modeling mechanism [2] of the unified modeling language (UML).

J. Abawajy et al. (Eds.): ATCI 2018, AISC 842, pp. 621–626, 2019.
https://doi.org/10.1007/978-3-319-98776-7_69

2 The Summary of the Design Process

2.1 The Demand Analysis

The project team and the XX company's staff made the requirement document together through communication and negotiation. In the process of our visit, we found the sales of various types of dealers are still using written documents, so this system lasts people's custom of records in the form of documents for every ins and outs. The record method is very convenient for after query and management, just the ticket number will be able to check out the consumption situation.

Personnel management can be divided into two aspects. One is the buyers' personal information, which records the information such as name, contact information, company. The other is the employee's personal information, which records the information such as position information.

2.2 Profile Design

Through the demand analysis, system design is divided into five modules: the basic information management, warehouse management, sales management, system management and statistics management. The system administrator can manage the information of procurement, the clerk, inventory, sales management and statistics analysis. See Fig. 1.

2.3 The Detailed Design

Information management is divided into three secondary modules: Commodity information management is to manage the basic information of the commodity's name, specification and grade. Shop assistant management is to manage the basic information of the shop assistant. Business management is to manage the buyer's name, the salesman's name and address, and to add or delete contact information.

Warehouse management is divided into two secondary modules: Procurement module, is about the company's warehouse entry. The module records the detailed information of the commodity's name, date, quantity and so on. Warehouse-out management, is to manage outbound of non-sale goods.

Marketing management is divided into two secondary modules: Sales management, the module is done for the company's sales record, which is about the information of commodity, quantity, sales assistant. Returns management is a module when buyers want to return the commodity to the company, which can realize the returns function through searching and modifying the receipt.

Statistical management is divided into two secondary modules: Inventory statistics, the module can realize detailed search and statistical functions of the inventory, inventory information and dispatch information. It is convenient for the company to manage its inventory. Sales statistics can realize the function of statistics and analysis of the sales situation. It can find the details of each sales single or return.

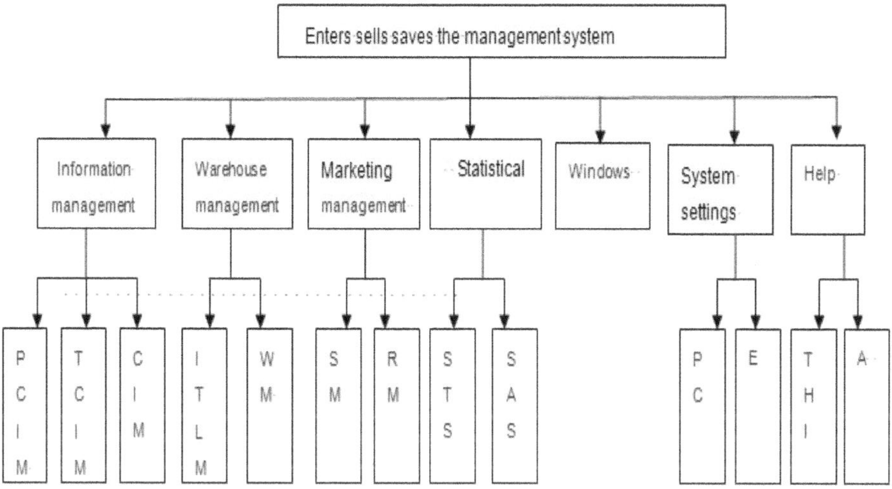

Mark:
PCIM: Purchasing commodity information management
TCIM: The clerk information management
CIM: Commodity information management
ITLM: Into the library management
WM: Warehouse management
SM: Sales management
RM: Returns management
STS: Storing statistical
SAS: Sales statistical
PC: Password change
E: Exit
THI: To help illustrate
A: about

Fig. 1. System diagram

2.4 The Database Design

Database designed eight tables, respectively is: Buyer (Purchaser information table), Employee (Shop assistant information table), Commodity (Commodity information table), InStorage (Warehouse), OutStorage (Outbound table), Sell (Sell Sales), Storage (Inventory table), Marketing (Sales record). The structure of Buyer information table and Commodity information table is explained in detail as an example. As follows in Tables 1 and 2.

2.5 The Choice of Tools

The development used JSP technology, which is used XML tags and scriptlets of Java programming language class. It is used to encapsulate processing logic of generating

Table 1. Buyers information

Field	Type	Null	Instructions
Buyer Num	Int		Purchaser number
Company	Nchar(40)	√	Buyer company name
Buyer Name	Nchar(8)		The salesman name
Tel	Nchar(12)	√	Fixed telephone
Field	Type	Null	Instructions
Mobile Phone	Nchar(11)	√	The mobile phone
Address	Nchar(50)	√	The company address
Birthday	Datetime	√	The salesman birthday

Table 2. Commodity information

Field	Type	Null	Instructions
Commodity Num	Int		Product id
Commodity Name	Char(20)		Trade name
Specification	Int		Specifications and quantity
Grade	Nchar(1)	√	Level
Start	Datetime		The date of production
Finish	Datetime		Shelf life
Yield	Nchar(50)	√	Origin
Material	Nchar(50)	√	The raw material
Volume	Int	√	Capacity
Code	Nchar(13)	√	Bar code
Suppliers	Int	√	Supplier

dynamic web pages. Pages also can be accessed through the tags and scriptlets in the server-side resource application logic2 [3]. JSP separates Web logic and Web design, supports reusable component-based design, makes the development of a web-based application quickly and easily [2]. JSP (JavaServer Pages) is a kind of dynamic page technology. Its main purpose is to separate presentation logic from the Servlet [4].

3 The Key Technology

(1) The commodity information for dynamic loads in the storage module, reveals all commodity information in the new form and in which you can choose goods.
(2) The implementation of the details of library documents shows in the outbound module, and then realizes the real change of goods quantity through the control method. By using a DataRow classes, so as to realize the effect they want.
(3) When choosing the form of goods, need to value his father's form after selecting the goods. Through the Internet and read some information, team members

eventually chose Show Dialog method to display and establish the relationship between the two form [5].

4 Application System Efficiency Analysis

(1) To improve the management efficiency and reduce labor costs
(2) To reduce the purchasing cost
(3) To make management more scientific and operation more standard
(4) To prevent old, dead account and strengthen the financial supervisor
(5) Intelligent analysis of business data, to promote efficient decision-making
(6) To improve the comprehensive utilization of warehouse capacity

5 Conclusion

This system adopts the latest JSP technology development. Using the database system, it manages a wide range of data and is a set of convenient and quick management system for users. A lot of problems arise in the process of development, at the same time, we have a deeper understanding of software development by solving the problems. The current system has shortcomings. We will try hard to improve the system and put it into effect. The students of common development really got a good exercise. In the following teaching process, I should as far as possible let the students involve in the actual project.

References

1. Watts, D.J.: Small Worlds: The Dynamics of Networks Between Order and Randomness. Princeton University Press, Princeton (1999)
2. Aleksy, M., Korthaus, A., Schader, M.: Use Java and the CORBA realization distribute type system. J. Pingxiang Coll. **4**, 104–105 (2005)
3. Titus, J.: ECN technical editor: the eclipse of stand. J. Zhongkai Agrotech. Coll. **19**(2), 32–35 (2006)
4. Clay Richardson, W., Donald, A.: The Java high class weaves a distance: JDK 5. Sci. Technol. Book Rev. **3**, 17–18 (2006)
5. Vuong, J.L.: A semantics-based routing scheme for grid resource discovery. In: E-Science: First International Conference on E-Science and Grid Computing, pp. 58–70, 90 (2005)
6. Duchowski, T.: Eye Tracking Methodology: Theory and Practice. Springer, New York (2003)
7. Zhu, Z., Ji, Q.: Eye and gaze tracking for interactive graphic display. Mach. Vis. Appl. **15**(3), 139–148 (2004)
8. Schnieders, D., Fu, X., Wong, K.: Reconstruction of display and eyes from a single image. In: Proceedings of IEEE International Conference on Computer Vision and Pattern Recognition (CVPR 2010), pp. 1442–1449 (2010)

9. Zhang, C., Chi, J., Zhang, Z., Wang, Z.: The research on eye tracking for gaze tracking system. Acta Autom. Sin. **36**(8), 1051–1061 (2010). (in Chinese)
10. Yoo, D., Kim, J., Lee, B., Chung, M.: Non-contact eye gaze tracking system by mapping of corneal reflections. In: Proceedings of IEEE International Conference on Automatic Face and Gesture Recognition. IEEE Computer Society, Washington, DC, pp. 94–99 (2002)
11. Yoo, D., Chung, M.: A novel nonintrusive eye gaze estimation using cross-ratio under large head motion. Comput. Vis. Image Underst. **98**(1), 25–51 (2005)
12. Han, K., Wang, X., Zhang, Z.L., Zhao, H.N.: A novel remote eye gaze tracking approach with dynamic calibration. In: Proceedings of 15th IEEE International Workshop on Multimedia Signal Processing (MMSP 2013), pp. 111–116 (2013)

Speed Identification of PMSM Based Least Square Algorithm for Closed Loop *I/F* Control

Yang Yu(✉), Leyao Cong, Jianbin Lu, Renjie Xie, and Xia Tian

North China Electric Power University, Baoding 071003, China
ncepu_yy@163.com

Abstract. In recent years, the application of PMSM is becoming more and more extensive and the speed control of a sensorless permanent magnet synchronous motor (PMSM) has gradually become a hot spot of current research. In order to accurately estimate the rotor speed and improve the control accuracy of PMSM, a speed identification algorithm based on recursive least squares for closed-loop *I/f* control is proposed in this paper. Based on the *I/f* control framework, this paper deduced the dynamic mathematical model of PMSM and established the identification model of least square method with genetic factors, so as to achieved the estimation of rotor position and speed. The simulation results show that the algorithm could quickly and accurately track rotor speed and keep track of rotor position in a very small error range, accomplished tracking at any speed.

Keywords: Speed Identification · Least square algorithm
Closed loop *I/F* control · PMSM

1 Introduction

1.1 A Subsection Sample

The simple structure has been configured for permanent magnet synchronous motor (PMSM) by cancelling external excitation device and providing an air-gap magnetic field with the included permanent magnet [1]. Additionally, due to its high power density and high torque inertia ratio, PMSM has been extensively applied into various industrial drive fields such as electrical vehicle, numerically-controlled machine tools and aerospace engineering.

In practical applications, in order to obtain accurate information of the rotor position and speed of the permanent magnet synchronous motor, various mechanical encoders are needed [2–4]. However, the use of the encoder improves the cost and complexity of the system and reduces the ability to resist interference. Therefore, the speed control of a sensorless permanent magnet synchronous motor (PMSM) has gradually become a hot spot of current research [5, 6].

The basic *I/f* control is an open loop control scheme. There are problems such as the current frequency and amplitude are not controlled, the speed is easily disturbed, the load is easy to lose step when the load is changed, and the efficiency is low.

J. Abawajy et al. (Eds.): ATCI 2018, AISC 842, pp. 627–633, 2019.
https://doi.org/10.1007/978-3-319-98776-7_70

The novelty in this paper, as an extension of the original *I*/*f* control, is to propose a speed identification of PMSM based least square algorithm for closed loop *I*/*f* control which to accurately estimate the rotor position and speed and achieve sensorless control. Based on the *I*/*f* control framework, this paper deduces the dynamic mathematical model of PMSM, and establishes the identification model of least square method with genetic factors, so as to achieve the estimation of rotor position and speed.

2 Modeling of SPMSM in Current Vector Oriented Coordinate System

As a complex nonlinear system, some ideal assumptions are made for surface-mounted PMSM (SPMSM) as follows: (1) Three-phase winding of stator is completely symmetrical; (2) Saturation of iron core is ignored, and losses of hysteresis and eddy-current are disregarded; (3) Air gap of motor is uniform, and magnetomotive force produced by stator winding and magnetic field in space are both distributed sinusoidally.

For stator current vector is, the synchronous rotating coordinate system d*q* for SPMSM is introduced and plotted in Fig. 1, where d* and q* axes are real and imaginary axes respectively and dq is the real rotating frame of rotor. Its amplitude is i_s located at q* axis. The included angle between q* axis and d axis is θ. The projections of is in d*q* frame to dq axes are denoted as i_d and i_q in dq frame individually.

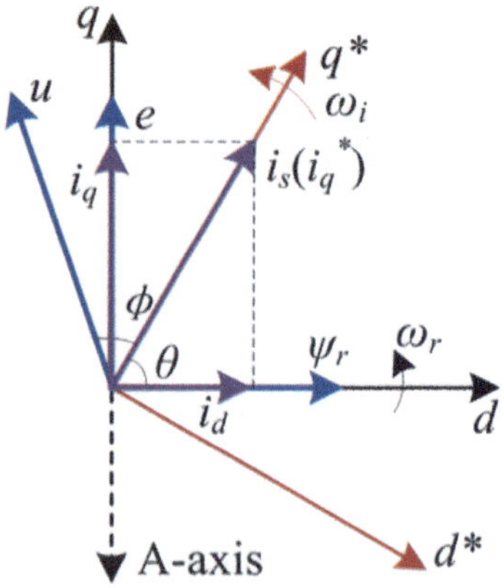

Fig. 1. Phasor diagram including dq and d*q* frames for SPMSM.

In d*q*o, equation of stator voltages can be listed as the following [7]:

$$u_{sd^*} = \frac{d\psi_{rd^*}}{dt} - n_p\omega_i L_s i_s - n_p\omega_i\psi_{rq^*} \tag{1}$$

$$u_{sq^*} = R_s i_s + L_s \frac{di_s}{dt} + \frac{d\psi_{rq^*}}{dt} + n_p \omega_i \psi_{rd^*} \tag{2}$$

where u_{sd*} and u_{sq*} are the d* axis and q* axis voltages of stator, R_s is the resistance of stator, Ls is the inductance of stator, ψ_{rd*} and ψ_{rq*} are the d* axis and q* axis flux linkages of rotor, ω_i is the rotating mechanical angular velocity of current vector i_s, n_p is the number of pole pairs for rotor.

Additionally,

$$T_e = \frac{3}{2} n_p \psi_{rd^*} i_s \tag{3}$$

where T_e is the magnetic torque.

According to Fig. 1, we can get:

$$\psi_{rd^*} = \psi_r \sin\theta \tag{4}$$

where θ is the angle between q* axis and d axis, which describes the situation current vector is pulling ψ_r.

Substitution of (4) into (3), we can obtain

$$T_e = \frac{3}{2} n_p \psi_r i_s \sin\theta \tag{5}$$

Equation of motion for rotor can be written as follows:

$$J\frac{d\omega_r}{dt} = T_e - T_{sp} - B\omega_r = \frac{3}{2} n_p \psi_r i_s \sin\theta - T_{sp} - B\omega_r \tag{6}$$

where J is the moment of inertia, ω_r is the mechanical angular velocity, T_{sp} is the torque of STS indicating the load torque, B is the viscous damping coefficient. Hence, the relation among the angle θ, ω_i and ω_r is

$$\frac{d\theta}{dt} = \Delta\omega = n_p(\omega_i - \omega_r) \tag{7}$$

Then, (1) and (2) can be converted as:

$$u_{sd^*} = -n_p \psi_r \omega_r \cos\theta - n_p \omega_i L_s i_s \tag{8}$$

$$u_{sq^*} = R_s i_s + L_s \frac{di_s}{dt} + n_p \psi_r \omega_r \sin\theta \tag{9}$$

Consequently, (6), (7), (8) and (9) construct the dynamic model of SPMSM in stator current vector oriented coordinated system.

3 Speed Identification Based On Least Square Method

3.1 Description of Least Square Method

Least squares identification method is commonly used in engineering, the principle is that when the identification system in operation, each made a new observation data, based on the results of previous estimates, using new observational data according to the recursive rules in the previous estimation result, the estimated value of the new parameters. To reduce the estimation error. In this way, with the continuous introduction of the new observation data, the estimation parameters are corrected until the parameter estimates reach a satisfactory degree of accuracy. Fig. 2 is an identification model of the least square method.

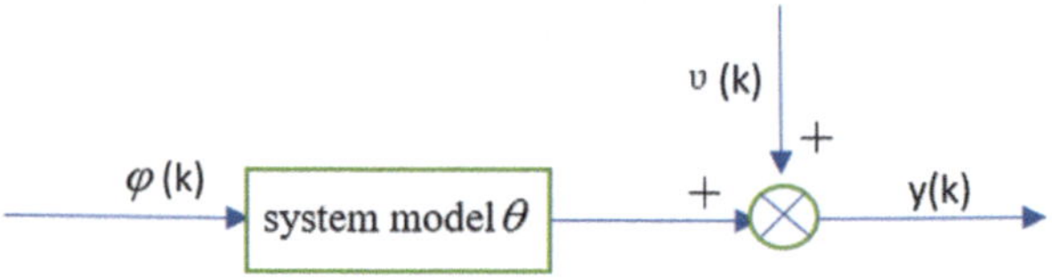

Fig. 2. Identification model of the least square method

Regression information vector $\varphi(k) = [\varphi_1(k)\,\varphi_2(k)\cdots\varphi_n(k)]^T$; System interference noise $v(k) = [v_1(k)\,v_2(k)...v_n(k)]$; The output vector measured by the system $y(k) = [y_1(k)\,y_2(k)...y_n(k)]$; parameter vector of the system $\theta = [\theta_1\,\theta_2...\theta_n]$.

The output of the model can be expressed as:

$$y(k) = \varphi^T(k)\theta + v(k) \tag{10}$$

The principle of least square identification is to use the input of the system to output the data, the two criterion function can be expressed as:

$$J(\theta) = \sum_{i=1}^{k} \left[y(i) - \varphi^T(i)\theta\right]^2 \tag{11}$$

$\hat{\theta}$ is defined as: when $J(\theta)$ takes the minimum value, the value of θ. The expression of the least square method can be derived from the extremum theorem:

$$\hat{\theta}(k) = \left[\varphi^T(k)\varphi(k)\right]^{-1}\varphi^T(k)y(k) \tag{12}$$

Let:

$$P(k) = \left[\varphi^T(k)\varphi(k)\right]^{-1} \tag{13}$$

The recursive formula of the least square method can be derived from (12) and (13):

$$\hat{\theta}(k) = \hat{\theta}(k-1) + L(k)\left[y(k) - \varphi^T(k)\hat{\theta}(k-1)\right] \tag{14}$$

where the gain vector *L*(*k*) and the covariance matrix *P*(*k*) can be expressed as:

$$L(k) = \frac{P(k-1)\varphi(k)}{1 + \varphi^T(k)P(k-1)\varphi(k)} \tag{15}$$

$$P(k) = P(k-1) - L(k)\varphi^T(k)P(k-1) \tag{16}$$

In order to make the least square method be able to identify time-varying parameters, the forgetting factor λ is introduced on the basis of general least squares method, which can be written as:

$$\hat{\theta}(k) = \hat{\theta}(k-1) + L(k)\left[y(k) - \varphi^T(k)\hat{\theta}(k-1)\right] \tag{17}$$

$$L(k) = \frac{P(k-1)\varphi(k)}{\lambda + \varphi^T(k)P(k-1)\varphi(k)} \tag{18}$$

$$P(k) = \frac{1}{\lambda}\left[P(k-1) - L(k)\varphi^T(k)P(k-1)\right] \tag{19}$$

where $0 < \lambda \leq 1$.

3.2 Least Square Speed Identification Algorithm for *I/F* Control Framework

This paper is based on the closed loop *I/f* control framework, the rotational speed of the PMSM rotor is selected as a parameter to be identified.

(9) can be written as:

$$\psi_r \omega_r \sin\theta = -L_s \frac{di_s}{dt} - R_s i_s + U_{sq^*} \tag{20}$$

The discrete-time treatment of (11) can be obtained:

$$(\psi_r \sin(\varepsilon(k)))\omega_r(k) = -L_s \frac{i_s(k) - i_s(k-1)}{T} - R_s i_s(k) + U_{sq^*}(k) \tag{21}$$

where

$$y(k) = -L_s \frac{i_s(k) - i_s(k-1)}{T} - R_s i_s(k) + U_{sq^*}(k) \tag{22}$$

$$\varphi(k) = (\psi_r \sin(\varepsilon(k))) \tag{23}$$

$$\hat{\theta}(k) = \omega_r(k) \tag{24}$$

The expression of the rotor speed of the PMSM rotor can be obtained by Substituting (22)–(24) into (17)–(19).

4 Simulation Tests

The proposed algorithm of speed identification is integrated into the closed-loop *I*/*f* control algorithm. The parameters of the motor are displayed as follows: stator resistance $R_s = 2.875\ \Omega$, stator inductance $L_s = 0.033$ H, number of pole pairs $n_p = 20$, permanent magnet flux linkage $\psi_r = 0.3$ Wb, inertia $J = 0.51\ \mathrm{kg \cdot m^2}$, viscous coefficient $B = 0.002$ N/m/s, the load torque $T_L = 2\ \mathrm{N \cdot m}$, forgetting factor $\lambda = 0.9$.

The mechanical angular velocity ω_r starts up from standstill to the rated speed in 1 s. When the speed references are selected as three different values: 1.047 * 6 rad/s (60 r/min), 1.047 * 15 rad/s (1500 r/min), and 1.047 * 60 rad/s (600 r/min), respectively; and the simulation results of motor speed are shown in Figs. 3, 4 and 5. It can be seen that whether at high speed, medium speed or low speed, the estimated values of the rotor speed can accurately estimate the actual value. The algorithm can achieve full-speed tracking of PMSM rotor speed.

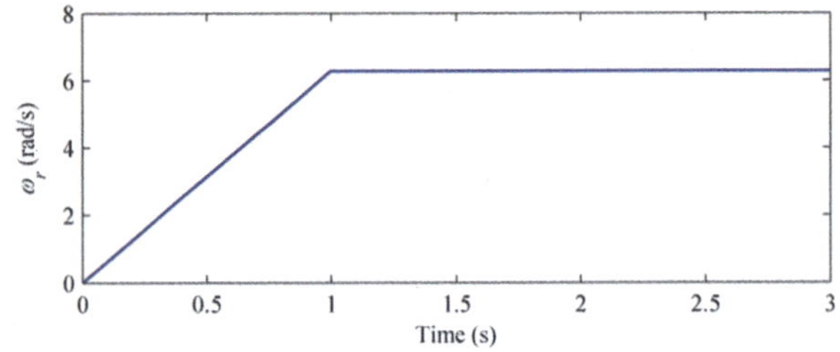

Fig. 3. Startup from standstill to 1.0470*6 rad/s (60 r/min)

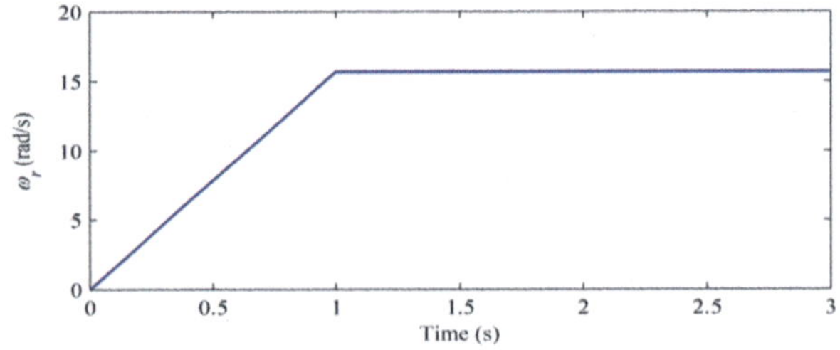

Fig. 4. Startup from standstill to 1.0470*15 rad/s (150 r/min)

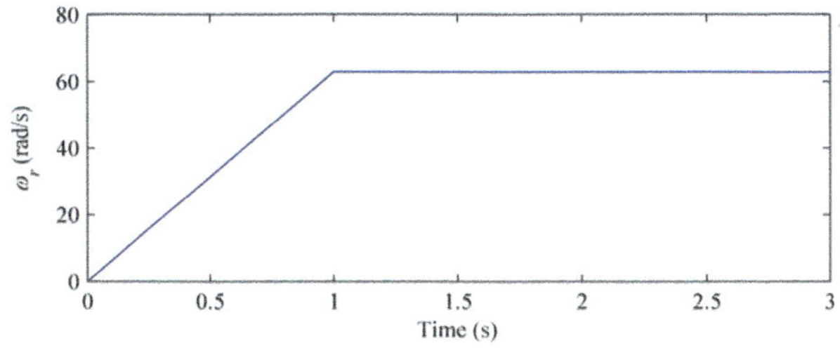

Fig. 5. Startup from standstill to 1.0470*60 rad/s (600 r/min)

5 Conclusion

In this paper, a speed identification algorithm based on recursive least squares for closed loop *I/f* control is proposed for the PMSM sensorless vector control system. The control structure is simple, and the computation is relatively small. It can track rotor speed quickly and accurately, and accomplish tracking the speed at different references. The correctness and effectiveness of the control scheme are verified by simulation, and it has a great significance for guiding the practical application.

Acknowledgements. This work was partially supported by the National Natural Science Foundation of China (grant no. 51407077), the Fundamental Research Funds for the Central Universities of China (grant no. 2017MS10) and the Science and Technology Project of State Grid Corporation of China Headquarter (Grant no. 5204BB16000F).

References

1. Yu, Y., Mi, Z.Q., Guo, X.D., Xu, Y.K.: Low speed control and implementation of permanent magnet synchronous motor for mechanical elastic energy storage device with simultaneous variations of inertia and torque. IET Electr. Power Appl. **10**(3), 172–180 (2016)
2. Arif, S.J., Asghar, M.S.J., Sarwar, A.: Measurement of speed and calibration of tachometers using rotating magnetic field. IEEE Trans. Instrum. Meas. **63**(4), 848–858 (2014)
3. Fang, M.C., Wang, P.: Initial rotor position detect method of PMSM for digital AC servo system. Electr. Mach. Contr. Appl. **36**(11), 27–29 (2009)
4. Mohabati, H.R., Moghani, J.S., Boroujeni, S.T.: Complex vector modelling and sequence analysis of the integrated three-phase rotating transformer for design of a symmetrical structure. IET Electr. Power Appl. **10**(7), 649–657 (2016)
5. Piippo, A., Hinkkanen, M., Luomi, J.: Adaptation of motor parameters in sensorless PMSM drives. IEEE Trans. Ind. Appl. **45**(1), 203–212 (2009)
6. Po-Ngam, S., Sangwongwanich, S.: Stability and dynamic performance improvement of adaptive full-order observers for sensorless PMSM drive. IEEE Trans. Power Electron. **27** (2), 588–600 (2012)
7. Fatu, M., Teodorescu, R., Boldea, I., Andreescu, G.D., Blaabjerg, F.: I-F starting method with smooth transition to EMF based motion-sensorless vector control of PM synchronous motor/generator. In: IEEE Annual Power Electronics Specialists Conference, pp. 1481–1487 (2008)

A Data Storage and Mining Algorithm Based on Distributed File System

Jian Nong(✉)

Guangxi Colleges and Universities, Key Laboratory of Professional Software Technology, Wuzhou University, Wuzhou 543000, China
hanbingtbm@yeah.net

Abstract. In order to improve control property and storage property of embedded metallurgy system, this paper raises a kind of data structure optimum storage algorithm based on embedded metallurgy system of optimum basic function adaptive matching in metallurgy control system data storage and summarizes the security challenges faced by data storage in cloud computing environment. The advantages and the protection measures should be taken into improve the data security architecture in the cloud computing environment. In addition, data file preprocessing and verification process of data files and dynamic operation details of data are described in detail. In view of the problems of data file storage access control in cloud computing environment, this paper proposes an access control scheme based on data storage algorithm by cloud computing. It's proved by stimulation outcome that it efficiently reduces data storage expense and enlarges data storage space by using method raised in this paper, which further improves control ability and storage ability of data interaction in embedded metallurgy system.

Keywords: Data storage · Data mining · Distributed file system
ThinkDB system · Schematic diagram

1 Introduction

Today people have the ability to use large-scale distributed computing resources in the network [1]. Uncontrollable data will inevitably bring users to store in the cloud data insecurity and cloud computing service provider's distrust [2, 3]. In the traditional IT service solution, the vast majority of application software and data information is running or stored in the user's local physical equipment and absolute controllable range [4]. In view of broad space for development and application prospects about the cloud computing, cloud computing data storage issues have been widespread concern scholars [5, 6]. Cloud computing service provider unified management of user data files and centralized storage and data security real-time monitoring and data failure timely processing [7]. However, all service maintenance and data are entrusted to the cloud computing service providers.

At present, although many scholars believe that cloud computing service providers to provide strong storage capabilities of the data center, but the storage problem is still

J. Abawajy et al. (Eds.): ATCI 2018, AISC 842, pp. 634–638, 2019.
https://doi.org/10.1007/978-3-319-98776-7_71

the main problem faced by people in the use of cloud computing data storage service [8, 9].

Cloud computing is regarded as an information technology to large-scale, intensive development trend [10]. Despite the widespread concern of scholars, the existing research on cloud computing data storage is still in its infancy, which has a large space for improvement and important applications and academic research value.

2 Key Technologies of Data Storage

In this paper, we studied the method of cloud data storage and built a data storage mechanism. The system adopts a multidimensional architecture with three layers of network. Schematic diagram of multidimensional system for data storage and the sketch of data storage multidimensional system based on cloud platform are used.

It is very important for us to design and implement a wide range of cloud authentication mechanism and a cloud data transmission storage technical solutions. To deal with this change, we also need to built the future system of our storage model with the direction of efforts.

The organization's information storage policy describes the importance of information storage indicating that the management's commitment to the organization and managing information storage methods that provide support.

3 Distributed File System

Distributed file system is designed as a distributed file system suitable for running on general purpose hardware. Distributed file system differs from other distributed file systems. Distributed file system can provide high throughput data access on the application. In addition this system can achieve the reading data file system. Distributed file system can relax requirements so that the flow can be achieved in the form of streaming access data file system. The distributed file system is consist of hundreds or thousands of servers that store file data fragments. The fault detection and automatic fast recovery is a core design goal of distributed file system. Applications running on distributed file system must stream their data sets, which are not ordinary programs running on common file systems. The focus is on data throughput and response time. The distributed file system can be written in Java, and any Java machine can run data node by using the ultra-portable Java language, which is easy to deploy distributed file system to a wide range of machines. The typical deployment is to run the name node software by a specialized machine. The architecture does not exclude instances of multiple data nodes running on a single machine, but the actual deployment will not be the case. Distributed file system supports traditional inherited file organization. The schematic logic diagram of distributed file system is shown in the Fig. 1.

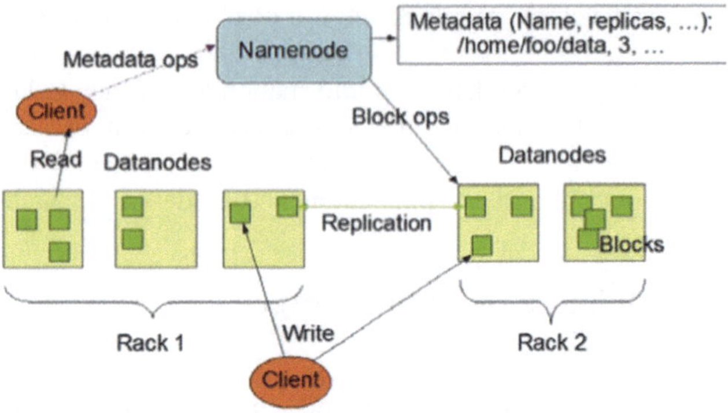

Fig. 1. The schematic logic diagram of distributed file system

From Table 1, we can see that the difference between two different systems by evaluating indicator.

Table 1. Evaluating indicator

Evaluating indicator	Conventional file system	Distributed file system
Design principles	Data integrity consistency	Temporal consistency of data
Management object	Static storage/random read	Continuous and orderly, compressed storage
Typical operation	Data crud relational data query	Data crud sequential scanning query
Data storage	Disk storage passive query	Time index cycle storage data compression

4 Data Storage and Mining Algorithm

Suppose that the file F is made up of strings, $F = b_1, b_2,\ldots b_n$. Among them, the character b_i can be considered as an integer in the range $[0,\ldots, 2^l - 1]$. l is the number of bits that a byte holds.

Step 1. The minimum prime value $p(p > 2^l - 1)$ is taken as the modulus of the character operation in the file, That is, all elements in the string are bounded by Z^p.
Step 2. Select a suitable integer m, so that

$$n = m + k\left(\frac{m}{n} < 1 + \varepsilon,\ \varepsilon > 0\right) \tag{1}$$

Step 3. Select the n vectors and these vectors are arbitrarily related to different m vectors. These m vectors are randomly extracted from (a_1, a_2,...a_n), and the possibility of nonlinear vector is very large. n vectors can be defined as

$$a_i = (a_{i_1}, a_{i_2}, \ldots a_{i_n}) \in Z_p^m, \ 1 \le i \le n.$$

Step 4. The matrix F is split and recombined using the selected vectors. Divide the file F into a sequence of length m, then

$$F = (b_1, b_2, \ldots b_m), \ (b_{m+1}, b_{m+2}, \ldots b_{2m}), \ldots \tag{2}$$

Let s_1 = (b_1, b_2,...b_m), s_2 = (b_{m+1}, b_{m+2},...b_{2m}),..., then

$$F = c_{i_1}, c_{i_2}, \ldots, c_{i_{n/m}}, \quad i = 1, 2, \ldots, n. \tag{3}$$

$$c_{i_k} = a_i \cdot s_k = a_{i_1} \cdot b_{k-1_{m+1}} + \ldots + a_{i_m} \cdot b_{k_m} \tag{4}$$

$$|F_i| = |F|/m \tag{5}$$

If the m block (F_1, F_2,...F_m) of the file F has been given, then you can rebuild the file F in the following steps.

Let A = a_{ij}, 1 $\le$ i, j $\le$ m is a $m \times m$ matrix, and a_i is the i line, then

$$A \cdot \begin{bmatrix} b_1 \\ \cdots \\ b_m \end{bmatrix} = \begin{bmatrix} c_{11} \\ \cdots \\ c_{m1} \end{bmatrix} \tag{6}$$

We can obtain formula (7) as follows

$$\begin{bmatrix} b_1 \\ \cdots \\ b_m \end{bmatrix} = A^{-1} \cdot \begin{bmatrix} c_{11} \\ \cdots \\ c_{m1} \end{bmatrix} \tag{7}$$

where $b_j = a_{i_1} c_{1_k} + \ldots + a_{i_m} c_{m_k}, \ 1 \le j \le N.$

The presented algorithm and simulation experiment in this paper show that, as a kind of key technologies the data storage and mining algorithm can improve effectively the ability of data storage in multidimensional system and distributed file system. Because the algorithm effectively connects the customer database and the component library this algorithm also can achieve data mining in distributed file system by using mining method of the data storage and mining algorithm.

5 Conclusion

Data for efficient storage, acquisition, and processing functions were implemented by using computer technology so as to establish a data storage system. It provided a certain data support for enterprises through the use of the system. With the development of the Internet of things, a large number of data generated were of diversity and complexity, and there were considerable difficulties in the processing of these data. The technology of data acquisition and storage were studied and discussed, thus the author established networking data based on the massive data storage system and the system had good query performance and very high storage efficiency.

Acknowledgment. This work was supported by Academic Research Project of Wuzhou University (Grant No. 2013D009) and Basic Ability Improvement Project of Young College Teachers of Guangxi (Grant No. KY2016LX366).

References

1. He, Z.-X.: Research and design of data security multidimensional protection system in cloud computing environment. J. Residuals Sci. Technol. **13**(6), 211–219 (2016)
2. Sookhak, M., Gani, A., Buyya, R.: Dynamic remote data auditing for securing big data storage in cloud computing. Inf. Sci. **380**(112), 101–116 (2017)
3. Xu, S.: A novel data storage model and method based on bayesian network. J. Residuals Sci. Technol. **13**(6), 121–124 (2016)
4. Jayapandian, N., Rahman, A.M.J.M.Z.: Secure and efficient online data storage and sharing over cloud environment using probabilistic with homomorphic encryption. Cluster Comput. **22**(3), 1–13 (2017)
5. Aga, F.G., Woo, J., Song, J.: Controllable quantized conductance for multilevel data storage applications using conductive bridge random access memory. Nanotechnology **28**(9), 115–123 (2017)
6. Zhang, Y., Wang, J., Chang, Y.: The key technologies of the internet of things storage system for massive real time data storage. J. Residuals Sci. Technol. **13**(8), 21–25 (2016)
7. Aswathy, G., Rajesh, C.S.: Multiplexing recording in nickel-ion-doped photopolymer material for holographic data storage applications. Appl. Opt. **56**(8), 1566–1575 (2017)
8. Zhao, Y., Cai, Y., Fan, G.: Dynamical behavior for fractional-order shunting inhibitory cellular neural networks. J. Nonlinear Sci. Appl. **9**(3), 4589–4599 (2016)
9. Zhao, Y.: Research on active control of the dynamic vibration for underwater robot. J. Balkan Tribol. Assoc. **22**(5), 770–779 (2016)
10. Qu, Z.-Y., Su, W.-Z., Zhao, L.: Distributed data storage method based on information decentralization algorithm. J. Comput. Appl. **26**(3), 1102–1105 (2016)

A Novel Ship Detector Based on Gaussian Mixture Model and K-Means Algorithm

Jingwei Yu[1], Qian Wan[1,2,3](✉), Qianqian Liu[5], Xinqiang Chen[4], and Zhibin Li[1]

[1] Southeast University, Nanjing 210096, China
1564656392@qq.com
[2] Hualan Design & Consulting Group, Nanning 530011, China
[3] Guilin University of Electronic Technology, Guilin 541004, China
[4] Guangxi University of Finance and Economics, Nanning, China
[5] Shanghai Maritime University, Shanghai 201306, China

Abstract. Ship detection in videos has become a popular research topic in maritime field due to its important role in the unmanned ship era. However, unknown environments, severe interference and high accuracy challenged existing ship detectors' performance. Now, more attention has been paid to detect ship in coastal or offshore areas. Less attention is being paid on detecting ships in inland waters as these areas can be monitored manually through existing surveillance videos. As we are entering unmanned era, less human-involvement is required in maritime related researches. Thus, to detect ships in inland waterways, we presented a novel framework for detecting ships based on Gaussian mixture model (GMM) and K-means clustering algorithm methods. Firstly, GMM is used to model general background in each ship frame. Thus, we can remove background pixels in each frame. In that way, we can obtain all ships pixels in each frame. Then, K-means clustering method is used to find each ship's position in ship-frames. Experimental results showed that our proposed ship detector achieved over 90% in both precision accuracy and recall rate. Findings of the research can provide essential and initial visual information for maritime authorities, vessel crews and other related traffic participants, and improve maritime safety.

Keywords: Ship detect · Multi-view · Gaussian mixture model K-means algorithm · Background suppression

1 Introduction

Waterway transportation is becoming importantly in our daily life. A large amount of goods has been shipped to other countries and continent due to its cost-effective characteristics. Hence, safety related issues have attracted a lot of attention by the field researchers due to its important role in reducing losses for human beings in the situation of maritime accidents. An efficient way of enhancing maritime safety capability is to regulate human beings' operations at sea. For instance, Chen proposed framework of checking, finding and reducing human beings' misconducts after marine accidents [1]. Christine Chauvin showed some important models and concepts in the psychology field

J. Abawajy et al. (Eds.): ATCI 2018, AISC 842, pp. 639–646, 2019.
https://doi.org/10.1007/978-3-319-98776-7_72

which help maritime researchers understand human beings' role in maritime safety deeply [2]. Some researchers also found that the misconducts caused by first-line operators (including captain or officials on watch) may lead to further and severe damages after maritime accident happened [3–5].

To reduce property loss caused by maritime accidents, more and more maritime practitioners and researchers tried many modern techniques to reduce maritime risk. In fact, development of computer vision technologies has provided us a new sight in solving the problem. A key problem in the field is to detect ships with a high accuracy and reliability. Ren used many methods to detect small targets in maritime surveillance videos. For instance, Ren extracted salience features of ships and thus obtain a satisfied detection result [6–8]. But, these methods mainly used visible-technology images to detect ships. Infrared based technologies have also attracted many researchers' interests. Lee proposed a novel multi-frame model for detecting moving ships [9]. Similar researches can also be found in [10].

As ship detection in videos provides visual tracking information which can be easily understandable by human beings, visual ship detecting has been an active topic in the field. A critical challenge for ship detecting is wave interference. That's to say, some ship detection algorithms may fail when encountering strong waves in background. To address that problem, we developed a novel ship detect model based on Gaussian Mixture Model (GMM) and K-means clustering algorithm. The remainder of paper is recognized as follows: Sect. 2 describes details of ship detecting framework. Section 3 shows some experimental detection results and conclusion is drawn in the last section.

2 Methodology for Ship Detecting

2.1 Framework Overview

Framework for the proposed ship detecting method includes background modelling and suppressing, targets clustering which are detailed in the following sub-sections. Specifically, the proposed framework tried to rule out background and clutters from image sequences such as water, wave and so on. Gaussian mixture model (GMM) is employed to model the dynamic background in our study. After background suppression process, K-means clustering method is used to cluster ships in the frame and each cluster represents a ship. Each cluster is considered as a detected ship in the image. The flowchart for the proposed ship detector is shown in Fig. 1.

2.2 Background Suppressing and Determining Region of Interest

2.2.1 Background Modeling and Suppressing with Gaussian Mixture Model

Background consists a large proportion in ship tracking image sequences. Thus there is a need of segmenting and suppressing clutter and background for better tracking performance. Gaussian mixture model and its variants had been studied in many background modeling and subtraction applications with satisfied results [11–14]. As our focus is not to acquire background suppression in pretty high accuracy, we employ

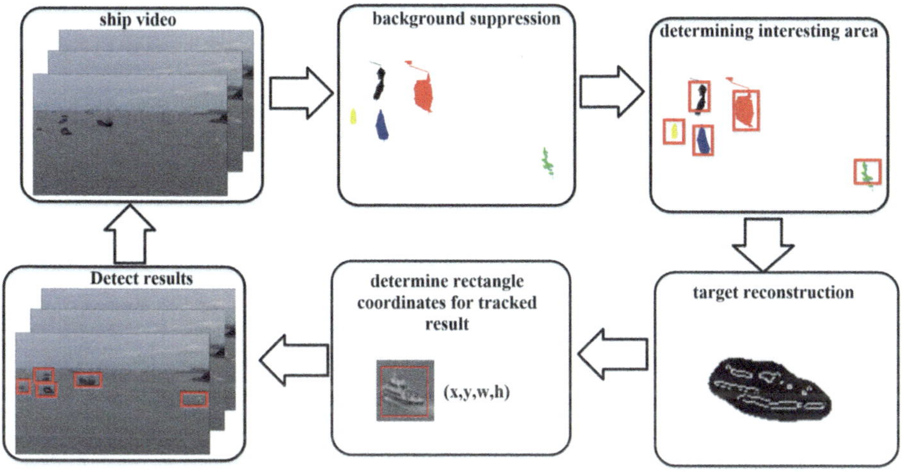

Fig. 1. Flowchart for the proposed ship detector.

classical GMM algorithm to model background in ship tracking videos. GMM collects and stores each pixel's intensity in consecutive frames which is called pixel process and Eq. (1) shows intensity set of given pixel (p_x, p_y) from first frame to nth frame [11, 12].

$$S_{intensity}(p_x, p_y) = \{i \in [1, n] | \mathrm{I}(p_x, p_y, i)\} \quad (1)$$

where $\mathrm{I}(p_x, p_y, i)$ is intensity of pixel at location (p_x, p_y) in the ith frame. $S_{intensity}(p_x, p_y)$ is intensity pixel set of (p_x, p_y) in image sequences. For simplicity, $\mathrm{I}(p_x, p_y, i)$ is abbreviated as $\mathrm{I}_{p_x, p_y, i}$

Noticed that intensity values of $S_{intensity}(p_x, p_y)$ are collected in successive frames, Possibility of obtaining pixel intensity from nth frame can be modeled in Eq. (2) by using mixture of K-Gaussian distribution.

$$P(\mathrm{I}_{p_x, p_y, n}) = \sum_{i=1}^{K} w_{i,n} * \Upsilon(\mathrm{I}_{p_x, p_y, n}; \mu_{i,n}; \Sigma_{i,n}) \quad (2)$$

In Eq. (2), $w_{i,n}$ represents weight of ith Gaussian in the mixture at nth frame. Parameters of $\mu_{i,n}$ *and* $\Sigma_{i,n}$ are the mean and variance values for the ith Gaussian distribution of nth frame respectively. $\Upsilon(\cdot)$ depicts corresponding Gaussian probability dense function which is formulated in Eq. (3). Parameter d of Eq. (3) represents image channel. In other words, current image is gray image if value of d equals 1. Symbol dis_{mah} is Mahalanobis distance which is formulated in Eq. (4).

$$\Upsilon(\mathrm{I}_{p_x, p_y, n}; \mu_{i,n}; \Sigma_{i,n}) = \frac{1}{(2\pi)^{\frac{d}{2}} |\Sigma_{i,n}|^{\frac{1}{2}}} e^{-\frac{1}{2} dis_{mah}^2} \quad (3)$$

$$dis_{mah} = \sqrt{\left(\mathrm{I}_{p_x,p_y,n} - \mu_{i,n}\right)^T \Sigma_{i,n}^{-1} \left(\left(\mathrm{I}_{p_x,p_y,n} - \mu_{i,n}\right)\right.} \tag{4}$$

It was inferred that covariance matrix $\Sigma_{i,n}$ in the diagonal form could achieve same modelling accuracy with lower computational complexity. Formulation of covariance matrix $\Sigma_{i,n}$ is depicted in Eq. (5) where $\sigma_{i,n}$ is standard deviation and I_{iden} is identity matrix.

$$\Sigma_{i,n} = \sigma_{i,n}^2 \mathrm{I}_{iden} \tag{5}$$

Based on above GMM model, we can assume n*th* frame intensity $\mathrm{I}_{p_x,p_y,n}$ can be modeled in GMM if Eq. (6) is satisfied. In Eq. (6), C is a constant with its value ranging from 2.5 to 3.5 approximately. Furthermore, background modeling is turned out to find optimal GMM performance by iteratively renewing parameters in equation set (7). If Eq. (6) cannot be met, it means all Gaussian distributions cannot match $\mathrm{I}_{p_x,p_y,n}$. Then values of $\mu_{i,n}$ and $\sigma_{i,n}$ remain same while $w_{i,n}$ is updated with $(1 - \beta)w_{i,n-1}$.

$$\left|\mathrm{I}_{p_x,p_y,n} - \mu_{i,n}\right| < C\sigma_{i,n} \tag{6}$$

$$\begin{aligned} w_{i,n} &= (1-\beta)w_{i,n-1} + \beta \\ \mu_{i,n} &= (1-\theta)\mu_{i,n-1} + \theta \mathrm{I}_{p_x,p_y,n} \\ \sigma_{i,n}^2 &= (1-\theta)\sigma_{i,n-1}^2 + \theta\left(\mathrm{I}_{p_x,p_y,n} - \mu_{i,n}\right)^T\left(\mathrm{I}_{p_x,p_y,n} - \mu_{i,n}\right) \end{aligned} \tag{7}$$

where β determines learning rate of GMM while θ determines learning factor.

2.2.2 Finding Ships with K-Means Clustering Algorithm

With help of GMM algorithm, we can get foreground pixels by removing majority background pixels such as water waves, shore-based buildings efficiently. Factually, the obtained foreground pixels may involve from different ships as a couple of vessels may appear in one frame. In another word, we need determine interesting area so that M2S2T can track ships in image sequences with higher efficiency and accuracy. Intuitively, pixels' distance from same ship in the same frame has smaller distance in comparison with their counterparts from other ships. In addition, each ship in the same frame varies in features such as shape, intensity, size etc. Such diversity provides potentiality of clustering same ship's pixels with clustering algorithm. Here, K-means method is employed to implement clustering process.

Suppose that we have acquired background-removed frame set $\mathrm{I}_{set-bkr}$, which has been converted to gray-scale image sequences. Each frame $\mathrm{I}_{set-bkr}(i)$, $i = 1, 2\ldots m$ in the frame set can be considered as two dimensional points. One dimension represents x-axis coordinate while another dimension for y-axis coordinate. For each frame, K-means algorithm divides its points into several clusters based on its grouping principle. Specifically, we set Cen_k as centroid of cluster cls_k and *p*(*j*) as the member of cls_k. Parameter *num* depicts total number of points in cluster cls_k. Min_{cls_k} represents minimum of squared error between *p*(*j*) and Cen_k. K-means algorithm obtains clusters by finding minimum Min_{cls} in Eq. (8) [15]. As background has been eliminated by background suppression process mentioned in above section, different clusters obtained by K-means method could be considered as different ships. Such clusters area is the region of interesting (ROI) that M2S2T will follow and track.

$$\text{Min}_{cls} = \sum\nolimits_{k=1}^{K} \text{Min}_{cls_k} \tag{8}$$

$$\text{Min}_{cls_k} = \sum\nolimits_{j=1}^{num} ||p(j) - \text{Cen}_k||^2 \tag{9}$$

3 Results and Analysis

Shanghai Port is one of busiest and largest port in China and huge amount of ships have been sailing and docking in Shanghai port every day. Hence, detecting ships taken from Shanghai port is meaningful for vessel surveillance and traffic control. Experiments were implemented with 8-GB RAM and 2.59-GHz processor. The source code was written in Matlab language. Our experiments included two cases, the case-a demonstrates our ship detector's performance on ships that have clear contours, while the case-b demonstrates our ship detector's performance on ships that have clear and ambiguous contours.

3.1 Measures of Goodness of Detection

We employed recall rate (Re) and precision rate (Pr) to evaluate performance of the ship detection models. Definitions of Re and Pr are shown in Eqs. (10) and (11), respectively. Re demonstrates ratio of real ship targets that detected by our detector. Larger value of Re represents better performance of our detector. The indicator Pr shows accuracy performance of our proposed ship detector. Similarly, larger Pr means that ship detector missed less targets during the detection procedure.

$$\text{Re} = \frac{T}{T + T_F} \tag{10}$$

$$\text{Pr} = \frac{T}{T + F_T} \tag{11}$$

where T the detected number where ships are the true targets. Parameter T_F represents detected ship number that ships have not been detected. The F_T is the detected ship number where targets have been detected as ships wrongly. The higher values of Pr and Re indicators show better performance.

3.2 Detection Results and Analysis

Generally, all of ship targets, in both large and small imaging sizes, have been detected correctly shown in Fig. 2. Firstly, targets detected in maritime images have shown that ships had clear shapes have been found successfully. That is to say, our detector can detect ships that far from our camera (i.e. closing to water-sky line). We found that although two ships were close to each other in the initial image, our detector have detected them separately. In addition, the size of detected ship, highlighted in the red rectangle, encircled the real-target ship was almost same as naked-eyes can recognize. In sum, under this situation, our detector has achieved satisfied performance with accurate detection result.

Fig. 2. Detection results on ships with clear contours.

Figure 3 showed another detection results when to-be-detected ships either have clear or ambiguous contours. It can be seen that two ships are closed to each other. Hence, the detector detects these two ships into a single ship. But, it is acceptable as the two neighboring ships are pretty close to each other. There is less significance for detecting these two ships into one ship. Another challenge in this ship-detecting image is that there is an ambiguous ship in the image. This ship is easily detected as a buoy or outlier due to its blurry image features. However, our proposed detector has removed background (such as buoys, waves etc.) successfully. The ship target with less clear contours was not classified as noises.

Quantitative results of detector performance of case-a and case-b are shown in Table 1. It can be shown that both case-a and case-b have obtained over 0.95 accuracy in the detecting situations. The recall rates of both cases are larger than 0.8. Both indicators of Re and Pr have verified the ship detector's performance. According to above analysis, it is trustable to say that our proposed ship detector showed trustable performance in challenging situations.

Fig. 3. Detection results on ships with clear and ambiguous contours.

Table 1. Statistical performance of the proposed detector

	Re (percent)		Pr (percent)	
Case-a	0.98	0.90	0.97	0.90
Case-b	0.97	0.83	0.95	0.81
Average	0.975	0.865	0.960	0.855

4 Conclusions

Ship detection is a prerequisite for many maritime related applications and the research still attracts many researchers' interests. This research proposed a novel framework for detecting ships in a robust and trustable way. Our framework included two steps which are background-modeling and suppressing, targets clustering. We have verified our detector's performance on many typical situations. We found that our detector not only works well in situations where ships have clear and observable contours. But, our proposed framework also worked well when ships' contours are ambiguous. In future, we will expand our researches in the following aspects. First, our detector requires good maritime weather conditions. That's to say, our detector cannot achieve satisfied detecting performance during terrible conditions. Besides, we would like to enhance our detector's performance under strong interference such as inland buildings.

Acknowledgement. The research is jointly supported by the Scientific Research Project of Ministry of Housing and Urban-Rural Development (2017-R2-032, 2017-K2-009), the National

Natural Science Foundation of China (51508122), and Guangxi Key Research and Development Program (GuiKe AB16380280).

References

1. Chen, H.S., Msa, W.: Analytic hierarchy process based SWOT analysis on passenger ship safety. China Maritime Safety (2015)
2. Chauvin, C.: Human factors and maritime safety. J. Navig. **64**(4), 625–632 (2011)
3. Er, Z., Çelik, M.: Definitions at human factor analysis for maritime safety management process. In: International Association of Maritime Universities-IAMU 5th Conference (2005)
4. Er, Z., Celik, M.: Definitions of human factor analysis for the maritime safety management process (2005)
5. Berg, H.P.: Human factors and safety culture in maritime safety (revised). Transnav Int. J. Marine Navig. Safety Sea Transp. **7**(3), 343–352 (2013)
6. Ran, X., Ren, L.: Small and weak target detection on sea surface based on visible light video image processing. J. Shanghai Maritime Univ. **2**, 004 (2010)
7. Ren, L., et al.: Saliency detection for small maritime target using singular value decomposition of amplitude spectrum. IETE Tech. Rev. **34**, 1–11 (2016)
8. Shi, C., et al. Architecture of vision enhancement system for maritime search and rescue. In: International Conference on ITS Telecommunications (2009)
9. Lee, H.J., Huang, L.F., Chen, Z.: Multi-frame ship detection and tracking in an infrared image sequence. Pattern Recogn. **23**(7), 785–798 (1990)
10. Crisp, D.J.: The state-of-the-art in ship detection in synthetic aperture radar imagery. Org. Lett. **35**(42), 2165–2168 (2004)
11. Stauffer, C., Grimson, W.E.L.: Adaptive background mixture models for real-time tracking. In: IEEE Computer Society Conference on Computer Vision and Pattern Recognition. IEEE (1999)
12. Mukherjee, D., et al.: Dual Gaussian mixture model with pixel history for background suppression. In: 2014 IEEE International Conference on Systems, Man, and Cybernetics, SMC. IEEE (2014)
13. Szpak, Z.L., Tapamo, J.R.: Maritime surveillance: tracking ships inside a dynamic background using a fast level-set. Expert Syst. Appl. **38**(6), 6669–6680 (2011)
14. Mukherjee, D., Wu, Q.J., Nguyen, T.M.: Multiresolution based gaussian mixture model for background suppression. IEEE Trans. Image Process. **22**(12), 5022–5035 (2013)
15. Jain, A.K.: Data clustering: 50 years beyond K-means. Pattern Recogn. Lett. **31**(8), 651–666 (2010)

Research on Intelligent Traffic Scheduling Based on Dijkstra Algorithm

Huiyan Shen(✉)

Lianyungang JARI Electronics Co., Ltd., Lianyungang, Jiangsu, People's Republic of China
ShenHYyeah@yeah.net

Abstract. With the development of intelligent card (IC), traffic monitoring, Radio Frequency Identification (RFID), Electronic Toll Collection (ETC) and other intelligent transportation technology, traffic managers can accurately measure of road traffic flow, which provide conditions for road traffic intelligent decision. This paper uses the city traffic network graph, and each road traffic flow measurement data, according to the length of road and road congestion condition, uses the Dijkstra algorithm, to provide the best travel route for the city's driver, which can effectively avoid the traffic congestion, improve the operation efficiency of city traffic.

Keywords: Dijkstra algorithm · Intelligent traffic
Radio frequency identification

1 Introduction

With the development of new technology, such as Internet of things, large data technology, artificial intelligence and cloud computing, traditional traffic engineering has entered the age of intelligence. The use of intelligent information collection and decision-making technology can reasonably manage city traffic, to avoid traffic congestion and reduce air pollution. Many existing Internet of things technology can achieve the collection function of traffic information, for example, broadband wireless access technology and mobile terminal technology can achieve real-time transmission of road conditions. In literature [1–3], the use of intelligent transportation networking system based on Internet of things can realize the real-time acquisition of the city road traffic network information, intelligent transportation based on Internet of things is to apply sensor technology, wireless communication technology, radio frequency identification (RFID) technology, network technology and global positioning system to traffic transportation management system, establish a real-time, accurate and efficient transportation management system and control network. In literature [1], radio frequency identification (RFID) technology is used to contact license plate number with RFID tag ID through the electronic license plate form, and stored into electronic tag; RFID detection device (Reader) arranged in the road network can collect vehicle identification information, location information and other the traffic information (such as the number of vehicles, vehicle location, the acquisition time), and can realize the automatic violation fines, illegal information records and other functions. RFID

J. Abawajy et al. (Eds.): ATCI 2018, AISC 842, pp. 647–651, 2019.
https://doi.org/10.1007/978-3-319-98776-7_73

technology is a non-contact automatic identification technology implemented by radio frequency communication [4–6]. Its characteristics are: non-contact, two-way communication, location, long time tracking management and all-weather information collection, and less affected by environment. RFID technology, combined with Internet, communication and other technologies, can achieve global object tracking and information sharing [7–9].

This paper assumes that the RFID readers are deployed in every road in the city traffic network, the car is equipped with RFID card, RFID reader can read RFID card information by car on the road, and statistic traffic flow and velocity, to quantify the degree of congestion of the road. The cost function is constructed based on the flow and velocity of the road and the length of the walking path. According to the principle of Dijkstra algorithm, we find out the shortest and most smooth route for the driver.

2 Problem Modeling

According to the connection of city traffic network, the traffic network topology map is set up with the intersection as the node and the road connecting the intersection as edge. This paper represents the intersection position by * according to the coordinates of the intersection. The line connecting the intersection represents the road and forms the city network traffic map (see Fig. 1).

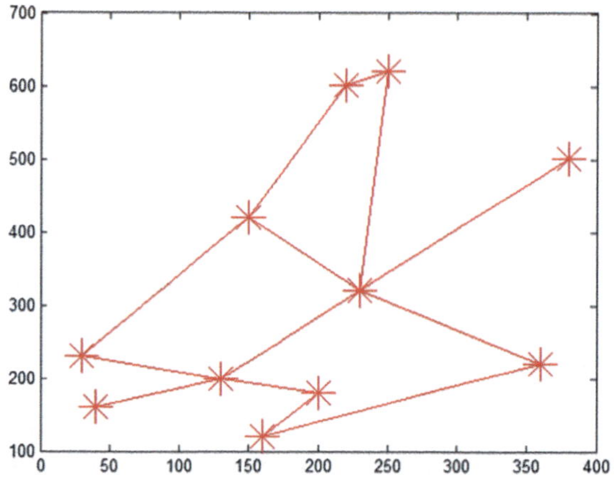

Fig. 1. City traffic network map

When calculating the cost of each road in the city network traffic map, there are mainly two factors: (1) the length of the road. That is the distance between the ends of the road, which is denoted by L; (2) the road congestion level. That is marked by the vehicle flow velocity by the RFID reader records, which is denoted by S. Considering the above two factors, the cost function of establishing the vehicle through the road is as follows:

$$f = \alpha * L + \beta * S \tag{1}$$

Among them, α, β are the weight values of road length and road congestion.

3 Dijkstra Algorithm Introduction

Dijkstra algorithm is proposed by Dijkstra E W in 1959, which produces the shortest path according to the increasing order of path length. It is considered to be a classical and effective algorithm to solve the shortest path problem between single source points.

When calculating the shortest path in a graph by the Dijkstra algorithm, you need to specify the starting point s (that is, the calculation starts from the vertex s). In addition, two sets of S and U are introduced. The role of S is to record the vertexes which have find out the shortest path and the shortest path length, and the role of U is to record the vertexes which have not yet obtained the shortest path (the distance from the vertex to the starting point s). At the beginning, there is only the starting point s in S; U is all the vertices of the s point, and the path of the vertex in the U is the path from the starting point s to the vertex. Then, find the vertex with shortest path from the U and add it to the S; then, update the vertices and vertices corresponding to the path in the U. Then, we find vertex with the shortest path from the U and add it to the S; then, update the vertex and the vertex corresponding to path in the U. Repeat the operation until all the vertices are traversed.

4 Algorithm Description

A RFID reader is set up on every road in city traffic network, and the flow rate of the motor vehicle is counted. Then the cost value of the road is calculated according to the formula (1), as the same, the cost value of all roads in city traffic network is given. Making use of the traffic network graph with cost value for each edge, combined with the Dijkstra algorithm, we search the optimal travel path for motor vehicle drivers based on the trade-off between the length of roads and the degree of road congestion. The specific operation steps are as follows:

(1) Use formula (1) to calculate cost for each edge of traffic network map, and establish the traffic network cost graph.
(2) At the beginning, S contains only the starting point s. U contains other vertices except s, and the distance of vertex in U is the distance from s to the vertex. If s is directly connected with vertices in U, then the distance of the vertex in U is the weight value from s to the vertex in network graph; otherwise, it is infinity.
(3) Select the vertex K with the shortest path from U and add the vertex K to S; meanwhile, the vertex K is removed from the U.
(4) Update the distance from each vertex in U to the starting point s. The reason for updating the distance of vertices in U is that K is the vertex with the shortest path to the starting point in the last step, so that K can be used to update the distance of other vertices, that is as follows:

$$\text{dis}[j] = \min\{\text{dis}[j], \text{dis}[k] + \text{dis}\{k, j\}\}.$$

(5) Repeat steps (3) and (4) until all the vertices are traversed.

5 Simulation Results and Analysis

In order to verify performance of algorithm, the simulation is carried out on the established city network traffic map (see Fig. 1). In Fig. 1, select the starting point and target point. When there is no congestion in the road, the speed of the motor vehicle is fast. The main factor that affects the road cost function is the length of the road. In this case, the travel path calculated by our algorithm is shown in Fig. 2, which is the optimal path in length and can assist the driver along the nearest path to the target.

Figure 3 shows a section of road congestion occurs on the shortest path, the RFID reader on the road records statistical information of vehicle velocity and feedback to cost function of the road, meanwhile increasing cost value of road congestion.

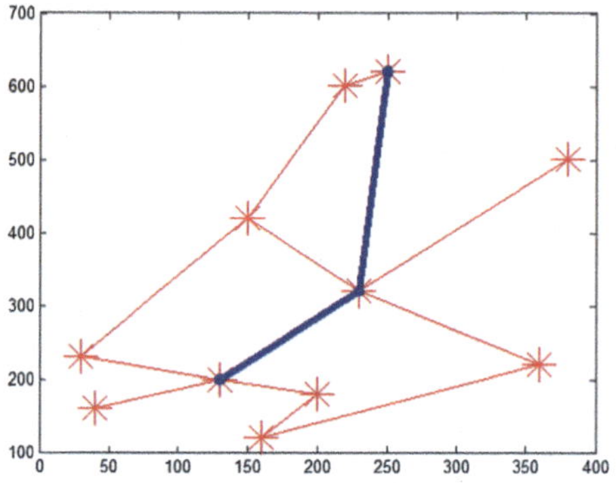

Fig. 2. The optimal path search under road is free

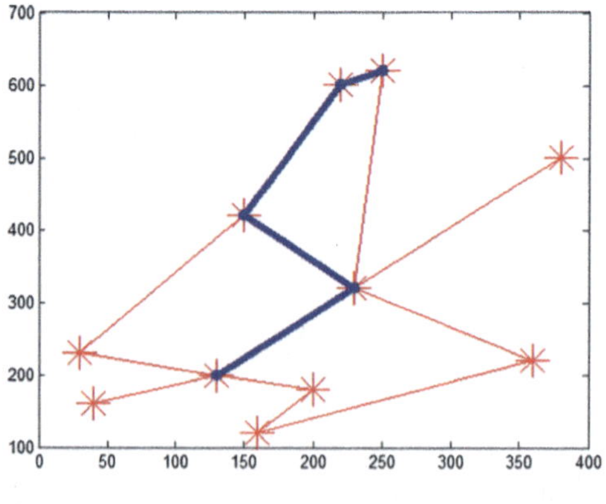

Fig. 3. The optimal path search under road is congestion

According to the updating network information, this article calculates the optimal path in Fig. 3, which avoids road congestion, and completes a good traffic scheduling function.

6 Conclusion

In this paper, the RFID reader is used to record vehicle flow velocity of each road in city network map. Considering vehicle flow velocity and length of each road, the road cost function is constructed. On the traffic network with assigned value, considering the path length and road congestion, the Dijkstra algorithm is used to provide the best travel route for the city's driver, which can avoid the traffic jam, and can provide a near-shortest travel route for drivers, improve the operating efficiency of the city.

References

1. Gan, Q., Zhou, J.: Application of RFID in realtime road traffic information acquisition system. Technol. Highw. Transp. **6**(3), 125–129 (2012)
2. Chen, X., Shi, J.: Context-awareness of vehicle smart terminal and the system implementation. J. Nanjing Univ. **51**(2), 243–250 (2015)
3. Yan, P., Zhang, X.: Smart home based on internet of things. J. Nanjing Univ. **48**(1), 30–34 (2014)
4. Xiao, W., Xu, R.: Intelligent urban public transportation system based on RFID-information collection and release system. Road Traffic Saf. **6**(9), 25–28 (2006)
5. Younghwan, Y., Jinhwan, K., Haosong, G., et al.: RFID reader and tag multi-hop communication for port logistics. Korea Information and Communication Society (2009)
6. Zheng, W.: RFID Application and Simulation Research of RFID Technology in Intelligent Transportation System. Dalian University of Technology, Dalian (2009)
7. Anagnostopoulos, C.N.E., Anagnostopoulos, I.E., Loumos, V., Kayafas, E.: A license plate-recognition algorithm for intelligent transportation system applications. IEEE Trans. Intell. Transp. Syst. **7**(3), 377–392 (2006)
8. Wen, Y., Lu, Y., Yan, J.: An algorithm for license plate recognition applied to intelligent transportation system. IEEE Trans. Intell. Transp. Syst. **12**(3), 830–845 (2011)
9. Dalida, J.P.D., Galiza, A.J.N., Godoy, A.G.O.: Development of intelligent transportation system for Philippine license plate recognition. In: Region 10 Conference, pp. 3762–3766 (2017)
10. Li, J., Ren, Q.: Study on supply air temperature forecast and changing machine dew point for variable air volume system. Build. Energy Environ. **27**(4), 29–32 (2008). (in Chinese)

Construction of the Network Learning Community of the New Generation of Migrant Workers

Jun Mo[1,2] and Chunmei Mao[3(✉)]

[1] School of Business, Hohai University, Nanjing 211100, China
[2] Wuxi Vocational College of Science and Technology, Wuxi 214028, China
[3] School of Public Administration, Hohai University, Nanjing 211100, China
maochm@hhu.edu.cn

Abstract. With the rapid development of information technology and the improvement of network learning function, network learning has put forward higher requirements for the positioning of the role of the new generation of migrant workers. The new generation of migrant workers are discussed between the network learning community and the network environment in this paper. It points out that the new generation of migrant workers are provided a new platform for the development of online learning community, and then the significance and the present situation of the migrant workers community network learning is analyzed under the network environment. Finally the approach of construction of network learning community for migrant workers is given under the network environment.

Keywords: Network learning community · New generation · Migrant workers ThinkDB system

1 Introduction

With the development of computer network, the concept of network learning community has been formed [1]. The e-learning community is a learning group composed of learners and their assistants, including teachers, experts and tutors in the virtual environment of the network [2, 3]. In the community, the learners can mutually communicate, exchange and cooperate each other based on the common learning willingness and specific learning tasks in the network environments and they can share all kinds of learning resources, learning experience and emotional study between learners and facilitators. In order to build a good interpersonal relationship, mutual influence and promotion each other are finished.

Every member of the network learning community of migrant workers is the receiver of knowledge, experience and skills. The learning achievement of the community has important guiding significance for the development of migrant workers and the actual learning activities. At the same time, communication in the e-learning community is basically generated in learning activities, and it brings strong practicality in the process of deliberative discussion. In fact, it is also a simulation of learning

J. Abawajy et al. (Eds.): ATCI 2018, AISC 842, pp. 652–656, 2019.
https://doi.org/10.1007/978-3-319-98776-7_74

activities. To a large extent, it will promote communication and communication between migrant workers.

2 ThinkDB System of Networked Learning Community of the New Generation of Migrant Workers

The ThinkDB system can meet the requirements of integrated data management in the field of wide area monitoring. ThinkDB system can carry on the traditional management mode based on relational data, and it will be online processing data requirements and features with real-time data acquisition to be integrated, so as to achieve the purpose of the data storage system developed through the use of innovative management mode. The following figure showed the characteristics of the application of the system, which can perform real-time data acquisition and processing, intelligent, Internet, realize the collection, processing, storage for high speed data etc, and can realize the fusion and management of real-time data, structured data and unstructured data.

Real time data acquisition is to collect the data of the external environment accurately and effectively, and to process the collected data in a certain period of time or a certain period of time. The data processed by the system were often only valid, such as the data generated by the equipment in the industrial field and the pressure and temperature of the sensor. Real time database was a system that can be used to process, store, manage and collect data in real time. Innovative real-time database system was mainly used in the field of control, measurement and control, military, space exploration, aerospace and other fields, because the system of the transaction and data had the characteristics of time. The database system was the traditional system which was developed on the basis of permanent data processing on, and emphasized the consistency and integrality of data and ignored the related data of the time limit. The innovation of real time data system has been improved to this disadvantage.

The system can not only provided data and factual data, combined with the correlation analysis, the subscription fusion application services, but also provided real-time data characteristics, historical data query, real-time data and history data publish subscribe analysis etc. And it was also able to provide access to a standard relational database. The data with the characteristics storage and in accordance with the traditional model of structured data management so as to provide technical support for enterprise's comprehensive data management, which can meet the needs of data storage in different fields and different industries.

Standard package API functions was provided by the ThinkDB system and the system will provide a standard version of the API interface so as to achieve the user in different language environment for the use of the system when designing the system architecture for data storage system. In addition to providing the above content, the system will also provide access to management data and relational data based on SQL. In order to achieve a transparent access to the database process, the speed of the development process of the database application was increased greatly and used a combination of different databases used commonly in the API function in the access system.

3 Fuzzy Cognitive Map Approach

In case of the learning FCM, each chromosome consists of n(n − 1) genes is defined as follows

$$CR = \left[w_{12}, w_{13}, \cdots, w_{1N}, w_{21}, w_{23}, \cdots, w_{2N}, \cdots, w_{N,N-1}\right] \tag{1}$$

Where, w_{ij} specifies a weight for an edge from ith to jth factor node, which are floating point numbers from the range [−1, 1], n is the size of the FCM model.

The fitness function f is defined as

$$f() = h(Error) \tag{2}$$

$$h(x) = \frac{1}{\alpha x + 1} \tag{3}$$

$h(x)$ is used for formulating the fitness in the range (0, 1). α is established experimentally, usually set to be 100 or 10.

Let $\widetilde{C^{(t)}}$ define the output vector by initial model W at time t. Suppose the given input data length is L. Based on the initial model W and the same beginning vector $C^{(0)}$, a series of output vector is generated as: $C^{(0)}, \widetilde{C^{(1)}}, \widetilde{C^{(2)}}, \cdots, \widetilde{C^{(t)}}, \cdots \widetilde{C^{(L-1)}}$. So $Error$ is the difference between $\widetilde{C^{(t)}}$ and $C^{(t)}$, defined as

$$Error = \beta \sum_{t=1}^{L-1} \sum_{i=1}^{n} \left| C_i^{(t)} - \widetilde{C_i^{(t)}} \right|^2 \tag{4}$$

$$\beta = \frac{1}{(L-1) \cdot n} \tag{5}$$

Where β is the parameter used to normalize the model.

The data collected by the system was constantly changing, especially the real-time data in the data, which was collected from the data generated by the instrument. In order to make the user more convenient, the system also had a two development framework that can be used for custom purposes. The following Fig. 1 showed the ad hoc network of data storage system. The numbers in Fig. 1 are all IP addresses.

The tool that can realize the real-time synchronization of data must be installed independently, because it was not part of the ThinkDB aware database, but was the component independent to the ThinkDB system. It can realize the function of data communication, data perception in the database and real-time data transmission. It can greatly improve the security of the data in real time synchronization by using the network physical isolation technology, and also is able to achieve a one-way transmission of the source database to the mirror database. The data in the ThinkDB database system and Server SQL was migrated so as to establish multi database integration system which can use data backup and recovery tools in two ways for the database system of the data stored to delete backup, renew the backup and restore the

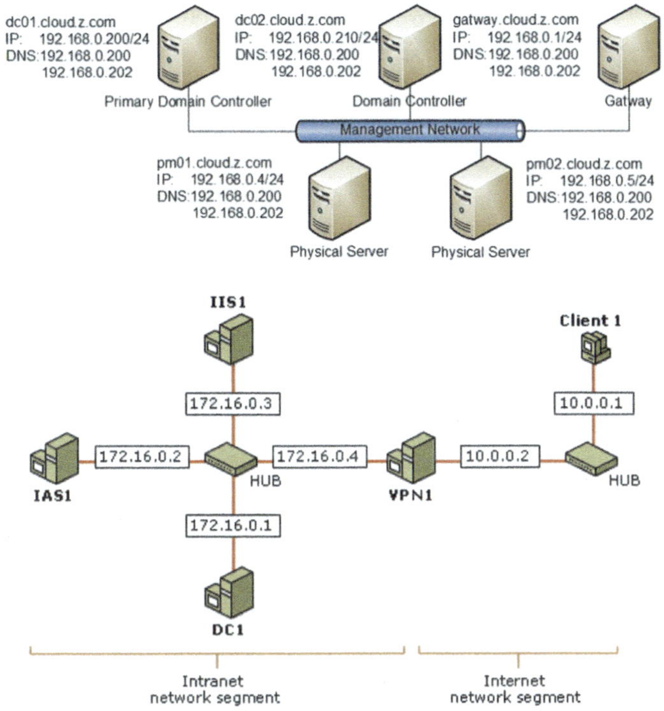

Fig. 1. The ad hoc network of data storage system

needed created data. One way is to achieve the full backup of the data through the command of the way, and the other way is to achieve data backup function by using the recovery and backup data management interface.

4 Conclusion

In conclusion, the development of web-based learning community surpasses the traditional knowledge transfer learning mode, including the experience of the new generation of migrant workers' practical problems, creating practical knowledge and enhancing the sense of mission of the new generation of migrant workers in lifelong learning. The development of the new generation of migrant workers will become the trend of the new generation of migrant workers in the future. The new generation of migrant workers learning community based on the network needs more effective support and guidance to achieve greater results.

Acknowledgment. This work was supported by Project of Social Education Planning for Jiangsu Province in 2017 (Grant No. JSS-C-2017001) and Fund Guide Project of Wuxi Professional College of Science and Technology in 2016 (Grant No. RJ1607) and Special Project of

Foreign Language Teaching of Philosophy and Social Sciences in Colleges and Universities of Jiangsu Province in 2016 (Grant No. 2016SJA740021).

References

1. Kelleret, P., Tchernev, N., Force, C.: Object-oriented methodology for FMS modeling and simulation. Int. J. Comput. Integr. Manuf. **10**(6), 405–434 (2017)
2. Anglarti, A., Grieco, A., Pacella, M.: Object-oriented modeling and simulation of flexible manufacturing systems: a rule-based procedure. Simul. Model. Pract. Theory **10**(10), 209–234 (2012)
3. Wolfgang, K., Kasper, O.: Beta SIM-A framework for discrete event modeling and simulation. Simul. Pract. Theory **6**(6), 573–599 (2008)
4. Bmzzone, A., Orsonl, A.: AI and simulation-based techniques for the assessment of supply chain logistic performance. IEEE Comput. Soc. **65**(21), 154–164 (2013)
5. Hwang, H.S.: An improved model for vehicle routing problem with time constraint based on genetic algorithm. Comput. Ind. Eng. **42**(2–4), 361–369 (2012)
6. Luis, G.: A result on projection for the vehicle routing problem. Eur. J. Oper. Res. **85**(3), 610–624 (2015)
7. Dimitris, J.B., David, S.: A new generation of vehicle routing research: robust algorithms, addressing uncertainty. Oper. Res. **44**(2), 286–304 (2016)
8. Lin, S.: Computer solutions of the travelling salesman problem. Bell Syst. Comput. J. **44** (10), 2245–2269 (2015)
9. Lin, S., Kernighan, B.W.: An effective heuristic algorithm for the travelling salesman problem. Oper. Res. **21**(2), 498–516 (2013)
10. Duhamel, C., Potvin, J.Y., Rousseau, J.M.: Tabu search heuristic for the vehicle routing problem with backhauls and time windows. Trans. Sci. **31**(1), 49–59 (2017)

Research on Visualization of Intelligent Substation Information Flow Based on Simulated Annealing Algorithm

Xian Zheng Feng(✉), XiaoJun Liao, HongJun Li, and Bin Zhao

State Grid Sichuan Electric Power Company Skill Training Center,
Chengdu, China
fengxianzheng@foxmail.com

Abstract. The smart substation SCD file is a configuration file that can reflect the system configuration information. It describes the instance configuration and communication parameters of all intelligent electronic devices in the substation, the communication configuration between the IEDs, and the primary system structure of the substation. The large number of SCD configuration files is not desirable due to artificial memory. At present, it is mainly used special software to read, and then manual search, this method is time-consuming and laborious and error-prone. The visualization method based on the simulated annealing algorithm uses the network topology diagram to read the information flow relationship between the IEDs of the entire station, so that the information interaction between the target IED and the associated IED can be conveniently viewed, and the debugging and maintenance efficiency is improved.

Keywords: Smart substation · SCD file · Simulated annealing algorithm
Information flow

1 Introductions

SCD (System Configured Description) file full name is the intelligent substation system configuration description file, that is, the only system configuration file of the entire station. This file describes the instance configuration and communication parameter information of all intelligent electronic devices (referred to as IED) of the smart substation, and between the IEDs. Contact information and substation configuration information such as system configuration [1]. In accordance with the SCD file management software provided by the manufacturer, SCD information flow diagrams are mostly displayed on a single device and cannot be used well for information flow display of the entire station. Therefore, the grid inspection department put forward the need for visualization of the entire station SCD file. Visualization of SCD files is represented by IED devices as nodes. The connections between IEDs are connected by a straight line. After the algorithm is deployed, the SCD information flow is uniformly distributed, the graphics are clear, and the intersections are few. It is easy for users to distinguish.

J. Abawajy et al. (Eds.): ATCI 2018, AISC 842, pp. 657–663, 2019.
https://doi.org/10.1007/978-3-319-98776-7_75

At present, the visualization methods for automatic layout of network diagrams mainly include simulated annealing algorithm [2], ant colony algorithm [3], and force-guided algorithm [4]. The ant colony algorithm has a long solving time. Although adjustment methods can be adopted, it is computationally time-consuming. Still cannot be solved, it is difficult to be accepted. The force-guided algorithm has a short solution time and can be used in scenes that require fast graphics. The force-guided algorithm does not optimize the graphical route. Therefore, it is necessary to study more rapid and optimized graphical visualization algorithms. Simulated annealing is actually a greedy algorithm, but its search process introduces random factors. The simulated annealing algorithm accepts a solution that is worse than the current solution with a certain probability, so it is possible to jump out of this local optimal solution and reach a global optimal solution. At present, it has been widely used in engineering, such as VLSI, production scheduling, control engineering, machine learning, neural networks, and signal processing and other fields. In this paper, the simulated annealing algorithm is improved, and the program is verified with Matlab software. It is used to visualize the information flow of intelligent power-transformed SCD files, and the results are analyzed. It is convenient for the debug personnel to view and improve the debugging and maintenance effectiveness.

2 Smart Substation Information Flow

In a smart substation system, it can be divided into a three-tier model and a two-tier network. The three-tier model is the station control layer, the bay layer, and the process layer. The two-tier network is the process layer network and the station control layer network. The process layer includes transformers, circuit breakers, disconnects, current/voltage transformers and other primary devices and their associated intelligent components as well as independent intelligent electronic devices. Spacer equipment generally refers to secondary devices such as relay protection devices, system measurement and control devices, and monitoring IEDs. It implements the function of using one interval of data and acts on the primary device of the interval, i.e., various remote input/output sensors. Communicate with the controller. The station control layer includes the automation station level monitoring and control system, station area control, communication system, and time synchronization system, etc., and realizes monitoring, control, alarming, and information interaction functions for the entire station equipment, and completes data acquisition and supervisory control (SCADA) and operation. Locking and synchronization pastor acquisition, electrical energy collection, protection information management and other related functions.

The information flow in the substation is divided into time-based messages, manufacturing message specifications, sampled value messages, and object-oriented substation event messages. Among them, the information transmission between substation and control mainly includes telemetry, remote signaling, remote control, and remote adjustment; the information transmission between substation devices is mainly the information of workstation switching and calling other devices; information between substation layer and bay layer Under normal circumstances, the transmission is mainly a small amount of information flow that is transmitted regularly. When a substation accident occurs, information on status changes of switches and switches, etc. caused by the fault must be transmitted preferentially, which will increase the flow of information. The information transmission between the equipment at the interval layer mainly includes adaptive relay protection, back-up protection and other information flow, and the flow rate is small. The characteristics are that the real-time performance is high and the probability of occurrence is small; the information transmission between the interval layer and the process layer. It is mainly the exchange of relevant information in the interval layer; the information transmission between the process layer devices mainly includes the SV message, the status of the primary device upload and diagnostic information, and the time synchronization message. The information flow is stable and continuous, with high traffic and high real-time performance. The schematic diagram of intelligent substation information flow transmission is shown in Fig. 1. This article focuses on the visual presentation of information flows between IEDs.

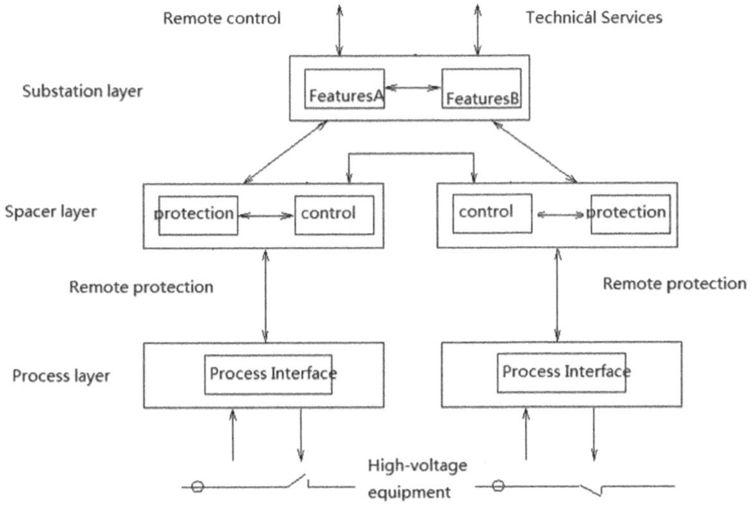

Fig. 1. Transmission diagram of intelligent substation information flow

3 Simulated Annealing Algorithm

3.1 Algorithm Introduction

Simulated annealing is actually a Greedy algorithm, but its search process introduces random factors. Simulated annealing algorithm accepts a solution with a certain probability to be worse than the current solution, so it may jump out of this local optimal solution. It is a stochastic optimization algorithm based on Monte-Carlo iterative solution strategy. Its starting point is based on the similarity between the annealing process of solid matter in the physics and general combinatorial optimization problems. The simulated annealing algorithm is based on the principle of solid annealing. The solid is heated to a sufficiently high temperature, and then allowed to cool slowly. When the temperature is increased, the internal particles of the solid become unordered with increasing temperature, the internal energy increases, and the particles gradually cool down. In an orderly manner, the equilibrium state is reached at each temperature, and finally the ground state is reached at room temperature, and the internal energy is minimized. According to the Metropolis criterion, the probability that a particle tends to balance at temperature T is $\exp(-\Delta E/(kT))$, where E is the internal energy at temperature T, ΔE is its change, and k is the Boltzmann constant. Using the solid annealing simulation combinatorial optimization problem, the internal energy E is modeled as the objective function value f, and the temperature T is evolved into the control parameter t, which is the simulated annealing algorithm for solving the combination optimization problem.

3.2 Algorithm Steps

Table 1. Algorithm steps

Step number	Algorithm content
1	Read the stream and save it in the adjacency table
2	Randomly distribute initial node positions
3	Determine if it is a fixed node. If it is not a fixed node, proceed to the next step
4	Given the initial state randomly, set a reasonable annealing strategy. (Select each parameter value, initial temperature T0, cooling rate, etc.)
5	Let $x' = x + \Delta x$ (Δx be a small uniformly distributed random disturbance), calculate $\Delta E = E(x') - E(x)$
6	If $\Delta E < 0$, accept x′ as a new state, otherwise accept x′ with probability $P = \exp(-\Delta E/(kT))$, where k is Boltzmann's constant. The specific method is to generate a random number a between 0 and 1, if $P > a$ then accept x′, otherwise reject x′, the system remains in the state x
7	Repeat steps 5 and 6 until the system reaches equilibrium
8	Reduce the temperature according to the law given in step 4 and re-execute steps 5–7 at the new temperature until $T = 0$ or a predetermined low temperature
9	Iterate n times until you reach the desired effect
10	Output visualization

The algorithm steps are as follows (Table 1).

4 Algorithm Verification

Using the algorithm proposed in this paper, the information flow of a smart substation SCD file is visualized. The number of IEDs in the substation is 131, and there are 400 information flows between IEDs. The initialization parameters are: IED number 131,

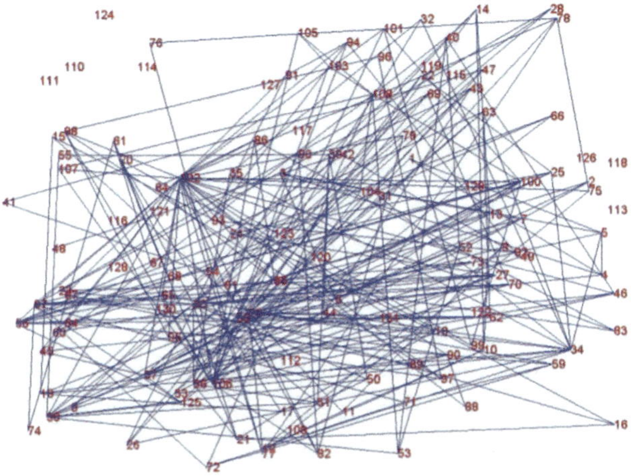

Fig. 2. The initial position before using the algorithm

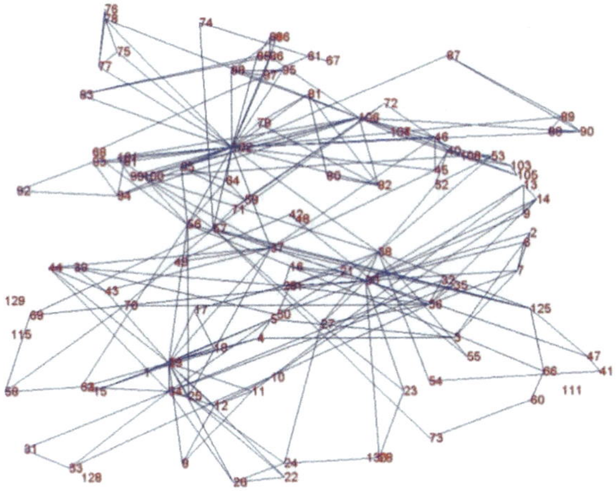

Fig. 3. Layout results of the simulated annealing algorithm

initial temperature = 13100, termination temperature = 0.001, temperature decay

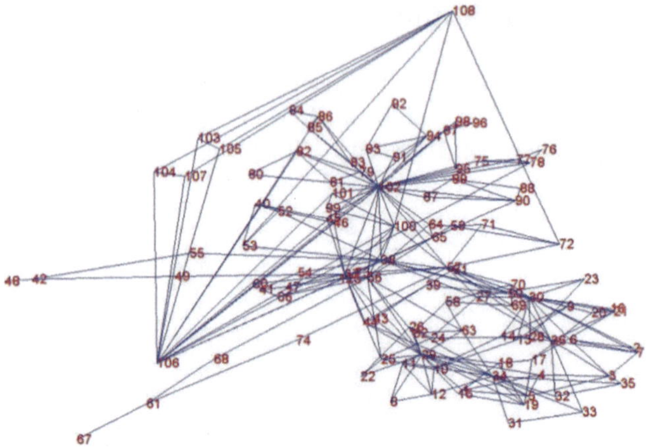

Fig. 4. Simulated annealing algorithm layout result after bus bar protection

rate = 0.98, internal Monte Carlo iteration number = 100. Figure 2 shows the initial position before the algorithm is used, Fig. 3 shows the layout result using the simulated annealing algorithm and Fig. 4 shows the layout result of the simulated annealing algorithm after the fixed bus protection device. The results show that the number of intersections of the edges and the distribution uniformity of the nodes are significantly reduced. After this algorithm is adopted, it is convenient to view the connection relationships between the nodes.

5 Conclusions

Using the simulation-based algorithm proposed in this paper, the intelligence substation information flow graph has fewer intersection points, clear structure, and convenient viewing. It takes less time to generate the graphs each time using the algorithm. In practical applications, the bus bar protection device, line protection device, and transformer protection device can be displayed. This is not possible with the visualization software Gephi. Set the initial parameters of the algorithm. Using this algorithm, the information flow between the target IED and the associated IED can be conveniently viewed, which facilitates debugging personnel's viewing and improves the efficiency of debugging and maintenance.

References

1. mc 61850–6 Communication Networks and Systems in Sub—Station Part 6: Configuration description language for communication in electrical substations related toreds [S3]
2. Wang, Y.: Study on Automatic Layout of Power Flow Diagram Based on Simulated Annealing. Hangzhou Dianzi University, Hangzhou (2009)

3. Lu, Z., Li, X.: Automatic layout of transmission line single-line diagram based on ant colony online line loss analysis. Electr. Power Syst. Autom. **35**(21), 74–77 (2011)
4. Zhang, J., Ye, Y., Chen, L., Chen, J., Zhan, Z., Zhuang, X.: Automatic layout of provincial power grid uniform wiring diagram based on new force guidance algorithm. Electr. Power Syst. Autom. **37**(11), 107–111 (2013)
5. Aleksy, M., Korthaus, A., Schader, M.: Use Java and the CORBA realization distribute type system. J. Pingxiang Coll. **4**, 104–105 (2005)
6. Titus, J., ECN Technical Editor: The eclipse of stand. J. Zhongkai Agrotechn. Coll. **19**(2), 32–35 (2006)
7. Richardson, W.C., Avondolio, D.: The Java high class weaves a distance: JDK 5. Sci. Technol. Book Rev. **3**, 17–18 (2006)
8. Vuong, J.L.: A semantics-based routing scheme for grid resource discovery. In: E-Science: First International Conference on E-Science and Grid Computing, pp. 58–70, 90 (2005)
9. Wang, Q.: Automatic control system regulation scheme of textile air conditioning. Sci. Technol. Inf. **22**, 395–396 (2010). (in Chinese)
10. Liu, J.: Advanced PID Control and MATLAB Simulation, 2nd edn. Electronics Industry Press, Beijing (2003). (in Chinese)
11. Li, J., Ren, Q.: Study on supply air temperature forecast and changing machine dew point for variable air volume system. Build. Energy Environ. **27**(4), 29–32 (2008). (in Chinese)

Cloud Service Discovery and Key Matching Methods in the Semantic Internet of Things

Jian Nong[1(✉)] and Pengyu Xie[2]

[1] Guangxi Colleges and Universities Key Laboratory of Professional Software Technology, Wuzhou University, Wuzhou 543000, China
jipbqd@yeah.net
[2] Guangxi Power Grid Co., Ltd., Nanning 530023, Guangxi, China

Abstract. The nature of the Internet of things is the Internet connected by objects and things, which is a bridge to realize the seamless link between the physical world and the information world. But the diversity of the object information and the lack of understanding of the main body can not make the data perceived by the Internet of things meet the needs of the information world, which hinder the development of the Internet of things. Advanced applications and services require explicit semantic knowledge, so the Internet of things requires the support of semantic technology. This paper gives the main framework of cloud service discovery method in semantic Internet of things, and puts forward the technical support of cloud service discovery method based on intelligent algorithm. Firstly, the ontology of cloud computing involved in the discovery process is built by referring to the mutual constraints in cloud services and obtaining common knowledge in the field of cloud computing. Secondly, the different matching methods are proposed for the matching of services under different circumstances. Finally, the evaluation method of query results is designed in detail.

Keywords: Cloud service discovery · Semantic internet of things
Matching methods · Cloud ontology

1 Introduction

With cloud computing attention, the number and types of cloud services provided by cloud computing will increase. How to quickly find the required cloud service becomes the key problems to be solved in the field of cloud computing. The traditional cloud service discovery method is based on the search of keyword. Users need to query their own cloud services manually. The query results will contain irrelevant information, which is easy to miss relevant services, and it can not meet users' advanced query requirements. In order to solve the existing problems in the process of the cloud service discovery, according to the different requirements the network cloud service discovery method based on ontology semantic object cloud is proposed to provide more precise information for users of cloud services.

J. Abawajy et al. (Eds.): ATCI 2018, AISC 842, pp. 664–669, 2019.
https://doi.org/10.1007/978-3-319-98776-7_76

2 Matching Method of Network Cloud Service Discovery Based on Ontology Semantic Object Cloud

2.1 Matching Process

Aiming at the different grammatical differences of the same semantic concepts in cloud services, we design the equivalent matching. Aiming at the good inheritance relationship and obvious hierarchical structure of cloud ontology, we design the inclusion matching. In addition, considering the general situation of matching time, we design the matching of services in different cases, such as the exact matching of concepts and the similarity matching between concepts. This paper proposes a variety of service matching methods in order to avoid missing the services related to user requests, and reduce the retrieval of services that are not related to user needs and the work of cloud service requesters in finding cloud services.

The matching stage of network cloud service discovery based on ontology semantic object cloud is shown in Fig. 1.

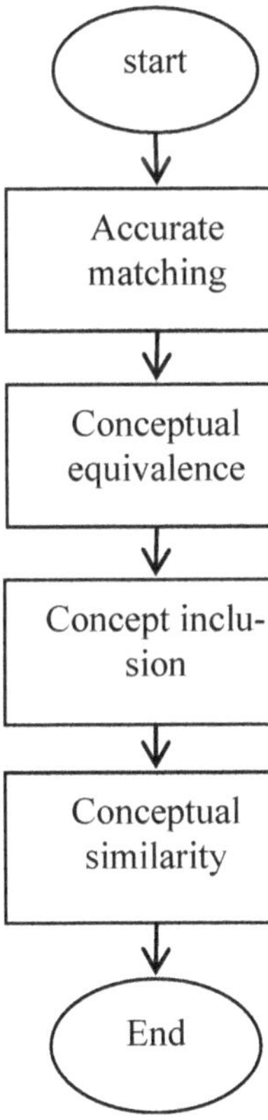

Fig. 1. Cloud ontology based cloud service discovery matching process in semantic Internet of things

2.2 Similarity Matching

The common nodes are calculated by statistics of two conceptual nodes. The similarity of concepts represents the commonality between concepts. The formula for calculating conceptual similarity is shown in Formula (1).

$$Sim_{conp}(x,y) = \rho\frac{|\alpha(x)\cap\alpha(y)|}{|\alpha(x)|} + (1-\rho)\frac{|\alpha(x)\cap\alpha(y)|}{|\alpha(y)|} \tag{1}$$

Among them, ρ is an influence degree and the range of value is [0, 1].

$\alpha(x)$ is a set of node x up to up-nodes, $|\alpha(x)\cap\alpha(y)|$ is the intersection of nodes x and y up to up-nodes, and is also a symbol of the commonality of concept x and concept y. $Sim_{conp}(x,y) = 0$ means that the concept x is not similar to y, when $Sim_{conp}(x,y) = 1$ means that the concept x and y are equal. However, the similarity reasoning can not compare the similarity between two adjacent nodes.

2.3 Equivalence Matching

The equivalence reasoning based on the label value is to determine the similarity between the concepts of adjacent nodes. For example, calculate the two versions of the software compatibility, and two adjacent node is used to represent different versions of the software. The high similarity will be obtained through similarity matching, however, if adjacent node represents the time sequence of the tag, the similarity will be different. The similarity between the two neighbor node concepts is shown by the formula (2).

$$Sim_{equ}(x,y) = Sim(x,y) + \frac{w^{|C_1-C_2|}}{n} \tag{2}$$

Among them, C_1, C_2 are the label values of concept x and concept y, and $Sim(x,y)$ is the similarity value obtained by similarity matching.

2.4 Similarity Matching of Data Types

The similarity reasoning of data types is used to determine the similarity between two data types. The similarity calculation of data types in the same domain is shown by the formula (3).

$$Sim_{data}(x,y) = 1 - \left|\frac{x-y}{C_{\max} - C_{\min}}\right| \tag{3}$$

$C_{\max}$ and $C_{\min}$ represents the range of values for digital attributes.

3 Evaluation Method

In this paper, the accuracy rate and recall rate are used to evaluate cloud service discovery system based on cloud ontology. On the one hand, evaluation is the accuracy rate of the service results of the system; on the other hand, it is the recall rate of the service results of the system. The most commonly used evaluation criterion for information retrieval is the rate of accuracy - recall. Accuracy is the ability of the system to exclude irrelevant information. Precision is the standard of measurement. Recall rate refers to the ability of the system to query relevant results, and recall is the standard of measurement.

The specific description of the evaluation of the accuracy of the service results can be defined as follows. The number of service results found by cloud service discovery system based on cloud ontology is R, and the number of services that satisfy user needs is T, so the precision of service P is

$$P = \frac{T}{R} \tag{4}$$

The specific description of evaluation of recall rate of service result can be defined as follows. The total number of hidden service results which is not found by cloud service discovery system based on cloud ontology is K, and the number of service results related to user needs is T, so the success rate of service results RP is

$$RP = \frac{T}{K + T} \tag{5}$$

But in many cases, users often focus on the results of ranking. Users will not click and browse for the results of ranking. Therefore, when evaluating the prototype system of cloud service discovery based on cloud ontology, considering the ranking results based on the accuracy rate recall rate, we use the $P@N$ evaluation method. P represents the value of the precision of the service result, which represents the requirement information for the user. Considering the location component of the search results, only the accuracy of the N results from the top ranking is detected. When the N increases gradually, the value of $P@N$ is gradually approaching precision. A specific description can be defined as follows.

After setting user requirements through the preprocessing stage of cloud ontology based semantic Internet of things cloud service discovery prototype system, M query words are constructed, and P_i represented the $P@N$ results of each query word. The $P@N$ formula for the result of the service is

$$P@N = \frac{\sum_{i=1}^{M} P_i}{M} \tag{6}$$

Compared with the cloud ontology based cloud service discovery method, the cloud service discovery method based on cloud ontology is more satisfied with the user's needs. Transportation is due to cloud ontology based cloud service discovery, which has strong filtering and reasoning functions. After the user input the requirement information, cloud service discovery method based on the cloud ontology can output the cloud service which is more in line with the user's requirements. The contrast object of the experiment is cloud service discovery method of unused cloud ontology, but there is a lack of relationship between concepts. The success rate and *P@N* rate of the above three methods are analyzed, and the relevant data are shown in Table 1.

Table 1. Comparison of *P@N* values based on the number of cloud services

The number of cloud services	Cloud service discovery method without cloud ontology	A method of using cloud ontology without inter conceptual relationships	Cloud service discovery method with cloud ontology
30	0.104	0.521	0.752
60	0.168	0.534	0.781
90	0.212	0.548	0.801
120	0.275	0.554	0.799
150	0.324	0.562	0.855

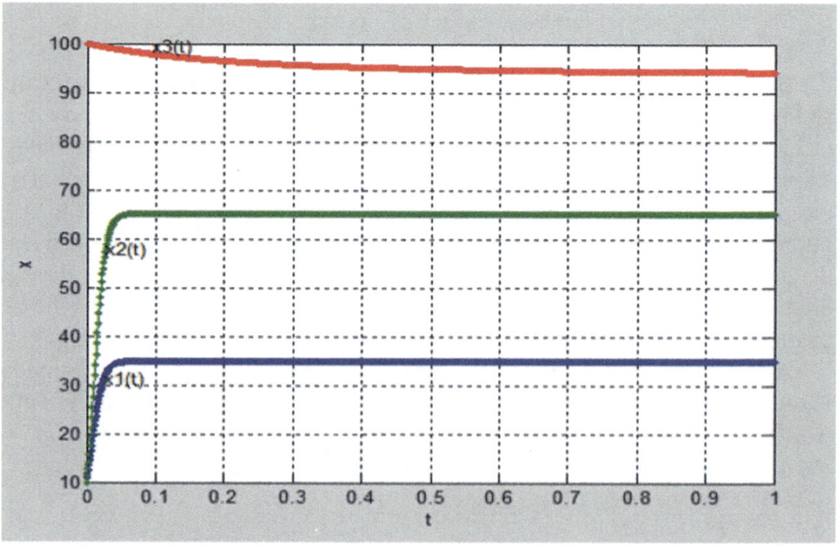

Fig. 2. Comparison of *P@N* values in three method

The comparison of *P@N* values in three method is shown in Fig. 2.

4 Conclusion

A cloud ontology of cloud service discovery method is proposed based on cloud ontology in this paper. The method is constructed in the process of the cloud found service and the use of cloud cloud ontology. Semantic reasoning machine is used to match semantic similarity between cloud service description information and user requirement information, and get the similarity degree between services, and efficiently find cloud services that meet users' needs, which is more conducive to people's understanding of cloud service discovery and related operations.

Acknowledgment. This work was supported by Academic Research Project of Wuzhou University (Grant No. 2013D009) and Basic Ability Improvement Project of Young College Teachers of Guangxi (Grant No. KY2016LX366).

References

1. Kelleret, P., Tchernev, N., Force, C.: Object-oriented methodology for FMS modeling and simulation. Int. J. Comput. Integr. Manuf. **10**(6), 405–434 (2017)
2. Anglarti, A., Grieco, A., Pacella, M.: Object-oriented modeling and simulation of flexible manufacturing systems: a rule-based procedure. Simul. Model. Pract. Theory **10**(10), 209–234 (2012)
3. Wolfgang, K., Kasper, O.: Beta SIM-A framework for discrete event modeling and simulation. Simul. Pract. Theory **6**(6), 573–599 (2008)
4. Bmzzone, A., Orsonl, A.: AI and simulation-based techniques for the assessment of supply chain logistic performance. IEEE Comput. Soc. **65**(21), 154–164 (2013)
5. Hwang, H.S.: An improved model for vehicle routing problem with time constraint based on genetic algorithm. Comput. Ind. Eng. **42**(2–4), 361–369 (2012)
6. Luis, G.: A result on projection for the vehicle routing problem. Eur. J. Oper. Res. **85**(3), 610–624 (2015)
7. Dimitris, J.B., David, S.: A new generation of vehicle routing research: robust algorithms, addressing uncertainty. Oper. Res. **44**(2), 286–304 (2016)
8. Lin, S.: Computer solutions of the travelling salesman problem. Bell Syst. Comput. J. **44**(10), 2245–2269 (2015)
9. Lin, S., Kernighan, B.W.: An effective heuristic algorithm for the travelling salesman problem. Oper. Res. **21**(2), 498–516 (2013)
10. Duhamel, C., Potvin, J.Y., Rousseau, J.M.: Tabu search heuristic for the vehicle routing problem with backhauls and time windows. Trans. Sci. **31**(1), 49–59 (2017)

Research on a Lightweight Security Communication Method Based on OTP and OTK

Zhao Donghao and Lu Yu(✉)

Shijiazhuang Branch of Army Engineering University,
Shijiazhuang 050003, China
zhaodonghao3@163.com

Abstract. In recent years, as a product of the new era, the Internet of Things technology has become more and more widely used in various aspects such as intelligent transportation, logistics management, and electronic tags. With the rapid development of Internet of Things technology, there are also many problems. In view of the current characteristics of the Internet of Things (IoT): the terminal devices are of uneven size, the storage and processing capabilities are inconsistent, and the connections are intermittent. In order to anti these characteristics, we come up with a low energy consumption, low computational communication project which doesn't need the third party and has the relatively high security. The project comprises an identity authentication scheme based on OTP (One Time-Password) and a key agreement scheme based on OTK (One Time-Key). This project can achieve the desired authentication and secure communication proved by the theoretical analysis, it doesn't have any security defect and has good security and credibility.

Keywords: IoT · Identity authentication · Key agreement

1 Introduction

The Internet of Things is a composite network composed of wireless sensor networks, RFID networks, and the Internet. It has the characteristics of small size of some terminal devices, weak storage capacity and computational processing capacity, and intermittent and intermittent connection. In view of these characteristics, this paper proposes a lightweight authentication scheme that is less computational, more efficient, and does not require third-party fairness compared to traditional schemes.

In 1981, Lamport first proposed the issue of secure communication in an insecure channel and solved it using a hash chain to generate a one-time password. At present, the research on one-time password authentication at home and abroad has also made some progress. Zhang Hong and Chen Zhigang et al. proposed to use software methods instead of challenge/response token hardware functions. Challenges were generated by the client to achieve authentication and secure communication between the two parties. However, during the authentication processes, both the client and the server must perform multiple Hash operations, the amount of calculation is large. Longyan Gong et al. proposed using

J. Abawajy et al. (Eds.): ATCI 2018, AISC 842, pp. 670–676, 2019.
https://doi.org/10.1007/978-3-319-98776-7_77

Hash functions and pseudo-random functions to generate mutually independent OTP to implement mutual authentication and resist various attack methods. However, the scheme did not encrypt the authentication information in order to reduce the amount of computation and left a security risk.

This paper summarizes and analyzes OTP-based identity authentication schemes in recent years, improves OTP authentication, and proposes a new key negotiation method OTK. This method avoids the use of complicated public key algorithms and can be performed simultaneously with OTP authentication. The calculation and communication costs are small, the authentication efficiency is high, and subsequent key negotiation problems can be solved. The calculated session key can be directly generated as the data encryption key for subsequent communication. The two parties can complete the authentication as well as the key agreement through OTP and OTK, and the secure communication can be smoothly performed.

2 Certification Scheme Design

2.1 Design Ideas

Based on the study of one-time passwords, the communication scheme proposes a one-time key agreement protocol. It also has the advantages of simple implementation, low cost, and no need for third-party authentication. OTP and OTK both use the unique seed (Seed) that is pre-stored in production to join the challenge parameter and use a one-way hash algorithm to generate a one-time dynamic password or key to complete the authentication and key agreement.

We usually divide the one-time password authentication into three types: time synchronization dynamic password, event synchronization dynamic password, and challenge/response based dynamic password. Due to the complexity of the Internet of Things communications environment, data transmission and processing on the network have a certain delay, although the time-based and event-based synchronization of dynamic passwords has high security but technically require both sides of the communication synchronization, the cost is higher, more difficult. It does not apply to the networking environment, so we use challenge/response based OTP authentication principles.

In the challenge/response mechanism, the uncertainty comes from the authentication server. When the client requests authentication, the server generates a random number and sends it to the user; the user uses a certain one-way function (this protocol uses a hash function) to hash the random number, converts it into a password, and sends it to the authentication server. Then the server will verify the validity of the user's identity in the same way (Fig. 1).

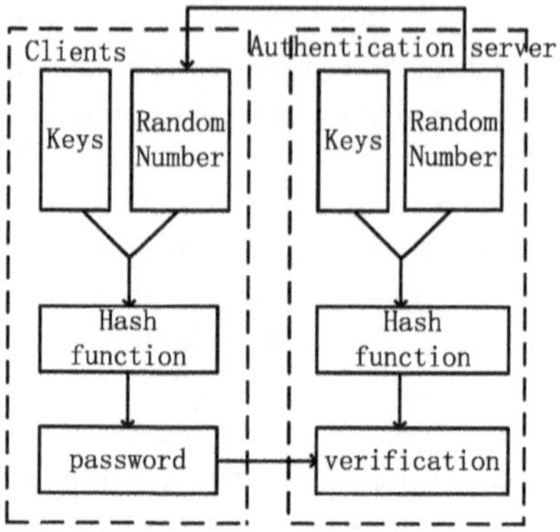

Fig. 1. The principle of OTP and OTK based on the challenge/response mechanism

2.2 Key Generation Method OTK

First assuming that both parties of the communication have completed the authentication, where X is the counter, time factor, or challenge information for successful OTP verification.

The two parties share a serial number SN and the Seed, and both parties also enjoy Seed2, and calculate Hash (Seed2, X) to obtain the session key.

When calculating the session key, in addition to using a hash function, you can also use a symmetric encryption algorithm that uses one or both of Seed, X, and Y as the key, and encrypt other variables in Seed, X, and Y, resulting as the dense session key.

The session key may be directly generated as a data encryption key for subsequent communications or may be generated as a data encryption key for subsequent communications in a certain transformation manner.

According to the above description, the key generation method OTK is simple to implement and has low energy consumption and is suitable for terminals of Internet of Things with inconsistent computing capabilities; the number of communication times between the two nodes in the authentication process is relatively small, and the communication connection may be intermittent. Finally, the method does not require third-party certification, to avoid the many problems posed by the third-party certification.

2.3 Scheme Description

The authentication scheme is mainly divided into two stages: initialization and authentication. The symbols and meanings involved in the scheme are shown in Table 1.

Table 1. Symbols and meanings involved in this article

Symbols	Meaning
K_{UR}	Node public key
K_{US}	Node private key
R	Random data string
A B	Both sides of the communication node
SN	Terminal device serial number
C_i C_j	The challenge value required to generate a one-time password
P_i P_j	OTP generated one-time password
K_i K_j	OTK-generated session keys
H()	Hash function

- **Initialization phase**

The initialization phase mainly completes the sharing of serial numbers and seeds of the communication nodes and the exchange of public keys.

In the production of an OTP device, a unique serial number and seed are generated for each OTP device and written to the OTP device, and the serial number and seed are usually written to the authentication server's database. Due to the problem that the communication parties cannot interact with the authentication server, we propose a new seed management method.

Seed distribution can be achieved through short-distance communication. After one party generates the seed and serial number and completes the exchange through Near Field Communication: NFC touch, audio coding, or two-dimensional code scanning, the two parties can implement security authentication and communication through the OTP mechanism.

The same is true for the exchange of public keys. Because of the risk of being unable to communicate with the authentication server, we use the same method as the serial number and seed sharing: Near-field communication such as NFC touch.

At this point, the serial number, seed, and public key of the two communication nodes have been exchanged and shared, initialization is completed, and initialization is performed only once before authentication. The basic flow chart in the initialization phase is shown in the Fig. 2.

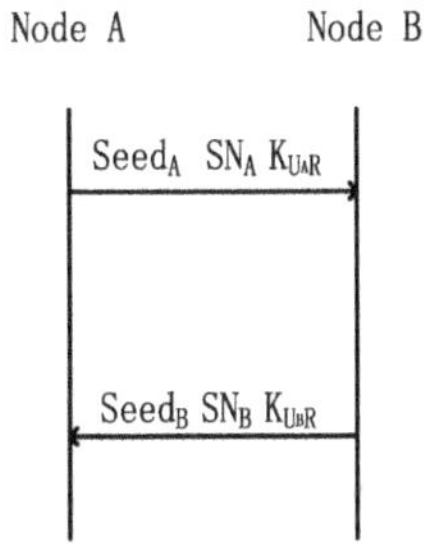

Fig. 2. Authentication initialization

- **Certification Stage**

  i. Node A randomly generates challenge values C_i; uses node B's public key to encrypt and send sequence number SN and challenge value C_i to node B;
  ii. Node B uses its own private key to decrypt the received encrypted information to obtain SN and C_i, generate a one-time password based on the challenge value C_i, generate session key using OTK, and generate challenge value C_j;
  iii. Node B uses node A's public key to encrypt and send a one-time password P_i and challenge value to C_j node A;
  iv. Node A uses its own private key to decrypt the encrypted information P_i and C_j received from Node B to obtain the sum. The OTP is used to verify the one-time password P_i. If the verification is successful, the OTK is used to generate the session key K_i based on the challenge value C_i, and the OTP is used to generate a one-time challenge value P_j. Sexual passwords and use OTK to generate session keys K_j based on challenge values C_j;
  v. Node A uses Node B's public key to encrypt and send a one-time password P_j to Node B;
  vi. Node B uses its own private key to decrypt the encrypted information P_j received from node A. OTP verifies the one-time password P_j. If the verification passes, the OTK is used to generate the session key based on the challenge value K_j;
  vii. Node B uses node A's public key to encrypt and return K_j to node A;
  viii. The authentication is completed and the final session key is $K_i + K_j$.

At this point, the authentication is completed and a symmetrical key is generated between the nodes as a subsequent session key. The basic flow chart in the authentication phase is shown below (Fig. 3).

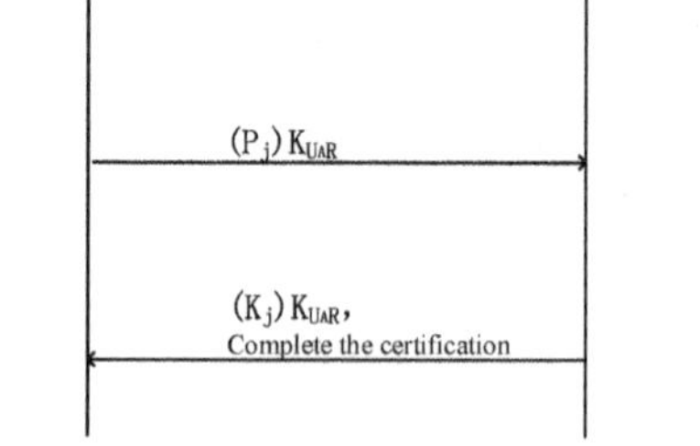

Fig. 3. Basic flow chart in the authentication phase

3 Program Performance Analysis

- Replay attacks

As the one-time password is joined by challenging stored in itself and the seeds using a hash algorithm to generate. Each time the generated challenge value is different, the calculated password is also different, which can effectively prevent replay attacks.

- Fractional attacks

Different from the S/Key scheme, the password information of the previous authentication in this scheme is not saved. The password of the latter authentication is also not related to the previous password, which can effectively prevent the decimal attack.

- Counterfeit attack

Different from the traditional one-way authentication, the two parties in the program will compare the received one-time password with their own calculated password. If they find something wrong, they will stop the communication and achieve two-way authentication, which can effectively prevent counterfeit attacks.

- Message tamper

In this scheme, the hash function is used to calculate the one-time password and the session key. The security features of the hash function can enable the two parties to judge the integrity of the message and effectively prevent the message from being tampered with.

- Denial of service attacks

In the process, both parties in the communication calculate a simple hash function and compare the result calculated by the other party with their own. This is because the characteristics of the one-time password do not store the data, which improves the efficiency and prevents denial of service attacks effectively.

4 Conclusion

With the development of the Internet of Things, more and more articles of life and work have also been added, among which there is no shortage of terminal equipment with a small amount of storage and calculation. This paper proposes a secure communication scheme based on OTP and OTK. It is optimized on the basis of the existing one-time password authentication. The amount of data calculation, transmission volume is small, the number of interactions between the two parties is small, and some storage and calculation are well solved. Small amount of Internet of Things terminals communication security issues. Through the analysis of its security, we can know that the program can resist replay attacks, fractional attacks, counterfeit attacks, message tampering and denial of service attacks. Through formal analysis, this solution can achieve expected authentication and secure communication without security flaws with good security and credibility.

References

1. Zhang, H., Chen, Z.: Design and analysis of a new one-time password authentication scheme. Comput. Eng. **30**(17), 112–113 (2004)
2. Xi, Y.: An improved OTP authentication system and its application in mobile commerce. Intell. Comput. Evolut. Comput. **180**, 615–624 (2013)
3. Yang, M., Peng, X.: A lightweight one-time password authentication scheme. Small Microcomput. Syst. **35**(8), 1808–1811 (2014)
4. Gong, L., Pan, J., Liu, B., et al.: A novel one-time password mutual authentication scheme on sharing renewed finite random sub-passwords. J. Comput. Syst. Sci. **79**, 122–130 (2013)
5. Wang, P., Wang, D., Huang, X.: Research progress of password security. J. Comput. Res. Dev. **53**(10), 2173–2188 (2016)
6. Deng, F., Zhu, Y.: Password-based client/server authentication protocol. Comput. Eng. Appl. **51**(20), 72–76 (2015)
7. Lei, X., Xue, R.: Logic Method of Cryptographic Protocol Analysis. Science Press, Beijing (2013)
8. Wang, Q.: Automatic control system regulation scheme of textile air conditioning. Sci. Technol. Inf. **22**, 395–396 (2010). (in Chinese)
9. Liu, J.: Advanced PID Control and MATLAB Simulation, 2nd edn. Electronics Industry Press, Beijing (2003). (in Chinese)
10. Li, J., Ren, Q.: Study on supply air temperature forecast and changing machine dew point for variable air volume system. Build. Energy Environ. **27**(4), 29–32 (2008). (in Chinese)

A Reasoning and Query Method of Equipment PHM Ontology Based on Jena

Jingsheng Liu(✉), Xu Li, Yuliang Lu, Hui Liu, and Litao Wang

Army Beijing Representative Office, No. 97, Heping West Road, Xinhua District, Shijiazhuang 050003, Hebei, China
764468402@qq.com

Abstract. With the continuous development of computer retrieval technology, people have more and more demand for information semantic retrieval, and the ontology reasoning and query method based on Jena can be researched to solve this problem effectively. Through the research of ontology theory, the ontology technology is applied to the field of equipment maintenance support, and the equipment PHM ontology is built and persisted into the database. The ontology reasoning rules of PHM ontology are defined, and the ontology is deduced by using Jena inference engine. The query of PHM ontology is realized by using SPARQL query language, and it lays the foundation for the next step to study the information system of equipment maintenance support based on ontology.

Keywords: Jena · PHM · Ontology reasoning · SPARQL · Persistence layer Inference rule

1 Introduction

Since the 1990s, computer network technology has developed rapidly. Examples of information retrieval in various fields and industries using computer technology are commonplace. However, traditional information retrieval systems mostly use keyword-based retrieval techniques and cannot retrieve information semantics. This phenomenon also exists in the field of equipment maintenance support [1]. The Semantic Web refers to an intelligent network that can recognize and process information semantics and realizes barrier-free communication between people and computers. In recent years, the development of Semantic Web technology is gradually filling the large gaps in people's description of information semantics. Ontology-based information retrieval technology has also developed along with it.

Based on the concept of extracting the domain, Ontology defines and expresses the relationship between concepts, forms a unified domain information description standard, and uses OWL (Web Ontology Language) to standardize the description of domain information, and implements information semantic representation [2, 3]. In this paper, based on Jena's equipment PHM ontology reasoning and query methods, explore and solve the problem of information semantic retrieval in the field of equipment maintenance support. For this reason, this paper constructs and equips PHM knowledge ontology, constructs PHM knowledge ontology reasoning rules based on

J. Abawajy et al. (Eds.): ATCI 2018, AISC 842, pp. 677–684, 2019.
https://doi.org/10.1007/978-3-319-98776-7_78

Jena's reasoning mechanism, studies ontology reasoning and query technology, and realizes equipment PHM knowledge ontology reasoning.

2 Ontology Inference Related Concepts

2.1 Ontology Reasoning Technology

Ontology inference is the key to realize the application of information semantics. On the basis of constructing domain knowledge ontology, according to the standardized knowledge description, the knowledge semantics are inferred according to certain rules. At present, the ontology reasoning problem is usually solved by an ontology reasoning machine, and the inference rules are different. The reasoning machine used is different. Reasoning based on basic description logic [4–6] uses Tableaux algorithm for reasoning. The classic inference engine system includes Pellet, Racer and Fact++. The rule-based reasoning method [7] infers the ontology OWL description file according to the custom OWL inference rules. The widely used inference engine system includes Jess and Jena. The method based on the first-order predicate prover uses the convenience of transforming OWL description statements into FOL (first-order logic) for OWL reasoning. The typical application is the Hoolet inference engine.

2.2 Jena Inference Engine

Jena is a Java-based application development package that handles ontologies on the Java development platform. It supports the application development of OWL, RDF, and RDFS. The included function interface calls can implement ontologies persistent storage, query, and reasoning [8–10]. Jena's main framework includes [11]:

(1) **RDF read and write.** RDF is developed on the basis of XML. It describes resources in the form of triples and is a standard that the W3C recommends to describe resources. The RDF API provided by Jena (com.hp.hpl.jena.rdf.model package) can handle the RDF file.
(2) **Ontology subsystem.** Ontologies can be expressed in a variety of description languages. The most common ones are OWL, DAML+OIL, and RDFS. The Ontology API (com.hp.hpl.jena.ontology) provided in the Jena framework supports common ontology description languages. It can extract ontology information in combination with the inference subsystem.
(3) **Ontology storage.** Jena can store persistent ontology OWL files in the database, using the com.hp.hpl.jena.db package.
(4) **Ontology inquiry.** Jena provides an ontology query engine, ARQ, to implement ontology queries based on SPARQL and RDQL.
(5) **Rule-based reasoning.** The Jena framework contains the ontology reasoning subsystem (com.hp.hpl.jena.reasoner.package). Users use the system's own inference rules and custom inference rules to implement RDFS and OWL file inference.

2.3 Jena Reasoning Model

Ontology inference usually uses the general rule inference model [12, 13] in Jena inference engine, as shown in Fig. 1:

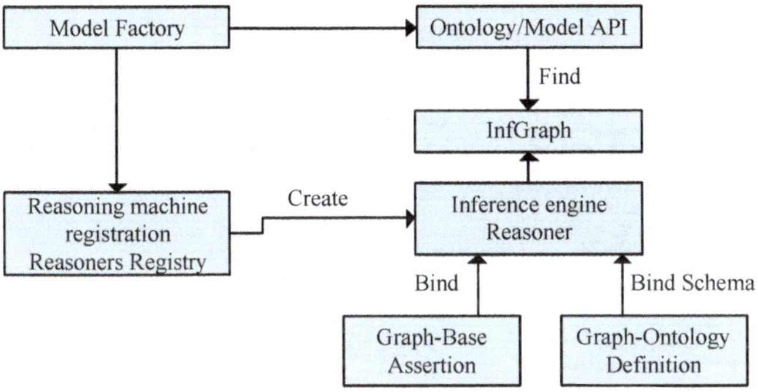

Fig. 1. Jena reasoning model

The process of using the Jena reasoning mechanism to reason about the PHM ontology is as follows: (1) Using Protege to construct the domain ontology, obtain the EPHMO.owl ontology file and store it in the database persistently; (2) Semantic annotation of the EPHMO.owl ontology file, Create PHM ontology model objects through Model Factory; (3) User-defined equipment with PHM inference rules and register to create inference engine; (4) Connect constructed PHM ontology files with Ontology/Model API to create ontology inference engine, Semantic information inference, InfGraph represents the model object for reasoning.

3 Equipment PHM Ontology Construction and Storage

3.1 Equipment PHM Ontology Construction

Based on the theoretical research on the field of equipment maintenance support, this paper uses Protege to build a PHM ontology knowledge base, extracts concepts from six aspects of equipment basic information, equipment testing, health status assessment, fault diagnosis, fault prediction and fault maintenance, and defines concepts. The object attributes and data attributes were used to instantiate an ontology with the background of the maintenance and management of a certain type of missile weapon system.

Use the protege to define the relevant concepts of the extraction concept, as shown in Fig. 2:

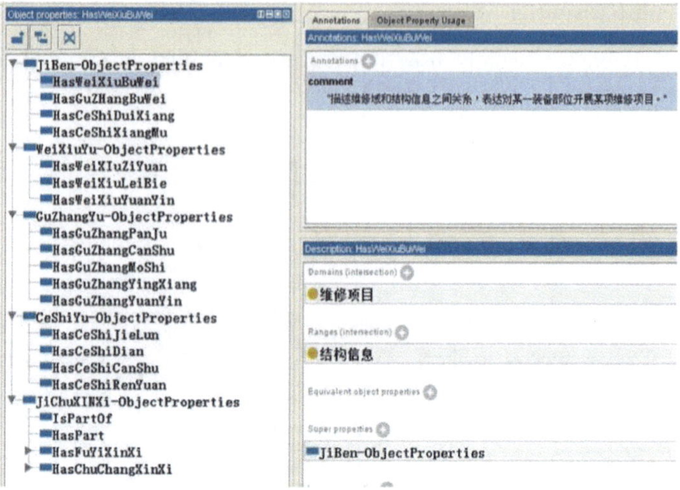

Fig. 2. Attribute definition

Using the OWL ontology description language to describe the equipment PHM

```
<owl:Class rdf:about="部组件">
    <rdfs:subClassOf rdf:resource="结构信息"/>
</owl:Class>

<owl:ObjectProperty rdf:about="#IsPartOf">
    <rdfs:subPropertyOf rdf:resource="#JiChuXinXi-ObjectProperties"/>
    <rdfs:domain rdf:resource="结构信息"/>
    <rdfs:range rdf:resource="结构信息"/>
</owl:ObjectProperty>

<owl:DatatypeProperty rdf:about="#HasEquipmentID">
    <rdfs:subPropertyOf rdf:resource="#ZhuangBei-DataProperties"/>
    <rdfs:domain rdf:resource="结构信息"/>
    <rdfs:range rdf:resource="&xsd;ID"/>
</owl:DatatypeProperty>
<owl:DatatypeProperty rdf:about="#HasEquipmentModel">
    <rdfs:subPropertyOf rdf:resource="#ZhuangBei-DataProperties"/>
    <rdfs:domain rdf:resource="结构信息"/>
    <rdfs:range rdf:resource="&xsd;string"/>
</owl:DatatypeProperty>
```

Fig. 3. Part of EPHMO.owl document schematic

information, the EPHMO.owl document is shown in Fig. 3:

3.2 Ontology Storage

In order to apply the PHM ontology, it is necessary to study ontology storage problems. Usually, the user stores ontology information directly in the OWL file, and obtains ontology information through parsing of the OWL file. However, in real life, the ontology contains a large amount of information, resulting in a relatively low parsing efficiency. The ontology storage development package included in the Jena framework provides a function interface for equipping the database with the PHM ontology persistence. In the text, these function interfaces are used to store the EPHMO.owl file using SQL Server 2005.

The database model is created by calling the ModelMaker function interface. After the program is run, the PHM body owl file is stored in the SQL Server database. The code for storing the EPHMO.owl file is as follows:

```
// Loading database driver
IDBConnection connection = new DBConnection (str.URL, str.User, str.Password, strDB);
Class.forName ("com.microsoft.sqlserver.jdbc.SQLServerDriver");
// Create a database named "equipment"
ModelMaker maker = ModelFactory.createModelRDBMaker (connection);
Model defModel = maker.createModel ("equipment");
// Read the ontology file
InputStreamReader in = null;
File file = new File("E:\EPHMO.owl");
// Ontology data storage
In = new InputStreamReader(new FileInputStream(file), "UTF-8");
defModel.read(in, "http://www.owl-ontology.com/EPHMO");
```

4 Equipment PHM Ontology Reasoning and Query

4.1 Equipment PHM Ontology Inference Rules Construction

Jena's own general reasoning rules are used to infer the basic relations of the classes and the basic properties of the attributes, including the hierarchical relationship of the same class, the mutual exclusion of different classes, the functionality of attributes, and transitivity [14]. General rules have nothing to do with the specific domain ontology, reasoning on the ontology public content. The general rules are stored in the Jena source file and the basic definitions are as follows:

```
# Conceptual level
Rdfs: (?a rdfs: subClassof ?b), (?b rdfs: subClassof ?c) —> (?a rdfs: subClassof ?c)
# Reversed (Inverseof)
Rdfs: (?P owl: Inverseof ?Q), (?X ?P ?Y) —> (?Y ?Q ?X)
# Transitivity (TransitivePropery)
Rdfs: (?P rdfs: type owl: TransitivePropery), (?A ?P ?B), (?B ?P ?C) —> (?A ?P ?C)
```

Based on the general inference rules, the inference rules need to be defined according to the actual conditions of constructing the ontology to implement implied relational inference and meet the actual needs of users. The fragmentation of the custom inference rules for the PHM Ontology equipment is as follows:

```
@prefix fa: <http://www.semanticweb.org/ontologies/ EPHMO.owl#>.
@include <RDFS>.
@include <OWL>.
[rule1: (?A IsPartof ?B) —> (?B HasPart ?A)]
[rule2: (?A IsPartof ?B), (?B IsPartof ?C) —> (?A IsPartof ?C)]
[rule3: (?A IsPartof ?B), (?A HasGuZhangXianXiang ?C) —>
(?B HasGuZhangXianXiang ?C)]
[rule4: (?A IsPartof ?B), (?A HasCeShiCanShu ?C) —> (?B HasCeShiCanShu ?C)]
......
```

Rule1 indicates that IsPartof and HasPart are symmetric relations; rule2 indicates the hierarchical structure of components; rule3 indicates that if component A is a substructure of component B, fault A occurs when fault A occurs, and fault B occurs when fault B occurs; rule4 It means that if component A is a sub-structure of component B, testing of A requires the testing of parameter C, and testing of B also requires testing of parameter C.

The custom equipment PHM ontology inference rules and general inference rules are saved in the form of ephmo.rules file, which is convenient for subsequent ontology reasoning.

4.2 Equipment PHM Ontology Reasoning and Query

Ontology inference means that the inference engine parses and infers the owl ontology document according to the ontology reasoning rule inference, and obtains the semantic information contained in the ontology. In the process of reasoning on the equipment PHM body, you need to use the Ontology/Model API to call the EPHMO.owl file and the ephmo.rules file. Some of the procedures are as follows:

```
// Create a registration inference engine and load inference rules into the inference engine
String ruleFile = "file:./EPHMO/ephmo.rules";
List rules = Rule.rulesFormURL(ruleFile );
GenericRuleReasoner reasoner = new GenericRuleReasoner(rules);
Reasoner.setOWLTranslation(true);
Reasoner.setDerivationLogging(ture);
Reasoner.setTransitiveClosureCaching(ture);
// Create the ontology model and read the ontology file
String ontoFile = "file:./EPHMO /EPHMO.owl";
Model model = ModelFactory.createDefaultModel();
Model.read(ontoFile);
OntModel ont = ModelFactory.creatOntologyModel (OntModelSpec.OWL_DL_MEM_RDFS_INF, model);
Resource configuration = ont.createResource();
Configuration.addProperty(ReasonerVocabulary.PROPruleModel, "hybrid");
```

Ontology inference display needs to be implemented with SPARQL query language. SPARQL is a query language developed specifically for querying RDF. The conditional clause for querying is similar to RDF and is expressed in the form of a triplet [15]. The query result is the result of matching the conditional triplet with the RDF data triplet. Using SPARQL to query the ontology mainly includes three aspects. One is to use InfModel to extend the conventional Model, it supports any related reasoning ability; the other is to use the QueryExecution interface to execute the query; the third is to use ResultSetFormatter to obtain the query result. Use SPARQL to query the part of the equipment PHM body code as follows:

```
String queryString =
"PREFIX rdfs: <http://www.w3.org/2000/01/rdf-schema#>\n " +
"PREFIX rdf: <http://www.w3.org/1999/02/22-rdf-syntax-ns#>\n " +
"PREFIX rdfs:< http://www.w3.org/2002/07/owl#>\n" +
"PREFIX rdfs:<http://www.w3.org/2001/XMLSchema#>\n " +
"SELECT ? subject ? object " +
"WHERE {? subject rdfs: subClassOf ?object} ";
 Query query = QueryFactory.create(queryString);
 InfModel inf = ModelFactory.creatInfModel(ruleFile, ontoFile);
 QueryExecution qe = QueryExecutionFactory.create(query, inf);
 ResultSet results = qe.execSelect();
 ResultSetFormatter.out(System.out, results, query);
```

The output of program execution is shown in Fig. 4:

subject	object
JieGouXinXi	ZhuangBeiJiChuXinXi
BuZuJian	JieGouXinXi
FenXiTong	JieGouXinXi
ShengChanChangJia	ShengChanXinXi
ChuChangRiQi	ShengChanXinXi
ShengChanXinXi	ZhuangBeiJiChuXinXi
ZhuangBeiCeShi	EPHMO

Fig. 4. SPARQL search results

5 Conclusion

In this paper, it is difficult to realize the semantic retrieval problem in the field of equipment maintenance support information. Based on Jena's equipment, PHM ontology reasoning and query methods are proposed. On the basis of analyzing and researching the ontology theory, using Protege software to construct and equip the PHM ontology, instantiating a certain type of missile weapon system as a background, and instantiating the equipment PHM ontology into the database. The ontology reasoning mechanism is used to customize the PHM ontology reasoning rules, and the PHM ontology file EPHMO.owl is used. Based on Jena's ontology reasoning technology, the basic reasoning of equipment PHM ontology is realized. Based on the research of this paper, we can further carry out the research on the ontology-based equipment maintenance support information system, improve the network and intelligence level of information systems, and achieve more effective information services.

References

1. Wang, X.: Research on Inquiry and Reasoning Based on Ontology and Their Applications in the Military Area. Chongqing University, Chongqing (2012)
2. Yuan, X., Tang, Q., He, J.: Construction of university knowledge ontology and its application research. Mod. Electron. Tech. **37**(16), 58–61 (2014)
3. Hu, J., Ji, Y., Meng, Y., et al.: Equipment support knowledge ontology construction based on protégé. Mod. Electron. Tech. **6**, 206–210 (2010)
4. Sun, P.: Research on Ontology Reasoning Based on Description Logic in Semantic Web. Jilin University, Jilin (2009)
5. Wen, K., Lu, Z., Wu, J., et al.: Design and implementation of reasoning system based on description logic. J. Chin. Comput. Syst. **29**(1), 57–60 (2008)
6. Tang, S., Cai, Z., Wang, J., et al.: Non-standard inferences in description logics. Pattern Recog. Artif. Intell. **23**(4), 522–529 (2010)
7. Liu, C.: Trustworthy Software Evolution Based on Ontology and Rule-Reasoning. Chongqing University, Chongqing (2010)
8. Ling, S., Huo, L., Wang, C.: Research on Chinese ontology application oriented semantic network. Comput. Technol. Dev. **24**(2), 194–197 (2014)
9. Jia, X., Lin, Z.: A research of large-scale remote sensing interpretation ontology storage and inference. In: CACIS 2009, China, 17 July 2009
10. Xiong, Z., Yu, J.: Comparative analysis of APIs for OWL ontology. Inf. Res. **2**, 72–76 (2015)
11. Yin, S.: Research on the Reasoner of Intelligent Question Answering System. Hunan University, Hunan (2013)
12. Tan, Y., Xiao, B., Chen, J., et al.: The suvery of Jena's reasoning and applying. J. Hebei Acad. Sci. **26**(4), 14–17 (2009)
13. Ouyang, L., Yin, S.: Design and implementation of multi-layer reasoner. Comput. Appl. Softw. **31**(10), 161–166 (2014)
14. He, L., Miao, J., Liu, L., et al.: Overview of ontology and Jena. J. Hangzhou Norm. Univ. (Nat. Sci. Ed.) **12**(5), 467–473 (2013)
15. Xie, G.: SPARQL-a new RDF query language. J. Xiangnan Univ. **30**(2), 80–84 (2009)

Computational Advertising Under the Convergence of Customized Communication Paradigm

Gang Li[1(✉)] and Jing Cao[2]

[1] Southwest Jiaotong University, Sichuan, China
13744354@qq.com
[2] Neusoft University, Sichuan, China

Abstract. Computational advertising is a new interdisciplinary discipline that emerged with the development of network technology. Its core is the use of big data technology and two-way interactive platform to achieve accurate matching of advertising, context, and consumers. Using text analysis and case analysis, this research found a totally different advertising and communication formats from the traditional advertising, that is customized recommendation, converged communication, intelligent adjustment, and programmatic buying. In the current rapid development of big data and "Internet+," data has become the basic production factor in the field of advertising marketing. It provides the most objective decision-making basis for advertising creativity, planning, delivery, and monitoring, so that it can maximize the precise matching of advertising. Computational advertising will surely bring about tremendous changes in the entire advertising industry chain.

Keywords: Computational advertising · Big data · Precise marketing

1 The Main Characteristics of Computational Advertising

Long before the concepts of computational advertising and big data marketing were proposed, the need for precise communication has already existed. But only when this need is superimposed on big data technology and a two-way interactive platform, it becomes a real industrial reality.

1.1 Precise Matching Is the Core Demand of Computational Advertising

Traditional brand advertising is based on planning, creativity, media resources and customer relationships, while computational advertising is data and computation driven.

The essence of the idea is to recommend the most suitable advertising information and product type to different users according to different preferences of different users in different time and space and in different contexts. In short, different people will see different advertisements for the same product. Even the same person sees different advertisements in different scenarios, that is, highly personalized advertising information. From this we can see that precise matching is the core of computational advertising, big data thinking and ability are the basis of technology, and building an

J. Abawajy et al. (Eds.): ATCI 2018, AISC 842, pp. 685–690, 2019.
https://doi.org/10.1007/978-3-319-98776-7_79

interactive platform for communication value creation is an important carrier for obtaining user data and achieving accurate matching.

1.2 The User's Participation of Building a Two-Way Interaction Platform Creates the Value of Advertising

As the "Internet+" universal technology has installed more smart elements and flexible screens into various terminal devices, the mediating trend of the terminal will become more apparent, resulting in a "everybody is media and everything is communication" of self-media, pan-media phenomenon. In this way, traditional media audiences are increasingly dissatisfied with the status of passively receiving information. They use their own hands to use all available equipment to actively participate in the process of content production and information dissemination. In this way, the boundaries between various actors such as producers, distributors, forwarders, receivers, and feedback providers gradually fade away, and more and more unidirectional audiences in the age of mass media have evolved into "Internet+" era's two-way interactive participants.

1.3 Data Thinking and Data Capabilities Improve Advertising Effectiveness

The Internet's general technology system makes the available data sources very rich and has even expanded from the sample to the whole. People can no longer obtain large amounts of data through sampling and can use advanced algorithms to perform deep mining and processing of these data at a lower cost. In this way, we can find specific consumer demand and media usage habits for each user, and targeted advertising and marketing communications, and can even invite consumers to participate in the entire process of producing, disseminating, and consumption.

2 A New Mode of Computational Advertising in a Customized Communication Paradigm

2.1 Customized Recommendation: Tailored Advertising Information

Consumers in the Internet era have changed from popular and segmented target audiences to micro-differentiated and personalized users. The definition of customers given by big data is no longer a group of abstract standardized faces, but a full, three-dimensional image. Therefore, for the enterprise, no matter whether it is the product production or the marketing communication, it is no longer continuing to meet the market demand, but has turned to meet individual pursuits, and the aggregated market concept has become a fragmented individual concept. With convenient and interactive communication methods, customized production and customized communication become possible. Consumers gradually become involved in the corporate R&D, design, production, supply, marketing, distribution, and after-sales of the company, and become "producers and consumers". Enterprise production and marketing intelligence, flexibility and customization.

Traditionally, brand ads placed on mass media are difficult to personalize and match precisely, until the emergence of search advertising has come up with tailor-made computational advertising prototypes. User input of keywords is equivalent to telling search engines what information they need, and search engines push specific ads to users through efficient algorithms. Adsense technology developed by Google further provides content-related advertising by analyzing the content of web pages viewed by users. These attempts have led to a more precise evolution of advertising, and when computational advertising has superimposed more technologies such as machine learning, big data, cloud computing, and e-commerce, the currently widely used form of computational advertising -the recommendation system has achieved rapid development. Tmall, JD.com, Amazon, etc. can analyze the massive browsing and purchasing data on their own websites, use data mining algorithms and comparison of consumer preferences, discover the potential needs of each consumer, and forecast and recommend products that may be of interest to them. In the future, with the deep development of computational advertising technology, customized advertising information will more accurately match the actual needs of everyone based on more comprehensive user's personal data.

2.2 Converged Communication: Multi-screen Communication in the Fusion Media Era

Media integration breaks the boundaries between different media, different departments, different types of jobs, and different terminals. With the extensive installation of smart chips and flexible screens and the rapid development of the Internet of Things, users no longer have only media such as newspapers, magazines, and televisions, but more and more fusion media such as smart devices with content elements and information service functions, smart phones, tablet computers, smart watches, smart glasses, smart refrigerators, smart cars, and intelligent robots. The fusion screen constitutes the infrastructure for communication between users and users, users and devices, and devices and devices.

With the large-scale use of this two-way communication of intelligent, interactive, and convergent screens, multi-screen propagation of advertisements and multi-screen viewing by users have become reality clearly. Coupled with breakthroughs in media convergence technology and big data technology, users are looking at the screen while the screen is also gaining insight into user behavior. The core value of computational advertising lies in achieving accurate matching. Since many users have multiple screens and their media habits are multi-screen and cross-screen, advertisers certainly hope to have insights into media use behavior and product consumption requirements, make their advertising information to be accurate, set personalized content for different screens, and even reach the target consumer across screens seamlessly, thereby optimizing advertising effectiveness and minimizing transmission costs. At the same time, content carrier & operators, platform operators, advertising companies, and advertisers have come to need to measure the effectiveness of multi-screen communication. Although current television screens, movie screens, computer screens, mobile phone screens, and other various screens each have a relatively independent computing and measurement system, the concept and practice of computational advertising gradually expand from online advertising to the entire advertising field. The spread of advertisements across screens will become an important format for computational advertising in a custom paradigm.

2.3 Intelligent Adjustment: Dynamic Strategy that Changes Instantly

In the era of integrated marketing communications, large and medium-sized enterprises formulate corresponding advertising budgets each year according to the overall goals of corporate development, and formulate annual, quarterly, monthly, and even weekly advertising marketing plans. These plans have uniformly stipulated media formats, creative ideas, interest appeals, promotional products, and price plans for advertising in a relatively long period of time. Once made, these plans seldom change. This is also the core principle of integrated marketing communication- all media and marketing channels send the same voice to all consumers in all time and space. However, in fact, the consumer's propensity to consume in different contexts is quite different. A unified integrated marketing communications strategy cannot meet the rapidly changing market competition and consumer demand in the "Internet+" era. Consumers are in touch with advertising information before and after the tendency of real-time buying behavior which is the most effective demand information, but it is difficult to accurately capture this kind of demand information with a traditional marketing strategy and instantly convert it into actual purchase behavior.

Under such circumstances, the dynamic computational advertising communication strategy becomes an inevitable choice for advertising communication under the converge & customized communication paradigm. It not only can obtain real-time consumer dynamics, but also can adjust advertisements in real time according to the specific consumer's historical data and current behavior. The ideal strategy for computational advertising communication should be to disseminate the most suitable advertising information to different consumers according to different needs or different needs of the same consumer under different time and space scenarios. That is, at the right time, in the right place, with the appropriate spokesperson, in a proper creative manner, with appropriate value claims, to convey the right price information to the right user. For example, Uber's smart algorithm will dynamically price and respond to local taxi demand at that time and will quickly feedback this price information to users to adjust the market supply and demand status. The price of a microwave oven on the Amazon website will change 9 times in a day. In the future, this kind of intelligent dynamic advertising marketing strategy will become more common through big data and cloud computing technologies.

2.4 Programmatic Buying: Smart Trading Ecosystem

In the initial stage of online advertising, because the number of online media and online advertising customers is relatively small, media can directly sell advertising resources to users. With the rapid development of the Internet industry and the rapid increase in the number of media and advertisers, an ad. mediation platform has gradually emerged to connect media and advertisers to complete advertising transactions. In recent years, media integration and the "Internet+" brought about by the trend of pan-mediating and universal e-commerce have led more and more individuals and organizations to become self-media, and more and more small and medium-sized enterprises have become more accurate. The demand for advertisers, the proliferation of supply and demand, further demanded an increase in the efficiency of advertising and reduced advertising transaction costs. The pattern of advertising exchanges to an online advertising transaction model such as "advertisers—agents—demand side platform—exchange-supply side platform".

In this model, the demand-side platform automatically performs the advertising media buying process on behalf of advertisers through a digital advertising exchange. This online advertising buying method is called programmatic buying. Programmatic buying is an important way to realize the core requirements of precision matching, which is emphasized by computational advertising. Its placement form, delivery time, and budget allocation are all more flexible and more precise. With the gradual development of the programmatic buying mode of personal computers, mobile programmatic purchases and multi-screen programmatic purchases have become topics of concern in the past two years. After the cross-screen and cross-device programmatic purchases are implemented, the advertisement dissemination will be able to target the various smart terminal screens of the same-origin users and their fragmented, multi-screen media usage habits to provide more precise advertising information.

3 Conclusion

The new technology and economic paradigm requires fusion and customized communication that matches it, and computational advertising came into being. In the current rapid development of big data and "Internet+," data has become the basic production factor in the field of advertising marketing. It provides the most objective decision-making basis for advertising creativity, planning, delivery, and monitoring, so that it can maximize the precise matching of advertising. The resulting industrial changes involve all stakeholders in the advertising industry chain, including companies, advertising agencies, third-party agencies, media, content producers, service providers, and users, and the company's core competitiveness and organization. The pace of the convergence of customized communication and computational advertising systems will soon become the future.

References

1. Qingzhen, L.: "Internet + Movie" in the perspective of the technology economy paradigm. Contemporary Film (2) (2015)
2. Guoming, Y.: "Using "Internet+" new normal to construct a new landscape. News and Writing (9) (2015)
3. Viktor, M., Kenneth, C.: The Age of Big Data: The Great Transformation of Life, Work, and Thinking. Zhejiang People's Publishing House, Hangzhou (2013)
4. Melanie, M.: Complex. Hunan Science and Technology Press, Shanghai (2011)

5. Duchowski, T.: Eye Tracking Methodology: Theory and Practice. Springer, New York (2003)
6. Zhu, Z., Ji, Q.: Eye and gaze tracking for interactive graphic display. Mach. Vis. Appl. **15**(3), 139–148 (2004)
7. Schnieders, D., Fu, X., Wong, K.: Reconstruction of display and eyes from a single image. In: Proceedings of IEEE International Conference on Computer Vision and Pattern Recognition (CVPR 2010), pp. 1442–1449 (2010)
8. Zhang, C., Chi, J., Zhang, Z., Wang, Z.: The research on eye tracking for gaze tracking system. Acta Autom. Sin. **36**(8), 1051–1061 (2010). (in Chinese)
9. Yoo, D., Kim, J., Lee, B., Chung, M.: Non-contact eye gaze tracking system by mapping of corneal reflections. In: Proceedings of IEEE International Conference on Automatic Face and Gesture Recognition, pp. 94–99. IEEE Computer Society, Washington, DC (2002)
10. Yoo, D., Chung, M.: A novel nonintrusive eye gaze estimation using cross-ratio under large head motion. Comput. Vis. Image Underst. **98**(1), 25–51 (2005)
11. Han, K., Wang, X., Zhang, Z.L., Zhao, H.N.: A novel remote eye gaze tracking approach with dynamic calibration. In: Proceedings of 15th IEEE International Workshop on Multimedia Signal Processing (MMSP 2013), 2013, pp. 111–116 (2013)

Network Security Management and Implementation Based on ACL

Mengqing Feng[1,2(✉)]

[1] College of Computer and Information Engineering, Zhengzhou University of Industrial Technology, Zhengzhou 451150, Henan, People's Republic of China
903901419@qq.com
[2] Machine Learning and Data Researching Institute, Zhengzhou University of Industrial Technology, Zhengzhou 451150, Henan, People's Republic of China

Abstract. At present, small and medium-sized enterprises demand for the confidentiality of the information network is more and more high, in view of the network protection company to provide professional information security technology and equipment of the disadvantages of high cost, stronger by using Access Control lists (Access Control List, ACL) technology as an independent security technology of network security management requirements in terms of design, focus on the router using ACL technology to protect sensitive data in the network, through the demand analysis to build the network topological structure of the simulation diagram, in the process of implementing the main command, finally based on the needs of specific testing network, more intuitive reflect the ACL network protection fast implementation in small and medium-sized enterprises and the characteristics of its low cost high practicability.

Keywords: Network security · Network attack · ACL · Flow control

1 Introduction

With the development of computer network, people's demand for the function of network is increasing gradually. While the network brings convenience to people, it inevitably produces various kinds of attacks. To achieve high standards of network security, data flow to achieve more accurate identification, separation and control, network devices such as routers, switches, must the data packet filtering, classify, different treatment, effectively protect the network security [1–4]. Access Control List (ACL) is produced and quickly and widely used in an environment where network service quality is increasingly demanding. ACL is a set of instruction set, according to the different requirements of network security, through the permutation and combination and orderly, effective list, the conditions of implementation of the classification of traffic filter, bag filter technology is the core of the ACL implementation. ACL can be used in network devices such as switches, routers and firewalls to better protect the network.

J. Abawajy et al. (Eds.): ATCI 2018, AISC 842, pp. 691–703, 2019.
https://doi.org/10.1007/978-3-319-98776-7_80

In the 1980s, access control technology was born in order to meet the need to manage Shared data authorization access on mainframe systems. With the development of computer network, the ideas and methods of access control technology are rapidly applied to various fields of information system [5–8]. Access control technology is the core technology to realize safe, is the security system of a solution, and ensure the confidentiality and integrity of key information technology, the research of access control has become one of the highlights in agro-scientific research in the computer science.

At present, ACL technology has become an independent security technology. How to implement ACL system and improve its performance is an important field in network security. Tang [9] is given through a router configuration ACL to enhance network security laboratory, Ruan [10] to study the application of ACL in the computer lab, Zhao [11] shows that the importance of ACL in campus network control technology, Chang [12] in Cisco router, for example, is how to implement network security strategy by using router ACL configuration are discussed. However, due to the diversity of network structure, most researches only summarize partial ACL configuration. This design by simulation experiments, with specific company needs as an example of network configuration and test, more intuitive reflect the ACL in the practical application of network protection of quick implementation and the flexibility of its application.

2 The Working Principle of ACL

An ACL is actually a traffic filter on a router that can identify specific types of packet traffic based on grouped attributes, such as IP addresses. ACL can perform specific operations after they are identified, such as preventing them from passing a particular interface. ACL are composed of access control entries (ACE), which are commonly referred to as ACL statements, and are created by a standard (such as source address, destination address, protocol number, and port number).

The ACL reads the following information from the packet header in the third layer: source IP address, destination IP address, ICMP message type; Read the following information from the fourth layer packet header: TCP/UDP source port number, TCP/UDP port number. Depending on the predefined rules, you decide which packets can be received and which packets need to be rejected to achieve the purpose of access control.

ACL provide a means to distinguish data packets from various types of packets, which can be used to control the user's access. The working principle of ACL is shown in Fig. 1.

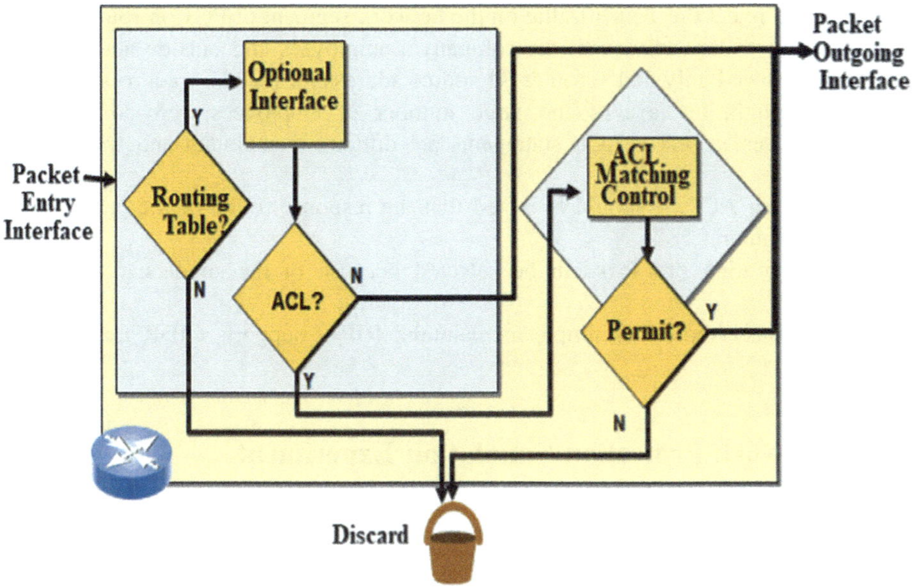

Fig. 1. Working principle of the ACL

3 Requirements Analysis of ACL Building Network Security Management Model

3.1 Functional Requirements

A company is building its own network. To ensure the security of the company's network and internal data, the following requirements are proposed for the network:

Use dynamic routing protocol to ensure that the whole network can ping.

Only the manager's host (PC0) can Telnet boundary network router (R2) and can manage the router R1 and R3 remotely.

Employees (PC1) and external network users are not allowed to access the company's TFTP server, but employees can access other servers (Web servers and DNS servers).

3.2 Performance Requirements

It is required that the whole network can ping, and the ICMP traffic should be released on the router.

Only the PC0 can Telnet boundary network router, while the router R1 and R3 can be remotely managed, and only the Telnet traffic of the source address of PC0 host address can be released on router R1.

You can also reject the Telnet traffic on the network segment of PC1 on router R1.

The user is not allowed to visit the company's employees and outside net TFTP server, can be allowed only on the router R1 source address for PC0 host address of the TFTP flow through, because of the large number of employees, refused to all employees to access a server ACL statements are difficult to maintain and waste of bandwidth.

When applying ACL, it should be noted that the response traffic of the packet is allowed on the router.

The RIPv2 routing protocol can be selected because of the small size of the company.

RIP protocol is simple and simple to maintain. RIPv2 supports CIDR for small networks.

4 ACL Network Protection Simulation Experiment

4.1 Simulation Topology

The solution topology diagram is shown in Fig. 2.

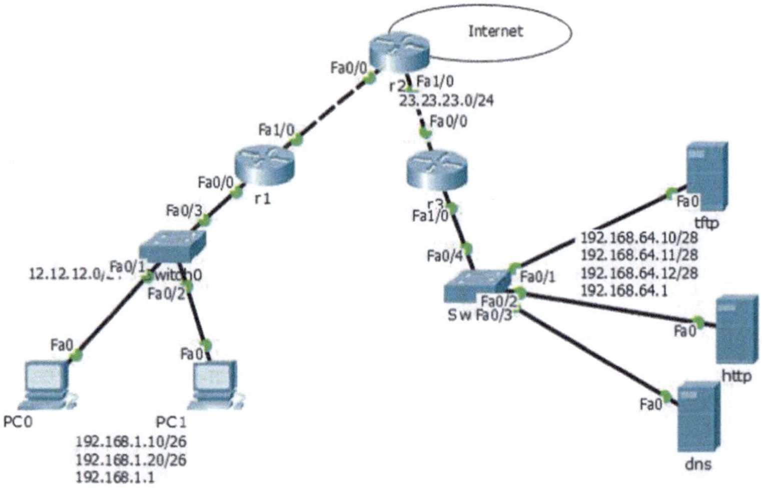

Fig. 2. Topology diagram of the solution

(1) ACL configuration on router R1.

ip access-list extended telacl

// create extended ACL, named telacl;

permit tcp host 192.168.1.10 any eq telnet

// Allows the source address to pass through Telnet traffic to any address at 192.168.1.10;

permit udp host 192.168.1.10 host 192.168.64.10 eq tftp

// Allow the source address to be 192.168.1.10 to the destination address for TFTP traffic of 192.168.64.10;

permit icmp any any

// Allow any ICMP traffic to be passed to facilitate testing of network connectivity;

permit udp any any eq domain

// Allows any DNS traffic to be passed, and the user can connect to the network via the DNS server;

permit tcp any any eq www

// Allows arbitrary HTTP traffic to be passed to connect to the network;

deny ip any any

// Implicitly reject all traffic;

interface FastEthernet0/0

ip access-group telacl in

// The ACL telacl is invoked in the inbound direction of the fastthernet0/0 interface;

（2）The ACL configuration on router R2.

```
ip access-list extended tftpacl
  // Create extension ACL name tftpacl;
deny udp any host 192.168.64.10 eq tftp
  // The reject host address is 192.168.64.10 for TFTP traffic, and the
external network user cannot access the internal TFTP server;
deny tcp any any eq telnet
  // Reject all Telnet traffic; The external network user cannot remotely
control the internal router of the company;
interface Loopback0
ip access-group tftpacl in
  // Call ACL tftpacl in the direction of the loopback L0;
```

（3）The ACL configuration on the router R3.

```
ip access-list extended internet
  // Create extended ACL with the name Internet;
permit icmp any any
permit udp host 192.168.64.10 eq tftp 192.168.1.0 0.0.0.63
  // Allow the TFTP traffic of the host to 192.168.64.10 to go to
192.168.1.0/26;
permit tcp host 192.168.64.11 eq www any
  // Allow HTTP traffic to the source address 192.168.64.11 to any address;
permit udp host 192.168.64.12 eq domain 192.168.1.0 0.0.0.63
  // Allow DNS traffic for the host to 192.168.64.12 to go to 192.168.1.0/26;
  interface FastEthernet1/0
ip access-group internet in
  // In the interface fast thernet1/0 interface is called;
```

4.2 Test

Connectivity Test
Ping Internet on PC0 (2.2.2.2.2), the test results are shown in Fig. 3.

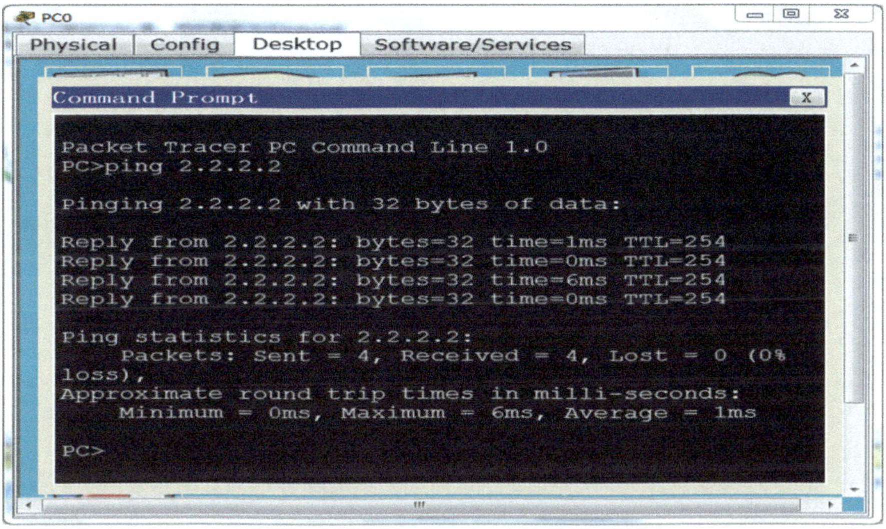

Fig. 3. Connectivity of PC0 to Internet is normal

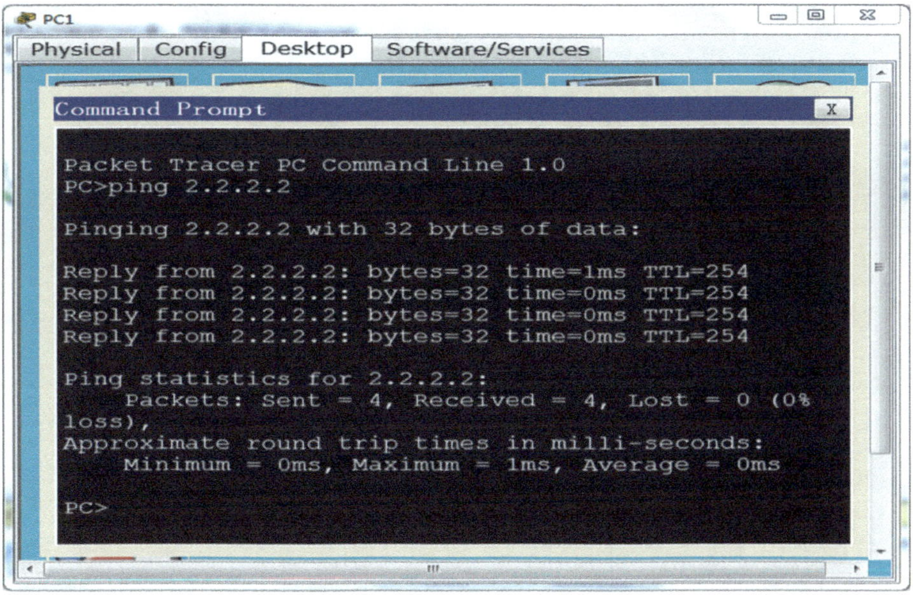

Fig. 4. The connectivity of PC1 to the Internet is normal

Ping Internet (2.2.2.2) on PC1, the test results are shown in Fig. 4.

Results analysis: Figures 3 and 4 show that the connectivity of the two PCS is normal, so that the network user can access Internet.

Flow Control Test

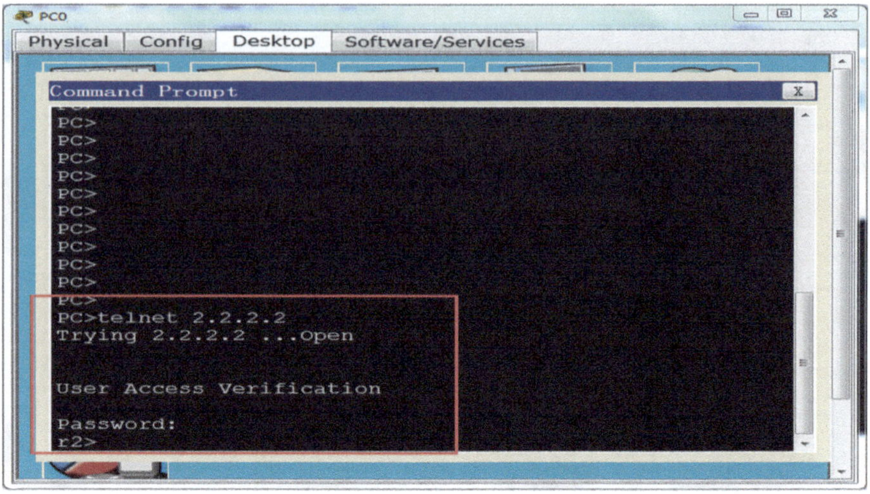

Fig. 5. PC0 can access Internet remotely

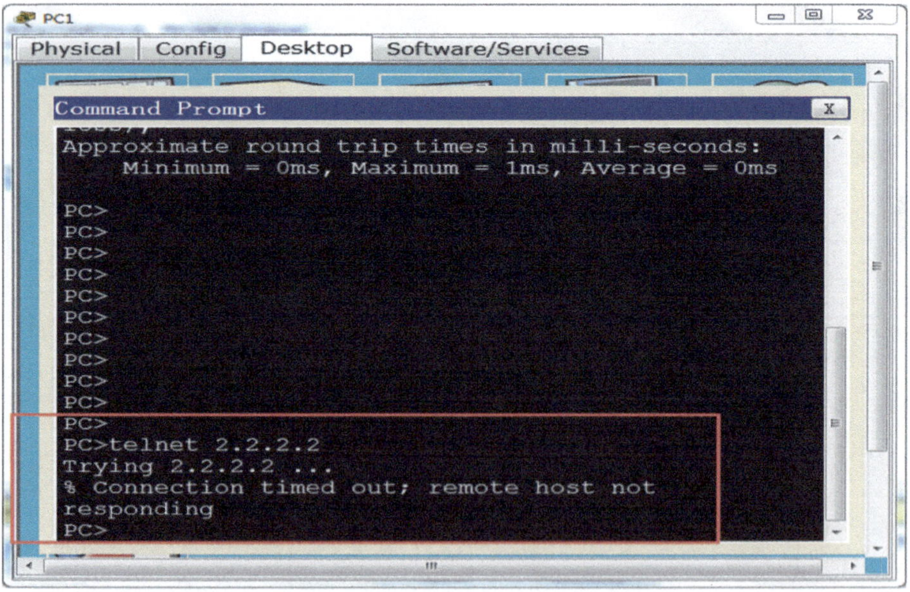

Fig. 6. PC1 cannot access Internet remotely

Telnet on PC0 Internet network, test results are shown in Fig. 5.

The test results are shown in Fig. 6.

Result analysis: according to the display information of Figs. 5 and 6, only the PC0 can Telnet boundary network router.

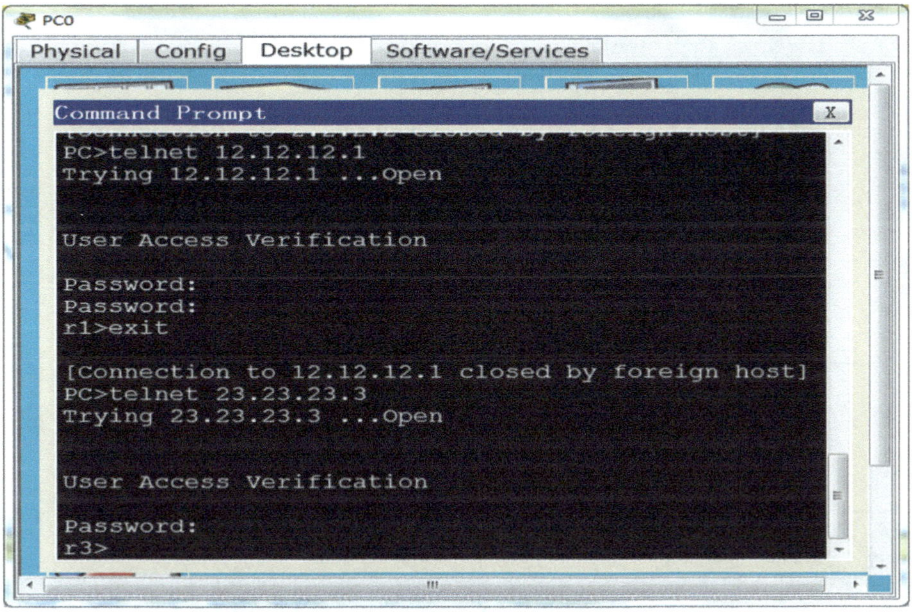

Fig. 7. PC0 remote control network router

The remote control of R1 and R3 was conducted respectively, and the test results were shown in Fig. 7.

Result analysis: Fig. 7 shows that PC0 can control the router R1 and R3 remotely. On router R1, only the Telnet traffic with the source address as the PC0 host address is allowed.

As shown in Fig. 8 on the DNS server, the domain name is: www.huxin.com.cn, and the mapping address is HTTP server address 192.168.64.11.

Connect the company's internal domain name on PC1, as shown in Fig. 9.

Result analysis: the display information of Fig. 10 shows that PC1 can access HTTP server and DNS server. Traffic on router R1 to HTTP server and DNS server; The flow of the release server to PC1 in the router R3.

Send packets to the TFTP server on PCO, as shown in Fig. 10.

Use the show IP access-lists telacl command on R1 to see the ACL match, as shown in Fig. 11.

Result analysis: the display information of Fig. 11 shows that the TFTP packet sent by PC0 matches the extended ACL named telacl on router R1.

Similarly, send packets to the TFTP server on PC1, as shown in Fig. 12.

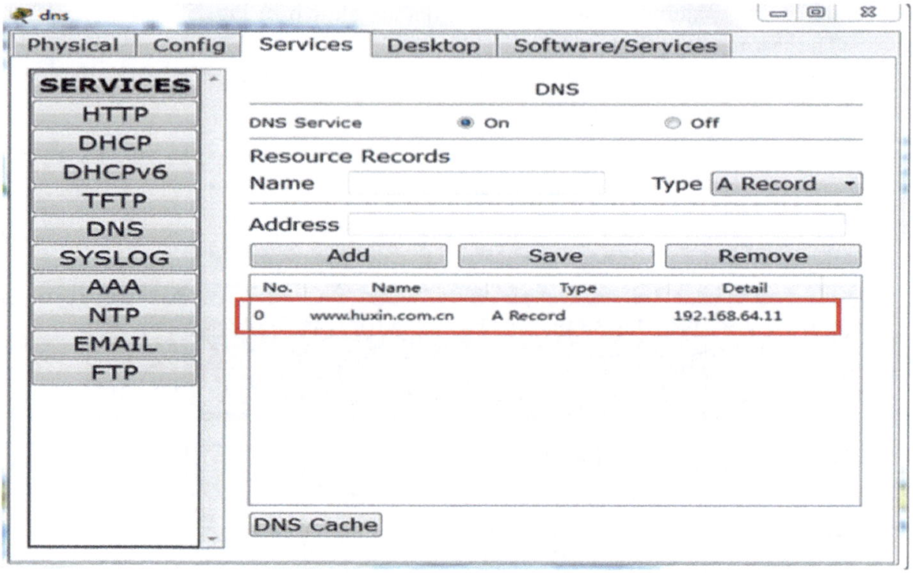

Fig. 8. Configuration of DNS server.

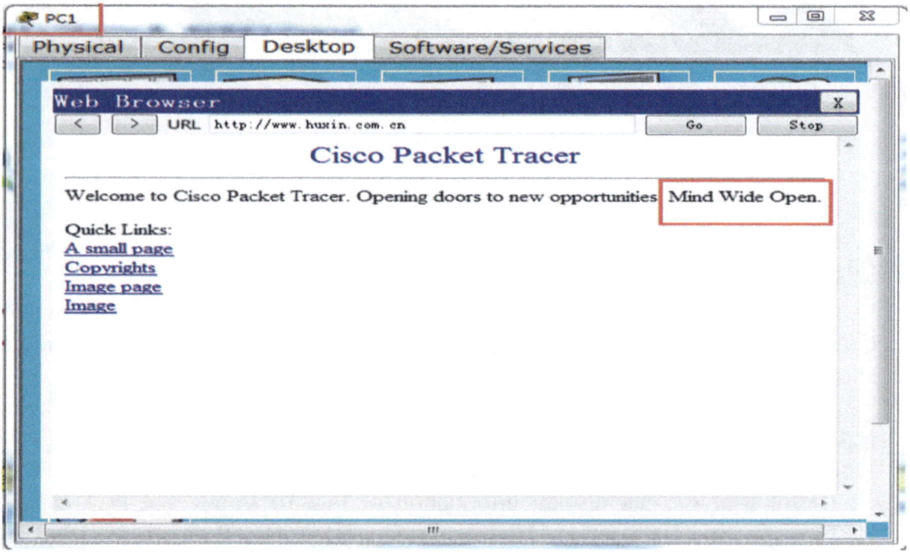

Fig. 9. PC1 access to the company's internal domain name server

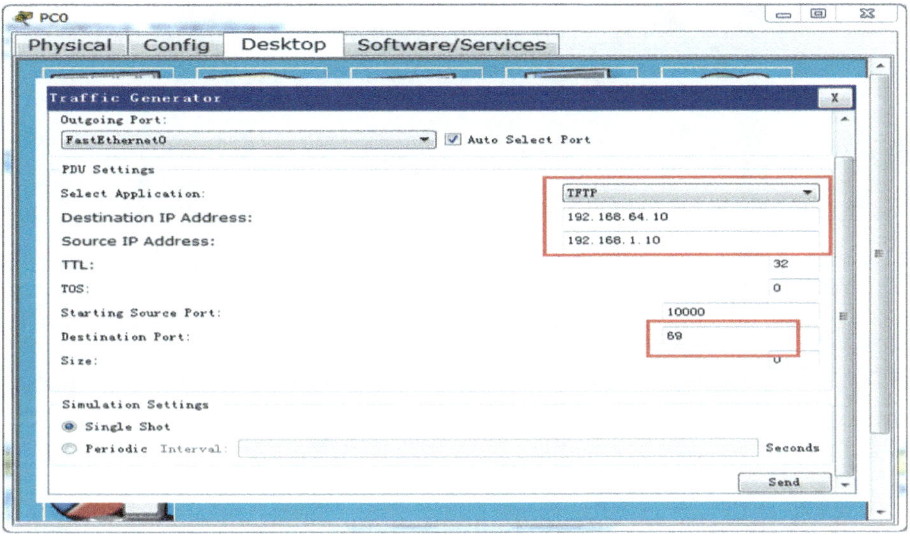

Fig. 10. Send TFTP flow package on PCO

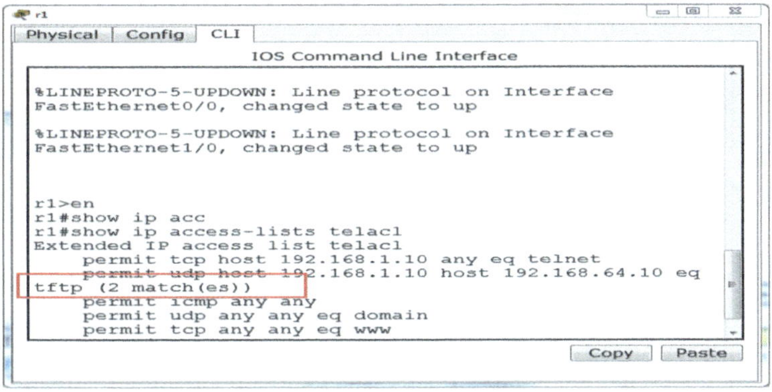

Fig. 11. The number of ACL matches on router R1

Check the ACL match on router R1, as shown in Fig. 13.

Results analysis: Fig. 13 shows that the matched number of packets did not change; The test results from Figs. 8, 9, 10, 11, 12 and 13 show that PC1 cannot access the TFTP server, but can access other servers of the company; You can also see that PC0 can access the TFTP server.

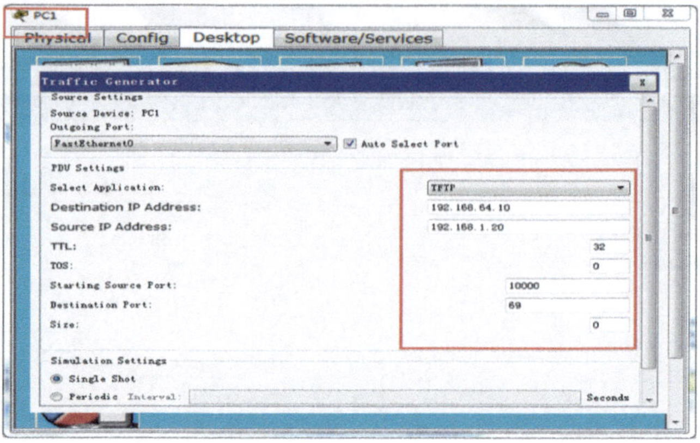

Fig. 12. Sends TFTP traffic packets on PC1

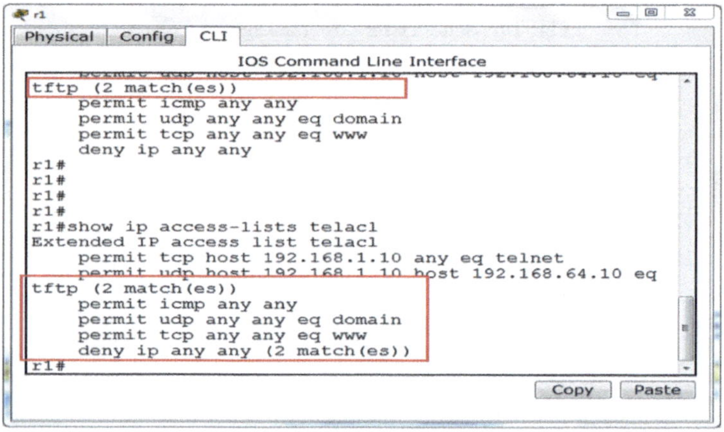

Fig. 13. The number of ACL matches on router R1

5 Conclusion

The computer network security problem is becoming more and more important with the development of the Internet, due to the openness of the computer network, the network is vulnerable to various attacks and damage, no matter in the LAN or wan, there are numerous factors, including natural and man-made form of potential security threats. In order to better solve the problem of network protection, through the demand analysis of structures, network topology structure of the simulation diagram, the main commands in the process of implementation of explanation, finally based on the needs of specific testing network, more intuitive reflect the ACL application flexibility and principles that should be paid attention to in practice and skill. ACL can implement effective authorization management for all users who have access to a particular resource. Although the ACL technology is relatively mature, to the realization of the flexible network user access control requirements of network security, but still have a lot of insufficient place, ACL in ACL is based on the information filtering in the packet header, and cannot provide a more detailed analysis, internal level can't identify the application layer, with some limitations.

References

1. Zhao, G.R.: Analysis of ways to protect the libraries' network information security. Shanxi Libr. J. (2010)
2. Conrad, E., Misenar, S., Feldman, J.: Chapter 5—Domain 4: Communication and Network Security (Designing and Protecting Network Security). Cissp Study Guide, pp. 219–291 (2016)
3. Casey, J.: Protect and survive: network monitoring tools, rather than traditional security measures of firewalls and IDSs, provide the strongest protection against 'denial of service' attacks - intrusion detection systems - monitoring network performance. J. Pediatr. **158**(4), 686 (2011)
4. Das, D., Nanda, S.: Securing computer networks by networking multiple OS kernels. Revisting network security: protecting computer networks from malwares. In: Internet Security, pp. 95–98. IEEE (2014)
5. Lee, Y.G., Kim, H.C., Kim, J.J., et al.: A design of home network security protocol using user authentication and access control technology. In: International Conference on Convergence and Hybrid Information Technology, pp. 30–34. IEEE Computer Society (2008)
6. Lakbabi, A., Orhanou, G., Hajji, S.E.: Network access control technology—proposition to contain new security challenges. Int. J. Commun. Netw. Syst. Sci. **5**(8), 505–512 (2013)
7. Yao, Y., Li, L., Yang, X.: Analysis and comparison on new network security access technology-NAC and NAP. In: International Conference on Computer Science and Network Technology, pp. 1736–1739. IEEE (2012)
8. Gao, H.C.: Application of access control list in network security. Comput. Knowl. Technol. (2014)
9. Tang, Z.J., Li, H.S.: Research on the application of ACL based network security management. J. Sichuan Inst. Technol. **22**(1), 48–51 (2009)
10. Ruan, J.: Application and research of ACL in computer lab. Inf. Comput. Theor. Ed. **11**, 17–18 (2011)
11. Zhao, Z.: Research and application of campus network control technology based on ACL. Agric. Netw. Inf. **2009**(9), 103–105 (2009)
12. Chang, J.: Network security strategy for Cisco router ACL configuration. Comput. Opt. Disk Softw. Appl. **6**, 198 (2014)

Indoor Navigation System Based on Sub High Frequency Sound Wave

Bocheng Liu and Xiaoman Wang(✉)

School of Software, Nanchang University, Nanchang, China
jsjjcjx@163.com

Abstract. In recent years, an on-board navigation system has emerged and gradually met people's needs for navigation to improve the current congestion of road vehicles, but it's not easy to be used in underground parking lots, large indoor venues, exhibitions and commercial centers. We use sub high frequency sound wave signals for acoustic communication. The mobile phone communicates with each transmitting device, measures the distance by the TOA ranging method and determines the position by the maximum likelihood estimation method. Once the location is determined, it is displayed in customized indoor map and navigated through the Dijkstra algorithm planning path. Indoor navigation system can suit China's national conditions and meet people's daily travel needs.

Keywords: Indoor navigation system · Acoustic wave communication
Dijkstra's algorithm

1 Introduction

In recent years, in order to improve the current congestion of road vehicles, an on-board navigation system has emerged and gradually met people's needs for navigation on the ground roads, greatly facilitating people's lives. The navigation software currently used in China includes Kay Rucker, Midea and Gao De. However, such navigation systems often use technologies such as GPS in the US and Beidou in China [1]. They can only be used outdoors, that is, where there are satellite signals, and they cannot navigate the room.

Under the current situation, underground parking lots and indoor navigation have become new exploration directions. Therefore, it is important to develop an indoor navigation system that suits China's national conditions and meet people's daily travel needs. It has important practical significance and economic value.

With the rapid development of all walks of life in society, people's demand for positioning and navigation is increasing. Especially in a complex indoor environment, it is often necessary to determine the location information of various facilities and items indoors. However, due to the constraints of positioning time, positioning accuracy, and complex indoor environments, relatively well-established indoor positioning technologies are still not well utilized. Therefore, experts and scholars have proposed many indoor positioning technology solutions such as GPS technology, infrared technology,

J. Abawajy et al. (Eds.): ATCI 2018, AISC 842, pp. 704–710, 2019.
https://doi.org/10.1007/978-3-319-98776-7_81

Bluetooth technology, radio frequency identification technology, Wi-Fi technology, ultrasonic technology, and so on.

GPS is currently the most widely used positioning technology. The disadvantage is that the positioning signal is weak when it arrives at the ground. It cannot penetrate buildings, and the cost of the locator terminal is relatively high. Infrared positioning technology can only spread within the line of sight, the transmission distance is short, and it is easily disturbed by the fluorescent lamp or the light in the room, which has limitations in the precise positioning [2]. The price of Bluetooth devices is relatively expensive, and for a complex space environment, the stability is disturbed by noise signals.

Although Wi-Fi, which is specifically used for positioning, is accurate in positioning, it is not easy to deploy, has high prices, and consumes a lot of energy. It is not practical for indoor applications where positioning accuracy is not too high. In view of the above problems, the system uses a mobile phone APP as a carrier with sub-high-frequency sound wave signals requires only a small sound wave transmitter rather than satellite signals, does not need to spend a lot of capital to install equipment, and it is lightweight and convenient without any space. The transmitter only needs to play sound with low power consumption [3]. The transmitter is hard to damage, easy to maintain and environmentally friendly. It can meet people's daily travel needs and has important practical significance and economic value.

2 System Analysis

2.1 System Technology Architecture

The system's Android client is developed using Java language. The iOS client is developed in Objective-C language. The mobile phone is used as the carrier to use commercial sound communication library named SinVoice Library, with the highest performance, the most stable performance and the most commercially-available method, which has been the first to be provided in the market. We use sub high frequency sound wave signals for acoustic communication. The mobile phone communicates with each transmitting device, measures the distance by the TOA ranging method and determines the position by the maximum likelihood estimation method, so as to achieve the positioning purpose. Once the location is determined, it is displayed in our customized indoor map and navigated through the Dijkstra algorithm planning path.

The server is developed using Python and provides various interfaces, including external service interfaces, map resource data, navigation paths, user management, shop management, and so on. The database uses open source and free MySql (Fig. 1).

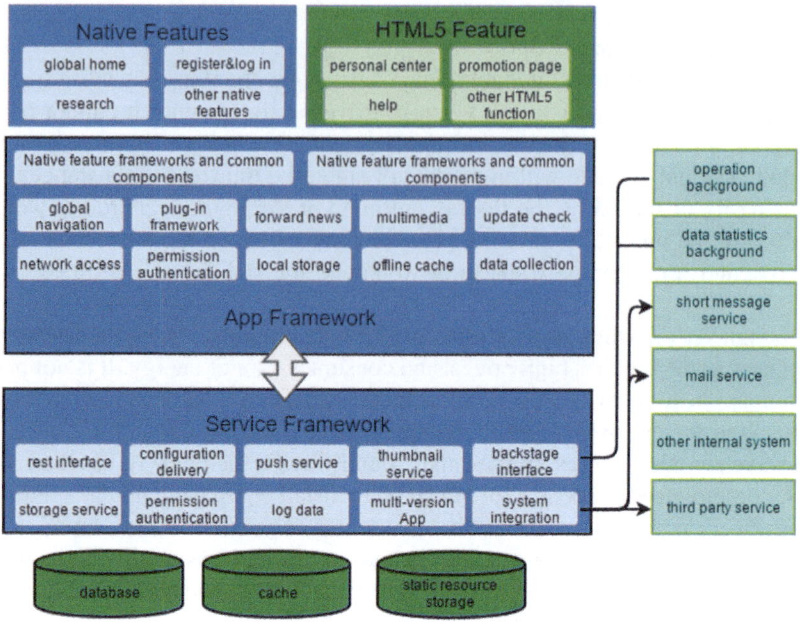

Fig. 1. System architecture diagram

2.2 Key Technology

(1) Sub High Frequency Sound Wave Coding and Decoding Technology

Acoustic wave communication encodes data with a single frequency sound signal. When these single-frequency sounds are played, the receiving party, after receiving the sound, recognizes the frequency and decodes the data according to the frequency [4].

The audio protocol for the sound wave communication comes from the technical documentation of chirp, which can realize sound wave information transmission between Android mobile phones or other various sounding devices. Chirp describes the technical details of smart devices relying on sound for data communication. However, the audio protocol of acoustic communication can be designed freely. For example, the sound in the chirp audio protocol can be changed to dual-tone or even multi-tone, so as to increase the capacity of information within a unit of time, thereby increasing the transmission speed.

(2) Sub High Frequency Acoustic Positioning Technology

The TOA ranging method is a method of calculating the distance between two nodes by measuring the transmission time of the signal between the transmitting node and the receiving node. TOA stands for Time Of Arrival, which means arrival time. This method generally requires that the transmission speed of the signal is small, the time between two measurement nodes is strictly synchronized, and the speed of signal propagation is known. The system uses TOA ranging method and maximum likelihood estimation method to obtain high positioning accuracy, even if there is some error in ranging.

(3) Path Planning Technology

The system uses Dijkstra's algorithm on the map to calculate the best route to the designated place and provide it to the user, so that the user can avoid the wrong path, take a detour, and walk the dead, and save a lot of time.

(4) QR Code Technology

The software adopts the two-dimensional code scanning technology to connect the QR code with the location points on the map, realize the function of automatic reading to realize the information automatic processing, and then records the location of the parking spot on the map to achieve the purpose of navigation.

3 System Implementation

3.1 System Function Module Design

The server function modules include a user management module, a shop management module, and a user cloud service module, and include three major databases. At the same time, it provides an external service interface to facilitate interconnection with other services.

(1) Positioning function: The system receives the acoustic signal, decodes the calculated distance, and displays it on the map after precise positioning.
(2) Navigation function: The system Provides positioning and navigation services for nearby parking lots, shopping malls, airports, schools, libraries, hospitals, scenic spots, theaters, museums, etc., and then the user selects a destination on the map. The system will automatically plan your itinerary and update it immediately using the path finding algorithm.
(3) Parking Service: If you already have a parking navigation record, there is a number and location for parking spaces. When you need to record the parking position after parking, click the stop button, the system will save the parking space information, and synchronize the information to the cloud.
(4) Shopping guide: When you want to go shopping after parking, the system will automatically recommend some shops nearby information, just a click, you can find the shop you want to go.
(5) Account management function: The system provides a user management center to support user registration and login. Users can modify and manage account information, passwords, and secret security information, as well as historical navigation records, map correction and other functions.

3.2 System Expansion Function Design

(1) Business Expansion

The system can cooperate with shops, shops can promote their own shops and products on the APP. The system can cooperate with other large venues such as movie theaters, train stations, etc (Fig. 2).

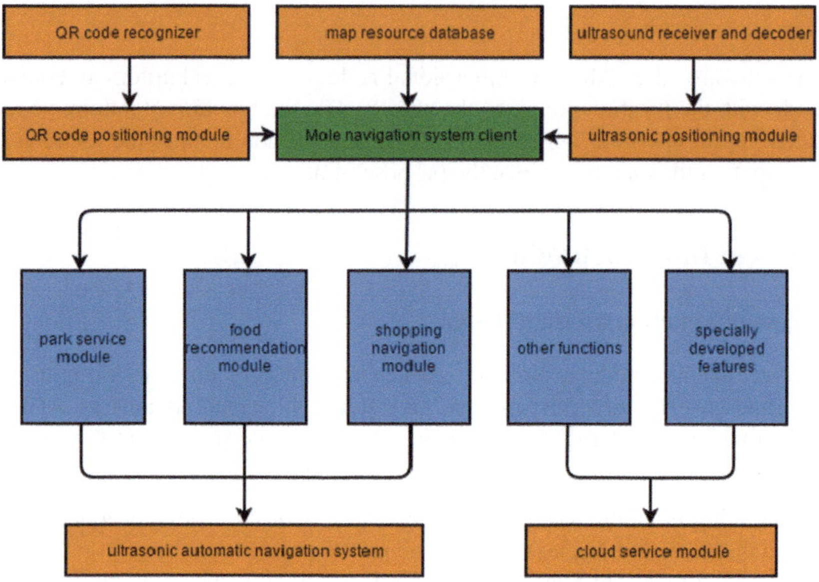

Fig. 2. Client architecture diagram

(2) User Service Expansions

Parking navigation features: The system quickly finds the nearest best parking lot and provides navigation services.

Parking reservation function: Users can reserve parking spaces to prevent parking spaces from being robbed.

Parking payment function: Users can use APP to pay for parking fees, which is quick and easy.

In large venues such as train stations, shopping malls, and entertainment centers, two users can use the mobile phone's secondary high-frequency sound wave navigation to navigate each other's location, thereby avoiding the trouble of not finding each other.

Through the main interface, you can see that the system provides location navigation services including parking lots, shopping malls, airports, schools, libraries, hospitals, scenic spots, theaters, museums, etc.

If you already have a parking navigation record, there is a parking space number and location. When you need to record the parking position after parking, click the stop button.

In addition, this software also provides the function of secondary navigation of previous navigation records. After clicking on the destination navigation, the system will automatically locate according to the collected sub high frequency sound wave signals, find your location, and the destination. The location is displayed. You only need to follow the directions to find the parking space. The system will save your parking space information.

When you want to go shopping after parking, the system will automatically recommend some nearby shops information to you, just with a click, you can find the shop you want to go.

When you are hungry, but do not know what is delicious nearby, the system will automatically recommend some of the nearby food for you, just with a click, you can enjoy your meal.

4 Conclusion

As an app that can be applied to various mobile phones, if it is able to collect maps of large commercial plazas across the country, then more and more people will use this software as a pure public welfare software designed to facilitate people's lives. If the software is successfully used in the market, it can also be used as a business value. For example, in the software landing interface implanted advertising.

Our software can be used as a closed-space precision navigation software that people are accustomed to using, just like today's Baidu maps and Gaode maps. At the same time, we can also do some advertisement promotion for the software, adding more member programs and so on. This software will have greater bright spots, thus create more economic value. When our system reaches a certain scale, it will be able to create more value if it can be incorporated into an existing large-scale map navigation system.

References

1. Rueppel, U., Stuebbe, K.M.: BIM-based indoor-emergency-navigation-system for complex buildings. Tsinghua Sci. Technol. **z1**, 362–367 (2008)
2. Jiang, Y., Fang, Y., Yao, C., et al.: A design of indoor and outdoor navigation system. In: 2011 International Conference on Communications Technology and Applications, pp. 1–5 (2011)
3. Guo, C., Hu, Q.: Indoor map service system based on Wechat two-dimensional code. In: International Asia-Pacific Web Conference, pp. 561–564 (2016)
4. Liu, J.-G., Shi, D.-M., Leung, M.K.: Indoor navigation system based on omni-directional corridor guidelines. In: 2008 International Conference on Machine Learning and Cybernetics, pp. 1271–1276 (2008)
5. Duchowski, T.: Eye Tracking Methodology: Theory and Practice. Springer, New York (2003)
6. Zhu, Z., Ji, Q.: Eye and gaze tracking for interactive graphic display. Mach. Vis. Appl. **15**(3), 139–148 (2004)
7. Schnieders, D., Fu, X., Wong, K.: Reconstruction of display and eyes from a single image. In: Proceedings of IEEE International Conference on Computer Vision and Pattern Recognition (CVPR 2010), pp. 1442–1449 (2010)

8. Zhang, C., Chi, J., Zhang, Z., Wang, Z.: The research on eye tracking for gaze tracking system. Acta Autom. Sin. **36**(8), 1051–1061 (2010). (in Chinese)
9. Yoo, D., Kim, J., Lee, B., Chung, M.: Non-contact eye gaze tracking system by mapping of corneal reflections. In: Proceedings of IEEE International Conference on Automatic Face and Gesture Recognition. Washington, DC, pp. 94–99. IEEE Computer Society (2002)
10. Yoo, D., Chung, M.: A novel nonintrusive eye gaze estimation using cross-ratio under large head motion. Comput. Vis. Image Underst. **98**(1), 25–51 (2005)
11. Han, K., Wang, X., Zhang, Z.L., Zhao, H.N.: A novel remote eye gaze tracking approach with dynamic calibration. In: Proceedings of 15th IEEE International Workshop on Multimedia Signal Processing (MMSP2013), pp. 111–116 (2013)

Smart Campus Information Security Analysis and Research

Hua Li[(✉)], Shiyu Zhang, and Laifeng Tang

Tianfu College of Southwestern University of Finance and Economics, Chengdu, Sichuan Province, China
553002949@qq.com

Abstract. With the popularization of smart campuses, the issue of big data security in smart campuses cannot be ignored. It is necessary to build a smart campus security protection system. This article systematically analyzes the problems faced by smart campus big data security, and proposes the smart campus big data application security strategy. By analyzing the security technologies used in big data, it proposes a method to build a smart campus big data security protection platform, in order to ensure the normal wisdom campus run.

Keywords: Smart campus · Information security · Large data security

1 Introduction

At present, with the development of the Internet and education, higher education is increasingly becoming open, international, and smart. Smart campuses are a large and complex system that integrates information technologies such as the IoT, cloud computing, and big data. Smart Campus promotes a new model of campus informationization planning, construction, management and service intelligence. With the development of smart campuses, the amount of data in the system grows exponentially in geometric progression. At the same time, the development of big data and cloud computing needs to increase the openness of the database. This open trend is the collection, transmission, and storage of massive data. The improper protection of big data in smart campuses has caused sensitive data to be stolen or modified in bad faith, resulting in information leakage. At the very least, it will cause chaos in school management, damage school honors, directly affect the future development of schools, jeopardize national security, and affect society stable. Inadequate security protection of big data systems may also result in malicious modification of teaching or research information, directly leading to disordered teaching and scientific research. How to better solve information security problems in the construction and operation of smart campuses is an important part of promoting the healthy development of smart campuses Question.

J. Abawajy et al. (Eds.): ATCI 2018, AISC 842, pp. 711–715, 2019.
https://doi.org/10.1007/978-3-319-98776-7_82

2 Smart Campus Information Security Risks

2.1 The Collection of Information

The method used for current information collection is mainly to use the network for data collection. Smart campus has many data sources and large data volumes. To conduct safety analysis and evaluation on the collected data, it is necessary to ensure the authenticity and reliability of data sources and prevent information leakage and ensure information security.

2.2 Information Processing

The processing of information includes the extraction of information and the cleaning of data. The extraction of information refers to the conversion of various types of data into relatively easy to process data. Cleaning data refers to the removal of data that is worthless. Various types of massive data in smart campuses need to be encrypted after extraction in the process of information transmission.

2.3 Information Storage

Big data storage uses multiple nodes for distributed storage. The storage of big data needs to be stored in the cloud using cloud storage technology, and the data is transferred to the cloud storage. The secure storage technology of the cloud is beyond the control of the smart campus security system, which will lead to data outside the security boundary of the campus security system. Loss of control increases the difficulty of data protection.

2.4 Auditing Aspects of Data

Smart campus data are complex, data structures are complex, and data types are diverse. It is difficult for auditors to determine the focus of audits. It is impossible to fully grasp and understand the connotation of data in a short period of time, which increases the difficulty of auditing.

3 Smart Campus Big Data Application Security Strategy

3.1 Strengthen and improve rules and regulations for big data security management, and ensure the application security of big data from the system. Scientifically and reasonably use big data within the legal framework to coordinate and handle the conflicts between the opening of big data and the protection of information security.

3.2 To standardize the construction standards and operating mechanisms of big data, standard construction standards and operating mechanisms can ensure that Big Data operates reliably under a unified safety standard.

3.3 Establish a security management platform system centered on heterogeneous data to ensure the secure access of big data through a unified management platform system. Under the premise of guaranteeing the use efficiency, the confidentiality, isolation, integrity, availability, and traceability of the data are realized, and wisdom management and application are realized.
3.4 Strict security control levels are implemented according to the level of confidentiality of big data and user requirements, and big data and user hierarchies are set to different levels. On the basis of security level protection, big data security access policies are formulated, active binding, role-based access control, and attribute-based access control are integrated so that different data can be safely mapped to different users.
3.5 The comprehensive improvement of security protection capabilities and the use of big data technology to analyze large amounts of data can uncover abnormal behaviors of the network and quickly and accurately find the source of the attack and quickly respond to security threats in a targeted manner.

4 Smart Campus Big Data Security Technology

4.1 Data Acquisition Security Technology

Data collection security technology Big data uses distributed technology to collect big data. During the collection process, there may be hidden dangers and threats such as data corruption, data loss, data leakage, and data theft. Authentication, data encryption, and data integrity protection should be used. Safety mechanisms to ensure the safety of the collection process.

4.2 Big Data Storage Security Technology

The storage of big data is mainly through the cloud storage technology to store various types of data resources. In the cloud environment, the encryption of big data must adopt the homomorphic encryption algorithm. After the big data is divided into 64 MB small data blocks by the cloud file system, the original encrypted data cannot be decrypted using the decryption key.

4.3 Data Transmission Security Technology

During data transmission, the confidentiality of the communication data must be detected, the data packets in the transmission are captured, and then the frames are parsed according to the protocol, and the unencrypted data is encrypted to ensure that the important data is wirelessly transmitted in ciphertext form. Data integrity and consistency can be detected. By parsing the captured data packets, the frames can be parsed to find the fields that guarantee the integrity of the data. The data that is missing or tampered can be accurately identified and discarded, requiring retransmission. The raw data ensures that the data received by the receiver is the original data.

4.4 Data Mining Security Technology

Data mining is the core of big data applications. Through data analysis and mining, various security incidents such as hacker attacks, illegal operations, internal and external threat sources are discovered from big data, and security alarms and dynamic responses are issued in a timely manner. For example, the multi-dimensional analysis of big data using deep learning algorithm and fuzzy algorithm can find all kinds of potential local features and overall features of attackers.

4.5 Security Audit Technology

Security auditing records, manages, categorizes, and stores data operations events, then analyzes, queries, and statistics the collected security events, and mines potential knowledge models behind the data for subsequent intrusion detection. Its core role is to accurately record the attack behavior, to trace security incidents, and to ascertain the ownership of the accident liability.

4.6 Prevent Advanced Persistent Threat Attacks

APT (advanced persistent threat) has the characteristics of strong concealment, strong pertinence, multiple attack methods, wide attack range, and high difficulty in prevention. To cope with APT attacks, there are currently solutions such as sandboxing, anomaly detection, full-flow auditing, deep-protocol resolution anomaly identification, and attack source tracing.

5 Smart Campus Big Data Security Protection Platform Construction

Based on the comprehensive analysis of big data storage, application, management, technology, and other security strategies, we established a smart campus big data security protection platform based on the three major modules.

(1) Data acquisition and transmission module: Responsible for data collection, information encryption and transmission, and hierarchical sharing, and at the same time standardize and manage the collected data.
(2) Data mining module: Responsible for the analysis and mining of big data, with more depth and breadth results, able to issue security alarms and make dynamic responses in a timely manner.
(3) Data security audit and maintenance module: Responsible for ensuring the information security and security platform operates normally and stably. On the basis of the three major modules, build a smart campus big data security protection platform architecture.

The security management in the big data security protection platform can obtain the corresponding security baseline by collecting the configuration information of the big data application platform, automatically monitor the big data platform, compare the monitoring results with the baseline, and quickly discover the system operation. The abnormal behavior can restore the original security configuration in time to ensure the security of the smart campus big data application platform.

6 Summary

The issue of big data security in smart campuses can not be ignored. It is necessary to build a smart campus security protection system. This article systematically analyzes the problems faced by smart campus big data security, puts forward a smart campus big data application security strategy, and analyzes the security technologies adopted by big data to build a smart campus big data security protection platform to ensure the normal operation of a smart campus.

References

1. Yang, J., Liao, Z., Feng, C.: A review of large data storage architecture and algorithms. Comput. Appl. **36**(9), 2465–2471 (2016)
2. Zhang, G., Li, C., Xing, C.: Core Technology Behind Large Data. Electronic Industry Press, Beijing (2017)
3. Chen, H., Bhargava, B., Fu, Z.: Multilabels-based scalable access control for big data applications. IEEE Cloud Comput. **1**(3), 65–71 (2014)
4. Li, L., Yu, X., Yang, Y.: Review on the research progress of homomorphic encryption. Comput. Appl. Res. **32**(11), 3209–3214 (2015)
5. Sun, Z.: The construction and Research on the network security of the intelligent campus of Higher Vocational Education. Heilongjiang Sci. Technol. Inf. **33**, 182 (2016)
6. Yu, D., Peng, L.: When does inferring reputation probability countervail temptation in cooperative behaviors for the prisoners' dilemma game? Chaos Solitons Fractals **78**, 238–244 (2015)
7. Wang, Q.: Automatic control system regulation scheme of textile air conditioning. Sci. Technol. Inf. **22**, 395–396 (2010). (in Chinese)
8. Liu, J.: Advanced PID control and MATLAB simulation, 2nd edn. Electronics Industry Press, Beijing (2003). (in Chinese)
9. Li, J., Ren, Q.: Study on supply air temperature forecast and changing machine dew point for variable air volume system. Build. Energy Environ. **27**(4), 29–32 (2008). (in Chinese)
10. Clay Richardson, W., Avondolio, D.: The Java high class weaves a distance: JDK 5. Sci. Technol. Book Rev. **3**, 17–18 (2006)

Benefit Evaluation of Power Grid Project Based on Gray Triangle Whiten Weight Function Model

Hui Kang[1], Dongxiao Niu[1], Yaru Han[1(✉)], Xianing Jin[2], Yi Song[2], and Xiao Wu[3]

[1] School of Economics and Management, North China Electric Power University, Beijing 102206, China
{Kangh, ndx}@ncepu.edu.cn, hfhmth@163.com

[2] State Grid Economic and Technological Research Institute Co., Ltd., Beijing 102210, China
{jinxianing, songyi}@chinasperi.sgcc.com.cn

[3] State Grid Tianjin Economic Research Institute, Tianjin 300171, China
wuxiao0117@126.com

Abstract. With the rapid development of the smart grid in China, it is becoming more and more important to make a scientific and rational benefit evaluation of the power grid project. In this paper, a comprehensive evaluation index system of general power grid projects is established. Based on the grey triangle whiten weight function theory and the variation coefficient method, a benefit evaluation model is established, and an empirical analysis of the project of a certain city power grid is carried out. The results show that the model has a strong practical value and provides ideas and methods for the benefit evaluation of related power grid projects.

Keywords: Triangle whiten weight · Power grid project · Benefit evaluation

1 Introduction

With the increasing complexity of China's power grid and the expansion of the investment scale, how to make a scientific and reasonable evaluation of the power grid project's production efficiency has become an important issue for power grid companies. Under the guidance of the "Thirteenth Five-Year Plan" development and construction goals of the State Grid, the company's development business process has been gradually optimized, the level of management leanness has been significantly improved, and higher requirements have been put forward for the benefit evaluation of grid projects.

Many existing literatures have carried out research on the benefit evaluation of grid projects. The literature [1] establishes the evaluation index system from two aspects of investment income and investment decision, and uses triangular fuzzy number theory to evaluate the investment benefit of distribution network project. The literature [2] studies the evaluation index system of power grid projects from five aspects: process,

J. Abawajy et al. (Eds.): ATCI 2018, AISC 842, pp. 716–723, 2019.
https://doi.org/10.1007/978-3-319-98776-7_83

technical performance, sustainability, investment efficiency, and impact. The literature [3] constructs an evaluation index system for power grid construction projects in terms of implementation process, economic benefits, environmental and social impacts, and grid performance. Although there are many literatures on the benefit evaluation of grid projects, there are still some problems in their research. For example, the statistical analysis data is not standardized, the index system constructed is not reasonable, and the evaluation method used is not applicable. Therefore, a benefit evaluation index suitable for general grid projects is constructed, and a scientific and reasonable evaluation method is used to evaluate the benefits. It is practical for power grid enterprises.

2 Project Benefit Evaluation Index System Construction

2.1 Project Benefit Evaluation Index System

This paper screens a large number of indexes from three aspects: economic benefits, social benefits, and environmental benefits, and constructs a general grid project evaluation index system, as shown in the following figure [4–8] (Fig. 1):

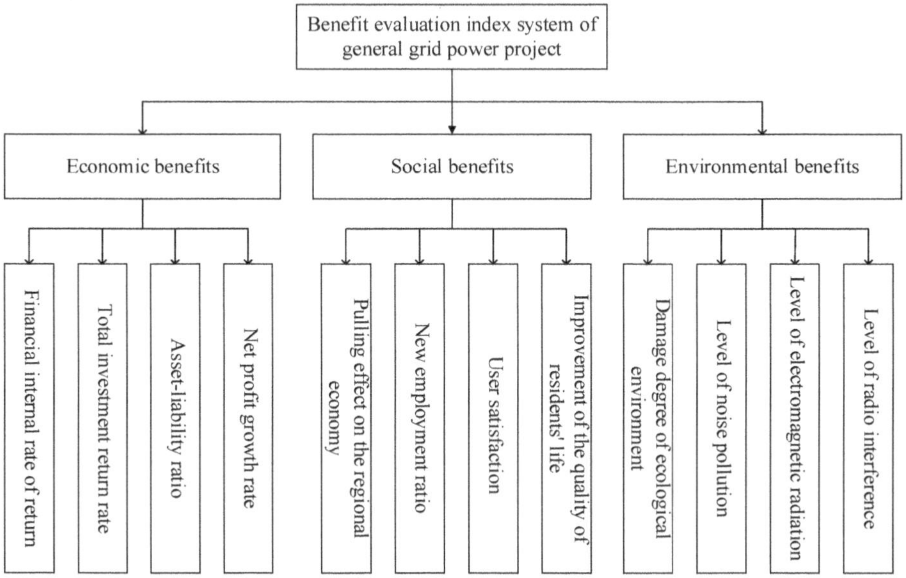

Fig. 1. Benefit evaluation index system of general grid power project

2.2 Evaluation Index Description

(1) Economic benefits
 (1) Financial internal rate of return
 Financial internal rate of return (FIRR) refers to the discount rate when the financial net present value of the project is equal to zero over the entire calculation period. Under normal circumstances, when the financial internal rate of return is greater than or equal to the industry benchmark rate of return, the project is feasible. It is a positive index.
 (2) Total investment return rate
 Total investment return rate refers to the percentage of EBITDA of the total investment of the project, which reflects the profitability of the project and is a positive index.
 (3) Asset-liability ratio
 Asset-liability ratio is the ratio of the total liabilities to the total assets. Under normal circumstances, the asset-liability ratio should be controlled within a certain standard, which is an interval index.
 (4) Net profit growth rate
 Net profit growth rate refers to the ratio of the increase of the net profit in this year to the total net profit of the previous year during the operation period of the project, which reflects the change in the overall operating capacity of the project and is a positive index.
(2) Social benefits
 (1) Pulling effect on the regional economy
 If the power grid construction can match the development situation of the area, it will play a promoting role in the improvement of the regional total economic volume, the adjustment of industrial structure, the attraction of investment volume and the improvement of infrastructure. The index is a qualitative index.
 (2) New employment ratio
 The index refers to the proportion of newly-increased employment of the project during the construction period and operation period to the total number of newly-increased employment in the area covered by the project every year, reflecting the improvement degree of the project on the local employment status, which is positive index.
 (3) User satisfaction
 User satisfaction refers to the ratio between the number of users who are satisfied with the power quality, service quality, etc. in the coverage area of the grid project and the total number of users. It is a positive index.
 (4) Improvement of the quality of residents' life
 The important goal of the power grid construction project is to improve the people's living standard and the quality of life. This index is a qualitative index.
(3) Environmental benefits
 (1) Damage degree of ecological environment

The index reflects the mutual adaptability between the power grid project and the natural environment, and it is a qualitative index.

(2) Level of noise pollution
Noise pollution includes the noise impact caused by construction and mechanical operation on the residents in the stage of project construction, as well as the audible noise pollution generated during transmission line and substation operation, which is a qualitative index.

(3) Level of electromagnetic radiation
The area passed by high voltage transmission lines will be affected by electromagnetic radiation to a certain extent. This index is a qualitative index.

(4) Level of radio interference
Due to the characteristics of their own operations, substations and overhead transmission lines will cause certain levels of interference to radio communications, radio broadcast television and radio navigation stations. This index is a qualitative index.

3 Construction of Comprehensive Evaluation Model of Project Benefits

3.1 Construction of Gray Triangle Whiten Weight Function Comprehensive Evaluation Model Based on Variation Coefficient Method

There are n objects, m evaluation indexes, s grey classes and the sample observation value of object i about the index j is $x_{ij}(i = 1, 2, \cdots, n;\ j = 1, 2, \cdots, m)$. The corresponding object i is evaluated according to the value of xij, and the specific steps are as follows [9, 10]:

(1) Determination of the evaluation grey class
According to the requirements of the evaluation, the number of grey classes is divided into M, and the range of the values of each index is also divided into M grey categories. For example, the range $[a_1, a_{M+1}]$ of the j index is divided into $[a_1, a_2], \cdots, [a_k, a_{k+1}], \cdots, [a_M, a_{M+1}], \cdots,$ in which the range of $a_k(k = 1, 2, \cdots, M)$ is determined according to the actual problems and qualitative research results.

(2) Determination of the evaluation index weight
The variation coefficient method is chosen here to determine the weight of the index. The specific methods are as follows:

(1) The standard deviation σ_k of each index is calculated to indicate the absolute variation of each index. The equation is as follows:

$$\sigma_k = \sqrt{\sum_{i=1}^{n} (x_{ik} - \bar{x}_k)^2 / n} \quad (k = 1, 2, \cdots, m) \tag{1}$$

Where: $\bar{x}_k$ refers to the average number of the k index.

(2) The variation coefficient c_k of each index is calculated, which reflects the relative variation of each index, and its calculation equation is:

$$c_k = \sigma_k / \bar{x}_k \tag{2}$$

Normalizing the coefficient of variation of each index, and the weight of each index is:

$$w_j = \frac{c_j}{\sum_{k=1}^{m} c_k} \tag{3}$$

Then the weight vector of each index is $W = (w_1, w_2, \cdots, w_m)$.

(3) Establishment of triangle whiten weight function
Assuming that the whiten weight function value of $\frac{a_k + a_{k+1}}{2}$ belongs to the k grey class is 1, and $\left(\frac{a_k + a_{k+1}}{2}, 1\right)$ is connected to the starting point a_{k-1} of the $k-1$ grey class and the end point a_{k+2} of the $k+1$ grey class. And then obtain the triangle whiten weight function of the j index about the k grey class is $f_j^k(x)(j = 1, 2, \cdots, m; k = 1, 2, \cdots, s)$. For $f_j^1(x)$ and $f_j^s(x)$, the number field continuation can be extended to the left and right to a_0, a_{s+2} respectively according to the specific circumstances.
The realization value x for the index j, the membership degree $f_j^k(x)$ of grey $k(k = 1, 2, \cdots, s)$ is calculated by

$$f_j^k(x) = \begin{cases} 0, & x \notin [a_{k-1}, a_{k+2}] \\ \frac{x - a_{k-1}}{\lambda_k - a_{k-1}}, & x \in [a_{k-1}, \lambda_k] \\ \frac{a_{k+2} - x}{a_{k+2} - \lambda_k}, & x \in [\lambda_k, a_{k+2}] \end{cases} \tag{4}$$

Where: $\lambda_k = \frac{a_k + a_{k+1}}{2}$.

(4) Calculation of comprehensive clustering coefficient
Calculating the comprehensive clustering coefficient of object i about grey $k(k = 1, 2, \cdots, s)$:

$$\sigma_j^k = \sum_{j=1}^{m} f_j^k(x_{ij}) \times w_j \tag{5}$$

Where: $f_j^k(x_{ij})$ is the whiten weight function of the object i belonging to the k grey class under the index j; x_{ij} is the value of the object i about the index j; w_j is the weight of the index j in the comprehensive clustering.
Judging from the $\max_{1 \le k \le s} \left\{ \sigma_j^k \right\} = \sigma_j^{k*}$ P, the object i belongs to the k* grey class.

3.2 Empirical Analysis

This paper takes an example of a general project of a city power grid as an example to carry out an empirical analysis. The detailed calculation process is as follows.

(1) Data standardization processing

The quantitative index value is the measured data of the project. For the qualitative index, the expert scoring method is used to assign the value, the assignment interval is from 0 to 100. All index values are standardized, and the values after treatment are between 0.4 and 1.

(2) Determination of the object of evaluation and evaluating the grey category

The project benefit in five years are five evaluation objects respectively, which can reflect the changes of the benefits of the project in every year. According to the actual situation and data characteristics of the project, this paper adopt 4 evaluation grey classes: "excellent", "good", "middle" and "bad". The range of each index is also divided into 4 grey categories, namely (1, 0.9), (0.9, 0.8), (0.8, 0.6), (0.6, 0.4).

(3) Determination of index weight

The weight of the evaluation index determined by the variation coefficient method is shown in the following Table 1.

Table 1. The table of index weight

Index	Weight	Index	Weight
Financial internal rate of return (C1)	0.11	User satisfaction (C7)	0.05
Total investment return rate (C2)	0.07	Improvement of the quality of residents' life (C8)	0.11
Asset-liability ratio (C3)	0.04	Damage degree of ecological environment (C9)	0.07
Net profit growth rate (C4)	0.17	Level of noise pollution (C10)	0.07
Pulling effect on the regional economy (C5)	0.07	Level of electromagnetic radiation (C11)	0.10
New employment ratio (C6)	0.09	Level of radio interference (C12)	0.05

(4) Evaluate the extension value of each index and the whiten weight function value of the corresponding grey class

Take $a_0 = 0.2$, $a_6 = 1.1$, and extend the number field continuation to the left and right, along to a_0 and a_6, to calculate the corresponding $f_j^1(x)$ and $f_j^4(x)$. According to the equation above, the extension value of each year's index and the whiten weight function value of the corresponding grey class are calculated.

(5) Calculation of comprehensive clustering coefficient

According to the equation above, the comprehensive clustering coefficients σ_j^k of 4 grey categories of "excellent", "good", "middle" and "poor" in each year are calculated, as shown in the following Table 2.

Table 2. The table of comprehensive clustering coefficient

Year	Excellent	Good	Medium	Poor	Grade evaluation
2012	0.3148	0.5053	0.3860	0.1690	Good
2013	0.1391	0.2833	0.4849	0.3719	Medium
2014	0.1417	0.6479	0.5823	0.1468	Good
2015	0.5296	0.5314	0.1555	0.0000	Good
2016	0.6908	0.1395	0.0061	0.0000	Excellent

From $\max_{1 \le k \le s}\left\{\sigma_j^k\right\} = \sigma_j^{k*}$, we can see that from 2012 to 2016, the level of comprehensive benefits of the project is good, medium, good, good and excellent, indicating that although the benefit of the project has declined slightly in 2013, it is developing towards an excellent direction.

4 Summary

In order to meet the requirements of grid development for the benefit evaluation of power grid projects, the benefit evaluation index system of general projects of power grid is constructed firstly in this paper according to the principle of index selection from three aspects: economic benefits, social benefits and environmental benefits. Secondly, based on the variation coefficient method and the grey triangular whitening weight function theory, a comprehensive evaluation model of project benefit is established. Finally, taking a general project of a city power grid as an example, the empirical analysis has been carried out to prove the practical value of the evaluation model built in this paper.

Acknowledgments. This work was one of the results of the science and technology project of State Grid Corporation headquarters called "Research and Application of Equipment Statistical Analysis System and Decision-making Model of Grid Development Based on Real-time Information" (Project No. B3441317K008), and it was supported by Natural Science Foundation of China (Project No. 71471059), "the Fundamental Research Funds for the Central Universities (2018ZD14)" and "the 111 Project (B18021)".

References

1. Li, X.J., Yu, T.K., Liu, X.Y.: Evaluation method of investment benefit of distribution network project based on triangular fuzzy number. Distrib. Util. **2015**(10), 62–66 (2015)
2. Yang, R.M., Jia, W., Ma, Q.L.: Research and application of post-evaluation system of power grid project. Zhejiang Electr. Power **2015**(4), 1–6 (2015)
3. Zhan, Z.H., Zhou, M., Wu, Q., et al.: Study on post-evaluation index system and evaluation method of power grid construction project. Hubei Electr. Power **39**(2), 63–66 (2015)
4. Li, J., Li, X.H., Liu, S.Y., et al.: Evaluation of investment benefit of distribution network based on ideal solution and grey relational degree. East China Electr. Power **2012**(1), 13–17 (2012)
5. Liu, S.L., Cao, Y., Feng, Y.L., et al.: Study and application of evaluation and decision model of distribution network investment benefit. Power Syst. Protect. Control **2015**(2), 119–125 (2015)
6. Zhu, X.R., He, Q.G.: Construction of evaluation index system for power grid investment. Product. Res. **2011**(2), 86–87 (2011)
7. Wang, F.: Discussion on the evaluation system construction of the economic benefits of power grid construction projects. Mod. Econ. Inf. **2012**(21), 226 (2012)
8. Song, L.L., Xu, Q.S., Wu, Y.W.: Weighted grid investment benefit evaluation and multi-objective optimization. J. Electr. Power **2013**(6), 503–506 (2013)
9. Zhao, H.R., Li, N.N.: The comprehensive risk assessment of urban power grid planning based on AHP-grey triangle whiten weight function. Shaanxi Electr. Power **41**(10), 14–18 (2013)
10. Wang, Y., Xu, M., Di, R.: The application of gray triangle whiten weight function in social responsibility evaluation of textile enterprises. Light Ind. Stand. Qual. **2015**(6), 57–58 (2015)

An Empirical Analysis of the Impact of Media Coverage on Stock Returns

Weiqing Wang, Ziyuan Li(✉), and Jiajun Fu

Donlinks School of Economics and Management,
University of Science and Technology Beijing,
30 Xueyuan Road, Haidian District, Beijing 100083, China
18813128949@163.com

Abstract. Behavioral finance believes that investor behavior will be influenced by psychological factors, while investor behavior has a direct impact on asset pricing. This paper selected the monthly data of Shanghai and Shenzhen Exchange A shares from January 2016 to December 2016 by simple random sampling. We defined media variable as the number of media coverage which is obtained from Baidu's news search engine and established a panel model. We tried to explore the influence of media coverage on investor sentiment and stock returns, and to see which effect is more significant among the stock groups with different degree of media coverage. A pretest of simple correlations reveals the number of media reports lagged behind the first phase negatively affects the stock return rate of this period. The result shows that there is an "attention driven effect" in the low- middle level of media coverage group, the "media effect" exists in the middle-high level of media coverage group and the "media effect" exists in Chinese stock market through single factor analysis of variance. Moreover, this paper classifies the samples with control variables and shows that the media effect is significant.

Keywords: Media reports · Stock returns · Single factor analysis
Multiple regression

1 Introduction

A large number of empirical studies have proven that media reports can affect asset pricing, and that repeated news also has an impact oń stock prices. This is in contrast to traditional finance that "asset prices reflect all known information and new information is immediately reflected in prices. The concept of causing fluctuations is contrary to the view. Increasing more market speculation has caused traditional financial theories to suffer, so that researchers have begun to pay attention to the psychological factors of investors. Based on this, this paper investigates the influence of media reports on investor sentiment and stock returns, as well as explores the relationship between media reports and stock returns.

J. Abawajy et al. (Eds.): ATCI 2018, AISC 842, pp. 724–731, 2019.
https://doi.org/10.1007/978-3-319-98776-7_84

2 Reference Reviews

Researchers have been generally conducting their works by collecting financial news on television or newspapers. Niederhoffer examined the stock market reaction through global news events reported by The New York Times [1]. John and Richard studied the Wall Street Journal's takeover rumors and found that the stock price began to rise significantly before the rumor appeared [2]. Later, researchers began to use online media reports to evaluate. Chan collects headline news from Dow Jones Newswires to study the monthly returns of sample companies and finds that news reports have "inertia" in stock returns, which means they maintain positive returns over the long-term [3]. Barber and Odean proposed the "attention-driven theory" that companies with high media focus have higher excess returns than those with low media [4]. Fang and Peress, by studying media reports and cross-sectional stock returns, found that the company's profitability in low-media coverage was significantly higher than that of high-media-reported companies, and proposed the "media effect" theory [5]. Tho and Chau used the 14 key macroeconomic indicators of the United States from January 2002 to April 2012 to test the return rates of the stock markets of 12 countries in Asia within two minutes of data release. It was found that Asian emerging markets are more responsive than Asian developed markets. Bigger, think that emerging markets rely more on US news information [6].

According to the existing researches, although researchers have different views on the relationship between media reports and stock returns, the study found that media reports have a certain impact on stock returns and even corporate external governance. In this kind of research, the definition and measurement of media factors are not the same-there is no uniform standard. Quantitative methods, news sources, selection periods, and whether there are emotions will lead to differences in results. This article selects the number of media reports as the media factor and explores the specific impact of media reports on stock returns.

3 Data Sources and Index Design

3.1 Data Sources

Until the end of December 2016, there were 3,299 listed A-shares in Shanghai and Shenzhen Stock Exchange. This article selected 80 stocks as samples by simple random sampling. For the research purpose, the sample was further processed: (1) 8 companies were listed after 2016; (2) 4 companies were excluded from the classification of 2012 securities industry class I companies; (3) 5 companies were excluded as ST or *ST; (4) Remove 10 companies that cannot obtain complete detailed data. At the same time, this article only studies the A shares issued by the company. After processing the original sample, we finally obtained 53 firms with the time interval ranges from January 2016 to December 2016. All data are from WIND financial terminal and CSMAR Database.

3.2 Index Design

Media refers to the medium of dissemination of information. At present, the quantification methods of media factors mainly include event research methods for conducting research before and after a specific news event, and agent variables method for constructing media proxy variables. This article chooses to use the number of media reports as proxy variables. On the one hand, the number of media reports can reflect the media's attention to the company to a certain extent. On the other hand, this method is technically easier to implement.

$$R_{i,t} - R_{f,t} = \alpha_{i,t} + \beta_1 MR_t + \beta_2 ln(MV_{i,t}) + \beta_3 BM_{i,t} + \beta_4 LI_{i,t} + \beta_5 ln(MN_{i,t-1}) + e_{i,t}$$

The variables and explanations chosen in this paper are shown in Table 1.

Based on Baidu's news search platform, this paper takes the listed company abbreviation as a keyword, and uses the train head data collector to capture the title of the web news, the time of reporting, the news source and the number of repeated news items as four batches. We count the number of news, so we can obtain the monthly coverage data of each sample firm from January 2016 to December 2016. The variables and explanations chosen in this paper are shown in Table 1.

Table 1. Variable summary table

Variable	Description
$R_{i,t}$	Stock returns, as explanatory variables
$R_{f,t}$	The risk-less rate of month *t*, and this paper selects the monthly interest rate of the 3 months fixed deposit base rate published by the People's Bank of China
$R_{m,t}$	The market returns of month t and this paper selects the monthly return rate of CSI 300 index
MR_t	The risk of market asset portfolio in month *t*, which is equal to $R_{m,t} - R_{f,t}$, as a control variable
$MV_{i,t}$	The stock market value of the stock *i* in the month *t*, and its logarithm, as a control variable
$BM_{i,t}$	Stock *i*'s book value ratio in month *t*, as a control variable
$LI_{i,t}$	The exchange rate of the stock *i* in the *t* month, as a control variable
$MN_{i,t}$	The number of media reports on the stock *i* in the *t* month, as an explanatory variable

4 Empirical Analysis

4.1 Descriptive Analysis of Media Reports

We take advantage of Baidu's news search platform to obtain statistics on the monthly media coverage of sample companies from January 2016 to December 2016. Descriptive statistics are shown in Table 2.

Table 2. Statistical characteristics of the monthly average number of media reports

Month	Mean	Median	Max.	Min.	Stan. dev.	Kurtosis	Skewness
Jan.	75	54	936	2	127	42	6
Feb.	25	15	142	1	25	8	2
Mar.	64	44	384	1	71	8	2
Apr.	65	47	341	5	61	9	3
May	36	25	177	1	35	4	2
Jun.	34	23	188	1	39	7	2
Jul.	29	22	97	1	24	0	1
Aug.	44	27	267	6	51	8	3
Sept.	31	22	315	1	46	28	5
Oct.	32	23	136	3	32	3	2
Nov.	28	19	120	1	29	2	1
Dec.	28	16	254	1	41	18	4

From a time-series perspective, the overall media coverage of sample companies from January 2016 to December 2016 has no apparent growth. We can see that the average news coverage was minimum in February, owing to the Chinese New Year holiday in February. The kurtosis values are all greater than 0, suggesting that most companies' media coverage data is concentrated around the average, with less extreme data. A skewness greater than 0 indicates that the distribution of media reports is skewed to the right, i.e. only a small number of company media reports exceed the average number, and most companies' media reports are at a low level. At the same time, we also find that if the average of the total number of media reports for the whole year is used as the boundary, the sample shares will be divided into "high report companies" and "low report companies", and the number of companies with the average number of reports will be within the entire sample period. Usually also above the average, if the stock below the average is reported to be above the average in a certain month, then this state will continue for at least one period like "inertia". From a qualitative point of view, we believe that the inertia of the media report in the next issue may be due to repeated reports or follow-up reports on the previous issue. As time passes, the information contained in the news gradually becomes "ineffective". Without the value of the media's re-reporting, the company has returned to a "low report" status. Therefore, we believe that the number of media reports in the previous period affects the number of media reports in the current period.

4.2 Single Factor Analysis of Variance

In order to study whether there is "inertia" in the stock returns under the influence of media reports, we evaluate the relationship between the number of media reports and the average rate of return in the current, the next, the next two and the next three months respectively. The specific experimental construction steps are: dividing the sample shares into three equal parts according to the number of different media reports, evaluating the equal weighted average return rate for each group of stocks as well as the equal weighted stock return rate difference between the low media coverage group and the high media coverage group stock. The factor is the degree of media coverage while the dependent variable is the stock return rate with a significance level of 5%. The results are shown in Table 3. And Table 4 shows the average number of each group's media coverage,

Table 3. Single factor variance analysis of media reports and stock returns

Category	Monthly average yield of stock (%)					
	Low	Middle	High	Low-high	F	P-value
Current	−1.92	3.50	−4.24	2.31	0.025	0.876
Next	−10.03	−11.41	0.89	−10.92	4.825	0.036
Next-two	−2.14	0.58	4.22	−6.37	1.327	0.258
Next-three	−6.55	1.49	−1.46	−5.10	0.532	0.471

Table 4. The average number of media coverage in each group

Average number of media coverage			
Low	Middle	High	Average
36	45	117	65
11	32	66	36
16	17	70	34
15	23	49	29

According to Tables 3 and 4, there is a large difference in the number of media reports between groups. Considering the average rate of return, the ranking from high to low are the media coverage group, the high media coverage group and the low media coverage group sequentially. If only discuss the low media group and the media group in the situation, we should consider the existence of "attention driven effect"; while if only discuss in media group and high media group in the situation, we should consider the existence of Fang and Peress proposed the "media effect" of [8], i.e. the number of media reports and stock returns are the inverted "U" type. This paper holds both these two effects existing in the stock market, and the difference is mainly due to the sample or time diversification.

When setting the number of media reports as the factor, there are no significant differences in the average return rates for the current month, the next two months, and the next three months for the stocks with different numbers of media reports this month. The average return rate for the next month is statistically significant at the 5% level ($p = 0.036$) and the monthly rate of return difference is −10.92%. Combined with the above, the number of media coverage in this period will not only affect the number of media reports in the next period, but also will also affect the stock returns of the next period. This is also understandable: behavioral finance breaks the hypothesis of rational man. Irrational investors will be influenced by media reports, and the types and quantities of stocks held by them will be adjusted through buying and selling. At first, investors may neglect or fail to notice the information contained in media reports, but with the increase of series or repeated reports in a short period, investors' attention has been brought to change their investment decisions. But if we want to cause changes in stock returns, we need a large volume of transactions and a certain amount of time to accumulate, which can explain why the number of media reports will affect the next stock returns. But with the increase of the number of times, the information contained in the current media reports has lost its timeliness and has been gradually reflected in the price, which has lost its value, making the difference between the high media coverage group and the low media coverage group is not significant.

4.3 The Impact of Media Coverageon Stock Returns

From univariate analysis, we believe that the number of media coverage will affect the next stock returns, but the number of media reports is not the only factor affecting stock returns. We need to add other variables to further confirm. The CAPM that a single portfolio of securities or assets relative to the price level response to market price fluctuations (beta) determines different stock returns; Fama-French found in the study of the stock price and the market value of the market portfolio, the risk factor, market value and book value ratio factor [7] to establish the FF three factor model; other scholars in the study also joined the liquidity factor [8] failed to further explain the three factor model to reflect the risk factors of compensation.

In order to fully describe the stock returns, this paper is based on the CAPM model, the introduction of market portfolio risk MR_t, market capitalization in regression $ln(MV_{i,t})$, the book market ratio $BM_{i,t}$, flow factor $LI_{i,t}$ and the media factor $ln(MN_{i,t})$ as explanatory variables, stock returns R_i as the dependent variable.

$$R_{i,t} - R_{f,t} = \alpha_{i,t} + \beta_1 MR_t + \beta_2 ln\left(MV_{i,t}\right) + \beta_3 BM_{i,t} + \beta_4 LI_{i,t} + \beta_5 ln(MN_{i,t-1}) + e_{i,t}$$

In which i is a sequence number, $i = 1, 2, 3\ldots$; t indicates the period of observation (month), $t = 1, 2, 3\ldots$; $\alpha_{i,t}$ are intercepts related to the I of a single strand; β_1, β_2, β_3, β_4 and β_5 are the coefficients; $e_{i,t}$ is the error term. By single factor variance analysis, the number of media reports and stock returns in an inverted "U" type, so the number of the media reports as the boundary set up three models were tested: (1) low media group and the media group, "attention driven effect"; (2) in the media group high media group, "media effect"; (3) the relationship between all the sample inspection reports and stock returns of stocks. Table 5 summarizes the results of the model regression.

In the single factor analysis, we can get the impact of the number of media reports on stock returns, but we cannot measure the size of this effect. From the coefficients of $ln(MNi, t-1)$ of Table 5, the "low-medium" combination is negative (−0.002) and the "medium-high" combination is negative (−0.005). We can find that after the low-middle group added control variables such as market risk factor MR, market value factor $ln(MVi,t)$, book market value ratio factor BMi,t, and flow factor LIi,t, the media The effect of the number of reports on stock returns is positively negative (β_5= −0.002,

Table 5. Media reports on the return of return on stock returns

	Low-middle		Middle-high		Low-middle-high	
Variable	Coeff.	t-statistic	Coeff.	t-statistic	Coeff.	t-statistic
MRt	−0.968	(−12.995***)	−0.859	(−15.919***)	−0.806	(−12.591***)
ln(MVi,t)	0.079	(1.378)	0.122	(3.917***)	0.137	(4.344***)
BMi,t	−0.194	(−1.879*)	−0.263	(−4.173***)	−0.320	(−4.992***)
LIi,t	0.001	(14.578***)	0.001	(21.382***)	0.001	(19.280***)
ln(MNi,t-1)	−0.002	(−0.883)	−0.005	(−2.247**)	−0.009	(−2.364**)
R-squared	0.779		0.805		0.797	
Adjusted R-squared	0.743		0.773		0.765	
F-statistic	21.794		25.400		25.190	
Prob.	0.000		0.000		0.000	
D.W.	1.904		2.060		1.966	

Note: * * *, * *, * indicated that the variables were statistically significant at the level of 1%, 5%, and 10%, and the level of significance in this article was 5%.

$t = -0.883$). That means, in the low-to-middle group and medium-high group, the higher the number of media reports, the lower the stock returns. From all the samples, we can see that the number of media reports negatively influences stock returns at a statistical level of 5%, which is consistent with the "media effect" theoretical result. That is, the higher the level of media coverage, the lower the level of stocks return rates. The high degree of media coverage makes investors more fully aware of the company's situation, reducing information asymmetry, which makes the company may be overestimated due to information opacity. In contrast, companies with low media coverage will receive a risk premium due to asymmetric information. In the single factor analysis, the stocks with low media coverage and the stocks with high media coverage have significant differences in the return rate of the stocks in the next period. In the multi-factor analysis, the media factors are significant at the statistical level of 5% except for the low-middle groups.

5 Conclusion

This paper first defines the media variables as the number of media reports, obtains the number of sample company news via Baidu's news search platform, and uses the method of single factor analysis of variance and the method of adding control variables to explore the relationship between the number of media reports and stock returns in different periods. The main conclusions are: (1) The number of lagging media reports negatively affects the current stock returns. (2) The single factor analysis of variance shows that there is a significant difference between the stocks with low media coverage and those with higher media coverage in the next period. (3) Under the circumstance of only discussing the low media report group and the Chinese media report group, we believe that there is a "driving-attention effect"; in the case of only discussing the Chinese media report group and the high media report group, this article believes there is a “media effect”. Based on this result, this paper adds control variables to classify the samples and shows that media effects are significant.

Traditional financial theory argues that asset prices reflect all known information, and duplication of information is “invalid,” but it is not always true in reality. All kinds of stock market reports will affect investor decision-making and thus affect asset prices. For stock investors, stocks with a high degree of media attention do not necessarily have a high return rate. They must learn to strengthen their ability to discriminate between media reports and their own actions to make rational decisions. For company managers, they must manage consciously. They should appropriately and timely spread company’s related news to the investors positive company, and establish a good corporate image, so as to attract investment and promote the healthy development of the company. For media practitioners, it is necessary to adhere to their own hearts, strictly abide by industry norms and self-regulation regulations, prohibition False reports or malicious manipulations of public opinion or other actions that disrupt market behavior. For regulators, they must pay attention to the influence of media reports on the stock market, strengthen supervision and management of the media, and prevent lawbreakers from using the media to maliciously manipulate the stock market, for prompting a smooth and orderly market.

Due to the limitation of this research period, the research has some limitations. Further research is needed to confirm whether the rule is universal.

References

1. Niederhoffer, V.: The analysis of world events and stock prices. J. Bus. **44**, 193–219 (1971)
2. John, P., Richard, Z.A.: Is possible in the crazy world of acquisition rumors. J. Finance **65** (4), 154–168 (1990)
3. Chan, W.: Stock price reaction to news and no-news: drift and reversal after headlines. J. Financ. Econ. **70**(2), 223–260 (2003)
4. Barber, B.M., Odean, T.: All that glitters: the effect of attention and news on the buying behavior of individual and institutional investors. Rev. Financ. Stud. **21**(02), 785–818 (2008)
5. Fang, P.: Media coverage and the cross-section of stock returns. J. Finance **59**(3), 2023–2052 (2009)
6. Tho, N., Chau, N.: Impacts of the US macroeconomic news on Asian stock markets. J. Risk Finance **15**(2), 149–179 (2014)
7. Fama, F., Investor, R., Firm, V.: J. Account. Res. **48**(1), 1–19 (1993)
8. Pastor, T.B., Stambaugh, L.M.: The role of the media in the internet IPO bubble. J. Financ. Quant. Anal. **44**(03), 657–682 (1996)
9. Yu, D., Peng, L.: When does inferring reputation probability countervail temptation in cooperative behaviors for the prisoners’ dilemma game? Chaos Solitons Fractals **78**, 238–244 (2015)
10. Jianwei, L., Qingchang, R.: Study on supply air temperature forecast and changing machine dew point for variable air volume system. Build. Energy Environ. **27**(4), 29–32 (2008). (in Chinese)

Research on the Countermeasures of Property Service Enterprises Constructing Smart Community

Yanqing Cui(✉) and Haoyuan Li

Dalian Neusoft University of Information, Dalian 116000, Liaoning, China
Summer_cuiyq@163.com

Abstract. The level of property management has a direct impact on the development of the enterprise. There are many problems in the property management of the community. Constantly improving the quality of property management is the driving force behind development. The rise of smart communities has provided new development opportunities for property service enterprises. This paper presents that the construction of a smart community should focus on the property and create an entire community service ecosystem. This article builds an intelligent community system model and architecture based on the four-tier structure of smart communities, which based on the Internet, Internet of Things, and Big Data. The model is the core of the property service enterprise, which is closely connected with the upstream retailers, logistics enterprises, service providers and the downstream community consumers, and all the roles constitute an intelligent community business ecosystem.

Keywords: Internet+ · Community E-commerce · Property management Smart community

1 Introduction

"Internet+" is the combination of the Internet and traditional industries, which aims to promote the development of traditional industries. On March 5, 2015, at the Third Session of the Twelfth National People's Congress, Premier Li Keqiang proposed to formulate an "Internet+" action plan. In the government work report, Li Keqiang proposed that "the "Internet+" action plan be formulated to promote the integration of mobile internet, cloud computing, big data, and internet of things with modern manufacturing, promote the healthy development of e-commerce, industrial internet, and internet finance, and guide the Internet The company expands its international market."

As a typical traditional service industry, property service companies are facing many problems, such as low service quality and single source of profits. Based on the existing problems, this paper proposes a method of reconstructing the property service model using "Internet+" technology, and builds a smart community business mode.

J. Abawajy et al. (Eds.): ATCI 2018, AISC 842, pp. 732–738, 2019.
https://doi.org/10.1007/978-3-319-98776-7_85

2 Smart Community Concept and Structure

The smart community is a new type of integrated community management operation mode that integrates owners, merchants, service providers, and information technology providers. Based on this model, the service provider takes the actual needs of the owners as its starting point, comprehensively considers its future needs. The smart community realizes informatization and intelligence of the community life through information technologies, such as the Internet, the Internet of things, and cloud computing. It brings safe, comfortable and convenient community life services to the owners. The structure of the smart community mainly includes the perception layer, network layer, application layer, and access layer, as shown in Fig. 1.

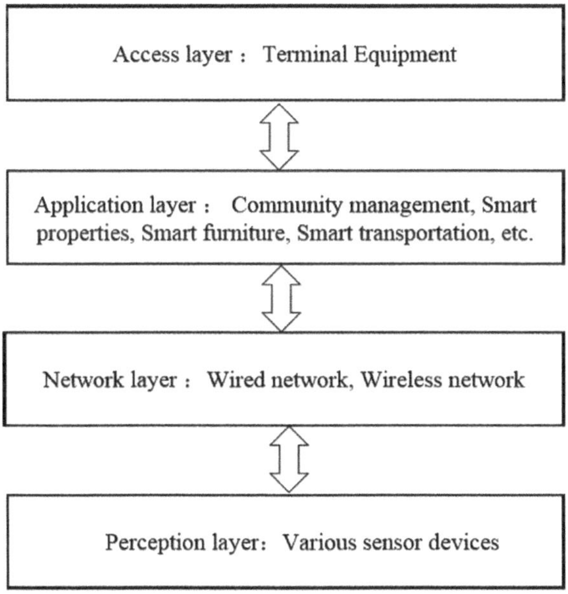

Fig. 1. The four-level structure of a smart community

3 Analysis of the Advantages of Property Service Enterprises Establishing a Smart Community

(1) Unique user resources: Each property company holds the information resources of its own community owners. This is a market that other Internet companies or other property companies cannot seize.
(2) Unique public resources: Property service companies have unique advantages in the use of community bulletin boards, elevator advertisements, and public parking lots within the entire property management area. Combining with the needs of the owners in the property area, targeted business activities such as advertising, service and merchandise agency are carried out.

(3) Right of negotiating with businesses around the community: The property company holds the functions of management and security of the community. Therefore, in the community smart community business model, the merchant and the property company are cooperative partners.

4 Smart Community Construction Strategies

4.1 Smart Community Model Construction

Through the study of the value chain of property service, the construction of smart community includes the property service system and the community electronic commerce system. The platform is dedicated to building convenient property services, community exchanges, and business district services. Through the integration of business resources around the community, it aims to build a smart community where property, business, and community residents converge.

The construction of a smart community will bring a win-win situation for both owners and property companies. On the one hand, it improves the efficiency of property management and reduces costs; on the other hand, it provides accurate and intelligent community services for owners. It contains four major functional modules:

(1) Property Self-service

Based on basic community services, community residents will be provided with daily services including on-line payment, online repairs, and notices.

(2) E-commerce service

According to the needs of residents of the community, the platform integrates community-related business services, provides online and offline e-commerce services for users, and creates new retail models. At the same time, combined with big data technology, it provides precision services and improves user experience.

(3) Social services

Build a "life circle" and "friend circle" based on neighborhood relationships to increase user stickiness through social networking. The community is an ecosystem, and the smart community is not only a consumer platform that meets the material needs for the owners, but also a spiritual home for the community owners.

(4) Property Management System

It is an important part of the smart community, and the basic system for internal management of enterprises. It improves the efficiency and response time of property workers.

5 Smart Community System Construction

5.1 The Overall Structure of the Smart Community System

The development of a smart community requires the integration of online and offline resources. It essentially improves the quality of its services and forms a virtuous circle of ecosystem with the owners. On this basis, it seeks new development directions and profit models. The platform covers property service systems and community business systems, as shown in Fig. 2.

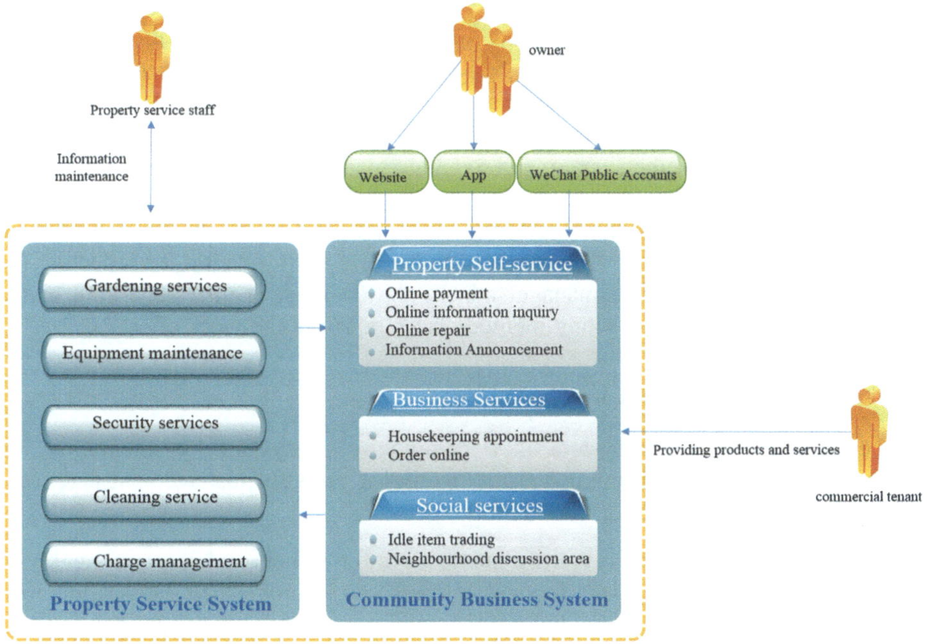

Fig. 2. The overall structure of the smart community system

5.2 Property Service System

The property service system standardizes the various projects served by the realty company, and the precise platform for property management is the basis for the property to enhance its core competitiveness. The system integrates five modules: engineering equipment, security, cleaning, and gardening and fee management. The system sets daily plans, service procedures, and execution standards for each vascular object, and automates it to achieve full life cycle management. Functions are as follows:

(1) Automatic work plan: The daily work plan for cleaning, horticulture, engineering equipment and security is automatically organized to generate a work plan. All

work plans are automatically scheduled by the system, and the system automatically triggers the work order and delivers it to the executor.

(2) Basic data management: Property service basic data maintenance, including basic information and status maintenance. For example, equipment maintenance consumables, maintenance work hours, fault records, etc.
(3) Payment Management: The automatic payment system can greatly reduce the company's labor costs. The system automatically sends the payment notification to the owner's mailbox or mobile phone. If the user pays through the online payment system, the payment status in the system can be updated. The online payment method supports multiple online payment methods such as PC and mobile.
(4) Employee self-service: Work plans, on-site work processes, job instructions can be directly transmitted to the mobile terminal. Employees can instantly transfer the on-site operation process and results to the system through the mobile terminal, realizing the flow of on-site office work and operations, real-time docking of the job site and the system, and dynamic management and control throughout the operation.
(5) Report function: Form the report of the completion of the maintenance work plan and the process records. The report is an important information for reflecting the effect of the work. The manager adjusts the decision in time according to the report, improves the working methods, and improves the work effect.

5.3 Community Business System

The community business system is a community ecosystem by integrating the resources of the community and realizes intelligent property management and commercial convenience. Functions are as follows:

(1) Property Self-service: Provide online services for owners and respond quickly to user needs. Through client services, owners can improve their satisfaction with basic property services, including payment, enquiries, complaints, repairs, and announcements.
(2) Community Business Services: By integrating the surrounding shopping districts, it provides various online and offline services such as catering, entertainment and housekeeping to meet the individual needs of the owners.

   - Convenient shopping process: Community residents can complete the shopping process anytime, anywhere through the mobile terminal, and the operation process is simple;
   - Lower decision-making costs: Every business in the platform is familiar and trusted by community residents, and the strong sense of trust established through daily consumption has greatly reduced the user's decision-making time;
   - Rapid response delivery service: The biggest advantage of the community life service e-commerce platform is the fast delivery of orders and convenience advantages that traditional e-commerce platforms do not have;

- Precision Marketing: Merchants publish product information and promotion information through the platform, and according to the user's spending habits, precision marketing, improve user experience.

(3) Social services: By setting up the owners' discussion area, the owner's sense of identity with the community is enhanced and the user's adherence to the platform is enhanced. Through analysis of information data which are shared and exchanged by the owners through the platform, Service provider can precisely match commercial services and improve service levels.

6 Conclusion

With the continuous upgrading of global Internet technology and the change of people's lifestyle, the traditional property service model can no longer meet the requirements of most consumers. Property service companies are facing a huge impact and urgently need to seek a development path. This article proposes a strategy for real estate service companies to build smart communities by integrating the resources of owners, businesses, and real estate service companies, and using Internet technology to upgrade the internal and external information systems of the company to improve the property service level and response speed. From the service model, it expands the single source of profit for traditional properties, integrates businesses around the community, and creates an online and offline interactive e-commerce ecosystem.

Fund. Liaoning Doctoral research start-up fund guidance program project, Project Number: 20170520320.

References

1. Ji Jiangming, F., Ge Yuping, S.: Research on the development strategy of shanghai smart community based on the concept of "three networks and one cloud". Urb. Dev. Stud. **23**(08), 119–124 (2016)
2. Gu Xiaohui, F.: Research on O2O mode of "social + personalization" community. Co-Oper. Econ. Sci. **16**, 143–145 (2015)
3. Zhang Yanguo, F., Zhu Shitao, S.: Internet+ community services: a new trend of smart community services. Jianghan Tribune **11**, 139–144 (2017)
4. Nie Lei, F.: The core and trend of community cloud services under the background of internet+. J. Shanghai Adm. Inst. **18**(06), 14–18 (2017)
5. Liu, S., Lu, F., Xu, X., Li, S., Guyue, T.: Development evaluation and wisdom management strategy of smart community in Shenyang City. Planners **33**(05), 14–20 (2017)
6. Yu, D., Peng, L.: When does inferring reputation probability countervail temptation in cooperative behaviors for the prisoners' dilemma game? Chaos Solitons Fractals **78**, 238–244 (2015)
7. Lipson, J., Vuong, S.: A semantics-based routing scheme for grid resource discovery. In: E-Science: First International Conference on E-Science and GridComputing, pp. 58–70, 90 (2005)

8. Qizhen, W.: Automatic control system regulation scheme of textile air conditioning. Sci. Technol. Inf. **22**, 395–396 (2010). (in Chinese)
9. Liu, J.: Advanced PID Control and MATLAB Simulation, 2nd edn. Electronics Industry Press, Beijing (2003). (in Chinese)
10. Li, J., Ren, Q.: Study on supply air temperature forecast and changing machine dew point for variable air volume system. Build. Energy Environ. **27**(4), 29–32 (2008). (in Chinese)

An Analysis of Film Subtitle Translation from the Perspective of Intercultural Communication

Jinhua Chen(✉)

Wuhan University of Science and Technology,
Huangjia Lake, Hongshan District, Wuhan, Hubei, China
cjhgzyx@foxmail.com

Abstract. With the development of film industry, more and more excellent foreign films are brought into China, meanwhile, many domestic films are going outside. In order to let both people at home and abroad understand each other, translators bear lots of responsibility to express the culture. This paper concentrates on the Intercultural communication, and how translators build the relationships between the native language and target language, in this paper i will talk about the translation strategies in the film subtitle from the perspective of Intercultural communication. The purpose of this thesis is to improve the future film subtitle translation and help people to know very well about the content and topics in the foreign films and understand foreign films culture in a better way.

Keywords: Intercultural communication · Film subtitle translation
Study strategy

1 Objective

In the past, we just saw Chinese movies, because people couldn't understand the foreign tune. Nowadays the situation is changing because of the internet. People can enjoy different types of movies online without any trouble. Just as what i have mentioned in the abstract that the intercultural communication has totally changed the situation. Due to a big team of translators and interpreters have pour into the film market. What's more, English becomes the most widely used language in the world. More and more America and European movies are very popular among Chinese audience. In order to let people have a much better view of the charisma of English films and know about the conventions and culture in foreign countries. The film subtitle translation plays an very important role.

The study of film subtitle translation is not only involves in two languages exchanging, but also involves in the game between various language cultures. Therefore, to study the film translation, it is necessary to understand the following problems.

(1) How does Intercultural communication influence people to choose to see the foreign films?

J. Abawajy et al. (Eds.): ATCI 2018, AISC 842, pp. 739–746, 2019.
https://doi.org/10.1007/978-3-319-98776-7_86

(2) How does education affect quality of watching movies?
(3) What are the implications of film subtitle translation and intercultural communication?

2 Methods

2.1 The Definition of Intercultural Communication

The basis of film subtitle translation is the study of globalization and localization. One scholar named Anglo pointed out that intercultural communication is the research object of the elliptical culture with the dual focus of culture and communication. The history of intercultural communication shows that all theories are only a relative solution, and each theory conflicts with other reality. So we feel the intercultural communication is not very reliable. In fact, the basis of intercultural communication is not something conceptualized, but something that needs to be created. It is the transmission of the practice clears a basic fact that the degree of social distance between different people, information exchange and communication, understanding and dialogue communication behavior is different, in order to merge together and overcome loneliness, self-awareness, cognitive environment, choice of social need, such as certain spread over a period of time, it will become relatively stable. Although just few scholars in China did the research on film subtitle translation theory, the translation of film can borrow some ideas from literature translation theory. However, because of the unique characters of film works translation, it is still a little bit different from the literal translation.

The study of subtitle translation also provides a substantial theoretical basis for intercultural communication. Some scholars begin to study the relevant issues from the perspective of intercultural communication. For example, Phyllis Ratline (2015: 25) studied some practical problems involved in the translation of theatre and film. Including the translators' self-cognition, translator and collaborators' work, play translation and dubbing and subtitle translation. Micheal Cronin (2011: 36) did a study about how cultural perfect combines through a case study of several typical American movies, probes into the development of film translation, the acceptability of different types of audiences, translation and cultural conflict and language difficulties, the plight of film translation in the context of globalization, etc. Through a large number of case studies, Carol (2015: 1) discussed the historical development of film translation, the media of film translation, subtitle translation and screen multilingual translation.

2.2 Study Theory and Methods

This study adopted the theoretical basis of descriptive translation. This is because from the previous studies on subtitle translation of domestic films, most of them are prescriptive research, focusing too much on some micro level. Such as subtitle translation features, the contrast between dubbing translation and subtitle translation, making the domestic film translation research stagnated and cannot reach a higher level academically. In addition, domestic film translation researchers rarely focus on this field for a long time, so research is not systematic and there is no continuity. However, the translation theory, especially the multivariate system has a strong explanatory power on the problems of the current film translation.

According to the above research questions and objects, this paper combines the research methods of quantitative research and qualitative research, specifically combining questionnaire survey and interview. Quantitative research, also known as quantitative research, is a measure and analysis of quantifiable aspects of things to test the research methods on certain theoretical assumptions about the subject. Qualitative research is, however, through the interaction between researchers and subjects to the in-depth, detailed, long-term experience, and then to the "quality" of things to get a more comprehensive explanatory understanding (姜飞, 2017: 8).

Quantitative research and qualitative research point out that these two research methods can be used in combination based on the research object, which can be divided into several kinds of data collection (陈晓丽, 2015: 18). According to the classification of the subject, the research on the film translation industry is more suitable for the qualitative research method, and the in-depth interview is more comprehensive understanding. Here we use both questionnaires and interview to illustrate the all the problems.

3 Data Analysis and Results

This questionnaire is mainly designed to understand the audience's views on the phenomenon of subtitle translation. The author believes that age, education degree, living area are the factors that will influence people's affection for film subtitle translation, So the questionnaire is divided into two parts, the first part is to understand the basic information of the respondents, including three questions: education degree, requirements for English and Chinese subtitles, requirements for dubbing in English and Chinese. The second part is to get the results for each questionnaire. The author believes that these factors may be related to the audience's understanding of the phenomenon of subtitle translation and intercultural communication.

The author chose different education background (primary, middle & high school, college, post graduate student), living area (first-tier cities, second-tier cities, third-tier cities, remote area), different age (12–18, 18 to 40, 40 to 60, over 60 years old) as samples:

Education background	Primary	Middle & high school	College	Post graduate
	12%	30%	78%	84%
Living area	First-tier	Second-tier	Third-tier	Remote
	78%	50%	36%	20%
Age	12-18	18-40	40-60	Over 60
	52%	64%	28%	10%

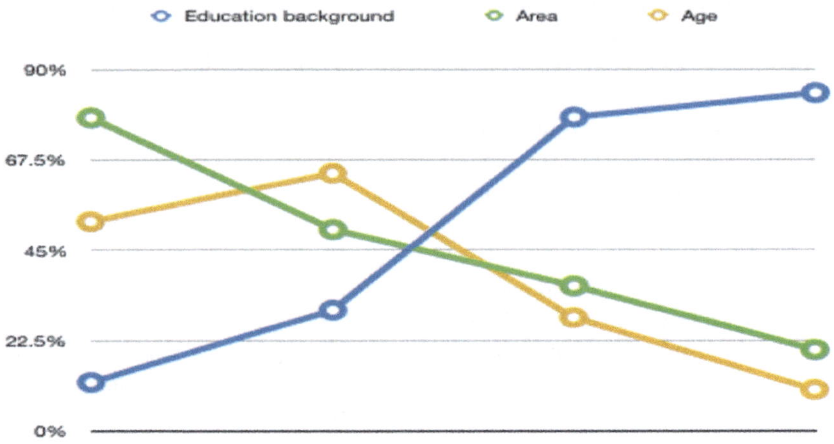

From this figure, we can get the following results:

1. The higher education background, the higher acceptance for English movie.
2. From middle school to college education, the acceptance rises deeply.
3. The acceptance in college and post graduate is of no big difference.
4. From first-tier city to forth-tier, the acceptance decrease gradually.
5. With the increase of age, the acceptance rate of English films increased first and then decreased, with the largest proportion between 40 and 60 years old.

We can see from the above figure that people got more education, live in more developed cities, they will accept the foreign culture easily and have more intercultural communication.

In order to know the people's requirements for English and Chinese subtitles, another 50 people were randomly selected as samples:

Requirements for Chinese subtitle	Requirements for English subtitle	Requirements for Chinese & English subtitle
36%	9%	55%

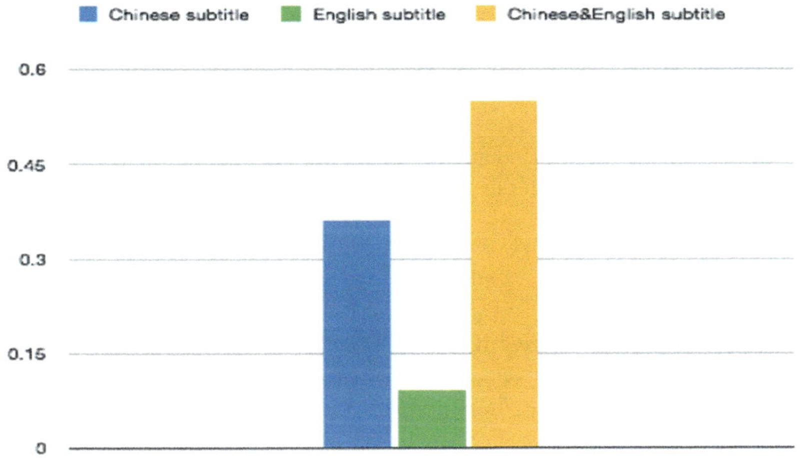

From this form, we got one result as follows.

Most people require Chinese and English subtitles.

There are few people who only need English subtitles, indicating that the reading ability of English is significantly higher than that of listening comprehension.

People who get more involved in intercultural communications will have less needs for film subtitle translation.

The last samples was made to see the requirements for dubbing in English and Chinese as follows:

Requirements for Chinese dubbing	Requirements for English dubbing
28%	72%

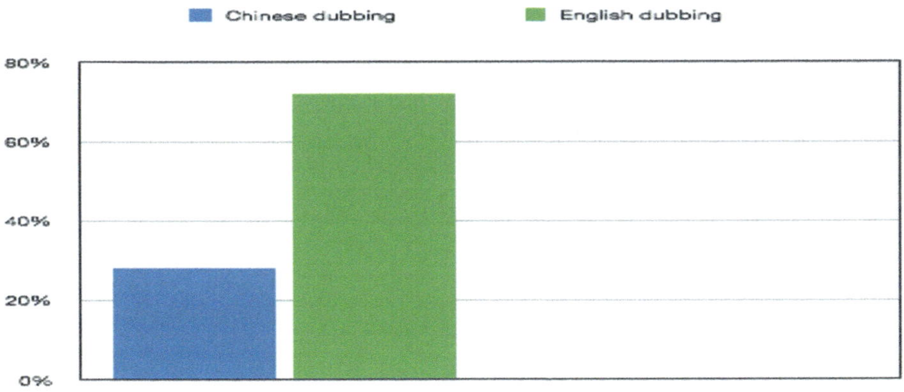

From this form, we got the third conclusion as follows.

(1) Most people like the original voice dubbing, and it is hard to accept the Chinese dubbing.
(2) For young people, listening to English dubbing can also improve their English listening ability.

Specifically, the degree of education led to the recognition of subtitle translation phenomenon, and the higher the level of education, the higher the quality requirement of subtitle translation.

3.1 Interview and Results

Interview is a way of qualitative research. The results of the questionnaire can only provide some quantitative answers, which are not sufficient for the discussion of the nature of the problem, while the interview can dig deeper and complement the questionnaire.

Interview Object
According to the research object, the author interviewed six people in the industry respectively, including four movie subtitle translators, one translation researchers and one who is related to international movie industry. The specific situation is shown in the following table:

Item No.	Gender	Age	Occupation	Research object type	Film subtitle translation and relevant work experience
1	Male	40	Professor (teaching British and America literature)	Translator	CCTV-10 documentary subtitle translation (English to Chinese) (part-time job, 10 years
2	Male	32	Professor (teaching translation)	Film subtitle translator	About 6 years for film subtitle translation
3	Female	21	Senior student	Film subtitle translator	About 2 years internet film subtitle translation experience (English to Chinese)
4	Male	35	Interpreter	Film subtitle translator	About three years CCTV-6 subtitle translation (part-time job, English to Chinese)
5	Female	36	Movie producer	Media	Have the movie producing working experience
6	Female	48	Professor (interpretation study)	Scholar	More than 20 years for translation teaching, study and practice experience

Interview Designing

Brinkmann pointed out that a scientific and rigorous interview process requires the following seven steps: determining research topics, design, interviews, transcripts, analysis, validation, and report (杨刘琴, 2017: 132). Based on this, the author designed the whole interview process, as shown in the following table:

Determining research topics	Based on previous research, the author puts forward the relevant hypothesis of Chinese film subtitle translation (1) The formation of this phenomenon is affected by various factors, such as the development of China's economy, and the promotion of Chinese culture to the outside world (2) This phenomenon may continue to deepen (3) This phenomenon has profound significance and influence on film translation and intercultural communication
Design	Plan to interview related field people, including film translators, scholars, media people and so on
Interview	Use detailed interview outline for personal interview, each interview for 30 min or so, and recording
Transcripts	The interview materials should be transcribed by words
Analysis	The types of interviewees were classified and all materials were analyzed in detail
Validation	Verify the validity of the interview
Report	Summarize the whole interview process and explain the phenomenon

From the analysis of the interview data and influence of film subtitles translation. We can got that translation mainly include dubbing process, audience, purpose, quality control, and participants in the process of dubbing. These factors are not isolated, but are interlinked, interrelated and inseparable. Based on the study of quantity, this paper mainly discusses the phenomenon of subtitle translation in China. The results answer the research questions in advance, explains the attitude of the involved parties to this phenomenon, the future development and enlightenment of film intercultural communication.

4 Conclusion

Nowadays, the international community is an agricultural development society, and the significance of intercultural communication is becoming more and more prominent. Under such background, various new cultural phenomena are more or less the result of the interaction between various factors and participants in the present society. This study was based on theory of intercultural communication through qualitative research methods, the phenomenon of sex, trying to explore the essence through the phenomena which provides a new understanding of this phenomena.

References

1. Andre, L.: Translation, Rewriting and Manipulation of Literary Fame. Shanghai Foreign Language Education Press, Shanghai (2010)
2. Anthony, P.: Exploring Translation Theories. Rout ledge, New York (2010)
3. Goteborg, H.: Subtitle: diagonal translation. Perspective **2**(1), 102 (1994)
4. Goteborg, H.: Screen Translation: Six Studies in Subtitle, Dubbing and Voice-Over. Hopi Service, Copenhagen (2012)
5. Aida, E.: Toward a Science of Translation. Shanghai Foreign Language Education Press, Shanghai (2001)
6. Gerber, D., Wilson, D.: Relevance: Communication and Cognition. Shanghai Foreign Language Education Press, Shanghai (2001)
7. Yu, D., Peng, L.: When does inferring reputation probability countervail temptation in cooperative behaviors for the prisoners' dilemma game? Chaos Solitons Fractals **78**, 238–244 (2015)
8. Qizhen, W.: Automatic control system regulation scheme of textile air conditioning. Sci. Technol. Inf. **22**, 395–396 (2010). (in Chinese)
9. Liu, J.: Advanced PID Control and MATLAB Simulation, 2nd edn. Electronics Industry Press, Beijing (2003). (in Chinese)
10. Li, J., Ren, Q.: Study on supply air temperature forecast and changing machine dew point for variable air volume system. Build. Energy Environ. **27**(4), 29–32 (2008). (in Chinese)

Classification of Neurons Based on Deep Learning and Feature of Image Gray Symbiosis Matrix

Tao Chen and Wenbin Liu(✉)

Department of Public Course, Wuhan Technology and Business University, Wuhan 430065, Hubei, China
812926096@qq.com

Abstract. In order to realize the automatic and efficient biometric neuron recognition analysis, an automatic classification method based on deep learning and image gray symbiosis matrix is proposed. The method is based on image preprocessing, and the category feature space is made up of the statistics which are the means and standard deviations of image gray symbiosis matrixes, then the deep learning is used to establish the mapping relationship between the feature space and neurons categories. The collected 537 biological neuron images were used to test the method, and the accuracy of the test dataset was 94.4%. The above results show that the automatic classification of biological neurons with deep learning and image gray symbiosis matrix has the high accuracy and reliability.

Keywords: Neurons · Classification · Deep learning
Image gray symbiosis matrix

1 Introduction

The study of spatial geometry of neurons is an important project in human brain planning, for the study of the geometric morphological features of neurons, different experts focus on the different indicators to depict. The most basic problem in understanding the morphological characteristics of neurons is the recognition of neurons. The science is still not well solved to identify different types of neurons. In the biological anatomy, the differentiation of neurons is mainly based on geometrical morphology and potential distribution, and the geometry of the neurons is mainly obtained by dyeing technology [1].

In recent years, many scholars have done a lot of work on neuron recognition. For the recognition of neurons in the micrograph of brain slices, Gurevich proposed a mathematical method, which combining the contour features of binary images [2–4]. Shang used the classifier "voting" composed of probabilistic neural network, BP neural network and fuzzy classifier to classify 7 kinds of neurons [5]. Myonglae implemented a pattern recognition system of the visual form of the nerve in the hardware level [6]. Lv established the classification model of neuron by analytic hierarchy process, graph

J. Abawajy et al. (Eds.): ATCI 2018, AISC 842, pp. 747–754, 2019.
https://doi.org/10.1007/978-3-319-98776-7_87

section and geometric similarity [7]. Carpenter identified the color vision neurons of primates based on the non-dimensional shard color vision system model [8].

The above methods of neuron recognition need more complex image analysis techniques, and it is not possible to achieve relatively simple and fast classification. The gray symbiosis matrix is a simple and fast effective technique to extract image features. By means of statistical image gray symbiosis matrix attribute value as its eigenvector, it can be regarded as a description of image texture features, which can be further used for classification, identification and retrieval. In its application, Lv studied the relationship between texture features and mineral phases in iron ore sintering based on gray symbiosis matrix [9]. Zhang characterized the defect of steel wire rope based on image gray symbiosis matrix [10]. Eleyan used the symbiosis matrix and its statistical characteristics as a new method for face recognition [11].

Now deep learning has become a mature and popular research method. Therefore, in this paper, the study of neuron classification based on the deep learning and the feature of image gray level co-occurrence matrix is proposed. The method is based on image preprocessing, and the category feature space is made up of the statistics which are the means and standard deviations of image gray symbiosis matrixes, then the deep learning is used to establish the mapping relationship between the feature space and neurons categories. Thus, it can help biology and medical researchers classify a large number of neuron images efficiently and accurately.

2 Classification of Neuron Images

The structure and function of different neurons involve multiple factors, and the geometrical features of neurons are the most important. The geometric morphological features of neurons constitute its spatial conformation. It consists of three parts, they are dendrites, which are used to receive information, cell body, for the information processing, and axon, for the information transmission. Because of the growth and change of these three parts, the neurons present various geometric forms, and this paper studies how to use the geometric features of neuron image to accurately identify it.

NeuroMorpho contains a large number of geometric morphological data of neurons [12], three representative types of neuron images are selected from this study (Fig. 1). These neurons, in turn, are basket cell, bipolarcell, and cone cell. The characteristics of these neurons are as follows:

(a) Basket cell (b) Bipolar cell (c) Cone cell

Fig. 1. Forms of nerve cells

(1) Basket cell: one of inhibitory neurons in the cerebella cortex, axons formed into fine branches, surrounded by flowers and baskets, thus designated Basket cells (Fig. 1a), hereinafter represented by the first class of neurons.
(2) Bipolarcell: Bipolar cells, receiving the signal input of photoreceptors, and the two ends of the cell are dendrite distribution (Fig. 1b). The second neuron represents the following.
(3) Cone cell: a photoreceptor cell, synaptic terminals, each cell has inside and outside, and an internal nuclear and all sorts of mitochondria, synaptic terminals form such as bipolar cells, cone is shorter but wide (Fig. 1c), below expressed in the third type of neurons.

3 Gray Symbiosis Matrix

The gray symbiosis matrix is an image processing method that describes texture by studying the spatial correlation property of image grayscale. The grayscale distribution occurs repeatedly in the spatial position of the image to form the texture. Therefore, there is a certain gray relation between the two pixels separated by a certain distance in the image space, that is, spatial correlation of image grayscale. The calculation of grayscale symbiosis matrix is as follows [13]:

(1) Take one point (x, y) in the image ($M \times N$) and deviate from it ($x + \Delta x$, $y + \Delta y$), then the gray value of the point is (g1, g2).
(2) The point (x, y) is moved in the whole image space, and the values of multiple (g1, g2) values are obtained. If the image gray scale is K, then the combination of (g1, g2) is the square of K.
(3) For the entire image, for each type (g1, g2), the number of occurrences is counted.
(4) The number of occurrences of (g1, g2) is normalized, and the probability P of (g1, g2) is obtained, and P is the gray symbiosis matrix of the image.

When taking the different values of the difference of distance (Δx, Δy), the different gray symbiosis matrixes are obtained. For fine texture, (Δx, Δy) can take (1, 0), (1, 1) and (2, 0), etc.

When $x = 0$, $y = 1$, the pixel is horizontal, which is 0° scanning.
When $x = -1$, $y = 1$, the pixel is opposite to the right diagonal, which is 45° scanning.
When $x = -1$, $y = 0$, the pixel is vertical, which is 90° scanning.
When $x = -1$, $y = -1$, the pixel is opposite to the left diagonal, which is 135° scanning.

In this way, the probability that the two pixel gray level occurs simultaneously will constitute the image gray symbiosis matrix.

To solve the rotation problem of neuron image acquisition, the mean and standard deviation of the contrast, correlation, energy and homogeneity of gray symbiosis matrix in different angles were calculated, which can be the characteristic parameters of the gray symbiosis matrix of this image.

In order to obtain the characteristic vector sample data of all kinds of neuron images, the steps of image preprocessing, gray level symbiosis matrix calculation and feature extraction are carried out. The extracted values will be the input parameters of the deep learning, so the feature dataset of the neuron image consists of the following eight characteristics:

(1) The mean value of the contrast of gray symbiosis matrix.
(2) The standard deviation of the contrast of gray symbiosis matrix.
(3) The mean value of the correlation of gray symbiosis matrix.
(4) The standard deviation of the correlation of gray symbiosis matrix.
(5) The mean value of the energy of gray symbiosis matrix.
(6) The standard deviation of the energy of gray symbiosis matrix.
(7) The mean value of the homogeneity of gray symbiosis matrix.
(8) The standard deviation of the homogeneity of gray symbiosis matrix.

4 Deep Learning

The training of intelligent method is mainly to train the artificial neural network, which is studied through the known input data and output results (categories). The traditional artificial neural networks are the BP neural network and the deep learning in recent years. The deep learning architecture consists of multi-layer nonlinear computing units. Each lower level output will be a higher level input, which can be the effective features from a large number of input data. Deep learning has many advantages [14]:

(1) The representation of complex higher-dimensional functions such as high variable functions;
(2) Reducing computational complexity;
(3) The best simulation of the human cerebral cortex;
(4) Reusing and sharing in similar tasks.

The typical deep learning model has convolution neural network, DBN and stack self-coding network.

5 Experimental Results

Deep learning is one of the common methods for pattern recognition, and it's made up of a number of neurons that are connected by weights. The value of the connection weight implies the ability to express knowledge. In order to obtain the output of deep learning network, it is necessary to assign weights to each neuron, and to calculate the corresponding excitation value for each neuron. In the deep learning network, nonlinear mapping functions can be applied to represent the mapping of feature space to classification space [15].

5.1 Deep Learning Model

The images of biological neurons are classified by deep learning toolbox. The deep learning used in the toolbox is multi-layer network, which is divided into input layer, multiple hidden layers and output layer. In the network, its input structure is 8 input neurons, the hidden layer neurons are multi-layered, and the three output neurons correspond to the three categories (Fig. 2).

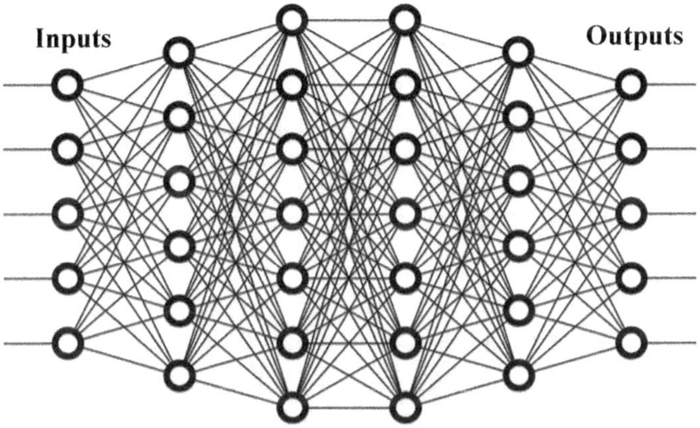

Fig. 2. Structure of deep learning

5.2 Model Training

Both the hidden layer neurons and the output layer neurons use the sigmoid function, and each output value of the deep learning represents the probability of the corresponding classification. The depth learning toolbox uses a quantified join gradient training function. The deep learning is using by the supervised learning method and the training set consisting of input characteristics is studied, which is based on the error back propagation method. Firstly, the initial weights are given for the deep learning network, and these weights are modified according to the error to minimize the error.

The experimental images are composed of 537 neuron images, and the numbers of these three categories are 201, 172 and 164. The training set, verification set and test set of deep learning are randomly divided into 323, 107 and 107. Through the constant testing and adjustment of the depth learning network layer, it is found that the network test results are the best when the hidden layer number is 6. With the monotonous decreasing of network error in the training process, the verification error can only be reduced to one point (Fig. 3). The minimum value of the verification error was obtained when the experiment was performed in step 102, and the verification error was 0.024.

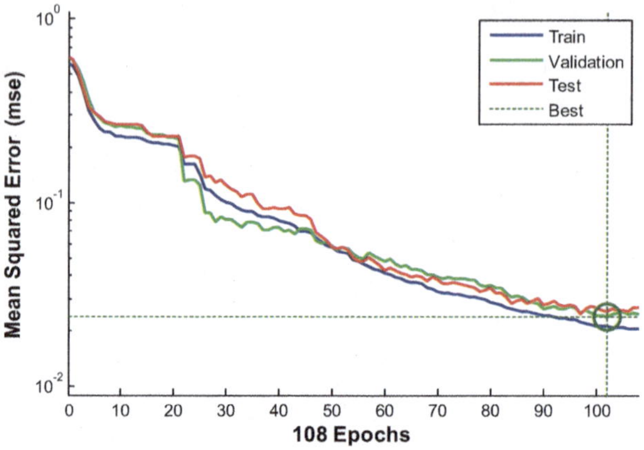

Fig. 3. Training errors

5.3 Results Analysis

The test set consisted of 107 images, 101 images (94.4%) were correctly classified by artificial neural network, and six (6.6%) error categories were classified. Table 1 is the results of the test classification: where columns represent the true classification and the rows indicate the depth learning classification.

Table 1. Test results

Deep learning classification	True classification		
	First	Second	Third
First	39	2	0
Second	1	31	2
Third	0	1	34

In the error classification, one of the first class of neuron errors was identified by deep learning as the second type of neuron, and three of them were identified as the first and third types of neurons, and two of the third neuron errors were identified as the second type of neuron. Therefore, the above classification results show that deep learning is better for the overall recognition effect of these three types of neuron images.

Based on the above experimental results, it is shown that the application of deep learning and image grayscale symbiosis matrix for biological neuron classification is of certain practical value, so it can be used as a technical reference for researchers of biological neurons. At the same time, it still has some experimental error, which may be due to the unequal number of neuron images trained in the model. The future work is how to better extract the morphological parameters of neurons to improve the training accuracy of deep learning. Thus, the accuracy of neuron image classification can be improved.

6 Conclusion

In this paper, an automatic identification method of biological neurons based on depth learning and image gray symbiosis matrix feature and deep learning is proposed. The method is based on image preprocessing, and the category feature space is made up of the statistics which are the means and standard deviations of image gray symbiosis matrixes, then the deep learning is used to establish the mapping relationship between the feature space and neurons categories. The collected 537 biological neuron images were used to test the method, and the accuracy of the test dataset was 94.4%. Based on the above experimental results, it is proved that the biometric neuron image recognition with deep learning and image gray symbiosis matrix has the certain practical value. This method may be further applied in the future: it can expand the classification of biological neurons, and can also be used for the interpretation of other properties of neuron images.

References

1. National Postgraduate Mathematic Contest in Modeling: Postgraduate mathematical modeling contest in 2010. http://gmcm.seu.edu.cn/00/45/c31a69/page.htm. Accessed 11 Dec 2017
2. Gurevich, I.B., Myagkov, A.A., Sidorov, Y.A.: A new method for automated detection and identification of neurons in microscopic images of brain slices. Pattern Recogn. Image Anal. **22**(4), 558–569 (2012)
3. Gurevich, I., Beloozerov, V., Myagkov, A.: Systems of neuron image recognition for solving problems of automated diagnoses of neurodegenerative diseases. Pattern Recogn. Image Anal. **21**(3), 392–397 (2011)
4. Gurevich, I.B., Myagkov, A.A., Trusova, Y.O.: On basic problems of image recognition in neurosciences and heuristic methods for their solution. Pattern Recogn. Image Anal. **25**(1), 132–160 (2015)
5. Shang, X.-J., Liu, X.-M., Li, C.-F.: Morphological classification of neurons. J. Jilin Univ. (Inf. Sci. Edit.) **29**(3), 237–244 (2011)
6. Chu, M., Kim, B., Park, S.: Neuromorphic hardware system for visual pattern recognition with memristor array and CMOS neuron. IEEE Trans. Ind. Electron. **62**(4), 2410–2419 (2015)
7. Lv, D., Kang, C., Tai, L.: The classification and recognition of neurons form. Math. Pract. Theory **41**(14), 156–164 (2011)
8. Carpenter, G.A., Chelian, S.E.: DISCOV (DImensionless Shunting COlor Vision): a neural model for spatial data analysis. Neural Netw. **37**, 93–102 (2013)
9. Lv, X., Bai, C., Qiu, G.: Relationship between texture features and mineralogy phases in iron ore sinter based on gray-level co-occurrence matrix. ISIJ Int. **49**(5), 709–718 (2009)
10. Zhang, D., Zhao, M., Zhou, Z.: Characterization of wire rope defects with gray level co-occurrence matrix of magnetic flux leakage images. J. Nondestruct. Eval. **32**(1), 37–43 (2013)
11. Eleyan, A., Demirel, H.: Co-occurrence matrix and its statistical features as a new approach for face recognition. Turk. J. Electr. Eng. Comput. Sci. **19**(1), 97–107 (2011)
12. George Mason University: NeuroMorpho.Org. http://neuromorpho.org/neuroMorpho/index.jsp. Accessed 17 Oct 2017

13. Baidu: Gray level co-occurrence matrix. http://baike.baidu.com/view/4293543.htm. Accessed 15 Nov 2017
14. Shi, F., Wang, X.-C., Yu, L.: MATLAB Neural Network Analysis of 30 Cases. Beijing University of Aeronautics and Astronautics Press, Beijing (2010)
15. Ruan, Q.-Q.: Digital Image Processing in the MATLAB. Tsinghua University Press, Beijing (2013)

The Application of Auto CAD Drawing Technique in Interior Design

Hongxing Yi(✉)

City College of WUST, Wuhan, China
lucyyhxc@126.com

Abstract. With the improvement of the people's living standards, the development of the interior design industry is becoming more and more orderly and regular. Interior designers use a variety of design software to design indoor space. The most commonly used design software is Auto CAD. Using Auto CAD drawing can not only make the picture succinct, the view is beautiful, but also be easy to modify, and the reuse rate is high. If designers are proficient in Auto CAD, they will greatly improve their productivity when designing and drawing interior space.

Keywords: Auto CAD · Technique · Interior design · Drawing · Set up

1 Drawing Preparation

Excellent interior designers should develop good drawing habits.

First, using the Auto CAD drawing, the drawing should be divided into the layers. This will provide convenient conditions for future revision. In the drawing of the interior design drawings, different layers should be set up respectively. For example: walls, doors, windows, furniture, dimensions, frame……

Second, drawing by color, distinguishing different Linetype and Lineweight by color will facilitate your design work. For example, White No. 7 is used to draw thick lines, and blue No. 4 is used to draw medium lines. The rest of the colors are used to draw thin lines. The above work is set in the Layer Properties Manager (Fig. 1).

2 To Draw a Design Drawing

Before using Auto CAD to draw the drawings of interior design drawings, we should first set the color of the wall layer as 7 white according to the drawing preparation work, and set the wall layer to the present, then we can draw the wall in the floor plan.

According to the regulations of the international drawing standard, the shape of the cutting surface should be drawn with thick lines. The plane's wall is section, the No. 7 white drawing, convenient the future when printing output shown as thick lines. By analogy, in accordance with the requirements of the drawing specification, the door and window are drawn one by one.

J. Abawajy et al. (Eds.): ATCI 2018, AISC 842, pp. 755–760, 2019.
https://doi.org/10.1007/978-3-319-98776-7_88

In this way, the output of the drawings will be printed in the future to show the difference between the thick and thin lines. It can meet the basic requirements of drawing standard.

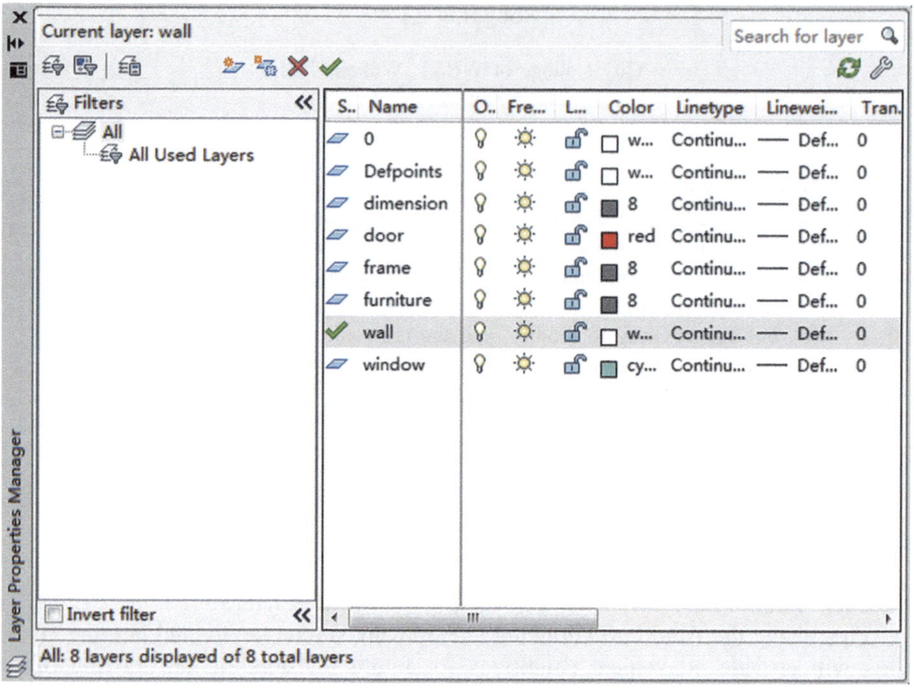

Fig. 1. Layer properties manager

3 Setting Skills of Dimension Style Manager

There are two things we need to do before marking the dimensions of walls, doors and windows. First, we should be determined the print size of the drawings of the plan or elevation, that is, the size of the drawings. In general, the common interior space design drawings are printed and outputted by A3 photocopy paper, and the residential space design drawings are usually printed out by A4 photocopy paper. Second, drawing the frame, determine the proportion of the print drawings. For example, the plane drawing of the living spaces are usually printed in the scale of 1: 100. To complete the above two things, we can set up the Dimension Style Manager.

Different drawings are printed in different proportions, and different styles of marking must be established respectively. For example, if the output of the plan is 1: 100, the output of the elevation is 1: 50, and the output of the profile is 1: 5, then you need to click on the new button to create three different marking styles and name them with different names.

The following is a detailed description of a drawing with a print ratio of 1:100.

- Click the New button to create a new dimension style.
- Change the new style name, and click the Continue button.
- Set the Fit menu. Set the Use overall scale of below the Scale for dimension features to 100 (Fig. 2).
- The rest of the menus can be set up according to your preferences.

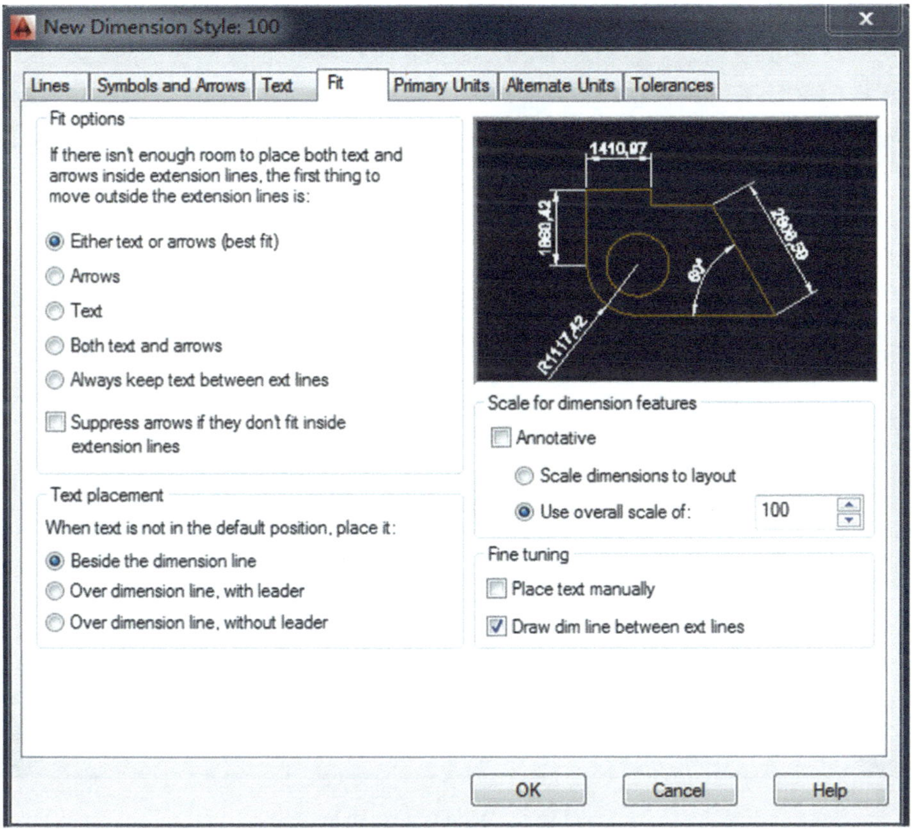

Fig. 2. Set the use overall scale of

In the process of setting up, one of the most important links is that the Scale for dimension features must be completely consistent with the proportion of the future output and printing of the drawing. For example, the three different dimension styles above, the Scale for dimension features are 100/50/5 respectively.

By analogy, more output ratios require more matching dimension styles. The goal is to make a set of design drawings look very regular and neat when printed in the future.

4 Text and Dimension Annotation

By setting up the Dimension Style Manager, we can mark the size of the design drawing, and describe the materials and colors used in the design. In the process of annotation, if you want to change the font, or change the word space, we can also modify the text style. For example, we can change the font, and adjust the Width Factor (Fig. 3), and so on.

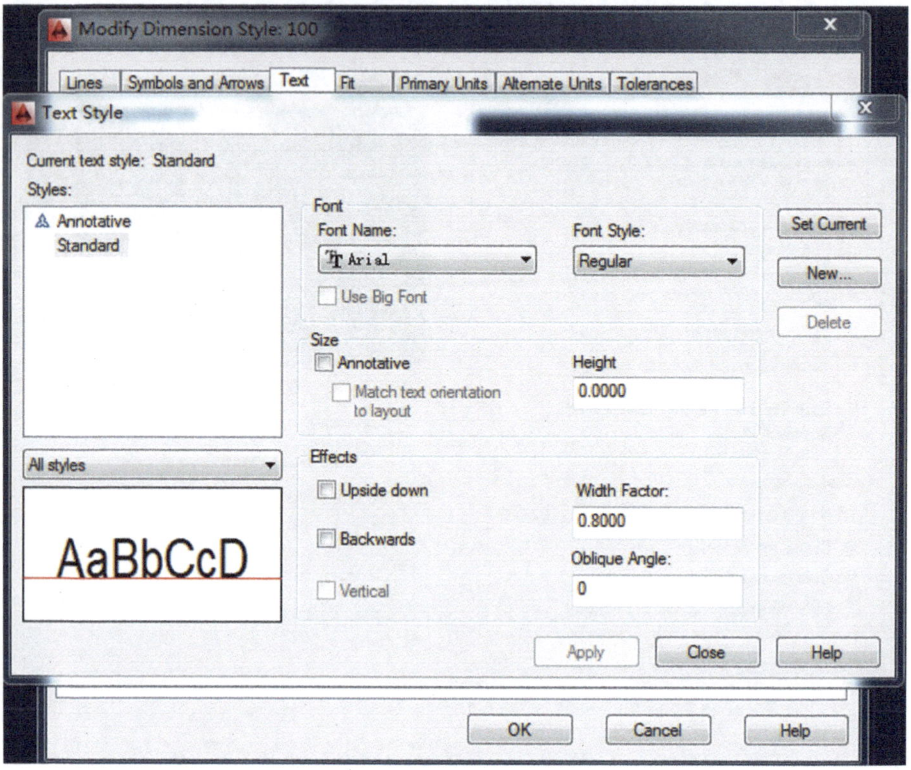

Fig. 3. Modify text style

5 Print Setup

A complete set of design drawings for indoor space includes planes, elevations, profiles, node details, etc. When all the drawings are drawn, we can print them out. Then we need to set up the Page Setup Manager.

The following are specific settings for the Page Setup Manager.

First, displays the New Page Setup dialog box, in which we can enter a name for the new page setup and specify the page setup to use as a starting point.

Second, select a printer, select Paper size, select Plot area, select Drawing orientation, select Plot style table and so on (Fig. 4).

Third, displays the Plot Style Table Editor, in which we can view or modify plot styles for the currently assigned plot style table (Fig. 5).

Fourth, press Ctrl key and P key at the same time, print the drawing.

In this way, we can finally get a set of specifications, beautiful, orderly interior design drawings.

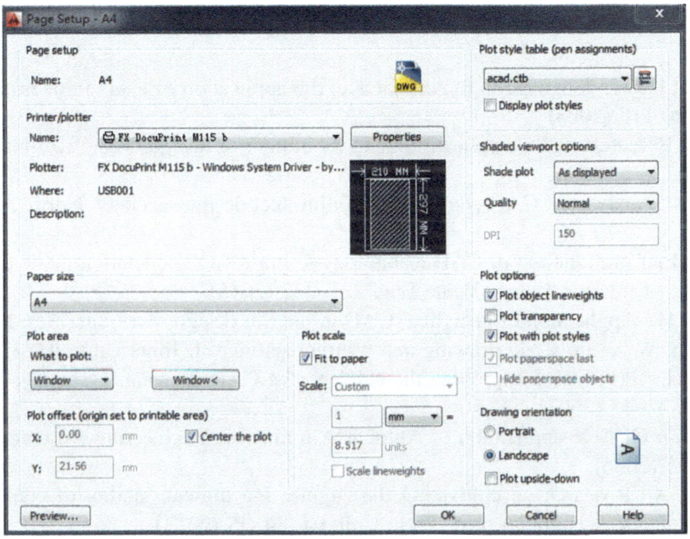

Fig. 4. Page setup

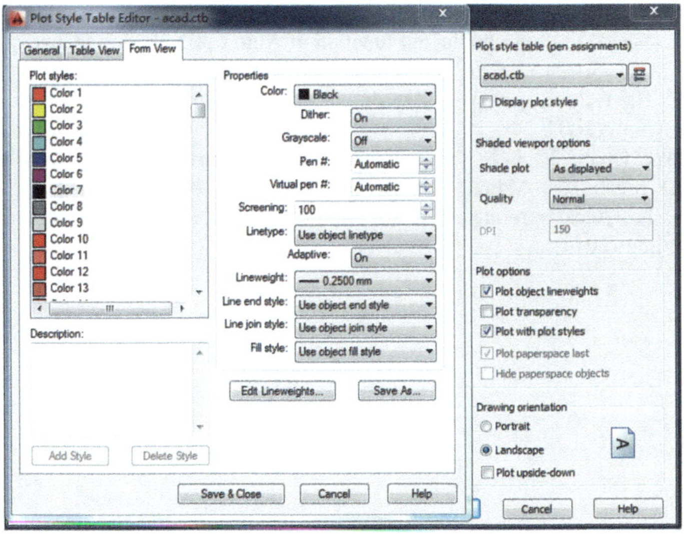

Fig. 5. Plot Style Table Editor

6 Summary

Nowadays, with the rapid development of science and technology, the computer hardware is updated and the computer software is developed. The work of our interior designer is becoming more and more efficient, especially Auto CAD, which has become an indispensable tool in the design work with many advantages.

References

1. Yina, H.: Design language in the digital age: the application of CAD in interior design. Art Panor. **06**, 118 (2008)
2. Yan, C.: Research on the application of CAD in interior design. Manuf. Autom. **15**, 33–50 (2010)
3. Pamin, N.: Analysis of CAD course setup for product design specialty. North. Art **02**, 73–74 (2010)
4. Jie, H.: Brief introduction of CAD technology to improve the operation guide and operation technique of interior design. Value Eng. 221–222 (2015)
5. Zhiying, H.: Application of analytical CAD in interior design. Art Educ. Res. **16**, 66 (2013)
6. Chengzhi, W.: Auto CAD drawing and printing settings. J. Jimei Univ. **04**, 37–42 (2002)
7. Xieying, L.: Practice of improving the efficiency of CAD drawing. J. Jiujiang Vocat. Tech. Coll. **01**, 42–43 (2007)
8. Chang, Y.: Flexible application of AutoCAD in drawing engineering drawings. Machinery **11**, 54–55 (2006)
9. Guiying, X.: Research on cultivating the engineering drawing ability of civil engineering major students. J. Harbin Vocat. Tech. Coll. **04**, 34–35 (2013)
10. Rene, P.: Some practical problems about AutoCAD dimensioning. Modul. Mach. Tool Autom. Manuf. Tech. **11** (2003)
11. Xiaofei, Z.: Interior Design Engineering Drawing Method and Its Example. China Architecture and Building Press, Beijing (2007)
12. Shiguo, L., Xiao, J.: Using the layout function of Auto CAD 2000 to realize quick drawing. CAD/CAM **11**, 30–31 (2005)
13. Hong, Z.: The technique of designing output drawing with AutoCAD layout. Res. Explor. Lab. **12**, 37–38 (2004)
14. Taigan, W., Huan, T., Lei, T.: Drawing skills of electrical drawing template based on AutoCAD. Mech. Des. Manuf. Eng. 81–83(2014)
15. Rongwei, T.: Quick Introduction to Furniture Design CAD Drawing. Chemistry Industry Press, Beijing (2014)

Fault Diagnosis for Uncertain Stochastic Nonlinear Systems via LMI Approach

Haiyi Sun[1], Ning Li[2(✉)], Shuxian Miao[1], and Jinbao Wang[1]

[1] College of Science, Shenyang JianZhu University, Shenyang 110168, China
[2] College of Sciences, Northeastern University, Shenyang 110819, China
lining80@163.com

Abstract. The problem of fault diagnosis observer design is researched for a class of stochastic nonlinear systems with uncertainties. The stochastic nonlinear systems with uncertainties model considered in this paper possesses the disturbances, lock-in-place (LIP) and loss of effectiveness (LOE). The proposed model is a more general LOE-LIP stochastic nonlinear model and covers several common actuator fault tolerant situations. Based on the theory of Lyapunov stability theory, Itô stochastic differential theory and linear matrix inequality approach a robust fault diagnosis observer is designed. When faults occur, the fault diagnosis observer can estimate the fault quickly and accurately. Finally, numerical simulation results show that the fault diagnosis observer can diagnosis the fault rapidly and accurately.

Keywords: Stochastic system · Nonlinear system · Fault diagnosis observer Linear matrix inequality (LMI)

1 Introduction

With the increasing complexity of the modern system and the increasing demands for its security, fault detection technology has gained more and more attention, and has achieved many important research results in recent decades [1–4]. In the actual system, the nonlinear phenomena of the object or process are universal and cannot be eliminated completely. The nonlinearity increases the complexity of the system and brings about the analysis and synthesis of the system. Furthermore, the vast majority of the real system will become the complex stochastic system affected by the random noise. The fault detection method based on deterministic system has great limitation and it is difficult to apply to stochastic system. Therefore, it is of great theoretical significance and application value to study the fault detection method of nonlinear random system [5–7].

Control delay is a factor needs to be considered in the reality control design. Delay may be caused by measurement of system variables, controller calculation, processes for actuators to build up the required control force, and etc. A small control delay may result in the instability of the control system. In [8, 9], the effect of control delays on the stability of controlled aeroelastic systems are studied systematically. The above methods arc applicable when the actuator has no faults, and they may become invalid when the faults exist.

J. Abawajy et al. (Eds.): ATCI 2018, AISC 842, pp. 761–767, 2019.
https://doi.org/10.1007/978-3-319-98776-7_89

Motivated by the above related work, in this paper, the problem of fault diagnosis observer is researched for a class of stochastic nonlinear systems with uncertainties. The stochastic nonlinear systems with uncertainties model considered in this paper possesses the disturbances, lock-in-place (LIP) and loss of effectiveness (LOE). The proposed model is a more general LOE-LIP model and covers several common actuator fault tolerant situations. Based on the theory of Lyapunov stability theory, Itô stochastic differential theory and linear matrix inequality approach a robust fault diagnosis observer is designed.

The letter is organized as follows. In Sect. 2, the mathematical model for the stochastic nonlinear systems with uncertainties model, LIP and LOP is presented. And some preliminaries are given. Section 3 gives the design of the fault diagnosis observer. Numerical simulations are given in Sect. 4. Section 5 briefly concludes the research work.

2 System Description and Preliminaries

We consider two types of actuator fault simultaneously, namely LOE and LIP, parameter uncertainties and stochastic disturbances. Hence, the uncertain stochastic nonlinear systems can be introduced as follow.

$$\begin{cases} dx(t) = [(A+\Delta A)x(t) + Bu(t-\tau(t)) + f(t,x) + Dv(t)]dt \\ \qquad\qquad + g(x)dw(t) \\ y(t) = Cx(t) \end{cases} \tag{1}$$

where ΔA is the parameter uncertainties; and $y(t)$ is the measured output. $w(t)$ is a one-dimensional standard Wiener process defined on the probability space. The parameters C and D are known matrices with appropriate dimensions. $\tau(t)$ represents the time-varying delay. There is constant η that satisfies $1-\dot{\tau}(t) \geq \eta^2$. $g(x)$ is unknown smooth nonlinear functions. w is an independent r-dimensional wiener process defined on a complete probability space with the incremental covariance $E\{dwdw^T\} = \sigma(t)\sigma^T(t)dt$. ΔA is assumed to satisfy the following matching condition of $\Delta A = M\Delta F(t)N$, where $\Delta F(t)$ is satisfied $\|\Delta F(t)\| < 1$.

The LOE and LIP fault models can be obtained with [10]

$$u(t) = [\phi(t) - \sigma(t)]r(t) + \sigma(t)]\tilde{r}(t) \tag{2}$$

where $r(t) \in R^p$ applied control vector; $\tilde{r}(t)$ is LIP fault input; and $\phi(t) \in R^{p\times p}$ diagonal matrix function that satisfies

$$\phi(t) = \begin{cases} diag(1,1,\cdots,1), & t < T_1; \\ diag(\phi_1,\phi_2,\cdots,\phi_p), & t \geq T_1; \end{cases} \tag{3}$$

where $\phi_i \in [0,1]$ is loss coefficient; T_1 is the time period at which the LOE fault is observed; $\sigma(t) \in R^{p\times p}$ is diagonal matrix function that satisfies

$$\sigma(t) = \begin{cases} 0, & t < T_2; \\ diag(\sigma_1, \sigma_2, \cdots, \sigma_p), & t \geq T_2; \end{cases} \tag{4}$$

where $\sigma(t) \in \{0, 1\}$ is LIP relation and coefficient; T_2 is time period at which the LIP fault is detected.

By substitution of this type of fault into Eq. (1), then we can obtain the stochastic wing flutter model including control delay, disturbances, external bounded disturbance, LOE and LIP the following form:

$$\begin{cases} dx(t) = [(A + \Delta A)x(t) + B[\phi(t - \tau(t)) - \sigma(t - \tau(t))]r(t - \tau(t)) \\ \qquad + B\sigma(t - \tau(t))\tilde{r}(t - \tau(t)) + f(t, x)]dt + g(x)dw(t) \\ y(t) = Cx(t) \end{cases} \tag{5}$$

Remark 1. Equation (5) is a more general LOE-LIP flutter model and covers several common actuator fault tolerant situations. If $\phi_i = 1$ and $\sigma_i = 0$, then the ith input is faultless, i.e., Eq. (5) degenerates into nominal system. If $\phi_i \in (0, 1)$ and $\sigma_i = 0$, then the Eq. (5) becomes the stochastic flutter model with LOE fault. If $\phi_i = 1$ and $\sigma_i = 1$, then the Eq. (5) becomes the stochastic flutter model with LIP fault.

Without loss of generality, the rational assumptions are given as following.

Assumption 1. It can be seen from the structure of $f(t, x(t))$ and $g(x)$ satisfies the Lipschitz condition $\|f(t, x(t)) - f(t, \hat{x}(t))\| \leq L_f \|e(t)\|$, $\|g(x)\| \leq L_g \|e(t)\|$, $L_f > 0, L_g > 0$.

Definition 1. The solution process $\{x(t), t \geq 0\}$ of stochastic system

$$dx = f(x)dt + g(x)dw \tag{6}$$

is said to be bounded in probability, if

$$\lim_{c \to \infty} \sup_{0 \leq t < \infty} P\{\|x(t)\| > c\} = 0.$$

Lemma 1. For any vectors $x, y \in R^n$ and positive definite matrix $Q \in R^{n \times n}$, the following matrix inequality holds:

$$2x^{\mathrm{T}}y \leq x^{\mathrm{T}}Qx + y^{\mathrm{T}}Q^{-1}y.$$

Lemma 2 [11]. Given constant symmetric matrix W, and matrices M, N with appropriate dimensions, then $W + N^T F^T(k) M^T + MF(k)N < 0$ holds for all $F(k)$ satisfying $F(k)^T F(k) \leq I$ if and only if there exists $\varepsilon > 0$ such that $W + \varepsilon^{-1} N^T N + \varepsilon M M^T < 0$.

Lemma 3 [12]. Consider the stochastic non-linear system (6). If there exists a positive definite, radially unbounded, twice continuously differentiable Lyapunov function $V : R^n \to R$, and constants $c_1 > 0, c_2 > 0$ such that

$$\ell V(x) \leq -c_1 V(x) + c_2$$

Then the system has a unique solution almost surely and the system is bounded in probability.

3 The Fault Diagnosis Observer Design

In this section, we design an observer which can provide the information of wing flutter states. Considering the form of the model (6), the following observer is designed:

$$\begin{cases} \dot{\hat{x}}(t) = A\hat{x}(t) + f(t, \hat{x}) + L[y(t) - \hat{y}(t)] \\ \hat{y}(t) = C\hat{x}(t) \end{cases} \tag{7}$$

Definite the error vector

$$e(t) = x(t) - \hat{x}(t) \tag{8}$$

Then we can obtain the error system.

$$\begin{aligned} de(t) = {} & [(A + \Delta A - Lc)e(t) + B[\phi(t-\tau(t)) - \sigma(t-\tau(t))] \cdot r(t-\tau(t)) \\ & + B\sigma(t-\tau(t))\tilde{r}(t-\tau(t)) + f(t,x) - f(t,\hat{x})]dt + g(x)dw(t) \end{aligned} \tag{9}$$

Theorem 1. If there exists a positive-definite matrix P, U, S and considering the estimation error Eq. (9), we suppose that the observer gains be satisfied with

$$\begin{bmatrix} \Omega_{11} & \Omega_{12} & \Omega_{13} & \Omega_{14} & \Omega_{15} \\ * & \Omega_{22} & \Omega_{23} & \Omega_{24} & \Omega_{25} \\ * & * & \Omega_{33} & \Omega_{34} & \Omega_{35} \\ * & * & * & \Omega_{44} & \Omega_{45} \\ * & * & * & * & \Omega_{55} \end{bmatrix} < 0 \tag{10}$$

where $\Omega_{11} = P(A - LC) + (A - LC)^T P + (\mu + 2L_f + 2L_g^2)P$, $\Omega_{12} = PM$, $\Omega_{13} = PB$, $\Omega_{14} = PB$, $\Omega_{15} = N$, $\Omega_{22} = -\varepsilon I$, $\Omega_{23} = \Omega_{24} = \Omega_{25} = \Omega_{34} = \Omega_{35} = \Omega_{45} = 0$, $\Omega_{33} = -S$, $\Omega_{44} = -U$, $\Omega_{55} = -\varepsilon I$. Then the estimation error dynamics system is stable.

Proof . According to the stochastic Lyapunov stable theory, we choose the following Lyapunov-Krasovskii functions candidate for system (9) as:

$$V(t) = e^{\mathrm{T}}(t)Pe(t) \tag{11}$$

By Itô's formula, along the state trajectory of systems (9), the stochastic derivative of $V(t)$ is obtained as follows:

$$\begin{aligned}\ell V_1(t) &= 2e^{\mathrm{T}}(t)P\dot{e}(t) + Tr\{g^T(x)Pg(x)\} \\ &= e^{\mathrm{T}}(t)[P(A-LC) + (A-LC)^T P]e(t) + 2e^{\mathrm{T}}(t)P\Delta Ae(t) \\ &\quad + 2e^{\mathrm{T}}(t)PB[\phi(t-\tau(t)) - \sigma(t-\tau(t))]r(t-\tau(t)) \\ &\quad + 2e^{\mathrm{T}}(t)PB\sigma(t-\tau(t))\tilde{r}(t-\tau(t)) \\ &\quad + 2e^{\mathrm{T}}(t)P[f(t,x) - f(t,\hat{x})] + 2Tr\{g^T(x)Pg(x)\}\end{aligned} \tag{12}$$

Since Assumption 1 and Lemma 1 and Lemma 2, we have the following inequalities.

$$e^{\mathrm{T}}(t)P[f(t,x) - f(t,\hat{x})] \leq e^{\mathrm{T}}(t)L_f Pe(t) \tag{13}$$

$$e^{\mathrm{T}}(t)P\Delta Ae(t) \leq e^{\mathrm{T}}(t)\left(\varepsilon PMM^T P + \varepsilon^{-1}NN^T\right)e(t) \tag{14}$$

$$\begin{aligned}2e^{\mathrm{T}}(t)PB[\phi(t-\tau(t)) - \sigma(t-\tau(t))]r(t-\tau(t)) &\leq e^T(t)PBS^{-1}B^T Pe(t) \\ &\quad + \|\phi(t-\tau(t)) - \sigma(t-\tau(t))S\tilde{r}(t-\tau(t))\|^2\end{aligned} \tag{15}$$

$$2e^{\mathrm{T}}(t)PB\sigma(t-\tau(t))\tilde{r}(t-\tau(t)) \leq e^T(t)PBU^{-1}B^T Pe(t) + \|\sigma(t-\tau(t))U\tilde{r}(t-\tau(t))\|^2 \tag{16}$$

Applying (13)–(16), it follows that

$$\begin{aligned}\ell V(t) &\leq e^{\mathrm{T}}(t)[P(A-LC) + (A-LC)^T P + 2L_f P + 2L_g^2 P + \varepsilon^{-1}PMM^T P + \varepsilon NN^T \\ &\quad + PB^T S^{-1}BP + PB^T U^{-1}BP]e(t) + \varepsilon\|\sigma(t-\tau(t))\|^2\|U\tilde{r}(t-\tau(t))\|^2 \\ &\quad + \varepsilon\|\phi(t-\tau(t)) - \sigma(t-\tau(t))\|^2\|Sr(t-\tau(t))\|^2 \\ &\leq -\mu V(t) + \delta + e^{\mathrm{T}}(t)\Omega e(t)\end{aligned} \tag{17}$$

where $\delta = \varepsilon\|\phi(t-\tau(t)) - \sigma(t-\tau(t))\|^2\|Sr(t-\tau(t))\|^2 + \varepsilon\|\sigma(t-\tau(t))\|^2\|U\tilde{r}(t-\tau(t))\|^2 \geq 0$, $\Omega = P(A-LC) + (A-LC)^T P + (\mu + 2L_f + 2L_g^2)P + \varepsilon^{-1}PMM^T P + \varepsilon NN^T + PB^T S^{-1}BP + PB^T U^{-1}BP$, $\mu > 0$.

Using Schur complement and matrix inequality (10), we can get $\Omega < 0$, i.e.,

$$\ell V(t) \leq -\mu V(t) + \delta \tag{18}$$

Using Lemma 3, from Eq. (18), we can get the error dynamics system Eq. (9) is stable. The proof is completed.

4 Numerical Simulations

In this section, numerical example of the uncertain stochastic nonlinear systems (1) is given to demonstrate the effectiveness of the observer (7). Consider the system (1) with the following parameters. Taking $A = \begin{bmatrix} 0.6 & 0.2 \\ 0 & 0.7 \end{bmatrix}$, $B = \begin{bmatrix} 0.3 & 0 \\ 0.2 & 0.3 \end{bmatrix}$, $C = \begin{bmatrix} 2 \\ -0.1 \end{bmatrix}^T$, $g(x) = \begin{bmatrix} 0.1\cos x & 0 \\ 0 & 0.3 \end{bmatrix}$, $f(x(t)) = \begin{bmatrix} 0.5x_1(t) \\ \sin(x_2(t)) \end{bmatrix}$, $\sigma(t) = \begin{cases} 1, & 40 \leq t \leq 80 \\ 0, & else \end{cases}$, $r(t) = 0$, $w(t) = \begin{bmatrix} rand[0,1] \\ 1.2rand[0,1] \end{bmatrix}$. The time-varying delay $\tau(t) = 0.5\sin t$. Consider the parameter uncertainty $\Delta A(t)$ in the general form. Applying Theorem 1 and solving the LMI (10) by using LMI toolbox of Matlab, it is found that LMI (10) is feasible, i.e., fault diagnosis observer is effective. The residual signal of the error system (9) is numerically demonstrated as the Fig. 1. It is shown that the designed fault detection observer can be effectively detected when the fault occurs.

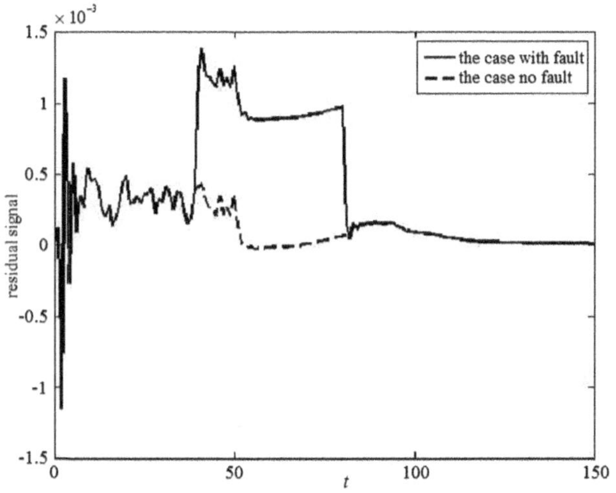

Fig. 1. The residual signal between the system (5) and the observer system (7).

5 Conclusions

In this paper, we study the fault diagnosis observer design of the uncertainties stochastic nonlinear systems. The stochastic nonlinear systems with uncertainties model considered in this paper possesses the disturbances, LIP and LOE. The proposed model is a more general LOE-LIP stochastic nonlinear model and covers several common actuator fault tolerant situations. Based on the theory of Lyapunov stability theory, Itô stochastic differential theory and linear matrix inequality approach a robust fault diagnosis observer is designed. When faults occur, the fault diagnosis observer can estimate the fault quickly and accurately. Finally, numerical simulation results show that the fault diagnosis observer can diagnosis the fault rapidly and accurately.

Acknowledgement. This work was supported in part by the National Science Foundation of China under Grant 61673100, in part by the Fundamental Research Funds for the Central Universities under Grant N180504009, in part by the Liaoning Province Doctor Starting Foundation under Grant 20170520283, in part by the Natural Science Foundation of Liaoning Province under Grant 20170540769, in part by the Liaoning BaiQianWan Talents Program under Grant 2017076, in part by the Fundamental Research Funds for the Central Universities under Grant N140504009, and in part by the Research Topic of Chinese Society of Academic Degrees and Graduate Education under Grant B-2017Y0802-139.

References

1. Fang, F.J., Ye, H., Zhong, M.Y.: Fault diagnosis of networked control systems. Ann. Rev. Control **31**(1), 55–68 (2007)
2. Pertew, A.M., Marquez, H.J., Zhao, Q.: LMI-based sensor fault diagnosis for nonlinear Lipschitz systems. Automatica **43**(8), 1464–1469 (2007)
3. Zhang, H.G., Li, M., Yang, Y.: Fuzzy model-based robust networked control for a class of nonlinear systems. IEEE Trans. Syst. Man Cybern. Part A Syst. Hum. **39**(2), 437–447 (2009)
4. Mao, Z.H., Jiang, B., Shi, P.: Observer based fault-tolerant control for a class of nonlinear networked control systems. J. Franklin Inst. **347**(6), 940–956 (2010)
5. Gao, H.J., Fei, Z.Y., Lam, J.: Fault detection for markovian jump systems with sensor saturations and randomly varying nonlinearities. IEEE Trans. Circuits Syst. I Regul. Pap. **59** (10), 2354–2362 (2012)
6. Liu, M., Shi, P., Zhang, L.X.: Fault-tolerant control for nonlinear markovian jump systems via proportional and derivative sliding mode observer technique. IEEE Trans. Circuits Syst. I Regul. Pap. **58**(11), 2755–2764 (2011)
7. Zhuang, G., Yu, X., Chen, J.: Fault diagnosis for uncertain stochastic nonlinear systems. Circuits Syst. Sig. Process **34**(9), 2839 (2015)
8. Huang, R., Hu, H.Y., Zhao, Y.H.: Designing active flutter suppression for high-dimensional aeroelastic systems involving a control delay. J. Fluids Struct. **34**(10), 33–50 (2012)
9. Huang, R., Qian, W.M., Hu, H.Y., Zhao, Y.H.: Design of active flutter suppression and wind-tunnel tests of a wing model involving a control delay. J. Fluids Struct. **55**(5), 409–427 (2015)
10. Hu, L.Z., Chen, F.Y., Jiang, B., Tao, G.: Control strategy for a quadrotor helicopter with state delay via improved guaranteed cost control and quantum adaptive control. J. Aerosp. Eng. **30**(4), 04017004-1–04017004-9 (2017)
11. Petersen, I.R.: A stabilization algorithm for a class of uncertain linear systems. Syst. Control Lett. **8**(4), 351–357 (1987)
12. Psillakis, H.E., Alexandridis, A.T.: Adaptive tracking control for stochastic uncertain nonlinear systems satisfying short- and long-term cost criteria. Int. J. Control **79**(2), 107–118 (2006)

Clustering Fuzzy Network Dynamic Data Simulation Research

Bo Tang(✉)

School of Mechatronic Engineering, Xi'an Technological University, Xi'an 710021, China
nwputang@163.com

Abstract. Traditional clustering analysis needs complete data set, but in some cases, data is incomplete missing. This brings difficulties to clustering analysis. Cluster analysis is a multivariate statistical analysis method for the study of classification problems. We use variables related to missing data of the auxiliary information, establish some relationships between data can reflect the model based on the relationship between these variables on the analysis to estimate the data, determine a reasonable compensation for value of missing data interpolation to the original position of missing data thus, to construct a complete data set for iterative clustering analysis. The experimental results show that the proposed method can accelerate the iterative parameter estimation of defect data rate, correction for large scale data defects, compared to the current method in execution time and speedup performance advantages.

Keywords: Dynamic network · Fuzzy network · Data defect correction Simulation

1 Introduction

With the rapid development of the Internet and people's increasing demand for network information sharing, dynamic network operations increasingly rely on data information. Missing data is divided into two cases of missing and arbitrarily missing blocks [1]. A missing block means that valid data appears continuously in the entire data sequence. An arbitrarily missing refers to a missing data sample that may appear anywhere in the data sequence. Missing data spectral analysis uses limited discrete data to estimate signal parameters. Because the continuity of the data phase is destroyed, generally the direct method based on FFT has a serious spectrum leak, and it becomes more complicated as the proportion of missing data increases [2]. Time domain missing data recovery is based on the assumption that missing data and valid data have the same spectral components. MUSIC, ESPRIT, and other high-resolution techniques can provide very accurate spectral estimates, but are more dependent on the model order, and have better data recovery accuracy for smaller data loss ratios. The earliest non-parametric method applied to the restoration of missing data of line model nuggets is the missing data adaptive phase estimation method (GAPES) proposed in the literature [3]. This method inherits the excellent statistical performance of the matched filter bank method, and is then extended to 2-D to recover the phase history data missing from the

J. Abawajy et al. (Eds.): ATCI 2018, AISC 842, pp. 768–773, 2019.
https://doi.org/10.1007/978-3-319-98776-7_90

angle diversity SAR. The data interpolation and spectral extrapolation of non-parametric methods are realized, the proportion of missing data in the block is increased to 65%, and the imaging effect is better than the CLEAN method [4]. Based on the GAPES method, literature 61 introduces the tuning factor into the missing data spectral estimation problem, compensates for the small deviation between the sample covariance and the true value, and confirms that it has higher estimation accuracy than the GAPES in the absence of small samples.

2 Data Defect Real-Time Correction Principle

The dynamic fuzzy network domain data for real-time defect correction process, parameter estimation of fuzzy field defect data using genetic optimization method, the optimal parameters, the optimal parameters based on the multidimensional fuzzy correlation domain data specific to different data sources of the approximate measure, constructing approximate to the fuzzy domain defect data as the center of the source, and the defects have been filled value as the observation value for subsequent fuzzy domain data real time defect correction process, the flaw data value sequence sensitive to fill, to complete the fuzzy domain data real-time correction of defects.

2.1 Parameter Estimation of Data Defect in Fuzzy Domain Based on Genetic Algorithm

The log likelihood function with parameters is regarded as the objective function. According to the dynamic network fuzzy domain data samples, the corresponding constraint conditions of the parameters are obtained, and the data defect estimation model of the dynamic network fuzzy domain is constituted by the objective function and the constraint conditions. The specific process is as follows:

We suppose that Y represents a data set of a dynamic network fuzzy domain, which contains a total of k variables $Y_1, \cdots, Y_K$. The data set contains defect data, which is represented as $Y = (Y_o, Y_m)$, Y_o is used to describe a set of data with an observed value, and Y_m represents a set of defective data. We use the genetic algorithm to estimate the parameters of the fuzzy domain data set Y_o. The upper and lower bounds of the variable $Y_1, \cdots, Y_K$ are obtained from the dynamic network fuzzy domain data set Y_o, which is marked as $[\min_l, \max_l]$. The logarithmic likelihood function of the mean and covariance matrix of the data variables in a dynamic network fuzzy domain is described as:

$$f(\mu, \Sigma) = -n'/2(2\pi) - -n'/2|\Sigma|(y_l - \mu_k)^T \tag{1}$$

In the equation, $\Sigma = (\sigma_{pq})$ represents the corresponding covariance matrix of the variable $Y_1, \cdots, Y_K$. y_l is used to describe vectors that record l corresponding variables in fuzzy domain data. n' represents the number of data records in a dynamic network fuzzy domain. The larger the logarithmic likelihood function, the more accurate the estimated parameters are. The original problem can be transformed into a single objective optimization problem of μ, Σ corresponding to the f maximization that satisfies all constraints. We describe its data model by using the next Formula:

$$\begin{cases} \max\limits_{\mu,\Sigma} f \quad s.t. \begin{cases} \min_1 \leq \mu_1 \leq \max_1 \\ \vdots \\ \min_K \leq \mu_1 \leq \max_K \end{cases} \end{cases} \tag{2}$$

In the equation, $\min_\iota$ and $\max_K$ represent the lower and upper limit of the ι variables of the dynamic network fuzzy domain respectively.

To determine the dynamic network to estimate the fuzzy domain data deletion model and parameter estimation of parameters, in populations of randomly generated parameters under constraint conditions, the population size can be based on fuzzy domain data loss rate to set the objective function of the f as the fitness function, e iteration.

The iterative cut-off condition of our population parameter estimation is the optimal parameter, and the corresponding fitness function value interval is less than a given value α, that is, $|f_\iota^* - f_{\iota-1}^*| < \alpha$. f_ι^* represents the corresponding objective function value of the optimal parameter after the iterative cycle ι. When all the termination conditions are satisfied, the iteration is terminated and the optimal estimate is obtained.

2.2 Defect Data Filling Sequence Decision and Defect Value Filling

For a tuple containing multidimensional missing data in a dynamic network fuzzy domain, it is treated as a tuple containing only one dimension missing data. It is assumed that the fuzzy field data set of N' data sources arrived at t time period contains n' defect data tuple, and the corresponding defect data source set is $MS = \{S_1, \cdots, S_{n'}\}$. The real-time correction of fuzzy domain data defects is to fill in the missing data of n' missing data sources. The mean Euclidean distance [5] is used to measure the multidimensional approximation between data sources in the fuzzy domain of dynamic network. Assuming that there exists data defect in the j' dimension property of the dynamic network fuzzy domain at present stage, the multidimensional approximation degree between data source S_ι and S_K is:

$$Sim_K = 1/(1 + \min\{d_{\iota K}^T, d_{\iota K}^S, d_{\iota K}^A\}) \tag{3}$$

Among them, $d_{\iota K}^T$, $d_{\iota K}^S$ and $d_{\iota K}^A$ are used to describe the distance between the data source of the dynamic network fuzzy domain and the time, space and attribute. Based on the nearest neighbor list of missing data sources, and the approximate information between the missing data sources and their neighboring nodes, the approximate graph centered on the defect data source is constructed. The distribution of all relative complete nearest neighbor nodes of each defect data source in the fuzzy domain is characterized by the weight of the vertex of the approximate graph:

$$w_{S_\iota} = 1 - \prod_{S_K \in N(s_\iota)} (1 - sim_{\iota K}) \tag{4}$$

In the formula, $N(s_i)$ represents the set of relative complete adjacent nodes of the missing vertex s_i in the approximate graph. For a dynamic network fuzzy domain with n' missing data, the corresponding *seq* can be transformed into a Bayesian network.

3 Experimental Results and Analysis

In order to prove the comprehensive effectiveness of the real-time correction method based on Genetic Optimization in dynamic network fuzzy domain, a simulation experiment is needed. The simulation experiment environment is: Windows7 system, Matlab R2010a, Intel (R) Core (TM) i7-3770 CPU, 2G memory. The published data set ASL is used as the experimental data set.

Figure 1 is a partial amplification curve for data correction in dynamic network fuzzy domain, where A represents the actual dynamic network data. B represents the correction result of method, and C indicates the correction result of this method [6]. It can be seen from the correction that the two methods have obtained good correction results for the case that the amount of defect data in the fuzzy domain is less. For the case of large amount of defect data, the visual error of this method is obviously smaller than the reference method from the degree of the fitting of the curve. This is mainly because the method of parameter estimation in fuzzy domain defect data is adopted in this paper, and the similarity degree between different data sources is measured by the unique multidimensional correlation of fuzzy domain data.

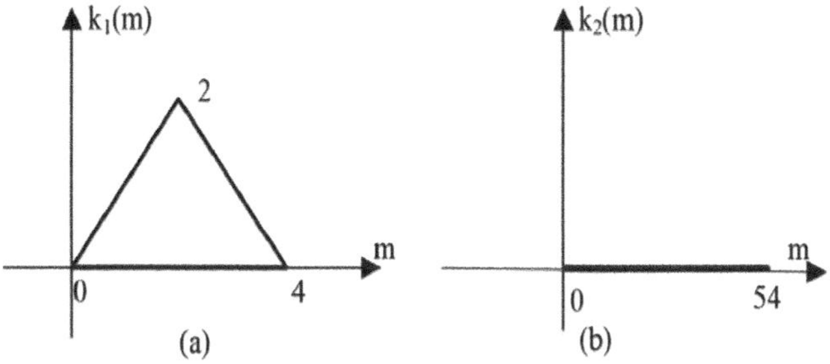

Fig. 1. Modifying defect data with different methods

From the results, it can be seen that this method is based on the result of parameter estimation, and the real time correction of defect data in fuzzy domain is very close to the actual value. It is proved that this method can correct the defect data accurately.

In the ASL dataset, different size subset is selected and executed on multiple PC machines. The ASL data set's defect records are all 40%. All attributes contain missing values. The execution time (s) on the data set and the execution acceleration ratio (%) are shown in Fig. 2.

Analysis of Fig. 2 shows that the corresponding data defect correction time is shorter, and the corresponding data defect correction is speeded up. The correction is real-time. Taking common data defects d_1, d_2 and d_3 as an example, when data defects

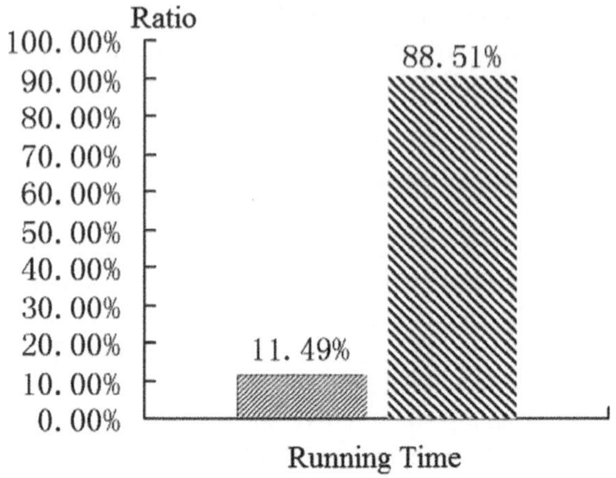

Fig. 2. Running time and speedup ratio comparison results

appear, the data defect testing items involved in the fuzzy domain are c_1, c_2, c_3 and c_4. The test data is normally expressed in "0", and the different commonly used "1" is expressed. This method is used to fill the filling, and the condition attributes of the other objects are better filled. This means that this method uses the multidimensional correlation characteristic of fuzzy domain data to measure the degree of approximation among different data sources. The approximate graph based on the fuzzy domain defect data source is constructed, and the filled defect value is taken as the observation value, which is used for subsequent fuzzy field data defect real-time correction process. It fills in the order sensitivity of the defect data missing value, and the data defect correction accuracy is high.

4 Conclusion

With the advent of the era of large data, the demand for dynamic network data quality has been gradually improved. The problem of data defect in fuzzy domain is common in dynamic network. Many data defects affect the integrity and authenticity of the overall data, which affects the post processing operation of dynamic network data. In the face of the high real-time requirement of dynamic network, it is necessary to speed up the iteration speed of data defect correction method and to estimate the data defect more accurately. In this paper, a method for real-time correction of data defects in dynamic network fuzzy domain based on genetic optimization is proposed. This method not only accelerates the estimation of the iterative speed of the data defect parameters of the fuzzy domain, but also improves the accuracy of the data defect correction.

Acknowledgements. This research is financially supported by the following projects.

1. Shaanxi Provincial Education Office natural special project: Atrial fibrillation digitalized model and less degree of freedom parallel mechanism system (NO. 2013JK1012).

2. Natural fund of Shaanxi provincial science and technology department: Atrial fibrillation signal analysis and ablation therapy (NO. 2013JQ4038).

References

1. Zareapoor, M., Shamsolmoali, P., Jain, D.K., Wang, H., Yang, J.: Kernelized support vector machine with deep learning: an efficient approach for extreme multiclass dataset. Pattern Recogn. Lett. (2017)
2. Deglint, J., Chung, A.G., Chwyl, B., Amelard, R., Kazemzadeh, F., Wang, X.Y., et al.: Photoplethysmographic imaging via spectrally demultiplexed erythema fluctuation analysis for remote heart rate monitoring. In: Multimodal Biomedical Imaging XI, vol. 9701, p. 970111. International Society for Optics and Photonics (2016)
3. Castells, F., Mora, C., Rieta, J.J., Moratal-Perez, D., Millet, J., et al.: Based on generalization center clustering, incomplete data set filling method. Small Microcomput. Syst. **38**(9), 2017–2021 (2017)
4. Zhang, X., Cheng, Y.: Filling method. Appl. Probab., Stat. Data Missing Value Compon. Random Based For. Model **33**(1), 102–110 (2017)
5. Calkins, H., Jais, P., Steinberg, S.: Based on sparse iterative covariance estimation, missing data spectral analysis and timely domain reconstruction method. J. Electron. Inf. **38**(6), 1431–1437 (2016)
6. Wang, Y., Wang, R., Chen, W., et al.: Based on genetic optimization, missing data filling algorithm of regulation system. Power Syst. Protect. Control **44**(21), 182–186 (2016)
7. Yu, D., Peng, L.: When does inferring reputation probability countervail temptation in cooperative behaviors for the prisoners' dilemma game? Chaos Solitons Fractals **78**, 238–244 (2015)
8. Juan, L.V.: A semantics-based routing scheme for grid resource discovery. In: E-Science: First International Conference on E-Science and Grid Computing, pp. 58–70, 90 (2005)
9. Wang, Q.: Automatic control system regulation scheme of textile air conditioning. Sci. Technol. Inf. **22**, 395–396 (2010). (in Chinese)
10. Liu, J.: Advanced PID Control and MATLAB Simulation, 2nd edn. Electronics Industry Press, Beijing (2003). (in Chinese)
11. Li, J., Ren, Q.: Study on supply air temperature forecast and changing machine dew point for variable air volume system. Build. Energy Environ. **27**(4), 29–32 (2008). (in Chinese)

An Improved Support Vector Machine of Intrusion Detection System

Mingyuan Xin(✉)

Heihe University, Heihe 164300, Heilongjiang, China
Xmy5686@163.com

Abstract. Intrusion detection system is an important part of network security defense technology, and an intrusion detection system is proposed to improve support vector machine. Analysis of support vector machine (SVM) algorithm, simulated annealing algorithm is proposed to optimize the parameters of the support vector machine (SVM), reduce the non-response rates and the rate of false positives, intrusion detection system design and simulation experimental results show that support vector machine (SVM) based on simulated annealing algorithm of intrusion detection performance than for improving intrusion detection.

Keywords: Simulated annealing algorithm · Support vector machine
Intrusion detection

1 Introduction

With the application of cloud computing and big data technology, the process of information age is further deepened. People pay more attention to the security of information. Information encryption and firewall technology are commonly used in the Internet to ensure the security of digital information. However, with the increasingly complex information security environment, the application needs cannot be satisfied. Is a kind of active defense technology, intrusion detection system can effectively make up for the inadequacy of the firewall, it through the study of the comparison and analysis of network data, so as to determine the security of network data, and make corresponding security defense technology. Network intrusion detection mainly consists of two detection methods: misuse detection and anomaly detection. Abnormal detection is based on the behavior detection method. The misuse detection technique is to detect abnormal behavior by matching the abnormal model, and it is usually unable to detect the unknown abnormality. How to improve the detection rate of intrusion detection system and reduce the false alarm rate is the key problem to improve the system performance, and it is getting more and more attention.

The processing capacity of massive amounts of data in the network puts forward higher requirements on the performance of intrusion detection system. Forrest to intrusion detection problem as put forward by the normal data and abnormal data classification problem to solve, so you can use the principle of pattern recognition to detect, including the intrusion detection technology based on neural network, immune model of intrusion detection technology and USES the data mining is the same train of

J. Abawajy et al. (Eds.): ATCI 2018, AISC 842, pp. 774–780, 2019.
https://doi.org/10.1007/978-3-319-98776-7_91

thought. Several methods above are based on a large number of complete audit under the data set is based on the previous sample, and that then collecting and processing large amounts of sample time consumption, the influence of real-time intrusion detection system. In this paper, we use simulated annealing algorithm to optimize support vector machine intrusion detection system, and optimize the problem of small sample set and large dimension data processing.

2 Support Vector Machine (SVM)

Support vector machine (SVM) is an effective classification algorithm, which mainly solves two categories of classification problems. Let's say the training sample is (x1, y1), ··· , (xi, yi) ∈ X × {±}. Where X represents non-null variables, a hyperplane is found in multidimensional space, and the classification hyperplane w • x + b = 0, And must be satisfied and must be satisfied.

$$y_i[(w \cdot x_i) + b] - 1 \geq 0 \tag{1}$$

For the nonlinear classification problem, the appropriate inner product function K (xi, xj) can be implemented, and the optimized objective function becomes:

$$Q(a) = \sum\nolimits_{i=1}^{m} a_i - \frac{1}{2}\sum\nolimits_{i,j=1}^{m} a_i a_j y_i y_j K\left(x_i \cdot x_j\right) \tag{2}$$

In the formula, And for all i = 1, …, m is true. And there are

$$\sum\nolimits_{i=1}^{m} a_i y_i = 0 \tag{3}$$

Considering the constraints of 1.3, the classification decision function is expressed as:

$$f(x) = sgn\left(\sum\nolimits_{i=1}^{m} a_i y_i \langle x, x_i \rangle + b\right) = 0 \tag{4}$$

After the initial problem is transformed into dual problem, the complexity of the calculation depends on the sample size, which is mainly the support vector number of the sample, which enables the support vector function to solve the high-dimensional problem.

3 Simulated Annealing Support Vector Machine

The core of support vector machine is the parameter selection problem, which is related to the recognition ability and generalization ability of the algorithm. Simulated annealing algorithm is a global random search algorithm proposed by Metropolis et al., which can effectively solve combinatorial optimization problems. Thought of the algorithm is of high temperature solid material in the physical cooling process, generally divided into heating process, isothermal process and cooling process, in which the key isothermal process, use based on the Metropolis criterion constraint algorithm converges to the global optimal solution. In this paper, using simulated annealing algorithm to optimize the parameter selection of support vector machines not only saves the model training time, but also improves the prediction rate of the model. Using simulated annealing algorithm, the parameters of vector machine are selected as follows: simulated annealing support vector machine.

Step1: Initialization. C and W are the upper and lower bounds of support vector machines. $1 < C < 100$ and $1 < W/1 < 100$, the initial values of two parameters are randomly generated, and the initial state value is calculated by K cross validation T_0. Any initial solution S_0.
Step2: To produce the new. The new solution S_{new} was generated by random movement of the two parameters according to formulas 5 and 6. The two parameters are fed back to the support vector machine model and the new status value T_{new} is obtained.

$$w\_next = w\_pre + w\_factorstep * wMAX * (1.0 * rand()/RAND\_MAX - 0.5) \tag{5}$$

$$C_{next} = C_{pre} + C_{factorstep} * CMAX * (1.0 * rand()/RAND\_MAX - 0.5) \tag{6}$$

Step3: Calculate the increment $df = f(S_0) - f(S_1)$ from S_0 to S_1, f is the target function.
Step4: Where, f is tested. Use formula 7 to determine whether a new solution is produced. Objective function. γ Random number on the interval (0, 1).

$$\begin{cases} acceptS_{new} : f(S_1) \leq f(S_0) \\ acceptS_{new} : f(S_1) > f(S_0)\, and\, \gamma < exp(f(S_1) - f(S_0)/T) \\ else\ deleteS_{new} \end{cases} \tag{7}$$

Step5: Cool down. The maximum number of iterations is set to 100β, β is the dimension of the problem. Suppose there are N parameters to determine the state of the system, and the maximum number of iterations is Nmax = N * 100. When a preset number of iterations is reached Nmax. The current temperature T is cooled. Use formula 4 to cool down. The range of K is $0.5 < K < 0.99$.

$$T = k * T \tag{8}$$

Step6: End the algorithm. If the current solution is better than the original solution or the maximum number of iterations, the algorithm ends, otherwise the steps 2 and 3 will be executed until the end of the algorithm is satisfied.

4 Simulation Experiment and Result Analysis

Design simulation experiment to analysis method, the experimental sample data from the KDD99 data set, the data set includes four types of abnormal DOS (denial of service attacks) and R2L (remote terminal of illegal access), U2R (local super user not granted rights to access) and surveillance and probing (port monitoring or scan). For example, ping-of-death, syn flood, smurfect password, buffer overflow attacks, port-scan, ping-sweep, etc. The records in the data set are represented in CSV format, with 41 feature items, including the network data feature data of TCP connection and traffic statistics. On the sample data randomly selected 4 times, each time repetition rate is less than 1/10 of the sample data, data of each extraction is 1/2 the training sample data, 1/2 is test sample data, in order to see improve vector machine the advantages of intrusion detection system, we use the same sample data sets to the traditional intrusion detection system has carried on the same test, experimental data as shown in Table 1.

Table 1. Random sample data set

Sample number	Sample data number	Abnormal data	Normal data
1	1000	190	810
2	1500	336	1164
3	2000	481	1519
4	2500	527	1973

The false alarm rate and false alarm rate are two important indexes to detect the performance of intrusion detection system, and the performance of the intrusion detection system should have the lowest possible false alarm and false alarm rate. The false alarm rate is defined according to formula 9, and the false alarm rate is defined according to formula 10. Often the data

$$FR = MA - MR/MA \tag{9}$$

$$ER = ME/MR \tag{10}$$

In the formula, the FR represents the omission rate, ER represents the false alarm rate, M_R indicates the abnormal sample number, M_A is the abnormal sample number, and the M_E representative is detected the normal data number in the abnormal data. In the process of experiment, we compare the detection time of two intrusion detection systems to determine the performance of real-time index of intrusion detection system. The experimental data are shown in Table 2.

Figures 1 and 2 are the results of the intrusion detection data of the basic intrusion detection and improvement support vector machine. According to Figs. 1 and 2 shows that improved support vector machine (SVM) intrusion detection regardless of sample size of data sets, it non-response rates and the rate of false positives than did not improve the intrusion detection, and with the increase of sample size, intrusion detection and support vector machine (SVM) has not occurred against the background of the oscillation phenomena.

Table 2. Experimental data

Methods	M_R	M_A	FR	M_E	M_N	ER
Basic intrusion detection	696	810	13.96	16	190	8.42
	1000	1164	14.12	26	336	7.57
	1321	1519	13.01	48	481	9.89
	1751	1973	11.23	47	527	8.90
Improved support vector machine intrusion detection	736	810	9.13	14	190	7.36
	1065	1164	8.50	22	336	6.54
	1393	1519	8.29	34	481	7.06
	1826	1973	7.45	46	527	8.72

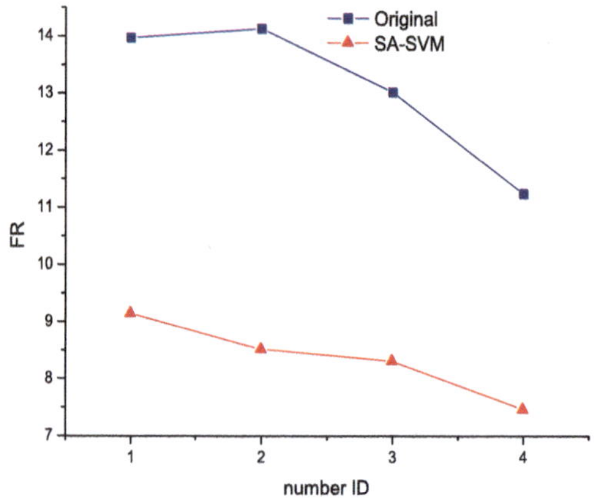

Fig. 1. Non-response rates

5 Conclusion

With the increase of network data processing and information security incidents, the performance of intrusion detection technology, as an active defense strategy, is more demanding. On the basis of the original algorithm, this paper puts forward using annealing algorithm to optimize the parameters of the support vector machine (SVM) choice, design the support vector machine (SVM) model of intrusion detection system improvement, and design simulation experiment to validate the model, and experimental data analysis, the conclusion is given. This experiment is carried out in the simulation environment, which is different from the actual intrusion detection system. The application of the intrusion detection system in the real network environment is the content of future research. In addition, due to the condition problem,

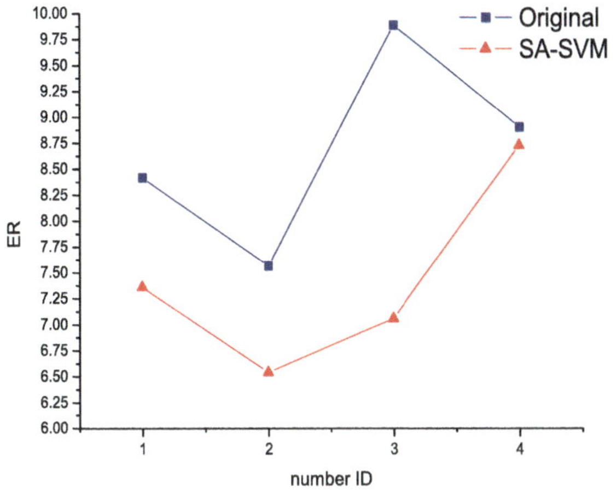

Fig. 2. The rate of false positives

the real-time verification of intrusion detection system is not accurate enough. It is only based on the estimation of the general time and will be improved in the future work.

Acknowledgements. The youth innovation talent training program of the general undergraduate colleges and universities in Heilongjiang Province, UNPYSCT-2017104.

References

1. Shivasankaran, N., Senthil Kumar, P., Venkatesh Raja, K.: Hybrid sorting immune simulated annealing algorithm for flexible job shop scheduling. Int. J. Comput. Intell. Syst. **8** (3), 455–466 (2015)
2. Du, G.L., Xue, N.: The research on mutil-objective location routing problem based on genetic simulated annealing algorithm. Appl. Mech. Mater. **543** (2014)
3. Ge, H.: Application and research of improved genetic algorithm in TSP problem (2016)
4. Zhang, H., Bai, G., Liu, C.: A broadcast path choice algorithm based on simulated annealing for wireless sensor network. In: IEEE International Conference on Automation and Logistics (2012)
5. Engelbrecht, A.: Particle swarm optimization. In: Proceedings of the Companion Publication of the 2015 on Genetic and Evolutionary Computation Conference (2015)
6. Tinoco, J.C.V., Coello, C.A.: hypDE: a hyper-heuristic based on differential evolution for solving constrained optimization problems. In: EVOLVE-A Bridge Between Probability, Set Oriented Numerics, and Evolutionary Computation II (2013)
7. Yu, D., Peng, L.: When does inferring reputation probability countervail temptation in cooperative behaviors for the prisoners' dilemma game? Chaos Solitons Fractals **78**, 238–244 (2015)
8. Vuong, J.L.: A semantics-based routing scheme for grid resource discovery. In: E-Science: First International Conference on E-Science and Grid Computing, pp. 58–70, 90 (2005)
9. Qizhen, W.: Automatic control system regulation scheme of textile air conditioning. Sci. Technol. Inf. **22**, 395–396 (2010). (in Chinese)

10. Liu, J.: Advanced PID Control and MATLAB Simulation, 2nd edn. Electronics Industry Press, Beijing (2003). (in Chinese)
11. Li, J., Ren, Q.: Study on supply air temperature forecast and changing machine dew point for variable air volume system. Build. Energy Environ. **27**(4), 29–32 (2008). (in Chinese)

Construction of China's Land Network Communication—Shandong Sub System

Chen Wu(✉), Tingting Guo, Hao Zhang, ChunShuai Lv, and JinLong Song

Seismological Bureau of Shandong Province, Jinan 250014, China
13953160911@163.com

Abstract. The state land network project is China's "11th Five-Year" period included in the national high-tech industrial development project of the 12 major national science and technology infrastructure construction projects, this paper on the background of the state land network project and the significance of the project and the state land network communication system based on seismic network industry are briefly introduced, and explain in detail the technical scheme the virtual network system is built on the basis of the GRE tunnel, and to the application of the system, the key technology of VPN package. In this paper, several key technologies are studied in detail, and they are applied to the communication system of terrestrial network. Especially, the construction scheme of Shandong land network communication system is explained in detail. Combined with the author's practical experience, this paper summarizes the problems encountered in the operation and maintenance of the communication system of Shandong land network, and provides a reference for the later construction and operation of the land network.

Keywords: Terrestrial network · GRE tunnel · Packaging technology VPN technology · SDH special line · Efficient operation and maintenance

1 Project Background

Mainland China tectonic environment monitoring network referred to as "lu state net", the project by the national development and reform commission (NDRC) in the "11th five-year plan" on the national major infrastructure construction projects of science and technology, and belongs to the national high and new technology industrial development projects. As the general responsible for the project by the China seismological bureau, the ministry of education, Chinese academy of sciences, China meteorological administration, the ministry of land and resources bureau of surveying and mapping, and ZongCan subsidiary construction. In after the completion of the project, our country has formed by the 260 base stations and 2000 continental tectonic environment monitoring network composed of regional station [2]. Project is the final request will be 260 benchmark station observation data, through the exclusive network platform in real-time transmission to the beijing-based China state lu network data center. The national earthquake industry data transmission task undertaken by China earthquake information network. The system consists of the national center in Beijing, more than

J. Abawajy et al. (Eds.): ATCI 2018, AISC 842, pp. 781–789, 2019.
https://doi.org/10.1007/978-3-319-98776-7_92

30 provinces, municipalities and autonomous regions in our country the provincial center and node of more than 300 seismic stations all over the country, and the communication network also includes a professional seismic data of satellite communication system [1], the system adopts the C band and Ku band, communication range can cover the and surrounding countries and regions in the country.

Shandong lu state using ground communication network base station, the ground communication system consists of the following main parts: channel planning, networking, security, network security, network bandwidth monitoring such as key content. The whole network system, relying on network platform based on the industry to build virtual network system, if the earthquake information network with the state of lu network system in the overlapping part of a communication channel, the use of the existing industry network channel transmission, it can greatly reduce the cost of input to the system construction and operation [5], and can realize the request of long channel run independently.

2 Communication System Design

The system by setting up a firewall, hierarchical deployment details of the security policy, to ensure that the state of lu network into network independently, and the protection means of network center and node of each benchmark station, complete the independence and security of the system, at the same time guarantee with the existing seismic industry network and Internet to realize logical isolation.

2.1 Design Thought

In GNSS reference station and meet system under the condition of continuous gravity station data transmission, network system based on industry, the ground transmission way to realize various reference station and the lu state between center and the reference station network connectivity to support a variety of professional services.

(1) In order to ensure the security of network and easy management, system management in the center of the office network under unified planning and construction; In order to avoid a single node fault causes the network interruption, to ensure smooth communication, should reduce transit node number and agency information network center to the site in accordance with the minimum number of nodes to achieve the channel access.
(2) To achieve ShengJu data center and the state of lu data sharing and exchange of network center, available ShengJu information network center and the state of lu SDH network center line. Similarly also available ShengJu data centre to data information and the entire province information node 2 m long SDH line, realize the GNSS reference station, continuous gravity station channel integration construction and sharing of [3].
(3) Using the China seismological bureau of industry network (VPN private network), realizes the communication line backup.

2.2 Design Implementation

According to the state of lu network project design requirements, the operation of the system to be independent as net, no exchange of visits between base station. Therefore, adopting the traditional mode of the GRE tunnel structures, a set of virtual network communication system, the implementation and industry network logic isolation.

(1) Land state network USES a class B network IPv4 network private address section, when the lu state network in access network equipment industry network industry network use special network IP address [3].
(2) ShengJu information network center distribution of three industry network IP address, used for VPN, IP telephone, monitoring server access; Base station USES a class B IP address, consisting of 28 subnet mask, 16 can assign IP addresses, all the disposable addresses.
(3) In the state of lu network center, ShengJu information network center and the reference station and equipped with a dedicated VPN equipment, use of certificate authentication, VPN equipment real-time online status and traffic statistics [3], at the same time also is attaining the unity of each datum station monitoring and various reference station and the state of lu one-stop point-to-point VPN network center docking.

2.3 Transmission Scheme

Lu Tai communication network USES the GRE tunnel mode to realize the channel transmission, also accord with the earthquake industry network status of both the land state network communication system design requirements. Tunnel protocol to other data frames or package to encapsulate and then sent through the tunnel, the Internet channel transmission load data is encapsulated, because the new frame head can provide routing information. Use tunnel passing data (or load) can be a different protocol data frames or package, because the tunnel technology is a kind of through the use of the Internet infrastructure and transfer data between network mode of transmission. Encapsulation tunnel through the earthquake industry Network state lu channel transmission Network system of data information, realize the Network center and each benchmark station node in the system of one-stop point-to-point VPN connection VPN (Virtual Private Network) is a Virtual Private Network. VPN can provide a high level of security protection, to prevent data theft and other unauthorized users contact with this data, you can use advanced encryption and identification protocol can prevent data snooping on [4] (Fig. 1).

Lu state ground network communication system by earthquake industry network channel virtual network across the country, the system and other systems are independent of each other operation, realize the isolation. Network node is in the center of the core layer and access layer, which is divided according to the function of the system and structure. National data center, the five data sharing subsystem formed network center, all the base station node belong to the access layer (1). System of data terminal equipment, is through the system - which is the benchmark station at the end of the node to access network lu state. The benchmark station according to the following two ways to access networks: one is the way to access seismic stations; 2 as a way to access

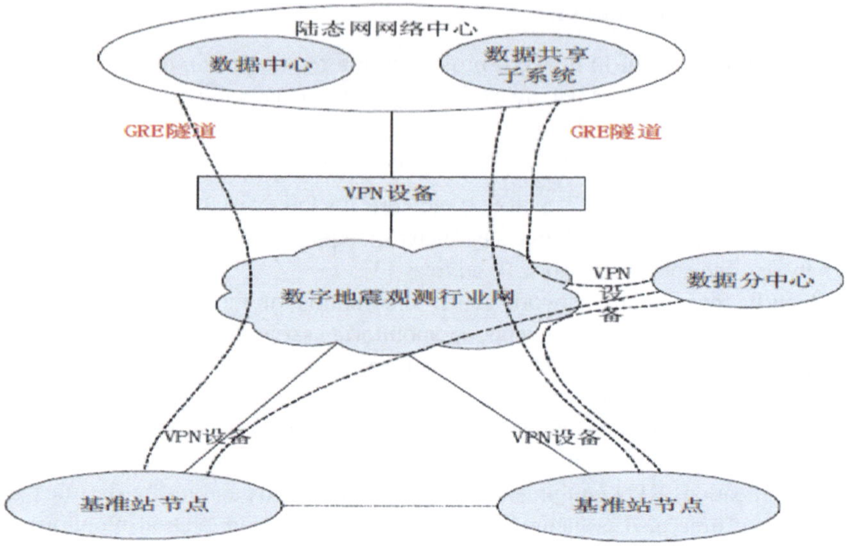

Fig. 1. Network base station channel access land state falls map

earthquake industry area center; Base station node device adopts the RJ45 Ethernet interface mode connection [7]. Base station can be divided by function type and GNSS reference station and gravity datum station, to make it at the same time to realize the remote transmission of data and local storage backup functions, and also for improving the security of system redundancy and data (2), each base station equipped with storage module at the same time.

3 Construction and Operations

ShengJu data center and each base station node communication channel bandwidth is not less than 2 m, each reference station node communication channel can provide no less than 4 m breaking flow ability. Data center unified integration construction is mainly composed of data exchange module, data management and processing module, data acquisition module, monitoring module and database management of Shared services and information release module platform, responsible for the GNSS reference station and the continuous gravity station database platform of management and storage and backup, in order to improve the efficiency and security of the data storage equipment, set up a data store for the professional services; Compressed storage and unified data management and data security backup.

3.1 Implementation Plan

Located in the shandong province lu eight GNSS reference station in the network project and a continuous observation of gravity station by seismological bureau of shandong province is responsible for the communication system of the construction, operation and maintenance. According to the state of lu network project of shandong nine stations (including eight GNSS reference station and 1) gravitational observatory site of geographical distribution and the composition of our province "11th five-year plan" earthquake industry network, for each network node situation analysis, solution to its connection with central site can use two kinds of communication solutions (Fig. 2).

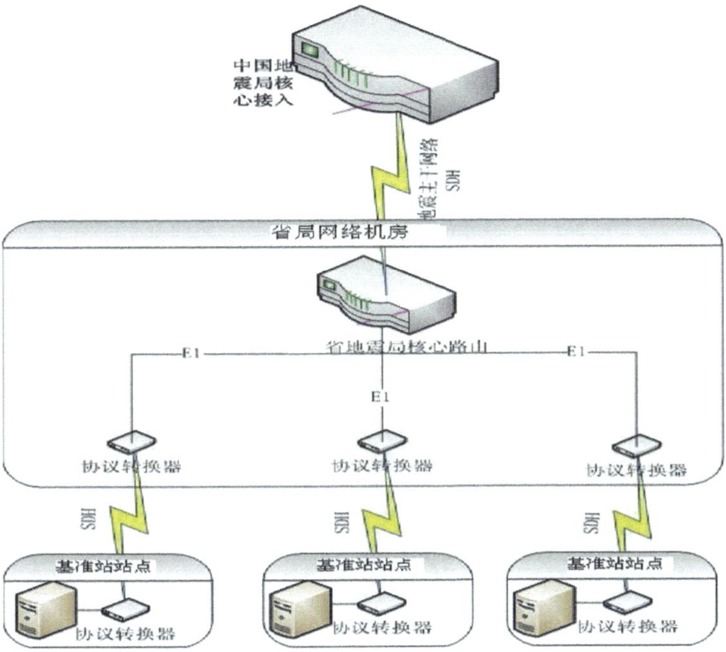

Fig. 2. Data transmission scheme of GNSS reference station network design

SDH is a kind of high efficiency based on time division multiplexing synchronous digital communication technology, shandong nine stations adopt the rent SDH line connection ShengJu center site. Because the SDH has automatic routing selection ability, maintenance, control and management function is strong, unified standards, and convenient circuit, so it as a new generation of ideal transmission system, easier to transmit high rate of business, and can well meet the needs of the rapid development of communication network [6] (Fig. 3).

Shandong nine with ShengJu industry web site is located city seismological bureau established industry dedicated channels, therefore, shandong 8 GNSS reference station and a gravity stations are permitted to use the ground of GNSS and gravity observation data transmission on the virtual private network. Eight GNSS reference station and a continuous gravity observation stations have four station is run by seismological bureau, the other five station is working with the ministries management. Overall, due to reasons such as management system and site conditions, managed by seismological bureau of site operations is significantly better than by the ministries and commissions outside management site (Fig. 4).

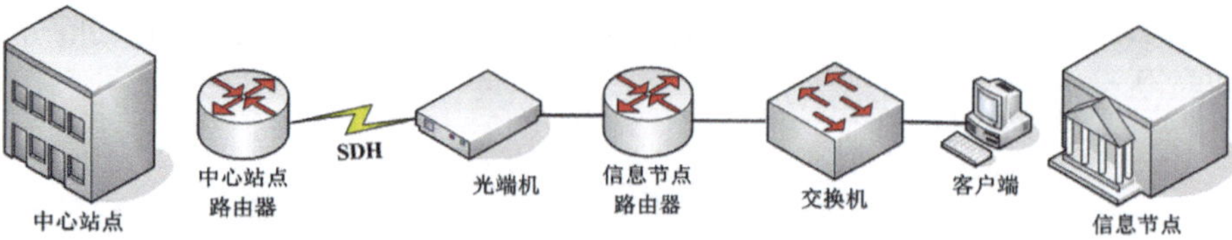

Fig. 3. SDH connection dedicated line access user map

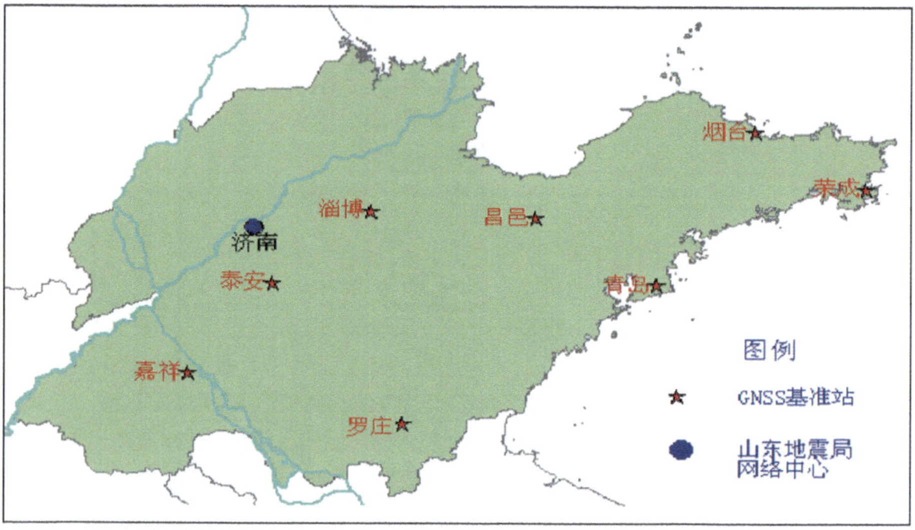

Fig. 4. GNSS reference station distribution in shandong province

3.2 Line Access Seismostation of Base Station Operations

Taian and RongChengTai jinghang canal, currently has 2 m SDH access ShengJu regional centers, and the GNSS reference station distance of the three places are three earthquake monitoring stations, so use by laying optical fiber network access stations to access the earthquake industry way that line access way of seismic stations (Fig. 5).

The three stations due to the earthquake stations on itself, so the operational environment is good, unless there is a massive power failure or operator link failure, few interruptions, the annual average network connectivity rate above 99%. So, the three stations should strive for a higher operating ratio in the future.

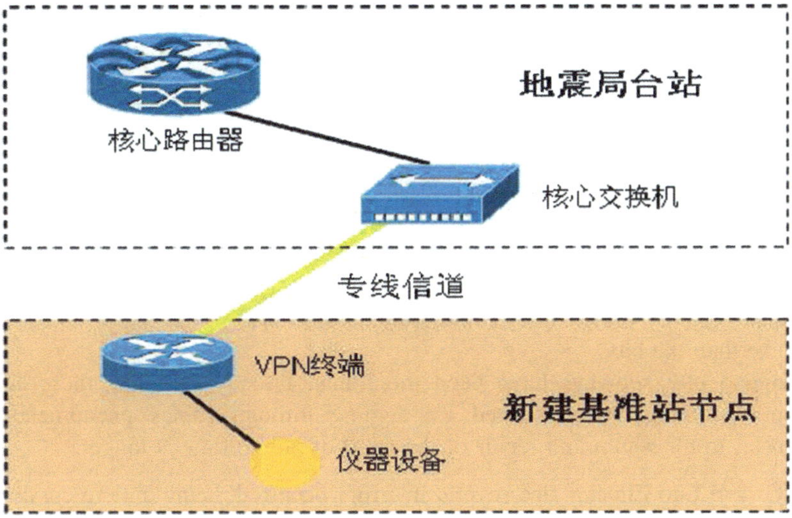

Fig. 5. Base stations through dedicated line access station design

3.3 Earthquake Area Center Access to the Base Station Operations

Because nearbu, yantai, zibo, Luo Zhuang and without seismic station near Qingdao base station and gravity station, so only by using SDH network line access to their home city earthquake industry network regional center city node, and then into the provincial regional center (Fig. 6).

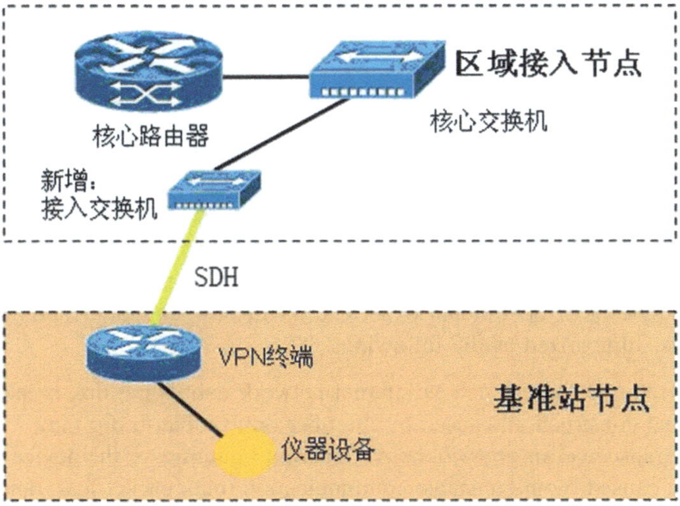

Fig. 6. Reference station access earthquake regional center

Nearbu, although is the site of seismological bureau management, construction site is not on station, but in the wild, so the number of interrupt significantly more than the other three sites, most of the interruption due to power outages, the annual average network connectivity rate can reach 98%. The remaining five managed by the ministries of site connected of the lowest rates (and the lowest in all nine site) is Qingdao GPS station and Qingdao gravity station, the two site is also the most bad treatment after failure. Mainly the following two reasons:

1. The two sites are managed by forces, every once in a while, head of the base station will be replaced, so after failure, not easy to contact, head and units in and out of the limited, gravity station in a remote mountainous area power supply security is far worse than the city.
2. Qingdao city subways have been more than three years time, there are other municipal construction caused by digging through cables phenomenon often appear, troubleshoot and repair of this kind of failure time is longer.

Zibo and Luo Zhuang also several interrupt occurred, being able to contact to the base station, so whether it's a power outage or operators fault caused by interruption can be processed in a timely manner. In all site operations is one of the best the yantai, reason can be summed up in the following two aspects:

1. Yantai seismological bureau network operating ratio is high, performance is good, in the province's 17 cities in 2013 made information science the data comparison "cities and counties in series" first.
2. Yantai site is located in the city bureau of meteorology, due to the similar nature of work units with our seismological bureau, from management to maintenance is more attention on the site, such as fault reasons such as power outages and operators can be processed in a timely manner.

4 Conclusion

Seismological bureau of shandong province of shandong province nine stations (including 8 reference station and a gravity station) the communication system of construction, installation, debugging. Due to the design and construction process are perfect, so all sites in the process of the communication system operations, does not appear larger technical problems. According to my years of state lu network communication system of operational experience, will now communication system common fault is summarized as the following.

1. The operator to causes such as internal network debugging link break;
2. Municipal construction, caused by the fiber optic cable to dig into;
3. Optical transceiver or protocol converter from lightning, or the device itself quality problem caused by the unstable communication (data packet loss, delay too long).
4. A receiver or network storage devices such as causes of data receiving is not successful.

When the station attendant after receive a call from a communication failure, can follow the steps below to troubleshoot:

(1) Check the station power supply is normal, the equipment without electricity.
(2) Such as power supply is normal check router indicator light is normal, do you have any red or orange light is flashing.
(3) Check the router's Ethernet port (so) if there is a loose connection.
(4) Check station (metro) room and optical transceiver at the ends of the base station or protocol converter with or without abnormal blinking, such as bright red or orange light (this can be as the basis of running business some troubleshooting).
(5) If the above checks all without exception, shall file a declaration to the local unicom fault, with unicom technical personnel after some troubleshooting.

In operational process, technical problems are not difficult, more outstanding management problems, especially working with the ministries ops site. Lu state in network operations in the future, how to coordinate relations with outside the ministries and commissions, distinguish responsibility, establish a good contact current is an urgent need to consider to solve the problem.

References

1. Liang, H.: On the construction of environmental monitoring network engineering in mainland China. J. Shaanxi Earthq. **135**(3), 24–26 (2008)
2. He, A., et al.: Application of wireless network based on VPN technology in seismic precursor network. J. Geodesy. Geodyn. **27**(6), 47–51 (2007)
3. Guo, F.R., et al.: Mainland China tectonic environment monitoring network system design. Jiangsu Modern Computer, pp. 36–40 (2014). Preceding (Top)
4. He, B.: IP Virtual Private Network Technology. People's Posts and Telecommunications Press, Beijing (2012)
5. Wu, C., et al.: The construction of the Shandong sub system for the construction of the network monitoring system for environmental monitoring in China. J. Geodesy. Geodyn. **32** (6), 138–140 (2012)
6. Wu, C.: The application of GPON optical fiber technology in Shandong seismic information network. J. Earthq. Disaster Prevent. Technol. **8**(1), 104–111 (2013)
7. Wu, C., et al.: Research on the high availability of Shandong seismic data center. J. Comput. Appl. Res. **32**(12), 19–20 (2015)
8. Yu, D., Peng, L.: When does inferring reputation probability countervail temptation in cooperative behaviors for the Prisoners' Dilemma Game? Chaos Solitons Fractals **78**, 238–244 (2015)
9. Liu, J.: Advanced PID Control and MATLAB Simulation, vol. 2. Electronics Industry Press, Beijing (2003). (in Chinese)
10. Li, J., Ren, Q.: Study on supply air temperature forecast and changing machine dew point for variable air volume system. Build. Energy Environ. **27**(4), 29–32 (2008). (in Chinese)

Research on Management Innovation of Equipment Manufacturing Industry—From the Perspective of Intelligent Production and Service Network

Xing Li, Liyan Jia(✉), and ZiHan Yang

College of Economy and Management, Longdong University, Gansu 745000, China
leebj@189.cn

Abstract. Equipment manufacturing industry is the foundation of domestic economic development. Taking substantial measures to promote innovative development of equipment manufacturing industry is of great significance. Research suggests that new-type industry innovation platform as an important tool for industry innovative development has functions not only including management and information service, but also covering individuation, module integration, and whole-course value chain supply strategy. It can be obtained through structural analysis method and function value analytical method that under the condition of intelligent production and service network, the basic composition of the platform includes decision-making system, information perception and transmission system, intelligent production and service organization system, and innovation support system. Further research has found that the platform will be "embedded" in the industrial organization system to "integrate" with the intelligent production and service network system. Its existence form has diverse characteristics.

Keywords: Equipment manufacturing industry · Intelligent production Service network · Innovation platform

1 Introduction

Currently, despite of the world first place of total output value of China's equipment manufacturing industry, there is still a long way to go before achieving the goal of becoming an equipment manufacturing power. Comparatively speaking, the industry is in urgent need of high-end development, which is highlighted in five major areas, namely, major complete equipment for traditional industry transformation and upgrading, energy-saving, environmental protection and resource development and utilization equipment, new energy vehicles and other advanced transportation equipment, high-end CNC machine tools and Intelligent manufacturing equipment, key basic parts, precision instruments and apparatus and intelligent control system, and special equipment for development requirements of emerging industries. There exist four common problems in equipment manufacturing industry upgrading development field:

J. Abawajy et al. (Eds.): ATCI 2018, AISC 842, pp. 790–795, 2019.
https://doi.org/10.1007/978-3-319-98776-7_93

Firstly, the level of industrial structure is low, the value chain is severely bottom-oriented, which has limited industry function extension; Secondly, low technology content and less high-value-added products have influenced product supply quality; Thirdly, the imbalance of factor structure and the sudden increase of costs have prevented industry from continuously upgrading; Fourthly, a lot of industry development environmental regulations and prevailing "policy opportunism" phenomenon have restricted innovation enthusiasm. Provide "point-to-point" services in place through full-course value chain, namely, analyze users' previous service needs with big data (Fig. 1) and cloud computing technology, and adopt the same service strategy based on the need of the same "point".

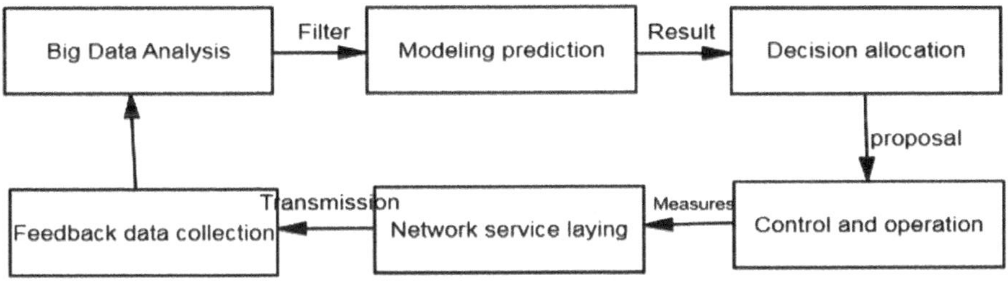

Fig. 1. Service network "point" layout based on big data

2 Research Design

2.1 Innovative Platform Is the Important Component of Industry Innovation System

From the industry level, the industry innovation platform can promote the organic integration of regional innovation system with industry innovation management. Overcoming technical problems by joint means of industrial base research, generic technology research, and cutting-edge technology research can not only meet industry development demands, but can also providing technical support for construction of regional innovation system. From the enterprise level, existing studies have shown that there exist predominant problems during equipment manufacturing companies' innovation process, such as the difficulty to concentrate talents, capital, knowledge and other innovation elements, which results in high innovation costs. Industry innovation platform can provide approaches to alleviate these problems. On one hand, innovative talents have the willingness to realize innovation value, but their implementation methods are often relatively simple. Platforms can be used as carriers to provide innovative communication channels for innovative talents by implementing effective management systems and standards to create an environment of fair competition for innovation; On the other hand, industry innovation platform can play a very prominent role in weakening the phenomenon of redundant construction and improving resource sharing. Meanwhile, in terms of improving the efficiency of market-oriented innovation and reducing corporate technology innovation risk, industry innovation platform has following advantages: (1) Platform participants share innovation risks together and disperse risk action points, which is particularly important for the flourishing SMEs. (2) Relieve the pressure of innovation result transformation generated by technological innovation and knowledge spillovers, form values quickly and efficiently, and promote the development of high-end industries. Statistical data shows that the yield rate of scientific results and technological transfer rate of scientific and technological achievements in China are low, and the environment of scientific and technological achievement transformation is poor. Innovation incubators, industrial alliances, industrial parks and other forms can effectively improve the transformation of scientific and technological achievements. As China's science and technology activities increase year by year, the investment in key talents such as innovative talents and innovation funds has gradually increased, scientific and technological innovation results generated by which urgently need more models to achieve rapid transformation. Practice shows that industry innovation platform has owned great advantages in gathering innovative resources, constructing the innovative achievement incubation mechanism, promoting integration of science and industry.

2.2 From the Perspective of Industry Evolvement

From the perspective of evolutionary economics, industry evolvement can be seen as a dynamic process with disequilibrium change. Under the framework of evolutionary analysis, inner-industry enterprises can be seen as microscopic subjects with different learning abilities and adaptability.

A large amount of empirical studies have shown that the industry will experience three general changes during the evolution process: industrial technology, industrial scale, and industrial organization. According to industry innovation dynamic evolution model and related theories (A-U model) by Abernathy and Utterback, product innovation is the focus in the early industry development stage; in the industry transition period, the emergence of dominant design makes product innovation turn to process innovation; When industry development tends to be stable, innovation activities are jointly dominated by incremental product innovation and process innovation. Under the conditions of intelligent production and service network, industry innovation platform in accordance with the A-U theory, can highlight function enhancement and industrial chain upgrading and regard the priority path for industry upgrading as "process upgrading → product upgrading → function upgrading → industry chain upgrading". According to the "smile theory", the design and marketing services in the industry chain can reflect additional values, with function upgraded to the extension of the upstream and downstream value chains, such as the expansion from processing and manufacturing to the areas of creative, design, marketing, and services. Industry chain upgrading depends on knowledge accumulation and innovation to spills over from a certain industry chain and to crosses over to four-dimension with higher value content, wider enterprise chain, longer supply and demand chain, and wider space chain. Following are its main manifestation forms: (1) Extend from the production area to the field of productive services, broaden the corporate chain by means of knowledge innovation; (2) Integrate supply and demand parties through mergers and reorganizations to lengthen supply and demand chains; (3) Explore smart production and service network portfolio, achieve global production and supply, and expand industry space chain; (4) Embed a global value chain (GVC), establish a national value chain (NVC) strategy in parallel with it, and build a high value chain.

Under the condition of intelligent production and service networks, module integration will become a new feature of the industry innovation platform. The module integration will make the innovation platform reform from "embedment" in industry organization system to "integration" in the intelligent production and service network system transformation. During the process, different forms of industrial innovation platform will be generated.

Several functional modules will carry out innovation activities simultaneously to create an ecosphere of competition and survival of the fittest. The most innovative module will gain opportunities and rights to achieve modular integration and eventually establish innovative products or technologies.

Based on referring to previous research, this paper believes that under the conditions of intelligent production and service network, the safeguard mechanism within the construction period of industry innovation platform includes: policy and market-oriented mechanism, risk mechanism, modularization coupling mechanism, and innovative ecological evolutionary mechanism. (1) Policy and market-oriented mechanisms.

Government-driven and market-driven are the two basic driving forces for platform operation. At different stages of platform growth, their function mechanisms are different. (2) Risk mechanism. It mainly includes platform external environmental risks and platform internal environmental risks, among which platform's external risks include the economic and social development environment, policy regulations, market risks, etc.; platform internal risks include strategic decisions, organizational leadership, and program control. (3) Modularization coupling mechanism. Under the conditions of intelligent production and service networks, in the new-type industry innovation platform with knowledge module as the carrier, self-disciplined and relatively independent modules will efficiently form coupling network clusters through certain rules. (4) Innovative ecological evolutionary mechanism. In the initial stage of platform establishment, the government can promote equipment manufacturing industry upgrading by providing financial support and policy attraction. During the platform development period, the government can purchase from platform enterprises with public funds and provide investment and financing guarantees to platform enterprises to promote equipment manufacturing product up-gradation; at the platform mature stage, market leading industry innovation platform with mature module integration can promote equipment manufacturing industry function upgrading mainly through market resource allocation, platform subject entry and exit, and penalty mechanism; During the platform reform period, the government will get involved again through abatement of fiscal tax and establishment of industrial breakthrough innovation system, to realize full-course value supply with market leading industry innovation platform and to promote the upgrading of the equipment manufacturing industry chain.

3 Conclusion

The industry innovation platform is an important component of the industry innovation system, as well as a significant method to promote the high-end development of equipment manufacturing industry. According to the study in this paper, the government-dominating type, market-driven type, and hybrid-oriented type industry innovation platforms have their own advantages and disadvantages. The platform management modes and operating mechanisms promoted by different forces also have their own advantages. To effectively promote the high-end development of traditional industries, it is necessary to select the construction model of an innovation platform based on the characteristics of the industry and the needs of the industry, and to give the platform innovative functions. Viewing from the process of industry evolvement, there exist a mutually evolutionary relationship between industry cluster and innovation platform governance. The governance model of innovation platform needs to be adjusted when characteristics of industry clusters change. In view of this, it is necessary to maintain certain policy flexibility when confirming the strategic positioning of the innovation platform. It can be considered from the following aspects.

(1) At the level of science and technology policy, guide capital flow and technology flow in categories, integrate scientific research resources, promote the rational distribution of scientific research resources, and restrain excessive concentration of resources.
(2) In terms of industry development, aim at the middle and high-end industries according to domestic and international economic and social development environment, adhere to policy guiding industry direction, promote the traditional industry optimization and updating with high-end industry clusters, meanwhile facilitate the construction of enterprise-centered market-oriented multiple innovation system, and improve industry independent innovation capability.
(3) From the perspective of platform subject constraint, when building innovative carriers in the government-oriented industry innovation platforms, binding documents such as inter-subject responsibility contract must be formulated to balance the interests between the subjects, and prevent cutthroat competition.
(4) Legally, legislative protection of patent achievements independently researched and developed by the industry innovation platform must be guaranteed. Analyze the inapplicability of the existing intellectual property rights system through legal interpretation, induction, and interpretation of legal application possibility. Supplement advice, regulations, and management methods that are helpful for industry innovation platform efficient operation, establish a legal system with the main contents of "science and technology policies, local scientific and technological regulations, and government provision".

References

1. Li, A., Donghua, Z., Zhongyu, X.: Study on service model of collaborative manufacturing industry cluster innovation platform. In: Proceedings of the Tenth Annual Conference on Science and Technology Policy and Management in China, Jilin, Changchun (2014)
2. Li, X., Qu, T., He, J.M., Hou, B.J.: Seek excellence on the hospital quality management competition. Basic Clin. Pharmacol. Toxicol. **119**(S4), HHME16-H24 (2016)
3. Communication Promoters Group of the Industry-Science e-search Alliance. Securing the future of german manufacturing industry-recommendations for implementing the strategic initiative INDUSTRIE 4 Final report of the Industrie 4.0 Working Group (2013)
4. Wahlster, W.: Industry 4, from the internet of things to smart factories, 3rd European Summit on future internet (2012)
5. Li, X., Qiu, W., Zhang, P.: Discussion on service innovation of cultural and creative park: cultural value perspective. Macroecon. Manag. **S1**, 34–36 (2017)
6. Yu, D., Peng, L.: When does inferring reputation probability countervail temptation in cooperative behaviors for the Prisoners' Dilemma Game? Chaos Solitons Fractals **78**, 238–244 (2015)
7. Vuong, J.L.: A semantics-based routing scheme for grid resource discovery. In: E-Science: First International Conference on E-Science and GridComputing, pp. 58–70, 90 (2005)
8. Wang, Q.: Automatic control system regulation scheme of textile air conditioning. Sci. Technol. Inf. **22**, 395–396 (2010). (in Chinese)
9. Liu, J.: Advanced PID Control and MATLAB Simulation, vol. 2. Electronics Industry Press, Beijing (2003). (in Chinese)
10. Li, J., Ren, Q.: Study on supply air temperature forecast and changing machine dew point for variable air volume system. Build. Energy Environ. **27**(4), 29–32 (2008). (in Chinese)

The Reform and Research on Training Program and Curriculum System of Computer Network Engineering Professionals in the Big Data Era

Hui Wang, Gang Liu(✉), Yizhuo Liu, and Mao Lu

College of Computer Science and Engineering,
Changchun University of Technology, Changchun, China
liugang@ccut.edu.cn

Abstract. With the advent of the big data era, in order to meet the needs of science and technology, in light of the shortcomings of the traditional computer network engineering professional talent training program and curriculum system, the computer network engineering professional talent training program and curriculum teaching reform implementation method have been studied and formulated. The curriculum system suitable for the training of network engineering professionals was formulated. It consists of a basic curriculum group, a professional curriculum group and professional development courses, the innovative teaching methods and an online teaching platform were introduced, including pre-class, in-class and after-class three. At last, the assessment and implementation plans for theoretical courses in network engineering were given. According to the development of big data, continuous innovation and reform, computer network professionals can better adapt to the requirements of the big data era and cultivate high-tech talents that meet the needs of the times and jobs.

Keywords: Big data · Computer network · Curriculum system

1 Introduction

With the growing development of mobile Internet, cloud computing and big data, big data technology has been applied to all aspects of the computer. The development of big data technology has a close relationship with computer networks. Therefore, the talent training program and training objectives for the direction of computer network engineering in local engineering colleges are adjusted in time can better meet the requirements of the development of the big data era. At present, the goal of computer network engineering students in the context of big data is to cultivate relevant knowledge about computer networks, mobile Internet, cloud computing, and big data, so as to ensure that students have strong practical ability and high computer network professionalism. With the general management capabilities under the computer cloud platform and the application and development capabilities of the computer network, they can become computer-skilled talents engaged in network design, network maintenance, network infrastructure development, computer cloud platform management, and big data application storage in computer network engineering.

J. Abawajy et al. (Eds.): ATCI 2018, AISC 842, pp. 796–801, 2019.
https://doi.org/10.1007/978-3-319-98776-7_94

2 Current Status of Talent Training Programs and Curriculum Systems for Computer Network Engineering in Local Engineering Colleges

2.1 The Single Mode of Professional Talents Training Scheme and Social Disconnect

Most of the current talent training programs for computer network engineering cannot match the needs of the talent market. In the big data era, the rise of big data centers and related industries will increase the demand for big data technology talents. However, in the current talent training program for computer network technology, the application of big data technology has not yet been demonstrated. It is important to note that most network talent training programs are still traditional network maintenance, security, and applications.

2.2 The Teaching Methods of Some Professional Teachers Are Behind

At present, the teaching process of computer network engineering courses in many local engineering colleges is mainly based on arranging teacher demonstrations or arranging confirmatory experiments. Theoretical teaching hours are far greater than practical ones. Confirmatory teaching can no longer meet the needs of big data era, resulting in students lacking good communication skills and innovative spirit. Innovative teaching should come into being.

2.3 Inadequate Teaching Environment and Teaching Equipment

The storage and analysis of big data research needs the support of computer cloud platform technology. At present, there are already several large data centers, large-capacity storage devices, and high-configured servers in the country. These devices require considerable capital investment. However, the amount of funds for scientific research and teaching in universities is limited, so there are few schools that have established a comprehensive platform for big data and cloud computing experiments. As a result, the training and improvement of students' ability to use big data is limited, and is divorced teaching content from actual needs.

3 Computer Network Engineering Professional Training Program and Curriculum Teaching Reform Implementation

3.1 Computer Network Curriculum Group Construction and Optimization

Optimize and integrate the current curriculum system in the talent training program, eliminate the setting of some basic courses or reduce the time of theoretical teaching, add some elective courses or professional limited selection courses related to big data processing, analysis and innovation and entrepreneurship. At the same time, it selects textbooks suitable for professional development and selects excellent textbooks for scientific calculations, algorithmic language design, big data processing analysis, and the cultivation of students' innovation and entrepreneurship. At the same time, practical training and practical courses can be increased. In addition, courses such as big data analysis, data visualization, data mining and other optional courses and practical application courses related to big data analysis, innovation and entrepreneurship are appropriately added. In the rational setting and optimization of courses, combined with teaching practice, some of the elaborate and profound basic theories involved in some courses can be described briefly, focusing on or adding knowledge about application and practice, and strengthening students' ability to apply innovation, big data processing analysis awareness and creative entrepreneurial thinking.

For the network engineering professional course system, the structure should be composed of basic course group, professional curriculum group, and professional development course in the order from the lower grade to the higher grade, and the corresponding course group is required for each stage. Basic courses are the basis of professional courses and applied courses. The focus is on cultivating knowledge acquisition capabilities and professional basic capabilities of students. The course focuses on students' cultivation of professional basic knowledge building capabilities. Professional courses focus on the development of students' developmental and creative abilities. They directly face the group of jobs.

At present, it focuses on three directions: website design direction, network engineering direction and the Internet of things direction. The website design direction is for website design and management, webpage production, network marketing, etc. The direction of network engineering is for network management and maintenance, system management, comprehensive wiring and testing; Internet of things engineering direction is for building and transportation intelligence, transportation and logistics, energy saving and environmental protection. The structure of the network engineering professional training model is shown in Fig. 1.

3.2 The Introduction of Innovative Teaching Methods

Under the background of big data, the classroom teaching of computer network engineering majors in local engineering colleges can introduce some innovative teaching methods based on their own characteristics. For example: Micro-curriculum, which provides students with detailed, mobile and new experience of the micro-class, it has short lesson times, fine content, modularization, contextualized features; The MOOC, massive open online course is linked to the micro-curricular and micro-curriculum system, and asks students to answer questions every three to five minutes. After the questions are answered correctly, they can continue to watch the follow-up courses content.

With the introduction of micro-teaching and the bisection teaching, based on the learning environment and features of network engineering students, the reformed teaching methods are introduced. As shown in Fig. 2, the online teaching and learning platform consists of three parts: pre-teaching, in-teaching and post-teaching.

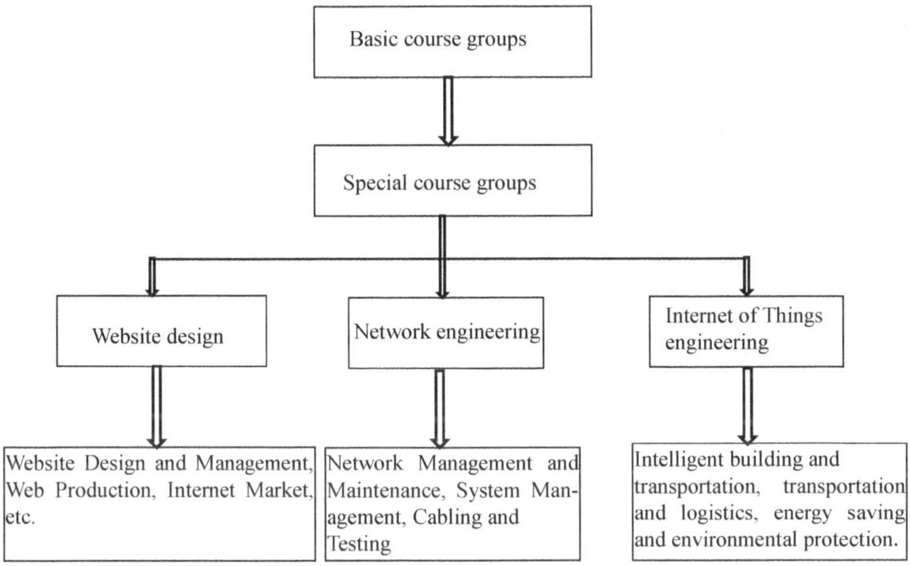

Fig. 1. The structure of the network engineering professional training model

First of all, pre-teaching is to download learning tasks online platform, students can actively learn according to the task, learning methods include watching online video, learning materials or instructions provided by the teacher. When students ask questions online, the teacher will focus on replying to answer questions of students; The students are given concentration response based on their mastery of the pre-teaching tasks, and the theoretical part can be taught using the MOOC model or the bisection method teaching. The experimental courses can be grouped according to the comprehensive cases and small projects arranged by the teacher. During the post-class, assignments and experimental reports can be submitted online on the platform. The teacher can make the assessment of the online platform.

3.3 Reform and Exploration of Teaching Evaluation and Assessment

Traditional computer network courses are mainly based on the combination of attendance, classroom performance, in-class or after-school homework assignments, mid-term and final exam evaluation. This will inevitably lead to blind spots and limitations. In the era of big data, all behaviors related to curriculum learning are regarded as basic data and then quantified and input into the database. The student's learning trajectory is recorded through related technologies, and the student's learning behavior can be restored to a visualized scene. The basis of the teaching evaluation is the overall data of this learning information, which can objectively demonstrate the complete image of the learning process, so as to achieve multi-dimensional comparison and scientific evaluation. For example, teachers and students can use micro-videos such as WeChat public number, online cloud classroom, teacher space, etc. Teachers can publish micro-videos of the course content before class to achieve pre-class understanding, teaching instruction in class, and application expansion after class.

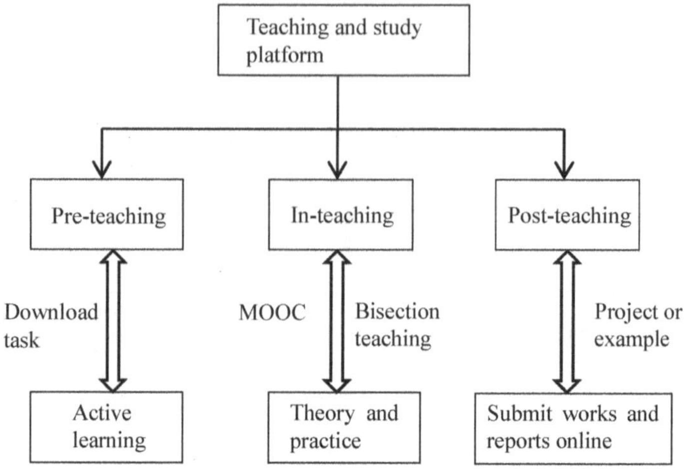

Fig. 2. Frame of the curriculum teaching model.

Table 1 shows the evaluation implementation plan for the theoretical course of network engineering. In accordance with the new talent training program, teaching model after the reform and pre-teaching content assessment on pre-class online teaching platform, regular assignments, regular experiments, and periodical comprehensive assessment methods, students are assessed to increase the pre-study content assessment and reduce the proportion of comprehensive assessments at the end of the period, so as to avoid The appearance of the examination paper as the focus of assessment of the course takes place, and the main process assessment is more

Table 1. Assessment plan for theory courses of network engineering.

Assessment summary	Examination type	Examination content	Assessment form and method	Proportion of results
Professional theory course assessment, including pre-school online learning, homework and final comprehensive assessment	Pre-teaching question implementation	Pre-class online learning on the online teaching platform	Record examination	20%
	Regular assignments and regular experiments	Basic knowledge, language ability and design ability	Submit online or reply in class	20%
	Comprehensive assessment at the end of the period	Report content or final examination paper	Design results or final examination papers	60%

important. According with the new talent training program and teaching mode, performance of students has been improved, and their practical ability has been enhanced. This is more conducive to future learning. The new talent training program and teaching model have improved student.

4 Conclusion

The development and application of mobile internet, cloud computing and big data technology brought new requirements and challenges to the training of computer network engineering professionals in local engineering colleges. In order to avoid the disconnection from the market demand, the computer network professional talent training program and curriculum system should follow the advanced computer technology such as mobile Internet, cloud computing and big data, and continue to carry out innovative research on network professional courses, and pay attention to practice, the development of curriculum and instructional materials, timely adjustment of talent training programs and syllabus and teaching plan to meet the actual requirements of social jobs.

Acknowledgements. The authors acknowledge the project of the Higher Education Institute of Jilin (Grant: JGJX2017D57).

References

1. Zhao, W.T., Xia, D.L.: The construction of course group and knowledge system of cloud computing based on the development of professional ability. Comput. Knowl. Technol. **12** (4), 167–169 (2016)
2. Bao, A.H., Chen, W.W., Liu, P., Wen, A., Wang, Z.: Content system construction and practice of cloud computing courses. Comput. Eng. Sci. **36**(A2), 42–45 (2014)
3. Lu, W.W.: The reform of teaching model based on learning analysis technology under the background of big data. New Curric. **15**(28), 169 (2015)
4. Li, L.J., Liao, X.H., Tan, L.: Study on teaching reform of C language programming course based on SPOC. Comput. Educ. **32**(1), 74–76 (2016)
5. Su, X.H., Zhao, L.L., Ye, L.: Exploration and Practice of Blended Teaching Based on MOOC+SPOC. China Univ. Teach. **12**(7), 60–65 (2015)
6. Yu, Y., Zhou, G.H., Li, H.Y.: The reform of C language programming practice teaching in CDIO mode. Comput. Educ. **10**(2), 122–126 (2016)
7. Yu, D., Peng, L.: When does inferring reputation probability countervail temptation in cooperative behaviors for the Prisoners' Dilemma Game? Chaos Solitons Fractals **78**, 238–244 (2015)
8. Liu, J.: Advanced PID Control and MATLAB Simulation, edn. 2. Electronics Industry Press, Beijing (2003). (in Chinese)
9. Li, J., Ren, Q.: Study on supply air temperature forecast and changing machine dew point for variable air volume system. Build. Energy Environ. **27**(4), 29–32 (2008). (in Chinese)
10. Clay Richardson, W., Avondolio, D.: The Java high class weaves a distance: JDK 5. Sci. Technol. Book Rev. **3**, 17–18 (2006)

Non-fragile H_∞ Controller Design for Linear Discrete-System with Quantization

Liang Zhang[1(✉)], Gang Luo[2], and Haoyan Zhang[1]

[1] Bohai University, Jinzhou, China
md18638@126.com
[2] Jinzhou Normal College, Jinzhou, China

Abstract. This paper considers the problem of non-fragile state feedback H_∞ control for liner discrete-time system. Attention is focused on the design of a no fragile state feedback H_∞ controller to mitigate the effect of controller parameters variation such that the closed-loop control system is asymptotically stable with a prescribed H_∞ performance. A logarithmic quantize was introduced to deal with quantitative problem in this paper. Finally, the Linear matrix inequality (LMI) was obtained by Lyapunov stability theory.

Keywords: Discrete-time system · Non-fragile · H_∞ state feedback controller LMI · Quantify

1 Introduction

H_∞ Control has been received the extensive attention in the past twenty years because in dealing with the advantages of structured uncertain system. It also has been successfully used in the robot, aerospace, electric system and other fields [1–3]. To closed-loop system parameters, especially, the stability is very sensitive to the uncertainty and change of controller, which will be called fragile or non-resilient [4]. Each controller may have the vulnerability for which the traditional optimal and robust controller design method through the example by literature [5]. When the controller parameters slightly offset will lead to the stability of the closed-loop system was damaged or performance degradation. Recently, the control of the fragile caused attention [6, 7]. At the same time, we consider the system which didn't have the H_∞ control and fragile [8, 9].

Based on the existing literature, a new kind of a fragile state feedback controller design method was put forward and on processing the signals of quantification quantify problem were introduced in this paper. A new controller of the linear matrix inequality (LMI) design conditions can be obtained by a new control law and the Lyapunov function which has a non-fragile and at the same time satisfy the performance index.

J. Abawajy et al. (Eds.): ATCI 2018, AISC 842, pp. 802–807, 2019.
https://doi.org/10.1007/978-3-319-98776-7_95

2 Problem Formulation

Consider the linear discrete-system equation for the state-space as follow.

$$\begin{aligned} x(k+1) &= Ax(k) + Bu(k) + Ew(k) \\ z(k) &= Cx(k) + Du(k) + Fw(k) \\ x(0) &= 0 \end{aligned} \tag{1}$$

Where $x(k) \in \mathbb{R}^n$, $u(k) \in \mathbb{R}^m$, $z(k) \in \mathbb{R}^q$, $w(k) \in \mathbb{R}^v$ is the system state variables, $u(k) \in \mathbb{R}^m$ is the system input variables, the controlled output variables and disturbance variables. $A \in \mathbb{R}^{n\times n}$, $B \in \mathbb{R}^{n\times m}$, $E \in \mathbb{R}^{n\times v}$, $C \in \mathbb{R}^{q\times n}$, $D \in \mathbb{R}^{q\times m}$, $F \in \mathbb{R}^{q\times v}$ is appropriate dimension coefficient matrix.

Consider the logarithmic quantizer form as follow, $q(x(t)) = [q_1(x_1(t)), q_2(x_2(t)), \ldots, q_n(x_n(t)),]^T$ the logarithmic quantize q is made up of $q_1, q_2, \ldots, q_n$, q_i is the components for x, $q_1, q_2, \ldots, q_n$ are independent of one another, quantizer q_i are defined as follow.

$$q_i(x_i(t)) = \begin{cases} q_i(l), & \frac{1}{1+\sigma_i} q_i(l) < x_i(t) \\ 0, & x_i(t) = 0 \\ -q_i(-x_i(t)), & x_i(t) < 0 \end{cases}$$

Quantize q_i satisfy the following form.

$$Q_i = \{\pm q_i(l) : q_i(l) = \sigma_i^l q_i(0), l = 0, \pm 1, \pm 2, \ldots\} \cup \{0\}, q_i(0) > 0, 0 < \sigma_i < 1$$

Each quantize for q_i is logarithmic where σ_i is the quantitation density of each q_i, we have $\Upsilon_i = \frac{1-\sigma_i}{1+\sigma_i}, i = 1, 2, \ldots, n$, $\sigma = [\sigma_1, \sigma_2, \ldots, \sigma_n]^T$, $q(x(t)) = (I + \bar{\Delta}x(k))$.

Consider the stochastic gain which have non-fragile State feedback control laws as following.

$$u(k) = (K + \Delta K)(I + \bar{\Delta}(t))x(k) \tag{2}$$

where $\Delta K^T \Delta K \le I, \bar{\Delta}^T(t)\bar{\Delta}(t) \le I$.

The closed-loop system can obtained as following.

$$\begin{aligned} x(k+1) &= \tilde{A}x(k) + Ew(k) \\ z(k) &= \tilde{C}x(k) + Fw(k) \\ x(0) &= 0 \end{aligned} \tag{3}$$

Where $\tilde{A} = A + B(K + \Delta K)(I + \bar{\Delta}(t))$, $\tilde{C} = C + D(K + \Delta K)(I + \bar{\Delta}(t))$.

The purpose of this paper is to design the controller (2), which makes the following two conditions are established.

(1) When $w(k) = 0$, the closed-loop system is asymptotic stability.

(2) $\gamma > 0$ was the given performance indexes for H_∞ control, where $x(0) = 0$, $w(k) \in L_2[0, \infty)$, the system output $z(k)$ to satisfy $\sum_{k=0}^{\infty} \|z(k)\|_2 < \gamma^2 \sum_{k=0}^{\infty} \|w(k)\|_2$.

The following lemma will be used in the derivation.

Lemma 1: For the given matrix Γ, Λ [10], M is the uncertainties and Φ is the symmetric matrix. If the inequality holds.

$$\Phi + \Gamma M \Lambda + \Lambda^T M^T \Gamma^T < 0$$

There need a constant scalar $\varepsilon > 0$ to satisfy the matrix as follow.

$$\Phi + \varepsilon^{-1} \Gamma \Gamma + \varepsilon \Lambda^T \Lambda < 0$$

3 Main Results

Theorem 1: Consider the closed-loop system (3), for the given scalar λ and $\gamma > 0$, if there is a matrix $P > 0$ and K makes the following conditions is established.

$$\begin{bmatrix} -P & 0 & A^T + (I + \bar{\Delta}(t))^T (K + \Delta K)^T B^T & C^T + (I + \bar{\Delta}(t))^T (K + \Delta K)^T D^T \\ * & -\gamma^2 I & E^T & F^T \\ * & * & -P^{-1} & 0 \\ * & * & * & -I \end{bmatrix} < 0 \tag{4}$$

So the closed-loop system (3) is asymptotically stable and the H_∞ control performance index $\gamma > 0$ can be guaranteed.

Proof: Choose a Lyapunov function as following.

$$V(x(k)) = x^T(k) P x(k) \tag{5}$$

define

$$J = \Delta V(\mathrm{k}) + z^T(\mathrm{k}) z(\mathrm{k}) - \gamma^2 w^T(\mathrm{k}) w(\mathrm{k})$$

The following formula can be obtained by the calculation as following.

$$\begin{aligned} J &= x^T(\mathrm{k}+1) P x(\mathrm{k}+1) - x^T(\mathrm{k}) P x(\mathrm{k}) + [\tilde{C} x(\mathrm{k}) + F w(\mathrm{k})]^T \times [\tilde{C} x(\mathrm{k}) + F w(\mathrm{k})] - \gamma^2 w^T(\mathrm{k}) w(\mathrm{k}) \\ &= x^T(\mathrm{k}) \tilde{A}^T P \tilde{A} x(\mathrm{k}) + x^T(\mathrm{k}) \tilde{A}^T P E w(\mathrm{k}) + w^T(\mathrm{k}) E^T P \tilde{A} x(\mathrm{k}) + w^T(\mathrm{k}) E^T P E w(\mathrm{k}) + x^T(\mathrm{k}) \tilde{C}^T C x(\mathrm{k}) \\ &\quad + x^T(\mathrm{k}) \tilde{C}^T F w(\mathrm{k}) + w^T(\mathrm{k}) F^T \tilde{C} x(\mathrm{k}) + w^T(\mathrm{k}) F^T F w(\mathrm{k}) - \gamma^2 w^T(\mathrm{k}) w(\mathrm{k}) = \xi^T(\mathrm{k}) \Xi \xi(\mathrm{k}) \end{aligned} \tag{6}$$

where $\xi(k) = \begin{bmatrix} x^T(k) & w^T(k) \end{bmatrix}^T$,

$$\Xi = \begin{bmatrix} \tilde{A}^T P \tilde{A} - P + \tilde{C}^T \tilde{C} & \tilde{A}^T P E + \tilde{C}^T F \\ E^T P \tilde{A} + F^T \tilde{C} & E^T P E + F^T F - \gamma^2 I \end{bmatrix} \quad (7)$$

When $\Xi < 0$, it is easy to prove the system is asymptotically stable by $w(k) = 0$, if $\Delta V(\mathrm{k}) + z^T(\mathrm{k}) z(\mathrm{k}) - \gamma^2 w^T(\mathrm{k}) w(\mathrm{k})$, need $\Xi < 0$, $J < 0$ and $\Xi < 0$ can be obtained by using Schur on (4). $\mathrm{J} = \sum_{K=0}^{\infty} V(x(k)) + \sum_{K=0}^{\infty} \|z(k)\|_2 - \gamma^2 \sum_{K=0}^{\infty} \|w(k)\|_2 < 0$ can be obtained sum over by J, because $\sum_{K=0}^{\infty} V(x(k)) > 0$, so we can obtained $\sum_{k=0}^{\infty} \|z(k)\|_2 < \gamma^2 \sum_{k=0}^{\infty} \|w(k)\|_2$.

Matlab can't solve directly because there has the nonlinear term and error in Theorem 1. So Theorem 2 can be obtained through congruent transformation and S-procedure technique.

Theorem 2: Consider the closed-loop system (3), for the given scalar ε, δ and $\gamma > 0$. If there is a matrix $\bar{P} > 0$ and $\bar{K}$ makes the following LMI is established.

$$\begin{bmatrix} -\bar{P} & 0 & \bar{P}A^T + \bar{K}B^T & \bar{P}C^T + \bar{K}D^T & \bar{P} & 0 & \bar{P} & 0 \\ * & -\gamma^2 I & E^T & F^T & 0 & 0 & 0 & 0 \\ * & * & -\bar{P} & 0 & 0 & B & 0 & B\bar{K}^T \\ * & * & * & -I & 0 & D & 0 & D\bar{K}^T \\ * & * & * & * & -\varepsilon I & 0 & 0 & \bar{P} \\ * & * & * & * & * & -\varepsilon^{-1} I & 0 & 0 \\ * & * & * & * & * & * & -\delta I & 0 \\ * & * & * & * & * & * & * & \delta^{-1} I - 2\delta^{-1} \bar{P} \end{bmatrix} < 0 \quad (8)$$

So the closed-loop system (3) is asymptotically stable and the H_∞ control performance index $\gamma > 0$ can be guaranteed. At the same time, the quantized error can be removed also the controller is non-fragile.

Where $P^{-1} = \bar{P}, P^{-1}K^T = \bar{K}KP^{-1} = \bar{K}^T, K = \bar{K}^T \bar{P}^{-1}, P = \bar{P}^{-1}$.

Proof: The controller gain error can be removed through (4) by using Schur and Lemma 1, and obtain the following expressions.

$$\begin{bmatrix} -P & 0 & A^T + (I + \bar{\Delta}(t))^T K^T B^T & C^T + (I + \bar{\Delta}(t))^T K^T D^T & (I + \bar{\Delta}(t))^T & 0 \\ * & -\gamma^2 I & E^T & F^T & 0 & 0 \\ * & * & -P^{-1} & 0 & 0 & B \\ * & * & * & -I & 0 & D \\ * & * & * & * & -\varepsilon I & 0 \\ * & * & * & * & * & -\varepsilon^{-1} I \end{bmatrix} < 0 \quad (9)$$

The quantized error can be removed through (9) by using Schur and Lemma 1, obtained (10).

$$\begin{bmatrix} -P & 0 & A^T + K^T B^T & C^T K^T D^T & I & 0 & I & 0 \\ * & -\gamma^2 I & E^T & F^T & 0 & 0 & 0 & 0 \\ * & * & -P^{-1} & 0 & 0 & B & 0 & BK \\ * & * & * & -I & 0 & D & 0 & DK \\ * & * & * & * & -\varepsilon I & 0 & 0 & I \\ * & * & * & * & * & -\varepsilon^{-1} I & 0 & 0 \\ * & * & * & * & * & * & -\delta I & 0 \\ * & * & * & * & * & * & * & -\delta^{-1} I \end{bmatrix} < 0 \quad (10)$$

Formula (11) is left multiply by J^T, right multiply by J, where $J = diag(P^{-1}, I, I, I, I, I, I, P^{-1})$. We can obtained $-P^{-1}\delta^{-1}P^{-1}I < \delta^{-1}I - 2\delta^{-1}P^{-1}$ from $(I - P^{-1})(I - P^{-1}) \geq 0$, obtain the following expressions.

$$\begin{bmatrix} -P^{-1} & 0 & P^{-1}A^T + P^{-1}K^T B^T & P^{-1}C^T + P^{-1}K^T D^T & P^{-1} & 0 & P^{-1} & 0 \\ * & -\gamma^2 I & E^T & F^T & 0 & 0 & 0 & 0 \\ * & * & -P^{-1} & 0 & 0 & B & 0 & BKP^{-1} \\ * & * & * & -I & 0 & D & 0 & DKP^{-1} \\ * & * & * & * & -\varepsilon I & 0 & 0 & P^{-1} \\ * & * & * & * & * & -\varepsilon^{-1} I & 0 & 0 \\ * & * & * & * & * & * & -\delta I & 0 \\ * & * & * & * & * & * & * & -P^{-1}\delta^{-1}P^{-1} I \end{bmatrix} < 0 \quad (11)$$

Suppose $P^{-1} = \bar{P}, P^{-1}K^T = \bar{K}$, and obtain $KP^{-1} = \bar{K}^T, K = \bar{K}^T\bar{P}^{-1}, P = \bar{P}^{-1}$, so (8) can be proofed.

4 Conclusions

This paper considers the problem of no fragile state feedback H_∞ control for liner discrete-time system. By the new control law, a sufficient condition for the design of a non-fragile state feedback controller, which based on linear matrix inequality (LMI) discrete-time system is given by using the Lyapunov stability theory. Made in the controller parameter changes or uncertain will still be able to guarantee the stability of closed-loop control system gradually and with the given H_∞ performance index, a logarithmic quantize was introduced to deal with quantitative problem. This paper only considering the gain of control devices by additive, multiplicative gain of non-fragile state feedback controller problem will be researched in the future.

References

1. Chang, X.H.: Robust Output Feedback H-Infinity Control and Filtering for Uncertain Linear Systems. Springer, Heidelberg (2014)
2. Zhao, X.D., Shi, P., Zheng, X.L.: Fuzzy adaptive control design and discretization for a class of nonlinear uncertain systems. IEEE Trans. Cybernet. **46**(6), 1476–1483 (2016)

3. Zhao, X., Yin, Y., Niu, B.: Stabilization for a class of switched nonlinear systems with novel average dwell time switching by T-S fuzzy modeling. IEEE Trans. Cybernet. **46**(8), 1952–1957 (2016)
4. Li, Z.C., Fei, Z.Y., Karimi, H.R.: New results on stability analysis and stabilization of time-delay continuous Markovian jump systems with partially known rates matrix. Int. J. Robust Nonlinear Control **26**(9), 1873–1887 (2016)
5. Yin, S., Gao, H.J., Qiu, J.B.: Descriptor reduced-order sliding mode observers design for switched systems with sensor and actuator faults. Automatica **76**, 282–292 (2017)
6. Hu, Q.L., Jiang, B.Y., Zhang, Y.M.: Observer-based output feedback attitude stabilization for spacecraft with finite-time convergence. IEEE Trans. Control Syst. Technol. (2017)
7. Kao, Y.G., Xie, J., Wang, C.H.: A sliding mode approach to H∞ non-fragile observer-based control design for uncertain Markovian neutral-type stochastic systems. Automatica **52**, 218–226 (2015)
8. Hu, Q.L., Jiang, B.Y.: Continuous finite-time attitude control for rigid spacecraft based on angular velocity observer. IEEE Trans. Aerosp. Electron. Syst. (2017)
9. Wang, Z.D., Ding, D., Shu, H.S.: Non-fragile H_∞ control with randomly occurring gain variations, distributed delays and channel fadings. IET Control Theory Appl. **9**(2), 222–231 (2014)
10. Petersen, I.R.: A stabilization algorithm for a class of uncertain linear systems. Syst. Control Lett. **8**(4), 351–357 (1987)

Military Text Named Entity Recognition Based on Deep Learning

Jianlong Li(✉) and Panqing Wang

Simulation Training Center, Army Engineering University, Zhengzhou, China
360689880@qq.com

Abstract. A named entity recognition method based on recursive neural network model is proposed for named entities in texts of military field. This method mainly uses the recursive neural network model to record the structural advantages of the context. At the same time, it adds the word feature vector to the input layer to complete the named entity recognition in the military text. Compared with the results of traditional methods, this method can complete the task of naming entity recognition in texts of the military field, and finally the F-value on test corpus can reach 81.32%.

Keywords: Deep neural network · Named entity recognition
Recurrent neural network

1 Introduction

1.1 A Subsection Sample

With the rapid development of information technology, a large amount of information is presented in the military's database. The military desperately needs the technology to handle these massive amounts of information. Named entity recognition is one of the key technologies in many information processing technologies [1]. Recent years, named entity recognition had a lot of research work in the field of science. It is increasingly being taken seriously by people. Usually, named entities include organization name, person's name, place name, time word, numeral, etc. Other types of named entities in different application areas need to be defined according to specific needs. Named entity recognition is an important technology for information extraction, question answering systems, grammar analysis, and machine translation. It has received very high attention at home and abroad [2]. Therefore, the study of named entity recognition has important theoretical and practical significance. Named entity recognition is not well resolved in military text. According to the characteristics of military texts, we proposes a named entity recognition method for military texts. It solved the problem of named entity recognition in texts in the military field.

J. Abawajy et al. (Eds.): ATCI 2018, AISC 842, pp. 808–816, 2019.
https://doi.org/10.1007/978-3-319-98776-7_96

2 Related Overview and Model Introduction

2.1 Military Text Overview

Military text is the collective name for all kinds of texts related to the military. The format of military texts is relatively uniform and fixed, with accurate, concise and confidential contents. Some military texts still have phenomena such as colloquialization and networking. There are a large number of military named entities in military texts. They have a unique grammatical structure and their composition is relatively complex. This article focuses on the research of named entity recognition for military texts. Military naming entity refers to the collective name of military-related named entities in military texts. It mainly includes ranks of military and military personnel, names of military agencies, army numbers, military equipment names, and military names.

2.2 Recurrent Neural Networks

In the field of deep learning, traditional multi-layer perceptrons have excellent performance and have achieved many successes. It has created records on many different tasks, such as hand-written digital recognition and target classification. Even so, most experts will reach a consensus: the functions that MLP can achieve are still quite limited [3]. The reason is that the human brain has astonishing computational functions, and the classification task is only a small part of it. We can not only identify individual cases, but also analyze the overall logical sequence between the input information. These information sequences are rich in a large amount of content, information has a complex temporal relationship with each other, and the information length is various. This is something the traditional MLP cannot solve. RNN officially came into being to solve this sequence problem [5]. The key point is that the hidden state of the current network will retain the previous input information and be used as the output of the current network. The structure of RNN is different from MLP, The input layer and the signal from the hidden layer of the previous element in the sequence act together on the current hidden layer, as shown in the Fig. 1:

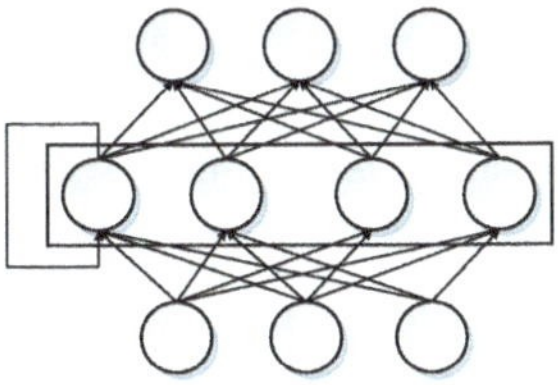

Fig. 1. Recurrent neural network structure diagram

RNN is usually used to describe the dynamic time behavior sequence, and the state is cyclically transmitted in its own network, and it can accept more extensive sequence sequence structure input. Unlike feed-forward deep neural networks, RNN places more emphasis on the feedback of the network [4]. Because there is a connection between the current state and the past state, the RNN can have a certain memory function.

3 Military Named Entity Recognition Based on Recurrent Neural Network

The main machine learning method adopted for the named entity recognition task is the conditional random field model. However, when the conditional random field model is used for entity recognition, the feature must be extracted first through the given training corpora. However, the identification of useful features will depend more on expert knowledge in specific areas, and not all features will have a positive effect on the outcome of predictions. Therefore, a large number of experiments are needed to design the template and filter out the optimal feature subset. However, when the corpus changes, it takes a high labor cost to repeat the previous feature engineering. In order to solve these problems, an efficient named entity recognition system was established. This article is based on the recursive neural network model for military domain text named entity recognition. Its structure is shown in Fig. 2. The first layer is the input layer, which is mainly responsible for vector mapping of the input words or words, and the input layer combines these vectors in order to obtain the input vector. The second layer is the hidden layer, through the recursive neural network to find useful, more advanced feature information. The third layer is the output layer, which is mainly responsible for the search of the optimal tag sequence for the input sentence.

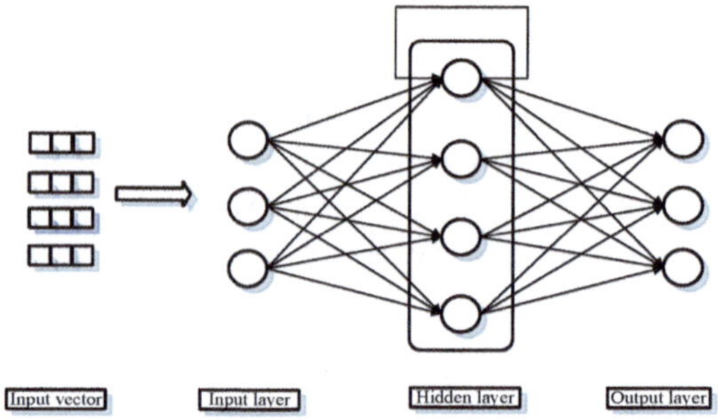

Fig. 2. Named entity recognition structure diagram based on recurrent neural network model

3.1 Representation of Distributed Vectors

In a named entity recognition task, each word after textual preprocessing is usually used as a feature, called a word feature. Word features can reflect the basic information of the original character set more completely. In addition, part of speech as a feature input model can significantly improve the performance of named entity recognition, called part-of-speech features. If traditional feature expressions are used for word features and part-of-speech features, then any two words or any two parts of speech are isolated and unrelated. A large number of studies have shown that distributed representation of the features of words can effectively improve the performance of natural language processing systems. A distributed representation of word features and part of speech features, Express each word or each part of speech tag as a low-dimensional real number vector, then the Euclidean distance between any two words or any two part of speech tags will be closer. The distributed representation of word features can solve problems such as dimension disasters and local generalization limitations in machine learning. Compared with the traditional feature representation, the intrinsic relationship between input data can be explored more deeply, and its internal syntax and semantic similarity can be captured.

In the sequence labeling task, the mark result corresponding to a word often has a more important semantic relationship with the local language environment in which it is located. Therefore, this paper chooses to combine the word vector and the word vector as input vectors. The joint vector of words is shown in the Fig. 3. The vector we choose to combine words is mainly based on the following two reasons: (1) Word-based NER cannot understand the meaning of Chinese characters. (2) NER based on words strongly depends on the quality of the word segmentation results and the completeness of the dictionary. If the word segmentation errors occur, it will directly affect the NER results. Therefore, we choose to combine NER with words. It can be effectively complementary to improve the recognition performance of the NER system. We use the Skip-gram model in the Word2vec tool to conduct unsupervised training of military corpora to generate word vectors and word vectors.

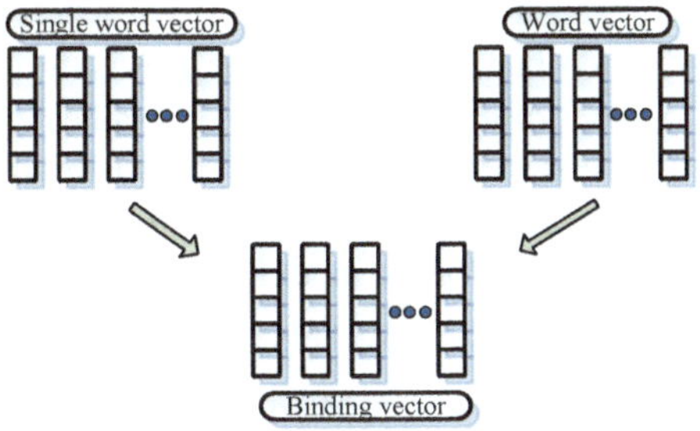

Fig. 3. Joint vector structure diagram

3.2 Recurrent Neural Network Model Structure

Due to the recursive connection of the RNN, it enables the recording and preservation of a wider range of contextual information. Getting stronger associative information, so that it has increasingly attracted the attention and interest of researchers. This article uses the Jordan-RNN, which is a typical RNN structure, as shown in the figure (Fig. 4).

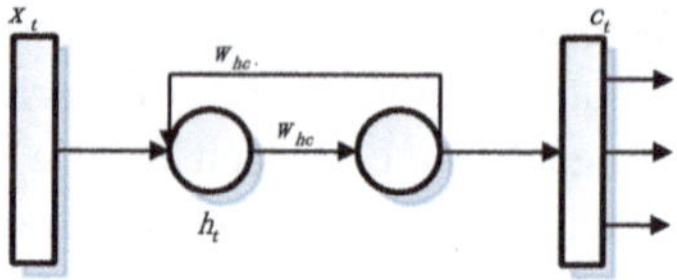

Fig. 4. Jordan-RNN structure diagram

In the Jordan-RNN, suppose the input vector at time t is x_t, When calculating the corresponding hidden layer representation, the predicted probability vector c_{t-1} of the output layer of the previous node is input as part of the calculated hidden layer, as shown in Eq. 3.1. We use h_t as the new eigenvector at time t to get the score of each marker after linear summation and input it into the sortermax classifier to normalize to get the probability c_t of each marker, as shown in formula 3.2. In the training process, if the correct marker at time t is y_t, the loss function can be obtained according to formula 3.3, and then the parameters of the entire neural network are updated in reverse using the stochastic gradient descent algorithm.

$$h_t = f(w_{xh} \cdot x_t + w_{hc\cdot} \cdot c_{t-1} + b_h) \tag{3.1}$$

$$c_t = g(w_{hc} \cdot h_t + b_c) \tag{3.2}$$

$$loss = -\log[c_t]_{y_t} \tag{3.3}$$

f represents the activation function, this article uses sigmoid function, the value of the function range (0, 1), the formula shown in 3.4. w_{xh} indicates the connection parameter between the input layer and the hidden layer, $w_{hc\cdot}$ indicates the connection parameter between the output layer and the hidden layer, and b_h indicates the offset vector. g denotes the probability prediction function for the input flag, as shown in Eq. 3.5. w_{hc} represents the parameter matrix connecting the hidden layer and the output layer, and b_c represents the offset vector.

$$f = \frac{1}{1 + e^{-x}} \tag{3.4}$$

$$g = \frac{e^{o_k}}{\sum_{k=1}^{K} e^{o_k}} \tag{3.5}$$

The difference from the single hidden layer MLP is that the hidden layer of the RNN is used as a memory element of the network to store information. The computation of the hidden layer also mainly depends on the current input and the hidden layer result of the previous word, and the hidden layer unit of each step contains the previous history information. Historical input information can be represented by hidden layer abstraction, and as the RNN unfolds, it continuously integrates new input textual information and retains and disseminates it layer by layer. Therefore, RNN has a structural advantage for named entity recognition tasks that require access to long-span context information.

3.3 Parameter Training

The essence of recurrent neural network training is to calculate the unknown parameters in the model on the training corpus. The unknown parameters mainly include several parameters of the hidden layer. It also contains the transformation matrix and bias matrix in the logistic regression layer. Training a neural network requires the use of a back propagation algorithm and a stochastic gradient descent algorithm. Specific parameters training process is:

(1) All parameters of the network are initialized randomly, including hidden layer and logistic layer parameters.
(2) Randomly pick a training sample. First forward propagation, the output of the hidden layer is passed to the logistic regression layer. Map the extracted highest level features to the corresponding tag information. The data mark value is used for supervised training of the model, and the connection weights are continuously adjusted to reduce the probability error between the model's target prediction mark and the actual mark.
(3) Reverse propagation. Calculate the conceptual error between the target prediction marker and the actual marker in the forward propagation process, propagate the error from the logistic regression layer to the hidden layer, and continuously adjust the hidden layer parameters.

4 Military Text Entity Identification Process

For military texts, the detailed entity identification process is shown in Fig. 5. It mainly includes data crawling, data processing, data processing, vocabulary distributed feature training, neural network model training, entity recognition and recognition result extraction. First of all, crawl the relevant documents of the military legacy system. The preprocessing of military texts includes the filtering of special symbols, manual annotation, word segmentation, and other operations. Corpus is divided into two parts: training set and test set. Put the training set into the model for training, and then use the training parameters to test the model recognition effect.

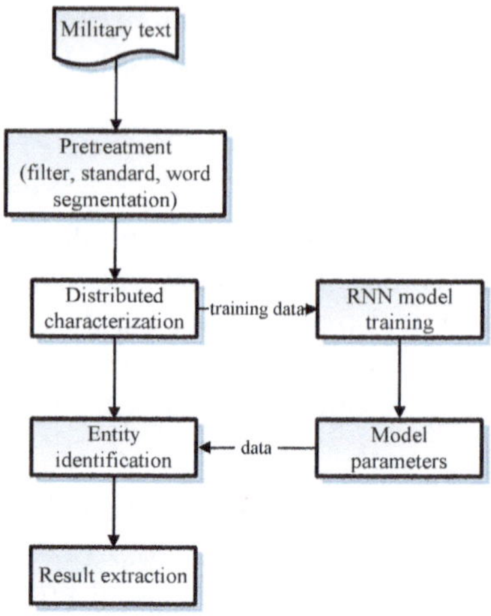

Fig. 5. Military text entity identification flow chart

5 Experimental Results and Analysis

5.1 Experiment Settings

This article uses Python to implement the relevant code and complete the construction and training of the entire model. As there is no comparatively authoritative and unified military corpus, the military text library is constructed using manual collection. The experimental sample includes a total of 142312 words. We chose 80% of the experimental samples as training corpus and the remaining 20% as test corpus. All vector representations use 200 dimensions, the input layer has a context window size of 5, the hidden layer number is 1, the gradient reduction learning rate is 0.001, and the number of iterations is 60. This article sets three groups of comparative experiments for military texts.

Using word features and part-of-speech features to build feature sets and training conditional random fields models on training corpus. The obtained conditional random field model was used to identify the named entity on the test corpus, and the recognition result was evaluated and recorded as CRF.

Unsupervised learning distributed representations of linguistic features and distributed representations of POS features on training corpusBuilding and training recurrent neural network models using distributed representations of word features and distributed representations of part-of-speech features, Named entity recognition was performed on the test corpus and the results were evaluated as Jordan-RNN1.

Unsupervised learning of the distributed representation of the features of words, the distributed representation of part of speech features, and the distributed representation of word features on the training corpus. Construct and train recurrent neural network models using input distributed vector representations, Named entity recognition was performed on the test corpus and the results were evaluated as Jordan-RNN2.

5.2 Experimental Results

To facilitate the comparison of the effectiveness of different methods of named entity recognition, we typically use recall, accuracy, and F-values to measure the performance of named entity recognition. Calculate the correct rate, recall rate, and F-value for the three groups of results. The results are shown in the Table 1.

Table 1. Experimental results

Experiment	Correct rate (%)	Recall rate (%)	F-value (%)
CRF	73.54	73.96	73.75
Jordan-RNN1	78.22	77.92	78.07
Jordan-RNN2	81.03	81.62	81.32

Experiment 1 uses the conditional random field model to identify named entities. In order to reduce the manual development of features, this experiment uses only word features and part-of-speech features for feature set construction. The accuracy, recall rate, and F-number were 73.54, 73.96, and 73.75.

In experiment 2, the recursive neural network was used to identify named entities. The accuracy, recall rate, and F value were 78.22, 77.92, and 78.07. Compared with the traditional CRF model, the average accuracy rate, recall rate, and F value of the named entity recognition method based on the recursive neural network model increased by 6.3%, 5.3%, and 5.9%.

In Experiment 3, based on the named entity recognition of recursive neural network model, the word feature vector was added in the input layer. The accuracy rate, recall rate and F value were 81.03, 81.62, 81.32. Compared with the result of experiment 2, the correct rate, recall rate, and F value of experiment 3 increased by 3.7%, 4.5%, and 3.4%.

In summary, comparing the results of Experiment 1 and Experiment 2, the named entity recognition method based on recursive neural network not only reduces the workload of manual formulation of features in the traditional methods, but also achieves the same effect as the conditional random field model in the naming entity recognition task of military texts. Comparing the results of experiment 2 and experiment 3, in named entity recognition based on recurrent neural network model, the combined input vectors of words effectively improve the effect of named entity recognition.

6 Conclusion

In this paper, we use pre-trained distributed vectors as input instead of manually formulating features to reduce labor costs. And the RNN can record the context of the advantages of the structure applied to the military domain name entity recognition task, while the input layer to increase the word feature vector, to a certain extent, improve the system of named entity recognition effect. The correct rate, recall rate, and F-value obtained last were 81.03, 81.62, and 81.32. Judging from the experimental results, the method adopted in this paper can well complete the identification of military text named entities.

How to further improve the effectiveness of named entity recognition is the focus of the next step.

References

1. Ouyang, L., Tian, Y., Tang, H., et al.: Chinese named entity recognition based on B-LSTM neural network with additional features. BMC Med. Inf. Decis. Mak. **17**, 67 (2017)
2. Ouyang, L., Tian, Y., Tang, H., et al.: Chinese named entity recognition based on B-LSTM neural network with additional features. In: Security, Privacy, and Anonymity in Computation, Communication, and Storage (2017)
3. Dernoncourt, F., Ji, Y.L., Szolovits, P.: NeuroNER: an easy-to-use program for named-entity recognition based on neural networks (2017)
4. Wei, Q., Chen, T., Xu, R., et al.: Disease named entity recognition by combining conditional random fields and bidirectional recurrent neural networks. Database J. Biol. Databases Curation (2016)
5. Park, G., Lee, H.G., Kim, H.: Named entity recognition model based on neural networks using parts of speech probability and gazetteer features. J. Comput. Theor. Nanosci. **23**(10), 9530–9533 (2017)
6. Nam, J.: Semi-supervised neural networks for nested named entity recognition (2014)
7. Chapram: Arabic named entity recognition using artificial neural network. J. Comput. Sci. **8**(8), 1285–1293 (2012)
8. Alfred, R., Leong, L.C,, On, C.K., et al.: A rule-based named-entity recognition for malay articles. In: Advanced Data Mining and Applications, pp. 288–299. Springer, Heidelberg (2013)
9. Yu, D., Peng, L.: When does inferring reputation probability countervail temptation in cooperative behaviors for the prisoners' dilemma game? Chaos Solitons Fractals **78**, 238–244 (2015)
10. Ge, X.: On how to improve the effectiveness of college english translation teaching. Engl. Teach. **18**, 5 (2015)

Differential Expression Analysis Based on Expression Data of Multiple Platforms

Ming Zheng(✉) and Mugui Zhuo

Guangxi Colleges and Universities Key Laboratory of Professional Software Technology, Wuzhou University, Wuzhou, China
370505375@qq.com

Abstract. Differential expression analysis is one of the basic objectives of the study of the transcriptional group, which plays an important role in revealing the function and regulation of genes, as well as the fluctuation of selective shear. Many researches have adopted many measuring platforms to get more accurate results. Research shows that fusion of multi-platform expression data can improve the accuracy and reliability of differential expression analysis. Most of the existing differential detection studies of fusion multi-platform expression data are mainly fusion of various types of gene chip expression data, and less RNA-Seq expression data are considered. Moreover, the existing methods ignore many useful information, such as measurement errors and the volatility produced by repeated experiments. In view of the problems existing in the existing methods, this paper proposes a differential detection model mpDE (multi-platform Differential Expression model), which combines multi-platform transcriptome data to find differentially expressed genes and isomers. The model integrates the technical measurement errors of the expression data and the expression level of the different experimental platforms into the model, and takes into account the fluctuation of the same platform in different conditions of biological repetition or technical repetition, thus improving the accuracy of the difference detection. In this paper, mpDE is applied to two human multi-platform expression data sets for differential expression detection. It involves the traditional 3' chip, exon arrays, HTA2.0 chip, and RNA-Seq four common transcriptional level measurement platform for Affymetrix.

Keywords: Data integration · Multi-platforms · RNA-Seq
Differential expression analysis · Gene expression data

1 Objective

The differential expression analysis of transcriptional groups is to identify genes or isomers of differential expression (DE) by analyzing the expression data of transcriptional groups under different conditions, which plays an important role in revealing the regulation of gene regulation or the change of selective shear of genes. With the development of sequencing technology, the cost of sequencing is decreasing. Many studies choose to use a variety of expression level measurement platform for the analysis of the transcriptional group, so the public database (such as NCBI GEO [1]) has accumulated

J. Abawajy et al. (Eds.): ATCI 2018, AISC 842, pp. 817–823, 2019.
https://doi.org/10.1007/978-3-319-98776-7_97

a large number of multi-platform expression data. Research shows that fusion of multi-platform data can increase the performance and reliability of differential detection [2].

Gene chip and RNA-Seq based on high throughput sequencing are the two high-throughput gene expression level measurement techniques in transcriptome research. Gene chip experiments usually hybridize probes with fluorescent labels and probes on chips. After completing the hybridization, we processed and scanned the chip to get the fluorescent image of the chip, and finally got the continuous image gray value, that is, the original data of the gene chip. The size of the gray value of the probe represents the concentration of the corresponding RNA fragments in the sample. In the RNA-Seq experiment, the samples were transformed into cDNA, and then sequenced after the bridge amplification. Then the sequences of the same base sequence (read) in the same sequence were sequenced, and the expression level of the gene was reflected by the reading segment count on the gene reference sequence. The expression level of corresponding gene can be calculated by using the gray value of probe produced by gene chip experiment and the discrete reading segment counted by RNA-Seq experiment. Gene chip appears early, repeated experiments and mature data analysis methods, but because of its inherent technical limitations, the background noise is large and difficult to eliminate. RNA-Seq sequencing technology has little background noise and is more sensitive to low expressed genes or isomers, but many experiments have less repeated experiments, which reduces the reliability of the experimental results. Therefore, it is a challenge to systematically analyze the multi-platform expression data and obtain valuable information from the features of these platforms, which has also prompted many fusion and clustering algorithms to emerge.

At present, the research methods of multi-platform differential expression analysis are mainly divided into two categories. One is to use meta-analysis [3] for differential detection. By merging and summarizing a number of similar research results, the total amount of samples can be increased, and the accuracy of detection and the consistency of statistical analysis results are improved. The two is to merge the multi-platform expression data into an expression matrix through the method of data conversion [4], and then use the corresponding differential detection method to carry out the subsequent differential expression analysis.

The existing multi-platform expression data analysis studies mainly through the fusion of different platforms to obtain the gene or isomer expression level, combined with the existing differential detection methods to find different genes or isomers. Most of the methods are based on the point estimates to calculate the expression level of genes or isomers, ignoring a lot of useful potential letters. Interest, for example, the technical measurement error of expression level under different experimental platform. In addition, due to sample collection, experimental environment and experimental platform, the same measurement platform will produce the volatility of technology repetition or biological repetition under different experimental conditions, which reduces the accuracy of differential expression analysis, but the existing multi-platform differential expression analysis methods do not take these problems into account. Aiming at these problems, this paper proposes mpDE (multi-platform Differential Expression model). The model is based on the multi-layer Bayesian model, and combines the characteristics of different gene expression level measurement platform to model the technical measurement error of the expression level and expression level of multiple platforms. At the same time, the volatility of the same measurement platform in different experimental conditions is considered. The probability distribution of expression value of gene or isomer based on platform and repeated experiments.

2 Method

The original data format of different experimental platforms is different. The goal of this paper is to design a general multi-platform expression data fusion model. So for different experimental data, we first use the original data processing methods of each platform to obtain the probability distribution of the continuous gene expression level, and then design the multi-layer Bayesian model. The probability distribution of gene expression level obtained by different platforms was fused to get the final platform independent gene expression level. It is usually assumed that the true logarithmic expression of each gene or isomer obeys the Gauss distribution, so the mpDE model assumes that the logarithmic expression of each gene or isomer in the observed data D obeys the distribution of the logarithm of the logarithmic expression of each gene or isomer. In order to improve the computational efficiency, the model adds a hidden variable to represent the real logarithmic expression value [5]. The mpDE model uses EM algorithm and variation method to estimate the posterior distribution of the parameters in the model, and uses PPLR [6] standard to identify differentially expressed genes or isomers.

The mpDE model combines the EM algorithm with the variation method to optimize the implicit variables in the model, optimizes the lower bounds of the function through iteration, and finally estimates that the posterior distribution of the posterior distribution model parameters is independent, then the posterior distribution Q of the implicit variables in the variation EM algorithm [7] is decomposed into the lower formula.

$$\mathrm{Q}(\theta) = \prod_{r}^{R}\prod_{p}^{P}\prod_{c}^{C} Q(x_{rpc}) \prod_{\mathrm{p}}^{\mathrm{P}}\prod_{\mathrm{c}}^{\mathrm{C}} Q(u_{pc}) \prod_{r}^{R}\prod_{c}^{C} Q(\lambda_{pc}) \prod_{c}^{C} Q(u_{c}) Q(\lambda) \tag{1}$$

When the EM algorithm converges, the mpDE approximates the posterior distribution by formula (1). It is known that each condition is independent of each other. It is easy to get the mean and variance of the expression values of the platform independent genes or isomers under each condition.

Since the algorithm is approximated to some posteriori, the results are distributed, and we can evaluate the genes or isomers that are significantly different in two experimental conditions. If the data set contains two experimental conditions, the expression level of conditional C1 is higher than that of conditional C2, then the probability of the differential expression distribution is called the PPLR value by the calculated probability value. The PPLR value is similar to the significant probability value p-value in the traditional statistical test. A threshold is set up to determine the genes or isomers of the difference in the rules, and the PPLR value can be converted to p-value [8] through the formula. The descent rule differences in genes or isoforms, can also be identified, only need to be changed $(-\infty, 0)$.

3 Results

3.1 MAQC Data Set

MAQC [9] comes from the Microarray Quality Control (MAQC) of the US Food and Drug Administration (US Food and Drug Administration), which is used to evaluate the gene expression level of high quality samples under different sequencing platforms. The project includes three parts: MAQC_I, MAQC_II and MAQC_III (Sequencing Quality Control, SEQC). MAQC dataset is widely used as a standard dataset to evaluate the performance of different methods on different platforms. The MAQC data set mainly uses two human samples, which are universal reference RNA (Universal Human Reference RNA, UHRR) and brain reference RNA (Human Brain Reference RNA). The experiment includes four experimental conditions. The experimental A contains only UHRR samples, and the experimental B contains only HBRR samples. The samples of experimental C and experimental D are composed of 3:1 and 1:3 in the samples of UHRR and HBRR, respectively. In this paper, we use the data set at the experimental conditions A and B, Human GenomeU133 Plus 2 Array (HGU133), Human Exon 1.0ST Array (HuEx1.0ST) and three kinds of chip data, in which 5 technical duplicates are contained in each experimental condition. 0 data, each experimental condition contains 3 technology duplicated chips.

3.2 SEQC Data Set

SEQC [10] is the latest experimental data set provided by MAQC project. Its main purpose is to evaluate RNA-Seq technology. Both SEQC and MAQC data sets use two samples of UHRR and HBRR, and the proportion of 3:1 and 1:3 is mixed to form samples C and sample D, but the sequencing platform used in the SEQC experiment is Illumina HiSeq 2000, which increases the sequencing depth and sequencing length. This paper mainly uses SEQC data concentration experimental conditions A and B RAN-Seq double terminal sequencing data, including 8 sequencing channels under each experimental condition, which can be repeated as 8 techniques. In addition, the SEQC data set provides more than 20000 isomers tested by the RT-qPCR (Reverse Transcription qPCR) experiment, and the LFC method is also used to screen the RT-qPCR data. The results can be used as the real difference between the isomers. The differential expression results of 1002 single isomers with high confidence were obtained through screening, including 643 DE isomers and 359 non-DE isomers.

3.3 Experiment

This section first validates the reasonableness of the mpDE model, and then verifies the accuracy of the mpDE model gene expression level calculation, and compares it with the single platform expression level calculation method. Then the performance of mpDE model difference detection was verified from two aspects of gene and isomer differential expression analysis, and compared with single platform differential detection results,

MRS and limma combination and RSP model. Finally, we evaluate the time and space complexity of the mpDE model, and compare it with MRS and RSP.

In this paper, a qRT-PCR verifying gene ENSG00000152583 was randomly selected from the MAQC data set. The expression of the gene was significantly different by qRT-PCR test. The expression level of the gene was low under the UHRR condition, and the expression level was higher under the HBRR condition. In this paper, the rationality of the model is verified by comparing the level distribution of the original expression level measured by different platforms of the gene and the expression level distribution of the platform independent (u_{PC}) and the platform independent (u_c). The results of the experiment are shown in Fig. 1. The HGU133, HuEx1.0ST, HTA2.0, and RNA-Seq in Fig. 1 represent the original expression data of the four platforms obtained by the gene expression level (expressed in thin lines); the same line with the four platforms, but the rough lines are the level of the platform related gene expression under each condition obtained by mpDE computing, that is, PC The posterior distribution, the thick line said mpDE obtained a platform independent gene expression level, C selection of the posterior distribution. It can be seen from the graph that the mpDE model excludes the influence of technical measurement error and repeated experiment volatility. By using the variation EM algorithm, the more reasonable platform correlation and the level distribution of the platform independent gene expression are calculated, so as to achieve the purpose of integrating the multi-platform experimental data. Finally, the platform independent expression level distribution can be applied to subsequent analysis of various gene expression, such as finding differential genes and clustering. The following part is applied to differential expression detection in transcriptome.

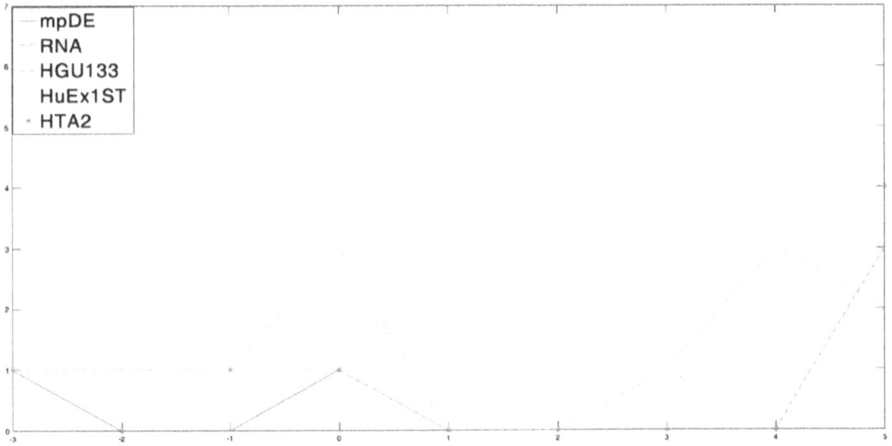

Fig. 1. The expression level distribution of gene ENSG00000152583 under the two conditions of UHRR and HBRR mpDE model.

The experimental results are shown in Fig. 1. Under each condition, the highest correlation coefficient is marked by bold and underline. As can be seen from Fig. 1, the accuracy of gene expression levels measured by different platforms is different, and

RNA-Seq is the most accurate in single platform measurement, while the accuracy of the mpDE model results is significantly higher than that of the three gene chips, and is higher than that of RNA-Seq. It can be seen that the fusion of multi-platform expression data can improve the accuracy of expression level calculation, and help to improve the accuracy of subsequent differential expression detection.

4 Conclusion

In this paper, a multi-platform transcriptional difference detection model (mpDE) is proposed to solve the existing problems in the differential expression analysis of multi-platform expression data. The model uses multi-platform to express the data, while considering the fluctuation of the technical measurement error and the same platform in different experimental conditions. It can better simulate the probability distribution of the true expression value of the gene or isomer, and improve the accuracy of the difference detection. The platform independent expression value obtained by the model not only can be applied to differential expression detection, but also can be applied to the analysis of other multi-platform data such as clustering, control network analysis and so on. Therefore, the application of mpDE model is very promising.

As a whole, the multi-platform data fusion model can significantly improve the accuracy of single platform differential gene detection, but it is difficult to calculate the isomer expression level in the differential isomer detection, especially the gene chip uses a limited number of probes to detect the abundance of the corresponding transcript. The diversity of selective shear is a great challenge to the performance of the technology, which results in the low accuracy of the detection of the isomers. In addition, the original data analysis method relies on the annotation information of genes, isomers and probes, and the improvement of annotation information has a certain influence on the calculation accuracy of the expression level. In contrast, the RNA-Seq technology has no restriction on the fixed probe, and the breadth and depth of the transcriptional detection are far beyond the gene chip, so it has higher accuracy in the selective shear study. Because of the limitations of gene chip technology in the measurement of isomer expression level, the advantages of multi-platform data fusion model are not significant compared to the RNA-Seq single platform in the aspect of isomer differential expression analysis. With the improvement of the chip experiment technology and the continuous improvement of the annotation information, the accuracy of the expression level of the isomers will also be improved. The accuracy of the multi-platform data fusion model in the heterogeneity analysis of isomers will be improved. Finally, finding differentially expressed genes is a basic goal of gene expression measurement experiments. Our current work mainly focuses on this aspect. On this basis, we will extend these research ideas to the detection of differential gene sets in future work, and improve the effectiveness of data analysis methods from the perspective of system biology.

Acknowledgements. This work was supported by grants from The National Natural Science Foundation of China (No. 61502343), the Guangxi Natural Science Foundation (Nos. 2017GXNSFAA198148, 2015GXNSFBA139262) foundation of Wuzhou University (No.

2017B001), Guangxi Colleges and Universities Key Laboratory of Professional Software Technology, Wuzhou University.

References

1. Barrett, T., Wilhite, S.E., Ledoux, P., et al.: NCBI GEO: archive for functional genomics data sets-update. Nucleic Acids Res. **41**(D1), D991–D995 (2013)
2. Kim, J., Patel, K., Jung, H., et al.: AnyExpress: integrated toolkit for analysis of cross-platform gene expression data using a fast interval matching algorithm. BMC Bioinf. **12**, 14 (2011)
3. Moradifard, S., Hoseinbeyki, M., Ganji, S.M., et al.: Analysis of microRNA and gene expression profiles in Alzheimer's disease: a meta-analysis approach. Sci. Rep. **8**, 17 (2018)
4. Heider, A., Alt, R.: virtualArray: a R/bioconductor package to merge raw data from different microarray platforms. BMC Bioinformatics **14**, 10 (2013)
5. Jeanmougin, M., de Reynies, A., Marisa, L., et al.: Should we abandon the t-test in the analysis of gene expression microarray data: a comparison of variance modeling strategies. PLoS ONE **5**(9), 9 (2010)
6. Liu, Y., Chiaromonte, F., Ross, H., et al.: Error correction and statistical analyses for intra-host comparisons of feline immunodeficiency virus diversity from high-throughput sequencing data. BMC Bioinf. **16**, 14 (2015)
7. Liu, Z., Song, Y.Q., Xie, C.H., et al.: Clustering gene expression data analysis using an improved EM algorithm based on multivariate elliptical contoured mixture models. Optik **125**(21), 6388–6394 (2014)
8. Brulard, C., Chibon, F.: Robust gene expression signature is not merely a significant P value. Eur. J. Cancer **49**(12), 2771–2773 (2013)
9. Schuierer, S., Carbone, W., Knehr, J., et al.: A comprehensive assessment of RNA-seq protocols for degraded and low-quantity samples. BMC Genom. **18**, 13 (2017)
10. Xu, J.S., Gong, B.S., Wu, L.H., et al.: Comprehensive assessments of RNA-seq by the SEQC consortium: FDA-led efforts advance precision medicine. Pharmaceutics **8**(1), 8 (2016)

Computer Network Security Technology Research

Shen Wei[1,2(✉)]

[1] The First Affiliated Hospital, Jilin University, Changchun, China
abyswabysw@163.com
[2] College of Computer Science and Technology, Beihua University, Jilin, China

Abstract. With the development of the information technology, computer information system has played an important role in modern society. As long as any break appears in computer information system, it takes a huge effect in whole society. Based on the analysis of network information safety factors, and then put forward common computer network information safety protection strategy, and the development of the network information safety was prospected and formed the network information safety protection system. There is an important, direct and realistic significance in research about computer system security. With the development of computer and communication technology, the computer net will penetrate every field of our social life as an important method of exchanging information. So, recognizing the frangibility and potential threatens of the net and some existent security problems objectively, taking effectively security strategy and ensuring the security of the net information are the things that every country, and group and every person must envisage. From the basic concept and existent problem of the net information security, some digital security technologies, such as information cryptography technology, digital abstract, digital signature, digital envelope and digital credentials have been analyzed in the paper.

Keywords: Computer network · Network information security · Security threats
Preventive measures

1 Introduction

Along with the rapid development of the computer technology and Internet [1], accelerated the process of global informatization, internet is being entered into thousands of households [2]. It plays an indispensable role in our production and life [3]. Network and network users grow exponentially, the scale of network is becoming larger and larger. Therefore, more and more network malicious acts occurred, how to effectively guarantee the normal operation of the network has already become a key question [4]. In order to prevent the destroy to the network and the loss of information resources caused by malicious damage, network managers urgently need to know the current state and development trend accurately, discover the attacks and harm behaviors timely [5]. Hereford, how to ensure the information is true, complete, effectively, legally transceiver, the problem of network security has become a hotspot [6]. As is shown in Fig. 1.

J. Abawajy et al. (Eds.): ATCI 2018, AISC 842, pp. 824–831, 2019.
https://doi.org/10.1007/978-3-319-98776-7_98

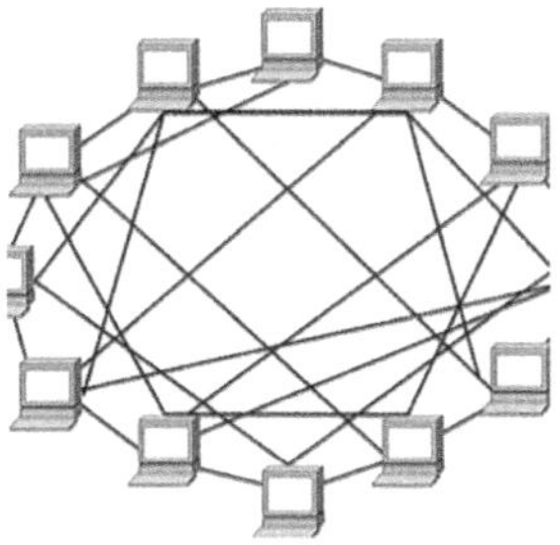

Fig. 1. Network topology structure

2 The Computer Network Security

Network security is refers to the protection of the hardware and software system of network system, and the data, factors from accidental or malicious destruction, so that the whole system can reliably run continuously, provide uninterrupted service [7]. In essence, the network security is to ensure the safety of information. Broadly, as long as involving the network information authenticity, confidentiality, integrity, availability, and controllable theory category of technology is the study of network security. Network security is a comprehensive discipline, usually involves the password techniques, information security technology, applied mathematics, computer science and other fields [8].

2.1 The Concept of Network Original Design

Early in the network development, people are more valued, emphasizes the network convenience and availability, but ignores the security of the network. At that time, the use of the network is limited to a small range, so the network security has not been seriously. But with the development of science and technology, and human needs updating, network to overcome the geographical restrictions, the distribution in the small scope of branch network, connect successfully, and all over the world [9]. At this point in transmission due to the correlation of network sensitivity information, information security is threatened by the great. At the same time, all kinds of network products are developed on the basis of network protocol, there are more or less are unsafe.

2.2 The Openness of the Network

The basic characteristics of the open Internet, the Internet is built on the basis of free and open [10]. Today's network resources is increasingly widely used, at the same time also provides an opportunity for hacking into. The contradiction between resource sharing and network security and edgy.

2.3 The Administrative Control Variation of the Network

Along with the network all over the world, private network as demand also inevitably linked with external public sites directly or indirectly [11]. Just due to network, the correlation of network attack one of the most weak link in the chain can cause the collapse of the whole security system. More complicated so that the operation of the network environment, controllable sharply reduced, the network security also becomes poor.

3 The Threats to Network Security

Online there are all kinds of security threats, common natural disasters, the vulnerability of network system itself, user error, artificial malicious attacks, computer viruses, spam and spyware, computer crime, etc. As is shown in Fig. 2.

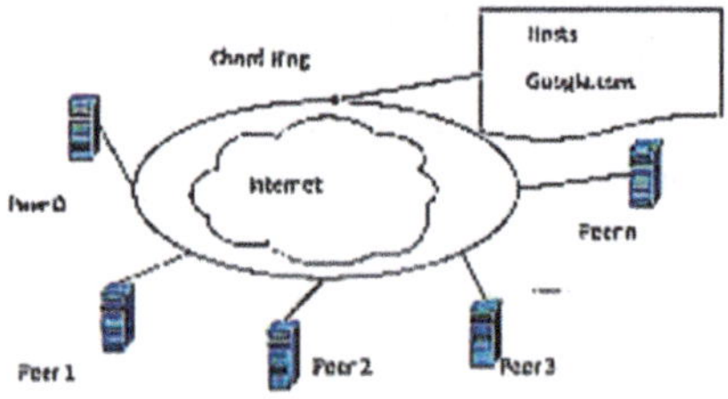

Fig. 2. The internet

3.1 Natural Disasters

Computer information system is just a smart machine, easily affected by natural disasters and the environment. At present, we use a lot of computer room is no shock, fire prevention, waterproof, lightning protection, electromagnetic leakage or interference measures, such as grounding system is lack of thoughtful consideration, the ability to resist natural calamities and accidents, was poor.

3.2 The Vulnerability of Network System Itself

The most significant advantages of the technology of the Internet is open [12]. However, this wide open, from a security point of view, thus become the vulnerable weakness. Along with the Internet rely on the safety of the TCP/IP protocol itself is not high, run the protocol of the network system is spoofing attacks, denial of service, such as data capture and data tampering with threats and attacks.

3.3 User Error

User security awareness is not strong, the user password is simple, the user will give their account to leak, etc., will be a threat to network security.

3.4 Artificial Malicious Attacks

This is the biggest threat to the computer network attack. Malicious attacks and can be divided into active attack and passive attack. Attack in various ways to selectively destroy the validity of the information and integrity; Passive attack is in normal working conditions, does not affect the network to intercept, steals, deciphering to obtain important confidential information. These two kinds of attacks can cause great harm to computer networks, and lead to a leakage of important data [13]. Now use the network software is more or less exist some shortcomings and vulnerabilities, network hackers often use intrusion into important means of information system, eavesdropping, obtain and attack into important information about the sensitivity, modify, and destroy the normal use of the information network, data loss or system paralysis, have significant political influence and economic losses to the country.

3.5 A Computer Virus

Computer viruses can be stored, executable and can be hidden in the executable programs and data files without being found that trigger the access control system after an executable program, it is contagious, latent, triggers and destructive sexual characteristics' computer virus is mainly transmitted by copying files, files, and run the program operation [14]. In the course of everyday use, floppy disk, hard disk, CD and network is the main way of spreading the virus. Computer virus after running light could reduce the system efficiency, or may damage files, delete files, even make the data loss, destruction of the system hardware, all kinds of unpredictable consequences. In recent years, the emergences of a variety of malignant viruses are based on the spread of the network, the computer network virus destructive is large, it brings the network a very severe losses.

3.6 Spam and Spyware

Some people use the E-mail address for commercial, religion, political activities, such as your E-mail forced into other people's email, forcing others to accept junk mail [15]. The main purpose of the different from computer viruses, spyware is not cause damage to the system, but steal system or user information, threat to user privacy and computer security, and possibly a small range of affect system performance.

3.7 Computer Crimes

Is usually used to steal passwords and other means of illegal invasion computer information system, the spread of harmful information, malicious damage computer system, the implementation of embezzlement, theft, fraud and financial crime activities.

4 Network Security Defense Measures

Network security is a complex project, which involves the technology, equipment, management, use and legislation system, and so on various aspects of factors [16]. Want to well realize information security, we must form a complete set of network information security system, technology, equipment, management, use and factors such as the legislative system of coordinated development, be short of one cannot.

4.1 The Traditional Technology

Firewall. Firewall is a kind of exclusive hardware, also can be set up a set of software on common hardware. Logically, a firewall is a separator, the limiter and analyzer, main effect is when the communication between two or more network, can have the scale of the access control of the firewall, filtering information. Common type of firewall has a packet filter, application proxy, network address translation NAT and monitoring. The use of a firewall can maximize the normal operation of the internal and external network, but cannot prevent attack from within the LAN; it has become the biggest limitations of firewall. At the same time, with the development of technology, some of the ways to decipher also makes the firewall has certain hidden danger, so in the firewall technology remains to be a better development. The commonly used fire Wall with sky net, rising; there are 360 firewall, etc.

Access Control Technology. Reference to access control technology, it has to be referred to design the access control to the three basic concept, subject, and object and access authorization. Subject: can apply to other entities action of active substance, also sometimes referred to as users or visitors, including users, user groups, terminal machine, card, even application service program, etc. The passive entity object: interview with other entities, it could be information, files, records, can also be a processor, memory, network contact, etc. Access authorization: the subject of object access rights, without the permission of the access authorization is given for each pair of subject and object [17]. Access control technology is by setting the access permissions to restrict access to the subject of access object access, consciously or unconsciously to prevent unauthorized users access to data. The technology is one of the main strategies of network security and protection, its main task is to ensure no illegal use of and access to network resources. The scope of the wider access control technology, including: network access control, network access control, inventory level control and a variety of means such as attribute controls.

Data Encryption Technology. Data encryption technology refers to the data transmission, the information some mixed addition, recoding, so as to achieve the function of the hidden information, make illegal users cannot read the information content, effectively protect the information system and data security, network security is one of the core technology. Data encryption technology can be realized in network OSI7 layer protocol of multilayer, look from the encryption technology application logic position, common data encryption are: link encryption (encryption) under the network layer, node encryption (protocol on the transport layer encryption), end-to-end encryption (more than the network layer encryption), etc. According to different functions, data encryption technology can be divided into: in the identification of data storage, data transmission, data integrity, and key management technology, etc.

4.2 The New Technology

Intrusion Detection System (IDS). Intrusion detection technology is the "firewall", a new generation of network security technology in recent 10 years, to a great extent, it made up for the inadequacy of the firewall. It is through several key points in the network or computer system to collect information and analyze, detect whether there is a violation of security policy, are under attack, can effectively find intrusion behavior legitimate user's misuse of the privilege, is a collection of test, record, alarm, response technology, can take the initiative to protect themselves from attack of dynamic network security policy, also is the core part of P2DR. Although there is a lot of "firewall" integrated intrusion detection module, but because of the limitation of technology and performance compared with the professional intrusion detection systems are not. Profession of intrusion detection system is a kind of active network security protection tool, provides for internal and external attacks and real-time protection, wrong operation before network system compromised intercept corresponding invasion [18]. Current intrusion detection system intrusion detection methods include: feature detection, anomaly detection, detecting state, protocol analysis, etc. Intrusion detection system also exists some problems, such as false positives; For example, when attacked by the delicate and organized, to find such a complex attack is difficult. In general speaking, the development trend of intrusion detection system can be summarized as a distributed intrusion detection and intrusion detection common intrusion detection framework (CIDF), application layer intrusion detection, smart layer, combined with other network security, intrusion detection based intrusion detection system evaluation system. Due to the insufficiency of intrusion detection system, already can't meet the demand of the development of the network, the future development of intrusion detection technology is mainly in the following aspects: intelligent intrusion detection, large-scale distributed intrusion detection, the application layer intrusion detection, a distributed intrusion detection system and common intrusion detection, etc.

Virtual Private Network (VPN) Technology. A virtual private network (Virtual Private Network, VPN) is a kind of based on public data network, give the user a feeling connected directly to the private LAN services. VPN technology is relying on the Internet service provider (ISP) and other Internet service providers (NSP), in the public network to establish a dedicated data communication network technology. In a virtual private network, the connection between any two nodes and no traditional end-to-end physical link needed for the private network, but the use of the resources of a public site of dynamic. A lot of the role of virtual private networks (VPNS), for instance: to realize network security, simplify network design, lower cost, easy to extend, flexible connection, complete control of the initiative, and so on. In the mainly discusses its security, virtual private networks (VPNS) transmission is mostly private information, so the safety of the user data is more concerned about. At present, the technology of the VPN mainly adopts four to ensure information security, tunneling, encryption, key management, the user identity authentication technology and equipment. Through these technologies can effectively resist criminals tamper, eavesdropping, intercept information ability, ensure the confidentiality of data.

Cloud Security. Cloud security is the latest embodiment of information security, network era as cloud computing becomes a hot topic in network development, puts forward the concept of "cloud security". Cloud computing is an emerging business computing model of distributed processing, parallel processing and the development of network computing, is a kind of super computing model based on the Internet. It USES high-speed Internet transmission capacity, the data processing from the PC or server to Internet super computer cluster. Cloud computing is a virtual computing resources pool, it through the Internet to provide users with resource pool of computing resources [19]. For users, cloud computing has access at any time, according to need to use, at any time extension, according to the advantages of paying. At the same time, cloud computing is characterized by good efficiency, management intensive, high utilization rate of equipment, saving energy and reducing consumption, service background, monetization, instant, elasticity and so on many characteristics.

5 Conclusion

With the rapid increasing of the size of the network, network heterogeneity and complexity, the hacking techniques are develop rapidly, which makes network security issues facing huge challenges, in order to protect the security of the network, we need to analyze network status information existed, then predict future network state, make a proper strategy according to the prediction result response, to reduce the losses caused by the network security issues. Network is safe, has affected the survival and development in all walks of life, and is also the future social development one of the important influence factors. All in all, only the good level of network security, network era of cyberspace will exert its maximum guarantee for human being, the action of the service.

Acknowledgement. The authors wish to thank the Education Department of Jilin Province [2015] (127th) and [2015] (182nd), Jilin science and Technology Bureau (201467006), Science and Technology Department of Jilin Province (20150204045NY). Under which the present work was possible.

References

1. Parikh, D., Chen, T.: Data fusion and cost minimization for intrusion detection. IEEE Trans. Inf. Forensics Secur. **3**(3), 381–389 (2008)
2. Gui, H.X.: Research of the intrusion detection system based on data mining. In: Proceeding of the International Conference on e-Education Entertainment and e-Management, pp. 190–192 (2011)
3. Zhang, B.Y., Chen, Z.Q., Tang, W.S., Fan, Q., Yan, X.A., Wang, S.L.: Network security situation assessment based on stochastic game model. Adv. Intell. Comput. **6838**, 517–525 (2011)
4. Lv, H.Y., Cao, Y.D.: Research on network risk situation assessment based on threat analysis. In: 2008 International Symposium on Information Science and Engineering, pp. 252–257 (2008)
5. Wei, Y., Lian, Y.F.: Network security situational awareness model based on log audit and performance correction. Chin. J. Comput. **32**(4), 763–772 (2009)
6. Acampora, G.: Exploiting timed automata based fuzzy controllers for designing adaptive intrusion detection systems. Soft. Comput. **16**(7), 1183–1196 (2012)
7. Li, X.Y., Wang, Q.X., Yang, L.: Research of network security situation index system and visualization technology. J. Commun. **32**(11), 109–118 (2011)
8. Hubballi, N., Biswas, S., Nandi, S.: Network specific false alarm reduction in intrusion detection system. Secur. Commun. Netw. **4**(11), 1339–1349 (2011)
9. Liu, N., Li, G., Liu, Y.: A method of network security situation prediction based on gray neural network model. Appl. Mech. Mater. **63–64**(6), 936–939 (2011)
10. Shifflet, J.A.: Technique independent fusion model for network intrusion detection. In: Proceeding of the Misdates Conference on Undergraduate Research in Computer Science and Mathematics, vol. 3, no. 1, pp. 13–19 (2005)
11. Wen, C.L., Wen, C.B.: The multistate sequential filter with multisensory data fusion systems and control in aerospace and astronautics. In: Proceedings of the 1st International Conference on Systems and Control in Aerospace and Astronautics (2006)
12. Dartique, C., Jang, H., Zeng, W.: A new data- mining based approach for network intrusion detection. In: Proceedings of the 7th Annual Conference on CNSR. IEEE Computer Society, Los Alamitos, pp. 372–377 (2009)
13. Conti, M.D., Pietro, R., Gabrielli, A., Mancini, L.V.: Secure topology maintenance and events collection in WSNs. Secur. Commun. Netw. **4**(7), 744–762 (2011)
14. Li, Z.T., Ma, Y.M., Wang, L., Ma, J.E.: A novel real- time aggregation method on network security events. Kybernetes **40**(5), 912–920 (2011)
15. Xiao, H.D, Li, J.H. Arithmetic of security data fusion under high performance computing environment. In: Proceeding of WSEAS2005 (2005)
16. Tian, J., Chen, J., Dou, L.H., Zhang, Y.H.: The research of test and evaluation for multisensory data fusion systems. In: Proceedings of Intelligent Control and Automation, vol. 3, no. 6, pp. 2104–2108 (2002)
17. Ritchey, R., Ammann, P.: Using model checking to analyze network vulnerabilities. In: Proceedings of 2000 IEEE Symposium on Security and Privacy, pp. 156–165. IEEE Computer Society Press, Los Alamitos (2000)
18. Wu, S.L.: A geometric probabilistic framework for data fusion in information retrieval. In: 10th International Conference on Information Fusion (2007)
19. Sheyner, O., Haines, J., Jha, S., et al.: Automated generation and analysis of attack graphs. In: Proceedings of 2002 IEEE Symposium on Security and Privacy, pp. 273–284. IEEE Computer Society Press, Berkeley California (2002)

Intelligent Property System of Science and Technology Park Based on GPON Technology

Chen Wu(✉), Dan Tian, Le Yang, Xiao Liang Fang, Jin Long Song, and Dong Jing Zhang

The Seismological Bureau of Shandong Province, Jinan 250014, China
13953160911@163.com

Abstract. The mitigation of earthquake disasters in Shandong province are briefly introduced composition and function of the intellectual property system, as well as the project background, the demand of construction of the system. Intellectual property system is introduced based on the basic principle of Gpon technology and the construction of the advanced mode. Combination of industry information construction and the development of future business planning, architecture design, from the network technology is introduced, system building, etc., for the construction of the intellectual property system was analyzed, and the passive optical network based intelligent property development direction for the construction of the proposed his own view.

Keywords: The network architecture · Pon technology · Network integration System construction · Intellectual property

1 Introduction

In order to better development of earthquake prevention and disaster relief cause of Shandong province, our bureau in Shandong province in 2011 began the construction of the mitigation of earthquake disasters. The park technology system includes: the new research activities building integrated wiring, data center machine room, data communication network, information platform, emergency command center, the intellectual property system, etc. The new intellectual property is a complex system, its operation is based on advanced network construction scheme of GPON technology, not only including security system, access control system, measuring system and the energy consumption of form a complete set of equipment (such as communication and storage system), includes redundant data communications connections, environmental control equipment, monitoring equipment and various safety devices [1]. Now the intellectual property system for this park plays a vital role, is bearing the weight of the daily operation of the entire park, provide us with timely, reliable and efficient service.

2 Construction Background

The mitigation of earthquake disasters park technology in the construction of system, seismological bureau of Shandong province in accordance with the "systematic, intelligent, modular, integration" the construction of ideas, fully research the functional

J. Abawajy et al. (Eds.): ATCI 2018, AISC 842, pp. 832–839, 2019.
https://doi.org/10.1007/978-3-319-98776-7_99

requirements and index requirements of the technical system. Function and performance index of the requirements to follow the principle of "practical, advanced, forward-looking", satisfy the operation and management of the existing business system needs, and meet the requirements of the development of future at least 10 to 15 years.

Intellectual property system for the entire park "neural network", is also a park can be the basis for normal operation of the platform. According to specification for design of intellectual property system, the system according to class A intelligent property system design, power equipment host and backstage support according to the redundancy design, air conditioning, etc., to ensure that the communication link uninterrupted. Characteristics of the intellectual property system is based on the optical network, efficient, convenient service for the integration of advanced system, can realize intensive, and scientific management, and compared with other kinds of public network security. The system can provide a for protecting against and mitigating earthquake disasters in Shandong province park intelligent management platform of high reliable, high performance, fully consider the network stability, compatibility and extensibility. Efficiency and security is the core of the system requirements, for that the construction of the system response to the specific requirements and comprehensive consideration of the development of network technology. According to the development of information technology and inform the actual situation of the construction of earthquake disasters: construction area is large, the information available, more professional, more requirements and practical application, and the park phase ii, the construction of the three periods of reserve enough space for development. Optical network using Ethernet switches and PON equipment hybrid networking way [2].

2.1 Property Management System Present Situation and the Problems We Are Facing

Intellectual property is a sunrise industry, all of the new creatures are all the same to follow the pace of the society, constantly advancing with The Times to obtain long-term development. With the continuous improvement of the social informatization degree, intellectual property is in the process of growth. Property management hierarchy of needs, categories, forms of management mode, need to gradually use information management, improve the work efficiency. To strengthen property management level. Intellectual property management system arises at the historic moment under the background of it is in this system, it is a set of object-oriented system, the distributed structure property [3].

In recent years, the progress of information technology and property management level gradually improve, especially in some cities, property management, intelligence information obtained rapid development. Due to fast development, and thus derived the property management software developers. Software developers are generally more professional computer professionals. As their involvement, the function of property management software, a great change. As the management software is based on the platform, online and other large stand-alone version of management software has become the mainstream in the stage of the product, tend to be more intelligent property management also. With the development of intelligent property, rapid popularization of network technology, computer technology in the future time will be more deeply and widely application popularization in property management, property management will realize seamless and intelligent information management system, intensive management become a reality. At this time of the intellectual property management system will be the application of the wide area network technology, office automation and intelligent systems, financial electronic system, the combination of reflects the development trend of modern property system [4].

2.2 The Design Idea and Implementation Approach

Intelligence property in the whole park is playing an increasingly important role in mitigation of earthquake disasters, the current intellectual property can meet the new demand is put forward under the new environment, improve the service ability and guarantee service of high reliability, high efficiency, reduce our operating costs, achieve the goal of green energy saving [5]. In order to adapt to the trend of the development of the property management system and the park construction and the needs of the development of the future to run all of the intellectual property system is based on the technology of GPON optical network platform.

The Introduction Of PON Technology. Called the PON English Passive 0 pticalnetwork, translated into Chinese for Passive optical networks, is refers to the optical distribution network does not contain any electronic devices and electronic power, optical distribution network (dn) 0 all consists of optical divider and other Passive components. A passive optical network including center office end device optical line terminal (0 lt) and the user terminal equipment (0 nu) optical network unit. 0 dn used to connect to 0 lt and ONU, including passive optical splitter, optical fiber and coupler. Hold the Ethernet, voice, video, high-definition can networks, such as business, from OLT ~ I | ONU is using a fiber-optic access, 0 nu can use a cable or Ethernet cable to the user equipment. PON technology belong to point to multipoint fiber according to the technology, is the mainstream of the world's broadband access technology, compared with the point-to-point active optical network technology, the cost of PON technology is not high, the maintenance is relatively simple, its transmission bandwidth is wide. With the number of Internet more and more, is becoming more and more high to network bandwidth requirement, PON technology to meet the rapid growth of the bandwidth requirements [1] (Fig. 1).

Network Platform to Build. Traditional tree structure is based on the binary tree structure, adopts vertical scaling method, by adding higher layer and the expansion of higher performance of switch devices. But the topology is difficult to overcome the inherent defects of traditional tree structure: the flow in the core collection root node, the pass is hot, the core node easily become the bottleneck of network performance; Network has such problems as serious overload [5]. In order to overcome the traditional network mode inconvenience we network using Ethernet switch when mixed with PON equipment networking model. Core layer adopts double machine core switching mechanism configuration; In the access layer of floor focused on access mode of double linked to the core switch, other floor adopts the PON equipment networking, OLT equipment placed in the core room, in each floor of each floor beam

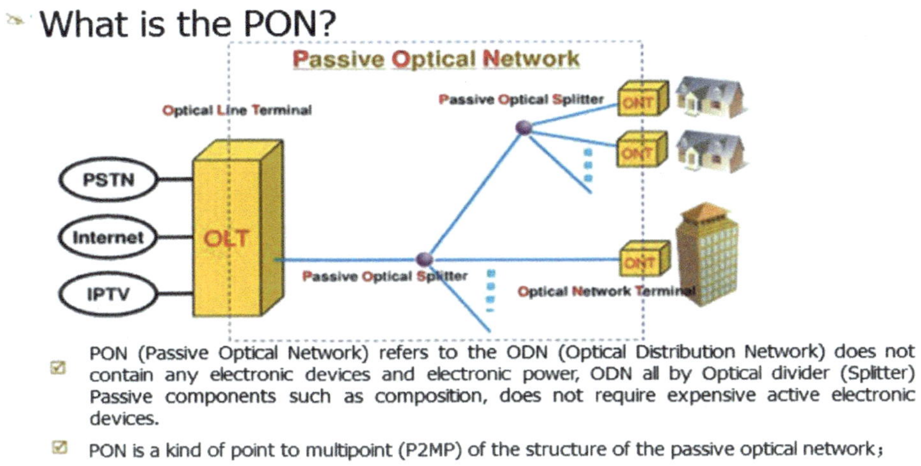

Fig. 1. Gpong technical brief sketch.

splitter based on the demand configuration, information on each node according to the specific needs configuration ONU equipment [6]. Network integration is development trend of network building, through fusion can achieve lower costs, reduce the management complexity, improve security advantages [7]. The intellectual property system construction goal is: to realize office automation, information sharing, and scientific management, and compared with other kinds of common network security. Traditional property management system model has not adapted to the development of The Times and realistic demand, the advanced intelligent property system to be able to provide mitigation of earthquake disasters in Shandong province park property management platform of a high reliable, high performance, fully consider the network stability, compatibility and extensibility of the system (Fig. 2).

Building Intelligent Property System. This system includes security system: the main courtyard and the building environment for video surveillance and perimeter, Access control system: mainly on the site's main entrance and underground car park entrance sets smart gate vehicle management, energy consumption Metering system: and expansion on the basis of energy metering system in the engine room, statistical analysis of electricity consumption on each floor.

Security System

(1) The construction requirements

Realize to park in the monitoring area of real-time video monitoring, image storage, historical images query; Monitoring network and content breadth of coverage, monitoring network should be fully covered in key areas, monitoring content video monitoring system should be adapted to the needs of the business (Fig. 3);

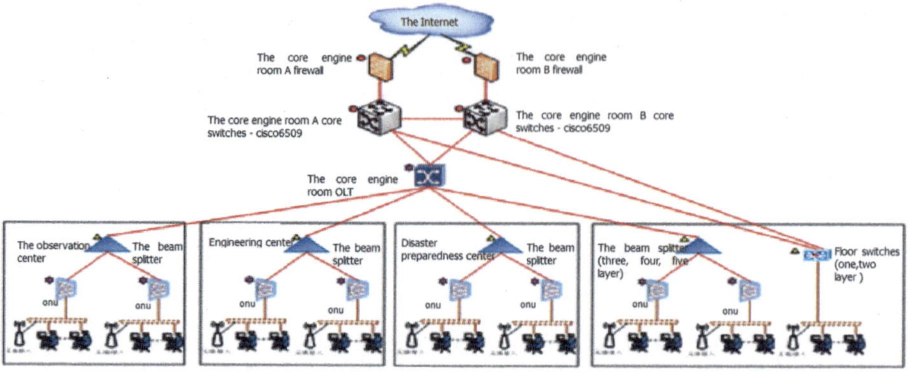

Fig. 2. Internet network topology diagram.

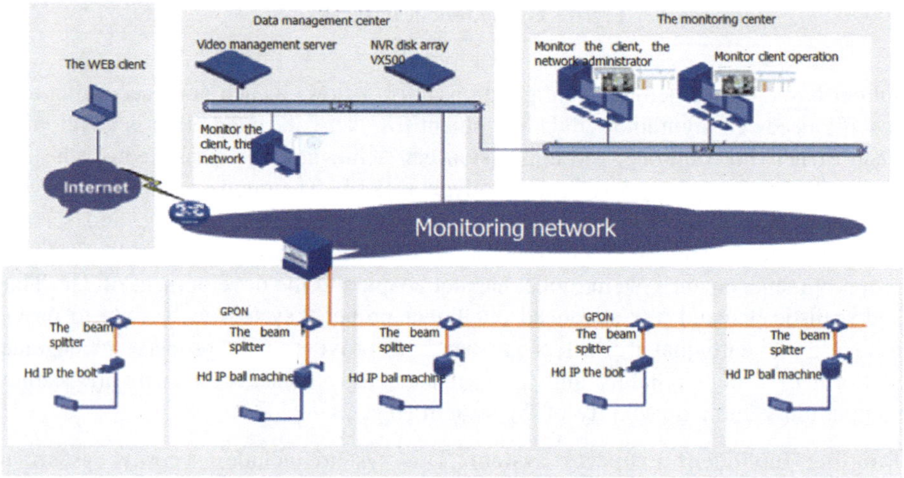

Fig. 3. Schematic diagram of security system.

(2) Equipment layout

Storage devices: according to the design requirements of the original design stored for 30 days, in the center of the third floor room set up three sets of yu VX500 storage devices, storage space for 42 t. Display equipment: in monitoring control room set up four client computer and four sets of 42 inch LCD TV, TV wall functions in the most practical and economical way. At the same time, through the HDMI transmission, the sound amplification functions. Front-end equipment: according to the design and change the file, the system set up 64 cameras (including: southern district 9 sets of comprehensive lab building 19, networks center 9 units, 15 units observation, three emergency center, north outside 9 units). Infrared correlation: according to the design requirements, this system has access to the infrared correlation equipment 12 (9 to the north, south to 3, and 6 to standby).

Minimum Energy-Control Systems

(1) The construction requirements

In response to a call to conserve energy, total into the box on each floor in the park, the total power consumption of monitoring, remote meter reading. This system platform for the east electronic InduView power equipment in the center of the three layer data environment monitoring center has been set up, this expansion was carried out on the platform, the debugging (Fig. 4).

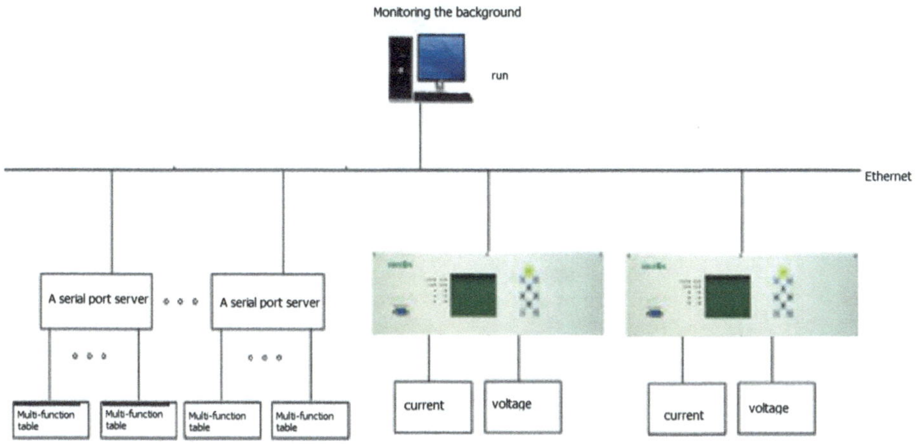

Fig. 4. Schematic diagram of energy consumption measurement.

(2) Equipment layout

Centralized power metering device: transformer room in the center of the observation and comprehensive lab building at level 5, respectively set up a centralized power metering device, the center and the comprehensive lab building electricity metering. And through the network system to transmit data to the data center. Intelligent monitoring instrument: in southern observatory, network center and disaster preparedness center each layer set a intelligent monitoring instrument, through a serial port server to transmit data to the data center. For the floor distribution of electricity metering. Ac transformer, transformer in the corresponding power distribution cabinet installation, implementation of three phase current, voltage, power monitoring, etc.

Gateway Control System

(1) The construction requirements

Entrance control system for the main entrance to the park and the underground parking garage management design, the main inlet and outlet, respectively, set brake, underground garage entrances to set a brake, meet in use at the same time. Inward and outward set card reader for the use of the worker and the official access over a long distance (Fig. 5).

(2) Equipment layout:

Simon luxury gate: the gate two sets, a set of standby. City: Simon, two sets of the door, two sets of backup. Two sets of card reader: Simon the doorway, two sets of backup. Remote controller: Simon, two sets of the door, two sets of backup. Network expander: Simon, two sets of the door, two sets of backup. Credit card machine: Simon is designed.

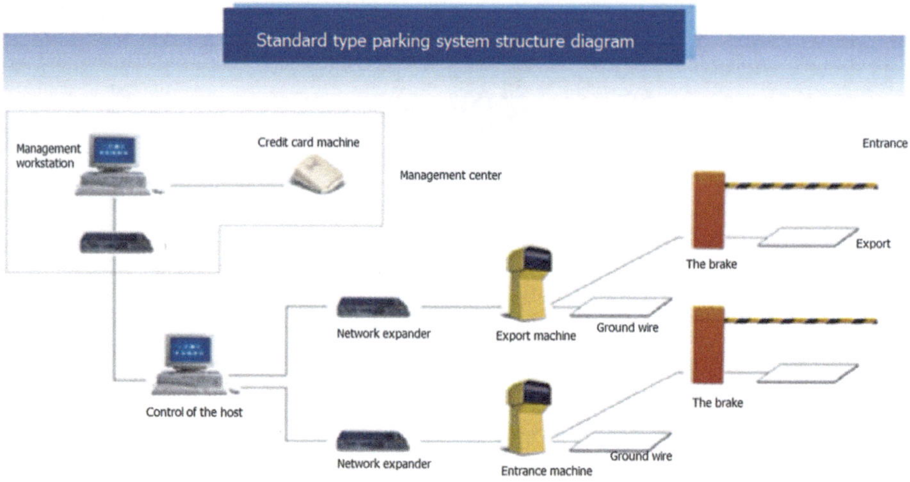

Fig. 5. Schematic diagram of the control system of the parking lot.

3 Conclusion

With the development of information technology, I use the existing mature technology, technology of earthquake disasters in Shandong province park construction support system of basic information platform. According to the concept of "intellectual property" advanced construction, on the basis of the original hardware server and storage resources, new hardware server and storage resources, and use the advanced optical network has been built, establish intelligent property background data processing and storage and unified supporting platform, provides the reliable infrastructure services, comply with the technology trend of development of the industry in the future. To sum up, the advanced intelligent property system based on optical network has been in protecting against and mitigating earthquake disasters in Shandong province park give full play to the advantage of its role, we also hope that through efficient operations of the system, to better serve our cause of protecting against and mitigating earthquake disasters. For brother ShengJu hope this article can be used for reference in construction of intellectual property system, also hope that we can in our information development on the path of earthquake prevention and disaster relief cause as a strength.

References

1. Wu, C., Wang, F., et al.: The virtualization technology in the application of seismic information network business. Earthq. Geomagn. Obs. Study **32**(3/4), 250–255 (2012)
2. Wu, C., Fang, X., et al.: The GPON optical fiber technology in the application of seismic information network in Shandong. J. Earthq. Def. Technol. **8**(1), 104–111 (2013)
3. Huang, Y.: The design of the intelligent property management system and research. J. Henan Sci. Technol. **36**(10), 80 (2013)
4. Li, G., Sun, J., et al.: The MPLS VPN in Tianjin earthquake monitoring system of high speed LAN applications. J. Earthq. Def. Technol. **7**(1), 92–99 (2012)
5. Kan, Y.: The era of cloud computing data center construction and development. J. Inf. Commun. **166**(6), 100–102 (2011)
6. Qu, Z.-J.: The light field network coding node model based on logic operation. J. Electron. **40**(7), 1304–1308 (2012)
7. Yu, D., Peng, L.: When does inferring reputation probability countervail temptation in cooperative behaviors for the prisoners' dilemma game? Chaos Solitons Fractals **78**, 238–244 (2015)
8. Aleksy, M., Korthaus, A., Schader, M.: Use Java and the CORBA realization distribute type system. J. Pingxiang Coll. **4**, 104–105 (2005)
9. Richardson, W.C., Avondolio, D.: The Java high class weaves a distance: JDK 5. Sci. Technol. Book Rev. **3**, 17–18 (2006)
10. Vuong JL. A semantics-based routing scheme for grid resource discovery. In: E-Science: First International Conference on E-Science and GridComputing, vol. 90, pp. 58–70 (2005)
11. Wang, Q.: Automatic control system regulation scheme of textile air conditioning. Sci. Technol. Inf. **22**, 395–396 (2010). (in Chinese)
12. Liu, J.: Advanced PID Control and MATLAB Simulation, 2nd edn. Electronics Industry Press, Beijing (2003). (in Chinese)
13. Li, J., Ren, Q.: Study on supply air temperature forecast and changing machine dew point for variable air volume system. Build. Energy Environ. **27**(4), 29–32 (2008). (in Chinese)

Iterated Particle Filter for Bearing Only Target Tracking

Fasheng Wang[1(✉)], Xucheng Li[2], Baowei Lin[3], and Junxing Zhang[1]

[1] School of Information and Communication Engineering, Dalian Minzu University, Dalian 116600, China
wangfasheng@dlnu.edu.cn
[2] Department of Software Engineering, Dalian Neusoft University of Information, Dalian 116600, China
[3] School of Information and Communication Engineering, Dalian University of Technology, Dalian 116024, China

Abstract. Among many of the existing nonlinear filtering methods, particle filter has drawn special attention due to its ability to deal with nonlinear/non-Gaussian state estimation problems. But it suffers from degeneracy problem. In order the suppress this problem, in this paper, we introduced an iterated particle filter which is applied to the bearing only target tracking problem. The proposed method is on the base of the iterated Kalman filter which is used for building a proposal distribution for the particle filter. The iterated Kalman filter can improve the estimation accuracy by iteratively incorporating the current observations which makes it a better candidate for building proposal distribution and easy to be integrated into the particle filtering framework. Simulation results demonstrate that the iterated particle filter shows superior performance to the other methods in bearing only target tracking.

Keywords: Particle filter · Bearing-only tracking · Iterative particle filter

1 Introduction

Bearing only sensors are commonly used to collect measurements of the object in many real-world applications, such as surveillance, under water weapon systems, sonar-based robotics navigation and submarine tracking. Tracking object of interests using the measurements gathered from bearing only sensors is the main task of the famous bearing only target tracking (BOT) problem. BOT is widely used in many military and civil applications [1]. It can be formulated as a nonlinear estimation problem and solved using the Kalman filter.

A famous method for dealing with bearing only target tracking is the extended Kalman filter (EKF). EKF is based on the first order terms of the Taylor series expansion of system models, which often introduces large estimation errors, especially when the system models are highly nonlinear. Another commonly used method for BOT is unscented Kalman filter (UKF) which does not depend on the Taylor series expansion. In UKF, the state distribution is specified by a set of deterministically chosen sample points. These sample points propagate through the nonlinear system

J. Abawajy et al. (Eds.): ATCI 2018, AISC 842, pp. 840–848, 2019.
https://doi.org/10.1007/978-3-319-98776-7_100

models and obtain accurate state estimation results. Despite its efficiency in nonlinear state estimation, the UKF cannot be applied to general non-Gaussian distribution. An alternative approach for BOT is the sequential Monte Carlo method (SMC), which is also named as particle filter.

The particle filter is a simulation-based approach for the recursive Bayesian filtering which is built on the concept of importance sampling [2]. Its basic idea is to compute the required expectations with respect to the posterior probability distribution using a set of particles (or samples) drawn from proposal distribution [3]. Each particle has an importance weight which is used to evaluate its contribution to build the posterior distribution. When the number of particles increases, the empirical distribution approximated by the particle set will become as close as possible to the true posterior probability distribution. Particle filter has become very popular due to its ability to handle highly nonlinear/non-Gaussian state-space models. Many researchers have applied this method to solve different nonlinear filtering problems emerging in different areas, such as finance, robotics, geophysical systems, communication, control, navigation and tracking [4], and signal processing [5], visual tracking [6, 7], and so on.

Although particle filter has been applied with success to many fields, it still suffers from two common problems, that are degeneracy phenomenon and impoverishment problem [8]. Researchers have done much work to conquer these two problems. Particle degeneracy phenomenon is such a problem that after several iterations, only one particle has large weights while the other particle weights are negligible. This phenomenon can be suppressed by designing better proposal distributions. The proposal distribution should be as close as possible to the true posterior distribution to ensure the particles drawn from this distribution can better simulate the true posterior probability distribution. The transition prior $p(x_t|x_{t-1})$ is a good candidate for proposal distribution [9]. But it does not take the current observation y_t into account, which yields inaccurate state estimation results. The unscented Kalman filter can be used to generate a Gaussian importance density function which yields the famous unscented particle filter (UPF). The UPF is adopted to solve many real-world problems such as target tracking, navigation, and so on. In [10] a mixture Kalman particle filter (MKPF) is proposed, which uses a mixture Kalman filter to generate the proposal distribution. Wang et al. [11] utilized an iterative version UKF (IUKF) to build the proposal distribution yielding an iterative UPF (IUPF). IUPF has been used to solve object tracking problem.

Introducing resampling step in the particle filtering framework [12] also can avoid degeneracy phenomenon. The idea of resampling is to remove the particles with low weights and to multiply the particles with high weights. However, resampling step could introduce another problem: particle impoverishment where the diversity of children particles is lost because most of the parent particles are removed because of their low weights. In recent years, many intelligent approaches are integrated in to the particle filtering framework to conquer the sample impoverishment problem, such as ant colony optimization [13, 14], particle swarm optimization [15], etc.

In this paper, we introduced an iterated particle filter for bearing only target tracking. The iterated particle filter uses an iterated Kalman filter to build proposal distribution, which can make better use of the current observation to improve the tracking accuracy. The paper is organized as follows: Sect. 2 gives the basic theory of particle filter, Sect. 3 shows the details of the iterated particle filter. Simulation results are given in Sect. 4 and the last section draws conclusion.

2 Particle Filter

The target tracking can be modeled using dynamic state space model. Following [9], we represent the state vector at time t as $x_t \in R^{\mathrm{dx}}$, and the target state is evolved through the following model,

$$x_t = f(x_{t-1}, u_t) \tag{1}$$

where u_t is the system process noise and it is independent of time;$f : \Re^{d_x} \times \Re^{d_u} \rightarrow \Re^{d_x}$ is the system transition function. The state transition function can be represented by a probabilistic form: $p(x_t|x_{t-1})$. At each discrete time step, an observation $y_t \in \Re^{d_y}$ is obtained corresponding to the state vector x_t using the following observation function.

$$y_t = g(x_t, v_t) \tag{2}$$

where $g : \Re^{d_x} \times \Re^{d_v} \rightarrow \Re^{d_y}$, and v_t is the measurement noise and it is independent of the system noise and time. The observation function can be represented by $p(y_t|x_t)$.

From the Bayesian perspective, all relevant information of the target state is embodied into the posterior distribution $p(x_t|y_{1:t})$ where $y_{1:t} = \{y_1, y_2, \ldots, y_t\}$ is a set of observations up to time step t. According to the Bayes' theorem, $p(x_t|y_{1:t}) = \frac{p(y_{1:t}|x_t)p(x_t)}{p(y_{1:t})}$. To compute the posterior probability distribution function recursively, applying the Chapman-Kolmogorov equations, we can obtain the following equations:

$$p(x_t|y_{1:t-1}) = \int p(x_t|x_{t-1})p(x_{t-1}|y_{1:t-1})dx_{t-1} \tag{3}$$

$$p(x_t|y_{1:t}) = \frac{p(y_t|x_t)p(x_t|y_{1:t-1})}{p(y_t|y_{1:t-1})} \tag{4}$$

Equation (3) is called prediction step, and Eq. (4) is called update step. We can see from Eq. (3) that integration operation is needed, thus it is difficult to obtain analytical solutions when the state space models are nonlinear and the system noises are non-Gaussian.

Particle filter is a simulation based approach for recursive Bayesian estimation, where the posterior distribution $p(x_t|y_{1:t})$ is approximated by a set of weighted particles or samples $\{\omega_t^{(i)}, x_t^{(i)}\}$, $i = 1, 2, \ldots, N$:$p(x_t|y_{1:t}) = \sum_{i=1}^{N} \omega_t^{(i)} \delta(x_t - x_t^{(i)})$, δ is the Dirac delta function.

Intuitively, the particles should be drawn directly from the posterior probability distribution, but it is impossible because the true posterior distribution is unknown. Thus, we must resort to the importance sampling method in which the particles are sampled from a proposal distribution (importance distribution) $\pi(x_t|x_{t-1}, y_{1:t})$. Then, the particle weight is considered as the ratio between posterior distribution and importance distribution and can be computed recursively using the following equation.

$$\omega_t^{(i)} = \omega_{t-1}^{(i)} \frac{p(y_t|x_t^{(i)})p(x_t^{(i)}|x_{t-1}^{(i)})}{\pi(x_t|x_{t-1}^{(i)}, y_{0:t})} \tag{5}$$

Resampling is needed when particle degeneracy occurred according to the effective sampling size: $ESS(\omega_t) = 1/\sum_i (\omega_t^{(i)})^2$. After resampling, statistics can be calculated to provide state estimations. The statistics could be the minimum mean-squared error (MMSE) or maximum a posteriori (MAP). In our work, we will use MAP statistic for target tracking.

3 Proposed Method

3.1 Iterated Kalman Filter

We assume all the relevant variables are Gaussian, and all the distributions are Normal distributions. Thus, we have the following,

$$p(y_t|x_t) = \frac{1}{\sqrt{(2\pi)^{d_y}|R_t|}} exp\left(-\frac{1}{2}(y_t - g(x_t))^T R_t^{-1}(y_t - g(x_t))\right) \tag{6}$$

$$p(x_t|y_{1:t-1}) = \frac{1}{\sqrt{(2\pi)^{d_x}|P_t|}} exp\left(-\frac{1}{2}(x_t - x_{t|t-1})^T P_t^{-1}(x_t - x_{t|t-1})\right) \tag{7}$$

where $||$ is the determinant, R and P are the covariance matrices of the measurement and system state. According to Eq. (4), the posterior probability distribution function can be represented as the product of (6) and (7). As maximizing $p(x_t|y_{1:t})$ is equivalent to minimizing its negative log, that is $\ell(x_t) = -log(p(x_t|y_{1:t})1)$, we can obtain the following equation.

$$\begin{aligned} \ell(x_t) =& \frac{1}{2}\{y_t - g(x_t)\}^T R^{-1}\{y_t - g(x_t)\} \\ &+ \frac{1}{2}\{x_t - x_{t|t-1}\}^T P_{t|t-1}^{-1}\{x_t - x_{t|t-1}\} \end{aligned} \tag{8}$$

Using Newton-Raphson method to iteratively minimize the above equation will yield an MAP estimate of x_t [16]. If the estimation of the jth iteration is $\hat{x}_t^j$, The second order Taylor expansion of $\ell(x_t)$ at $\hat{x}_t^j$ is as follow.

$$\ell(x) = \ell^j + [\ell_x^j]^T(x - \hat{x}_t^j) + \frac{1}{2}(x - \hat{x}_t^j)^T \ell_{xx}^j (x - \hat{x}_t^j) \tag{9}$$

where $\ell^j = \ell|_{x=\hat{x}_t^j}$, $\ell_x^j = \nabla_x \ell|_{x=\hat{x}_t^j}$, and $\ell_{xx}^j = \nabla_x \nabla_x^T \ell|_{x=\hat{x}_t^j}$. ℓ_x^j and ℓ_{xx}^j are the gradient and Hessian of ℓ with respect to x_t. Let Eq. (9) to be zero, we can obtain the next iteration value of x.

$$x^{j+1} = x^j - (\ell_{xx}^j)^{-1} \ell_x^j \tag{10}$$

According to Eq. (8), the gradient of ℓ is,

$$\ell_x^j = -g_x^j(\hat{x}_t^j) R_t^{-1} \cdot (y_t - g(\hat{x}_t^j)) + P_{t|t-1}^{-1}(\hat{x}_t^j - \hat{x}_{t|t-1}^j) \tag{11}$$

and the Hessian of ℓ is consequently computed as,

$$\ell_{xx}^j = [H_t^j]^T R_t^{-1} H_t^j + P_{k|k-1}^j \tag{12}$$

where $H_t^j = g_x(\hat{x}_t^j)$. Then, the inverse matrix of ℓ_{xx}^j can be obtained.

$$[\ell_{xx}^j]^{-1} = P_{t|t-1} - P_{t|t-1}[H_t^j]^T \cdot \{H_t^j P_{t|t-1}[H_t^j]^T + R_t\}^{-1} \cdot H_t^j P_{t|t-1} \tag{13}$$

Here, $[\ell_{xx}^j]^{-1} \triangleq P_t^j$. Substituting (11) and (13) into Eq. (10), the value of the $(j+1)^{th}$ iteration is obtained.

$$\hat{x}_t^{j+1} = \hat{x}_t^j + P_t^j [H_t^j]^T \cdot R_t^{-1}\{y_t - g(\hat{x}_t^j)\} - P_t^j P_{t|t-1}^{-1} \cdot [\hat{x}_t^j - \hat{x}_{t|t-1}] \tag{14}$$

The iteration process state from $j = 0$, and the one-step-ahead prediction of the state is used for the beginning point, that is $\hat{x}_t^0 \triangleq \hat{x}_{t|t-1}$.

3.2 Iterated Kalman Particle Filter

In this section, we incorporate the iterated Kalman filter into the particle filtering framework to build the proposal distribution. The particle set of time $t-1$ is $\{x_{t-1}^{(i)}, \omega_{t-1}^{(i)}\}_{i=1}^N$. At time step t, the particles will be propagated through system model (1) to obtain the one-step-ahead prediction:

$$x_{t|t-1}^{(i)} = f(x_{t-1}^{(i)}, u_t) \tag{15}$$

The Jacobians of the system model (1) can be computed as:

$$F_t^{(i)} = \frac{\partial f(x_t)}{\partial x_t}\Big|_{x_t = x_t^{(i)}} \tag{16}$$

$$G_t^{(i)} = \frac{\partial f(u_t)}{\partial u_t}|_{u_t=\bar{u}} \tag{17}$$

Then, one-step-ahead prediction of the covariance is obtained using the following equation,

$$P_{t|t-1}^{(i)} = F_t^{(i)} P_{t-1}^{(i)} [F_t^{(i)}]^T + G_t^{(i)} \tag{18}$$

The iterative procedure begins by computing the Jacobians of the observation model (2),

$$H_t^{(i)} = \frac{\partial g(\hat{x}_t^j)}{\partial \hat{x}_t^j}|_{\hat{x}_t^j = x_{t|t-1}^{(i)}} \tag{19}$$

$$V_t^{(i)} = \frac{\partial g(v_t^j)}{\partial v_t^j}|_{v_t^j=\bar{v}} \tag{20}$$

According to the previous section, we detail the proposed iterated Kalman particle filter as in Fig. 1.

3.3 Experiments

We evaluate the performance of the proposed method by applying the method to bearing-only target tracking problem and compare the proposed method with Kalman filter (KF), generic particle filter (PF) and Kalman particle filter (KPF) [17]. Suppose the target is moving in the 2D (x, y) plane, and the observer is keep still at the coordinate origin. The state transition model is defined as follow.

$$x_t = \Phi x_{t-1} + \Psi u_t \tag{21}$$

The object state at time step t is represented as $(x, \dot{x}, y, \dot{y})_t^T$ where x and y are the position of the target, $\dot{x}$ and $\dot{y}$ are the velocities at x and y directions. The system noise u_t is assumed to be zero mean Gaussian white noise with covariance matrix $Q = qI_2$, where I_2 is a 2D identity matrix. q is set as 0.001. The matrix Φ and Ψ are as follows:

$$\Phi = \begin{pmatrix} 1 & 1 & 0 & 0 \\ 0 & 1 & 0 & 0 \\ 0 & 0 & 1 & 1 \\ 0 & 0 & 0 & 1 \end{pmatrix} \tag{22}$$

$$\Psi = \begin{pmatrix} 0.5 & 0 \\ 1 & 0 \\ 0 & 0.5 \\ 0 & 1 \end{pmatrix} \tag{23}$$

Algorithm 1: Iterated Kalman Particle Filter

Input: Number of particles N;initial particle set $x_0^{(i)}$
Output: State estimation results $\bar{x}$.

Initialization: $t = 0$;
for i=*1,2...,*N **do**
  Draw initial particles $x_0^{(i)}$ from the prior $p(x_0)$;
end
for $t = 1, 2, ...$ **do**
  Importance Sampling:;
  for i=*1,2,...,*N **do**
    Compute the Jacobians of the system model using equation (16) and (17);
    Propagate the particles through the system model to yield the one-step-ahead-prediction $x_{t|t-1}^{(i)}$ (15);
    Compute the one-step-ahead prediction of the covariance $P_{t|t-1}^{(i)}$ using equation (18);
    for $j = 1, 2, ..., M$ **do**
      Compute the Jacobians of the observation model using equation (19)-(20);
      Update the Covariance using equation (13);
      Update the state using equation (14);
    end
    Set $x_t^{(i)}= x_t^M$, $P_t^{(i)}=P_t^j$;
  end
  Calculate the particle weights using equation (5) and normalize the weights;
  Resampling: remove the particles with low weights and multiply the particles with high weights;
end

Fig. 1. The proposed iterated Kalman particle filter.

As the observer is at coordinate origin, the observation is the noisy bearings of the target.

$$y_t = \tan^{-1}(y_t/x_t) + v_t \tag{24}$$

The observation noise is assumed to be zero mean Gaussian white noise with covariance matrix $R = rI_1$ where I_1 is a 1D identity matrix. r is set as 0.005. The initial state vector of the target is $x_0 = (-0.03, -0.001, 1.6, -0.055)^T$. At time $t = 0$, we assume that the initial state vector has a Gaussian distribution with mean $\bar{x}_0 = (0, 0, 1.4, -0.05)^T$ and covariance P_0.

$$P_0 = \begin{pmatrix} \sigma_1^2 & 0 & 0 & 0 \\ 0 & \sigma_2^2 & 0 & 0 \\ 0 & 0 & \sigma_3^2 & 0 \\ 0 & 0 & 0 & \sigma_4^2 \end{pmatrix} \tag{25}$$

The four parameters are set as: $\sigma_1 = 0.5$, $\sigma_2 = 0.005$, $\sigma_3 = 0.3$ and $\sigma_4 = 0.01$. In this experiment, total time step is $T = 24$, particle number $N = 4000$, iteration number $M = 5$ and the number of Monte Carlo run is 50.

The estimated target trajectories are shown in Fig. 2. The initial position of the target is unknown to the methods. We can see from the estimated trajectories that IKPF is superior to the other three methods. Table 1 summarizes the mean square errors (MSEs) of position (x, y) estimation. KF generates the highest MSE. The MSE mean of PF is $(0.0293, 0.247)$, while KPF is $(0.0226, 0.269)$, and IKPF generates the lowest MSE $(0.0221, 0.243)$.

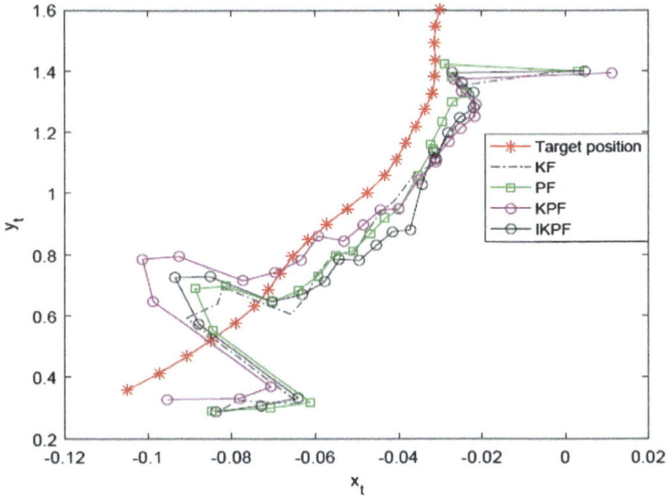

Fig. 2. Estimated target trajectories.

Table 1. The mean and variance of the MSE.

Methods	Items	MSE mean	MSE variance
KF	x position	0.0342	0.00183
KF	y position	0.421	0.1191
PF	x position	0.0293	0.000929
PF	y position	0.247	0.0058
KPF	x position	0.0226	0.00368
KPF	y position	0.269	0.0395
IKPF	x position	0.0221	0.00496
IKPF	y position	0.243	0.0385

As for the time cost, IKPF costs 3.272 s for each Monte Carlo run, which is much longer than the other three methods (KF: 0.04, PF: 0.076, KPF: 1.197) since IKPF iteratively incorporates the observations to improve the estimation accuracy.

4 Conclusion

In this paper, we have proposed an iterated particle filter for bearing-only target tracking. To build an appropriate proposal distribution to draw particles, we introduced iterated Kalman filter as the proposal distribution. The improved particle filter can make better use of the current observation, which leads to more accurate estimation results. As the proposed method costs more computational time compared to traditional methods, we will consider to reduce the computational time in our future work.

References

1. Sarkka, S.: Bayesian Filtering and Smoothing. Cambridge University Press, Cambridge (2013)
2. Wang, F., Lu, M., Zhao, Q., Yuan, Z.: Particle filtering algorithm. Chin. J. Comput. **37**(8), 1679–1694 (2014)
3. Djogatovic, M., Stanojevic, M., Mladenovic, N.: A variable neighborhood search particle filter for bearings-only target tracking. Comput. Oper. Res. **52**(1), 192–202 (2014)
4. Yu, Y., Cheng, Q.: Particle filters for maneuvering target tracking problem. Sig. Process. **86** (1), 195–203 (2006)
5. Doucet, A., Johansen, A.: A tutorial on particle filtering and smoothing: fifteen years later. In: Crisan, E.D., Rozovsky, B. (eds.) Handbook of Nonlinear Filtering, pp. 656–704. Oxford University Press, Oxford (2011)
6. Wang, F., Lin, B., Zhang, J., Li, X.: Object tracking using Langevin Monte Carlo particle filter and locality sensitive histogram based likelihood model. Comput. Graph. **70**, 214–224 (2018)
7. Wang, F., Xiao, Z., Chen, W., Li, X., Lu, M.: Tracking objects in video-based education using an enhanced particle filter. J. Intell. Fuzzy Syst. **31**(5), 2573–2581 (2016)
8. Li, T., Sun, S., Sattar, T., Corchado, J.: Fight sample degeneracy and impoverishment in particle filters: a review of intelligent approaches. Expert Syst. Appl. **41**, 3944–3954 (2014)
9. Van der Merwe, R., Doucet, A., Freitas, N., Wan, E.A.: The unscented particle filter. In: Advances in Neural Information Processing System (2000)
10. Wang, F., Zhao, Q.: A new particle filter for nonlinear filtering problems. Chin. J. Comput. **31**(2), 346–352 (2008)
11. Wang, F., Li, X. Lu, M.: Improving particle filter with better proposal distribution for nonlinear filtering problems. In: Proceedings of International Conference Wireless Algorithms, Systems, and Application, pp. 1–12 (2013)
12. Li, T., Bolic, M., Djuric, P.M.: Resampling methods for particle filtering: classification, implementation, and strategies. IEEE Signal Process. Mag. **32**(3), 70–86 (2015)
13. Xu, B., Chen, Q., Zhu, J., Wang, Z.: Ant estimator with application to target tracking. Sig. Process. **90**, 1496–1509 (2010)
14. Wang, F., Lin, B., Li, X.: An ant particle filter for visual tracking. In: Proceedings of International Conference on Computer and Information Science, pp. 417–422 (2017)
15. Zhao, J., Li, Z.: Particle filter based on particle swarm optimization resampling for vision tracking. Expert Syst. Appl. **37**, 8910–8914 (2010)
16. Havlik, J., Straka, O.: Performance evaluation of iterated extended Kalman filter with variable step-length. J. Phys.: Conf. Ser. **659**(012022), 1–12 (2015)
17. Wang, F., Guo, Q., Lin, Y.: A Kalman particle filter for bearing-only target tracking. J. Comput. Inf. Syst. **7**(15), 5628–5635 (2011)

An Improved Strategy of Distributed Network Crawler Based on Hadoop and P2P

Hongsheng Xu[1,2(✉)], Ke Li[1,2], and Ganglong Fan[1,2]

[1] Luoyang Normal University, Luoyang 471934, China
85660190@qq.com
[2] Henan Key Laboratory for Big Data Processing and Analytics of Electronic Commerce, Luoyang 471934, China

Abstract. Crawler technology is a technology that uses web crawlers to capture information on a web page to get valuable information. The web crawler is a program that automatically extracts web pages. It downloads Web pages from WEB for search engines. It is an important part of the search engine. This paper introduces the architecture of distributed Hadoop platform, and by the principle of the web crawler puts forward design scheme of distributed crawler based on Hadoop and P2P technology, and ultimately determine the system layout of distributed crawler, module partition and distributed crawler flow control and power can be realized.

Keywords: Hadoop · Network crawler · P2P · Data processing
Distributed network

1 Introduction

Today, the web crawler is mainly divided into two kinds of general crawler and focused crawler. The general search engine is targeting a possible coverage of the network; the general search engine provides keyword based retrieval. Most focused crawler is a program to automatically download page, it crawl under the established goals, have a choice to access the World Wide Web page and related links, to obtain the required information. And the general crawler (general purpose web crawler), a focused crawler does not pursue large coverage, but aim to capture related to a specific topic ", prepare data resource for topic oriented queries [1]. At the same time, it is according to the semantic information of support the query.

The distributed web crawler system Google is a central host and three units responsible for the crawler machine, and the three machines only with the central host communication. The central host reads from a URL file system, and put them to the other machine crawler process. Crawler uses asynchronous I/O and gets data from three hundred sites all the crawler will download page compressed and stored on disk. Then the indexing process from these HTML pages will extract the URL stored in a disk file in the URLResolver process of reading this store linked files, which will link relative conversion (link to browse through a folder button means local links) for absolute super link (link, a precise location to touch the files of the file can be stored in a file server, the web or a company intranet Then, it is provided to the host [2]. The drawback is that

J. Abawajy et al. (Eds.): ATCI 2018, AISC 842, pp. 849–855, 2019.
https://doi.org/10.1007/978-3-319-98776-7_101

once the central host fails, the whole system will stop working, and the URL allocation module of the central host will always become the bottleneck of the whole system performance.

Here a context diagram way design a focused crawler, they named it Context Graphs Focused Crawler. The learning web citation relation method proved the efficiency is not high, but it was also an important means of innovation. The system will use the reverse link to the page of the web service to find and establish a reference in the relationship between the two layer ", thereby establishing a focused crawler crawling path. The search parameters by the user changes, reference of each page will establish a corresponding, finally they merge. In this process, the classifier will determine the hierarchy, they determined, page links will join the queue thus, all need to extract the web crawlers.

This strategy is generally adopted when web crawlers are captured on Web pages for two reasons: First, the World Wide Web, the actual depth can reach the maximum of 17 layers "between connections, so there is from one page to the shortest path problem in another page. If using the depth first, there may be from a very low PageRank web crawling to a real PageRank high", is not convenient calculation of PageRank (personal understanding).

Second, breadth first strategy is beneficial for parallel crawling of multiple crawlers. This multi crawler cooperative crawler usually grabs the link within the station first, and then crawls out of the connection outside the station, and the seal is very strong.

This design is mainly based on eclipse software, the design process first access to server "URL, access to URL has been formed from the client access to the URL, and two methods, analytic website to obtain the updated URL, then according to the URL connection of the web page to access web page source code, access to web pages the contents of the web page source code, remove redundant spaces, labels, and finally the formation of XML document, the server also includes access to IP agent.

The working principle and working principle of traditional web crawler distributed web crawler of node similarity between a pluralities of nodes through the collaboration on the climb from the page, so that the efficiency of distributed web crawler is much higher than that of traditional web crawler for distributed web crawler. Typically, it is each node not only from no access but also from other nodes to accept URL. Access to URL library and the Web page are itself.

2 Optimization Analysis of Distribution Network Crawler Based on P2P

Because of the characteristics of Web information is distributed, so it will take the way of distributed web crawler design can greatly improve the efficiency of crawling data. The web crawler can be obtained by means of distributed network free resources ordinary PC users, can reduce the cost of data crawling; reduce the burden on the network. The design of a distributed first understand the web crawler of traditional web crawler climb to take the principle of data, improved and optimized on the basis of it.

General web crawler from one or several of the initial "URL" on the initial list; in the process of crawling in the continuous crawling queue from the current page selected new URL into the system, until the stop condition, crawling strategy can be divided into depth first, breadth first and best first three. Depth first in many cases will lead to the problems in the current common crawler, the breadth first and best first method.

Depth first, that is, starts from the start page, select a URL, enter, analyze the URL in this webpage, and choose a reentry. Such a link, a link to go deep tracking, after processing a route, then deal with the next route.

One example is: in the feudal system, the feudal imperial succession system is the eldest grandson of the highest priority, then the second. If the eldest son died, it is so the second higher priority than the sun.

The crawler design is relatively simple. However, the depth first web crawler has a problem: the portal site provides links are often the most valuable, PageRank is also very high, and each depth level, “the value and PageRank will be correspondingly decreased. This indicates that an important page usually distances closer to seed, and over deeply the crawled web pages but the value is very low, as is shown by Eq. (1) [3].

$$e_i = \exp\left\{ -\frac{1}{2} v_i'(k) S^{-1}(k) v_i(k) \right\} \quad (1)$$

The network uses the fully distributed unstructured topology is the P2P structure between the nodes is equal, each node is both the client and the server, the utility model has the advantages of good overall stability and failure will not impact on the system performance of a large node in the system, thus better fault tolerance [4]. But once the scale of the network large inter node communication efficiency is low.

Uniform resource locator (URL, English abbreviation Uniform Resource Locator) is also called the web address, is an Internet standard resource address. It was originally developed by Tim Berners and Li Faming used as a web address. Now it has been the World Wide Web Consortium for the preparation of the Internet standard RFC1738.

In the history of the Internet, uniform resource locator is a very basic step. URL syntax is general, scalable, part of it using the ASCII code to represent the Internet address. General uniform resource locator marks the start of the use of a computer network the agreement.

JSON is a lightweight data interchange format, easy for machines to parse and generate, but also convenient for people to read and write. Json is used to store and exchange of text messages, have similar functions and XML [5]. We don’t need any other data package can deal with JSON data in JavaScript, because JSON is a native JavaScript format.

Distributed web crawler structure based on P2P, a distributed system composed of multiple nodes, each node has its own unique ID number, all the nodes have a control node for each node is assigned ID, dynamic monitoring management of the system, the remaining nodes for a crawler node. A special coordination center node crawler node, is responsible for coordinating the communication between nodes, once the center node exit or collapse, other nodes will select the remaining nodes in ID, the largest center node.

3 An Improved Strategy of Distributed Network Crawler Based on Hadoop and P2P

The network server framework, the social platform of real-time information service website based on FriendFeed and the development of polymerization. It is used in the FriendFeed can be extended to non blocking web server and related tools of the source version. The Web framework looks somewhat like the web. py or Google webapp, but in order to effectively use non blocking server environment the Web framework also contains some useful tools and optimization.

Web crawler module DNS parser module, download module and process control module, web crawler is the core part of nodes, it communicates directly with the Web server. The Web page through to the thread control module sends the URL download, because the focus of this paper is the focus of the distributed module, module is not discussed.

Distributed file system is the underlying implementation of large-scale distributed computing support. If there are multiple servers in the LAN, and also distributed in different shared folders on the server, this is not conducive to the management of the administrator and user access while using a distributed file system, system administrators can put on the different servers shared folders together, built into a tree [6]. In the point of view of users, all shared files stored only in one place, only need to access a DFS root share, as is shown by Eq. (2), where it will be able to access distribution on the network file or folder, and do not need to know the physical location of these documents. In view of this, Hadoop implements a a similar to the GoogieGFS distributed file system HDFS.

$$J = \sum_{i=1}^{n} [y_i - (b_1 x_{1i} + b_2 x_{2i} + \cdots + b_m x_{mi} + a)]^2 \quad (2)$$

To control the crawler data source: the data source including the crawling of the crawler, add and delete, and the configuration of the crawl site (for example, change HTML code through the HTMLParser keyword matching feature captures the corresponding types of news). The data source configuration requires regular updating and maintenance systems need to provide a good interface and as far as possible to reduce the difficulty of the configuration and maintenance workload.

The advantages of breadth first strategy lie in its design and implementation is relatively simple, and the basic idea of this strategy is: with the seeds within a certain distance of "important degree, consistent with the actual situation [7]. In the application of focused crawler, breadth first strategy can be combined with Web filtering technology, i.e. with the breadth first crawl pages, then the irrelevant filter out. But the disadvantage of this method is that with the increasing of web crawling algorithm, efficiency becomes low.

The initial URL concentration corresponding to the Web page in accordance with certain rules of dividing task subset, task specific subset from the central node of the corresponding node to the division. In the process of collection if node discovery does not belong to their own range of information collection, feedback the information to the central node, a central node re assigned to the corresponding node. Centralized allocation is completed through the center node task. This approach often needs more forwarding to central node URL, easy to cause performance bottlenecks.

Hadoop is composed of a multi machine distributed system, the cooperation between nodes using master-slave mode, namely in the Hedoop system, there will be a Master and a slave. (1) Maste; NameNode is mainly responsible for the work and the work of JobTraeker. JobTracker's main duty is to start tracking and scheduling each Slave task execution.

4 Experiments and Analysis

In a distributed web crawler system node never access URL download from the web pages in the URL library, from web pages retrieved from a new URL, using the Hash function to calculate the Hash value, will belong to the node URL library into the unused access queue, send URL will not belong to the node to its belonging to the node [8]. The working principle of cooperation between nodes distributed web crawler system.

This paper implements a multi thread crawler program based on breadth first bias algorithm through JAVA language. By implementing this crawler program, we can collect URLs of a site at fixed point. If we need to collect other information, we can parse URLs and get corresponding information at the same time.

Tornado and now the mainstream Web server framework (including the framework of most Python) have obvious differences: it is a non blocking server, but the speed is quite fast. Thanks to the non blocking mode and the use of epoll, Tornado per second can handle thousands of connections, which means for real-time Web service, Tornado is an ideal Web framework [9].

$$\hat{x}^{(1)}(k+1) = \left(x^{(0)}(1) - \frac{b}{a}\right) \cdot e^{-ak} + \frac{b}{a},\ k = 0,\ 1\ldots,\ n \tag{3}$$

Because of the wide area network design and implementation of web crawler is more complex, so the design of a web crawler crawler system on the LAN. Web crawler refers to the crawler all the distribution in the same LAN, as is shown by Eq. (3), where it is the communication among the nodes is connected by means of a high speed. In order to avoid duplication in the search for web crawler and the load balance of each node, we must choose the reasonable allocation strategy.

Manual acquisition and incremental acquisition (add timing task). The essence of both is the open web crawler to the data source site to perform data acquisition tasks according to grasp configuration, the difference is to perform a specific task manually collected by the administrator after the trigger, as is shown by Fig. 1. The incremental acquisition will periodically update the part of the high frequency of web crawler in order to achieve real-time updates of the database server.

There is a commonly used in web search strategy of focused crawler. The best strategy best first priority and it is in accordance with a web page analysis algorithm to predict the similarity of candidate URL and the target page, relevance or theme, and

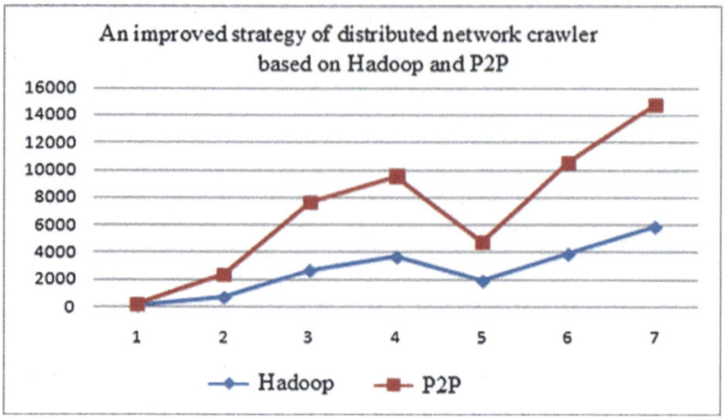

Fig. 1. An improved strategy of distributed network crawler based on Hadoop and P2P

select the evaluation of one or several of the best URL into the crawling step [10]. The defects of this strategy are possible there will be a lot of related web pages are ignored, but relatively, this method can reduce the number of independent "30%–90%.

According to the function of the participle provided by the analysis module, the content of the capture is divided into sub - processing, and Word sequence after word segmentation to index module index module. Inverted index according to the segmentation results and capture the text, the index into the index database for users Hadoop distributed file system on the retrieval. Then according to the segmentation results, the establishment of an index rate analysis key frequency. When the user wants to query is by keyword query interface when processed by the query processing. Query processing by the user search keywords word segmentation module, it will form a set of keywords. These keywords set is passed to the indexer and searcher to index database retrieval, retrieval results by sorting, returned to the user. In the realization of the process in the web (erawler), index (top dexer), and retrieval (Query) is MaPReduce framework of realization of distributed computing based on P2P and Hadoop.

5 Summary

The web crawler is designed in this paper; a distributed crawler follows P2P and Hadoop technology to design the web crawler. When visiting a site, it will first check whether robots.txt exists, the site root directory if it exists, the robot will search range according to the contents of the file to determine access; if the file does not exist then search robot along the link to crawl. Site administrators can define the web crawler can be access permissions by it.

Acknowledgements. This paper is supported by Henan key Laboratory for Big Data Processing & Analytics of Electronic Commerce, and also supported by the science and technology research major project of Henan province Education Department (17B520026, 15A120012), Key scientific research projects in Henan province universities (17A880020).

References

1. Zhu, X.: Design and implementation of P2P based distributed theme Crawler system. Information Management Department of Nanjing University, Institute of Multimedia Information Processing, Nanjing (2014)
2. Salakhutdinov, R., Mnih, A.: Bayesian probabilistic matrix factorization using Markov chain Monte Carlo. In: Proceedings of ICML, pp. 880–887 (2013)
3. Xu, H., Zhang, R.: Semantic annotation of ontology by using rough concept lattice isomorphic model. Int. J. Hybrid Inf. Technol. **8**(2), 93–108 (2015)
4. Jin, Y.: Design and Implementation of Distributed Web Information Acquisition System. Harbin University of Technology, Institute of Measurement and Control Technology and Communication Engineering, Heilongjiang (2013)
5. Herlocker, J.I., Konstan, J.A., Borchers, A., Riedl, J.: An algorithmic for performing collaborative filtering. In: Proceedings of the Conference on Research and Development in Information Retrieval (SigIR 1999), pp. 230–237 (2015)
6. Wu, L., Liu, N.: Design and Implementation of Distributed Network Crawler. Hubei School of Computer Science, Wuhan University, Key Laboratory of Ministry of Information and Network Security, Shanghai (2011)
7. Xu, H., Zhang, R.: Novel approach of semantic annotation by fuzzy ontology based on variable precision rough set and concept lattice. Int. J. Hybrid Inf. Technol. **9**(4), 25–40 (2016)
8. Bai, H, Wang, J.L. et al.: Research and Implementation of Distributed Multi-theme Web Crawler System, Graduate School of Chinese Academy of Sciences, National Network New Media Engineering Technology Research Center, National Institute of Acoustics, Chinese Academy of Sciences, Beijing (2009)
9. Li, X.: Design and implementation of distributed crawler system. In: China Science and Technology Information, (2014)
10. Salakhutdinov, R., Mnih, A.: Probabilistic matrix factorization. In Proceedings of NIPS (2007)
11. Yu, D., Peng, L.: When does inferring reputation probability countervail temptation in cooperative behaviors for the prisoners' dilemma game? Chaos, Solitons Fractals **78**, 238–244 (2015)

Campus Network Control Management System Based on Python

Qiang Miao(✉), Hui Li, and Zonggang He

Dalian Neusoft University of Information, Dalian 116023, Liaoning, China
miaoqiang@neusoft.edu.cn

Abstract. In order to make it easy to maintain and manage the campus network equipment, on the basis of depth analysis of current campus network deployment and the support features for a variety of network protocols of Python, the design will establish remote connection with network equipment to realize the remote centralized control of network communication equipment with the remote Telnet function of router and switchboard. The design uses PHP to realize the graphical campus network management interface, and finally presents a complete campus network management system through Web page. The system will bring convenience on network management and maintenance for the administrator.

Keywords: Campus network · Python · Management system

1 The Brief Introduction of Python Language

Python is a kind of object-oriented interpreted computer programming language [1]. Some famous universities have set the Python courses owing to the conciseness, readability and strong extensibility. In view of Python is completely free, a variety of open source scientific computing libraries provide the Python calling interfaces. The users can install Python and most of the extension libraries freely on any computers. Python language is fully support inheritance, overload, derive and multiple inheritance which can enhance the reusability of the resource code. When Python is executed, first the resource code of .py file is compiled into the byte code of Python, and then the Python Virtual Machine will execute the compiled byte code. Python can also realize system programming task, the provided API which is the ideal programming tool of many system administrators can provide system maintenance and management easily. Python standard library includes some functions on the system maintenance:

(1) Text processing which includes text formatting and regular expression matching.
(2) Network communications which include the function of nested word, SSL encrypted communication, asynchronous network communication.
(3) Network protocols which support varieties of network protocols such as Telnet, HTTP, FTP, SMTP, POP, and provide the frame of writing network server.

J. Abawajy et al. (Eds.): ATCI 2018, AISC 842, pp. 856–865, 2019.
https://doi.org/10.1007/978-3-319-98776-7_102

2 The Present Situation Analysis of Campus Network

The topological structure of the campus network is complicated, and there is a huge number of student users and teacher users. On the basis of the enormous demand of network communication on teaching, scientific research and daily network access, as well as the network environment deployment being relate to interconnection of different teaching areas and communication of different functional departments, the network capacity, network security, network redundancy and network extensibility of network formation and maintenance should be fully considered.

Take the local campus network environment as an example, the whole campus network is built with Ethernet technology. The access layer of the office buildings and teaching buildings, and the links from access layer to distribution layer are all twisted pair. The distribution layer is accessed to the core equipment which is CE12808 in the computer room through optical fiber. The access layer is hundreds of megabytes of bandwidth. From distribution layer to core layer is gigabit bandwidth. The core layer is ten thousand of megabyte bandwidth. The wireless network is also accessed to the campus network core through CE12808. The access layer and distribution layer are all two-layer link, and the equipment of central computer room provide the routing addressing of the campus network communication. The central computer room uses five core equipment which are CE12808, ME60, Teachier8512, Lib8512 and Server8512 to connect the whole campus network as well as provide Internet exits. The following part will show the network topology connection taking teaching building, dormitory building and wireless network connection as examples. The access layer and distribution layer network topology of teaching building A2 (see Fig. 1), the network topology of dormitory building 1 (see Fig. 2) and the network topology of wireless user access area (see Fig. 3) are shown below:

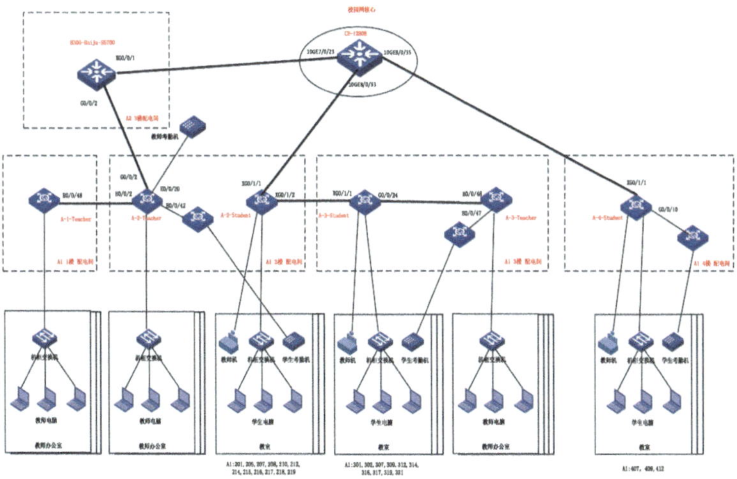

Fig. 1. Access layer and distribution layer sketch map of teaching building.

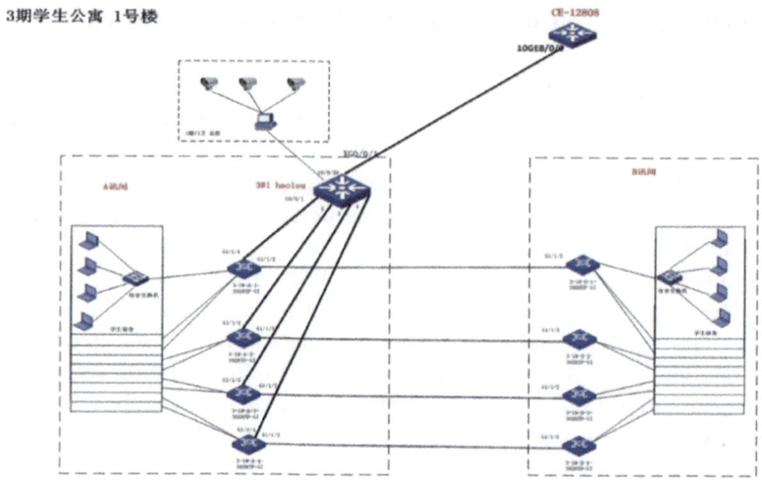

Fig. 2. Access layer and distribution layer sketch map of dormitory building.

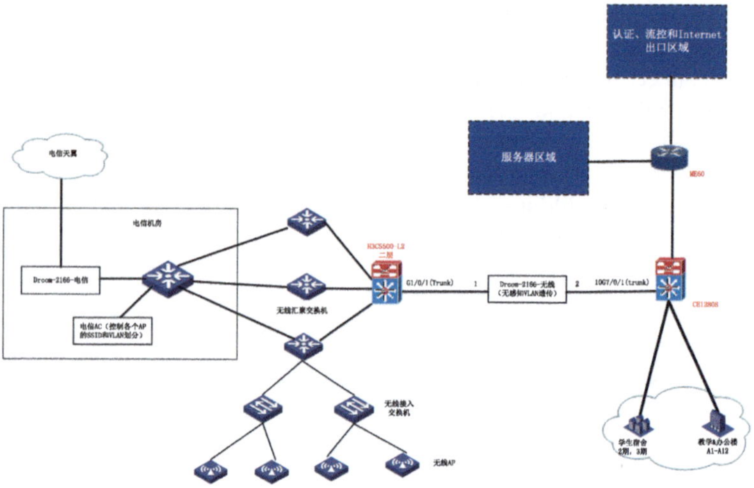

Fig. 3. Wireless network access topological graph of central computer room.

The Vlan information distribution and management strategy of the unified planning and management of the campus network are shown below:

(1) Managing Vlan and business Vlan separation.
(2) The managing Vlan of the network equipment are 1021, 1022 and 1023 corresponding to Internet equipment management of student dormitories, classrooms and offices.

(3) The specific user will use the specific Vlan. Vlan could span multiple switches. For example, the Vlan of all teachers' office is 800, and the Vlan of all teachers' computers in the classrooms is 500.
(4) The students use different Vlan in different classrooms. For example, Vlan of classroom A201 is 101, Vlan of classroom A205 is 215.
(5) The student dormitory uses Vlan isolation. For example, in student dormitoryies 1–18, a single room uses a single Vlan, and there are isolations between rooms. And in student dormitories 18–25, a single access point uses a single Vlan, and there are isolations between access points.
(6) The QinQ technology is used in the distribute router of student dormitories in order to manage and assign IP addresses easily.

The following Table 1 gives the Vlan management information.

Table 1. Managing Vlan of access layer and distribution layer.

Number	Purpose	IP address segment	Gateway
1021	Internet equipment of dormitories managing Vlan	172.17.248.0/23	172.17.249.254
1022	Internet equipment of classrooms managing Vlan	172.18.248.0/24	172.18.248.254
1023	Internet equipment of offices managing Vlan	172.19.248.0/24	172.19.248.254

Users in campus network could access campus network through wired mode or wireless mode. The teaching buildings, administration buildings, dormitories and library acquire IP addresses through PPPoE to access campus network with wired mode. Central computer office room acquires IP addresses through DHCP to access campus network in laboratories of building A2–A3. Users could access campus network through WiFi signal in the campus with wireless mode. Both PPPoE and wireless mode should be through authentication, and the authentication and charging function are completed by three Radius servers.

3 Design of Network Control Management System

3.1 Design of Management

The main functions of campus network control management system are obtained through analyzing the specific function and demand. The main functions include network equipment parameters query and setting, monitoring task and management, data management, system remote control and maintenance, user management (see Fig. 4). The detail information is shown below:

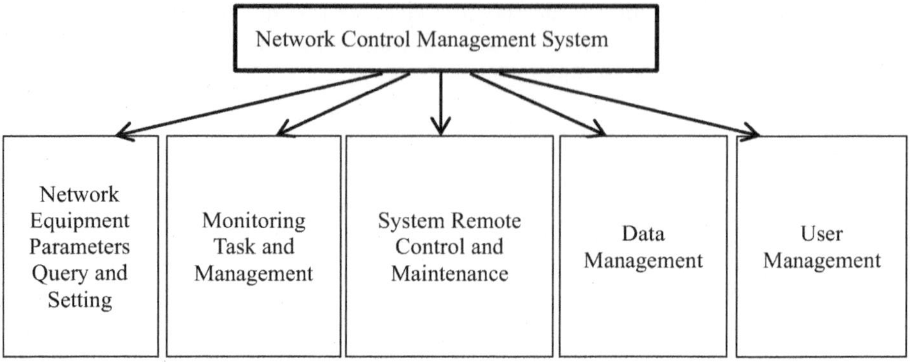

Fig. 4. Main function of campus network control management system.

(1) Network equipment parameters query and setting include the basic configuration information of switches and routers, device port parameters configuration and information query.
(2) Monitoring task and management include condition monitoring, network performance monitoring and flow monitoring.
(3) Data management will complete the query and statistics of the monitoring data, as well as analysis and management in the form of figures and Tables
(4) System remote control and maintenance include the copies of monitoring data, download and transmission, remote control, network equipment interface management.
(5) User management include crud and permission settings of the users. Permission settings include different levels of administrators and different teacher users.

3.2 The Use Process of the Management System

All users are provided a unified entrance in the Web page to login the system, the system will identity and verify the login information. Different system resources access are according to different permission roles. Permission role mechanism is assigned through user role by the highest level administrator. Users could operate equipment with allowed permission through remote access information. First the area management adds corresponding campus, area and buildings, and then locate the equipment to the specific buildings and classrooms. System users could get the information of any equipment in the campus. The address module is used to manage the addresses in campus. The campus network design strategy is considered, and the design also includes the management of all the addresses in the campus with the corresponding Vlan information.

3.3 The Implementation of Network Energy-Saving Through Remote Control

The task of network environment energy-saving is also considered during the design. In the current campus network environment, layer two switches in access layer are commonly used in laboratory network connection and multi-media classroom network connection. In order to provide convenient network connection, the switches are always in the electrified state. In the situation, the network equipment such as switches still consume power even if the laboratories and classrooms are not used. Normally laboratories and classrooms are unused from 20:00 to 8:00. The testing results show that the network equipment in unused state will consume a large quantity of electricity. The power consumption of the equipment is about 90% of the full load state [2]. The 220 V/24 W switch as an example, power consumption is 0.024° per hour, and 0.288° of 12 working hours. On the basis of all the campus have realized complete coverage of the wireless network, and a classroom with the capacity of 90–100 people owning 4–5 switches, the whole power consumption of a night will be 1.2°. And the campus owes hundreds of such classrooms and laboratories, the whole power consumption will be hundreds of degrees according to the above content. There will be large amount of waste electricity on monthly calculation.

In this network environment, in order to realize the network energy-saving, the design will carry out controllable management to the network resources. The specific measure is to shut down the unused equipment and ports, and to restart them necessarily in order to avoid wasting more power. In the current network structure, the consumption power of the link is independent with its utilization [3], which means the power consumption will not change no matter what the utilization of the link is. In fact, even if it is during idle phase with no data frame transmission, the link will still send frames with no information content consistently in order to keep synchronized. So the power consumption of link is mainly depend on the design capacity of the link, and has nothing to do with the actual link flow [4]. So shutting down the unused links can save the unnecessary power consumption.

4 The Implementation of Network Control Management System

4.1 The Implementation of Remote Connection and Control in Python

First configure the managing Ip, username and password for login of the physical equipment in the network environment. This step can implement the remote equipment management. The second step is to access the network equipment through remote connecting mode, then execute the commands to get the return value, and then implement the interaction with network equipment as well as obtain the storing data in Python. Finally, after getting the parameters and data information of the network equipment through Python, the user could tell different command according to command identifier, and to implement different query or control function. The core script of login is shown below:

```
#!/usr/bin/env  python
import pexpect
address = '172.22.248.42'
userName = 'tel_user'
password = 'Tel_user@123'
cmd = 'telnet ' + address
prompt = '<DPTECH>'
cmd2='show ip interface brief'
child = pexpect.spawn(cmd)
#print (cmd)
index = child.expect(['Login',pexpect.EOF,pexpect.TIMEOUT],timeout=5)
if index == 0:
        child.sendline(userName)
        index = child.expect('Password:',timeout=5)
        child.sendline(password)
        child.expect(prompt,timeout=5)
        #print ("login successful!")
        child.sendline(cmd2)
        child.expect(cmd2,timeout=5)
        child.expect(prompt,timeout=5)
        #child.expect('More ----',timeout=10)
        str1 = child.before
        print str1
        #sstr1=str(sstr1)
        #sstr2='/r'
        #sstr1=sstr1[sstr1.find(sstr2) +1:]
else:
       print ('expect "login",but get EOF or TIMEOUT')
child.close()
```

4.2 The Implementation of System Interface in PHP

After getting the parameters and data information of the network equipment through Python command, all the data should be imported into the management control system platform written in PHP, and be displayed in a page through the graphical interface. The design uses the built-in function exec of PHP. In order to display the data in a better form, the design uses regular expressions of PHP, and reorganize the return information. Part of the core code is shown below:

```
$re=(preg_grep( '/^vlan/' , $jieguo));
    echo 'Interface physical Protocol Description IP Address Second Address';
    foreach ($re as $value){
        ?>
        <ul>
        <li> <?php echo $value;?> </li>
        </ul>
        <?php
        }
        ?>
```

In the main interface of system management (see Fig. 5), the user can click every part of the left list to view the information, and to implement the independent management of every classroom. The interface of the classroom management (see Fig. 6) could show the switch information, router information, protocol and the switch state record. The menu column also includes equipment management record and user management interface record.

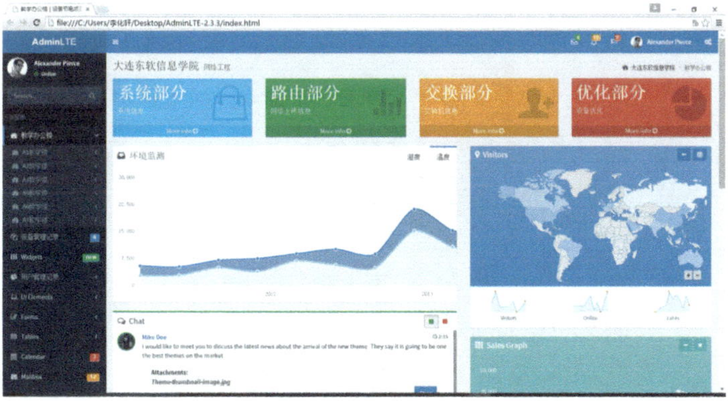

Fig. 5. Main Interface of system management.

Fig. 6. Interface of classroom management.

Remote access and configuration of the network equipment are very important to the control and management for the campus network. Some basic commands such as ping could be used to check that the equipment is in up-state or down-state. The administrator could judge whether the equipment is operating normally through the feedback information. The system uses a settled time interval to poll the equipment which need maintenance and management in the campus network to discover problems and email the problem information to the system administrator timely. The administrator could manage remote access network equipment through a central management site, check the corresponding operation parameters and management information in the form of graphics, and do management operation according to the specific demand.

The system testing and operating include the application testing of network equipment parameters, equipment interface information, network state and flow management in the campus network. The analyzing tests of protocol and application used by data flow (see Fig. 7) and the analyzing tests of data flow direction (see Fig. 8) are shown below:

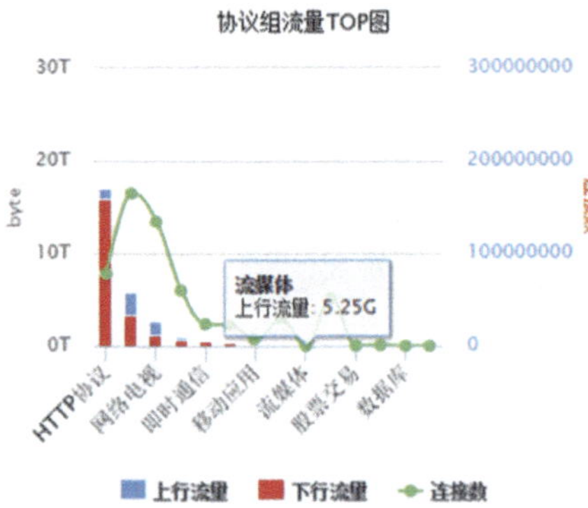

Fig. 7. The protocol and application used by data flow.

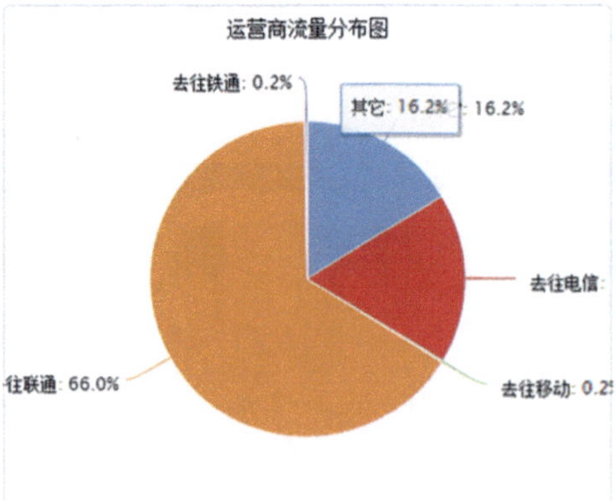

Fig. 8. The data flow direction.

5 Conclusion

The application tests of the design include the function of network equipment parameters, network state and flow management. The testing results show that all function of the system is normal, and the operation of the system has no harmful effects on the network communication. The design achieves the goal that the administrator could manage campus network conveniently. With the operation of specific script events, the design implements network equipment energy-saving. The network security module will be further implemented in the subsequent development.

Acknowledgements. This work was supported by the second batch of Cooperative Education Collaborative Education Project in 2017 of Department of Higher Education of the Ministry of Education. The project name is Construction of Practice Teaching System for Network Engineering with Deep Integration of Industry-University-Research. The Project Number is 201702043013.

References

1. https://www.python.org/
2. Heller, B., Seetharaman, S., Mahadevan, P., et al.: Elastic tree: saving energy in data center networks. In: Proceedings 7th USENIX Conference Networked Systems Design and Implementation, p. 17 (2010)
3. Hlavacs, H., Da Costa, G., Pierson, J.: Energy consumption of residential and professional switches. In: Proceedings CSE 2009, vol. 1, pp. 240–246 (2009)
4. Bianzino, A.P., Chaudet, C., Rossi, D., et al.: A survey of green networking research. IEEE Commun. Surv. Tutor. **14**(1), 3–20 (2012)
5. Yu, D., Peng, L.: When does inferring reputation probability countervail temptation in cooperative behaviors for the prisoners' dilemma game? Chaos, Solitons Fractals **78**, 238–244 (2015)
6. Aleksy, M., Korthaus, A., Schader, M.: Use Java and the CORBA realization distribute type system. J. Pingxiang Coll. **4**, 104–105 (2005)
7. Titus, J.: ECN technical editor: the eclipse of stand. J Zhongkai Agrotech. Coll. **19**(2), 32–35 (2006)
8. Richardson, W.C., Avondolio, D.: The Java high class weaves a distance: JDK 5. Sci. Technol. Book Rev. no. 3, pp. 17–18 (2006)
9. Juan, L., Vuong, S.: A semantics-based routing scheme for grid resource discovery. In: E-Science: First International Conference on E-Science and GridComputing, pp. 58–70, 90 (2005)
10. Wang, Q.: Automatic control system regulation scheme of textile air conditioning. Sci. Technol. Inf. **22**, 395–396 (2010). (in Chinese)
11. Liu, J.: Advanced PID Control and MATLAB Simulation, 2nd edn. Electronics Industry Press, Beijing (2003). (in Chinese)

Image Recognition Technology in Motion Trajectory Planning Based on Intelligent Robot

Zhang Yu(✉)

School of Mechano-Electronic Engineering, Xidian University, Xi'an, China
JohnZhang322@hotmail.com

Abstract. To achieve higher requirements for industrial robots in practical applications, the image pre-processing algorithms, key feature extraction and recognition algorithms, and trajectory planning and generation algorithms for wheel hub images were studied to realize image feature of wheel hub and pre-processing before recognition. Then, the problem of misunderstanding in corner detection was solved. The identification of wheel hub images, the planning and generation of the processing trajectory of wheel hub were completed. Finally, this research was applied to the industrial robot image recognition automatic processing system. The experiment verified the practical feasibility of the industrial robot trapped image recognition automatic processing system. It is concluded that the future intelligent robot can better serve human beings.

Keywords: Industrial robots · Image recognition · Moving trajectory planning Algorithm

1 Introduction

On December 19, 2016, the State Council issued the *Thirteenth Five-Year Plan for the Development of Strategic and Emerging Industries of the State*, which further promoted the development of the robotics industry as a national strategy and highlighted the importance of the robot industry [1].

As a high-end field of equipment manufacturing industry, the industrial robot is a strategic emerging industry with a focus on development in China. In recent years, thanks to the influence of China's internal and external environment, the pulling of traditional industries, and the support of major national industrial policies, the industrial robot industry in China has developed rapidly and has become the world's largest robot market for three consecutive years. The level of industrial robot technology has also become a key factor in determining the manufacturing industry's nuclear record and competition. It has become an important tool and method to reshape the competitive advantages of China's manufacturing industry, and has become a pragmatic choice for accelerating the transformation and upgrading of China's traditional industries.

At present, industrial robots are widely used in the automotive and parts industries, electronic 3c industries, food manufacturing and pharmaceutical industries, logistics and warehousing industries, and rubber chemical industries. Among them, the

J. Abawajy et al. (Eds.): ATCI 2018, AISC 842, pp. 866–873, 2019.
https://doi.org/10.1007/978-3-319-98776-7_103

application of automobiles and their components is the most successful and the most extensive. With the rapid development of China's automotive and parts industries, the production capacity of automobiles and their parts has increased hundreds of times. As an alternative to manual repetitive labor or automated equipment in hazardous and harsh environments, industrial robots are becoming more and more widely used [2–4]. However, there is still a huge gap between the level of industrial robots used in the automotive and parts industries and the developed countries in the United States, Germany, Japan, and South Korea. It is understood that the use density of industrial robots in China is 49 units per 10,000 in 2016, only 1/8 of that in developed countries. Therefore, it is imperative to conduct research on the application of industrial robots in relation to the specific industry needs of the automotive and parts industries.

2 Image Pre-processing Technology of Industrial Robot Image Recognition System

The study of the visual system shows that more than 75% of the information that humans obtain from the outside world is obtained through the visual system. Industrial robot image recognition system is to complete various measurements and judgments by using industrial cameras and related equipment for image acquisition and processing instead of the human visual system in the externally trapped image system of industrial robots. The process flow of the image recognition system of industrial robots can be roughly divided into five modules: The relationship between the modules is shown in Fig. 1.

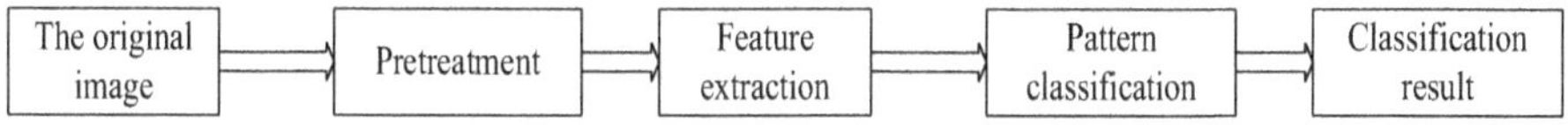

Fig. 1. Image recognition system processing flow.

To a certain extent, the original image will inevitably be damaged by all kinds of noise pollution in the collection process. Therefore, it cannot be directly used in the image recognition system, and the original image must be pre-processed before feature extraction and pattern classification.

2.1 Image Denoising Process

To obtain a high-quality digital image, it is necessary to reduce or eliminate the noise information in the image. The essence of image denoising is to use the type and characteristics of image noise to remove unwanted noise information from the image while maintaining the main feature integrity of the original image as much as possible.

Gaussian noise is also known as positive noise. Gaussian noise originates mainly from changes in the humidity or frequency of an electronic device, and its probability of distribution of grayscale values conforms to a normal distribution. Gaussian noise is the most widely used noise model in the image. It is also easier to process in the spatial and frequency domains. The PDF is shown in formula (1):

$$p(z) = \frac{1}{\sqrt{2\pi}\sigma} e^{-(z-\mu)^2/2\sigma^2} \tag{1}$$

According to the formula, in fact, the pdf of the gray value z is the probability density function of the normal distribution. Therefore, the gray value distribution of the Gaussian noise follows the normal distribution of the mathematical expectation μ and the variance σ^2.

Rayleigh noise is a noise that is subject to Rayleigh distribution. Its PDF is shown in formula 2:

$$p(z) = \begin{cases} \frac{2}{b}(z-a)e^{-(z-a)^2/b}, & z \geq a \\ 0, & z < a \end{cases} \tag{2}$$

The main value μ and variance σ^2 are $\mu = a + \sqrt{\pi b/4}$.

Gamma noise is the noise that obeys the distribution of gamma curves and often appears on laser and radar imaging. Its PDF is shown in formula 3:

$$p(z) = \begin{cases} \frac{a^b z^{b-1}}{(b-1)!} e^{m}, & z \geq a \\ 0, & z < a \end{cases} \tag{3}$$

In the formula, parameter a > 0, and b is positive integer.

Exponential noise is a special case when the gamma noise parameter is b = 1. Its PDF is shown in the formula:

$$p(z) = \begin{cases} ae^{-ax}, & z \geq a \\ 0, & z < a \end{cases} \tag{4}$$

In the formula, the noise is displayed in the image as bright spots (a < b) and dark spots (a > b) according to the magnitude relationship between the grayscale values a and b. In addition, if the probabilities of the gray values a and b are similar and not zero, the impulse noise value will produce bright and dark noise points scattered on the image. According to the characteristics displayed on the image, the noise in this case is also called salt-and-pepper noise. Salt-and-pepper noise is generally the result of impulse interference. Since the wheel hub is a metal object, the reflection intensity on the surface is different, and noise similar to salt-and-pepper noise is also formed.

2.2 Image Enhancement Processing

Image enhancement refers to the enhancement of the useful information of the observer in the image according to the requirements, so that the visual effect of the image is better and it is easier to handle the operation afterwards.

Gray-scale transformation method: Image gray-scale transformation is an image enhancement method based on image pixel point operation. It converts the gray value of the pixel in the image into a new gray value according to a mathematical transformation formula to improve the contrast of the digital image.

Histogram equalization method: The histogram of the image is a statistical report graph that reflects the distribution of gray values of image pixels, while the histogram equalization is an operation that adjusts the number or the proportion of each grayscale pixel in an image to be equal by a certain algorithm. The effect is to expand the number or the proportion of gray value pixel points, and compress the number or the proportion of gray value pixel points, thereby improving the overall image contrast and gray change. Therefore, the image is displayed more clearly [5].

2.3 Image Binarization

The target image processed in this paper is obtained under the condition that the background illumination is uniform, and the image hierarchy is clear. Therefore, this paper comprehensively considers the global binarization processing with simple selection and processing, and solves the global threshold of the image through iterative method. The specific iterative process is shown in Fig. 2.

3 Wheel Hub Image Feature Extraction of Industrial Robot Image Recognition System

Aiming at feature extraction and recognition problems in the image recognition process of automobile wheel hub images, how to accurately extract key features from the wheel hub image and complete the description and recognition of the overall characteristics of the vehicle hub is discussed [6]. As the main research contents, the basic principle of edge detection of automobile wheel hub images is first introduced in detail, and the edge features of wheel hub images are extracted on the basis of edge differential operator detection algorithms. On this basis, by analysing the symmetry characteristics of the contour circle, the hub contour circle feature method based on the symmetric Hough circle transformation and the hub image contour point feature extraction method based on the curvature scale space are proposed. There two methods can extract the key features of the hub image. Finally, based on the key features, the feature vector of the hub image is constructed by combining the invariant moment description method of the image to complete the hub recognition [7].

3.1 Hub Edge Detection Based on Differential Operator

The hub image in this article is an image taken on a laboratory processing console. The background of the image is uniform and the gray value at the edge generally changes in price. Therefore, based on the differential operator detection theory, the local maxima of the amplitude values are calculated and the edge points are extracted. The definition and use of several commonly used differential operators are as follows: Roberts operator, Prewitt operator, Sobel operator, Laplacian operator, Canny operator.

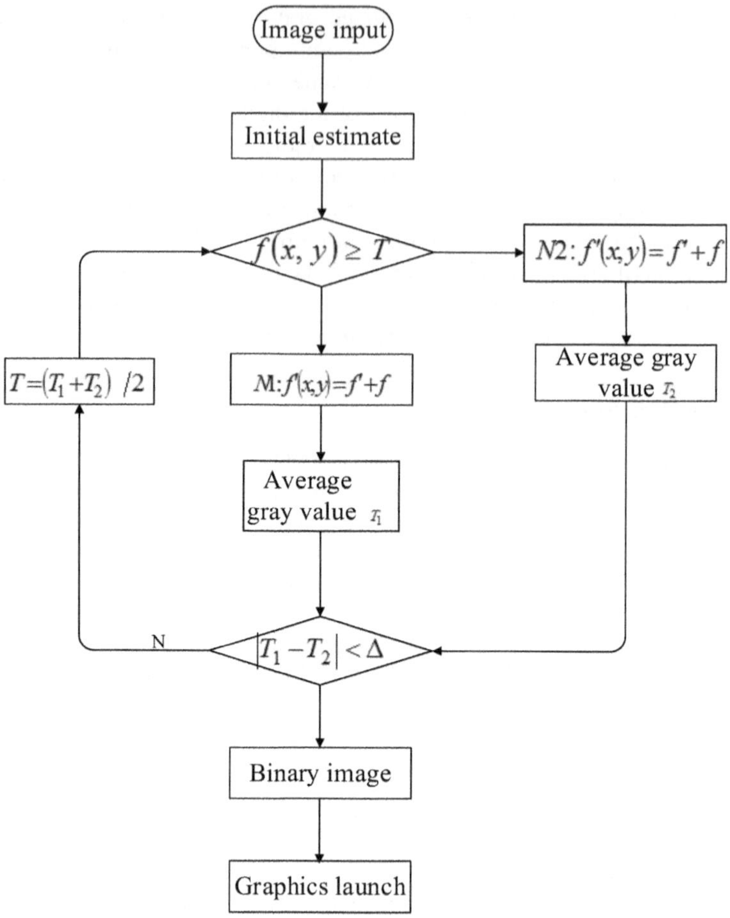

Fig. 2. Global binarization threshold iteration process

The Canny operator has the best edge detection in the experiment. It not only can detect the edge more accurately but also has certain resistance to noise. Therefore, this chapter uses the Canny operator to perform edge detection on the hub image.

3.2 Hub Image Feature Vector Construction and Recognition Based on Invariant Moments Description

To verify the applicability of image moment invariant in the hub image, the original hub image is panned, scaled, mirrored and rotated at different angles to obtain images after different processing. Then, the eigenvalues of the invariant moments of the hub images with different transformations are calculated. The resulting invariant moment eigenvalues are shown in Table 1.

Table 1. Characteristics of invariant moments of hub images

Invariant moment features	Original image	Panning	Reduced in half	Acoustic image	Rotation 45°	Rotation 90°
Φ_1	2.8511	2.8511	2.8511	2.8511	2.8511	2.8511
Φ_2	8.0890	8.0891	8.0942	8.0891	8.0908	8.0891
Φ_3	11.8206	11.8204	11.8211	11.8204	11.8224	11.8204
Φ_4	12.9720	12.9722	12.9724	12.9723	12.9719	12.9726
Φ_5	25.5434	25.5437	25.5452	25.5437	25.5501	25.5432
Φ_6	17.1695	17.1696	17.1698	17.1698	17.1665	17.1706
Φ_7	−25.4968	−25.4969	−25.4967	−25.4970	−25.4828	−25.4982

From the data in Table 1, it is known that the invariant moment feature has strong robustness to various transformation of the wheel hug image, and is suitable for identification as a whole feature of wheel hub detection in practical application.

4 Trajectory Planning of Industrial Robots Based on Key Features of Hub Image

The robot trajectory planning is based on the kinematics of the robot. According to the description of the trajectory, a certain method is used to obtain a procedure that can control the robot to execute the corresponding trajectory [8]. This belongs to the robot floor plan. Trajectory planning can be performed either in the robot joint space or in Cartesian space. In practical applications, planning methods should be selected according to different application purposes and scenarios. However, regardless of the method chosen in the end, the goal is to plan a continuous and smooth trajectory curve so that the end effector moves smoothly without large fluctuations.

Joint space trajectory planning: At present, the trajectory planning is performed in the joint space, and the joint trajectory interpolation method is generally used. The commonly used interpolation methods include the cubic polynomial interpolation method, the high-order polynomial interpolation method, and the parabola transition type linear interpolation method.

Trajectory planning in Cartesian space. The currently used Cartesian space trajectory planning methods include linear interpolation and circular interpolation.

5 Industrial Robot Image Recognition Automatic Processing System Construction and Experiment

To verify that the method presented in this paper can effectively identify the key features of automobile wheel hub and realize trajectory planning and automatic processing on industrial robot systems, an industrial robot image recognition automated processing system is built based on the original industrial robot platform by adding industrial cameras, light sources and computers.

5.1 Hardware Integrated Design of Industrial Robot Image Recognition and Processing System

The workflow of the system is shown in Fig. 3. Through the industrial camera and the lens shooting target object under the ring light source, the digital image signal is transmitted to the computer through the gigabit net port for processing. After the computer completes a series of processing activities such as image recognition and trajectory planning, an industrial robot executable program is generated. After the robot control cabinet accepts the foregoing program, it releases the operation instruction to the industrial robot, and the robot accepts the operation instruction and completes the corresponding processing operation.

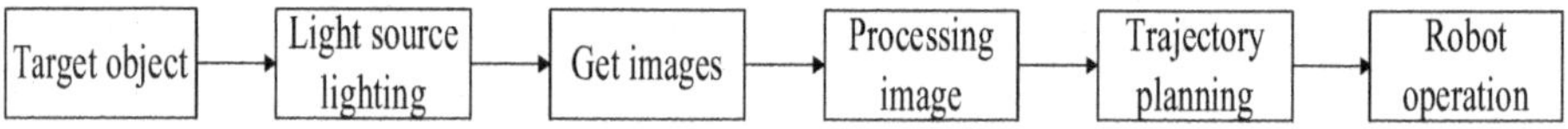

Fig. 3. Process flow chart of the system.

5.2 Software Development of Industrialized Human Image Recognition Processing System

The function module of automatic process system for industrial robot image recognition includes: Image acquisition module, image preprocessing module, feature extraction module and classification recognition module.

Through the actual operation in the laboratory industrial robot image recognition automatic processing system, the key feature extraction, model identification and generation of the trajectory program for the hub are completed. Due to manufacturing errors and installation errors of industrial robots and CCD industrial cameras, and the influence of vibration, light source, and noise of the working environment on the system, these factors should be controlled as much as possible in the recognition process. In practical applications, compensation should be made according to the corresponding operation environment.

6 Conclusion

In this paper, the trajectory planning technology of industrial robot based on image recognition is studied and applied. In order to solve the current automation problems in the hub process, image recognition technology and industrial robot technology are used to complete the application of the hub automatic processing system. It mainly includes the research of related theory of image preprocessing, the design and optimization of image feature extraction and pattern recognition algorithm theory, the research of industrial robot trajectory planning method, and the establishment of an automatic processing system based on industrial robot image recognition technology.

References

1. Kim, Y., Kim, B.K.: Time-optimal trajectory planning based on dynamics for differential-wheeled mobile robots with a geometric corridor. Trans. Ind. Electron. **64**(7), 5502–5512 (2017)
2. Yazici, A., Kirlik, G., Parlaktuna, O., et al.: A dynamic path planning approach for multirobot sensor-based coverage considering energy constraints. Trans. Cybern. **44**(3), 305–314 (2017)
3. Zhu, H., Guan, Y., Chen, S., Su, M., Zhang, H.: Single-step collision-free trajectory planning of biped climbing robots in spatial trusses. Robot. Biomim. **3**(1), 1 (2016)
4. Ruolong, Q.I.: Obstacle avoidance trajectory planning for gaussian motion of robot based on probability theory. J. Mech. Eng. **53**(5), 93 (2017)
5. Xu, J., Yang, H., Liu, M., Tian, J., Liu, B.: Research on winding trajectory planning for elbow pipe based on industrial robot. Int. J. Adv. Manuf. Technol. **93**(1–4), 1–9 (2017)
6. Roozegar, M., Mahjoob, M.J., Esfandyari, M.J., et al.: XCS-based reinforcement learning algorithm for motion planning of a spherical mobile robot. Appl. Intell. **45**(3), 1–11 (2016)
7. Jahanpour, J., Motallebi, M., Porghoveh, M.: A novel trajectory planning scheme for parallel machining robots enhanced with NURBS curves. J. Intell. Rob. Syst. **82**(2), 257–275 (2016)
8. Park, J., Karumanchi, S., Iagnemma, K.: Homotopy-based divide-and-conquer strategy for optimal trajectory planning via mixed-integer programming. IEEE Trans. Robot. **31**(5), 1101–1115 (2017)
9. Yu, D., Peng, L.: When does inferring reputation probability countervail temptation in cooperative behaviors for the prisoners' dilemma game? Chaos, Solitons Fractals **78**, 238–244 (2015)
10. Jianwei, L., Qingchang, R.: Study on supply air temperature forecast and changing machine dew point for variable air volume system. Build. Energy Environ. **27**(4), 29–32 (2008). (in Chinese)

Realtime 3D Reconstruction with Mobile Devices

Xiaomeng Wang(✉)

Hefei University of Technology, Hefei, China
wxm_030@163.com

Abstract. Real-time 3D reconstruction is a core issue in computer vision. Mobile devices provide a convenient tool for 3D reconstruction and stereo vision. First, this paper introduces the technical parameters and depth measurement principle of the tango phone (Lenovo PHAB2 Pro). Secondly, the algorithm flow of the 3D reconstruction system based on mobile devices is described in detail. Finally, it is implemented on the mobile device, and several sets of reconstruction experiments are given. Experimental results show that the reconstruction system can reconstruct objects and large scale 3D scenes conveniently in real time.

Keywords: Tango · ToF · 3D reconstruction · TSDF

1 Introduction

Real-time 3D reconstruction is a core issue in computer vision. Currently, mobile devices integrate a variety of sensors, and as their data acquisition and processing capabilities continue to increase, they provide a convenient tool for rapid 3D data acquisition, 3D scene understanding, and effective information extraction.

Project Tango is Google's public research in 2014. The purpose is to make the mobile device have human-like spatial awareness and motion capture capabilities. Tango device is equipped with a depth sensor, which provides a new mobile platform for active 3D reconstruction. Currently, Tango only supports several hardware devices, including the Tango "Peanut" prototype phone, Tango "Yellowstone" tablet, Lenovo PHAB2 Pro and Asus Zenfone AR.

There have been some related work on 3D reconstruction based on Tango devices. Dryanovski [1, 2] and others developed CHISEL software to acquire depth data through Tango device and use active vision to achieve large-scale 3D scene reconstruction. Based on CHISEL [1], Dzitsiuk et al. [3] added geometric prior in the 3D reconstruction process to reduce the noise on the planar structure and to some extent fill the holes on the planar structure. The depth sensor provided by Tango does not apply to bright outdoor scenes, therefore, Schöps [4, 5] uses the devices's GPU to compute depth maps via plane sweep stereo. Hong [6] studied the SLAM algorithm based on Tango equipment and optimized the 3D reconstruction results of Tango API.

CHISEL proposed by Dryanovski is based on two devices: one is the Tango "Yellowstone" tablet and the other is the Tango "Peanut" mobile phone. The experimental

J. Abawajy et al. (Eds.): ATCI 2018, AISC 842, pp. 874–882, 2019.
https://doi.org/10.1007/978-3-319-98776-7_104

equipment used in this article is Tango's first consumer phone Lenovo PHAB2 Pro. Based on this phone, Tango API can also be used to achieve real-time 3D reconstruction of indoor scenes, but the source code is not open. So, this paper builds a 3D reconstruction system based on Lenovo PHAB2 Pro. First, this paper introduces the technical parameters and depth measurement principle of the Lenovo PHAB2 Pro. Secondly, the system flow of the 3D reconstruction based on mobile devices is described in detail. Finally, it is implemented on the mobile device, and several sets of reconstruction experiments are given.

2 Lenovo PHAB2 Pro

Lenovo PHAB2 Pro (see Fig. 1) has a Qualcomm Snapdragon CPU and 4 GB of RAM. It contains a low-cost inertial measurement (IMU). In addition, it also contains three rear camera: wide-angle fish-eye camera, high-resolution color camera, depth camera. With Lenovo PHAB2 Pro, depth data (Fig. 1b) and texture data (Fig. 1c) of the scene can be obtained. The technical parameters of Lenovo PHAB2 Pro are listed in Table 1.

a. hardware b. 3D point cloud c. color image

Fig. 1. Hardware and raw data of Lenovo PHAB2 Pro

The depth measurement of Lenovo PHAB2 Pro uses ToF technology. For each pixel, the 3D image sensor chip can perceive the time it takes for infrared rays to be projected from the depth camera to the object. Calculate the distance according to the speed of light to obtain depth information (Fig. 2).

3 Mobile Device-Based Real-Time 3D Reconstruction System

This paper builds a real-time 3D reconstruction system with mobile devices based on CHISEL and Tango API. Figure 3 is a flowchart of the overall algorithm.

The detailed algorithm flow is as follows:

(1) Data acquisition and preprocessing

First, get the original depth data and color data through the Tango APIs. The depth data provided by Tango is a 3D point cloud in the depth camera coordinate system. Therefore, a point cloud projection method is required to obtain a 2D depth image.

Table 1. Technical parameters of Lenovo PHAB2 Pro

Items	Specification
Color camera resolution	1920 × 1080
Color data frequency	30 Hz
Depth camera resolution	224 × 172
Depth accuracy	2 cm
Depth effective value	0.5–4 m
Depth data frequency	3–5 Hz
Screen size	2560 × 1440
Memory	64 GB ROM, 4 GB RAM

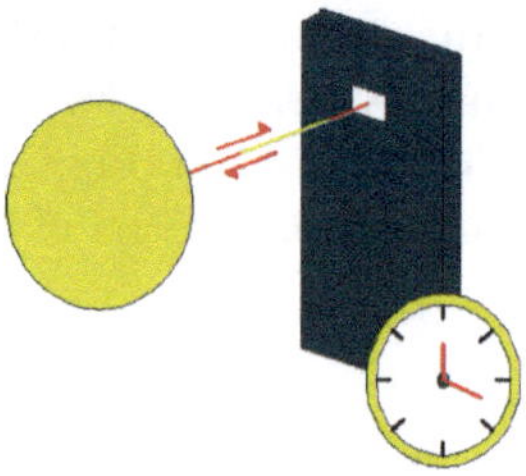

Fig. 2. Schematic diagram of distance measurement using Time of Flight

The correspondence between a space point (x, y, z) and its pixel coordinates $[u, v, d]$ (d refers to the depth value.) in the image is as follows:

$$\begin{aligned} u &= \frac{x}{z} * f_x + c_x \\ v &= \frac{y}{z} * f_y + c_y \\ d &= z \end{aligned} \tag{1}$$

where f_x, f_y, c_x, c_y is the camera intrinsics.

In addition, since the depth data and the color data are not in the same coordinate system, registration is also required. The registration results are shown in Fig. 4.

(2) Single-frame color point cloud acquisition

The module takes color image and depth image as input, and generates a single-frame color point cloud by means of projection. Figure 5 is a schematic diagram of a single-frame color point cloud.

(3) Pose Estimation Through VIO

Figure 6 shows the four single-frame point clouds collected at different times. Only when the position and posture of each point cloud is known, the point clouds can be merged. Lenovo PHAB2 Pro is equipped with a 30 Hz built-in visual inertial ranging

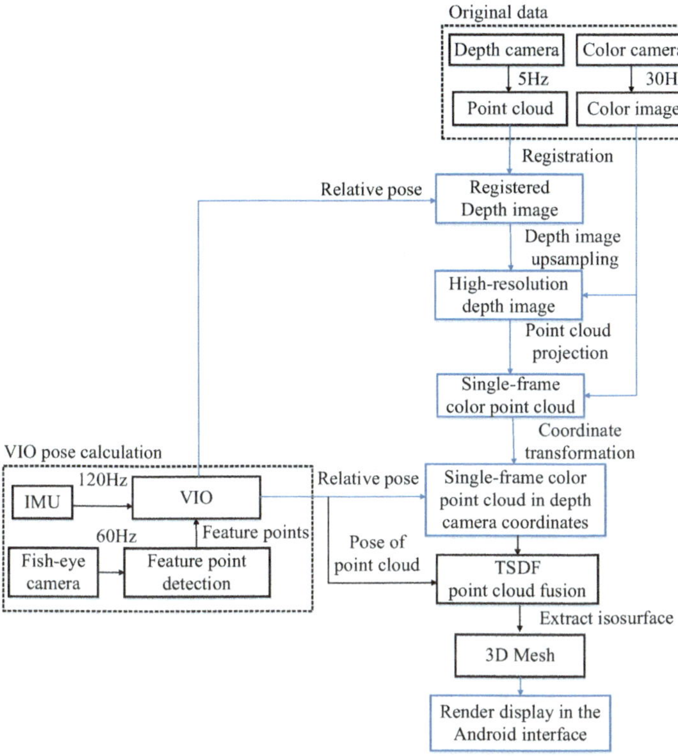

Fig. 3. Realtime 3D Reconstruction System Based on Lenovo PHAB2 Pro

Fig. 4. Registration results of depth data and color data

module and this module has been implemented in the Tango API. This article directly reads pose information from the Tango API for registration and point cloud fusion. Figure 7 shows the gradual integration of four gesture point clouds.

The VIO module used in this paper is an open-loop system. As the trajectory increases, the pose measurement will produce a significant deviation (see Fig. 8).

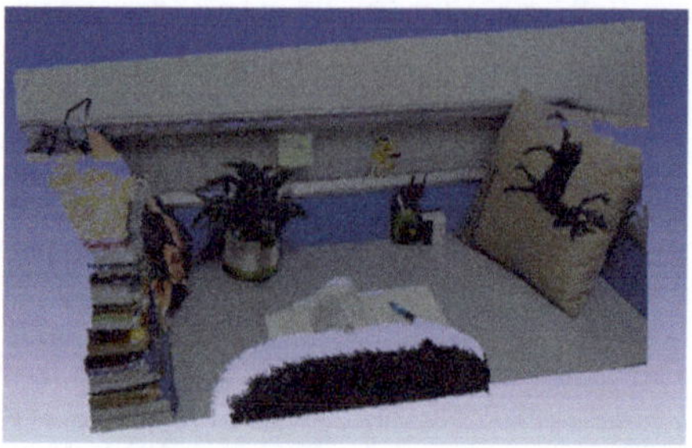

Fig. 5. The 3D color point cloud of single frame

Fig. 6. 4 single - frame color point clouds

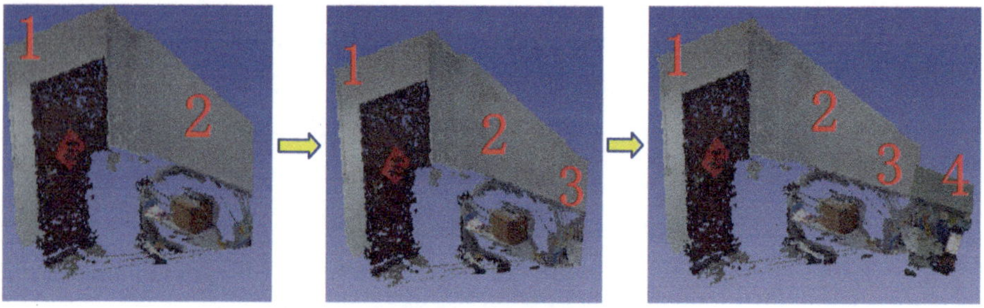

Fig. 7. The merging process of point clouds

(4) 3D Point Cloud Transformation

The module completes the coordinate system transformation of the point cloud. The purpose is to make the point-cloud pose read from the Tango API and the current color point cloud correspond to each other, so as to carry out the next point cloud fusion operation.

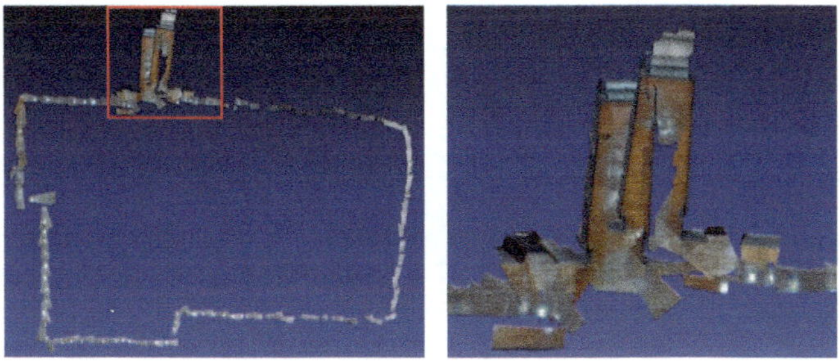

Fig. 8. System pose drift in a long corridor

(5) TSDF Cloud Integration

The module's input is the transformed color point cloud and the point cloud pose. Using TSDF to achieve the fusion of multi-frame point cloud data, that is, the point cloud calculated by each frame will be integrated into the global model.

After the depth data is aligned with the color data, each depth value corresponds to a color value. For simplicity, the RGB color space is used to represent color consistency under different lighting conditions. The voxel storage information is:

```
struct Voxel
{
    float  sdf ;               // TSDF
    uchar  colorRGB[3] ;       // the color
    uchar  weight ;            // the weight
};
```

If no color information is stored in the voxels, a non-textured 3D model is obtained. Figure 9 shows a comparison of models with and without textures.

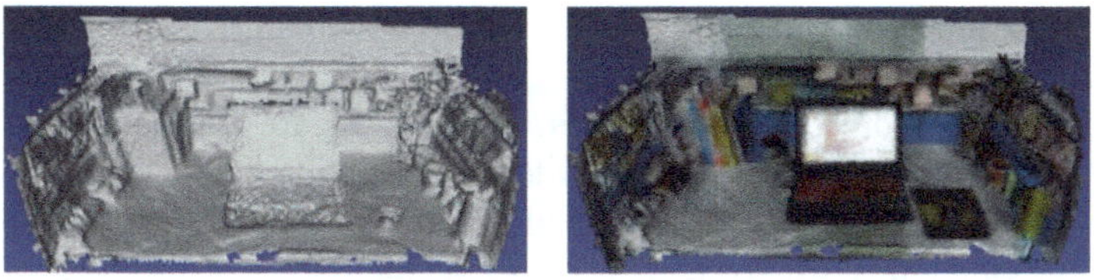

Fig. 9. 3D Mesh model of the table scene

(6) Mesh generation

The module uses the Marching Cubes algorithm to extract the iso-surface from the stored TSDF to generate a 3D mesh model.

(7) Rendering display

This module mainly displays the 3D mesh model on the screen through OpenGL ES. Figures 10 and 11 show the coordinate transformation process.

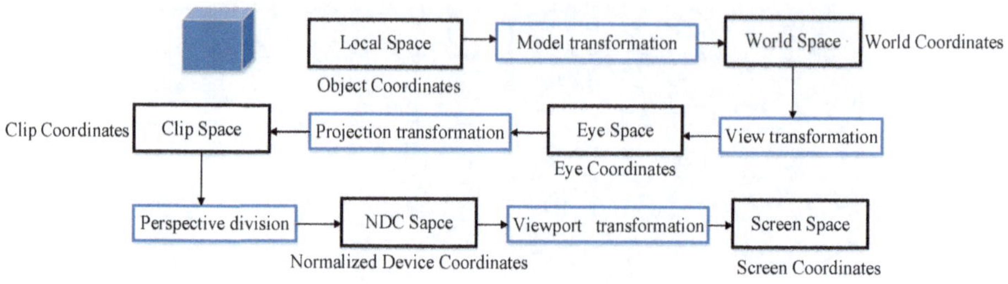

Fig. 10. Coordinate transformation process

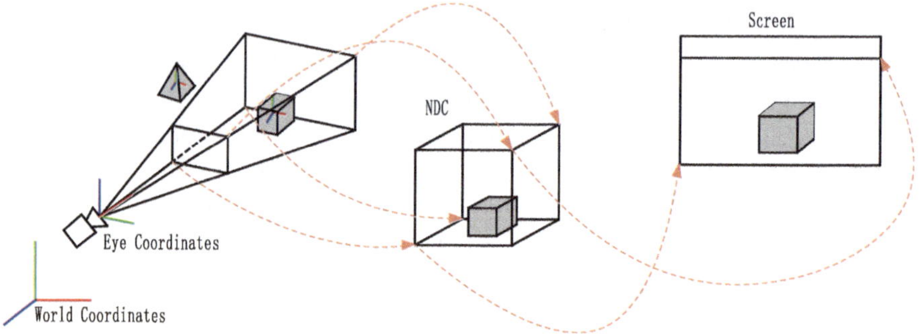

Fig. 11. Schematic diagram of the coordinate transformation

4 System Implementation and Experimental Results

System Implementation
According to the flow of the 3D reconstruction system described above, it is implemented on Lenovo PHAB2 Pro. The interface is shown in Fig. 12. The operating environment is: 2.3 GHz Intel Core i5, 12 GB RAM, Ubuntu16.04 operating system. The development tool is: Android Studio 2.2.

Figure 12a shows object reconstruction interface. Figure 12b shows indoor scene reconstruction interface. The designed Android application can switch between three viewing modes: first person perspective, third person perspective and top perspective; The status information is displayed at the top of the interface which contains the current available memory, the maximum measurement distance, the voxel grid resolution and the power amount information. The blue line in the reconstruction scene is the camera trajectory recorded in real time; In addition, the application also has the functions of the start and stop of the reconstruction, the saving of the model and the screen clearing.

Experimental Results
As shown in Figs. 13 and 14, the real-time 3D reconstruction system proposed in this paper can create and display 3D models of different resolutions in real time on mobile devices.

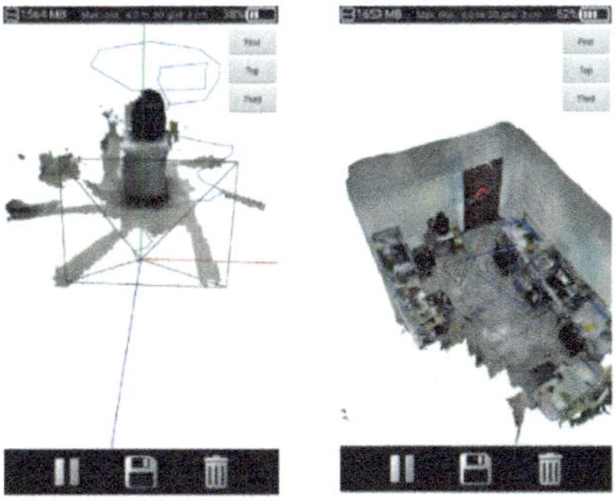

a. b.

Fig. 12. Interface of the real-time 3D reconstruction system on Lenovo PHAB2 Pro

a. 5mm b. 1cm c. 3cm

Fig. 13. Table scene reconstruction results under different resolution

a. 5mm b. 1cm c. 3cm

Fig. 14. Reconstruction results of schoolbags under different resolution

5 Conclusion

Based on Lenovo PHAB2 Pro, this paper implements a 3D reconstruction system with mobile devices, which can complete real-time 3D reconstruction of objects and 3D scenes. Due to the drift of the VIO module, further work can be done to correct the trajectory using methods such as bundle adjust.

References

1. Dryanovski, I., Klingensmith, M., Srinivasa, S.S., Xiao, J.: Large-scale, real-time 3D scene reconstruction on a mobile device. Auton. Robots **41**(6), 1423–1445 (2017)
2. Klingensmith, M., Dryanovski, I., Srinivasa, S.S., Xiao, J.: Chisel: real time large scale 3D reconstruction onboard a mobile device using spatially hashed signed distance fields. In: Proceedings of Robotics: Science and Systems (2015)

3. Dzitsiuk, M., Sturm, J., Maier, R., Ma, L., Cremers, D.: De-noising, stabilizing and completing 3D reconstructions on-the-go using plane priors. In: 2017 IEEE International Conference on Robotics and Automation (ICRA), pp. 3976–3983 (2017)
4. Schöps, T., Sattler, T., Häne, C., Pollefeys, M.: 3D Modeling on the go: interactive 3D reconstruction of large-scale scenes on mobile devices. In: Proceedings of 2015 International Conference on 3D Vision, pp. 291–299 (2015)
5. Schps, T., Sattler, T., Hne, C., Pollefeys, M.: Large-scale outdoor 3D reconstruction on a mobile device. Comput. Vis. Image Underst. **157**(C), 151–166 (2017)
6. Hong, C.: SLAM Research based on Project Tango. Computer School, Wuhan University, Wuhan (2017). (in Chinese)

Three Dimensional Finite Element Analysis on the Failure Process of a Reinforced Concrete Arch Structure by Damage Theory

Shaohui Xiong[1(✉)] and Xuesong Tang[2]

[1] China Merchants Chongqing Communications Research and Design Institute Co., Ltd., State Key Laboratory of Bridge Engineering Structural Dynamics, 33, Xuefu Avenue, Nan'an District, Chongqing 400067, China
381234694@qq.com, xiongshaohui@cmhk.com

[2] School of Civil Engineering and Architecture, Changsha University of Science and Technology, Changsha 410114, Hunan, China

Abstract. A damage constitutive model with 3 independent parameters is derived from the two-order form of the general theory for elastic damage problems. Based on this damage constitutive model and by using damage mechanics – additional load – finite element method, a three dimensional finite element analysis computer program is established. The damaging process and failure of a reinforced concrete arch structure which was removed from a real old bridge is simulated numerically. The damage parameters of concrete are determined through a targeting calculation process. The values of element damage and the nodal displacements along three directions under different load level are obtained. The theoretical results of the failure mode, displacements are in good agreement with the experimental observations. Numerical results reveal the complicated behavior of concrete in a damaging process. The reinforced concrete structures exhibit pronounced nonlinearity at large loading levels.

Keywords: Reinforced concrete · Arch structure · Damage mechanics
Three dimensional finite element method · Numerical simulation
Nonlinearity

Damage mechanics was proposed by Kachanov in 1958, the core of which is shrinkage of effective loading area, namely effective stress. Classic damage theory developed on the basis of the damage model of Kachanov was deficient in its theory (references [1–5]). A general theory of elastic isotropy and anisotropic damage was built up by second law of thermodynamics in references [3–5].

Damage mechanics is a good way to analyze the concrete structure with flaws [6–8]. In reference [9], elastic damage theory based on temperature effect was built up through the basic laws of thermodynamics. In reference [10], nonlinear loading process of the concrete girder was simulated through subsection curve damage model, where the damage level was also predicted. In reference [11], the dynamic performance of concrete structure was studied through damage caused by crack. In reference [12], the damage process of arch structure was analyzed and calculated without considering spatial effects. In this paper, two-order form of the general theory for elastic damage and damage mechanics – additional load – finite element method [13] are adopted to

J. Abawajy et al. (Eds.): ATCI 2018, AISC 842, pp. 883–893, 2019.
https://doi.org/10.1007/978-3-319-98776-7_105

make a more precise three dimensional finite element analysis on a reinforced concrete arch structure.

1 Concrete Damage Model Based on Elastic Damage

Concrete structure is easy to occur elastic-brittle damage. According to the second law of thermodynamics, the general form of constitutive relationship of stress-strain is [4]:

$$\sigma_{ij} = 2\mu\varepsilon_{ij}M_{\mu}(D) + \lambda\varepsilon_{kk}\delta_{ij}M_{\lambda}(D) \tag{1}$$

where λ, μ is elastic constant of Lame, δ_{ij} is tensor of Kronecker, σ_{ij}, ε_{ij} are stress and strain tensor respectively; D is isotropic damage variable; $M_{\lambda}(D)$, $M_{\mu}(D)$ are two damage effect functions, representing the effect of damage on two elastic constants. The general form of dual variable of damage Y (i.e. thermodynamic dual variable of damage variable D) is:

$$Y = -\frac{1}{2}\lambda M'_{\lambda}(D)(\varepsilon_{kk})^2 - \mu M'_{\mu}(D)\varepsilon_{ij}\varepsilon_{ij} \tag{2}$$

where the apostrophe is derivation. The general form of damage evolution equation is:

$$\dot{D} = AY^m = A\left[-\frac{1}{2}\lambda M'_{\lambda}(D)(\varepsilon_{kk})^2 - \mu M'_{\mu}(D)\varepsilon_{ij}\varepsilon_{ij}\right]^m \tag{3}$$

where A and m are damage evolution parameters of concrete structure.

In practice, series of load is given by damage effect functions $M_{\lambda}(D)$ and $M_{\mu}(D)$:

$$\sigma_{ij} = 2\mu\varepsilon_{ij}\left[1 - \sum_{n=1}^{N}\beta^{(n)}D^n\right] + \lambda\varepsilon_{kk}\delta_{ij}\left[1 - \sum_{n=1}^{N}\alpha^{(n)}D^n\right] \tag{4}$$

$$Y = \frac{1}{2}\lambda\sum_{n=1}^{N}\alpha^{(n)}nD^{n-1}(\varepsilon_{kk})^2 + \mu\sum_{n=1}^{N}\beta^{(n)}nD^{n-1}\varepsilon_{ij}\varepsilon_{ij} \tag{5}$$

$$\dot{D} = A\left[\mu\varepsilon_{ij}\varepsilon_{ij}\sum_{n=1}^{N}n\beta^{(n)}D^{n-1} + \frac{1}{2}\lambda(\varepsilon_{kk})^2\sum_{n=1}^{N}n\alpha^{(n)}D^{n-1}\right]^m \tag{6}$$

where $\alpha^{(n)}$, $\beta^{(n)}$ (n = 1, 2,…, N) are dimensionless positive damage parameter of concrete structure. Make $M_{\lambda}(D) = M_{\mu}(D)$, meaning $\alpha^{(n)} = \beta^{(n)}$, where the effect of damage on Poisson Ratio is ignored. Strain tensor from the second order approximation of formula (4)–(6) (i.e. N = 2) is divided into volumetric strain and shear strain, then

$$\sigma_{ij} = \left(1 - \alpha_1 D - \alpha_2 D^2\right)\left(2\mu\varepsilon_{ij} + \lambda\varepsilon_{kk}\delta_{ij}\right) \tag{7}$$

$$Y = (\alpha_1 + 2\alpha_2 D)\left[\frac{3}{2}\mu\varepsilon_{eq}^2 + \left(3\mu + \frac{9}{2}\lambda\right)\varepsilon_m^2\right] \tag{8}$$

$$\dot{D} = AY^m = A(\alpha_1 + 2\alpha_2 D)^m\left[\frac{3}{2}\mu\varepsilon_{eq}^2 + \left(3\mu + \frac{9}{2}\lambda\right)\varepsilon_m^2\right]^m \tag{9}$$

where ε_{eq} is equivalent strain of von Mises, the effect of shear deformation on damage evolution; ε_m is average normal strain, the effect of volumetric deformation on damage evolution. There are 3 independent damage parameters in the damage constitutive model of concrete, α_1, A and m, $1 + a_2 = 1$. α_1 and a_2 depend on the concrete property. Parameter A presented overall damage speed and deformation capacity of concrete. Parameter m presented an exponential relationship between damage evolution and the state of strain of concrete. These 3 independent damage parameters can be defined by targeting calculation process and test results.

2 Finite Element Numerical Method for Damage Mechanics Problems

Element stiffness equation derived from principle of virtual work for any element i of second-order form in elastic damage constitutive model is:

$$\left(1 - \alpha_1 D - \alpha_2 D^2\right)[K_e]_{(i)}\{\delta\}^e = \{F\}_{(i)} \tag{10}$$

where $[K_e]_{(i)}$ is a common element stiffness matrix, $\alpha_1 + a_2 = 1$, when $D = 1$ and the concrete structure was in failure, effective elastic constant decreased to 0. $\{F\}_{(i)}$ is a common element equivalent node load vector. Item with damage variable D on the left of Eq. (10) was moved to the right, and additional loading of damage $\{F\}_{(i)}^{(D)}$ was defined as:

$$\{F\}_{(i)}^{(D)} = \left(\alpha_1 D + \alpha_2 D^2\right)[K_e]_{(i)}\{\delta\}^e \tag{11}$$

Equation (10) can be written as

$$[K_e]_{(i)}\{\delta\}^e = \{F\}_{(i)} + \{F\}_{(i)}^{(D)} \tag{12}$$

Equation (10) is equal to Eq. (12), which means damage effect is able to lower element stiffness. If element stiffness is invariable, element loading increases after damage.

An overall finite element equation was obtained by assembling a global stiffness matrix with each element stiffness matrix:

$$[K]\{\delta\} = \{F\} + \{F\}^{(D)} \tag{13}$$

where $[K]$ is the global stiffness matrix of structure, $\{\delta\}$ is the total nodal displacement vector, $\{F\}$ is the total equivalent nodal load vector formed by external load, $\{F\}(D)$ is additional load vector of structural damage.

Quasi-static loading procedure can be divided into many small loading steps. In each loading step, damage velocity can be approximated as damage increment ΔD, then Eq. (9) can be changed as:

$$\Delta D = A(\alpha_1 + 2\alpha_2 D)^m \left[\frac{3}{2}\mu\varepsilon_{eq}^2 + \left(3\mu + \frac{9}{2}\lambda\right)(\varepsilon_m)^2\right]^m \tag{14}$$

The computational format of damage mechanics-additional load-finite element is:

(1) Divide the total loading procedure into n loading steps firstly. Take the initial value, $D = D^{(0)}$ (initial damage field), $\{F\}_0^{(D)} = 0$, $\mu_i^{(0)} = 0$, $\varepsilon_{ij}^{(0)} = 0$. Within the first loading step, work out displacement field $\mu_i^{(1)}$ and strain field $\varepsilon_{ij}^{(1)}$ through the finite element Eq. (13).
(2) Calculate the damage degree increment $\Delta D^{(1)}$ of each element produced within the first loading step according to Eq. (14). The damage degree of each element is:

$$D^{(1)} = D^{(0)} + \Delta D^{(1)} \tag{15}$$

(3) Overall additional loading vector $\{F\}_{(1)}^{(D)}$ of the concrete structure was obtained by calculating overall additional loading vector for each element under the current damage field through Eq. (11). Work out displacement field $\mu_i^{(2)}$ and strain field $\varepsilon_{ij}^{(2)}$ within the second loading step through finite element Eq. (13).
(4) Calculate step (2) and step (3) until achieving ultimate loading $P_{\max}$. When the damage degree D of an element achieved or exceeded 1, made $D \equiv 1$ with no change. This element was in failure without bearing capacity. Though element stiffness factor decreased to 0 and stress value was 0, displacement and strain would change.

3 Numerical Result and Analysis

A Static load failure test was performed on a reinforced concrete arch rib from an old bridge for 28 years in the structural testing center of Changsha University of Science and Technology. The concrete mark is C25, bar diameter is 14 mm. Some of its geometrical parameters are: arch height $H = 3.073$ m, arch length $L = 19.72$ m, arch thickness $T = 20$ cm, the inclination of arch axis at the arch base $\beta = 35.5°$, the cross section of the arch is rectangle, its height $h = 28$ cm. Both ends of the arch were clamped. There were three loading points which were in the 1/4 span, middle span and 3/4 span of the arch. There were 18 levels of loading. In every level of loading, the action of these three loading points were the same, that was $P_1 = P_2 = P_3 = P$. The ultimate load $P_{\max} = 65$ kN. After the test, the left arch base broke up, the bar at the left arch base shielded and produced large plastic deformation and displacement, shown in Fig. 1. There were macro-crack at the right arch base, concrete in the 1/8 span and 7/8 span was in serious damage, a large and deep hole formed on the bottom of the 1/8 span of one arch, shown in Fig. 2. No failure occurred on the top of the arch.

Fig. 1. Photo of reinforced concrete arch after failure

The arch structure was not within the perpendicular plane totally, the projection was a slight S shape. The spatial effect should be considered to make sure accuracy. In the spatial finite element model, geometry of arch was build up according to test data. The finite element grid was divided into 6 layers of element along the height of the arch cross section, the two second layers of element from the upper and lower of the finite element grid are equivalent reinforced element. The concrete parameters are: elastic modulus of cement concrete E_1 = 30 Gpa, Poisson Ratio v_1 = 0.17, elastic modulus of bar E_2 = 210 Gpa, bar diameter d = 14 mm. Elastic modulus of equivalent reinforced element were defined according to the principle of equivalent axial stiffness, E^* = 43.57 GPa, Poisson Ratio v^* = 0.17. It was observed there were some damage at the arch base and in the places of 1/8 and 7/8 span before test. So the initial damage in these places is 0.1, the initial damage in other places is 0. It was only observed corrosion on some parts of the bar at the arch base after the test. There was initial damage in the equivalent reinforced element at the arch base. The loading process was divided into 100 steps. Three damage parameters α_1, A and m were adjusted to perform targeting calculation according to ultimate load and failure mode of arch from the test. Parameters of the damage model were defined, $\alpha_1 = 0.05$, $\alpha_2 = 0.95$, $A = 1.0 \times 10^{-5}$, $m = 0.54$.

Distribution and evolution of damage field of arch structure of loading process are shown in Fig. 3. Evolving curve of damage degrees in different parts of arch structure is shown in Fig. 4. There was basically no damage on the arch structure under small loading level. When P/P_{max} = 83%, failure began on the upper edge of the left arch base. The left arch base broke with large plastic deformation in the bar under ultimate load. When P/P_{max} = 90%, failure occurred in some elements of the upper edge of the right arch base. When P/P_{max} = 95%, failure occurred in some elements of 1/8 span and the lower edge of the 7/8 span. No failure occurred in elements of vault. Basic law presented in Figs. 3 and 4: damage was slowly developed at the beginning of loading.

Fig. 2. A big hole formed in the 1/8 span

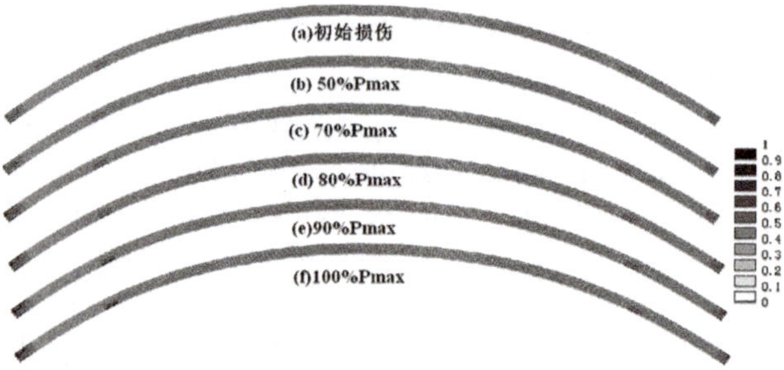

Fig. 3. Distribution and evolution of damage field of arch structure

The damage degree increased to 1 when close to the ultimate loading. It showed elastic and brittle performance of concrete.

In vertical displacement, the downward is positive and upward is negative. When P/P_{max} = 90.88%, the largest vertical displacement was in the middle span and its test value was 56.43 mm. When P/P_{max} = 90%, the results from damage theory and linear elastic finite element (irrespective of damage of concrete) are 58.75 mm and 46.25 mm respectively. The result from linear elastic finite element (irrespective of damage of concrete) is smaller with big error. Vertical displacement curves in different places of the arch structure are shown in Figs. 5, 6, 7, 8 and 9. There are 3 curves in each figure: test curve, curve calculated through damage theory and curve calculated through linear elastic finite element without regard to damage.

In Fig. 5, the curve calculated through damage theory is close to the test curve. Notably there is a step near P/P_{max} = 70% for the test curve in Fig. 5, which is reflected from the curve calculated through damage theory. These curves in Fig. 7 are consistent because the damage in the vault was always small.

In Fig. 8, the test curve is a complicated curve, which is in the same direction compared to the curve calculated through damage theory. These two curves are close

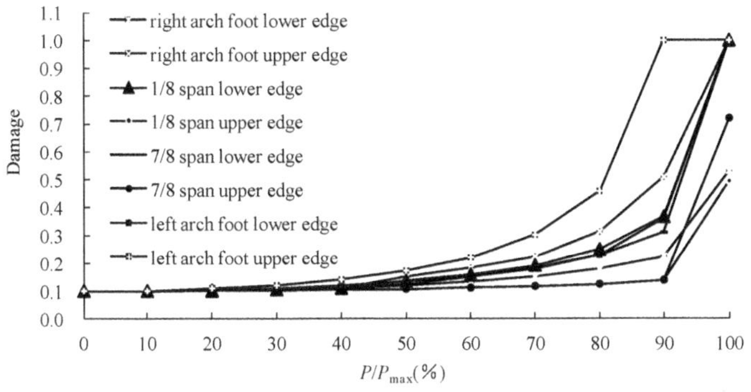

Fig. 4. Evolving curve of damage degrees in different parts

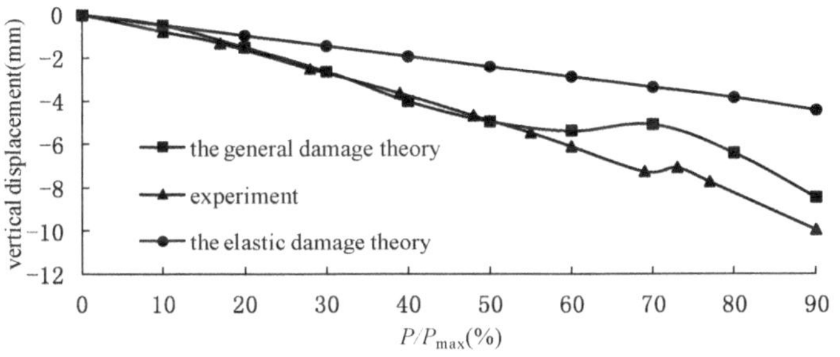

Fig. 5. Vertical displacement in 1/8 span

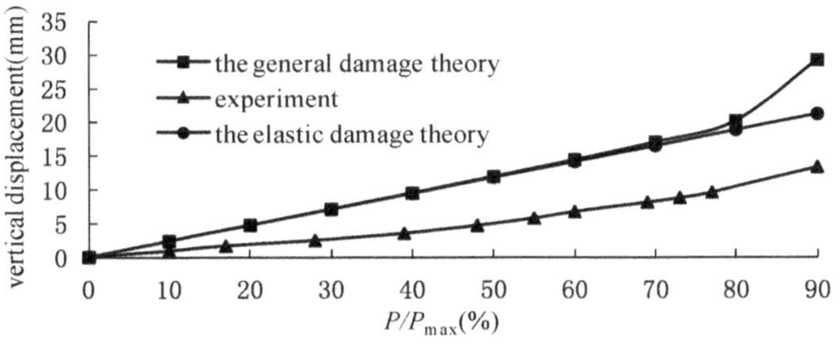

Fig. 6. Vertical displacement in 1/4 span

when P/P_{max} = 90%. When P/P_{max} = 83%, failure occurred in two places of the left arch base and 7/8 span, stress field and strain field redistributed with the evolution of damage field. Displacement curve in the 3/4 span is very complicated since it is sensitive to these two failure places. In Fig. 9, the curve calculated through damage theory is close to the test curve.

The results of linear elastic finite element irrespective of damage are in straight line. In all, the results from damage theory are better than the results from linear elastic finite

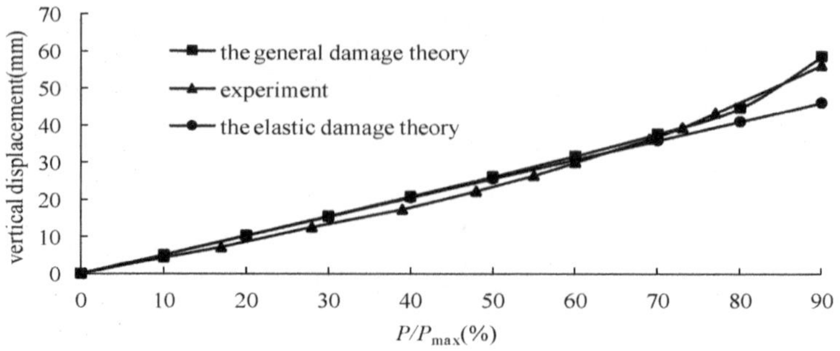

Fig. 7. Vertical displacement in 1/2 span

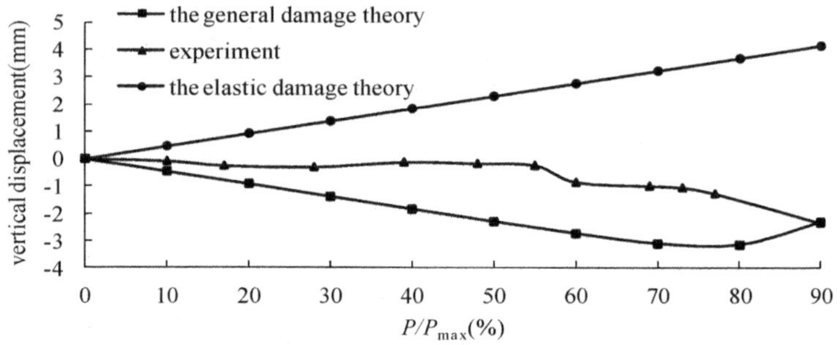

Fig. 8. Vertical displacement in 3/4 span

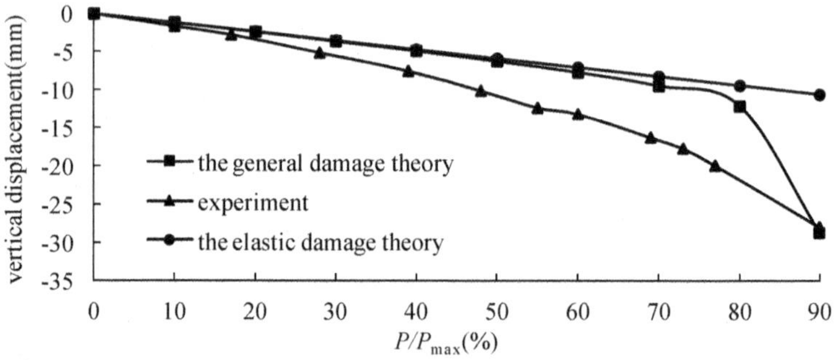

Fig. 9. Vertical displacement in 7/8 span

element irrespective of damage, which can reflect nonlinear behavior of concrete during loading.

In lateral displacement, north direction is positive. When $P/P_{max} = 90.88\%$, the lateral displacement of the middle span is 2.9 mm. If the arch rib was tend to produce lateral displacement during test, lateral friction would occur in the loading point which was opposite to the lateral movement direction. Add 0.036 P of lateral force in the 1/4 loading point, −0.024 P in the 1/2 loading point, −0.032 P in the 3/4 loading point (P is the vertical load of this loading point, the direction is in line with the coordinate axis). Then results from damage theory agree with the test results. Lateral displacement curves in different places of arch structure are shown in Figs. 10, 11, 12, 13 and 14. There are 2 curves in every figure: test curve and curve of results from damage theory.

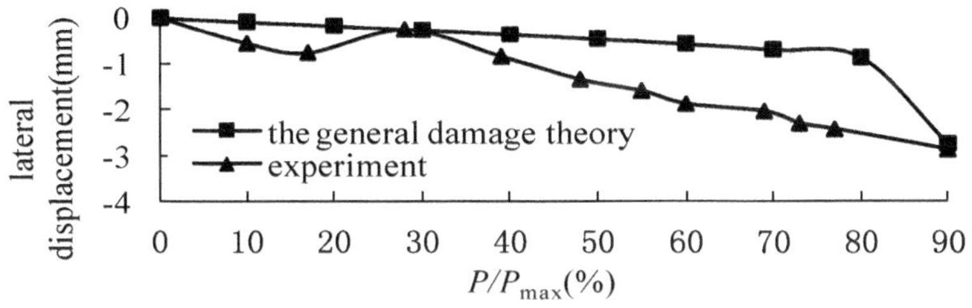

Fig. 10. Lateral displacement in 1/8 span

It is shown in Fig. 10 that the curve begins to present obvious nonlinear changes when P/P_{max} = 70% and is close to the test curve when P/P_{max} = 90%. There is a step on the curve between P/P_{max} = 10% and P/P_{max} = 30%, caused by personal error.

It is shown in Fig. 11 that error is large during loading process even though calculated results and test results agree well when P/P_{max} = 90%. This error may come from modeling error of finite element. There was a loading point in 1/4 span. The lateral force added was positive, which had a large effect on nodal displacement of 1/4 span. At the beginning of loading of the test curve in Fig. 12, the vault waved. When P/P_{max} = 40%, test results and calculated results agree well. It is shown in Fig. 13 that the error between calculated results and test results is big in loading process, caused by the difference of elastic modulus of concrete in 3/4 span and elastic modulus of concrete of the finite element model in damage theory. In Fig. 14, calculated results and test results agree the best.

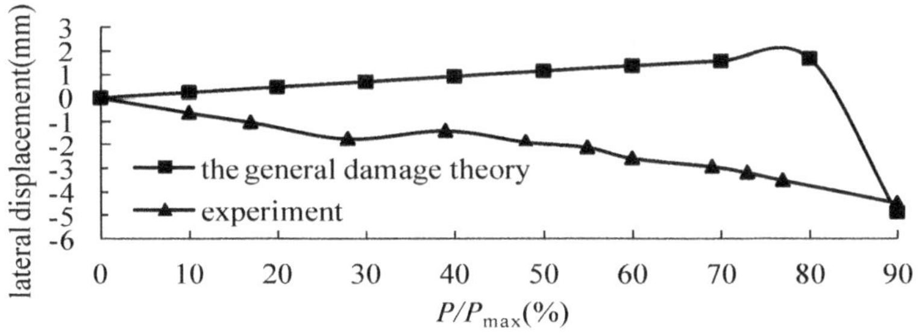

Fig. 11. Lateral displacement in 1/4 span

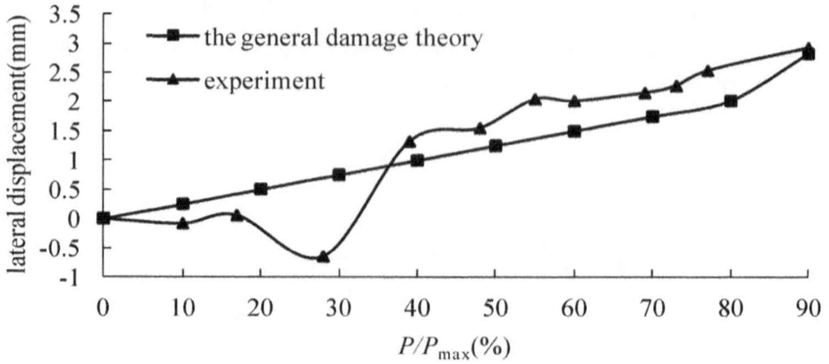

Fig. 12. Lateral displacement in 1/2 span

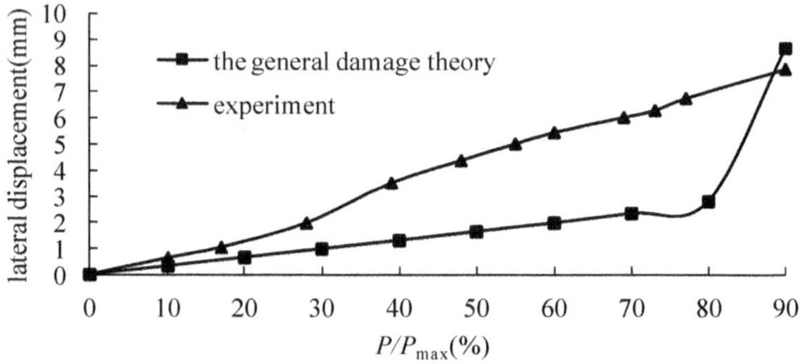

Fig. 13. Lateral displacement in 3/4 span

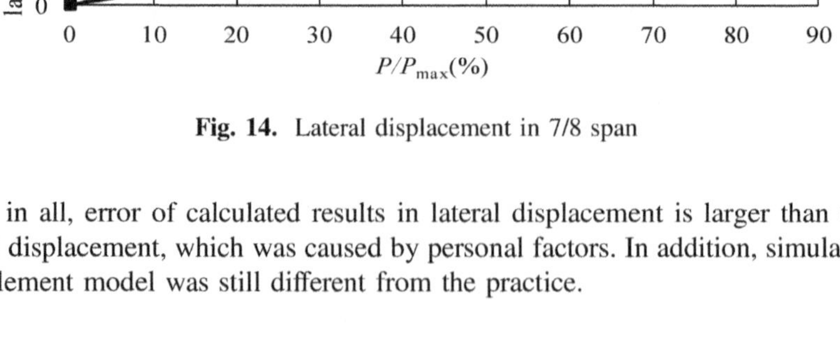

Fig. 14. Lateral displacement in 7/8 span

All in all, error of calculated results in lateral displacement is larger than that in vertical displacement, which was caused by personal factors. In addition, simulation in finite element model was still different from the practice.

4 Conclusions

(1) Numerical results showed that calculated results through damage theory were better than that through linear elastic finite element irrespective of damage. Damage constitutive model in this paper reflected nonlinear mechanics behavior of concrete materials during failure process.
(2) Damage failure process of reinforced concrete structure could be simulated precisely and damage degree could be forecasted quantitatively according to finite element numerical analysis on the basis of damage theory.
(3) Damage effect of concrete materials should be taken into consideration under large load to improve the accuracy of structural analysis.

Acknowledgement. This research was financially supported by **the National Basic Research Program of China (973 Program) (2015CB057702)**.

References

1. Rabier, P.J.: Some remarks on damage mechanics. Int. J. Eng. Sci. **27**(1), 29–54 (1989)
2. Gao, W., Zheng, Q., Yu, S.: Elastic isotropic damage expressed by double scalar. Chin. J. Theor. Appl. Mech. **28**(5), 542–549 (1996)
3. Tang, X., Jiang, C., Zheng, J.: A general form of isotropic elastic damage constitutive equation. Appl. Math. Mech. **22**(12), 1317–1323 (2001)
4. Tang, X., Zheng, J., Jiang, C.: Continuum Damage Theory and Application. China Communications Press, Beijing (2006)
5. Tang, X.S., Jiang, C.P., Zheng, J.L.: Anisotropic elastic constitutive relations for damaged materials by application of irreversible thermodynamics. Theor. Appl. Fract. Mech. **38**(3), 211–220 (2002)
6. Zhang, W., Cai, Y.: Continuum Damage Mechanics and Numerical Applications. Springer, Heidelberg (2010)
7. Zhu, W., Zhao, Q., Tang, C., Zhuo, J.: Mechanical model and numerical simulation of fracture process of concrete. Adv. Mech. **32**(4), 579–598 (2002)
8. Yin, S.: Fracture Damage Theory and Application. Tsinghua University Press, Beijing (1992)
9. Tang, X., Zheng, J., Jiang, C.: A general continuum theory of elastic damage coupled with thermal effect. Chin. J. Appl. Mech. **19**(1), 1–4 (2002)
10. Yu, H., Tang, M., Wu, J.: Research on capacity of beam based on damage constitutive model for concrete. Chin. J. Appl. Mech. **28**(4), 388–392 (2011)
11. Chen, L., Peng, H., Chen, Z.: Effect of crack face contact and friction on dynamic stress intensity factors for a double-edge cracked plate. Chin. J. Appl. Mech. **28**(3), 299–303 (2011)
12. Tang, X.S., Zhang, J.R., Li, C.X., Xu, F.H., Pan, J.: Damage analysis and numerical simulation for failure process of a reinforced concrete arch structure. Comput. Struct. **83**(31–32), 2609–2631 (2005)
13. Tang, X., Yang, J., Jiang, C., Zhang, X.: Damage mechanics-additional load-finite element method for fatigue life prediction of axisymmetrical structural members. Chin. J. Aeronaut. **23**(2), 97–101 (2002)

Study on the Function and Evolution Mechanism of Urban Innovation System

Yan Ningning(✉)

School of Management and Economics, Jingdezhen Ceramic Institute, Jingdezhen 333403, China
lsemail2008@163.com

Abstract. On the basis of the theoretical research achievements of innovation and innovation system at home and abroad, the concept of urban innovation system is defined by drawing on the definition of national innovation system and regional innovation system and combining the characteristics of urban innovation, and the elements of urban innovation system are analyzed, and its structure model and innovation ability are put forward. Culvert, the evolution mechanism of urban innovation system is studied by using the theory of super-cycle, and the open principle of urban innovation system evolution, the principle of self-stability, the principle of emergence and the principle of selection and evaluation are revealed.

Keywords: Urban innovation system · Function and evolution mechanism Theory of hyper-cycle

1 The Meaning of the Urban Innovation System

The organization for economic cooperation and development believes that the national innovation system is a system network composed of a series of public institutions (national laboratories, universities) and private institutions (enterprises). The activities and interconnections and effects of these institutions determine the ability of a country to spread knowledge and sound the innovative performance of the country. The national innovation system is mainly aimed at "system failure". The failure of the system mainly lies in the cooperation and lack of knowledge flow between enterprises and scientific research institutions, enterprises and universities as well as enterprises and enterprises. The basic research direction of the state funding is not matched with the application and Development Research of the industry, and colleges and universities cannot cultivate talents with strong innovation and entrepreneurial consciousness, which are urgently needed by the enterprises. The intermediaries, such as avoidance of innovation, venture risk and technology transfer, do not play their due role in promoting knowledge flow. Enterprises cannot effectively absorb the information of digestion innovation in time and so on. These all seriously affect the efficiency of a national innovation system. In the national innovation system, the most important is the interrelationship between different participating agencies. Without connection, there will be no innovation system.

J. Abawajy et al. (Eds.): ATCI 2018, AISC 842, pp. 894–900, 2019.
https://doi.org/10.1007/978-3-319-98776-7_106

Regional innovation system is a network system, which is composed of enterprises, universities, research institutions and governments participating in the development and diffusion of new technology, and the network system for creating, storing and transferring the interaction of knowledge, skills and new products. It embodies the hierarchical characteristics of the national innovation system. The elements of regional innovation system consist of main elements, functional elements and environmental elements. The main elements include enterprises in the region, universities, scientific research institutions, intermediary service institutions and local governments. The functional elements include institutional innovation, technological innovation, management innovation and service innovation, and environmental factors include system, mechanism, government or legal regulation, infrastructure construction and guarantee conditions. The relationship between the regional innovation system and the national innovation system is mainly manifested in the following: the national innovation system is composed of various open regional innovation systems, and the regional innovation system is the basic support and component of the national innovation system. The coordinated operation between the regional innovation system and the regional innovation system is the prerequisite and guarantee for improving the quality and efficiency of the national innovation system. The regional innovation system emphasizes the bureaucratic structure of local governments, the long-term development of local special industries, the differences between the core and periphery of the industrial structure, and the innovation performance.

The city is a part of the national economy. Therefore, the urban innovation system is also an integral part of the national or regional innovation system. The innovation system of important large cities is a place where national innovation resources are highly concentrated, and is a strategic fulcrum of the national innovation system. However, the research on urban innovation and urban innovation system is far less than the national innovation system because of the close blending of the urban innovation system and the national innovation system and the regional innovation system, the complexity of the urban innovation activities and the development of the urban economics and the theory of urban sociology. The system of regional innovation is mature.

Drawing on the definition of national innovation system and regional innovation system and combining the characteristics of urban innovation, the connotation of urban innovation system can be understood as follows: the urban innovation system is a subsystem of the national innovation system, which is the support of the development of the city in the national innovation system, the demand of the market and the optimization of the strategic layout of the country (region). A system network that develops into a dynamic, undertaking the function of innovation source in regional technological innovation and institutional innovation activities. It directly determines the linkage, knowledge flow and innovation performance of institutions in a country or region. Urban innovation system is the basis and carrier of urban innovation activities, and is the basic guarantee for implementing urban innovation activities.

2 The Function of Urban Innovation System

Innovation is the soul of a nation's progress and an inexhaustible motive force for national prosperity. In the face of the global scientific and technological revolution and innovation wave, the comprehensive strength of the country is gradually embodied as the economic strength supported by scientific and technological innovation. The competition between countries and countries is the competition of scientific and technological strength and the competition of talents. In the final analysis, the competition of independent innovation ability is in the final analysis. City is the collective carrier of innovation strategy and the base of innovation. Building an innovative city is the embodiment and specialization of building an innovative country. Looking at the development process of the advanced cities in the world, the emergence and development of the advanced cities ultimately depend on the technological innovation ability of the city.

At present, China's overall ability of independent innovation is still weak, and there is a big gap between the competitiveness of cities and those of developed countries. In order to conform to the trend of the development of the international city and seek a new round of the dominant position in the international competition pattern, it is of great practical significance to carry out the development strategy of the innovative cities in the major cities of our country.

The traditional quantitative economic growth mode relies on high input and high consumption to promote economic growth. In recent years, the city is facing such problems as shortage of resources and serious environmental pollution. This consumption growth mode can no longer meet the needs of urban development in China. It is gradually replaced by the quality based economic growth mode, that is, promoting economic growth through technological progress and the improvement of labor quality. The core of the construction of an innovative city is to improve the independent innovation ability of the city, adjust the investment structure by the independent technical achievements and increase the investment to the information industry, which is the essence of the new road of industrialization. The city must rely on the innovation of science and technology, gradually realize the transformation from the factor driven growth to the innovation driven growth, and make the scientific and technological progress an important way to change the way of economic growth.

Urban competitiveness is unique and difficult to be imitated in the city. Compared with other cities, it can gather more factors of production, achieve better allocation of resources, and create more social wealth. Urban sustainable development and lasting competitive advantage come from its core competitiveness, that is, the high technology and high efficiency of the city. Innovation is the source of high technology. Only by increasing the investment in the high-tech industry and improving the preferential policies of technological innovation, can the city maintain the core technical advantages of the continuous innovation. Therefore, building innovative cities, strengthening and improving industrial innovation, institutional innovation and innovation input are the fundamental guarantee for enhancing urban competitiveness.

The interaction of various elements of the urban innovation system accelerates the flow of technology, information, talent and capital flows in the system. There are many ways and means to interact with each other. There are differences in connection form and way between different elements. This link includes the links between enterprises and enterprises, enterprises and universities, enterprises and scientific research institutions, government departments and scientific research institutions, as well as between enterprises, universities and scientific research institutions. The way of interaction between different elements is different, and their connection forms are also different. The general ways of interaction among the elements in the urban innovation system include cooperation research, technical cooperation, cooperation patents, cooperative papers, technology transfer, technology share, information sharing, personnel flow, joint research institutions, cooperative enterprises, joint ventures and so on. Through various forms of interaction, the rapid flow of knowledge, technology and personnel in enterprises, universities and scientific research institutions will help to improve the overall efficiency of the urban innovation system.

The optimization activities of urban innovation system are closely related to the factors of science and technology, economy, society and politics in the city. Therefore, the analysis of the macro factors related to the development of the urban innovation system is of great significance to the direction and goal of the innovation of the system and the formulation of the scientific optimization strategy and countermeasures.

The efficiency of innovation in the urban innovation system is low from the incoordination within the innovation system, but this incoordination is no longer the incoordination of the factors of production, but the incoordination of the development of technology, knowledge, system and service in the innovation system. The urban innovation system may not meet each other's internal factors, making the whole innovation system inefficient. With the serious lag of one or several factors innovation, this incompatibility and incoordination becomes more and more intense, which has a strong hindrance to the innovation of other factors and the formation of a bottleneck effect. In turn, the force of other factors also accelerates the renewal of the bottleneck factor, making it a certain period and a certain area. The most active and critical factor in the new system. Once the renewal is completed, the urban innovation system will be in a state of short coordinated operation, which is most conducive to the creation of innovative results and economic development. But at the same time, there will be new incompatibility within the innovation system. The urban innovation system is spiraling up and developing in the process of "uncoordinated, coordinated and uncoordinated".

The relationship between urban innovation system and its subsystems is objective and not based on human will. Through human intervention and control measures, people can try to shorten the process of the system from incoordination to coordination, prolong the time of the system in a coordinated state and increase the output rate of the innovation results, thus accelerating the economic development of the city.

3 The Evolution Mechanism of Urban Innovation System

The super cycle links the cycle with development, and believes that nature is developing and developing in circulation. Cycle is not always going round and round, nor is it a simple repetition of the same process. If the cycle is carried out in an open state away from equilibrium, it will gain autonomy and direction. Hyper cycle is a hierarchical circular development system. It shows that circulation is a basic form of development, and complexity is derived from a simple cycle. The cycle can be divided into three levels, the reaction cycle, the catalytic cycle and the super cycle, which correspond to the characteristics of the metabolism, self-replicating and mutagenicity of the life system.

The metabolism of urban innovation system is based on the unity of opposites between assimilation and alienation. The urban innovation system continuously obtains resources from outside to provide nutrition for the life of urban innovation system. On the one hand, the urban innovation system assimilates the resources into the innovation elements within the city through various internal cycles, which is the assimilation process. On the other hand, the urban innovation system consumes the input resources and produces new products and technologies needed by the market, as well as the formation of new urban state, which is the process of alienation. Assimilation and alienation are mutually conditional and indispensable. Once the transformation stops, the life of a city's innovation system will also cease.

The urban innovation system is self-replicating. Through self-accumulation and continuous innovation, its technical level and personnel quality are constantly improved. The scale of the urban innovation system is continuously expanded, the structure is continuously optimized and the ability is constantly enhanced. It has the original traces of the original urban innovation system, and the reproduction and reproduction of the urban innovation system, and the continuous reproduction and reproduction of the urban innovation system. The self is new, and the new self is a spiral and higher self.

The system of urban innovation is also mutagenicity. The changes in economic policy, the occurrence of scientific and technological leaping, the change of raw material supply, the success of technological innovation, the change of human resources quality, the change of competition situation, the change of the demand of the users and so on, will make the operation of the urban innovation system qualitative change.

The super cycle theory points out that the super cycle structure evolves in a stable and unstable contradictory movement, and the stability depends on the self-replicating capacity of the self-replicating unit and the coupling strength between the members, and the instability is derived from the errors in the copying process and the inherent randomness caused by the interaction of complexity. The disturbance of the external environment. The super cyclic structure will change gradually without breaking the original structure, and will change abrupt on the basis of gradual change, so that the old structure will be disintegrated and the new structure will be established. Based on these viewpoints, the open principle, self-stabilization principle, emergent principle and selection evaluation principle of urban innovation system optimization are summarized.

The principle of openness. We should take the initiative to take measures to control the development of the city, promote its healthy development, and delay its aging and timely transform. At the same time, we need to dynamically control the optimization and implementation of the urban innovation system, make the most effective use of innovative resources, and guide the healthy development of the urban innovation system.

The principle of self-stability. It refers to the system that reintroduces the output information and data in the running process to the system, compares the deviations between the output value and the target value, and corrections and controls the deviations, thus realizing the process of the system's predetermined target. Self-replicating is a more complex system of system population stability. It realizes the interaction between various elements and subsystems through the catalytic cycle, and realizes self-replication and continues the life of the system.

The principle of emergence. The optimization of urban innovation system should be good at controlling the emergence and promoting the optimization of the urban innovation system so as to enable it to operate at a higher level. In the process of optimizing the urban innovation system, it is necessary to grasp the opportunity and direction in the process of the optimization of the urban innovation system. By establishing an efficient dynamic mechanism, using the subjective initiative of the people to restrain or promote some emergence, the system will evolve according to the purpose of the people.

Select the principle of evaluation. In the process of evolution, the beneficial variation of survival will be preserved, and the harmful variation of survival will be eliminated. This is natural selection. The principle of selection and evaluation of super cycle is similar to Darwin's natural selection principle. It is the principle that the system is formed in the interaction of the group and the group and the environment on the evaluation of the system, the evolution and the evolution, and the choice of the evolutionary path.

The selection and evaluation principle of urban innovation system requires the combination of competition and synergy. In the process of system evolution, there is not only the competition of "natural selection and survival of the fittest", but also the "synergy" and "integration". Urban innovation systems of different types and regions are not only competitive but also synergistic. They are isolated and integrated, and thus choose and evolve.

Acknowledgements. This research was financially supported by modeling and optimization of tacit knowledge diffusion partners in R&D teams under the informal network environment, science and technology project of Jiangxi provincial education department in 2017 (GJJ170784).

References

1. Lou, S., Feng, Y., Zheng, H., Gao, Y., Tan, J.: Data-driven customer requirements discernment in the product lifecycle management via intuitionistic fuzzy sets and electroencephalogram. J. Intell. Manuf. 1–16 (2018)
2. Wang, Y., Li, J., Wang, H.H.: Cluster and cloud computing framework for scientific metrology in flow control. Clust. Comput. 1–10 (2017)
3. Weiss, L.: Collection and connection: the anatomy of knowledge sharing in professional service. Organ. Dev. J. **17**(4), 61–72 (1999)
4. Peng, L., Fang, W.: Heterogeneity of inferring reputation of cooperative behaviors for the prisoners' dilemma game. Phys. A: Statis. Mech. Appl. **433**, 367–378 (2015)
5. Ge, X.: On how to improve the effectiveness of college English translation teaching. Engl. Teach. (18) (2015)

6. Chen, T.: Effective measures of english teaching translation. Overseas Engl. (20) (2015)
7. Wang, H.: On college english translation teaching models in the internet environment. Engl. Square (2) (2016)
8. Fan, M.: The cultural elements in english translation teaching. Engl. Square (20) (2014)
9. Wang, H.: The problems and countermeasures of english translation teaching in college of traditional Chinese medicine. Overseas Engl. (20) (2014)
10. Watts, D.J.: Small worlds. The dynamics of networks between order and randomness. America: 56A2579BB6, 13 May 1999
11. Aleksy, M., Korthaus, A., Schader, M.: Use Java and the CORBA realization distribute type system. J. Pingxiang Coll. (4), 104–105 (2005)
12. Titus, J.: ECN technical editor: the eclipse of stand. J. Zhongkai Agrotech. Coll. **19**(2), 32–35 (2006)

Research on the Autoloader Device' Fault Diagnostic Mechanism Based on the Current Detection

Guobao Ding[1(✉)], Dan Li[1], and Fanghuai Gou[2]

[1] Army Academy of Armored Forces, Changchun, China
15304459101@189.cn
[2] Unit 31694, Benxi, China

Abstract. Automatic loader is one of the important part of the artillery autoloader device the system. The agency's work is normal or not that directly related to the whole system operation state, and even the development of entire situation. It has a very important significance that classifies and recognizes the fault through timely finding faults and monitoring the state. It can also avoid unnecessary trouble and loss and provide a powerful guarantee for the victory of war. The work environment of automatic loader is very bad and the vibration is also obvious, which leads to the background noise in the signal is quite large. A part of the useful signals may be covered by strong noise. The traditional signal analysis methods may not reach the expected effect of the diagnosis. Here introduced current signal analysis, as MCSA. It uses induction motor stator current as a signal analysis the point, and studies the characteristics and the corresponding relation of failure. As a new method of fault diagnosis, it used in some other means inconvenience or easily found the problem the implementation of the occasion. This paper's processing method is real-time monitoring the four sets of dc servo motor operation by hall current sensor. And it can also transfer the electrical signals to DASP signal collection system for post processing. In the later processing, it can classify all kinds of signal under the condition, and extract current signal characteristic parameters which are related. Finally it can use BP neural network to classify and diagnosis for fault.

Keywords: Autoloader · Fault diagnosis · MCSA

1 Introduction

In the modern war, the artillery must complete the high intensity combat task in a short time, and the automatic loading system is an important part of the artillery. In the period of the war, the automatic loading system of the artillery automatically supplies the mechanical failure of the transmission mechanism, which will affect the smooth completion of the task and even affect the whole task. The situation of the war situation is unthinkable. At present, all kinds of military equipment use the integrity of combat readiness to measure the level of the combat effectiveness, mainly based on the accumulated experience and statistics, and lack of scientific and accurate [1]. From the above facts, we can see the importance and urgency of the fault diagnosis technology

J. Abawajy et al. (Eds.): ATCI 2018, AISC 842, pp. 901–906, 2019.
https://doi.org/10.1007/978-3-319-98776-7_107

of automatic loading system for automatic loading system. Therefore, a fault diagnosis method based on current parameter integrated detection automatic loading system is developed. It is very necessary to find out the cause of the fault as soon as possible, reduce the degree of its danger and reduce the economic loss. Yes. The purpose of mechanical equipment fault diagnosis is to ensure that the mechanical equipment can achieve the desired results within the permitted scope [2]. Therefore, the fault diagnosis system should effectively obtain and utilize the diagnosis information of mechanical equipment, thus effectively identify the working state of the mechanical equipment, find out the possible faults or cause the causes of these failures, and even predict the development trend of the equipment state [3].

2 Current Signal Analysis Method

The current signal analysis method, in short MCSA, is in essence the use of the stator current of the induction motor as the entry point of the signal analysis. The corresponding relationship between the characteristics and the fault is studied. As a new method of fault diagnosis, it can be used in some complicated fields, such as some conventional means are difficult to implement or ask. An occasion that is more difficult to find. This method is sensitive to the torque fault of the transmission system, because it is essentially a conversion from torque to electrical energy. There is a very close relationship between the electromagnetic torque (air gap torque) of the induction motor and the load torque of the system, and the fluctuation of the load torque directly affects the air gap torque, and the change of the air gap torque can cause the stator current change through the stator flux. Therefore, in order to meet the requirements of monitoring motor load mechanical failure, we can measure the stator current to monitor the change of load torque of the system. Through the research and Discussion on the fault diagnosis method of the automatic loading system part of the artillery, the current signal analysis method is introduced into the fault diagnosis of the analysis of the motor load equipment (automatic transmission mechanism), which will greatly broaden the method of fault diagnosis and the ability to analyze the obstacle [4].

This paper mainly classifies and identifies the fault states of the actuator part of the artillery automatic loading system. The faults often occur in the automatic loading system of the artillery, such as the unlocking of the brake, the non rotation of the projectile, the unfolding of the mechanism, the non push of the projectile and so on. First, the common fault forms of the automatic transmission mechanism are understood, and the mechanism of the failure is analyzed. In addition, the method of signal analysis and feature extraction commonly used in the fault diagnosis of automatic loading system of artillery is also understood. Due to the firing process of the artillery, great noise will be produced. The traditional method is difficult to achieve the desired results. Therefore, it is applied to the acquisition of Holzer current sensor by the learning current signal analysis method. In the electrical signal, the processed data are then diagnosed by BP neural network. The results show that the processed data diagnosis results are consistent with the type of fault set in advance, and the ideal diagnosis results are obtained [4].

3 Fault Principle of Current Diagnosis for Automatic Loading System

In the actual application of artillery automatic loader, the common faults are mainly caused by two aspects: control device failure and actuator failure. Here we mainly study the mechanical part of the automatic loading system, which is the actuator. The actuator is mainly composed of a rotating projectile, a turning mechanism and a projectile mechanism, which is mainly driven by four DC motors to actuate the corresponding action. The brake of the bomb transport machine is unlocked. After the transmission motor is slowed down by transmission, the projectile is driven to rotate and the cartridge cup is transported to the loading line. Under the action of the overturned motor, the projectile mechanism is turned over to the loading angle. Under the action of two projectile motors, the projectile is pushed into the bore. The overturn motor and the push motor are done. In this series of actions, the working load of the four DC servo motors changes obviously and forms part of the working current curve of the actuator, so that the working state of the artillery automatic loader can be judged according to the fluctuation of the working current curve [5].

4 The Relationship Between Fault Types and Current Changes in Actuator of Automatic Loading System

The maintenance and diagnosis of mechanical faults in military equipment directly affect the exertion of self propelled artillery in war. The self-propelled artillery units of our army also draw on the maintenance support system of the traction artillery and armored soldiers, but the current equipment maintenance and support work is mainly realized by the mode of replacement and post maintenance, and this is mainly dependent on the long-term experience accumulation and the statistical law of the traditional artillery. This will inevitably lead to a lack of accuracy and scientificity. Therefore, it is very necessary to develop a fault diagnosis system that can find out the cause of failure as soon as possible, reduce the degree of damage and reduce economic losses. The main components of the automatic loading system actuator are motor, gear, transmission device and so on. The fault diagnosis of the system is mainly based on the monitoring values of various components and the non measurable signals observed by the on-site inspectors. The fault components and the reasons are located according to the experience accumulated by the experts and the structure function of the system [6]. At present, the detection of automatic loading system parameters is mainly a large number of fault diagnosis parameters which are collected by various types of sensors. These parameters mainly refer to various operating parameters of the automatic loading system, such as current, vibration, voltage, speed and so on. However, the changes of these parameters are often used to distinguish a kind of fault. It provides a reliable basis. After analysis, it is found that the failure of the mechanical part of the automatic loading system is related to the working current of the four related DC motors. Therefore, some rules can be summed up. By the change of a certain period of time, the failure of the mechanical part, that is, the rotary projectile, the mechanism and the projectile mechanism, can be judged [7].

5 Fault Diagnosis Model

It can be seen from the upper model that the central control unit can store the important data in the storage unit and send the corresponding data to the fault diagnosis unit after preprocessing the collected signals. In order to judge the working condition and failure of the equipment, the result can be displayed not only in time, but also in time to know the maintenance and storage of the maintenance personnel. For historical data [8] (Fig. 1).

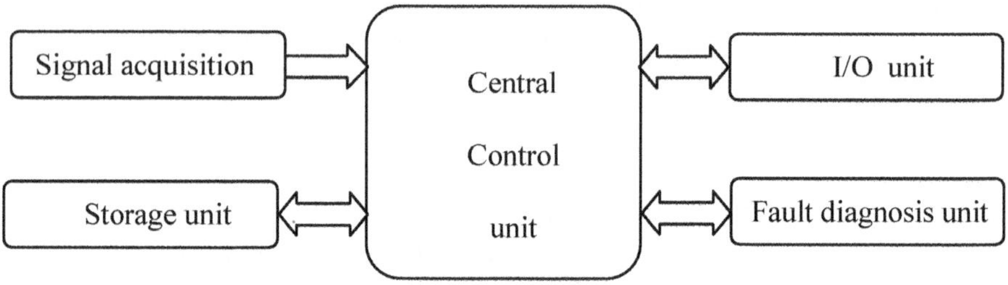

Fig. 1. Model diagram of fault diagnosis

6 Automatic Loading System Fault Diagnosis System

The fault diagnosis system of automatic loading system is aimed at innovation and practicability, including the lower computer and the upper computer. The lower computer is responsible for the real-time collection of the field data of the automatic loading system, and the simple fault diagnosis. The upper computer uses the BP neural network to form the automatic fault intelligent diagnosis center of the automatic loading system and the automatic loading system database. Combined with comprehensive diagnosis of fault [9] (Fig. 2).

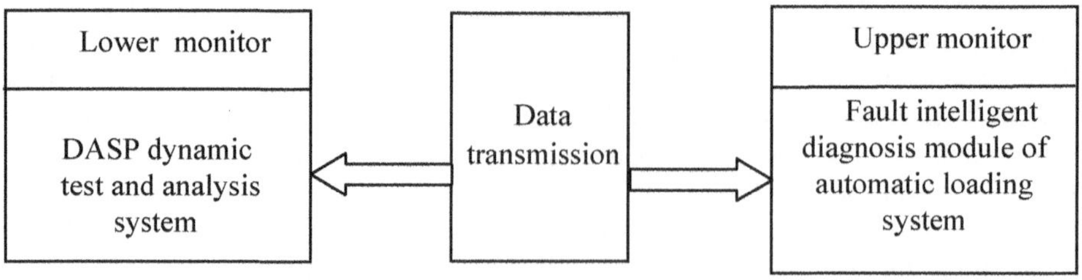

Fig. 2. Fault diagnosis system composition

The intelligent diagnosis module of upper computer fault: Based on the characteristic values of different current curves, combined with the database of automatic loading system, the intelligent diagnosis of automatic loading system fault is carried out by BP neural network.

The DASP dynamic test and signal analysis system of the lower computer: the current parameters and vibration parameters of the automatic loading system are collected by the Holzer current sensor and the piezoelectric acceleration sensor, and the current value and current curve are displayed in real time, or further analysis is carried out in the data management of the upper computer.

The fault diagnosis system of automatic loading system has a good man-machine interface. It uses the Holzer current sensor to collect the current of the DC servo motor in real time. Using the DASP dynamic test and signal analysis system, the current value and the curve trend of each DC servo motor are displayed on the notebook computer. The value of the collected current is stored in the computer. The data collected in the field are entered into the fault intelligent diagnosis center of the automatic loading system of the upper computer, and the characteristic values of the current curve are extracted. The BP neural network fault diagnosis module and the automatic loading system database are combined to identify the various obstacles of the automatic loading system, which can be automatically installed. The working state of the mechanical structure of the automatic feeding mechanism is displayed, and the working mode of the automatic loading system is displayed in the form of text and graphics. When abnormal working state occurs, the system sends out alarm information.

7 Conclusion

The actuator part is an important part of the artillery automatic loading system, and the frequency of its failure is also high, so it is of great significance to carry out state monitoring and fault diagnosis. Because of the artillery in the launching process, it will bring a huge shock wave, the noise is unavoidable, the analysis of the traditional fault signal will not reach the ideal diagnosis effect. Therefore, this paper applies the current signal analysis method to the troubleshooting process of the actuator part of the artillery automatic loading system. Next: the common fault states and failure mechanism of artillery automatic loading system are analyzed, and the failure mechanism of actuator part is emphatically analyzed. On the basis of a simple understanding of the common method of current signal analysis, feature extraction and fault identification, the electrical signals collected by the Holzer current sensor are analyzed and processed in the normal and fault conditions of the artillery automatic loading system to prepare for fault identification. Finally, the characteristic value of the current signal is extracted. As the input vector of the neural network, the fault diagnosis is carried out by BP neural network. The result shows that the working state can be identified accurately.

References

1. Fornberg, B., Lehto, E.: Stabilization of RBF-generated finite difference methods for convective PDEs. J. Comput. Phys. **230**(6), 2270–2285 (2011)
2. Degang, C., Suyun, Z.: Local reduction of decision system with fuzzy roughstes. Fuzzy Sets Syst. **161**(13), 1871–1883 (2010)
3. Liu, Y., Sun, J.: The concept of approximation based on fuzzy dominacerelation in decision-making. In: Wang, G., Liu, Q., Yao, Y., Skowron, A. (eds.) Proceedings of RSFD-GrC 2003, pp. 382–385. Springer, Heidelberg (2003)
4. Barreto, M.A., Tygel, M., Rocha, A.F.: Automatic Downhole Card Generation and Classification. In: SPE 33605
5. Lee, J., Tsai, J.: On-line fault detection using integrated neural networks. Appl. Artif. Neural Netw. **17**(9), 436–446 (1992)
6. Sanchez, V.D.A.: Advanced support vector machines and kernel methods. Neurocomputing **55**, 5–20 (2002)
7. Tan, S.C., Lim, C.P., Rao, M.V.C.: A hybrid neural network model for rule generation and its application to process fault detection and diagnosis. Eng. Appl. Artif. Intell.: Int. J. Intell. Real-Time Autom. **20**, 203–213 (2007)
8. Lee, K.C., Cho, H.R., Kim, J.S.: An expert system using an extended AND–OR graph. Knowl.-Based Syst. **2**, 38–51 (2008)
9. Kengpol, A., Wangananon, W.: The expert system for assessing customer satisfaction on fragrance notes: using artificial neural networks. Comput. Ind. Eng. **51**, 567–584 (2006)
10. Zhang, B., Lang, Z.Q., Billings, S.A., et al.: System identification methods for metal rubber devices. Mech. Syst. Signal Process. **39**, 207–226 (2013)

Big Data-Driven Innovation: An Empirical Research on Sustainable Development of Business Model

Yucong You(✉)

Guangzhou College of Business and Technology, Guangzhou 51000, Guangdong, China
stoneyyc@163.com

Abstract. With the emergence of new information technologies and application models, the global data volume has exploded and humanity has entered the "big data" era. The hidden business value behind the huge data is bound to bring more innovation opportunities to firms. How to fully tap the commercial value of big data so as to shape the core competitiveness has become a key issue that firms urgently need to address. In view of this, this paper draws on existing researches at home and abroad, combining resource-based theory, dynamic capability theory, together with manager cognitive theory, to discuss: How does big data drive business model innovation? By the research object of 180 businesses, this study obtained relevant data through questionnaire survey; SEM method is adopted to empirically test the theoretical model and related hypotheses; conclusions are drawn: resource integration ability bears a significant positive impact upon business model Innovation; Real-time insight and forecasting capabilities exert a significant positive impact on business model innovation; In-depth analysis capabilities cast a significant positive impact upon business model innovation. This paper reveals the relationship between the impact of big data capabilities on business model innovation and its internal mechanism through empirical research. It opens the "black box" of the relationship between big data and business model innovation; it renders a new perspective for the research of big data and business model innovation. Based on the research results, this study puts forward some suggestions providing certain reference for the practice of businesses.

Keywords: Data-driven · Sustainable development
Business model innovation

1 Introduction

1.1 Research Background and Significance

The huge amount of data in the era of big data poses new challenges to enterprises. The original business model will inevitably no longer adapt to the development of enterprises, and innovation is bound to become an inevitable choice. Manyika, Chui, Brown et al. (2011) clearly pointed out that in the era of big data, data is one of the corporate production materials, and big data is the frontier of the next innovation, competition, and

J. Abawajy et al. (Eds.): ATCI 2018, AISC 842, pp. 907–914, 2019.
https://doi.org/10.1007/978-3-319-98776-7_108

productivity enhancement. Big data is most likely to bring lower costs to businesses, giving birth to more opportunities for innovation. Big data is triggering profound technological changes in the world and providing conditions for business model innovation. The value of big data comes from three main domains: data, know-how and mindset. With the advancement of data processing technology, the realization of data business value is diversified, which tends to provide enterprises with a good opportunity to realize business model innovation. The amount of information the company obtains will grow substantially, leading to a better opportunity for mining the business value of big data.

1.2 Research Questions

Based on the aforementioned realistic and theoretical background, this study carried out relevant literature and theoretical research to explore the relationship between big data drive and business model innovation, and try to analyze its internal mechanism and causes; based on this, the research question is: In the era of big data, what drive business model innovation? What is the mechanism by which big data driving capabilities affect business model innovation?

2 Literature Review

Current academic circles on business model innovation are based on a resource-based-view-practice-model of "resource-competitive-competitive-predominance-strategy" (Bamey 1991; Grant 1991). Peteraf (2003) pointed out that the key for enterprises to obtain excess profits and maintain their competitive advantages lies in their unique resources. The key to an enterprise's competitive advantage is internal rather than external factors, that is, its unique resources and its ability to form on the basis of such resources. Zott and Amit (2008) further supplements the definition of business model based on a system perspective, pointing out that the business model is, as a matter of fact, a business activity system composed of enterprises and their trading partners. MuE (2014; 2015) contends that the management cognitive process is divided into three Steps: Scanning, in which managers selectively focus on certain decision-relevant information; interpretation, interpretation of the selected information and a certain meaning; information application, manager decision based on perception of information. Most scholars have not yet proposed the concept of big data capabilities. The impact of big data on business model innovation still needs to be reflected in its combination with other corporate resources to the ability level and impact on business model innovation (Pateli 2014; Lavalle 2016; You 2017). The foundation of big data capabilities is resource integration. Lavalle (2011; 2017) further pointed out that data is a strategically-resourced resource other than human, financial, and material resources. Sustained competitive advantage comes from valuable, scarce, and difficult to imitate alternative resources, but resources are the capabilities of the enterprise. You (2018) pointed out that how to update, integrate and utilize these key big data resources in the dynamic environment is an important part of the enterprise's big data capabilities.

To sum up, scholars have carried out a whole list of studies on business model innovation, yet the studies on the relations between business model innovation and big data drive is quite limited, especially those researches with a dynamic perspective and of empirical manner are scarce. In view of this, this paper is to discuss the question: How does big data drive business model innovation?

3 Empirical Research

3.1 Theory Framework of the Model

This study contends that data itself, as a resource, does not necessarily drive companies to innovate their existing business models. The value of big data lies in the value of other companies' other resources that are integrated into big data capabilities. Big data drive capability is the driving force. Big Data helps enterprises to further study the external environment, and can promote the enterprise to re-examine the internal management process, triggering improvement and innovation of the enterprise's production model and cooperation model to achieve the purpose of the innovative business model.

3.2 Variables and Relational Hypotheses

From the above, we can see that, both theoretically and practically, the driving effect of big data capability on the business model of the enterprise is quite apparent. Therefore, the following hypothesis is proposed in this paper:

H: Big Data drive capability exerts a positive impact on the business model innovation

H1: resource integration capability exerts a positive impact on business model innovation;

H2: in-depth analysis capability exerts a positive impact on the business model innovation;

H3: insight and prediction capability exerts a positive impact on business model innovation (Fig. 1).

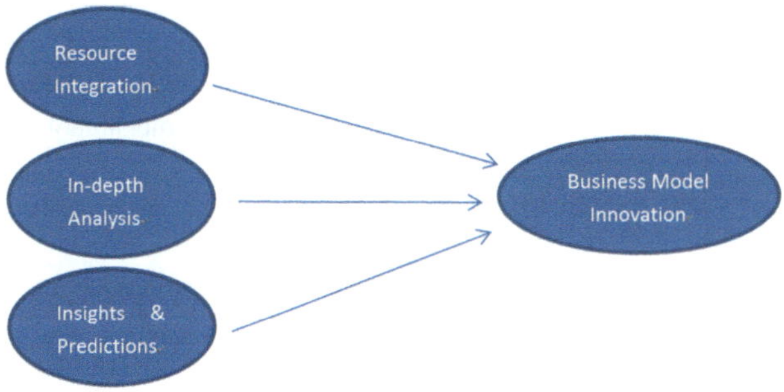

Fig. 1. Theoretical model

3.3 Questionnaire Design and Variable Measurement

The overall structure, content and composition of the questionnaire are related to the research purpose and theoretical basis. This questionnaire focuses on this research topic, and designs related questions based on the core research variables and related theoretical knowledge. Its purpose is to provide effective data for this research. The questionnaire is mainly composed of the following components: Basic information of the company: the establishment time, location, sales, assets, ownership, and number of employees; Capacity of big data Table; Business Model Innovation Scale.

This article mainly studies the path and mechanism of the impact of big data capabilities of commercial enterprises on business model innovation. In view of the deepening transformation and upgrading of commercial enterprises in recent years, big data has become more and more significant. The target object of this study is aimed at 180 regional commercial enterprises. It covers manufacturing, power, heat, gas and water production and supply, information transmission, software and information technology services, etc. The reason why these industries are chosen is because the development of these industries is most closely related to the development of information technology, with typical representativeness and a certain degree of generality. Sampling method for this research survey is random sampling. The questionnaire is mainly issued through three ways: on-site distribution and recycling; email; social platform. In this survey, 200 questionnaires were distributed, 188 were recovered, 28 invalid questionnaires were removed, 160 valid questionnaires were obtained, and the effective recovery rate was 80%. Based on previous research, this study divides big data capabilities into resource integration capabilities, in-depth analysis capabilities, and real-time insight and prediction capabilities as shown by Table 1.

Table 1. Variable measurement of big data capabilities

Variable	Code	Measurement item
Resources integration	RI1	Continuous, real-time access to various data within and outside the firm
	RI2	Big data technologies can continuously be learn and updated in the firm
	RI3	Big data resources are effectively coordinated according to business goals
In-depth analysis	IA1	Value information from big data can quickly be assorted
	IA2	A variety of unstructured data can be analyzed real time
	IA3	Most corporate decisions rely on the support of big data analytics
Insights & prediction	IP1	Real-time insight into the market can be realized based on big data
	IP2	Customer behaviour can be comprehended and predicted based on big data
	IP3	Prediction for competitors can be realized based on big data

It is illustrated by Table 2 the Variable Measurement of Business Model Innovation as follows.

Table 2. Variable measurement of business model innovation

Variable	Code	Measurement item
Business model innovation	BMI1	New products, services, information
	BMI2	A large number of new customers
	BMI3	Diverse participants
	BMI4	Irreplaceable existing business model
	BMI5	New operating process
	BMI6	New transaction method

4 Data Analysis and Results

4.1 Variable Reliability and Validity Analysis

Sample data was verified with CDF verification factor analysis. According to You (2016), the AMB indicator is selected. As shown in Figs. 2 and 3, the data has high reliability and validity, being suitable for the next analysis.

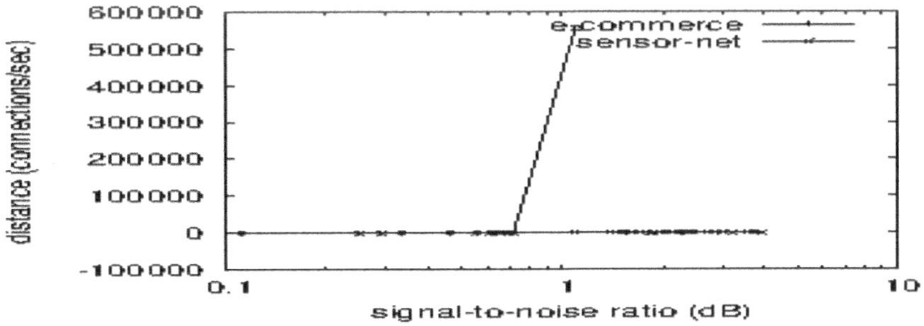

Fig. 2. Ajowan mean bandwidth

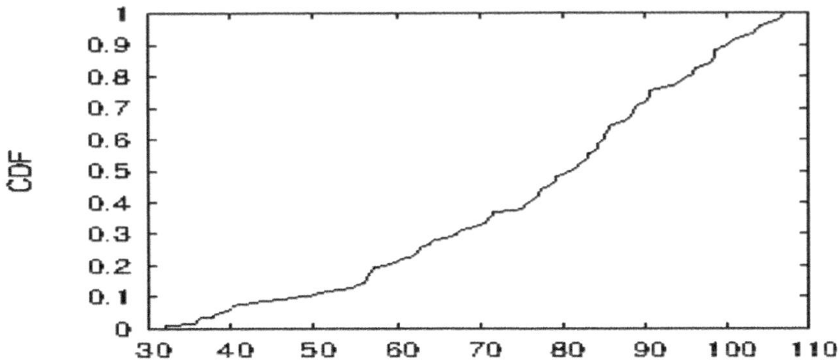

Fig. 3. CDF validity index

4.2 Hypothesis Testing

Prior to hypothesis testing, this paper first performed correlation analysis and descriptive analysis of variables. This study uses SPSS to analyze the major latent variables involved in the text—big data capabilities, decision-making rationality, and business model innovations. The results are shown in Table 3. It can be seen that each variable is significantly related at a significant level of 0.01, indicating that continue with hypothesis testing analysis.

Table 3. Inter-variable correlation coefficient

	1	2	3	4
RI	0.799			
IA	0.678**	0.826		
IP	0.525**	0.689**	0.898	
BMI	0.509**	0.478**	0.601**	0.801

Note: * indicate $p < 0.05$; ** indicate $p < 0.01$

As shown in Fig. 4, Lamport Clocks Discriminant Validity Test, the correlation data and coefficients of this variable are marked with significant stability and higher validity.

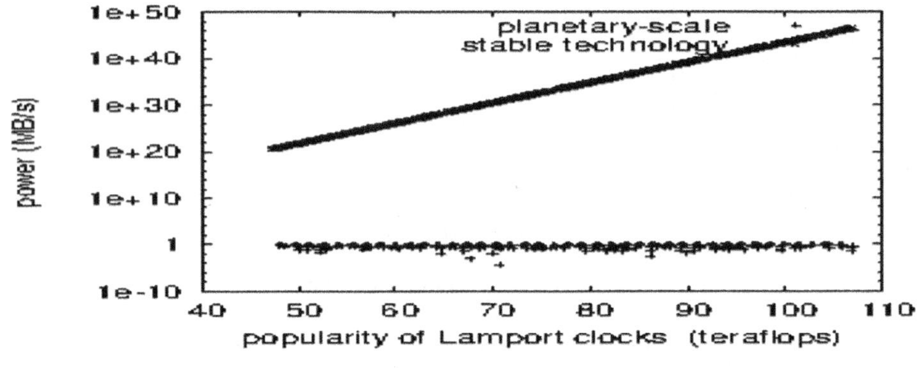

Fig. 4. Lamport clocks discriminant validity test

Full model analysis was performed based on the theoretical model. The specific results are shown in Table 4 and Fig. 5. The hypotheses of the three path relationships are significant, namely H1, H2, and H3.

Table 4. Path coefficient

Hypothesis	Path	Coefficient	T-value	Conclusion
H1	RI → BMI	0.55	3.52	Significant
H2	IA → BMI	0.48	3.02	Significant
H3	IP → BMI	0.37	4.00	Significant

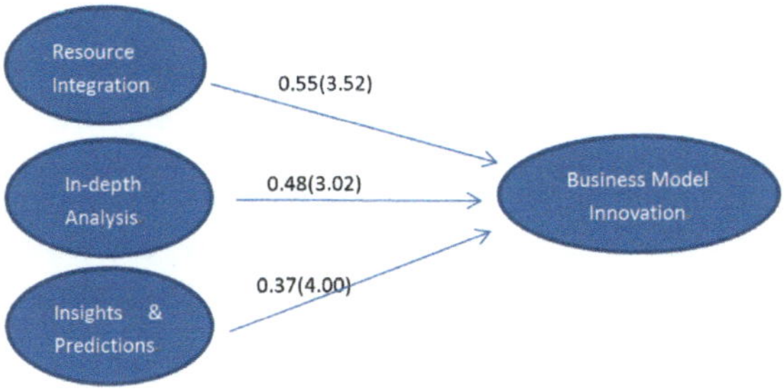

Fig. 5. Model with path coefficient

Therefore, based on the above test results, the following conclusions can be drawn: The hypothesis that the big data capability has a significant direct positive impact on business model innovation is supported. Among them: resource integration capability has a significant direct positive effect on business model innovations of 0.55 (3.52), so H1 is supported; in-depth analysis capabilities cast a significant direct positive effect on business model innovations of 0.48 (3.02), so H2 is supported; insight and prediction ability has a significant direct positive effect on business model innovations of 0.37 (4.00), so H3 is supported.

5 Conclusion and Suggestions

Based on the data analysis results, we can see that big data capabilities mainly affect the business model innovation from the following aspects. For enterprises in the era of big data, data is an emerging strategic resource that follows people, finance, and property. Big data resources need to be integrated with other resources and rise to capacity level. Combined with other capabilities of enterprises, they can improve information flow, value flow and capital flow within the enterprise, and make resources circulate within the enterprise, thereby improving the company's own value chain and realizing products. The real-time insight and forecasting ability based on big data ensures that the company predicts market dynamics and industry competition on the basis of known data, and strengthens the connection between the company and customers, suppliers, and other stakeholders. Recognizing the opportunities and challenges of the environment has increased the probability of companies finding niche markets. Faced with the pressure of competition in the industry, if companies want to maintain their core competitiveness, they need to promote their own business model innovation.

This article provides important implications for how business enterprises can use business data model to realize business model innovation. First, enterprises should focus on the cultivation of big data capabilities. The empirical research results of this paper show that the ability of big data has a partial driving effect on business model innovation. The ability of resource integration real-time insight and forecasting has a direct driving effect on business model innovation, but in-depth analysis capability does not exist, but it does not mean that the cultivation of in-depth analysis ability is not important. Second, in order to fully tap the business value of big data, enterprises should adapt to development and realize business model innovation. Third, business managers should respect and make full use of data and information, reduce cognitive bias, and improve the quality of decision-making. Research results show that decision-making rationality plays a partial intermediary role in the relationship between real-time insight and predictive ability and business model innovation, that is, decision-making. Combining with business decision-making, new requirements are put forward, and data-driven enterprise decision-making will take into account both data use and expert judgment.

Acknowledgements. This research was financially supported by the Guangdong Province Philosophy and Social Sciences "13th Five-Year Plan" 2016 annual academic projects (Guangdong foreign trade "supply side reform" driven development path analysis, project number GD16XYJ30) and Guangzhou philosophy and social science development "13th five" plan 2017 Research on Resource Allocation and Innovation Driving Path of Guangzhou Foreign Trade Enterprises under the View of Structural Reform of Supply Side, Project No. 2017GZGJ20).

References

Lynch, C.: Big data: how do your data grow? Nature **9**, 29–32 (2008)

Manyika, J., Chui, M., Brown, B., et al.: Big data: the next frontier for innovation, competition and productivity. McKinsey Glob. Inst. **2**, 15–21 (2011)

Mu, E., Kirsch, L.J., Butler, B.S.: The assimilation of enterprise information systems: an interpretation systems perspective. Inf. Manag. **5**, 18–25 (2015)

McAfee, A., Brynjolfeson, E.: Big data. Harv. Bus. Rev. **10**, 35–37 (2012)

Lavalle, S., Lesser, E., Shockley, R., et al.: Big data, analytics and the path from insights to value. MIT Sloan Manag. Rev. **2**, 21–38 (2011)

Cong, Y., Xia, Y.: An empirical study on the mechanism of institutional innovation driving foreign trade competitiveness. Adv. Hum. Soc. Sci. Res. **1**, 28–35 (2017)

You, Y., Yi, L.: An empirical study on the relationship between innovative enterprises and the cultivation performance of innovative talents in universities with CIA mode. Adv. Comput. Sci. Res. **2**, 18–25 (2017a)

You, Y., Yi, L.: An empirical study on the driving path of optimizing the resources allocation of foreign trade enterprises from the perspective of energy supply-side reform. Earth Environ. Sci. **9**, 90–95 (2017b)

You, Y., Yi, L.: An empirical study on the performance of regional innovation resource allocation system from the perspective of supply side reform. Adv. Soc. Sci. Res. **5**, 90–95 (2016)

Yi, L., Yo, Y.: An empirical study: CDA-based data innovation-driven intelligent traffic control system model of canton city. Adv. Comput. Sci. Res. **3**, 10–15 (2017)

Research on the Application of Block Chain Technology in Military Supply Chain

Yingjin Lu[1], Zhangyi Xu[1(✉)], and Xisheng Le[2]

[1] Wuhan University of Technology, Wuhan 430063, China
lyjingl35@163.com

[2] Hubei Provincial Communications Investment Logistics Group CO., LTD., Wuhan 430000, China

Abstract. Supply chain is generally composed of suppliers, manufacturers, distributors and users. As the military production is on schedule, and the difference with the general supply chain is that there is barely no distributors, the military supply chain covers the parts procurement, manufacturing, assembly, quality inspection, transportation and other aspects. Based on block chain technology, four kinds of supply chain models of different core-control centers are presented by the type and quantity of military products procurement in this paper. And it analyzes the typical applications of block chain technology in the military supply chain further, namely, the entire process logistics management, risk management in supply chain, cost management in supply chain.

Keywords: Military supply chain · Block chain technology
Supply chain payment risk management · Supply chain cost management

1 Introduction

As the military products have highly strict requirements on technology and security, participators of the supply chain to participate in the operation is of a higher threshold and the participators are relatively stable and interdependent, which results in confidentiality requirements in the production information, product information, customer information, etc. [1–3]. Due to the confidentiality requirements, and unsound supply chain information management system or platform, the main information flow includes order information, production information, quality inspection information, inventory information, transportation information, delivery information [4]. At present, each member mainly organizes the procurement, production and transportation business by following the contract [5]. It is difficult for the member enterprises in the supply chain to grasp the upstream and downstream information, which does not really play the role of supply chain coordination. Owing to the information asymmetry, competitive and large-scale core enterprises often pass capital risk to upstream and downstream SMEs through arrears, requirements of prepayment and bulk procurement in the military supply chain [6, 7]. On the other hand, with the poor financing capacity and huge financial pressure, those SMEs may result in cooperative fatigue, friction or conflict to the whole supply chain, once the delivery or payment extension happened.

J. Abawajy et al. (Eds.): ATCI 2018, AISC 842, pp. 915–920, 2019.
https://doi.org/10.1007/978-3-319-98776-7_109

2 Principles and Characteristics of Block Chain Technology

2.1 Principles of Block Chain Technology

Each block in the block chain represents a node in the chain, and each block structure consists of a block header and a block [8]. The block header mainly functions as a message marker and inter-block link, and block header information contains the previous block address, the block time stamp, the function value of the current block and the random number, while the block information mainly contains the number of transactions for the current block, and all transaction records generated during creation process of the block [9].

The block chain is a data structure in which blocks are ordered to link up, and it records the information of the nodes of the whole chain. It is mainly composed of block header information to achieve the link by pre-setting block chain technology to introduce the verification mechanism. When an external node finds the solution to meet the requirements, the results of the node will be broadcast to the node, and the other nodes will receive and verify its correctness to form a chain structure.

Block chain uses an asymmetric encryption algorithm, which has two keys: public key and private key. Public key and private key are a couple. When using publicly encrypted data, only the private key can decrypt, and vice versa. The node encrypts the written information with a private key and operates with a data signature, and the other member nodes can verify the authenticity of the data by a public key.

2.2 Features of Block Chain Technology

According to the link way of block structure and block, the block chain technology has the following characteristics [10], decentration, intelligent contract, security and trustworthy, common maintenance.

Decentration means the absence of a center in the chain can control the entire chain or network, and that is to say chain or network nodes have the same rights and obligations. When the information of each node changes, other node information also does the corresponding adjustment at the same time. So if any node is in abnormal situations, the entire chain or network operation will not be effected, which is different from the traditional supply chain with the existence of the core business or control center to organize the operation of the entire supply chain.

The definition of smart contract is intelligent contracts generated by various types of scripting language and algorithm through block chain. In the intelligent contracts, the key terms of the contract, the behavior of the trigger conditions are set up to determine whether the implementation of the contract is ordered, and the implementation of the smart contract is mandatory, which improves the efficiency of contract performance and reduces the risk of default.

Secure and trustworthy is defined as the transparency of node entrance or information change, which means other nodes can not affect its implementation, can only be a common certification, and help to redundant backup to reduce the possibility of data tampering, with the use of time stamp technology making the block chain traced.

Common maintenance refers to the equality of members or nodes that block chain emphasizes, and because there is no unified command and management agencies, the distributed accounting management is implemented, which requires members or nodes to conduct self-management, self-regulation and system operation.

3 Model Selection of Military Supply Chain Based on Block Chain Technology

According to the procurement quantity and operation and control requirements of the military supply chain, the military supply chain can be divided into the alliance chain model and the private chain model based on the block chain technology. Alliance chain is the block chain of part-decentration or multi-center, which is applied to multiple entities organization or alliance, and the consensus process is mainly controlled by a group of nodes. The private chain is a fully centralized block chain that applies to the internal data management and auditing of a particular organization and is entirely controlled by the central organization [2].

The cooperative state of the military supply chain W and the cooperative state within each subsystem W_i can be measured by the cooperative state parameter $X = (x_1, x_2, x_3, \ldots x_9)$ given by formula 1 and formula 2, and the cooperative state of each system is linearly related to the state parameter. That is

$$W = \omega_{x_1}x_1 + \omega_{x_2}x_2 + \cdots + \omega_{x_{11}}x_{11} \tag{1}$$

$$W_i = \omega_{ix_1}x_1 + \omega_{ix_2}x_2 + \cdots + \omega_{ix_9}x_9 \tag{2}$$

Among them, the vector $(\omega_{x_1}, \omega_{x_2}, \ldots \omega_{x_9})$ and $(\omega_{ix_1}, \omega_{ix_2}, \ldots \omega_{ix_9})$ respectively represent the coordination state parameter weighted value of the military supply chain and the coordination state parameter weighted value of the i-th subsystem of the military supply chain. The specific solution process is as follows

$$\omega_i^* = \frac{1}{m}\sum_{i=1}^{m} \omega_{x_i} \tag{3}$$

$$\omega_j^* = \frac{1}{f}\sum_{j=1}^{m} \omega_{x_j} \tag{4}$$

Among them, f is the number of the military supply chain nodes that are included in the j-th theme.

4 Applications of Block Chain Technology in Military Supply Chain

4.1 Full Management of Logistics Operations

The military supply chain logistics includes orders issued, product procurement, product quality inspection, product assembly, transportation, acceptance delivery, payment and etc.

In the military supply chain model mentioned in the previous, whether the alliance chain model or the private chain mode, once the core-control center is determined and logistics operation of each node is completed, the block chain information of the other nodes on the chain is updated, the model organized the production, and the information is fed back to the other nodes, as shown in Fig. 1.

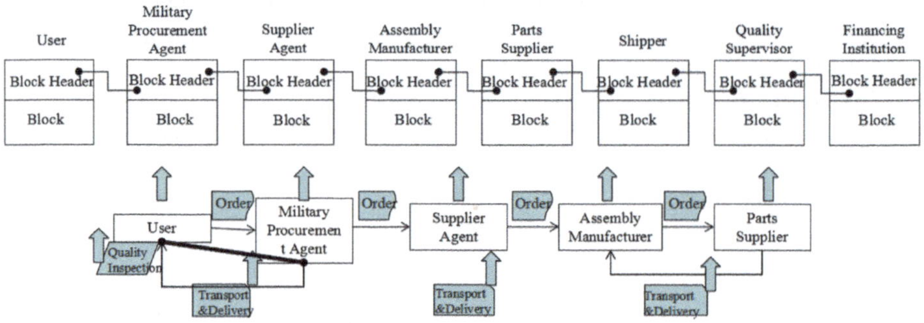

Fig. 1. Full management of logistics operations based on block chain technology

4.2 Payment Risk Management in Military Supply Chain

Basically, military procurement agent or supplier agent is the core-control center in the military supply chain. The original payment model may require downstream enterprises to afford for the production costs in advance and even defer payment, which brings higher financial risk or financial management costs to downstream enterprises. Under the new payment model, the parts suppliers only charge the sales of components, the assembly manufacturers only charge the total assembly cost without purchase cost of components, the carriers only collect the transportation service fees, and the military agent and the supplier agent only charge the transfer difference. As shown in Fig. 2, as the core-control center of the supply chain, the military procurement agents are tied together with financial institutions. When the work of the node on the chain is completed, it will trigger the use of block chain technology in advance which sets the payment conditions and smart contracts. Under this condition, the financial institutions will pay money to the node until the product delivered to the end users' hands. The core-control center of the supply chain can achieve the capital control and use of the entire chain, greatly reducing the overall risk of capital, as shown in Fig. 2.

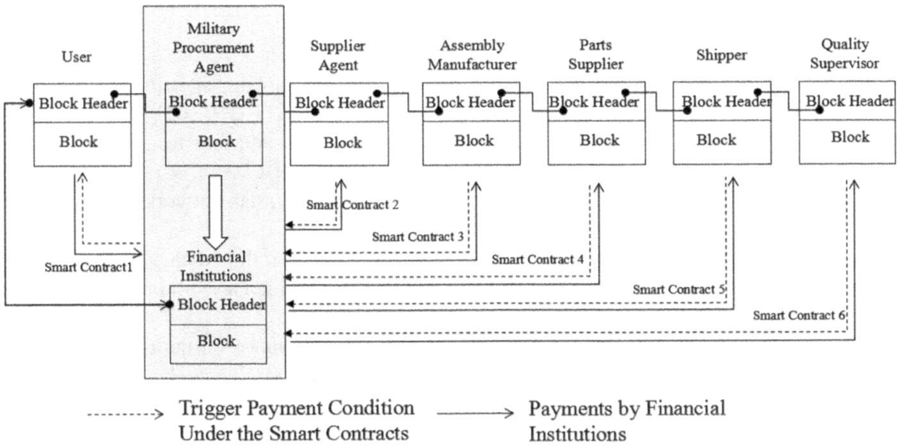

Fig. 2. Payment risk management in military supply chain based on block chain technology

4.3 Cost Management in Military Supply Chain

As a result of the use of block chain technology, the information in the supply chain can achieve real-time sharing, which eliminates the bullwhip effect, each node enterprise does not has to handle the risk of information asymmetry by redundant inventory and increasing the number of suppliers, and according to the production cycle and order cycle, the enterprises can organize production, achieve on-time production and zero inventory, significantly reduce the cost of raw materials, finished products squeeze, and also increase the marginal efficiency of enterprises. The supply chain's core-control center carries on centralized management and control use by mastering transaction costs, transportation costs, production costs on the supply chain, which can not only solve the financing problem of small and medium-sized enterprises, but also reduce the financial cost of the enterprises in each node of the supply chain, and finally realize the cost optimization management.

5 Conclusion

Owing to the characteristics of decentralization, intelligent contract, trustworthy, no tampering, trace-ability etc., it not only has bright prospect of the application in the financial industry, digital encryption currency and so on, but also has great application value in the supply chain, which can effectively reduce the logistics cost and improve the logistics agility. Based on the analysis of the characteristics of the military supply chain, this paper constructs four military supply chain models based on block chain technology, and analyzes the entire process management, risk management and cost management of military supply chain by applying block chain technology. In addition to the military supply chain, block chain technology also has meaningful application value in pharmaceutical supply chain, drug trace-ability, and multimodal transport.

References

1. He, Z.-X.: Research and design of data security multidimensional protection system in cloud computing environment. J. Residuals Sci. Technol. **13**(6), 211–219 (2016)
2. Sookhak, M., Gani, A., Buyya, R.: Dynamic remote data auditing for securing big data storage in cloud computing. Inf. Sci. **380**(112), 101–116 (2017)
3. Xu, S.: A novel data storage model and method based on bayesian network. J. Residuals Sci. Technol. **13**(6), 121–124 (2016)
4. Jayapandian, N., Rahman, A.M.J.M.Z.: Secure and efficient online data storage and sharing over cloud environment using probabilistic with homomorphic encryption. Cluster Comput. **22**(3), 1–13 (2017)
5. Aga, F.G., Woo, J., Song, J.: Controllable quantized conductance for multilevel data storage applications using conductive bridge random access memory. Nanotechnology **28**(9), 115–123 (2017)
6. Zhang, Y., Wang, J., Chang, Y.: The key technologies of the internet of things storage system for massive real time data storage. J. Residuals Sci. Technol. **13**(8), 21–25 (2016)
7. Aswathy, G., Rajesh, C.S.: Multiplexing recording in nickel-ion-doped photopolymer material for holographic data storage applications. Appl. Opt. **56**(8), 1566–1575 (2017)
8. Zhao, Y., Cai, Y., Fan, G.: Dynamical behavior for fractional-order shunting inhibitory cellular neural networks. J. Nonlinear Sci. Appl. **9**(3), 4589–4599 (2016)
9. Zhao, Y.: Research on active control of the dynamic vibration for underwater robot. J. Balkan Tribol. Assoc. **22**(5), 770–779 (2016)
10. Qu, Z.-Y., Su, W.-Z., Zhao, L.: Distributed data storage method based on information decentralization algorithm. J. Comput. Appl. **26**(3), 1102–1105 (2016)

Research on the Application of Computer Aided Audit Technology

Leping Jiang(✉)

College of Modern Economics and Management,
JUFE, Nanchang 330013, China
174599736@qq.com

Abstract. The rapid development of enterprise informatization has made the large-scale popularization of the accounting computerization, and the accounting computerization has changed the audit object, the audit content, the audit method and the audit scope. The traditional manual audit has not been able to meet the needs of the modern social information audit. Computer aided audit technology is the innovation and combination of computer technology and audit technology, and its use and promotion depend to some extent on the development of audit. At present, the construction of audit information in China has made great achievements. It not only lays a foundation for the introduction of computer aided standards and legal norms of the system, but also provides a broad platform for the application of computer aided audit technology. This article is based on the practice, through the analysis and study of the application of computer aided audit technology, to find out the existing problems in the application and promotion of computer aided audit technology, and put forward the corresponding countermeasures and suggestions, hoping to provide guidance for the development of computer aided audit technology in China to a certain extent.

Keywords: Audit technology · Computer technology · Auditors
Computer aided audit technology · Audit informatization

1 Research Background and Significance

1.1 Research Background

With the wide application of computer technology, in recent decades, computer technology has penetrated into various fields of natural science and social science, and audit, as a profession that has the function of supervising economic and social functions, inevitably accepts the huge challenge of the wave of information. The rapid development of enterprise informatization has made the large-scale popularization of the accounting computerization, and the accounting computerization has greatly changed the audit object, the audit content, the audit method and the scope of audit. As an important means of maintaining economic order and supervising economic activities, audit is bound to be deeply influenced by computer technology, especially the rapid development of audit technology and methods under the impetus of computer technology.

J. Abawajy et al. (Eds.): ATCI 2018, AISC 842, pp. 921–927, 2019.
https://doi.org/10.1007/978-3-319-98776-7_110

1.2 Research Meaning

At present, our country's audit enterprise is in the transition period from the traditional audit to the modern audit. Based on the practice, this paper analyzes and studies the application status of computer aided audit technology, finds out the problems existing in the application and popularization of computer aided audit technology, and tries to find out a reasonable solution. The development of mobile audit technology has the most practical significance.

2 Research Ideas and Innovations

2.1 Research Ideas

First of all, the theory of computer aided audit technology is summarized. Secondly, through investigation and research, the application status of computer aided audit technology in China is analyzed and the existing problems are studied. Finally, some reasonable suggestions are put forward in combination with the previous theoretical analysis and empirical research.

2.2 Innovation

Starting from the computer aided audit technology, this paper focuses on the application of computer aided audit technology in China, because the computer aided audit technology is a hot and focus problem in the research of current audit technology. Although there are many research on computer aided audit in China, it involves computer aided. The research on the application of audit technology is not very common. Therefore, this paper studies the application status of computer aided audit technology in China, focuses on the analysis of the problems existing in the application of computer aided audit technology and tries to find a reasonable solution.

3 Overview of the Theory of Computer Aided Audit Technology

The popularization and application of information technology has caused great changes in enterprises. In the wave of enterprise information management, the accounting information system inevitably has been greatly impacted. Accounting information system is a comprehensive, systematic and comprehensive reflection of financial activities and management process in the form of financial data and provides decision-making basis for the decision makers of the enterprise. According to the application status of computer aided audit technology, it can be divided into three categories: system oriented, data oriented and other new technologies. Figure 1 is a classification of computer aided audits.

With the vigorous development of accounting computerization, audit technology has developed from manual work to manual and information technology. In recent years, the types of audit projects have been increasing, and audit is closely linked with

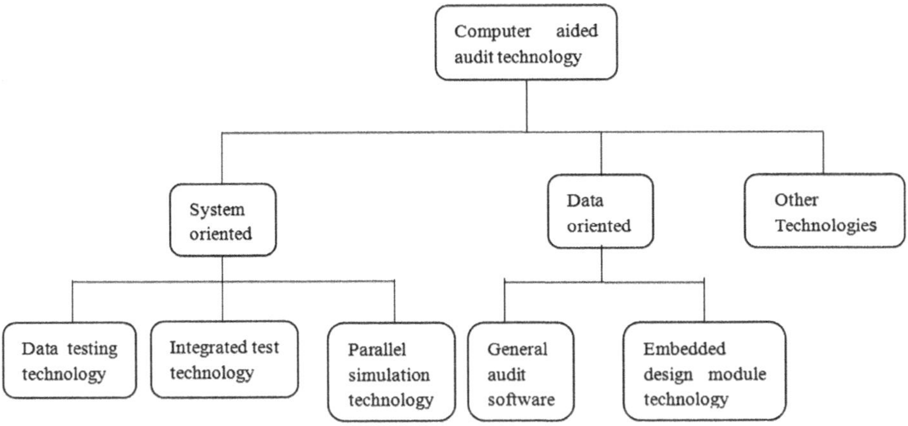

Fig. 1. Classification of computer aided audits

many industry departments (such as finance, real estate, water conservancy, customs, tax and so on). The audit object is no longer only financial data. These changes are urgently required to change the audit technology and methods.

4 The Application of Computer Aided Audit Technology in China

4.1 Statistical Test of Questionnaire

Information Degree Analysis. Reliability test is mainly used to measure the items in the questionnaire have high internal consistency and stability. The high stability indicated that the survey subjects had no significant difference in the same measurements at different time and space. According to the main contents of this study and the design of the questionnaire, this paper uses Cronbach's α coefficient to test the internal consistency of the scale. It is generally accepted that the α coefficient above 0.7 is acceptable. The results of the reliability test in this paper are shown in Table 1.

Table 1. Reliability test results of questionnaire (Cronbach's α coefficient)

Groups	Cronbach's α coefficient
Part one	0.774
Part two	0.791

Validity Analysis and Statistical Analysis of Data. The KMO value of the principal component factor is 0.767, indicating that the data are suitable for factor analysis. The specific results are shown in Table 2. Statistical analysis of the application of computer aided audit technology.

Table 2. Contact with accounting information system or management system

	Option	Frequency	Percentage	Effective percentage	Cumulative percentage
Effective	Regular	242	68.0	68.0	68.0
	Sometimes	70	19.7	19.7	87.6
	Very seldom	28	7.9	7.9	95.5
	Hardly	16	4.5	4.5	100.0
	Total	356	100.0	100.0	

Table 3. Whether computer aided audit technology is used to audit the information system of the audited unit.

	Option	Frequency	Percentage	Effective percentage	Cumulative percentage
Effective	Regular	158	44.4	44.4	44.4
	Sometimes	148	19.7	19.7	86.0
	Very seldom	42	11.8	11.8	97.8
	Hardly	8	2.2	2.2	100.0
	Total	356	100.0	100.0	

As shown in the above table, Tables 2 and 3 show the basic situation of Computer Assisted Auditing technology used by our auditors in their daily work. As can be seen from Table 2, 87.6% of the auditors who have been investigated have a high frequency of contact with the accounting information system or management system, and only 12.4% of the auditors have little or almost no contact with the accounting information system or management system in the audit work. From Table 3, we can see that among the respondents, the proportion of people who often or sometimes use computer aided audit technology to audit the information system of the audited units is up to 86%, the proportion of people rarely used is 11.8%, and only 2.2% of those who do not use it. From these two scales, we can see that at present, the auditors have a great opportunity to contact the accounting information system or the management system, and the proportion of the personnel to the information system is also higher.

Table 4 shows that 62.4% of the respondents often use the information system of the audited units to obtain relevant information, and 25.8% of the people sometimes use the information system of the audited units to obtain relevant information, and only 11.8% of the personnel with few or almost no use are used. This phenomenon reflects that it is common to use audit information system to obtain audit related information. The audit work environment makes the auditors more exposed to computer technology during their work. The following descriptive statistical analysis is carried out on several common computer aided audit techniques, such as Table 5.

Table 4. Using the information system of the audited unit to obtain relevant information

	Option	Frequency	Percentage	Effective percentage	Cumulative percentage
Effective	Regular	222	62.4	62.4	62.4
	Sometimes	92	25.8	25.8	88.2
	Very seldom	30	8.4	8.4	96.6
	Hardly	12	3.4	3.4	100.0
	Total	356	100.0	100.0	

Table 5. Description of statistics

Type of technology	N	Minimum value	Maximum value	Average	Standard deviation
General design software	356	1.00	5.00	2.2022	1.18374
Spreadsheet	356	1.00	4.00	1.1236	1.39204
Database technology	356	1.00	5.00	2.3820	1.17251
Integrated test technology	356	1.00	5.00	3.3652	1.19450
Parallel simulation technology	356	1.00	5.00	3.3146	1.21117
Data testing technology	356	1.00	5.00	3.0225	1.18657
Embedded audit module technology	356	1.00	5.00	3.1404	1.25033
Effective N	356				

As you can see from Table 5, in addition to the spreadsheet technology, some of the respondents are unfamiliar with the remaining 6 common computer aided audit techniques. As you can see from the mean, the use of spreadsheet technology is high, between regular use and sometimes use; general audit software technology and database technology are between sometimes used and rarely used; and integrated test technology, parallel simulation technology, data testing technology and embedded audit module technology are rarely made Between use or almost no use.

4.2 Analysis of the Results of Questionnaire Survey

To sum up, it is quite common for our audit staff to contact or utilize computer aided audit technology. However, most auditors are skilled in using one of these technologies, such as spreadsheet technology, general audit software, and other common computer aided audit techniques. There are two difficulties in the application of computer aided audit technology in China.

The Limitations of Computer Aided Audit Technology. Most auditors are not professional computer experts. What they need is simple, easy to operate tools. The use of audit software to assist audit is a more commonly used computer aided audit method.

Although the existing audit software has replaced a large number of manual audit work, the auditors are only routinely operated according to the procedural standards of the audit software. It is impossible to judge the authenticity and legality of data, so it is difficult to detect hidden doubts in data.

Lack of Unified Computer Aided Auditing System and Legislative Guarantee. The International Auditing Standards No. 16 has made relevant regulations on how to carry out computer aided audit in computer environment. However, based on the basic national conditions of our country, it still needs to formulate the computer aided audit technical standards for our country. At present, our country's audit information is still in the construction, the research on computer aided audit technology is still in the exploratory stage, without specific criteria, standards and norms as reference. When using computer aided audit, auditors will have many problems due to unclear direction and vague concept.

5 Summary

The research method of this paper is to conduct an empirical study by questionnaire, and discuss the application status of computer aided audit technology in China. The results of the survey depend entirely on the objective and fair answers of the respondents. The questionnaire type is relatively single, which also limits the subjective idea of the respondents. This is the aspect that this article definitely needs further improvement. At present, the application methods of computer aided audit technology are varied, and its future development direction includes two aspects.

The first is to integrate information technology with other industries. Modern audit involves many fields. The contents of the audit are not only financial data, but also related to other industry data, such as engineering data in engineering audit, coordinate data in land audit, and so on. The acquisition of these information needs the data information of the industry.

The second is the combination of artificial intelligence technology. Artificial intelligence technology is one of the focuses of modern information technology research. If we combine it with audit technology, we will greatly reduce the workload of auditors.

Acknowledgements. Funded projects: 2017 Jiangxi provincial university humanities and social sciences research project (Youth Project). Project name: From the perspective of regional ecological economy under the government environmental accounting research – in the Ganjiang River area as an example. Subject number: JC17207.

References

1. Schultz, E.E.: Sarbanes–Oxley—a huge boon to information security in the US. Comput. Secur. **23**(6), 353–354 (2014)
2. AICPA. Statement on Auditing Standards No. 99: Consideration of fraud in a financial statement audit, SAS 99. Auditing Standards Board of the American Institute of Certified Public Accountants (AICPA), vol. 10, pp. 5–25 (2012)

3. Knechel, W.R., Payne, J.: Additional evidence on audit report lags. Audit. J. Pract. Theory **20**(1), 137–146 (2011)
4. Duan, H: Research on the Impact of Risk Preference in Risk Oriented Audit Mode. Doctoral degree thesis of Southwest Jiao Tong University, vol. 12, pp. 11–14 (2016)
5. France, G.: Security and Risk Management of E-commerce, pp. 08–70. Huaxia Publishing House, Beijing (2011)
6. Elliott, R.: 21st century assurance. AAA Audit. Sect. Mid-Year Meet. **1**, 12–15 (2011)
7. Selima, G., Mc Namee, D.: The risk management and internal auditing relationship: developing and validating a model. Int. J. Audit. **3**, 159–174 (2009)
8. Zhang, Renshou: On the limitation and improvement of audit risk model. Guangdong Audit **6**, 8–14 (2010)
9. Lv, B.: What does joining WTO mean to CPA? China Financ. Econ. News. **9**, 14 (2012)
10. Cameron, R.: Modern Auditing and Assurance Service, pp. 54–55. John Wiley & Sons Australia Ltd, Milton (2014)

Improved Hybrid Collaborative Filtering Algorithm Based on K-Means

Xiaofeng Li[1(✉)] and Dong Li[2]

[1] Department of Information Engineering, Heilongjiang International University, Harbin 150080, China
mberse@126.com

[2] School of Computer Science and Technology, Harbin Institute of Technology, Harbin 150001, China

Abstract. Due to the rapid development of e-commerce, both users and items are increasing, which leads to the problems of low efficiency and inaccurate recommendations when using collaborative filtering recommendation algorithms. In order to solve this problem better, this paper proposes an improved hybrid collaborative filtering algorithm based on K-Means. Firstly, use K-Means user clustering to divide users into different clusters according to the user's behavioral habits, and then use item similarity to solve the sparseness of the scoring matrix. Finally, use collaborative filtering recommendation algorithm based on this to first determine which users belong to which clusters, calculating user similarities within a cluster. Experiments show that the algorithm proposed in this paper can avoid computing similarity on the entire data set to find the nearest neighbors, minimize the amount of computation, and improve the efficiency of the recommended algorithm.

Keywords: K-means · Collaborative filtering algorithm · Clustering

1 Introduction

With the continuous increase of network data, the 21st century has entered the era of the Internet and the era of big data [1, 2]. In this era of information overload, how to quickly and accurately find the products that users are interested in the massive data is imminent. At this point, data mining is not only widely studied in academia but also widely used in industry. Data mining is mainly to search for some features of information from massive data [3]. Data mining has a variety of algorithms, generally based on the characteristics of the data on the classification of the data or finds the correlation between the data to be classified, and when using the collaborative filtering recommended algorithm is based on the relationship between the users to find the user's nearest neighbor. Therefore, we can combine the data mining algorithm and the collaborative filtering algorithm, and use the data mining algorithm's classification method to classify the associated data into the same category [1, 5]. This can also avoid the inaccuracy of finding the user's neighbors in the collaborative filtering algorithm. This paper adopts k-means algorithm to cluster data. Firstly, it introduces the core idea of k-means algorithm, then analyzes the advantages and disadvantages of the method and

J. Abawajy et al. (Eds.): ATCI 2018, AISC 842, pp. 928–934, 2019.
https://doi.org/10.1007/978-3-319-98776-7_111

adjusts it appropriately to combine with collaborative filtering algorithm. Finally, analyze the experimental results to prove the efficiency of the algorithm.

2 K-Means Algorithm

The k-means algorithm, also known as the K-means algorithm, is a relatively simple clustering method. K, as the input parameter of the algorithm, assumes that there are n objects in a set, which are divided into K clusters. The similarity in the cluster is higher than that between clusters K-Means. The K-means algorithm used for partitioning, the average of each object in the cluster as the centroid of the cluster. K-means partitioning algorithm is shown in Algorithm 1.

Algorithm1 K-means partitioning algorithm

Input: Data set D: there are n objects, K: the number of clusters to be divided
Output: The set of k clusters
1. Select K objects randomly in D as the initial centroid of the cluster;
2. Calculate the distance from other objects in D to the centroids of K clusters, and assign each object to the nearest cluster;
3. Recalculate the centroid of each cluster based on the results in (2);
4. Repeat step 2, step 3, until no more changes occur.

Suppose there is a set of objects distributed in the space represented by the rectangle in Fig. 1(a). Set K = 2, that is, the user wants to divide these objects into 2 clusters (distinguished by two colors, red and blue). According to the above algorithm, 2 objects are randomly selected as the initial centroid of the cluster, and the red and blue two-color crosses are used to mark, it is shown in Fig. 1(b).

By comparing each object's distance to the centroid of each cluster, it is assigned to clusters that are closer together. This distribution is displayed in different colors in Fig. 1(c). Next, there is a need to calculate the mean value of each cluster as the centroid of the cluster based on the current distribution of each object. Use these new cluster centroids as shown in Fig. 1(d). The distance between the object and the center of mass of the cluster is repeatedly calculated to determine the cluster to which the object belongs and the updated cluster centroid, until finally no longer changes.

The k-means method requires K-values to be entered in advance. When the K-value cannot be determined, it will have a large impact on the results of the algorithm. We can choose the most suitable one by repeatedly trying different K comparison results. However, the algorithm is easy to understand, the calculation speed is fast, and the effect is also very good. The experimental results will be used to prove the advantages of the k-means method.

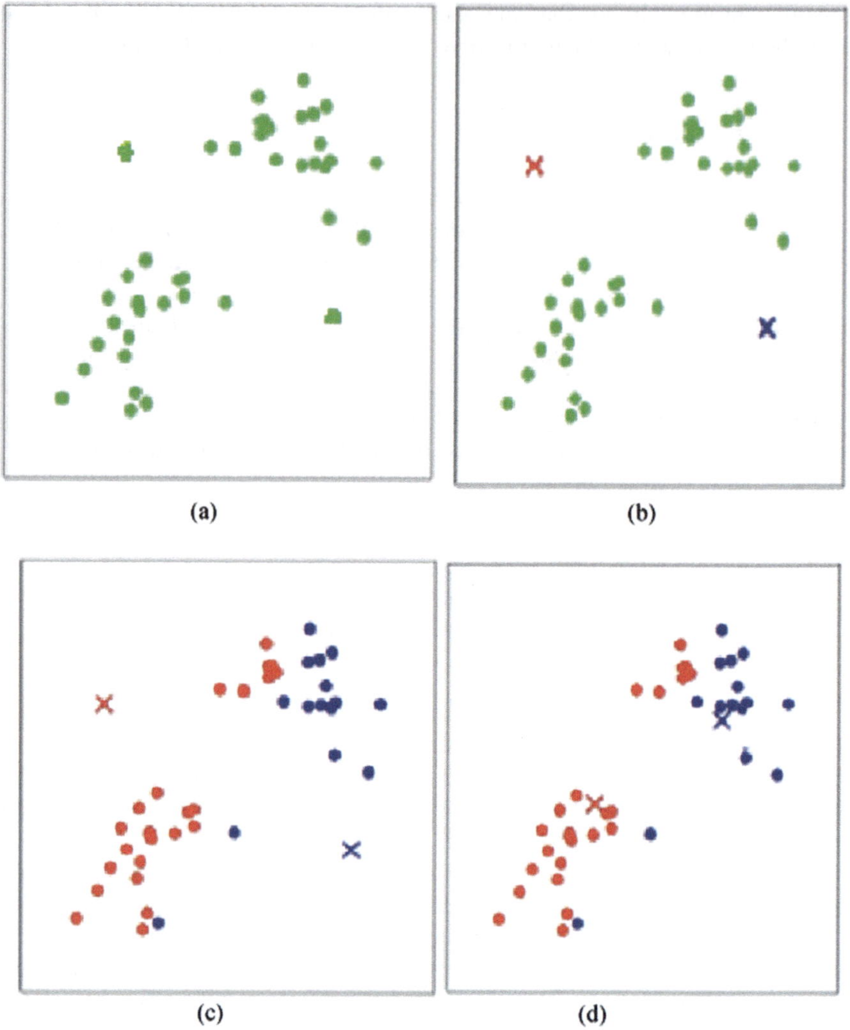

Fig. 1. Clustering process

3 Improved Hybrid Collaborative Filtering Algorithm Based on K-Means

3.1 K-Means User Clustering

With the continuous increase of users, the types of goods, etc., are also constantly increasing. The traditional collaborative filtering algorithm has been unable to accurately and quickly find the interests of users from big data. Data mining is also constantly evolving. Data mining mainly digs out the needed information in big data, so we can combine data mining and collaborative filtering algorithms. Firstly, relevant data are mined in big data, and then users are really interested in the relatively small related data to recommend them to users. At this point, what data mining method is used becomes the key. Clustering algorithm is a widely used data mining algorithm [5]. Clustering is mainly based on similarity or distance and other methods to classify data into different classes, making it in the same class. The distance between data is relatively close or similar. The k-means method is a relatively simple and effective algorithm in the clustering algorithm, and this algorithm is also constantly improving and optimizing. In this paper, there is a need to combine k-means algorithm and hybrid collaborative filtering algorithm.

K-means user clustering is used to classify data sets based on the distance between users, so that user data with a short distance is in the same cluster, and the distances are far in different clusters. This method can reduce the time efficiency of searching for the nearest neighbor, and divide the users into different clusters. When searching for the nearest neighbor, it is only necessary to search in the cluster to which the user belongs. Therefore, this not only solves the sparseness of the scoring matrix to some extent, but also avoids calculating the similarity between users in big data, improving the efficiency and the accuracy.

In this paper, the Euclidean distance measurement formula is used, because the data set (movielens) used by the user to score the movie is used. Therefore, the distance between users should be judged according to the user's rating of the item. Due to the scarcity of the user-item scoring matrix, the Euclidean distance measurement formula needs to be revised. The revised formula is as follows:

$$D_{mn} = \sqrt{\sum\nolimits_{i \in I_{mn}} (V_{mi} - V_{ni})^2 / count(I_{mn})} \tag{1}$$

Where, D_{mn} represents the Euclidean distance between user m and user n. I_{mn} represents a collection of scores for the same item by two different users. $count(I_{mn})$ represents the number of elements contained in this collection.

3.2 Algorithmic Description

The k-means user clustering algorithm is mainly proposed to reduce the amount of computation and improve the recommendation quality. Below we combine the k-means user clustering and hybrid collaborative filtering algorithm, the actual recommendation idea has not changed. Only when using hybrid collaborative filtering for recommendation, the input data is changed from the original user-item rating data to the output based on the k-means user clustering algorithm. The basic hybrid collaborative filtering algorithm calculates the similarity between users and the similarity between items on the entire user-item scoring matrix. However, the algorithm used in this chapter needs to first determine which cluster the user belongs to, and then calculates the user's similarity between the two, so the main difference is in the search of the nearest neighbor. The specific steps can be described as Algorithm 2.

Algorithm2 Hybrid collaborative filtering algorithm based on K-means

Input: K represents the number of clusters that the dataset is divided into by the k-means method; t represents the number of users' nearest neighbors, and the user-item rating matrix $R(u \times i)$.

Output: The target user's N item recommendation

1. K-means user clustering. Calculate the Euclidean distance between users according to Equation (1) and k-means algorithm implementation steps to divide the data set into K clusters $U = \{U_1, U_2, U_3, \ldots, U_k\}$.
2. Fill in the user-score matrix. Similarity is calculated in each class, and the user-item rating matrix is populated based on the similarity predicted scores.
3. Find the cluster to which the target user belongs. Calculate the Euclidean distance between the target user and each cluster's centroid and compare its size, where the smallest representative target user belongs to the cluster.
4. Calculate the similarity between users. The similarity between users is calculated in the cluster to which the target user belongs and saved as a similarity matrix.
5. Nearest neighbor. Compare the similarity with the target users and sort them in descending order. Select the first t nearest neighbor.
6. Predicted score. Referring to the target user's nearest neighbor's rating of the item, the target user's rating of the item is predicted.
7. Recommendation. Sort the predicted scores in descending order, and select the top N as the final recommended item set.

4 Experimental Results and Analysis

The selection of K in the k-means method has a crucial influence on the result. When the K value is large, the k-means clustering consumes time and takes up memory in use. When the value of K is small, the efficiency will be affected when the user similarity is calculated to find the nearest neighbor of the user. Therefore, the appropriate K value has a crucial influence on the experimental results. The results of multiple experiments show that K = 15 has the best effect. Therefore, the value of K in the paper experiment is selected as 15.

In order to prove that the proposed algorithm improves the efficiency of the recommendation, sample hybrid collaborative filtering, traditional user-based collaborative filtering, and hybrid collaborative filtering algorithm combined with k-means user clustering perform performance comparisons on the data set. Select the number of nearest neighbors of the target user as 10–50. Compare the difference between the predicted score and the test set score. Use the MAE value to measure the merits of the three algorithms. When different values are selected for K, the MAE values of the two algorithms are shown in Fig. 2.

Figure 2 shows the variation of MAE when the three recommended algorithms take different values for k. The following conclusions can be drawn based on the curve changes in the figure: The MAE of the hybrid collaborative filtering recommendation

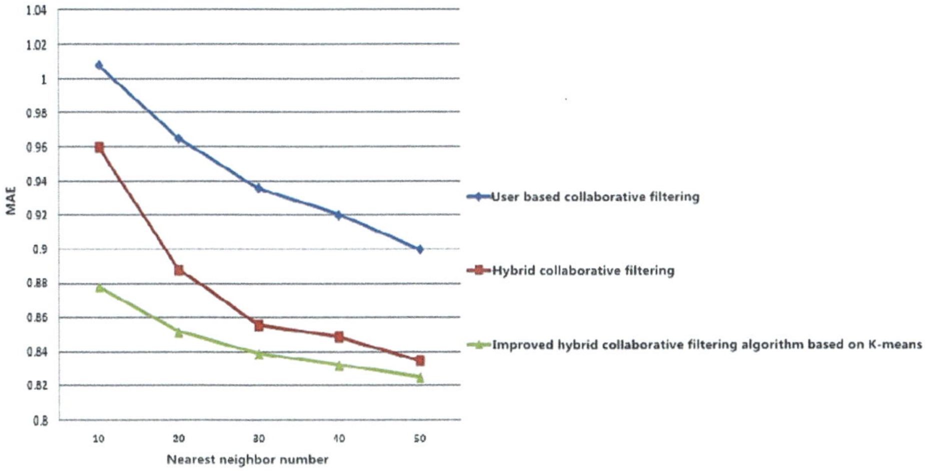

Fig. 2. The MAE values of three recommended algorithms under different K values

algorithm based on k-means user clustering is the smallest regardless of the value of k. It is proved that the algorithm is optimized to some extent. When the value of k, that is, the number of nearest neighbors gradually increases, the MAE value of the hybrid collaborative filtering recommendation algorithm based on k-means user clustering is less, and the trend of change is relatively stable. This is due to the fact that the algorithm is first based on the user's clustering, placing users in close proximity in the same cluster, and has used project similarity to solve the problem of sparseness of the scoring matrix, so the search is more accurate when searching for nearest neighbors. When the value of k is gradually increased, the impact on finding the nearest neighbor is not great, and the final MAE value does not change much so as to prove the accuracy of the algorithm in theory.

5 Conclusion

This paper briefly describes the core idea of the k-means algorithm and analyzes its advantages and disadvantages. According to the characteristics of the k-means algorithm and the collaborative filtering algorithm, it is decided to use the Euclidean distance measurement formula in the k-means clustering. The sparseness characteristics of the dataset make appropriate adjustments to the Euclidean distance formula, giving detailed steps of the algorithm. Finally, the experiment was performed on the data set. The MAE value was used as a measure to compare the quality of the recommended algorithm and analyze the experimental results and the reasons for the experimental results. The accuracy of the theoretical analysis and the effect of the algorithm optimization were proved.

References

1. Wei, Z., Nan, L., Ying, H., Hongtao, Z.: An improved K-means clustering based collaborative filtering algorithm. J. Anhui Univ. (Nat. Sci.) **40**(2), 32–36 (2016)
2. Wan, J.: Collaborative filtering recommendation algorithm based on user's comprehensive information particle swarm optimization and K-means clustering. In: Proceedings of 2015 International Conference on Industrial Informatics, Machinery and Materials (IIMM 2015), pp. 77–80. Science and Engineering Research Center (2015)
3. Katarya, R., Verma, O.P.: A collaborative recommender system enhanced with particle swarm optimization technique. Multimed. Tools Appl. **75**(15), 9225–9239 (2016)
4. Zhi, H.J.: Application of Hybrid Algorithm Based on Content and Collaborative Filtering in Recommender System. Donghua University, Changning (2016)
5. Zhai, D.H., Jiang, Y.U., Gao, F., et al.: K-means text clustering algorithm based on initial cluster centers selection according to maximum distance. Appl. Res. Comput. **8**(12), 20–28 (2014)
6. Peng, L., Fang, W.: Heterogeneity of inferring reputation of cooperative behaviors for the prisoners' dilemma game. Physica A **433**, 367–378 (2015)
7. Aleksy, M., Korthaus, A., Schader, M.: Use Java and the CORBA realization distribute type system. J. Pingxiang Coll. **4**, 104–105 (2005)
8. Titus, J.: ECN technical editor: the eclipse of stand. J. Zhongkai Agrotech. Coll. **19**(2), 32–35 (2006)
9. Richardson, W.C., Avondolio, D.: The Java high class weaves a distance: JDK 5. Sci. Technol. Book Rev. **3**, 17–18 (2006)
10. Li, J., Vuong, S.: A semantics-based routing scheme for grid resource discovery. In: E-Science: First International Conference on E-Science and Grid Computing, pp. 58–70, 90 (2005)
11. Qizhen, W.: Automatic control system regulation scheme of textile air conditioning. Sci. Technol. Inf. **22**, 395–396 (2010). (in Chinese)
12. Jinkun, L.: Advanced PID Control and MATLAB Simulation, 2nd edn. Electronics Industry Press, Beijing (2003). (in Chinese)
13. Jianwei, L., Qingchang, R.: Study on supply air temperature forecast and changing machine dew point for variable air volume system. Build. Energy Environ. **27**(4), 29–32 (2008). (in Chinese)

Efficient Parallel Clustering Spectral Algorithm Based on Hadoop

Ling Ma(✉)

Department of Information Engineering, Heilongjiang International University, Harbin 150080, China
Lingmaml@sina.com

Abstract. This paper presents an efficient parallel spectral clustering algorithm. First, the *kd*-tree technique was introduced to sparse the similarity matrix. Then, when calculating the eigenvectors, the Laplacian matrix was stored in the Hadoop file system, and the eigenvectors were obtained by the distributed Lanczos operation. Finally, efficient parallel k-means clustering was used to process the transposed matrix of eigenvectors to obtain clustering results. By using different parallel strategies for each step of the algorithm, the entire algorithm achieves a linear increase in speed. Experiments show that the clustering speed increases almost linearly with the expansion of the processing data size. The proposed parallel spectral clustering algorithm is suitable for massive data mining.

Keywords: Clustering algorithm · K-means · Hadoop

1 Introduction

By using a large number of inexpensive computer clusters instead of expensive servers, data processing costs can be greatly reduced. The large-scale applications based on cloud computing platforms all have features such as heterogeneous, large-scale, and distributed, and are suitable for data-intensive applications and processing [1, 2]. How to speed up spectral clustering algorithm has attracted wide attention of many scholars. These research work can be summarized into four categories: spectral clustering based on sampling techniques, spectral clustering based on low rank approximation, spectral clustering based on preprocessing methods, and Spectral clustering of parallel computing techniques.

Meng et al. [3] proposed the method of matrix sparse-t-nearest neighbors, and finally implemented the parallelization of spectral clustering using the matrix sparsified by the nearest neighbor method, and verified the algorithm by an empirically-learned document data. This proves that this algorithm can effectively handle the big data problem. In this paper, the similarity matrix is sparsely processed firstly. When calculating the eigenvectors, the Laplacian matrix is stored in the Hadoop file system, and the eigenvectors are obtained by the distributed Lanczos operation. Finally, efficient parallel k-means clustering was used to process the transposed matrix of eigenvectors to obtain clustering results. Different parallel strategies were used to parallelize each step of the algorithm, and the spectral clustering algorithm's operating speed was greatly improved.

J. Abawajy et al. (Eds.): ATCI 2018, AISC 842, pp. 935–941, 2019.
https://doi.org/10.1007/978-3-319-98776-7_112

2 Related Concepts and Descriptions

Spectral clustering algorithm was proposed in 1973. It is based on spectrogram theory in graph theory. It is a kind of point pair clustering algorithm. Its essence is to transform the clustering problem into the optimal division problem of the graph. Due to the combined nature of graph partitioning problems, the optimal partitioning of graphs is an NP-hard problem. In general, the steps for a standard serial spectral clustering algorithm are as follows:

Algorithm1 Spectral clustering algorithm

1. Calculate to get the similarity matrix $S = \exp(\frac{\| x_i - x_j \|^2}{2\sigma^2}) \in R^{n \times n}$;
2. The construction of angle matrix $D = d_i = \sum_{j=1}^{n} w_{ij}$;
3. The construction of angle matrix $D = d_i = \sum_{j=1}^{n} w_{ij}$;
4. Normalized Laplacian Matrix L Calculation;
5. The k smallest eigenvectors of the matrix L are calculated and form the matrix $Z \in R^{n \times k}$, and the k smallest eigenvectors are used as the matrix Z
6. Standardize Z into $Y \in R^{n \times k}$
7. K-means algorithm is used to gather data points $y_i \in R^k (i = 1, \ldots, n)$ into k clusters

3 Hadoop-Based Parallel Spectral Clustering Algorithm Design

3.1 Parallel Computing Similarity Matrix

When building a sparse similarity matrix, most of the relevant sample data points are saved by the t-nearest neighbor method, and irrelevant ones are removed. In fact, most uncorrelated data points are noise data. Calculating distances between data points and then sorting the distance matrix for these noises will result in high computational costs and waste of time. Here, we introduce the *kd*-tree indexing technique proposed in the literature [4]. The *kd*-tree is a data structure used for the rapid search of multidimensional data, first proposed by Bentley. In a *kd*-tree, each tree node stores a d-dimensional data point. The construction process of a *kd*-tree is a process of circularly using each dimension of data to divide a data set. It is shown in Fig. 1.

Literature [4] designed a nearest-neighbor interpolation method to evaluate lost data in wireless sensor networks, and used a *kd*-tree structure to speed up the search. In order to find neighbors, the nearest neighbor search algorithm elaborates on the concept of distance matrix. The general distance matrix uses Euclidean distance, which is defined as Eq. (1).

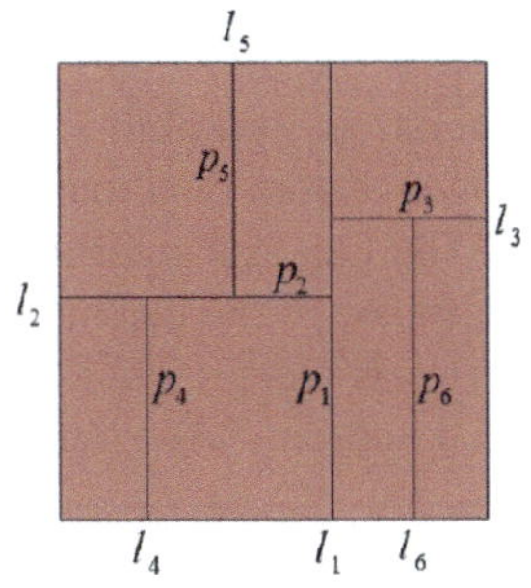

Fig. 1. *kd*-tree for storing two dimensional data

$$d(x_i, x_j) = \sqrt{\sum_{k=1}^{K} \left(x_i^{(k)} - x_j^{(k)}\right)^2} \tag{1}$$

Since the *kd*-tree performs linear segmentation on the space, if the Euclidean distance is used, the search range determined by the center and the radius is spherical, and it is relatively complicated to determine the inclusion and intersection of the sphere and the cube. Equation (1) has been modified to use the following form of distance definition. It is shown in Eq. (2).

$$d(x_i, x_j) = \sqrt{\sum_{k=1}^{K} \lambda^k \left(x_i^{(k)} - x_j^{(k)}\right)^2} \tag{2}$$

According to [5], we defined the similarity between data points. It is shown in Eq. (3).

$$w_{ij} = \exp\left(-\frac{d^2(x_i, x_y)}{\xi_i \xi_j}\right) \tag{3}$$

After sparse processing of data, the sparse similarity matrix is parallelized according to literature [6]. The process is shown in Fig. 2.

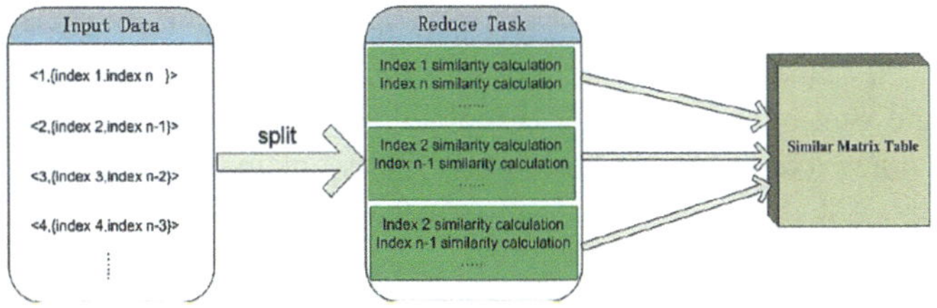

Fig. 2. Parallelization of similar matrix

3.2 Parallelization Calculates the k Smallest Eigenvectors

The Lanczos algorithm is an iterative algorithm invented by Cornelius Lanczos to calculate the eigenvalues and eigenvectors of a matrix, or the singular value decomposition of a rectangular matrix. The Lanczos algorithm is very effective particularly in the face of very large, sparse matrices. In the calculation of the largest (or smallest) k eigenvectors of the matrix, Lanczos is more suitable because it only needs to iterate k times to find the k eigenvectors. Lanczos's algorithm see Algorithm 2.

Algorithm2 Spectral clustering algorithm

1. $v_1 \leftarrow norm$ is a random vector of 1

2.Iteration: for j=1,2,…,m

$w_j \leftarrow Lv_j - \beta_j v_{j-1}, a_j \leftarrow (w_j, v_j)$

$w_j \leftarrow w_j - \alpha_j v_j$

$\beta_{j+1} \leftarrow \| w_j \|$

$v_{j+1} \leftarrow w_j / \beta_{j+1}$

3.Return

3.3 Parallelization of K-Means Clustering

1. MapC Function Design with Combine

The input to the Map function is the center of all the records that need to be clustered and the iterations of the previous round. Its task is to calculate the distance from each record to the center point, and mark the new attribution category for it again. The input record <key, value> pair is in the form of <line number, record line>. Each Map function reads the cluster center description file. For each record entered, the Map function calculates the nearest center of the class and marks the record as belonging to the new class. For the intermediate result output by the Map function, the <key, value> pair is in the form of <cluster category ID, record attribute vector>. The Combine operation introduced reads the output of the Mapper process on each node and computes the center point of the local data.

2. ReduceC Function Design with Combine

The intermediate result of Map C function output is used as input to the Reduce function. Its task is to calculate a new clustering center for use by the next Map Reduce Job. The data <key, value> pair input to the Reduce function is in the form of <cluster class ID, {record attribute vector set}>. All records with the same key value (i.e. records with the same class ID) are sent to a Reduce. To add the sum of the key data points and the sum of the recording components to obtain the average value of each component, a new cluster center description file is obtained. The output of the reduce function <key, value> is in the form of <cluster category ID, mean vector>.

4 Experiment and Result Analysis

In this lab, we used 10 computers to build a Hadoop cluster. Among them, 8 sets are dual-core 2.6 GHZ CPUs, 4 GB of memory, and the operating system is Ubuntu 10.04; 2 sets are quad-core 2.8 GHZ CPUs, 8G of memory, and Ubuntu 10.04 operating system is also installed. The Hadoop version is 0.20.2, and Gigabit Ethernet cards are used between each machine to connect through switches.

The experiment uses the classical data set Data Set1 provided by the KDD Cup'99 to test the correctness of the proposed parallel spectral clustering algorithm; 10000 (data set DS1), 50,000 (data set DS2), 100000 (data set DS3), 1000000 (data set DS4), 5000000 (data set DS5) data to verify the superiority of the proposed parallel algorithm, data samples, such as the Literature [5] data structure to construct multidimensional data. In the experiment, acceleration ratio and scalability were used as evaluation indicators.

4.1 Correctness Verification

Table 1 shows the clustering result of the data set Data Set1 in stand-alone and the parallel spectral clustering algorithm mode mentioned in this paper. The test results show that both the standard serial spectral clustering algorithm and the parallel spectral clustering algorithm mentioned in this paper achieve high accuracy, and their error rate is lower than 2%. This shows that in these two modes, the parallel spectral clustering algorithm proposed in this paper is correct. Table 1 is the comparison of the clustering accuracy rate between the standalone mode and the parallel algorithm mode proposed in this paper.

Table 1. Comparison of clustering accuracy between single machine mode and parallel algorithm pattern proposed in the paper

Type	The amount of data	The standard model		The parallel mode in this paper	
		Correct recognition number	Error recognition number	Correct recognition number	Error recognition number
Normal	17811	17828	365	17811	290
U2r	256	254	5	256	3
Dos	16877	17512	275	17441	188
R21	3673	3765	100	3801	87
Probe	4766	4576	101	4601	73
Average error rate		1.97%		1.46%	

4.2 Acceleration Ratio Test

The speedup ratio is an important indicator for verifying the performance of parallel algorithms and is defined as Eq. 4. Where T represents the running time of the serial algorithm and p is the running time of the parallel algorithm. In general, the greater the speedup, the less time it takes for parallel computations to achieve higher efficiency. In the experiment, we tested the acceleration of the conventional spectral clustering algorithm and the parallel spectral clustering algorithm proposed in this paper when dealing with different data sizes. Table 2 shows the running time of datasets DS1 to DS5 under different cloud platform cluster numbers.

$$S_A = \frac{T_S}{T_P} \tag{4}$$

As can be seen from Table 2 and Fig. 3, as the data set increases in size, the algorithm is getting better and faster than the performance. The main reasons are as follows: (1) The <key, value> pairs in the Map and Reduce stages of the parallel spectral clustering algorithm are relatively reasonable; (2) We added Combine operations after the Map phase and the Reduce phase, which greatly reduced the communication cost between the named node (primary node) and the data node (slave node). Thus, as the amount of data evolves to a large-scale level, the acceleration-to-performance ratio increases dramatically.

Table 2. Running time comparison

Data volume	Machines	Similar matrix (s)	Eigenvector (s)	K-means (s)	Total time (s)
DS1	1	0.338	0.482	0.156	1.025
(10000)	2	0532	1.089	3.485	5.054
DS2	1	7.254	9.314	2.954	19.514
(50000)	2	9.841	12.741	4.147	26.874
DS3	1	19.225	23.984	8.574	51.784
(100000)	2	11.453	15.201	4.111	30.043
DS4	1	7621.780	9702.565	3411.543	20757.73
(1000000)	2	34.220	46.034	16.342	100.312
DS5	1	31802.602	190.558	80.870	88145.31
(5000000)	2	150.876	97.905	39.215	421.276

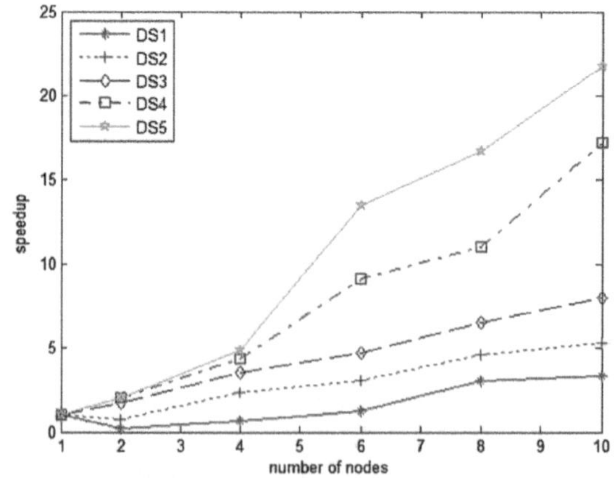

Fig. 3. Speed up performance test

5 Conclusion

In this paper, the traditional spectral clustering algorithm is studied, and an efficient parallel spectral clustering algorithm is designed. The strategy of parallelized spectral clustering algorithm is to improve the distance matrix and similarity matrix, and introduce the *kd*-tree technology to sparse the similarity matrix. In the calculation of eigenvectors, Laplacian matrices are stored on the Hadoop file system, and feature vectors are obtained through distributed Lanczos operations. Finally, we use efficient parallel k-means clustering to process the transposed matrix of eigenvectors to obtain clustering results. By using different parallel strategies for each step of the algorithm, the entire algorithm achieves a linear increase in speed. Experiments show that with the expansion of data processing scale, the clustering speed has almost linear growth, and the proposed parallel spectral clustering algorithm is suitable for massive data mining. The research results can provide research basis for designing a better large-scale data-efficient parallel clustering algorithm later.

References

1. Pai, Z., Lizhen, C.: The data layout strategy for data intensive applications in the cloud computing environment method. Comput. J. **33**(8), 1472–1480 (2016)
2. Peng, W., Dan, M.: Advances in data intensive computing programming models. Comput. Res. Dev. **47**(11), 1993–2002 (2012)
3. Chao, M.: The Research of Spectral Clustering Algorithm in the Distributed Environment. Beijing University of Posts and Telecommunications, Beijing (2013)
4. Li, Y.Y., Parker, L.E.: Nearest neighbor imputation using spatial-temporal correlations in wireless sensor networks. Inf. Fusion **15**(SI), 64–79 (2014)
5. Zelnik, M.L., Perona, P.: Self-tuning spectral clustering. In: Proceedings of Advances in Neural Information Processing Systems, pp. 1601–1608 (2015)
6. Cheng, W.: The Realization of Parallel Clustering Algorithm on MapReduce, pp. 27–38. Zhejiang University, Zhejiang (2012)
7. Peng, L., Fang, W.: Heterogeneity of inferring reputation of cooperative behaviors for the prisoners' dilemma game. Physica A **433**, 367–378 (2015)
8. Aleksy, M., Korthaus, A., Schader, M.: Use Java and the CORBA realization distribute type system. J. Pingxiang Coll. **4**, 104–105 (2005)
9. Richardson, W.C., Avondolio, D.: The Java high class weaves a distance: JDK 5. Sci. Technol. Book Rev. **3**, 17–18 (2006)
10. Vuong, J.L.: A semantics-based routing scheme for grid resource discovery. In: E-Science: First International Conference on E-Science and Grid Computing, vol. 90, pp. 58–70 (2005)
11. Qizhen, W.: Automatic control system regulation scheme of textile air conditioning. Sci. Technol. Inf. **22**, 395–396 (2010). (in Chinese)
12. Jinkun, L.: Advanced PID Control and MATLAB Simulation, 2nd edn. Electronics Industry Press, Beijing (2003). (in Chinese)
13. Jianwei, L., Qingchang, R.: Study on supply air temperature forecast and changing machine dew point for variable air volume system. Build. Energy Environ. **27**(4), 29–32 (2008). (in Chinese)

A New Real-Time Assessment Method and System Design for Closed-Loop Simulation Experiment

Yu Wang[1(✉)], Dao-Wei Liu[1], Ming-Fei Liao[1], and Yu-Feng Du[2]

[1] Luoyang Electronic Equipment Test Center, Luoyang 471003, China
253389416@qq.com
[2] Tianjin University of Technology, Tianjin 300384, China

Abstract. The experiment assessment is an important part in closed-loop simulation experiment research job. The conventional post assessment method belongs to a kind of static assessment, wasting time, and the conclusion is not comprehensive. As an improvement, real-time assessment system based on behavior component is designed and implemented using technologies such as assessment behavior components modeling, multiple cooperative engine coupling, etc. It can not only acquire instant and process data of simulation experiment, but also realize real-time analysis, information contrast and index dynamic analysis. It provides a new method and tool support to effectively solve the problem of closed-loop simulation experiment assessment. Verified by the experiment and application, the technology is feasible and the system can adapt to different situations and run stably, when adjusting the simulation speed ratio, adding simulation entities, changing experimental scenarios, expanding scale of simulation. The system has an extensive application prospect in the field of closed-loop simulation experiment assessment.

Keywords: Behavior components · Real-time assessment
Closed-loop simulation experiment

1 Introduction

The conventional post assessment method hasn't been able to meet the need of real-time assessment, which is required in the experiment activities such as experiment index analysis and confrontation result assessment. Research of real-time assessment has carried out in various fields [1–5]. With the development of the evaluation means, methods, equipment conditions, real-time evaluation technology and system have been applied more widely in training and equipment test [6]. According to the circuit composition and time characteristics, training simulation and test simulation commonly used real-time or lacking-real-time simulation adopting person or equipment in circuit for meet the demands of command training, equipment operation and equipment test. The technology of real-time assessment has been applied commonly [8, 9]. On the other hand, analysis simulation, based on the closed-loop simulation experiment, have many technological difficulties such as man not in loop, ultra-real-time data rate, bad

J. Abawajy et al. (Eds.): ATCI 2018, AISC 842, pp. 942–952, 2019.
https://doi.org/10.1007/978-3-319-98776-7_113

cooperative coupling of cause and effect sequence. There are more difficulties and research needs to research and apply real-time assessment technology in closed-loop simulation.

Aim at the shortages of post assessment technology in closed-loop simulation experiment such as assessment method, process, conclusions, etc. the conventional solution way is to modify the simulation engine. As a result, it changes the functions of simulation platform, break out the closed-loop simulation running style. The advantage of the conventional solution way is to provide the human-computer mutual assessment function of man in loop. Its weakness is to break out the closed-loop simulation organization and influence the stability of simulation platform. Its influence and indetermination of the conventional solution way still be unknown [10]. For now it is clear that the conventional solution way can reduce simulation experiment speed and efficiency, extending the manufacturing period, enhancing the cost.

We put forward a new solution way of real-time assessment in closed-loop simulation experiment according to the behavior components. The real-time assessment system has been designed and implemented using technologies such as assessment behavior components modeling, multiple cooperative engine coupling, MVC (Model-View-Controller) structure mode, etc. The advantages of the system include keeping current technology organization of simulation platform, following closed-loop simulation experiment technology way. The key technology to realize real-time assessment in closed-loop simulation experiment includes assessment behavior modeling, and adjustment experiment link. The system provides a new method and tool support to effectively solve the problem of closed-loop simulation experiment assessment.

2 Closed-Loop Simulation Experiment and Assessment

2.1 The Closed-Loop Inference and Characteristics

By means of simulation model and system based on simulation modeling technology, closed-loop simulation experiment can obtain large amounts of simulation results data. And on this basis, comprehensive analysis of the simulation results is carried out, using post-processing methods such as statistical analysis and regression analysis, etc. It can discover and understand how the different factors change the experiment results behind the complex phenomenon, as well as the interaction relationship between factors [11]. As a result, the experiment analyst can find the rules that different factors following, and obtain satisfactory solution and understanding of experiment. Closed-loop simulation experiment is mainly used in analysis simulation and experiment prediction, and provides the support for the study of feasible scheme, optimized decisions, equipment system application scheme. Its characteristics include closed-loop circuit structure, ultra-real-time, large samples, numerous entity, and more complicated calculation model. The requirement of the simulation system analysis count capacity is high. Generally it implements three stages according to experiment design, experiment implementation and experiment assessment.

2.2 Post Assessment Method and Insufficient

Experimental assessment is an important part in closed-loop simulation experiment research job. Closed-loop simulation experiment produce lots of data at ultra-real-time. It has great significance for simulation experiment to study how to acquire data around assessment index, implement assessment, and make conclusions from these huge amounts of data. The currently conventional method is post assessment offline at the end of closed-loop simulation experiment work. It carries out post analysis and process of the experimental data recorded in the process of simulation experiment, and obtains visualization of the statistical information. Compared with real-time assessment, post assessment method mainly has the following deficiencies:

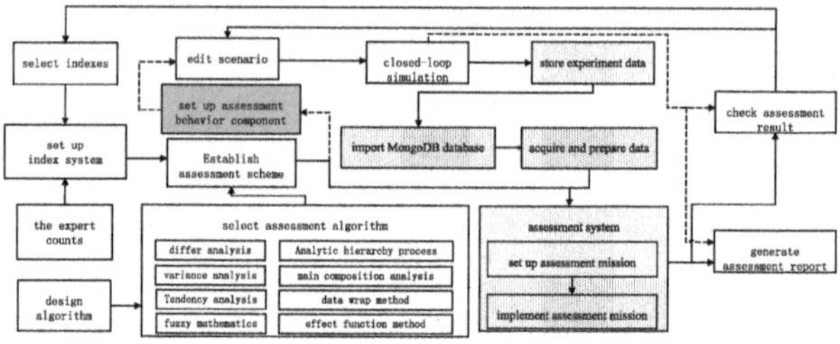

Fig. 1. Closed-loop simulation experiment post assessment process compared with real-time evaluation

First, post assessment belongs to a kind of static evaluation method. Although it is easy to implement, but the evaluation methods, such as the index assessment method, analytic hierarchy process, index calculation method, etc. less consider the counter environment. The process of index weight assignment, mutual relationship, and unitary aggregation and so on, is indistinct implication. And the result will be influenced severely by analyst personally. It regards simulation end as time section of assessment results. The assessment results can't be able to output real-time with simulation running. The assessment index can't be able to monitor dynamically and real-time analysis in the process of experiment. It is relatively complicated to retrieval assessment index change process of large samples and numerous entities.

Second, post assessment process takes longer. Experimental analyst spends a lot of time to check and contrast data of situation, motor, detection, communication, etc. after the experiment. Moreover, the analyst needs to complete assessment data collection table, to understand the key data, analysis and judgment combining with the situation of repetition. The whole process is time-consuming, low efficiency, boring operation, low automation level and trivial. As showed in Fig. 1. This insufficient will affect the further experiments. In addition, with the expansion of the simulation extent, closed-loop simulation required longer and the conventional post assessment takes a long time with modify of the experiment content each time [14]. If experiment implementation and assessment are simultaneously, experiment analyst can control real-time efficiently the simulation process and assessment results. It will enhance the efficiency of project research and establishment, save economically experiment time. This new idea will make it possible to apply closed-loop simulation experiment in extensive analysis simulation.

Last, assessment conclusion of the conventional post assessment method is not comprehensive. As post assessment only can be carried out after the simulation implementation, and focus on the results value of the assessment index, the changing information of the assessment index need further analysis. Both jobs can't be carried out at the same time. This dimension decline process of information will lead to the loss of assessment information, inaccurate assessment results. It is lack of overall evaluation to the rationality of experiments problems such as the deployment manners, coordinated timing, power application.

3 Real-Time Assessment System Design

Real-time assessment system of closed-loop simulation experiment is designed by system modeling thought. Different functions represent different modules. These modules support data analysis and experiment assessment according with simulation synchronously. Real-time assessment system is made of data collecting module, real-time analysis module, information contrast module, index analysis module and report generation module. It can not only acquire instant and process data of closed-loop simulation experiment, but also fulfill real-time analysis, information contrast, and index analysis dynamically. Its data process is showed in Fig. 2. The main function and design of the modules describe as follows.

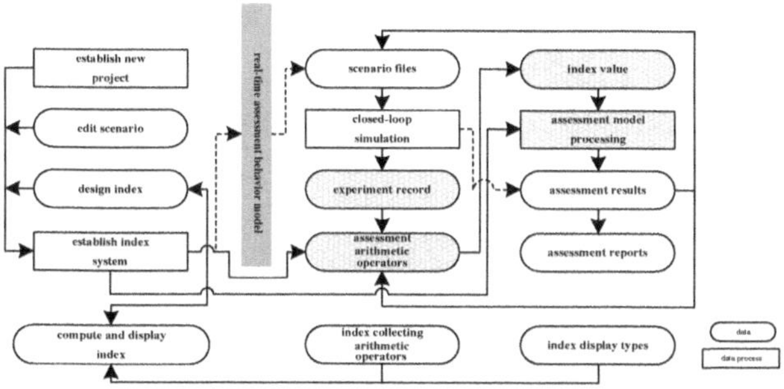

Fig. 2. Real-time assessment data process of closed-loop simulation experiment

3.1 Data Collecting Module

This module run in the real-time assessment behavior component, also provide human-computer interface. It support instant and process data collecting from all entities, mainly include instruction receive unit, arithmetic operators implement unit, data collecting unit, synchronous output unit, data storage unit, etc. According to the choice, settings and requirements of the assessment analyst, the module make different assessment index for different entities. According to the simulation step running, the module real-time collects diriment data for different entities.

3.2 Real-Time Analysis Module

This module provide human-computer software interface in real-time assessment system. It mainly contains instant information real-time analysis unit, process information real-time analysis unit, selection control unit and information storage unit. It can read and process raw data provided by data collecting module, display information, online real-time statistical analysis and contrast process information according to the choice and settings of the assessment analyst. It provides detail information of simulation process information and instant information. This module supports the dynamic control in the simulation process, such as scaling of information detail degree, movement operation. It can complete real-time data reception, information analysis, information display, scaling back, information read and the storage, etc. work. As a result, the researcher can comprehensively grasp the closed-loop experiment process combined with scenario playback tool (Fig. 3).

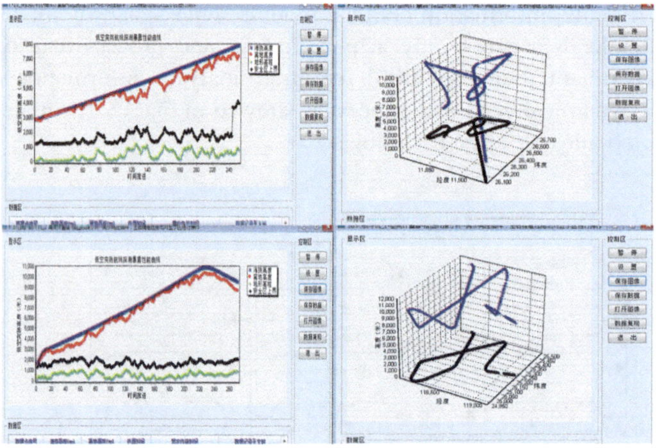

Fig. 3. The process interface of real-time analysis module with simulation running

3.3 Information Contrast Module

The module mainly consists of information display unit, selection control unit and information storage unit. It can real-time contrast and display multi-type and irregular instant information of all entities via vector element, vector array and vector matrix. It makes the analyst fully grasp the transient change of the experiment information by setting the scope and contrast item of the entity and the assessment criteria thresholds, changing the comparative perspective view (Fig. 4).

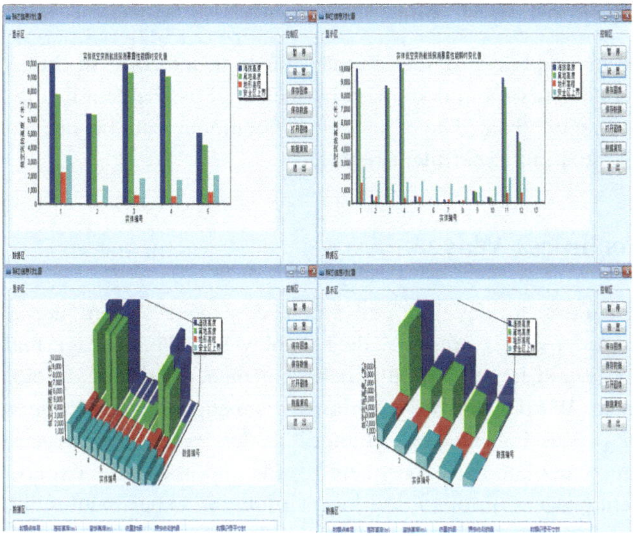

Fig. 4. The process interface of information contrast module with simulation running

3.4 Index Analysis Module

Mainly including dynamic analysis display unit, assessment result calculation unit, selection control unit and the result storage unit, this module can edit coefficient of assessment criteria, in accordance with the rule. Assessment results are calculated real-time through the assessment criteria and raw data. The module help analyst grasp the instant conditions and change process of the assessment index by dynamic display. These are used for comprehend crucial problem of experiment, such as strength assistance, application occasion, strength deployment (Fig. 5).

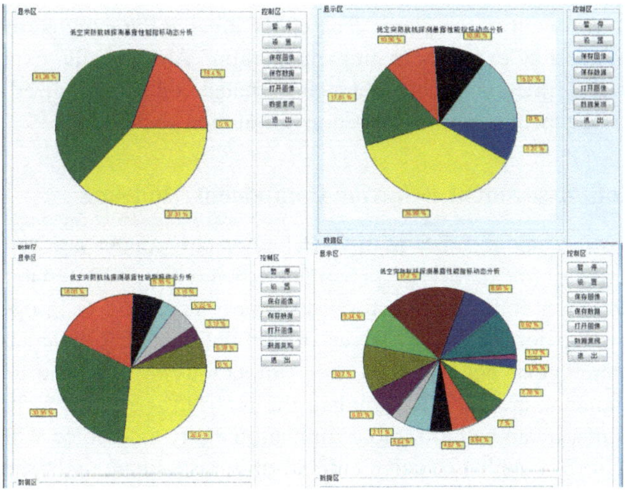

Fig. 5. The process interface of index analysis module with simulation running

3.5 Report Generation Module

Report generation module provides human-machine interface in the system, to automatically generate assessment document by means of assessment results, process chart, and the experimental data. The assessment document can be used as the basis of artificial edit simulation experiment report.

4 Main Technical Way

The main difficulties to develop real-time assessment system are data collecting, information contract, index analysis, data display and data storage real-time dynamically of high rate and large sample experiment data, produced by closed-loop simulation experiment. We put forward that the key is using behavior component model and real-time data processing method to monitor the experiment process, analyze the experiment data, calculate the experiment results and assess the experiment results. In the development process of the closed-loop simulation experiment real-time evaluation system, we concluded the following key technologies that have been over fulfilled and implicated:

4.1 The Closed-Loop Simulation Experiment Real-Time Assessment Technology

With scale of simulation system becoming larger and larger, the model and data are more and more complex. The conventional post assessment technology can't meet the requirement of closed-loop simulation experiment. Now it is crucial to research closed-loop simulation experiment real-time assessment technology. Essentially the technology solves the problem of assessment by model technology. The technical way of thinking is to develop assessment behavior component model before the closed-loop simulation experiment implementation. The model are built and assembled into an entity model to form entity templates. The entity joins in the simulation running in the closed-loop simulation experiment implementation. As a result, experiment implementation and experiment assessment are coordinated and synchronized so as to realize the real-time assessment of the closed-loop simulation experiment.

4.2 Real-Time Assessment Behavior Component Modeling

Behavior is abstract about entity ability. Behavior component is a reflection and the simulation carrier of the entity or personnel's behavior characteristic. It is used to describe entity action control, behavioral decision, implement task, operation cooperation such as behavior. It is the driver of simulation entity independent decision-making and intelligent behavior. In the simulation process real-time assessment behavior components are used to simulate the real-time assessment behavior for the experiment results. It can be built into a simulation entity assembled with other mission behavior component together to carry out real-time assessment understanding behavior for experiment results.

Assessment component model and equipment components, behavior components, auxiliary components are assembled together. As the entity implement tasks using the component model in closed-loop simulation experiment, assessment component will give the results of action and experiment. So the experiment implementation and assessment implement at the same time. As shown in Fig. 6 is the model interaction process among entities in real-time assessment progress.

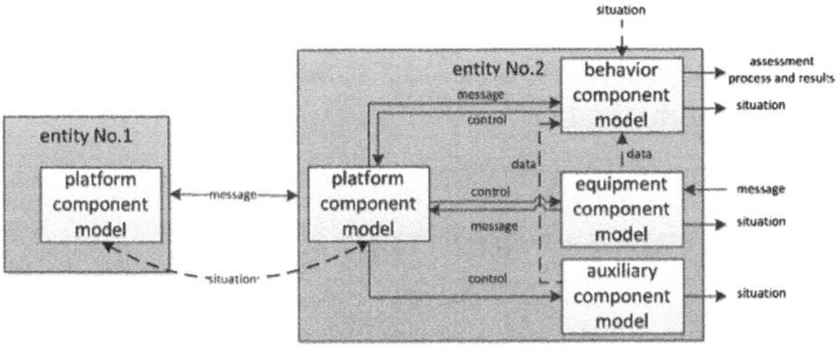

Fig. 6. The model interaction process with real-time assessment in progress

4.3 Multiple Cooperative Engine Coupling

Using the modular modeling technology to solve the problem of real-time assessment, the most difficult is realize the synchronization among the closed-loop simulation engine, data real-time processing engine and the index dynamic display engine. It should avoid not only prevent data loss and reading-writing error due to speed sampling, experiment delay due to slow simulation running speed, but also data overflow due to high data rate. By using multi-threaded calculation processing technology, the real-time characteristic of experiment data acquisition, processing, reading-writing is improved. By using shared GPU memory to improve the communication transmission speed among GPU thread, the problem of the indicators display dynamically is solved. By minimizing time-consuming of the calculation and display, multiple cooperative engines are coupled together. Even if the assessment data quantity increases suddenly, real-time processing and dynamic display engine is still effective. Practical experiments have shown that all the engines can operate stably in the case of an ultra-real-time speed ratio at 100.

4.4 Based on the MVC Function Module

MVC framework provides abstract interface of three kinds of element that is model, view and controller. The application module are independent, so as to improve the reusability and extensibility of the module. Each function module of closed-loop simulation experiment real-time assessment system is designed and developed Based on MVC pattern. The model input, processing, output of all entities interacted in closed-loop simulation experiment are separated. Each module can be divided into model layer, view layer, control layer three levels. The logical structure of functions is clear and system can be flexible configuration, meet the demand of the expansion of the subsequent simulation applications at the same time.

4.5 Assistant Generating Assessment Report

By template technology, assistant generating assessment document can shorten the evaluation process to improve work efficiency. The steps include the template design and document generation. Template design consist static content and dynamic content design respectively. Defining the style, bookmark logo and regular expression can implement dynamic output function of the template. The assessment document will generate according to the logo to locate charts, data, text and other evaluation result position in the document, match the style and the results. According to the required format created by the template, closed-loop simulation experiment assessment documents are generated automatically.

5 System Implementation and Application

The experiment of system application is designed as flight path optimization closed-loop simulation experiment about how aircrafts by low altitude penetration breakthrough the radar warning-detect network. We use the velocity and acceleration as aircraft instant motor performance assessment index, and use low altitude penetration route detection exposure information as aircraft process penetration performance assessment index, to measure the detection probability of routes in different height layer. We regard aircraft entity detection exposure instant change value as assessment index, to analysis of different entities low-altitude penetration altitude, the height from the ground, terrain elevation, the differences between the upper and lower bounds of the safety aspects. As shown in Fig. 7 from real-time assessment system process interface, experimental assessment data can output and update synchronously along with simulation running and experiment situation changes. The analysts can not only real-time

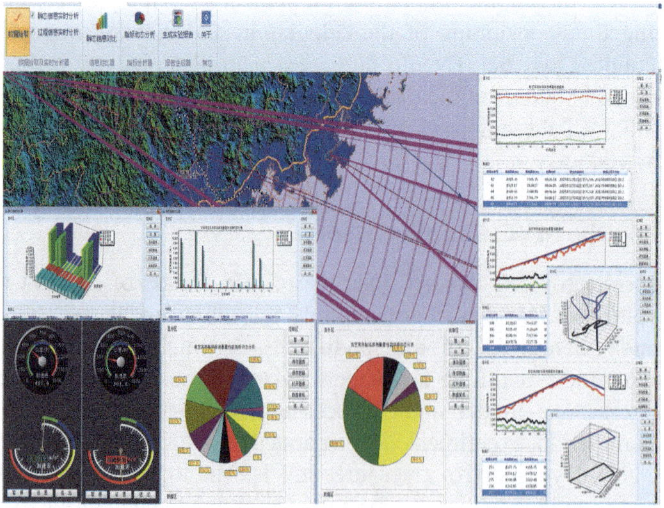

Fig. 7. The process interface of closed-loop simulation experiment real-time assessment

control experiment situation and circumstance, but also real-time assessment experiment scheme by means of the data flow and dynamic graph of assessment results. By the system application and experiment verification, the system still can adapt to different situations and operation stably as changing the simulation speed ratio, increasing the simulation entities, changing experiment scenarios, expanding simulation scale.

6 Conclusions

The conventional post assessment method in closed-loop simulation experiment has poor flexibility and takes longer. This paper uses the behavior components modeling technology to build real-time assessment system. It provides a new thought and tool to research and solves the problem of closed-loop simulation experiment assessment. Verified by the experiment and instance application, system functions are comprehensive. It can realize real-time assessment and dynamic monitoring to closed-loop simulation experiment. It makes the closed-loop simulation running combined with simulation real-time assessment synchronously and effectively, achieves the purpose of implementation between experiment and assessment in the same time. It has higher accuracy and less time cost than the conventional post assessment method, and makes experiment analyst extricated from heavy and complicated post artificial assessment. The new method ensures the efficiency of closed-loop simulation experiment work, and can well meet the needs of large-scale analysis simulation experiment.

Real-time assessment system based on behavior components still have following deficiencies. Firstly, the evaluation behavior components need to be modified and rerun the simulation experiment when assessment purpose and rules change. Secondly, to carry out research and development need have a more comprehensive and in-depth understanding about simulation experiment design, modeling, implementation, assessment, etc. process technology when using modeling method to solve the problem of assessment.

Assessment process is usually a kind of complex data process, and assessment data processing method differ according to evaluation purposes. In order to meet the requirement of two aspects of complexity and variability, the system requires improvement in the follow-up job. Especially, it should adopt flexible model assessment, efficiency arithmetic operator design and so on methods and patterns that reflect assessment flexibility. Strengthening the reusable, extensible and reconfigurable characteristics of the real-time assessment behavior components, the system will have an extensive application prospect in the complex closed-loop simulation experiment assessment.

References

1. Chen, C.: Modeling of air combat based on effectiveness evaluation. Syst. Eng. Electron. **1**, 79 (2015)
2. Dan, S.: Based on the DSP implementation of real-time effect evaluation. J. Mod. Electron. Technol. **11**, 46 (2015)

3. Zhou, Z.: The research on damage effect real-time assessment of aerial target intercept-on by medium-range or long-range ship-to-air missile. J. Proj. Rocket. Missiles Guid. **4**, 61 (2012)
4. Wang, J.: Model of efficiency real-time evaluation of artillery firepower damage based on DEA. Ordnance Ind. Autom. **8**, 39 (2010)
5. Wan, L.: Real-time evaluation method for flight manipulation in air countering. Fire Control Command Control **7**, 31 (2012)
6. Huo, Z.: Real-time analysis system of flight test vibration data based on POWER PC + FPGA architecture. Mod. Electron. Tech. **7**, 26 (2011)
7. Zhu, X.: A real-time assessment method for space testing based on multimodels. Meas. Control Technol. **1**, 153 (2015)
8. Sun, J.: Design and implementation of real-time assessment system for a kind of radar countermeasure simulation training. Electron. Technol. Softw. Eng. **14**, 198 (2015)
9. Lv, K.: Research into the evaluation method of simulation training operation of shipborne radar countermeasure equipments. Shipboard Electron. Countermeas. **3**, 88 (2011)
10. Fu, D.: Research of assessment method for operation capability and combat effectiveness. Mil. Oper. Res. Syst. Eng. **4**, 35 (2006)
11. He, Q.: Comparison of different platforms for analytic combat simulation. J. Syst. Simul. **7**, 94 (2011)
12. Ye, P.: Static evaluation algorithm in game-tree searching. Comput. Eng. Design **7**, 25 (2004)
13. Liu, Z.: Research on missile-borne computer closed loop simulation system based on six-degree-of-freedom model. **7**, 346 (2005)
14. Qiao, Y.: The pilot-aircraft closed loop simulation method investigation and the application. Sci. Technol. Innov. Her. **34**, 26 (2014)

Application and Research of Blockchain Technology in P2P Network Distributed Data Storage

Sun Peng(✉)

[1] Sanya Aviation & Tourism College, Sanya 572000, Hainan, China
sun_pengl@hnair.com

Abstract. With the rapid development of information technology, people have entered the era with information explosion. The cooperation, competition and trust between people take the information as the carrier, and information resources are undoubtedly the most valuable resources. Once the information is lost, the loss is heavy. The traditional way of information recording cannot prove the belonging of the information, information is easily leaked or lost. People live in an environment where information is not completely credible, and the legitimate rights and interests are threatened. In this paper, a distributed data storage strategy based on P2P network is designed to ensure that the user backup text information will not be lost. The signature verification technology based on elliptic curve series algorithm is adopted to realize the ownership authentication of the information, and secure encryption and decryption technology is adopted to ensure the security of the user's information.

Keywords: Blockchain · P2P · Network · Information safety

1 Introduction

Through the design and implementation of this system, we can explore the implementation of blockchain technology with better performance and more favorable usage scenarios. Through the application and innovation of blockchain technology, it helps to explore more application direction of blockchain and expand its research field. Through integrating and refining its technology branches, we can promote the technological development and improve user experience. The research and design of the blockchain technology and its application can help to summarize the development experience of decentralized products, learn and make use of advanced technological ideas, build decentralized and self-made computer systems and solve the problem of mutual trust in the world.

J. Abawajy et al. (Eds.): ATCI 2018, AISC 842, pp. 953–960, 2019.
https://doi.org/10.1007/978-3-319-98776-7_114

2 Research on Key Technology of Blockchain and Data Storage

2.1 Blockchain

Blockchain is the underlying technology of the encrypted currency represented by bitcoin. It is a new distributed infrastructure and computing paradigm that uses block chained data structure (a chain data structure composed of data blocks in a sequential way according to the time) to validate and store data, use distributed node consensus algorithm to generate and update data, use cryptography to ensure the security of data transmission and access and uses the intelligent contract composed of automated script code to program and manipulate data [1–3].

The organic integration of the various mechanisms and modules of the blockchain makes such a decentralized accounting system become possible. Because the code and community of the blockchain system are all open source, users who pay attention to or use this system will protect the healthy development of the blockchain system to the greatest extent. The number of honest nodes continues to grow, and the performance of nodes improves continuously, which make the blockchain system more stable, safe and reliable. The ecological system of blockchain can be simply divided into protocol layer, extension layer and application layer. They are independent and inseparable from each other and provide good decentralized services together. The following is an introduction for the functions of each layer, and Fig. 1 is the schematic diagram of the block chain system architecture.

2.2 Cryptography Technology

(1) Public key and private key system
The typical components of modern encryption algorithm include encryption and decryption algorithm, public key and private key. When encrypting, encrypting plaintext by encryption algorithm and public key, and obtaining ciphertext. When decrypting, decrypting ciphertext by decryption algorithm and private key, and obtaining plaintext. According to whether the public key and private key are consistent, they can be divided into symmetric encryption algorithm and asymmetric encryption algorithm. These two models are applicable to different usage scenarios and form complementary relationship, which can be combined to form a combination mechanism in many cases according to the actual situation of the business [4, 5].

(2) Elliptic curve series algorithm
The elliptic curve refers to the plane curve determined by Weierstrass equation $y^2 + a_1xy + a_3y = x^3 + a_2x^2 + a_4x + a_6$. If F is a domain, $a_i \in F, i = 1, 2, \ldots, 6$. The number pairs (*x*, *y*) satisfying this equation is called the point of the elliptic curve E on F domain. The F domain can be a rational number domain, and it also can also be a finite field $GF(\text{Pr})$. The elliptic curve is usually expressed by E. In addition to all the points of the curve E, we need to add a special point O called infinite point.

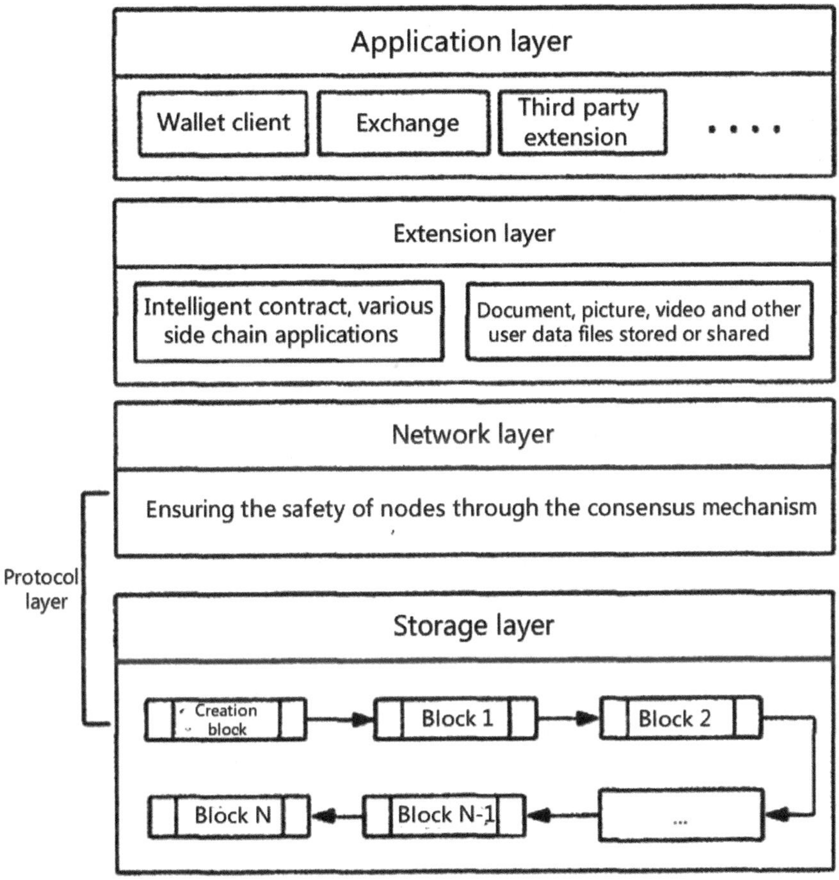

Fig. 1. The schematic diagram of the block chain system architecture

In the encryption of elliptic curve, a special form of elliptic curve is used, that is, an elliptic curve defined on finite field, and the equation is: $y^2 = x^3 + ax + b(\bmod p)$. Where, P is a prime number, a and b are two non-negative integers that less than P, which satisfy: $4a^3 + 27b^2(\bmod p) \neq 0$. Where, $x, y, a, b \in F_p$, then the point (x, y) and the infinite point O satisfying the condition form the elliptic curve.

3 Application of Blockchain Technology in P2P Network and Distributed Storage

3.1 P2P Network

Blockchain uses a network architecture based on P2P. P2P means that each computer in the same network is fair and equal, each node provides network services together and without any special (central) nodes. Each node in the block chain is connected to each other to form a blockchain network, which ensures the security of the whole blockchain system [6].

The node related functions in the blockchain network include:

(1) Each node receives the message that the transaction is broadcast to the whole network, collects the new transaction into the node memory and assembles it into blocks.
(2) Each node attempts to get the qualification of the creating block, and then broadcast the legal and valid effective data to the whole network.
(3) After the transaction in the block is verified to be legal and effective and the proof of work is verified, each node identifies the block's effectiveness and writes a new block at the end of the blockchain, so that the whole blockchain can be extended continuously.
(4) All nodes default the longest chain is the only correct blockchain main chain, and on this basis, continue to complete the above process of extending blockchain.

3.2 Distributed Storage

The decentralized storage mode is more or less faced with the problem of information security and permanent storage, and distributed cloud storage based on block chains will be the best solution to solve this problem. Different from the cloud storage space provided by decentralized, distributed cloud storage based on blockchain technology can not only be used for storage, but also can prove that the data is true and trustworthy, and will never be modified, which can be said to be an encrypted distributed cloud storage [7, 8].

3.3 Consensus Mechanism

In any decentralized architecture, because the equal status of each node involved in the system, a reasonable consensus is needed to be reached to maintain the health and stability of the blockchain system when there are disagreements in decision. As one of the core technologies of the blockchain, the consensus mechanism realizes a series of strategy algorithms that encourage users to participate in and maintain the system. It can not only solve conflicts when the cooperation between nodes appears problems, but also encourages users to participate in the maintenance work of the blockchain system and improve the activity and stability of system. The consensus mechanism is constantly being innovated under the impetus of the community, which makes the application field of the blockchain more extensive and promotes the sound development of the block chain ecosystem. The following is the introduction of the consensus mechanism of the blockchain.

3.4 Research on Design Difficulties

The design and development of the system mainly include the following difficulties:

(1) The design of consensus mechanism
In the blockchain system, different applications have different choices for consensus algorithms. This system has no mining process (such as bitcoin mining),

and the participation of the user for system maintenance is not high, but the degree of decentralized is higher, a certain number of quality service nodes must be guaranteed, so the system adopts the consensus mechanism based on DPoS. In the process of design and development, it is necessary to clear the process of the trustee's production and the specified process of block creation qualification, as well as the operation need to be completed in the whole reward cycle, block cycle, and voting cycle [2, 9]. In order to guarantee the service quality of the trustee, it is necessary to conduct credit evaluation for the trustee and provide the basis for voting of user. For those nodes with poor service quality, they can vote against to reduce the level of the node. Each small module has a more complex business process, and it is difficult to design and code, so before the development, we need to sort out the ideas and write the design documents to avoid missing the key steps.

(2) Block data update
The main difficulty of this part lies in how to verify the missing block, loading process of block data and how to verify and solve the problem of block bifurcation after loading. For these questions, we mainly refer to the process of block data synchronization and verification of other blockchain applications and apply it to this system to design and implement related specific processes.

(3) Transaction data extension
Digital currency application based on blockchain in the early days simply takes blockchain as cashbook to record some transfer information, and then packing and storing it. But now, we need to store different forms of transaction data to facilitate users to classify queries and extract them, and we can clearly see the change records of information. This part needs to design a reasonable data table structure, so that transactions can be linked to extended information. In the code, it is displayed in the transaction with the asset field.

3.5 System Function Module

This system mainly includes the following modules: P2P network module, transaction module, block module, user module, consensus mechanism module, and encryption information management module. P2P network module contains node creation and network connection. Transaction module mainly includes transaction generation and transaction information expansion. Block module mainly includes the creation, verification and synchronization of blocks. User system module includes the registration and login of user and user public and private key management. Consensus mechanism module mainly includes the reward cycle and the entrustment cycle. Encryption information management module mainly includes user text basic management, text encryption and decryption process and text information transmission function. The function and structure of each module are shown in Fig. 2. Among them, the transaction module is one of the core functions of the blockchain. In this system, the transaction records the valuable operation of the user in the system, but it is not necessarily the transfer of tokens.

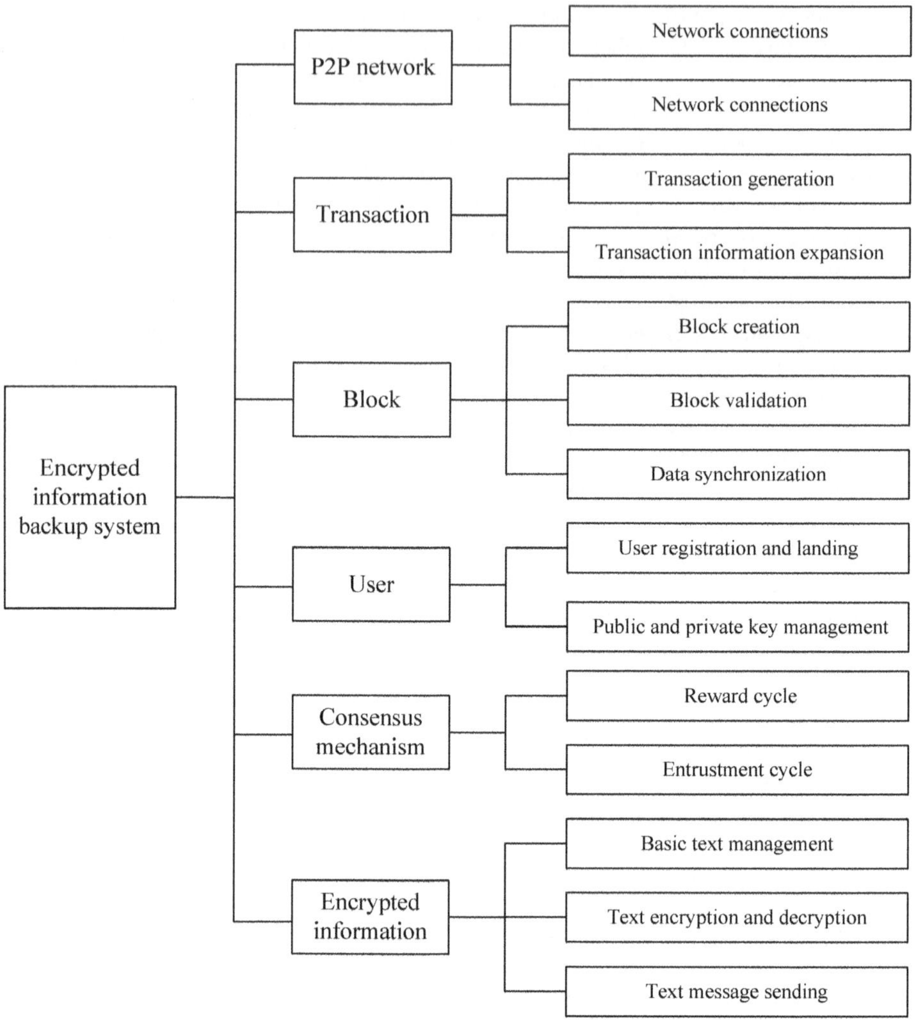

Fig. 2. System function module diagram

3.6 System Deployment

The deployment of this system is very simple. After the system is developed, configuring a server that basically meets the requirements of the operation as an initial node, and then connecting the initial node or other user node through the node created by the user to complete the P2P network connection. After forming a multi node network, the initial node can be deleted at any time, and it will no longer affect the normal operation of the whole system. In this way, a blockchain network built entirely by users themselves are preserved, each node holds a data backup, and each data backup records the current online node information. Figure 3 is the deployment of the system.

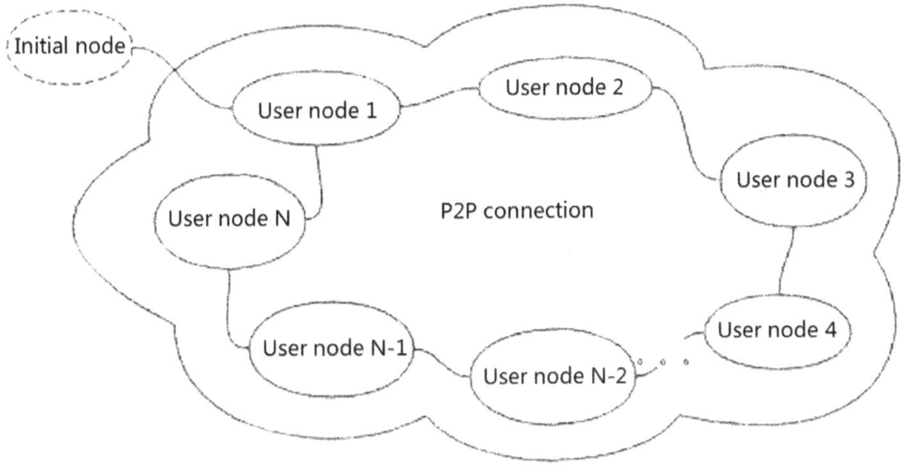

Fig. 3. The deployment of the system

4 Conclusion

This paper mainly studies the application of blockchain technology in the P2P network distributed data storage, analyzes the key technologies related to the encryption of blockchain, puts forward a solution for the design difficulties, and finally gives the outline design, which has certain practical significance and reference value for the future related research.

References

1. Peters, G.W., Panayi, E.: Understanding Modern Banking Ledgers Through Blockchain Technologies: Future of Transaction Processing and Smart Contracts on the Internet of Money. Social Science Electronic Publishing. arXiv:1511.05740 (2016)
2. Zhang, Y., Wen, J.: The IoT electric business model: using blockchain technology for the Internet of Things. Peer-to-Peer Netw. Appl. **10**(4), 983–994 (2017)
3. Szebeni, S.: Method and system for handling of group sharing in a distributed data storage, particularly in p2p environment (2017)
4. You, P., Huang, Z., Peng, Y., et al.: Towards a delivery scheme for speedup of data backup in distributed storage systems using erasure codes. J. Supercomput. **7**, 1–15 (2015)
5. Yli-Huumo, J., Ko, D., Choi, S., et al.: Where is current research on blockchain technology—a systematic review. PLoS ONE **11**(10), e0163477 (2016)
6. Peters, G.W., Panayi, E.: Understanding modern banking ledgers through blockchain technologies: future of transaction processing and smart contracts on the internet of money. In: Banking Beyond Banks and Money. Springer International Publishing (2016)
7. Mainelli, M., Smith, M.: Sharing ledgers for sharing economies: an exploration of mutual distributed ledgers (aka blockchain technology). J. Financ. Perspect. **3** (2015)

8. Atzori, M.: Blockchain technology and decentralized governance: is the state still necessary. **6** (1) (2016). Social Science Electronic Publishing, New York
9. Tian, F.: An agri-food supply chain traceability system for China based on RFID and blockchain technology. In: International Conference on Service Systems and Service Management, pp. 1–6. IEEE (2016)

Simulation Research of HVDC Control Strategy Based on Matlab/Simulink

Fang Zhao(✉) and Haiyun Zhu

Zhengzhou Technical College, Zhengzhou 450121, China
zhaojifang216@126.com, 33645975@qq.com

Abstract. There are deficiencies in HVDC transmission technology, According to the concept of vector control, the paper proposes the use of the secondary side of the transformer leakage inductance to replace the role of the reactor in the main circuit, and built a Matlab/Simulink simulation model. The simulation results show that even if there is no commutation reactor and filter, the voltage and current waveforms on the primary side of the transformer are still very good sine waves. This control method saves costs.

Keywords: HVDC transmission · Transformer · Vector control

1 Introduction

Since the first high-voltage DC transmission line in the world was put into commercial operation in 1954, as an increasingly mature technology, HVDC has been widely used in long-distance, high-power, submarine cable transmission, and asynchronous communication between two AC systems. However, at both ends of the current HVDC transmission circuit, a commutation reactor must be used to suppress the rising speed of the DC over-current and filter out the characteristic harmonics generated by the inverter. The cost of the converter reactor is very high. The ratio of the cost of the converter reactor and transformers is 0.5:1 to 1:1. This paper studies the role of the transformer's leakage reactance in the transmission system instead of the reactor, as a result, expensive reactors can be eliminated and the cost of HVDC transmission can be reduced. This technology will have important practical significance and economic significance in the field of HVDC transmission as well as other power system fields.

2 HVDC Principle and Control Strategy

2.1 HVDC Circuit Structure

HVDC includes two components: a converter station and a pair of ground cables. The converter station uses voltage source converter (VSC) technology, its output is regulated by the VSC control system. Its operating mechanism is shown in Fig. 1 [3].

The VSC is used for the transmitter and receiver converters. The two converters have the same structure. The converter consists of a converter bridge, a converter reactor, a DC capacitor and an AC filter. Each bridge arm of the commutation bridge is

J. Abawajy et al. (Eds.): ATCI 2018, AISC 842, pp. 961–966, 2019.
https://doi.org/10.1007/978-3-319-98776-7_115

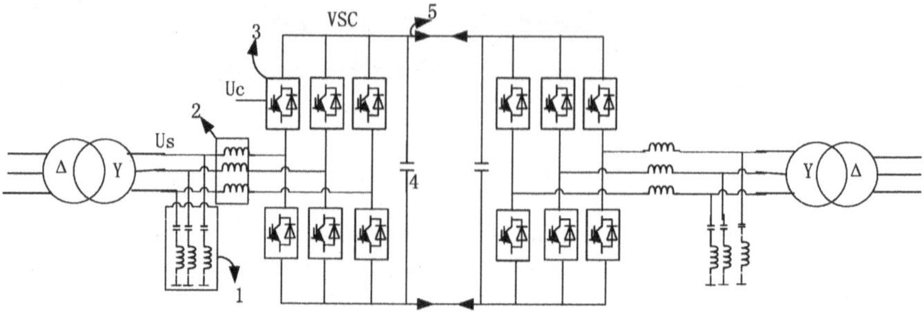

Fig. 1. HVDC schematic

made up of IGBTs in series. The commutation reactor is the link between the VSC and the AC side energy exchange, and also acts as a filter. The function of the DC capacitor is to provide voltage support for the inverter, buffer the inrush current when the bridge arm is off, and reduce the DC side harmonics [1, 2]. This paper proposes to remove the commutation reactor from the system and directly control the IGBT's turn-on and turn-off generation of PWM wave by vector control of the secondary side of the transformer.

2.2 HVDC Control Strategy

Transformer is high-order, nonlinear, strongly coupled multivariable system, According to the vector control idea, establishing transformer simulation model in two-phase rotating coordinate system. Take Y/Y-connected symmetrical three-phase transformers as an example.

u_A, u_B, u_C is respectively the primary three-phase voltage; i_A, i_B, i_C is respectively the primary three-phase currents; ψ_A, ψ_B, ψ_C is respectively the primary three-phase winding total flux linkage; $u_a^{'}$, $u_b^{'}$, $u_c^{'}$ is respectively the three-phase voltage converted to the secondary side of the primary side respectively; $i_a^{'}$, $i_b^{'}$, $i_c^{'}$ is respectively the three-phase currents of the secondary side after conversion; $\psi_a^{'}$, $\psi_b^{'}$, $\psi_c^{'}$ is t respectively he total magnetic flux of the three-phase winding after secondary conversion; $R_A = R_B = R_C = R_1$ is respectively the primary three-phase resistance; $L_A = L_B = L_C = L_{l1} + L_m$ is respectively the primary three-phase inductors; L_{l1} is the leakage inductance of the primary winding, L_m is the primary inductance of the primary winding, $M_{AB} = M_{BC} = M_{CA} = -L_m/2$ is the mutual inductance between the three-phase windings on the primary side, $M_{Aa}^{'} = M_{Bb}^{'} = M_{Cc}^{'} = L_m$, $M_{Ab}^{'} = M_{Bc}^{'} = M_{Ca}^{'} = -L_m/2$ is the mutual sense between the primary and secondary sides after the secondary side is translated, $R_a = R_b = R_c = R_2^{'}$ is respectively the three-phase resistance after the secondary conversion; $L_a^{'} = L_b^{'} = L_c^{'} = L_{l2}^{'} + L_m$ is respectively the three-phase self-sensing of the secondary side after conversion, $L_{l2}^{'}$ is the leakage inductance of the secondary winding; $M_{ab}^{'} = M_{bc}^{'} = M_{ca}^{'} = -L_m/2$ is the mutual inductance between the three-phase windings on the secondary side after conversion. According to Fig. 2, the rules of the primary and secondary positive directions are all based on the motor law, and the voltage equation can be established as follows. The voltage balance equation for the primary windings can be written as:

$$\begin{cases} u_A = i_A R_1 + \dfrac{d\psi_A}{dt} \\ u_B = i_B R_1 + \dfrac{d\psi_B}{dt} \\ u_C = i_C R_1 + \dfrac{d\psi_C}{dt} \end{cases} \tag{1}$$

The voltage balance equations for the secondary windings can be written as:

$$\begin{cases} u_a^{'} = i_a^{'} R_2 + \dfrac{d\psi_a^{'}}{dt} \\ u_b^{'} = i_b^{'} R_2 + \dfrac{d\psi_b^{'}}{dt} \\ u_c^{'} = i_c^{'} R_2 + \dfrac{d\psi_c^{'}}{dt} \end{cases} \tag{2}$$

Through the coordinate transformation, the three-phase transformer's AC quantity in the three-phase ABC coordinate system is converted into the DC quantity in the d-q-0 coordinate system.

Assume $p = \frac{d}{dt}$, Written in matrix form is:

$$\mathbf{U} = \mathbf{R} \cdot \mathbf{I} + p\psi \tag{3}$$

$$\begin{bmatrix} u_A \\ u_B \\ u_C \\ u_a^{'} \\ u_b^{'} \\ u_c^{'} \end{bmatrix} = \begin{bmatrix} R_1 & & & & & \\ & R_1 & & & & \\ & & R_1 & & & \\ & & & R_2 & & \\ & & & & R_2 & \\ & & & & & R_2 \end{bmatrix} \begin{bmatrix} i_A \\ i_B \\ i_C \\ i_a^{'} \\ i_b^{'} \\ i_c^{'} \end{bmatrix} + \mathrm{p} \begin{bmatrix} \psi_A \\ \psi_B \\ \psi_C \\ \psi_a^{'} \\ \psi_b^{'} \\ \psi_c^{'} \end{bmatrix} \tag{4}$$

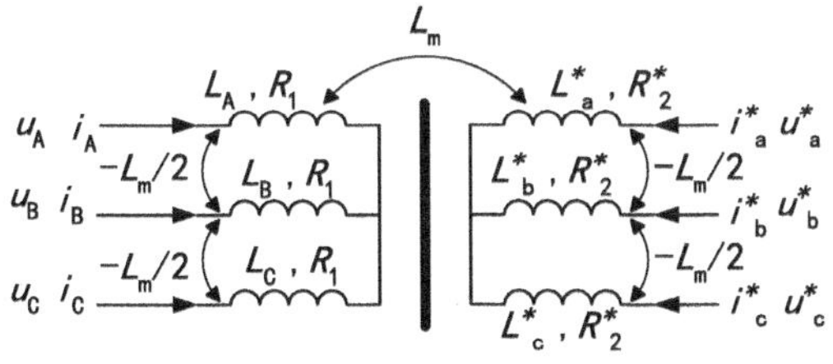

Fig. 2. Schematic diagram of Y/Y connected three-phase transformer

For a three-phase symmetric system, the zero-sequence component may not be considered, transformer voltage equation expression in d-q-0 coordinate system is:

$$\begin{bmatrix} U_{d1} \\ U_{q1} \\ U_{d2} \\ U_{q2} \end{bmatrix} = \begin{bmatrix} R_1 & & & \\ & R_1 & & \\ & & R_2 & \\ & & & R_2 \end{bmatrix} \begin{bmatrix} I_{d1} \\ I_{q1} \\ I_{d2} \\ I_{q2} \end{bmatrix} + p \begin{bmatrix} L_1 & & M & \\ & L_1 & & M \\ M & & L_2' & \\ & M & & L_2' \end{bmatrix} \begin{bmatrix} I_{d1} \\ I_{q1} \\ I_{d2} \\ I_{q2} \end{bmatrix} + \begin{bmatrix} & -\omega L_1 & & -\omega M \\ \omega L_1 & & \omega M & \\ & -\omega M & & -\omega L_2' \\ \omega M & & \omega L_2' & \end{bmatrix} \begin{bmatrix} I_{d1} \\ I_{q1} \\ I_{d2} \\ I_{q2} \end{bmatrix}$$

3 Matlab/Simulink Simulation Model and Results

3.1 HVDC Simulation Model

Building the simulation system based on the coordinate transformation relationship described above is generally composed of three major components: Single-phase AC input 220 V and full-wave rectifier by diode; Three-phase IGBT inverter bridge converts the rectified DC power into three-phase AC power; transformer. Simulation block diagram shown in Fig. 3.

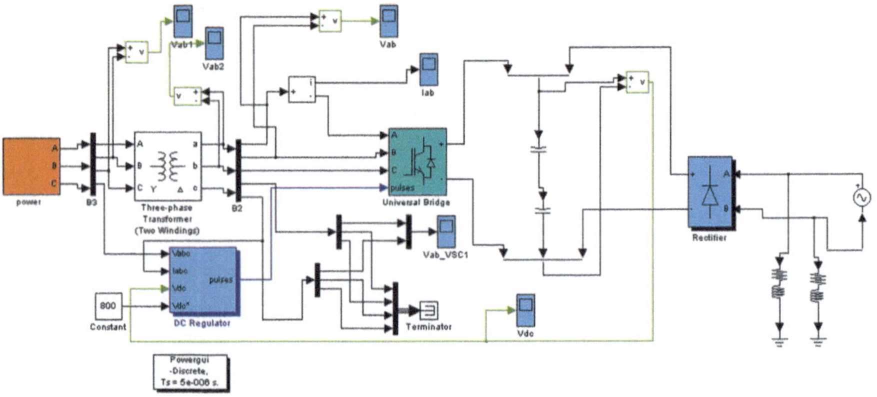

Fig. 3. DC transmission system simulation block diagram

3.2 Simulation Results

The transformer used in this system is Δ/Y connection, the original and secondary parameters are: voltage 380 V/190 V, winding group 0.02 Ω/0.02 Ω, inductance 0.05 H/0.05 H, capacity 1.5 KVA, frequency 50 Hz. Using this system is for simulation, There is no commutation reactor and filter between the transformer and the inverter unit. After adopting the vector control of the transformer, the expensive commutation reactor and the filter are omitted, however, the inverted voltage and current waveforms have the same effect as the prior art waveforms. The voltage waveform is shown in Fig. 4, the current waveform is shown in Fig. 5.

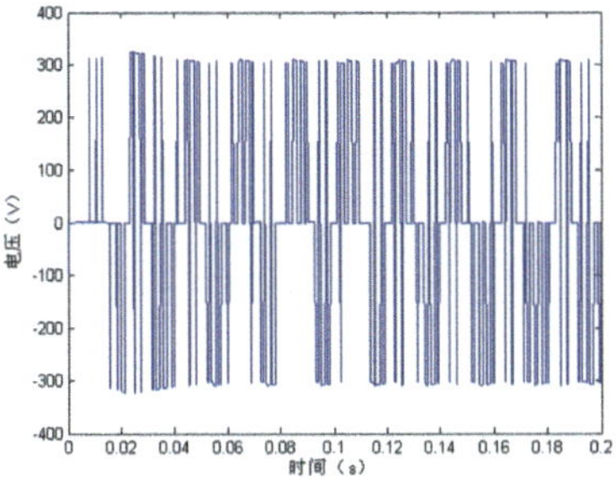

Fig. 4. Inverted voltage waveform

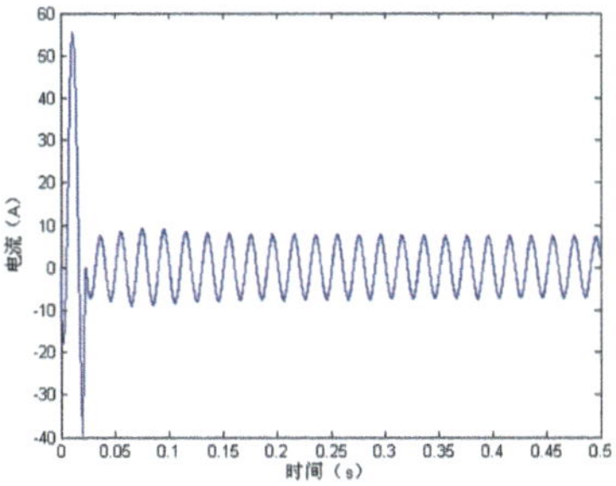

Fig. 5. Inverted current waveform

The line voltage waveform of the secondary side of the transformer is the same as in Fig. 4. The line voltage waveform of the primary side of the transformer is shown in Fig. 6. The inverted voltage and current are fed back to the grid. The current waveform fed back to the grid side is shown in Fig. 7.

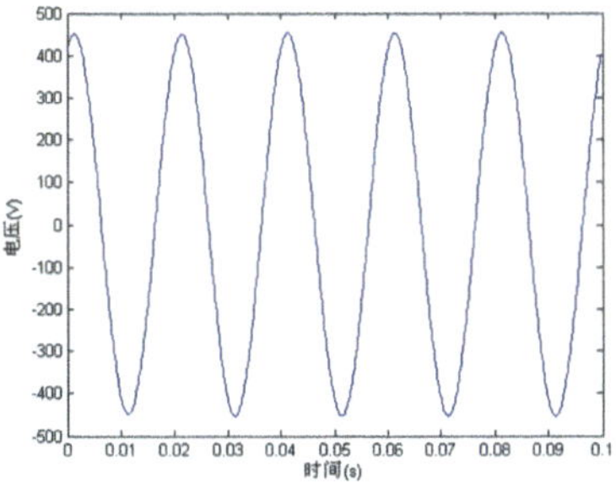

Fig. 6. Transformer primary voltage waveform

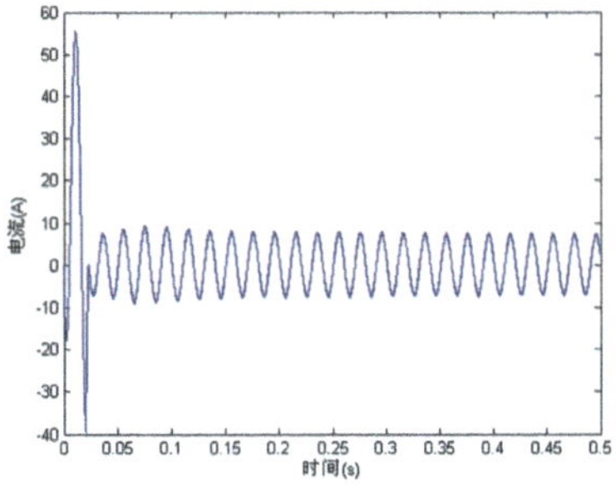

Fig. 7. Current waveform fed back to the grid side

4 Conclusion

The simulation results prove that the vector control of the transformer can be used, and the leakage reactance of the transformer can replace the current reactor and the filter, so that the cost of the HVDC system, FACTS system and reactive power compensation (SVG, ASVG) is greatly reduced, this technology can also flexibly control the direction of power transmission.

Acknowledgements. 2017 Henan Provincial Department of Education Key Research Project. Number: 17B413009.

References

1. Milind, S., Ramanarayanan, V.: Design and analysis of a linear type electromagnetic stirrer. IAS2004 8486-5/04/, 188–194
2. Hörle, N., Eriksson, K., Maeland, A., Nestli, T.: Electrical supply for offshore installations made possible by use of VSC technology; In: Cigre 2002 Conference, Paris, August 2002, 14(6), pp. 15–20
3. Castro, A.D., Ellström, R., Häffner, Y.J., Liljegren, C.: Co-ordination of parallel AC–DC systems for optimum performance. In: Power Delivery Conference, September 1999, 18(6), pp. 30–37 (1999)

A New DV-Hop Location of Wireless Sensor Networks Based on Firefly Algorithm

JianYin Lu(✉) and DaQiong Zhao

Chao Hu University, Hefei 238000, Anhui, China
893238956@qq.com

Abstract. Aiming at the accumulation of average Localization error in the wireless sensor network, resulting in the lack of location accuracy, a new DV-Hop Localization Algorithm of Wireless Sensor Networks based on Firefly Algorithm (FADV-Hop) is proposed. The new Localization Algorithm uses the fitness function to represent the sum of the error of the unknown node and the n standard node. By theoretical reasoning and simulation, the results show that the FADV-Hop Localization algorithm reduces the location error and improves the Localization accuracy.

Keywords: Firefly algorithm · DV-Hop · Localization algorithm

1 Introduction

With the rapid development of wireless sensor network [1] technology, it has been widely used in military, intelligent home, intelligent old age, and geographical environment monitoring. And become one of the hot spots in the study of intelligent city construction. Among them, wireless sensor network location technology is a key problem to be solved based on location service. At present, the Wireless Sensor Network location algorithm is mainly based on distance measurement and no distance measurement. Among them, the Wireless Sensor Network location algorithm based on distance measurement requires information such as relative angle or arrival time of the wireless device in the region to locate the unknown node and locate the result more accurately. But at the expense of a great deal of human, material and financial resources. At present, the wireless sensor network location algorithm based on ranging is mainly Received Signal Strength Indicator [2], Time of Arrival [3], Angle of Arrival [4]. The distance independent wireless sensor network location algorithm is not required to invest a lot of equipment and manpower, the positioning effect mainly depends on the connectivity of unknown nodes and beacon nodes in the wireless sensor network. As well as obtaining jump distance information from the neighbor nodes to help achieve the goal of positioning the unknown nodes, the cost is low and the positioning accuracy is not high. But in many cases, the positioning effect can meet people's needs. Therefore, the location algorithm without distance measurement has been widely implemented.

The DV-Hop location algorithm [5] is a Wireless Sensor Network location algorithm widely used at present, which is easy to implement. In view of the large

J. Abawajy et al. (Eds.): ATCI 2018, AISC 842, pp. 967–974, 2019.
https://doi.org/10.1007/978-3-319-98776-7_116

accumulation error and the shortage of positioning accuracy in DV-Hop algorithm, many researchers have put forward many improvement measures: literature [6]. An improved algorithm is proposed based on the number of jumps, but combined with the RSSI technology, the complexity of network communication technology is increased. The document [7] proposes an optimization scheme based on particle swarm algorithm, and the positioning accuracy is improved, but the hardware overhead is increased by introducing signal measuring instrument. The document [8] proposes an optimization measure based on the optimal node communication radius, and the calculation is not increased, but the network convergence speed is not flexible enough. Therefore, based on the firefly algorithm, this paper proposes an improvement scheme FADV-Hop, transforming the original problem of lack of positioning precision into a multidimensional optimization problem and solving it.

2 Analysis of DV-Hop Algorithm

DV-Hop positioning algorithm is a Wireless Sensor Network location algorithm based on distance vector routing, which is simple in structure, even restricted by natural environment and other factors. The DV-Hop algorithm is widely used in different environments because it has little impact on positioning performance and low cost. DV-Hop algorithm process is briefly described as three steps: calculating the average jump distance; Calculate the minimum number of jumps; Calculates the initial value of an unknown node.

2.1 Implementation of the Un-Hop Positioning Algorithm

First, each beacon node in the network broadcasts to each other to obtain the location information of the other beacons in the network and the minimum number of jumps to be reached. Calculate the average distance per jump, such as (1), based on the number of jumps and the position information of other beacon nodes.

$$HopSize_i = \frac{\sum_{i \neq j} \sqrt{(x_i - x_j)^2 + (y_i - y_j)^2}}{\sum_{i \neq j} hop_{ij}} \tag{1}$$

Formula (1), beacon node $i(x_i, y_i)$, beacon node $j(x_j, y_j)$, hop_{ij} represents the minimum number of jumps between the first and the second j beacons.

Second, each unknown node n gets an average *HopSize* for each jump, The minimum jump hop_i calculates the distance d_n from the other beacon nodes, Such as formula (2):

$$d_n = HopSize * hop_i \tag{2}$$

Finally, the actual value of the unknown node is calculated. (x_1, y_1), (x_2, y_2), $\cdots$, (x_n, y_n) represent the value of the beacon node 1 to n, and the distance between the unknown node and the beacon node $1, 2, \cdots, n$ is $d_1, d_2, \cdots, d_n$. Before $n - 1$ formula subtracts the result of the formula n, And linear agenda group: $AL = b$, And

$$A = \begin{bmatrix} 2(x_1 - x_n) & 2(y_1 - y_n) \\ 2(x_2 - x_n) & 2(y_2 - y_n) \\ \vdots & \vdots \\ 2(x_{n-1} - x_n) & 2(y_{n-1} - y_n) \end{bmatrix}, b = \begin{bmatrix} x_1^2 - x_n^2 + y_1^2 - y_n^2 - d_1^2 + d_n^2 \\ x_2^2 - x_n^2 + y_2^2 - y_n^2 - d_2^2 + d_n^2 \\ \vdots \\ x_{n-1}^2 - x_n^2 + y_{n-1}^2 - y_n^2 - d_{n-1}^2 + d_n^2 \end{bmatrix},$$
$$L = (x, y)^T \tag{3}$$

The coordinates of the unknown node, such as formula (4), can be calculated.

$$L = (A^T A)^{-1} A^T b \tag{4}$$

2.2 Question

In this analysis, we can find the coordinates of an unknown node, but the coordinates of the unknown node obtained by the above method, due to the error of the mean jump, and the accumulation of multi hop error, etc. The high positioning accuracy is not very accurate, But every element in the observation formula (3) appears d_n, which represents the distance between the unknown node and the beacon node, so long as the d_n is kept closer to the true value, Finally, the more accurate the actual value of the unknown node is, the closer it is to reality. Therefore, it is difficult to achieve the desired positioning effect by simplifying the process of positioning at the expense of positioning accuracy. Therefore, based on the location algorithm in the firefly wireless sensor network, the problem is proposed, which can be solved by converting the solenoid linear solution to a two-dimensional combinatorial optimization problem. A defect of low precision of hop positioning algorithm.

The distance between the standard node i and the unknown node is $\overline{d}_i$. Calculate the distance d_i according to the average distance of each jump and the number of jumps, Then the location error is as (5):

$$f(x,\ y)_i = \overline{d}_i - d_i = \sqrt{(x - x_i)^2 + (y - y_i)^2} - d_i \tag{5}$$

The complexity function *Fitness*() represents the sum of the error of the unknown node and the n standard node, such as (6).

$$Fitness(x,\ y) = \sum_{i=1}^{n} f(x,\ y)_i \tag{6}$$

As can be seen from above, the smaller the value of the complexity function *Fitness*(), the more accurate the positioning effect is.

3 DV-Hop Algorithm Based on the Optimization of Firefly Algorithm

According to the above analysis, the problem is transformed from a linear equation group to a multi-dimensional optimization problem.

3.1 Algorithm Thinking of Fireflies

In 2008, University of Cambridge scholar Yang Xinshe proposed the firefly algorithm Firefly, Algorithmic [9, 10], which is a group intelligent algorithm based on natural inspiration. It is an intelligent bionics algorithm that imitates the behavior of fireflies in nature, and fireflies attract their mates to reproduce by luminescence. Some researchers believe that firefly luminescence is also useful for prey. But in either case, firefly luminescence follows the following rules:

(1) Every firefly has the same chance of attracting other fireflies. The luminous fireflies will be attracted by the shiny individuals and fly toward them;
(2) The light emitted by fireflies will be weakened to a certain extent by the air and other media, that is, as the distance between fireflies increases, the attraction and brightness will decrease gradually.
(3) The brightness of fireflies is limited by the target function in the smart algorithm, and if there are no individuals brighter than themselves in the visible area of fireflies, fireflies will implement the free flight strategy.

3.2 FADV-Hop

The individual brightness of fireflies at each location is fixed, and the current individual brightness at other locations decreases as the distance increases and is influenced by the absorption rate of the transmission media. In this topic, due to the low range of coding values, it is necessary to reduce the decay of brightness in the propagation process to a certain extent. The current individual at the location of the distance r, $I(r)$ using formula (7):

$$I(r) = \frac{I_0}{1 + \gamma r^2} \tag{7}$$

In the formula (7), I_0 represents the initial brightness of fireflies, γ is the absorption rate of the transmission media. r is the distance between two individuals.

The attraction between individuals β is also relative, its size is limited by the distance r and the absorption rate γ of the transmission media, and the current individual's attraction at the distance r position is such as formula (8).

$$\beta(r) = \beta_0 e^{-\gamma r^2} \tag{8}$$

In formula 8, β_0 is the attraction of the light source (r = 0), and γ is the absorption rate of the transmission media.

When the current individual i finds a brighter individual j in the visual range, the individual position is updated using formula (9):

$$x_i' = x_i + \beta_0 e^{-\gamma r^2}(x_i - x_j) + \alpha\left(R - \frac{1}{2}\right) \tag{9}$$

In formula (9), x_i' is the updated position of individual i towards individual j, which is brighter than itself, x_i is the current position of the individual, $\beta_0 e^{-\gamma r^2}$ is the attraction of individual j to current individual i. α is the step factor, a constant between [0, 1], R is a n dimensional random factor between [0, 1], subject to uniform distribution. The average positioning error is defined as formula (10).

$$AvgError = \frac{\sum_{i=1}^{n} \sqrt{(x - x_i)^2 + (y - y_i)^2}}{n \times r} \tag{10}$$

In formula (10), n is the number of unknown nodes in the region, (x_i', y_i') is the coordinates of the unknown node from the new algorithm, (x, y) is the actual coordinate of the unknown node as originally specified in the experiment, and r is the communication radius of the node.

4 Simulation and Result Analysis

The region of the FADV-Hop positioning algorithm network simulation is 100 m * 100 m, and the node communication radius r is set to 30 m.

According to previous scholars' research experience, the initial value of each parameter in the firefly algorithm is set as follows: the initial attraction β_0 is set to 1.0 and the random parameter α is set to 0.9. The absorption rate of the transmission media γ is set to 0.009, which is set to 50 iterations by several experiments because of the large coding range. In order to prove the superiority of the FADV-Hop algorithm in wireless sensor network positioning, the DV-Hop algorithm is used as an experimental comparison.

4.1 Algorithm Process

There are m communication nodes in the simulation area, including n beacon nodes, so the number of unknown nodes is $(m - n)$, the location information of the beacon nodes is known, and the firefly algorithm is introduced into DV-Hop algorithm, the unknown nodes are located as follows:

(1) Use formula (1) to calculate the average hop distance per beacon node.
(2) Calculate the fitness function according to formulas (7) and (8).

(3) Initialize a population, setting parameters such as iterations $t = 0$, I_0, α, β_0, and population size, maximum iterations $T\max$, etc. For the current individual, if there is a brighter individual than yourself, the current position is updated according to formula 9, otherwise free flight. $t = t + 1$, if $t = T\max$, the algorithm ends.

4.2 Performance Analysis

See Figs. 1, 2, 3 and 4.

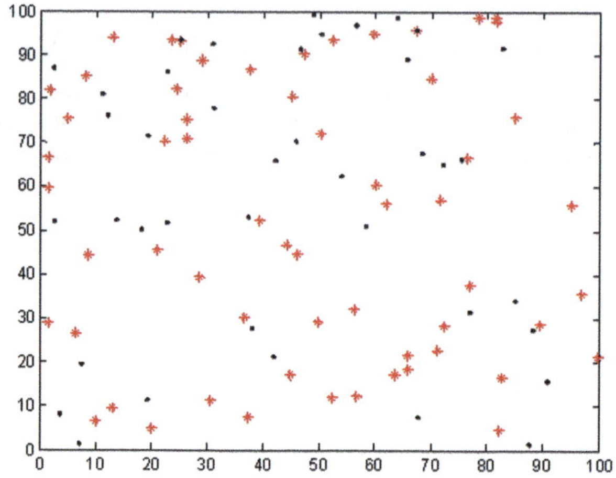

Fig. 1. Network topology of the beacon node and the unknown node (ratio 2:3), the asterisk indicates the beacon node and the dot represents the unknown node.

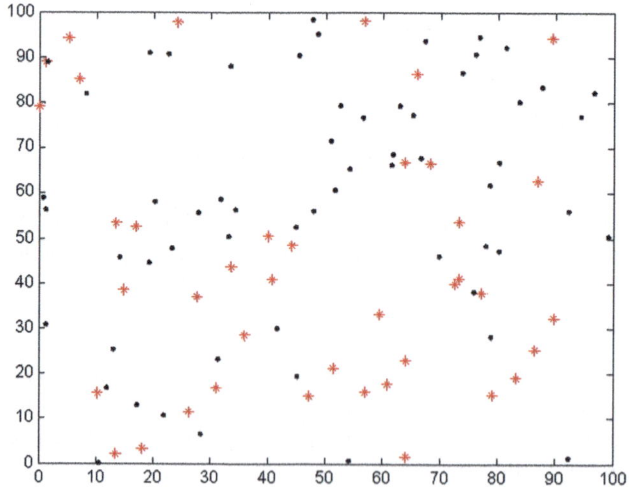

Fig. 2. Is the topology of the beacon and unknown nodes, (ratio 3:2). The asterisk indicates the beacon node and the dot represents the unknown node.

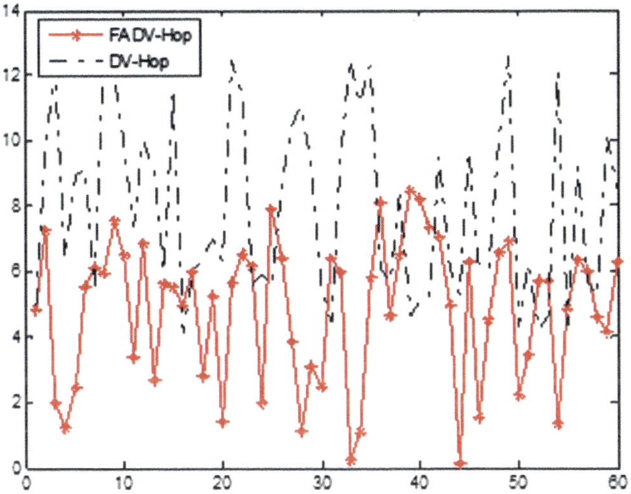

Fig. 3. The average location error (ratio 2:3), the difference between the density of different nodes 3:2 and the average location error is calculated by formula 5. The average location error in the DV-Hop positioning algorithm is 8.16 14, and the average location error in the FADV-Hop positioning algorithm is 3.24 15.

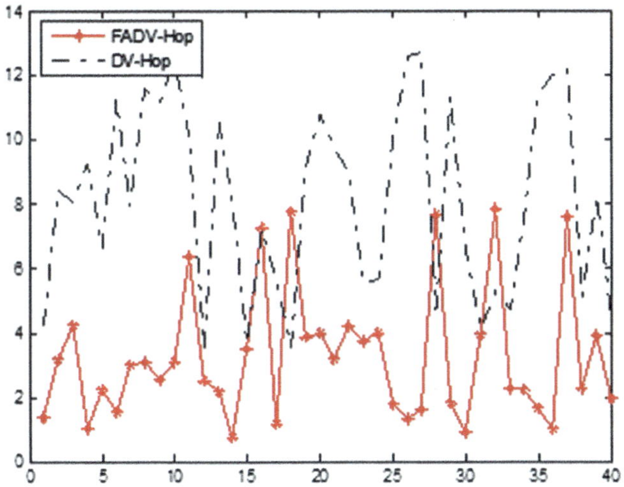

Fig. 4. Is a comparison of the average location error of FADV-Hop and DV-Hop in the case of 3:2 distribution ratio between the beacon node and the unknown node. The average location error of DV-Hop is 7.6639, and the average location error of FADV-Hop is 4.6353.

5 Conclusion

Based on the DV-Hop location algorithm, this paper presents a FADV-Hop location algorithm based on firefly algorithm, aiming at the average location error caused by the calculation of the average hop distance. The specific improvement strategy, theoretical reasoning and experimental proof are given. Under the same network environment, the positioning accuracy of FADV-Hop location algorithm is improved significantly. The research of location algorithm, which combines routing and positioning, is the focus of the next step.

Acknowledgements. The research presented in this paper was supported by Chao Hu University, Hefei, Anhui.

Funding. The authors acknowledge a key project from the Natural Science Foundation in Anhui Province, China (Grant: KJ2014A096).

References

1. Sajid, M., Wahid, A., Pervaiz, K., et al.: SMIC: sink mobility with incremental cooperative routing protocol for underwater. Wirel. Sens. Netw. **44**(4), 883–892 (2016)
2. Mardini, W., Khamayseh, Y., Almodawar, A.A., et al.: Adaptive RSSI-based localization scheme for wireless sensor networks. Peer-to-Peer Netw. Appl. **9**(6), 991–1004 (2016)
3. Bassi, P., Clark, G., Rossi, B.: Distribution of arrival times of air shower particles. Phys. Rev. **92**(2), 441–451 (2016)
4. Peng, B., Kürner, T.: Three-dimensional angle of arrival estimation in dynamic indoor terahertz channels using a forward–backward algorithm. IEEE Trans. Veh. Technol. **66**(5), 3798–3811 (2017)
5. Das, B.B., Ram, S.K.: Localization using beacon in wireless sensor networks to detect faulty nodes and accuracy improvement through DV-Hop algorithm. In: International Conference on Inventive Computation Technologies, pp. 1–5. IEEE (2017)
6. Kanmaz, M., Aydin, M.A.: Indoor localization for wireless sensor network and DV-hop. PressAcademia Procedia **5**(1), 88–95 (2017)
7. Zhan, W.S., Chen, H.W., Chin, C.S. et al.: Cluster heads distribution of wireless sensor networks via adaptive particle swarm optimization. In: Fourth International Conference on Computational Intelligence, Communication Systems and Networks, pp. 78–83. IEEE Computer Society (2012)
8. Tao, Q., Zhang, L.: Enhancement of DV-Hop by weighted hop distance. In: Advanced Information Management, Communicates, Electronic and Automation Control Conference, pp. 1577–1580. IEEE (2017)
9. Zhao, C., Wu, C., Chai, J., et al.: Decomposition-based multi-objective firefly algorithm for RFID network planning with uncertainty. Appl. Soft Comput. **55**, 549–564 (2017)
10. Sánchez, D., Melin, P., Castillo, O.: Optimization of modular granular neural networks using a firefly algorithm for human recognition. Eng. Appl. Artif. Intell. **64**, 172–186 (2017)

Co-saliency Detection via Weighted Low-Rank Matrix Fusion

Song Yu and Yuan Xie(✉)

School of Software, Central South University, Changsha, China
154711035@csu.edu.cn

Abstract. In order to improve efficiency, most saliency and co-saliency detection models are built at the superpixel level. In this paper, we propose a new co-saliency detection based on weighted low-rank matrix fusion. Firstly, a multi-scale superpixel pyramid are segmented by different thresholds using superpixel segmentation algorithms, and we use a weighted low-rank matrix recovery model to generate corresponding saliency map at each scale. Then, a reliable fused saliency map is generated by using adaptive fusion multi-scale saliency map. Finally, a co-saliency prior based on Gaussian Mixture Model (GMM) was introduced to detect the co-saliency objects in multiple images. The experiment results in the standard datasets show that the saliency map obtained by the proposed method can accurately highlight the all of saliency area and effectively suppress the background noise, which have higher accuracy and better recall rate.

Keywords: Saliency · Co-saliency · Low-rank matrix recovery
Multi-scale superpixel pyramid

1 Introduction

With the rapid development of many imaging devices, saliency detection has become a hot topic in the field of computer vision. Numerous saliency detection models have been proposed to reveal biological vision mechanisms and explain human cognitive processes.

The single image saliency detection model was originally based on a method of global contrast and local contrast. Cheng et al. [1] calculated the salient values of the image pixels using the color statistics histogram contrast, and used the difference in spatial consistency to evaluate the global contrast. Yan et al. [2] proposed a method for detecting saliency, setting up three image layers at different scales, and the significant measure of each layer has different local contrasts. Hou et al. [3] proposed a frequency domain residual method to obtain a saliency map by calculating the average logarithmic spectral difference between the logarithm spectrum of the image and a series of natural images. Achanta et al. [4] proposed a frequency tuning algorithm to directly assign pixel significance. Harel et al. [5] calculated the significance by using random walks on their designed graph model architecture. Yang et al. [6] applied a popular sorting algorithm to a closed-loop graph model with image superpixels as nodes, and

J. Abawajy et al. (Eds.): ATCI 2018, AISC 842, pp. 975–987, 2019.
https://doi.org/10.1007/978-3-319-98776-7_117

ranked them based on background and foreground sequence similarities, thereby obtaining a saliency map. Recently, low-rank matrix recovery algorithms have been found to be useful in saliency detection. Shen et al. [7] proposed a unified model that integrates the underlying visual features and high-level visual features. In this model, an image can be decomposed by low-rank matrix recovery, where the salient regions can be treated as sparse noise, but it does not take into account the problems associated with single-scale superpixel segmentation. Moreover, the above methods are only applicable to single image saliency detection and do not consider the consistency information contained in multiple images.

The purpose of the co-saliency detection is to detect a common, salient foreground region in a pair or group of images. In recent years, as a new research branch of visual significance, the detection of synergistic significance targets has attracted the interest of a large number of scholars. Jacobs et al. [8] first proposed and defined cooperative saliency detection as the discovery of special objects in image groups with similar backgrounds. Chen et al. [9] proposed a distribution-based representation method to quantify sparse features of an image. Based on the feature distribution of a pair of images, it presents a progressive process that enhances forward feedback, and can identify the synergetic effect of image pairs. Sexual objects, co-saliency maps are obtained by merging different saliency measures. Li et al. [10] linearly combine the saliency maps of the three single images with the two co-saliency maps by using a complex cooperative multi-layer image approach. However, these two methods can only be used for a pair of images and cannot be extended to multiple image groups. Chang et al. [11] generated a saliency map for each image, and used the most prominent part of the image as the final co-salient objects. Fu et al. [12] used clustering to learn global consistency information from image groups. They used three types of bottom-up saliency and co-saliency cues (contrast cues, spatial cues and corresponding cues) to measure the clusters. Conspicuous area. However, these methods merely fuse various saliency maps or saliency measures with fixed weights by simple summation or multiplication, which ignores the essential intrinsic link of multiple saliency measures.

Because in most computer vision applications, saliency detection is often used as a pre-processing step, rapid saliency detection is particularly important. In order to improve efficiency, the existing saliency detection methods are based on the superpixel level. However, there are significant differences in the results of saliency detection at different superpixel scales. This paper proposes a co-saliency detection method based on weighted low-rank matrix fusion. Firstly, different segmentation thresholds are set, superpixel segmentation is performed on the original image to form a multi-scale superpixel pyramid, features are extracted on the superpixel segmentation map of each scale, and a novel weighted low-rank matrix recovery model is adopted to each scale superpixel segmentation map to generate a single-scale saliency map. Then, the saliency maps of all scales are adaptively blended to obtain the reliable fused saliency map. Finally, the single image saliency detection method is extended to introduce a co-saliency prior based on GMM model to the completion of the detection of co-salient objects. The experimental results show that the proposed method is superior to other state-of-the-art methods. The main contributions of this paper are:

- We use superpixel segmentation method on the image at different scales and establish a multi-scale superpixel pyramid to achieve multi-scale significance detection.
- We propose a weighted low-rank matrix recovery (WLRMR) model to optimize the traditional low-rank matrix recovery method in saliency detection.
- We propose a co-saliency prior based on Gaussian Mixture Model (GMM) to extend the saliency detection of single images to co-saliency detection of multiple images.

2 Low-Rank Matrix Recovery

Recently, low-rank matrix analysis methods have shown great potential in the detection of significance. [13] decompose the image into redundant information and saliency information. Redundant information shows a high degree of regularity, while saliency information shows abruptness. In [7], the entire decomposition process is considered as a low-rank matrix recovery problem, defined as follows:

$$\min_{L,S}\left(rank(L) + \lambda\|S\|_0\right) \quad s.t.\ F = L + S \tag{1}$$

where $F = [f_1, f_2, \ldots, f_n]$ is a feature matrix by stacked N feature vectors, L is recovered low-rank matrix which models background and S is sparse matrix which models saliency.

Since the above problem is NP, the alternative convex surrogate to solve the above problem is

$$\min_{L,S}\left(\|L\|_* + \lambda\|S\|_1\right) \quad s.t.\ F = L + S \tag{2}$$

where $\|L\|_*$ denotes nuclear norm of matrix L, and $\|S\|_1$ is the l_1 norm of S.

3 Proposed Co-saliency Detection Model

As shown in Fig. 1, multi-scale saliency fusion is divided into several steps. Firstly, we introduce how to use weighted low-rank matrix recovery model to compute the single scale saliency map in Sect. 3.1. Then we explain how to fuse the multi-scale saliency maps in Sect. 3.2. Finally, the fused saliency map is smoothed by considering the relation of the neighborhoods in Sect. 3.3. Furthermore, we extend the single image detection to co-saliency detection of multi images in Sect. 3.4.

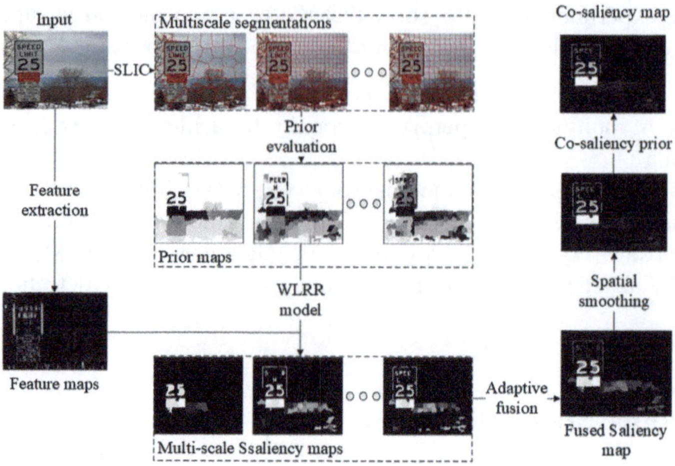

Fig. 1. Framework of proposed algorithm

3.1 Weighted Low-Rank Matrix Recovery Model

Given an input image *I*, we first partition it into *N* nonoverlapping superpixels $P[P_1, P_2, \ldots, P_n]$ by using the SLIC algorithm [14]. For each superpixel P_i, a D-dimension feature vector is extracted and denoted as $f_i \in \mathbb{R}^D$, then the image feature matrix *F* can be obtained by assembling all the feature vectors, denoted as $F = [f_1, f_2, \ldots, f_n] \in \mathbb{R}^D$. The problem of salient object detection is to design an effective model to decompose the feature matrix *F* into a low-rank matrix *L*, which represents the redundant background information, and a sparse matrix *S*, which represents the salient object. To solve the issues as mentioned in Sect. 5, we propose a novel background prior weighted low-rank matrix recovery (WLRR) model for salient object detection as follows:

$$\min_{L,S}\left(\|L\|_* + \lambda\|S\|_1\right) \quad s.t.\ W \circ F = W \circ L + S \tag{3}$$

where $\circ$ denotes element-wise multiplication of two matrixes, and *W* is the background prior weighting matrix, which we will introduce in the following part of our letter. $\lambda > 0$ is a tradeoff parameter to balance *L* and *S*.

Our WLRR model extends the classic matrix recovery model by taking the background reliability of an image into consideration, which strengthens the low-rank and sparsity of *L* and *S*, respectively. As the weighting matrix assigns smaller weights to most of salient object feature vectors in feature matrix *F*, the l_1 norm of the corresponding vectors in the recovered matrix sparse matrix *S* are inclined to be small. Thus, salient objects can be highlighted more effectively.

In this letter, we employ color, location, and boundary connectivity, and fuse them to generate a high-level background prior map, then this map is transferred to a weighting matrix.

(1) *Color prior*: As discussed in [7], warm colors such as red and yellow are more pronounced in our daily life, here, we directly use the color prior proposed in [7] to get our background color prior. Suppose, we have already obtained the color prior for superpixel P_i, denoted as C_i, then the corresponding background color prior is defined as

$$CP(i) = 1 - C_i \tag{5}$$

(2) *Location prior*: Generally, objects which locate near the image center are more attractive to people [15]. On the contrary, objects that are far away from image center have a greater probability of belonging to the background. For each superpixel P_i, we calculate the distance from its mean position to image center location c, denoted as $d(p_i, c)$. Then the background location prior of P_i can be defined as

$$LP(i) = 1 - \exp(-d(p_i, c)/\sigma^2) \tag{4}$$

where σ is the standard deviation that controls the width of the Gaussian "bell.

(3) *Boundary connectivity prior*: Inspired by the work in [16, 17], we know that background regions have very high probability of connecting with the borders of the image, while very few object regions connect to image borders. Here, we use the length of intersection between a superpixel P_i and image boundary superpixels to quantify how heavily P_i is connected to the image borders. Our boundary connectivity prior of background is defined as

$$BCP(i) = \frac{1}{1 + \exp(-\frac{|P_i \cap B|}{\Re(i)})} \tag{6}$$

where $|\cdot|$ denotes the length of intersection, B is the set of boundary superpixels, and $\Re(i)$ represents the pixel number of P_i. Note that boundary superpixels are those border superpixels and superpixels that are connected with image border superpixels on an undirected graph. Each superpixel represents a graph node and the connectivity of a superpixel to the border superpixel is calculated by Dijkstra's Algorithm [18]. If a superpixel is connected with boundary superpixels, it will have a bigger value in the boundary connectivity prior map.

After we get the above three background priors, the final background prior of superpixel P_i can be obtained by

$$W(i) = CP(i) \cdot LP(i) \cdot BCP(i) \tag{7}$$

For the convenience of representation and computation, we ensemble $W(i)$ into the weighting matrix W as follows:

$$W = \begin{pmatrix} W(1) & \cdots & W(N) \\ \vdots & \ddots & \vdots \\ W(1) & \cdots & W(N) \end{pmatrix} \in \mathbb{R}^{D \times N} \tag{8}$$

In order to avoid the trivial solution, we normalize the elements of W to [0, 1].

Our WLRR model is obviously a convex optimization problem, it can be efficiently solved by the proximal alternating direction method of multipliers (ADMM) [19]. We first introduce a Lagrangian multiplier Y, then the solver of WLRR model can be transferred to minimizing the following augmented Lagrangian function:

$$\begin{aligned} G(L, S, Y, \mu) = & \|L\|_* + \lambda \|S\|_1 \\ & + Tr(Y^T (W \circ (F - L) - S)) \\ & + \frac{\mu}{2} \|W \circ (F - L) - S\|_F^2 \end{aligned} \tag{9}$$

where $Tr(\cdot)$ denotes the trace of a matrix, $\mu > 0$ is a positive constant that controls the penalty for violating the linear constraints.

After decomposing the image feature matrix into the low-rank matrix L and the sparse matrix S and optimizing them, the saliency value of superpixel P_i can be obtained by calculating the l_1 norm of the elements of the i-th column of S. The specific calculation formula is as follows:

$$SP(i) = \frac{\|S(:, i)\|_1}{\sum_i \|S(:, i)\|_1} \tag{10}$$

3.2 Multi-scale Saliency Fusion

As mentioned in Sect. 3, detecting saliency in a single scale always fail in capturing the intrinsic salient object. To obtain a reliable solution for saliency detection, we propose a multi-scale fusion method. At first, we construct a multi-scale superpixel pyramid with different segmentation scales. Then the single scale saliency map is computed as mentioned in Sect. 3.1. Finally, we compute the fused saliency map by the adaptive weights that takes all scales saliency results into account.

To compute the adaptive fusion weights, we need to reassign the saliency value that obtained at each scale to the finest over-segmentation scale. The saliency value of a superpixel is the mean saliency value of the pixels contained in it. By placing the saliency value of the superpixels of a single scale into a row vector. Then stacking all vectors together forming a saliency indication matrix Z. The ideal situation is that the saliency detection results are consistent in all scale, so the rank is one. We can model this problem as low-rank matrix recovery:

$$\min_{L,E} \left(\|L\|_* + \lambda \|E\|_1 \right) \quad s.t.\ Z = L + E \tag{11}$$

The optimal solution E^* represents the disparity of the multi-scale salient object detection results. Summing up the E's absolute value by rows, then we get a vector $[E_1^*, E_2^*, \ldots, E_n^*]^T$, where n is the number of scales. Large E_i^* represents the i-th saliency map are not consistent well with others. So the corresponding saliency map should be endowed with a small weight. The adaptive weights are:

$$w_i = \frac{\exp(-E_i^{*2})}{\sum_i \exp(-E_i^{*2})} \tag{12}$$

Then, the fused saliency maps can be obtained by

$$S_{map}^{fuse} = \sum w_i \cdot S_{map}^i \tag{13}$$

3.3 Spatial Smooth

Due to the super pixel segmentation process in the image, it is not considered that the saliency regions of the images should be clustered together in space, and the smoothness between adjacent super pixel points is ignored. Inspired by the relevant work in [20], in order to further optimize the fused saliency map, a loss function is established:

$$E = \sum_i \omega_i^{bg} * s_i^2 + \sum_i \omega_i^{fg}(s_i - 1)^2 + \sum_{i,j \in Nei(i)} \omega_{ij}(s_i - s_j) \tag{14}$$

where s_i denotes the saliency value of the superpixel P_i, ω_i^{bg} is the background probability, ω_i^{fg} is the foreground probability (in our method, $\omega_i^{fg} = S_i^{fuse}$), $Nei(i)$ denotes the neighbors of superpixel P_i. The weight ω_{ij} is defined as

$$\omega_{ij} = \exp(-\frac{d_{app}^2(p_i, p_j)}{2\sigma^2}) \tag{15}$$

where $d_{app}^2(P_i, P_j)$ denotes the l_2 distance of their average colors in the CIE-LAB color space, σ is 10 in our experiment.

3.4 Co-saliency Detection

Co-saliency detection is discovering the similar object in a group of images. The co-saliency objects should be saliency in their own image and have high similarity with certain features (e.g., intensity, color, texture or shape) in a set of images [21]. We show our co-saliency detection flowchart in Fig. 2.

Firstly, given a set of images $I_{set} = [I_1, \ldots, I_i, \ldots, I_n]$, we compute the single saliency map S_i for i-th image as we proposed in Sects. 3.1, 3.2 and 3.3. Then we segment the saliency map S_i into binary mask M_i with adaptive threshold defined as:

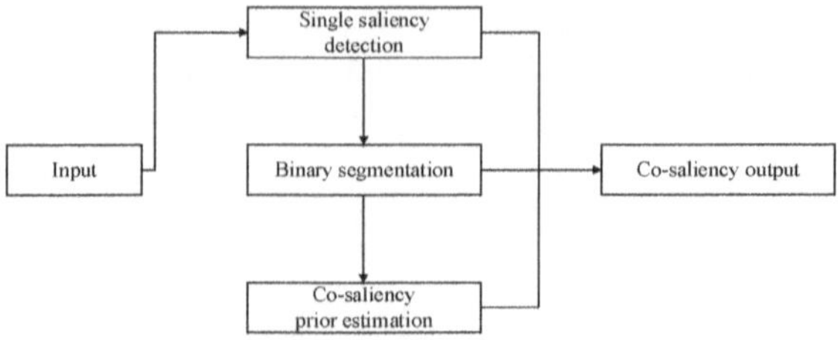

Fig. 2. The flowchart of co-saliency detection

$$T_i = \alpha \cdot mean(S_i) \tag{16}$$

where $\alpha = 2$ in our experiment. The pixel (or superpixel) with the saliency value which is bigger than the adaptive threshold is identified as foreground, otherwise it is identified as background.

The key problem is how to find the co-salient pattern for a given image set. To tackle this problem, we employ GMM to build the co-saliency prior. The GMM with 5 Gaussian distributions is used to build foreground color models G_i for the foreground pixels for i-th image. We use G_i to estimate the foreground probability of the pixels in a mask M_j of the j-th image. For each image we get n estimations. By averaging them, we get the co-saliency prior for each image.

4 Experiment and Analysis

In order to verify the effectiveness of the proposed algorithm, this paper selects the internationally published datasets MSRA [22] and ESCCD [2] for experiments. Both datasets contain 1000 natural images and manually labeled binary images. The image background in MSRA is simple and contains a single object, while the images in the ESCCD have the more complex background. In addition, this paper also carried out experiments on the dataset Image Pair [10] to verify the practicality of co-saliency detection. The dataset Image Pair includes 105 pairs of natural images, total 210 images.

4.1 Qualitative Analysis

In order to evaluate the method proposed in this paper, we compared seven existing methods for saliency detection, including GS [17], HDCT [23], LR [7], MR [6], HS [2]., SF [24] and RBE [16]. The results are shown in the Fig. 3. The method of this paper has a good overall effect. The contour of the saliency map obtained is closest to the true value, and the details are also relatively clear. It can accurately distinguish the foreground and background areas, and can clearly highlight the entire area of salient objects.

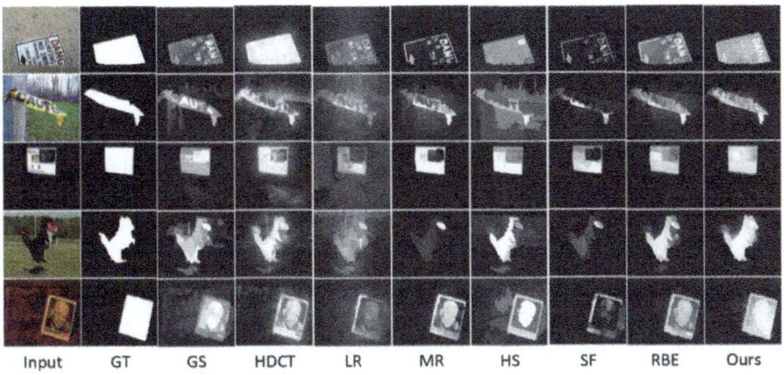

Fig. 3. The comparison of saliency detection methods on MSRA dataset

In addition, in order to evaluate the effectiveness of the proposed method for co-saliency detection, three kinds of collaborative saliency detection methods were selected for comparison, including Liu [25], Fu [12], and Li [10]. The results are shown in the Fig. 4. Compared with other algorithms, the proposed co-saliency detection method of this paper can accurately separate the foreground and background, and can accurately extract the internal information.

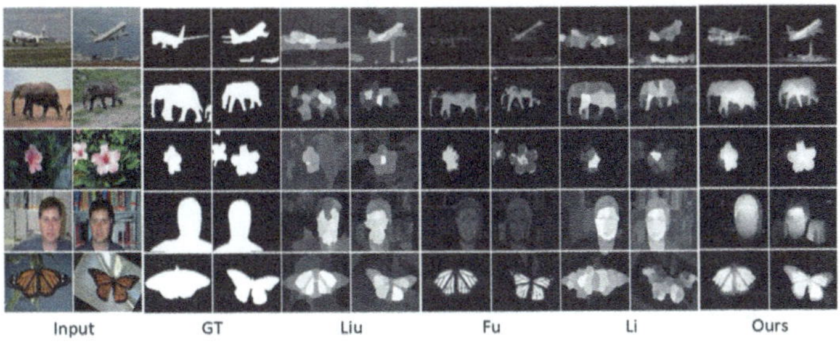

Fig. 4. The comparison of co-saliency detection methods on image pair dataset

4.2 Quantitative Analysis

In order to further verify the effectiveness of the proposed algorithm, in this paper, we use the PR curve, ROC curve, MAE, F-measure to evaluate the performance for quantitative analysis. This evaluation criterions are related to precision and recall, the formulas for precision and recall are as follows:

$$precision = \frac{Num(S_B \cap G)}{Num(S_B)} \tag{17}$$

$$recall = \frac{Num(S_B \cap G)}{Num(G)} \tag{18}$$

where S_B is the binary segmentation of the saliency map, G is a real binary map, $Num(S_B)$ and $Num(G)$ represent the number of pixels in which the value is 1.

The F-measure value is calculated as follows:

$$F\text{−measure} = \frac{(1+\beta^2) * \text{precision} * \text{recall}}{\beta^2 * \text{precision} + \text{recall}} \quad (19)$$

We conduct comparative experiments on the MSRA data set, as shown in the Fig. 5. The PR curve and ROC curve (red in the figure) of the proposed method are higher than other methods and have the lowest MAE score.

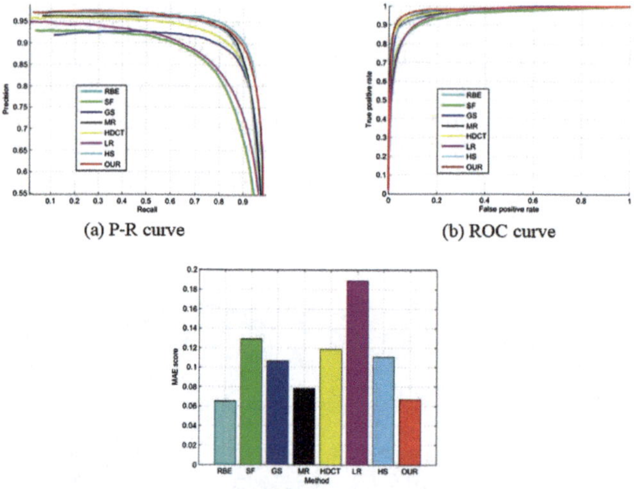

(a) P-R curve (b) ROC curve

(c) MAE score

Fig. 5. The quantitative comparison of saliency detection method on MSRA dataset

We conduct comparative experiments on the ESCCD data set, as shown in the Fig. 6. Compared to MSRA datasets, the saliency detection method is more challenging because the background of the images in the ESCCD dataset is more complex. However, the method proposed in this paper still performs well. The PR curve and ROC curve are located at the top, and the MAE score is the lowest.

In Fig. 7, we show quantitative comparisons of the results of the proposed method with various synergic saliency detection methods on MAE and Precision, Recall, and F-measure results. From Fig. 7, Fu's detection effect is the worst. Compared with other methods, the proposed method has better F-measure and MAE.

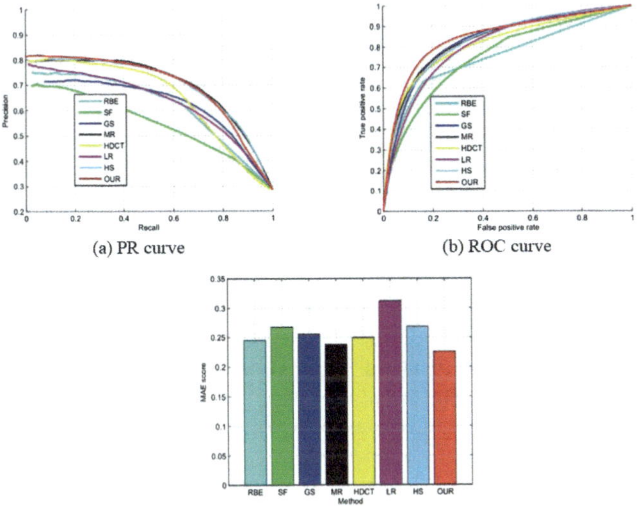

(a) PR curve (b) ROC curve

(c) MAE score

Fig. 6. The quantitative comparison of saliency detection methods on ESCCD dataset

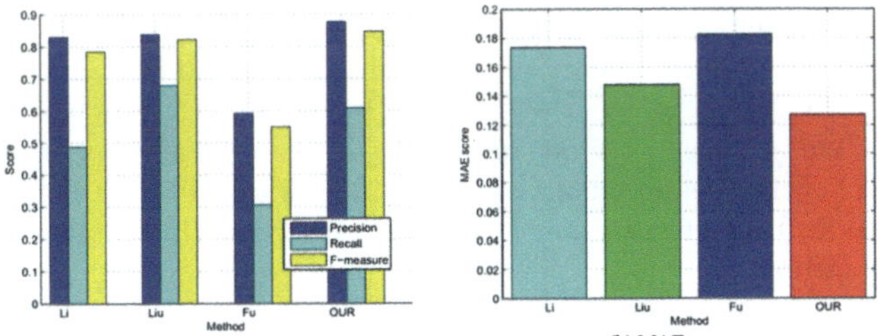

(a) F-measure score (b) MAE score

Fig. 7. The quantitative comparison of co-saliency detection methods on image pair dataset

5 Conclusion

In this paper, we have proposed a co-saliency detection method based on weighted low-rank matrix fusion, which obtains more reliable saliency maps through different scale superpixel segmentation combined with low-rank matrix analysis. We also propose a co-saliency prior based on Gaussian Mixture Model (GMM), which extends the single image saliency detection method to multiple images and then detects co-salient objects from multiple images. Moreover, the proposed method in this paper can achieve parallel computing. Therefore, in fact, the complexity of the proposed method is equivalent to the single-scale detection method. The experimental results on the public dataset also verify the effectiveness of the proposed method and is superior to the state-of-the-art methods.

References

1. Cheng, M.M., Zhang, G.X., Mitra, N.J., Huang, X.L., Hu, S.M.: Global contrast based salient region detection. In: IEEE Conference on Computer Vision and Pattern Recognition, pp. 409–416. IEEE (2011)
2. Yan, Q., Xu, L., Shi, J.P., Jia, J.Y.: Hierarchical saliency detection. In: IEEE Conference on Computer Vision and Pattern Recognition, pp. 1155–1162. IEEE (2013)
3. Hou, X.D., Zhang, L.Q.: Saliency detection: a spectral residual approach. In: IEEE Conference on Computer Vision and Pattern Recognition, p. 2280. IEEE (2007)
4. Achanta, R., Hemami, S., Estrada, F., Susstrunk, S.: Frequency-tuned salient region detection. In: IEEE Conference on Computer Vision and Pattern Recognition, p. 1597. IEEE (2009)
5. Harel, J., Koch, C., Perona, P.: Graph-based visual saliency, pp. 545–552 (2007)
6. Yang, C., Zhang, L., Lu, H., Ruan, X., Yang, M.: Saliency detection via graph-based manifold ranking. In: IEEE Conference on Computer Vision and Pattern Recognition, pp. 3166–3173 (2013)
7. Shen, X., Wu, Y.: A unified approach to salient object detection via low rank matrix recovery. In: IEEE Conference on Computer Vision and Pattern Recognition, pp. 853–860 (2012)
8. Jacobs, D., Goldman, D., Shechtman, E.: Cosaliency: where people look when comparing images, pp. 219–228 (2010)
9. Chen, H.: Preattentive co-saliency detection. In: 2010 17th IEEE International Conference on Image Processing, ICIP, pp. 1117–1120 (2010)
10. Li, H., Ngan, K.N.: A co-saliency model of image pairs. IEEE Trans. Image Process. **20**, 3365–3375 (2011)
11. Chang, K., Liu, T., Lai, S.: From co-saliency to co-segmentation: an efficient and fully unsupervised energy minimization model, pp. 2129–2136. IEEE (2011)
12. Fu, H., Cao, X., Tu, Z.: Cluster-based co-saliency detection. IEEE Trans. Image Process. **22**, 3766–3778 (2013)
13. Lang, C., Liu, G., Yu, J., Yan, S.: Saliency detection by multitask sparsity pursuit. IEEE Trans. Image Process. **21**, 1327–1338 (2012)
14. Achanta, R., Shaji, A., Smith, K., Lucchi, A., Fua, P., Suesstrunk, S.: SLIC superpixels compared to state-of-the-art superpixel methods. IEEE Trans. Pattern Anal. Mach. Intell. **34**, 2274–2281 (2012)
15. Judd, T., Ehinger, K., Durand, F., Torralba, A.: Learning to predict where humans look. In: IEEE International Conference on Computer Vision, pp. 2106–2113 (2009)
16. Zhu, W., Liang, S., Wei, Y., Sun, J.: Saliency optimization from robust background detection. In: IEEE Conference on Computer Vision and Pattern Recognition, pp. 2814–2821 (2014)
17. Wei, Y., Wen, F., Zhu, W., Sun, J.: Geodesic saliency using background priors. In: Fitzgibbon, A., Lazebnik, S., Perona, P., Sato, Y., Schmidt, C. (eds.) Lecture Notes in Computer Science, vol. 7574, pp. 29–42 (2012)
18. Dijkstra, E.W.: A note on two problems in connexion with graphs. Numer. Math. **1**, 269–271 (1959)
19. Ren, X., Lin, Z.: Linearized alternating direction method with adaptive penalty and warm starts for fast solving transform invariant low-rank textures. Int. J. Comput. Vis. **104**, 1–14 (2013)

20. Zhu, W.J., Liang, S., Wei, Y.C., Sun, J.: Saliency optimization from robust background detection. In: IEEE Conference on Computer Vision and Pattern Recognition, pp. 2814–2821. IEEE, New York (2014)
21. Li, H., Meng, F., Ngan, K.N.: Co-salient object detection from multiple images. IEEE Trans. Multimed. **15**, 1896–1909 (2013)
22. Liu, T., Yuan, Z., Sun, J., Wang, J., Zheng, N., Tang, X., Shum, H.: Learning to detect a salient object. IEEE Trans. Pattern Anal. Mach. Intell. **33**, 353–367 (2011)
23. Kim, J., Han, D., Tai, Y., Kim, J.: Salient region detection via high-dimensional color transform and local spatial support. IEEE Trans. Image Process. **25**, 9–23 (2016). A Publication of the IEEE Signal Processing Society
24. Perazzi, F., Kraehenbuehl, P., Pritch, Y., Hornung, A.: Saliency filters: contrast based filtering for salient region detection. In: IEEE Conference on Computer Vision and Pattern Recognition, pp. 733–740 (2012)
25. Liu, Z., Zou, W., Li, L., Shen, L., Le Meur, O.: Co-saliency detection based on hierarchical segmentation. IEEE Signal Process. Lett. **21**, 88–92 (2014)

Research on Product Form Design Based on Computer Technology

Zhang Liang(✉) and Zhao Jian

Qingdao Binhai University, Qingdao 266555, China
76201691@qq.com

Abstract. Objective: The new concept of consumption requires that products not only meet the technical requirements such as the use of functions and structures, but also satisfy people's aesthetic needs through novel designs, reasonable layouts, beautiful lines, and detailed handling; Methods: The combination of industrial design modeling theory and Computer Aided Industrial Design technology has created computer-aided modeling methods based on computer platforms, enabling product modeling to quickly and intuitively perform 3D rendering, virtual assembly, animation demonstration and simulated working state; Results: Before the product is put into production, a complete three-dimensional three-dimensional digital model can be presented in the computer for the designer to compare and analyze. The designer can arbitrarily modify and perfect the industrial product material, color and other modeling elements, so as to reduce the complex design process, shorten product development cycle, significantly increase product performance and market competitiveness; Conclusion: Digital models using computer-aided software design can output data directly for rapid prototyping and manufacturing, So, as an important application technology of form design, computer-aided design (CAD) technology should meet the new requirements of form design.

Keywords: Computer · CAID · Form design · Software

Since the beginning of the 21st century, with the improvement of living standards, people's concept of consumption has changed greatly. People have begun to focus on individual spiritual needs after meeting material needs. The new concept of consumption requires that products not only meet the technical requirements of using functions and structures, but also require novel products, reasonable layout, beautiful lines, and detailed treatment, and have certain aesthetic value [1].

The rapid development of digital technology has brought revolutionary changes to industrial design. The emergence of computer platforms has provided a new creative environment for industrial design. The combination of industrial design theory and CAD technology produces Computer Aided Industrial Design CAID based on computer platform [2]. The involvement of computer-aided industrial design software, such as Rhino 3D, enables product modeling to quickly and intuitively perform 3D rendering, virtual assembly, animation demonstration, and simulation of working conditions. Before the product is put into production, a complete three-dimensional spatial digital model can be presented in the computer for designers to compare and analyze. The designer can modify the industrial product material, color and other modeling

J. Abawajy et al. (Eds.): ATCI 2018, AISC 842, pp. 988–996, 2019.
https://doi.org/10.1007/978-3-319-98776-7_118

elements arbitrarily [3]. Digital models using computer-aided software design can output data directly for rapid prototyping and manufacturing, thereby reducing many complicated design processes, shortening the product development cycle, and significantly increasing the product's expressiveness and market competitiveness.

1 Basic Theory of Form Design for Industrial Design

Product shape design requires a comprehensive consideration of a series of internal constraints and external control factors and their relations. The constraints include the function, structure, material and man-machine relationship of the product. The control factors include the color, texture, decoration, appearance, and so on [4]. From the point of view of industrial design, the form design of products should deal with the relationship between the use function, man-machine relationship and aesthetic principles of products. Among them, the use of aesthetic rules to meet the technical and man-machine constraints in the design of product form is the key to form design. The geometric elements of product form include dots, lines, planes and bodies. The relationship between these elements constitutes the basis of product form design. Designers use the aesthetic laws to select and organize these geometric elements to design the product form to enhance the artistic effect of the product form.

The basic theories of morphological design of industrial design mainly include formal beauty rules, formal psychology and perception of form, morphological composition design, form proportional design, and morphological linear design. Formal beauty, form psychology and perception belong to domain knowledge, while form composition design, proportional design and linear design are form design methods. The form structure is designed to design the basic layout of the product modeling, and the shape proportion design is used to coordinate the relationship between the parts of the body. The morphological linear design is used to process the line between the boundary and the joint surface of these shapes. These three aspects ultimately determine the shape of the product.

1.1 Surface Segmentation of Shape

The morphological design of a product is to use a body to contain some single bodies that contains technical, functional and structure, and then to form a product's overall image by organizing together a plurality of single bodies according to a certain form of body composition laws and methods. How to organically combine multiple single forms to form a unique and beautiful shape is the purpose of art modeling, and is also the core issue of form composition.

The division of the surface of the body can be understood as the area division of the two-dimensional facade of the product shape, which can be used to design the facade (appearance) of a certain view of the product form. The segmentation of surface (elevation or shape) is to obtain different area zoning with a certain number ratio relation, which makes it more beautiful in structure, line type, color area or position arrangement, Among them, the segmentation of the characteristic rectangle is the most characteristic. The segmentation of the feature rectangle is shown in Fig. 1.

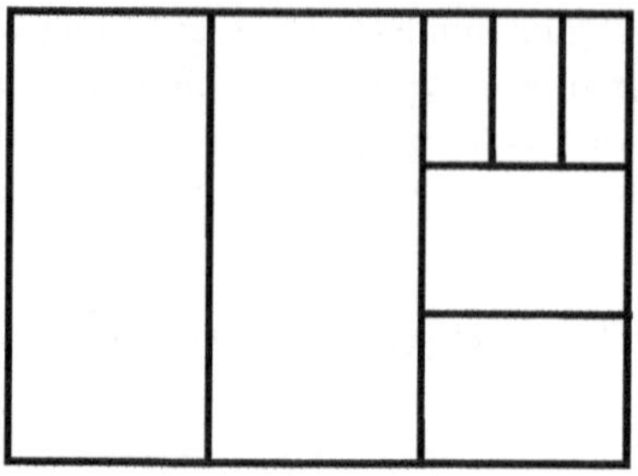

Fig. 1. The segmentation of the feature rectangle

1.2 Proportionality Design of Form

Proportional design is a very important aspect of form design and should be treated with caution. The proportion of products is designed to solve the proportion coordination problem of product shape with proportional knowledge, and the specific size of the product form is adjusted to make the proportion of each size conform to the proportion of root mean square, the proportion of golden section, the proportion of the middle value, the proportion of the integer and so on. Excellent design always induces the relatively independent dimensions of any unit to a clear overall size, changing the proportions of the unit size to make a perfect relationship with each other [5].

The accuracy of the form ratio is shown in Fig. 2. The ratio of the upper edge of the spout to the height of the pot, the upper edge of the pot handle and the height of the pot, the width of the pot handle, and the width of the pot body are integer ratios.

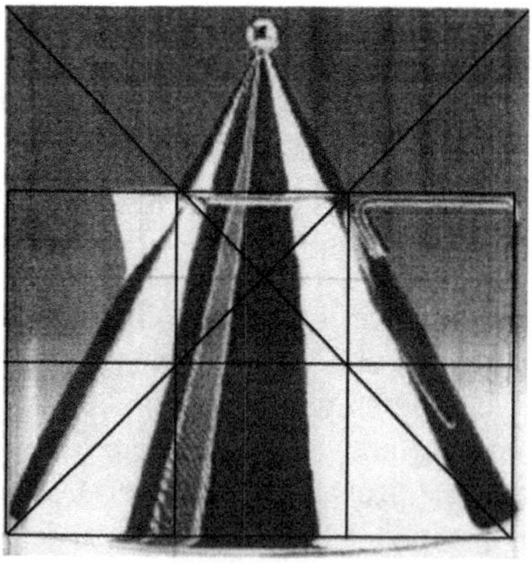

Fig. 2. The accuracy of the form ratio

1.3 Shape Design of Line Type

A successful product design is always closely related to the correct selection and organization of its linetype. The linear design of the form is an important part of the final determination of the shape, because the effect of overall harmonization can be achieved by rationally processing the arrangement, penetration, transition, and transition of each part of the line type. The main task of linetype design is to select and organize linetype. When choosing a line type, the designer should consider the perception of the line, as shown in Table 1.

Table 1. Description of the characteristics of various lines

Classification	Name	Line description
Slow curve	Stability curve	The overhanging position is in an equilibrium position and it cannot be flattened
Velocity curve	Parabola	It should not be symmetrical and adapt to some large organic shapes
Direction curve	Drooping curve	It does not have straight edges and is slightly curved at the edges
Independent curve	Spiral curve	It is difficult to combine with other curves, but there are many potential features

2 The Basic Application of Computer-Aided Design Technology

The computer-aided form design technology of industrial design is a technology that assists designers in the design of form using the design method of industrial design under the computer environment, and provides technical support for the application of the form design method of industrial design on the computer [6]. In general, the study of computer-aided morphological design techniques for industrial design can start from two aspects: on the one hand, it studies how to apply the relevant methods of industrial design morphological design to computer-aided morphological design, such as the application of aesthetic principles in morphological design; On the one hand, it is to study the computer's auxiliary methods, such as the use of handwriting on the sketch input, advanced rendering technology.

2.1 Curve Drawing and Organization

The form and linetype design of industrial design mainly refers to the selection and organization of linetype. In terms of line type selection, the computer can assist in providing feature curve templates, especially those commonly used in modeling design (such as track lines, parabolic lines, draping lines, etc.); Considering the linear organization, computers can help designers associate and coordinate free curves, such as associating a group of free curves into a family curve or a proportional curve [7]. Scribbles are often used in form, and their organization is an important part of linear organization. The family curves have the common law of curvature change, and they can be coordinated and unified. If a group of free curves in a shape are arranged in the same family curve, the alignment of the form is easy to achieve a coordinated visual effect.

2.2 The Segmentation and Simplification of Form

When the designer from the initial form of design to further design, he will find that the morphological elements of the product may be more complex, regardless of the primary and secondary, the focus is not prominent, is not conducive to surface segmentation, it is necessary to simplify the target screen. Through a certain simplification process, the designer can clarify which elements in the facade should be considered and which elements can be simplified.

The commonly used simplified methods are as follows: (1) The thickness of the line indicates the near and far, strong and weak effect; (2) Using atomization to detect the layout characteristics of the morphology; (3) Individually display contour lines, horizontal structure lines, and vertical structure lines separately. The lines in the form can be divided into contour lines, horizontal structure lines, vertical structure lines, etc. In the computer environment, we can use separate display contour lines, horizontal structure lines, vertical structural lines and other operations to deepen the impression of the designer. The shape contour shows the flow chart, as shown in Fig. 3. Of course, shape simplification can also be handled by far near effect or atomization effect, and this method can also be used to check the advantages and disadvantages of surface segmentation scheme.

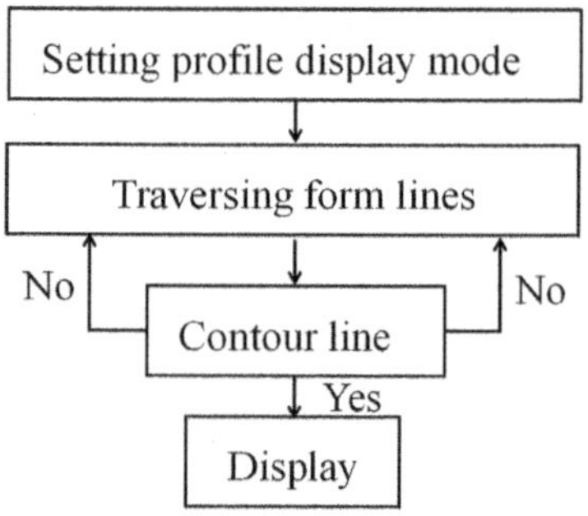

Fig. 3. Shape profile display flow chart

3 Comparison and Selection of Modeling Platform

3.1 The Characteristics of 3dsmax and Rhino

There are many application softwares suitable for industrial design. Here we mainly introduce two commonly used softwares: One is 3dsmax, and the other is Rhino [8]. According to the current situation, both of these softwares are suitable for industrial design aided design applications. 3dsmax has been developing for nearly 20 years. It is the most popular professional 3D software. It is not developed for industrial design, only later, it has gradually been applied to industrial design. 3dsmax has a variety of modeling methods such as lofting, Boolean operations, patches, meshes, polygons, NURBS surfaces and so on; Its modeling method is flexible and free, each modeling method provides a powerful command group, which brings great convenience to the modeling; 3dsmax is adaptable and capable of building a variety of special free surfaces.

Rhino is a relatively young professional 3D software, but it has a great potential for development, it gradually becomes the new darling of industrial design with its own powerful NURBS surface modeling function, which is loved by many industrial designers [9]. Although the function of Rhino may be too single compared to 3dsmax, but its powerful function of NURBS surface modeling is beyond doubt. Although 3dsmax also provides the modeling method of NURBS, but it is far from Rhino, and Rhino's modeling method is more rigorous and accurate than 3dsmax. In addition, NURBS itself has a very excellent definition of curves and surfaces, coupled with the support of powerful software commands, making Rhino also become a strong force to build complex freeform surfaces.

3.2 Similarities and Differences Between 3dsmax and Rhino in the Application of Industrial Design

The main modeling method of 3dsmax is network format, the principle is to divide a geometry or a surface into an appropriate number of grids (the quantity can be adjusted), and to edit the elements such as vertices, line segments, and polygons on the grid. So as to edit and adjust the shape, which is somewhat similar to the clay fabrication method in real life. This modeling method can be controlled without strict dimensional accuracy. It mainly depends on the operator's perceptual grasp, and it completes the modeling process by building the image through the mastery of the morphology and the spatial imagination [10]. This process is the same for both forward engineering and reverse engineering.

Because 3dsmax does not have enough dimensional accuracy, it is not suitable for workflow in the later stage of design, such as mechanism design or assembly design. But the modeling method, which is similar to clay sculpture, has a great degree of freedom, and more emphasis on the designer's perceptual understanding, and the entire process is fast, free, unconstrained, which is very beneficial to the free play of the designer. Therefore, we often use 3dsmax for rapid modeling and performance in the early stage of industrial design, that is, the design stage of the scheme, and make full use of the virtual three-dimensional effect to realize the exploration of the shape modeling and the effect of the scheme [11].

Rhino is based on NURBS and its core. The key to applying Rhino for model building is to draw and build curves. Because of the excellent and flexible precision control of Rhino, it is very easy to draw standard or free curves. n addition to curves, Rhino also provides commands for directly generating surfaces or bodies, and provides control points similar to curves for shape adjustment. This seemingly free modeling function seems to be equivalent to the flexibility of 3dsmax's grid modeling, but in fact, Rhino needs some rational analysis or logical analysis to complete modeling, and it is not natural to have a curve to build a surface [12].

Rhino's application needs to have a reasonably rational understanding. For the product form design, this state is more suitable for the later stage, because at the later stage, designers need to organize their own design programs, and at the same time begin to approach the engineering design stage, which is quite necessary for rational analysis, so that the downstream link can be better connected. Although at this stage there are also more advanced engineering software such as Pro-e that can be used for modeling processing and analysis, and it can be done in one step. However, relatively speaking, Rhino is easy to use, much simpler, and more suitable for product form designers.

4 Research on the Design Method of Product Modeling Based on Rhino

The original idea of product depends on sketching, while sketch conception is polysemous. Using Rhino can help designers to study styling through the ambiguity of sketches. Taking LED lamps as an example, the designer, based on the study of the LED lighting technology, hopes that the luminaire can obtain uniform illumination by increasing the lighting area, and use the touch sensor to adjust the motor to accurately set the angle of the light source, and integrate the simple sculptural modeling into the interior environment. The initial idea is shown in Fig. 4.

Fig. 4. The initial idea

After customer recognition scheme innovation, we need further optimization based on this model. Using Rhino can effectively extract the original creative styling features and scrutinize the details. According to the "principle of overall imagery priority", the main feature that affects product identification is the sideline [13]. After extracting the side view curves in Rhino, optimizing the styling on this basis can ensure the main features of the original styling. Shape optimization, as shown in Fig. 5.

On the basis of maintaining the main characteristic curve of the side view, the shape can be boldly changed, as shown in Fig. 6.

It can be seen that Rhino with strong modeling ability can be used as the creative tool and product expression tool of industrial designers.

Fig. 5. Shape optimization

Fig. 6. Shape optimization

5 Conclusions

The application of science and technology in manufacturing determines the development of computer-aided industrial design. The completion of the industrial design task is transformed by a single designer to be completed by the design team, and the computer aided industrial design ensures the effective division of labor and cooperation among the different professionals. The main purpose of industrial designers using computer-aided industrial design technology is to better achieve the unification of technology and art. The modeling ability of computer aided industrial design is already very powerful. And the computer model from sketch to three-dimensional is more convenient than the traditional method, but it has further development space.

References

1. Weng, C., Ai, X.: Product Form Design. Peking University Press, Beijing (2016)
2. Zhang, T.: Design and research of computer aided industrial design system. J. Liaoning Inst Sci. Technol. **18**(4), 6–8 (2016)
3. Chen, W.: Application of virtual reality technology in industrial design. Dev. Innov. Mech. Electr. Prod. **19**(05), 122–123 (2006)
4. Ye, D.: Discussion on the design method of the improved product. J. Guilin Univ. Electr. Technol. **26**(4), 276–277 (2006)
5. Zhang Wenli. Research on virtual environment-based intend-driven product form design. Nanjing University of Aeronautics and Astronautic (2006)
6. Xi, T.: Discussion on VR gulding industrial design development. Packag. Eng. **28**(7), 125–126 (2007)
7. Ai, P.: Research on NURBS curve quality control method in computer-aided industrial design. J. Mach. Des. **30**(6), 101–102 (2013)
8. Li, Q.: Discussion on 3D modeling of computer aided industrial design. Inner Mongolia Sci. Technol. Econ. **143**(13), 60–61 (2007)
9. Xi, S.: Design of computer aided product modeling based on Rhinoceros. Mech. Manag. Dev. **122**(4), 61–62 (2011)
10. Zhang, Q.: Application of CAID technology in industrial design. Packag. Eng. **28**(5), 123–124 (2007)
11. Zhang, L.: Virtual reality technology in the application of transport design process. Yanshan University (2011)
12. Ren, G., Cui, Y., Yang, J.: Industrial product modeling design based on Rhino 3D. Packag. Eng. **29**(1), 124–125 (2008)
13. Cheng, Y.: Humanized design of automobile modeling based on computer aided industrial design. J. Guizhou Univ. (Nat. Sci.) **31**(1), 123–124 (2014)

Product Quality Prediction Based on BP Neural Network and Rough Set Theory

Yingying Su(✉) and Lianjuan Han

Mechanical Engineering School, Shenyang University,
Shenyang 110044, Liaoning, China
syyle2012@163.com

Abstract. Product quality is critical to the survival and development of enterprises. Quality prediction is an important step in the field of quality control. In this paper, based on the study of BP neural network and rough set theory, the data mining model and algorithm for the combination of BP neural network and rough set theory are designed to give full play to their advantages and overcome the existing problems. Combined with the actual manufacturing process of an enterprise, algorithm solution and simulation analysis are carried out to verify the feasibility of the method and improve the accuracy of product quality prediction.

Keywords: Quality prediction · BP neural network · Rough set

1 Introduction

Excellent product quality is a necessary condition for satisfying consumer demand and one of the important indicators of competition among enterprise, which can enables enterprises to take a favorable position in the market and enhance the image of the enterprise. With the rapid development of society and the maturity of the market, product quality has aroused extensive attention from all walks of life, and quality requirements are also improving. Therefore, quality control is particularly important. Quality prediction is a key part of quality control. Product quality prediction plays a positive role in solving the problem of product disqualification.

At present, the traditional methods for product quality prediction are mechanism method, regression analysis method, least squares estimation method and so on. Nowadays, during the increasingly diverse development of economy, the lag in traditional methods can no longer satisfy the needs of contemporary products. Since the era of artificial intelligence, more and more methods have been used in quality control. Time series analysis and neural networks are widely used in prediction methods. Artificial neural network has the ability of self-learning, self-organization, self-adaptation, strong nonlinear function approximation and high fault tolerance, which can realize simulation, binary image recognition, prediction and fuzzy control. However, neural network has the problem of local convergence and other issues. Rough set is a new mathematical method to deal with inaccurate, uncertain and incomplete data. Through data analysis and reasoning, the hidden relationship of rough set is discovered, useful data is found, and knowledge reduction is performed. In order to eliminate the

J. Abawajy et al. (Eds.): ATCI 2018, AISC 842, pp. 997–1004, 2019.
https://doi.org/10.1007/978-3-319-98776-7_119

problem of local convergence and non-agile processing information of artificial neural networks, neural networks and rough sets are combined to build a comprehensive model based on synthesis algorithm of BP neural network and rough set for product quality prediction to improve the accuracy and speediness of product quality prediction.

In the process of part production, the welding factor parameter in the welding process of the component is one of the key factors affecting the product quality. Combination of artificial neural network and rough set method is presented to predict product quality in this paper based on the selection of influence factors of the welding process to the product quality, so as to solve the problem of unqualified welding.

2 Model Construction of Product Quality Prediction Based on BP Neural Network and Rough Set

2.1 BP Neural Network

Artificial neural network is a set of neuron processing units, where each connection is associated with a weight [1]. Redundant information can be dealt with in artificial neural network because of its weight. Self-organizing ability is relatively strong. Back-propagation (BP) neural networks are the most widely used method of artificial neural networks. Because of gradient descent, data can converge to local minimum. BP neural network consists of input layer, hidden layer and output layer.

The core of BP neural network to achieve its function is the algorithm. It is a kind of multilayer feed-forward network trained according to error backward propagation, and the most widely applied in neural network. Because of better mapping results of the multilayer perceptron in BP algorithm, the most common application is the single hidden layer network, which is a three-layer perceptron.

In the three-layer perceptron, the input vector is $X = (x_1, x_2, \ldots, x_i, \ldots, x_n)^T$, the hidden layer output vector is $Y = (y_1, y_2, \ldots, y_j, \ldots, y_m)^T$, the output layer output vector is $O = (o_1, o_2, \ldots, o_k, \ldots, o_l)^T$, the desired output vector is $d = (d_1, d_2, \ldots, d_k, \ldots, d_l)$, the weight matrix between the input layer and the hidden layer is $V = (v_1, v_2, \ldots, v_j, \ldots, v_m)^T$, and the weight matrix between the hidden layer and the output layer is $W = (w_1, w_2, \ldots, w_k, \ldots, w_l)$. The three-layer BP network is shown in Fig. 1.

The mathematical relationship between different layers of signals is shown below.

In output layer, the relationship is shown by formulas (1) and (2).

$$o_k = f(net_k) \quad k = 1, 2, \ldots, l \tag{1}$$

$$net_k = \sum_{j=0}^{m} w_{jk} y_j \quad k = 1, 2, \ldots, l \tag{2}$$

In hidden layer, the relationship is shown by formulas (3) and (4).

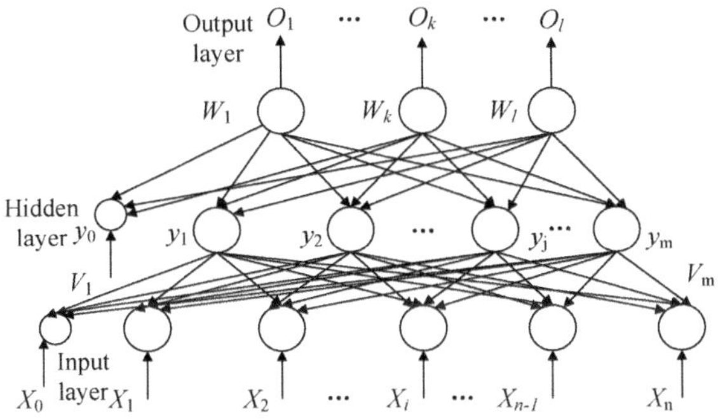

Fig. 1. The three-layer BP network

$$y_j = f(net_j) \quad j = 1, 2, \ldots, m \tag{3}$$

$$net_j = \sum_{i=0}^{n} v_{ij} x_i \quad j = 1, 2, \ldots, m \tag{4}$$

In the two formulas above, $f(x)$ is a transfer function, which is a mono sexual Sigmoid function and shown by formula (5).

$$f(x) = \frac{1}{1 + e^{-x}} \tag{5}$$

$f(x)$ is continuous and conductive, which is shown by formula (6).

$$f' = f(x)[1 - f(x)] \tag{6}$$

Bipolar Sigmoid functions (or hyperbolic tangent functions) can also be applied base on needs, which is shown by formula (7).

$$f(x) = \frac{1 - e^{-x}}{1 + e^{-x}} \tag{7}$$

The above formulas constitute a three-layer perceptron mathematical model [2].

2.2 Rough Set Theory

The Rough Set theory is a data analysis theory proposed by Polish mathematician Z. Pawlak in 1982 [3]. The data can be directly analyzed and reasoned in rough set theory based on mathematical knowledge. Prior knowledge is not be required, and conditional attributes are optimized and combined to achieve significant success in data decision-making and analysis, pattern recognition, data mining, etc. As one of the core contents of the rough set theory, knowledge reduction is to delete the irrelevant or unimportant knowledge under the condition of keeping the classification ability of the knowledge base unchanged [1]. Indefinite, inaccurate, and verbose data is found and deleted by means of reduction to extract rules. There are the following concepts for reduct and core in knowledge reduction.

Suppose $Q \subseteq P$, if Q is independent and $ind(Q) = ind(P)$, then Q is a reduction of P. Obviously, there are a variety of reductions in P. The set of all necessary relations in $P$is called the core of P, which is written as $core(P)$ [3].

2.3 Establishment of Synthesis Algorithm of BP Neural Network and Rough Set

BP neural network has strong self-learning, self-adaptability, fault tolerance and generalization ability. However, it will fall into the problem of local minimization and slow convergence in the process of solving the complex nonlinear problem for BP neural network. Therefore, the other methods are combined to exert their respective advantages and find the optimal value. The rough set theory with complementary BP neural network is selected for data processing.

Firstly, the original sample data is preprocessed by BP neural network. The data with larger errors are deleted, and then the new data set is arranged. Secondly, the new data set is discredited and rough set theory is applied to reduce the discrete new data. Finally, relevant rules are extracted. The algorithm flow is shown in Fig. 2.

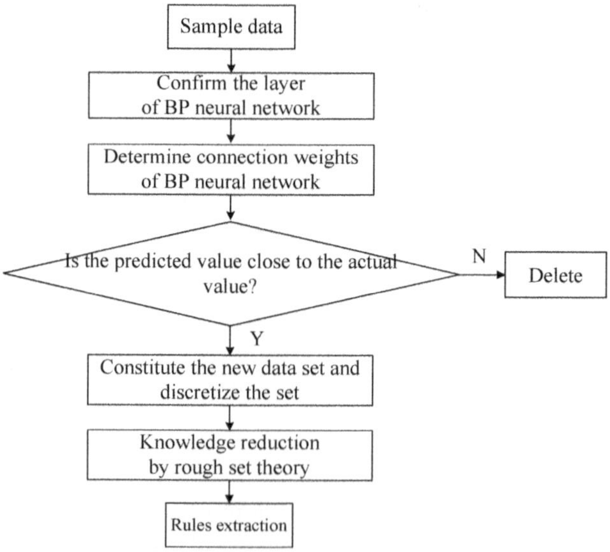

Fig. 2. The algorithm flow

3 Construction of Quality Evaluation Index System for Welding Products

Welding quality refers to the degree of welding products meeting the requirements of design technology and has a great influence on product performance, life, personal and property safety. Therefore, evaluation system is established in order to improve the quality of welding products. Under normal circumstances, the evaluation of welding quality is divided into appearance and internal defects. Appearance defects generally refer to the formation and surface defects of welding, namely welding width, evenness, surface crack and so on. Nondestructive testing of internal defects is necessary such as air holes and slag inclusion. As welding width is a key factor influencing the welding quality, the width and its influence parameters of a certain component are analyzed in the welding process. Specific indicators include welding current, welding voltage, welding speed, welding wire diameter and extension length of welding wire. The quality evaluation system of welding products is shown in Fig. 3.

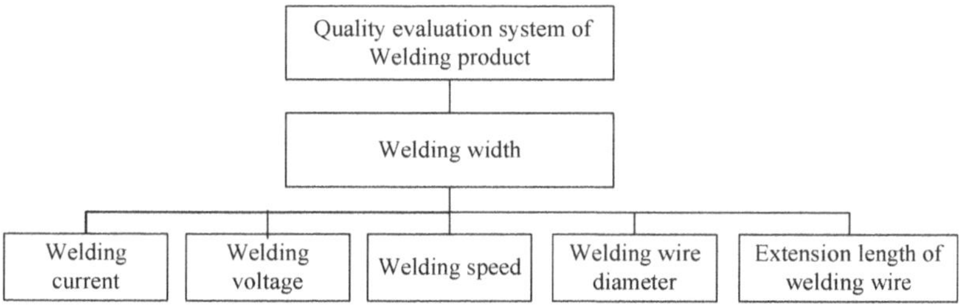

Fig. 3. Quality evaluation system of welding products

4 Case Analysis

The quality evaluation system of welding products is constructed in Fig. 3. Five parameters of the secondary indicators are selected as indicators for analysis. 120 data are selected, in which the sample information is the first 100 groups to extract rules, and the rest 20 groups are used in verification. Through the rule information determined in the welding process, the relationship between the evaluation indicators is found out to select the appropriate welding parameters.

4.1 Preprocessing of Sample Data

In the processing of the raw data, the three layer BP neural network is selected. The condition attributes of C and D in the decision table are respectively the welding parameters and the width of the weld. The x_1, x_2, x_3, x_4, x_5 is used to represent the condition attribute C, namely $C = \{x_1, x_2, x_3, x_4, x_5\}$; and the decision attribute $D = \{y\}$. Some of the data in the decision table is shown in Table 1.

Table 1. Partial data of the decision table

C					D
x_1	x_2	x_3	x_4	x_5	y
…	…	…	…	…	…
210	162	22.1	3	16	7.12
200	184	22	3.5	15	7.42
190	182	23	3	15	7.15
220	165	22.4	2	18	7.26
190	207	23.1	2.5	13	6.49
180	197	23.6	3	14	7.82
170	160	22.5	3.5	15	6.48
240	171	24	2.5	17	6.49
230	190	23.6	2.5	12	7.94
200	197	22.3	2	14	7.86
…	…	…	…	…	…

Three layers of BP neural network are used for preprocessing, in which there are 6 neurons in the input layer, 1 neuron in the output layer and 9 neurons in the hidden layer. The selected training samples are the first 50 in the decision table and the prediction samples are the last 50. The model of BP neural network is established to predict, the inconsistent data are deleted, and the remaining data sets are composed to perform the calculation of the next step of rough set theory.

4.2 Rule Extraction of New Data Set

In step 1, rough set theory is applied to discretization.

Table 2. Discretized decision table of partial data

C					D
x_1	x_2	x_3	x_4	x_5	y
…	…	…	…	…	…
2	5	1	0	4	5
1	2	2	2	4	3
3	5	5	0	0	0
4	5	0	2	5	5
5	5	5	2	2	0
5	0	4	2	4	5
3	0	2	2	5	5
3	0	1	0	2	3
5	4	3	0	2	0
…	…	…	…	…	…

The important processing step in rough sets is discretization processing, by which the final result will be affected. Equidistance discretization processing is adopted and results are shown in Table 2.

In step 2, rules are extracted based on rough set theory reduction. In order to eliminate verbose and inaccurate data, the results by reduction and extract rule information are shown as follows.

Rule1: $x_1(1)ANDx_2(3)ANDx_3(2)ANDx_5(3) \Rightarrow \mathrm{y}(2)$
Rule2: $x_1(0)ANDx_2(1)ANDx_3(3)ANDx_5(0) \Rightarrow \mathrm{y}(4)$
Rule3: $x_1(4)ANDx_2(4)ANDx_3(0)ANDx_5(5) \Rightarrow \mathrm{y}(5)$
Rule4: $x_1(2)ANDx_2(3)ANDx_3(4)ANDx_5(1) \Rightarrow \mathrm{y}(2)$
Rule5: $x_1(1)ANDx_2(2)ANDx_3(2)ANDx_5(4) \Rightarrow \mathrm{y}(3)$
Rule6: $x_1(3)ANDx_2(5)ANDx_3(5)ANDx_5(0) \Rightarrow \mathrm{y}(0)$
Rule7: $x_1(5)ANDx_2(5)ANDx_3(5)ANDx_5(2) \Rightarrow \mathrm{y}(0)$
Rule8: $x_1(3)ANDx_2(0)ANDx_3(2)ANDx_5(5) \Rightarrow \mathrm{y}(5)$
Rule9: $x_1(3)ANDx_2(0)ANDx_3(1)ANDx_5(2) \Rightarrow \mathrm{y}(3)$
Rule10: $x_1(5)ANDx_2(4)ANDx_3(2)ANDx_5(0) \Rightarrow \mathrm{y}(0)$
Rule11: $x_1(1)ANDx_2(5)ANDx_3(0)ANDx_5(5) \Rightarrow \mathrm{y}(3)$
Rule12: $x_1(3)ANDx_2(5)ANDx_3(0)ANDx_5(5) \Rightarrow \mathrm{y}(5)$
Rule13: $x_1(5)ANDx_2(4)ANDx_3(1)ANDx_5(2) \Rightarrow \mathrm{y}(1)$
Rule14: $x_1(0)ANDx_2(4)ANDx_3(2)ANDx_5(5) \Rightarrow \mathrm{y}(2)$
Rule15: $x_1(5)ANDx_2(1)ANDx_3(0)ANDx_5(4) \Rightarrow \mathrm{y}(1)$
Rule16: $x_1(0)ANDx_2(0)ANDx_3(0)ANDx_5(1) \Rightarrow \mathrm{y}(4)$
Rule17: $x_1(3)ANDx_2(4)ANDx_3(3)ANDx_5(5) \Rightarrow \mathrm{y}(2)$
Rule18: $x_1(3)ANDx_2(5)ANDx_3(3)ANDx_5(2) \Rightarrow \mathrm{y}(2)$
Rule19: $x_1(4)ANDx_2(5)ANDx_3(1)ANDx_5(2) \Rightarrow \mathrm{y}(1)$
Rule20: $x_1(3)ANDx_2(0)ANDx_3(0)ANDx_5(4) \Rightarrow \mathrm{y}(3)$
Rule21: $x_1(2)ANDx_2(5)ANDx_3(3)ANDx_5(3) \Rightarrow \mathrm{y}(1)$
Rule22: $x_1(5)ANDx_2(3)ANDx_3(4)ANDx_5(5) \Rightarrow \mathrm{y}(2)$
Rule23: $x_1(0)ANDx_2(4)ANDx_3(4)ANDx_5(1) \Rightarrow \mathrm{y}(5)$
Rule24: $x_1(2)ANDx_2(1)ANDx_3(1)ANDx_5(3) \Rightarrow \mathrm{y}(5)$
Rule25: $x_1(2)ANDx_2(5)ANDx_3(0)ANDx_5(0) \Rightarrow \mathrm{y}(2)$
Rule26: $x_1(1)ANDx_2(3)ANDx_3(5)ANDx_5(0) \Rightarrow \mathrm{y}(3)$
Rule27: $x_1(3)ANDx_2(0)ANDx_3(1)ANDx_5(0) \Rightarrow \mathrm{y}(1)$
Rule28: $x_1(4)ANDx_2(0)ANDx_3(2)ANDx_5(3) \Rightarrow \mathrm{y}(3)$
Rule29: $x_1(4)ANDx_2(0)ANDx_3(2)ANDx_5(1) \Rightarrow \mathrm{y}(0)$
Rule30: $x_1(0)ANDx_2(2)ANDx_3(3)ANDx_5(1) \Rightarrow \mathrm{y}(4)$
Rule31: $x_1(0)ANDx_2(5)ANDx_3(2)ANDx_5(5) \Rightarrow \mathrm{y}(3)$

According to the above processing results, the parameters are reduced according to the method of equidistant division area to obtain the relationship between the selection of the welding parameters and welding width. The meaning of rule 1 is that When the selected interval of welding current is (167.5, 175.5), the interval of welding voltage is (22.99, 23.32), the interval of welding speed is (186.66, 199.99), the interval of extension length of welding wire is (15, 16), and the interval of welding width is (7.17, 7.50).

It was found that 19 groups of data are in accordance with the rules extracted by synthesis algorithm of BP neural network and rough set algorithm by applying the 20 groups of data to verify the extracted rules. The method of using the BP neural network can improve the accuracy of the rules information of the rough set theory. According to the rules, the parameters affecting welding width in the welding process are welding current, welding voltage, welding speed and extension length of welding wire. In the quality analysis of welding products, the unqualified products can be deleted first through the rule extraction to improve the product quality.

5 Conclusions

Aiming at the problem of local convergence of BP neural network, a synthesis algorithm based on BP neural network and rough set are presented to better extraction of planning and more accurate of quality prediction. Welding width is selected to analyze the influence parameters of a component in the welding process, and the quality evaluation system of welding products is established. The synthesis algorithm of BP neural network and rough set are used to predict the influence factors of welding width of the component, which proves the feasibility of the method, improves the accuracy of the product quality prediction, makes the data mining results better and satisfies the requirement of the development of modern manufacturing quality.

References

1. Shao, F.: Data Mining Principles and Algorithms, 2nd edn. Science Press, Beijing (2009)
2. Han, L.: Artificial Neural Network Theory, Design and Application, 2nd edn. Chemical Industry Press, Dongcheng (2007)
3. Zhang, W., Wu, W., Liang, J.T.: Rough Set Theory and Method. Science Press, Beijing (2001)
4. Yang, J.: Practical Tutorial of Artificial Neural Network. Zhejiang University Press, Zhejiang (2001)
5. Zhang, S., Sun, J., Zhang, M.: Rough sets and BP neural network for power distribution network fault diagnosis. Electr. Qual. **362**(5), 40–43 (2017)
6. Yuan, N., Yang, L.: Safety evaluation of construction technology based on rough sets-neural network. Saf. Environ. Eng. **19**(1), 60–64 (2012)
7. Li, G., Wang, K., Wenqiang, T.: Research on product requirements analysis method based on Big Data and rough set. Chin. J. Eng. Des. **23**(6), 522–529 (2016)
8. Yang, Z., Zhang, C., Wu, Y.: Research on rough sets and RBF neural network based transformer fault diagnosis method. Electr. Meas. Instrum. **51**(21), 34–39 (2014)
9. Peng, L., Fang, W.: Heterogeneity of inferring reputation of cooperative behaviors for the prisoners' dilemma game. Physica A **433**, 367–378 (2015)
10. Clay, W., Richardson, D. A.: The Java high class weaves a distance: JDK 5. Sci. Technol. no. 3, 17–18 (2006)

Research on the Innovation of Industrial Economy Under the Background of Big Data

Xing Li(✉), ZiHan Yang, JianMing He, and Bingjie Huo

College of Economy and Management, Longdong University, Qingyang 745000, Gansu, China
leebj@189.cn

Abstract. With the advent of the era of big data, the issue of "big data" has arisen rapidly from the technological level in entrepreneurial to the top agenda of national strategies. This paper intends to give a basic demonstration of the concept of big data as a starting point, and exams the strategic model of entrepreneurial technological innovation in the context of big data in China through the analysis of the characteristics of big data.

Keywords: Entrepreneurial technological innovation · Big data Strategy

1 Introduction

The advent of "big data" era marks the indication that traditional data mining methodology is no longer able to adapt to the constantly-changing environment, and is inadequate in the aspects such as data collection, data storage, data analysis and visualization. Meanwhile, there has been a distinct growth of dependence on data from all walks of life; the quantitative analysis method based on data now intends to be replaced with qualitative analysis method based on experts which requires a large amount of devotion to time and energy. It is under these circumstances that traditional entrepreneurial technological innovation faces a big challenge. As traditional data mining method and entrepreneurial technology gradually fail to adjust to the large-scale data which emerges quickly in the field of entrepreneurial technology and realize timely processing and analysis of such data, it loses the data analysis scheme and ability that help make decision. Therefore, this paper intends to put forth the effective way of facing the opportunities and challenges in the "big data" environment, responding to the radical changes brought about by big data to the technological innovation management of traditional entrepreneurial, from the level of thinking to technology, and realizing the goal of decision-making innovation of the technological innovation management in China. It is not only the foothold of this paper but the problems it anticipates to solve.

J. Abawajy et al. (Eds.): ATCI 2018, AISC 842, pp. 1005–1009, 2019.
https://doi.org/10.1007/978-3-319-98776-7_120

2 The Basic Concept of Big Data

Big data is not the aggregation of a large amount of data in traditional sense, but a set of information processing food technology that entails new data processing entrepreneurial technology, transferring entrepreneurial technology and applying entrepreneurial technology. It is different from mass data in essence, so its characteristics and connotations cannot be explained by mass data. The international information consultation company—Gartner—believes that big data "has surpassed the existing information entrepreneurial technology to process extreme information resources." On Wikipedia, big data is defines as "the dataset that cannot extract, analyze and process its content through traditional information entrepreneurial technology." NSF defines it as a large-scale, diversified and complicated dataset that is generated by scientific instrument, E-mail, sensor, Internet of things and video and audio soft wares. Mckinsey suggests that big data is a dataset whose collecting, storing, managing and analyzing abilities surpasses those of typical database soft wares. This definition has two meanings: (1) the size of dataset that is consistent with the standard of big data is changeable, and will grow over time and with the advancement of entrepreneurial technology (2) there are some differences among departments in the size of dataset that is consistent with the standard of big data (Fig. 1).

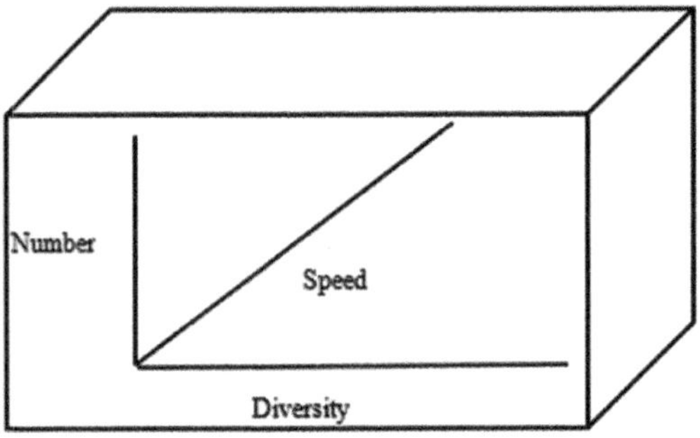

Fig. 1. The 3D growth of data

3 The Characteristics of Big Data

Big data is of 3Vs, namely, volume, variety and velocity. By volume, it means data volume increases at an unprecedented speed. Big data, larger than the former MB data, is a huge data cluster whose unit transforms to TB and PB. Variety means the types of big data and data source, including structured, semi-structured and unstructured data from traditional and untraditional information source like weibo, texts, journals, sensor data and video and audio profiles. Velocity suggests the timeliness of big data. As big data has time effect as regards generation, analysis and application, data stream must be applied to procedures and decision-makings so as to achieve the effect. Volume indicates the value of big data. It is the ultimate characteristic of big data and is capable of procuring insight and value. It was due to the entrepreneurial technology including artificial intelligence, data mining entrepreneurial technology and machine learning that the advent of big data came into emergence. Entrepreneurial technology based on big data can transform huge volumes of data to information and make it convenient to make decision and take actions (Fig. 2).

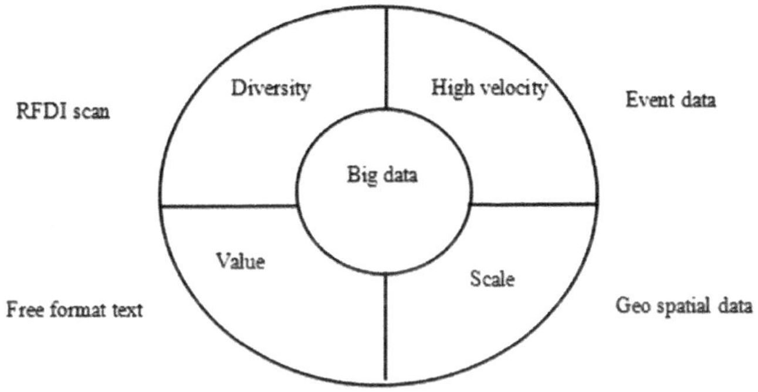

Fig. 2. The characteristics and components of big data

4 The Strategic Model of Entrepreneurial Technological Innovation in the Context of Big Data in China

"Enhance the innovation ability and build a innovation-driven country" has been the core thought of the future development strategy since the 17th CPC National Congress. In the 18th National Congress, it came up with the great idea of "implementing the innovation-driven development strategy". Under such circumstances, how to achieve the "innovation-driven development" in entrepreneurial technological innovation management is an urgent issue demanding to be considered by science and technology workers in China.

Entrepreneurial technological management entails the whole process from the conceiving, designing, developing, producing of new products and crafts, as well as the developing, recognizing and applying of the market to commercialization. It is a dynamic unity of economic and social coordinated development of food company technology, which is composed of interwoven subsystems of input, process, feedback and output. Managing the system requires maintaining and sharpening competitive edges through meeting the increasing and changing demands of customers, so as to achieve the goal of the current and long-term economic benefits and best integrate the inner elements and related resources of companies (Fig. 3).

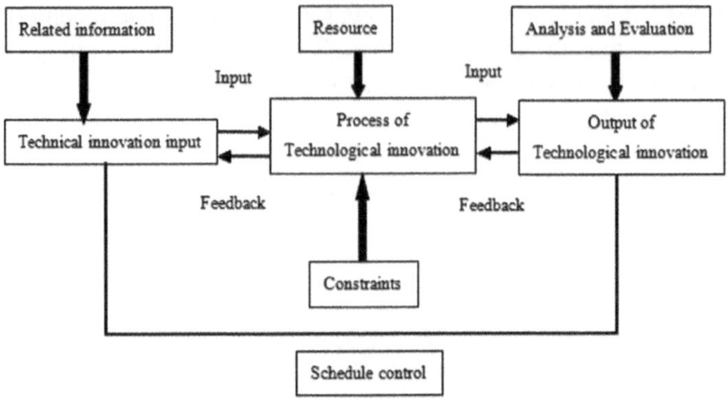

Fig. 3. The macro model of entrepreneurial technological innovation management

5 Management of Entrepreneurial Technological Innovation Resources

Entrepreneurial technological innovation resources mainly include fund resource of research and development, instrument and equipment resource related to entrepreneurial technological innovation, human resource of entrepreneurial technological innovation, and information resource of entrepreneurial technological innovation. The management of these innovation resources is demonstrated through the source, collection scheme, realization method and the specific prescription and measures of various resources; the accumulation, reservation, update and storage of various resources; the optimal allocation and usage of various resources during the implementation of entrepreneurial technological innovation; the duties and responsibilities of the departments and staff of various resources; the monitoring and the evaluation of effect while using various resources.

6 Decision-Making Management of Entrepreneurial Technological Innovation

The decision-making of technological innovation is the key link to entrepreneurial technological innovation, which directly affects the risks and final outcome of entrepreneurial technological innovation. To ensure the scientificity and reasonality of decision-making, companies must: (1) establish a democratic system to ensure the democratic centralism during the process of decision-making and the scientificity of the method of decision-making; (2) set up a program evaluation organization including internal and external experts; (3) issue a debate about the feasibility of entrepreneurial technological innovation from the following aspects: market demand, the maturity of entrepreneurial technology, innovation ability, the scheme of implementing entrepreneurial technology, and the benefits and risks of innovation.

The process of entrepreneurial technological innovation is an kind of economic activity of high risks, such as the risks of entrepreneurial technology, market, intellectual property, credit, and time. Risk management means to project and analyze the risks during the process of entrepreneurial technology; make warning as well as evaluation; monitor and control risks; resolve and avoid risks; put forth the remedy measures and scheme; identify the responsibility and reasonable undertaking of risks. The system management of entrepreneurial innovation technology contains the following three systems: (1) the specific operation system of various links and procedures during the process of entrepreneurial technological innovation. As entrepreneurial technological innovation is a dynamic process of integrating the achievements of researches and market demand, it is bound to have complicated links and procedures. Therefore, it is imperative to systematize and standardize the specific operation and functioning of them to ensure the final innovation outcome; (2) the management system of coordinating the involved parties of entrepreneurial technological innovation.

7 Conclusion

All the countries, companies and research institutions are standing at the same starting line in regard to the research of big data. How to seize the opportunity, face the challenges and obtain a leading position in this fast-changing complex competition environment is the ultimate goal of this research team and all the scientific entrepreneurial technological innovation workers in China.

References

1. Author, F.: Article title. Journal **2**(5), 99–110 (2016)
2. Edwards, T.: Innovation and organizational change developments towards an interactive process perspective. Technol. Anal. Strateg. Manag. **124**, 445–464 (2002)
3. Li, X., Qu, T., He, J.M., Hou, B.J.: Seek excellence on the hospital quality management competition. Basic Clin. Pharmacol. Toxicol. **119**(S4) (2016). HHME16-H24
4. Dougherty, D., Hardy, C.: Sustained product innovation in large mature organizations: overcoming innovation to organization problems. Acad. Manag. J. **39**, 1120–1153 (1996)
5. Rothwell, R.: Towards the fifth—generation innovation process. Int. Mark. Rev. **11**, 7–32 (1994)
6. von Hippel, E.: The dominant role of users in the scientific instrument innovation process. Res. Policy **5**, 212–239 (1976)
7. Boschi, R.A.A., Balthasar, H.U., Menke, M.M.: Quantifying and forecasting exploratory research success. Res. Manag. **125**, 14–21 (2002)
8. Li, X., Jianmin, H., Hou, B.J., Zhang, P.: Exploring the innovation modes and evolution of the cloud-based service using the activity theory on the basis of big data. Clust. Comput. (2017)
9. Peng, L., Fang, W.: Heterogeneity of inferring reputation of cooperative behaviors for the Prisoners' Dilemma game. Physica A **433**, 367–378 (2015)
10. Titus, J.: ECN technical editor: the eclipse of stand. J. Zhongkai Agrotech. Coll. **19**(2), 32–35 (2006)

Software and Hardware Design of Data Safety System Under Power Architecture

Lei Yao(✉)

Chengdu Polytechnic, Chengdu 610041, Sichuan, China
565169780@qq.com

Abstract. Information is the core of the application of all kinds of data systems. The maintenance and operators of IT systems tend to pay more attention to the security and availability of the server system, and easily ignore the importance of the security of the information itself. The degree of informatization in China is also increasing, and the localization of hardware and software has become an important part of safeguarding the national security, especially when it comes to the use of domestic software in the core areas of the national economy and the people's livelihood, which is of great significance to national security. This project studies the date oriented guarantee system. The self-developed data safety software is installed on the domestic server based on Power architecture, which can not only guarantee the security of government data in all aspects, but also realize the independence and controllability of the localization of core hardware and software so as to ensure the security of national data.

Keywords: Power framework · Data safety · Safety technology

1 Purpose and Importance

Information is the core of the application of all kinds of data systems, and the normal operation of all data systems is based on the premise of information security. However, it is worth noting that the IT system maintenance and operators tend to pay more attention to the security and availability of the server system, and easily ignore the importance of the security of the information itself. In fact, the core of all kinds of information application and business system is all kinds of information. When special cases occur, the necessary hardware resources, such as switches, servers, and storage devices, are easy to recover or reconfigure quickly, however, if the information is damaged or lost, it will be difficult to recover. Therefore, the real-time protection of information security has become the top priority of information security.

With the progress of the times, the degree of informatization in China is also improving, and the localization of software and hardware has become an important part of safeguarding national security. Previously, the occurrence of "Prism" event sounded the alarm for the protection of national information security. At the same time, it also made domestic software manufacturers more deeply aware of the importance of the development of the information industry and the important responsibility of protecting the national information security.

J. Abawajy et al. (Eds.): ATCI 2018, AISC 842, pp. 1010–1017, 2019.
https://doi.org/10.1007/978-3-319-98776-7_121

Especially for the aspect that whether the core area of national economy and people's livelihood uses domestic software, it is of great significance to national security. The fermentation of "Prism" event has made the information security much concerned by the society. In addition, the domestic IT manufacturers need to break the monopoly of the foreign IT giant in order to obtain their own market space, so since 2013, the IT system has entered the acceleration period of the domestic substitution. We do not simply abandon foreign equipment, but under the same conditions, we will give priority to the choice of domestic equipment and services, so as to continuously promote the healthy development of the domestic IT industry.

In 2015, General Office of the State Council released "Guidance Concerned with the Coordinated Development of E-government", which points out that we will take five years or so to unify and normalize and realize the overall completion of the network of national e-government and also strengthen substantially the safety capacity for network information. At the same time, they will intensify the research and spread of technical products with security and reliability, and request national e-government network and mission-critical system to take priority to the domestic hardware products as they purchase.

In September of the same year, Prime minister Li Keqiang presided over executive meeting of the State Council and passed "Executive Summary concerning Promoting the Development of Big Data", which demonstrates that government information and public data should be connected with each other and shared publicly. Meanwhile, this paper emphasizes that they will intensify the construction of information safety system protecting private data.

In a word, the situation of information safety of the "Internet plus" era is quite hard. Therefore, the significance of autonomous control is increasingly prominent. The development of core based software and hardware, which means critically getting rid of others' control for core technique, is the route one must take to make sure network safety even country safety. Meanwhile, the 13th 5-year plan suggestion puts forward that through the overall and efficient online service, we can realize the organized and efficient power operation, thus promoting the connection between government and mass further, continuously enhancing the intellectualized level of government for dealing with all kinds of affairs and impressible problems, and carrying out the big data strategy and network power strategy. Therefore, with the driving of this policy, localizing information safety products will become the heatedly political points in some years.

Sichuan province also pays more attention to the development of information safety and takes it as their great deployment of accelerating the cultivation of five great high-end and growth industries, which is also under the macroscopic background of the new and normal state and also takes it as a great measure of giving full impetus to the development of science and accelerating healthy development in Sichuan province. Since the definition of five high-end and growth industries in July of last year, Sichuan province pushed the development of information safety industry with strong power and released "Inform Concerning the Coordinated and Balanced Advance of 56 Great and Important Industrial Project and the Complete Construction of Five High-End and Growth Industries" and other related papers and set up the propulsion and development group in charge of the five great high-end and growth industry, led by provincial leaders, which all demonstrates their determination to pushing the development of the five great high-end and growth industry. And information safety industry lies on the first place of five and has great importance, which is no need saying.

This project research faces the safety system of data and installs the data safety software developed independently on the homemade server based on Power architecture, which can not only guarantee the overall data safety of e-government and also realize the autonomous control of localization of core hardware and software and protect national data safety.

This project based on the research and development of Power structure, the data deduplication technology, Hadoop cluster storage technology and automatic generation of KVM on Power virtual machine and other technologies, supports the continuous protection of data and emergent operation and maintain when some failures happen. And it automatically takes over the server business in 3 min through the disaster tolerance virtual machine to ensure continuity of things and solve data security problems systematically, which is very suitable for the e-government system with high demand for data safety and also enhances the overall stability and security in the field of relative data application substantially.

2 Technical Routes

This project is based on independent research and development and independent intellectual property rights. The concept of rapid prototyping method and agile development is taken as the leading factor in the project development, and the corresponding technical route is as follows (Fig. 1).

The first step is to build a hardware platform based on Power architecture. This project selects the RedPower server of Wuxi Zhongtai.

The second step is to port the code of the company's existing core product Heifang data backup and recovery system (V6.0) to the RedHat 7.1 of the Power platform, and form a Beta1 version of the multidimensional linked list CDP government data guarantee system based on the Power architecture, and then combine the software to the RedPower server to produce the prototype for trial sale.

The third step is to develop the arbitrary time point round-trip CDP function of the multidimensional linked list. Using this technology, we can achieve continuous protection for government data, data recovery and instant mount at any time point, and simulation exercise and emergency takeover.

The fourth step is to use the CAPI + FPGA of Power architecture to optimize the duplicate data deleting module of the product, use Hadoop as a backup storage, and use Erasure Codes technology on the Power architecture to reduce the storage redundancy consumption from 3 times to 1.4 times.

The fifth step is to study the automatic creation, deletion and editing technology of the KVM on Power virtual machine, which is used for emergency operation and maintenance function of business.

The sixth step is to integrate the key technologies to tackle problems to upgrade the product, and form the Beta2 version of the multidimensional linked list CDP

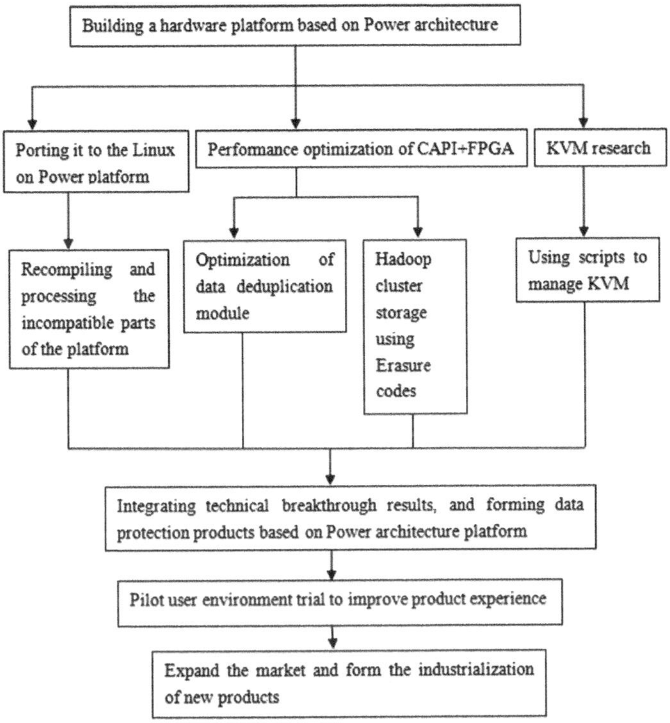

Fig. 1. Schematic diagram of technical route

government data guarantee system based on the Power architecture, and test its performance.

The seventh step is to find pilot users, continue to test and optimize in the actual environment, and improve user experience based on user feedback.

The eighth step is to put the multidimensional linked list CDP government data support system based on Power architecture into market operation and form industrialization.

The key points of the technology are as follows:

The first is the multidimensional linked list arbitrary time point round-trip CDP algorithm.

A TimeStamp and a occupant Unit Number are added in each data block of logical volumes to mark {Time, Unit Number}. Unit number represents which storage unit is occupied, and Time indicates when it is occupied. Here we need to set up two tables separately, one is LMT (logic mapping table) and the other one is IRT (io recording table). LMT records where the data block is pointing at the current time, and LMT always has data of the current volume size. IRT records when the data block is occupied by anyone and the IRT table is the largest, which is the size of the backup set data (See Fig. 2).

This algorithm can quickly reconstruct the data of any time point according to the block index, and automatically recover the space occupied by the volume real-time

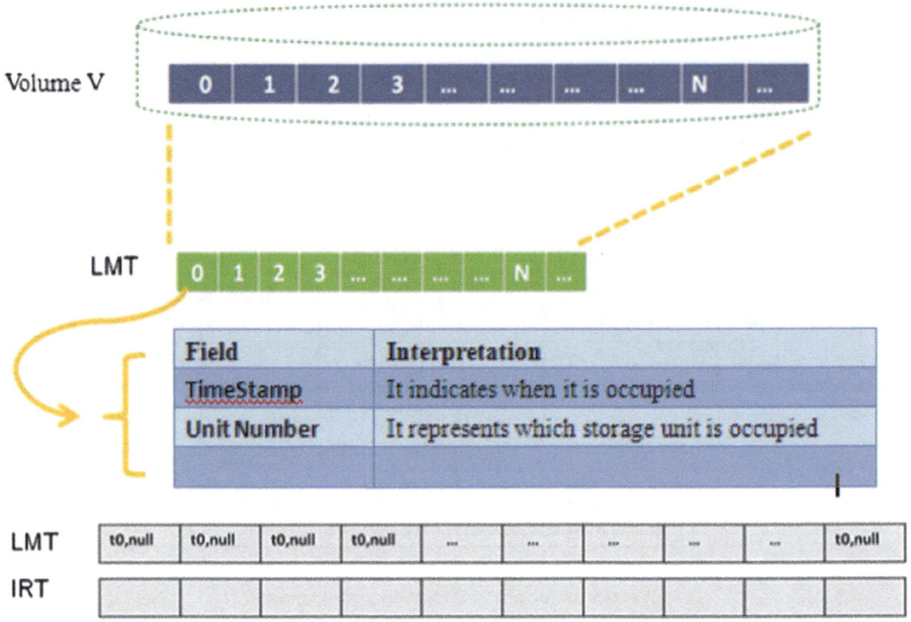

Fig. 2. The arbitrary time point round-trip CDP algorithm for multidimensional linked list

protection data according to the reserved time strategy of the set of volume real-time protection data.

The second is to optimize the duplicate data deleting module.

The method of duplicate data deletion module for judging duplicate data is: The SHA-1 digital signature algorithm is used to calculate the hash value of each data block, which is called the data block fingerprint, and the fingerprint is used as the only identification of the data blocks to index each data block. The fingerprint index table is then searched for the fingerprint of each data block. If the same fingerprint is found in the fingerprint index table, it is considered that the data block is repeated. If the same fingerprint is not found in the fingerprint index table, it is considered a new data block. On the X86 platform, when the amount of data is getting larger and larger, the CPU consumed by extracting fingerprints and fingerprint search is also getting larger and larger, resulting in slow operation of the system. The optimization on the RedPower platform using the CAPI + FPGA scheme can greatly improve the deduplication performance and support the deduplication of the larger amount of data (See Fig. 3).

The third is to study the use of Hadoop cluster storage as backup storage.

Using Hadoop as backup storage, Hadoop storage usually has 3 copies of data files, and 200% of storage space is wasted. The way to reduce waste is mainly to reduce the number of copies, however, when the number of copies is reduced to less than 3, the risk of data loss will be very high. Hadoop RAID uses Erasure Codes technology to reduce the storage redundancy consumption from 3 times to 1.4 times, however, on the X86 platform, the calculation of Erasure Codes will make CPU usage rate reach 99%,

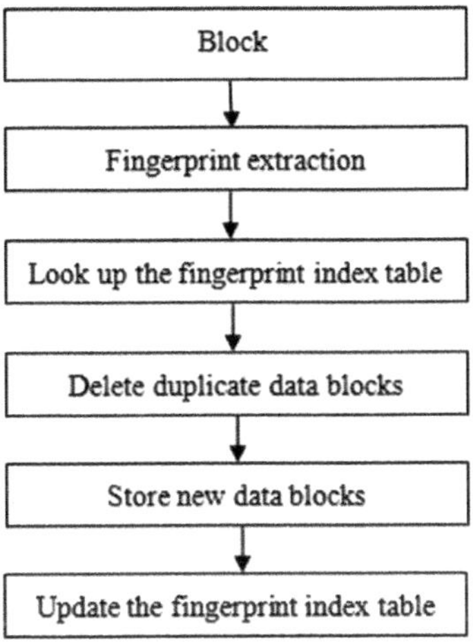

Fig. 3. Optimization of data deduplication module

and on the Power architecture platform, FPGA can be used to calculate Erasure Codes to release CPU utilization rate.

The fourth is the automatic generation of the KVM on Power virtual machine.

The virtual machine is automatically generated through the CPU, memory, network code mode, disk (Virtual disk volume raw device is mapped to virtual machine), and so on of the script customizable virtual machine for emergency operation and maintenance, and the virtual machine can start in 3 min using the latest time point version data. When the number of virtual machines is less than 6, the CPU occupancy rate of the host of the X86 architecture increases with the increase of the number of virtual machines. When the number of virtual machines is 5, it reaches 80%. When the host of the Power architecture adds a full load virtual machine, the occupancy of the host CPU only increases by 5%. If the comparison is carried out according to 80% load performance, the ability of Power to open the virtual machine is about 3–4 times that of X86.

3 The Corresponding Hardware System

The final product of the project is a hardware and software integrated equipment. The hardware configuration is recommended according to the different client-side size, and the server-side software is pre installed in the RedPower server when leaving factory. Users can work only by accessing the network cable or optical fiber line.

The specific product form is like Fig. 4 below:

The main content of the test:

The first point is to support the regular backup of files and databases under Windows, Linux and AIX systems. The backup data is stored in the Hadoop storage, the data deduplication at the target end starts to be used, and the storage space occupied by the backup version is reduced.

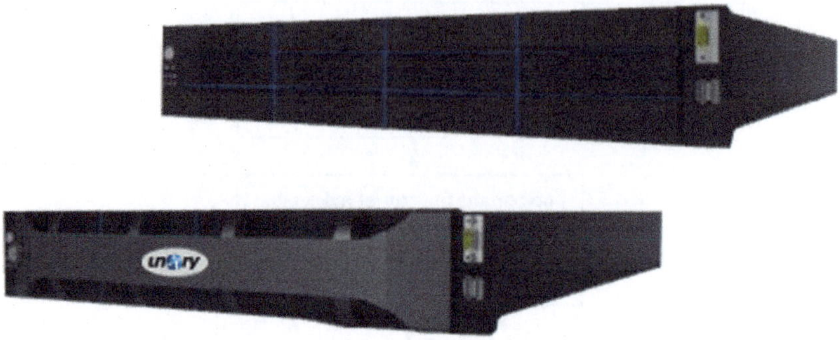

Fig. 4. Hardware system

The second point is the real time protection of the logical volumes of the operating system above Windows 2003 and Linux 2.6, and the support of the protection of the amount of data above 1 PB.

The third point is to support the backup version that selects any time point in seconds to generate virtual disk, and to mount the virtual disk through the ISCSI/FC protocol in 1 min to get the data of the corresponding time point.

The fourth point is the virtualization emergency operation and maintenance. Virtual servers can take over the e-government server business in 3 min using the version data of the latest time point.

In the process of demonstration, we verify the stability and performance of the product, and collect customer opinions for improvement.

4 The Technical Level Expected by the Project

The first is to support the online regular backup protection of files and mainstream databases under Windows, Linux and AIX systems.

The second is to support the real-time protection of logical volumes of the operating system above Windows 2003 and Linux 2.6, and to support the protection of the amount of data over 1 PB.

The third is to support the global data duplication features, save only one copy of the same data in all backup data, and minimize the use of backup storage space to the utmost extent, and the maximum deduplication data ratio is up to 95%.

The fourth is to support Hadoop cluster as backup storage, and use Erasure Codes technology to reduce the storage redundancy consumption from 3 times to 1.4 times.

The fifth is that one device of this project can support less than 25 disaster tolerant virtual machines, supports emergency operation and maintenance when the server fails, and can automatically take over the server business online within 3 min by the disaster recovery virtual machine to ensure the continuity of application. When the server is restored to normal, it can support the recovery of the new business data after the takeover of the disaster tolerance virtual machine to the related servers.

Acknowledgement. Fund project: Scientific research project funded by the Sichuan Provincial Education Department "Research on Data Safety System Based on Power Architecture" (18ZA0171).

References

1. Yang, D., Wang, J.: Path selection of "13th five-year" e-government transformation and development. E-Goverment, 20 April 2015
2. Anhui, Sichuan, and Henan launched new strategy for high growth industries. Inf. Deciders Mag., 01 December 2014
3. Su, Y., Lu, J.: A CDP backup scheme based on volume level. J. Jiangsu Univ. Sci. Technol. (Natural Science Edition) 2016-05-23 **1**
4. Li, M.: What to see after the internet conference. Innov. Time 2014-12-20
5. Jia, X., Gu, Q.: Research on mobile internet security. Mob. Commun. 2011-05-30
6. Xiao, Z.: A review of mobile internet research. Telecommun. Sci. 2009-10-15
7. Peng, L., Fang, W.: Heterogeneity of inferring reputation of cooperative behaviors for the Prisoners' Dilemma game. Phys. A **433**, 367–378 (2015)
8. Aleksy, M., Korthaus, A., Schader, M.: Use Java and the CORBA realization distribute type system. J. Pingxiang Coll. **4**, 104–105 (2005)
9. Titus, J.: ECN technical editor: the eclipse of stand. J. Zhongkai Agrotech. Coll. **19**(2), 32–35 (2006)
10. Richardson, W.C., Avondolio, D.: The Java high class weaves a distance: JDK 5. Sci. Technol. Book Rev. (3), 17–18 (2006)
11. Li, J., Vuong, S.: A semantics-based routing scheme for grid resource discovery. In: E-Science: First International Conference on E-Science and Grid Computing, pp. 58–70, 90 (2005)

A Solution for Quality Monitoring System of Indoor Distribution System

Wenjun Su(✉)

Department of Electronic Information, Guangzhou Civil Aviation College, Guangzhou 510000, China
suwenjun@139.com

Abstract. There are many problems in the indoor distribution system, which cannot be directly monitored. There are a lot of problems that cannot be found in time in the current network because of the lack of effective problem detection and problem location tool. The system uses the passive wireless answering system technology to detect link status quickly, which can locate and exclude problems, improve coverage and user perception. The system can monitor the quality state of the indoor distribution system in real time and make the network operation and maintenance active.

Keywords: Indoor distribution system · Fault diagnosis · Monitoring Manageable control

1 Introduction

The passive indoor distribution system has the following problems and pain points:

(1) All passive indoor distribution systems cannot be directly monitored. A large number of failures cannot be found in time.
(2) The structure of the indoor distribution system is complex and the fault is stealth which is difficult to judge directly. The workload of fault detection and location is large and the efficiency is low because of lack of effective tools and Many restrictive factors.
(3) Indoor distribution system has many devices and is distributed in different areas, so it is difficult to inspect the whole system. The maintenance work can only take into account the smaller part, and the inspection efficiency is low.
(4) The maintenance of the system is in the passive triggering way. After the user's complaint feedback, it starts the troubleshooting in the corresponding indoor area, which is not conducive to the improvement of the user's perception and the promotion of the brand image.

In order to solve these problems, the quality monitoring system of indoor distribution system is developed.

J. Abawajy et al. (Eds.): ATCI 2018, AISC 842, pp. 1018–1022, 2019.
https://doi.org/10.1007/978-3-319-98776-7_122

2 Working Principle

The diagnostic system is mainly composed of an external probe or probe antenna, a combiner, a monitoring host, a network management platform and a APP. The schematic diagram of the system deployment structure is shown in Fig. 1.

The functions of each part are described as follows. (1) Monitoring host: According to the instructions, the detection signal is transmitted, and then the probe signals and information at the end of the indoor distribution system are detected, collected and transmitted back. (2) External probe: Receive and monitor the monitoring signal of the host and answer it and paste it in the existing antenna housing. (3) Probe antenna: Integrating probes and antennas to receive and monitor the host's monitoring signals and respond. (4) Combiner: Access the monitoring host to the indoor distribution system.

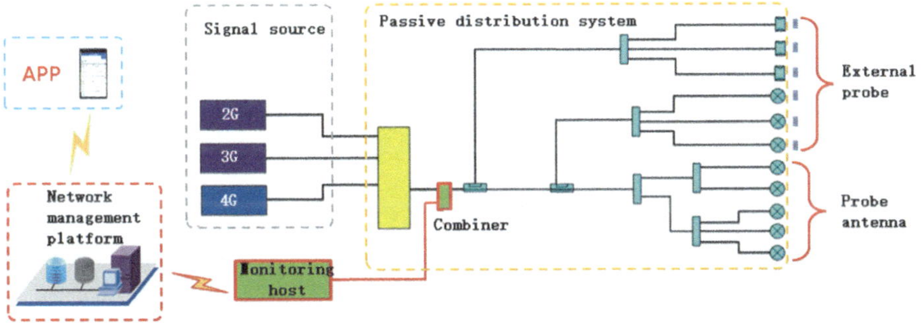

Fig. 1. Schematic diagram of the quality monitoring system

A monitoring probe is installed on the antenna cover of indoor distribution system, and a new combiner is added to the host of the original system to access the monitoring host. The external probe at the antenna receives the signal from the monitoring host and responds. The monitoring host detects the probe signal and collects the information of the feedback to the network management platform.

In the case of fault diagnosis, the monitoring host sends out the detection signal which enters the RF distribution system through the combiner and transfers to the antenna at the end of the link. The information is analyzed by the network management platform, and the link state of the radio frequency indoor distribution system is obtained which is pushed to the client to display. And the fault diagnosis of the chamber system is realized. In this way, the fault diagnosis of radio frequency distribution system is realized.

The functions of the network management platform include network element statistics, alarm statistics, site management, monitoring unit management, link state detection, alarm management, polling management, and room data query.

3 Operation Effects

(1) The link status is detected quickly, and the problems can be located and eliminated.

Using the quality monitoring system, difficulties of problems location in distributed system can be solved. After the deployment of the device, the detection of the distributed system takes about 3 min a time. After the test is finished, the problems can be displayed directly in the topology map of APP. It is convenient, simple and efficient that network signal is not interrupted and service is not affected. The quality monitoring system is used for statistical analysis of the distribution system problems which is shown in Fig. 2.

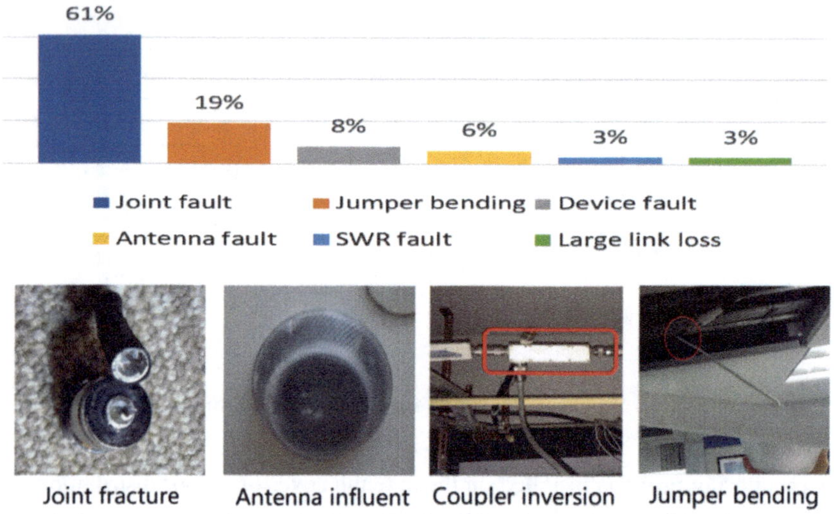

Fig. 2. Fault diagnosis

(2) The coverage rate and the user perception are improved.

After troubleshooting, the signal coverage rate increased by 10–60%, and user perception improved significantly. The case of a typical sites is like Table 1.

Table 1. Comparison of coverage before and after system deployment

Site and floor	Proportion greater than −70 dBm		
	Before	After	Improvement
27th floor of Site A	41.40%	54.20%	12.80%
7th floor of Site B	8.69%	68.15%	59.46%
6th floor of Site C	53.44%	81.72%	28.28%
5th floor of Site D	6.78%	32.61%	25.83%

(3) Traffic is promoted and Income is increased.

The indoor distribution system equipped with the quality monitoring system can improve coverage by fast locating and troubleshooting, and the traffic can be increased by 20–30%. Taking the A site as an example, the monthly traffic is increased by about 150 GB. According to the 50 yuan per GB traffic, the monthly income can be increased by 7500 yuan (Fig. 3).

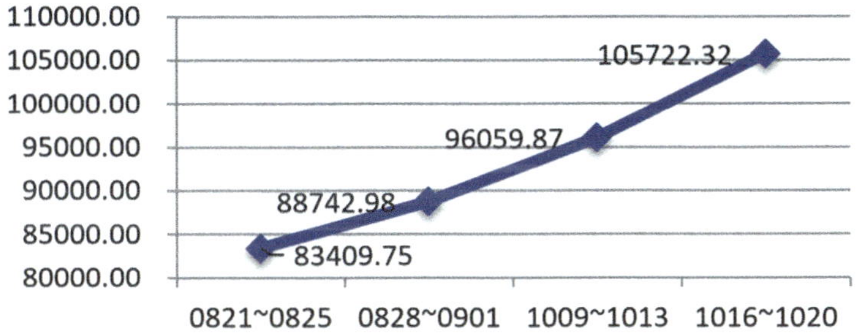

Fig. 3. Before and after the installation of the system (Unit: MB)

4 Application Prospects

(1) The efficiency of indoor distribution system maintenance and inspection is improved.

Usually, the maintenance and inspection of the traditional indoor distribution system mainly aims at the active equipment. It uses indoor CQT to inspect the link state indirectly during inspection, and it takes a long time to check the link state of the system.

The quality monitoring system can reduce the workload of maintenance and inspection of the indoor distribution system, provide work efficiency, increase the number of single patrol stations, shorten the time of single site inspection, and reduce the workload of manual inspection without reducing the maintenance requirements.

(2) The system helps to change the maintenance of indoor distribution system from passive to active mode.

Usually, the maintenance of the indoor distribution system is in a passive way. After the user complaints, the maintenance department start the corresponding troubleshooting and troubleshooting process, which is not conducive to the timely detection and processing of the fault.

Using the quality monitoring system, we can detect all links in the indoor distribution system according to the needs and find out the faults in the system link in time. It is beneficial to eliminate the faults in the indoor distribution system in time, guarantee the network coverage, improve the user perception and enhance the brand image.

(3) Network assets are cleared up.

Through the application of the quality monitoring system, the topology of the indoor distribution system is promoted by the topology map, which can assist the verification and correction of the assets of the indoor distribution system, and help the indoor distribution system to be regulated.

(4) Effective auxiliary means are provided in the acceptance of works.

The application of this system in the new indoor distribution system can easily and quickly discover the quality problems in the construction process, and provide a good auxiliary means for the acceptance of the project. It can improve the quality of the intersection, reduce the difficulty of operation and maintenance, realize fast completion, quick opening, quick acceptance and rapid intersection.

5 Conclusion

By installing the quality monitoring system, the network coverage is perfected, the business performance is fully played, the business income is increased. And the maintenance and inspection efficiency of the room system is improved while the maintenance and inspection of the indoor distribution system is reduced. The system helps maintenance of the indoor distributed system change from passive to active, and actively and timely detect and eliminate the problems in the link of the indoor distributed system. It can help clarify network assets and provide effective means for engineering acceptance.

References

1. Wenjun, S.: Design method of LTE indoor distributed antenna system. Telecommun. Netw. Technol. **03**, 69–72 (2017)
2. Dongyang, X.: Indoor distribution system fault location scheme and case analysis. Telecommun. Technol. **11**, 85–88 (2013)
3. Lei, W.: Monitoring and management analysis of indoor distribution system. Telecommun. Technol. **03**, 41–42 (2010)
4. Peng, L., Fang, W.: Heterogeneity of inferring reputation of cooperative behaviors for the Prisoners' Dilemma game. Physica A **433**, 367–378 (2015)
5. Ming, F.: The cultural elements in English translation teaching. Engl. Sq. **20** (2014)
6. Honglian, W.: The problems and countermeasures of English translation teaching in college of traditional Chinese medicine. Overseas Engl. **20** (2014)
7. Watts, D.J.: Small worlds. The dynamics of networks between order and randomness. America: 56A2579BB6, l999-05-13
8. Aleksy, M., Korthaus, A., Schader, M.: Use Java and the CORBA realization distribute type system. J. Pingxiang Coll. **4**, 104–105 (2005)
9. Titus, J.: ECN technical editor: the eclipse of stand. J. Zhongkai Agrotech. Coll. **19**(2), 32–35 (2006)
10. Richardson, W.C., Avondolio, D.: The Java high class weaves a distance: JDK 5. Sci. Technol. Book Rev. **3**, 17–18 (2006)
11. Li, J., Vuong, S.: A semantics-based routing scheme for grid resource discovery. In: E-Science: First International Conference on E-Science and Grid Computing, pp. 58–70, 90 (2005)

The Development Dilemma and Countermeasures of China's Personal Credit Industry in the Era of Large Data

Ying Hu and Zhenxing Ge(✉)

College of Technology and Art Jingdezhen Ceramic Institute, Jingdezhen, Jiangxi Province, China
huhujdz@163.com, 181795184@qq.com

Abstract. The wide application of large data technology has brought unprecedented opportunities for development in all walks of life, and also provided a new processing mode for the booming personal credit industry in China. The development of large data personal credit will improve the financial ecological environment, which is conducive to the establishment of an honest and trustworthy market economic environment, and has an irreplaceable role in promoting the healthy development of Internet finance. Based on literature analysis, investigation and summary and other methods, this paper holds that as a emerging thing, there are still many difficulties in the development of big data personal credit, and there are still a long way to go. Strengthening the independent innovation in the field of personal credit, speeding up the legislative process, making industry standards, improving the consciousness of the whole people, and attaching importance to the training of professional talents will help to create a good market environment for the large data personal credit collection, and the development path of the personal credit industry can be more smoothly in the era of large data.

Keywords: Large data · Personal credit · Dilemma · Countermeasures

1 The Current Situation and Characteristics of China's Personal Credit Industry in the Era of Large Data

1.1 Development Status

In 2004, the central bank of China launched the personal credit information system. In January 2006, the personal credit information system was officially completed and operated nationwide. In 2015, the central bank issued the "Notice on Preparations for Personal Credit Reporting Business", which requires eight private credit agencies, such as Zhima Credit and Qianhai Credit, to prepare for personal credit business. It marks the gradual formation of a new type of credit collection system in China, which is dominated by the central bank and supplemented by the market. These eight institutions have highlighted the characteristics of "large data" and "internet credit" in the specific planning of personal credit business, which aims to carry out personal data collection. The gate of the personal credit industry of large data opens.

J. Abawajy et al. (Eds.): ATCI 2018, AISC 842, pp. 1023–1030, 2019.
https://doi.org/10.1007/978-3-319-98776-7_123

In February 22, 2018, central bank of China announced that Baihang Credit Co., Ltd. was awarded the first personal credit business license. It is a market oriented personal credit institution, which is led by the China Internet Financial Association and co funded by the 8 market institutions mentioned above. The emergence of Baihang Credit integrates the information collection of 8 shareholders and more Internet institutions, covers the blind spot of traditional financial credit, and forms a huge database of credit information. It will solve the problems of information segmentation in the field of internet finance to a great extent.

1.2 Development Characteristics

1.2.1 A Wide Range of Coverage

Represented by the central bank credit center, the traditional personal credit information mainly comes from the credit datas of commercial banks and the records of public service system. Because the threshold of bank loan or credit card is too high, the number of people covered by it is relatively small. Data shows that nearly 900 million people have their credit information in the central bank's personal credit center, but 500 million of them lack credit records. However, the coverage of large data collection is more extensive. Based on the internet behavior data records, web browsing records, online shopping evaluation, etc., the application of large data technology can take them as a factor of credit evaluation, more people will be recorded by the large data personal credit.

1.2.2 Data Sources Are Diverse and More Timeliness

With the development and progress of social communication, electronic commerce and third-party payment, more and more records are produced on the Internet, and all the characteristics and risk judgments of consumers can become credit data. Figure 1 shows the composition of the traditional personal credit data represented by the central bank's credit center. It is clear that these personal credit data sources are single and incomplete. The personal credit information of large data includes not only the historical data of traditional credit, but also the data of user's trading behavior and interpersonal relationship. These data come from the network, is massive and

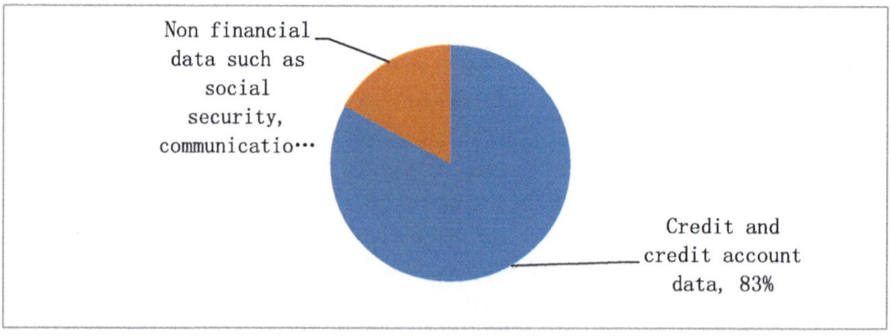

Fig. 1. The data composition of the central bank's personal credit center

comprehensive, and can be updated and processed in time that shows more effectively and conveniently.

1.2.3 Credit Evaluation Is More Comprehensive and Application Scene Is Richer

The personal records produced from the internet are more trivial and detailed, and to some extent they reflect the true credit level of individuals. Compared with traditional credit, the personal credit of large data evaluates objectively in many dimensions, such as credit history, behavior preference, performance ability and so on. Any aspect will affect the result of the evaluation.

The establishment of personal credit, in addition to the use of traditional financial activities, will also be integrated into our life aspects, in more scenes, and play a more important role. For example, reservations for hotels, renting houses and renting a car usually need to pay a deposit to prevent a lack of credit, but in the form of personal credit of large data, the deposit will be cancelled once personal credit is established. Online direct processing of visas, waiting for temporary access to fast track, these are also life convenience which large data personal credit can bring us.

1.2.4 The Link of Personal Credit Is Obvious Closeness

Different from the three segment development of the United States, the individual credit system, which is to be set up by China, is a combination of information collection, data analysis, credit evaluation and credit business in an attempt to form a complete ecological closed loop. Personal credit agencies are intended to create their own competitive advantages in each link, complete the credit evaluation work within

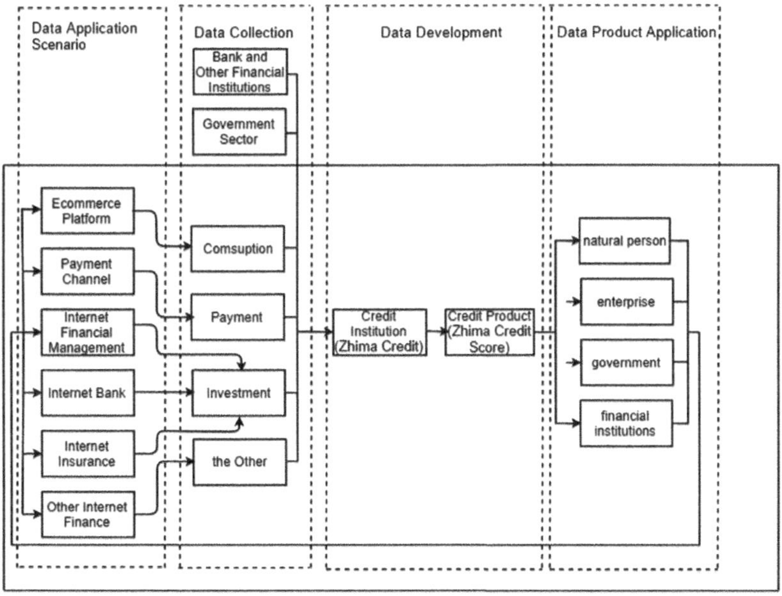

Fig. 2. Model of personal data collection for large data

the system. The credit collection is no longer an independent business, it is integrated into the specific application scene. Figure 2 illustrates the closeness of this credit system by taking the Alibaba ecosystem as an example. The left side includes e-commerce, internet finance and other fields that provide personal credit data. In the middle, the credit agency Zhima Credit carries out data development to get Zhima credit scores. On the right, it is the user of Zhima credit score. But the user is locked to the left application scene, thus there is a closed application ecology.

2 There Are Many Difficulties in the Development of China's Personal Credit Industry in the Large Data Era

2.1 Personal Privacy Risks

"Regulations on the management of credit information" of China issued in 2013 states that "collection of credit information should be consent by oneself, and not allowed to collect without one's permission." However, in the era of large data, information technology is changing the way of collecting and using personal information. It makes "all data credit" possible, personal information is massively collected. Large data enterprises ignore the foregoing terms, unauthorized collection and use of data are abound, personal privacy has been violated. On the other hand, hacker attacks and network viruses may put personal privacy in danger, and the diffusivity and destructiveness of network risks affect the public's trust in information security.

2.2 Non Compliance with the Principle of Independent Third Party

The independent third party means that it neither directly compete with the providers and users of the credit information, nor participate in the competition of the personal credit market. Personal credit institutions of large data are usually set up by internet companies. These institutions not only mine data, but also evaluate and use data. This is equivalent to being both a referee and a contestant which does not conform to the principle of an independent third party. The data of large data credit agencies are mainly applied to their own business systems, and the validity of such personal credit evaluation reports still needs to be discussed. When data information is shared to other institutions, fairness and objectivity must be discounted based on self interest and market position.

2.3 The Effectiveness of the Credit Evaluation Model Needs to Be Verified

At present, the personal credit institutions in the market are basically based on large data collection and credit evaluation models, which collect personal behavior data, such as social communication and e-commerce consumption data. However, these institutions can not dock with the central bank credit center, the credit records of financial institutions are not included in the statistics category of large data collection. Therefore, the authority of large data personal credit in China is not enough and the quality is difficult to control. Moreover, the credit evaluation models are developed by various companies themselves. Different institution with different models and data sources, the result of credit evaluation for the same person will be also different. The accuracy and effectiveness of these credit evaluation models are yet to be verified. To develop different models for different scenes, the assessment model will be more complex and difficult to explain.

2.4 Information Sharing Mechanism Is not Sound

Each personal credit rating institution in the market wants to form its own business closed loop and take advantage of the market because of its own interests. These measures objectively divide the information chain of the market, bringing about the negative effects of partial information and lack of effective products. Just as Ali's e-commerce data is impossible to be provided to the Tencent, Tencent's social data will not share with Zhima Credit, the phenomenon of "information island" is becoming more and more serious. So that the credit institutions face the problems of obtaining data difficult and high cost. In addition, private credit agencies are difficult to obtain the financial data from central bank or public data from government, and there is data barrier. It is one of the problems to be solved urgently in the development of big data personal credit to smooth the channels of sharing credit information and acquire more diversified data dimensions.

2.5 Laws and Regulations Are not Perfect, and the Level of Supervision Needs to Be Improved

At present, in China, laws and regulations to standardize personal credit reporting mainly include "Regulations on the Administration of Credit Industry" and "Administrative Measures of Credit Institutions". As a new way of credit reporting, large data personal credit is a useful supplement to the personal credit system, and it should also be subject to legal constraints. However, there are many new problems, such as the range of data collection, the operation specification of the business, the way of credit evaluation, these have changed a lot. To a certain extent, it is beyond the scope of supervision of the current legal system, and it is imperative to improve the supervision mechanism.

3 The Countermeasures for the Development of Large Data Personal Credit

3.1 Speed Up the Legislative Process and Improve the Regulatory Mechanism

First of all, the government should formulate the law of personal information protection. Although China's Internet credit agencies claim that they will not disclose personal credit information, there are still some loopholes in the actual operation. Therefore, according to the characteristics of large data personal credit, it is necessary to clear the scope and standard of the collection of credit data in the laws and regulations, improve the use of credit information and standardize access channels of personal credit information. To achieve the balance between personal information protection and sharing, the personal credit of large data can be stable and sustainable development.

Secondly, we should strengthen the management of personal credit institutions. The institutions must be classified, and the employees must have clear qualifications. There must be an irregular rush inspection to security system of the persona institutions. The act of disclosing and selling information must be prohibited. The third party organizations which are not good enough to protect credit data must be rectified and punished. We should develop a regulatory mechanism suitable for personal credit reporting of large data.

3.2 Encouraging Technology Innovation of Large Data Personal Credit, Which Is Mainly Market Driven

Mastering large data technology is the core of developing large data personal credit. Large data personal credit requires a variety of technical synergies in data collection, induction and processing. Cloud computing makes the processing and analysis of massive data more simple and convenient. It effectively reduces the time of credit data processing. Machine learning will further optimize the performance of credit model and improve the efficiency of personal credit evaluation. The requirement for technology of large data personal credit is strict. When the personal credit agencies have technical innovation, learn advanced foreign science and technology, construct a reasonable credit evaluation index system and model (Table 1), develop personal credit mode and technical means that is suitable for the national conditions of our country, and research the products suitable for market, the personal credit system in China will be more reasonable and perfect.

Table 1. Evaluation index system for large data personal credit

Index system composition	Character	Identity information	Occupation
		Educational level	Information stability
	Ability	Level of consumption	Consumption preference
		Consumption stability	Credit card amount
	Capital	Vehicle information	Real estate information
		Investment and financial information	
	Behavior	Network search record	Love donation
		Performance of the contract	Blacklist of law enforcement
		Blacklist of Internet consumption	
	Social Contact	Social breadth	Social evaluation
		Social activity	Effect
		Social blacklist	

3.3 Take the Government as the Leading Factor, Establish and Improve the Credit Information Sharing Mechanism

First, the government should play the leading role in the development of personal credit system, actively discuss with the leading credit institutions to formulate scientific and reasonable standards for large data personal credit industry. It will improve the processing capacity of data. The second is to connect the credit data of the market credit agencies which reach the industry standards to the central bank's credit system. Information barrier is broken and a smooth channel for the sharing of credit information will realize. The last is to build a unified information platform for large data personal information, to encourage individual credit agencies to break the phenomenon of information island, and to form a complementary situation with the personal credit system of the central bank. Building information sharing mechanism to jointly promote the development of large data personal credit, so that credit data will circulate in different institutions or fields.

3.4 Strengthen Personal Credit Information Publicity, and Promote People's Awareness of Credit

At present, individuals in China have weak awareness of credit protection. They are lack of relevant theoretical knowledge and policies which always makes them in a weak position. In order to establish a healthy social credit atmosphere and standardize the social credit order, it is necessary to increase the publicity and popularization of personal credit knowledge, and to improve the awareness of the individual's recognition of personal credit. Only by establishing the social overall environment of honesty and trustworthiness, promoting the self-protection consciousness of the credit subject, strengthening the credit consciousness and establishing the mechanism of "honesty encourage and credit punishing", can we promote the formation of social credit culture and establish a sound social credit system.

3.5 Pay Attention to the Training of Talents and Provide Intellectual Support

The modeling of large data credit is a complex task. It requires the technicians with high professional level that includes a strong understanding of personal credit business and powerful mathematical modeling, data mining skills. Talent is the core of the development of science and technology. At present, the large data acquisition personnel in China are relatively scarce. The government should create a good training environment, establish a good training platform and improve the sound training mechanism. Chinese institutions of higher learning can be encouraged to create some major related to large data personal credit. Cultivating management and technical personnel, supporting the research of related issues such as large data personal credit, expanding the number of experts and talents, and encouraging enterprises to carry out training for large data credit personnel, the number of large data talents and advanced technicians will both increase. It will provide intellectual support for the development of large data personal credit.

References

1. Youyuan, L., Gang, K.: Challenges and countermeasures of large data credit reporting in China. Large Data **01**, 27–34 (2017)
2. Yong, W., Pusheng, W.: Privacy risks and coping strategies of large data credit reporting. Study Dialect. Nat. **07**, 118–122 (2016)
3. Guanlin, C.: Personal credit industry in the era of large data. China Finance **07**, 90–91 (2015)
4. Yixia, C., Shuwen, G.: Large data promote the booming development of personal credit market. Shanghai Informatiz **03**, 16–19 (2016)
5. Miller, M.: Credit reporting systems around the globe: the state of the art in public and private credit registries. World Bank, Washington (2000)
6. Peng, L., Fang, W.: Heterogeneity of inferring reputation of cooperative behaviors for the Prisoners' Dilemma game. Physica A **433**, 367–378 (2015)
7. Ming, F.: The cultural elements in English translation teaching. Engl. Sq. **20** (2014)
8. Honglian, W.: The problems and countermeasures of English translation teaching in college of traditional Chinese medicine. Overseas Engl. **20** (2014)
9. Watts, D.J.: Small worlds. The dynamics of networks between order and randomness. America: 56A2579BB6,l999-05-13
10. Aleksy, M., Korthaus, A., Schader, M.: Use Java and the CORBA realization distribute type system. J. Pingxiang Coll. **4**, 104–105 (2005)
11. Titus, J.: ECN technical editor: the eclipse of stand. J. Zhongkai Agrotech. Coll. **19**(2), 32–35 (2006)
12. Richardson, W.C., Avondolio, D.: The Java high class weaves a distance: JDK 5. Sci. Technol. Book Rev. **3**, 17–18 (2006)
13. Li, J., Vuong, S.: A semantics-based routing scheme for grid resource discovery. In: E-Science: First International Conference on E-Science and Grid Computing, pp. 58–70, 90 (2005)

An Improved Apriori Algorithm Based on Data Mining

Chen Luan(✉), Hong Li, and Ying Qin

Heilongjiang Institute of Architectural Technology, Harbin 150080, China
chenluancl2017@sina.com

Abstract. For multi-core architectures, making full use of GPU's high-floating-point computing resources to achieve load balancing among multiple cores. This paper presents the way to use an efficient GPU graphics processing unit to speed up frequent pattern mining algorithms (GPFPM). This paper uses a parallel GPU process allocation mechanism to determine the number of GPU threads required based on the number of sorted items. Using GPU to calculate the threshold can reduce the time spent in support calculations, decrease the times of checks and comparisons, improve the candidate set confirmation time, and return the results to the CPU for the next-order operation. The experimental results show that in the multi-core platform and the hybrid platform of multi-core and multi-GPUs, the algorithm performs well, which excavates frequent set association rules with high efficiency and accuracy, satisfies the demand for efficient data mining of multi-core and heterogeneous platforms, and validates the effectiveness and feasibility of data mining algorithms based on multi-core and multi-graphics processing units.

Keywords: Data mining · GPU · Apriori algorithm

1 Introduction

There are many methods proposed for the association mining of frequent item set patterns, and they can be divided into two categories: the generation and testing of candidate items and the integration of frequent items. The candidate project generation and testing is represented by the Apriori algorithm; and the frequent item integration method is represented by the FP-tree algorithm. FP-growth uses a tree structure to store data in an FP-tree, which eliminates the need to generate candidate item sets to reduce the frequency of repeated access to the database, so as to accelerate its computational performance and achieve better performance in a single core operation [1]. However, when transplanting to a multi-core platform, there are problems such as cutting of the tree structure and the difficulty of load balancing, which is not conducive to parallel operations. Meanwhile, the Apriori algorithm uses a step-by-step approach to generate a set of candidate items for each level. The method is easy to make parallel data segmentation, and because it needs to generate and check whether the candidate project cluster is a high-frequency project, it uses the data matrix as the main storage method, which is convenient to facilitate the cutting and parallel operation. When the number of computing cores is increasing, the advantages of the Apriori algorithm will become more pronounced [2, 3].

J. Abawajy et al. (Eds.): ATCI 2018, AISC 842, pp. 1031–1037, 2019.
https://doi.org/10.1007/978-3-319-98776-7_124

In recent years, the development of GPU is getting faster and faster, and it can replace some of the functions of CPU. Therefore, the research and improvement of data mining algorithms using GPU has gradually become mainstream. Traditional Apriori algorithms are based on CPU architecture for improvement and research. As pattern mining has been widely used, it is necessary to use faster algorithms to speed up the calculation of data due to the amount of data in the database is increasing. In the previous chapters, two types of Bitmap and GPU-FPM were mentioned. Bitmap requires a large amount of memory space for storage, and since the AND operation is stored using Bit, it is difficult to calculate the number 1. The GPU-FPM has different computational complexity because of different candidate projects so the load balance is poor [4]. As the GPU operation needs to wait for all the calculations to be completed before it is returned to the main memory of the CPU, a lot of waiting time will be added and the operation time will be too long when the load balancing is not good [5].

In general, with the increasing amount of data, the speed of data mining gradually slows down. Since data mining is so important to us in the era of big data, an efficient mining algorithm has become an urgent need to achieve high efficiency by improving the shortcomings of the Apriori algorithm and using the parallel processing features of the Multi-core and GPU hybrid architecture so that algorithm can still quickly excavate the association rules between the large number of computing tasks. Therefore, it is significant to develop parallel processing Apriori algorithm based on multi-core and multi-graphics processing units.

2 An Efficient GPU Graphics Processing Unit to Speed Up Frequent Pattern Mining Algorithm

Apriori algorithm can be mainly divided into two parts; one is to confirm whether the candidate item meets the threshold value and the other is the high frequency item is merged into the candidate items of the next order. At the beginning of the operation, it is necessary to convert Apriori's TID format and delete items that are not belonging to high frequency items and then convert Table 1 of Tid to the Tid Value and Tid Index used by GPU-FPM, as shown in Fig. 1, and pass it to the GPU.

Table 1. Database conversion into TID form storage

Item	TID value		
A	T1	T2	
B	T1	T2	T3
C	T1	T4	

GPFPM will cut candidate items into smaller units and run parallel to the GPU's core. A GPU Thread will only be responsible for the Tid Value's check of a candidate item. GPU only needs to check if the number is equal. When the number is more, then

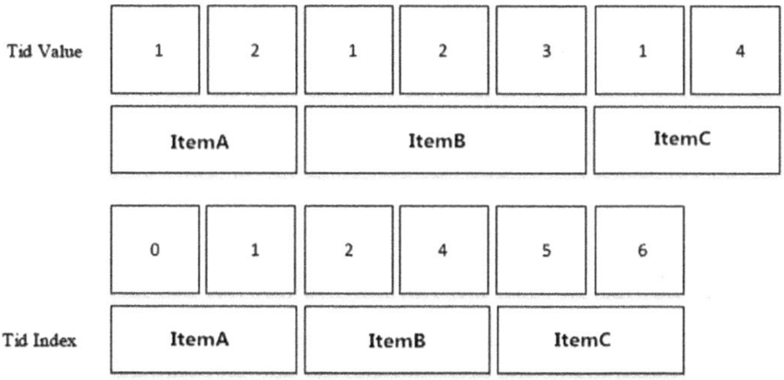

Fig. 1. Tid Value and Tid Index form

it jumps out of the Thread operation time, and returns 0; when the equal situation occurs, it returns 1 as shown in Fig. 2. The second-order candidate items set {AB}, {AC}, {BC}, as shown in Table 2, Tid Values as shown in Fig. 1. Therefore, after passing information to GPU, each thread of the GPU only needs to check the corresponding item Tid Value and store the result in the Output matrix. Each GPU Thread has a corresponding Output matrix to store the result, for example: GPU Thread ID 0, TID Value 1 assigned to item A, and the completed TID Value of item B, GPU Thread ID1, TID Value 2 assigned to item A, and the completed TID Value of item B, GPU Thread ID 2, the TID Value 1 assigned to item A, and the completed TID value of item C. Each GPU Thread checks its own data for the same number, then, outputting 1 if there is, 0 if not. For example, if items A and B both have TID Values of 1 and 2, then both GPU Thread ID 0 and GPU Thread ID 1 will output 1 while GPU Thread ID 3 will output 0 because Item A and Item C don't have a common transaction list 2. The matrix using this output is sent back to the CPU for statistics.

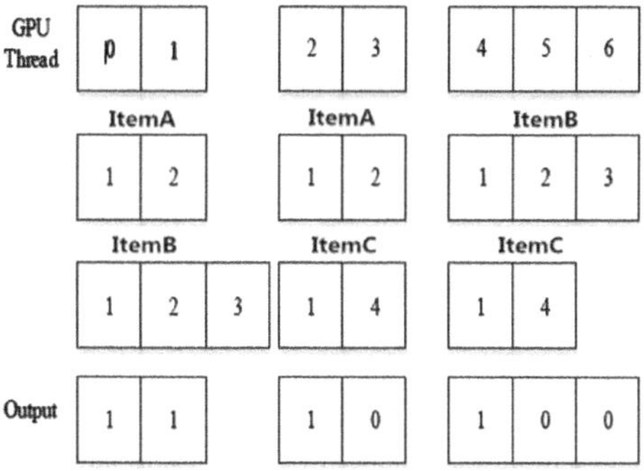

Fig. 2. Schematic diagram of calculation method with candidate projects by GPU

When an operation is completed, for example, the matrix with the operation result of 1110100 in Fig. 2 is returned, since the Tid Value of the item A is 2, the Tid Value of the item B is 3, so the segment of the Output matrix is 22 3, and the 0–1 matrix is the

Table 2. Calculation of candidate item number

Item	Number of times
AB	?
AC	?
BC	?

candidate item {AB}, the 2–3 candidate item {AC}, the 4–6 candidate item {BC}. Only the quantity of 1 is required to obtain the same number of orders for the candidate item, as shown in Table 3.

Table 3. Calculation of candidate item number

Item	Number of times
AB	2
AC	1
BC	1

3 Multi-GPU Utilizes GPFPM Computing Architecture

A single GPU can already achieve better computing speeds. If more display chips calculate at the same time, it is able to achieve better results. As the GPFPM method operates internally on the GPU, each GPU Thread does not need to communicate with others, so adding more GPUs can also work properly. However, it is not suitable to use the CPU's Thread mode to switch, as shown in Fig. 3. Because the Thread switch takes time and it needs to wait for the operation to switch over a period of time, but the time is not easy to control, for example, it cannot be switched during the data transfer process to avoid errors, then using a single CPU to control multiple Thread switching is difficult, and it can't play the best computing performance.

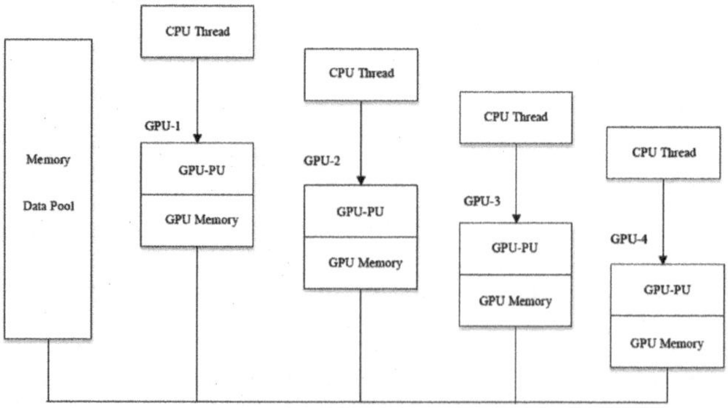

Fig. 3. Using the CPU Thread control Multi-GPU computing hardware architecture

Using OpenMP in related research can make use of multi-core operations, and utilizing different cores to control the corresponding GPUs, then, transferring data and collecting operation results in a one-to-one manner, as shown in Fig. 4, under this circumstance, it can not only reduce the waste of computing resources, but achieve better computing performance. Due to the multi-GPU operation mode, the original GPFPM algorithm also needs to be modified, as shown in Fig. 4. Since the GPU environment cannot be established at the same time with others, the GPU environment needs to be completed step by step, including the compilation process of Kernel code. The parts modified by OpenMP are: Due to the large number of candidate projects and the need for more GPU-assisted operations, the Threads of each GPU are gradually filled, GPU operations are started and then completed, then, the GPU operation results are reclaimed via each CPU Core, and calculate the number of occurrences of each candidate.

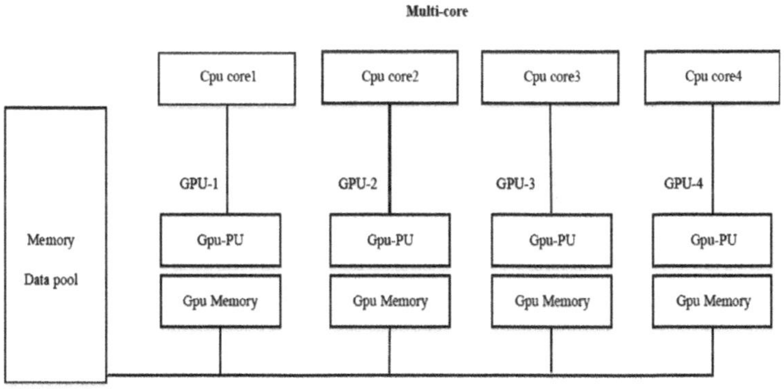

Fig. 4. The Multi-Core CPU control of Multi-GPU operations on Multi-core hardware architecture

Multi-GPU GPFPM is different from GPFPM in that it incorporates OpenMP allocation. Since only using OpenCL cannot control multiple GPUs, OpenMP is used to create a CPU core to control a GPU. For this purpose, only part of the algorithm flow, such as OpenMP controls different core CPU allocations to the GPU and retrieves the results, needs to be modified, while the working part of the GPU does not need to be modified just as GPFPM, which is the Multi-GPU GPFPM algorithm. As for Bit-map, since each GPU Thread is just the same amount of computation and is calculated with the AND method, each order data needs to correspond to each other when using the AND operation. If there is missing situation or the order does not contain this item, it needs to be stored as 0 so that to achieve the order of goods. Normal data are mostly sparse trading orders, so they waste a lot of memory to store 0, while the GPU-FPM only needs to retain high-frequency items, and it can save memory. However, because the length of each candidate is different, the results are in poor load balance.

4 Experiments and Results Analysis

In order to ensure the effectiveness of this study (GPFPM) and the proposed GPU-FPM algorithm, the same hardware and software platform is used as a platform for computational evaluation and will be compared with the previous algorithm. The database used for comparison is shown in Table 4.

Table 4. The comparison of database

Database name	T	D	N	Type
T10I4D100KN100K	10	100000	100000	Synthesis
T10I4D1000KN1000K	10	1000000	1000000	Synthesis
Retail	10.3	88154	16422	Truth/sparsity
Chess	35	3176	76	Truth/sparsity

In this experiment, the same hardware and software architecture as the paper is used. The input data was T10I4D100K, and the threshold value is 200 (0.2%), as shown in Fig. 5. GPFPM effectively improved the execution speed of GPU-FPM, and it obtain speed-up ratio of 2.6 times. As the GPU's core speed is relatively slow, it is less suitable for high computational calculations.

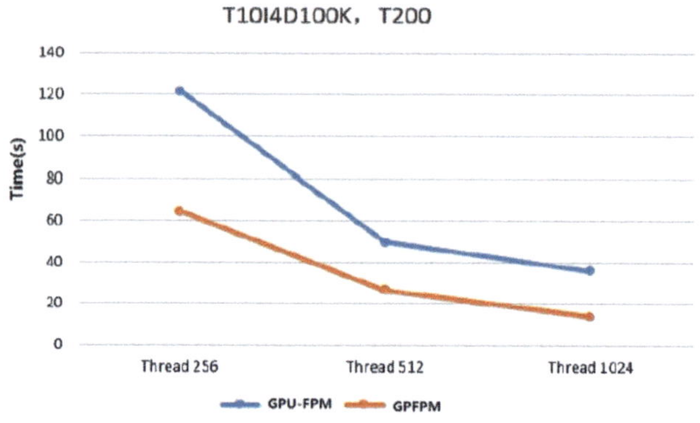

Fig. 5. The operations results of T10I4D100K

5 Conclusion

This paper proposes combination advantages of Bitmap and GPU-FPM, so as to achieve faster Apriori algorithm. Meanwhile, according to OpenCL platform features, using Multi-Core CPU to control Multi-GPU operations, can improve faster computing efficiency, and solve the core computing speed limit of a single GPU's speed of operation. Finally, OpenMP is used to control the Multi-Core CPU in order to control different GPUs and complete the Multi-GPU implementation.

References

1. ShouJian, Y., Yi, Z.: Improvement of Apriori algorithm based on prefix itemset. Comput. Appl. Softw. **34**(02), 290–294 (2017)
2. Yingying, B., Chen, S.: Improved Apriori algorithm based on association rule mining. Electron. Technol. Softw. Eng. (03), 203–204 (2017)
3. Wenhui, S., Jianling, G.: Improvement of Apriori algorithm based on matrix. Comput. Technol. Dev. **26**(06), 62–64 (2016)
4. Yu, L., Sha, L.: An improved Apriori algorithm based on weight. J. Xi'an Post Telecommun. Univ. **22**(04), 95–100 (2017)
5. Zhu, X.: An improved Apriori algorithm for association analysis. In: Science and Engineering Research Center, Proceedings of 2016 International Conference on Education, Training and Management Innovation (ETMI 2016), pp. 5–10. Science and Engineering Research Center (2016)
6. Peng, L., Fang, W.: Heterogeneity of inferring reputation of cooperative behaviors for the Prisoners' Dilemma game. Physica A **433**, 367–378 (2015)
7. Titus, J.: ECN technical editor: the eclipse of stand. J. Zhongkai Agrotech. Coll. **19**(2), 32–35 (2006)
8. Aleksy, M., Korthaus, A., Schader, M.: Use Java and the CORBA realization distribute type system. J. Pingxiang Coll. **4**, 104–105 (2005)
9. Richardson, W.C., Avondolio, D.: The Java high class weaves a distance: JDK 5. Sci. Technol. Book Rev. **3**, 17–18 (2006)
10. Li, J., Vuong, S.: A semantics-based routing scheme for grid resource discovery. In: E-Science: First International Conference on E-Science and GridComputing, pp. 58–70, 90 (2005)

Research on the Influence of Brand Fit on Brand Emotion and Brand Loyalty

Wenguang Liang, Yi Ye(✉), and Man Bao

Wuzhou University, No. 82, Fuminsan Road, Wuzhou,
Guangxi Zhuang Autonomous Region, China
136725703@qq.com, 2587537726@qq.com

Abstract. Based on theoretical analysis, brand fit may have an impact on consumer brand emotion, and the latter variable influences brand loyalty further. The results of the analysis of 258 questionnaires show that brand fit has a significant positive effect on consumer brand emotion and brand loyalty. Moreover, brand emotion has a significant positive impact on brand loyalty and plays a mediating role between brand fit and brand loyalty. Therefore, enterprises need to accurately position and design the brand so that their products can meet the needs of consumers and enhance the level of consumer loyalty.

Keywords: Brand fit · Brand loyalty · Brand emotion

Consumer brand loyalty can provide a guarantee for long-term profit of enterprises and ideal marketing effect [1]. Therefore, how to establish and cultivate consumer brand loyalty is a major problem faced by the current practice of enterprises, and the forming mechanism that reveal the consumer brand loyalty is also in the ascendant in academia.

The establishment of consumer brand loyalty is a complicated process. Some consumers may prefer the brand because the design style of the brand is more in line with their own preferences, and some consumers may draw up a more positive evaluation and get fond of the brand ultimately because of the performance of the product (such as practicality, convenience, etc.) after contacting and using the product. As a consequence, consumer loyalty to a particular brand may depend on the differences in individual factors or the differences in situational factors. Some researches detect that when the brand image of an enterprise has a high level of consistency with the consumer's image or personality, consumers may have an affinity for the brand, which is called brand fit. Brand fit will prompt consumers to have emotional preference for the brand, and further translate into their loyalty to the brand. Therefore, based on the perspective of emotion, the research will investigate the mediating role of brand emotion in the process that brand fit influences brand loyalty, hoping that the findings can provide meaningful guidance for management practice.

1 Theoretical Framework and Research Hypothesis

1.1 Brand Fit and Brand Loyalty

Brand fit refers to consumers' psychological assessment of they themselves as well as the matching and consistency of the brand [2]. When there is a high degree of

J. Abawajy et al. (Eds.): ATCI 2018, AISC 842, pp. 1038–1045, 2019.
https://doi.org/10.1007/978-3-319-98776-7_125

consistency between the brand and consumers' self-concept [3], consumers can utilize the brand to demonstrate and express self-concept [4]. Consumers' consumption process can also be regarded as a process of self-concept construction to some extent. Brand loyalty refers to the attitudes and behaviors of consumers who are partial to a brand and repurchase the same brand in the long run because of the influence of a variety of factors such as product quality, price and so on after consumers contact and use the brand. As a result, consumers' emotional state of the brand can affect their preference for the brand and their loyalty to the brand further. When the functions of a brand can build the consumers' self-image, consumers will trust and evaluate the brand positively. The trust and positive evaluation means that enterprises can provide products with standard quality for consumers and meet the needs of consumers' long-term self-concept, which helps consumers to establish long-term psychological contract with enterprises and drive dependence and needs for the brand. Based on the viewpoint, the research proposes the following hypothesis:

H1: Brand fit has a significant positive effect on brand loyalty.

1.2 Brand Fit and Brand Emotion

Brand fit allows consumers to have a positive evaluation of the brand, which leads to more positive emotions. Affect refers to the "valence state" [5] of an individual, and emotions and moods are usually external forms of affect. Brand emotion is defined as the potential of a brand that enables consumers to generate positive emotional responses [6]. The brand love proposed by Carroll and Ahuvia [7] includes three aspects, namely, consumer's trust, passion for the brand and attachment to the brand. Consumers have a good impression on a certain brand, and if the brand has a higher degree of matching and consistency with the self-image of the consumers, they will have positive psychological emotions. When these positive emotions can be accumulated and precipitated, the brand will establish a close emotional connection with the consumers, and even allow consumers to have the psychological state of attachment. Eventually the consumers have the desire to purchase the brand and become loyal users of the brand. On the basis of the viewpoint, the research puts forward the following hypothesis:

H2: Brand fit has a significant positive effect on brand emotion.

1.3 Brand Emotion and Brand Loyalty

Brand loyalty can be classified into two types: attitude loyalty and behavior loyalty. The attitude behavior refers to the consumers' belief in the brand, and the behavior loyalty refers to the consumers' love for the brand with the actual repurchase behavior. In modern society, consumers' evaluation of the products has been changed from the traditional quality evaluation in the past to the evaluation that whether to meet the individual's preferences. Therefore, enterprises need to cater to the change of consumers' needs and improve their positive emotional effects on the brand by means of technological innovation. In addition, with the coming of the experience economy, consumers pay more attention to the experience of the brand of the enterprise. Consequently, the design concept of a brand produced by the enterprise needs to be more compatible with the feeling and experience of the consumers so that the consumers can be impressed by the products and establish an attachment relationship with the brand. This emotional attachment is the precondition for the consumer loyalty to the brand [8]. Based on the point of view, the research proposes the following hypothesis:

H3: Brand emotion has a significant positive effect on brand loyalty.

1.4 The Mediating Role of Brand Emotion

Based on the above hypotheses and combined with the basic principles of intermediary effects, the research holds that brand emotion plays a mediating role between brand fit and brand loyalty, that is to say, brand fit will influence the consumers' emotional state of the brand first, and this emotional state will further influence the level of consumer brand loyalty. In view of the viewpoint, the research raises the following hypothesis:

H4: Brand emotion plays a mediating role between brand fit and brand loyalty.

2 Research Methodology

2.1 Samples and Data Collection

The research collected data in the form of questionnaires. A total of 300 questionnaires were distributed, and 270 were recovered. According to certain criteria, after looking over and checking up the questionnaires, 12 invalid questionnaires were excluded, and 258 valid questionnaires were eventually obtained. In terms of gender, there are 94 males and 164 females; in terms of age, there are 18 people under the age of 20 and 187 people between the ages of 20 and 30, accounting for 72.48% of the total, and 53 people over the age of 30; in terms of education degree, there are 45 people with a high school education or below, 77 people with a college degree, 109 people with a bachelor degree and 27 people with a master's degree or above; in terms of monthly income, there are 17 people in 1000–2000 yuan, 69 people in 2000–3000 yuan, 58 in people 3000–4000 yuan, 83 people in 4000–5000 yuan and 31 people in 5000 yuan.

2.2 Variable Selection

The scale involved in the research mainly adopts the measurement scale developed by western scholars. Among them, the brand fit employs the scale compiled by Karen [9], which contains 6 items, for example, "this brand is consistent with my opinion on myself". When using 5-point Likert scoring method, the Cronbach's α value of this scale is 0.869; brand emotion uses the scale compiled by Chaudhuri and Holbrook [7] scale, which contains 3 items, such as "I feel good when using the product of the brand" and so on, and when using 5-point Likert scoring method, the Cronbach's α value is 0.863; the brand loyalty is based on the scale compiled by Elena and Jose [10], which includes three items, such as "I will choose this brand when I buy this kind of product next time" and so on, and when using 5-point Likert scoring method, the Cronbach's α value is 0.843. Moreover, in order to exclude some other variables that may have adverse effects on the results, the research chooses gender, age, education degree and monthly income as control variables, and controls them in regression analysis.

2.3 Validity Test

In order to ensure that the variables involved in the research have high internal consistency and validity, it is indispensable to test the validity of the variables. The average variation of extraction and composite reliability are demonstrated in the research. As shown in Table 1, the average extraction capacity is above 0.5, and the composite reliability value is above 0.7. Therefore, the reliability degree of the measurement scale adopted in the research is relatively high, and the data can also meet the needs of follow-up research and analysis.

Table 1. Test of convergence validity of scale

Variables	Measurement items	Factor loading	Average extraction capacity	Composite reliability
Brand fit	BF1	0.632	0.536	0.873
	BF2	0.745		
	BF3	0.880		
	BF4	0.709		
	BF5	0.797		
	BF6	0.602		
Brand emotion	BA1	0.744	0.682	0.864
	BA2	0.857		
	BA3	0.870		
Brand loyalty	BL1	0.618	0.536	0.837
	BL2	0.783		
	BL3	0.860		
	BL4	0.724		

3 Research Results

3.1 Test of Common Method Biases

Since the data used in the research are derived from the answers of the same subjects, the common method biases may have a potential impact. The method adopted in the research is the combination of program control and statistical control. In the aspect of program control, before the subjects fill in the questionnaires, the survey questionnaires are accompanied by the relevant filling instructions to affirm that personal privacy will not be involved in the collected data and the questionnaire data is limited to academic research but not for commercial purposes so that the subjects can feel relieved and fill in the questionnaires carefully. In the aspect of statistical control, the Harman single-factor method is used for testing, and the factors of brand fit, brand emotion and brand loyalty are analyzed at the same time. The results manifest that the first factor explains

20.57% of the total variance, which is far less than half of the total variance explained (65.74%). As a result, it is believed that there is no serious problem of common method biases in the data collected by the research, and the conclusion is convincing.

3.2 Correlation Analysis

The correlation analysis of brand fit, brand emotion and brand loyalty is shown in Table 2. As is known from Table 2, brand fit is positively correlated with brand emotion and brand loyalty; there is a significant positive correlation between brand emotion and brand loyalty.

Table 2. The mean, standard deviation and correlation coefficient of each variable

	Mean	Std.D	2	3	4	5	6	7
1 Gender	1.33	0.454						
2 Age	26.53	4.652	—					
3 Educational background	2.67	0.384	0.733	—				
4 Monthly income	3015	346.323	0.343^{*}	0.199	—			
5 Brand identity	3.267	0.978	0.018	0.014	-0.035	—		
6 Brand trust	3.245	0.949	0.105	0.027	0.154	0.449^{**}	—	
7 Brand loyalty	3.545	0.978	0.013	0.059	0.012	0.453^{**}	0.523^{**}	—

3.3 Hypothesis Testing

The results of correlation analysis indicate that there is a strong correlation between brand fit, brand emotion and brand loyalty. In order to test the hypothesis between brand fit, brand emotion and brand loyalty, the research combines the testing method recommended by Baron and Kenny [11] with the SPSS21.0 software to test the main effect and the mediating effect. The results of regression analysis are displayed in Table 3.

From the results of the regression analysis in Table 3, it can be noted that brand fit has a significant positive impact on brand loyalty ($\beta = 0.369$, $P = 0.000$), and the hypothesis H1 is supported by the data; brand fit has a significant positive impact on brand emotion ($\beta = 0.412$, $P = 0.000$), and the hypothesis H2 is supported by the data; on the basis of controlling brand emotion, the regression coefficient of brand use to brand loyalty is still significant after adding the brand fit ($\beta = 0.353$, $P = 0.000$), but the coefficient decreases obviously, whereas the regression coefficient of brand emotion to brand loyalty has reached a significant level ($\beta = 0.218$, $P = 0.000$). Combined with the three-step test of mediating role, it can be judged that brand emotion plays a part

Table 3. The results of regression analysis

	Brand trust	Brand loyalty	
Gender	0.135	0.122	0.130
Age	0.004	0.018	0.026
Education degree	0.154	0.107	0.104
Monthly income	0.024	0.124	0.103
Brand identity	0.412**	0.369**	0.353**
Brand trust			0.218**
R^2	0.372**	0.482**	0.445**
F	224.463	211.045	120.98

mediating role in the relationship between brand fit and brand loyalty, and the hypothesis H3 and H4 are supported by the data.

4 Conclusions and Discussion

4.1 Conclusions

Based on 258 consumers' questionnaires, the research explores the influence of brand fit on consumer brand loyalty and its influencing mechanism. It is found that the higher the matching degree between consumers and brands is, the higher the brand loyalty is; secondly, brand loyalty is also influenced by brand emotion, and brand emotion serves as a bridge between brand fit and brand loyalty.

4.2 Implications in Practice

The Enterprise Can Strengthen and Enhance the Brand Loyalty of the Consumers by Means of Brand Fit. The scientific and reasonable brand design contributes to the enhancement of brand fit and brings out brand emotion of the consumers, which can transformed into brand loyalty of the consumers further. The results reveal that brand fit has a positive impact on brand loyalty, so enterprises need to understand consumers' self-concept in the process of marketing and integrate consumers' image into the process of product design. The product can express the meaning of the brand by displaying the consumers' self-concept through shape design and color matching. The meanings of the brand help to express the image of the brand, and at the same time, it will be more consistent with the self-image of the consumers. Hence, the consumers can express the self-concept accurately when they purchase and use the brand, which is the premise of strengthening consumer brand loyalty.

The Enterprise Can Strengthen the Brand Emotion of the Consumers through the Brand Fit. Brand fit not only helps consumers' cognition of self-concept and self-image, but also helps consumers establish the correct self-image in the process of social interaction. Related researches on consumers' behavior have demonstrated that consumers purchase products of a certain color or style mainly because the color or style design of these products is conducive to the improvement of customer's satisfaction. When consumers contact with and use products that are consistent with their self-concept and self-image, these brands can often bring pleasure and comfort to them, and they will have strong attachment to the brand. Besides, consumers will express their self -image in social interaction by purchasing and using a particular brand, which requires the brand to play the role of conveying information. Therefore, brand fit helps consumers to understand social cognition and self-cognition, which will enhance their brand emotion.

The Enterprise Can Provide Consumers with Brand Personalization Strategy to Strengthen Consumer Brand loyalty. Brand is of great value, and it can be considered as a valuable intangible asset of the enterprise. However, whether the brand of an enterprise can be accepted by the majority of consumers and whether it can provide the target consumers with distinctive brand experience and self-concept and self image depend on the behavior of an enterprise, so the enterprise also need to give the brand distinctive personality. Only if the brand with its own personality is able to be differentiated from other brands can the enterprise gain profit constantly and establish long-term competitive advantages. Consequently, enterprises should focus on the psychological needs of the consumers during the process of marketing and try to design and create the brand that is consistent with consumers' personalities. Furthermore, it is essential to highlight differentiated characteristics of the brand and form a relatively perfect brand personality.

4.3 Limitations

The limitations of the research are mainly manifested in the following aspects: firstly, although the choice of measurement scale has been verified by the results of numerous empirical researches, the content is not enough, hence there may be some flaws in the validity of the measurement; secondly, cross-sectional data is adopted in the research, and future research should employ longitudinal study to collect data so as to obtain more accurate conclusions about the relationship between variables.

Foundation Program. This article is the result of the 2015 Wuzhou University's research project "The Impact of Brand Brand Community Value on Community Identity, Brand Trust, and Loyalty" (2015A003).

References

1. Sirgy, M.J., Grewal, D., Mangleburg, T.F., et al.: Assessing the predictive validity of two methods of measuring self-image congruence. J. Acad. Mark. Sci. **25**(3), 229 (1997)
2. Swartz, T.A.: Brand symbols and message differentiation. J. Advert. Res. **23**(5), 59–64 (1983)
3. Swaminathan, V., Page, K.L., Gürhancanli, Z.: "My" brand or "our" brand: the effects of brand relationship dimensions and self-construal on brand evaluations. J. Consum. Res. **34** (2), 248–259 (2007)

4. Erevelles, S.: The role of affect in marketing. J. Bus. Res. **42**(3), 199–215 (1998)
5. Chaudhuri, A., Holbrook, M.B.: The chain of effects from brand trust and brand affect to brand performance: the role of brand loyalty. J. Market. **65**(2), 81–93 (2001)
6. Carroll, B.A., Ahuvia, A.C.: Some antecedents and outcomes of brand love. Market. Lett. **17** (2), 79–89 (2006)
7. Fournier, S., Yao, J.L.: Reviving brand loyalty: a reconceptualization within the framework of consumer-brand relationships. Int. J. Res. Mark. **14**(5), 451–472 (1997)
8. Miller, K.W.: Investigating the idiosyncratic nature of brand value. Aust. Market. J. **15**(2), 81–96 (2007)
9. Delgado-Ballester, E., Munuera-Alemán, J.L.: Brand trust in the context of consumer loyalty. Eur. J. Mark. **35**(11/12), 1238–1258 (2001)
10. Baron, R.M., Kenny, D.A.: The moderator-mediator variable distinction in social psychological research: conceptual, strategic, and statistical considerations. J. Pers. Soc. Psychol. **51**(6), 1173 (1986)

Characteristic of W3C' Standardization and Its Enlightenment to China's Association Standardization Work

Qing Xu(✉)

China National Institute of Standardization, No. 4 Zhichun Road, Haidian District, Beijing 100191, China
xuqing@cnis.gov.cn

Abstract. W3C is an important international standard developing organization in Web field. W3C's standards effectively promote the compatibility of Web technology and have played a fundamental supporting role in the development and application of Internet technology. This paper analyzes the association standardization work mode of W3C. The standard developing principle, the main stages of the standard developing procedure of W3C are studied, and on this basis, the enlightenment to the development of association standard in China is put forward, in order to provide reference and help for the social organizations in China to carry out the association standardization work.

Keywords: W3C · Standard developing procedure · Association standards Standard developing organization

1 Introduction

Since March 2015, China promulgated <The Plan for Furthering the Standardization Reforms> (Reform Plan for short) by the state council is put forward to develop association standards, China's association standards grow up briskly. The reform plan is a milestone in the development of standardization. As the highlight and important content of China's standardization system reform, the association standard is made clear in the normative documents issued at the state council level for the first time. The group standard was born in the fertile soil of the market economy, and its inherent advantages such as voluntary nature, rapid response to market and innovation demand have created its unique development advantages. On September 5, 2017, the party central committee and the state council issued <the central committee of the communist party of China under the state council about to carry out the action guidance for quality improvement>, and points out that the standard supply system reform, push convert standard from production to consumption of consumer goods, service, speed up the cultivation standards development group. This is China's first quality work programing document issued under the name of the CPC central committee and the state council. It is of great significance and will have a profound impact on China's quality development. On October 14, 2017, the Chinese theme of the world standards day is to promote the quality improvement of standardization. On October 15, 2017, director of

J. Abawajy et al. (Eds.): ATCI 2018, AISC 842, pp. 1046–1053, 2019.
https://doi.org/10.1007/978-3-319-98776-7_126

SAC Tian Shihong pointed out that stresses the quality improvement, standardization should be direct point in China's reform and development of the round table conference. Only by effectively playing the basic, leading and strategic role of standardization in quality improvement can the quality be promoted and promoted. We need to increase market standards and develop efforts to guide the development of association standards, stimulate market participants to participate in standardization efforts, and build a standard diversified supply system. On November 4, 2017, after years of revision of the standardization law of the People's Republic of China by the 30th session of the twelfth session of the standing committee of the National People's Congress voted, President xi jinping publish signed A presidential order NO. 78, to come into force on January 1, 2018. New clear the standardization law of the state shall encourage society, association, chamber of commerce, federation, social group such as industry technology alliance to coordinate related market main body common group standards for meet the needs of the market and innovation, for the first time gives association standard legal status, to foster development groups standard provides a strong legal protection, the landmark. The standardization law revision in the standardization work for nearly 30 years of comprehensive summary on the basis of fully absorbing standardization reform results and practical experience, formed the new standard system, management system and operational mechanism, to promote the development of standardization of innovation is of great significance and far-reaching influence. The procedures of standard developing in the ISO/IEC guide rule that applied by our country including nine stages: conceptual stage, proposal stage, drafting stage, committee stage, voting stage, approval stage, publication stage, review stage and withdrawal stage in order to coordinate with the international standard development [1]. The goal of setting up this nine stages is to serve the standard-developing procedures, which indicates consensus, justice and equity, openness and transparency [2]. This paper analyzes the association standardization work mode of W3C. The standard developing principle, the main stages of the standard developing procedure of W3C are studied, and on this basis, the enlightenment to the development of association standard in China is put forward, in order to provide reference and help for the social organizations in China to carry out the association standardization work.

2 Introduction of W3C

World Wide Web Consortium, the most authoritative and influential international neutral technical standardization organization in the web field, which founded in MIT computer science lab in October, 1994, is devoted to accomplish the mission of utilizing the full use of web. The most important task of W3C is developing the web criterion (also known as W3C Recommendations), which depicts web communicating protocols (such as HTLM and XHTML) and other structural modules. W3C membership mechanism opened to all kind of organizations (including commerce, academy, government entity, profit-drive or not) and individuals. These entities could become W3C members via making membership agreements. W3C, who have 3 coordination groups, 169 community groups and 3 business groups, have set up 5 technical preserves, explored 23 standard plan up to March, 2004.

2.1 Organization Structure

Without typical form, the management style and the program style can be considered at least as two respects of organization structure of W3C. From the management point, four W3C hosts, which located in MIT, America, ERCIM, Europe, Keio University, Japan and BeiHang university, China, can maintain its operation jointly, the employees can be lead by board of directors and admin directors. From the process documents, the process document, member agreement and patent policy clearly defined the obligation and profit of every participant of W3C. The key ingredients of W3C are Advisory Committee, Advisory Board, Technical Architecture Group, TAG, the admin director, chartered groups and so on. Every W3C member delegate one individual to form the advisory committee to occupy the role of censorship and appoint the advisory and technical architecture group. The advisory committee appoint the advisory group as advice organization. And the technical architecture group have the duty that seek the principle of document network frame. The obligation of board directors and admin directors in W3C have the duty to appraise the consensus made in the range of W3C. The chartered groups are formed with delegated members and charted specialists who made the most deliverables according to the procedures guided by W3C program.

2.2 Technical Organization

The technical organization of W3C including working group, interest group, coordination group and incubation group The working group generally formulates available documents. The interest group is a group of people who want to evaluate potential network technologies and policies to communicate ideas in the form of BBS. The coordination group manages dependencies within and outside the W3C to facilitate communication with other groups. Incubators develop new internet-related concepts (including unformed specifications, guidelines, and innovative ideas for applications) in a year and within a year. The working group and interest group are the core of the whole standard development. The working draft of the working group, as a prototype of the W3C standard, is finally become recommendations via standard revision process. The managers of the working group will organize standard revision and voting procedures and have the right to invite outside experts to participate in the work of the working group.

3 Standard Developing Principles

The mission of W3C is to enable the world wide web to reach its full potential by developing protocols and guidelines to ensure long-term web development. The basic principles of W3C are:

3.1 Open Standards Principles

Five major global organizations signed an agreement recognizing and abiding by a set of principles supporting modern standards in 29, August 2012; An open and collective empowerment model will help fundamentally improve the way people around the world develop new technologies and innovate for humanity.

3.2 Design Principles

The operation of W3C are guided by the following design principles:

- Web for All, The social value of the network is that it enables human beings to share knowledge in communication, business and opportunity.
- Web on Everything, The number of devices available to access the Web has increased dramatically. Mobile phones, smart phones, personal digital assistants, interactive television systems, voice response systems, kiosks, and even some household appliances are available online.

4 W3C's Standard Developing Procedures

In the process of releasing a new standard at the W3C, the specification is gradually established as a recommendation by following the following strict procedures: Submissions—Notes—Working Groups (WG)—Working Draft (WD)—Candidate Recommendation (CR)—Proposed Recommendation (PR)—Recommendation (REC)—Abolish or Tab.

Document types at each stage: proposal submission, publication of records of proposals described, draft work proposed, candidate recommendations, draft recommendations, W3C recommendations. As shown in Table 1:

Table 1. W3C standard development phase and corresponding files.

Project Phase	Relative files	
	Name	Ab.
Submission Phase	Submit position paper	Submission
Notes Phase	Publish a record of the scanning of the proposal	Notes
Working Group Establishing Phase	Establishing the working group	WG
Working Draft Phase	Publish the proposed work draft	WD
Candidate Recommendation	Candidate recommendation	CR
Proposed Recommendation	Draft recommendation	PR
Recommendation	W3C recommendation	REC

5 Standard Developing Stages

The stage of specific formulation of W3C standard and work content as follows:

(1) W3C Submissions

W3C members submit their proposals to the alliance for which hope to become web standards.

Any W3C member can submit a proposal to the alliance that wants to become a Web standard. Most W3C recommendations originate from a submission to the federation. The W3C will decide whether to initiate improvements to the proposal if a project is submitted to the W3C's work area (or charter).

(2) W3C Notes

Members submit a record to the W3C to refine the description of the proposal, which the W3C USES only to discuss.

Typically, a W3C recommendation will be a record. A record is a description of a proposal refined as a public document. The content of the record is edited by the member who submitted it, not the W3C, and the record can be updated, replaced, or discarded at any time. The release of the record does not represent an endorsement of it, nor does it indicate that the W3C has started any work related to this record.

(3) W3C Working Groups (WG)

After the proposal is approved by the W3C, it can form working groups, including members and other interest groups. The working group will define the timetable and issue a draft relate to the proposal.

(4) W3C Working Draft (WD)

W3C working drafts will be released on the W3C web site as usual. Working drafts will be explain working in operation, its contents can be updated replaced or abandoned at any time which should not be used for any reference material. The work draft released by the working drafts issued by W3C is documents which can be reviewed by community, including W3C members, public and other technical organizations.

(5) W3C Candidate Recommendation (CR)

The candidate proposal is a document that meets the technical requirements of the working group and is widely reviewed. Is more complicated than other specification, need more money, more time and more testing of these specifications can be recommended as a candidate to distribute, candidate recommendation is also a kind of "work in progress", the document can be updated at any time, replace, or abandoned, also should not be used as reference material.

(6) W3C Proposed Recommendation (PR)

The proposed recommendation means the final stage of the work of the working group. The proposed recommendation is also a "ongoing work". This document can be updated, replaced, or discarded at any time. But even if it doesn't mean any official endorsement from the W3C, the proposed recommendation is very close to the final recommendation, both in content and time.

(7) W3C Recommendation (REC)

The W3C recommendation has been reviewed by W3C members and has been formally approved by the W3C director. The W3C recommendation is a stable document that can be used as a reference material.

(8) Reverse and Abolish about W3C Recommended Standard

The W3C recommends that the release of the errata should be updated, and a new version will be released when sufficient substantial revisions are accumulated. If the version after the recommendation contains many mistakes and conflicts or affect implementation and cannot be resolved, the working party may request the competent cancel suggestion recommended, or competent can directly cancel recommendation are put forward.

From the draft stage to the recommendation stage, the working relationship of each stage is shown in Fig. 1. Each stage also includes the possibility of returning to the early stage, repeating the current stage, terminating the project, and entering the next stage. From WD to REC, the maturity level is higher and higher, and the person in charge (Director) can refuse to develop to a higher maturity stage, or to return to the low maturity stage.

6 The Enlightenment of the W3C Standard Developing Procedures to the Standardization Work of Chinese Social Organizations

Due to the W3C standard setting up the strict and normative standard procedure, the standard has high quality and wide acceptance. In the concrete research analyses the principles and standards in its standard procedures, standardization of our country, especially the association standardization work has the certain enlightenment and reference, mainly embodied in the following aspects:

6.1 Extensive Freedom to Solicit Proposals

Standard-setting organizations are based on standardized requirements generated within a certain range of practices, and standards are developed. Therefore, in the first step of the standard formulation, which is the proposal stage, the source of the proposal is the key. When a W3C solicitation proposal is available to any member of the W3C, this ensures that the standard proposals it receives come from within the broadest range. Only the standard proposals come from the broadest range, and their standardization requirements are the best to reflect the needs of all stakeholders, thus ensuring the value of the standards set. Many social groups to carry out the standardization work in China is just beginning, in the standard system revision project solicitation, want to notice to solicit proposals widely freedom, ensuring that all stakeholders have an opportunity and right to standard proposal is put forward, so as to ensure the most widely recognised standards.

6.2 Improve the Supervision Mechanism Process and Improve the Standard Quality

To establish a full-process monitoring mechanism, we can draw lessons from foreign standard procedures, such as FGDC in the United States, and establish the supervision and management mechanism of the whole process. Give and implement a standard drawing board, standardization technology committee, and on the table, and the competent department of intelligent supervision and management functions, such as in the different stages of standards, the arrangement of fixed organization management, and work rules of time constraints, promote rapid standards program. Improve the standard maintenance mechanism, institutionalize standard maintenance work, and institutionalize standard maintenance on the basis of existing standard review work. First, define the standard maintenance management body and establish its responsibility. Secondly, to establish a standard maintenance system with strong operability, it is recommended to review the standard at least every five years, and it can be submitted by the bid committee or the sub-standard committee. Finally, the procedures of standard maintenance work are clarified, and the results of maintenance are defined and explained to improve the efficiency and quality of standard maintenance.

6.3 Innovate Standardized Working Methods and Meet New Challenges

As the extension of standardization role, standardization increasingly expanding coverage, particularly in the smart city, energy efficiency, the Internet of things, emerging field of nanotechnology and so on standardization, and services to the standard requirements. In China, attaches great importance to the standardization work in the field of emerging, in November 2013, the national standards committee joint department such as national development and reform commission has issued a "strategic emerging industries standardization development plan" expanding accompanied with standardized field is standardization work way and the method of innovation and improvement. In the face of new tasks and needs of economic and social development, the pattern and methods of standardization activities also need to be constantly transformed and updated. Standardization activities require cooperation, coordination and efforts across departments. Famous standardized expert Chuntian Li thinks, starting from the traditional industrialization era, standardization in today is difficult to deal with new problems, the progress of science and technology presents a comprehensive trend of more and more obvious, and the increasing of large systems engineering, cross-industry, synergy across different departments, professional standards. The standardization of the whole process will become the dominant mode of standardization, and standardization should go to the track of "system management". Therefore, in the process of promoting standard modernization, China should actively innovate its working methods and methods to meet the needs of the new situation of economic and social development.

7 Conclusion

By establishing a scientific standardization working mode, W3C ensures that its standard development follows the basic principles of standardization activities well, and thus guarantees the quality of W3C standards. Since March 2015, China has cultivated and developed association standards for three years, and more and more social organizations begin to carry out association standardization work. Many achievements have been achieved, but there are also many problems. In the actual standard developing work, China's social organizations should make full use of W3C's standardization work mode, pay more attention to extensive freedom to solicit proposals, improve the supervision mechanism process and improve the standard quality, innovate standardized working methods and meet new challenges.

Acknowledgment. This research was financially supported by Standardization Administration of the People's Republic of China project <Research on association standardization> (Project Number: 572018B-6052).

References

1. GB/T 16733-1997, Stage division and code of procedure for development of national standards
2. Wang, Z.: Standardization of basic knowledge of knowledge, 113 (2010)
3. Ge, X.: On how to improve the effectiveness of college english translation teaching. Engl. Teach. **18** (2015)
4. Chen, T.: Effective measures of english teaching translation. Overseas Engl. **20** (2015)
5. Wang, H.: On college english translation teaching models in the internet environment. Engl. Sq. **2** (2016)
6. Fan, M.: The cultural elements in english translation teaching. Engl. Sq. **20** (2014)
7. Wang, H.: The problems and countermeasures of english translation teaching in college of traditional Chinese medicine. Overseas Engl. **20** (2014)
8. Watts, D.J.: Small worlds. The dynamics of networks between order and randomness. America: 56A2579BB6, l999-05-13
9. Aleksy, M., Korthaus, A., Schader, M.: Use Java and the CORBA realization distribute type system. J. Pingxiang Coll. (4), 104–105 (2005)
10. Jon, T.: ECN technical editor: the eclipse of stand. J. Zhongkai Agrotech. Coll. **19**(2), 32–35 (2006)

On the Prospect of the Intelligent Computer-Aided Translation

Changhong Zhai(✉)

Wuhan Polytechnic University, Wuhan 430028, Hubei, China
594561792@qq.com

Abstract. Computer-aided Translation (CAT) technology, which plays an increasingly important role in the modern translation career, is employed to assist the translation by means of computer software, hardware, network equipment etc. This paper holds that the overwhelming trend of intelligentialize will surely promote intelligent Computer-aided Translation.

Keywords: Computer-aided Translation · Automation
Intelligent Computer-aided Translation

1 Introduction

Computer-aided Translation (CAT) technology, which plays an increasingly important role in the modern translation career, is employed to assist the translation by means of computer software, hardware, network equipment etc. Before the emergence of Computer-aided Translation software, translation was very difficult. In 1990, TRADOS -the German translation company put its terminology management tool MuhiTerln into the market. Therefore, the Computer-aided Translation technology has been commercialized and developed rapidly in all countries of the world. Since then, the working form of the translator has been qualitatively changed. The extensive application of Computer-aided Translation software in the international translation field has promoted the research and development of Computer-aided Translation software by domestic software development company. And some achievements have been made, such as the high reputation of ya xin CAT, etc. Therefore, with the rapid development of information technology, the development of intelligence in all aspects of human social life is inevitable, which is full of the significance of the times for the research on the intelligent development of Computer-aided Translation or Intelligent Computer-aided Translation.

In particular, there are a large number of repetitive or similar words, sentences in the process of translation. If it is a manual translation, you need to write each of the repeats several times. Computer-aided Translation technology solves this problem very well. Computer-aided Translation technology has the function of automatic memory and search. It can selectively store some translation content according to the user's needs. When the user encounters repeated translation, the system will automatically search for matching translation results. Translation references and suggestions will be given for similar translation. That is, complete matching and fuzzy matching. So we

J. Abawajy et al. (Eds.): ATCI 2018, AISC 842, pp. 1054–1058, 2019.
https://doi.org/10.1007/978-3-319-98776-7_127

know that the core of Computer-aided Translation technology is translation memory technology, and another important component is term management. The user can save the terms as needed. The saved set of terms forms a terminology database. Providing fast translation services for current and future translation tasks which can not only improve work efficiency, but also ensure the implementation of the principle of consistency in translation.

2 An Introduction About Intelligentialize

"Intelligent" is a dynamic word, it refers to a collection of modern communication and information technology, computing network technology, industry technology, intelligent control technology for a certain application. In today's society, with the development of information technology, digitization, networking and information are increasingly melting into people's lives, thus the concept of intelligence gradually permeates into the square aspects of people's lives. For example, smart phone, smart residential area is closely related to people's life. Intelligence is a breakthrough development of automation.

3 The Features of Computer-Aided Translation Automation

Computer-aided Translation is a highly automated computing machine to assist in human translation. It is the necessary stage of the development of intelligent Computer-aided Translation, So it is necessary to understand the automatization of Computer-aided Translation for the analysis of the trend of Computer-aided Translation Intelligence. Find a breakthrough from the middle. To promote the realization of its intellectualization. Computer-aided Translation technology is a change of the automation of translation work. As mentioned above, the application of Computer-aided Translation technology has improved the efficiency of the councillor. It is a great progress in the history of human translation. However, Computer-aided Translation automation still exists in some areas to be improved. For example, the combination of Computer-aided Translation technology and translation theory is not ideal. The translation results are more rigid. It is primarily a literal translation.

3.1 Integration with Translation Theory

As mentioned above. Computer-aided Translation (CAT) is a kind of technology that applies many kinds of computer automation equipment to translation work, and its work is accomplished mainly by means of translation memory technology and terminology database. In the process of memory, copying and pasting, the position of translation theory will be neglected. In this way, the position and function of translation theory is likely to be marginalized and stealthy as the translator's subjectivity is what it was before the culture turned. However, the position and role of translation theory can never be ignored. It has a fundamental guiding role in translation practice. It is an inseparable aspect of translation practice. It helps people to have a better understanding of translation activities. From the study of the original translation, the original text to the translator and the reader, to the study of the translator's subjectivity after the cultural transformation. Help improve the quality of translation from a deeper level.

3.2 Integration with Translation Practice

Due to the diversity of style and the variability of word meaning. The translator should be flexible in the process of translation. Do not blindly apply the previous translation results. In terms of style, it mainly refers to the translation of literary and advertising language, which is subjective and sentimental and emotional. For example, translation of Nike's advertising language "Just do it". According to the literal meaning of the match, the result may turn to "just do it", "just do it. There are many examples of the meanings of words. For example, the translation of "Never trouble trouble until trouble troubles you". The first and fourth "trouble" is a verb. It means "trouble". The second and third "trouble" is a noun. It means "difficulty". If just let the computer automatically match the translation result, it will make a big joke. Automatic matching saves time and effort, but these shortcomings can not be ignored. Therefore, to improve the level of intellectual ability of computer aided translation is also the demand of human civilization development.

3.3 Learning Profile

Nowadays, the need of Computer-aided Translation technology is becoming more and more urgent. This has also promoted training in related translation tools and language service management. But, as a new thing, the teaching of Computer-aided Translation is only in its initial stage. There are many problems in many aspects, the complexity and imperfection of Computer-aided Translation technology have made it more difficult for teachers and students to learn. For example, although the software is efficient in extracting the terms of a single word, it is not effective to extract multiple words, and the results of the alignment tool are mixed in quality. still need artificial proofreading, more labor cost, the urgent needs of the people to learn Computer-aided Translation technology and the needs of the development of Computer-aided Translation itself in promoting the process of intelligent Computer-aided Translation.

4 Development Trend of Intelligent Computer-Aided Translation

Computer-aided Translation technology is one of the advanced tools of human society development. The goal is to constantly improve the automatic level. Gradually translate the computer into the main body of the translation project. Realize the intelligence of Computer-aided Translation. Saving manpower and improving human productivity, the author thinks that this development can be called Intelligent Computer-aided Translation. The conception of intelligent development of Computer-aided Translation can be divided into following three aspects: First, improve the combination with translation theory. Theory is the foundation of practice. Computer-aided Translation technology should be closely combined with translation theory, and the combination of Computer-aided Translation and translation theory can be used in the selective operation. That is to establish a classified terminology database based on translation theory, and the user selects the corresponding translation theory according to the original text before the text is translated. The translation of literary works such as poetry can choose Xu Yuan's theory of "three beauties" The translation of non-literary works can be used to select the Dynamic functional equivalence theory of Eugene Nida. Or choose literal translation, paraphrase, assimilation and alienation according to the situation and customer needs. In simple terms, translation theory is taken as the general standard of classification, and its branch is named after each specific translation theory. At the same time, At the same time, because translation theory is dynamic development and continuous improvement, some translation methods should be added directly as category. Then, the translation of words, phrases and texts should be summarized under each category. Finally, in the practice of translation, the precise selection of translation results through a specific program identification system will theoretically guarantee the accuracy of the translation and thus improve the quality of the translation.

Secondly, expand the source of the terminology database. To achieve the accuracy of translation, you can also start from expand the source of the terminology database. The terminology database is the self-collection and generalization of the terms in translation practice, which determines the size of a person's terminology database depending on the translation workload and time of translation. However, in today's society, professional translation has become more and more inclined to part-time translation, so it is difficult for a single translator to form a systematic and comprehensive terminology database. It can be seen that the existence of the terminology database cannot depend solely on the individual translator. It is wise to extend the source of the terminology database from point to face.

Based on the protection and respect of individual labor results. The realization of this change requires a paid mechanism to help the completion, which can be understood by adding a term library intelligent user identification system in the Computer-aided Translation software. The terminology database can be imported at will, but it can only be exported in a particular form, a particular program, when the first translator and the Computer-aided Translation software development company reach an agreement. The first translator paid share the terminology database with the Computer-aided Translation software department. The Computer-aided Translation software development company paid shares the terminology database with the second translator required for the term. But the terminology database not directly Shared by the second translator to other translators except the first translator. In this way, the Computer-aided Translation software development department acts as a shared intermediary in assuming the role of the terminology database set. In a sense, the Computer-aided Translation software development company is the terminology database of all translators. This provides resource support for intelligent Computer-aided Translation.

Finally, simplify the operating procedures. One of the requirements of intellectualization is simple operation. Simple operation is not only the embodiment of high intelligence level. It is also the key to the widespread adoption of the technology. For example, the simplification of computer typing software. Initial typing software required users to master five input methods. Then become master pinyin input method.

And now it's already possible to make a hand-written blur on a desktop computer. This improvement in typing software greatly increases the age range of computer users. Because the gender of the translator is dominated by women, taking into account the influence of family, interests and other factors. The first stage of the Computer-aided Translation software operating procedures for many translators is somewhat to be simplified. To simplify the operation procedure of the software and the rational division of the translation tasks, and promotes the benign development of the translation cause.

5 Conclusion

To sum up. The intelligent development of Computer-aided Translation is not only a recognition of the powerful function of Computer-aided Translation. It is also the bold presupposition of future development trend. It is believed that the intelligent development trend of Computer-aided Translation software can solve the major problems such as translation memory bank and terminology database. To realize the transformation of saving memory and the terminology database.

References

1. Bo, Y.: Preliminary study on the evaluation of computer-aided translation software. Sci. Educ. Inf. **3** (2015)
2. Yao, X.: Research on the application of computer aided translation in English translation practice teaching. Overseas Engl. **13** (2012)
3. Zhao, J.: An analysis of intelligent technology in the automation control of electric process. Sci. Technol. Outlook **25**(35) (2015)
4. Jin, G.: The present situation and development trend of computer-aided translation technology. J. Shen Yang Eng. Coll. (natural science edition) **6**(3) (2010)
5. Feng, Z.: Machine translation studies. China Translation and Publishing Company, Beijing (2004)
6. Peng, L., Fang, W.: Heterogeneity of inferring reputation of cooperative behaviors for the prisoners' dilemma game. Physica A **433**, 367–378 (2015)
7. Fan, M.: The cultural elements in english translation teaching. Engl. Sq. **20** (2014)
8. Wang, H.: The problems and countermeasures of English translation teaching in college of traditional chinese medicine. Overseas Engl. **20** (2014)
9. Watts, D.J. Small worlds. The dynamics of networks between order and randomness. America: 56A2579BB6, l999-05-13
10. Aleksy, M., Korthaus, A., Schader, M.: Use Java and the CORBA realization distribute type system. J. Pingxiang Coll. **4**, 104–105 (2005)
11. Titus, J.: ECN technical editor: the eclipse of stand. J. Zhongkai Agrotech. Coll. **19**(2), 32–35 (2006)
12. Clay Richardson, W., Donald, A.: The Java high class weaves a distance: JDK 5. Sci. Technol. Book Rev. **3**, 17–18 (2006)
13. Vuong, J.L.: A semantics-based routing scheme for grid resource discovery. In: E-Science: First International Conference on E-Science and Grid Computing, vol. 90, pp. 58–70 (2005)

Fault Recognition Method Based on Deep Learning

Jingjie Yang(✉)

Liaoning Jianzhu Vocational College, Liaoyang 111000,
Liaoning Province, People's Republic of China
331857956@qq.com

Abstract. This paper applies the deep learning method to fault recognition. In order to improve the performance of deep learning method, short-time Fourier transform, first-order difference method and principal component analysis methods are applied to data treatment and the main features are extracted. On this basis, the deep belief network (DBN) classifier is used to carry out the on-line fault recognition. For the discrete state space model of the three tank water systems under random disturbances, recognition experiments are made in the open-loop control, and closed-loop control, which realize the on-line recognition of failure. Simulation results show the effectiveness of the proposed algorithm.

Keywords: Fault recognition · Machine learning · Deep belief network
DTS200 three-tank system

1 Introduction

Process monitoring and fault recognition are of great significance to improve the safety and reliability of industrial processes. Fault recognition technology can be achieved by qualitative and quantitative methods [1–3]. When control system tends to be complicate, the system will tend to show a time-varying, nonlinear and variable coupling, modal characteristics, and it will be more difficult to get an accurate mathematical model. While the only way for fault recognition is to extract the key characteristic information of data generated in the running process. Data-driven fault recognition method based on machine learning such as Bayesian classifier [4] probability graph model, support vector machine (SVM) [5], artificial neural network algorithm, can extract effective features from mass and high dimensional data. Sometimes, the wavelet packet decomposition method is used to decompose the input of the failure frequency into a narrow bandwidth signal, and the fault recognition is made by the empirical mode decomposition method and the error inverse propagation algorithm [6, 7]. Paper [8] firstly trains the extreme learning machine parameters and then identifies the faults of the wind turbine. Deep learning method is a new field of machine learning research, which is based on neural network and imitates the human brain mechanism to identify the target and perceive information [9].

This paper applies the deep learning method to fault recognition. Firstly, the normal working state and all kinds of fault states of the controlled object in industrial process, are defined as different labels (category). Industrial process data is collected according

J. Abawajy et al. (Eds.): ATCI 2018, AISC 842, pp. 1059–1066, 2019.
https://doi.org/10.1007/978-3-319-98776-7_128

to the characteristics of the object, and some important features are extracted to create a new sample set from the source data set by Fourier transform, the first-order difference method and principal component analysis method. Secondly, the new sample sets are applied to train deep belief network. Thirdly, input the time series of the characteristics of the sample, and realize the real-time separation of the fault. Fourth, the DBN output value is transformed and the threshold is set to realize the accurate identification of the unknown fault.

2 Feature Extraction

In the fault recognition technology, After the controlled object reaches steady state, the measured value of its variables can be approximated by the fixed value function. However, after the failure, the measured value of its variables has a strong time self-correlation, and the range of data varies greatly. It is difficult to depict them in simple function form. Therefore, the fault data has no typical distribution characteristics, which challenge the application of DBN for fault regression and classification. That is the reason why some effective feature extraction steps are required before the classification step.

Firstly, the Feature Extraction is made by First-order Difference. The observed variable value of the controlled object is not stable, which is related to the discrete sampling time k, therefore, the common characteristics of high stability in the time series of the observed variables should be obtained. The first order difference operation with a step length 1 is applied for each feature vector in each operation mode of the controlled object, and the effective characteristics are extracted.

Secondly, Short-time Fourier Transform is used for further feature extraction The sample data in the time domain is an aperiodic signal, which can be extended as a periodic signal through replication. Then, the discrete time domain Fourier transform method is applied to map the characteristic data of the sample in the time domain to the frequency domain. The short - time Fourier transform can be realized by using window functions such as the Hamming window in time. Analyzing the local frequency domain characteristics of the signal can guarantee the real time fault recognition.

Finally, the principal component analysis is used, the original sample data is projected into a new space corresponding to the eigenvalue of a linear independent group in the original sample, which can realize the correlation between data characteristics. When the dimension of this space is smaller than the sample characteristic, the original data feature is compressed and reduced.

3 Deep Learning Fault Recognition Model

3.1 Deep Learning Principle

Deep learning model with restricted Boltzmann machine is a kind of improvement of Boltzmann machine [10]. Boltzmann machine is a kind of random networks and its network training process is slow because of the interconnected relations among the units in the same layer. Hinton introduced a restricted Boltzmann machine (RBM) as basic structure for a single network, and the training of DBN is realized by training layer by layer. The validity of this method is proved by the automatic coding machine, which laid concepts and methods of DBN. A restricted Boltzmann machine includes a visual layer, a hidden layer, and there is no interconnection between the units in the same layer. This is why the RBM is highly efficient.

Deep learning is a machine learning that aims at establishing a deep structural model, and the model contains at least three hidden layers. The depth belief network (DBN) proposed in literature [11, 12] is a typical deep structure. Suppose that a DBN has n hidden layers, and g^i represents the vector of the ith implicit layer, and then the DBN model can be represented as:

$$p(x, g^1, g^2, \ldots g^n) = p(x|g^1)p(g^1|g^2)\ldots p(g^{n-1}|g^n) \tag{1}$$

where, the conditional probability is:

$$p(g^i \middle| g^{i+1}) = \sigma(-b_j^i - \sum_{k=1}^{n^{i+1}} W_{kj}^i g_k^{i+1}) \tag{2}$$

Where is the point in the layer, is the weight matrix of the layer, and is the as the following:

$$\sigma(x) = {}^1\!/_{1} + \exp(-x) \tag{3}$$

3.2 On-Line Failure Recognition

The characteristics data of the controlled object under different operating mode are used to train the offline DBN. In order to realize on-line monitoring, two alarm variables are introduced; 1. Monitoring length W: if the current sample is classified as a fault sample, the test results of followed W sampling time points will be monitored continuously. 2. Alarm threshold: if there is more than T data in the monitored W sample, the failure mode will be reported. In order to improve monitoring efficiency, the forecast label of W sampling points is stored in storage space. The system firstly discriminating the operation mode of the $k+1^{th}$ sampling point. If its label already exists, the system matches it directly to save calculation time.

In fact, the pre-established label spaces cannot cover all possible failure types. When a new fault occurs, the traditional classifier will misjudge it as a known fault sample. In this paper, a complete fault space is assumed, which will avoid the error classification for "new failure".

After an off-line training is completed, one particular type of new fault sample is put into the classifier, the classifier will output a predict value as $\hat{y} = [\hat{y}_A, \hat{y}_B, \cdots]$, then label value will be transformed into an positive domain by power operation, if it meet at this time:

$$\forall \hat{y}_i \in \hat{y},\ \hat{y}_i \geq T_r \tag{4}$$

The classifier outputs an abnormal value, which is a fault of new type. On the contrary, it is judged that the classifier performs well and can judge correctly.

4 Experimental Analysis

4.1 Modeling and Simulation of DTS200 Three-Volume Tank

Firstly, a dynamic equation of the liquid level of the DTS200 three-volume tank was established, then it is simulated in the MATLAB environment. After the three-volume water tank comes to a stable open loop operation, the PID closed-loop control algorithm is added to realize a new stable station in disturbed environment. Further, the actual leakage failure was quantified, then the failure was injected into the open loop and closed-loop condition, the validity of the controller was verified, and the stability of the control system was verified.

The working principle of the DTS200 three-volume tank is shown in Fig. 1. The water tank T1 and T2 are connected with two control pumps P_1 and P_2. Each tank has two manual operated valves, and the lower part of the three-volume tank is a sink. Proper adjustment of the opening of V4, V5 and V6 can simulate the leakage failure of tank T1, T3 and T2.

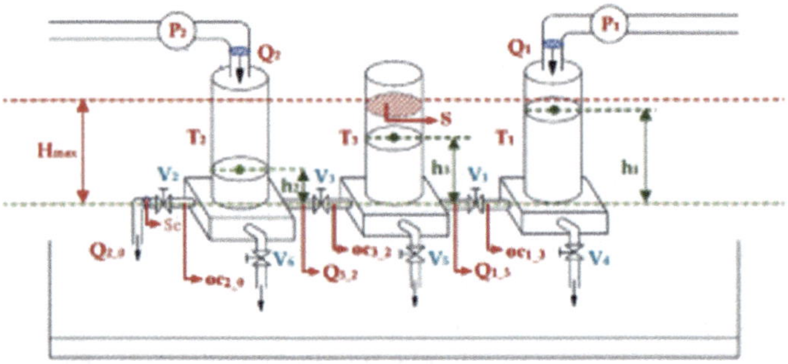

Fig. 1. Simulation structure of three tank water tank system.

Table 1 shows the parameters of the three-volume water tank model. The flow velocity in the opening is defined as Q, and a_z is the outflow coefficient. According to the generalized Torriori law, the unknown velocity quantity Q1_3, Q3_2, Q2_0 are shown in Eq. (5)

$$\begin{cases} Q_{1\_3} = oc_{1\_3} S_c \mathrm{sgn}(h_1 - h_3)\sqrt{2g|h_1 - h_3|} \\ Q_{3\_2} = oc_{3\_2} S_c \mathrm{sgn}(h_3 - h_2)\sqrt{2g|h_3 - h_2|} \quad \xi = \frac{S_c\sqrt{2g}}{2S} \\ Q_{2\_0} = oc_{2\_0} S_c \sqrt{2g|h_2|} \end{cases} \tag{5}$$

It is assumed that the disturbance factor $d(t)m^3/s$ and the water tank failure $f(t)m^3/s$ are both external inputs in a continuous time systems. For the tank water level, a dynamic equation is established. After linearization and discretization, the space model of the three water tank control system in discrete time forms is as formula (6):

$$\begin{cases} \dot{x}(t) = A^L x(t) + B_u^L u(t) + B_d^L d(t) + B_f^L f(t) \\ y(t) = C^L x(t) \end{cases} \tag{6}$$

Where Q_i (i = 1, 2) is the amount of water flow into tank T_i, Q_{i_j} (i = 1, 2, 3; j = 2, 3) is the flow velocity in the pipe between water tank T_i and T_j; $\boldsymbol{Q_{2\_0}}$ is the velocity flowing out of T_2; $\boldsymbol{Sc}$ $(5 \times 10^{-5}\,(m^2))$ is the cross sectional area of the pipe and $\boldsymbol{S}$ $(1.54 \times 10^{-2}\,(m^2))$ is that of the tank; $\boldsymbol{oc_{1\_3}}$ (0.450289) is the outflow coefficient between tank T_1 and T_3; $\boldsymbol{oc_{3\_2}}$ (0.461526) is the outflow coefficient between tank T_2 and T_3, and $\boldsymbol{oc_{2\_0}}$ (0.611429) is that of T_2; h_i (i = 1, 2, 3) is the liquid level of water tank T_i; $\boldsymbol{H_{max}}$ (0.61 ~ 0.63(m)) is the highest liquid level; $\boldsymbol{Q_{1MAX}}$ $(0.097(m^3/s))$ is the maximum of water flow into tank T_1 and $\boldsymbol{Q_{2MAX}}$ $(0.097(m^3/s))$ is that of T_2.

When water is pumped into the tank, its liquid level fluctuated in a range of [−1, 1] cm, therefore, the range of the random perturbation in Mat lab simulation experiment is:

$$\begin{cases} d_1(k) = 7.5 \times 10^{-6} \times \sin(\frac{k}{10}) \times (rand(1) - 0.5) \\ d_2(k) = 8 \times 10^{-6} \times \sin(\frac{k}{10}) \times (rand(1) - 0.5) \\ d_3(k) = 2 \times 10^{-6} \times \sin(\frac{k}{10}) \times (rand(1) - 0.5) \end{cases} \tag{7}$$

According to the actual situation, the leakage fault type and its corresponding valve stroke and leakage amplitude are defined as Table 1.

Table 1. Leakage fault

Fault type	The valve stroke	Leakage amplitude $f_i(t)$ $(i = 1, 2, 3)$
Normal	0°	$0\,(m^3/s)$
Small leak	$(0^\circ, 30^\circ]$	$(0, 34.09] \times 10^{-3}\,(m^3/s)$
Medium leakage	$(30^\circ, 75^\circ]$	$(34.09, 50.45] \times 10^{-3}\,(m^3/s)$

4.2 Modeling and Simulation of DTS200 Three-Volume Tank

At any time after the three-volume water tank system reaches steady state, a leakage failure is injected into one of the water tanks. After the failure lasts for a certain time, the three tank system finally reaches a new steady state. The sampling interval is 0.1 s, and all sampling points during the time interval between the failure and the new steady-state are used as sample data. The height of the three tanks is used as characteristic h1, h2 and h3. The fault recognition under closed loop control has three steps as the follows. The first step is DBN training and test for fault recognition accuracy. After adding the disturbance to the three-volume water tank, all the characteristic data in the steady state are no longer stable, especially the control quantity. After two feature extraction methods, the original eigenvector is multiplied by 15 dimensions, where the sampling window is Fourier transform, and the window function is the Hamming window with 25 sampling points. Then, the principal component analysis (PCA) is applied to select a training set and testing set according to the importance of the eigenvector, the final number of feature extraction F_e is shown in Table 2, which also shows the classification accuracy and DBN parameters in two cases. Case A has two fault type and Case B has three fault types. Take DBN parameter 8-4-2-1 for example, it is a DBN network with two hidden layer structure (RBM1 and RBM2), 8 input neurons in the visible layer, RBM1 and RBM2 contains 4 and 2 neurons respectively, the output layer of DBN has only 1 neuron. All the learning rate of RBM is set to 1 and the number of iterations is 1000 times.

Table 2. DBN parameters and test result

Case	Label space (Fault type)	F_e	DBN parameters	Test accuracy (%)
A	Label: only medium leakage in T_3	8	8-4-2-1	94.50
	Label 2: T1, T2, T3 are normal	10	10-7-3-1	98.62
	Label 1: only small leakage in T_1	15	15-10-7-3-1	98.38
B	Label 1: only small leakage in T_1	8	8-4-2-1	98.92
	Label 2: only medium leakage in T_2	10	10-7-3-1	98.08
	Label 3: T1, T2, T3 are normal	15	15-10-7-3-1	97.43

After the feature extraction, augmented resolvation and dimensionality treatment, the classifier identification accuracy is higher. The classification performance will be reduced when the classification task is faced with disturbances, multi-classification and minor faults.

The second step is on-line fault recognition. In the case of disturbance, the monitoring results of two typical faults are tested. In case 1, there are two fault labels as "Label 1, T1, T2, T3 are normal" and "Label 2: only medium leakage in T2", its test accuracy is 89.17%. Set the fault monitoring length parameter W is 50, and the fault alarm threshold parameter T is 25. In case 2, there are two fault labels as "Label 1, T1, T2, T3 are normal" and "Label 2: only small leakage in T3". The small leakage fault in the non-controlled water tank, which is more difficult to be accurately identified, and the characteristic data has a poor stability and a wide range of variation. The fault alarm is delayed, but the fault type can still be correctly identified.

The third step is the fault recognition process in an incomplete labeling space. Four operating modes of three-tank water tank is defined as the follows: Mode 0: Tank T1, T2, T3, are all in closed loop under normal work; Mode 1: only minor leaks occur at T1; Mode 2: only moderate leaks occur at T2; Mode 3: only minor leaks occur in T3. Under this assumption, a new operation mode data is entered into the over-limit learning machine, and the recognition accuracy is shown in Table 3.

Table 3. Recognition accuracy in an incomplete labeling space

Type of new fault	Recognition accuracy
1: Take Mode 3 as new fault	99.63%
2: Take Mode 1 as new fault	99.32%
3: Take Mode 2 as new fault	98.44%

The experiment shows that the system can identify the failure "belongs to the unknown category" with more than 98% accuracy. After adding the output index transformation and threshold judgment program, the system false alarm rate of fault detection will be greatly reduced.

5 Conclusion

This paper proposes a three-water-tank fault recognition method based on DBN. Firstly, two feature extraction methods, short-time Fourier transform and first-order difference method, are used to multiply the original eigenvector, where the sampling window is Fourier transform, and the window function is the Hamming window; Then, the principal component analysis (PCA) is applied to determine the final number of feature extraction; finally, a deep belief network is trained and applied in the fault recognition, especially on-line fault recognition in incomplete labeling space. A simulation experiment is carried out on the DTS200 three-volume tank system. The results show that the method can achieve the expected targets and the fault recognition is reliable.

References

1. Chang, C., Chen, J., Liew, A., et al.: Fuzzy expert system for fault recognition in power systems. Int. J. Eng. Intell. Syst. Electr. Eng. Commun. **5**(2), 75–81 (1997)
2. Lu, Y., Wang, F., Jia, M., et al.: Centrifugal compressor fault recognition based on qualitative simulation and thermal parameters. Mech. Syst. Signal Process. **81**, 259–273 (2016)
3. Liu, Y., He, X., Wang, Z., et al.: Optimal filtering for networked systems with stochastic sensor gain degradation. Automatica **50**(5), 1521–1525 (2014)
4. Muralidharan, V., Sugumaran, V.: A comparative study of Naive Bayes classifier and Bayes net classifier for fault recognition of monoblock centrifugal pump using wavelet analysis. Appl. Soft Comput. **12**(8), 2023–2029 (2012)
5. Widodo, A., Yang, B.S.: Support vector machine in machine condition monitoring and fault recognition. Mech. Syst. Signal Process. **21**(6), 2560–2574 (2007)
6. Bin, G., Gao, J., Li, X., et al.: Early fault recognition of rotating machinery based on wavelet packets—empirical mode decomposition feature extraction and neural network. Mech. Syst. Signal Process. **27**, 696–711 (2012)
7. Huang, G., Huang, G., Song, S., et al.: Trends in extreme learning machines: a review. Neural Netw. **61**, 32–48 (2015)

2 Overview of Equipment Support Information Network

The information network of equipment support refers to the network that provides information guarantee for the equipment support service. In modern warfare, relying on equipment to ensure the information environment and achieve comprehensive security of equipment are the basis for winning a modern war [2]. The following summarizes the equipment support information network from two aspects: the goal of equipment support and the analysis of the needs of the equipment support information network.

2.1 Target of Equipment Support

The fundamental purpose of equipment protection is to fully integrate existing equipment resources and minimize the cost to realize the equipment readiness requirements. According to the actual situation and future development trend of our army's equipment support, the target system of our army's equipment support can be embodied in three aspects: support force, support capability and safeguard action. The ultimate goal is the comprehensive protection equipment to ensure the realization of equipment. The protection force is based on information, relies on military-civilian integration, and meets the requirements of the combination of peacetime and warfare, as well as professional integration.

2.2 Demand Analysis of Information Network Equipment Support

The equipment support information network is the prerequisite for the efficient operation of the equipment support system, and it is the basis for the smooth development of our military equipment support work. In the information age, a large number of informatization equipment is flooded in the equipment support system. It is necessary to ensure the efficient development of the equipment support business, and also to deal with the security threats in the security process. This requires an analysis of the needs of the equipment support information network from both security and mission aspects.

In terms of security requirements, with the increasingly diversified means of attack, the security threat facing the equipment support information network is becoming more and more serious. The traditional equipment support system has adopted corresponding safety precautions, but these security measures can only be passive and single protection against specific security problems, and can not coordinate all kinds of security protection mechanisms to fundamentally solve network security problems. At the same time, the security defense mechanism of the key nodes of the network is basically solidified and can not be dynamically adjusted according to the actual situation of the network operation. Therefore, it is necessary to consider the security protection ability of the network building a new equipment support information network.

In terms of task requirements, with the continuous acceleration of the replacement of ordnance equipment, various models of ordnance equipment are installed in the military, resulting in various types of maintenance services, maintenance and other security services becoming more and more complex. At present, the equipment support business mainly adopts the special equipment support system to guarantee, which leads to repeated construction of various soft and hard support resources, and ensures that the resource data format is not uniform. This problem causes the equipment support information sharing, processing and utilization very inconvenient, and can not adapt to the demand of joint operational command under the information war environment. Therefore, it is necessary to consider the processing of equipment support information so as to improve the efficiency of the equipment support service in the construction of new equipment support information network.

The security problem is the first problem to be considered in the whole network construction process because of the particularity of the application scene of the equipment support information network. In order to better serve the equipment support service, efficient information sharing and processing ability is also a problem that needs to be considered in the process of network construction. Therefore, in the process of building the new equipment support information network, we should consider not only the safety of the network but also the efficiency of the equipment support business.

3 Overview of Equipment Support Information Network

The equipment support information network under the cloud environment is a complex network system, and the research on its construction must be guided by the scientific methodology. Therefore, the basic structure of the information network is constructed on the basis of combining the features of the equipment support information network under the cloud environment. The detailed construction method is expounded in three aspects consist of the security control center, data center and equipment support service.

3.1 Basic Framework of Equipment Support Information Network

According to the characteristics of the Cloud Architecture and the equipment support information network, the information network architecture of the equipment support under the cloud environment is shown in Fig. 1. It includes the infrastructure layer, the

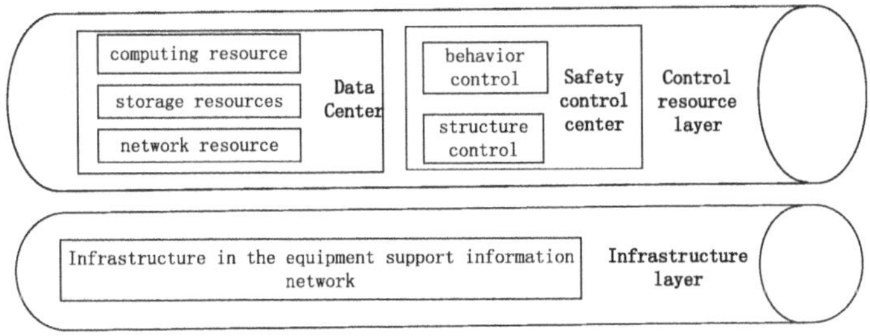

Fig. 1. The architecture of equipment support information network under cloud environment

control resource layer, the platform service layer, and the application software layer. Compared with the traditional three-tier cloud architecture, the equipment support information network architecture in the cloud environment adds a layer of control resources. The control resource layer is consist of data center and security control center. The data center provides resource guarantee for equipment support business and solves the problem of limited resource nodes in the equipment support information network. The security control center solves the security problem of the equipment support environment and satisfies the equipment guarantee.

3.2 Detailed Construction Method

In the basic framework of equipment support information network in cloud environment, the control resource layer is the key to realize the function and safety requirement of equipment support business. Therefore, the detailed construction of the equipment support information network in the cloud environment in this section mainly refers to the construction of the control resource layer. The construction of control resource level is expounded from two aspects: safety control center and data center.

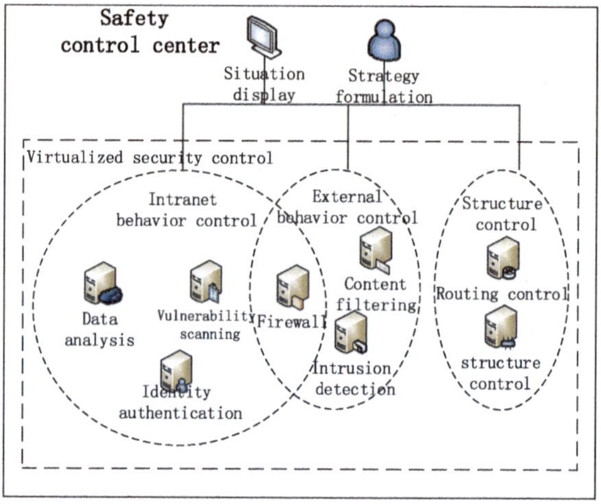

Fig. 2. The composition of the safety control center

Construction of Safety Control Center

The safety control center is a key part of the equipment support information network in the cloud environment to achieve safety protection. It is the network security control center, and its basic structure is shown in Fig. 2. The security control center is a collection of many security components. The security components in the security control center are all coordinated and controlled, which is different from the independent security components in the traditional network system [3]. The specific principle of the work is that the security state of the components in the network can be collected and displayed by the situation display mechanism. The strategy formulation mechanism collects the security status according to the situation and formulates an effective security control strategy, and feeds it back to the corresponding security component. Security components are protected against the network according to the security control strategy and its own security mechanism. In this way, the security control loop is formed to achieve the goal of comprehensive and active defense to the network.

The virtualized security control can be divided into two types: behavior control and structure control. The behavior control mainly solves the problems of network security. According to the actual operation of the network, the security measures can be divided into two parts: the internal network behavior control and the external network behavior control. The behavior control of the external network is mainly aimed at the access of the external network nodes to the internal network. The internal network behavior control is mainly aimed at the running status of the internal network. The implementation of various security mechanisms in behavior control is based on virtualized security component virtual machine, supported by trusted hardware foundation, and constructed by virtual management interface based on security control strategy. Structure control aims at the application scene of the network to solve the problem that the traditional network structure is not adapted to the requirements of the new environment. The realization of structural control is mainly based on the technology of Software Defined Network (SDN). SDN technology is used to control the phase separation of the network and manage the logical topology of the network, so as to achieve structural control.

Construction of Data Center

The data center is the basis for the equipment support information network to realize business functions in the cloud environment. The computing, storage and network services of equipment support business can be realized by the virtualization of the hardware and software resources in the network by using the distributed computing, parallel computing and network computing functions of cloud computing. Its basic structure is shown in Fig. 3.

The cloud data center is mainly composed of computing resources, storage resources and network resources, as well as corresponding application services. Computing resources mainly provide computing processing and computing resource management for equipment support services, which are mainly composed of CPU and memory. The services provided by storage resources mainly include distributed storage, data backup, resource allocation, data migration, etc., which are mainly composed of storage units and databases [4]. The services provided by network resources mainly include the backup and migration of nodes, the backup and migration of network, the load balance of network, etc., which are mainly composed of network node devices and network connection devices. Virtualized resources and corresponding cloud services access the network transport layer through the virtual network interface. The construction of data center is flexible to provide resources for network services, which greatly improves the efficiency of security.

Design and Simulation of Optical Wireless Communication Device Based on SCM

Pengfei Liu[1(✉)], Jinzhu Wang[1,2], Xiaodong Zhou[1], and Yuliang Lu[2]

[1] Army Engineering University, Shijiazhuang 050003, Hebei, China
418650692@qq.com
[2] Army Beijing Representative Office, Shijiazhuang 050003, Hebei, China

Abstract. Wireless laser communication is a cutting edge technology combining laser communication and wireless communication. The author proposes a laser wireless communication solution based on single chip microcomputer. The system adopts STC12C5A60S2 MCU as the main control chip and encodes laser as communication carrier to realize short-range wireless optical communication. Through Optisystem simulation, the line-of-sight communication within 7.2 m can be achieved without blocking at a laser power of 5 mW and a Beam divergence of 170 mrad.

Keywords: Optical wireless communication · SCM · Laser

1 Introduction

At present, with the rapid development of electronic technology and computer technology, people are constantly pursuing automation and intelligence in the fields of life and production. More and more microelectronic device data transmission channels are also more convenient. Wireless signals are used to achieve wireless. Data transmission as a new wireless communication technology has great research and application value [1, 2].

Optical wireless communication (OWC) utilizes high-speed flashing optical signals to realize modulation and transmission of information, and photo-electricity conversion of signals is achieved through photosensitive detection technology, and finally transmission and reception of information are completed. At present, the mobile Internet industry is booming, and humanity has entered the era of “big data and cloud computing” with huge demand for information. Radio frequency communication is an important wireless communication technology currently being used. Radio frequency wireless communication is widely used in various fields of production and life. It is currently the most mature communication technology and has changed people’s production and living habits to a great extent. In the future development of wireless communications, RF communications will continue to play an important role. Compared with radio frequency communication, laser-based wireless optical communication has the advantages of strong confidentiality, large information capacity, and low cost. With the rapid development of science and technology and the continuous progress of society, wireless optical communication technology based on semiconductor lasers will surely gain deep and wide application with its own unique advantages.

J. Abawajy et al. (Eds.): ATCI 2018, AISC 842, pp. 1074–1081, 2019.
https://doi.org/10.1007/978-3-319-98776-7_130

2 System Design

The optical wireless communication system includes a signal source, a control module, a driving module, a transmission window, an OWC channel, a reception window, a photoelectric conversion module, and an information processing module. as the picture shows.

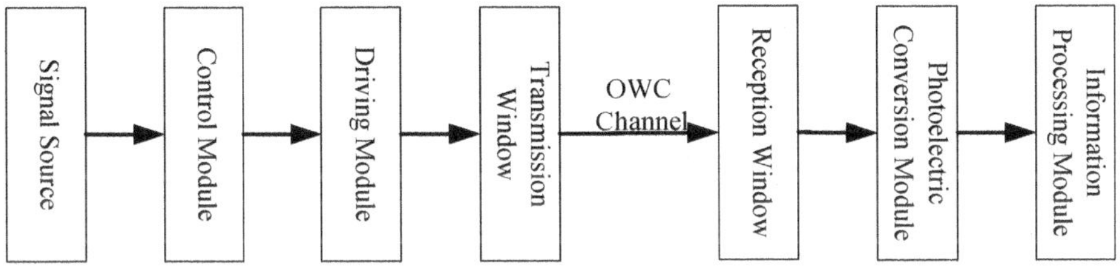

Fig. 1. System of optical wireless communication

2.1 Signal Source

Used for system terminal input and display sender's sending information.

2.2 Control Module

The MCU module is a super-impulse, high-speed, low-power STC89C52RC micro-controller minimum system control module that can use the data sent from the signal source to control and drive the laser driver circuit.

2.3 Drive Module

The module is mainly composed of a laser driving circuit and a laser array light source in the periphery of the MCU. Using the binary signal value transmitted by the MCU module, the optical signal is converted into an optical signal by using light intensity modulation to realize modulation and send out visible light data. In laser driving circuit, considering that the operating frequency of the operational amplifier is relatively low, and a single transistor can complete the operation from one hundred megahertz to one GHz, the laser array driver device in the system selects a single transistor.

2.4 Launch Window

The launch window uses a three-group three-piece structure consisting of three single lenses. The beam waist of the incident Gaussian beam is on the back focal plane of the collimator lens, and the sagittal plane beam waist can be imaged on the back focal plane of the collimator lens through a weak lens. In this way, collimation and aberration correction of the laser information beam emitted from the semiconductor laser are realized, and emission to the receiving end space is performed in an extremely strong direction in the atmosphere (Fig. 1).

2.5 Optical Wireless Communication Channel

Figure 2 shows a linear baseband transmission model of optical wireless communication system.

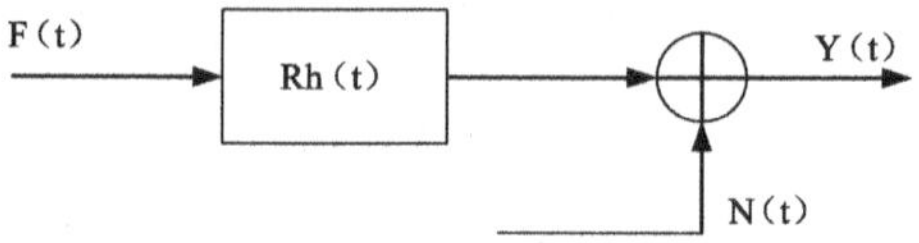

Fig. 2. Linear baseband transmission model of optical wireless communication system

As shown in the figure, F(t) is the emission light input from the launch plate, R is the response efficiency of the photosensitive detection element of the receiving plate, h(t) is the impulse response of the baseband channel, N(t) is the optical noise, Y(t)) is the output photocurrent and its expression is:

$$Y(t) = RF(t)\,h(t) + N(t); \tag{1}$$

Y(t) is proportional to the integral of the instantaneous optical power received on the surface of the photosensitive detector. Indoor lighting and sunlight in the system channel can all be regarded as optical noise. Optical filters and condensers can be used to effectively attenuate the optical noise. The convex lens can enhance the focus of the emitted light, and can expand the detection range of the photosensitive detection module. At the same time, the use of optical filters to filter out stray light can further improve the channel transmission quality and extend the communication distance.

2.6 Receive Window

The receiving window uses a single lens to receive the laser beam. The optical power collected on the focal plane is equal to the power collected by the receiving optical window aperture area from the received light field, and the received solid angle of view is:

$$\Omega_r = \frac{\pi}{4}\left(\frac{d}{f_c}\right)^2 = \frac{A_r}{f_c^2} \tag{2}$$

In the formula, A_r is the detection area; d is the detector diameter; f_c is the lens focal length.

2.7 Photoelectric Conversion Module

The photosensitive detection module receives the laser data and uses the photodiode to convert the optical signal into an electrical signal. Photosensitive detection module uses direct detection technology, the received optical signal is reduced to electrical signals through the photodiode. The light intensity received by the photodiode of the core device is proportional to its own effective receiving area. In the line-of-sight link, the receiving end can increase the effective receiving area of the photodiode by reducing the lens focal length or increasing the refractive index of the lens.

2.8 Information Processing Module

The core device of this module is the minimum system of STC89C52RC one-chip computer, can deal with the signal that the photoelectric receiving module produces, and restore it to original data information.

3 Simulation Analysis

The Optisystem software is used to simulate the communication performance of the system [3]. The transmitter uses a CW Laser as the light source. A Pseudo-Random Bit Sequence Generator pulse generator (NRZ) and a NRZ Pulse Generator are used to generate pseudo-random NRZ digital signal. Amplitude Modulator, is used to perform OOK modulation on the laser signal. The modulated signal passes through the atmospheric channel to the receiving end, receives the signal through the PIN detector and the TIA amplifier, and uses the BRE Analyzer to test its error performance. Architecture of the simulation system shown in Fig. 3.

The system device and atmospheric channel parameter settings are shown in Table 1.

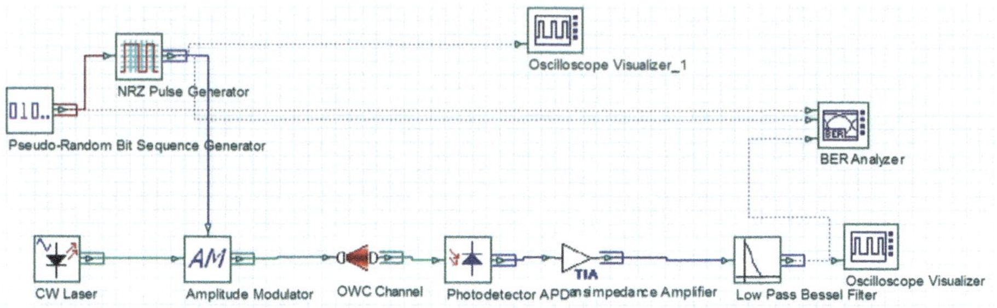

Fig. 3. Architecture of the simulation system

Table 1. Main parameters of the simulation system

Devices	Parameters	Settings
CW laser	Frequency	1550 nm
	Power	5 mW
Pseudo-random bit sequence generator	Bit rate	4 kb/s
	Mark probability	0.5
OWC channel	Range	5–11 m
	Attenuation	30 dB/km
	Beam divergence	170 mrad
	Transmitter aperture diameter	5 cm
	Receiver aperture diameter	4 mm
Photodetector APD	Responsivity	0.7 A/W
	Dark current	10 nA
	Thermal noise	1e−22 W/Hz
Low pass bessel filter	Cutoff frequency	3 kHz

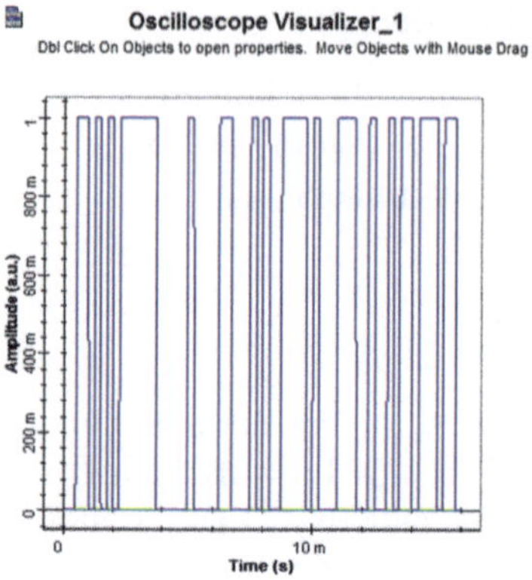

Fig. 4. Launched signal waveforms

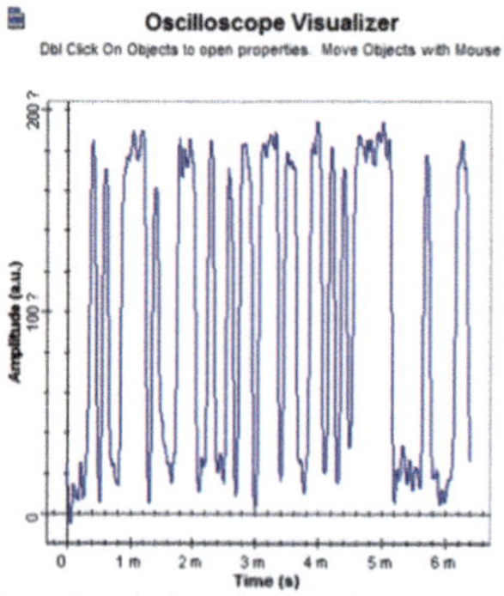

(a) Received signal waveforms at 5m

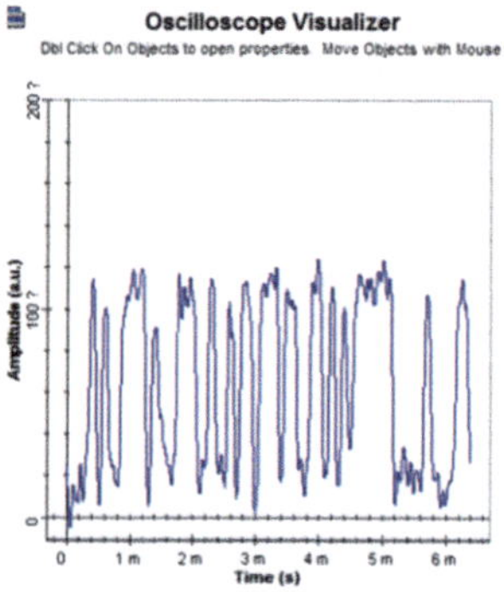

(b) Received signal waveforms at 7m

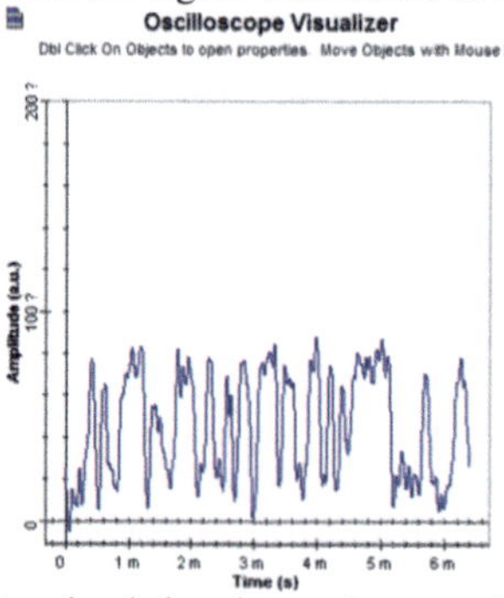

(c) Received signal waveforms at 9m

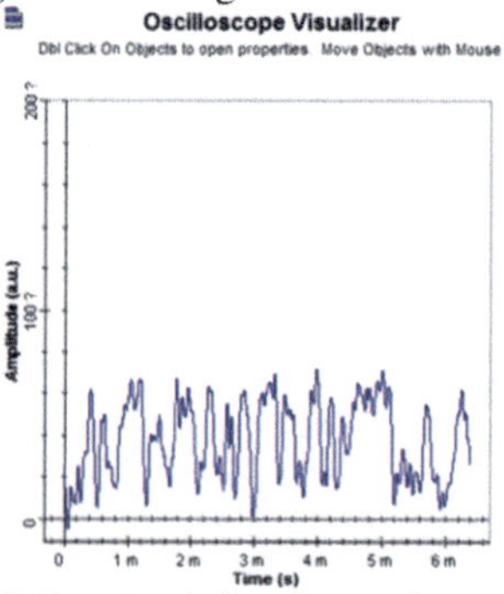

(d) Received signal waveforms at 11m

Fig. 5. Received signal waveforms

Under the above parameter settings, the transmitter emission signal is shown in Fig. 4, and Fig. 5 shows the signal waveforms received by the system at 5, 7, 9 and 11 m, respectively. As can be seen from the figure, the received signal is basically the same as the transmitted signal on the waveform, and the transmitted digital information can be extracted from the waveform to achieve signal transmission. The level of the received signal at 9 and 11 m drops significantly, mainly due to the large divergence angle, which results in attenuation of the received signal with distance transmission.

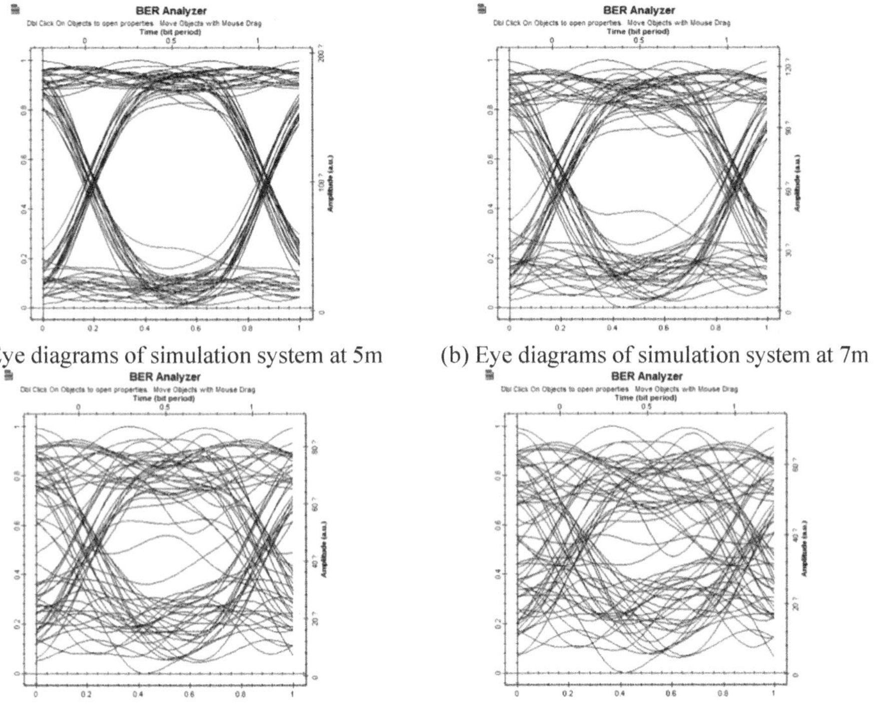

(a) Eye diagrams of simulation system at 5m (b) Eye diagrams of simulation system at 7m

(c) Eye diagrams of simulation system at 9m (d) Eye diagrams of simulation system at 11m

Fig. 6. Eye diagrams of simulation system

Figure 6 shows the eye diagrams for signal reception at 5, 7, 9, and 11 m. From Fig. 9(a) and (b), it can be seen that the Q values at both the 5 m and the 7 m reach 10.1124 and 5.70641 respectively, indicating good communication performance. The system's bit error rate can reach 10^{-9} orders of magnitude. From the figure, it can be seen intuitively that the eyes are open, the eye diagrams are clear, and the quality is good. This means that the inter-code crosstalk is small and the communication performance is good. From Fig. 9(c) and (d), it can be seen that the Q values at both 9 and 11 m are 3.443 and 0, respectively, and the communication performance is already poor at 9 m, and it is completely impossible to communicate at 11 m. Table 2 shows the main performance parameters of the system.

Figure 7 shows the error rate as a function of distance. It can be seen from the figure, the system error rate increases as the distance increases. In the case of a

Table 2. Performance parameters of system

Parameters (m)	Max. Q factor	Min. BER	Eye height
5	10.1124	2.40316e−24	0.000111666
7	5.70641	5.69721e−9	4.17199e−5
9	3.443	0.000283938	6.69645e−6
11	0	1	0

transmission power of 5mW, and Beam divergence of 170 mrad, the BER of less than 10^{-9} is used as a measure, and its effective communication distance can be 7.2 m.

The noise of communication system mainly includes the noise of amplifier, the undercurrent noise of photodiode the flashing of background light, and the flashing noise of the letter itself. Amplification of the noise of the circuit and from the power or other external electrical interference emission and receive optical window.

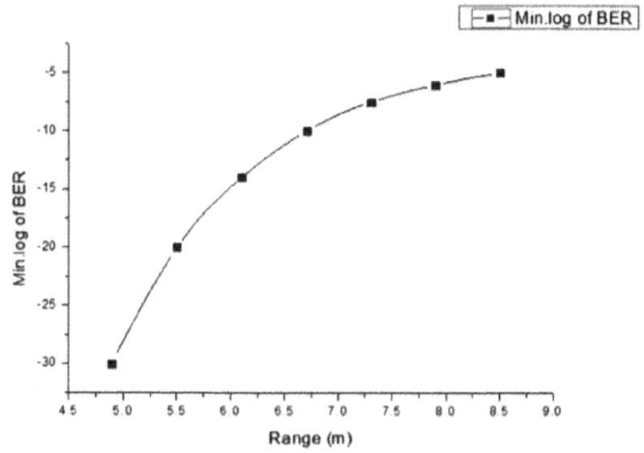

Fig. 7. Min.log of BER vs. range

4 Conclusion

SCM wireless communication network has a very close relationship with people's daily life. Therefore, the completion of the design work of the system is not only to meet the needs of its own use, but also to adapt to the development of the entire society. Based on the design and implementation of wireless optical communication device based on single-chip microcomputer, laser transmission technology, laser receiving technology and single chip microcomputer technology are combined, and through the establishment of communication protocol, the communication reliability of the system is enhanced, and the bit error rate is reduced. A reliable transmission of the communication is completed. The laser communication can be realized under the condition that the 7.2 m internal emission power is 5 mW, and Beam divergence is 170 mrad.

References

1. Zhang, J.: Design and implementation of wireless communication system based on single chip microcomputer. **36**(1), 83–85 (2017)
2. Fan, X.: Design of wireless optical communication device based on MCU. **32**(5), 67–68 (2015)
3. Li, F., Hou, Z., Wu, Y.: Optisystem experimental simulation of FSK optical wireless communication system. **9**(5), 42–47 (2011)
4. Peng, L., Fang, W.: Heterogeneity of inferring reputation of cooperative behaviors for the Prisoners' Dilemma game. Phys. A: Stat. Mech. Appl. **433**, 367–378 (2015)
5. Watts, D.J.: Small worlds. The dynamics of networks between order and randomness. America: 56A2579BB6, l999-05-13
6. Aleksy, M., Korthaus, A., Schader, M.: Use Java and the CORBA realization distribute type system. J. Pingxiang College, No. 4, pp. 104–105 (2005)
7. Titus, J.: ECN technical editor: the eclipse of stand. J. Zhongkai Agrotech. Coll. **19**(2), 32–35 (2006)
8. Aleksy, M., Korthaus, A., Schader, M.: Use java and the CORBA realization distribute type system. J. Pingxiang College, No. 4, 104–105 (2005)
9. Richardson, W.C., Avondolio, D.: The Java high class weaves a distance: JDK 5. Scientific & Technology Book Review, No. 3, pp. 17–18 (2006)
10. Lipson, J., Vuong, S.: A semantics-based routing scheme for grid resource discovery. In: E-Science: First International Conference on E-Science and GridComputing, pp. 58–70, 90 (2005)

Fitting of Nonlinear Function Based on Elman Neural Network

Yong Shan, Wenbin Liu(✉), and Tao Chen

Department of Public Course, Wuhan Technology and Business University, Wuhan 430065, Hubei, China
812926096@qq.com

Abstract. In order to solve some complex nonlinear systems in the practice application, a nonlinear function equation method based on Elman neural network is proposed. A limited number of input and output data of nonlinear function equations were used to train the Elman neural network, so that the network could express the nonlinear function. Finally, the output of nonlinear functions is predicted by using a trained neural network. The results show that the nonlinear function based on Elman neural network has a good fitting effect.

Keywords: Nonlinear function · Fitting · Elman neural network

1 Introduction

In the application of scientific practice, some more complex nonlinear problems are encountered, and it is difficult to express with more precise mathematical equations. Based on the above situation, the artificial neural network can be used to solve these problems [1]. Artificial neural networks can treat unknown systems as a black box. Through the finite input and output data of nonlinear problems, Artificial neural networks can be trained to express unknown nonlinear functions. Finally, the nonlinear problem is solved by using a trained artificial neural network [2].

The most common neural network is BP neural network, and which is a feedforward neural network, which has nothing to do with the previous output of the network. The input of the feedback neural network includes the feedback of delayed input or output data, so the research is better [3].

In this paper, a more common post - feed neural network - Elman neural network is used. At the same time, in order to simulate the solution of such nonlinear problems, a specific nonlinear function is used herein for the method test of the Elman neural network. The specific formula of the nonlinear function is as follows, and the graph is shown in Fig. 1.

$$y = 2x_1^2 + 3x_2^2 \tag{1}$$

J. Abawajy et al. (Eds.): ATCI 2018, AISC 842, pp. 1082–1087, 2019.
https://doi.org/10.1007/978-3-319-98776-7_131

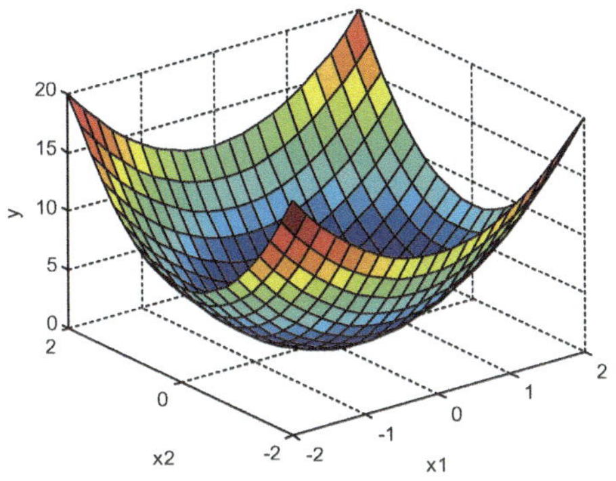

Fig. 1. Graph of nonlinear function

2 Elman Neural Network

Elman neural network is a common feedback neural network, which was proposed by Elman in 1990 [4]. The common BP neural network is a feedforward network whose output is determined by the current input and weight matrix, but not the previous output of the network. The input of the feedback neural network includes the feedback of delayed input or output data, and because of the input of feedback, it is a feedback dynamic system. The learning process of the system is the change process of the neuron state, which will eventually reach the stable state of a neuron, marking the end of the learning process [5]. Elman neural network model adds a layer in the hidden layer of feedforward network, as a one-step delay operator to achieve the purpose of memory. Thus, the system has the ability to adapt to time-varying characteristics and can directly reflect the characteristics of the process system [6].

The Elman neural network is generally divided into four layers: the input layer, the hidden layer (the middle layer), the receiving layer and the output layer (Fig. 2). The input layer, the hidden layer and the output layer are similar to the feed forward network. The unit of the input layer only serves as the signal transmission function, and the output layer unit plays a linear weighting function [7]. The transfer function of the

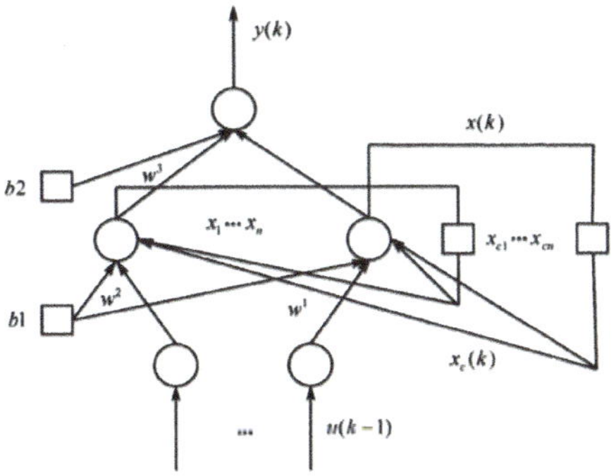

Fig. 2. Structure of Elman neural network

hidden layer unit can be linear or non-linear, and the receiving layer is also called the state layer. It is used to remember the output value of the previous moment of the hidden layer unit and return the input to the network [8].

As shown in the picture above, the nonlinear state space expression of Elman network is as follows.

$$y(k) = g\left(w^3 x(k)\right) \tag{2}$$

$$x(k) = f\left(w^1 x_c(k) + w^2 u(k-1)\right) \tag{3}$$

$$x_c(k) = x(k-1) \tag{4}$$

In the formula, y is the input node vector of m dimension, x is n dimension middle layer node unit vector, u is r dimension input vector, and x_c is n dimensional feedback state vector. The w^3 is the connection weight of the middle layer to the output layer, and w^2 is the connection weight of the input layer to the middle layer, and w^1 is the connection weight of the receiving layer to the middle layer. The g(*) is the transfer function of the output neuron, which is a linear combination of the middle layer output. The f(*) is the transfer function of middle layer neurons, and S function is often used. The Elman neural network adopts BP algorithm to make weight correction, and the learning index function takes the error sum of squares and functions [9].

3 Experimental Results

First, the nonlinear function equation in Eq. (1) is used to generate several groups of input and output data randomly. The data is then processed into two parts: training set and test set. The training set will be used for the training of Elman neural network. Finally, the fitting effect of Elman neural network is tested with the test set.

3.1 Data Preprocessing

The 400 sets of input and output data are randomly generated using a non-linear function in Eq. (1). In order to prevent the single variable from affecting other variables, and to obtain a comparable range of values, linear transformation is performed on all data, and the range of variable values after transformation is [0.01, 0.99]. The 400 groups of processed data were randomly divided into training sets and test sets, thus the data volume of the training set is 320, and the data of the test set is the other 80 groups.

3.2 Elman Neural Network Model

According to the input and output data of the nonlinear function, the established Elman neural network structure is 2-5-1 (Fig. 3). The two input neurons represent x_1 and x_2 in the equation, and the number of hidden layer neurons is set to 5, one output neuron corresponds to the y value in the equation finally [10].

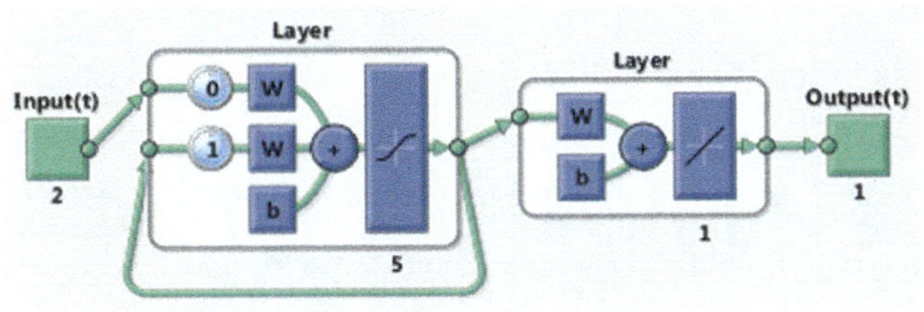

Fig. 3. Structure of 2-5-1

The experimental data consists of 400 preprocessed set of data, and each set of data has two input values and an output value corresponding to it. The training set and test set of Elman neural network were randomly divided into 320 groups and 80 groups. The data of the 320 training sets is used to train Elman neural network, with the gradual decrease of network error in the training process, the training error tends to be stable after more than 2000 training (Fig. 4).

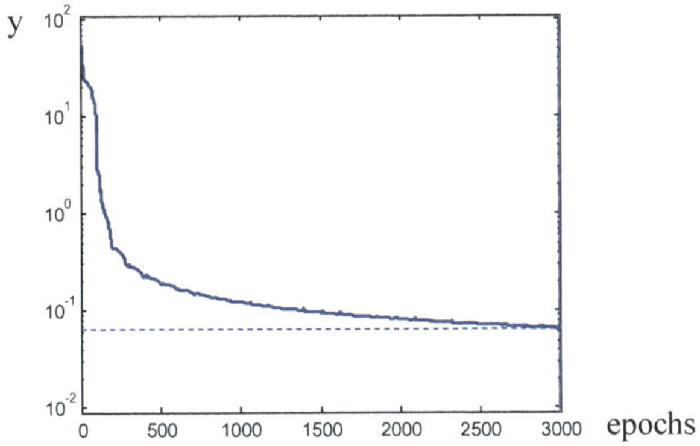

Fig. 4. Training errors

3.3 Result Analysis

The trained Elman neural network was used to predict 80 sets of test set data that did not participate in the exercise, then the prediction output and expected output of Elman neural network are shown in Fig. 5.

As can be seen from Fig. 5, the prediction output of Elman neural network is better fitting the real value of the nonlinear function. The average relative error of the 80 test set data was less than 3.4%, so the prediction accuracy of Elman neural network is high. There is still some error in the small amount of test data, which indicates that there are some problems in the fitting of nonlinear function based on Elman neural network, it is necessary to improve the training accuracy of neural network in the future.

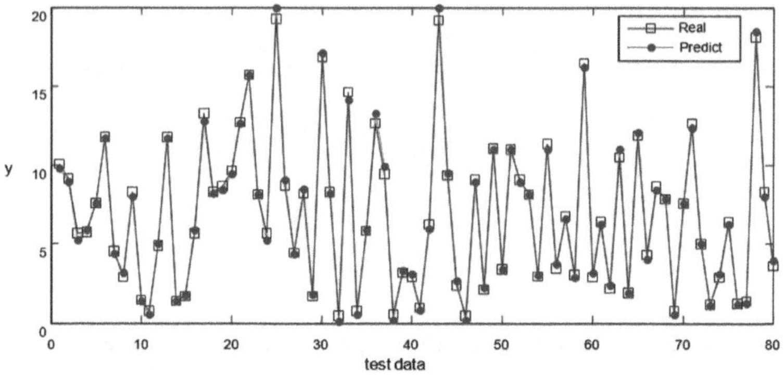

Fig. 5. Prediction of Elman neural network

4 Conclusion

In this paper, a method based on the Elman neural network is for nonlinear function fitting for related research. Methods by nonlinear function equation of 2000 groups of input and output data of Elman neural network training, enables the network to express the nonlinear function. Through the use of trained Elman neural network to predict the output of the nonlinear function, its prediction relative error less than 3.4%, this indicates that the nonlinear function fitting based on Elman neural network is feasible. In the future, training data can be increased to improve the training accuracy of neural network.

References

1. Abdulkarim, M., Shafie, A., Razali, R.: Function fitting for control source electro-magnetics data using Elman neural network. In: IEEE, International Conference on Computer and Information Science 2012, pp. 453–457 (2012)
2. Sun, J.: Application of Elman neural network in GPS elevation fitting. Geospat. Inf. **6**(6), 109–113 (2013)
3. Sutrisno, I., Abu Jami'in, M., Hu, J.: An improved Elman neural network controller based on quasi-ARX neural network for nonlinear systems. IEEJ Trans. Electr. Electron. Eng. **9**(5), 494–501 (2014)
4. Zhihong, L.U., Xiaodong, W.U., Guo, L.: Prediction of the number of rodent community composition species based on Elman neural network in Alasan desert. Ecol. Environ. Sci. **18** (2), 92–98 (2015)
5. Wang, X., Zhang, H.J., Wang, F.: The improved Elman neural network application in X-ray detector. Nucl. Electron. Detect. Technol. **17**(5), 194–199 (2015)
6. Yang, B., Ma, X.: Temperature prediction and analysis based on BP and Elman neural network for cement rotary kiln. Proc. Spie **79**(2), 357–364 (2010)
7. Huang, Y.C., Tian, J.P., Yang, H.L.: Thermal error modeling for machine tool based on genetic algorithm optimization Elman neural network. Modul. Mach. Tool Autom. Manuf. Tech. **15**(3), 82–89 (2015)

8. Tian, S., Chen, J.: An evaluation model for tailings storage facilities using improved neural networks and fuzzy mathematics. J. Appl. Math. **32**(6), 95–100 (2014)
9. Zhang, X.L., Shao-Qing, L.I., Tian, L.Y.: Research of flatness pattern recognition based on the Elman neural network. Caai Trans. Intell. Syst. **16**(1), 45–49 (2010)
10. Shi, F., Wang, X.-C., Yu, L.: MATLAB neural network analysis of 30 cases. Beijing University of Aeronautics and Astronautics Press, Beijing (2010)

Research on Information-Based Teaching Model of Computer Aided Design in Big Data Era

Wenjun Wang(✉)

City College of WUST, Wuhan, China
15827425957@163.com

Abstract. Computer aided design in college is a required course for the environmental design major. The traditional teaching model has obvious limitations in the background of the current big data era. In order to realize the modernization of education and make full use of big data and information technology to provide a newer and more efficient learning method for computer-aided design learning. The construction of the information-based teaching model is the basis of university teaching in the era of big data. Therefore, this paper focuses on the path of computer aided design on the construction of the information-based teaching model.

Keywords: Big data · Information · Computer aided design

With the continuous development of information technology and the popularity of various terminal devices, we have entered a new era of big data. The arrival of the era of big data is affecting all areas of humanity. It brings not only massive data, but also changes our traditional thinking patterns and behavioral habits. In the field of education, especially the computer-aided design education of the environmental design major, it is influenced by the background of this era because of its teaching content, teaching methods and audience groups.

At the same time, colleges and universities have also followed the pace of the times in the teaching philosophy and launched various information-based teaching models. The exploration and research will be adapted to the update of traditional education and teaching models, and explore more scientific, efficient and sustainable teaching methods and teaching models.

1 University Computer Aided Design Teaching Status

Lack of Motivation for Learning. The younger students have very limited understanding of the role and importance of computer-aided design courses in practical work, and at the same time, they are less familiar with professional design software. Therefore, many students lack the interest and motivation to learn.

Professional Orientation is Not Strong. Computer-aided design in the environment-design major aims to achieve universal applicability to students, but it emphasizes the

J. Abawajy et al. (Eds.): ATCI 2018, AISC 842, pp. 1088–1092, 2019.
https://doi.org/10.1007/978-3-319-98776-7_132

lack of specialization of the profession. As a result, it takes a period of time for the students to match what they have learned with actual work.

Students' Ability in Computer Aided Design is Uneven. As a required course for environmental design majors, because of the unbalanced level of education previously received by students, after entering colleges and universities, students' computer skills and capabilities have a significant gap.

2 Research of the Teaching Model

Traditional computer-aided design mainly depends on the teaching materials and the experience. Under the background of big data era, teaching resources will no longer stick to this. More text, pictures, and video teaching resources will enable students to acquire computer-aided design knowledge in different ways. These more vivid teaching resources can not only stimulate students' interest, but also enable them to have more choices in learning time and space. Solving the problem of individual differences in learning interest and student ability.

Big Data Makes Teaching Model More Scientific and Reasonable. Traditional computer-aided design teaching models need innovation and change. With the aid of the big data platform, the big data diagnostics teaching will be fully implemented. The system will judge the easiest points to improve according to the student's big data. Then the teacher will carry out an accurate explanation of the related knowledge points and provide relevant exercises to ensure students' understanding. And then a few days later, equipped with corresponding exercises for diagnosis and consolidation. After some practice, the teaching effect has been significantly improved.

Teachers should adopt new and more scientific teaching models and design the advantages of big data to the teaching, so that computer-aided design teaching effects can be significantly improved. Make full use of MOOC, micro-classes, and flipped classrooms. These new teaching modes allow students to use various resources to learn relevant courses. On the other hand, teachers can make related learning materials and send the data and information to the platform for student reference and study.

Big Data Helps Precision Teaching. Education informatization is a concrete application of the combination of hardware and software. The main purpose of computer aided design education informatization is to explore new teaching models and improve teaching quality. According to the requirements for the cultivation of innovative talents, the use of information technology, exploration of new modes of education, promotion of modernization of education, and improvement of students' overall quality. Jumping out of the boring descriptions and numbers on the paper, in the process of information-based teaching in the era of big data, intuitively find typical cases based on learning effect analysis charts, analyze detailed results of teaching results, after-school assignments, etc. for collective communication, and then follow the system. Collect lessons by mastering the knowledge of statistics.

Based on cloud computing, big data, etc., data collection, real-time analysis, and learning diagnosis are used to realize the accuracy of teaching management and teaching evaluation, and to promote teaching in accordance with individual aptitude and individualized learning.

Smart Matching. In the first stage of the class, based on the preset sequence of the system, the teacher concentrated on the typical topics marked during lesson preparation and prioritized the common problems in class testing. In the second stage, the system matched the student's answer to the database intelligence and automatic recommendation. After giving each student different learning resources, the teacher views the situation in real time, feeds back the high-frequency commonality problems that the student responded to, and personalized counseling. In the final stage, the teacher dynamically adjusts the typical topics, once again focus on common issues, combined with the answers to high-frequency wrong answers in a timely manner to explain, and consolidate. After the end of a class, students' individual doubt points have been effectively solved, and the score rate has also been greatly improved.

During the teaching process, each student will receive a customized and personalized operation. At the same time, the system will also generate a record of each student's practice, intelligent matching to the relevant practice after the formation of personalized advanced exercises, students can preview in advance according to relevant tips. Behind the big data system is a full range of data warehouses. Teachers have some references when arranging exercises; on the other hand, each student's learning resources are tailored to the system based on their own error-prone points.

Taking full advantage of the Flipped Classroom, supplemented by the most advanced on-line distance education platform and online technology, teachers are not traditional knowledge initiator but rather a mind-leader who allows students to get their own answers.

In the age of the Internet, people's demand for education changing the training mode that focuses on skills and obedience in the industrial age. It has put forward a wide range of demands for individuals who are highly individualized, considerate, self-aware, and capable of exploring.

Flipped Classroom. Readjust the time inside and outside of the classroom. And give the students the initiative to learn. In this classroom mode, students use video resources or information data in the platforms to learn about the content of their lectures without having to use valuable classroom time to learn basic knowledge. Classrooms serve as a place for teachers and students to communicate and interact, including questions and answers. The teaching model of the Flipped Classroom can play an active role in the computer aided design courses.

MOOC. It is an online teaching mode based on big data technology. Publish resources, learn management systems, and integrate learning management systems with more open network resources. In this mode, teachers can publish curriculum materials on the Internet's data platform. Students use the materials in these interactive platforms to learn computer-aided design content on the Internet. Teachers can set test practices in the course to allow students to self-check learning results. In the learning data platform, everyone can discuss with each other about some problems in the learning process, thereby improving learning interest and learning efficiency. The data platform also records the weak points of student learning. Accumulation of these data provides teachers with more accurate and difficult points for lesson preparation and teaching. Therefore, teachers will focus on explaining the weak points of students when preparing lessons and teaching.

Micro Lesson. Mainly using video as the main carrier, it records the difficult points in the process of computer aided design teaching in and out of the classroom. These more detailed, specialized, and subjective teaching content with more specific and targeted. Micro Lessons can significantly improve the efficiency of student learning related difficulties. Reduce student learning pressure.

Evaluation System. The application of the new teaching mode has made it difficult for the traditional assessment system of computer aided design to meet the needs of reality. Therefore, new standards need to be established to analyze and evaluate student learning outcomes and teacher teaching outcomes.

For students, one of the most important traditional assessments of attendance is no longer applicable. Instead, teachers can create independent learning files based on students' participation in learning. Record each student's history information such as registration, discussion and practice in relevant courses in real time, and use this as a basis for evaluation. To change the traditional students' passive learning methods with the main purpose of obtaining scores, to effectively improve students' learning enthusiasm and learning efficiency.

For teachers, because of the major changes in the teaching model, as a support for the realization of the curriculum model, the instructors of computer aided design courses need to be able to answer students' questions in the self-learning process at any time. At the same time, teachers also need to have excellent courseware production capabilities and Internet knowledge in order to produce a large number of computer aided design curriculum learning data and apply it to the public interactive platform for students to learn and reference.

Strengthen the construction of teachers. In order to meet the needs of improving the ability of instructors in the context of the big data era, computer aided design instructors need to improve their professional knowledge and Internet capabilities and keep pace with the times. Education management departments and schools should also provide teachers with more opportunities for on-the-job training and even opportunities for international exchange.

3 Conclusion

In the era of big data, we should make full use of big data informatization methods to make complex content easily to understand, enrich the teaching model, and simplify assessment methods for students' learning outcomes. Computer-aided design relying on the Internet for informatization can not only improve students' enthusiasm for learning, but also greatly improve students' learning efficiency.

References

1. Zhao, J.Y.: Research on the informative teaching mode of university under the environment of big data. Inf. Sci. **34**(1), 92–95 (2016)
2. Yang, X.M., Wang, H.L.: Education big data application model and policy recommendations. E-educ. Res. **09**, 56–63 (2015)
3. Wang, Y.X.: Study on flipped classroom teaching mode based on low cost platform. In: Fujian Normal University, pp. 52–63 (2015)
4. Chen, L., Chen, Y.H.: The modern path of education is analyzed by informatization. Educ. Res. **11**, 114–118 (2013)
5. Cai, Z.Q.: Analyze the role of teachers in flipped classroom. J. Bingtuan Educ. Inst. **04**, 51–53 (2014)
6. Lou, Z.H.: Case study of teaching structure reform under information environment. In: Shanghai Normal University, pp. 32–43 (2013)
7. Gen, J.: Computer aided instruction software design and development. In: Shandong Normal University, pp. 28–33 (2003)
8. Zhong, X.L.: Research on teaching design based on flipped classroom concept in information environment. Open Educ. Res. **01**, 60–66 (2013)
9. Luo, S.F.: The comparison and change of flipped classroom and traditional teaching mode. Orient. Enterp. Cult. **15**, 54–56 (2014)
10. Peng, M.C.: The enlightenment of flipped classroom on teacher training in vocational colleges. J. Changzhou Coll. Inf. Technol. **06**, 55–56 (2014)
11. Liu, J.Z., Wang, D.: A review on the research and practice of flipped classroom at home and abroad. Theory Pract. Contemp. Educ. **02**, 72–75 (2014)
12. Fan, X.Y.: Study on flipped classroom teaching model in America. In: Northeast Normal University, pp. 17–23 (2015)

Mixed Feature Selection Method Based on SVM

Hairui Lei(✉) and Xiufeng Gao

Shijiazhuang Campus of Army Engineering University,
Shijiazhuang 05003, China
larry134@163.com

Abstract. Aiming at feature selection in network traffic classification, a mixed feature selection algorithm based on SVM was proposed. This method uses the eigenvalues based on the ratio of detection rate and false alarm rate as the evaluation index for feature screening. First filter the noise and irrelevant features and reduce the feature dimension by using the F-score and the information gain in the filtering mode. Then, based on the selected cross-features subsets, use the sequential back search algorithm in the wrapping mode select the optimal feature subset in combination with SVM. The simulation results show that the optimal feature subset selected by this algorithm has good classification ability, and it also has better performance in reducing modeling time and testing time.

Keywords: Feature selection · SVM · F-score · Sequence back search

1 Introduction

Feature selection is an important research content in the field of pattern recognition, data mining and other fields. It selects the feature subsets in the original feature set to form feature subsets to achieve the purpose of reducing the data dimension while maintaining or improving the system classification performance. Different from feature extraction, feature selection preserves the original physical features. Therefore, it can actually reduce storage requirements, measurement requirements, computational costs, etc.; while features extracted by feature extraction are linear combinations of original features, and each feature-reduced feature is related to the original, so the feature extraction feature retained under no physical meaning. In addition, feature extraction cannot reduce storage requirements, measurement requirements, computational costs, etc., and does not have the advantages of feature selection [1]. Feature selection plays an important role in building a concise and efficient classification system. It not only reveals the hidden patterns and laws hidden in the system, but also makes it possible to visualize the classification results. Therefore, feature selection, especially feature selection based on SVM (Support Vector Machines) has attracted much attention.

According to whether the evaluation criteria depends on the learning algorithm, the feature selection algorithm can be divided into filtering mode and wrapping mode. Filter is independent of the learning algorithm, low calculation amount, and the noise

J. Abawajy et al. (Eds.): ATCI 2018, AISC 842, pp. 1093–1100, 2019.
https://doi.org/10.1007/978-3-319-98776-7_133

characteristics can be effectively removed. Wrapping mode needs to determine the classifier algorithm in advance and then use the classifier to evaluate the feature set, which has good classification performance, but large computation cost. At present, how to combine filtering mode and wrapping mode to further improve the classification performance is the focus of feature selection algorithms. For filtering mode, an algorithm based on Maharanis distance feature ranking and exhaustive search was proposed to improve the classification accuracy of intrusion detection systems [2]. And, a method for filtering noise and redundant features using information gain was proposed to reduce the difficulty of dimension reduction and data processing [3]. For wrapping mode, a method that combines Kappa coefficients and fuzzy neural networks is proposed to select the optimal feature set [4]. And, a wrapping algorithm based on Bayesian network classifier was proposed to achieve the purpose of improving system detection performance [5].

This paper proposes a mixed feature selection method based on SVM, which combines filtering mode and wrapping mode. By constructing a feature evaluation value based on the ratio of detection rate and false alarm rate, it is used as a feature selection evaluation index. Firstly, the F-score and the information gain are used to filter the noise and irrelevant features respectively, and the feature subsets are cross-combined. Then based on the selected cross feature subsets, a sequential back search algorithm is used to select the optimal features in combination with the SVM.

2 Mixed Feature Selection Method

2.1 Method Flow and Related Definition

Given the original feature vector $F = \{F_1, F_2..., Fi..., F_m\}$, where m is the feature dimension and F_i is the ith feature in the feature vector. The fisher score of F_i in the feature vector F is represented by FS_i, and the feature subset calculated and filtered by fisher score is represented by FS_Sub. The information gain value of F_i in the feature vector F is represented by IG_i, and the feature subset calculated and filtered by the information gain is represented by IG_Sub. A feature evaluation value is the ratio of the detection rate to the false positive rate of a feature or group of features, introduced by F_DS.

$$F\_DS_i = DR_i/FR_i \tag{1}$$

Above formula (1), i represents an ith dimension feature or an ith group feature; DRi represents the detection rate of the ith dimension feature; FR_i indicates the false positive rate of the ith dimension feature. If the detection rate of a feature is higher, the false positive rate is lower, the feature evaluation value is larger, and the classification ability is better.

The overall process for feature selection is demonstrated as in Fig. 1.

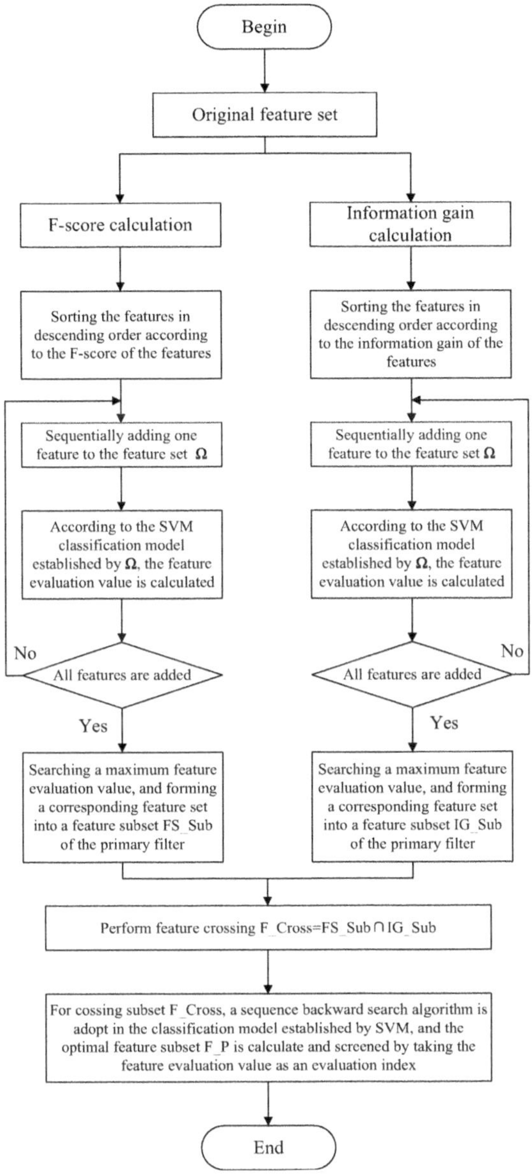

Fig. 1. Process of the mixed feature selection

2.2 F-Score

F-score maximizes the distance between samples of different classes, while minimizing the distance between samples of the same class, and uses the ratio of the two as the fisher score of the feature. FS_k represents the fisher score for the kth dimensional feature.

$$FS_k = \frac{S_{b,k}}{S_{w,k}} = \frac{(\overline{m_{1,k}} - \overline{m_k})^2 + (\overline{m_{2,k}} - \overline{m_k})^2}{\delta_{1,k}^2 + \delta_{2,k}^2} \tag{2}$$

Above the formula (2), $S_{b,k}$ is the class-to-class dispersion, indicating the distance between different class samples; $S_{w,k}$ is the intra-class dispersion, indicating the distance between similar samples; $\overline{m_{1,k}}$, $\overline{m_{2,k}}$, and $\overline{m_k}$ represent the mean of the kth dimensional features of the positive, negative, and all samples. $\delta_{1,k}^2$, $\delta_{1,k}^2$ represent the variance of the kth dimensional features of positive and negative class samples. The larger fisher score of features, the stronger classification ability of features and the greater contribution to classification.

2.3 Information Gain

Information gain is an effective filtering feature selection algorithm based on sample information, which can be used to measure the correlation between a feature and a category. The greater the information gain value of the feature, the higher the correlation between the feature and the category, and the stronger the category distinguishing ability of the feature. The information gain of feature a is defined as the difference between the original information demand and the new demand and expressed by IG(A) [6].

$$IG(A) = Info(D) - Info_A(D) = -\sum_{i=1}^{m} p_i \log_2(p_i) - \sum_{j=1}^{v} \frac{|D_j|}{|D|} \times Info(D_j) \tag{3}$$

Above the formula (3), $p_i = |D^*|/|D|$, p_i represents the probability that any sample belongs to the class C_i; $|D_i|$ is the number of samples belonging to class C_i; $|D|$ is the total number of samples; m is the number of sample categories; v is the number of partitioned subset D_j.

2.4 Feature Filtering and Feature Cross-Combining

In the mixed feature selection method, fisher score and information gain are combined to select feature subsets according to their own characteristics, and then these feature subsets are crossed. Since the method of feature filtering using fisher score and information gain respectively is the same, only the feature filtering method for fisher score is given below.

(1) Setting an empty feature set Ω;
(2) For the original feature vector F, the fisher score FS_i of each dimension feature F_i is calculated using formula (2);
(3) Sorting fisher scores FS_i of all features in descending order to form a feature sequence FS_Sort;
(4) From fs_sort, sequentially remove a feature and add it to Ω;
(5) According to the feature subset in Ω, a classification model is established by SVM, and the feature evaluation value F_DS_i is calculated;

(6) Repeat steps (4) and (5) until all features in FS_Sort are added to Ω;
(7) Searching for the maximum F_DS_i value;
(8) The feature subset corresponding to the maximum F_DS_i value is taken as the filtered feature subset to form the feature subset FS_Sub.

In the same way, for the original feature vector F, the information gain value IG_i of the feature is tested and calculated using formula (3), and the feature subset IG_Sub is screened and formed by the feature evaluation value.

Feature cross-combining is to combine the feature subsets screened by different feature selection algorithms together, and select the same features existing in different feature subsets to form cross-feature subsets. Features in the cross-feature subset have the ability to classify features from different angles. In the meanwhile, the features in the cross-feature subset will have better adaptability when facing data sets with different characteristics. A subset of cross- feature is represented by F_Cross, $F\_Cross = FS\_Sub \cap IG\_Sub$.

2.5 Feature Quadratic Optimization Selection

The cross-feature subset has removed noise features, but it still has some redundant features. Based on the cross-feature subset, the feature optimization selection uses the sequence backward search algorithm, uses SVM as classifier to establish the classification model, and uses the feature evaluation value as evaluation index to further perform the feature optimization selection. The method is shown in Table 1.

Table 1. Wrapper method based on sequence backward search and SVM algorithm

Input: F_Corss = {F1, F2,..., Fi,..., Fk}, K is the number of features
Output: Optimized feature subset F_P
Step1: The SVM model is established by F_Cross, and the classification accuracy is tested, and the value is given to F_DS_Max
Step2: for i = 1:k, The feature Fi is remove from F_Cross to obtain a feature subset F_Cross'
Step3: SVM model is established by F_Cross', and the classification accuracy F_DSi is tested. If F_DSi is larger than F_DS_Max, F_DSi is assigned to F_DS_Max
Step4: Returning to step 2, when i is equal to k, step 5 is executed
Step5: Assigning the feature subset F_Cross' of the maximum classification accuracy F_DS_Max to F_Cross and F_P, and returning to step 1 if there is a change in F_P, otherwise outputting BF as the optimal feature subset

3 Simulation Experiment

Simulation experiments are carried out on a 10% training subset of KDDcup99 [8] data set. In that experiment, a feature selection algorithm is implemented in the MATLAB R2011B environment, LibSVM (libsvm-mat-3.1) is used as a training and teste tool. The kernel function adopts RBF kernel function in C-SVM, and its parameters are set as follows: penalty coefficient c is 1.2, and kernel radius g of RBF function is 2.8.

3.1 Experiment on Feature Filtering and Feature Cross Combination

Feature filtering is mainly used to remove noise and extraneous features. In the test, about 6000 records were randomly selected from 10% KDDcup99 training subset at a ratio of normal to abnormal data ratio of about 4:1 as training set, and more than 2300 records which did not intersect were randomly selected as test set. According to the original 41-dimensional features, fisher scores and information gain values of different features are calculated and sorted in descending order respectively. Based on the sorted fisher sub-feature sequence and information gain feature sequence, the features are selected to form a feature subset in turn, and the classification model is established by SVM, and the feature evaluation value is calculated. The calculation results are shown in Figs. 2 and 3.

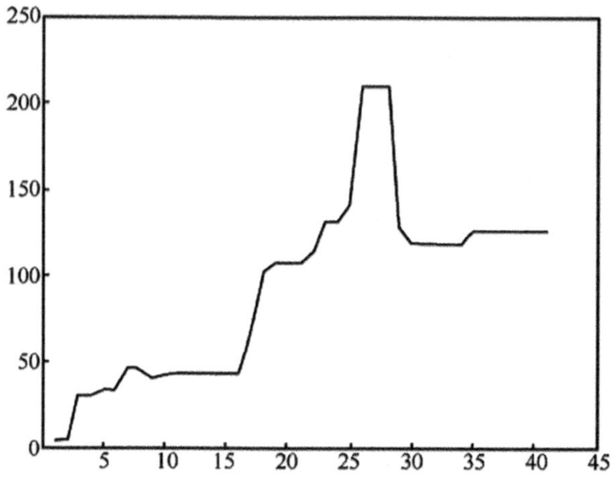

Fig. 2. Result of feature evaluation value calculated by fisher score

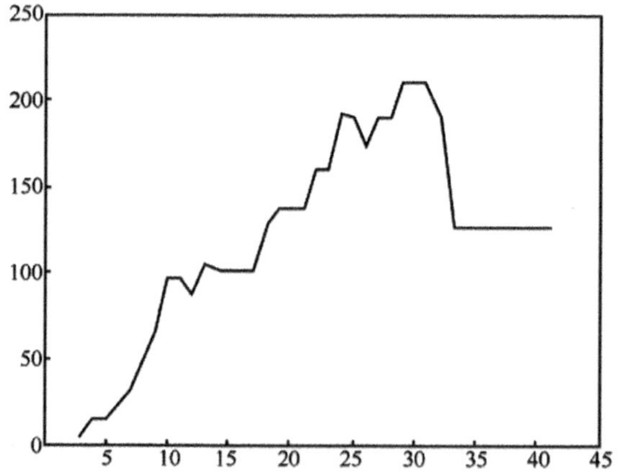

Fig. 3. Result of feature evaluation value calculated by information gain

Based on the feature subsets FS_Sub and IG_Sub, feature cross merging is performed, and the intersection is selected, the results are shown in Table 2.

Table 2. The subset of features selected in the experiment

Characteristic set	Characteristic dimension	Included features
FS_Sub	26	336 33 23 12 24 4 29 34 2 27 40 28 41 39 26 37 10 35 25 38 30 22 5 32 14
IG_Sub	28	5 3 23 24 33 36 2 32 4 12 29 10 34 39 26 37 27 38 25 40 28 22 41 35 30 6 1 11
F_Cross	25	2 3 4 5 10 12 22 23 24 25 26 27 28 29 30 32 33 34 35 36 37 38 39 40 41

3.2 Experiment of Feature Quadratic Optimization Selection

The feature secondary optimization selection is to further optimize the feature subset F_Cross by sequence backward search, establish a classification model by SVM, sequentially select features to form the feature subset, take the feature evaluation value as an evaluation index, remove redundant features, and obtain an optimized feature set. In the same manner, approximately 9000 records were randomly selected as the training set and over 2,300 records were selected as the test set. In the experiment, several sets of different data are used to test, and the corresponding feature subsets are obtained. Through the further test and evaluation of these feature subsets, the optimal feature subset F_P is determined and selected, which contains 13-dimensional features, namely 2, 3, 4, 10, 12, 22, 23, 28, 32, 33, 34, 37, 39, respectively.

According to the characteristics of this paper, in the same experimental environment, compared with GFR method, F-Score method and IIG method, the modeling time, test time and classification accuracy of the test algorithm are compared. The test results are shown in Table 3. GFR is a method of wrapping by sequential backward search; F-Score is an improved fisher algorithm; IIG is an improved algorithm for information gain. In the test, also from 10% KDDcup99 training subset, randomly selected about 26,000 records as the training set, with more than 12,000 records that do not intersect the training set as the test set.

Table 3. Comparison test results of feature selection algorithm

Feature selection method	Characteristic dimension	Modeling time (s)	Test time (s)	Classification accuracy (%)
GFR	19	5.10	2.26	98.64
F-score	19	7.20	2.90	98.46
IIG	12	5.84	2.28	98.3
All features	41	7.63	4.89	99.27
The method in this paper	13	4.48	1.73	99.29

4 Result Analysis

In this method, the feature evaluation value is used as the evaluation index of feature selection, and the classification ability of the algorithm is improved by secondary feature selection. Compared with the feature selection using fisher score or information gain alone, the quadratic feature optimization of the proposed method ensures that the selected feature subset has better classification ability. Compared with the feature selection using backward search only, the primary feature filtering of this method can effectively remove the noise features and avoid the influence of noise features in the subsequent feature selection. The test results in Table 3 show that in the same test environment, compared with other algorithms, the proposed method has better classification ability, and its modeling time, test time and classification accuracy have better performance.

References

1. Zhang, Y., Yang, A., Xiong, C., et al.: Feature selection using data envelopment analysis. Knowl.-Based Syst. **64**(64), 70–80 (2014)
2. Zhao, Y., Zhang, Y., Tong, W, et al.: An improved feature selection algorithm based on MAHALANOBIS distance for network intrusion detection. In: International Conference on Sensor Network Security Technology and Privacy Communication System, pp. 69–73. IEEE (2013)
3. Tesfahun, A., Bhaskari, D.L.: Intrusion detection using random forests classifier with SMOTE and feature reduction. In: International Conference on Cloud and Ubiquitous Computing and Emerging Technologies, pp. 127–132. IEEE (2014)
4. Araújo, N.V.D.S., Oliveira, R.D., Ferreira, E.T. et al.: Kappa-fuzzy ARTMAP: a feature selection based methodology to intrusion detection in computer networks. In: IEEE International Conference on Trust, Security and Privacy in Computing and Communications, pp. 271–276. IEEE (2013)
5. Balamurugan, S.A.A., Rajaram, R.: Effective and efficient feature selection for large-scale data using Bayes' theorem. Int. J. Autom. Comput. **6**(01), 62–71 (2009)
6. Mahalingam, P.R., Vivek, S.: Predicting financial savings decisions using sigmoid function and information gain ratio. Procedia Comput. Sci. **93**, 19–25 (2016)
7. Maldonado, S., Weber, R.: A wrapper method for feature selection using support vector machines. Inf. Sci. **179**(13), 2208–2217 (2009)
8. KDDCup99KDDdataset[EB/OL]. http://kdd.ics.uci.edu/databases/kddcup99/kddcup99.html (2011)
9. Peng, L., Fang, W.: Heterogeneity of inferring reputation of cooperative behaviors for the prisoners' dilemma game. Phys. A: Stat. Mech. Appl. **433**, 367–378 (2015)
10. Lipson, J., Vuong, S.: A semantics-based routing scheme for grid resource discovery. In: E-Science: First International Conference on E-Science and GridComputing, pp. 58–70, 90 (2005)

Research on Key Technology and Application of Big Data Integrated Machine

Ming Zheng(✉) and Mugui Zhuo

Guangxi Colleges and Universities Key Laboratory of Professional Software Technology, Wuzhou University, Wuzhou, China
370505375@qq.com

Abstract. In recent years, China's big data industry has already had a good foundation and faced with a rare opportunity for development. This paper analyzes and compares the current situation of big data integrated products at home and abroad, and summarizes the key technologies of big data integrated machine. In view of the imperfect support system of large data industry in China, the problem of security and development of information security system has not been established, and a large data machine based on domestic chip is studied and proposed, and the related hardware, software and application demonstration are introduced in detail, in order to solve the massive storage, continuous expansion, and Yun Wei. Deployment, data personalized processing and other problems. By comparing the research and design of large data machine at home and abroad, the large data machine based on domestic chip has the characteristics of data processing security, high concurrent processing efficiency and stable operation.

Keywords: Domestic chip · Big data integrated machine · Parallel processing Data security · Gene analysis

1 Objective

In recent years, China's big data industry has already had a good foundation, faced with a rare opportunity for development [1], but the level of open and sharing of data resources is low, the ability of technological innovation and support is not strong, the application level of large data is not high, especially the support system of large data industry is not perfect, and the management of safety and development is not yet established yet. And information security system. At the same time, with the rapid growth of data and the diversification of data sources, traditional data processing, storage and analysis have shown a shortage in actual production. How to obtain high performance, high availability, and fast response capability of complex queries at lower cost has become a major challenge in the modern IT field. At the same time, the safety specification is one of the principles of the development of the big data industry [2]. We should focus on the formation of safe and controllable technology products in the large data base software and hardware, and form a batch of products, solutions and services to meet the needs of major applications.

J. Abawajy et al. (Eds.): ATCI 2018, AISC 842, pp. 1101–1106, 2019.
https://doi.org/10.1007/978-3-319-98776-7_134

The main challenges we face in this process are the difficult management of mass data, difficult system performance and difficult support for complex analysis. The specific solution is to use hardware acceleration, expand data cache, use high-speed inline network, enhance the linear expansion of the system, use hierarchical storage, compressed storage. In the same time, with the in-depth application of the information system and the increasing dependence of the information system on the information system, the concurrent processing capability of the hardware and software platform, the capacity of mass data processing, the response speed of the system, the stability of the hardware and software platform, the software and hardware, and the hardware and software are also used. The ability of platform scalability and other aspects has higher requirements and is on the rise. Due to the limitations of early IT technology and concept, many complex "shaft type" applications have been formed in the process of information development, which brings great challenges to information management. By simplifying the IT infrastructure, improving the utilization of hardware resources, reducing the cost of investment, reducing the energy consumption of equipment and the cost of operation and maintenance, it has become a common understanding [3].

Based on the above problems and solutions, the development of high-performance and big data integration machine is imminent. The big data Big Data Machine solves the problem of infrastructure demand in the era of big data. At present, many manufacturers at home and abroad have developed the large data integrated products with their own characteristics. According to type classification, the current big data products can be subdivided into data warehouse integrated machine, database integrated machine, middleware integrated machine and so on.

From the key technologies of hardware architecture, data storage, data processing and data security, the current domestic and foreign mainstream products are compared. On the hardware architecture, the similarity of each one is higher. For example, in the database architecture, the 4 products have adopted the Share-nothing architecture, and the Oracle Exadata and HUAWEI FusionCube also adopt the Share-Disk + Share-nothing hybrid architecture technology. In server architecture, MPP technology is mostly adopted. In the internal network communication, InfiniBand hardware network architecture is adopted. However, there are great differences between big data platform support and data storage. They are selected or optimized according to their key application scenarios.

2 Method

Currently, the common commercial server architecture can be roughly divided into the Symmetric Multi-Processor (SMP) architecture, the Non-Uniform Memory Access (NUMA) [4] architecture and the Massive Parallel Processing (MPP) [5] architecture on the server's overall architecture.

SMP is widely used in the X86 platform (single node). The system architecture has the following obvious characteristics: each CPU is symmetrical, CPU has no primary and subordinate relationships; all CPU sharing system resources (CPU, memory, I/O, etc.); the access of each CPU to the bus is at the same level, and the time required to visit any address in memory is the same. The task queue of all CPU parallel execution systems has good concurrency ability, but as the number of CPU increases, the memory access conflict will increase rapidly, and the scalability is greatly limited. SMP is the most common system architecture [6].

Hadoop Distributed File System (HDFS) [7] is based on the GFS design and development of Google, a distributed file system running on general hardware, and management of files deployed on multiple independent physical machines. HDFS is a highly fault-tolerant file system, which provides large throughput data access, supports TB-PB level data storage, and is suitable for application on large data sets. It is an essential part of the large data machine.

There are several main concepts in HDFS: File Name: in file system, file name is used to locate storage location. Metadata (Metadata): data that saves file attributes, such as file name, file length, file group, file storage location and so on. Data block (Block): the smallest unit of storage file. The storage media are divided into fixed areas, which are allocated according to these areas when used. File system: represents a way to store and organize computer data, which simplifies the access and lookup of data. Distributed file system is a subset of distributed system, which solves the problem of data storage. In other words, they are storage systems across multiple computers. Data stored on distributed file system is automatically distributed on different nodes.

MapReduce [8] is designed and developed based on Google distributed computing framework MapReduce, which is used for parallel computing of large data sets (larger than 1 TB). MapReduce is a computing model, framework and platform for large data parallel processing. It is a high performance parallel computing platform based on cluster, and also a parallel computing and transportation. Software framework is also a parallel programming model and method. MapReduce is easy to program, has good expansibility (by adding nodes), and ensures high fault tolerance by computing migration or data migration.

The main idea of MapReduce comes from functional programming language and vector programming language: Map (mapping) and Reduce (reduction) [9]. First, a Map function is determined by software implementation, and then a set of key value pairs is mapped to a set of new key value pairs by specifying a concurrent Reduce function through the Map function to ensure that all the key value pairs of all the maps share the same key group. MapReduce can effectively handle chaotic large data sets. The process is as follows: The Map function will parse each data in a disorganized large data set, extract the characteristics of each data, get key and value, and then the data will pass through the Shuffle phase (the data is summed up), and the final arrival of the Reduce stage is all the data that has been arranged. On the basis of summarizing the good data, we can further analyze and get the desired results [10].

3 Result

By integrating the host, storage, network, management software, data warehouse, middleware, or virtualized software in the traditional architecture, the integrated solution is formed to reduce the total cost (TCO) and improve the overall performance, as shown in Fig. 1. Nowadays, for large-scale data processing, we should get rid of two aspects of storage and computing power. In order to solve this problem, the current large data processing platform often uses multi core server cluster parallel computing architecture, and uses such tools as Cache, memory, auxiliary storage, distributed storage and other tools.

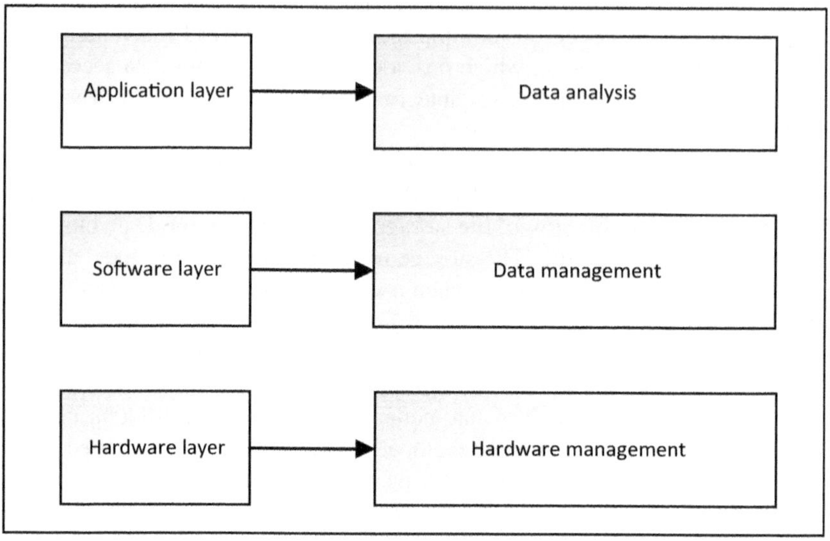

Fig. 1. Big data integrated architecture

The main architecture is divided into the following aspects: the hardware support layer: the hardware compression part of the large data machine based on the domestic chip, which is a large data support platform based on the Hadoop. To deal with parallel problems, multiple levels of parallel computing are needed to speed up processing, and at the same time, we need some storage support. The high density assembly processing platform of the domestic processor uses SMP/CC-NUMA multi-core multi-channel nodes, which is connected into a cluster form through high-speed network, and forms the storage of each node in a parallel cluster.

Software level: the software level of large data integrated machine based on domestic chip includes distributed data acquisition layer, mixed distributed storage layer and processing layer based on memory computing, integrated resource and system management layer. The storage structure of the large data integrated machine consists of distributed external storage, structured/semi-structured database and distributed memory 3-layer storage architecture, including distributed file system and relational database based on distributed file system, non-relational database and memory database. The data processing layer uses a hybrid processing architecture based on distributed memory computing, including memory computing engines and batch processing, graph computing, interactive processing and flow processing engines based on memory computing. And use the integrated management platform to manage the big data integrated machine, including centralized Web console and distributed cluster resource management system.

Application level: applied to the detection, diagnosis and analysis of cancer gene. The big data integration machine filters and compares the gene sequences of cancer patients, and generates automatic test reports by SNP analysis and cross validation. Through the analysis of the results of the report, we can obtain the characteristics of the cancer gene classification, classification and staging of the cancer, in order to achieve the purpose of the analysis of cancer genes. According to the results of the test, this result is compared with the previous experience of similar cases, and the corresponding analysis is given. This opinion is for reference for doctors and patients.

4 Conclusion

This paper introduces the background of the large data integration machine, discusses the necessity and urgency of the integration machine, and compares the current status of the mainstream integrated machine products, and introduces the key technology to realize the large data machine from the hardware and software level. The integrated product based on the domestic chip is put forward to realize the big data bearing platform based on container virtualization. In the application development mode, application run time, online development and delivery, dynamic scheduling, service quality assurance and so on, the application service development environment system architecture based on micro service is researched, and the large data machine for typical fields is developed. With the increase of the amount of data, the load of the system is getting bigger and larger, and the storage and query performance of the data is also decreased. In this case, the maximum performance of the system is played without increasing the cost of hardware.

The integrated machine has its unique advantages, and has different achievements in database, network and hardware. However, with the development of science and technology in China, many domestic manufacturers are going on. HUAWEI, dawning, wave and other domestic enterprises have gradually become the international leading level in the development of one machine. In the coming years, with the advent of the era of artificial intelligence, more powerful data storage and computing power is needed, and large data and cloud computing are still the focus of research. The development and optimization of the large data machine is a long-term project. The development of the big data integrated machine will benefit from the national government departments, equipment manufacturers, the first-line Internet Co holding massive data and hundreds of millions of people in our country. It is a long way to go to realize the industrialization of big data in areas closely related to finance, education, medical treatment, and life in China.

Acknowledgements. This work was supported by grants from The National Natural Science Foundation of China (No. 61502343), the Guangxi Natural Science Foundation (No. 2017GXNSFAA198148, 2015GXNSFBA139262) foundation of Wuzhou University(No. 2017B001), Guangxi Colleges and Universities Key Laboratory of Professional Software Technology, Wuzhou University.

References

1. Zhang, L.X., Wang, H.B., Li, Q.Z., et al.: Big data and medical research in China. BMJ **360**, 3 (2018)
2. Badia, J.G., De Puelles, P.G., De Lecuona, I.: Big data and information security. Aten. Primaria **50**(1), 3–5 (2018)
3. Kalid, N., Zaidan, A.A., Zaidan, B.B., et al.: Based real time remote health monitoring systems: a review on patients prioritization and related "big data" using body sensors information and communication technology. J. Med. Syst. **42**(2), 30 (2018)
4. Guo, X.M., Han, H.Y.: A Good Data Allocation Strategy on Non-uniform Memory Access Architecture. IEEE, Cambridge (2017)
5. Huang, W., Meng, L.K., Zhang, D.Y., et al.: In-memory parallel processing of massive remotely sensed data using an apache spark on Hadoop YARN model. IEEE J. Sel. Top. Appl. Earth Obs. Remote Sens. **10**(1), 3–19 (2017)
6. Fu, H.Y., Li, C., Fu, Y.Z., et al.: A parallel CNC system architecture based on symmetric multi-processor. In: Proceedings of 2016 Sixth International Conference on Instrumentation and Measurement, Computer, Communication and Control, pp. 634–637. IEEE, Cambridge (2016). <Go to ISI>://WOS:000390578900137
7. Begum, A.A., Chitra, K.: Formation of Single and Multinode Clusters in Hadoop Distributed File System. IEEE, Cambridge (2017)
8. Ramirez-Gallego, S., Fernandez, A., Garcia, S., et al.: Big Data: tutorial and guidelines on information and process fusion for analytics algorithms with MapReduce. Inf. Fus. **42**, 51–61 (2018)
9. Ezatpoor, P., Zhan, J., Wu, J.M.T., et al.: Finding top-k dominance on incomplete big data using map reduce framework. IEEE Access **6**, 7872–7887 (2018)
10. Xu, L.J., Dou, W.S., Zhu, F., et al.: Characterizing and diagnosing out of memory errors in MapReduce applications. J. Syst. Softw. **137**, 399–414 (2018)

Research on the Gene Regulation Network Construction Algorithm

Shen Wei[1,2(✉)]

[1] The First Affiliated Hospital, Jilin University, Changchun, China
abyswabysw@163.com
[2] College of Computer Science and Technology, Beihua University, Jilin, China

Abstract. With the completion of Human Genome Project, the discovery of the gene function has become the new research interest in the Post-genome times. Using the microarray techniques, it is possible for scientists to discover the regulatory relationships between genes. The gene regulatory network, in definition, is the simulation or reconstruction of the mutual relations among expressed genes. Gene regulatory network helps us to understand in organisms which, where, when and how genes are expressed though observing visual model of gene expression. In this way, gene regulatory network has been widely applied in the research on relations between genes and diseases or drug target designs. The algorithms of reconstructing gene regulatory networks based on Bayesian network are the main research interest of this paper. We study some models which are often used in reconstructing gene regulatory networks, weighed matrices model, Boolean networks model, mutual-information networks model, linear combination model, directed graphs and undirected graphs.

Keywords: Gene regulation network · Correlation analysis · Bayesian network
Search scoring

1 Introduction

Bayesian network can analyze probabilistic dependence degree of different variables with the form of probability, use graphical model in stating causal relationship of the variables, and dispose and analyze the incomplete data set by combining with the prior knowledge, so it can be widely used in the research on bioinformatics [1]. In Bayesian network, gene is expressed with node, and intergenic regulation function is expressed with directed edge, so it is suitable for the construction of gene regulation network in gene expression data [2].

The method of constructing gene regulation network based on Bayesian network was firstly proposed by Friedman [3]. Pe'er used the improved method in discretizing the gene expression data and BIC scoring function in seeking the optimal gene regulation network. Segal constructed Bayesian network structure based on the module network model [4]. Wang constructed a new gene regulation network algorithm combining K2 algorithm with mutual information method based on Bayesian network, and he firstly built undirected graph with mutual information, then selected the optimal node sequence with joint information entropy, and finally optimized the network

J. Abawajy et al. (Eds.): ATCI 2018, AISC 842, pp. 1107–1112, 2019.
https://doi.org/10.1007/978-3-319-98776-7_135

structure based on K2 algorithm [5]. Yin proposed to learn Bayesian network based on genetic algorithm, and he realized parallelization of the genetic algorithm to reduce the algorithm operation time simultaneously [6]. The researches above are all established on the single gene expressing data set whose number of samples is always insufficient, so accuracy of the gene regulation network constructed shall be affected seriously [7].

Based on the problems above, this paper proposed the algorithm of learning distributed data set based on Bayesian network: this algorithm can avoid the direct use of local data and realize individual privacy protection; simultaneously, it combines search scoring method with correlation analysis method and can realize full use of the information with concentration of distributed data [8].

2 MFD-GRN Algorithm

MFD-GRN algorithm refers to the process of learning various Bayesian network structures concurrently and intensively from various isomorphism data sets firstly, and this process is just the local learning [9]. At this moment, there are less samples contained in the data set, so K2 algorithm with search scoring can be used in learning the local structure [10]. Next, we can fuse local structures and form global network structure, and this process is the global learning [11]. As the integrated global data set contains huge number of samples, the method based on correlation analysis shall be selected when fusing local structures. Concretely speaking, there are several data sets defined which are distributed on s nodes of computer [12]. In local learning, we can obtain the local Bayesian network structure it adopts the method of correlation analysis and determines whether the edge in undetermined edge set can conform to conditions and be added in the edge set of global structure [13] (Fig. 1).

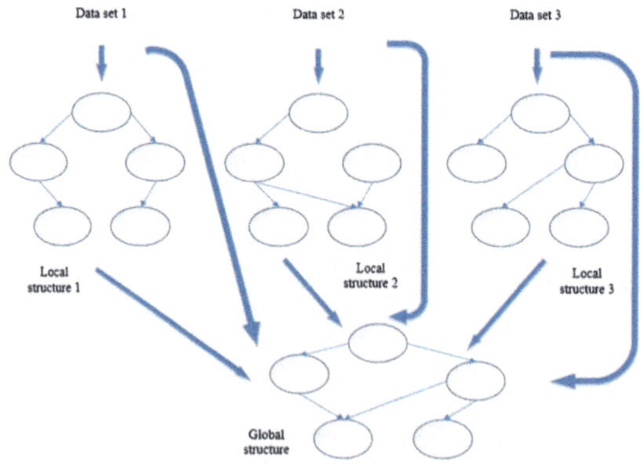

Fig. 1. MFD-GRN algorithm flow chart

2.1 Local Learning

In local learning, the K2 algorithm based on search scoring can be used, and s Bayesian network structures $G1 = (V, E1)$, …, $Gs = (V, Es)$ can be obtained in the synchronous and concurrently learning. K2 algorithm needs to specify upper bound μ of father node and input sequence of nodes, and it is essentially a greedy heuristic search method, with the purpose of obtaining Bayesian network structure. On the nodes with different sequence in a same data set, we can obtain different network structures by learning with K2 algorithm, for K2 algorithm is very sensitive to the input sequence of nodes. Firstly, we can give data set D containing N variables, and specify sequence of N nodes and upper bound μ of father node. Afterwards, we can find Bayesian network B matching with data set D with search scoring method, and adopt BDE scoring function [14].

$$SCORE(G|D) = \frac{P(G)P(D|G)}{P(D)} \tag{1}$$

2.2 Global Learning

Global learning is based on mutual information inspection method, with the purpose of fusing local structures to global structure. Mutual information of variables X and Y can be expressed as:

$$I|(X,Y) = \sum_{x,y} P(x,y) \log \frac{P(x,y)}{P(x)P(y)} \tag{2}$$

Next, it gives description to the simple steps of finding the minimum cutest of (A, F) node in Fig. 2, in which A refers to initial node and F refers to target node.

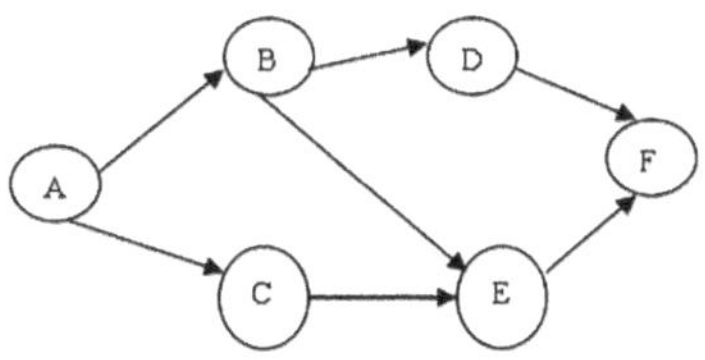

Fig. 2. A directed graph of the network structure

3 MFD-GRN Algorithm Testing with Alarm Network

Alarm network has 37 variables and total 46 directed edges, and it is the benchmark test set which can be widely used in learning Bayesian network structure. We can generate simulation data by utilizing Alarm network and produce 10 data sets which contain

5,000, 10,000, 15,000, 20,000, 25,000, 30,000, 35,000, 40,000, 45,000 and 50,000 samples points respectively. Then, we can give random non-uniform distribution to the various data sets on the five computer nodes, conduct experimental test, and compare experimental results of the distributed learning of MFD-GRN algorithm with learning results of K2 algorithm and TPDA algorithm in centralized learning. Table 1 gives concrete test results of partial data sets. The analysis indicates that: when number of samples exceeds 5,000, learning results of MFD-GRN algorithm are very close to those of centralized learning. The two curves in Fig. 3 refer to specific value of the computing time of K2 algorithm and the MFD-GRN algorithm and specific value of the computing time of TPDA algorithm and the MFD-GRN algorithm respectively, namely speed-up ratio. As indicated by the curve, the speed-up ratio shows near-linear growth along with increment of the number of samples [15].

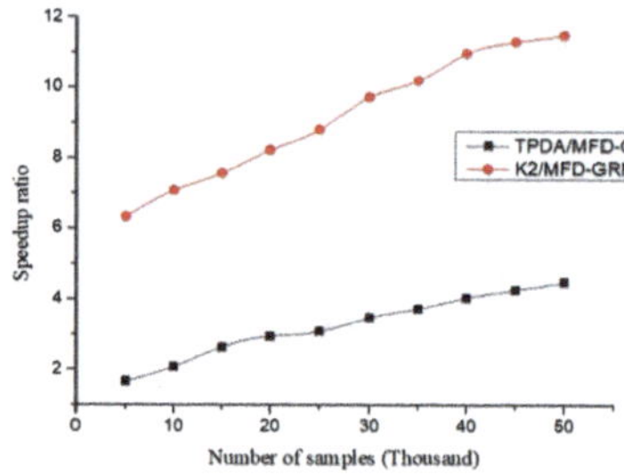

Fig. 3. Ratio of running time of alarm online concentrated learning and that of distributed learning

4 Experiment of Arabidopsis Gene Data Set

The gene expression data set based on stress response of Arabidopsis is used in testing MFD-GRN algorithm, and we can obtain experimental data from ATGen Express database of Arabidopsis and receive 9 static gene expression data sets in different

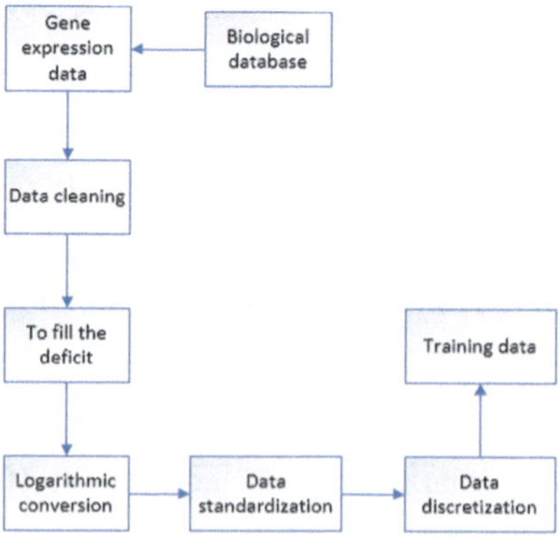

Fig. 4. Data preprocessing process

experimental conditions, including 24, 24, 20, 20, 18, 18, 15, 15 and 15 samples points respectively, and each data set contains 30 same genes [16]. To make the gene expression experimental data obtained adapting to the distributed learning algorithm, it needs to give pretreatment to the experimental data before learning, as shown in Fig. 4.

5 Conclusion

This paper gives research on construction of the gene regulation network based on Bayesian model by learning from the distributed data set, and proposed MFD-GRN algorithm [17]. In local learning, it can learn local structure respectively, independently and concurrently from various computer nodes with K2 algorithm; global learning is based on correlation analysis method, and global structure can be obtained by fusing local structures. MFD-GRN algorithm avoids direct acquisition of local individual data, thus individual privacy can be protected effectively, and it also gives full use to the information with concentration of distributed data.

Acknowledgment. The authors wish to thank the Education Department of Jilin Province [2015] (127th) and [2015] (182nd), Jilin science and Technology Bureau (201467006), Science and Technology Department of Jilin Province (20150204045NY). Under which the present work was possible.

References

1. Cooper, G.F., Herskovits, E.: A Bayesian method for the induction of probabilistic networks from data. Mach. Learn. **9**(4), 309–347 (1992)
2. Folter, S.D., Kieffer, M., Pařenicová, L., Weigel, D., Busscher, M., Kooiker, M.: Comprehensive interaction map of the arabidopsis mads box transcription factors. Plant Cell **17**(5), 1424–1433 (2005)
3. Gevaert, O., Vooren, S.V., Moor, B.D.: A framework for elucidating regulatory networks based on prior information and expression data. Ann. N. Y. Acad. Sci. **1115**(1), 240 (2007)
4. Gómez-Mena, C., De, F.S., Costa, M.M., Angenent, G.C., Sablowski, R.: Transcriptional program controlled by the floral homeotic gene agamous during early organogenesis. Development **132**(3), 429 (2005)
5. Pe'er, D., Regev, A., Elidan, G., Friedman, N.: Inferring subnetworks from perturbed expression profiles. Bioinformatics **17**(Suppl 1), S215 (2001)
6. Peeters, R.L.M., Zeemering, S.: Sparse gene regulatory network identification. In: International Conference on Knowledge Discovery and Emergent Complexity in Bioinformatics, vol. 4366, pp. 171–182. Springer (2006)
7. Wang, R., Jing, Z., Chen, L.: Modelling periodic oscillation in gene regulatory networks by cyclic feedback systems. Bull. Math. Biol. **67**(2), 339–367 (2005)
8. Wang, X., Sun, J.T., Chen, Z., Zhai, C.X.: Latent semantic analysis for multiple-type interrelated data objects. In: International ACM SIGIR Conference on Research and Development in Information Retrieval, pp. 236–243. ACM (2006)
9. Chen, R., Sivakumar, K., Kargupta, H.: Collective mining of Bayesian networks from distributed heterogeneous data. Knowl. Inf. Syst. **6**(2), 164–187 (2004)

10. Cheng, J., Greiner, R., Kelly, J.: Learning Bayesian networks from data, an information-theory based approach. Artif. Intell. **137**(1–2), 43–90 (2002)
11. Friedman, N., Linial, M., Nachman, I., Pe'er, D.: Using Bayesian networks to analyze expression data. J. Comput. Biol. **7**(3–4), 601 (2000)
12. Fujimoto, S.Y., Ohta, M., Usui, A., Shinshi, H., Ohme-Takagi, M.: Arabidopsis ethylene-responsive element binding factors act as transcriptional activators or repressors of GCC box-mediated gene expression. Plant Cell **12**(3), 393 (2000)
13. Lam, W., Segre, A.: A Distributed Learning Algorithm for Bayesian Inference Networks. In: IEEE Educational Activities Department (2002)
14. Nakamura, E.F., Loureiro, A.A.F., Frery, A.C.: Information fusion for wireless sensor networks, methods, models, and classifications. ACM Comput. Surv. **39**(3), 9 (2007)
15. Segal, E., Shapira, M., Regev, A., Pe'er, D., Botstein, D., Koller, D.: Module networks, identifying regulatory modules and their condition-specific regulators from gene expression data. Nat. Genet. **34**(2), 166 (2003)
16. Sterritt, R., Adamson, K., Shapcott, C.M., Curran, E.P.: Parallel data mining of Bayesian networks from telecommunications network data. Lecture Notes in Computer Science, vol. 37, no. 8, pp. 415–422 (2000)
17. Yin, L., Huang, C.H., Rajasekaran, S.: Parallel data mining of Bayesian networks from gene expression data. In: Parallel and Distributed Processing Symposium, International, vol. 18, p. 30156a. IEEE Computer Society (2004)

Prediction of Auricularias Bacterial Package Density Based on Improved BP Neural Network

Chunyu Mao, Zhuojuan Yang(✉), and Chao Gao

Jilin Engineering Normal University, Changchun, China
290414003@qq.com, 450818386@qq.com, 1319593820@qq.com

Abstract. For the BP algorithm learning slow convergence, easy to fall into the local minimum and can't get the global optimal solution defects, in the density forecasting of automatic processing of auricularias bacterial package used in the production line, this paper proposes to use genetic algorithm to optimize BP neural network, namely GA-BP algorithm to achieve reliable and rapid prediction of auricularias bacterial package density. The simulation results show that the algorithm has high prediction accuracy and short response time, and can provide a reliable basis for the production and operation of enterprises.

Keywords: Auricularias · BP algorithm · Neural network · Improved Prediction

1 Introduction

The prediction of the density of bacterial package is an important part of the production and processing of bacterial package. Due to the existence of many influencing factors in the processing of the auricularias bacterial package. When the equipment has been determined, the density of the bacterial package is affected by various factors such as material temperature, material moisture, material density, feed rate and feed pressure of the bacterium. If any one of these factors changes, it will affect The density of bacteria. The influence of these factors on the density is mostly non-linear, and often interacts with each other. It is very difficult to establish the mechanism model of the actual process, and it is difficult to obtain ideal results. The neural network has the ability of self-learning and self-adaptation. Through the provision of a set of mutually corresponding input-output data, analysis and master the potential laws between the two, and finally based on these laws, using the new input data to estimate the output results, thus achieving density prediction, reliability and accuracy Very high sex.

Combining the advantages of neural network and genetic algorithm, this paper proposes a hybrid training method GA-BP algorithm based on the combination of improved genetic algorithm and neural network. The genetic algorithm is used to optimize the neural network initial weight threshold, and then the neural network algorithm is finally used. Based on the processing data of bacterial package, neural network methods were used to study the relationship between various factors such as material temperature, material moisture, material density, feed rate, and feed pressure

J. Abawajy et al. (Eds.): ATCI 2018, AISC 842, pp. 1113–1120, 2019.
https://doi.org/10.1007/978-3-319-98776-7_136

of bacilli, and the density of bacterial package. The density of the package was predicted.

2 Summary of the Neural Network

2.1 BP Neural Network

BP network is a one-way communication of multilayer feedforward network communication, the structure shown in Fig. 1, has more than three or three-layer neural networks, including the input layer, a middle-tier (hidden) and the output layer. Full connection between upper and lower, and no connections between each neuron. When a pair of learning samples available to the network, and neuron activation values from the input layer of the middle-tier to the output layer spread in the output layer neurons receive input response of the network. Next, follow the direction reduced the target output and the actual error, from the output layer weights after the middle layer-by-layer to amend the connection, and finally returned to the input layer [3].

BP network is the most widely used as feedforward network, but the drawbacks, mainly has following several aspects. First of all, as a result of learning rate is fixed, the network convergence speed is slow, require longer training time. Secondly, the BP algorithm can make the convergence to a certain value, but can't guarantee that the error surface global minimum. Again, networks hidden layer of layer number and cell number selection, there is no theory, are generally based on experience or by repeated experiments to determine. Finally, the network of learning and memory have inaccuracies [4].

2.2 Genetic Algorithm

Genetic algorithms is a genetics and evolutionary processes at bio-Mimetic and established a search algorithm, reflects the "struggle for existence and inferior lost, survival of the fittest" the competition mechanism.

Genetic algorithm operation process can be described as follows:

(1) Randomly generated initial population.
(2) In the fitness function evaluation of chromosomes.
(3) Select a chromosome into the next generation of high fitness.
(4) Through genetic, mutation operation to generate new chromosomes.
(5) Repeat (2)–(4) steps until the book evolution of algebra. Genetic algorithms of basic processes such as shown in Fig. 2.

2.3 GA-BP Algorithm

BP algorithm has the advantage of simplicity and plastic, but is based on BP algorithm of gradient method, this convergence is slow and often suffering from local minimum point; GA at global search for lack of accurate local search capabilities. Combination of genetic algorithm and BP algorithm, complementary implementation, accelerate the

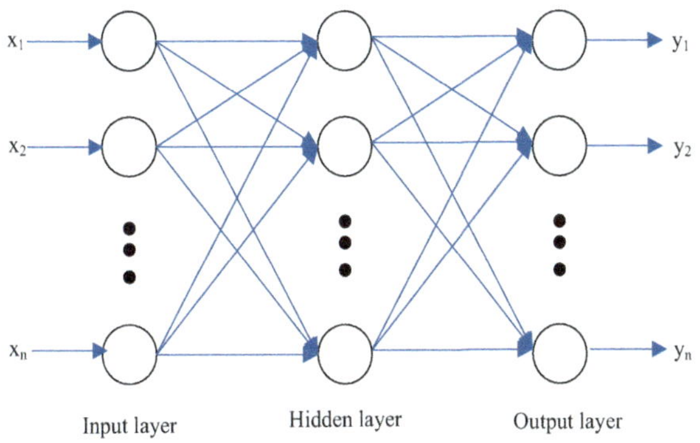

Fig. 1. The structure of BP neural network

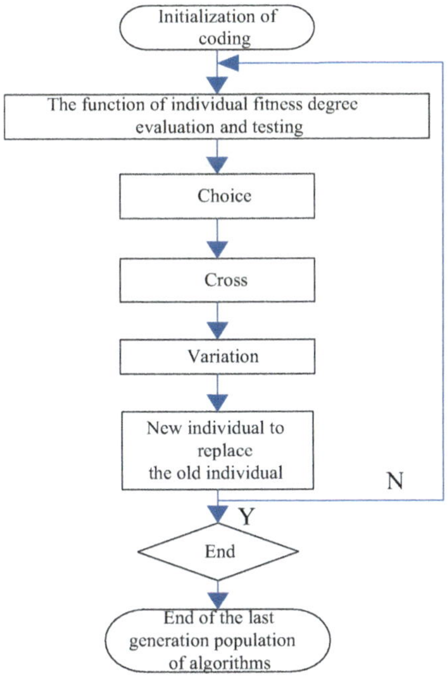

Fig. 2. Basic processes of genetic algorithm

speed of network learning, integrated to improve the approximation ability and generalization ability of learning in the process as a whole to help solve the problem [5][1].

GA-BP algorithm the basic idea is: lists may be present all the neurons in the neural network, the weights of all possible connections of neurons encode the binary code series or real digital represent individual, randomly generated the code groups, for

[1] Identify applicable sponsor/s here. *(sponsors)*

General genetic algorithms for optimization of calculations. Code decoded form neural network, and regular training of BP algorithm, resulting in predictions, using genetic algorithm optimizing received initial weights of a neural network.

The concrete steps:

(1) Randomly generated initial population, in which each individual hidden node number with a decimal integer.
(2) Calculate the fitness of each individual in the initial population;
(3) The initial population of the hidden nodes in the same individual, so that the composition of the different groups;
(4) M times for each small group evolution, evolutionary process to follow the general genetic algorithm (corresponding to the selection, crossover and mutation operations, the reproduction process of the best individual retention policies), and then select the best individual in each small group to form main groups;
(5) The main groups of individuals in accordance with the above described genetic operators are crossover and mutation operations, and then elect one of the outstanding individuals for BP training;
(6) Calculating the fitness of each individual in the main group;
(7) If you reach the pre-set the mean square error of the evolution of algebra or the network has reached a given value, then the algorithm ends, otherwise the main groups of individuals to return to the small groups, and then turn to Step (4).

3 Neural Network Optimization to Achieve Yield Forecast

3.1 Predictive Neural Network Model

This paper determines the use of a three-tier network structure to establish a neural network structure for density prediction, as shown in Fig. 3. There are 5 input nodes, which are the material temperature, material humidity, material density, feed rate, and feed pressure of the bacteria. The number of output layer nodes is 1, that is density. The number of hidden layer nodes is 8.

3.2 Predict the Training of the Neural Network Weights

The equations are an exception to the prescribed specifications of this template. You will need to determine whether or not your equation should be typed using either the Times New Roman or the Symbol font (please no other font). To create multileveled equations, it may be necessary to treat the equation as a graphic and insert it into the text after your paper is styled [6].

Training Algorithm. Given neural network input and output sample set; Determine the encoding of the network weight coefficients, select genetic manipulation and set of genetic parameters; Set the population size N randomly generated initial population; Each position strings from the coding population to obtain the Group N network weight coefficients, the N network with the same structure; By the input sample set using the forward propagation algorithm to obtain the Group N network weights corresponding

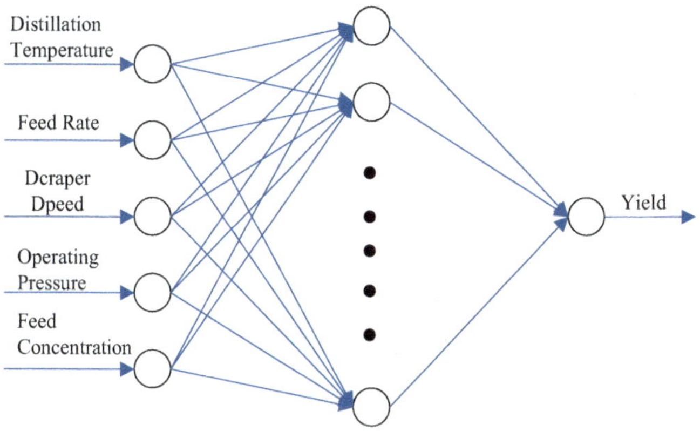

Fig. 3. Predictive neural network

to the N network output; Set the objective of the network function that will convert it to a fitness function to evaluate the N network; Selecting Options in the choice of space, according to the fitness; Get a new group based on that the selected crossover and mutation, and related algorithms, the parameters for the corresponding operation; Return to step, a set of optimized weight coefficient until the meet the performance requirements [7]:

(a) P of initial population, including cross-scale, cross-Pc, mutation probabilities of probability, as well as the rights to one of the values initialized; Using real-coded in the coding:
(b) Calculate each individual's evaluation function, and sort, select the network of individuals, according to the following probability:

$$p_s = \frac{f_i}{\sum_{i=1}^{N} f_i} \tag{1}$$

Where fi is the adaptation value of individual i, which can be used to measure for the error sum of squares E, Namely:

$$f(i) = \frac{1}{E(i)} \tag{2}$$

$$E(i) = \sum_{P}\sum_{K}(V_K - T_K)^2 \tag{3}$$

Where i = 1, …, N is the number of individuals; k = 1, …, m is the output layer nodes; p is the number of learning samples; the TK as the teacher signal [8];
(c) Two individual crossover operation to generate new individuals, using Probability Pc, no cross-operating individuals that is a direct copy;
(d) Using probability Pm mutations produce new individuals;

(e) Insert new individuals into the population P, evaluation of new individuals and to computation function.
(f) The end If you get the satisfaction of the individual, otherwise go to step (c). Reach the required performance will be optimized in the final group of individuals decoding can be optimized neural network connection weights;

Training Error. In order to test the ability of GA-BP algorithm to approximate the signal, this paper chooses the parameter of bacterial package density produced by Jilin Jilong Technology Development Co., Ltd. in July of 2017 as historical data. The absolute error index of the bacterial density is defined as [9]:

$$e_i = y_i - \hat{y}_i \tag{4}$$

Of which: $\hat{y}_i$ is the known bacterial density, y_i is the corresponding predicted bacterial density, e_i is the absolute error value [10].

3.3 Improved BP Neural Network to Predict the Simulation Results

Set the bacteriostatic density absolute error $e_i < 0.04\%$, take 50 samples, of which 20 as training samples and 30 as prediction samples, set up input and output sample sets. The measured and predicted values of the bacterial bag density are shown in Fig. 4, and the absolute error of the bacterial bag density is shown in Fig. 5. Figure 5 shows that with this algorithm, the bacterial package density prediction error $P > 0.48\%$ and the bacterial package density prediction absolute error $e_i < 0.04\%$ are within their permissible range. This algorithm satisfies the bacterial package density actual production purity prediction Needs [11].

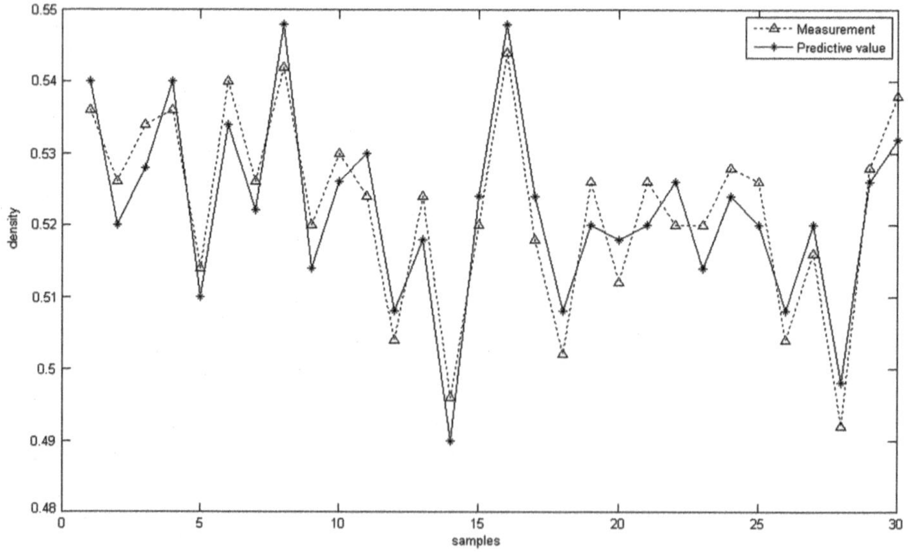

Fig. 4. Yield of essential oil of measured value and predicted value

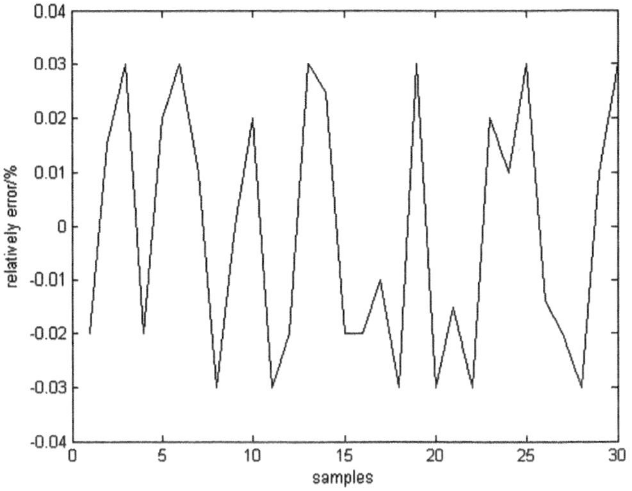

Fig. 5. Essential oil yield absolute error

4 Conclusion

In this paper, the combination of genetic algorithm and BP neural network, on the basis of speeding up the learning speed of neural network, improves the ability of approximation and generalization of the whole learning process, and realizes the complementary advantages of the two algorithms. By using the GA-BP algorithm to predict the bacterial package density, the change trend of the bacterium density can be grasped promptly and accurately, the feed rate and the feed pressure of the bacterium bag can be controlled, the waste of resources can be reduced, and the economic benefit and information management level of the product can be improved, has a wide range of application prospects.

Acknowledgments. Thanks IRTJLENU for the funding and equipment support provided in this article.

References

1. Jiao, X., Jing, B., Huang, Y., et al.: Research on fault diagnosis of airborne fuel pump based on EMD and probabilistic neural networks. Microelectron. Reliab. **75**, 296–308 (2017)
2. Yu, J., Yi, Z., Zhou, J.: Continuous attractors of Lotka Volterra recurrent neural networks with infinite neurons. IEEE Trans. Neural Netw. (2010)
3. Zhao, H., Zeng, X., He, Z.: Low-complexity nonlinear adaptive filter based on a pipelined bilinear recurrent neural network. IEEE Trans. Neural Netw. (2011)
4. Ahmad, M.T., Kumar, N., Singh, B.: Fast multilayer perceptron neural network-based control algorithm for shunt compensator in distribution systems. IET Gener. Transm. Distrib. **10**(15), 3824–3833 (2016)

5. Yang, S.E., Huang, L.: Financial crisis warning model based on BP neural network. Syst. Eng. Theory Pract. **25**(12), 12–19 (2005)
6. Zhang, L., Wu, K., Zhong, Y., et al.: A new sub-pixel mapping algorithm based on a BP neural network with an observation model. Neurocomputing **71**(10–12), 2046–2054 (2008)
7. Zhang, K., Huang, Q.: Pulse recognition method based on PSO-BP neural network. Mod. Electron. Tech. (2018)
8. Li, J.C., Zhao, D.L., Ge, B.F., et al.: A link prediction method for heterogeneous networks based on BP neural network. Physica A **495**, 1–17 (2018)
9. Ahmed, R., Sayed, M.E., Gadsden, S.A., et al.: Automotive internal-combustion-engine fault detection and classification using artificial neural network techniques. IEEE Trans. Veh. Technol. **64**(1), 21–33 (2015)
10. Khan, Y.: Partial discharge pattern analysis using PCA and back-propagation artificial neural network for the estimation of size and position of metallic particle adhering to spacer in GIS. Electr. Eng. **98**(1), 1–14 (2015)

Authenticated Identity-Based Private Key Sharing Scheme and Its Application

Weiqing Wan(✉)

Jiangxi University of Engineering, Xinyu 338029, China
wanwqing@163.com

Abstract. In order to solve the private sharing problem in the authenticated identity-based scheme, combined with (t, n) threshold verifiable secret sharing scheme, a new private key sharing scheme is proposed, and with the existing private key sharing scheme, a comparative analysis is made, analysis shows that this scheme have a greater advantage in terms of efficiency and it's also safety. Combined with authentication identity-based encryption scheme, a security electronic document interchange model in the government network is provided, to achieve the rapid exchange of electronic documents and to provide the security guarantee, through analysis shows that the scheme is safe and high efficiency.

Keywords: Identity-based encryption · PKI · BDH problem
Electronic official document · Secret sharing

1 Introduction

In recent years, the e-government of our country has been rapidly developed. Due to the requirement of confidentiality and cyber security [1] in the transmission of electronic official documents, the development of e-government has been affected. Therefore, to speed up the development of e-government, we must first resolve an important issue is how to exchange electronic documents safely and efficiently. In the literature [2] PKI is given in the e-government applications, these programs are better to solve the identity authentication, electronic document security delivery and other issues, but because of PKI's own shortcomings affect the efficiency of electronic document transmission, transmission the cost is also higher. Therefore, this paper introduces identity-based authentication and encryption scheme [3], to some extent, to solve the deficiencies in the PKI scheme to ensure the safety of electronic official document exchange at the same time also improves the operational efficiency.

However, the identity-based encryption scheme [4] requires a PKG centralized management system private key, which poses the risk of compromise and over-concentration of rights over the PKG. Therefore, this paper proposes a new private key sharing scheme [5] using the threshold-based secret sharing scheme (t, n), which reduces the risk of secret leakage in the identity-based encryption scheme and avoids the problem of excessive concentration of rights.

J. Abawajy et al. (Eds.): ATCI 2018, AISC 842, pp. 1121–1128, 2019.
https://doi.org/10.1007/978-3-319-98776-7_137

2 Identity-Based Authentication Encryption Scheme

2.1 Definition 2.1 [4]

Gives an additive cyclic point group G and a multiplicative cyclic point group.

G_1 of two prime numbers. We then call a mapping $e : G \times G \rightarrow G_1$ bilinear if we have $e(g^a, h^b) = e(g,h)^{ab}$ $e(ab,h)$, $e(ab,h) = e(g,h)^a = e(g,ah)$, for all $g, h \in G$ and $a, b \in F_q$.

2.2 Definition 2.2

BDH Problem [4]: $e : G \times G \rightarrow G_1$ is a prime number for a bilinear map $|G| = |G_1| = q$ and is defined as follows: Given $g, g^a, g^b, g^c \in G$, Calculated $e(g,g)^{abc}$, g is the generator, a, b, c randomly selected. Said an algorithm to solve the problem of BDH ε, if any

$$\Pr\left[A\left(g, g^{a,} g^b, g^c\right) = e(g,g)^{abc}\right] \geq \varepsilon \tag{1}$$

An identity-based encryption scheme [4] consists of the following four algorithms: Setup, Extract, Encrypt, and Decrypt. In general, Setup generates a public, distributed system parameter and a master key. Extract generates a private key corresponding to a given original ID. Encrypt encrypts a message with this given ID, and Decrypt decrypts the secret with a private key Text.

We often define the message space as $M = \{0,1\}^*$, unless otherwise stated. We need these algorithms to satisfy the standard conformance constraints, defining that d is the private key generated by the algorithm Extract given the ID, A is the public key, and then

$$\forall M \in Decrypt(param, A, C, d) = M, \quad \text{Here } C = Encrypt(\text{par}am, A, M).$$

2.3 Identity-Based Authentication Encryption Scheme [3]

With Authentication-Encryption: Enter a message, a (sender's) private key, a (recipient's) ID, and a ciphertext.

With authentication-decryption: Enter a ciphertext, a (sender's) ID, a (recipient's) private key, and output the corresponding plaintext.

We require both algorithms to meet the standard consistency constraints. In addition, for all messages M, the private keys corresponding to ID_A and ID_B are d_A, d_B

$$Authenticated - Encrypt(M, d_A, ID_B) = Authenticated - Decrypt(M, ID_A, d_B) \tag{2}$$

Specific programs are as follows: from the Setup, Extract, Authenticated-Encrypt and Authenticated-Decrypt four parts.

Setup: Enter $k \in Z$. The BDH parameter generator is run and a description of two groups G, G_1 and bilinear maps $e : G \times G \rightarrow G_1$ is output in k-polynomial time. $|G| = |G_1| = q$ is a prime number. Choose a random $a, \in F_q$ and a random g, $\in G$. Pick a cryptographic hash function (for some n)

$$H_1 : F_q \times G_1 \rightarrow \{0,1\}^n, H_2 : \{0,1\}^* \rightarrow G$$
$$H_3 : \{0,1\}^* \times \{0,1\}^* \rightarrow F_q, H_4 : \{0,1\}^n \rightarrow \{0,1\}^n \quad (3)$$

For proof of security, we consider the hash function as a random oracle model.

Output: Master key a and parameters

$$param = <e, G, G_1, g, g^a, H_1, H_2, H_3, H_4> \quad (4)$$

Extract: Enter one ID_A, output $d_A = H_2(ID_A)^a$. Suppose we have a symmetric cryptosystem that is semantically safe. We use E_k, D_k to denote the encryption and decryption functions of key K, and assume the key space $k \in \{0,1\}^*$.

Authenticated-Encrypt: Enter a message $M \in \{0,1\}^*$, a private key d_A, a ID_B and system parameters. Choose a random $\sigma \leftarrow \{0,1\}^n$, calculate $r = H_3(\sigma, M)$ and $s = e(d_A, H_2(ID_B))$

Then output ciphertext

$$C = <r, \sigma \oplus H_1(r,s), E_{H4(\sigma)}(M)> \quad (5)$$

Authenticated-Decrypt: Enter a ciphertext $<U, V, W>$, ID_A, *private key* d_B. *Calculate*

$$s = e(H_2(ID_A), d_B), \sigma = V \oplus H_1(U,s), M = D_{H_4(\sigma)}(M) \quad (6)$$

Check $U = H_3(\sigma, M)$ is equal. If not, reject the ciphertext, otherwise the plaintext M is output.

Consistency is clear because of the bilinear nature of

$$e(d_A, H_2(ID_B)) = e(H_2(ID_A), H_2(ID_B))^a = e(H_2(ID_A), d_B) \quad (7)$$

Identity-based authentication encryption scheme at the time of encryption because a less power and no dot product, so encryption faster than ordinary encryption, but also the existence of the key escrow problem, there is also the risk of disclosure, this article can be used to verify (t,n) threshold secret sharing scheme [5], a new private key sharing scheme is proposed, which reduces the risk of secret disclosure in identity-based encryption scheme and avoids the problem of excessive concentration of rights. The following describes the new private key sharing program.

3 Private Key Sharing Program

3.1 The Private Key d_A Is Distributed to $\{K_1, K_2, \ldots K_n\}$ N Trusted Centers as Follows

(1) User A randomly selects a number $S_{ID} \in Z_q^*$ and calculates $P = g^{S_{ID}}$, $Q = H_2(ID_A)^{S_{ID}} \bullet d_A$, and broadcast $<P, Q>$.

(2) The n trusted centers can verify whether equation $e(H_2(ID_A), P_{PKG} \bullet P) = e(Q, g)$ holds. because

$$\begin{aligned} e(H_2(ID_A), P_{PKG} \bullet P) &= e(H_2(ID_A), g^a \bullet g^{S_{ID}}) \\ &= e(H_2(ID_A), g^{a+S_{ID}}) = e\left(H_2(ID_A)^{a+S_{ID}}, g\right) \\ &= e\left(H_2(ID_A)^a, H_2(ID_A)^{S_{ID}}, g\right) \\ &= e\left(H_2(ID_A)^{S_{ID}} \bullet d_A, g\right) = e(Q, g) \end{aligned} \tag{8}$$

If the equation holds, then P and Q are legal.

(3) User A shares (t, n) with n verifiable $S_{ID} \in Z_q^*$ threshold secret sharing schemes [5] for n trusted centers.

First, choose a random $t-1$ degree polynomial $f(x) = a_0 + a_1x + a_2x^2 + \cdots + a_{t-1}x^{t-1}$ Make $a_0 = S_{ID}$. The other coefficients $a_1, a_2, \ldots a_{T-1}$, were randomly selected from Z_q^*. Then broadcast $< g^{a_0}, g^{a_1}, \ldots g^{a_{t-1}} >$, where $g^{a_0} = g^{S_{ID}} = P$.

User A secretly sends each of his trusted centers part of his private key $S_{ID} = f(i) \text{mod}_q$, each of which can be verified by equation $g^{S_{ID}} = g^{\sum_{j=0}^{t-1} a_j \bullet i^j}$.

For example: take $i = h, 0 \leq h \leq t-1$, $S_{ID_K} = f(h) \bmod q = a_0 + a_1h + a_2h^2 + \cdots + a_{t-1}h^{t-1} \bmod q$ $g^{S_{ID_K}} = g^{a_0 + a_1h + a_2h^2 + \cdots + a_{t-1}h^{t-1}} = g^{\sum_{j=0}^{t-1} a_j \bullet h^j}$

(4) Any t partial private key $S_{ID_{t_1}}, S_{ID_{t_2}}, \ldots S_{ID_{t_t}}$ can recover the private key d_A as follows.

Recovery of Temporary Private Key with Lagrange Formula S_{ID}:

$$S_{ID} = \sum_{j=1}^{t} \frac{\prod_{k \neq j, k=1}^{t} i_k}{\prod_{k \neq j, k=1}^{t} i_k - i_j} \bullet f(i_j) \bmod q \tag{9}$$

Recalculate $d_A = H_2(ID_A)^{-S_{ID}} \bullet Q$, restore the private key d_A.

Combined with identity-based encryption scheme and threshold cryptosystem, key escrow problem in identity-based encryption scheme can be effectively solved. However, existing solutions all use bilinear pairing to apply the secret sharing scheme to the identity-based threshold secret sharing scheme, so that a large number of bilinear pairing occurs in all aspects of the identity-based threshold secret sharing scheme Calculation. At present, there is no efficient and efficient method to calculate bilinear pairing. Therefore, the efficiency of identity-based threshold secret sharing scheme is not improved compared with PKI-based threshold cryptography.

The private key sharing scheme in this paper combines verifiable (t, n) threshold secret sharing scheme [5], which transforms the sharing problem of user private key into the secret sharing problem of integers, which reduces a lot of double-pair computing.

3.2 The Secret Key Sharing Program Ideas are Summarized as Follows

First random number generated by the user as a temporary private key $S_{ID}S_{ID}$. S_{ID} is then distributed to n trusted centers using a verifiable (t, n) threshold secret sharing scheme.

Need to restore the user private key d_A, only to recover the temporary private key S_{ID}, through the calculation $S_{ID} = \sum_{j=1}^{t} \frac{\prod_{k \neq j, k=1}^{t} i_k}{\prod_{k \neq j, k=1}^{t} i_k - i_j} \bullet f(i_j) \bmod q$ available.

Then through the calculation $H_2(ID_A)^{-S_{ID}} \bullet Q$, available private key d_A.

In the literature [5], the security proof of the integer secret sharing scheme is given, and it is clear that the security of the private key sharing scheme in this paper can be drawn.

3.3 Below the Calculation of the Amount of Private Key Sharing Program Analysis, the Results Shown in Table 1

At the same time, we compared the private key sharing scheme with some existing schemes, and compared and analyzed the private key distribution phase, the partial private key verification phase and the private key recovery phase. The comparative operations include: Bilinear pairing operation E, modulus index operation S, and Lagrange interpolation L, the other operations are not considered here due to the small amount of computation. The results are shown in Table 2.

Table 1. Private key sharing scheme computing amount of analysis

Operating	Private key distribution	Partial private key verification	Private key recovery
Bilinear pair e operation	0	2	0
Hash operation	1	1	1
Lagrange interpolation	0	0	1
Modulus index calculation	t	t	t

Table 2. The private key sharing scheme in this paper is compared with several schemes

Program	Private key distribution phase	Part of the private key verification phase	Private key recovery phase
Yu-Kong Program [6]	$(4n + 2t + 1)S$	$(2n + t+2)S$	$2nS + 1L$
Literature [7] program	$(3n + t)E$	$2E + (t + 1)S$	$tE + (t-1)S + 1L$
Dehkordi-MH regimen [8]	$3nS$	tS	$(nt) S + 1L$
This program	tS	$2E + tS$	$tS + 1L$

4 Based on Identity Authentication Encrypted Electronic Security Fast Exchange Model

Based on the authentication encryption scheme, the encryption scheme is less than one power and no dot product, so the encryption speed is faster than that of ordinary encryption. Meanwhile, the private key sharing scheme of this paper combines the verifiable (t, n) threshold secret sharing scheme, proposes a new private key sharing scheme, and converts the sharing problem of the user's private key into an integer secret sharing problem, which reduces the number of double lines to calculate.

The following will be based on the identity of the cryptographic cryptography solution for the electronic message security exchange. The results are shown in Fig. 1.

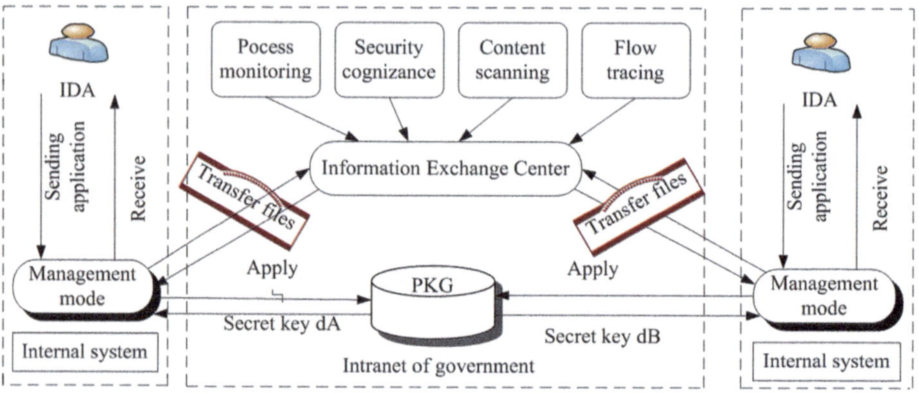

Fig. 1. Based on identity authentication encrypted electronic security fast exchange model

PKG runs the initialization algorithm based on the identity encryption scheme, and the random $a \in F_q$ is the main key, and $params = \langle e, G, G_{1,}g, g^a, H_1, H_2, H_3, H_4 \rangle$ is exposed.

The identity of the sender A is ID_A, which first submits an application to the management module, and the sender gets the identity of the recipient B from the electronic document information exchange center ID_B, Encrypting the electronic official document $M \rightarrow E_{H_{4(\sigma)}}(M)$ to be sent by randomly generated $\sigma \leftarrow \{0,1\}^n$ and symmetric encryption function E_k, and then encrypting the symmetric key with public key, and combining it into a ciphertext $C = <r, \sigma \oplus H_1(\mathrm{r}, s), E_{H_{4(\sigma)}}(M)>$ to send the file, ensuring the security of the file content.

Sender will send encrypted file and send to apply for A to belong to internal system management module, after get the license, by its internal system to join the system data in the encrypted file signature, through the digital signature and management module of process control, ensures the sender sent A identity authentication and non-repudiation.

Encrypted electronic official document sent to C exchange center of information, and then through the PKG to confirm the signature, at the same time according to the C and the private key d_A, to the electronic document transmission process flow tracking, and according to the actual need to scan the content of the electronic official document to be transmitted, sent to the receiver after determine the correct B.

Electronic documents the receiver's internal system B receives the encrypted files, through the digital signature of the sender of A system for identity authentication, certification can record the basic information of the encrypted file, after B forwarded to the recipient, at the same time to send the receipt to information exchange center. The receiver B applies the identity $\mathrm{ID_B}$ to PKG for private key d_B, by calculation $s = e(H_2(ID_A), d_B), \sigma = V \oplus H_1(U, s), M = D_{H_4(\sigma)}(M)$, check $U = H_3(\sigma, M)$. If not, refused to cipher, otherwise output plaintext M. Then enter the internal system to start the related business of electronic document.

Among them in the internal system, the user A and B to the private key used in this paper, the private key of Shared hosting solution to share n trusted center, effectively solve the problem of key escrow, the most important thing is managed is just A temporary private key, real private key is not easy to be n trusted center of multiple collusion attack, so security is guaranteed.

5 Security Analysis

The model is established in the government Intranet and related internal systems, and there is a certain physical isolation between the network and the external network, which makes the operating environment of the model higher. The PKG, which is used to generate the private key of the sender and receiver, is set up in the information exchange center, which is easier to manage and monitor.

In the literature [9], proof of the integrity and security of authentication encryption scheme based on identity has been given. The conclusions are as follows:

Theorem 1. Assumes that A is a polynomial time bounded attacker can forge cipher with the advantage of ε and can perform Q, H_2 query and Q_D decryption query at most. Then there is a polynomial time bounded algorithm B to solve the BDH problem with at least ${}^{\varepsilon}/_{Q_D}\binom{Q}{2}^2$.

Theorem 2. Assumes that A is a polynomial bounded AID-CCA attackers, the advantage is ε, and the number of queries in H_2 is defined by Q, and the number of H_1 queries is defined by Q_1. Also, suppose that our solution is cipher cannot be falsified. Then there is a polynomial time bounded algorithm B to solve the BDH problem with at least ${}^{\varepsilon}/_{Q_1}\binom{Q}{2}^2$.

Therefore, the security and identity authentication of the IBE scheme only used BDH hypothesis and random prediction model, which can resist the adaptive selection of cipher attack in the random prediction machine model.

6 Conclusion

According to the above analysis, identity-based authentication and encryption scheme is secure, so the electronic document exchange model in the government intranet guarantees the integrity and security in sender encryption, receiver decryption, and private key distribution. And the IBE program encryption faster than the average IBE encryption to ensure the efficiency of electronic document delivery. At the same time, a new private key sharing scheme is proposed to solve the key escrow problem of IBE scheme. Through comparative analysis, we can see that the private key sharing scheme is more efficient and secure, making the system more convenient to use.

References

1. Zhang, X.: Research on the security of electronic document system. Digit. Technol. Appl. **02**, 1007–9416 (2013)
2. Yang, S., Zhou, L., Chen, M.: PKI technology and its application in E-government. J. Univ. Electron. Sci. Technol. China (Soc. Sci. Ed.) **9**(4), 1008–8105 (2007)
3. Lynn, B.: Authenticated ID-ID-based encryption cryptology. ePrint Archive Report 072.2002 (2002)
4. Xu, L.: Research and Application Based on Identity and Password System. Shan Dong University, Jinan (2007)
5. Feldman, P.: A practical scheme for non-interactive verifiable secret sharing. In: Proceedings of FOCS 1987. ACM Press, New York (1987)
6. Yu, J., Kong, F.Y., Hao, R.: Publicly verifiable secret sharing with enrollment ability. In: 8th ACIS International Conference on Software Engineering, Artificial Intelligence, Networking, and Parallel/Distributed Computing, 2007 (SNPD 2007), pp. 194–199. IEEE Computer Society, Qingdao (2007)
7. Li, D.: Identity-based secret sharing and its application. Nanjing University of Posts and Telecommunications, Nanjing (2011)
8. Dehkordi, M.H., Mashhadi, S.: New efficient and practical verifiable multi-secret sharing schemes. Inf. Sci. Int. J. **178**(9), 2262–2274 (2008)
9. Vuong, J.L.: A semantics-based routing scheme for grid resource discovery. In: E-Science: First International Conference on E-Science and Grid Computing, pp. 58–70, 90 (2005)

Research on Topic Crawler Strategy Based on Web Page Extension and Best Priority Search Algorithm

Hongsheng Xu[1,2(✉)], Yanqing Lv[1,2], and Ganglong Fan[1,2]

[1] Luoyang Normal University, Luoyang 471934, China
85660190@qq.com
[2] Henan Key Laboratory for Big Data Processing and Analytics of Electronic Commerce, Luoyang 471934, China

Abstract. The topic crawler filters the topic independent links according to a certain web page analysis algorithm, keeps the topic related links and puts them into the URL queue to be fetched. Then, according to a certain search strategy, the next page URL is selected from the queue and repeated until a certain condition of the system is reached. The best priority search strategy predicts the similarity between candidate URL and target pages according to a certain web page analysis algorithm, and selects one or more URL which is the best evaluation to grab. The crawler algorithm based on Web page extension is to evaluate the web page or website with indirect relation through known web pages or data. The paper presents research on topic crawler strategy based on web page extension and best priority search algorithm.

Keywords: Topic crawler · Web page extension · Best priority search URL · Related links

1 Introduction

As an important part of search engine, web crawler design directly affects the quality of search engine. The web crawler is a special web page and downloads from the program analysis web page. It will download the web pages and collected web information stored in a local database for the use of search engines. The working principle of the web crawler is from one or several initial page link to start and get a link queue. With web crawler constantly from the crawled pages from the new link into the link queue, a stop condition until the crawler full system.

However, JumpStation and WWW Worm only arrange search results in the order in which the search tool finds matching information in the database. So there is no correlation. And RBSE was the first engine to introduce the concept of keyword string matching in search results, the earliest modern search engine, in July 1994, when Michael Mauldin introduced John Leavitt. Access to its index program and it is to create what is now known as Lycos [1]. April of the same year, David Filo, a Ph.D. student at Stanford University, and Yang Zhiyuan Gerry Yang, a Chinese American, co-founded a superdirectory index, Yahoo, and succeeded in bringing the concept of

J. Abawajy et al. (Eds.): ATCI 2018, AISC 842, pp. 1129–1136, 2019.
https://doi.org/10.1007/978-3-319-98776-7_138

search engines to the fore. Since then, search engines have entered a period of rapid growth. There are hundreds of name-named search engines on the Internet, and they retrieve as much information as ever before. Google, for example, has more than 3 billion pages in its database these days.

The breadth-first strategy is to perform the next level of search during crawling, after completing the current level of search, in order to cover as many pages as possible. In general, the breadth-first search algorithm is used. There are also many studies which apply the breadth-first algorithm to the focused crawler. The basic idea is that there is a high probability of topic correlation with the web page with the initial URL within a certain link distance.

Depth first starts with the initial URL page, selects a URL from the initial URL library through certain rules, analyzes the other URLs in this page, and selects one to re-enter. Until one line is processed and then the next URL. Depth-first policy is designed to be simple, but the importance of the connections provided in a portal decreases as the hierarchy deepens. Therefore, the depth first strategy has a direct impact on the grab efficiency and grab hit ratio.

Because the traditional web crawler is implemented in a centralized way, and because of its poor scalability and fault tolerance, in this era with big data as the background, it can no longer meet the needs of users for data. Distributed technology can search a large number of web pages as quickly as possible, so distributed web crawlers have gradually become a development direction of web crawlers.

2 Systematic Analysis of Topic Web Crawler

Web crawler is a special web page and downloads from the program analysis web page. It will download the web pages and the collected information is stored in the local database for the use of search engines [2]. The working principle of the web crawler is from one or several initial page link to start and get a link with the queue. "Crawl again from the crawled web pages to extract new links into the link queue, stop until the crawler program satisfies some condition system. It is an important part of the search engine.

The topic crawler filters the topic independent links according to a certain web page analysis algorithm, keeps the topic related links and puts them into the URL queue to be fetched. Then, according to a certain search strategy, select the next page URLL from the queue and repeat the above process until a certain condition of the system is reached. All web pages crawled by the web crawler will be stored by the system. Some analysis, filtering and indexing are carried out. For the topic crawler, the analysis results obtained by this process may also provide feedback and guidance to the subsequent crawling process.

Mercator is a web crawler for the Alta Vista search engine. It is written entirely by JAVA. Mercator is very extensible, and you can add, subtract or replace modules to achieve different functions. With only limited memory, most of the data structures are accessed in the hard disk. Mercator establishes a cache for the recently accessed URL. The hit ratio of the cache has reached 85. Mercator has proved that using JAVA language can achieve higher performance.

PageRank and HITS are the most common link analysis algorithms. PageRank algorithm considers the randomness of user access behavior and the existence of Sink pages, but neglects the purpose of most users. Aiming at this problem, two key concepts are put forward: authorityof authoritative web page and hubby of central web page.

Definition 1. If the hyperlink l points to the web page t, then the page t is called the subpage, also known as the target webpage. The basic idea of the theme web crawler is to divide the hyperlink and the downloaded page content according to the given theme in advance. Predict the relevance of the next URL to be grabbed and the topic of the current web page, ensure that as many crawls as possible, download pages related to the main, download as few irrelevant pages as possible. If the page p contains a hyperlink, p is called the parent page of link l.

The distributed web crawler is implemented on the basis of the traditional network crawler. Each node of the distributed network crawler can be regarded as a traditional network crawler. All nodes are managed through a central management node. So distributed network crawlers have the following advantages over traditional web crawlers: 1. Reliability: when a node crashes, Broadcast through a central management node to keep other nodes working. Extensibility: because distributed web crawlers are composed of multiple nodes and managed by a central management node, when new nodes are added, it cannot affect the work of the whole system, thereby enhancing the efficiency of the system, as is shown by Eq. (1).

$$\frac{f'(x_1)}{2x_1} = (b^2 + a^2)\frac{f'(x_2)}{4x_2^3} = \frac{\ln\frac{b}{a}}{b^2 - a^2}x^3 f'(x_3) \tag{1}$$

The biggest difference between a vertical search engine and a general search engine is that the vertical search engine is oriented to a certain area, so the vertical search engine web spider only collects the topic-related pages, and the topic-independent pages are discarded. This kind of web spider is called the theme spider [3]. After the theme spider downloads the web page to the local place, it needs to use the content-based theme discriminant method to calculate the theme relevance value of the web page. Web pages with a topic correlation degree lower than a certain threshold are discarded, and the methods of calculating the topic correlation are Boolean model and vector space model.

Search engines use certain strategies to collect, discover, understand, extract, organize and process information on the Internet, and provide search services for users [4]. The search engine system can be divided into three categories: directory search engine, robot search engine and meta search engine. Indexer, searcher and user interface are composed of four parts.

3 Web Page Extension and Best Priority Search Algorithm

In the breadth-first search algorithm, the expansion of the solution tree node is along the "fault" of the node depth, that is, the expansion of the node is carried out according to the degree of their proximity to the starting node. First, the first layer of nodes is generated. At the same time, check whether the target node is in the generated node or, if not, expand all the first layer nodes one by one to get the second layer node. And check whether the second layer node contains the target node… Before extending any node with length n 1, you must consider every possible state of the node with length n. Therefore, for the same layer of nodes, the value of solving problems is the same, and we can extend them in any order.

Crawler: its function is to roam the Internet, discover and collect information. It is often a computer program that runs day and night. To improve the speed of information is discovery and update. Indexer: its function is to understand the crawler search for information, extracted from the index items, an index table used to represent documents and generate document libraries. Index tables generally use some form of inverted table [5].

The title of a web page is a highly summary of a page. In general, the title of the home page is the official name of the site, and the title of the content page of the article in the website is the title of the article. The title of the home page is usually the title of the column [6]. Of course, this general principle is not fixed, it may change in practice, but in any case, and the rule will still be followed in general.

Retrieval: its function is to quickly check out the document in the index library according to the user's query, evaluate the correlation between the document and the query, and sort the results to be output. There are four kinds of information retrieval models: set theory model, as is shown by Eq. (2), where a is algebraic model, probability model and mixed model.

$$a_n'(l+1) = a_n(l)\left[\frac{v_i(l+1) - v_{\min}}{v_{\max} - v_{\min}}\right]^q; \quad q > 1, \quad i \in U_1 \tag{2}$$

The node communication module is composed of system node table, node parameter and log record, in which the system node table records the information of all crawler nodes, including the ID number of all nodes, IP address, port number, whether the central node, or not. In order to ensure the consistency of the views in the distributed environment, the system node table information of each node must be the same, and the system node table in each node should be updated when the node in the system enters, exits or crashes [7]. When the central node collapses or exits, all nodes can be expressed as consensus through the system nodes to elect a new central node table according to certain election rules.

In one page, there are often multiple links to other pages, only some of which point to the subject-related page or indicate its importance based on the link anchor text of the page. However, in the PageRank and HITS algorithms, these links are not distinguished from each other, so they often interfere with noisy links such as advertising. Level the basic idea of the link analysis algorithm is to divide the web pages into different blocks through the VIPS page segmentation algorithm, and then set up a page page for these blocks. Page link matrix is Z and X in page [8]. The PageRank on the page diagram at the page block level is W, X × ZG. BlockTo is The BlockRank on the block diagram is W. Some people have implemented block level PageRank and HITS algorithms, and the experimental results show that the efficiency and accuracy of these algorithms are better than those of the traditional corresponding algorithms.

The best priority search steps are typically:

(1) Take a node from the queue header; check whether it can be extended according to the expansion rule, and if it can produce a new node.

$$r_{ij} = \frac{\sqrt{r_i^2 + 2r_ir_j\cos\theta_{ij} + r_j^2}}{\sin\theta_{ij}} \geq \frac{xr_i + yr_j}{\left|x\overrightarrow{n_i} + y\overrightarrow{n_j}\right|}, \quad i \neq j; \quad i,j = 1,2,\ldots,4. \tag{3}$$

(2) Check whether the newly generated node already exists in the queue, if the new node has already appeared in the queue, as is shown by Eq. (3), where r is abandon the node and return to No. 1. Otherwise, if the new node has not appeared in the queue, add it to the end of the queue [9].

At the same time, the allocation of huge crawling tasks to crawler system needs to ensure the load balance of each node. The function of task allocation module is designed to ensure the solution of these problems, and the specific allocation algorithm is explained in detail in the following contents.

This paper discusses how to divide the IP addresses of different hosts and servers under the same domain name into two types: domain name and IP address. At the same time, according to the distribution of different files on each site, the document map is constructed, and the document map is obtained by combining with SiteRank distributed computing. It is proved that using distributed SiteRank computing, It not only greatly reduces the algorithm cost of single site, but also overcomes the shortcoming that single site has limited network coverage. One of the advantages is that it is difficult for common PageRank to cheat on SiteRank.

4 Topic Crawler Strategy Based on Web Page Extension and Best Priority Search Algorithm

The system composition of topic crawler is to filter the page, unlike the common crawler to deal with all the links of the page, the correlation between the page and the restricted domain is analyzed first. The link in the page is processed only when its subject relevance meets the requirements, because if the page is more relevant to the domain, it contains a greater probability of links and domain-related, which improves crawling accuracy. Although a few pages will be omitted, the overall effect is satisfactory [10]. Therefore, the analysis of theme correlation is the key to the design of theme reptiles.

When using the breadth-first search algorithm to solve the problem, we need to construct a data structure that shows the relationship between state characteristics and different states. This kind of data structure is called node. Different problems need to be described with different data structures.

The expansion rules of nodes are determined. According to the conditions given by the problem, one or more new nodes can be generated from one node. This process is often called an extension, as is shown by Eq. (4).

$$\mu_j(k) = P(M_j(k)|Z^k) = \frac{P[Z(k)|M_j(k), Z^{k-1}]P[M_j(k)|Z^{k-1}]}{\sum_{i=1}^{r} P[Z(k)M_i(k), Z^{k-1}]P[M_i(k)|Z^{k-1}]} \tag{4}$$

The relationship between nodes can generally be expressed as a tree. The search process of the search algorithm is actually to construct a solution tree according to the initial conditions and the extended rules and to find the node in accordance with the target state, The improved topic crawler algorithm is as follows:

Step 1: Get topic (String path) Segment.java
Step 2: Compulate topicweight (String topic) JDBCTest.java
Step 3: sortAndDelRepeat(int[]count) Ping.java
Step 4: delRepeat (String[] segment) Relative.java
Step 5: delRepeat (Vector url) mainF.java
Step 6: getParser (String url)
Step 7: String titleStr = p.getURLtitle

The general practice is to grab this page to compare the data with the previous one, if five such comparisons have not changed, will later crawl the page twice as long as the original; If this comparison changes five times in a row, then the time taken to crawl the page will be reduced to 1/2 of the original. In addition, as is shown by Fig. 1, and where the crawler does not need to recrawl the page when updating the content of the page. Just judge some of the properties of the page, and if it is the same, you do not need to update it.

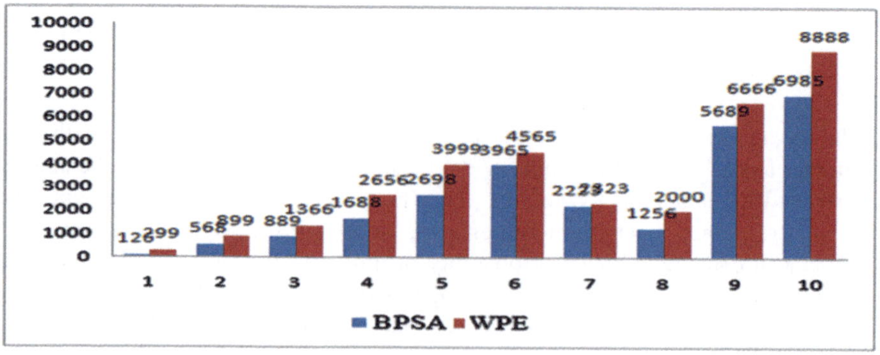

Fig. 1. Topic crawler strategy based on web page extension and best priority search algorithm

The development purpose of the theme web crawler proposed in this paper is to automatically extract web pages through web crawler technology, and realize the search engine to download web pages from the web it wants to visit. Then continue to visit other pages according to the downloaded pages and download them until they meet the user's needs.

The website size of resource discovery and management strategy is better than size is more simple and effective. The key point lies in the size of the crawler web site classification and site level (SiteRank) calculation method to calculate the SiteRank and PageRank is similar, but the website links between the need for a certain degree of abstraction, and calculate the link weights in the model.

ExtJS can be used to develop rich Internet applications RIA: rich Internet applications, it is mainly used to create front-end user interfaces, is a back-end technology-independent front-end AJAX framework. This algorithm is widely used in real-time search, including: page granularity analysis algorithm, website granularity analysis algorithm and web page block granularity analysis algorithm.

Tests on crawlers: enter an illegal URL. Be able to write error types to log files. Enter all types of URLs, only other links to http and https links are treated as exceptions. System speed tests: after many times. Running calculation average value. Get the system running efficiency is not very high. An average of about 3 pages per second download. Test portability: porting the system to other linux systems to run most of the lack of openssl library caused the system to go wrong.

5 Summary

The system composition of topic crawler is to filter the page, unlike the common crawler to deal with all the links of the page, the correlation between the page and the restricted domain is analyzed first. The link in the page is processed only when its subject relevance meets the requirements, because if the page is more relevant to the domain, it contains a greater probability of links and domain-related, which improves crawling accuracy. Although a few pages will be omitted, the overall effect is satisfactory.

Acknowledgements. This paper is supported by Henan key Laboratory for Big Data Processing and Analytics of Electronic Commerce, and also supported by the science and technology research major project of Henan province Education Department (17B520026, 15A120012), Key scientific research projects in Henan province universities (17A880020).

References

1. Wutka, M., Moffet, A., Mittal, K.: Sams teach yourself JAVA ServerPages 2.0 with Apache Tomcat in 24 Hours. Sams Publishing, Indianapolis (2003)
2. Wan Yuan, W., Wang, D.: A design port of URL allocation algorithm in a parallel Crawler system. Comput. Eng. Appl. (2016)
3. Tsui, F.F.: JSP: a research signal processor in Josephson technology. IBM J. Res. Dev. **24** (2), 243–252 (2013)

4. Cho, J., Garcia-Molina, H.: Effective page refresh policies for Web crawlers. ACM Trans. Database Syst. (TODS) **28**, 390–426 (2013)
5. Yu, J., Shi, S.: Distributed web page sorting algorithm and its transmission mode analysis. Comput. Eng. Appl. (2014)
6. Zhang, M., Gao, J., Ma, S.: Web information retrieval based on link description text and its context. Comput. Res. Dev. (2014)
7. Zhou, X., Wu, C.: Text knowledge discovery: text Mining based on information extraction. Comput. Sci. (2013)
8. Xu, H., Zhang, R.: Novel approach of semantic annotation by fuzzy ontology based on variable precision rough set and concept lattice. Int. J. Hybrid Inf. Technol. **9**(4), 25–40 (2016)
9. Chen, H., Luo, C., Wang, J.: Design and implementation of FTP search engine based on Web. Comput. Appl. 100–106 (2014)
10. Xu, H., Zhang, R.: Semantic annotation of ontology by using rough concept lattice isomorphic model. Int. J. Hybrid Inf. Technol. **8**(2), 93–108 (2015)

An Improved Speech Enhancement Algorithm Based on Combination of OMLSA and IMCRA

Hui Li(✉), Qiang Miao, and Hua Shen

Dalian Neusoft University of Information, Dalian 116023, Liaoning, China
lihui@neusoft.edu.cn

Abstract. Speech is the acoustic performance of language, it also is the most natural, most effective and most convenient communication ways for people. The existing classic speech enhancement algorithms include statistical model algorithm, signal subspace algorithm and spectral subtraction algorithm. This paper presents an improved speech enhancement algorithm of combination of OMLSA (Optimally Modified Log-Spectral Amplitude Estimator) and IMCRA (Improved Minima Controlled Recursive Averaging). Comparing with the classic algorithms above, the proposed algorithm have several advantage such as effective suppression of short wave noise for the voice interference, low distortion of the speech signal after processing, high naturalness of the speech, avoiding residual noise effectively. Simulation of the combined algorithm of OMLSA and IMCRA include analysis and design of the algorithm, algorithm modeling and simulation verification. The simulation show that the improved algorithm could obtain better results in speech enhancement.

Keywords: Speech enhancement · OMLSA · IMCRA

1 Introduction of Improved Speech Signal Enhancement Technology

As the basic part of the speech signal processing, speech enhancement technology is widely used in the pre-processing part of speech processing system with the purpose reducing the background noise [1]. In the two-state model, the combination algorithm of OMLSA and IMCRA has obvious advantage comparing with the other classic algorithms in noise reducing. The OMLSA algorithm which combines probability of speech existence and two-state model is used in the speech estimation. The OMLSA speech estimator has low complexity and better speech reduction effect. The IMCRA noise estimator includes two iterations of noisy speech signal power spectrum smoothing and minimum statistics. And the algorithm is applicable for the non-stationary noise situation.

J. Abawajy et al. (Eds.): ATCI 2018, AISC 842, pp. 1137–1144, 2019.
https://doi.org/10.1007/978-3-319-98776-7_139

2 Noise Spectrum Estimation Algorithm

2.1 Introduction of Noise Spectrum Estimation Algorithm

The estimation of noise power spectrum estimation is the important composed part of single channel speech enhancement system. The accuracy of noise power spectrum estimation is very important to the quality of speech enhancement. The noise overestimation leads to the speech distortion, and the noise underestimation leads to large amount of noise. Classic noise power spectrum estimation algorithms commonly use VAD (Voice Activity Detection) in the non-speech estimation and update noise power spectrum. In the stable noise environment such as white noise, the VAD algorithm can obtain satisfactory results. But the main defect of this way is the dependence on the accuracy of the voice activity detection. The voice activity accuracy is very low in low SNR. And the noise spectrum is changing in the noisy environment. So the noise estimation should not be in the non-speech, and should be update continuously in the voice activity period. In order to make up for the shortcomings of the classic algorithms, some new noise spectrum estimation algorithms which have no relationship with VAD and can update noise spectrum estimation during voice activity period are proposed.

2.2 The IMCRA Algorithm

The IMCRA algorithm is a widely used noise spectrum estimation algorithm in the single channel speech enhancement algorithm [2]. IMCRA gets the noise power spectrum through first-order recursion smoothing for noisy speech signal, then does the minimum search for the smoothing output signal, and then solve the probability of speech existence through a noise compensation factor, last solve the probability of non-speech by the second smoothing and minimum search according to the probability of speech existence. IMCRA estimates the noise with the time recursive averaging algorithm. Normally the influence of speech spectrum by noise is inhomogeneous which is used to estimate and update the noise spectrum by frequency band. Both of the two iterations include power spectrum smoothing and minimum statistics. The first iteration is to detect speech activity roughly for each frequency point. The smoothing operation of the second iteration is to eliminate the relatively strong speech presentence, and to use the short window in the minimum statistics.

3 Speech Presence Probability (SPP) Estimation Algorithm

3.1 Introduction of Speech Presence Probability Estimation Algorithm

Most of the speech enhancement algorithms are on the base of speech presence probability estimation and noise power spectrum estimation. The speech presence probability estimation algorithm is used to update the gain coefficient of the de-noising filter. The classic speech presence probability estimation algorithms include spectral subtraction algorithm and Wiener filtering algorithm [3].

3.2 The OMLSA Algorithm

OMLSA algorithm uses the speech estimation algorithm. The first step of OMLSA is to get the speech conditional probability through the estimation of prior probability of silence and prior SNR, and then realize speech estimation by calculating the speech effective gain G [4]. The OMLSA has some obvious advantage comparing with the classic method which are: minimum mean-square error of the log spectra for speech signals under signal presence uncertain, adaption in many kinds of noise environments, good noise reduction effect in most severe noise environment, good effect in avoidance of residual noise and reservation of week speech component especially in low SNR and non-stationary noise situation. The gain formula of OMLSA algorithm depends on the SPP and noise power spectrum estimation, and it has low complexity and wide application.

4 Design of Improved Speech Enhancement Algorithm of Combination of OMLSA and IMCRA

The improved algorithm includes several steps (see Fig. 1) which are discrete speech signal input, noise adding, speech noise spectrum estimation, speech spectrum estimation, speech enhancement and speech noise reduction.

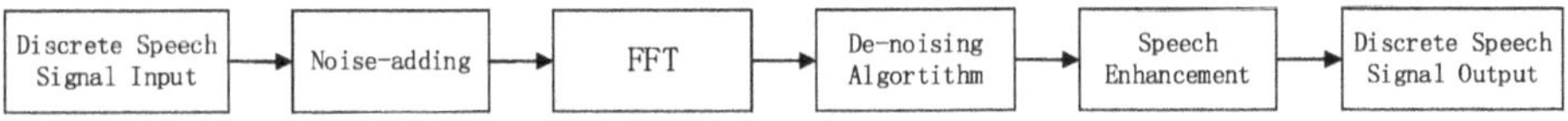

Fig. 1. Main flowchart of speech enhancement algorithm.

The algorithm flowchart (see Fig. 2) of the Matlab Gui design is shown below:

4.1 The Noise-Adding of Speech Signal

In the design, two kinds of noise are added to the pure speech signal.

(1) White Gaussian Noise

The probability distribution of Gaussian is normal formula, and the white noise has irrelevant second moment as well as constant first second. The White Gaussian Noise is an ideal model of analyzing channel additive noise. The power spectrum density of the White Gaussian Noise is with uniform distribution, and the amplitude distribution is with Gaussian distribution. The definition of the power spectrum density of the White Gaussian Noise is shown below:

$$S_n(f) = \frac{N_0}{2} \tag{1}$$

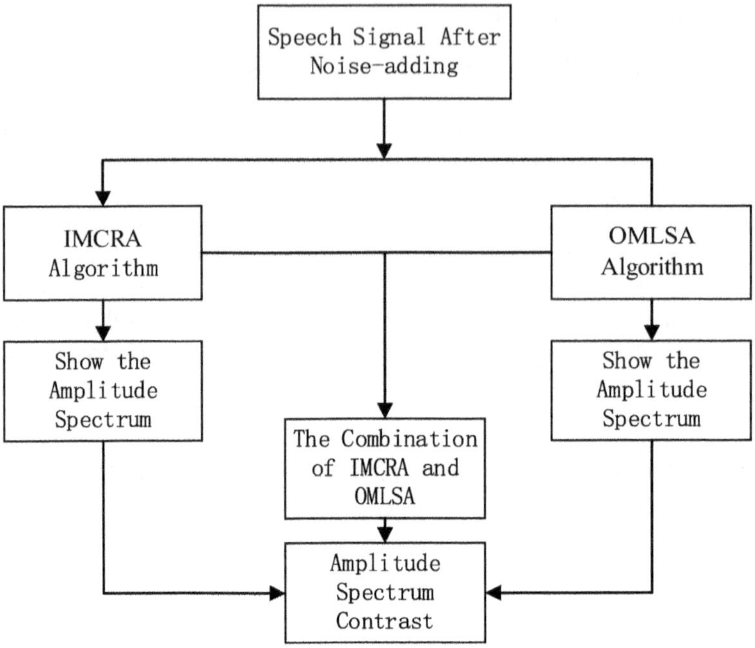

Fig. 2. Algorithm flowchart.

The formula $\frac{N_0}{2}$ stands for bilateral power spectrum density. Rhe theory of signal analyzing shows that the power spectrum density is the Fourier Transform of its autocorrelation formula which is shown below:

$$R_n(\tau) = \frac{N_0}{2} \cdot \delta(\tau) \tag{2}$$

(2) Colored Noise

Colored noise is a kind of noise whose power spectrum density is not flat. The power spectrum density in the infinite frequency range of the white noise is a constant which means some colored noise is existed, or in the other way means the power spectrum density is not flat. Most of the audio noise such as moving vehicle noise, the fan noise of computers, electric drill noise and the people walking noise are non-white low frequency spectrum. The white noise through channel will change into colored noise on the effect of channel frequency change.

4.2 Design of OMLSA Algorithm

OMLSA algorithm is on the base of the following formulas:

$$H_0(\lambda, k) : Y(\lambda, k) = D(\lambda, k) \tag{3}$$

$$H_1(\lambda, k) : Y(\lambda, k) = X(\lambda, k) + D(\lambda, k) \tag{4}$$

The formulas $Y(\lambda, k)$, $X(\lambda, k)$ and $D(\lambda, k)$ stand for the noise signal, pure speech signal and the short-time Fourier Transform of the noise signal. The formulas $H_0(\lambda, k)$ and $H_1(\lambda, k)$ stand for the hypothesis of the speech presence and no presence. The parameter λ stands for the index of time, and k stands for the index of frequency. After short-time Fourier Transform of the noisy speech, the noise signal and speech signal are all complex Gaussian variables. The presence probability of the speech signal is $P(H_1(\lambda, k))|Y(\lambda, k) = p(\lambda, k)$. If using the Bayesian rule, then the realization formula is shown below:

$$\left\{p(\lambda, k) = 1 + \frac{q(\lambda, k)}{1 - q(\lambda, k)}(1 + \xi(\lambda, k)) \times \exp^{(-v(\lambda, k))}\right\} \tag{5}$$

$$v(\lambda, k) = \frac{\gamma\xi}{(1 + \xi)} \tag{6}$$

The formula $q(\lambda, k) = p(H_0(\lambda, k))$ stands for the no presence probability of speech on the prior conditions, $\gamma(\lambda, k)$ and $\xi(\lambda, k)$ stand for the SNR of the posteriori and prior. And the calculation formula of $\xi(\lambda, k)$ on the prior condition if shown below:

$$\xi(\lambda, k) = \alpha G_{H_1}^2(\lambda - 1, k)\gamma(\lambda - 1, k) + (1 - \alpha)\max\{\gamma(\lambda, k) - 1, 0\} \tag{7}$$

According to the relevance of the continuous frames, the soft decision is completed by prior SNR based on the time domain distribution, and the spectral entropy VAD hard decision is used to complete the verification of the prior speech no presence probability $q(\lambda, k)$.

The OMLSA gain is obtained through the described method above, and the calculation formulas are shown below:

$$G(\lambda, k) = \{G_{H_1}(\lambda, k)\}^{p(\lambda, k)} G_{\min}^{1 - p(\lambda, k)} \tag{8}$$

$$G_{H_1}(\lambda, k) = \frac{\xi(\lambda, k)}{1 + \xi(\lambda, k)} \exp\left[\frac{1}{2}\int_{v(\lambda, k)}^{\infty} \frac{e^{-t}}{t} dt\right] \tag{9}$$

4.3 Design of IMCRA Algorithm

The realization of IMCRA algorithm is shown below:

The first step of the algorithm is to get the noise power spectrum through first order recursion smoothing of the noisy speech signal, the specific formula is shown below. In the formula, α_d stands for the fixed smoothing factor of the silent segment.

$$\bar{\lambda}_d(k,l+1) = \left[\alpha_d, \bar{\lambda}_d(k,l) + (l-\alpha_d))\right]|Y(k,l)| + \bar{\lambda}_d(k,l)p(k,l) \quad (10)$$

Next step is the noise power spectrum compensation, and the formula is shown below:

$$\bar{\lambda}_d(k,l+1) = \beta\bar{\lambda}_d(k,l+1) \quad (11)$$

The last step is to calculate the value of $p(k,l)$ from the following formula:

$$p(k,l) = \left\{1 + \frac{q(k,l)}{1-q(k,l)}(1-\varepsilon(k,l)\exp(-v(k,l)))\right\}^{-1} \quad (12)$$

In the above formula, the factor $q(k,l)$ is the prior probability of speech absence. To calculate $p(k,l)$, the first step is the smoothing of the noisy signal, the realization is the following formula.

$$S_f = \sum_{i=-w}^{w} b(i)|Y(k-i,l)|^2; S(K,l) = a_i S(k,l-1) + (1-a_i)S_f(k+l) \quad (13)$$

In the above formula, the normalized window is $\sum_{i=-w}^{w} b(i) = 1$, and the length of $b(i)$ is $2w+1$. The Hanning window is used in the normalization. The minimum search of the normalized output signal will get the next result:

$$S_{\min}(k,l) = \min\{S(k,l)|l-D+1 \le l \le l\} \quad (14)$$

In the above formula, D stands for the length of the window, so a compensation factor is obtained from the following formulas which is $B_{\min}$.

$$E\{S_{\min}(k,l)|\varepsilon(k,l) = 0\} = B_{\min}^{-1}\bar{\lambda}_d(k,l) \quad (15)$$

$$\gamma_{\min} = \frac{|Y(k,l)|^2}{B_{\min}S_{\min}(k,l)}, \xi(k,l) = \frac{S(k,l)}{B_{\min}S_{\min}(k,l)} \quad (16)$$

The next step is the calculation of the speech presence probability.

$$I(k,l)\begin{cases} 1; \gamma_{\min}(k,l) < \gamma_0, & \xi(k,l) < \xi_0 \\ 0, & else \end{cases} \quad (17)$$

Then perform the smoothing again and minimum search according to the calculation speech probability. The specific steps are shown below:

$$\bar{S}_l \begin{cases} \sum_{i=-w}^{w} b(i)I(k-i,l)|Y(k-i,l)|^2 / \sum_{i=-w}^{w} b(i)I(k-i,l); & \sum_{i=-w}^{w} b(i)I(k-i,l) \neq 0; \\ S_f(k,l-1); & else \end{cases} \tag{18}$$

$$\bar{S}(k,l) = a_i S(k,l-1) + (1-a_i)\bar{S}(k,l) \tag{19}$$

$$\bar{\gamma}_{\min} = \frac{|Y(k,l)|^2}{B_{\min}\bar{S}_{\min}(k,l)}, \bar{\xi}(k,l) = \frac{S(k,l)}{B_{\min}\bar{S}_{\min}(k,l)} \tag{20}$$

The minimum search process is the same with the first smoothing, and the probability of the speech absence is obtained from the following method.

$$\begin{array}{c} 1; \bar{\gamma} \leq 1, \bar{\xi}(k,l) \leq \xi_0 \\ (\gamma_1 - \bar{\lambda}_{\min}(k,l)/(\gamma_1 - 1); 1 < \bar{\lambda}_{\min}(k,l) < \gamma_1, \bar{\xi}(k,l) \leq \xi_0 \\ 0, else \end{array} \tag{21}$$

5 Simulation of Improved Speech Enhancement Algorithm of Combination of OMLSA and IMCRA

The simulation of the speech enhancement algorithm of combination of OMLSA and IMCRA is based on the Matlab Gui. The simulation system includes OMLSA simulation, IMCRA simulation and the improved combination algorithm simulation. The

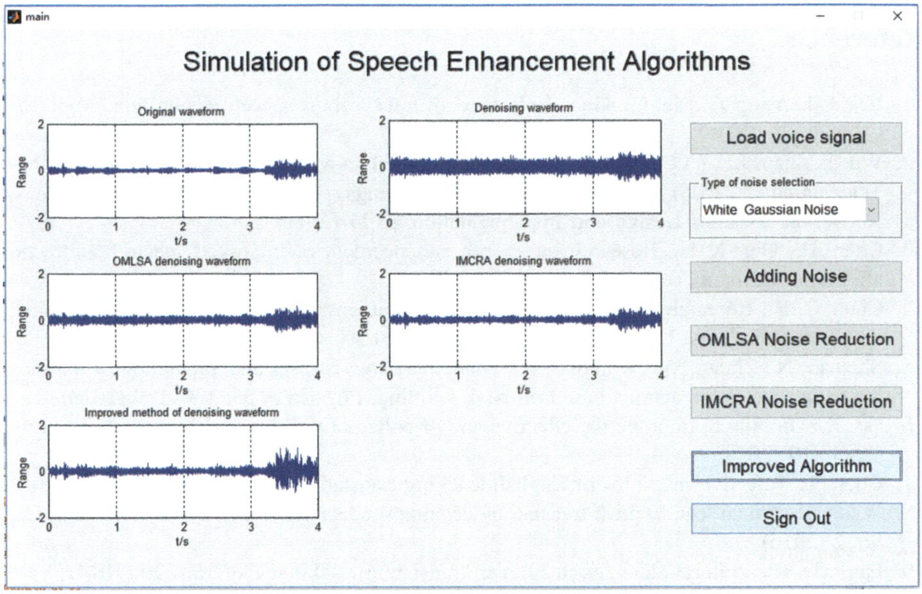

Fig. 3. Simulation results.

simulation interface (see Fig. 3) could show five power spectrum images which are the original speech signal waveform, the adding noise speech signal waveform, the waveform of OMLSA noise reduction, the waveform of IMCRA noise reduction, and the waveform of improved combination algorithm. The buttons on the right side include loading, noise choosing (White Gaussian Noise or colored noise) and adding, OMLSA noise reduction, IMCRA noise reduction, the improved algorithm simulation and sign out.

The simulation uses a segment of speech signal, and add the White Gaussian Noise. The simulation results show that the single de-noising algorithm can eliminate noise to a certain extent. But the improved speech enhancement algorithm of combination of OMLSA and IMCRA is much better than single de-noising method. More speech information is reserved and more noise is eliminated in the last image.

6 Conclusion

The improved Speech Enhancement Algorithm of Combination of OMLSA and IMCRA could obtain a better speech enhancement effect than single method. The algorithm is applicable to any noisy speech to obtain the enhancement effect. The simulation system uses the mixed voice and noise speech to test, and the test results show that the de-noising effect and speech enhancement effect are perfect.

Acknowledgements. This work was, in part, supported by the Liaoning Province Natural Science Fund (Grant No. 20170540052); Liaoning Province Science Fund for Research and Development Program (Grant Nos. 2017225078, 2017225079).

References

1. Lei, J.-J., Yang Z., Liu, G., Guo, J.: Review of noise robust speech recognition. Appl. Res. Comput. **26** (2009)
2. Wu, J., Zhao, J., Li, Q.: The IMCRA algorithm based on wiener filtering. J. Xian Univ. Posts Telecommun. **22**(05), 73–77 2017. In: Proceedings of 7th USENIX Conference on Networked Systems Design and Implementation, p. 17 (2010)
3. Chen, H., Qiu, X.-H.: Research on speech enhancement of improved spectral subtraction algorithm. Comput. Technol. Dev. **4**, 107 (2014)
4. Chen, C.-B.: Research and development in the home environment for speech enhancement system. South China University of Technology (2015)
5. Shizhao, N., Zheng, W., Wangmo, P., Yuan, N., Peng, L.: Big data prediction of durations for online collective actions based on peak's timing. Physica A **55**, 130–139 (2018)
6. Ge, X.: On how to improve the effectiveness of college English translation teach-ing. Engl. Teach. **18** (2018)
7. Chen, T.: Effective measures of English teaching translation. Overseas Engl. **20** (2015)
8. Wang, H.: On college English translation teaching models in the internet environment. Engl Sq. **2** (2016)
9. Fan, M.: The cultural elements in English translation teaching. Engl. Sq. **20** (2014)
10. Wang, H.: The problems and countermeasures of English translation teaching in College of Traditional Chinese Medicine. Overseas Engl. **20** (2014)

Research on Virtual Guzheng Based on Kinect

Shuyao Li, Kuangyi Xu, and Heng Zhang(✉)

College of Computer and Information Science,
IoE-AI Industrial Innovation Center, Southwest University,
Chongqing 400700, China
dahaizhangheng@163.com

Abstract. There are a lot of researches on virtual instruments, but there are few on classical Chinese instruments, and the techniques used are very limited. This paper uses Unity 3D and Kinect camera combined with virtual reality technology and gesture recognition method to design a virtual playing system of Guzheng, a traditional Chinese musical instrument, with demonstration function. In this paper, the real scene obtained by Kinect camera is fused with virtual Guzheng in Unity 3D. The depth data obtained by Kinect and the Suzuki85 algorithm are used to recognize the relative position of the user's right hand and the virtual Guzheng, and the hand gesture of the user is recognized by Kinect.

Keywords: Unity 3D · Kinect sensor · Virtual reality · Virtual performance
Gesture recognition

1 Introduction

Guzheng is one of the unique and important national musical instruments in China. It has beautiful timbre, wide range of timbre, rich playing skills and strong expressiveness, which is loved by the broad masses of people. Guzheng is an ancient national musical instrument born and raised on this fertile loess ground with a long history of Chinese culture. Zither music is elegant and subtle, and melodies are smooth and graceful. It can express many aspects of things, improve people's aesthetic ability, purify the soul, and cultivate sentiment. At the same time, Guzheng music can often express the author's ideal of life, artistic pursuit, value judgment, ethical attitude and so on. It can purify the appreciator's heart while spreading music, the moral sentiment of the appreciating person can be sublimated. Due to the historical relationship and the persistence of people in the pursuit of interests in the material society, we have gradually become indifferent to the consciousness of protecting traditional culture, inheriting traditional culture and multiplying traditional culture. Carrying forward the national musical instruments can also make the younger generation understand, know and love the cultural and musical traditions of their own nation, and carry on the excellent national culture.

In traditional musical instruments, people make wonderful music by playing all kinds of real instruments. However, due to difficulties in purchasing, playing skills and other reasons, many people do not have direct access to instruments to play. Now, with the rapid development of high and new technology, more and more musical instruments appear on our mobile devices. People who learn musical instruments, or those

J. Abawajy et al. (Eds.): ATCI 2018, AISC 842, pp. 1145–1156, 2019.
https://doi.org/10.1007/978-3-319-98776-7_140

without instruments or those who haven't learned playing musical instruments, can play virtual instruments wherever they are equipped with mobile devices. With the rapid development of VR and machine learning, people are no longer satisfied with simply operating virtual instruments on mobile devices. People hope to be able to feel the real feeling of playing virtual instruments and play virtual instruments. It may even innovate the performance art of Guzheng, feel the charm of the art from a new angle of view and appreciation.

This paper discusses the reconstruction process of virtual Guzheng playing scene. By using the popular VR technology and machine learning algorithm, using Kinect camera to obtain the user's hand position and identify the hand movement, the static art is transformed into dynamic. Without holding a real instrument, the performer can feel the touch of the real performance, improve the artistic expression ability, increase the artistic expression form, and carry forward the excellent Chinese national musical instrument, so that people who know Guzheng can practice Guzheng anytime, anywhere, so that people who can not play Guzheng can also play Guzheng, it can not only attract more people to learn how to play national musical instruments, but also contribute to the protection and inheritance of national intangible cultural heritage.

2 Related Work

At present, a lot of work has been done in the field of virtual instrument performance at home and abroad, and the research methods and techniques are different.

As early as 2000 and before, Paradiso [1] Wanderley and Battier [2] proposed to play music through electronic interfaces and gesture controllers. In 2001, the research on virtual musical instruments was innovated. A PC virtual reality system based on data gloves was realized, which was used to simulate the performance of piano, trumpet, percussion instrument and so on. This article focuses on the creation of virtual reality scenes, and does not deal with gesture recognition in detail. In 2005, a multi-functional virtual guitar [4] was Created which used color gloves and infrared cameras as input devices, formed a tone by confirming the distance between the gloves and synthesizing the hand movements of the player. In the musical instrument playing virtual reality, the most influential is the "air guitar" game frame. It uses gestures to interact, developing a powerful platform that provides seamless real-time interaction, intuitive playability, and coherent sound output. This game framework lays the foundation for virtual reality games, and many virtual games have learned from the framework and theoretical ideas of "air guitar", such as virtual piano playing system [5], virtual bell playing system [6] and so on.

Later, the monocular camera was used to obtain the gesture information of the user, to simulate the movement of music playing in 3D environment, and to issue the basic scale tone color [7]. Or using a high-definition camera captured the user's gestures to simulate virtual guitar play [8]. But they simplified guitar gestures, mapping chords to 12 different gestures, and in order to detect gestures, the operator had to place his left hand in a specific area of detection. Although the processing of gesture details had been added, this kind of performance method was divorced from the real guitar playing method, which maked the virtual performance appear unreal. Since the advent of

Kinect, there have been many technologies based on Kinect. For example, Virtual performance of three musical instruments (drum, guitar, spider king) has been realized [9] based on Kinect, Or using Kinect to sense 3D gesture space and top-down projector as visual feedback, with MIDI piano keyboard to achieve enhanced performance [10]. The system is a combination of virtual playing and keyboard hardware. Gesture control is used to enhance the performance effect, not to use gesture recognition for virtual performance, so there is no part of virtual instrument. Professor Pan Zhigeng of Zhejiang University of China took Erhu as an example, using Kinect camera combined with augmented reality technology and gesture recognition method to realize a virtual performance system of national bow string musical instrument [11]. He fused reality scenes and virtual instruments captured by Kinect into augmented reality scenes. The user's left hand is segmented from the depth data by the Bayesian skin model obtained by Kinect and a new image is created on the enhanced image again, which solves the problem of virtual and real occlusion in virtual playing scene. A 3D virtual gesture fitting method based on reverse dynamics and Markov model is used to recognize the left hand gesture and combined the movement of the right hand to complete the virtual performance of the instrument. This provides a good reference for the study of this project.

3 Instruments

3.1 Kinect

Kinect is a 3D external somatosensory camera developed by Microsoft Xbox 360. It also introduces dynamic capture, image recognition, microphone input, speech recognition, community interaction, etc. Users do not need any props, just use their own limbs to control the game. It uses the deep image based human skeleton tracking algorithm developed by Microsoft Research Institute of Cambridge, while the deep image is generated by Range Camera technology provided by PrimeSense. In addition, Kinect uses a 4-Mac array to recognize 3D stereo speech.

Kinect for windows SDK2.0 has the following characteristics:

- using XBox One Kinect sensors
- wider depth and color view in horizontal and vertical directions
- full HD Color Picture (full HD Color)
- Independent Illumination Infrared (Lighting Independent Infrared, 30 fps)
- depth fidelity (Depth fidelity, 3x)
- improved microphone (zero balance)
- 6 complete 25 skeletal sites (25 for each)
- thumb tracking, end of hand tracking, hand opening and closing gestures
- enhanced high quality operating range (close to 0.5 m, far 4.5 m)
- multiple applications can use sensors at the same time.

A. Kinect sensing hardware

The Kinectv2 has a color camera and a depth sensor as shown in Fig. 1. RGB camera is on the left, used to capture 1920 × 1080 color images, a maximum of 30 frames per second. On the right is the Depth sensor. Kinectv2 uses the "Time of Flight (TOF)" way to obtain Depth information from the time returned from the projected infrared reflection. The Depth sensor can't see the appearance, but next to the Color Camera is the infrared camera (the left) and the projective pulse modulated infrared projector (the right). Below the Kinect is a set of quaternion microphone arrays for sound source location and speech recognition.

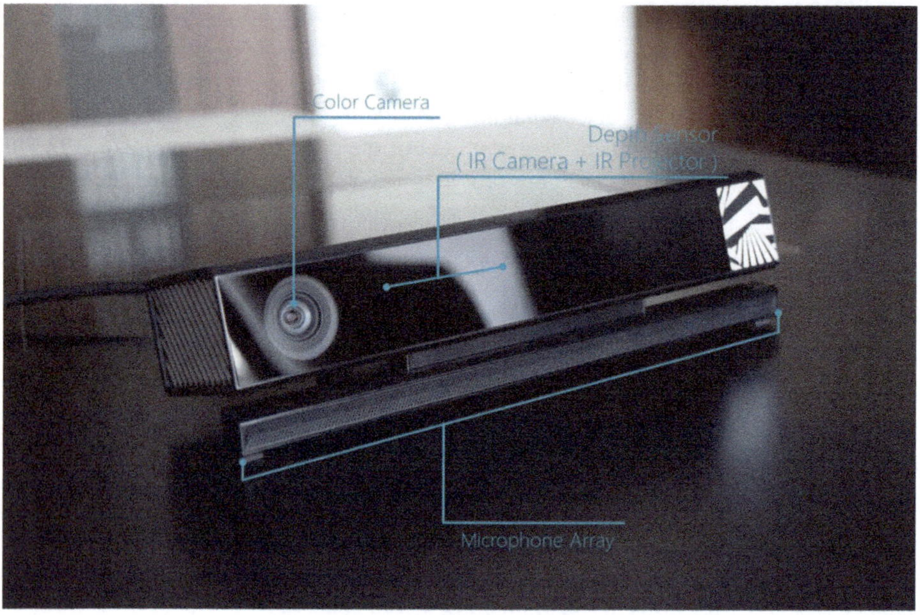

Fig. 1. Kinect V2 camera

Microsoft describes Kinect v2 as having "three eyes" and "two ears".
Three eyes: color camera, depth camera, infrared projector

- Infrared camera: used to capture color video images in a range of viewing angles.
- Infrared projectors: active projection of near infrared spectrum, irradiating rough objects, or penetrating glass, will form random speckles, called speckles, and can be read by infrared cameras.
- depth camera: analyze infrared spectrum and create depth images of bodies and objects in visual range.

Four ears: quaternion linear microphone array

- the sound is collected from 4 microphones, digital signal processing (DSP) is built in, background noise is filtered at the same time, the direction of sound source can be located.

B. Kinect software tools

Microsoft provides developers with Kinect for Windows SDK development tools that allow developers to develop applications in languages such as C or C#. There are also many software interfaces and toolkits in SDK, such as Unity 3D interface, virtual gesture creation tool and speech recognition tool. In addition to Microsoft, there are many third-party frameworks, such as United 3D Matla and OpenCV. Microsoft's Kinect provides support for Unity3D and has provided a third-party toolkit, Kinect. unitypackage.

This article uses unity3D to develop.

3.2 Unity3D

Unity3D is a comprehensive game development tool developed by Unity Technologies, which makes it easy for players to create interactive content such as 3D video games, architectural visualization, real-time 3D animation and so on. It is a fully integrated professional game engine. Unity3D attaches great importance to the efficiency of development and platform application promotion, it provides perfect documentation and many model scripts for beginners to quickly start.

Since 2016, researchers and IT manufacturers have attached great importance to virtual reality. More and more research departments and IT enterprises have invested a lot of money to deploy virtual reality technology research and development. Unity 3D is a free virtual reality development tool, and has been used in many games and websites. According to statistics, nearly 80% VR developers choose to use Unity3D engine for VR application system development.

4 Virtual Playing System

This paper constructs a virtual Guzheng playing system. In the process of playing Guzheng, the relative position of Guzheng and human body need to be suitable, and the right hand and the left hand need to play the correct scale with the correct fingering. Due to the limitation of the number of test equipment, the right hand is only considered in this paper. The reconstruction of the performance scene needs real Image captured by Kinect overlay Virtual instrument Model. The processing route architecture of the system is shown in the Fig. 2.

4.1 Modeling

In many studies of virtual instrument playing, the authenticity of virtual instrument modeling is ignored. For example, only a few lines are used to simulate the strings or only the contour of the instrument, not the texture, which results in poor performance of the system and the user can not tell which instrument he is playing.

In this paper, we use 3Ds MAX modeling software to model Guzheng, add texture and material, export standard fbx file, then read model file in Unity3D. The texture and

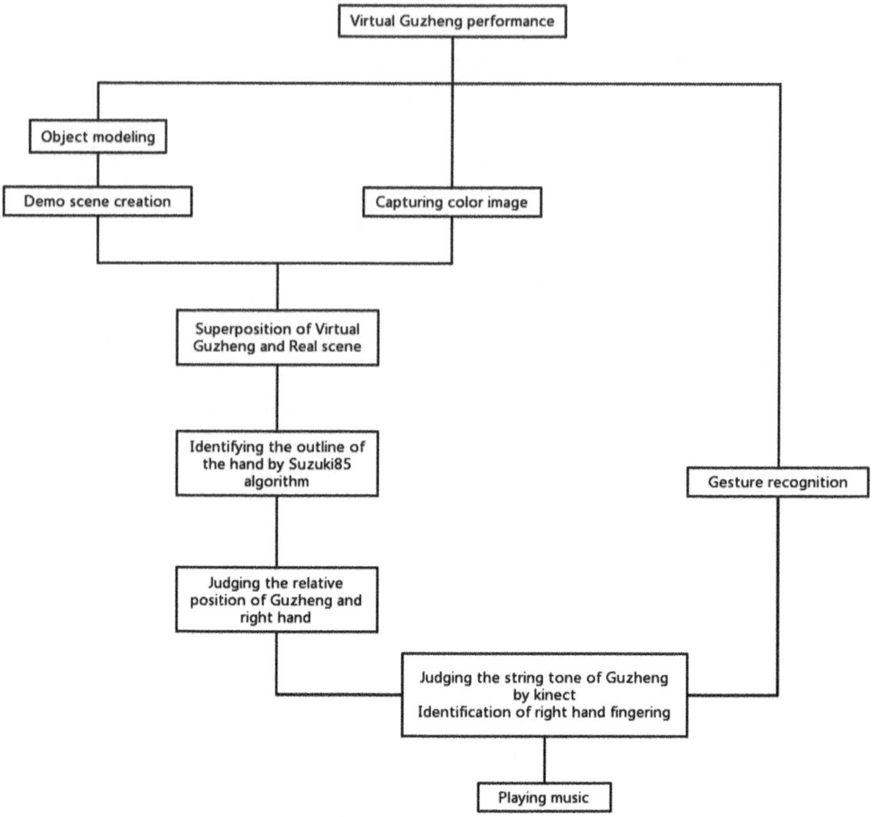

Fig. 2. The system architecture diagram

material of the model and the lighting of the scene are set to make the musical instrument model more realistic, as shown in Fig. 3.

In addition to modeling musical instruments, we also model the characters and other materials needed in the demo scene. In this paper, we make characters to play Guzheng animation in 3Ds MAX. Finally, we import all the prepared materials into Unity 3D, set up the terrain, lighting, sky box and other environmental conditions, and create a good demonstration scene, as shown in Fig. 4.

4.2 Interact

In the research of virtual instrument playing, the most important thing is how users interact with virtual environment.

This virtual playing system chooses to construct virtual environment in Unity3D, and uses Kinect camera to acquire user's body and hand position to interact with virtual environment. This article uses Carnegie Mellon's intermediate plug-in to develop Kinect with Unity3D. This paper mainly solves two problems: first, the relative position of Guzheng and user is reasonable, and it can move along with user. Second, it

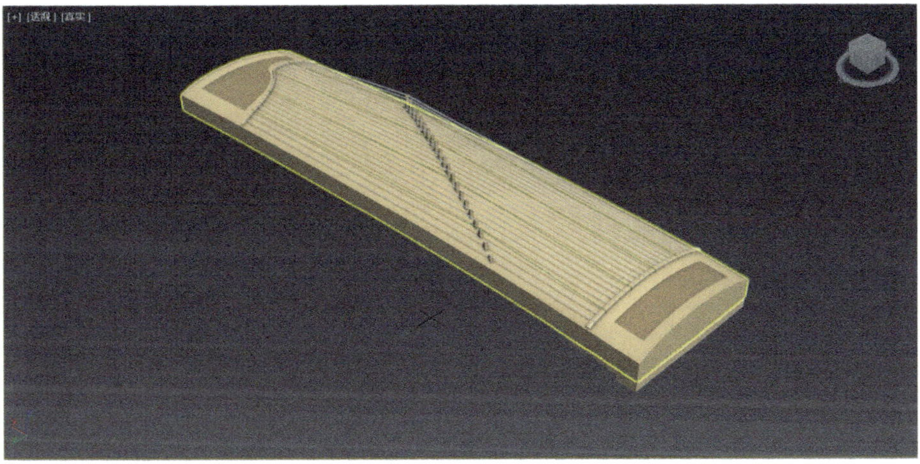

Fig. 3. Guzheng model

Fig. 4. Demonstration scene

is able to capture the user's right hand position and determine which string the right hand is playing.

The first is to determine the reasonable relative position of the Guzheng and the user. By observing the relative position of the real Guzheng and the player, it is found that the Guzheng is parallel to the waist at the height, and the former Yue Shan of the Guzheng in the left and right positions is probably level with the right shoulder. Referring to Carnegie Mellon's example project, Overlay Demo, we replaced the green ball with the Guzheng and the Kinect right hand position with the user's right shoulder and waist position. Then according to the real situation to determine the relative position of the virtual Guzheng and user and let Guzheng follow the user movement (Figs. 5 and 6).

For the second question, because during the playing process, people's gestures are constantly changing. If we test the position of each gesture in relation to the string of the zither, the amount of work will be very large, and the error will be caused by the error of the gesture. If it's just a gesture to determine the position, there is an error that makes the user not able to play the right sound. So we need accurate hand positioning to improve the user experience.

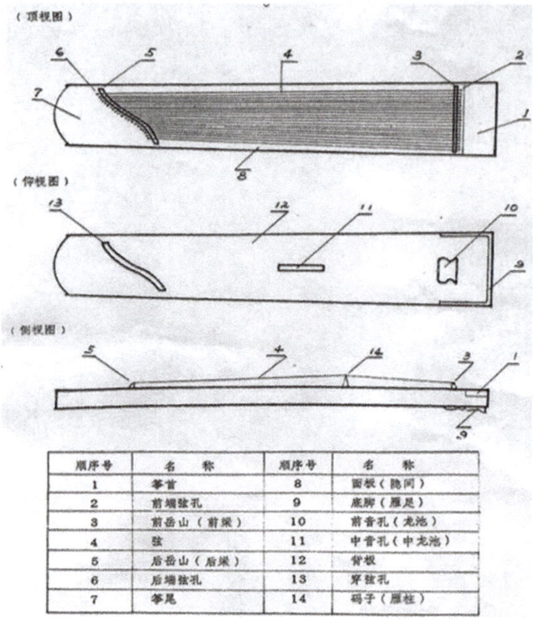

Fig. 5. Guzheng structure map

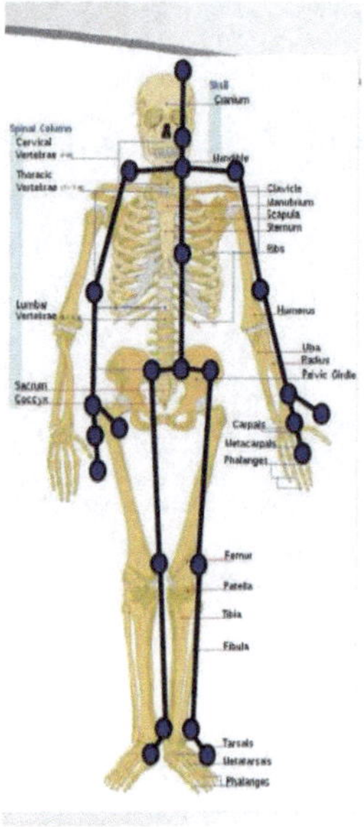

Fig. 6. Kinect V2 joint map

This paper uses the following steps to locate the hand:

1. Detect the profile of the right hand. This paper uses Suzuki85 algorithm to identify the right hand outline. By using OpenCV's FindContours function, we can extract the contour from a binary image. When the hand image in Fig. 7(a) is given, the image of the hand contour will be illustrated. The center of the palm is the center of the hand.
2. Find all fingertip candidates. From the center to the edge of the contour of the hand, there is a farthest point in each long projection. To find all fingertip candidates, we calculated the distance from each point to the center of the hand and compared it with the nearest, consecutive 40 points. The farthest point of the 41 points was chosen as a fingertip candidate. The fingertip candidates to be fond are shown in Fig. 7(b).
3. Find out the true fingertip from the fingertip candidates. This step is mainly to remove the fingertip candidate point near the wrist. Finally, the real fingertips to be selected are shown in Fig. 7(c).

After processing, combined with Kinect to locate the relative position of right hand and virtual Guzheng, the real-time Guzheng string can be judged, and then the virtual Guzheng can be played by gesture recognition.

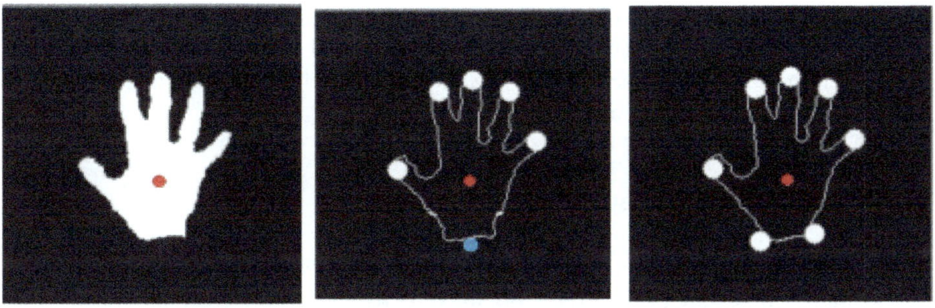

(a) (b) (c)

Fig. 7. The process of fingertip detection

4.3 Gesture

When it is able to judge the real-time Guzheng string, how to let the user play strings? We need to recognize the user's gestures and judge the Guzheng fingering used by the user. In this paper, Kinect is used to define six basic fingering of Guzheng: Tuo, Pi, Gou, Mo, Da, Ti, corresponding to thumb outward, thumbs inward, middle finger inward, index finger inward, ring finger inward and small finger inward.

The Kinect V2 can only recognize three gestures: five fingers open, five fingers tight and other gestures. So we need to create our own gestures. We use the Visual Gesture Builder (VGB) tool to do this.

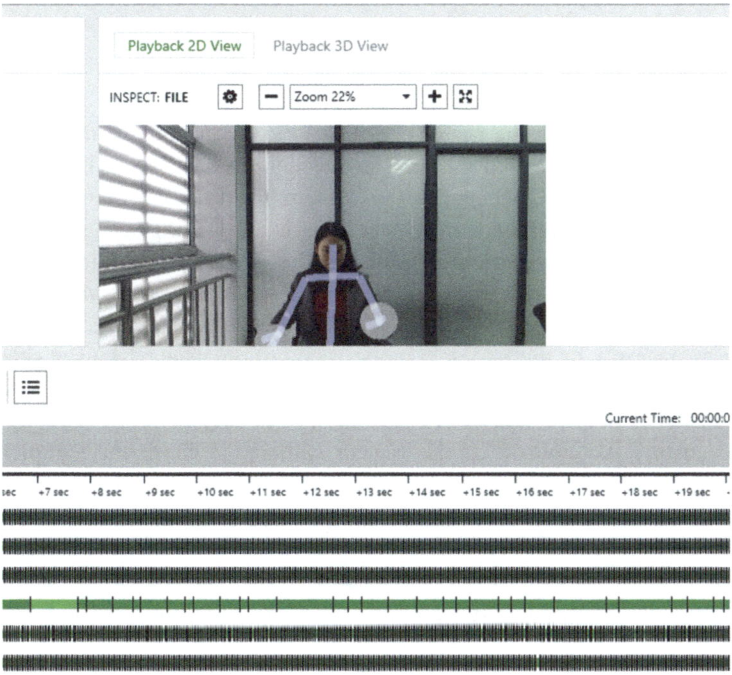

Fig. 8. Use Kinect Studio v2.0 to collect training samples

We first used Kinect Studio v2.0 to collect the Guzheng finger gestures of many people as training samples (Fig. 8), and then we trained them in Gesture Builder. By training all the samples of each finger, find the most matching gesture, and then call the training file in the program, as shown in Fig. 8.

5 System Implementation

In the Windows 10 operating system, the 3D model is created by using 3Ds MAX, and the position and gesture of the user is obtained by using the Kinect camera in Unity 3D. The system is realized by using the Visual Studio 2015 development tool and the C# programming language. After the system starts running, the Kinect camera automatically judges the nearest person to the camera as the user, gets the user's body position and adds the Guzheng. The location of the Guzheng is determined by the user's waist and right shoulder joint. Guzheng can be moved with the user, when the user faces the camera, the user can use his right hand to play virtual Guzheng. The Kinect camera combined with the Suzuki85 algorithm can judge the strings played in the right hand, and judge the gesture of the right hand by the hand gesture database. Use recording equipment to record the sound of each string, form a timbre library, and add to the system resources. The right hand gesture recognition and motion state need to notify the system to switch to the corresponding timbre resource immediately. Here we use the QT signal and slot communication mechanism. The gesture and position of the right hand are defined as a signal, and the signal is triggered when the state changes, and the related slot function will be executed immediately.

6 Summary

At present, the research on the performance of virtual instruments in China, especially in the application of traditional musical instruments, is still less. This paper studies a virtual playing system of Guzheng. In this system, the color image of the real scene is obtained by using the depth camera Kinect, the virtual playing scene is constructed by combining the musical instrument model, and the gesture recognition function of the Kinect camera is used to recognize the gestures used by the user. Although there are some innovations, there are still many areas for improvement.

There is an error in the shoulder node recognition when the user side is on the camera, and the position of the virtual instrument will deviate, which needs to be further improved. If there is an error in training the fingerprinting sample set, it will affect the gesture recognition of the Kinect camera. Because the sample set is not complete, there will be some errors. In this regard, consider using other hardware devices instead of Kinect camera to recognize gestures.

This research will bring more convenience and interest to music lovers. After some improvement and perfection, it can also be used in the teaching of musical instruments and other aspects. At the same time, we also promote the protection and inheritance of intangible cultural heritage, which is of great significance.

References

1. Paradiso, J.: Electronic music interfaces: new ways to play. IEEE **34**(12), 18–30 (1997)
2. Wanderley, M., Battier, M. (eds.): Trends in Gestural Control of Music. Ircam - Centre Pompidou (2000)
3. Cheng, X.M., Hao, Q., Wang, Y.T., Yan, D.Y.: The virtual band on virtual reality technique. Modern Sci. Instrum. **4**, 38–40 (2001). https://doi.org/10.3969/j.jssn.1003-8892.2001.04.011
4. Mäki-Patola, T., Laitinen, J., Kanerva, A., Takala, T.: Experiments with virtual reality instruments. In: Proceedings of the 2005 Conference on New Interfaces for Musical Expression. National University of Singapore, Singapore, pp. 11–16 (2005)
5. Liang, S.L.: A Virtual Performance Based on Computer Vision Gesture Recognition. North China University of Technology, Beijing (2007)
6. Hu, W.J.: Research and Realization of Chime-Bell Playing System Driven by Gesture Discrimination. Zhejiang University, Hangzhou (2007)
7. Wu, Z.: Research on Gesture Driving Virtual Performance System. Kunming University of Science and Technology, Kunming (2009)
8. Yang, G., Sung, T.P., Tsai, W.T.: Air guitar on Altera DE2-70 FPGA architecture. In: Proceedings of the 1st Asia-Pacific Workshop on FPGA Applications, Xiamen, pp. 56–63 (2012)
9. Hsu, M.H., Kumara, W., Shih, T.K., Cheng, Z.X.: Spider king: virtual musical instruments based on microsoft Kinect. In: Proceedings of the 2013 International Joint Conference on Awareness Science and Technology and Ubi-Media Computing (iCAST-UMEDIA). Tsuruga, pp. 707–713 (2013)
10. Yang, Q., Essl, G.: Augmented piano performance using a depth camera. Ann Arbor (2012). http://wwweb.eecs.umich.edu/nime2012/Proceedings/papers/203_Final_Manuscript.pdf
11. Zhang, S.C., Chen, G.F., Zhang, M.M., Zhao, P.T., Pan, Z.G.: Kinect-based virtual performance system of national bowed string instruments. Ruan Jian Xue Bao/J. Softw. **26** (Suppl. 2), 198–207 (2015). http://www.jos.org.cn/1000-9825/15030.html. (in Chinese)
12. Shin, J., Minkim, C.: Non-touch character input system based on hand tapping gestures using Kinect sensor. IEEE Access **99**, 1 (2017)
13. Leite, D.Q., Duarte, J.C., Neves, L.P., de Oliveira, J.C., Giraldi, G.A.: Hand gesture recognition from depth and infrared Kinect data for CAVE applications interaction. Multimed. Tools Appl. 1–33 (2016)
14. Kim, D., Kim, D.-H., Kwak, K.-C.: Classification of K-pop dance movements based on skeleton information obtained by a Kinect sensor. Sensors **17**(6), 1261 (2017)
15. Chabra, R., Ilie, A., Rewkowski, N., Cha, Y.-W., Fuchs, H.: Optimizing placement of commodity depth cameras for known 3D dynamic scene capture. In: Virtual Reality, pp. 157–166. IEEE (2017)
16. Barbulescu, A., Garciay, M., Begaultz, A., Boissieuxx, L., Cani, M.-P., Portazk, M., Viand, A., Duleryyy, R., Heinishzz, P., Ronfardxx, R., Vaufreydaz, D.: A system for creating virtual reality content from make-believe games, pp. 207–208. In: Virtual Reality. IEEE (2017)
17. Tredinnick, R., Boettchery, B., Smithz, S., Solovyx, S., Ponto, K.: Uni-CAVE: a Unity3D plugin for non-head mounted VR display systems. In: Virtual Reality, pp. 393–394. IEEE (2017)
18. Tariqul Islam, A.B.M., Scheel, C., Pajarola, R., Staadt, O.: Robust enhancement of depth images from Kinect sensor. In: Virtual Reality, vol. 71, no. 4, pp. 197–198. IEEE (2015)
19. Espitia-Contreras, A., Sanchez-Caiman, P., Uribe-Quevedo, A.: Development of a Kinect-based anthropometric measurement application. In: Virtual Reality, pp. 71–72. IEEE (2014)

20. Livingston, M.A., Sebastian, J., Ai, Z., Decker, J.W.: Performance measurements for the Microsoft Kinect Skeleton. In: Virtual Reality Short Papers & Posters, vol. 285, no. 1, pp. 119–120 (2012)
21. Satyavolu, S., Bruder, G., Willemsen, P., et al.: Analysis of IR-based virtual reality tracking using multiple Kinects. In: Virtual Reality Short Papers & Posters, vol. 3, no. 1, pp. 149–150 (2012)
22. Belinda, L., et al.: Interactive game-based rehabilitation using the Microsoft Kinect. In: IEEE Virtual Reality, pp. 171–172. IEEE Computer Society (2012)
23. Burba, N., et al.: Unobtrusive measurement of subtle nonverbal behaviors with the Microsoft Kinect. In: Virtual Reality Short Papers & Posters, pp. 1–4 (2012)
24. Hackenberg, G., Mccall, R., Broll, W.: Lightweight palm and finger tracking for real-time 3D gesture control. In: IEEE Virtual Reality Conference, pp. 19–26. IEEE Computer Society (2011)
25. Pan, Z., et al.: A real-time multi-cue hand tracking algorithm based on computer vision. In: IEEE Virtual Reality Conference, pp. 219–222 (2010)
26. John, C., Schwanecke, U., Regenbrecht, H.: Real-time volumetric reconstruction and tracking of hands and face as a user interface for virtual environments. In: Virtual Reality Conference, 2009, pp. 241–242. VR IEEE (2009)
27. Bott, J.N., Crowley, J.G., LaViola, J.J.: One man band: a 3D Gestural Interface for collaborative music creation. In: Virtual Reality Conference, pp. 273–274 (2009)
28. Pagliari, D., Pinto, L.: Calibration of Kinect for Xbox One and comparison between the two generations of Microsoft Sensors. Sensors **15**(11), 27569–27589 (2015)
29. Zhou, L., et al.: Posture reconstruction using Kinect with a probabilistic model. In: ACM Symposium on Virtual Reality Software and Technology ACM, pp. 117–125 (2014)
30. Napoli, A., et al.: Performance analysis of a generalized motion capture system using Microsoft Kinect 2.0. Biomed. Signal Process. Control **38**, 265–280 (2017)
31. Suzuki, S., Be, K.: Topological structural analysis of digitized binary images by border following. Comput. Vis. Graphics Image Process. **30**(1), 32–46 (1985)
32. Han, J., et al.: Enhanced computer vision with Microsoft Kinect sensor: a review. IEEE Trans. Cybern. **43**(5), 1318 (2013)

Source Extraction with Linear Predictor

Yongjian Zhao(✉)

Shandong University, Weihai 264209, China
zhaoyj@sdu.edu.cn

Abstract. This paper addresses the source extraction problem from noise measurements. A proper cost function is presented through extending the results from noise-free case. The function can remove the effect of noise to a great extent. A flexible learning algorithm with a linear predictor is introduced by minimizing the normalized mean square prediction error. Computer simulations are performed to confirm the validity and better performance of the proposed method.

Keywords: Predictor · Noise · Extraction · Factor · Correlation Mixture

1 Introduction

Recently there is an increasing interest in blind signal processing (BSP) due to its potential application in the domain of speech enhancement, wireless communication, biomedical signal processing, etc [1–3]. Indeed, BSP is a technique with purpose of estimating the original sources from all kinds of their mixtures, without the need of prior information about the mixing process and the sources themselves. The BSP approach is performed in a semi-blind context, since its achievement often requires certain information on the nature of the original sources. In contrast to correlation-based transformations such as principal component analysis (PCA) and factor analysis (FA), the BSP technique not only decorrelates the original sources (second order statistics) but also reduces higher order statistical dependence, attempting to make the recovered signals as independent as possible [4, 5].

Figure 1 depicted a block diagram of the blind signal processing process. Here n original sources $s_1(k), s_2(k), \cdots, s_n(k)$ pass through an unknown mixing system with additive noise $v(k)$. Subsequently, we can from m sensors obtain the mixed signals $x_1(k), x_2(k), \cdots, x_m(k)$.

Generally speaking, the BSP approach can be roughly divided into two classes: the simultaneous blind source separation (BSS) approach and one by one blind source extraction (BSE) [6, 7]. The BSS approach utilizes a neural networks and relative learning algorithm carries out separation of all source signals simultaneously. The separated signals are subject to the ambiguities of permutation and scaling. The BSE approach aims to extract original sources sequentially by eliminating the already extracted original sources from their mixtures through deflation techniques. Indeed, the BSE technique can be considered as a special class of BSS. The BSE approach has several advantages (such as global convergence and relative low complexity of learning

J. Abawajy et al. (Eds.): ATCI 2018, AISC 842, pp. 1157–1163, 2019.
https://doi.org/10.1007/978-3-319-98776-7_141

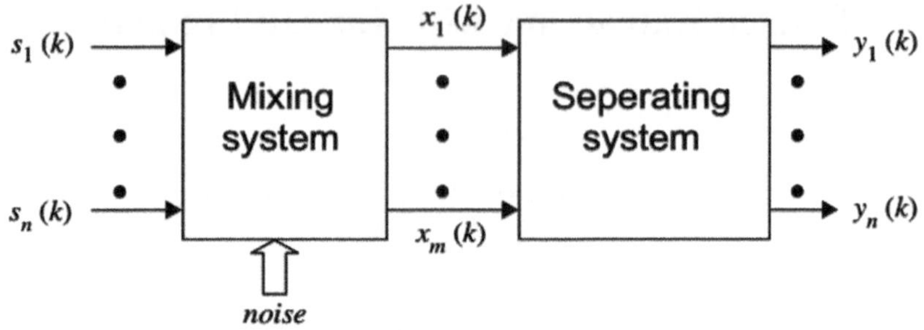

Fig. 1. Block diagram of the blind signal processing problem.

algorithm) over the simultaneous BSS approach. It must be noted that extraction of only one or a few desired signals may save much computation load when the number of original sources is large [8, 9]. In contrast to the ordinary BSS for multiple sources, BSE provide us with more freedom in signal separation.

Almost all of the existing BSE approaches treat simple cases where no measurements noise is taken into account. However, in realistic situations such as biomedical signal processing, it is common that certain measurement noise is added after the desired original source pass through an unknown system. Traditional BSE approach performs worse when the noise increases and it is very difficult to generate meaningful outcomes. There are several methods which handled BSE problem when additive noise is presented. Leong presented a noisy component extraction algorithm through combining standard BSE and a cascaded nonlinear adaptive estimator [6]. To achieve efficient component pursuit from noisy mixtures, the algorithm removes the effects of noise directly from the objective function. By yielding an asymptotically efficient estimator in the presence of an unknown observation noise, this algorithm can be viewed as a robust approach to noisy component extraction.

In this paper, the problem of source extraction from noisy measurements is considered in the domain of second-order statics. By extending the linear predictor criteria from the noise-free case, a proper cost function is introduced which has the same generic form as that for the noise-free case. The effect of noise is removed by specially rearranging the cost function on the basis of an estimate of the variance of the noise. Based on standard gradient principle and the standard stochastic approximations, a flexible learning algorithm is proposed which can be suitable for source extraction in noisy case. Computer simulations are presented to illustrate the performance of the proposed method.

2 Problem Formulation and the Proposed Algorithm

In the BSS/BSE model from noisy measurements, there are l source which can be expressed $s(n) = [s_1(n), \cdots, s_l(n)]^T$, which have passed through an unknown mixing system with added noise; by m sensors, we can obtain the observed mixed signals.

In the instantaneous mixing model, the mixtures are modeled as weighted sums of individual sources without dispersion or time delay, i.e.,

$$x(n) = As(n) + v(n) \tag{1}$$

where $A \in R^{m \times l}$ is an unknown non-singular mixing matrix, and the vector $v(n)$ is Gaussian noise. For convenience, we further assume that the original sources are zero mean and the elements of $v(n)$ are independent about the source signals.

Without knowing the original sources and the mixing matrix, the main purpose of component pursuit is to estimate a specific original component through the observed data $x(n)$ and some statistical properties of source signals. A proper criterion to exploit this property is to utilize a linear predictor within the component pursuit structure, where the weighted sum $y(n) = w^T x(n)$ passes through a linear predictor with coefficient b and length p. The normalized mean square prediction error (MSPE) can be expressed by

$$e(n) = y(n) - b^T y(n) \tag{2}$$

where

$$\begin{aligned} b &= [b_1, b_2, \cdots, b_p]^T, \\ y(n) &= [y(n-1), y(n-2), \cdots, y(n-p)]^T. \end{aligned} \tag{3}$$

Figure 2 shows the typical structure of source extraction with a linear predictor. Developing a component pursuit algorithm even in noise-free case is a difficult task. For simplicity, we make certain assumptions on the unknown original components, such as uncorrelatedness and different temporal structures, which can be given by

$$R_{ss}[1] = E\{s[n]s^T[n]\} = diag\{\rho_1[1], \cdots, \rho_l[1]\} \tag{4}$$

with $\rho_m[1] = E\{(s_m[n] \cdot s_m[n])\}$, $m = 1, \cdots, l$ where $E\{\cdot\}$ expresses the statistical operator and

$$R_{ss}[\Delta n] = E\{s[n]s^T[n - \Delta n]\} = diag\{\rho_1[\Delta n], \cdots, \rho_l[\Delta n]\} \tag{5}$$

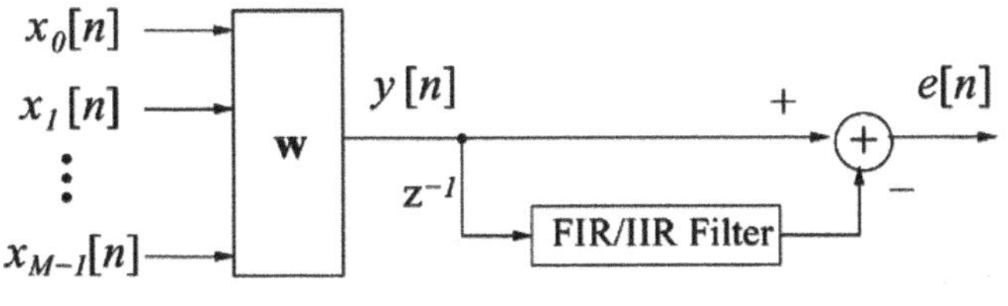

Fig. 2. The structure of source extraction.

with $\rho_l[\Delta n] \neq 0$ for some nonzero delay Δn. We further assume that the noise $v[n]$ is uncorrelated with the original component and has a correlation matrix expressed by

$$R_{vv}[\Delta n] = E\{v[n]v^T[n-\Delta n]\} = \begin{cases} 0, & for\ \ \Delta n \neq 0 \\ R_n, & for\ \ \Delta n = 0 \end{cases}. \tag{6}$$

We further assume $R_v = \delta_v^2 I$, where I denotes the identity matrix and δ_v^2 expresses the variance of the noise.

The normalized mean square prediction error (MPSE) of the structure from Fig. 2 can now be described as

$$\begin{aligned} E\{e^2[n]\} &= E\{y^2[n]\} - 2E\{y[n]b^T y[n]\} + E\{b^T y[n] y^T[n] b\} \\ &= w^T R_{xx}[0] w - 2\sum_{p=1}^{P} b_p w^T R_{xx}[p] w + \sum_{p,q=1}^{P} b_p b_q w^T R_{xx}[q-p] w \end{aligned} \tag{7}$$

where $R_{xx}[\Delta n]$ denotes the correlation matrix of the observed mixtures. After account for the effect of noise from (1), we can deduce

$$\begin{aligned} R_{xx}[\Delta n] &= AE\{s[n]s^T[n-\Delta n]\}A^T + E\{\xi[n]\xi^T[n-\Delta n]\} \\ &= AR_{ss}[\Delta n]A^T + R_{vv}[\Delta n]. \end{aligned} \tag{8}$$

Therefore, $E\{e^2[n]\}$ can be divided into two parts

$$E\{e^2[n]\} = e_s^2 + e_v^2. \tag{9}$$

The term e_s^2 is associated with the original components and given by

$$\begin{aligned} e_s^2 &= w^T A R_{ss}[1] A^T w - 2\sum_{p=1}^{P} b_p w^T A R_{ss}[p] A^T w + \sum_{p,q=1}^{P} b_p b_q w^T A R_{ss}[q-p] A^T w \\ &= w^T A \tilde{R}_{ss} A^T w \end{aligned} \tag{10}$$

where $\tilde{R}_{ss}$ is a diagonal matrix given by

$$\tilde{R}_{ss} = R_{ss}[1] - 2\sum_{p=1}^{P} b_p R_{ss}[p] + \sum_{p,q=1}^{P} b_p b_q R_{ss}[q-p]. \tag{11}$$

The term e_v^2 is deduced by the additive noise, which can be expressed as

$$e_v^2 = w^T R_v w + \sum_{p=1}^{P} b_p^2 w^T R_v w = \sum_{p=0}^{P} b_p^2 w^T R_v w \tag{12}$$

where b_0 is defined as $b_0 = 1$. Without loss of generality, different original components may have different prediction errors for certain linear predictor. Therefore, the error values are greatly dependent on the magnitude of components, so the prediction error is not convenient to utilize as an objective function. We introduce the normalized mean squared prediction error of the desired component $y(n)$ as the objective function, given by

$$J(w) = \frac{E\{e^2[n] - e_\xi^2}{E\{y^2[n]\} - w^T R_\xi w} = \frac{w^T A\tilde{R}_{ss} A^T w}{w^T A R_{ss}[1] A^T w}. \tag{13}$$

Applying the standard gradient stochastic optimization principle to $J(w)$, a learning algorithm can be obtained as

$$W(n+1) = w(n) - \frac{\mu}{\delta_y^2(n)}\left(e(n)\tilde{x}(n) - \frac{\delta_e^2(n)}{\delta_y^2(n)} y(n)x(n)\right), \tag{14}$$

where μ is the learning rate,

$$\begin{aligned} \tilde{x}(n) &= x(n) - \sum_{p=1}^{P} b_p x(n-p), \\ \delta_e^2(n) &= \beta_e \delta_e^2(n-1) + (1-\beta_e)e^2(n), \\ \delta_y^2(n) &= \beta_y \delta_y^2(n-1) + (1-\beta_y)y^2(n). \end{aligned} \tag{15}$$

Here β_e and β_y are called the corresponding forgetting factors, which are subject to $0 \leq \beta_e, \beta_y < 1$.

3 Computer Simulations

ABio7 is a benchmark proposed by Barros containing seven typical biological original components. We adopted three original components as shown in Fig. 3, which can be found in the ICALAB toolbox [10].

The coefficients of the linear predictor were random generated with a length of $P = 3$, and given by $b = [0.4787 \quad 0.2677 \quad -0.1562]$. A 4×3 mixing matrix was randomly generated and given by

$$A = \begin{bmatrix} -0.4328 & 0.1788 & -0.2165 \\ -0.0718 & 0.2655 & 1.1164 \\ 1.2167 & 0.0877 & -1.0968 \\ -0.8746 & 0.1299 & -0.0987 \end{bmatrix}. \tag{16}$$

To demonstrate the performance of the proposed technique, we set variance of the noise as $\delta_v^2 = 0.01$. By minimizing the normalized MSPE, the original component with the smallest normalized prediction error will be estimated. In this simulation, the forgetting factors were $\beta_e = \beta_y = 0.86$ and the step size $\mu = 0.01$. After convergence, the third signal in Fig. 3 is estimated as shown in Fig. 4.

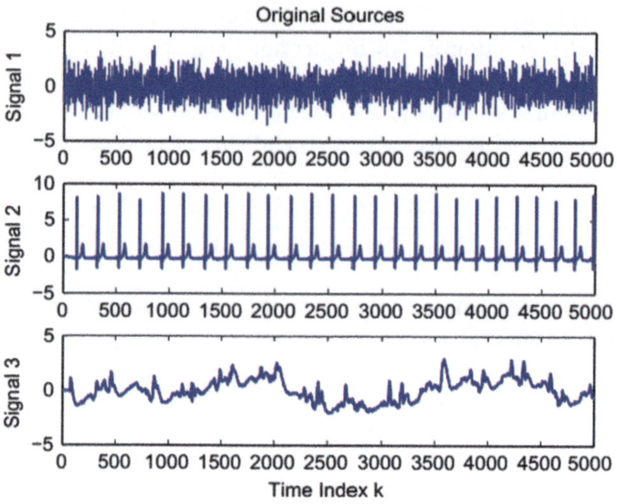

Fig. 3. Three original components used in the simulation.

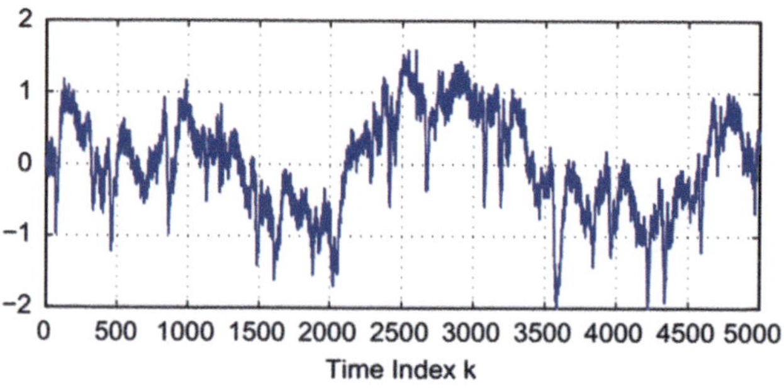

Fig. 4. The estimated signal in the simulation.

4 Conclusions

In this paper, we consider the signal extraction problem under blind environment when Gaussian noise is present. A proper cost function is introduced which can remove the effect of noise to a great extent. Applying standard gradient principle and the standard stochastic approximations, a flexible learning algorithm is developed by minimizing the normalized mean square prediction error with several different factors. Preliminary results shown here and illustrated with biological original sources illustrate that the proposed technique may be promising for blindly source extraction under noise environment.

Acknowledgement. This paper is partially supported by Shandong Provincial Natural Science Foundation, China (ZR2017MA046) and the National Science Foundation of China (11473019). The authors gratefully acknowledge the helpful comments and suggestions of the reviewers, which have improved the presentation.

References

1. Cichocki, A., Amari, S.: Adaptive Blind Signal and Image Processing. Wiley, New York (2003)
2. Barros, A.K., Cichocki, A.: Extraction of specific signals with temporal structure. Neural Comput. **13**(9), 1995–2003 (2001)
3. Hyvärinen, A., Karhunen, J., Oja, E.: Independent Component Analysis. Wiley, New York (2001)
4. Hyvärinen, A.: A unifying model for blind separation of independent sources. Sig. Process. **85**, 1419–1427 (2005)
5. Zhao, Y.J., Liu, B.Q., Wang, S.: A robust extraction algorithm for biomedical signals from noisy mixtures. Front. Comput. Sci. China **5**(4), 387–394 (2011)
6. Leong, W.Y., Mandic, D.P.: Noisy component extraction (NoiCE). IEEE Trans. Circ. Syst. **57**(3), 664–671 (2010)
7. Lu, W., Rajapakse, J.C.: ICA with reference. Neurocomputing **69**(16–18), 2244–2257 (2006)
8. James, C.J., Hesse, C.W.: Independent component analysis for biomedical signals. Physiol. Meas. **26**(1), 15–39 (2005)
9. Liu, W., Mandic, D.P.: A normalized kurtosis-based algorithm for blind source extraction from noisy measurements. Sig. Process. **86**(7), 1580–1585 (2006)
10. Cichocki, A., Amari, S., Siwek, K., Tanaka, T., Phan, A.H.: ICALAB Toolboxes. http://www.bsp.brain.riken.jp/ICALAB

Research on Reasonable Time Limit of Project Based on Monte-Carlo Computer Simulation

Zheng Huimin(✉) and Chen Xuan

Qingdao Binhai University, Qingdao 266555, Shandong, China
zhmhappy@126.com

Abstract. There are many uncertain factors in the actual project. The duration of the process is used as a decision variable, the multi objective function of time, cost and quality is described fuzzy, and the fuzzy equilibrium optimization model based on the three parties is established in this paper. The precision of the time limit obtained by the traditional PERT algorithm is low, In combination with engineering practice, The Monte-Carlo simulation method is used to solve the model more efficiently and provide the basis for determining the reasonable time limit of the project.

Keywords: Project duration · PERT method · Monte-Carlo simulation

1 Introduction

Project management problems can be solved by CPM (critical path method), program evaluation and review technique, etc. By the CPM, the duration of the project based on empirical data is determined as a definite value, and then the critical route is searched [1]. The project review technology is calculated by three time points of pessimistic time, most likely time, and optimistic time. This method mainly focuses on the evaluation of the project plan [2]. M Bevilacqua and H Steyn proposed the critical chain method and apply it to the project management system [3, 4]. Yin Qixin and others fully considered the indirect cost of the project, and established a nonlinear optimization model between construction cost and quality. Genetic algorithm is used to solve the problem, but the probability of completing the project within the predetermined time and cost is not given [5]. Zhang Lianying carried out fuzzy equilibrium optimization for the duration cost quality and Pareto solution is given [6].

However, there are many uncertainties in the actual project management. By comparing the simulation results with the PERT algorithm, the advantages and disadvantages of the two algorithms are compared, and then the basis for the analysis of the reliability of the construction period is provided [7].

2 Fuzzy Algorithm Model for Time Limit of Project

For part of the work task, query quota information or calculation can get work hours. However, due to changes in the external environment or personnel and material allocation, the working hours may be affected in practice. The PERT method is based on

J. Abawajy et al. (Eds.): ATCI 2018, AISC 842, pp. 1164–1171, 2019.
https://doi.org/10.1007/978-3-319-98776-7_142

the time required for the uncertainty of work and makes reasonable calculation. The three point time estimation method is used to determine the working hours.

a: The fastest possible time (optimistic time)
m: The most likely time (likely time)
b: The slowest possible time (pessimistic time)

The three time a, m and b are used to calculate the expected working hours of each job:

$$\mathrm{t(i,j)} = \frac{a + 4m + b}{6} \tag{1}$$

The Variance working hours of each job:

$$\sigma^2 = \left(\frac{b - a}{6}\right)^2 \tag{2}$$

3 Probability Analysis and Calculation

The network diagram presents a network with both sequential and directional directions, and its practical significance is to represent a planned task. The double code network diagram is widely used in project management, and it is difficult to draw compared with the horizontal map and the single sign network diagram. The core of network optimization is to improve and adjust the initial plan of the network in terms of time limit, cost and resources, so as to achieve the best results.

Cost-objective function is:

$$\min\tilde{C} = \sum_{i \in u} \left[\tilde{b}_{io} + \tilde{r}_i(\tilde{T} - \tilde{T}_{i0})^2\right] \tag{3}$$

$\tilde{T}_{i0}$: The longest fuzzy time
$\tilde{b}_{i0}$: The cost of completion
$\tilde{r}_i$: Fuzzy incremental factor of marginal cost

The relationship between the period of time and the quality is:

$$\tilde{q}_i = \frac{1}{1 + \left|\frac{\tilde{T}_i - T_{io}}{a}\right|^{2b}}, i = 1, 2, \ldots, \mathrm{n} \tag{4}$$

The Monte Carlo simulation method, also known as random simulation, random sampling, or statistical test, aims to estimate the distribution of the result variables depending on the input variables of a number of probabilities. Based on the theory of probability and statistics, random sampling is the main method. Through long-term engineering practice, it is found that the duration of tasks is similar to the triangular distribution. In the analysis of project duration and sensitivity analysis, the probability distribution of the duration of a process can be set to a triangular distribution. By setting the fastest completion time a, the slowest completion time B and the most likely completion time m. In this way, all process durations are defined according to the probability distribution.

The following is the network diagram of the project. The most likely completion time of the process is marked below the process, through the analysis, Table 1 is calculated (Fig. 1):

Table 1. Time schedule table

Procedure	Preceding activities	Minimum time	Medium time	Maximum time	Start time	Duration	Finish time
A	–	2	3	4	0.00	3.00	3.00
B	–	2	2	2	0.00	2.00	2.00
C	–	3	3	5	0.00	3.00	3.00
D	B, C	2	3	4	3.00	3.00	6.00
E	C	3	4	6	3.00	4.00	7.00
F	A	5	6	8	3.00	6.00	9.00
G	A, D	2	3	4	6.00	3.00	9.00
H	D, E	4	5	6	7.00	5.00	12.00
I	G	4	5	7	9.00	5.00	14.00
J	G, H	2	2	3	12.00	2.00	14.00
K	F	4	4	4	9.00	4.00	13.00
L	I	2	2	2	14.00	2.00	16.00
M	K, L, J	1	1	1	16.00	1.00	17.00
线路1	14.00						
线路2	14.00						
线路3	16.00						
线路4	17.00						
线路5	14.00						
线路6	12.00						
总工期	17.00						

According to the double code network diagram, the earliest start and start time of each process can be listed, and the total time difference is calculated. The calculation table of the double code network diagram is shown in Table 2.

Using formulas (1) and (2) to calculate the average working hours of each work and calculate the total time difference R, fill in Table 2. It can be seen from the table that the total float of procedure CDGILM is zero. So the key line is C → D → G → I → L → M.

The total duration is 17.5 weeks according to the expected time calculation.

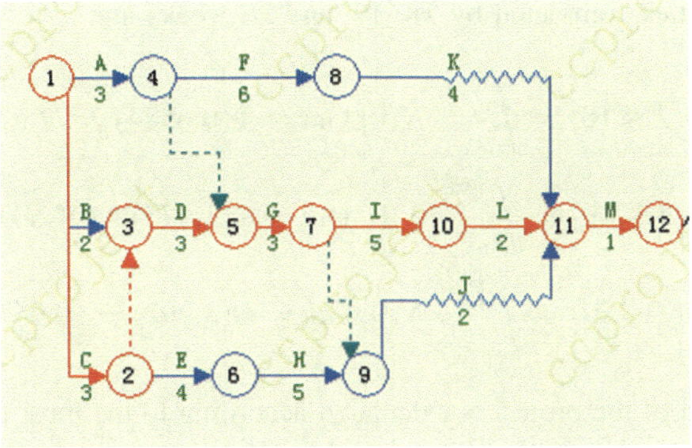

Fig. 1. Double code network diagram

Table 2. Double code network diagram calculation table

Procedure	Preceding activities	Start time	Duration	Finish time	Variance	ES (early start)	LS (last start)	Total float
A	–	0	3.00	3.00	0.33	0	3.33	3.33
B	–	0	2.00	2.00	0.00	0	1.33	1.33
C	–	0	3.33	3.33	0.33	0	0.00	0.00
D	B, C	3.33	3.00	6.33	0.33	3.33	3.33	0.00
E	C	3.33	4.17	7.50	0.50	3.33	5.17	1.84
F	A	3.00	6.17	9.17	0.50	3.00	6.33	3.33
G	A, D	6.33	3.00	9.33	0.33	6.33	6.33	0.00
H	D, E	7.50	5.00	12.50	0.33	7.50	9.33	1.83
I	G	9.33	5.17	14.50	0.50	9.33	9.33	0.00
J	G, H	12.50	2.17	14.67	0.17	12.50	14.33	1.83
K	F	9.17	4.00	13.17	0.00	9.17	12.50	3.33
L	I	14.50	2.00	16.50	0.00	14.50	14.50	0.00
M	K, L, J	16.50	1.00	17.50	0.00	16.50	16.5	0.00

Because the key work is CDGILM, so:

$$\sqrt{\sum \sigma^2} = \sqrt{\sigma_c^2 + \sigma_D^2 + \sigma_G^2 + \sigma_I^2 + \sigma_L^2 + \sigma_M^2}$$

$$= \sqrt{0.33^2 + 0.33^2 + 0.33^2 + 0.5^2 + 0^2 + 0^2} = 0.7594$$

The probabilities completed by 18, 19, and 20 weeks are:

$$p(T \leq 18) = \int_{-\infty}^{\frac{18-17.5}{0.7594}} N(0,1)dt = \Phi(0.6584) = 74.5\%$$

$$p(T \leq 19) = \int_{-\infty}^{\frac{19-17.5}{0.7594}} N(0,1)dt = \Phi(1.975) = 97.60\%$$

$$p(T \leq 20) = \int_{-\infty}^{\frac{20-17.5}{0.7594}} N(0,1)dt = \Phi(3.292) = 99.95\%$$

If the duration of the project is calculated according to the most probable time, the duration of the calculation is 17 weeks. According to a certain risk taking ability analysis, the risk will be greater in the eighteenth week. The risk of completion in the seventeenth week is greater. If the construction period can be fixed for 19 weeks, the risk will be greatly reduced.

4 Monte-Carlo Computer Simulation Algorithm for Time Limit Reliability

In the analysis of Crystal Ball by computer simulation software, the definition of "hypothetical unit" is to define the input of uncertainty. In the analysis and sensitivity analysis of project duration, it is necessary to define them as hypothetical variables because of the uncertainty of the time required for each process. The "prediction unit" is defined as the value calculated by the formula in the process of user specified simulation analysis. In the risk analysis of this project, the duration of each line should be set as "prediction unit", the longest line is the key line, and its length is the duration of the whole project.

Through Monte Carlo simulation, in order to get the most objective results, 1000 random numbers are generated by the simulation method. The probability distribution is shown as shown in Fig. 2.

Through simulation and analysis of three graphs, the probability that the project can be completed in the 18 weeks is 49.93%, the probability that the 19 weeks can be completed is 84.42%, and the probability of completing the 20 weeks is 97.33%.

Through the sensitivity analysis, it is found that the process I is the most important factor for the construction period, followed by C, D and G, and the more sensitive procedures should be focused on the adjustment and improvement. As shown in Fig. 2, the probability of completion of the project in 18 weeks is 49.93%, and there is a certain risk of delay. From Fig. 3., we can see that the effect of I process is 43.8%. If we want to shorten the duration of the project and reduce the risk of delay, we should start with I process.

By comparing the two methods, it is obvious that the advantages of simulation analysis relative to probability analysis. Through comparative analysis, it can be seen that using probabilistic network graph time parameter calculation, it is assumed that the distribution of formulas in each process is approximately normal distribution. This is

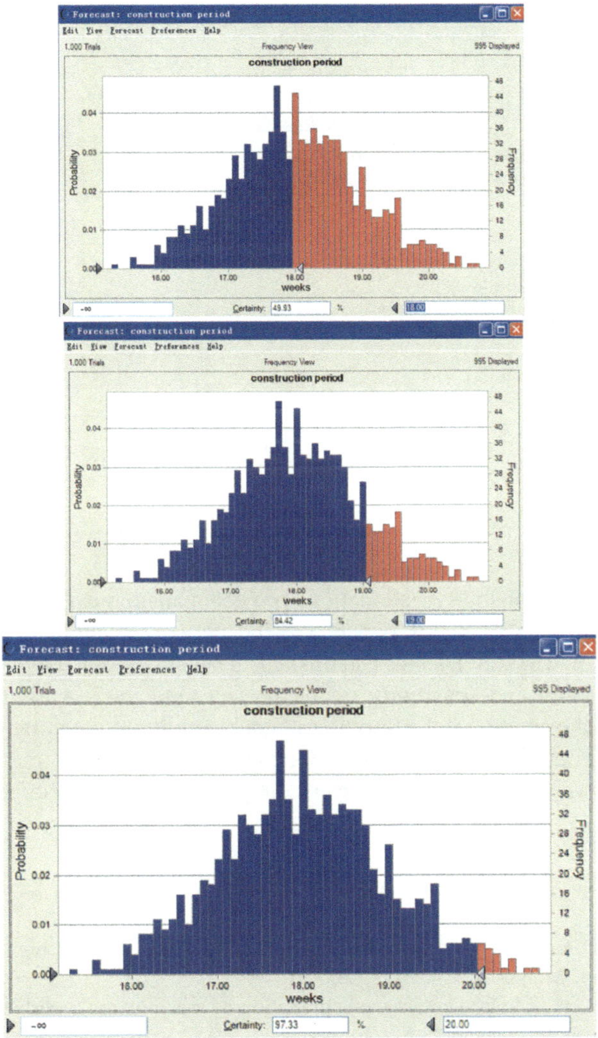

Fig. 2. Crystal Ball prediction probability distribution map

also in line with the actual situation, that is, in the completion of each process, the smoothest and the most unfavorable situation is relatively small, more in the most likely time to complete. But in the actual project, the most smooth and unfavorable conditions are random, and the generation of a set of random numbers is more in line with the completion time of the actual project. From this perspective, the completion probability of Crystal ball simulation is more in line with the actual situation. In addition, in the process of computer simulation, not only the probability of completion of the project can be completed in a given time, but also the influence of different factors on the time limit can be displayed intuitively, so the sensitivity analysis can be carried out.

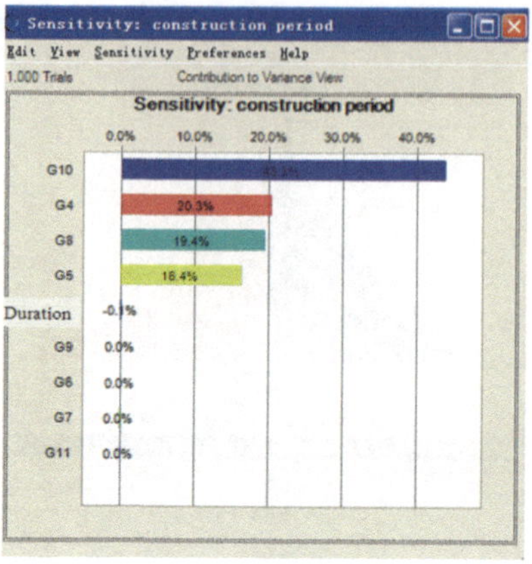

Fig. 3. Sensitivity analysis of time limit

5 Conclusion

The knowledge of probability calculation, project management, computer simulation and risk analysis is used in this paper, and the network planning project simulation analysis based on Crystal Ball is established. The simulation parameters are set to analyze the duration and sensitivity of the project, and the sensitivity of the risk of completion and the change of the duration of each activity to the duration of the project is obtained. In order to improve the overall benefit of the project, a reasonable means is used to adjust and optimize the activities of the project in the project period evaluation.

References

1. El-Rayes, K., Kandil, A.: Time-cost-quality trade-off analysis for highway construction. J. Constr. Eng. Manag. **131**(4), 477 (2005)
2. Xie, X., Chan, H.: Monte-Carlo based simulation of machine tool overhaul project duration and cost risk analysis. Manuf. Autom. **38**(5), 76–80 (2016)
3. Jiang, L., Xiaogang, L.: Risk analysis method for large-scale aerospace research and development projects based on Multi-Agent technology. J. Natl. Univ. Defense Technol. **34** (6), 148–152 (2012)
4. Guo, Y., Liu, E.L.: Using Monte Carlo method to improve project cost risk analysis. J. Tianjin Univ. Natl. Sci. Eng. Technol **35**(2), 199–202 (2002)
5. Gao, J.: Quantitative analysis technology of research projects based on Monte Carlo simulation technology. Sci. Technol. Bull. **30**(13), 31–35 (2012)
6. Zhang, L., Ying, Y.Y.: Fuzzy balanced optimization of time limit cost quality and its Pareto solution. J. Tongji Univ. (Natl. Sci. Edn.) **2**, 303–308 (2013)
7. Chen, W., Shi, Y., Teng, H., Lan, X., et al.: An efficient hybrid algorithm for resource
8. Shizhao, N., Zheng, W., Wangmo, P., Yuan, N., Peng, L.: Big data prediction of durations for online collective actions based on peak's timing. Physica A **55**, 130–139 (2018)
9. Jon, T.: ECN technical editor: the eclipse of stand. J. Zhongkai Agrotech. Coll. **19**(2), 32–35 (2006)

10. Markus, A., Axel, K., Martin, S.: Use Java and the CORBA realization distribute type system. J. Pingxiang Coll. **4**, 104–105 (2005)
11. Richardson, W.C., Avondolio, D.: The Java high class weaves a distance: JDK 5. Sci. Technol. Book Rev **3**, 17–18 (2006)
12. Vuong, J.L.: A semantics-based routing scheme for grid resource discovery. In: E-Science: First International Conference on E-Science and GridComputing, pp. 58–70, 90 (2005)

The Application of Big Data in the Analysis of Security Work Information

Gao Chai(✉)

Political College of National Defence University, Xi'an, China

Abstract. The security information often uses big data to conduct intelligence analysis. The paper studies the traditional information analysis system in China, and finds out the shortcomings in the traditional information analysis system, and then puts forward a new optimization idea, in order to provide support and help to the analysis of large data information in the security work of our country.

Keywords: Big data · Security work · Intelligence analysis

1 The Information Analysis of Big Data

In 2012, the United States federal government formally developed a new "large data research and development project". After this data R&D project was opened, how to maximize the use of data has become the focus of academic research. In data research, it is the most important task to carry out reasonable and orderly and systematic data storage and management, and to fully excavate data resources, and to share information decision-making in many aspects under the premise of guaranteeing large data.

The process of information analysis is very complicated and cumbersome. It needs to excavate a large number of data, that is, from a large number of incomplete, incomplete, or even fuzzy random data, to extract some useful information hidden in a variety of messy information. Using big data technology to analyze useful information, known as "knowledge discovery". Generally speaking, large data information analysis is mainly divided into seven steps: (1) Cleaning up mass data. (2) Screening massive data. (3) Collate and synthesize massive data. (4) Changing and deforming the data after the collation. (5) In depth mining of useful information. (6) The model evaluation of the data is carried out. (7) The display of detailed information knowledge.

2 The Defects of Traditional Large Data in Information Analysis

2.1 Backwardness of Traditional Algorithms

Large data in the application of information analysis in the application of statistics, probability and other principles, its specific analysis methods generally have the following several: (1) genetic algorithm, large data in the process of intelligence analysis of the survival of the survival of the survival of the fittest, self adaptation of the

J. Abawajy et al. (Eds.): ATCI 2018, AISC 842, pp. 1172–1177, 2019.
https://doi.org/10.1007/978-3-319-98776-7_143

evolution of the law. This genetic algorithm optimizes the probability search algorithm. It finds simple and convenient algorithm process in the process of calculation, and can be used in clustering analysis and rough set method. It is mainly used in incomplete and uncertain data analysis work, and classifies all kinds of inaccurate and incomplete information in data information. And extracting useful knowledge; (2) the decision tree method, after sorting data information, using the test function to set up the information data into different branches, in the subsets of each branch, set up the nodes and branches of the lower layer again, and finally deal with and transform the decision tree. This algorithm is mainly used for classification and excavation of security information.

2.2 The Contradiction Between the New Form of Crime and the Traditional Security Technology

In our security work, the technology of using large data algorithms to deal with information cases has been gradually developed. In the process of information collection, the traditional algorithm is the main method. But we should also realize that some defects in traditional algorithms can not satisfy the intelligence requirements in real life. The social transformation of our country not only has the adjustment of the structure of economic structure, but also the change of social relations and interest pattern. It is the time stage of social contradiction, which can easily induce all kinds of unsettled factors. In the traditional large data analysis, it is a difficult problem to secure the information situation of the whole database under the support of each item set, and it is a difficult problem to secure the security work in the case of the association rules. At present, the number of crimes in the society is increasing in general, but some traditional methods of crime are disappearing. The traditional data algorithm can not constantly update the criminal behavior database constantly, so as to increase the new type of crime and not be able to respond to the new form of crime in time. Therefore, we must find new association rules in time and guide our daily security work with new data intelligence collection technology. For example, the use of police face recognition client can realize remote face retrieval based on wireless communication, truly combine prevention and monitoring, improve security initiative and attack crime accurately. In addition, the traditional large data analysis only takes into account the frequency that the analysis project appears in the whole database, without taking into account the significance of the different project itself and the unique importance. The unique meaning of each item is ignored, the information of special value is regarded as a project of the same value, so that some of the less frequency but very valuable information is ignored in the past, which makes us often neglect some rare but serious circumstances and great social harmfulness in the security process. Criminal acts.

3 The Idea of Large Data Optimization Based on New Security Risks

3.1 Pay Attention to the New Security Role

In order to make better use of large data to help the security work intelligence work, it is necessary to strengthen the enhancement of screening technology for the sensitivity and relevance of new crime projects. In the reception and issue of information on security activities, there are many new information. For example, in the process of carrying out land expropriation, relocation, resettlement and civil disputes, a large number of data information is needed to help the security work to make decisions. In the process of controlling and combating the infiltration and destruction of the Falun Gong Cult Organizations, it also needs large data to provide guidance and guidance to the work. The new large data analysis can monitor the issue of the dynamic information, find the hidden danger in the security situation in time, and determine the target and focus of the attack and prevention. For example, in public security emergencies, if the social masses take photos and upload the network to criminals who are committing a crime, the case handlers can enter the images collected on the network into the face retrieval system to realize the quick confirmation of the identity of the criminals and ensure the first time to capture them. In order to disrupt the threat of community stability, we should grasp the trend of the development of the overall crime in a timely manner so that we can make the first move against the enemy. In the process of information supply, large data analysis can master the whole law and prevent and control the case. In the process of restricting the control of the other side, we should strive to exceed the criminal speed, improve the ability of the security forecast, establish the security early-warning mechanism, and actively strive for the initiative.

3.2 Optimization of Security and Large Data Algorithm

In order to better realize the role of big data in intelligence analysis, we must optimize the traditional big data algorithm. The algorithm of frequent itemsets is improved, which not only enhances the sensitivity of the individual meaning of each item in the large data information analysis, but also improves the efficiency of the data mining. Dynamic data in big data applications such as online services and web applications require precise information data. For the change of association rules that happen occasionally or frequently, we need to pay attention to data research, ensure the integrity of information and have independent value and significance.

The new data algorithm reform should break through the platform period that the traditional database can not achieve the association rules, increase the scale of the database and increase the efficiency of data information extraction. When the large database is updated, we should make full use of the previously acquired database, establish a new mining algorithm for association rules, and establish a new frequent item set according to the previously frequently appearing item set and the new dynamic data in the database. And it is constantly increasing in order of time.

In the new time period, we should pay more attention to the sensitivity of the new project. In the introduction of the parameter C ($1 \leq C \leq \infty$), it is necessary to keep the frequent items which are originally larger than or equal to the traditional ones. At the same time, when the new data set is considered, the scale of the database is continuously enlarged to ensure that the project is more stable by ten. We also need to scan old data that supports less than traditional frequent itemsets. Assuming that the data set in the original security problem record database is recorded as A, the newly added data set is recorded as a, and the entire record database is (A + a).

The basic idea of optimizing intelligence analysis is to assume that the security data item set A can be larger than or equal to the support degree of each item set in the original frequent itemset M′(A) and M′(a). Count can represent the total number of transactions in the whole security project. Countall (if there are two transaction sets A, and Az, for an item set 1, it is in A, is a frequent itemset, and in Az is not a frequent item set, then Countall = A is a common frequent itemset of (A + Az). Then: Support = (count (A) + count (a)). In the course of this calculation, the value of count can be constantly changed. In general, the greater the count value, the more the original information data is retained, and on the contrary, the weight of the newly increased intelligence data is increasing.

When the value of count is 1, it represents the process of security without considering the traditional information data. It is proved that the old unsettled factors in the process of social development have been eliminated by the times, and more time and energy should be used in the face of new emergencies. Set the minimum support degree to 0.5, set the count value to 2, and (minimum support degree C) = 0.1, and continue to inherit the above research, set the new information data as a, and set the traditional information data to A, then the data collection time of the specific security hidden danger is shown as shown in Fig. 1.

Tid	Itemset
1	A,D
2	A,B
3	B,C
4	A,B
5	A,B,C
6	B,C
7	A,D
8	A
9	B,C
10	B,C,F

Itemset	Support
A	6/10
B	7/10
C	5/10
D	2/10
A,B	3/10
A,D	2/10
B,C	5/10

Itemset	Support
A	10/16
B	11/16
C	7/16
A,B	5/16
B,C	7/16

Fig. 1. Traditional intelligence data analysis project set

In order to better join the new unsettled factors, we must continue to add new data information, assuming that it is E, then the optimized data collection pattern is shown in Fig. 2.

According to the analysis of traditional intelligence data, after introducing the new project parameter E, the security intelligence data algorithm can be optimized. This

Itemset	Support
A	10/16
B	11/16
C	7/16

→

Itemset	Support
A	10/16
B	11/16
C	7/16
E	4/16

Fig. 2. Improved information data analysis project set

optimization can balance all kinds of new unsettled factors through the proportion of different parameters in large information data, and use the proportion of security information to design a more efficient and reasonable application mode of information data, and then find the association rules in security. For example, 3000 common cases occur every month in an urban area, of which there may be only 30 major cases. In the process of traditional large data intelligence analysis, possible major events will be ignored. However, whether from the social influence or the nature of the case, or from the loss to the people and the state, the major cases are far more than ordinary ordinary cases. At this time, the association rules of the traditional algorithms have obviously not been able to adapt to the requirements of our social security. Therefore, we need to improve traditional computing methods and introduce new and more important new data functions for different security problems. In this way, each item and its frequency and its components are taken into consideration, and the weight is used to balance the analyzed information and establish a new information analysis index.

4 Conclusion

Large data has played a very effective role in the collection of security information work. It can track and monitor security risks well, so as to make a visual analysis, better protect the privacy of citizens, and maintain social stability. At present, the security work in China is still facing many challenges. It is necessary to adjust the calculation method of large data analysis in time to ensure the social application of the analysis of large data information with a more efficient and faster way of work.

References

1. Sageman, M.: Understanding Terror Networks. University of Pennsylvania Press, Philadelphia (2004)
2. Borthakur, D.: HDFS architecture guide (2012). http://hadoop.apache.org/common/docs/r0.20.0/hdfs_de-sign.pdf
3. Big data (2012). http://en.wikipedia.org/wiki/Big_data
4. Apache Struts (2012). http://struts.apache.org/index.html
5. Beaver, D., Kumar, S.: Finding a needle in haystack: Facebook's photo storage. In: Proceedings of OSDI (2010)
6. Shizhao, N., Zheng, W., Wangmo, P., Yuan, N., Peng, L.: Big data prediction of durations for online collective actions based on peak's timing. Physica A **55**, 130–139 (2018)

7. Titus, J.: ECN technical editor: the eclipse of stand. J. Zhongkai Agrotech. Coll. **19**(2), 32–35 (2006)
8. Markus, A., Axel, K., Martin, S.: Use Java and the CORBA realization distribute type system. J. Pingxiang Coll. **4**, 104–105 (2005)
9. Richardson, W.R., Avondolio, D.: The Java high class weaves a distance: JDK 5. Sci. Technol. Book Rev. **3**, 17–18 (2006)
10. Vuong, J.L.: A semantics-based routing scheme for grid resource discovery. In: E-Science: First International Conference on E-Science and GridComputing, pp. 58–70, 90 (2005)

Research on Signal Intensity and Distance in Cold Chain Logistics Data Acquisition

Rui Luo, ZiHong Zhang(✉), and Wei Xiong

Honghe University, Mengzi, Yunnan, China
35660520@qq.com, 56090733@qq.com, 459353359@qq.com

Abstract. This paper analyzes the software and hardware requirements of the cold chain logistics quality perception, constructs the overall framework and hardware of the cold chain logistics quality perception base on the Internet of things, designs and implements the hardware system and software system of the cold chain logistics quality perception, and tests the communication distance, data acquisition and monitoring, link quality and power consumption, which can effectively realize the real-time collection and transmission of the cold chain logistics quality perception, and provide the most basic sensing data collection and transmission software and hardware for the quality perception base on the Internet of things technology.

Keywords: Cold chain logistics · Link quality · Internet of things Data acquisition

1 Introduction

Cold chain logistics can be divided into three types of cold chain, living agricultural products and frozen food. It includes cold chain pretreatment, pre cooling, cold storage, loading, cold chain transportation, unloading and sales and many other cold chain links, and the cold chain logistics depends on the various factors of cold chain logistics environment, the diversity of quality parameters and the information asymmetry of the process. These characteristics eventually lead to the uncertainty in cold chain logistics process.

In view of the problem of data perception in the process of food cold chain logistics, combined with the actual demand of cold chain logistics monitoring and quality control, this paper uses wireless sensor network technology to monitor the quality perception and data acquisition of food in cold chain logistics, and takes into account the factors such as cold chain environment parameters, cost, volume, computing power and reliability, and the layout and optimization of integrated circuits. In this paper, the hardware and embedded software system of wireless sensor nodes and converging nodes of the Internet of things are designed and realized, and the communication distance, data acquisition and monitoring, link quality and power consumption are tested. It can improve the stability of Internet of things hard and soft data collection.

J. Abawajy et al. (Eds.): ATCI 2018, AISC 842, pp. 1178–1184, 2019.
https://doi.org/10.1007/978-3-319-98776-7_144

2 The Overall Architecture Design of Cold Chain Logistics Quality Perception

In this paper, the quality perception characteristics of cold chain logistics and the actual requirements of the monitoring system are synthetically analyzed. The wireless sensor network technology is used as the monitoring method of the cold chain logistics quality perception, and the cold chain environment of cold chain logistics is combined with the practical requirements. We use the simple and convenient wireless sensor network topology structure to complete the design of quality aware Internet of things network structure for cold chain logistics.

The overall architecture of the system is shown in Fig. 1. The data acquisition architecture based on the Internet of things mainly consists of two parts: wireless sensor network, cold chain logistics quality perception and IOT monitoring service center.

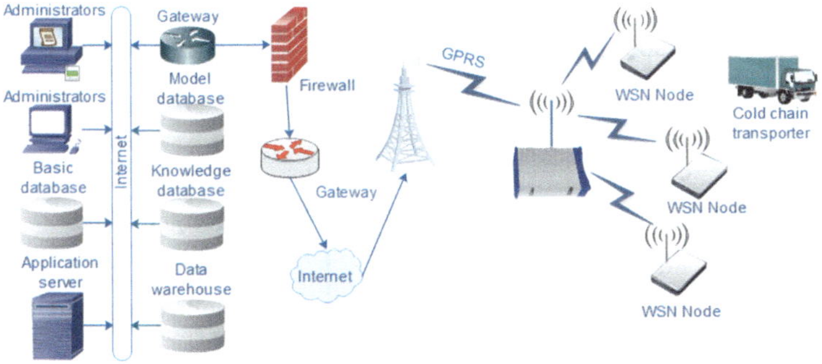

Fig. 1. The overall design block diagram of the data collection based on the Internet of things for quality perception in cold chain logistics

2.1 Design and Implementation of Wireless Sensor Nodes Based on Internet of Things

Wireless sensor nodes are mainly responsible for real-time perception, collection and transmission of cold chain environment information in cold chain logistics process. The wireless sensor nodes are mainly composed of main control module, radio frequency module, sensor module, battery power supply module, peripheral oscillator and JTAG interface. The principle block diagram of wireless sensor nodes is shown in Fig. 2.

2.2 Design and Implementation of the Convergence Node of IOT

As shown in Fig. 3, the interconnection node of the IOT is composed of a wireless sensor network coordinator and an ARM module. The wireless sensor network coordination is consistent with the collection module and radio frequency module chip model used by the wireless sensor nodes. It is responsible for the establishment, maintenance and communication of the cold chain logistics wireless sensor, and

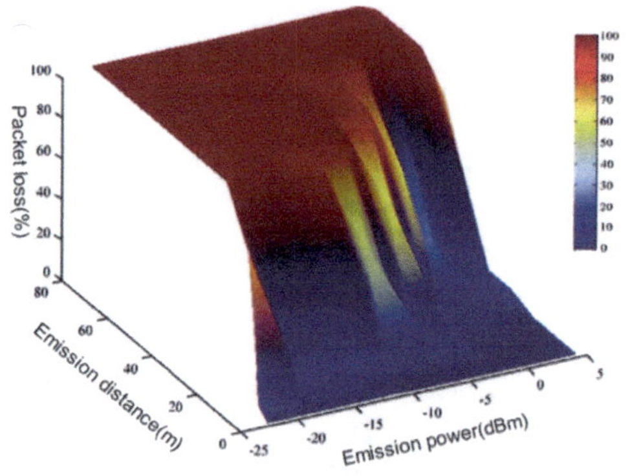

Fig. 6. Surface map of packet loss rate with different distances and transmitting power

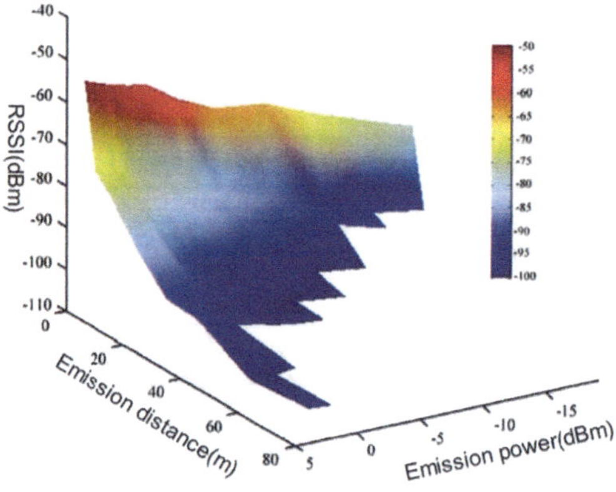

Fig. 7. Surface map of RSSI with different distances and transmitting power

communication distance between the nodes of the Internet of things is also larger, the packet loss rate is reduced, and the RSSI value becomes larger. When the transmission power is at 1 dBm, the effective communication distance of the Internet of things nodes is about 30 m, the data packet loss rate is about 4.4%, reaching to 0, and the RSSI value is about −90 dBm. The data communication between the nodes of the Internet of things is normal, its packet loss rate is low, and the effective communication distance tested can meet the data collection needs within the cold chain logistics range.

The paper tested the average RSSI value of wireless sensor nodes for the fresh grape in the actual cold storage at 0 °C and −18 °C for 24 h, and the quality of the communication link between the wireless sensor nodes and the IOT converging nodes at different locations in the cold storage, the average RSSI value between the converging nodes is shown in Fig. 8.

By comparing the average RSSI values of the wireless sensor nodes in the storage environment from 0 °C and −18 °C, the overall RSSI of each wireless sensor node in the constant temperature storage process (−18 °C) is lower than that under the constant

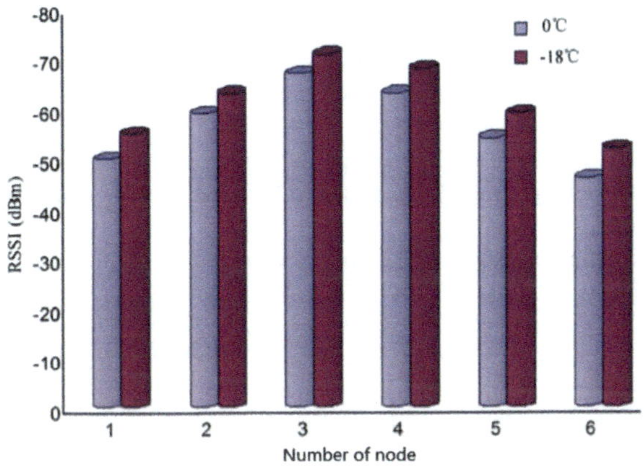

Fig. 8. RSSI comparison chart of wireless sensor nodes in constant temperature

temperature environment (0 °C), and the link quality of each node is relatively weak at the same distance. The RSSI value of the wireless sensor nodes will be affected by the corresponding temperature, and the lower the temperature, the RSSI value will be relatively low, but the results do not affect the normal communication of the collection data.

5 Conclusions

The greater transmission power of the IOT nodes, the effective communication distance between the IOT nodes will also increase. The packet loss rate of data will decrease and the RSSI value will increase. When the transmission power is 1 dBm, the effective communication distance of the IOT node is about 30 m. The Internet node can effectively realize the data collection and monitoring of the key parameters of the environment in the cold storage logistics process, and accurately reflect the fluctuation of the cold chain environment parameters, and the minimum temperature of the node can reach −18 °C. The high relative humidity can reach 70%, to meet the requirements of data acquisition in the range of cold chain logistics.

The longer the communication distance between the wireless sensor node and the IOT node, which due to the lower the corresponding RSSI value, and the greater the fluctuation of the temperature, it can due to the weaker the RSSI value, and the lower the temperature. The design and development of the Internet of things node has relatively stable RSSI value and relatively low work power, which can meet the practical needs of cold chain logistics.

Acknowledgment. The research was supported by scientific research fund of the Yunnan Provincial Education Department No. 2018JS476.

References

1. Xiao, X., Li, Z., Matetic, M., et al.: Energy-efficient sensing method for table grapes cold chain management. J. Clean. Prod. **152**, 77–87 (2017)
2. Xiao, X., He, Q., Li, Z., et al.: Improving traceability and transparency of table grapes cold chain logistics by integrating WSN and correlation analysis. Food Control **73**, 1556–1563 (2017)
3. Caione, C., Brunelli, D., Benini, L.: Compressive sensing optimization for signal ensembles in WSNs. IEEE Trans. Ind. Inf. **10**(1), 382–392 (2014)
4. Tzamalis, P.Q., Panagiotakos, D.B., Drosinos, E.H.: A 'best practice score' for the assessment of food quality and safety management systems in fresh-cut produce sector. Food Control **63**, 179–186 (2016)
5. Shizhao, N., Zheng, W., Wangmo, P., Yuan, N., Peng, L.: Big data prediction of durations for online collective actions based on peak's timing. Phys. A **55**, 130–139 (2018)
6. Ming, F.: The cultural elements in English translation teaching. Engl. Sq. **20** (2014)
7. Wang, H.: The problems and countermeasures of English translation teaching in college of traditional Chinese medicine. Overseas Engl. **20** (2014)
8. Watts, D.J.: Small Worlds. The Dynamics of Networks Between Order and Randomness. Princeton University Press, Princeton (2004)
9. Aleksy, M., Korthaus, A., Schader, M.: Use Java and the CORBA realization distribute type system. J. Pingxiang Coll. **4**, 104–105 (2005)
10. Titus, J.: The eclipse of stand. J. Zhongkai Agrotech. Coll. **19**(2), 32–35 (2006)
11. Clay Richardson, W., Avondolio, D.: The Java high class weaves a distance JDK 5. Sci. Technol. Book Rev. **3**, 17–18 (2006)
12. Li, J., Vuong, S.: A semantics-based routing scheme for grid resource discovery. In: E-Science: First International Conference on E-Science and Grid Computing, pp. 58–70, 90 (2005)

Data Communication Method in Collaborative Process Planning

Yingying Su(✉) and Lianjuan Han

Mechanical Engineering School, Shenyang University, Shenyang 110044, Liaoning, China
syyle2012@163.com

Abstract. Collaborative process planning is the application of computer supported cooperative Work (CSCW) in the field of CAPP, which provides a collaborative work environment for workers through computer, network and collaborative tools. Real-time updating and synchronization is the key to realize collaborative process planning. The key technology of real-time updating and synchronization is the communication mechanism. The data communication method and flow of collaborative process planning system are determined, and communication mechanism is established in this paper based on the study of collaborative process planning and network communication to ensure synchronous and real-time transmission of collaborative process planning.

Keywords: CSCW · Collaborative process planning
Data communication method

1 Introduction

The technology of CSCW is applied to the process planning, so that the personnel involved in process planning can overcome the obstacles of distance, time and foreign computer equipment. It is of great significance to improve the quality and speed of the process planning for the parts, especially the complex parts [1, 2].

In process planning, the geographically distributed several technologists synchronously browse the process documents and collaborative process planning, which need the support of Synchronous collaboration. Good interactivity is the key to the success of collaborative process planning system. If the user needs a long time to wait for each operation, the process planning work cannot be carried out normally. Therefore, real-time and efficient collaborative data exchange is the key to realize the real-time collaborative process planning in the distributed environment [3, 4].

Synchronous and real-time update of collaborative process planning is realized through communication mechanism, which is divided into the communication between the cooperative user and the server and the communication between the cooperative users. The former communicates are according to the general request/response mode, and the main purpose is to process and store data. The communication between the cooperative users is the key problem. On the basis of analyzing and determining the network transmission protocol and network communication mode of collaborative

J. Abawajy et al. (Eds.): ATCI 2018, AISC 842, pp. 1185–1194, 2019.
https://doi.org/10.1007/978-3-319-98776-7_145

process planning system, the communication method and flow are determined to ensure the synchronous and real-time transmission of collaborative process planning.

2 Network Communication Transmission T Protocol

TCP/IP (Transmission Control Protocol/Internet Protocol) is a network communication protocol, which regulates all communication devices on the network, especially the data interchange format and transmission mode between one host and another host. The TCP/IP protocol is organized into four conceptual layers, including application layer, transmission layer, internet layer and network interface layer. The reference model is shown in Fig. 1.

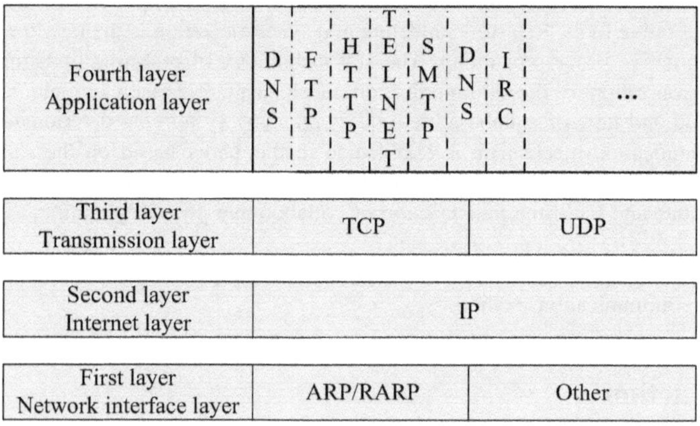

Fig. 1. Four layering reference model of TCP/IP

Socket is a set interface, which is the basic unit of TCP/IP network communication protocol. As the network programming interface of TCP/IP, it provides a convenient environment for the development and operation of network communication based on Windows. In the TCP/IP network application, the communication parties interact with each other, and the client/server mode is adopted to establish the connection process, which is shown in Fig. 2.

Server (first start):

(1) Open a communication channel and inform the client to accept the request of the customer on a certain port (such as HTTP: port 80).
(2) Waiting for the client to arrive at the port.
(3) Receiving and processing the request, sending the response data to the client.
(4) When the service is completed, the connection is closed.
(5) Return to step 2, waiting for the next customer request.

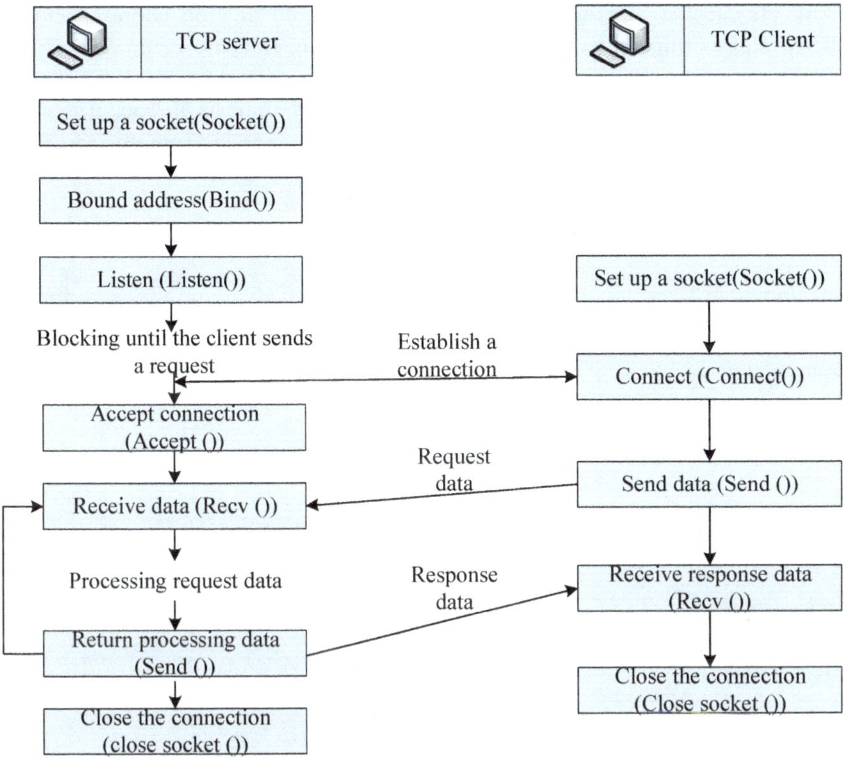

Fig. 2. Flow of socket communication between client and server

Client:

(1) Open a communication channel and connect to the specific port of the host on which the server is located.
(2) If the server accepts the connection request, it sends a service request message.
(3) Waiting and receiving the response.
(4) The communication channel is closed after the request finishes.

3 Network Communication Mode

At present, there are three network communication mode, which includes unicast, broadcast and multicast. The host communications with an IP address for one-to-one (unicast), one-to-many (multicast) and one-to-all (broadcast).

3.1 Unicast

The "one-to-one" communication mode is between the hosts, and the receiving and transmission of information is only between two nodes. The router and switch in the network choose the transmission path according to destination address, and the IP multicast data is transmitted to designated destination. Because of timely response to each customer, all web browsing is now based on IP unicast protocol. For example, the host H1 sends a request page to the server H2, and the unicast IP packet is shown in Fig. 3. However, because the server sends data streams to each client, the server will be overwhelmed if the number of customers is large.

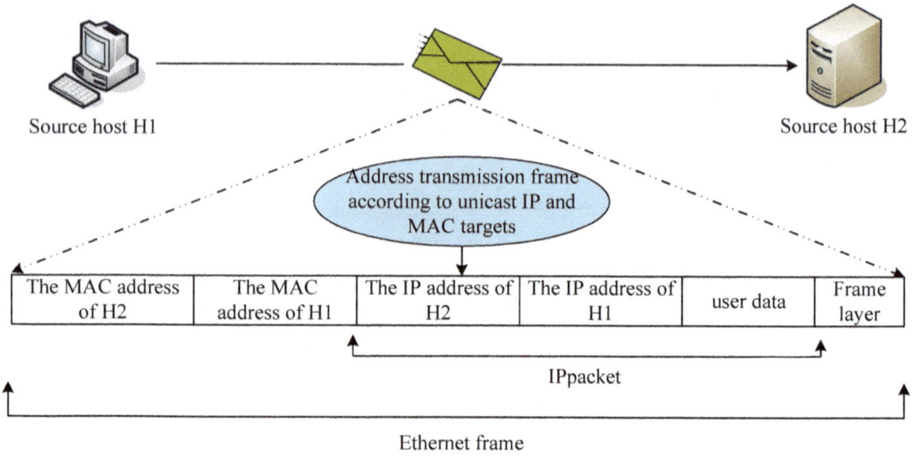

Fig. 3. Unicast IP packet

3.2 Multicast

Communication mode of multicast is "One-to-one group" between the hosts. Host joined the same set can accept all the data in this group. Switches and routers copy and forward the data to the demanders needed. The host can request or exit a group to the router. The routers and switches in the network have the choice of replicating and transmitting data. Only the data in the group are transferred to the host that joins the group. Multicast in IP networks is usually implemented through multicast IP addresses, and Fig. 4 shows a multicast IP packet.

Source device send packets to a set of devices for multicast addresses. Equipment belonging to the multicast group will be assigned an IP multicast group address, and multicast MAC addresses are headed with 6 sixteen values, and the remaining 6 sixteen bits are converted to the last 23 bits of the IP multicast group address.

3.3 Broadcast

Signals are sent to each hosts unconditionally and forward in the "one to all" communication mode between the hosts, and all the hosts can receive all the information (whether the client needs), because of its no path selection. So the network cost can be very cheap, and cable network is a typical radio network. Broadcast is also allowed in the data network, but it is restricted to the LAN scope of the two layer switch, and the

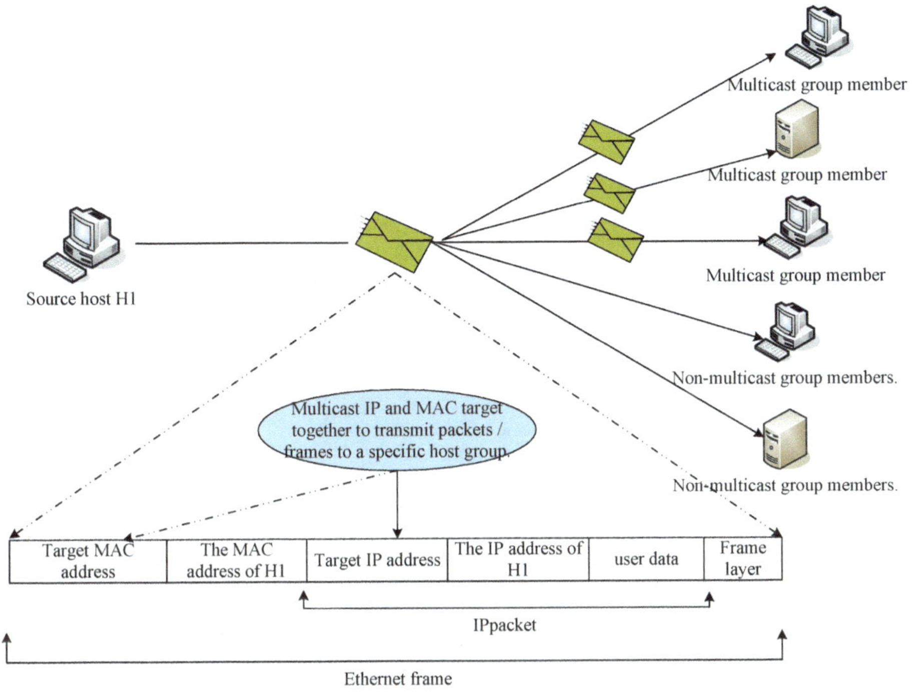

Fig. 4. Multicast IP packet

broadcast data is prohibited through the router to prevent the broadcast data from affecting the large area of the host.

Figure 5 shows a broadcast IP packet that must contain the broadcast MAC address corresponding to the broadcast IP address in Ethernet frames. In Ethernet, the length of the broadcast MAC address is 48 bits and its hexadecimal is denoted as FF-FF-FF-FF-FF-FF.

In collaborative process planning system, in order to achieve collaboration between multiple users, the data needs to be transmitted to multiple collaborative users, so multicast communication is required. As long as the communication client joins the same multicast group, the same information only needs one packet to be transmitted to all groups of cooperative users, which greatly improves the bandwidth utilization of the network.

4 Message Based Data Communication Mechanism

The synchronization of collaborative process planning system is realized through message communication among collaborative users [5–7]. Message is the application layer encapsulation of user data, and the basic communication unit of collaborative process planning system. According to the different contents of messages, communication

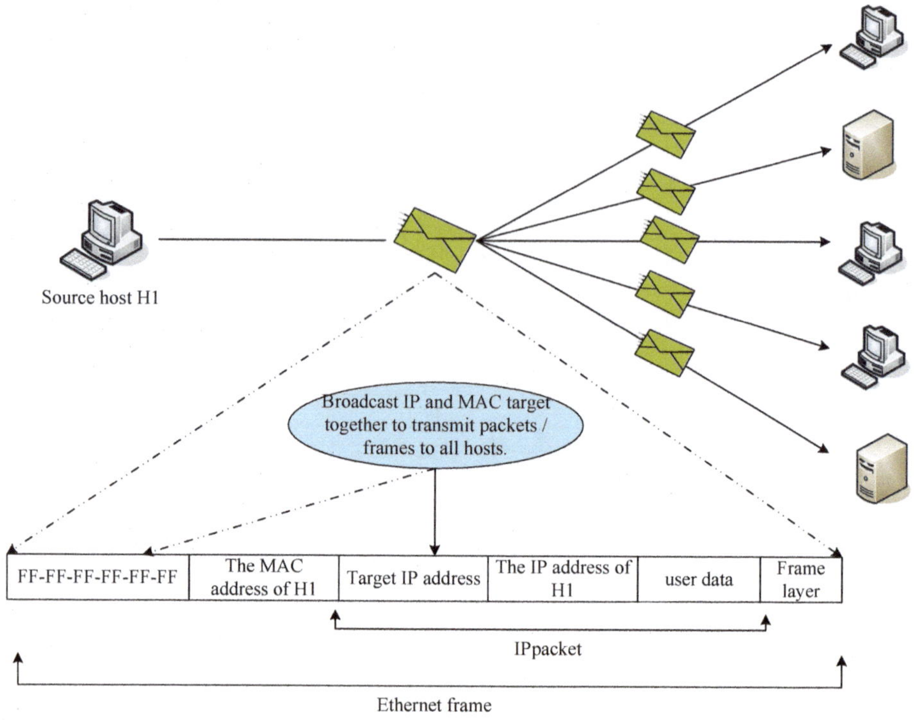

Fig. 5. Broadcast IP packet

messages of collaborative process planning system are divided into two kinds: instant messaging and collaborative application messages.

Instant communication message is related to the function of coordination layer communication message in collaborative process planning system. It defines the collaboration site to complete the task of mutual coordination communication language and has nothing to do with specific collaborative applications, which includes user authentication, shared whiteboard, multimedia audio and video communication.

Collaborative application messages are related to the function of application layer communication message in collaborative process planning system. The synergy between the application levels of the application function modules associates with the application of collaborative, so collaborative application message communication problems is mainly studied in this paper.

4.1 The Basic Structure of Network Messages

In the communication method of collaborative process planning system, the meaning of message envelope, message header and message body in network message is as follows.

Message Envelope. The message envelope packages the message body and message header, whose parameters and meanings are shown in Table 1.

Table 1. Main parameters and meanings of message envelope in collaborative process planning system

The main parameters of the message envelope	Meanings
To	Host address of the collaborative user for receiving message
From	Host address of the collaborative user for sending message
Message ID	Communication message identifier
Timestamp	The timestamp of the communication message
Encoding	The character encoding of the message header and message body
Encrypted	Encryption of message header and message body
Message	Message header

Message Header. The header represents information that describes the body data, such as property information of message body. The header is optional, and if the header is selected, the next element must be in the message envelope label. The message header contains parameters and meanings as shown in Table 2.

Table 2. Main parameter and meanings of message header in collaborative process planning system

Main parameters of message header	Meanings
Sender	Collaborative user identity for sending messages
Receiver	Collaborative user identity for receiving messages
Type	Message type
Performative	Type of communication behavior
Ontology	Message body ontology
Language	The expression language of the message body
Content	Message body

Message Body. The message body contains the active part of the message. In general, the data in the message body is given to the final message receiver. The representation of the message body is defined by the application itself, and the following is an exploration of collaborative application messages.

4.2 Collaborative Application Message

The collaborative application message is the communication message related to the application layer function of the collaborative process design system, and the collaboration of process planning is realized. The communication mode is shown in Fig. 6. Collaborative application messages rely on IP multicast to deliver and maintain the consistency of client operations within each design group.

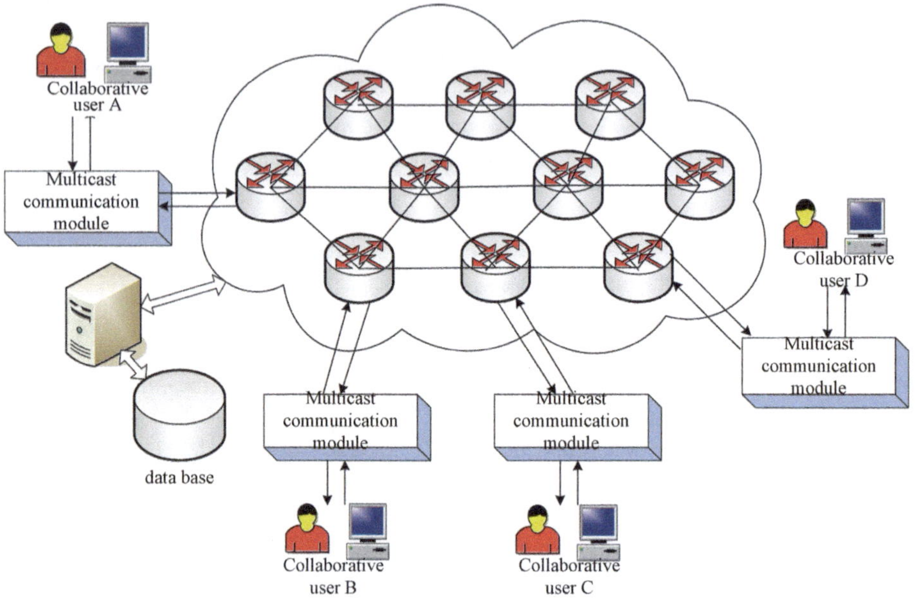

Fig. 6. Communication model of collaborative process planning

5 Realization Flow of Data Communication Method in Collaborative Process Planning

According to the characteristics of collaborative process planning, the message expression of the collaborative process planning system for multicast communication is designed. Firstly, serializable message base is user defined, including news target IP, transmission message type and message body attributes. Then specific network message classes are derived from the message base class. Thirdly, XML is encapsulated in serialization to network flow. Finally, data is sent with IP multicast. When the collaborative application messages are received by other clients in the collaborative group, exchange messages of the collaborative application are obtained and desterilized. The message is analyzed by class information according to the rules. The implementation flow is shown in Fig. 7.

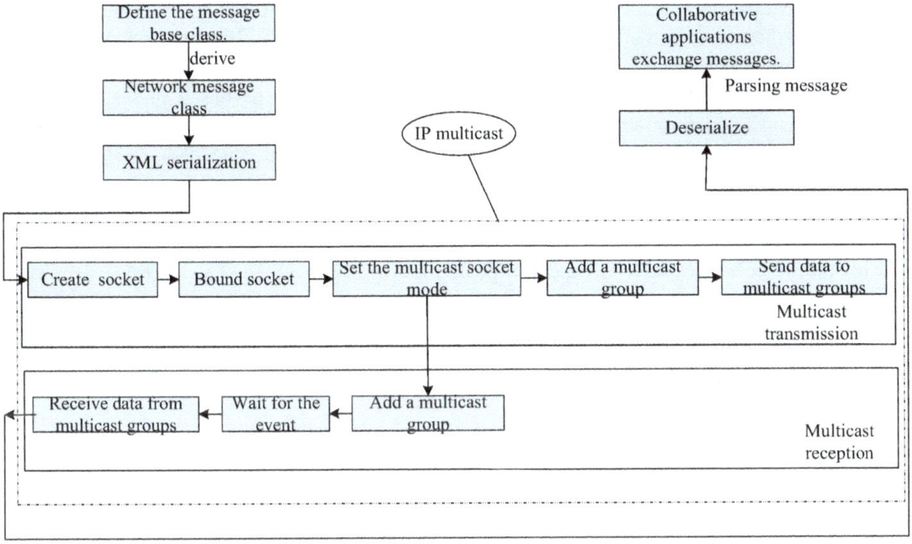

Fig. 7. Implementation flow of application message in collaborative process planning

6 Conclusions

In the face of the needs of enterprises, CSCW technology is introduced into the process planning form collaborative process planning. Synergetic collaboration of process planning is the key to the realization of collaborative process planning system. The most important data communication method in synchronous collaboration is studied. A method of data communication based on message and implementation flow are presented to ensure the implementation of synergetic coordination in process planning in this paper, which is of great significance to shorten the product development cycle, reduce product cost and improve product quality.

References

1. Milosevic Mijodrag, F., Lukic Dejan, S., Antic Aco, T.: e-CAPP: a distributed collaborative system for internet-based process planning. J. Manuf. Syst. **42**, 210–223 (2017)
2. Ming, X.G., et al.: Collaborative process planning and manufacturing in product lifecycle management. Comput. Ind. **59**(2), 154–166 (2008)
3. Rambabu, D., Lenin, A., Bhaskar, G.B.: Gap analysis and optimization of process involved in product design and development by integrating enterprise resource planning & product lifecycle management. Proc. Eng. **64**, 983–992 (2013)
4. Zhou Ziqiang, F., Dai Guohong, S.: Research of web-based 3D collaborative CAPP system - research of web-based 3D collaborative CAPP system. Comput. Aided Draft. Des. Manuf. **22**(04), 69–73 (2012)

5. Arshad Muhammad, F., Ullah Zahid, S., Ahmad Naveed, T.: A survey of local/cooperative-based malicious information detection techniques in VANETs. EURASIP J. Wirel. Commun. Netw. **2018**(1), 62–89 (2018)
6. Talipov Elmurod, F., Hojung Cha, S.: Communication capacity-based message exchange mechanism for delay-tolerant networks. Comput. Netw. **55**(15), 3408–3422 (2011)
7. Sahoo, J., Wu, E.H.L., Sahu, P.K., Gerla, M.: Congestion-controlled-coordinator-based MAC for safety-critical message transmission in VANETs. IEEE Trans. Intell. Transp. Syst. **14**(3), 1423–1437 (2013)
8. Shizhao, N., Zheng, W., Wangmo, P., Yuan, N., Peng, L.: Big data prediction of durations for online collective actions based on peak's timing. Phys. A **55**, 130–139 (2018)
9. Clay Richardson, W., Avondolio, D.: The Java high class weaves a distance: JDK 5. Sci. Technol. Book Rev. **3**, 17–18 (2006)
10. Li, J., Vuong, S.: A semantics-based routing scheme for grid resource discovery. In: E-Science: First International Conference on E-Science and Grid Computing, pp. 58–70, 90 (2005)

Analysis and Research of Fatigue Driving Detection Algorithm

Jun Wang[1] and Qiang Liu[1,2](✉)

[1] Innovative Research Team of Jilin Engineering Normal University, Jilin Engineering Normal University, Changchun 130052, China
78675985@qq.com, 719598353@qq.com
[2] State Key Laboratory of Automotive Simulation and Control, Jilin University, Changchun 130025, China

Abstract. Fatigue driving is an important cause of traffic accidents, the paper designs a real-time monitoring algorithm for fatigue driving, through the camera Obtain the eye image, Using the cumulative difference image processing technique to detect and track the eyes. According to the closing time of the human eye at a certain time, calculate the fatigue degree and take corresponding measures. The experiment result shows that: the algorithm is simple, fast, accurate and cost-effective, the research has laid a good foundation for the application and productization of the fatigue driving real-time monitoring system.

Keywords: Fatigue driving · Human eye to identify · Eye location

In recent years, with the increase of the number of car ownership in economic growth, the problem of traffic accidents has caused people's high attention. Fatigue is one of the important reasons for the traffic accident, effective fatigue detection system can be avoided by the traffic accidents caused by fatigue driving, reduce casualties and property losses. Domestic and foreign scholars have done a lot of research on this, such as the electronic sober belt developed by Japan and the steering wheel monitoring device developed by the United States.

Domestic based on driver's facial features of video signal fatigue driving detection algorithms are studied, but also stay on the theory and method of exploration, not standard and perfect solution of the system. This paper puts forward the fatigue driving detection algorithm based on facial features, with the help of infrared light of face to locate, find the face region, to accurate positioning of the eyes, and analysis of the status of evaluation eyes, judge whether the driver fatigue.

1 Fatigue Driving Test Method

Driving fatigue refers to the phenomenon that drivers continue to drive when their physical, mental and driving skills are reduced because of long driving. Fatigue driving will bring vehicle driver inattention, insensitive sensation, perception, thinking slow, misjudgment, weak willpower and so on many aspects of influence, resulting in road traffic accidents.

J. Abawajy et al. (Eds.): ATCI 2018, AISC 842, pp. 1195–1199, 2019.
https://doi.org/10.1007/978-3-319-98776-7_146

The monitoring of fatigue driving is divided into active subjective and passive objective.

The subjective monitoring method is mainly based on the subjective questionnaire, the driver's self-report form, the sleep habit questionnaire and the Stanford sleep scale, etc., to measure the fatigue degree of the driver.

There are two types of objective monitoring:

One is to use the relationship between physiological characteristics and fatigue of human fatigue, and to monitor the fatigue status of the test subjects by detecting these physiological characteristics. Such as electroencephalogram (EEG), EOG Electrooculography, electromyogram, breathing with the sound of the nasal airflow sensor measurement, breathing effect with the chest measurement sensor, etc.

The other is to use the relationship between driver's driving behavior and fatigue, to monitor the driver's fatigue state by detecting the driver's driving behavior.

1.1 Comparison of Various Monitoring Methods

Using electroencephalogram (EEG), EOG Electrooculography, electromyogram to detect fatigue is the "gold standard" for measuring sleep. Although these methods more accurate measurement results, but because of the need to be worn on the head when measuring electrode, and in driving indoor installation of the instrument is not realistic, nor into practical use (Table 1).

Table 1. Comparison of monitoring effects of various methods.

Detection method	Veracity	Real-time	All-weather	Anti-interference	Comfort	Practicability
Pulse	Ordinary	High	High	Ordinary	High	High
Electroencephalogram (EEG)	46.30%	High	High	High	Low	Low
Mouth closed	Ordinary	High	Low	Low	High	High
Head position	52.90%	Ordinary	Ordinary	Low	High	Ordinary
Steering wheel monitoring	Low	Ordinary	High	High	High	Ordinary
Reaction time tester	Low	Ordinary	High	Ordinary	Low	Ordinary
Blink test	48%	Ordinary	Ordinary	Low	Ordinary	Ordinary
PERCLOS(P80)	87.80%	High	High	High	High	High

1.2 Determination of Monitoring Methods

Can be seen from the comparison above all the methods in different degree, it can reflect the driver's fatigue, and PERCIOS P80 standard can better monitor the driver's fatigue state. Therefore, even considering the particularity of fatigue monitoring of the driver, the monitoring method should meet the following requirements:

① Non-contact, that is to do not interfere with the driver's driving behavior.
② Real-time, can detect the driver fatigue physiological characteristic parameters in real time.
③ All weather, can work in the dim light or **dark** and the driver wear glasses.

As far as the present situation is concerned, it is undoubtedly the best choice in the field of fatigue driving monitoring to adopt the appropriate improved method. Therefore, this design plan adopts this method and improves it.

2 DSP Monitoring System Based on Improved PERCLOS Method

2.1 The Composition of the Detection System

This design adopts PERCLOS method, using infrared light source, difference image and Kalman filter. System is mainly composed of infrared light source, COMS photo like, control main board and the corresponding software components.

The camera is mounted on the front of the driver without affecting the driver's vision. In the selection of the camera, the CCD (charged coupled device) and CMOS (complementary metal-oxide-semiconductor) are commonly used as sensors. Compared with the CCD, CMOS sensor does not require high voltage, but only requires a single-power low logic level voltage, the power consumption is equivalent to 1/10 of the CCD. In addition, CMOS sensor can integrate chip timing, noise suppression, A/D converter and DSP on an IC. This will help the system eventually integrate into the on-board equipment. At the same time, CMOS sensor does not require complex processing, which directly converts the electrons generated from the image transistor into a voltage signal, so the number of images generated is very high, up to 400 to 2000 frames/s. So this design chooses CMOS sensor.

Different wavelengths of infrared light have different retinal emissivity, the human eye can reflect 90% of the incident light at 850 nm wavelength, the human eye can reflect 40% of the incident light at 950 nm wavelength, so in the same illumination condition, use two cameras to capture images of the human eye, one image is infrared light with a wavelength of 850 nm, the other one is infrared light with a wavelength of 950 nm, the result of two images subtracting, just the image of the position of the retina, the size and location of the retina are then analyzed.

To get two images of different wavelengths of light, use two separate cameras, 90° crossover. When the image passes through a beam splitter, the two cameras are divided into two cameras respectively, and the corresponding infrared images are obtained by using the filter with the wavelength of 850 nm and 950 nm respectively. As a result, two images with different retinal images at the same time were obtained.

Because the use of infrared light source, on the one hand will not affect the operation of the driver, on the other hand, can effectively meet the requirements of the weather.

The collected image signal is differential mode treatment by the control board and the pupil image is obtained. At the same time, the Kalman filter assisted by neural network was used to predict the pupil position in real time. Practical application process, through the monitoring of potential difference, with reference to the distance to calculate the pupil size of absolute value, so as to solve the due to the driver's head movement (forward or backward) caused by the impact of the image size changes.

The first image data subtracts the second image data and gets the pupil image. At the same time, Kalman filter is used to predict the pupil position in real time. In practice, monitoring of pupil distance, use the pupil distance as the reference value to calculate the pupil size absolute value, Thus, the effect of image size change caused by driver's head movement (forward or backward) is solved.

2.2 Algorithm and Improvement of PERCLOS

A large amount of data was collected in this study, including blink time and half closed eyes. Data from eye tracker can be analyzed. These data can be used for the following analysis:

- Identify some parameters for blinking and closing eyes.
- Disconnect from lane and close eye time of relation.

Based on this study, the ratio of the width of the eyes to the diameter of the pupil (V/H) could have the desired effect. Fatigue scores are usually low when you start driving. With the driving time, scores are getting higher and higher, and they reach the highest point before the traffic accident. PERCLOS method to detect fatigue is generally in 10 s before the accident, to the driver's reaction time is too short, proposed under the condition of the fixed half closed eyes, to the eye pupil of horizontal and vertical diameter ratio V/H algorithm to detect fatigue, experiments show that generally occurs at least 2–3 min before the accident but combining the two methods compared to other method is more accurate and stable.

2.3 Eye Localization

Hough transform is used in this study to detect the pupil of the human eye. Hough transform is a method used to describe the shape of the region boundary, and is often used for the detection of straight lines, circles and ellipses. The basic idea is to transform the spatial domain of the image into the parameter space and describe the curve (regional boundary) in the image with some form of parameter satisfied by most boundary points. Hough transform detection technology based on local measurement parameters to calculate the boundary curve, therefore, and other goals for regional boundary noise or cover and cause of border, has the very good fault tolerance and robustness.

The shape of the eye is precisely round, so if you can find the center of the circle on the edge of the face image, you can pinpoint the person's eye. However, the presence of eyelashes and upper eyelids will block the edges of the circle. Hough transform has the advantages of noise and edge discontinuity, so it is a very effective method to test the pupil contour.

3 Conclusion and Prospect

This paper mainly introduces the research scheme of driver fatigue monitoring system, which is summarized as follows:

(1) A brief introduction to the overall design;
(2) PERCLOS (percentage of eyelid closure over the pupil overtime) Fatigue judgment method, namely determining fatigue parameters, detailed description of how to achieve fatigue;
(3) The positioning of the eye directly influences the identification and judgment of the next driver fatigue. In this paper, Hough transform is used to detect the pupil of the human eye.

The solution of this series of problems is determined, and a clear idea is established for the future hardware design and software programming.

Acknowledgments. This work was supported by the 13th five-year science and technology project of Ji Lin education department. 《Fatigue driving detection system based on human eye recognition and FPGA technology》 (JJKH2017166KJ).

References

1. Lu, X.: Research on Fatigue Driving Detection Based on Image Processing. Guangdong University of Technology, Guangzhou (2013)
2. Liu, C.-S.: Study on Facial Feature Detection and Fatigue State Recognition. Shandong University, Jinan (2012)
3. Sahayadhas, A., Sundaraj, K., Murugappan, M.: Detecting driver drowsiness based on sensors: a review. Sensors **12**, 16937–16953 (2012)
4. Zhang, X.B., Cheng, B.: Real-time detection of driver drowsiness based on steering performance. J. Tsinghua Univ. Sci. Technol. **50**(7), 1072–1076 (2010)
5. Ming, F.: The cultural elements in English translation teaching. Engl. Sq. **20** (2014)
6. Wang, H.: The problems and countermeasures of English translation teaching in college of traditional Chinese medicine. Overseas Engl. **20** (2014)
7. Watts, D.J.: Small Worlds. The Dynamics of Networks Between Order and Randomness. Princeton University Press, Princeton (2004)
8. Aleksy, M., Korthaus, A., Schader, M.: Use Java and the CORBA realization distribute type system. J. Pingxiang Coll. **4**, 104–105 (2005)
9. Titus, J.: The eclipse of stand. J. Zhongkai Agrotech. Coll. **19**(2), 32–35 (2006)
10. Clay Richardson, W., Avondolio, D.: The Java high class weaves a distance: JDK 5. Sci. Technol. Book Rev. **3**, 17–18 (2006)
11. Li, J., Vuong, S.: Semantics-based routing scheme for grid resource discovery. In: E-Science: First International Conference on E-Science and Grid Computing, pp. 58–70, 90 (2005)

Research on Computer Network Information Security and Preventive Countermeasures

Guang Li[1], Shuang Yu[1], and Xiaohui Chai[2(✉)]

[1] College of Resource and Environment, Anshun University, Anshun 561000, China
[2] School of Computer and Information Technology, Shangqiu Normal University, Shangqiu 476000, China
460818640@qq.com

Abstract. The collection, transmission and use of information are inseparable from the network. Precisely because the broad applications of computer network information security vulnerability and complexity. Makes computer networks more and more attacks, security has also become increasingly vulnerable, so that information security is growing more difficult. This article from network information security vulnerabilities, network security for common network attacks and countermeasures, to prevent the main construction principles of network security technologies, such as analysis of the various aspects of the current network information security issues, and solutions to common network attacks raised the technical level of, and the principles of network security. At its core are key to preventing network security technologies, and the comparison of several key technologies.

Keywords: Network information security · Firewall · Data encryption

1 Introduction

Network information security mainly means that the hardware, software and data in the network system are protected from being accidentally or maliciously destroyed, changed or leaked. It is undeniable that the Internet has changed people's ways of learning, working and living [1].

This paper analyzes the main problems of current network information security from various aspects, such as the vulnerability of network information security, the main technologies for preventing network security, common network attack methods and countermeasures and the principles of building network security, and proposes solutions of the common network attacks from the technical level and the principles that should be followed in building network security.

2 Overview of Network Information Security

The 21st century is an information-oriented era. Its important feature is to obtain information and exchange information through the Internet. With the rapid development of the information industry, the security problems of the Internet and network

J. Abawajy et al. (Eds.): ATCI 2018, AISC 842, pp. 1200–1205, 2019.
https://doi.org/10.1007/978-3-319-98776-7_147

information also increase rapidly. The annual economic losses caused by the fragility of information systems have been increasing year by year in the world, and the security problems have become increasingly serious. In the face of such cruel reality, the issue of network security has received unprecedented attention [2].

The threats to network security and confidentiality come from many sources, which will change over time. We roughly divide them into four aspects: unsafe factors of network components, unsafe factors of software, unsafe factors of personnel and unsafe factors of the environment.

3 Common Network Attacks

Computer network information can be logged in remotely, which gives intruders an opportunity. If an intruder steals or deciphers someone's account or password on the Internet, he or she can obtain authorized access to other people's networks and realize attempts to steal information resources. The main attacks used by intruders are roughly the following [3]:

(1) Eavesdrop on the contents of communication services and identify communication parties to understand the nature and content of the information transmitted in the communication network.
(2) Eavesdrop on data services and identify fake words, and access and utilize communication networks based on these communication words to understand the data exchanged in the network.
(3) Analyze traffic flows and infer key information. Through the analysis of the traffic in the communication network, the capacity, direction and time window information of the communication can be understood, which is crucial in military networks.
(4) Repeated or delayed transmission of information cause uses' confusion.
(5) Change the flow of information. The communication information in the network is modified, deleted or reordered so that users will make wrong responses.
(6) Block the network and inject a large amount of useless information into the communication network to occupy the channel to block the transmission of useful information.
(7) Deny access and prevent legitimate network users from performing their functions.
(8) Counterfeit routing and attack the network's switching equipment and direct network information to the wrong destination.
(9) Tamper with programs and disrupt operating systems, communications and application software, such as using computer viruses, "worm" programs, "Trojan horse" programs and logic bombs to perform software attacks.

4 Network Attacks Preventive Strategies

4.1 Improve Safety Awareness

Improving security awareness is mainly aimed at preventing and killing viruses, repairing system vulnerabilities and setting passwords. Common methods are as follows: Preventively open E-mails or files of unknown origin; letters and numbers shall be used as password; do not use simple numbers or letters as password and do not use name or birthday as password. It is best not to download unknown software or game programs from the Internet. Fix system vulnerabilities in time. On HTML-backed BBSs, be sure to look at the source code first if there are submitting warnings. Use pirated software with caution [4].

4.2 Back Up Important Data

Protect important personal data and develop the habit of backing up data at any time. If an unexpected event occurs, such as sudden power failure or electromagnetic interference or sudden crash, the file may be deleted and the hard disk may be formatted. So always save and back up files.

4.3 Install Antivirus Software

It is necessary to install antivirus software on computer, because most of the current viruses can be killed by antivirus software. After being installed, antivirus software must be updated and upgraded. Because the virus is developing very quickly, the corresponding antivirus software must also be added with the feature code of the new virus.

4.4 Improve the Browser's Security Level

The default security level of Internet Explorer is "medium", which can be set to "high". Active download can also be manually disabled in advanced options of Internet Explorer and even Java programs can be disabled.

4.5 Install System Protection and Repair Tools

When the computer's IE program or system file is modified, in addition to modifying the registry, special tools can also be used to repair, such as Super Rabbit magic settings, Windows optimization master. In addition, Google Toolbar and Baidu Toolbar are also good tools.

5 The Main Technologies of Network Security

The computer network security technologies are mainly divided into: firewall technology, host security technology, cryptography technology, access control technology, identity authentication technology, security management technology and security audit technology. This paper mainly introduces three technologies: firewall technology, cryptography technology and access control technology.

5.1 Firewall Technology

A firewall is a network device or group of devices that can be used to enhance access control between two or more networks, that is to build a protection on the layer interface between the internal network and the external network. This protection layer forces all connections to pass through it, and checks and connects. At the same time, we require that only authorized communications can pass through this protection layer so as to prevent the illegal invasion of internal network resources [5].

One of the most effective tools in network security strategies is the firewall which is widely applied to the Internet. Traditionally, firewalls fall into two broad categories: application-layer firewalls that use application gateways and network-layer firewalls that use filtering routers. Its structural model can be divided into two parts: strategy and control. The strategy refers to whether to grant the service requester the corresponding access rights and the control refers to the control of the resource access of the authorized visitor.

5.2 Cryptography Technology

With the widespread use of computer networks, the application of cryptography has also expanded. The most commonly used data encryption technologies at present are symmetric encryption and asymmetric encryption. Symmetric encryption technology means that the data encryption key and the decryption key are the same. The asymmetric encryption technology is different, which means that the encryption key and the decryption key are not the same, that is, it has a pair of keys, in which the public one is called "public key" and the private one is called "private key". These two keys need to be paired together before they can achieve their functions. In other words, a public key encrypted file must be decrypted with the corresponding person's private key, just like a key for a lock or a password for a bank card [6].

5.3 Access Control Technology

Access control is an extremely important part of computer security defense measures. It controls the proposed resource access request according to the validity of the identity on the basis of identity authentication. The purpose of access control is to protect the resources of the host computer and the legal use of the resources. Users can only access the system resources according to the specific permission they have been granted, which cannot exceed the scope of the permission or they will be rejected because of insufficient permissions. At the same time, access control is also a prerequisite for accounting and auditing.

6 The Principles of Network Security

In an open network environment, both secure and insecure elements, trusted and untrusted users exist at the same time. In order to protect the security of system communication and protect the interests and privacy of legitimate users, an open network environment requires security protection measures. At the same time, in order to guarantee the compatibility and information integrity of the system, reduce the time cost and space cost overhead, facilitate the use of users and the management of the system, the security protection finally follows several principles: Set up security measures in applications close to users. The number of trusted third parties is minimized. Separating the implementation of the security strategy from the specific security implementation means that the security management is convenient and versatile and the specific method of security implementation is flexible and diverse. The implementation of a security policy is a holistic concept. The characteristics of the original network and the versatility of the system shall be maintained as far as possible, and the principle of open network shall not be violated by the newly added security measures. Adopt a unified internal structure and external management interface to achieve the necessary security measures [7].

7 Conclusion

This paper mainly ensures network information security in three aspects, which are the firewall technology, data encryption technology and access control. It specifically focuses on firewall technology and data encryption technology. For the firewall, packet filtering technology and proxy server are mainly submitted. The former is quite simple since it does not require the support of user software. However, the configuration is relatively complicated. Access control techniques are proposed to protect the legitimate use of host resources of computer networks, but this section only puts forward some brief introductions. Because there are many common attacks in networks, including information collection; password attacks; Trojan horses; network eavesdropping; attacks on routers or servers and e-mail attacks, different attacks are naturally correspond to different defense measures. This paper only makes corresponding defense measures against the proposed attacks.

How to ensure legal users' legitimate access to resources on the Internet? From the demand brought by the development of network economy to network security products, antivirus software, intrusion detection, information encryption, security authentication and firewalls will have a huge market prospect. The main shares of the market are firewalls and high-end antivirus software; At the same time, the technology of passive defense of the Internet will gradually develop toward active detection and defense. The intrusion detection system is an active network security protection measure that adapts to this trend of development. Therefore, it is expected that the demand will increase rapidly.

References

1. Mendez, D.J., et al.: System and method for globally and securely accessing unified information in a computer network. U.S. Patent No. 9,361,603, 7 June 2016
2. Guo, J., et al.: An approach to network security evaluation of computer network information system with triangular fuzzy information. J. Intell. Fuzzy Syst. **28**(5), 2029–2035 (2015)
3. Sun, Q.L.: Information under the network environment using computer information security technology. In: 2015 International Conference on Intelligent Transportation, Big Data and Smart City (ICITBS). IEEE (2015)

4. Ab Rahman, N.H., Choo, K.-K.R.: A survey of information security incident handling in the cloud. Comput. Secur. **49**, 45–69 (2015)
5. Porras, P.A., Skinner, K.M., Dawson, S.M.: Security actuator for a dynamically programmable computer network. U.S. Patent No. 9,571,523, 14 February 2017
6. Shropshire, J., Warkentin, M., Sharma, S.: Personality, attitudes, and intentions: predicting initial adoption of information security behavior. Comput. Secur. **49**, 177–191 (2015)
7. Zhang, Y.: Research on the computer network security evaluation based on the DHFHCG operator with dual hesitant fuzzy information. J. Intell. Fuzzy Syst. **28**(1), 199–204 (2015)
8. Titus, J.: The eclipse of stand. J. Zhongkai Agrotech. Coll. **19**(2), 32–35 (2006)
9. Aleksy, M., Korthaus, A., Schader, M.: Use Java and the CORBA realization distribute type system. J. Pingxiang Coll. **4**, 104–105 (2005)
10. Clay Richardson, W., Avondolio, D.: The Java high class weaves a distance: JDK 5. Sci. Technol. Book Rev. **3**, 17–18 (2006)
11. Li, J., Vuong, S.: A semantics-based routing scheme for grid resource discovery. In: E-Science: First International Conference on E-Science and Grid Computing, pp. 58–70, 90 (2005)

The Influence of the Opening Community on the Traffic

Bingjun He[1], Yi Cheng[2(✉)], Han Li[2], and Shi Kang[2]

[1] Department of Mathematics, Bohai University, Jinzhou 121013, China
362360915@qq.com
[2] College of Mathematics and Physics, Bohai University, Jinzhou 121013, China
645924181@qq.com, 15566527260@163.com,
2367230337@qq.com

Abstract. This paper mainly studies show to improve the traffic by opening up completed residential and unit courtyards under the mechanism of no longer building closed residential areas. Is it possible to improve road capacity and traffic conditions? With this problem, we have established a mathematical model based on saturation, traffic congestion and other conditions with statistics and investigation data to get the specific impact of opening community on road traffic.

Keywords: Traffic saturation · Remittance saturation · Road congestion rate
Closed cell

1 The Establishment of the Model and Solution of the Question

In order to evaluate the influence of the opening of community to the traffic of the surrounding roads, we should analyze the traffic conditions in a reasonable, objective and comprehensive way, and take a multi objective principle as a principle to make a comprehensive evaluation of the degree of traffic impact, so as to get the influence of the opening community to the surrounding traffic.

2 Remittance Saturation

Road saturation is an important index to reflect road level. The higher the saturation is, the lower the service level of the road is. Therefore, this paper selects traffic saturation and inflow saturation as indicators to evaluate the level of peripheral traffic service.

Symbols and Their Meanings

$C_{汇}$—Maximum traffic flow rate of cross road;
$C_{通}$—The maximum traffic flow rate of one way at the intersection;
s—Number of intersections;
n—Number of one-way lanes at intersections;
C_i—The maximum traffic flow rate of each single way;

J. Abawajy et al. (Eds.): ATCI 2018, AISC 842, pp. 1206–1209, 2019.
https://doi.org/10.1007/978-3-319-98776-7_148

i—Lane number;
j—Number of imported lanes;
C_0—Theoretical capacity of 1 lanes;
$\beta_{条}$—Lane reduction coefficient;
$\beta_{交}$—Intersection reduction coefficient;
$\beta_{车道}$—Lane width reduction coefficient.

Build the Following Sodium Calculate Formula

$$C_{汇} = \sum_{s=1}^{n} C_{通};C_{通} = \sum_{i=1}^{j} C_i$$

$$C_s = C_0 \times \beta_{条} \times \beta_{交} \times \beta_{车道}$$

First, let v = 40 km/h, m = 600, b = 3.5, n = 1, by formula? We can get

$$C_{通} = 1640 \times 1 \times 0.83 \times 1 = 1361.2$$

$$C_{通} = C_s = 1361.2$$

Then, also by v = 40 km/h, m = 600, b = 3.5, n = 3, j = 1, we have

$$C_{S1} = 1640 \times 1 \times 0.83 \times 1 = 1361.2 = C_{s2} = C_{s3}$$

$$C_{汇} = C_{S1} + C_{s2} + C_{s3} = 4083.6$$

The rest of the algorithm is similar to that.

Establishment of Weight Value. The serial number of the main road is 1, the serial numbers of each traffic road in the community are 2, 3, 4 … n. We now take a community with three routes. It is assumed that five hundred drivers were randomly selected to investigate their selection of these four roads. Through statistical results, we obtain that the weight of road 1 is 0.6, the weight of road 2 is 0.1, the weight of road 3 is 0.2, and the weight of road 4 is 0.1.

Phase and Road Congestion Rate. The phase refers to the right of way at the same time in a certain direction of traffic flow, or the time period of distribution of the right of way. Road congestion rate is an important index to evaluate the number of vehicles on the same road. It is an evaluation standard in traffic control, that is, standardized traffic volume (Fig. 1).

The road congestion rate α can be expressed as follows:

Road congestion rate of phase 2: $\alpha_2 = \max(P_{23}/T_3, P_{25}/T_5)$
Road congestion rate of phase 4: $\alpha_4 = \max(P_{41}/T_1, P_{45}/T_5)$
Road congestion rate of phase 6: $\alpha_6 = \max(P_{63}/T_3, P_{61}/T_1)$
Road congestion rate of import and export: $\alpha = \alpha_2 + \alpha_4 + \alpha_6$
General road congestion control between 1 and 1.4 will not cause traffic jams.

Through the actual survey of the same time period (evening rush peak), the same location in the open area, the traffic volume of each lane, and then find the road congestion rate.

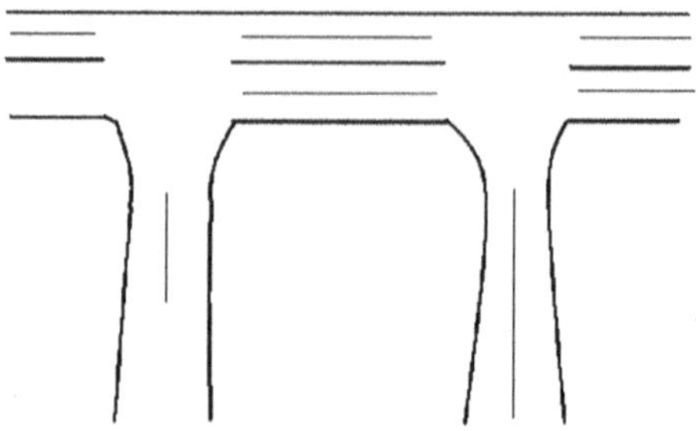

Fig. 1. The thumbnail of the import and export of the district

Before the opening of the community:

The flow of traffic into the lane 3, 5 is $P_{63} = 6100$ and $P_{45} = 3600$.
The flow of traffic into the lane 4, 6 is $P_{63} = 6100$ and $P_{45} = 3600$.
The saturated flow of 3, 5 vehicles is $T_3 = 5500$ and $T_5 = 3400$.
Send into the formula, we get $\alpha_4 = 1.059$ $\alpha_6 = 1.109$.
So $\alpha = 2.168$.

So the rush hour at night will cause congestion.
After the opening of the community:

The flow of traffic into the lane 1, 3, 5 is $P_{41} + P_{61} = 2400$, $P_{23} + P_{63} = 5800$ and $P_{25} + P_{45} = 2400$.

The flow of traffic into the lane 2, 4, 6 is $P_{23} + P_{25} = 4500$, $P_{41} + P_{45} = 2400$ and $P_{61} + P_{63} = 3700$.

So, $P_{23} = 3200$, $P_{25} = 1300$, $P_{41} = 1300$, $P_{45} = 1100$, $P_{63} = 2600$, $P_{61} = 1100$.
The saturated flow of 1, 3, 5 vehicles is $T_1 = 3800$, $T_3 = 5500$, $T_5 = 3400$.
Send into the formula, we can get $\alpha_2 = 0.582$, $\alpha_4 = 0.342$, $\alpha_6 = 0.473$.
So $\alpha = 1.379$.

Therefore, the open area will relieve the congestion phenomenon at rush hours. The opening of the community has a relieving effect on the traffic pressure on the surrounding roads. To a certain extent, the traffic congestion is solved, and a part of the traffic flow has been taken, which breaks the overall closure of the urban district.

3 Improving Traffic Advice

It is suggested that the safe separation measures between vehicle lane and non motor vehicle lane and pedestrian passage should be taken in the construction, so as to prevent accidents and ensure the safety of residents in the community.

It is worthwhile to suggest that open channels should be opened in the community without disturbing the normal work and life of the residents, so as to ensure the width of the driveway out of the entrance and exit, not to cause congestion, and facilitate the convenience of the residents and the traffic of the past vehicles.

As the district is close to the intersection in the project, it is suggested that the relevant departments should set up more traffic lights, or open the overpass or underground passage to ensure the smooth traffic of the main road.

References

1. Li, X.P.: Countermeasures for urban traffic congestion – study on closed district traffic opening. Full text database of Chinese Academic Dissertations, China, p. 238 (2014)
2. Cai, R.D.: Research on opening strategy of residential area based on vulnerability of urban road network, No. 7, p. 35 (2016)
3. Guo, X.J., Liu, K.: A Preliminary Study on Road Traffic Safety Planning in Residential District. Architectural Skills, China, p. 63 (2009)
4. Zhang, F.H.: A Brief Discussion on the Planning of Service Facilities in Residential Quarters. Private Science and Technology, China, p. 492 (2011)
5. Wang, W.: Guide to Urban Traffic Management Planning, p. 4. People's Transportation Press, Beijing (2003)
6. Gao, H.L.: Traffic Capacity of Highway Intersection, p. 43. Science Press, Beijing (2000)
7. Jiang, Q.Y., Xie, J.X.: Selected Case Selection of Mathematical Modeling, p. 63. Higher Education Press, Beijing (2006)
8. Wang, Q.: Automatic control system regulation scheme of textile air conditioning. Sci. Technol. Inf. **22**, 395–396 (2010). (in Chinese)
9. Liu, J.: Advanced PID Control and MATLAB Simulation, 2nd edn. Electronics Industry Press, Beijing (2003). (in Chinese)
10. Li, J., Ren, Q.: Study on supply air temperature forecast and changing machine dew point for variable air volume system. Build. Energy Environ. **27**(4), 29–32 (2008). (in Chinese)

An Improved K-means Clustering Algorithm Based on Distance Evaluation Function and Entropy

Peng Li[1(✉)], Xiaoyun Gong[2], and Qinghua Hao[3]

[1] Department of Information Engineering, Heilongjiang International University, Harbin 150080, China
610984970@qq.com

[2] Department of Computer Science and Technology, Heilongjiang College of Business and Technology, Harbin 150025, China

[3] College of Electronic Engineering, Harbin Institute of Information Technology, Harbin 150030, China

Abstract. In view of the low quality and efficiency of the traditional H-K (Hierarchical K-means) clustering algorithm, and its sensitivity to noise, this paper proposes the concept of distance evaluation function optimizing to the H-K clustering algorithm in the K value, treats the concept of information entropy as the similarity measurement standards to reduce the noise sensitivity of H-K clustering algorithm, and on this basis puts forward an improved H-K clustering algorithm: HKDE (Hierarchical K-Means Clustering Algorithm Based on Distance Evaluation Function and Entropy) algorithm. Based on the results of the simulation experiment, this paper has verified the effectiveness of HKDE algorithm, and has made a corresponding comparison between the traditional H-K clustering algorithm and the newly proposed algorithm in clustering quality, efficiency of the algorithm, and the algorithm of multidimensional data processing. The experimental results show that the algorithm proposed in this paper for the HKDE algorithm can be better applied in data clustering. It has a higher clustering accuracy and efficiency, and can deal with multidimensional data better.

Keywords: Clustering · Hierarchical K-means · Distance evaluation function

1 Introduction

At present, we are in a data era. All aspects of our lives are inseparable from data, from mobile phones, QQ to supermarkets, shopping malls, and banks. At any time, a large amount of data will be generated waiting for analysis and processing. Therefore, data mining technology has also achieved unprecedented development and improvement in this era. From the national economy, the people's livelihood to the production, the daily life, it is inseparable from database mining technology [1, 2].

In related applications of the database, Data Mining is called by many people as knowledge discovery (knowledge discovery in the database). Intelligent data mining finds valuable information knowledge from the database, which not only reflects the development of the past data, but also can predict the future development trend of the

J. Abawajy et al. (Eds.): ATCI 2018, AISC 842, pp. 1210–1218, 2019.
https://doi.org/10.1007/978-3-319-98776-7_149

data, provide decision support information for enterprises, and so on. It is the most commonly used and most important data processing process in data mining related processing technology, and it is also an important data processing tool. Through cluster analysis, valuable laws and information can be extracted from the existing huge data set. Clustering technology has been well developed. There are many kinds of data clustering algorithms, which are widely used in important fields such as customer classification, similarity search, trend analysis, and pattern recognition [3, 4].

2 The Disadvantages and Shortcomings of Traditional Clustering Algorithms

The traditional H-K clustering algorithm uses the advantage of the K-means algorithm and the hierarchical clustering algorithm. It highlights its unique algorithm advantages in the data processing of many data objects. However, with the deepening of data analysis and the increase of the complexity of the data itself, the H-K clustering algorithm also shows some deficiencies and defects in the clustering of some data. For example: (1) The algorithm is relatively sensitive to noise points or outliers, which affects the clustering quality of the algorithm. (2) Value selection of K has a great influence on the clustering quality of the algorithm, so it is necessary to first determine the K value and the initial clustering center, so if you do not determine the K value and the initial clustering center then the time complexity of the algorithm will be higher [5]. (3) As the complexity of the data increases, the clustering algorithm requirements for multi-dimensional and even high-dimensional data processing capabilities become more and more strict and important. However, the traditional H-K clustering algorithm is slightly insufficient in the processing capability of high-dimensional data, and there are major flaws in the time complexity of processing multi-dimensional data. (4) With the arrival of the data era, production and accumulation of large amounts of data directly lead to the massive growth of data. How can we effectively cluster big data in the vast ocean of data? It is very important to quickly and effectively obtain good clustering results. However, the traditional H-K clustering algorithm has a big flaw in this aspect of processing. Even in some special cases, the clustering result is invalid. Aiming at the defects and deficiencies of the traditional H-K clustering algorithm on the data clustering process mentioned above, we can effectively improve the H-K clustering algorithm so that the H-K clustering algorithm can be more effective and applicable to the current cluster and analysis of data, which becomes more important and more meaningful [6, 7].

3 Clustering Algorithm Based on Distance Evaluation Function and Entropy-HKDE Algorithm

In this paper, based on the study of a large amount of literature, we study the problems in the hierarchical K-means clustering algorithm, which is the traditional H-K clustering algorithm, such as: high time complexity, low clustering accuracy and the noise-sensitive problem. And a new clustering algorithm based on distance evaluation function and entropy is proposed: HKDE (Hierarchical K-Means Clustering Algorithm Based on Distance Evaluation Function and Entropy) algorithm.

3.1 Range of K Value

In the clustering algorithm, the range of the optimal number of clusters for data objects can be set to [K_{min}, K_{max}] [8].The range of optimal clustering numbers for data objects formed in this paper will be set to $[2, \sqrt{n}]$ (n is approximately equal to 4).

Definition 1. In the clustering processing space of the data object, the range of the optimal clustering number K is set as:

$$2 \leq K \leq \sqrt{n} \tag{1}$$

3.2 Distance Evaluation Function

In the clustering algorithm, the key to the quality of clustering lies in how the evaluation criteria of the algorithm clustering effect are selected. According to the evaluation rules of the distance evaluation function, we can say that the clustering criterion function selected by such a clustering algorithm is the best one that converges.

$$E = \sum_{i=1}^{k} \sum_{p \in C_i} |p - m_i|^2 \tag{2}$$

In formula (2) E is the sum of the squared errors of all data objects, p is the data object, and it is the average of the clusters.

Definition 2. In the clustering space of the clustering algorithm, the distance between the class and the class is expressed as the sum of the distance from the center of all the classes formed by the clustering algorithm to the center of this clustering space:

$$Dis_{out} = \sum_{i=1}^{k} |m_i - m| \tag{3}$$

In formula (3), Dis_{out} denotes the distance between classes,m denotes the mean of data objects, m_i denotes the mean of all data objects in the class of C_i, and distance calculation uses the Euclidean distance.

Definition 3. In the clustering space of the clustering algorithm $I = \{S, K\}$, the value of the intra class distance is the sum of the interior distances of all the classes formed by the clustering algorithm, and the distance inside the class is the sum of the data inside each class and the distance from the object to the geometric center:

$$Dis_{in} = \sum_{i=1}^{k} \sum_{p \in C_i} |p - m_i| \tag{4}$$

As shown in formula (4), Dis_{in} represents the distance within the class; p represents an arbitrary data object; m_i is the average of all data objects in the class of C_i.

Definition 4. In the clustering space of the clustering algorithm $I = \{S, K\}$, when $Dis_{out} = Dis_{in}$, the K value obtains the best value, so the distance evaluation function is expressed as:

$$F(S,K) = \left|\frac{Dis_{in}}{Dis_{out}} - 1\right| = \left|\frac{\sum_{i=1}^{k}\sum_{p\in C_i}|p - m_i|}{\sum_{i=1}^{k}|m_i - m|} - 1\right| \tag{5}$$

As we have described in detail above, when using the distance evaluation function to evaluate the clustering results of the clustering algorithm, the following conclusion can be drawn: When the distance evaluation function obtains the minimum value, the clustering effect of the clustering algorithm is optimal. We can say that the clustering algorithm chosen by this clustering algorithm is the best one that converges.

That is, the best choice for K is: $Min\{F(S,k)\},\ K = 2, 3, \ldots, n$.

3.3 Information Entropy

In information theory, information entropy is a measure of uncertainty, and the amount of information becomes a decrement of uncertainty, which also makes the concept of entropy, a measure of uncertainty, in current social production and scientific research has a very wide range of applications [4]. Information entropy theory is used to measure uncertainty. For discrete variables, information entropy is defined as:

$$H(x) = -\sum_{i} p(x_i)\log p(x_i) \tag{6}$$

Where $p(x_i)$ is the probability of x_i.

For continuous variable x, information entropy H is defined as:

$$H(x) = -\int f(x)\log f(x)dx. \tag{7}$$

3.4 The Application of Information Entropy in HKDE Algorithm

For cluster i, the probability that a data object in cluster i belongs to a class j is: $p_{ij} = \frac{n_{ij}}{n_i}$, where n_i is the number of data objects in the cluster i, k is the number of classes, and n_{ij} is the number of data objects in the cluster i that belong to the class j. Cluster C_i entropy calculations follow the following formula:

$$E(C_i) = -\sum_{j=1}^{k} p_{ij} \log_2 p_{ij} \tag{8}$$

3.5 HKDE Algorithm

The HKDE algorithm includes two stages: a partitioning clustering algorithm and a hierarchical clustering algorithm.

In dividing the clustering stage, the main purpose is to obtain the K value. First of all, according to the results of a large number of data clustering algorithms and the experience in the clustering algorithm research, the K value is limited to $2 \leq K \leq \sqrt{n}$, which limits the clustering results of the clustering algorithm to a certain extent. Secondly, we chose the data structure of k-d-tree in the selection of the data structure of the algorithm. Its purpose is to improve the efficiency of the algorithm. Finally, the K value is optimized. In the optimization stage, we use the k-means algorithm in the partitioning clustering algorithm. At the same time, in order to obtain the optimal K value, we select the corresponding distance evaluation function. This can not only improve the efficiency of the algorithm, but also can define a range of K values, and use the distance evaluation function to evaluate the distance between data objects and classes, so as to achieve the purpose of optimizing the K value so that the class process can obtain the optimal K value.

In the hierarchical clustering stage, we obtained the optimal K value in the partition clustering stage. At this stage, we use the optimal K value to perform agglomerative hierarchical clustering on the data objects. In the process of clustering, entropy is used as a guide to agglomerate the similarity measure of the hierarchical clustering process. That is, the current data object is added to all clusters, and the increment of the entropy value of each cluster, after the current data object is added, is calculated. By comparing the entropy increments of each cluster, the current data object is added to the cluster with the smallest increment of entropy to ensure that the current data object has the best cluster ownership. At the same time, a clustering threshold ε is introduced to define whether to calculate the entropy increment after the current data object is added to a cluster. The introduction of the probability threshold ε is used to reduce the calculation of the entropy increment. The introduction of entropy reduces the sensitivity of the clustering algorithm to noise points and improves the accuracy of data object clustering. This can make the algorithm get the best clustering quality and the best clustering accuracy. The introduction of the probability threshold ε reduces the running time of the algorithm from a certain extent, which can improve the overall algorithm's operating efficiency.

3.6 Algorithm Description

In the partition clustering stage, based on the application of the K-means algorithm, the distance evaluation function is added, and the K value is optimized by the distance evaluation function. The Algorithm 1 is described as follows:

Algorithm1: Get-K

Input: Data set S

Output: K

(1) Determine the best range of K values: $2 \le K \le \sqrt{n}$

(2) According to the range of K given in (1) and the K-means algorithm, the data objects are clustered.

(3) Construct a k-dimensional from the data objects in the data set S.

(4) Use the distance evaluation function to calculate the $F(S, K)$ value when K takes a different value.

(5) Get the $F(S, K)$ minimum value and record the K value.

(6) Output the optimal cluster number K

In the hierarchical clustering stage, entropy is used as a measure of cluster consolidation, and a probability threshold is introduced to define whether to calculate the entropy increment of the current data object after adding a cluster. The Algorithm 2 is described as follows:

Algorithm2: HKDE

Input: data set S, distance threshold ε

Output: K clusters

(1) Invoking the algorithm Get-K to obtain the K value;

(2) Each data object in S is numbered as a single cluster;

(3) while (the number of clusters is K) do

(4) Randomly select clusters C_i from S; $1 \le i \le n$

(5) Get clusters with distances less than or equal to ε from C_i to the centers of other clusters;

(6) Join C_i to the clusters obtained in (5) separately and calculate the increment of entropy after C_i is added to each cluster;

(7) Find the cluster $C_{\min}$ with the smallest increment of entropy;

(8) Assign C_i to the cluster $C_{\min}$ and renumber it as the cluster $C_{\min}$ number;

(9) end while

(10) return K clusters

3.7 Comparison Analysis of Algorithm

Compared with the traditional HK clustering algorithm, the traditional HK clustering algorithm has a large number of defects both in the clustering quality of the algorithm and the clustering effect of the algorithm. And the efficiency of the algorithm is low with the noise-sensitive problem. The HKDE algorithm proposed in this paper is mainly improved from the following aspects:

1. In the partition clustering stage, the HKDE algorithm introduces a distance evaluation function in the evaluation of the clustering effect of the k-means algorithm. Through this method, the clustering quality in the partition clustering stage is improved accordingly, and then the HKDE algorithm can reach the purpose of optimizing the K value in the H-K clustering algorithm.

2. In the hierarchical clustering stage, the HKDE algorithm introduces the concept of entropy as a metric for similarity to guide the implementation of the agglomerative hierarchical clustering of the algorithm. In the classification of data objects of the agglomerative hierarchical clustering, determine the ownership of data objects through the entropy incremental values. In the clustering of algorithms, the degree of density between data objects and the handling of noise points and outliers are concerned with entropy.

4 Simulation Experiment and Result Analysis

4.1 Data Set

The data set used is UCI's iris, breast-cancer, vehicle, and letter. The experiments mainly compare the clustering efficiency and clustering quality of HKDE algorithm and traditional HK algorithm. The specific parameters of the data set used in the experiment are shown in Table 1.

Table 1. Dataset parameter

Dataset	Name	Attribute number	Data size	Class number
Dataset1	Iris	12	160	2
Dataset2	Breast-cancer	26	920	3
Dataset3	Credit-g	19	6570	6
Dataset4	Letter	35	22186	23

4.2 Experiment Analysis and Comparison

Algorithm Clustering Quality. The accuracy and recall rate of the HKDE algorithm and the traditional HK algorithm's clustering quality were compared and analyzed. The data statistics of the HKDE algorithm and the data statistics of the H-K algorithm are on the Dataset1. The data statistics of HKDE algorithm is on Dataset2, the data statistics of H-K algorithm are shown in Table 2.

In order to compare the differences in clustering quality between the HKDE algorithm and the traditional HK algorithm in depth, the following cluster quality assessment method Q_{APC} was used to evaluate the clustering quality of the two algorithms in the datasets Dataset1, Dataset2, Dataset3 and Dataset4. As shown in Fig. 1.

It can be seen from Fig. 1 that the HKDE algorithm values on the four datasets are significantly higher than the HK algorithm. Therefore, it can be shown that the HKDE

Table 2. Statistical analysis of clustering results of HKDE algorithm on Dataset1 and Dataset2

Dataset	Algorithm	Cluster	Classification 1(91)	Classification 2(67)	Noise Point (2)	Precision	Recall Rate	
Dataset1	HKDE algorithm	1	88	1	1	0.978	0.967	
		2	3	66	1	0.942	0.985	
		合计	91	67	2			
	HK algorithm	1	85	2	2	0.955	0.934	
		2	5	65	1	0.915	0.970	
		合计	90	67	3			
Dataset	Algorithm	Cluster	Classification 2(375)	Classification 3(329)	Noise Point (5)	Precision	Recall Rate	Classification 1(211)
Dataset2	HKDE algorithm	1	205	4	2	1	0.972	0.972
		2	3	369	2	3	0.979	0.984
		3	3	2	323	2	0.979	0.982
		合计	211	375	328	6		
	HK algorithm	1	201	6	3	1	0.953	0.953
		2	5	364	2	6	0.966	0.971
		3	3	5	321	3	0.967	0.976
		合计	209	375	326	10		

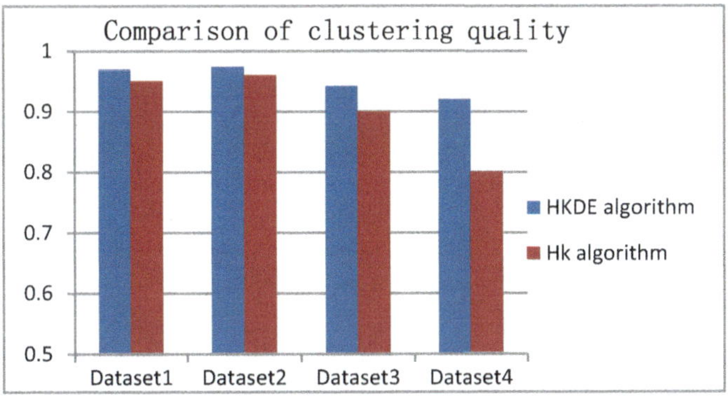

Fig. 1. Comparison of clustering quality

algorithm is higher in cluster quality than the traditional HK clustering algorithm. In addition, we can find that as the amount of data from the data set continues to increase, both algorithms have a declining trend in the clustering quality, and the HK clustering algorithm has a more obvious downward trend, while the HKDE algorithm has a certain degree of decline, but the magnitude of the decline is not very large and maintain good clustering quality. Therefore, the HKDE algorithm is relatively stable in terms of the stability of the clustering quality. In general, the HKDE algorithm has higher clustering quality and relatively stable clustering quality compared to the HK algorithm.

By explaining the environment, data set, clustering effect evaluation methods and other aspects of the algorithm simulation experiment and comparing with the traditional HK algorithm in the aspects of clustering quality, algorithm efficiency and so on, we get the conclusion that the HKDE algorithm has higher clustering quality and algorithm efficiency. The experiment shows that the HKDE algorithm can obtain the best clustering quality in the clustering process.

References

1. Ruiyu, J., Zhen, L.: Hierarchical clustering algorithm with information gain. Microelectron. Comput. 86–88 (2016)
2. Bo, F., Wenning, H.: Optimization of initial clustering center selection by K-means algorithm. Comput. Eng. Appl. 182–185 (2013)
3. Zhiming, X., Peng, W., Yan, H.: Research on the improvement of spectral clustering algorithm for multidimensional data K-means algorithm. Comput. Technol. Dev. 60–64 (2017)
4. Guozhu, W.: Cluster analysis and algorithm research. Mod. Bus. Ind. 105–107 (2017)
5. Liming, B.: A dynamic K-means clustering algorithm for determining K values based on multi-branch trees. Comput. Technol. Dev. 41–45 (2017)
6. Hongxing, H., Yuquan, Z.: Hybrid dynamic clustering algorithm based on partition and hierarchy. Comput. Appl. Res. 51–53 (2011)
7. Li, Z., Weidong, C.: Hybrid clustering algorithm based on partition and hierarchical methods. Comput. Eng. Appl. 127–129 (2010)
8. Ruiyu, J., Jianlin, S.: Determination of optimal clustering number of k-means based on clustering center optimization. Microelectron. Comput. 62–66 (2016)
9. Ge, X.: On how to improve the effectiveness of college English translation teaching. Engl. Teach. **18** (2015)
10. Chen, T.: Effective measures of English teaching translation. Overseas Engl. **20** (2015)
11. Wang, H.: On college English translation teaching models in the internet environment. Engl. Sq. **2** (2016)

Design of LQR Controller Based on Particle Swarm Optimization Algorithm for Aircraft Test Simulation Attitude Control System

Jian Fang(✉) and Wei Li

JiLin Engineering Normal University, Changchun, Jilin, China
757314739@qq.com

Abstract. A mathematical model is established for the self-designed vehicle test simulation attitude control system. On the basis of designing LQR controller based on linear quadratic optimal control method, Particle swarm optimization is used to optimize the parameter matrix of LQR controller. In order to stabilize the attitude control of the aircraft, reduce the response time and frequency of the system. Simulation experiment proves: The stability of the system is improved by the optimized LQR controller using particle swarm optimization, got a good control effect, it has high practical application value.

Keywords: Optimization problem · Particle swarm optimization
Aircraft test system · Attitude control

1 Introduction

The aircraft test simulation attitude control system is a multi-variable and strongly coupled nonlinear system. In this paper, the aircraft as a whole is studied. To abstract it into a ball with a 360° flexible motion, on a horizontal control platform, through the research on the control algorithm of the aircraft test simulation system, it is expected to provide a solution to improve the performance of the vehicle control system.

In this paper, a self-designed simulation attitude control system for aircraft is studied. The mathematical model of the system is established by Lagrange equation. For the mathematical model of the system, design of LQR controller based on particle swarm optimization, the Q and R matrix of LQR controller are optimized, the global optimal solution of Q and R matrix in LQR controller is obtained, then the optimal state feedback control matrix K is obtained. Through the comparison of simulation experiment, the method has better control effect.

2 The Aircraft Test Simulation Attitude Control System Physical Modeling

The structure of the simulation attitude control system for aircraft test is shown in Fig. 1.

J. Abawajy et al. (Eds.): ATCI 2018, AISC 842, pp. 1219–1226, 2019.
https://doi.org/10.1007/978-3-319-98776-7_150

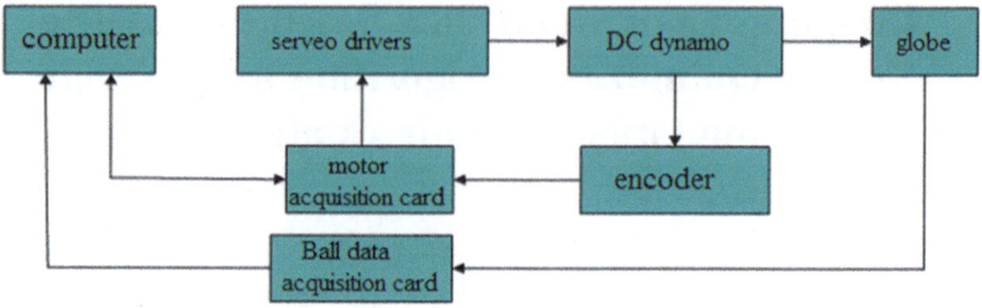

Fig. 1. Working schematic diagram of simulation attitude control system for aircraft test

The process of attitude control of the aircraft test simulation system is as follows: the position of the ball position acquisition card, the motor data acquisition card collects the data of two motor tail encoders, by calculating the displacement of the slant angular displacement and the sphere, angular displacement differential angular, the control quantity is calculated according to the control theory. Velocity, the displacement difference can be obtained, the control quantity is calculated according to the control theory, control to the servo driver, to control the motor, the dc motor is controlled by two servo drives. To control the ball position on the plate, the motor tail encoder is connected to the drive to form a closed loop, thus, the positioning of the ball on the plate and track tracking can be achieved.

Taking the support point of the plate as the origin of coordinates, establish the coordinate system as shown in Fig. 2.

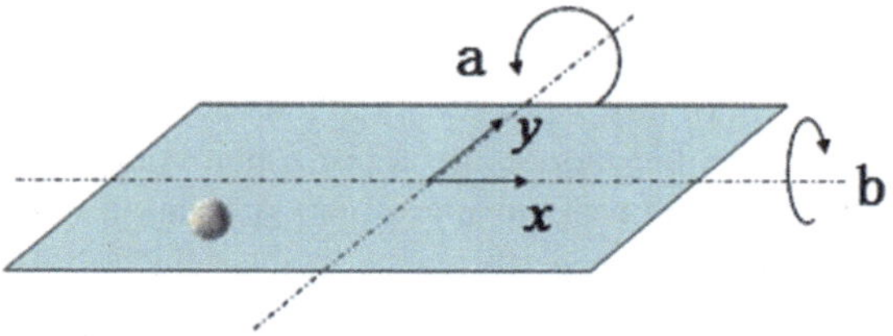

Fig. 2. Coordinate of Ball&Plate system

From the diagram, we can see that the system has four degrees of freedom: The position of the ball (x, y) and the tilt Angle of the plate (a, b). Because the two groups of variables are interrelated, only one of them is independent, make the following assumptions:

Ignore all the friction; Do not consider the area of the board and the rotation Angle limit. In any case, the ball is in contact with the plate; The plate is symmetric in the direction of x and y axis; the ball does not slide on the board and rotate around its vertical axis. The mathematical model of the system is established by Lagrangian mechanics method.

1. The kinetic energy of the sphere includes its own rotational kinetic energy and translational kinetic energy on the plate.

$$T_b = \frac{1}{2}m(\dot{x}^2 + \dot{y}^2) + \frac{1}{2}J_b\left(\omega_x^2 + \omega_y^2\right) \tag{1}$$

Because the sphere doesn't slide during the motion, thus

$$\dot{x} = r\omega_x, \dot{y} = r\omega_y \tag{2}$$

Bring (2) into (1) formula.

$$T_b = \frac{1}{2}\left(m + \frac{J_b}{r^2}\right)(\dot{x}^2 + \dot{y}^2) \tag{3}$$

2. The kinetic energy of the plate includes its own rotational kinetic energy and the rotational kinetic energy of the ball around the plate support.

$$T_p = \frac{1}{2}J_P(\dot{a}^2 + \dot{b}^2) + \frac{1}{2}\left(J_b + m|\Omega \times s|^2\right)(\dot{a}^2 + \dot{b}^2) \tag{4}$$

again $|\Omega \times s|^2 = \frac{(X\dot{a} + y\dot{b})^2}{\dot{a}^2 + \dot{b}^2}$, Plug in to (4)

$$T_p = \frac{1}{2}(J_b + J_p)(\dot{a}^2 + \dot{b}^2) + \frac{1}{2}m(x\dot{a} + y\dot{b})^2 \tag{5}$$

3. Select the zero potential energy point of the coordinate origin, and the potential energy of the plate is zero, and the potential energy of the sphere is:

$$V_b = mg(x \sin a + y \sin b) \tag{6}$$

4. The generalized force is provided by the torque generated by the servo system:

$$Q_X = T_X, Q_y = T_y \tag{7}$$

The combined formula (4), (5) and (6) can get the Lagrangian function:

$$\begin{aligned} L &= T_b + T_p - V_b \\ &= \frac{1}{2}\left(m + \frac{J_b}{r^2}\right)(\dot{x}^2 + \dot{y}^2) + \frac{1}{2}(J_b + J_p)(\dot{a}^2 + \dot{b}^2) + \frac{1}{2}m(x\dot{a} + y\dot{b})^2 - mg(x \sin a + y \sin b) \end{aligned} \tag{8}$$

For Lagrange equation, we take the following four variables:

$$q_1 = x, q_2 = y, q_3 = a, q_4 = b$$

$$Q_1 = Q_2 = 0, Q_3 = Q_x, Q_4 = Q_y$$

And the Eqs. (5) and (6) are brought in $\frac{d}{dt}\left(\frac{\partial L}{\partial \dot{q}_i}\right) - \frac{\partial L}{\partial q_i} = Q_i, \quad i = 1, 2, 3, \ldots, n$. The system of equations are as follows:

$$\begin{aligned} &\mathrm{x:}\ \left(\mathrm{m} + \frac{\mathrm{J_b}}{\mathrm{r}^2}\right)\ddot{x} - m\left(x\dot{a}^2 + y\dot{a}\dot{b}\right) + mg\sin a = 0 && \text{(a)} \\ &\mathrm{y:}\ \left(\mathrm{m} + \frac{\mathrm{J_b}}{\mathrm{r}^2}\right)\ddot{\mathrm{y}} - m\left(x\dot{b}^2 + x\dot{a}\dot{b}\right) + mg\sin b = 0 && \text{(b)} \\ &\mathrm{a:}\ \left(J_b + J_p + mx^2\right)\ddot{a} + m\left(xy\ddot{b} + \dot{x}y\dot{b} + x\dot{y}\dot{b} + 2x\dot{x}\dot{a}\right) + mgx\cos a = T_x && \text{(c)} \\ &\mathrm{b:}\ \left(J_b + J_p + my^2\right)\ddot{b} + m\left(xy\ddot{a} + \dot{x}y\dot{a} + x\dot{y}\dot{a} + 2x\dot{x}\dot{b}\right) + mgy\cos b = T_y && \text{(d)} \end{aligned} \quad (9)$$

Where (a) and (b) two equations describe the motion of the sphere on the plate, the relation between the acceleration of the sphere and the tilt angle of the plate and the angular velocity; (c) and (d) two equations show how the dynamics of the slanted plates are influenced by external forces and the position and velocity of the sphere.

In fact, the input of the system is not a torque T_X and T_y, and this is the angle a and b, this is because the load torque does not affect the position of the motor, based on this assumption, Eq. (9) Chinese (c), (d) can be ignored.

In a steady state, the tablet should be in a horizontal position, and then both angles are equal to 0, because the plate rotation angle range is not big (plus or minus 6°), the sine function can be replaced by its independent variable, the sina $= a, \sin b = b$. In this way, the differential Eq. (9) in the vicinity of the stable state (a), (b) is linearized as type 10(a) and (b).

$$\begin{aligned} &\mathrm{x:}\ \left(m + \frac{J_b}{r^2}\right)\ddot{x} + mga = 0 && \text{(a)} \\ &\mathrm{y:}\ \left(m + \frac{J_b}{r^2}\right)\ddot{\mathrm{y}} + mgb = 0 && \text{(b)} \end{aligned} \quad (10)$$

Take $x, \dot{x}, a, \dot{a}, y, \dot{y}, b, \dot{b}$ is the state of the system, the angular acceleration of the plate is controlled.

Make $x_1 = x, x_2 = \dot{x}, x_3 = a, x_4 = \dot{a}, x_5 = y, x_6 = \dot{y}, x_7 = b, x_8 = \dot{b}$ have following equation

$$\begin{cases} \dot{x}_1 = x_2 \\ \dot{x}_2 = \ddot{x} = -\frac{mg}{(m + J_b/r^2)}a \\ \dot{x}_3 = x_4 \\ \dot{x}_4 = \ddot{a} \\ \dot{x}_5 = x_6 \\ \dot{x}_6 = \ddot{\mathrm{y}} = -\frac{mg}{(m + J_b/r^2)}\mathrm{b} \\ \dot{x}_7 = x_8 \\ \dot{x}_8 = \ddot{b} \end{cases}$$

Make $x = [x_1, x_2, x_3, x_4, x_5, x_6, x_7, x_8]^T = [x, \dot{x}, a, \dot{a}, y, \dot{y}, b, \dot{b}]^T, y = x, u = [\ddot{a}, \ddot{b}]^T$, the state space description equation of the system is

$$\begin{cases} \dot{x} = Ax + Bu \\ y = x \end{cases} \tag{11}$$

Take g = 9.81 m/s^2, the moment of inertia of a sphere $J_b = \frac{2}{5}mr^2$, make $x_x = [x_1, x_2, x_3, x_4]^T, x_y = [x_5, x_6, x_7, x_8]^T$, get the following state-space formula

$$\dot{x}_x = \begin{bmatrix} \dot{x}_1 \\ \dot{x}_2 \\ \dot{x}_3 \\ \dot{x}_4 \end{bmatrix} = A_x x_x + b_x u_x \quad y_x = [1, 0, 0, 0]x_x, \quad A_x = \begin{bmatrix} 0 & 1 & 0 & 0 \\ 0 & 0 & -7.007 & 0 \\ 0 & 0 & 0 & 1 \\ 0 & 0 & 0 & 0 \end{bmatrix}, \quad b_x = \begin{bmatrix} 0 \\ 0 \\ 0 \\ 1 \end{bmatrix}$$

$$\dot{x}_y = \begin{bmatrix} \dot{x}_5 \\ \dot{x}_6 \\ \dot{x}_7 \\ \dot{x}_8 \end{bmatrix} = A_y x_y + b_y u_y \quad y_y = [1, 0, 0, 0]x_y, \quad A_y = \begin{bmatrix} 0 & 1 & 0 & 0 \\ 0 & 0 & -7.007 & 0 \\ 0 & 0 & 0 & 1 \\ 0 & 0 & 0 & 0 \end{bmatrix}, \quad b_y = \begin{bmatrix} 0 \\ 0 \\ 0 \\ 1 \end{bmatrix}$$

The aircraft test simulation system is linearized at the horizontal position, x, y direction is decoupled and structurally consistent, so you just have a qualitative analysis of the x direction. Give the system a unit step signal, system analysis in MATLAB environment, the position of the sphere and the tilt angle of the system is unstable, plate are both divergent, the system is unstable.

3 Optimization of LQR Controller Based on Particle Swarm Optimization Algorithm

This paper adopts LQR control aircraft (ball) attitude, then realize the control of ball balance position output, the LQR optimal control problem is to seek a control quantity $u^*(t)$, the optimal control problem of LQR is to seek a control quantity, so that the system can reach a stable state and ensure that the performance index J takes the minimum value, among

$$J = \int_0^\infty (X^T QX + u^T Ru)dt \tag{12}$$

the Q and R matrices in Eq. (12) are mutually restricted. So the key to design is to find the right matrix Q and R, as long as the Q and R matrices are determined, the state feedback matrix K is determined. But usually in LQR controller design process, the selection of matrix Q and R are rely on experience and the method of trial and error, with strong subjectivity, resulting in the controller design is not perfect, has affected the control effect.

Aircraft simulation test attitude control system state variable $[x_1, x_2, x_3, x_4]^T$ (that is $[x, \dot{x}, a, \dot{a}]^T$), as $[x_1, x_2, x_3, x_4]^T$ particle, the initial value of the particle's position and velocity is randomly generated within a certain range. Fitness function is an important step to optimize the particle swarm optimization, it is the standard for iterative evolution of the whole particle swarm algorithm, this is because we are designing a quadratic optimal control regulator, so we use linear quadratic performance indicator (12) as the fitness function. Because of Q is a 4 by 4 symmetric semi-positive definite matrix. R is a constant positive definite matrix. In order to make the weighted matrix have definite physical significance, we choose the weighted matrix Q as the diagonal matrix, This performance metric can be expressed as:

$$J = \int_0^\infty (q_1 x_1^2 + q_2 x_2^2 + q_3 x_3^2 + q_4 x_4^2 + Ru^2)dt \tag{13}$$

the q_1, q_2, q_3 and q_4 are the weights of the position, velocity, inclination and angular velocity of the sphere. In the objective function R is the square weighting of the control quantity u. According to the principle of particle swarm optimization, in D dimensions, X particles form a group, the position and velocity of the some (i) particle in space:

$$x_i = (x_{i1}, x_{i2}, \ldots, x_{iD}), v_i = (v_{i1}, v_{i2}, v_{i3}, \ldots, v_{iD}), \quad i = 1, 2, \ldots, m,$$

the best position of the some (i) particle is recorded $p_b(i)$, the best place to remember all the particles in a group is $g_b(i)$. The whole particle swarm updates its speed and location by tracking single extremum and optimal values. The particle search process is shown in Eqs. (14) and (15).

$$v_{id}^{t+1} = \omega v_{id}^t + c_1 \alpha (p_{id}^t - x_{id}^t) + c_2 \beta (p_{gd}^t - x_{id}^t) \tag{14}$$

$$x_{id}^{t+1} = x_{id}^t + v_{id}^t \tag{15}$$

Among: ω is the inertial weight; c_1 is the weight coefficient of the optimal value of the particle tracking; c_2 is the weight coefficient of the optimal value of the particle tracking group; α and β is the random number that changes in [0, 1]. p_{id}^t is the individual optimal solution after the t th iteration; p_{gd}^t is the global optimal solution of group in t sub-iteration, in order to make the algorithm more accurate search scope, The velocity of the particle is limited between $[-v_{\max}, v_{\max}]$.

The search process is as follows:

(1) Initialize the particle group. Initialize the particle group to set the acceleration coefficient c_1, c_2, Maximum Generation *gen*, Population size *pop*, The location and speed of the initial search point, the p_b value of each particle is the current position;
(2) Calculate the fitness value of each particle $F_{id}[i]$;

(3) For each particle, use its fitness value $F_{id}[i]$ and individual extremum $p_b(i)$ compare if $F_{id}[i] > p_b(i)$, the $Fit[i]$ replace p_b(i)
(4) For each particle, use its fitness value $Fit[i]$ and global extremum g_b compare, if $F_{id}[i] > p_b(i)$ the $Fit[i]$ replace g_b;
(5) According to formula (14) and (15) update the v_i velocity and place x_i of the particle;
(6) if the ending condition is satisfied (the error is good enough or the maximum number of cycles is reached), return (2) step.

4 Experiment and Conclusion

The initial state of the simulation attitude control system of the aircraft is obtained $[x, \dot{x}, a, \dot{a}]^T = [0, 0, 0, 0]^T$, select $c_1 = c_2 = 1.2$ to experiment, at the same time, the inertia weight formula is selected:

$$\omega^k = \omega_{\max} - \frac{k}{gen}(\omega_{\max} - \omega_{\min}) \tag{16}$$

among them, $\omega_{\max} = 1$, $\omega_{\min} = 0.3$, $gen = 25$ represents the iterative value of algorithm evolution. K is the current evolutionary algebra. The global optimal solution is obtained by programming the above particle swarm algorithm:

$$Q = \begin{bmatrix} -13.0384 & 0 & 0 & 0 \\ 0 & -13.605 & 0 & 0 \\ 0 & 0 & 49.7401 & 1 \\ 0 & 0 & 0 & 10.22166 \end{bmatrix}, \quad R = 1.8682 \tag{17}$$

with the help of MATLAB function $K = lqr(A, B, Q, R)$. The optimal feedback matrix is solved $K = [-13.0384, 50.5125, -13.6605, 10.2216]$.

The dynamic response curves of LQR controller and LQR controller based on particle swarm algorithm are shown in Figs. 3 and 4.

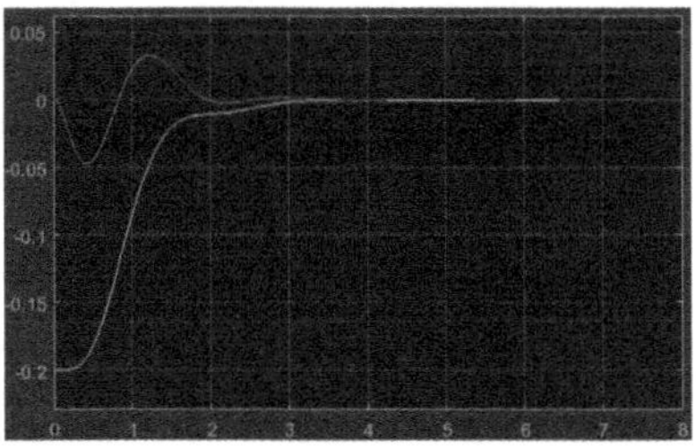

Fig. 3. Displacement and inclination output under conventional LQR controller

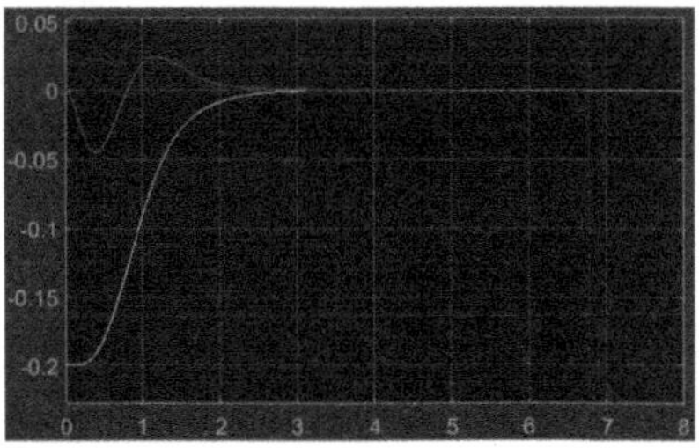

Fig. 4. Displacement and inclination output under LQR controller based on particle swarm optimization

The simulation curves of the controller in Figs. 3 and 4 can be obtained, the simulation curves of the controller in Figs. 3 and 4 can be obtained. LQR control and LQR control of particle swarm optimization algorithm make the system stable. However, the latter algorithm has the advantages of short stable time, small overshoot and low oscillation frequency.

By using particle swarm optimization to select weighted matrix Q, R can overcome the blindness of traditional LQR optimal control for Q and R. This paper is based on the linear model of aircraft simulation test attitude control system, Intelligent search using particle swarm algorithm, progressive optimization. It is not easy to fall into local optimum and easy to implement. Obtain the global optimal solution of Q R, to achieve optimum LQR controller design, and through MATLAB simulation experiment found that design of the optimal LQR controller design response speed faster, less overshoot, also can keep the steady state error is zero, the control effect is better.

Acknowledgments. Jilin provincial education department project «Automatic brake device design of automobile intelligent anti – error» (project number: JJKH2017162KJ). Supported by Program for Innovative Research Team of JiLin Engineering Normal University.

References

1. Zhao, J., Ruan, X.: The LQR control and design of dual-wheel upright self-balance robot. In: 7th World Congress on Intelligent Control and Automation, WCICA 2008, pp. 4864–4869 (2008)
2. Wang, Z., Sun, J., Yue, H.: The inverted pendulum based on LQR optimal control system research. Ind. Instrum. Autom. **3**, 6 (2005). (in Chinese)
3. Gao, Q., Yuan, L., Wu, J.: The four-rotor UAV attitude control based on the new LQR. J. Electron. Opt. Control **0405**(10), 46–52 (2014)
4. Prasad, L.B., Tyagi, B., Gupta, H.O.: Modelling and simulation for optimal control of nonlinear inverted pendulum dynamical system using PID controller and LQR. In: Proceedings of the 6th Asia Modelling Symposium, Bali, Indonesia, pp. 138–143. IEEE (2012)
5. Wu, X.: Double Inverted Pendulum Based on particle swarm optimization control research. Qufu Normal University, Shan-dong (2012). (in Chinese)
6. Aleksy, M., Korthaus, A., Schader, M.: Use Java and the CORBA realization distribute type system. J. Pingxiang Coll. **4**, 104–105 (2005)
7. Titus, J.: ECN Technical editor: the eclipse of stand. J. Zhongkai Agrotech. Coll. **19**(2), 32–35 (2006)
8. Clay Richardson, W., Avondolio, D.: The Java high class weaves a distance: JDK 5. Sci. Technol. Book Rev. **3**, 17–18 (2006)
9. Vuong, J.L.: A semantics-based routing scheme for grid resource discovery. In: E-Science: First International Conference on E-Science and Grid Computing, pp. 58–70, 90 (2005)

Website Fingerprinting Attack Method Based on DNS Resolution Sequence

Kai Wang(✉), Liyun Chen, and Xingkai Chen

Shijiazhuang Branch of Army Engineering University, Shijiazhuang, China
Wangkai8a8@gmail.com

Abstract. Aiming at the problem that the traditional fingerprinting attack feature extraction and attack method of HTTPS encrypted websites are difficult to effectively apply when using the CDN (Content Delivery Network) caching technology. In this paper, a fingerprinting attack method which uses the DNS resolution sequence in CDN as fingerprint and compares the fingerprint similarity of the websites using the longest common sub-sequence algorithm is proposed. The experimental results show that the average accuracy of the proposed method is 85.6%, and the standard fingerprint remains valid for a certain period of time, which can effectively identify HTTPS encrypted websites in CDN cache.

Keywords: Website fingerprinting attack · Encrypted network traffic
DNS resolution · HTTPS protocol

1 Introduction

With the development of the internet and the increasing awareness of people's privacy, more and more websites use HTTPS protocol based on SSL/TLS to protect privacy and personal information security. HTTPS protocol has good compatibility, and existing web applications can be seamlessly migrated to HTTPS [1]. On the other hand, the application of encryption technology makes the visibility of the network reduced, which brings difficulties to network management, traffic analysis and network prediction.

Although the encryption mechanism can hide the transmission content between the client and the server, it cannot effectively hide the size, timing, direction and other information of the data packet, and the website fingerprint attack can use this information to guess the website visited by the client [2].

2 Background

The traditional website fingerprinting attack object is the response data all from the same server, and the packets have the same source address or destination address. However, with the emergence of a large number of real-time internet applications and the need to improve the user access experience, Content Delivery Network (CDN) is widely used. The CDN streaming media service uses a domain name resolution system

J. Abawajy et al. (Eds.): ATCI 2018, AISC 842, pp. 1227–1233, 2019.
https://doi.org/10.1007/978-3-319-98776-7_151

(DNS) to redirect users to a nearby mirror server to obtain high-bandwidth, low-latency performance and provide users with high-quality media services [3].

There have been some studies on fingerprinting techniques for websites this years. Kim [4] and others proposed a method to identify the service by using the SessionID of the SSL/TLS packet header and the IP address of the server as a feature. This method divides the service into SSL/TLS traffic detection, signature-based traffic identification and ID-IP-based service identification is a three-level process. In this experiment, the method has achieved 95% accuracy for SSL/TLS encrypted traffic classification and 90% accuracy for SSL/TLS encryption service identification. However, after using of CDN technology, the resources of the website come from a number of different servers, and it is difficult to establish an accurate correspondence between ID-IP and services. Husák et al. [5] use the list of cipher suites contained in the ClientHello packet in the SSL/TLS protocol handshake process as the fingerprint, establishes the SSL/TLS cipher suite list and the HTTP User-Agent relationship dictionary, and uses this method in a real-time network environment. Under the experiment, the recognition rate of the network client reached 95.4%, but different encryption methods in the cipher suite supported by different software environments are different. This method is greatly affected by different operating systems and browsers, and does not apply to most environments. At the same time as the development of offensive methods, some scholars have also conducted research on countermeasures against fingerprints, focusing on methods such as maximum packet filling [6], traffic shaping [7], and HTTP obfuscation [8]. These methods have filled encrypted data packets. Processing, deformation, or disguise to prevent attacks, but for HTTPS encrypted traffic, the clear text negotiation process before the establishment of the HTTPS encryption channel is not encrypted.

3 Threat Model

In this section, we first introduce the principles of website fingerprinting using DNS resolution request sequences as fingerprints, further analyze the impact of the fingerprint system's impact on the fingerprint accuracy, and give a method for evaluating the accuracy of fingerprinting on a website.

3.1 Threat Principle

When accessing a website using the HTTPS protocol, the communication message is encrypted, but the DNS resolution process before handshake of the encryption protocol uses clear text communication. The CDN distributes site usually stored resources (such as pictures, CSS, and JavaScript files) in servers which in different geographic locations. When users access web pages, they load resources from different servers to improve access speed and optimize user experience. As shown in Fig. 1, when the client accesses the webpage, the browser requests the resources of the different domain name CDN servers to load resources in the rendering order, and the process generates a large number of DNS query messages.

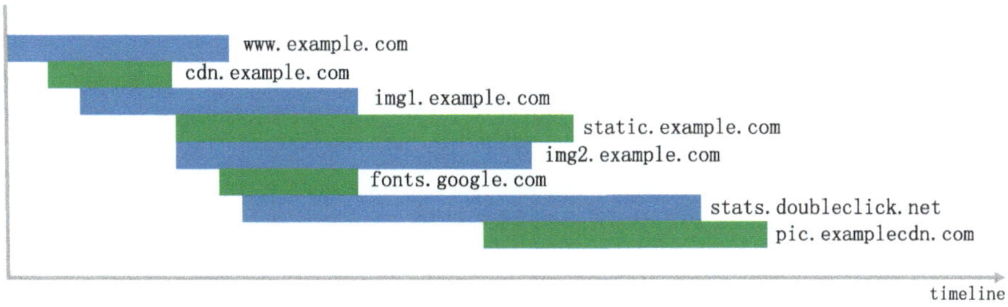

Fig. 1. DNS resolve sequence

During the page access process, the client generates a DNS query sequence, and it can be page fingerprint, which can be expressed as:

$$S = (d_1, d_2, \cdots, d_n) \tag{1}$$

Each page of the website is loaded with different resources, resulting different DNS requests, and the link location in source code (such as linked CSS file, JavaScript file code) written in the Document Object Model (DOM) [9] is different, and the loading order when the browser renders pages is different, so each web page has a relatively unique DNS request sequence.

3.2 Fingerprint Analysis

Although the DNS request sequence of the website page is relatively unique, often multiple software in the operating system run simultaneously to generate other DNS requests and form noise information. Network speed differences and different browser kernels cause DNS request packets to be out of order for different page rendering methods. The lack of data packets caused by caches at various levels in the network has a certain influence on the accuracy of fingerprints. The following specific analysis is performed.

Noise. The client often has multiple applications running at the same time and sending network data packets continuously. Other applications also generate DNS query messages when they connect to the server to exchange data. In the fingerprinting attack based on the DNS query sequence presented in this paper, it belongs to noise information. Expressed as:

$$S_1 = (d_1, d_2, \ldots, v, \ldots, d_n) \tag{2}$$

DNS Packets are Out of Order. In this paper, the DNS query sequence is used as a fingerprint, that is, an uplink data packet. Different browser kernels have different processing modes for rendering pages and resource loading methods, which leads to different sending order of resource request data packets, and results in out-of-order DNS query data packets. Expressed as:

$$S_2 = (d_1, d_2, \ldots, d_{i+1}, d_i, \ldots, d_n) \tag{3}$$

DNS Cache Impact. DNS caches exist widely in browsers, systems, ISP servers, etc. The cache time varies from site to site and affects DNS requests. TTL (Time to Live) is an important factor affecting the DNS cache time. In the CDN network, a smaller TTL (Time to Lives) is usually set for DNS response to achieve load balancing [10]. Due to the impact of caching, some DNS resolutions do not send query requests to the DNS server, resulting in missing DNS query sequences. Expressed as:

$$S_3 = (d_1, d_2, \cdots, d_{i-1}, d_{i+1}, \cdots, d_n) \tag{4}$$

3.3 Fingerprint Comparison

Fingerprint comparison is to determine the most likely identity class of an unknown fingerprint by measuring the similarity between an unknown fingerprint and a known fingerprint [11]. Through the above analysis, under the influence of noise, out-of-order and cache, the selected sequence of the DNS request sequence fingerprints still remains in order. The Longest Common Subsequence (LCS) algorithm can quickly and efficiently compare the website fingerprints. The similarity of defining the DNS query sequence x and y is:

$$similarity(x, y) = \frac{|lcs(x, y)|}{\max(|x|, |y|)} \tag{5}$$

where $|lcs(x, y)|$ represents the length of the longest common subsequence of x and y, $\max(|x|, |y|)$ represents the maximum length of the sequence x and y.

3.4 Accuracy Evaluation

In order to accurately reflect the accuracy of this feature and attack method, precision is used to represent the accuracy of page recognition. The precision rate is defined as:

$$precision = \frac{TP}{TP + FP} \tag{6}$$

4 Experimental Evaluation

In order to evaluate the fingerprinting attack method proposed in this paper, according to the model deployment experiment shown in Fig. 2, data collection is performed in the Internet environment.

The experiment uses a Python program to collect DNS data during requests. On account of the asynchronous loading technology used by sites, we use selenium and Phantomjs schemes to access the page, simulating the use load the full page resource.

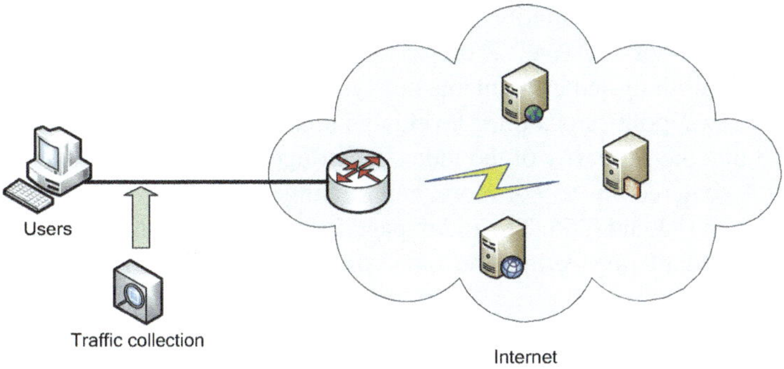

Fig. 2. Fingerprinting attack model

Within a month, continuous random access to the selected secondary web pages under HTTPS-encrypted sites. The number of single page visits was between 150 – 1100 times. The extracted DNS resolve sequences and is cleaned to form the dataset used in this experiment.

4.1 Experimental Results

Page Fingerprint Matching. The similarity of the attacked page fingerprint and the standard fingerprint is calculated by formula (5). For reference, the page fingerprint is compared with other secondary pages under the site. The effect on yahoo.com is shown in Fig. 3.

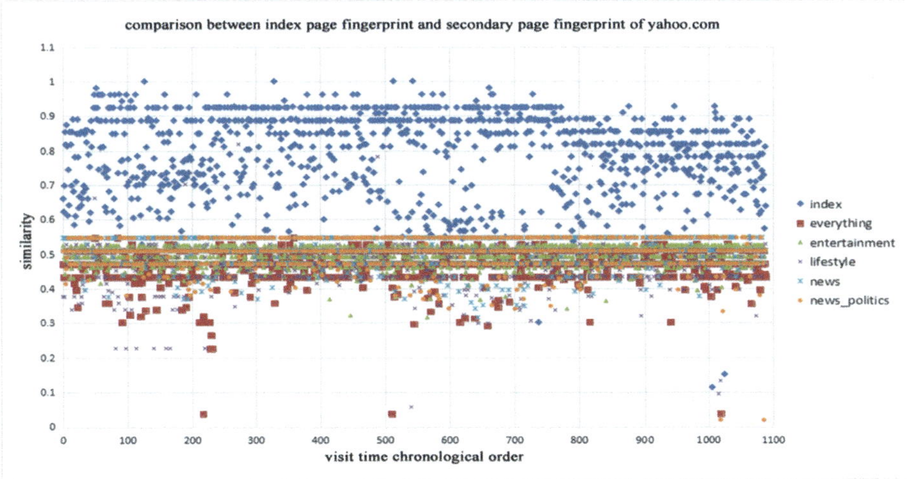

Fig. 3. Yahoo.com index page fingerprint and secondary page fingerprint similarity comparison

Figure 3 shows the similarity of yahoo.com between index page standard fingerprint and the rest of the five secondary pages, where index is the home page of yahoo.com, everything, entertainment, lifestyle, and news are secondary pages under yahoo.com, news_politics is a third-level page under yahoo.com. It can be clearly seen from Fig. 3 that the similarity of the index page fingerprint and the standard index page fingerprint is centered on 0.55 or more, and the fingerprint similarity with other second page is between 0.3 and 0.55. The index page fingerprint has a high similarity with the index page standard fingerprint, and other page fingerprints have a higher degree of discrimination.

The horizontal axis of Fig. 3 is sorted according to the actual access time. During the one month of the experiment, the similarity between the index page fingerprint and the standard index page fingerprint is not significantly reduced, and the fingerprint similarity can still be quite different from other pages.

Fingerprinting Attack Accuracy. Comparing the fingerprint extracted from the data set with the standard fingerprint, and calculating the similarity of all page fingerprints under the same site. The accuracy was calculated using the formula in (6). The experimental results obtained are shown in Table 1. The results show that this method has an excellent recognition rate for some websites and the average of the precision rate calculated for all experiments shows that the method has achieved an average accuracy of 85.6%.

Table 1. The accuracy of website fingerprinting attack

Website	Subpage number	Visited times	Average precision
stackoverflow.com	3	3268	0.919804102
cn.bing.com	4	618	0.79876636
open.163.com	8	1368	0.818522202
play.google.com	5	3267	0.835629017
snapchat.com	4	3399	0.931470588
wordpress.com	5	6519	0.722831843
xueqiu.com	6	1008	0.858134921
yahoo.com	5	5445	0.83030303
youtube.com	3	3267	0.998775635
mafengwo.com	5	802	0.855126175

5 Conclusion

According to the characteristics of CDN distribution process using DNS redirection and the HTTPS protocol is not encrypted during DNS resolution. A web site fingerprinting attack method based on DNS resolution sequence is proposed. The longest common subsequence algorithm is used to compare the similarity of page fingerprint. Experiments show that the attack method can effectively identify the attacked pages, the success rate reaches 85.6%, and the standard fingerprint can maintain the attack accuracy rate for a long time.

In the next step, based on this method, further research will be conducted to further improve the accuracy of the attack, user identification after network address translation, and fingerprinting attack technology corresponding to the user.

References

1. Pan, W.B., Cheng, G., Guo, X.J., et al.: Review and perspective on encrypted traffic identification research. J. Commun. **37**(9), 154–167 (2016)
2. Cai, X., Nithyanand, R., Wang, T., et al.: A Systematic approach to developing and evaluating website fingerprinting defenses. In: ACM SIGSAC Conference on Computer and Communications Security, pp. 227–238. ACM (2014)
3. Qin, Z., Zhou, F., Li, L.M.: The Impact of DNS on CDN streaming performance. J. Univ. Electron. Sci. Technol. China **42**(4), 577–580 (2013)
4. Kim, S.M., Goo, Y.H., Kim, M.S., et al.: A method for service identification of SSL/TLS encrypted traffic with the relation of session ID and Server IP. In: Network Operations and Management Symposium, pp. 487–490. IEEE (2015)
5. Husák, M., Čermák, M., Jirsík, T., et al.: HTTPS traffic analysis and client identification using passive SSL/TLS fingerprinting. Eurasip J. Inf. Secur. **2016**(1), 30 (2016)
6. Dyer, K.P., Coull, S.E., Ristenpart, T., et al.: Peek-a-boo, I still see you: why efficient traffic analysis countermeasures fail. In: IEEE Symposium on Security and Privacy, pp. 332–346. IEEE Computer Society (2012)
7. Wright, C.V., Coull, S.E., Monrose, F.: Traffic morphing. An efficient defense against statistical traffic analysis. In: Network and Distributed System Security Symposium, NDSS 2009, San Diego, California, USA, pp. 237–250. DBLP (2009)
8. Luo, X., Zhou, P., Chan, E.W.W, et al.: HTTPOS: sealing information leaks with browser-side obfuscation of encrypted flows. In: Network and Distributed System Security Symposium, NDSS 2011, San Diego, California, USA. DBLP (2011)
9. W3C. Document Object Model [EB/OL], 19 January 2005. https://www.w3.org/DOM/
10. Ager, B., Smaragdakis, G., Uhlig, S.: Comparing DNS resolvers in the wild. In: ACM SIGCOMM Conference on Internet Measurement 2010, Melbourne, Australia, pp. 15–21. DBLP (2010)
11. Gu, X.D., Yang, M., Luo, J.Z., et al.: Website fingerprinting attack based on hyperlink relations. Chin. J. Comput. **38**(4), 833–845 (2015)

Transformer Fault Diagnosis Model Based on Discrete Hopfield Neural Network

Shuo Ding(✉), Zhongyu Cheng, Qinghui Wu, Fang Zhang, and Youlin Yang

College of Engineering, Bohai University, Jinzhou 121013, China
dingshuo2004@sina.com

Abstract. To overcome the low accuracy of traditional transformer fault diagnosis methods, a new method is proposed combining analysis of dissolved gas (DGA) in oil with discrete Hopfield neural network (DHNN). The eigenvectors of transformer fault diagnosis were obtained via improved Rogers three-ratio method, then a fault diagnosis model of the discrete Hopfield neural network was established. To further test the generalization performance of the established model, a simulation test was done by taking examples of 4 major types of fault diagnosis of the main transformer in a certain substation. The simulation result shows that the fault diagnosis model based on discrete Hopfield neural network has high accuracy, fast speed and good generalization performance, and the practicability and validity of the proposed method are also verified.

Keywords: Discrete Hopfield neural network
Improved Rogers three-ratio method · Transformer · Fault diagnosis

1 Introduction

Transformer is an important equipment of any power system, and its operation state directly affects the security and stability of the whole system. Traditionally transformer fault diagnosis adopts the method of chromatographic analysis of dissolved gas in oil, but the dissolved gas analysis (DGA) in oil has two deficiencies in coding impairment and critical value criterion [1–4]. With the development of neural network and data mining technology in recent years, improved Rogers three-ratio method provides a new solution to the problem of fault diagnosis of transformer based on neural network. Among various artificial neural network types, BP neural network (BPNN) is the most typical and most widely applied neural network model [5–7]. References [8–11] all judged transformer fault states based on BPNN automatic diagnosis system, and the diagnosis accuracy was improved to some extent. However, the topology of BPNN is difficult to determine, and its convergence speed is lower, also it is easy to fall into local minima [12, 13].

Because of the strong associative memory of Hopfield neural network, it has been widely used in the field of pattern recognition with much success. Therefore, in this paper discrete Hopfield neural network (DHNN) is adopted to diagnose 5 states of a main transformer in a substation, and improved Rogers three-ratio method is also used

J. Abawajy et al. (Eds.): ATCI 2018, AISC 842, pp. 1234–1239, 2019.
https://doi.org/10.1007/978-3-319-98776-7_152

to obtain the input vectors to simplify the complex process of model establishment and low accuracy in diagnosis. At the same time the effectiveness of the DHNN diagnosis model is also tested.

2 Improved Rogers Three-Ratio Method

Presently, the most widely used method for transformer internal fault diagnosis is dissolved gas analysis (DGA) in oil. It can recognize the internal or potential failure of a transformer by analyzing the composition and concentration of dissolved gases in an oil-immersed transformer under load condition and fault condition. Traditional DGA methods include the characteristic gas judgment method, the Rogers three-ratio judgment method, the Rogers four-ratio judgment method and the improved Rogers three-ratio judgment method. Because the low accuracy of the former three methods in practical use, in this paper, the one with the highest accuracy, improved judgment Rogers three-ratio method is adopted to get the input feature vectors of neural network fault diagnosis model by analysis of three pairs of ratio between five different gases dissolved in the transformer oil.

3 Basic Principles of Discrete Hopfield Neural Network Algorithm

Hopfield neural network is a single-layer feedback network. It can be divided into a continuous type and a discrete type according to different activation function. DHNN uses binary type activation function, thus often applied in associative memory, whereas continuous Hopfield neural network is often applied in calculation optimization. Because the essence of transformer fault diagnosis is mode classification, this paper adopts DHNN to identify the fault state of a transformer.

The input sample matrix of n-order DHNN is set to be $X = (X_1, X_2, X_3,\ldots, X_n)^T$, Each neuron node in DHNN has two states, 1 or −1. The calculation of input weight $l_k(t)$ of the kth neuron at time t is as formula (1):

$$l_k(t) = \sum_{k=1}^{n} w_{kj} \cdot x_k(t) - q_k \quad j = (1, 2, \ldots, n) \tag{1}$$

w_{kj} is the connection weights between neurons; q_k is the threshold value of neuron k; $x_k(t)$ is the state of the kth neuron at time t. Then the calculation of $x_k(t + 1)$, the state of the kth neuron at the next moment is as formula (2):

$$x_k(t+1) = \mathrm{sgn}(l_k(t)) = \begin{cases} 1, & l_k(t) \geq 0 \\ -1, & l_k(t) < 0 \end{cases} \tag{2}$$

Let $z_k(t)$ be the output value of the kth neuron. Make $z_k(t)$ feedback to the input end. The output value of the network at the next moment, $z_k(t + 1)$ can be obtained. After several iterations according to formula (3), the network can finally reach a stable state.

The network output value meets formula (4), and $u_j(j \neq i)$ in formula (3) is the output value of the jth neuron.

$$\begin{cases} z_k(t+1) = \mathrm{sgn}(l_k(t)) = \mathrm{sgn}\left[\sum_{k=1}^{n} w_{kj} \cdot u_j(t) - q_k\right] \\ z_k(0) = x_k(0) \end{cases} \tag{3}$$

$$z_i(t+1) = z_i(t) \tag{4}$$

4 Establishment of DHNN Transformer Fault Diagnosis Model

Diagnosis of a substation main transformer in a city in China under normal state and 4 types of fault states, namely overheat in low temperature, overheat in high temperature, low energy discharge and arc discharge were taken as examples to check the feasibility of using DHNN for transformer fault diagnosis. 12 groups of data were measured in each state, and 60 groups of data were actually measured. The training sample set of DHNN (Some Samples) is shown in Table 1.

Table 1. Training sample set of DHNN (some samples)

Number	C_2H_2/C_2H_4	CH_4/H_2	C_2H_4/C_2H_6	Fault code
1	0.0311	0.7660	0.7647	I
2	0.0057	0.2509	0.2571	I
3	0.0928	0.8604	0.6647	I
4	0.0000	1.4475	1.6829	II
5	0.0012	2.3333	1.5000	II
6	0.0000	1.8125	1.9091	II
7	0.0046	6.9655	8.2797	III
8	0.0075	6.1071	9.6667	III
9	0.0050	7.1233	8.5714	III
10	1.2700	0.1763	8.5000	IV
11	1.1197	0.2400	8.3571	IV
12	1.0476	0.1698	9.0001	IV
13	3.4165	0.3891	12.4647	V
14	5.3480	0.4746	11.7778	V
15	4.9828	0.6287	12.7603	V

4.1 Setting Memory Standard Mode of DHNN

The code table of 5 states was obtained by the improved Rodgers three-ratio method, as shown in Table 2. First, the presorted mode was encoded and the coding rules are as follows: the original scope of diagnostic limits suggested by improved Rodgers three-ratio method was further divided according to the actual special distribution of the sample set, and the coding rules of 5 states were obtained as shown in Table 2. If one of the sample index values falls into the corresponding ranges of one or several grades in Table 2, the corresponding neuron state is set to be 1, otherwise it is set to be −1. '•' indicates that a sample data is within the specified range, while 'o' indicates otherwise. Then the memory standard model was obtained when 1 and −1 are taken as the values. The diagnosis results could be classified into 5 types and there were all together 3 evaluation index, so evaluation result was taken as the vertical axis, and evaluation indices were taken as the horizontal axis. A DHNN composed of 15(5 × 3) neurons was formed. The standard memory modes of DHNN diagnosis model is shown in Fig. 1. Fault code I stands for normal state; Fault code II stands for overheat in low temperature; Fault code III stands for overheat in high temperature; Fault code IV stands for low energy discharge; Fault code V stands for arc discharge.

Table 2. Coding rules for 5 states

Fault code	C_2H_2/C_2H_4	CH_4/H_2	C_2H_4/C_2H_6
I	<0.1	<1.0	<1.0
II	<0.1	>1–<3	>1–<3
III	<0.1	≥ 3	>3–<10
IV	>1–<3	≥ 3	>3–<10
V	≥ 3	<1.0	≥ 10

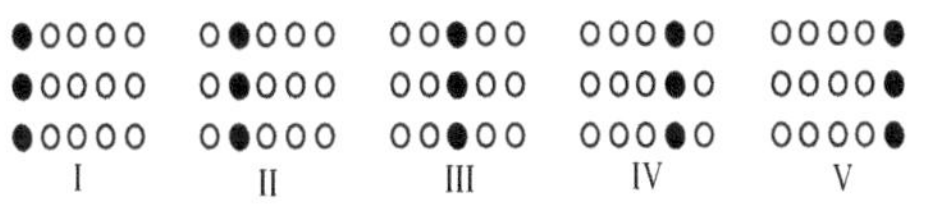

Fig. 1. Standard memory modes of DHNN diagnosis model.

4.2 Weight Determination of DHNN

First set up DHNN through using the function *newhop* in MATLAB R2014a, then trained it to get its memory weight matrix $\boldsymbol{w}$ and its threshold vector q.

4.3 Initialization of Input Samples

The training sample index value in Table 1 was converted into a pattern that could be recognized by DHNN according to the method described in Sect. 4.1, that is to encode the original data. The training sample was converted into a binary mode to be put into DHNN by using the function *sim* in the toolbox. After many times of iteration, the output state of DHNN no longer changed, that is, the network finally achieved a stable state, which indicates that the diagnostic model had been established.

5 Simulation Test

Tested the practical diagnosis ability of the established DHNN model. Among 15 samples in each state, one sample was randomly selected, and the test sample set (a total of 5 samples) was established. Then, the five samples were sorted by category, and the test sample set of DHNN was shown in Table 3.

Table 3. Test sample set of DHNN

Number	C_2H_2/C_2H_4	CH_4/H_2	C_2H_4/C_2H_6	Fault code
1	0.0132	0.6258	0.5621	I
2	0.0004	1.8644	1.6973	II
3	0.0057	6.7320	8.8393	III
4	1.1458	0.1954	8.6191	IV
5	4.9584	0.4975	12.3342	V

According to the coding rules described in Sect. 4.1, the test sample index values in Table 3 were converted into a two-value pattern recognized by DHNN. The coding of the diagnostic indices of 5 test samples were obtained as shown in Fig. 2.

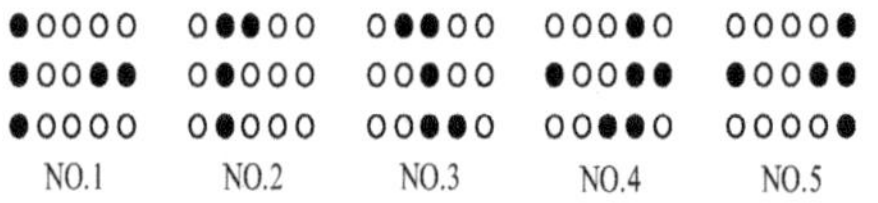

Fig. 2. Coding of the diagnostic indices of 5 test samples

Then simulation function *sim* was employed. DHNN finally converged after 20 times of iteration, and the diagnosis results of DHNN model is shown in Fig. 3. As can be seen from Fig. 3, the diagnosis results are completely consistent with the actual results, indicating that the established DHNN classification model has strong classification ability and generalization ability. Simulation experiment also finds that if the decrease in the number of evaluation index can reduce the complexity of the network model, but this will result in impaired classification accuracy and generalization ability. Therefore, to further improve the classification accuracy of the model, we can increase the number of evaluation index.

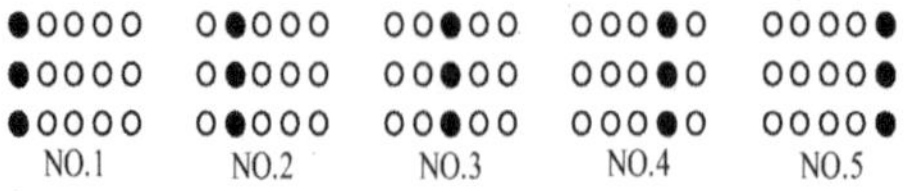

Fig. 3. Diagnosis results of DHNN model

6 Conclusion

In this paper, a model of transformer fault diagnosis was constructed based on DHNN, combining the improved Rogers three-ratio judgment method. And the model was used to diagnose the test data of the main transformer in a substation. The simulation experiment results show that the diagnosis accuracy of this model is high, and its structure is simple. Besides, it is also easy to implement, one can simply type the sample data into a binary memory model without normalization processing in advance, which greatly shortens the network training time, reduces the iteration times and improves the modeling efficiency. Above all, it is feasible to apply DHNN to fault diagnosis of power transformers.

Acknowledgements. This work is supported by National Natural Science Foundation (NNSF) of China under Grant 61401044 and Natural Science Foundation of Liaoning province under Grant 201602012.

References

1. Wu, H., Zhou, N., Wang, C.: Research for transformer fault diagnosis and MATLAB simulation based on RBF neural networks. Sci. Technol. Eng. **10**(5), 47–51 (2010)
2. Ding, S., Chang, X., Wu, Q., et al.: Study of transformer fault diagnosis based on GRNN and DGA. Electron. Meas. Technol. **37**(5), 142–146 (2014)
3. Hu, Q., Tan, X., Akella, M.R.: Finite-time and fault-tolerant spacecraft attitude control with torque saturation. AIAA J. Guid. Control Dyn. **40**(10), 2524–2537 (2017)
4. Zhao, X., Yang, H., Karimi, H.R., et al.: Adaptive neural control of MIMO nonstrict-feedback nonlinear systems with time delay. IEEE Trans. Cybern. **46**(6), 1337–1349 (2016)
5. Yin, S., Yang, H., Kaynak, O.: Sliding mode observer-based FTC for Markovian jump systems with actuator and sensor faults. IEEE Trans. Autom. Control **62**(7), 3551–3558 (2017)
6. Li, Z., Xu, Y., Fei, Z., Misra, S., et al.: Stability analysis and stabilization of Markovian jump systems with time-varying delay and uncertain transition information. Int. J. Robust Nonlinear Control **28**(8), 68–85 (2018)
7. Li, B., Hu, Q., Yu, Y., Ma, G.: Observer based fault tolerant attitude control for rigid spacecraft. IEEE Trans. Aerosp. Electron. Syst. **53**(5), 2572–2582 (2017)
8. Hu, Q., Shao, X., Guo, L.: Adaptive fault-tolerant attitude tracking control of spacecraft with prescribed performance. IEEE/ASME Trans. Mechatron. **99**, 1 (2017)
9. Zhao, X., Shi, P., Zheng, X.: Adaptive tracking control for switched stochastic nonlinear systems with unknown actuator dead-zone. Automatica **60**, 193–200 (2015)
10. Hu, Q., Li, L., Friswell, M.I.: Spacecraft anti-unwinding attitude control with input nonlinearities and velocity constraint. AIAA J. Guid. Control Dyn. **38**(10), 2042–2050 (2015)
11. Yin, S., Qiu, J., Gao, H., et al.: Descriptor reduced-order sliding mode observers design for switched systems with sensor and actuator faults. Automatica **76**, 282–292 (2017)
12. Gong, R., Ma, L., Zhao, Y., et al.: Fault diagnosis for power transformer based on quantum neural network information fusion. Power Syst. Prot. Control **39**(23), 79–84, 88 (2011)
13. Ding, S., Wu, Q.: Research on inverse model based on ANN and analytic method for induction motor. Autom. Control **5**(4), 356–370 (2011)

Spindle Fault Prediction Based on Improved Fuzzy Neural Network Algorithm

Guangwen Zhou, Chunyu Mao, and Mei Tian(✉)

Jilin Engineering Normal University, Changchun, China
zgw_zyl@sohu.com, maochunyu_2005@126.com,
32593829@qq.com

Abstract. The early mechanical failure of the spindle in the CNC machining center has the characteristics of concealment and complexity. In order to quickly and accurately identify the potential mechanical failure of the spindle, this paper combines PSO with strong global search capability, fast searching speed, and strong fault tolerance of the fuzzy neural network. With the characteristics of strong adaptability, an early mechanical failure prediction method based on fuzzy logic, RBF neural network and particle swarm optimization is proposed. In order to optimize the structure parameters of the fuzzy neural network, the speed updating formula and inertia weight of the particle swarm were improved, and an improved PSO optimization fuzzy neural network model for early failure prediction of the main shaft was established. The experimental results show that this model has higher accuracy and generalization ability for the prediction of early mechanical failure of spindle.

Keywords: Mechanical failure · CNC · PSO · Fuzzy neural network

1 Introduction

CNC machining centers is a high degree of automation, the complex structure of advanced processing equipment, is a knowledge-intensive, complex mechanical and electrical products. CNC machining center spindle is a major part of the production process, it was found CNC machining center spindle has a mechanical problem mis-diagnosed, long maintenance cycle and other issues in production [1]. To be able to find fault in advance to avoid large spindle failure, the paper HL1100 Vertical motorized spindle CNC machining centers for the study of the vibration machining centers, noise and temperature detected by detecting various signals and trends forecast, can grasp the operational status of equipment, we found that variation, predict the likelihood of early failure of the spindle to prevent timely detection of abnormal operation signal, so that timely adjustments.

2 Acquisition Method of Early Fault Signal

CNC machining center spindle when due early mechanical failure, various signals relatively weak, which gives the signal acquisition and processing has brought a lot of difficulties. Emerson's company of PeakVue is a new diagnostic technique for rolling

J. Abawajy et al. (Eds.): ATCI 2018, AISC 842, pp. 1240–1248, 2019.
https://doi.org/10.1007/978-3-319-98776-7_153

bearings and gear fault diagnosis development, it is Emerson's patented technology. By PeakVue technology can find abnormal mechanical vibration signal potential especially in the early and subtle abnormal signal. And early this abnormal signal energy is generally low, through the conventional vibration spectrum analysis, often hidden in the vibration spectrum of the underlying background energy which can't be found [2, 3]. PeakVue can be used to separate these signals low energy periodic damage, the causes of these signals are mostly early wear of rolling bearings and gears, and can capture and retain the true impact of these signals, it is possible to damage the roots and show a more accurate response severity of the problem.

3 Method of Early Failure Prediction

3.1 Agency of Fuzzy Neural Network

Fuzzy neural network and fuzzy logic and neural networks with their respective advantages, has great advantages in dealing with non-linear, fuzziness [4]. Taking into account the RBF neural network can overcome the BP neural network training for a long time, easy to fall into local optimum shortcomings, and in function approximation and classification ability, learning speed and other aspects are better than BP network, RBF fuzzy logic and neural networks establish fuzzy RBF neural network, the structure shown in Fig. 1.

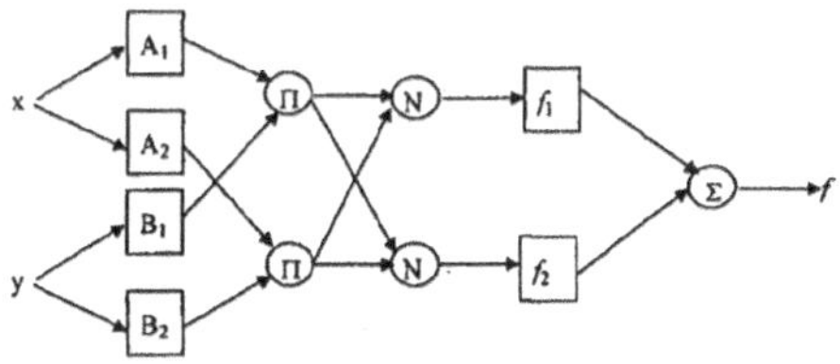

Fig. 1. Structure of fuzzy RBF neural network

(1) Layer1: fuzzy input variables, the output is:

$$o_i^1 = \mu_{A1}(x) \tag{1}$$

According to the form of the membership function selected, the corresponding parameters can be set, called the antecedent parameters. For example, select common Gaussian membership function, the formula (2) in all $\{\sigma_i, d_i\}$ value is composed of former members parameter sets, the learning process in the sample, adjust each parameter fuzzy RBF neural network capable of adaptive [5].

$$\mu_{A1}(x) = \exp\left[-\frac{\|x - d_i\|^2}{\sigma_i^2}\right] \tag{2}$$

(2) Layer 2: the output is the product of the input signal, in the sense that the rules of sample activation strength.

$$o_i^2 = \varpi_i = \mu_{Ai}(\sigma)\mu_{Bi}(y) \tag{3}$$

(3) Layer 3: normalized intensity of activation of each rule.

$$o_i^3 = \varpi_i = \frac{\omega_i}{\omega_1 + \omega_2} \tag{4}$$

(4) Layer 4: deblurring compute nodes, each node to calculate the corresponding Output Rules. It consists of all {ai bi ci} parameter set consisting of pieces called after parameter set.

$$o_i^4 = \varpi_i f_i = \varpi_i(a_i x + b_i y + c_i) \tag{5}$$

(5) Layer 5: Calculate the output of all the rules sum.

$$o_i^5 = f = \sum_i \varpi_i f_i = \frac{\sum_i \varpi_i f_i}{\sum_i \varpi_i} \tag{6}$$

Structure of Fuzzy Neural networks can be seen from this structure contains the fuzzy rule inference, clarity three steps, with a clear sense of fuzzy logic [6].

3.2 Improved Particle Swarm Optimization

PSO has strong global search capability, and the algorithm is simple and easy to implement, less parameters need to be adjusted, it is widely applied in various fields [7]. The basic principle is: first initialize the D solution space the size of a group of the population is m particles, wherein the velocity and position of the i-th particle are $v_i = (v_{i1}, v_{i2}, \ldots, v_{iD})$ and $x_i = (x_{i1}, x_{i2}, \ldots, x_{iD})$, and then by the fitness function to calculate the fitness value Pf particles, in order to determine the population's personal best (P_b) and groups of extreme value (g_b). In the optimization process, the particles precisely keep track of P_b and g_b of the speed (v_i) and position (x_i) to update, the update formula is:

$$v_{ij}(t+1) = wv_{ij}(t) + c_1\varepsilon_1\left(p_{bj}(t) - x_{ij}(t)\right) + c_2\varepsilon_2\left(g_{bj} - x_{ij}(t)\right) \tag{7}$$

$$x_{ij}(t+1) = x_{ij}(t) + v_{ij}(t+1) \tag{8}$$

Where i = 1, 2, …, m; particle represents the i-th; j = 1,2, …, D; the j-dimensional space; c_1, c_2 learning factor; t is the number of iterations; ε_1, ε_2 uniform distribution between 0 and 1 random number; w is the inertia weight.

The PSO shows that in the late insufficient particle algorithm local digging ability, easy to fall into local optimum, and therefore make the following improvements:

(1) Improved Particle Swarm Own Cognitive Strategies
After the optimization process of the particles, the particles themselves cognitive component only consider their own personal best, in order to better learn from the experience of other optimization particles, the average of all the individual extremes instead of individual extreme [8, 9], improvements speed iterative formula is:

$$v_{ij}(t+1) = wv_{ij}(t) + c_1\varepsilon_1\left(p_{meanj}(t) - x_{ij}(t)\right) + c_2\varepsilon_2\left(g_{bj} - x_{ij}(t)\right) \tag{9}$$

$$p_{meanj}(t) = \left(p_{1j}(t) + p_{2j}(t) + \ldots + p_{mj}(t)\right)/m \tag{10}$$

(2) Improved Inertia Weight
Particle swarm algorithm, inertia weight selecting a serious impact on the convergence rate, in order to increase the particle swarm preliminary global ability to explore and local ability to explore later, the use of S-shaped function, so that the particles in the initial optimization, the middle and late 3 stages with different flight speed: the higher the initial V_i, V_i rapid decline in the mid-late V_i constant and complete the final convergence. Formula is as follows:

$$w = (w_{max} - w_{min})/(1 + e^{(2e*t/tmax-e)}) + w_{min} \tag{11}$$

3.3 Improved PSO Fuzzy RBF Neural Network Algorithm

RBF FNN in the application, is easy to fall into local optimal and there is no reasonable method for determining the fuzzy rules, prone redundant rules, which restricts the wide application [7] of the network. PSP is to be optimized. First, we set the number of particles of PSO η, and the inertia weight w, learning factor c1, range c_2, particle velocity and position of the range, the maximum number of iterations $t_{max} = 100$. Second, determine the particle optimization fuzzy RBF neural network parameters c_{ij}, b_{ij} and mapping relations w_{ko}; and then, the speed and position of the random initialization particle; then, the definition of fitness function f(x) = E(x), calculate individual fitness value, coming in on the fuzzy neural network training set generated mean squared error (MSE) is configured to adapt to the function of formula (12) to calculate its fitness value. Where: y_i is the target value; $\widehat{y}_i$ is an output value of the model; η is the number of samples. Finally, the calculated values of the fitness required to decide whether to update. If necessary, according to the formula (10), (11) to update the particle velocity and position, until its optimization times reaches t_{max} or training error is less than E_{min}; later if the training is completed error does not reach E_{min} continues to iterate until E_{min} or t_{max} when the algorithm ends, optimal output fuzzy RBF neural network model [10–12]. Specific processes shown in Fig. 2

$$\mathrm{MSE}(\widehat{y}_i) = \frac{1}{\eta}\sum_{i=1}^{\eta}(y_i - \widehat{y}_i)^2 \tag{12}$$

4 CNC Machining Centers of Motorized Spindle Tool Fault Diagnosis Based on Improved PSO Fuzzy Neural Network

Since the CNC machine tool structure is more complex, the failure in the form of diversification, and therefore HL1100 model CNC machining center motorized spindle fault diagnosis system as an example to verify the proposed method is feasible and effective.

4.1 Failure Mode and Failure Analysis

Fault Feature E = {E1, E2, E3, E4} there are: E1 amplitude at the critical speed becomes large, a peak, E2 vibration frequency of the second harmonic of the rotation frequency, there are also multiple frequency vibration; maximum vibration mainly in misalignment couplings on both sides of the bearing, vibration increases with load; spectrum showed radial second harmonic and fourth harmonic vibration significantly. E3 spindle is weak vibration or noise, E4 and the fundamental frequency of vibration signals have a certain relationship. From the above, the main reason for these failures led to F = {F1, F2, F3, F4} there: F1 spindle system unbalanced, F2 spindle system

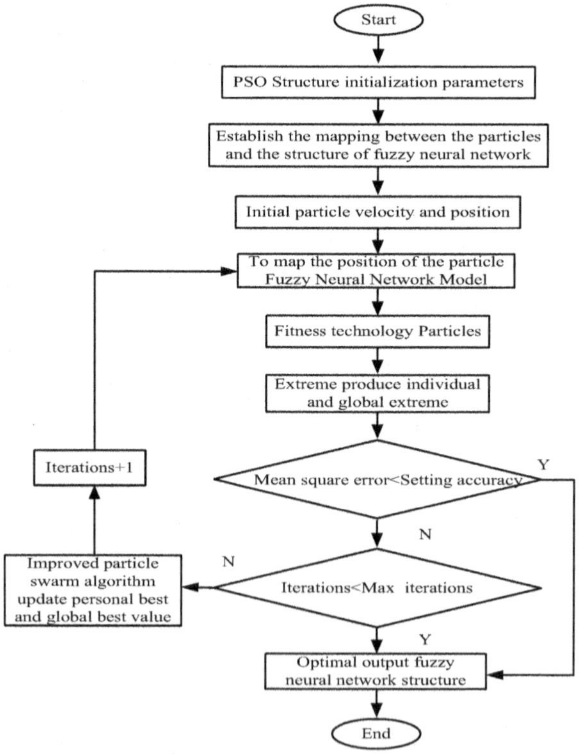

Fig. 2. Improved PSO fuzzy RBF neural network algorithm processes

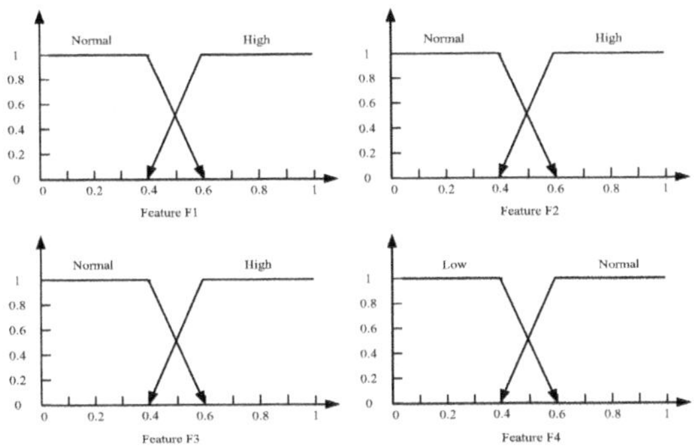

Fig. 3. Membership function distribution

misalignment, F3 spindle cracks, F4 rolling bearing fault. Moreover, the relationship between these symptoms and causes of complex, non-linear and obvious coupling.

4.2 Fault Fuzzy and Training Samples Obtained

The improved PSO optimized fuzzy RBF neural network model for fault diagnosis of CNC machine tools, first determine the degree of importance according to the fuzzy rule and failure characteristics and reasons for data representation blur. Fault features F1–F3 with the "normal" and "high" to describe the two fuzzy sets, "low" and "normal" characteristics of the two modes combined with F4 set to represent, according to experts' experience, the use of trigonometric functions as an input variable membership function, membership function for each input variable distribution shown in Fig. 3.

Combined with relevant expertise, obtain a degree of presence Cause fuzzy description (Table 1); thereby obtaining fault characteristics and causes of fuzzy rules corresponding library, the training samples. To obtain more diagnostic model generalization ability, its fuzzy rules should be as comprehensive as possible (Table 2).

Table 1. Fuzzy description.

The degree of presence cause	Membership
Must	0.8–1
Possible	0.6–0.8
Unclear	0.4–0.6
Not too possible	0.2–0.4
NO	0–0.2

4.3 Confirm the Diagnosis Model Structure

By the actual fault characteristics and causes of apparent input layer nodes diagnosis model is 4, corresponding to four fault characteristic; output layer 4 nodes, respectively, with corresponding cause of the malfunction, the output value in the range [0, 1]

Table 2. Symptom fuzzy relationship with the corresponding

Fault no.	E1	E2	E3	E4	F1	F2	F3	F4
1	0	0	0	0	0	0	0	0
2	0.9	0	0	0	1	0	0	0
3	0	0.9	0	0	0	1	0	0
…	…	…	…	…	…	…	…	…
29	0	0	0.9	0.9	0	0	0	1
30	0	0.9	0	0.9	0	0	1	1

representatives credibility of failure; and a fault fuzzy membership grade 5, and therefore determine the structure of the fuzzy RBF neural network diagnostic model for 4-20-35-4, thereby determining the dimensions of each particle in the particle group $20 \times 2 + 25 \times 4 = 140$-dimensional [12], particle swarm the remaining parameters are set as follows: the number of population m = 30, 100 times the number of iterations, the precision is 0.002, c1 = c2 = 2; particle velocity range is [−0.1, 0.1], the position of the range [−1, 1], wmax = 0.9, wmin = 0.4.

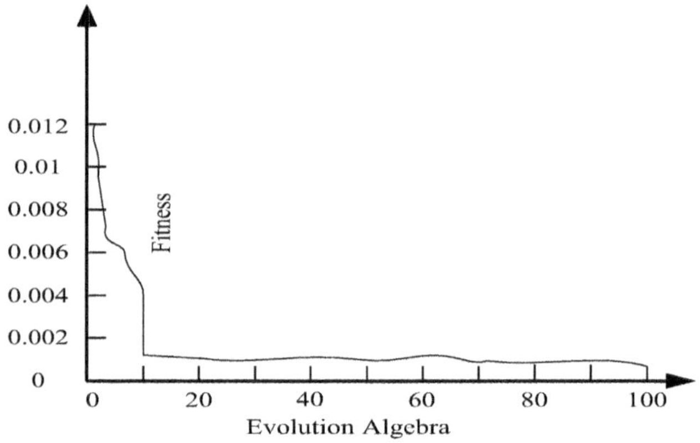

Fig. 4. Iterative process adaptation curve

Table 3. Diagnostic model test output and expected output

Fault no.	Test output				Expected output				Fault feature
	C1	C2	C3	C4	F1	F2	F3	F4	
1	0.93	0.12	0.20	0.26	1	0	0	0	E1
2	0.29	0.90	0.11	0.23	0	1	0	0	E2
3	0.22	0.19	0.92	0.16	0	0	1	0	E3
4	0.23	0.29	0.32	0.96	0	0	0	1	E4
5	0.13	0.23	0.95	0.96	0	0	0	1	E3, E4
6	0.27	0.92	0.32	0.96	0	1	0	1	E2, E4

4.4 Comparison of Results and Model Training

Training model, randomly selected group of samples in Table 2 for 30 in 20 groups as training samples, 10 groups as a test sample, the model in Fig. 2 Improved particle swarm optimization process RBF fuzzy neural network is trained until the end of the algorithm. Practical Fault Diagnosis Model Training Improved Particle Swarm algorithm iterations fitness curve is easy to know from Fig. 4, the proposed method particle swarm after 10 iterations it converges to 0.001; and improved fuzzy RBF PSO optimized Neural network converges fast, high precision, after 23 iterations precision step training fuzzy RBF neural network will reach the goal set by the accuracy of 0.001; ultimately, output six groups of test samples with the expected results of such Table 3. Table 3 can know from the table, only test output number 4, a set of fault category can't be accurately obtained, and the remaining five groups fault can be accurately identified [13].

5 Conclusion

Fuzzy Neural Network combines the advantages of fuzzy logic and neural networks, neural networks overcome the subjectivity and black box of fuzzy logic rules, enabling accurate identification of NC machine tool failure. The text of the improved PSO algorithm from the particle velocity iteration formula and inertia weight update strategy has been improved, and thus the structural parameters of fuzzy RBF neural network optimization, the establishment of an improved particle swarm optimization fuzzy RBF neural network CNC machine tool spindle servo system fault diagnosis model, to achieve an accurate identification of the spindle servo system failure. Experimental results show that the improved PSO early fault diagnosis of CNC machining center spindle fuzzy RBF neural network knowledge to speak clearly and quickly training, to more accurately determine the cause of failure to achieve optimization.

References

1. Zhang, B., Duan, L.L.: CNC machine tool fault diagnosis and repair technology. Hunan Agric. Mach. (2013)
2. Li, D., Yao, P., Bianchi, T.S., et al.: Fault diagnosis and recovery for a CNC machine tool thermal error compensation system. J. Manuf. Syst. **19**(6), 428–434 (2000)
3. Lee, J.-H., Yang, S.H.: J. Virol. **85**(6), 2723–2732 (2011)
4. Wu, D., Shao, J.: Based on machine fault diagnosis CNC ant colony algorithm and neural network. Mech. Des. Manuf. **2013**(1), 165–167 (2014)
5. Ming, L.: Research on CNC machine fault diagnosis based on the GSO-BP neural network. Mach. Des. Res. (2014)
6. Duan, M.: Based on PSO optimized fuzzy RBF neural network learning algorithm and its application. Contemp. Educ. Theory Pract. **2010**(1), 101–104 (2010)
7. Qin, P.: Establishment and application of Chlorella growth model based on optimizing BNN by genetic algorithm. J. Anhui Agric. Sci. (10) (2011)
8. Yang, D.M., Chen, M., Qing-Chao, W.U.: Research on CNC machine fault diagnosis based on KPCA-PSO-RBF neural network. Mach. Des. Manuf. (2016)

9. Wang, J., Wu, Q., Qin, B., et al.: Prediction with oxygen. PSO support vector machine BOF. Foundry Technol. **8**, 1806–1809 (2014)
10. Zhang, C.: On improved PSO algorithm based on fuzzy neural network. Harbin Institute of Technology, Harbin (2014)
11. Fu, G.X.: CNC machine fault diagnosis strategy and maintenance. Adv. Mater. Res. **722**, 228–232 (2013)
12. Ma, C.G., Wang, J.G., Li, Y.Z.: The remote monitoring management and fault diagnosis system of the distributed open numerical control system. Key Eng. Mater. **589–590**, 746–751 (2014)
13. Zhang, L., Zhai, J.: Fault diagnosis for oil-filled transformers using voting based extreme learning machine. Clust. Comput. 1–8 (2018)

Application of Big Data Analysis in the Automation Security of Power Grid Dispatching

Li Zi(✉)

Shenyang Institute of Technology, Fushun, Liaoning, China
121384978@qq.com

Abstract. This paper puts forward the problems of the power dispatching automation in China, and expounds the application of big data technology in China's power dispatching automation system. In this paper, the way to construct the security of the power dispatching automation system is also proposed, and the security of physical layer and network layer are elaborated in detail.

Keywords: Dispatching automation · Big data · Security

1 Introduction

The rapid development of today's society is accompanied by the rapid development of the electric power industry. The development of electric power industry is inseparable from the updating of the power dispatching automation technology, and the amount of data in power dispatching automation system also increases. The dispatching automation systems used by our electric power enterprises are different, therefore, the information exchange between them is very complex. For the total dispatching of provincial network, the traditional data management mode can no longer satisfy the current system requirements. With the emergence of big data, the dispatching automation system can effectively carry out sustainable development based on it. Big data can be processed from information collection, storage, analysis, mining and other aspects, so as to fundamentally realize the automatic control of power dispatching and the rapid development of electric power enterprises.

The electric power industry in China is combined with the automation and informatization in the process of today's social development. The intelligent power dispatching system carries the functions of information flow, power flow and business flow, which makes the electric power industry in our country have the best advantage to develop in the big data age.

J. Abawajy et al. (Eds.): ATCI 2018, AISC 842, pp. 1249–1254, 2019.
https://doi.org/10.1007/978-3-319-98776-7_154

2 The Security Problems of Power Dispatching Automation System in China

2.1 Firewall Security Problem

When installing the network firewall, the power dispatching automation system needs to intercept some malicious software, so as to ensure the safety of the system. However, in the process of firewall installation, some electric power enterprises did not fully consider the needs of the power dispatching automation system, which had a great impact on the daily use of the customers. In addition, some electric power enterprises did not set up according to the actual situation of the customers in the process of setting up the firewall. In the process of use, there will be some malicious program fire hacker attacks and so on. These problems will directly affect the use safety of the power dispatching automation system.

2.2 System Security Problems

At present, there are still some shortcomings in the process of development of the electric power industry in China and the use of dispatching automation system. The system structure is rather chaotic and lacks a certain integrity, which is very difficult to realize the safety management of power dispatching automation. At the same time, in the process of constructing the power dispatching automation system, its security vulnerabilities are not well considered and lack complete safety defense devices.

2.3 Management Problems

There are still many shortcomings in the process of building power dispatching automation system in China's power enterprises. For example, when constructing the network security of power dispatching automation system, the power company does not understand the system, which leads to some problems such as incomplete and unreasonable system in the process of construction. In that way, the power dispatching automation system can not solve these security problems. In addition, some electric power enterprises are not in place for safety management, and there is no high sense of system security, which leads to problems in the actual operation process, which will threaten the life safety and property safety of professional staff.

Techniques of Dealing with Big Data

2.4 Normalizing Data Structure

If you want to do well in the safety of the power dispatching automation system, it is indispensable to build a unified data platform, as shown in Fig. 1. In the process of dealing with data, the normalized data structure is needed. In the process of normalizing data structure, several points are demanded as following:

(1) Under the operating mode, the data information in the power dispatching automation system should be unified, and the information types of each transformer substation should be corresponded to data.

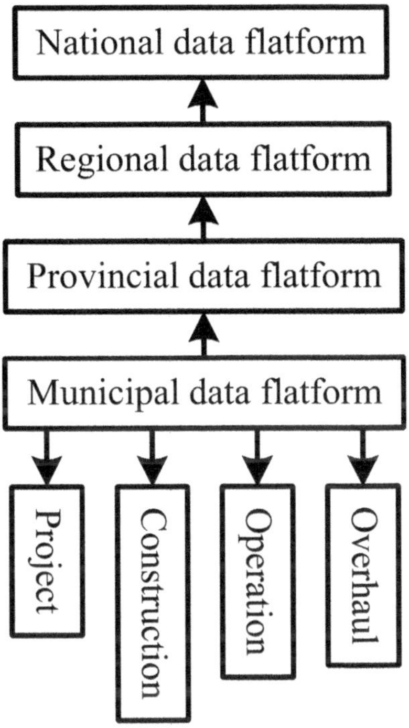

Fig. 1. Unified data platform

(2) Ensure that in the process of uploading data information, all the power companies' needs are met and the data should be unified with the data in State Grid Corporation of China.
(3) Under the current mode, the data information should be programmed, the big data age should be used to combine operating data of power dispatching with power planning data and the supporting business in the power dispatching automation system and its production and run should be integrated to ensure the utilization safety of this system and make the power dispatching automation system widely applied in people's life.

After the achievement of constructing data platform, it is necessary to transform the data information in various professional departments to improve the reporting and analysis of data information, so as to ensure the security of the system.

2.5 Application of Data Mining Technology

In the process of mining the power dispatching automation system, the modern learning system can be used to stare the development, mining and use of the data, and connect the traditional data information to ensure the accuracy of the data information and ensure the safety of the dispatching automation system. In view of the characteristics of the power dispatching automation system, the expansion method is used to deal with the system, and the analysis and calculation is made by combining the central processing unit and the graphics processor structure, so as to ensure the safety of the power dispatching automation system and the safety of the people's daily use of electricity, which will fundamentally improve the social and economic benefits of our country, improve the people's life and promote the rapid development of the power industry.

In addition, the semi supervised and semi learning forms are used to ensure the accuracy of the use of the data. At the same time, the data information structure can be combined to make the visualization analysis of the data information so as to achieve the final data mining effect.

3 The Construction of the Security of the Power Dispatching Automation System

In the era of big data, the security and defense system for the power dispatching automation system is an overall structure. Under the process of constructing this system, people need to design according to the power dispatching automation system. If some shortcomings are encountered in the process of designing, they need to improve them in time in order to make sure its security when the automatic system is used. There are some ways to construct secure power dispatching automation system as following:

3.1 Realizing the System Security

Solving the security problems of the power dispatching automation system becomes an inevitable trend in power industry in the future. The security and defense system for the power dispatching automation system should be improved and its safety utilization should be also guaranteed. The shortcomings, defects and chaos existing in the operation system should be solved in order to make sure its integrality to build a marvelous foundation for its use later on.

3.2 Realizing the Security of Its Physical Layer

The physical layer of the power dispatching automation system is mainly composed of the equipment security and the communication security. People are able to examine the security and reliability of the connection route of the power dispatching automation system by using the communication equipment. What's more, operators need to guarantee the security of transmission medium and the accuracy of operating software. They need to guarantee the safety of the communication devices in use and the power dispatching automation system in use.

3.3 Realizing the Network Safety

We are supposed to do our work in protecting the power dispatching automation system in network security layer to reduce the malicious and disease attack, so as to ensure its safety utilization and accelerate the development speed of power industry.

In the process of constructing the network safety and defense system, it is necessary to design a scientific and reasonable defense system in the purpose of ensuring its safety utilization to keep the network safe. In the process of transmission, access, utilization, analysis and search of data information in the power dispatching automation system, this system is expected to control data information strictly, in order to ensure the security of data information and material in utilization and also in sharing and avoid the invasion of spiteful program and virus and guarantee the security of the power dispatching automation system. In this way, the safe and stable environment of using electricity can be provided, thus promoting the development of our power industry.

3.4 Realizing the Security of Management

In the process of managing the power dispatching automation system, we need to keep the equipment in use safe. If some problems are discovered when examined, we are allowed to take some efficient measures. We need to implement the safety management system of the power dispatching automation system, and manage the related departments in the power dispatching automation system. Besides, the effective strategies need to be designed and also people need to make use of human resources to construct the power dispatching automation system in a better way.

With the development of today's social economy, there is actually potential safety hazard when the power dispatching automation system is utilized. Only when this problem is resolved can our power industry have a healthy development. Meanwhile, we should enhance the network construction of the power dispatching automation system and distribute the power circuit resources reasonably, so as to meet the electricity use needs of people. What's more, taking advantage of big data is an effective way to effectively carry out the comprehensive engineering of power dispatching automation system, and ensure that power data have certain visualization application in the process of use. The big data can become important backbone of this system and reserve more power energy for this system, thus accelerating the development of power industry and enhancing social and economic benefits.

4 Conclusions

The safety and implementation of power dispatching automation under the construction of big data can effectively guarantee the people's safety of using electricity. I hope our professional and technical personnel will strengthen the research on the safety and implementation of power dispatching automation under the construction of big data.

References

1. Sun, G.: The development and application of power grid dispatching. Electron. Technol. Softw. Eng. **2**(5), 167 (2016)
2. Li, G., Wang, X.: Development and design of integrated power dispatching automation system based on data warehouse. Autom. Instrum. **10**(10), 26–29 (2016)

3. Shizhao, N., Zheng, W., Wangmo, P., Yuan, N., Peng, L.: Big data prediction of durations for online collective actions based on peak's timing. Phys. A **55**, 130–139 (2018)
4. Fan, M.: The cultural elements in English translation teaching. Engl. Sq. (20) (2014)
5. Wang, H.: The problems and countermeasures of English translation teaching in college of traditional Chinese medicine. Overseas Engl. (20) (2014)
6. Watts, D.J.: Small Worlds. The Dynamics of Networks Between Order and Randomness. America: 56A2579BB6,l999-05-13
7. Aleksy, M., Korthaus, A., Schader, M.: Use Java and the CORBA realization distribute type system. J. Pingxiang Coll. **4**, 104–105 (2005)
8. Titus, J.: ECN technical editor: the eclipse of stand. J. Zhongkai Agrotech. Coll. **19**(2), 32–35 (2006)
9. Clay Richardson, W., Avondolio, D.: The Java high class weaves a distance: JDK 5. Scientific & Technology Book Review, No. 3, pp. 17–18 (2006)
10. Li, J., Vuong, S.: A semantics-based routing scheme for grid resource discovery. In: E-Science: First International Conference on E-Science and Grid Computing, pp. 58–70, 90 (2005)

Application of Big Data Mining Technology in Intelligent Safe Production on Cloud Computing Platform

Hongsheng Xu[1,2(✉)] and Ganglong Fan[1,2]

[1] Luoyang Normal University, Luoyang 471934, China
85660190@qq.com
[2] Henan Key Laboratory for Big Data Processing and Analytics of Electronic Commerce, Luoyang 471934, China

Abstract. Clustering and classification are two of the most important technology in data mining; they were respectively used to detect data and decision space to feature space dependent. The paper describe analysis of intelligent association rule extraction technology based on big data mining method firstly, then discusses application of cloud computing and Internet of things technology in intelligent security. The paper presents application of big data mining technology in intelligent safe production on cloud computing platform. The final experiment shows that the proposed method of big data mining is very effective.

Keywords: Big data mining · Cloud computing · Intelligent safe production Clustering · Association rule

1 Introduction

At present, our country in the field of production safety supervision is still facing basic data preparation is not sufficient, the problem of database construction to be perfect, mainly contains the following three aspects: (1) The basic data preparation is not sufficient, the database construction needs to be improved; (2) the lack of data analysis tools, information disclosure is not enough; (3) the talent is not sufficient, lack of professional talents [1].

Data mining technology as a new subject, starting in twentieth Century eighty, compared with the current utility, there have been differences in the nature of the study, at first, big data is mainly for the development of artificial intelligence technology, in other words, technical requirements of data mining is relatively high, is the process of innovation and transformation of data that is, since the goal of data hiding, and mining are potential value, to help the operator to dig out the data information effectively from a pile of miscellaneous, fuzzy and random database; data mining application in the commercial aspects of the law is to find out the information contained in the data and the value of information from a large amount of data, to provide intellectual credentials very important for decision makers.

Big data based modeling of complex problems often have extremely complex structure, which requires high data mining algorithms can in accordance with the

J. Abawajy et al. (Eds.): ATCI 2018, AISC 842, pp. 1255–1262, 2019.
https://doi.org/10.1007/978-3-319-98776-7_155

requirements of the task automatically or human-computer interaction from large data extraction and organization as has the structure of multi-layer/multi local feature, and reasoning can be carried out in the complex structure, in order to achieve the expected goal of mining.

The present information system: production safety supervision and management system, online monitoring system of tailing dam, occupational health management system (bureau of construction), which production safety supervision and management system is the State General Administration of troubleshooting information pilot, including: enterprise accounting, hidden investigation, security checks, credit evaluation, performance appraisal, intermediary services, knowledge base subsystem.

Practical work in production safety supervision personnel many times is a rush on the outside. This is the time to understand the safety production status information only through telephone consultation or communication, with the popularity of mobile Internet applications and smart phones, in smart phones are mobile law enforcement and moving supervision has become possible.

The project in safety production comprehensive supervision system based on expanding mobile law enforcement system and the mobile supervision system, law enforcement personnel through the mobile law enforcement box or smart phone to achieve on-site execution site to submit data for enforcement, ensure the execution of data real-time and avoid individual enterprises about human, race relations, bring a lot of resistance to the supervision and administration of production safety work; and mobile monitoring system for leaders at all levels can learn anytime, anywhere within the jurisdiction of the safety in production, mobile monitoring system dynamically generated all kinds of safety production statistics, query of multi angle, in order to all levels of leaders from time to time to inquire.

2 Analysis of Intelligent Association Rule Extraction Technology Based on Big Data Mining Method

Data is divided into structured data and unstructured data from the point of view of data organization form. Structured data, with a certain law, you can use the two-dimensional table structure to represent, and stored in the database, such as the academic management system data, bank transaction data generated.

Because most users do not have the required professional knowledge to accurately describe the demand for a route to the service, usually only fuzzy demand to express, for example, response speed, a relatively clear picture, safety higher, but these needs are often difficult to be ISP recognition and for the user to assemble the required service routing on the basis of it [2]. Therefore, it is necessary to user requirements fuzzy mapping for the ISP to use the exact information about the mechanism.

Clustering and classification are two of the most important technology in data mining; they were respectively used to detect data and decision space to feature space dependent. Data of these two kinds of characteristics were also known as data clustering and classification.

Definition 1. *The score of the sample space of phi r = {0, 1,... n of them said, |r|}, unknown scoring. With X_a, B said r from the sample space on the user item matrix X in a random variable, a, {1,..., K}, B, {1,... M}.*

$$X = [u_1, \ldots, u_K]^T, u_k = [x_{k,1}, \ldots, x_{k,M}]^T, \quad k = 1, \ldots, K \tag{1}$$

Big data can be used to calculate the challenging of ideas is to use the data sampling technique, by sampling the data smaller, in order to take advantage of existing techniques for data analysis [3]. However, this idea may suffer from two aspects of the question.

Rather than structured data, it is impossible to express and store the data in the database, such as sound, video and pictures, as is shown by Eq. (2), which are stored in the database [4]. The growth rate of the current unstructured data is much faster than the structured data.

$$x'_{ij} = \frac{x_{ij} - \overline{x}}{s_j} (i = 1, 2, \ldots, n; \ j = 1, 2, \ldots, m) \tag{2}$$

Big data is mainly formed by a series of large number of data sets. The size of the data is not normally collected by the human being, which is beyond the normal range of human beings.

Correlation exists between different data, data association because of the existence of the hidden, it is difficult to be human through simple discrimination method of treatment, but the need according to the relevant information of the relation, formulate corresponding data association scheme, in order to complete with a purpose of information processing tasks, this analysis method is applicable in complex tasks, information processing requirements of information management.

Well/makes the hybrid of mining the massive amount of data mining algorithms, approximate solution properties, mainly from the users of the mining process, mining the timeliness requirements, data of the growth of online mining technology put forward severe challenges, and traditional small data sets and learning, big data, uncertainty and high noise of independent and identically distributed hypothesis "failure to pursue the issue of optimal exact solution becomes almost impossible, forcing us to find problems with approximate solution.

And in the medical detection, ECG, EEG, with ultrasound, X ray, computed tomography (CT), magnetic resonance imaging (MRI), positron emission tomography (PET), single photon emission computed tomography (SPECT) and functional magnetic resonance imaging (FMRI) and other modal information is mutual the difference and complement each other, the different modal information appropriate integration has become an urgent demand for clinical diagnosis and treatment of disease.

Definition 2. *If a node number J in s t (K_1,..., K_M), is a set {IK_1,..., Ik_m, count of number of Ij} for t s (K_1,..., K_M).* Count.

administrative permission cannot from the administrative licensing system acquisition and so on.

With the rapid development of the Internet, the production of large data, classification as one of the most practical technology in data mining, has been widely used. At present, the common classification methods are decision tree, artificial neural network, SVM, Bayes, KNN and so on.

And communication needs of is users of a kind of applications is often in many aspects, and every aspect of the demand of customer service experience and the degree of influence is different [8]. Therefore, it is how to effectively and quickly analyze the user experience for the different needs of the dependence of and it are but also the ISP to achieve routing of service customization.

For example, in the financial field of big data, the decision task may be oriented to the national level, regional level, or local level. Even for a bank; may also be different for deposits and loans, or financial products. This makes the task of mining may also for different levels and different aspects.

Cloud computing platform to support the use of cloud computing technology, coordinate the use of computing resources, storage resources, network resources, information resources, application support resources and conditions of existing, unified construction and government sectors to provide infrastructure and support software, application function, information resources, operation security and information security services.

System provides the management of users, groups and roles, can conveniently adjust the relationship between departments, customers and between department's subordinate relationship and associated system roles and departments and users. In general it is tree authorization mechanism, were authorized to different staff, departments and role. On the basis of unified user management, to achieve between the various departments of the graded authorization configuration in order to achieve full coverage of the Department (Fig. 1).

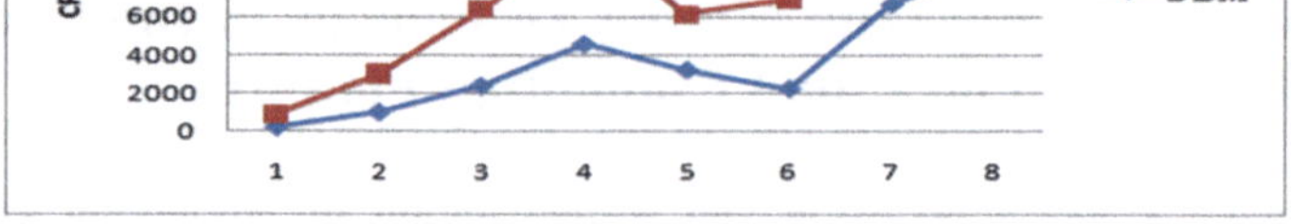

Fig. 1. Performance comparison of Intelligent safe production on cloud computing platform by BDM and TBD diagram.

At present, the system of supervision and administration of safe production in our country is: the combination of the integrated supervision and the supervision of the industry, the combination of state supervision and local supervision, the pattern of the combination of government supervision and other supervision.

Among them, the safety supervision information service exchange center is the hub of the exchange center, the main function is to provide data management, sharing and distribution of data management and sharing and distribution services for each unit of use and data management units.

The paper presents mine chase which is a big responsibility of data in the successful cases applied in the field of safety production, 2010 U.S. Internet users in the use of large coal mine network accountable by Macy's, Inc., a subsidiary of another coal mine Ruby safety supervision, investigate and deal with the data analysis found in advance the mine also at any time "explode" possible.

5 Summary

The safety production supervision is more scientific, more efficient and more optimized, the more information, the stronger analysis ability will promote the scientific nature of decision making. The paper presents application of big data Mining Technology in Intelligent safe production on Cloud Computing platform. But in the face of massive data, the traditional ETL tools are often completely ineffective, the main reason is the cost of data format conversion is too large, cannot meet the needs of massive data acquisition performance. Internet companies of massive data acquisition tools, Facebook's open source scribe, LinkedIn open source of Kafka, Taobao open source Timetunnel, Hadoop chukwa etc. can meet hundreds of MB per second log data acquisition and transmission requirement and upload these data to the Hadoop central system.

Acknowledgements. This paper is supported by Henan key Laboratory for Big Data Processing & Analytics of Electronic Commerce, and also supported by the science and technology research major project of Henan province Education Department (17B520026, 15A120012), Key scientific research projects in Henan province universities (17A880020).

References

1. Thanchalatudom, S., Assawamekin, N.: Using big data technology for information management in hybrid learning system. RNIS **12**, 179–182 (2013)
2. Lei, J., Ye, Q., et al.: Big data platform research, computer research and development based on open source ecosystem. **541**, 80–93 (2017)
3. Xu, H., Zhang, R., Lin, C., Ma, Y.: Novel approach of fault diagnosis in wireless sensor networks node based on rough set and neural network model. Int. J. Future Gener. Commun. Netw. **9**(4), 1–16 (2016)
4. Zhang, D., Xin, M.: Research on technical development of big data in smart grid. Chin. J. Electr. Eng. **1**, 2–12 (2015)

5. Cheng, X., Jin, X.: Big data system and analysis technology overview. J. Softw. **25**(9), 1889–1908 (2014)
6. Xu, H., Zhang, R.: Novel approach of semantic annotation by fuzzy ontology based on variable precision rough set and concept lattice. Int. J. Hybrid Inf. Technol. **9**(4), 25–40 (2016)
7. Silva, R.R., Hirata, C.M., Lima, J.D.C.: Computing BIG data cubes with hybrid memory. JCIT **11**(1), 13–30 (2016)
8. Hongsheng, X., Wang, L., Gan, W.: Application of improved decision tree method based on rough set in building smart medical analysis CRM system. Int. J. Smart Home **10**(1), 251–266 (2016)
9. Wang, H.: On college English translation teaching models in the internet environment. Engl. Sq. (2) (2016)
10. Fan, M.: The cultural elements in English translation teaching. Engl. Sq. (20) (2014)
11. Wang, H.: The problems and countermeasures of English translation teaching in college of traditional Chinese medicine. Overseas Engl. (20) (2014)

Research on Emotional Feature Mapping of Artificial Intelligence Based on Brain Wave Images

Hangming Zhang(✉)

College of Instrumentation and Electrical Engineering, Jilin University, Changchun, China
stefanzhang0427@sina.com

Abstract. To improve the ability of artificial intelligence to recognize human emotions and lay the foundation for real human-computer interaction, the biological characteristics of brain waves were introduced. Also, the visual features of images were taken into consideration in order to complete the mapping of the image visual features to the emotional features of images. Firstly, people's brainwaves were collected when they were enjoying the pictures, and the emotional vectors generated by the visual stimulation of the images were obtained through the calculation of the emotional matrix. Then, the mapping rules of visual features to emotional semantics were established, and the mapping of the image visual features to the emotional features of the images was well achieved. Finally, the mapped emotional features were verified. The verification results showed that the proposed mapping mechanism was effective and highly accurate. So, it has practical value in the field of artificial intelligence and emotional interaction.

Keywords: Artificial intelligence · Emotional feature · Brain wave
Image vision · Mapping rule

1 Introduction

With the rapid development of human-computer interaction system, image emotional information processing has drawn more and more attention. It has become one of the most interesting research topics in the field of artificial intelligence. Some studies have shown that an image contains a lot of emotional information, and that different images can arouse different human emotions. We hope that the computer can effectively describe and express the emotional information [1]. Therefore, effective mapping between image features and emotions is a new and challenging frontier topic in the field of image emotion recognition and emotional calculation.

At present, Japan, the United States, South Korea, Italy and other countries have started to explore and study the field of emotional information processing such as computing and identifying emotions. However, there are still few studies in this field in China. Existing research has found that when people browse through the images, the images will be delivered to the brain through the visual system. Under the guiding ideology of using human visual perception to improve the ability of computer vision

J. Abawajy et al. (Eds.): ATCI 2018, AISC 842, pp. 1263–1270, 2019.
https://doi.org/10.1007/978-3-319-98776-7_156

recognition, we use human brain's biological information induced by people's appreciation of image files to achieve highly intelligent, highly accurate, semantic and perceptual retrieval function [2]. It has always been a challenge and a hot topic in the field of image information to make use of the underlying visual features and high-level semantic features in order to make up for the lack of understanding of computer graphics. A good mapping mechanism can make up for the problem of "semantic gap" to a great extent. According to the records of early scholars, linear mapping is the main method of image emotional mapping. Due to the complexity of the relationship between image features and emotional space, linear mapping cannot describe their relationship well. On the basis of using linear mapping, some scholars also use regression analysis method to adjust the weight of each feature, making the mapping technology have a larger improvement [3]. With the development of technologies such as SVM, neural networks and genetic algorithms, the model of the relationship between the image features and the emotional space is made more diverse and closer to the human emotions through user training [4].

2 Methodology

2.1 Extraction of Brainwave Emotional Features

10 electrodes are used to record brain waves, and the relative number of ${}_{10}C_2 = 45$ correlation coefficient can be obtained. If we calculate the correlation coefficients of θ wave (5–8 Hz), α wave (8–13 Hz) and β wave (13–20 Hz) at different cycles, we can get a set of 45 × 3 = 135 coefficients. This set of values can be used as the eigenvalues of brain waves. From a mathematical point of view, a point on the 135-dimensional space represents a state of the mind, so that the 135 coefficients form a vector known as the "state vector". It is easy to deduce four basic emotional factors: joy, tension/anger, sadness and relaxation because human beings have four basic states of mind, such as pleasure, anger, sadness and joy. Of course, the 135-dimensional state vector also contains other kinds of information which are different from mental states, but we can consider using the learning method to deduce their intrinsic eigenvalues through the basic elements of "pleasure, anger, sadness and joy". The correlation coefficient c (α; j k) that has been recorded by each electrode is calculated and it's between the potential and the collected electrode j; k is the alpha band and c (α; j k) can be expressed as:

$$c(\alpha; jk) = \frac{\sum_{\alpha} X_f(f_n) X_K^*(f_n)}{\sqrt{\sum_{\alpha} |x_{j(f_n)}|^2 \sqrt{\sum_{\alpha} |X_k(f_n)|^2}}} \tag{1}$$

$X_j(f_n)$ is an electroencephalogram Fourier transform at the j-th electrode pole and the n-th frequency-domain point, and the sum exceeds the frequency-domain point of the α-band. In these interrelated coefficients, the four emotional factor eigenvalues of human beings can be captured (anger, sadness, joy, relaxation). The level strengths of the four emotional factors are respectively set as z1, z2, z3, z4, and the

135-dimensional state vector is y = {y1, y2,..., y135}. d is the weight vector; z is the emotional vector; c is the emotional matrix, so the relationship among them is:

$$\begin{bmatrix} c_{1,1} & c_{1,2} & \cdots & c_{1,135} \\ c_{2,1} & c_{2,2} & \cdots & c_{2,135} \\ \cdots & \cdots & \cdots & \cdots \\ c_{4,1} & c_{4,2} & \cdots & c_{4,135} \end{bmatrix} \begin{bmatrix} y_1 \\ y_2 \\ \cdots \\ y_{135} \end{bmatrix} + \begin{bmatrix} d_1 \\ d_2 \\ \cdots \\ d_4 \end{bmatrix} = \begin{bmatrix} z_1 \\ z_2 \\ z_3 \\ z_4 \end{bmatrix} \quad (2)$$

2.2 Image Emotional Mapping

BP neural network is a multi-layer feedforward neural network. Due to its simple structure, multiple adjustable parameters, multiple training algorithms and good manoeuvrability, BP neural network has obtained a wide range of practical applications [5]. This paper uses BP neural network to realize the semantic mapping between the image bottom feature and the emotion of images. The reason is that image emotion embodies the various emotional changes caused by the communication process between people and images. And the neural network can simulate the function and structure of human brain well. Also, the process in which various emotions are generated during the interaction between the image and the human body is quite complex. However, the neural network is very suitable for solving the problem of complex internal structure. That's because it has a very strong ability to learn, promote and generalize. In addition, the learning function of BP neural network is very powerful. Its connection weight and connection structure can be learned through learning, and it has strong fault tolerance and robustness.

BP network is a kind of neural network with three or more neurons. It consists of input layer, middle layer (hidden layer) and output layer. There is a full connection between the upper and lower layers, but there is no connection between the neurons of each layer [6]. Since the input is an image, the number of nodes in the input layer is the dimension of the input feature vector. Better experimental results can be obtained if the effective feature vector is trained by network. When the feature vectors are selected, the ones which can express the image features should be chosen. Appropriate features should be distinguishable, reliable, independent and small in number. At the same time, we also have to effectively eliminate the information and data that are not processed in the data source in order to ensure the valid input. After determining the data source, the number of nodes in the input layer is fixed. The dimension of the output layer can be determined according to the requirements of the user. If the number of the class patterns is m and the number of the mode types is small, the number of output layer nodes is equal to the number of the mode pattern m. When there are many types of patterns, the coding of the output nodes indicates the type of each pattern, that is, $\log_2 m$.

The number selection of neurons in the hidden layer is a very complex issue that often needs to be determined on the basis of the designer's experience and multiple experiments, so there is no ideal analytic representation [7]. For a BP network, any continuous function in a closed interval can be approximated by a BP neural network with a single hidden layer. Therefore, a three-layer BP network can accomplish any mapping from n-dimension to m-dimension. The number of neurons in the hidden layer is mainly related to the requirements of the problem and the number of the input and output units. If there are too many hidden units, there will be problems. For examples, the learning time will be too long. Also, the error will not always be the smallest, and the fault tolerance may be poor. Currently, the reference formula that can be used to select the best number of hidden units is as follows:

$$\begin{cases} \sum_{i=0}^{n} C_{n_i}^{i} > k \\ n_1 = \sqrt{n+m} + a \\ n_1 = \log_2 n \end{cases} \tag{3}$$

The initial value has a close relationship with the convergence of learning and the minimum of the local. When the initial weight is added to the input, the state value of each neuron is close to zero, and the weight is generally a random number with a relatively small value. The input samples also should be normalized so that those larger inputs still fall on the place where the gradient of the transfer function is large.

2.3 Self-Organizing Feature Mapping Network

Self-organizing feature mapping networks is also known as Kohonen networks, or Self-Organizing Feature Map (SOM) networks [8]. During the training phase, the network randomly inputs the training set sample to a corresponding specific input mode, and the output layer will have the maximum chance of winning a node at the same time. At the very beginning of the training, we do not know which input node will affect the input. Whenever the input category changes, the winning node also changes. At the same time, the node in the neighborhood of the winning node are also affected by the lateral excitations. Therefore, the winning node and the domain node's weight vector are all adjusted to the input sample in varying degrees. Meanwhile, both the winning node and its neighbors adjusted to the input: the closer to the winning node, the stronger the dynamics, and the farther the distance, the smaller the strength. When the network is self-organized, a large number of samples need to be used to train the weights of the network, and finally the neurons in the output layer can be sensitive to any specific mode. And when the features of the two pattern classes are close, the nodes representing the two classes are also close in position. Thus, an ordered feature map that can reflect the distribution of the sample pattern class is formed.

3 Results and Discussion

3.1 Brainwave Emotional Feature Extraction

This article selected Corel image database, including a total of 51138 images of various types, among which 100 representative images were chosen as learning samples. The brain wave acquisition device recorded the change of the potential value of each electrode. The brain wave acquisition device recorded the numerical changes of the potential of each electrode, and took the correlation coefficient between the i-th and the j-th of the 10 electrodes as the eigenvalue. In other words, if every two of the 10 electrodes were combined, 45 can be obtained. In these combinations, θ (5–8 Hz), α (8–13 Hz) and β (13–20 Hz) were used respectively to obtain correlation coefficient values (y1, y2, Λ, y135), in which we could capture four human emotional feature values (tension/anger, sadness, joy and relaxation). In the experiment, people who were mentally trained were chosen and then made to imagine nervousness and anger. The same treatment was also used for the other three basic emotions (sadness, joy and relaxation). At that time, the value of the emotion vector was determined as z = (1, 0, 0, 0), (0, 1, 0, 0), (0, 0, 1, 0), (0, 0, 0, 1), equivalent to the four basic emotional states of tension/anger, sadness, joy and relaxation. At the same time, each person's brainwave signal corresponding to each emotion was collected according to different basic emotional states. When the four basic state values z = (1, 0, 0, 0), (0, 1, 0, 0), (0, 0, 1, 0), (0, 0, 0, 1) and 135 interrelated system values (obtained by the electroencephalogram electrodes) of each verifier were substituted into the formula, we obtained a balanced emotional matrix that tended to be balanced through the different substitutions of several people.

The acquisition of brainwave signals was completed by 20 candidates from graduate and undergraduate students. Considering that girls are usually more sensitive to emotion than boys, we selected 16 girls and 4 boys with good health and without visual or psychological related diseases. Their eyesight or corrected vision was all normal.

A total of 40 typical representative pictures were selected, corresponding to the four basic emotions of intense/anger, sadness, joy and relaxation, and there were 10 pictures for each basic emotion. Pictures were chosen at random, and the examinees' brain waves were collected respectively when they were viewing the pictures. Also, the values corresponding to the respective electrodes were recorded. According to the emotional calculation formula:

$$C \cdot y + d = z \tag{4}$$

by taking the emotion matrix previously obtained and the 135 correlation coefficient values collected as well as a proper adjustment of the weight vector, we got the examinees' emotional vectors generated when they were observing the pictures.

3.2 BP Neural Network Mapping Experiments

The system used three layers of BP neural network: the network input layer, the hidden layer and the output layer. This paper mainly mentioned four kinds of emotions, namely, tension/anger, sadness, joy and relaxation. Also, the training network mapped the four emotional values through the colour and texture features of the image. According to the previous hidden layer design experience formula as well as the actual situation of this example, the number of hidden neurons in the network that could solve this problem should be between 19 and 25. The simulation of these cases was implemented in MATLAB, and the results were compared. The experimental results are shown in Table 1.

Table 1 showed that after 1000 trainings, the BP network with a hidden layer of 21 neurons had the best approximation to the function, and the training speed of the

Table 1. Network training error

The number of neurons	19	20	21	22	23	24	25
Network error	0.2297	0.0967	0.0740	0.1027	0.1164	0.2674	0.2980

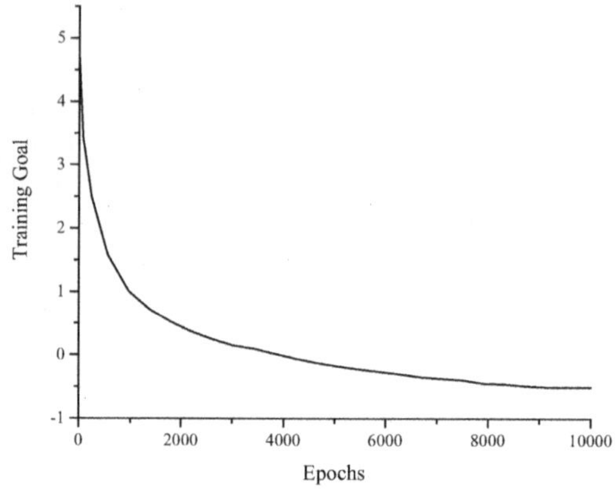

Fig. 1. Training error curve

network was very fast. Thus, the BP neural network with a hidden layer of 21 neurons should be chosen. Figure 1 showed the experimental results of the hidden layer with 21 nodes.

To sum up, the network established was a three-layer BP neural network with 432 dimensional features as the input end, 21 hidden nodes and 4 output nodes.

40 representative pictures were selected and their colour, texture and emotional features were extracted respectively. After the above steps, we trained these 40 pictures and input the colour and texture features of the image. The expected output was our induced brain emotional features of the image extracted. The neural networks were trained through MATLAB. The trained neural network was stored well, and then the emotional mapping of the image whose emotional features had not been extracted yet was carried out. Firstly, 50 images with unknown emotional features were randomly selected from the Corel database as the test image of the trained BP neural network. Secondly, the colour and texture features of the 50 test images were input respectively into the trained BP neural network and the mapping of the underlying features of the images to the emotional features was performed.

3.3 Comparison Between Electric Heating and Fire Exposure

In order to improve the speed of retrieval, we used the Self-Organizing Feature Mapping algorithm to retrieve the similarity distance by using the image emotional features. The main purpose of this method was to check whether the emotional features mapped by the trained BP neural network were reliable and accurate. The retrieval results were fed back to the users as shown in Table 2.

Table 2. Search results analysis table

	Euclidean distance	SOFM based on users space	SOFM retrieval based on BP neural network
Accuracy	54%	89%	95%
Success rate	59%	91%	95%

The essence of this algorithm was to organize the emotional features of images through a one-dimensional self-organizing mapping network, and the linear search method was basically used in the field of retrieval. Linear search was basically used within a class or neighborhood. According to the results of image query, 95% of the users were satisfied. From this point of view, we can see the mapping mechanism from image visual features to image emotional features not only can be realized by using a BP neural network but also has a high accuracy.

4 Conclusion

The emotional semantic of brain waves was increased on the basis of the semantic features of the image visual features. First of all, a number of people with mental imaging foundations were used. They imagined four kinds of basic emotions such as tension/anger, sorrow, joy and relaxation respectively, and obtained emotional vectors expressing the four basic emotions. And then 20 students were selected to observe the pictures and the electroencephalogram electrodes produced by them while they observing the pictures were collected to get the emotion matrix and the emotion vector. Then the BP neural network was used to establish the learning process of the initial visual features to the emotional vector. Finally, the self-organizing feature mapping method was used to verify the emotional features mapped by the trained BP neural network. The verification results showed that the proposed mapping mechanism was effective and highly accurate. So, it is of practical value in the field of artificial intelligence and emotional interaction.

References

1. Zhang, H.B., Huang, T.J., Xiu, Y., et al.: Fabric image emotional semantic analysis based on color and texture features. J. Tianjin Polytech. Univ. **32**(4), 26–32 (2013)
2. Wang, K.C.: The feature extraction based on texture image information for emotion sensing in speech. Sensors **14**(9), 16692 (2014)
3. Chau, C.J., Wu, B., Horner, A.: The emotional characteristics and timbre of nonsustaining instrument sounds. J. Audio Eng. Soc. Audio Eng. Soc. **63**(4), 228–244 (2015)

4. Liu, Y., Mougenot, C.: A study of emotional expression through tactile and gestural interactions. J. Multimed. **6**(1), 56–65 (2015)
5. Fisher, A.C., Rushby, J.A., Mcdonald, S., et al.: Neurophysiological correlates of dysregulated emotional arousal in severe traumatic brain injury. Clin. Neurophysiol. Off. J. Int. Fed. Clin. Neurophysiol. **126**(2b), 314 (2015)
6. Lee, H.J., Shin, D.I., Shin, D.K.: A study on the emotion classification as well as the algorithm of the classification applying EEG-data. In: Lecture Notes in Electrical Engineering, vol. 309, no. 1, pp. 515–521 (2014)
7. Campanella, F., Shallice, T., Lus, T., et al.: Impact of brain tumour location on emotion and personality: a voxel-based lesion–symptom mapping study on mentalization processes. Brain A J. Neurol. **137**(9), 2532–2545 (2014)
8. Du, R., Lee, H.J.: Brain wave patterns under emotional affect. Adv. Sci. **6**(1), 62–66 (2014)
9. Liu, J.: Advanced PID Control and MATLAB Simulation, 2nd edn. Electronics Industry Press, Beijing (2003). (in Chinese)
10. Li, J., Ren, Q.: Study on supply air temperature forecast and changing machine dew point for variable air volume system. Build. Energy Environ. **27**(4), 29–32 (2008). (in Chinese)

SAML Improvement Scheme in Cloud Environment

Zhang La(✉) and Lu Yu

Shijiazhuang Branch of Army Engineering University,
Shijiazhuang 050003, China
15612959442@163.com

Abstract. In order to cope with the continuous increase in the number of network applications in the cloud computing environment, the load of server operations based on SAML-based identity authentication interruptions requires a further increase in computing performance. This paper analyzes the characteristics of the cloud computing environment, studies the structure of the SAML framework, uses a combination of computational task outsourcing and cloud computing, and designs an improved solution to achieve better performance while reducing the workload of the asserting party and reducing input costs.

Keywords: Cloud computing · SAML · IDP

1 Introduction

With a unified identity authentication method based on SAML (Security Assertion Markup Language), a user can access the service resources of all network applications in the trust domain with only one authentication and does not need to re-authenticate when switching between applications. This is not only user-friendly, it also reduces the security risk of authentication information and reduces the cost of certification and maintenance. Currently SAML has been supported by major vendors such as Google, IBM, BEA, and Oracle. However, in recent years, the number of network applications and the number of users have doubled. When the SAML-based unified identity authentication method faces a growing number of user visits, the key assertion party has a large amount of load on the user. Diverting the calculated pressure and reducing the burden on the server has become a very urgent need.

2 Improved SAML Scheme

To solve the problem of too much pressure in the asserting party, this paper combines the cloud computing to improve the SAML-based unified identity authentication scheme. In recent years, cloud computing has developed rapidly. Cloud computing services are characterized by ultra-large scale, virtualization, high scalability, and extremely low cost. Assertive parties can purchase computing power from cloud service providers according to their own needs. The main feature of the improved solution is that the computing task of the asserting party is delivered to the cloud computing

J. Abawajy et al. (Eds.): ATCI 2018, AISC 842, pp. 1271–1275, 2019.
https://doi.org/10.1007/978-3-319-98776-7_157

service provider. The cloud computing service providers here mainly refer to the third-party cryptographic service providers in the cloud.

2.1 Overall Structure

The improved authentication scheme includes the following four entities (see Fig. 1).

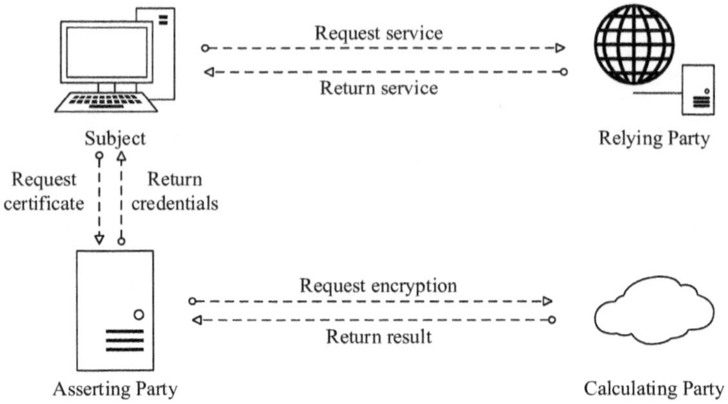

Fig. 1. Entity relationship diagram

Subject. The subject usually refers to a user, or it can be a program such as a web browser, and is a party that requests a certain service or accesses a resource. This article uses web browsers for analysis.

Relying Party. The relying party usually refers to the server accessed by the user. It is a service provider (service provider) and is denoted as SP.

Asserting Party. The asserting party usually refers to a third-party trust organization, and is an identity provider that provides a correct assertion of identity information identifying the principal, and is denoted as IDP.

Calculating Party. The computing party is the new entity proposed in this paper and is a cloud computing provider (Cloud Computing Provider) that provides third-party encryption services in the cloud. The Computation Party is responsible for sharing the IDP signatures, decryption, and other complex computing tasks in this article, and records them as CCP.

2.2 Specific Scheme

There are two common implementation frameworks for the SAML-based unified identity authentication method. After the comparative analysis, this paper adopted the Browser/Post framework with less pressure on IDP (see Fig. 2).

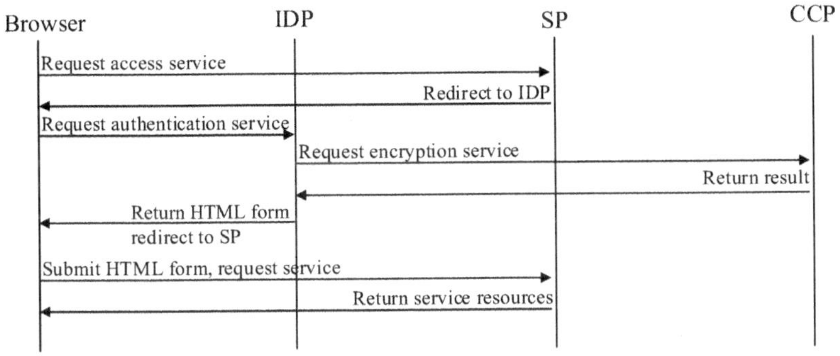

Fig. 2. Interaction flow chart

Figure 2 outlines the authentication process for the first request to access the service if the browser is not authenticated.

First, the browser sends an access service request to the SP;
Second, SP checks whether the browser is authorized to log in. Because there is no information related to identity authentication, that is, there is no relevant authorization, the user is redirected to the IDP for authentication;
Third, the browser submits identity information (a password or an X.509 certificate, etc.) to the IDP for authentication;
Fourth, IDP authenticates the identity of the browser, generates a SAML assertion, and places it in the HTML form. Passes the encrypted channel to the CCP. Digitally signs the HTML form;
Fifth, CCP completes the signature calculation and returns the result to IDP;
Sixth, IDP redirects the browser to the SP and returns the HTML form containing the SAML assertion;
Seventh, the browser again initiates an access service request to the SP and submits an HTML form containing a SAML assertion;
Eighth, the SP verifies whether it is included in the list of web applications authorized to log in in the HTML form. If it is included, the SAML assertion in the HTML form is interpreted. If the assertion is valid, the browser is authorized to log in and the service requested by the browser is provided.

After the browser is successfully authenticated for the first time, it can directly access many web applications in the HTML form without having to perform another authentication. That is, unified identity authentication is achieved.

3 Performance Analysis

3.1 Validity Analysis

Rich in Resources to Facilitate Large-Scale Promotion. The ultra-large-scale features of cloud computing ensure that it can meet the needs of IDP use, and the computing capacity will far exceed the self-purchased equipment. IDP can completely strip all complex computing tasks, especially digital signatures and private key decryption, from its original work, greatly reducing the workload of IDP servers and improving service capabilities.

Low Prices, Effectively Reduce Costs. The low cost of cloud computing determines that the ability to purchase commercial public cloud computing is enhanced or the cost of adding IDP hardware equipment is low. In addition, the high scalability of cloud computing ensures that IDP can adjust the amount of computing power purchased in real time according to different requirements for computing power at different times, such as holidays or days and nights, so as to achieve more reasonable real-time optimization of scale and reduce costs.

Convenient and Quick, No Geographical Restrictions. The virtualization features of cloud computing make IDPs adopting this improved scheme not have to consider geographical issues. As long as networking is possible, IDPs will not be affected even if they are hundreds of kilometers away from the CCP. And can save the space required by the site than the method of purchasing hardware devices.

3.2 Security Analysis

With the Original SAML-Based Authentication Security. The improved solution only spun off the pure computing part and sent it to the CCP. It did not change the logic of the SAML-based unified authentication protocol. Will not change the original anti-replay attacks, anti-hacking attacks, anti-man in the middle of attacks and other anti-attack features. Guaranteed security no less than the security of the original SAML-based unified authentication protocol.

Related to the Safety of CCP. There is a large amount of information transfer between IDP and CCP. The security of the delivery process is bound to be closely related to the security of the CCP providing encryption computing services. Therefore, the security improvements proposed in this paper will not be higher than the security of current third-party commercial cloud computing services.

4 Conclusion

The improved scheme proposed in this paper can simplify the task of the service provider in SAML, reduce the operating load, effectively improve the service ability and reduce the operating cost. It is of great significance for SAML to fully integrate into the increasingly mature cloud computing environment. Current deficiencies are mainly focused on the security aspects of cloud computing service providers. The next step will focus on how to separate security from the security of cloud computing service providers.

References

1. Zhou, C., Tian, X., Zhang, N., Yang, Y., Li, D.: Research on identity authentication technology in cloud computing. Comput. Sci. **43**(s1), 339–341 (2016)
2. Chen, H., Wang, G., Fu, Z.: Cloud services cross-domain authentication protocol formal analysis and verification based on fragment model check. Comput. Sci. **43**(4), 140–144 (2016)
3. Li, F., Wang, Q.: An identity authentication model for cloud computing. Math. Pract. Theory **47**(6), 116–126 (2017)
4. Gao, H., Xiao, T.: Improvement on SAML-based single sign-on model. Comput. Eng. Des. **32**(3), 827–829 (2011)
5. Wang, G., Zhang, B., Fei, X., Xiong, H.: SAML cross-domain single sign-on authentication protocol based on convertible proxy signcryption. Comput. Sci. **42**(4), 106–110 (2015)
6. Watts, D.J.: Small Worlds. The Dynamics of Networks Between Order and Randomness. America: 56A2579BB6, 13 May 1999
7. Aleksy, M., Korthaus, A., Schader, M.: Use Java and the CORBA realization distribute type system. J. Pingxiang Coll. **4**, 104–105 (2005)
8. Titus, J.: ECN technical editor: the eclipse of stand. J. Zhongkai Agrotech. Coll. **19**(2), 32–35 (2006)
9. Richardson, W.C., Avondolio, D.: The Java high class weaves a distance: JDK 5. Sci. Technol. Book Rev. **3**, 17–18 (2006)
10. Li, J., Vuong, S.: A semantics-based routing scheme for grid resource discovery. In: E-Science: First International Conference on E-Science and GridComputing, pp. 58–70, 90 (2005)

Improvement of Semantic Annotation Method in Semantic Internet of Things Based on Conceptual Ontology Model

Hongsheng Xu[1,2(✉)], Luosheng An[1,2], and Ganglong Fan[1,2]

[1] Luoyang Normal University, Luoyang 471934, China
85660190@qq.com
[2] Henan Key Laboratory for Big Data Processing and Analytics of Electronic Commerce, Luoyang 471934, China

Abstract. With the development of semantic Web technology, it provides a better solution to solve this problem, as the core of the semantic Web; ontology is a conceptual model of a clear specification. It provides a strong support for semantic similarity computation. This paper describes the semantic annotation method in the semantic web. This paper analyzes the construction of conceptual ontology in the semantic web. The paper presents improvement of semantic annotation method in semantic Internet of things based on conceptual ontology model.

Keywords: Ontology · Internet of things · Semantic web
Semantic annotation

1 Introduction

In recent years, due to the heterogeneous and distributed data in the Internet, in order to solve the interoperability problem of the Internet of things, the semantic Web technology is introduced into the Internet of things to form a semantic Internet of things.

The status of ontology construction can be in the early stages of software engineering development, and is still in the stage of manual workshop of individuals (or small groups). Establish the ontology mainly adopt artificial way [1]. Ontology is a kind of artistic activity and far from becoming a kind of engineering activity, each ontology development groups have construction principles, design criteria and different development stage of their own, so it is difficult to realize ontology sharing and reuse and interoperability.

Ontology in recent years has been widely applied in the field of Internet of things, through the representation of the semantics of domain concepts and relations, on the one hand, the concept of standardized labels; on the other hand, comprehensively describes the knowledge in this field, to understand the computer provides the basis. Ontology, as the core technology, provides support for the massive, heterogeneous resources search and service development of the Internet of things based on semantic.

The former is refers to the content of information only for thing's information, said must be standardized coding (UID or EPC); the latter refers to the first information

J. Abawajy et al. (Eds.): ATCI 2018, AISC 842, pp. 1276–1282, 2019.
https://doi.org/10.1007/978-3-319-98776-7_158

stored in attachment to entity of electronic tags (usually RFID electronic tag). Then none contact type machine, which is usually an RFID reader reading upload information stored on the Internet (specific in Internet information server). It can be used to write the model of Internet of things described in the form of set.

Therefore, the research on semantic interoperability of the semantic Internet of things mainly from the three levels of analysis [2]. This section mainly analyzes the construction strategy of semantic ontology based on rough concept lattice, and discusses the construction method of the core ontology in Internet of things.

In the semantic analysis of the existing networking: mainly concentrated in the IOT of ontology construction of it, the ontology of things is based on experts in the field of artificial construction, more authoritative a W3C semantic sensor networks incubator building SSN ontology, in addition to CSIRO, MMI, cesn, A3ME, OntoSensor etc. more than a dozen things networking domain ontology.

2 Research Construction of Conceptual Ontology in Semantic Web

The name of the Internet refers to the computer network that connects the whole world, and it can also refer to the information sharing which is realized on the physical network. The latter is the real Internet of things refers to the Internet of things.

The European Esprit project is KACTUS to research on the technology system life cycle process knowledge reuse, KACTUS method to develop the ontology and the engineering model, controlled by the development and application, on a per application have corresponding representation required for the application of ontology [3]. These ontologies can be reused by other ontology, and can also be integrated with subsequent applications. KACTUS provides an interactive environment that supports browsing, editing and management of ontology.

Definition 1. *Class C property P1 type definition for T1, q is a subclass of C, the set C1 in property p1 on the value of V, V type T2 if T1 and T2, the report there are inconsistent thunder attribute type.*

So the based on statistical parsing of Stanford depth NLP to extract knowledge tuples. They are using the extraction method based on the NLP model, but the model analysis of length is limited in a single sentence, ignoring the influence of lexical semantic context information in a sentence.

The process of constructing concept lattice is essentially the process of concept clustering [4]. In the process of application of concept lattice, is first solved is lattice structure problem, then combined with improved rule acquisition algorithm, in attribute reduction based on the background of the form is further simplified, because after attribute reduction of formal context may also contain redundant information, so the value reduction makes the form background information more simplified.

At present, with the rapid development of the Internet of things, networking equipment types is different, a variety of networking industry application platform for emerging, things have faced increasing isomerization problem, including data isomerization and application platform for the isomerization and equipment and the application of the heterogeneous.

In the development of the Internet of things and the development of the Internet of things, the user's demand is also increasing; the user begins to show the need for multi device linkage and multi data fusion. In order to meet the needs of users, networking applications of IOT platform resources and processing ability put forward higher requirements, the platform's expected to accurate intelligence to provide the application requirements of data resources and equipment, simplify the application processing, as is shown by Eq. (1) [5].

$$\lambda p_n = n\mu p_{n+1},\, p_{n+1} = \frac{\lambda}{n\mu} p_n = \frac{p^{n+1}}{n \cdot n!} p_0 \tag{1}$$

The annotation information of XML format can be generated based on the rules of semantic annotation. This system is part of the Parmenides project. Parmenides project has been applied to the labeling of GENIA biomedical corpus. Text mining based on Slot constraint is used in Parmenides to provide the analyst with annotation information.

Manually defined rules can alleviate the problem of manual annotation to a certain extent, and improve the efficiency of labeling. However this method exist following problems: need for different types of documents or in different application document to define different tagging rules; rules define when the need for artificial characteristics of the data analysis, summary information annotation rule, which makes defines the rules may be valid only in the part of the data, difficult in all data has good labeling effect.

Definition 2. *Feature item T, for a word in the document after the type of processing to mark the type of identification. The characteristics of sequence elements: the characteristic sequence of elements of a by a single feature T, namely a = (T), abbreviated as t. Based on this definition, the feature sequence element set can be used to represent the feature item set I. Feature sequence: the characteristic sequence s is a sequence of elements on the I of the feature item set, denoted as <a1a2... ar>.*

Internet of things as a new type of Internet applications, the difference is mainly reflected in the information uploaded to the Internet before the information. For example, the need to monitor and analysis of an ecological system in some kind of disease, not only need to sensor networks to return to the system in the current incidence, also need to get the system of climate change, such as temperature, precipitation, humidity and geographic information.

All of these data are returned by the mutual independent sensor network, and they use different application standards and service standards because of different fields. In recent years, the semantic similarity computation based on ontology has been in natural language processing, information extraction, information retrieval, data mining, machine translation, semantic annotation, ontology learning, and so on.

3 Analysis of Semantic Annotation in Semantic Web

METHONTOLOGY the method is used in the development of the University of Madrid Artificial Intelligence Library. This method is a more general method of ontology development, which is combined with the skeleton method and GOMEZ-PEREZ method.

However, the ontology mapping function KACTUS range of applications in the actual Co., there are still many problems, but KACTUS ontology development method provides a kind of thought, that is cut ontology in order to adapt to the specific application of ontology reuse [6]. This thought has an important significance in the actual development of ontology.

In the semantic network domain ontology construction research, the manually constructed ontology is time-consuming and laborious and error prone, current automation research construct ontology is still in the initial stage, project in the field has similar relation between the unit body, using ontology merging method based on concept lattice isomorphic generating, these units ontology merging, finally UNSPSC Ontology meta model as the core ontology, combined with expert knowledge in the field of electronic commerce, the core ontology semi automated extraction and e-commerce product ontology generation.

The result of this part is a natural language form of ontology specification document is defined as the following formula (2).

$$\begin{cases} x_t = \phi_0 + \phi_1 x_{t-1} + \cdots + \phi_p x_{t-p} + \varepsilon_t - \theta_1 \varepsilon_{t-1} - \cdots - \theta_q \varepsilon_{t-q} \\ \phi_p \neq 0, \theta_q \neq 0 \end{cases} \tag{2}$$

Equation (2) is an extensible knowledge model. User can redefine the representation of the system used. File output format can be customized. The internal representation of Protege2000 can be converted into a variety of forms of text representation, including XML, RDF (S), OIL, DAML, DAML + OII, OWL, and other languages [7]. User interface can be customized. To provide an extension of the API interface, the user can replace the Proteg2000 user interface display and data acquisition module to adapt to the new language.

Definition 3. *On the background of K = (G, M, R, G) as the object set, M attribute set to M = n, M = {M1, m2,... Mn}. For G β, attribute word Wg = b1b2... BN.*

$$[Initialize]\, Set\, 1 \rightarrow i,\ 1 \rightarrow j,\ V[i] = 0,\ \Pr = 0 \tag{3}$$

In ontology semantic similarity measure in the study, the current research lack of in-depth study on structural similarity, the problem will lead to ontology concept only reflect the linguistic characteristics of lexical and conceptual and not the vocabulary in specific scenarios, semantic characteristics.

Internet of things is a kind of Internet application that makes the information of the global share, and it is also a collection of information based on the Internet. According to the things of nature and its three properties, three key points of information and to store don't from other Internet applications.

Have the authors propose oriented networking sensing data attribute semantic tagging method for semantic web technologies of IOT data attributes based on annotation methods. Through the ontology hierarchical structure for attribute data of semantic description, attribute data from relational database abstraction, exists independent of the specific application of the attribute data.

According to the requirements of specific applications, networking and data attribute oriented data model, data calls in given data rich semantic attribute information, improve the physical meaning of the data networking and use value, and semantic tagging process for annotation of the semantics of semantic web of things, the intrinsic properties of the data is the contents of the data semantic annotation, embodies the physical meaning of the data.

Properties of the data more rich, physical meaning of data expression ability is stronger, the greater the value may be used. According to the characteristics of sensor data in the Internet of things, we will attribute data points for the dynamic attributes and static attributes and semantic annotation of content.

4 Improvement of Semantic Annotation Method in Semantic Internet of Things Based on Conceptual Ontology Model

In this paper, the calculation of the correlation, from two aspects of consideration is an instance document related degree on the other aspects is instances for semantic environment.

The level of networking architecture based on, from the point of view of the reference model, function model and function frame construct semantic architecture of things, will support networking applications of Semantic Functional decomposition, modular and mapped to the business layer, in order to meet the networking application of combined network platform semantic competence needs.

And improve the ontology based semantic similarity calculation method, explore a calculation method based on the similarity, the method can be used in the context of a single ontology and ontology, and aims to provide a efficient, simple and reliable semantic similarity calculation method in the semantic network analysis.

The specific development process is as follows: the application of the description: to provide the application environment and the application model of the required components. A preliminary design of related Ontology, as is shown by Eq. (4), and where X is search for existing ontology, extract relevant, reusable ontology. Related ontology extraction mainly from three aspects: task types, problem solving methods and domain types [8].

$$\sum_i \sum_j \left(X_{ij} - bi - \left(\bar{X}_j - b\frac{n+1}{2} \right) \right)^2 \tag{4}$$

Information sources are the semantic similarity between concepts is calculated by using the semantic similarity of Chinese words provided by CKIP. The semantic similarity between concepts is measured by the distance between the vocabulary and the WordNet dictionary.

All annotation tools are using DAML, RDFS, OIL, DAML + OIL, SHOE, or FLogi. As an ontology description language, the markup language uses XML, RDFs, DAML ten OIL, DAML or SHOE, and there is no tool to support the latest Web ontology description language OWL.

Based on sensor information in ontologies for the semantic network semantic annotation method, mainly including: (1) document with the concept of ontology is corresponding to the word marked, as a concept corresponding to the examples, usually in the form of RDF, RDF (resource) said. Using classification learning model to identify the semantic categories of documents (similar to text categorization), as is shown by Fig. 1 and then using statistical learning methods to identify named entities in Web pages.

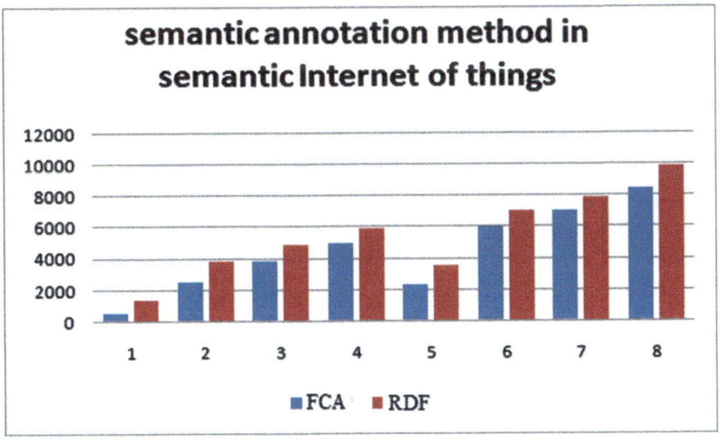

Fig. 1. Performance comparison of semantic annotation method in semantic Internet of things based on FCA and RDF diagram.

Concept lattice is from the given data extract implicit relationships between concepts and all, to construct the conceptual model of concept lattice), and use the computer to handle the symbolic form of performance and ontology and formal concept analysis from philosophy field, is a description of the relationship between the concept and the concept.

Finally, aiming at the reduction of formal context, concept lattice construction algorithm and improved rule acquisition algorithm combination is proposed based on semantic variable precision rough set concept lattice of value reduction algorithm is constructed to construct the concept lattice semantic.

5 Summary

Abstract model is based on domain feature, on field data attributes and conceptual relations of abstracted and classified, and to establish the relationship between the concepts of ontology attribute description and concept. Data attributes and relationships is expression of the physical meaning of the data, the data use value reflected in. This layer according to the characteristics of specific areas defined ontology concepts, attributes, and relationships, for example in the field of building energy-saving applications.

Acknowledgements. This paper is supported by Henan key Laboratory for Big Data Processing & Analytics of Electronic Commerce, and also supported by the science and technology research major project of Henan province Education Department (17B520026, 15A120012), Key scientific research projects in Henan province universities (17A880020).

References

1. Barnaghi, P., Wang, W., Henson, C., et al.: Semantics for the Internet of Things: early progress and back to the future. Int. J. Semant. Web Inf. Syst. **8**(1), 11–21 (2012)
2. Xu, H., Zhang, R., Lin, C., Ma, Y.: Novel approach of fault diagnosis in wireless sensor networks node based on rough set and neural network model. Int. J. Future Gener. Commun. Netw. **9**(4), 1–16 (2016)
3. Lee, C.S., Kao, Y.F., Kuo, Y.H.: Auomated ontology construction for unstructured text documents. Data Knowl. Eng. **60**, 547–566 (2007)
4. Xu, H., Zhang, R.: Novel approach of semantic annotation by fuzzy ontology based on variable precision rough set and concept lattice. Int. J. Hybrid Inf. Technol. **9**(4), 25–40 (2016)
5. Kang, X., Li, D., Wang, S.: Rough set model based on formal concept analysis. Inf. Sci. **222**, 611–625 (2013)
6. Xu, H., Zhang, R., Lin, C., Gan, W.: Semantic annotation of ontology by using rough concept lattice isomorphic model. Int. J. Hybrid Inf. Technol. **8**(2), 93–108 (2015)
7. Han, L., Chen, G.: AASA: a method of automatically acquiring semantic annotations. J. Inf. Sci. **33**(4), 435–450 (2007)
8. Xu, H., Wang, L., Gan, W.: Application of improved decision tree method based on rough set in building smart medical analysis CRM system. Int. J. Smart Home **10**(1), 251–266 (2016)
9. Jung, J.J.: Semantic annotation of cognitive map for knowledge sharing between heterogeneous businesses. Expert Syst. Appl. **39**(5), 5857–5860 (2012)
10. Ge, X.: On how to improve the effectiveness of college English translation teaching. Engl. Teach. (18), (2015)
11. Chen, T.: Effective measures of english teaching translation. Overseas Engl. (20) (2015)

Research on Attitude Control of Aircraft Test Simulation System Based on LQR Algorithm

Ming-yue Li(✉) and Jian Fang

JiLin Engineering Normal University, Changchun, Jilin, China
254295092@qq.com

Abstract. The aircraft test simulation system is a multi-variable, uncertain and strongly coupled nonlinear system, this article through the analysis of spacecraft testing simulation system of constructing, using Lagrange equation method, the mathematical model is established, in a linear quadratic optimal control theory as the theoretical basis for design the controller and the system simulation, at the same time LQR weighting matrix Q and R in the experimental study and analysis, find out the dynamic response of aircraft test simulation system between matrix Q and R and follow the basic law of the experimental results proved that the conclusion is correct and practical.

Keywords: Linear quadratic optimal control · Aircraft test system Attitude control

1 Introduction

The aircraft test simulation system is a multi-variable, uncertain and strongly coupled nonlinear dynamic system. The aim is to control the movement of the vehicle to a specified position or to move according to a given trajectory. Based on the whole vehicle as the research object, the abstract as a flexible movement of the ball can be 360°, at a level of control platform, through the study of the control algorithm of vehicle testing simulation system of research, expect for the flying car such as high risk control system find a solution in the field of aerospace.

2 Theory of Aircraft Test Simulation System

The overall structure of the aircraft test simulation system is shown in Fig. 1.

Vehicle testing simulation system of principle: sphere position sphere position data acquisition card to collect to the touch screen, motor data acquisition card, two tail motor encoder data collected motor connected to the tablet through the connecting rod, so the plate is obtained by calculation can tilt angular displacement and the displacement of the sphere angular displacement difference to angular velocity, displacement difference to speed, and then according to the algorithm to calculate the control quantity in automatic control. Control by a computer via USB data line to the servo drive, the drive of motor control and motor encoder is connected to the rear drive form a closed loop, which can realize positioning plate on the sphere, and trajectory tracking.

J. Abawajy et al. (Eds.): ATCI 2018, AISC 842, pp. 1283–1291, 2019.
https://doi.org/10.1007/978-3-319-98776-7_159

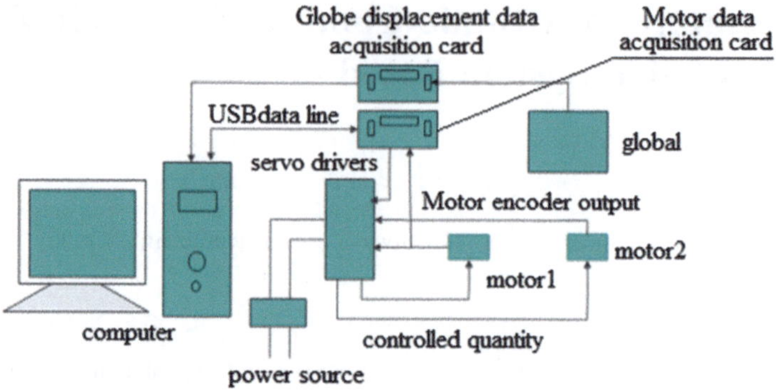

Fig. 1. Aircraft test simulation system overall structure diagram

The attitude control target of the aircraft test simulation system is two, one is the positioning, namely the control sphere movement to the specified position; The second is trajectory tracking, which is to control the ball to move according to a given trajectory, such as a circle and a rectangle. The whole process is dynamic, and it shows that the motion of the plate up and down is the positive and negative rotation of the motor.

3 Aircraft Test Simulation System Physical Modeling

Take the support point of the plate as the origin of coordinates, and establish the coordinate system as shown in Fig. 2.

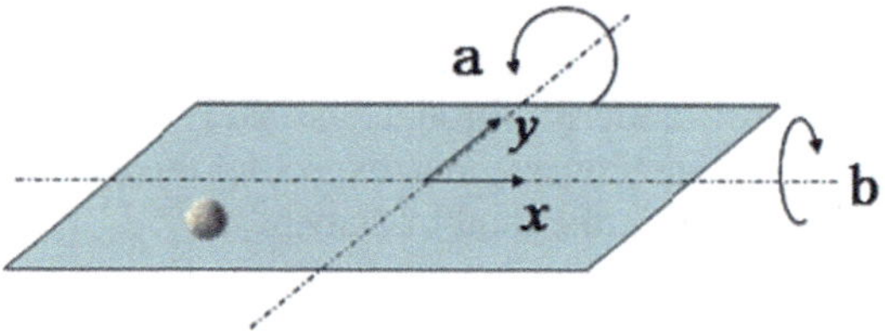

Fig. 2. Coordinate of Ball & Plate System

From the diagram, we can see that the system has four degrees of freedom: the position of the ball (x, y) and the tilt angle of the plate (a, b). Because the two groups of variables are interrelated, only one of them is independent. The following assumptions are made:

Ignore all the friction; In any case, the ball is in contact with the plate; do not consider the area of the board and the rotation Angle limit; The ball does not slide on the board and rotate around its vertical axis; The plate is symmetric about its support point in the x and y directions.

The mathematical model of the system is established by Lagrangian mechanics method. In the process of modeling, the parameters are as follows:

Quality of the sphere: m (kg) value 0.025
Radius of the sphere: r (m) value 0.14
Spherical horizontal and vertical displacement: x, y (m)
Ball level and vertical inclination: a, b (rad)
The ball position: s (m)
Ball level, rotation angular velocity in vertical direction: ω_x, ω_y (rad/s)
Plate rotation angular velocity: ω (rad/s)
The moment of inertia of the sphere: J_b ($kg \cdot m^2$)
The moment of inertia of the plate: J_p ($kg \cdot m^2$)
Flat and vertical torque: T_x, T_y $\left(kg \cdot m^2/_{s^2}\right)$

1. The kinetic energy of the sphere includes its own rotational kinetic energy and translational kinetic energy on the plate.

$$T_b = \frac{1}{2}m(\dot{x}^2 + \dot{y}^2) + \frac{1}{2}J_b\left(\omega_x^2 + \omega_y^2\right) \tag{1}$$

Because the sphere doesn't slide during the motion, thus

$$\dot{x} = r\omega_x, \dot{y} = r\omega_y \tag{2}$$

Bring (4) into (3) formula.

$$T_b = \frac{1}{2}\left(m + \frac{J_b}{r^2}\right)(\dot{x}^2 + \dot{y}^2) \tag{3}$$

2. The kinetic energy of the plate includes its own rotational kinetic energy and the rotational kinetic energy of the ball around the plate support.

$$T_p = \frac{1}{2}J_P(\dot{a}^2 + \dot{b}^2) + \frac{1}{2}\left(J_b + m|\Omega \times s|^2\right)(\dot{a}^2 + \dot{b}^2) \tag{4}$$

again $|\Omega \times s|^2 = \frac{(x\dot{a} + y\dot{b})^2}{\dot{a}^2 + \dot{b}^2}$, Plug in to (4)

$$T_p = \frac{1}{2}(J_b + J_p)(\dot{a}^2 + \dot{b}^2) + \frac{1}{2}m(x\dot{a} + y\dot{b})^2 \tag{5}$$

3. Select the zero potential energy point of the coordinate origin, and the potential energy of the plate is zero, and the potential energy of the sphere is:

$$V_b = mg(x \sin a + y \sin b) \tag{6}$$

4. The generalized force is provided by the torque generated by the servo system:

$$Q_X = T_X, Q_y = T_y \tag{7}$$

The combined formula (4), (5) and (6) can get the Lagrangian function:

$$\begin{aligned} L &= T_b + T_p - V_b \\ &= \frac{1}{2}\left(\mathrm{m} + \frac{J_b}{r^2}\right)(\dot{x}^2 + \dot{y}^2) + \frac{1}{2}(J_b + J_p)(\dot{a}^2 + \dot{b}^2) + \frac{1}{2}\mathrm{m}(x\dot{a} + y\dot{b})^2 - mg(x \sin a + y \sin b) \end{aligned} \tag{8}$$

For Lagrange equation, we take the following four variables:

$$\begin{gathered} \mathrm{q}_1 = x, \mathrm{q}_2 = y, \mathrm{q}_3 = a, \mathrm{q}_4 = b \\ \mathrm{Q}_1 = \mathrm{Q}_2 = 0, \mathrm{Q}_3 = \mathrm{Q}_x, \mathrm{Q}_4 = \mathrm{Q}_y \end{gathered}$$

And the Eqs. (5) and (6) are brought in $\frac{\mathrm{d}}{dt}\left(\frac{\partial L}{\partial \dot{q}_i}\right) - \frac{\partial L}{\partial q_i} = Q_i, \quad i = 1, 2, 3\ldots, n$. The system of equations are as follows:

$$\begin{aligned} &\mathrm{x}: \left(\mathrm{m} + \frac{\mathrm{J_b}}{\mathrm{r}^2}\right)\ddot{x} - m(x\dot{a}^2 + y\dot{a}\dot{b}) + mg \sin a = 0 \quad \text{(a)} \\ &\mathrm{y}: \left(\mathrm{m} + \frac{\mathrm{J_b}}{\mathrm{r}^2}\right)\ddot{\mathrm{y}} - m(x\dot{b}^2 + x\dot{a}\dot{b}) + mg \sin b = 0 \quad \text{(b)} \\ &\mathrm{a}: (J_b + J_p + mx^2)\ddot{a} + m(xy\ddot{b} + \dot{x}y\dot{b} + x\dot{y}\dot{b} + 2x\dot{x}\dot{a}) + mgx \cos a = T_x \quad \text{(c)} \\ &\mathrm{b}: (J_b + J_p + my^2)\ddot{b} + m(xy\ddot{a} + \dot{x}y\dot{a} + x\dot{y}\dot{a} + 2x\dot{x}\dot{b}) + mgy \cos b = T_y \quad \text{(d)} \end{aligned} \tag{9}$$

Where (a) and (b) two equations describe the motion of the sphere on the plate, the relation between the acceleration of the sphere and the tilt Angle of the plate and the angular velocity; (c) and (d) two equations show how the dynamics of the slanted plates are influenced by external forces and the position and velocity of the sphere.

In fact, the input of the system is not a torque T_X and T_y, and this is the Angle a and b, this is because the load torque does not affect the position of the motor, based on this assumption, Eq. (9) Chinese (c), (d) can be ignored.

In a steady state, the tablet should be in a horizontal position, and then both angles are equal to 0, because the plate rotation Angle range is not big (plus or minus 6°), the sine function can be replaced by its independent variable, the sina $= a, \sin b = b$. In this way, the differential Eq. (9) in the vicinity of the stable state (a), (b) is linearized as type 10 (a) and (b).

$$\begin{aligned} &\text{x:}\left(m+\frac{J_b}{r^2}\right)\ddot{x}+mga=0 \quad \text{(a)}\\ &\text{y:}\left(m+\frac{J_b}{r^2}\right)\ddot{\text{y}}+mgb=0 \quad \text{(b)} \end{aligned} \tag{10}$$

take $x, \dot{x}, a, \dot{a}, y, \dot{y}, b, \dot{b}$ is the state of the system, the angular acceleration of the plate is controlled.

Make $x_1 = x, x_2 = \dot{x}, x_3 = a, x_4 = \dot{a}, x_5 = y, x_6 = \dot{y}, x_7 = b, x_8 = \dot{b}$ have following equation

$$\begin{cases} \dot{x}_1 = x_2 \\ \dot{x}_2 = \ddot{x} = -\dfrac{mg}{\left(m + {J_b}/{r^2}\right)}a \\ \dot{x}_3 = x_4 \\ \dot{x}_4 = \ddot{a} \\ \dot{x}_5 = x_6 \\ \dot{x}_6 = \ddot{y} = -\dfrac{mg}{\left(m + {J_b}/{r^2}\right)}\text{b} \\ \dot{x}_7 = x_8 \\ \dot{x}_8 = \ddot{b} \end{cases}$$

make $x = [x_1, x_2, x_3, x_4, x_5, x_6, x_7, x_8]^T = \left[x, \dot{x}, a, \dot{a}, y, \dot{y}, b, \dot{b}\right]^T, y = x, u = \left[\ddot{a}, \ddot{b}\right]^T$, the state space description equation of the system is

$$\begin{cases} \dot{x} = Ax + Bu \\ y = x \end{cases} \tag{11}$$

Take g = 9.81 m/s^2, the moment of inertia of a sphere $\text{J}_\text{b} = \frac{2}{5}mr^2$.

Make $x_x = [x_1, x_2, x_3, x_4]^T, x_y = [x_5, x_6, x_7, x_8]^T$, get the following state-space formula

$$\dot{x}_x = \begin{bmatrix} \dot{x}_1 \\ \dot{x}_2 \\ \dot{x}_3 \\ \dot{x}_4 \end{bmatrix} = A_x x_x + b_x u_x \quad y_x = [1,0,0,0]x_x, \quad A_x = \begin{bmatrix} 0 & 1 & 0 & 0 \\ 0 & 0 & -7.007 & 0 \\ 0 & 0 & 0 & 1 \\ 0 & 0 & 0 & 0 \end{bmatrix}, b_x = \begin{bmatrix} 0 \\ 0 \\ 0 \\ 1 \end{bmatrix}$$

$$\dot{x}_y = \begin{bmatrix} \dot{x}_5 \\ \dot{x}_6 \\ \dot{x}_7 \\ \dot{x}_8 \end{bmatrix} = A_y x_y + b_y u_y \quad y_y = [1,0,0,0]x_y, \quad A_y = \begin{bmatrix} 0 & 1 & 0 & 0 \\ 0 & 0 & -7.007 & 0 \\ 0 & 0 & 0 & 1 \\ 0 & 0 & 0 & 0 \end{bmatrix}, b_y = \begin{bmatrix} 0 \\ 0 \\ 0 \\ 1 \end{bmatrix}$$

The aircraft test simulation system is linearized at the horizontal position, x, x and y direction is decoupled and the structure is consistent. So you just have a qualitative analysis of the x direction. Give the system a unit step signal, System analysis in MATLAB environment, the position of the sphere and the tilt Angle of the plate are both divergent, then, the system is unstable.

The design of the controller must be based on the controllability of the system, Therefore, it is necessary to analyze the controllability of the system before design. According to the complete controllability judgment of linear constant continuous system. $S = \left[B, AB, A^2B, \ldots, A^xB\right]$ there are $rank(b_x, Ab_x, A^2b_x, A^3b_x) = 4$.

And so you know, the system is controllable. Therefore, the LQR controller can be designed for the aircraft simulation test system, stabilize the system.

4 LQR Controller Design and Simulink Simulation

Optimal control is the core of modern control theory, according to the optimal control theory, the design of the optimal controller, the design of the optimal controller, the goal of optimal control is to make $J \rightarrow$ min, and the essence of it is, with little control to keep the small error, therefore, it achieves the optimal goal of energy and error synthesis.

$$J = \int_0^\infty \left(X^T QX + u^T Ru\right) dt \tag{12}$$

The state feedback control rules are minimized using Eq. (12).

$u^* = -R^{-1}B^T Px = -Kx$, and P is the solution of the Riccati algebraic equation $PA + A^T P + Q - PBR^{-1}B^T P = 0$, where Q and R are the weight matrix for the measurement of state quantity and input quantity, they are mutually restricted, the size of Q is directly proportional to the anti-interference ability of the system. Increase the Q, the anti-interference ability of the system is enhanced. The adjustment time of the system is shortened, but it also strengthens the system, energy consumption increases, he increase in R, educes the energy consumed by the system, but adjustment time increases. So the key to design is to find the right matrix Q and R, as long as we determine the Q and the R matrix, he state feedback matrix K is uniquely identified. In the MATLAB control toolbox, the command to solve the equation has been included:

$$K = \text{lqr}(A, B, Q, R)$$

For the orbiter simulation control system, system A, B has been determined, so just pick Q, R two weights, and you get K. Select the weight matrix to measure the input $R = [2]$, in order to satisfy the requirement, for this design instance, the design of Q is as follows:

$$Q = diag(Q_{11}, Q_{22}, Q_{33}, Q_{44}) \tag{13}$$

Q_{11}, Q_{22}, represents the actual value of the horizontal axis of the sphere and the velocity weight of the set value and the spherical horizontal axis respectively, Q_{33}, Q_{44} the inclination Angle of the horizontal axis and the angular velocity weight of the horizontal axis are respectively represented. The LQR control block diagram of system state feedback is built under Simulink. The simulation model is as follows (Fig. 3).

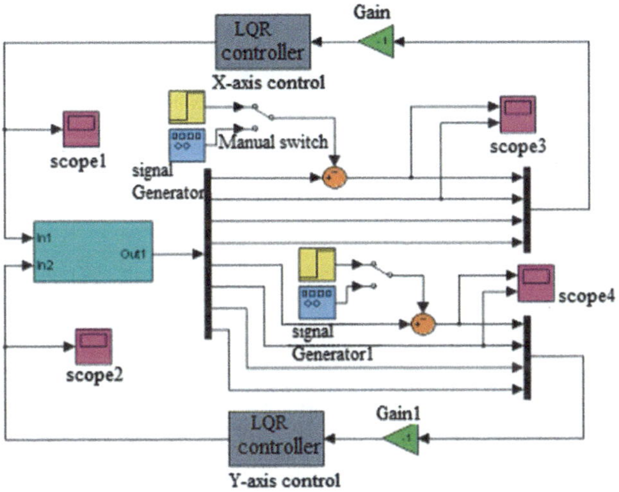

Fig. 3. LQR controller simulation model

The repeated simulation results of different Q values are found, Q11 = 300, Q22 = 4, Q33 = 1000, Q44 = 10, to get the K = [12.2474,12.925, −45.5128, −9.7993], the simulation results are shown in Fig. 4.

Controlled quantity u $= -kx = 12.2474x + 12.6925\dot{x} - 45.5128\theta - 9.7993\dot{\theta}$.

Figure 4 the upper part of the sphere is the horizontal displacement step response curve of the sphere, the x-coordinate is response time t, the unit is in seconds, the y-coordinate is the sphere deviation from the equilibrium position x, the units are divided into meters. The lower part represents the slope angle a step response curve of the horizontal axis platform board, the x-coordinate is response time t, the unit is in seconds, the ordinate is the tilt Angle a of the platform board, the units are radians.

The simulation curve shows that: when the sphere is given a disturbance, the system basically achieves the new equilibrium in the horizontal axis in 3.5 s, it is indicated that the LQR controller has good stability and rapidity in the control of aircraft attitude.

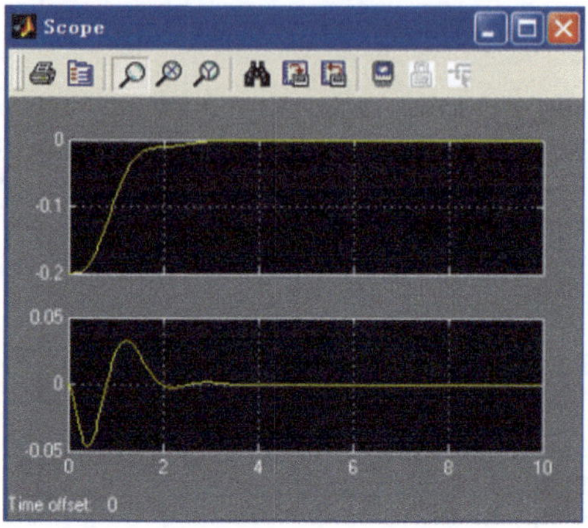

Fig. 4. LRQ control simulation curve

5 Conclusion

The linear quadratic optimal control theory is used in this paper, to realize the control of attitude and stability in the aircraft test system, but there are problems in the control process, for example, the selection of Q and R is based on experience and trial and error, it's more subjective, therefore, the controller design is not perfect. It affects the control effect. Such as, in the initial period of control, the zero point coordinates of the platform are higher, like the initial state cannot guarantee the horizontal position, it will exert a great influence on the control effect. In a word, the aircraft test simulation system is a challenging and practical subject, it provides some reference for the development of aerospace control.

Acknowledgments. Jilin provincial education department project 《Automatic brake device design of automobile intelligent anti – error》 (project number: JJKH2017162KJ).

Supported by Program for Innovative Research Team of JiLin Engineering Normal University.

References

1. Gao, Q., Yuan, L., Wu, J.: The four-rotor UAV attitude control based on the new LQR. J. Electron. Optics Control **46–52**(10), 0405 (2014)
2. Prasad, L.B., Tyagi, B., Gupta. H.O.: Modelling and simulation for optimal control of nonlinear inverted pendulum dynamical system using PID controller and LQR. In: Proceedings of the 6th Asia Modelling Symposium, Bali, Indonesia, pp. 138–143. IEEE (2012)
3. Balandat, M., Zhang, W., Abate, A.: On infinite horizon switched LQR problems with state and control constraints. Syst. Control Lett. **81**(4), 464–471 (2012)
4. Chen, D.J., Zheng, S.J., Wang, H.T.: Genetic algorithm based LQR vibration wireless control of laminated plate using photostrictive actuators. Earthq. Eng. Eng. Vib. **1**, 83–90 (2012)

5. Zhang, J., Zhang, W.: LQR-based adaptive control strategy for the planar double inverted pendulum. Electr. Inf. Eng. Mechatron. **1**(14), 117–124 (2011)
6. Aleksy, M., Korthaus, A., Schader, M.: Use Java and the CORBA realization distribute type system. J. Pingxiang Coll. no. 4, 104–105 (2005)
7. Titus, J., ECN Technical Editor: The eclipse of stand. J. Zhongkai Agrotechn. Coll. **19**(2), 32–35 (2006)
8. Clay Richardson, W., Donald, A.: The Java high class weaves a distance: JDK 5. Scientific & Technology Book Review, no. 3, pp. 17–18 (2006)
9. Vuong, J.L.: A semantics-based routing scheme for grid resource discovery. In: E-Science: First International Conference on E-Science and Grid Computing, pp. 58–70, 90 (2005)

The Attitude Control of Flying Car Test Simulation System Based on the Variable Theory Domain Fuzzy Algorithm

Yue Wu(✉) and Jian Fang

Jilin Engineering Normal University, Changchun, Jilin, China
28841898@qq.com, 757314739@qq.com

Abstract. The fuzzy controller has some limitations that lack of stability analysis and low control accuracy. And the stability of flight attitude is not ideal in the study of the optimal control of flight attitude stability of flying vehicles aiming at complex flight control for the system. According to the above problems and difficulties. This paper brings up the fuzzy control optimization algorithm for variable domain controlling by designing the variable domain on the basis of establishing an accurate mathematical model. Its purpose is finding a stable solution for the control system of the flying car domain. The simulation results show that the control precision is ideal by MATLAB platform for simulation.

Keywords: Fuzzy control · Variable domain fuzzy control
Simulation based on MATLAB

1 Introduction

The flying car is a vehicle that can travel both in the sky and on land. It can switch between the land driving mode and the flight mode. During the flight, the attitude is easily disturbed by factors such as external airflow, which leads to instability of flight posture and affects the quality of the preset flight task. The flight motion is characterized by multivariate, strong coupling and nonlinear, so it is obvious that design good attitude control system exerts great important role [1]. At present, the control methods of flying automobile system are summarized that PID control, fuzzy control, adaptive fuzzy control, etc.

PID control can only be applied to some simple controlled objects and the limitation of fuzzy control lies in the high degree of dependence on control rules. Adaptive fuzzy control has improved the self-adaptability and stability of fuzzy control, but it can not meet the precise requirements of industry. Therefore, in the fuzzy control, the expansion factor is introduced to adjust the field of input and output variables in real time. That is, in the premise of keeping the control rules unchanged, the domain is scaled with the change of the error, when the error increases, the domain is enlarged, whereas the error variable hour is reduced. The effect of domain shrinkage is the same as that of increasing control rules, which is equivalent to increasing the number of interpolation points, so the control accuracy of the controller is higher [2].

J. Abawajy et al. (Eds.): ATCI 2018, AISC 842, pp. 1292–1299, 2019.
https://doi.org/10.1007/978-3-319-98776-7_160

2 The Attitude Control System of Flying Car Test Simulation Physical Modeling

The simulation test attitude control system of flight vehicle is shown in Fig. 1.

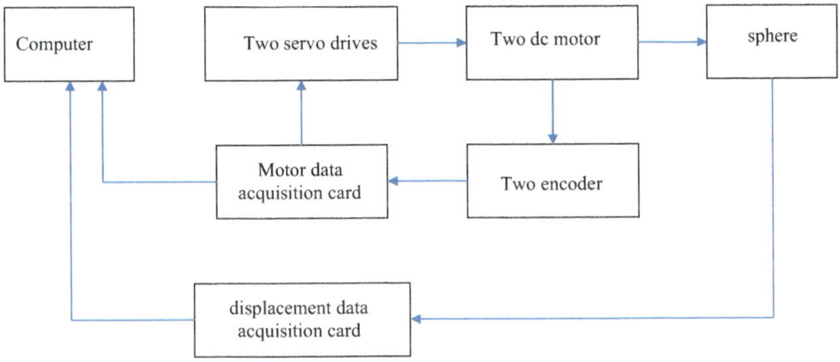

Fig. 1. The working principle diagram of the flying car simulation test attitude control system

The attitude control principle of flight vehicle test simulation system is as follows: The flying car as a whole is the research object. It is abstracted into a sphere of 360 degree motion, placed on a horizontal control platform. The ball position acquisition card collects the ball position data on the touch screen. The motor data acquisition card collects two motor tail encoder data. The motor and the plate are connected by connecting rod. Therefore, the tilt angle displacement of the plate and the displacement of the pellet can be obtained by calculation, angular displacement differential angular velocity, displacement difference can be obtained speed, and then according to the theory of automatic control in various transformation algorithm to calculate the control quantity. The computer is sent the control quantity controlled by the actuator to the motor to the servo driver via USB data [3]. And the motor rear encoder is connected to the drive to form a closed loop. Thus, the positioning of the ball and track tracking can be achieved.

Taking the physical model as reference, the support point of the drawing board is the coordinate origin, and the coordinate system as shown in Fig. 2 is established.

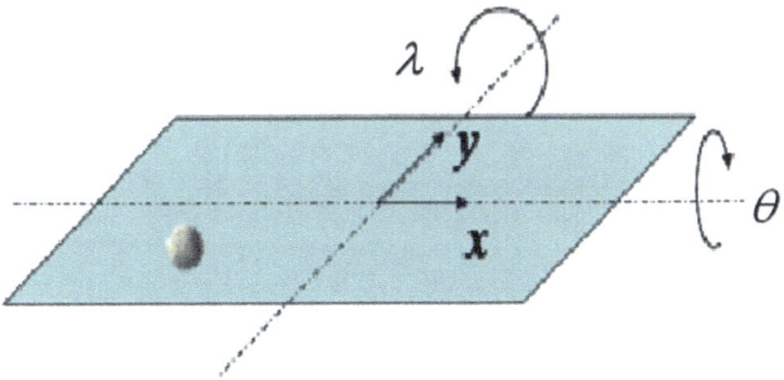

Fig. 2. Coordinate system diagram

This system has four degrees of freedom from the diagram: The position coordinates of the flying car (x, y). The tilt Angle of the plate (λ, θ). Because the two groups of variables are interrelated, only one of them is independent. The following assumptions are made:

- Ignore all the friction;
- Cars and tablets are in contact in any case;
- The area of the plate and rotation Angle are not considered;
- The flying car is symmetrical in the axial direction of the board.

The mathematical model of the system is established by Lagrangian mechanics method. In the process of modeling, the parameters used in the modeling are shown in Table 1.

Table 1. The meaning of the parameters

Parameter	Meaning	Unit
m	Quality of lying car	kg
r_a	Flying car radius	m
x, y	x, y direction displacement of the flying car	m
λ, θ	x, y direction inclination of the plate	rad
l	Position of flying car	m
ω_x, ω_y	x, y angular velocity rotation of flying car	rad/s
ω_b	Turns angular velocity of the flying car	rad/s
J_a	The moment inertia of the flying car	kg.m^2
J_b	The moment inertia of the plate	kg.m^2
M_x, M_y	The direction torque of the flying car	kg.m^2/s^2

The kinetic energy of the ball includes its own rotational kinetic energy and translational kinetic energy on the plate.

$$T_a = \frac{1}{2}m(\dot{x}^2 + \dot{y}^2) + \frac{1}{2}Ib(\omega_x^2 + \omega_y^2) \tag{1}$$

Because the car did not slip during the movement, so

$$\dot{x} = r_a\omega_x, \dot{y} = r_a\omega_y \tag{2}$$

Take (2) type into (1) type,

$$T_a = \frac{1}{2}(m + \frac{I_b}{r_b^2})(\dot{x}^2 + \dot{y}^2) \tag{3}$$

The kinetic energy of the plate includes its own rotational kinetic energy and the rotational kinetic energy of the ball around the plate.

$$T_b = \frac{1}{2}J_b(\dot{\lambda}^2 + \dot{\theta}^2) + \frac{1}{2}(J_a + m|\omega_b \times l|^2)(\dot{\lambda}^2 + \dot{\theta}^2) \tag{4}$$

and $|\omega_b \times l|^2 = \frac{(x\dot{\lambda} + y\dot{\theta})^2}{\dot{\lambda}^2 + \dot{\theta}^2}$, Plug in to

$$T_b = \frac{1}{2}(J_a + J_b)(\dot{\lambda}^2 + \dot{\theta}^2) + \frac{1}{2}m(x\dot{\lambda} + y\dot{\beta})^2 \tag{5}$$

The origin of the coordinates is zero potential energy, and the potential energy of the plate is zero, and the potential energy of the ball is:

$$V_a = mg(x \sin\lambda + y \sin\theta) \tag{6}$$

The generalized force is provided by the torque generated by the servo system:

$$Q_x = M_x, Q_y = M_y \tag{7}$$

Combined with (4), (5) and (6), the Lagrangian function is:

$$\begin{aligned} L = T_a + T_b - V_a &= \frac{1}{2}(m + \frac{J_a}{r_a})(\dot{x}^2 + \dot{y}^2) + \frac{1}{2}(J_a + J_b)(\dot{\lambda}^2 + \dot{\theta}^2) \\ &+ \frac{1}{2}m(\dot{x}\lambda + \dot{y}\theta)^2 - mg(x \sin\lambda + y \sin\theta) \end{aligned} \tag{8}$$

For Lagrange equation, the following four variables are obtained:

$$q1 = x, q2 = y, q3 = \lambda, q4 = \theta$$
$$Q1 = Q2 = 0, Q3 = Qx, Q4 = Qy$$

And the Eqs. (5) and (6) are brought in to $\frac{d}{dt}(\frac{\partial L}{\partial \dot{q}j}) - \frac{\partial L}{\partial qj} = 0,\ j = 1, 2, 3\ldots, n$, the system of equations is

$$x : (m + \frac{J_a}{r_a^2})\ddot{x} - m(x\dot{\lambda}^2 + y\dot{\lambda}\dot{\theta}) + mg \sin\lambda = 0 \tag{9}$$

$$y : (m + \frac{J_a}{r_a^2})\ddot{y} - m(y\dot{\theta}^2 + x\dot{\lambda}\dot{\theta}) + mg \sin\theta = 0 \tag{10}$$

$$\lambda : (J_a + J_b + mx^2)\ddot{\lambda} + m(xy\ddot{\theta} + \dot{x}y\dot{\theta} + x\dot{y}\dot{\theta} + 2x\dot{x}\dot{\lambda}) + mgx \cos\lambda = M_x \tag{11}$$

$$\theta : (J_a + J_b + mx^2)\ddot{\theta} + m(xy\ddot{\lambda} + \dot{x}y\dot{\lambda} + x\dot{y}\dot{\lambda} + 2x\dot{x}\dot{\theta}) + mgy \cos\theta = M_y \tag{12}$$

The two Eqs. (9) and (10) describe the motion of the ball on the plate, which shows the relation between the acceleration of the ball and the Angle of tilt of the plate and the angular velocity [4]. And (11) and (12) two equations show how the dynamics of slanting are influenced by external driving forces and the position and velocity of the ball.

Simplified Model of the Flying Car System. In fact, the input of the system is not the torque *MxMy*, but the Angle λ and θ. This is because the load torque does not affect the position of the motor. In this assumption, Eqs. (9)–(12), (11) and (12) can be ignored.

Under steady state, the plate should be in a horizontal position, while two angles are equal to zero, because the plate rotation angle range is not big (plus or minus 6°), The sine function can be replaced by its independent variable. That is to $\sin\lambda \approx \lambda$, $\sin\theta \approx \theta$, In this way, the differential Eqs. (9)–(12) of the stable state is obtained, and (10) is linearized. For example (13) and (14).

$$x : (m + \frac{J_a}{r_a^2})\ddot{x} + mg\lambda = 0 \tag{13}$$

$$y : (m + \frac{Ja}{ra^2})\ddot{y} + mg\theta = 0 \tag{14}$$

When $\mathbf{x}_x = [x_1, x_2, x_3, x_4]^{\mathbf{T}}$, $\mathbf{x_y} = [x_5, x_6, x_7, x_8]^{\mathbf{T}}$,

$$\dot{\mathbf{x}}_x = \begin{bmatrix} \dot{x}_1 \\ \dot{x}_2 \\ \dot{x}_3 \\ \dot{x}_4 \end{bmatrix} = A_x x_x + b_x u_x \quad y_x = [1,0,0,0]\mathbf{x}_x, \quad A_x = \begin{bmatrix} 0 & 1 & 0 & 0 \\ 0 & 0 & -7.007 & 0 \\ 0 & 0 & 0 & 1 \\ 0 & 0 & 0 & 0 \end{bmatrix}, b_x = \begin{bmatrix} 0 \\ 0 \\ 0 \\ 1 \end{bmatrix}$$

$$\dot{\mathbf{x}}_y = \begin{bmatrix} \dot{x}_5 \\ \dot{x}_6 \\ \dot{x}_7 \\ \dot{x}_8 \end{bmatrix} = A_y x_y + b_y u_y \quad y_y = [1,0,0,0]\mathbf{x}_y, \quad A_y = \begin{bmatrix} 0 & 1 & 0 & 0 \\ 0 & 0 & -7.007 & 0 \\ 0 & 0 & 0 & 1 \\ 0 & 0 & 0 & 0 \end{bmatrix}, b_y = \begin{bmatrix} 0 \\ 0 \\ 0 \\ 1 \end{bmatrix}$$

The vehicle simulation system is linear in horizontal position, the direction is decoupled and the structure is consistent, so it is necessary to conduct qualitative analysis of the direction.

3 Variable Theory Domain Adaptive Fuzzy Controller

3.1 The Scaling Factor of the Variational Domain

Single Input Single Output Adaptive Fuzzy Controller Scaling Factor. Given the fuzzy controller, the input and output fields are respectively denoted as $X = [-E, E]$

and $Y = [-U, U]$. The rules E and U are positive real Numbers. X and Y is the initial range relative to the variable domain.

α is defined as a scaling factor on the domain. For any $x \in X$, $X(x) = [-\alpha(x)E, \alpha(x)E]$. It's called the variable domain. Consequently, the same can be said of the scaling factor and the variation domain. In this paper, we give two practical application of single input single output fuzzy controller.

$$\alpha(x) = (|x|/E)^{\tau}, \tau > 0 \tag{15}$$

$$\alpha(x) = 1 - \lambda e^{-kx^2}, k > 0, \lambda \in (0, 1) \tag{16}$$

Expansion factor of dual input single output controller. $X = [-E, E]$,$Y = [-D, D]$ for the domain of input variables x and y respectively. $Z = [-U, U]$ is the domain of output variables. When they are independent of each other. The scaling factors $\alpha(x), \beta(y), \gamma(z)$ of X, Y, Z can be obtained by type 15 or 16 respectively. For example, if you represent the error domain, it will be often used to represent the error rate theory domain. Set $D = EC$ that is $Y = [-EC, EC]$ And β is defined on $X \times Y$, that is $\beta = \beta(x, y)$. In the case of domain scaling factor, we can take one of the following expressions:

$$\beta(x, y) = \left[\frac{|x|}{E}\right]^{\tau_1} \left[\frac{|y|}{EC}\right]^{\tau_2} \tag{17}$$

$$\beta(x, y) = 1 - \lambda e^{-k_1 x^2 - k_2 y^2} \quad k_1, k_2 > 0, \lambda \in (0, 1) \tag{18}$$

Among them,$0 < \tau_1$, $\tau_2 < 1$, the analysis type 1.17 and 1.18 are easy to obtain both can guarantee the change of the theory field is less than or equal to the initial scope. It is necessary to satisfy the boundary of the domain, so as to conform to the idea of the theory of change [5].

4 Design of Adaptive Fuzzy Controller for Flying Car System

The adaptive fuzzy controller is designed on the basis of fuzzy control, and the domain scaling factor is

$$\alpha(er) = 1 - 0.8e^{er^2}, \alpha(ec) = 1 - 0.78e^{er^2 - ec^2},$$
$$\beta(er) = 1 - 0.7e^{-er^2}, \beta(ec) = 1 - 0.7e^{-0.8er^2 - ec^2},$$

The output of fuzzy controller is $U = \frac{1}{2}(1 - 0.7e^{-er^2})u + \frac{1}{2}(1 - 0.7e^{-0.8er^2 - ec^2})u$, and u is the output of the fuzzy controller.

5 The Simulation Model Building of Flying Car System

The following model is used to illustrate the establishment of simulation platform in Simulink environment. We get the differential equations of the system by physical modeling (13) and (14).

Take $g = 9.81/s^2$, $J_a = \frac{2}{5}mr_a^2$ and bring it into (13) and (14).

$$x : \ddot{x} - \frac{5}{7}(x\dot{\lambda}^2 + y\dot{\lambda}\dot{\theta}) + 7.007\sin\lambda = 0$$

$$y : \ddot{y} - \frac{5}{7}(y\dot{\theta}^2 + x\dot{\lambda}\dot{\theta}) + 7.007\sin\theta = 0$$

The simulation results are shown in Figs. 3 and 4.

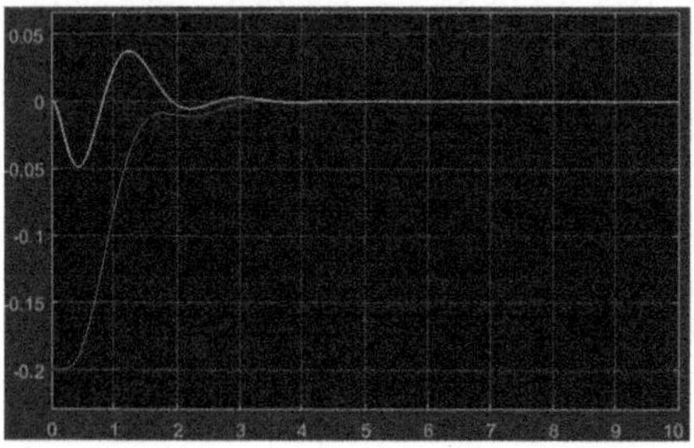

Fig. 3. Fuzzy simulation curve

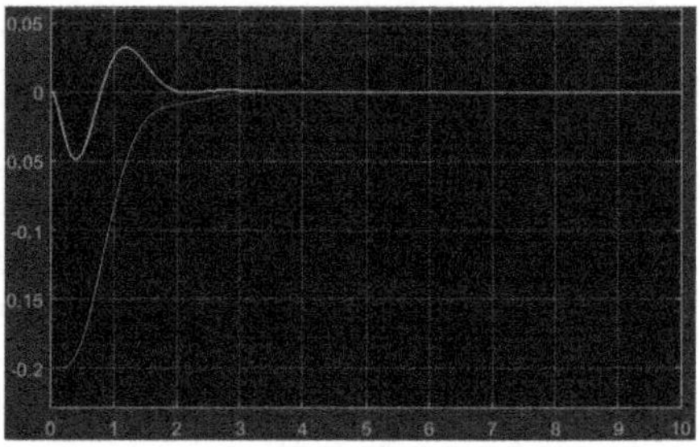

Fig. 4. Variable theory domain fuzzy control

Through the simulation curve we can see, the variable theory domain fuzzy control is more ideal than fuzzy control. It can be stable in a short time, with a small overshoot and strong robustness and adaptability to external disturbances.

6 Conclusion

According that the control parameters of the fuzzy controller are difficult to adjust automatically and the stability of flight attitude is not ideal. In this paper, the fuzzy control optimization algorithm is studied, which is compared with the conventional fuzzy controller. The simulation results of MATLAB show that the algorithm has better control effect. At the same, it lays a foundation for the application of Variable theory domain fuzzy controller in the field of flying car.

Acknowledgements. Jilin provincial education department project 《Automatic brake device design of automobile intelligent anti – error》 (project number: JJKH2017162KJ).

Supported by Program for Innovative Research Team of JiLin Engineering Normal University.

References

1. Li, F., Li, T., Wang, Y.: Uav four-rotor flight attitude stability control optimization. Comput. Simul. **10**, 43–47 (2016)
2. Cao, F., Me, H., Lei, X.: The development status and prospect of flying cars. Top. Forum Rev. **2**, 89–94 (2015)
3. Guo, H.: Several new methods of adaptive fuzzy control for variable domain. Dalian University of Technology (2013)
4. Chen, W.: Variable domain fuzzy PID control in dc motor. Northern Polytechnic University (2014)
5. Several improved control algorithms for uav flight attitude. Guangxi Normal University (2014)
6. Aleksy, M., Korthaus, A., Schader, M.: Use Java and the CORBA realization distribute type system. J. Pingxiang Coll. **4**, 104–105 (2005)
7. Titus, J.: ECN technical editor: the eclipse of stand. J. Zhongkai Agrotech. Coll. **19**(2), 32–35 (2006)
8. Richardson, W.C., Avondolio, D.: The Java high class weaves a distance: JDK 5. Sci. Technol. Book Rev. **3**, 17–18 (2006)
9. Vuong, J.L.: A semantics-based routing scheme for grid resource discovery. In: E-Science: First International Conference on E-Science and Grid Computing, vol. 90, pp. 58–70 (2005)

The Design of Vehicle Intelligent Anti-False Stepping System Based on Sensor Fusion

Wei Li(✉) and Jian Fang

Jilin Engineering Normal University, Changchun, Jilin, China
28841898@qq.com, 757314739@qq.com

Abstract. In the case of drivers who mistakenly use the accelerator pedal as the brake pedal during the driving process of the car, causing serious traffic accidents, from the perspective of intelligence, automation, STC12C5A60S2 single-chip microcomputer is used as the control core of data acquisition system, multi-sensor fusion technology and the fuzzy control algorithm is adopted to design a smart control system for the vehicle. The vehicle accelerator pedal information is collected by means of ranging, angular acceleration and pressure. Use the upper computer monitor system to collect the sample data and set the threshold. If the throttle is misused, the high speed motor will be used for emergency braking. The experimental results show that the system has low cost, simple structure, high accuracy and reliability, and can assist the driver effectively to ensure the safety of driving.

Keywords: Automobile active safety · Mistakenly step on the accelerator Multi-sensor fusion · Fuzzy algorithm

1 Introduction

With the rapid increase of car ownership and production, automobile accidents are becoming more and more serious, and the safety of automobiles is becoming more and more important to car designers, manufacturers and users. In the process of the driving, drivers often mistake the accelerator pedal as a brake pedal for tension, leading to serious traffic accidents. Therefore, it has become a research hot issue of many scholars to prevent the misuse of accelerator pedaling technology as a kind of vehicle active safety technology. At present, the research on the accelerator system and its key technology is still in its initial stage, so the key technology of preventing the mistake of stepping on the accelerator has to be perfected in many areas [1]. There are single sensor, misdiagnosis rate is high. In this paper, a new automatic brake system is proposed, using multi-sensor fusion technology, fuzzy control technology, the pressure sensor, angular acceleration sensor, acquisition throttle of voids and acceleration pedal pressure, determine whether to step on the accelerator. The fuzzy algorithm is used to obtain the fuzzy control rules and control the response speed of the high speed motor to realize the flexible braking.

J. Abawajy et al. (Eds.): ATCI 2018, AISC 842, pp. 1300–1306, 2019.
https://doi.org/10.1007/978-3-319-98776-7_161

2 System Overall Scheme Design

The system is composed of signal acquisition part, control unit and control mechanism. Signal acquisition part includes the pedal pressure, speed, acceleration, and the distance between the vehicle and the obstacle in front of the vehicle, and transmit the detection information in real time. The control unit adopts STC12C5A60S2 single-chip micro-computer with high speed, low power consumption and strong anti-interference as the core of the data acquisition system. The angular acceleration value of the pedal is known by A/D conversion, sampling and differential. The collected data is input into a central control unit composed of STM32 main control board, and the data is analyzed and processed in data detection, signal processing and control unit, and by using fuzzy algorithm and setting of threshold comparison, integration analysis, if the discriminant pressure, velocity and acceleration over a distance range setting threshold, the operating mechanism is given signal, control the brake pedal brake, When brakes the pedal brake, alarm will voice [2]. The system structure diagram is shown in Fig. 1.

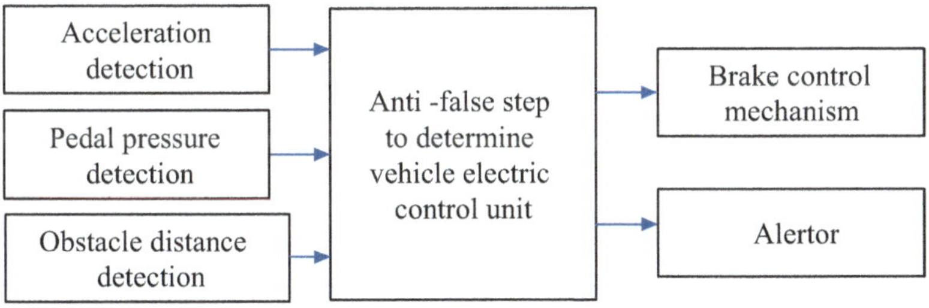

Fig. 1. Anti-false step system structure diagram

3 Research on the Key Technology of Anti-False Accelerator System

3.1 The Analysis of the False Step on the Determination of Accelerator

Acceleration Information Detection. In case of an emergency, the driver depress brake pedal faster than usual several times or even more, accidentally step on the gas, the pedal movement velocity and acceleration is much bigger than the normal step on the accelerator, so the sensor can be installed at the bottom of the pedal, when the sensor detects the accelerator pedal movement velocity or acceleration exceeds the preset threshold, will be judged as the accelerator is misstepped.

Detection of Pressure Information. In the normal driving process of the vehicle, the force of the pedal is uniform. Under a few dozen newtons, the force applied on the brake pedal could reach several hundred newtons in the event of a sudden emergency, so an RFP thin-film pressure sensor can be installed at the accelerator pedal. This sensor is thin, easy to bend, cheap, practical advantages, When stepping on the accelerator, stepping on the accelerator is greater than the preset force, and will be judged to be the wrong step on the accelerator [3].

Distance Detection. Set a safe distance and compare the actual distance, given a measuring range, using SW-L2DS50A laser ranging module for ranging. When detect the actual distance is in the scope of security, the automatic will start the automatic braking system. In this way, the automatic braking system will react quickly with a forced brake and stop the car when the driver does not have time to brake.

3.2 Study of Automatic Braking Method

The braking system is the key to the safety of the car. The driving safety decision is based on whether the distance of the workshop is less than the safe distance. And when the distance of the workshop is less than the safe distance, the automatic braking judgment is necessary [4]. The distance of the workshop can be monitored in real time by the vehicle ranging radar, and the safe distance must be determined by the system according to the relevant parameters collected. Since safe distance is a very important control parameter in system control, how to determine the size of safe distance is critical to system control. Automatic dynamic control should be determined according to the braking characteristics of the vehicle and the specific driving conditions of the vehicle.

Determination of Absolute Safety Distance. Absolute safe distance definition. Starting from the driver's braking, the car will slow down to the maximum braking force and reduce the speed until the vehicle is fully parked. The distance that the car passes is the minimum braking distance $S_{\min}$. When A stops behind B, there should be A distance between the two cars, which is called the minimum allowable workshop distance of D_0. The absolute safe distance S_{safe} is the sum of the minimum braking distance $S_{\min}$ and the minimum allowable distance D_0, namely:

$$S_{safe} = S_{\min} + D_0 \tag{1}$$

Determination of minimum allowable workshop distance. When the vehicle stops, it is necessary to keep a small distance between the two cars at a distance of D_0. According to the minimum distance of the actual vehicle, D_0 = 0.6 m is required to meet the safety requirements [5].

Analysis of Automobile Braking Process. Normal automobile braking process is divided into: driver reaction process, mechanical system reaction process, complete braking process. Reporter:

Driver reaction time t_1;
Mechanical system response time t_2;

The speed of the vehicle is V, the driving distance of the driver's reaction process and the mechanical system reaction process is S_1 and S_2 respectively, and the braking distance is S_3, which is as follows (Fig. 2).

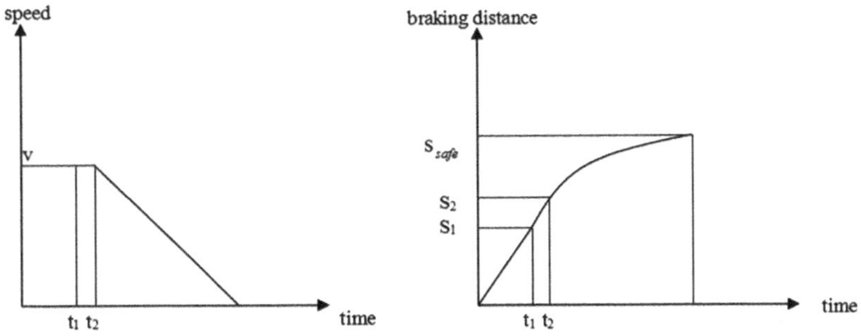

Fig. 2. Analysis of automobile braking process

Calculation of absolute safe distance:

$$\begin{aligned} S_{safe} &= S_{\min} + D_0 = S_1 + S_2 + S_3 + D_0 \\ &= V \cdot (t_1 + t_2) + V^2/2ug + D_0 \end{aligned} \tag{2}$$

Where u is the friction coefficient of wheel and ground, g is gravity acceleration (9.8 m/s^2).

4 System Software Design

4.1 Program Design

The control unit mainly program the operation program of the single chip computer, including the data collection and analysis of the sensor, the judgment of the wrong step on the accelerator, the motor output and the voice alarm. After the system is started, the initialization is first performed and the actuator is restored to its original position. Then, read and analyze the acceleration sensor, speed sensor, pressure sensor and distance detected data, determine whether to step on the accelerator, if you step on the accelerator, the drive motor action, make the car stop. Otherwise, read and analyze the sensor data. Software flow chart 5 is as follows (Fig. 3).

4.2 Experimental Data

Use LABVIEW to prepare the system monitoring software, when the driver encounters the emergency, the pedal angular velocity changes as shown in Fig. 4; The force changes of the pedal are shown in Fig. 5 (Table 1).

Fuzzy Control Rules. Using fuzzy control algorithm to describe accelerator pedal acceleration, pressure and obstacle distance, brake control system is adopted.

The relative velocity error and relative distance error are selected as two input control signals. The expected acceleration of the car is the output control signal. Set these signals to 7 fuzzy sets: NB = Negative direction of large deviation (Negative Big), NM = Negative direction of the deviation (Negative Medium), NS = Negative

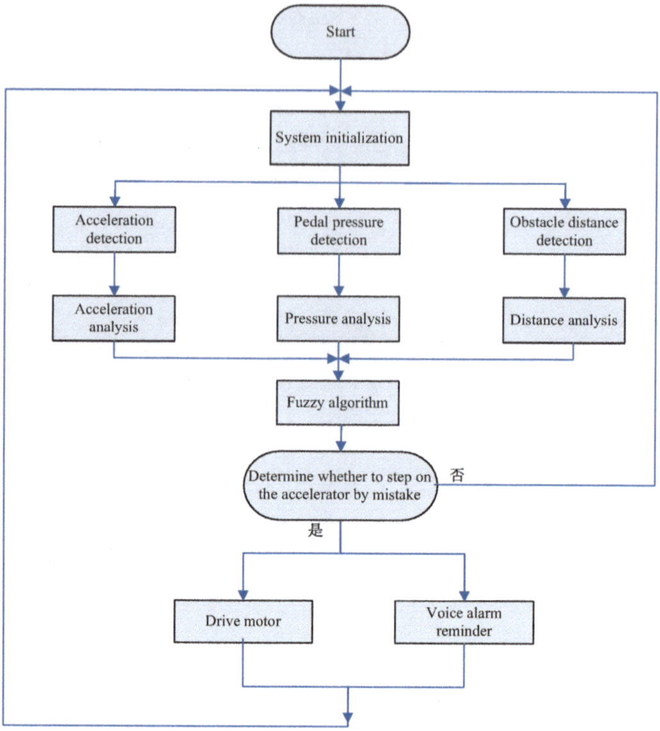

Fig. 3. Software flowchart

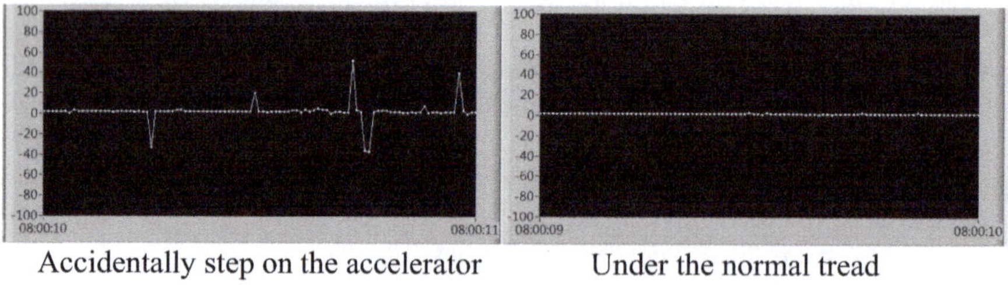

Fig. 4. Acceleration pedal angular velocity detection system

direction Small bias (Negative Small), ZO = Zero deviation (Zero), PS = Positive direction Small bias (Positive Small), PM = The deviation in the positive direction (Positive Medium), PB = Positive direction large deviation (Positive Big), which can form 16 fuzzy rules, because the motor will not reverse, Therefore, the fuzzy subset is not negative, and in this paper, the data result is directly reflected as zero (ZO), positive small (PS), positive medium (PM), and positive big (PB).

By using these fuzzy sets, fuzzy rules can be used to realize the fuzzy control of the driver's actual driving experience. On this basis, when the system determines that it should be decelerated, the angle acceleration and pedal pressure of the accelerator pedal can be used to further determine whether or not to tread correctly [6] (Table 2).

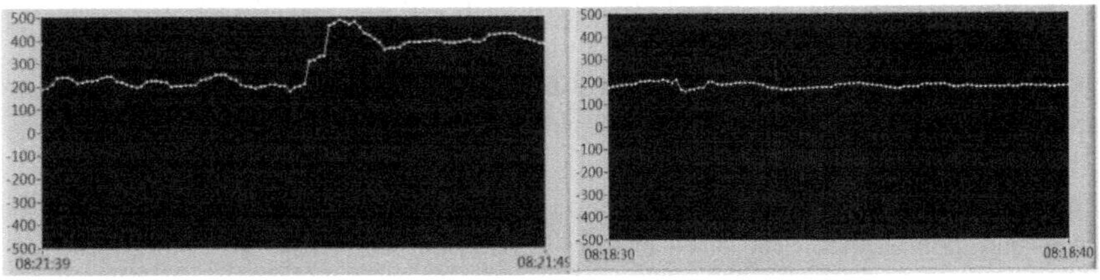

Accidentally step on the accelerator | Under the normal tread

Fig. 5. Acceleration pedal pressure detection system

Table 1. Experimental data

Variable	Tread situation		
	Normal tread	Emergency tread	Stepping on the accelerator by mistake
Angular speed (Rad/S)	−3–+3	−8–+10	−40–+50
Stress (N)	100–200	150–300	300–500

Table 2. Fuzzy control rules

Acceleration	Pressure			
	ZO	PS	PM	PB
ZO	ZO	PS	PM	PM
PS	PS	PS	PM	PB
PM	PM	PM	PB	PB
PB	PM	PB	PB	PB

The experimental data show that the acceleration of pedal angle and pedal pressure are generally higher than that of normal tread, and the measurement data can be used to judge whether or not to step on the accelerator pedal. If you step on the accelerator, causing alarm and emergency braking (Table 3).

Table 3. Test results

Project	Test number	Number of successful	Number of failures	Precision rate
Normal pedal	1000	1000	0	99.8%
stepping on the pedal by mistake	1000	996	4	

5 Conclusion

The intelligent accelerator system, as described above, can be used to analyze the pedal acceleration, pedal pressure and the distance between the car and the obstacle in front of the vehicle, and determine whether to step on the accelerator. The system realize when the driver accidentally step on the accelerator pedal, equivalent to step on the brake pedal. It is of great significance to protect the safety of people's life and property.

Acknowledgments. Jilin provincial education department project 《Automatic brake device design of automobile intelligent anti – error》 (project number: JJKH2017162KJ).

Supported by Program for Innovative Research Team of JiLin Engineering Normal University.

References

1. Jia, J.: Design concept of anti-misstep accelerator device. Automot. Electron. **8**, 9–10 (2006)
2. Mei, Z., Li, X.: Auto throttle control step on emergency brake device. J. Mech. Des. Manuf. **10**, 201 (2010)
3. Zhenqi, H., Zhu, C., Li, J.: Car and step on the accelerator system development. J. Automob. Eng. **8**, 713 (2011)
4. Mingji, W., Niu, H., Zhang, Y.: Auto throttle control step on intelligent control system design. J. Electron. Des. Eng. **2**, 57–62 (2017)
5. Wang, L.: Based on ARM + DSP control system of automobile rear-end collision prevention study. J. Jing Gang Shan Univ. Nat. Sci. Edit. **36**(1), 79 (2015)
6. Honghao, H.: The automotive anti-collision system based on the theory of grey fuzzy simulation study. J. Syst. Simul. **25**(11), 2791–2792 (2013)
7. Xiaojing, G.: On how to improve the effectiveness of college English translation teaching. Engl. Teach. **18** (2015)
8. Tairong, C.: Effective measures of English teaching translation. Overseas Engl. **20** (2015)
9. Hao, W.: On college English translation teaching models in the internet environment. Engl. Sq. **2** (2016)
10. Ming, F.: The cultural elements in English translation teaching. Engl. Sq. **20** (2014)
11. Honglian, W.: The problems and countermeasures of english translation teaching in College of Traditional Chinese Medicine. Overseas Engl. **20** (2014)
12. Watts D.J.: Small worlds. The dynamics of networks between order and randomness. America: 56A2579BB6, 1999-05-13

The TPMS Module in the Vehicle Positioning and Safety Warning System

Fang Shao(✉) and Yue Wu

JiLin Engineering Normal University, Changchun, Jilin, China
182089853@qq.com

Abstract. In this paper, we complete all the designs of vehicle active safety warning system and the design of tire pressure detection module; In the software algorithm, getting rid of the conventional fault location methods fundamentally, using positioning technology for traffic parameters, vehicle to vehicle's communication based on wireless communication technology, It also can analysis running state parameters of the relative velocity, acceleration, relative distance comprehensively, predicting vehicle's rear end collision dangerous degree under the environment of the running automobile, and the abnormal alarm; we can realize the anti collision model simulation system through software simulation, validating the model under extreme conditions can still effectively preventing rear end collision warning system. The system can meet the requirement of the vehicle positioning and safety warning system.

Keywords: Active safety · Positioning · TPMS module
Anti collision simulation

1 The Design of TPMS Sensor Module

Tire pressure monitoring used for real-time monitoring of tire temperature, air pressure, and providing the two main parameters of the specific values and the change trend in real-time, judging whether a fault exists in tire for the driver, to remind the driver to take corresponding measures in advance, to prevent aggravation of the risk. Tire pressure detection using the SCM technology, sensor technology and wireless radio frequency technology, is not only an important branch of vehicle active safety, but also one of the important topics of intelligent transportation research, so the study of fetal pressure detection has important significance [1].

2 Classification and Working Principle of TPMS

Compared with the comparison of various tire pressure detection scheme can be obtained according to the analysis method of tire pressure data and the degree of accuracy is divided into two major categories of indirect tire pressure detection system, which is based on the speed difference between the tire pressure detection system (Wheel-Speed Based TPMS, WSB TPMS); direct tire pressure monitoring system, which is based on the sensor the tire pressure monitoring system (Pressure-Sensor

J. Abawajy et al. (Eds.): ATCI 2018, AISC 842, pp. 1307–1314, 2019.
https://doi.org/10.1007/978-3-319-98776-7_162

Based TPMS, PSB TPMS). Two kinds of direct tire pressure detection system can be divided into active and passive, the classification of TPMS as shown in Fig. 1.

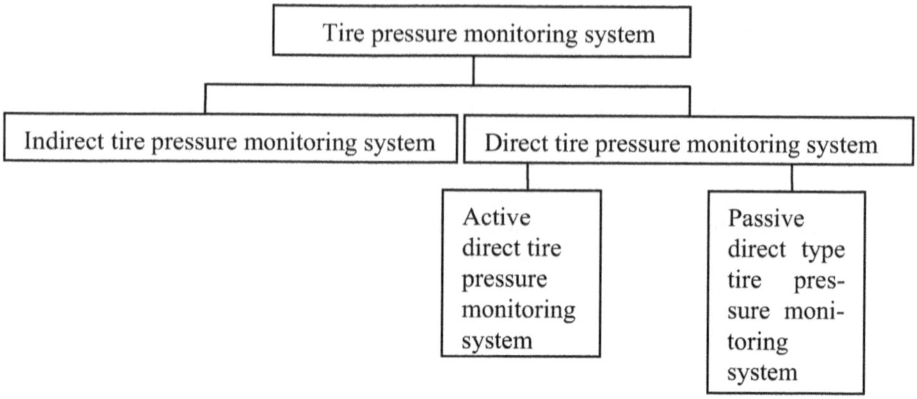

Fig. 1. Classification of tire pressure monitoring system.

2.1 Indirect Tire Pressure Detection System

The working principle of the indirect tire pressure detection system: the automobile tire has certain elasticity, the inner side of the tire is subjected to pressure, the outside is under tension, and the change of the radius of the tire is caused. A change of the tire radius leads to a certain speed difference exists between the wheel of the car. When the car running the automobile anti lock braking system (ABS) speed sensor measured the speed of the four wheels, the vehicle central controller through the integrated comparison and analysis is obtained indirectly four tires relative pressure conditions. The indirect tire pressure detection system is actually based on the measurement of the relative inflation pressure of the tire [3].

Indirect tire due to the pressure of different sizes lead to four tires with different radius, vehicle operation, will inevitably lead to four wheels rotate at different rotating speed, through a comparative analysis of the diagonal of the automobile wheel speed sum to judge whether there is pressure drop.

The formula for calculating the formula of the indirect type is as follows:

$$\omega = \upsilon/\gamma \tag{1}$$

$$T = \left|\frac{(\omega_{1f} + \omega_{ff}) - (\omega_{rf} + \omega_{1r})}{v}\right| = \left|\left(\frac{1}{r_{1f}} + \frac{1}{r_{rr}}\right) - \left(\frac{1}{r_{rf}} + \frac{1}{r_{1r}}\right)\right| \tag{2}$$

The formula of the indirect type is calculated by the above formula T absolute value and the system to set the safety threshold of T comparison, when T > T, the car central controller to control the alarm device.

2.2 Direct Tire Pressure Detection System

Direct tire pressure monitoring system, the pressure sensor, temperature sensor and other auxiliary hardware directly installed in the automobile tire, pressure sensor, temperature sensor measurement every tire pressure and temperature values, hardware assisted the measurement of pressure and temperature value according to certain modulation in wireless radio frequency (RF) are sent out and installed in car dashboard near the signal receiving device signal demodulation and be displayed. At the same time signal receiving device to determine whether to issue a warning information. Figure 2 wireless signal transmission schematic diagram of direct type tire pressure detection system.

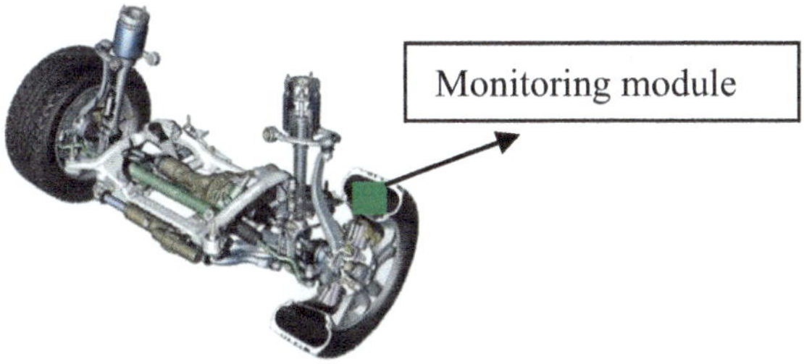

Fig. 2. Classification of tire pressure monitoring system

Active direct tire pressure monitoring system is will battery device is embedded in the detection module and mounted directly to the tire, drive system pressure sensor, temperature sensor, A/D converter, micro controller (micro control unit, MCU, RF module and digital interface work to complete the launch of tire parameter detection and detection result. The schematic diagram of the system is shown in Fig. 2.

Active direct tire pressure sensor testing system is mainly based on MEMS technology, according to the working principle of the pressure sensor in the detection system is different, mainly divided into two types: monitoring system based on capacitance variation and monitoring system based on the acoustic surface waves effect of Si (Fig. 3).

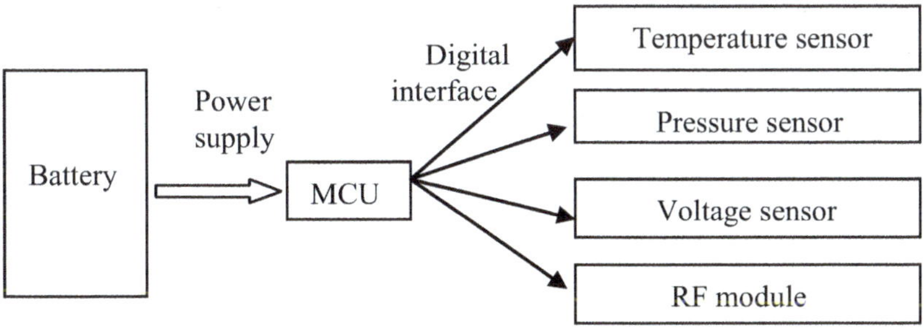

Fig. 3. The detection module of the active direct tire pressure detection system

Based on the acoustic surface waves effect of direct tire pressure monitoring system is using sensor based on acoustic surface waves effect of semiconductor GUI made direct measurement of internal pressure of the tire and its working principle: when under pressure, the pressure sensitive element produces the corresponding strain deformation, due to changes in the GUI as a result of the acoustic surface waves effect of the silicon semiconductor conductivity, causing Megumi stone bridge resistance changes, resulting in the loss of balance bridge and the output voltage, and according to the voltage measurement of change in pressure.

Based on capacitance variation of direct tire pressure monitoring system is using pressure capacitance measurement of change in pressure, working principle: using MEMSVV technology will pressure sensitive components are encapsulated into a capacitor, when the role of external pressure on the plates, the sensitive element deformation cause the capacitance change, causing variation in voltage between the two electrode plates, according to the change of voltage measuring pressure.

Based on the acoustic surface waves effect of direct tire pressure monitoring system is using sensor based on acoustic surface waves effect of semiconductor GUI made direct measurement of internal pressure of the tire and its working principle: when under pressure, the pressure sensitive element produces the corresponding strain deformation, due to changes in the GUI as a result of the acoustic surface waves effect of the silicon semiconductor conductivity, causing Megumi stone bridge resistance changes, resulting in the loss of balance bridge and the output voltage, and according to the voltage measurement of change in pressure.

2.3 Performance Requirements for TPMS

The automobile active safety warning system designed in this paper is mainly used in small cars and private cars, so the pressure test range of the tire pressure detection module in the system should include the standard tire pressure value of the small car. The standard tire pressure of a small car is generally around 210 KPa. Because of the own vehicle design, load and road condition is different, the tire pressure will fluctuate in a certain range, but not beyond the normal range 180 KPa–280 KPa. So the measuring range of the system detection module must to contains this range. According to the people's Republic of China on the automobile tire air pressure monitoring system of the standard system detection module measurement range should be selected as 100 KPa–450 KPa.

Tire pressure detection module is installed in the small space inside the tire, the monitoring module size, will drive module installation and tire maintenance work; quality of modules will tire eccentricity, when the car was traveling at a high speed lead to high frequency vibration, damage to the vehicle suspension mechanism and effect of driving comfort. So the volume and quality of the test module must be small enough. The monitoring module is embedded in the tire, the module is very inconvenient for the disassembly and maintenance, and it is necessary to require the detection module to work stably under the complex and bad working conditions, and prevent the detection module from running.

Each vehicles in general will have four monitoring module, each module are working in the same frequency band and each module emission RF signal sequence and time can not determine, is bound to cause the mutual disturbance and conflict between the four modules; when the installation device which is the same as two car is close to the distance, will cause conflicts and of interactions between the two sets of system detection module. So the system must be able to accurately identify whether the detection module belongs to its own system and quickly eliminate the collision between modules.

2.4 Selection of Chip Type for Tire Pressure Monitoring

Tire pressure monitoring function of the chip selection including the selection of the chip selection and module of the chip selection. Detection module chip selection must be reliability considering the volume, quality, chip and the chip working conditions. In addition, it is necessary to take into account, inter Alia, the chip of low power characteristics; receiving module of chip selection can do not consider the chip size, quality and low power consumption, but must consider the RF signal receiver sensitivity and fault tolerance.

Detection Module Chip Selection. The first stage and the second stage of the chip as discrete components, each a chip occupied printed circuit board area is with constant and each with a chip of the auxiliary circuit to occupy space on the circuit board, resulting in an increase in the volume and quality of detection module, and also increase the module complexity and hardware cost; the lines of communication between a plurality of discrete chips by the printed circuit board wiring connections, and chip pin and go line dry pick easy to fall off, cause the module reliability decreased. Considering the factors of volume, quality, reliability and development cost of the system, the third stage of the chip is selected as the monitoring module of the system. Currently, Infineon, general electric and Freescale three companies provide the third stage TPMS special sensor chip solutions, respectively SP3x series, NPX2 and mpxy8300.

MPXY8300 standby current is the smallest, and has the highest measurement accuracy. MPXY8300 select sensor as detection module. The performance parameters of MPXY8300 are shown below:

1. 10 bits precision pressure sensor, temperature sensor, X axis and Z axis acceleration sensor;
2. detection of battery voltage by A/D converter with 10 bits precision;
3. there are 4 10 bits precision A/D converters;
4. S08 kernel with SIM, interrupt and debug monitor of the 8 bit MCU, and there are FLASH 8 K;
5. the internal integration of the 315/434 MHz based RF PLL transmitter output;

   - amplitude shift keying and frequency shift keying modulation function;
   - programmable data baud rate generator;
   - support Manchester data encoding;
   - 128bit RF data buffer, RTS/CTS supports authentication function;
   - Working voltage charge of millet.

6. low frequency detector for differential input;

7. 4 ordinary I/O pins, support on pull/drop function and interrupt wake-up function:
8. a low frequency oscillator driver with a low power consumption wake-up timer and a support cycle reset;
9. low voltage detection, low voltage reset.

Chip Selection of the Tire Pressure Receiving Module. The radio frequency receiving module of the receiving module belongs to the central information processing system, which is mainly responsible for the signal receiving and demodulation. Therefore, the receiver chip can not have low power consumption, but it must have a high signal reception sensitivity and anti-jamming performance. In the system, we choose MC33596 as the radio frequency signal receiving chip, which is used by the Freescale company to produce the special Yu Chao high frequency radio frequency signal. The performance parameters of MC33596 are as follows:

1. the working frequency band is 915 MHz–304 MHz;
2. receiving OOK and SOK modulated radio frequency signals;
3. working voltage 3.6 V–2.1 V and 5 V optional;
4. SPI digital communication interface;
5. decoding the radio frequency signal based on Manchester, the highest data acceptance rate is 20 Kkbps;
6. the receiving sensitivity of −106.5 dBm;
7. Automatic wake-up function;
8. the working mode can be switched;
9. Receive Indicator (Signal Strength, RSSI) to support the digital and analog signal reception.

2.5 Software Design Requirements

Tire pressure detection module performance directly affects the whole system of tire pressure and temperature pre propagation effect, and module of the software design directly determines the module performance. Therefore, the detection module software design must be through many aspects into account.

Low Power Design. The tire pressure detection module is restricted by various factors, which can not be disassembled, installed and maintained regularly. At present, for most active tire pressure monitoring system, the battery life is the most important factor to limit the life of the module. Increase the capacity of the battery and reduce the overall power consumption of the module can solve the problem, and to increase the battery capacity is bound to increase the volume and weight of battery, thus reducing the overall power consumption of the module has become the key to increase the life module.

Reducing the overall power consumption of the module can take the following measures:

(1) using the appropriate clock frequency. Module clock frequency is directly determined by the crystal oscillator frequency, and module of the clock frequency and the system power consumption is proportional to; crystal oscillator start-up

time also directly affects the system's power consumption, startup time is long, the power consumption of the system.

(2) reduce the normal mode of operation of module MCU, using the mechanism of sleep wake up. Mpxy8300 belongs to high integrated on-chip system, has the normal operation mode (run mode, running in the background mode (active background mode) mode (wait mode) and wait stop mode stop mode) four working modes, the stop mode and stop mode, stop mode, stop mode 3 and stop mode 4 state, table 3.3 mpxy8300 four stop mode comparison.

(3) the control of the length of the launch data, the number of times the launch of the RF open and close. When the automobile is in a low speed state or stop state, the tire parameter information is forbidden, and the RF module is closed, and the high frequency oscillator is closed. When the vehicle speed is higher than 30 km/h, the measurement of tire parameters and emission measurement results, with the increase of the speed and the frequency of the measurement.

Tire Pressure Compensation. The standard inflation pressure of the automobile tire is under the condition that the environment temperature is 18. In the case of continuous operation of the vehicle, the tire is alternately deformed and the temperature of the tire

Table 1. MCU's stop mode comparison

Stop mode	Stop mode1	Stop mode2	Stop mode3	Stop mode4
LFO oscillator and PWU	Always open the work and have a clock			
MFO oscillator	Select Open	Select Open	Select Open	Select Open
HFO oscillator	Off state	Off state	Off state	Off state
MCU	Off state	Off state	Off state	Off state
RAM	Off state	Open state	standby mode	standby mode
Parameter register	Select Open	Select Open	Select Open	Select Open
FLASH	Off state	Off state	standby mode	standby mode
TIMER/PWM	Off state	Off state	Off state	Off state
DIGITAL I/O	Off state	Off state	standby mode	standby mode
PCI INTERFACE	Off state	Off state	standby mode	Select Open
Pressure sensor	Off state	Off state	standby mode	Select Open
Analog interface ACI	Off state	Off state	standby mode	Select Open
Acceleration sensor	Off state	Off state	standby mode	Select Open
Temperature sensor	Off state	Off state	Off state	Select Open
Normal temperature reset	Select Open	Select Open	Select Open	Select Open
Voltage reference	Off state	Off state	Select Open	Select Open
LFR detector	Select Open	Select Open	Select Open	Select Open
LFR DECODER	Open state	Open state	Open state	Open state
RFX charge pump	Select Open	Select Open	Select Open	Select Open
RFX data buffer	Open state	Open state	Open state	Open state
RFX transmitter	Select Open	Select Open	Select Open	Select Open
SPI	Off state	Off state	standby mode	standby mode
Wake up mode	Interrupt reset	Interrupt reset	Interrupt reset	Interrupt reset

deviates from the standard temperature, which causes the change of the tire pressure. Experiments show that when the temperature of the tire at 0–24 °C, tire inflation pressure has little change, approximate to a constant value without modifying; when the tire temperature is greater than or 24 °C is less than 0 °C, tire inflation pressure changes obviously and need to modify the tire pressure.

The British Michelin (Michelin) Company on the tire test, obtained the −40 0 °C - 25 C and 45 C at different temperatures on the tire pressure correction of 25 - 45 - 3.8° when the tire pressure correction (Table 1).

3 Concluding Remarks

The classification of tire pressure monitoring system and the working principle of the system are analyzed, and the development stage of the direct type of tire pressure sensor is introduced in detail. According to the requirement of the tire pressure monitoring system, the sensor of the detecting module is determined according to the comprehensive comparison of the current mainstream sensors. The hardware circuit of the tire pressure detection module, especially the antenna design, finally gives the design requirements of the software.

Acknowledgements. Education Department of Jilin province in 2017: Research on Parking Guidance System Based on Ultrasonic-Ranging.

Supported by Program for Innovative Research Team of JiLin Engineering Normal University

References

1. Kalinin, V., Lohr, R.: SAW based TPMS-theory and practice. In: Tire Technology EXPO 2004, Stuttgart, Germany (2004)
2. Benver, T., Kandler, M.: Solution for tire pressure monitoring systems. In: International Conference on Advanced Micro Systems for Automotive Applications (2003)
3. He, Z.J., Niu, Y.Q.: The new technology of automobile TPMS. Res. Dev. **7**, 60–62 (2014)
4. NHTSA Federal Motor Vehicle Safety Standards; Tire Pressure Monitoring Systems; Controls and Displays
5. ECE R64 Revision 1-Amendment 2 Uniform provisions concerning the approval of vehicles with regard to their equipment which may include: a temporary use spare unit, run flat-tyres, and/or a run flat-system, and/or a tyre pressure monitoring system
6. Qizhen, W.: Automatic control system regulation scheme of textile air conditioning. Sci. Technol. Inf. **22**, 395–396 (2010). (in Chinese)
7. Jinkun, L.: Advanced PID Control and MATLAB Simulation, 2nd edn. Electronics Industry Press, Beijing (2003). (in Chinese)
8. Jianwei, L., Qingchang, R.: Study on supply air temperature forecast and changing machine dew point for variable air volume system. Build. Energy Environ. **27**(4), 29–32 (2008). (in Chinese)
9. Peng, L., Fang, W.: Heterogeneity of inferring reputation of cooperative behaviors for the prisoners' dilemma game. Physica A **433**, 367–378 (2015)
10. Shizhao, N., Zheng, W., Wangmo, P., Yuan, N., Peng, L.: Big data prediction of durations for online collective actions based on peak's timing. Physica A **55**, 130–139 (2018)

Study on Risk-Benefit Quantification of Floodwater Resources Utilization

Lixiang Zhu[1(✉)] and Xianfeng Huang[2]

[1] Tongyu River North-Extended Water Supply Project Management Division of Lianyungang, Lianyungang 222005, China
lygzlx@163.com

[2] College of Water Resources and Hydropower Engineering, Hohai University, Nanjing 210098, China
xfhuang@hhu.edu.cn

Abstract. This paper combines the risk factors of floodwater resources utilization to quantify the risk-benefit of floodwater resources after adjusting limit water level. The limit water level is a key level involved the flood risk and benefits of the reservoir. After adjustment of limited water level, each risk rate correspond to each water level is obtained. Proportional distribution of increasing water content after adjusting the flood control level. Risk-benefit is comprehensive considered including industry, agriculture, life and ecology. The four benefits are quantified by C-D production function, the sharing coefficient method, water price based on energy and the energy theory. This paper took Shilianghe reservoir of Lianyungang City as the research case. Risk rates and benefits of each limit water level are calculated. Under the conditions of minimum risk and maximum benefits, the result that the control range of the limit water level is 23.9 m–24.9 m at Shilianghe reservoir can be obtained according to the relationship between the limit water level, risk rate and risk benefit.

Keywords: Flood utilization · Risk · Benefit · C-D production function
Energy analysis method

People's lives and property are threatened with frequently flood disaster in China, therefore, more and more scholars began to get involved in the field of flood resources utilization, and made some achievements in the utilization of flood resources and the calculation of benefit (Wang et al. 2004, 2009). However, there is a risk of flood utilization, how to control risk and improve efficiency, which is a particularly important issue in the process of flood resources utilization.

Foreign scholars pay more attention to the prevention of flood risk, but seldom mention the benefits of flood utilization. Roy and van Remco (2004) studied benefit evaluation research for the flood management policy in Holland, and he used CBA (Cost-Benefit Analysis) and MCA (Multi-Criteria Analysis) two methods to calculate the benefits. Due to Holland is more emphasis on flood control and disaster reduction, the study is aimed at the benefit of flood control policy, but the advanced benefit calculation method can be used as reference. Xu and Li (2009) established the evaluation index system and model of economic benefit about rainwater resources. Feng et al. (2013) built a social benefit evaluation model, which can be used to quantitatively evaluate the social benefits of flood resources. Fang et al. (2014) combined the dynamic control of the

J. Abawajy et al. (Eds.): ATCI 2018, AISC 842, pp. 1315–1324, 2019.
https://doi.org/10.1007/978-3-319-98776-7_163

reservoir limited water level with the short-term rainfall forecast, and then the utilization of flood water resources is calculated with the method for the long-serial regulation.

At present, there are many ways to transform the flood into resources, and the most common is through the adjustment of limited water level of reservoir. The limit water level is a critical level which can affect the reservoir flood risk and benefit, so we must correctly deal with relationship between risk and benefit (Feng and Han 2007). Considering the limit water level, this paper calculated flood risk rate and benefit value, and elucidated the quantitative analysis method for industrial, agricultural, life and ecological benefits.

1 Research Methods

1.1 Risk Analysis on Flood Resource Utilization

Flood resources as one of unconventional water resources, should be used on condition that the project is safety, and has no bad effect on ecological environment. In order to meet these conditions, flood resource utilization risk identification must be carried out. By identifying risks, the main risks are the limit water level rising, the adverse ecological impact, reservoir upstream uncertainty, water quality uncertainty and so on (Huang et al. 2015). The risk produced by adjusting limited water level is considered here. The major influence factors include the risk rate of dam safety and the risk rate that discharge flow overtakes the safety standard.

Using the method of mathematical statistics, the statistical data of the upstream reservoir duration as the basis for flood routing, find out the flood that made the reservoir reach the design standard water level, and take the probability of flood occurrence as the dam safety risk rate. The formula for calculation is as follows:

$$f_1(H_i) = \frac{n}{N+1} \times 100\% \tag{1}$$

where, $f_1(H_i)$ is the risk rate of dam safety while the limit water level is H_i, N is the total number of previous floods in the data series, n is the number of floods that the reservoir water level exceed design standard water level.

The risk rate that discharge flow overtakes the safety standard is calculated by using fuzzy mathematics method. The formula for calculation is as follows:

$$f_2(H_i) = \sum\nolimits_{V_i \in A} \mu_A(V_i) \times p(V_i) \tag{2}$$

where, $f_2(H_i)$ is the risk rate of excessive discharge while the limit water level is H_i, $p(V_i)$ is the probability of discharge belongs to the range of V_i, $\mu_A(V_i)$ is membership degree corresponding to V_i.

To estimate the comprehensive risk rate in follow equation

$$f(H) = w_1 f_1(H_i) + w_2 f_2(H_i) \tag{3}$$

where, w_1 is the weight of the risk rate of dam safety, w_2 is the weight of the risk rate of excessive discharge.

1.2 Benefit Quantization Method on Flood Resources Utilization

1.2.1 The Benefit of Industrial Water Supply

There are some commonly methods used to quantify the benefits of industrial water supply, including the residual method, economic contribution method, sharing coefficient method and water loss method. But in practice, these methods have some deficiencies, for example, economic contribution method produces larger results than true value. Considering the advantages and disadvantages of each method, the paper uses the Cobb Douglas (C-D) production function method to calculate the benefit of industrial water supply. This method considers investment returns of the non-water elements in industrial production. At the same time, through convert the elasticity of water demand, the influence of technological progress and government policy on the gross industrial output value is eliminated to make result more accurate (Xu and Li 2009).

1. C-D production function construction
   When calculating the benefits of industrial water supply, only consider the impact of industrial water on industrial production. Therefore, the water can be put into the production function as a combination of third inputs. Refer to the form of C-D function, the formula for calculation is as follows

$$Q = A \cdot K^{\alpha} \cdot L^{\beta} \cdot W_1^{\lambda} \tag{4}$$

   where, Q is total industrial output value, A is efficiency coefficient, K is industrial fixed assets, L is industrial labor force, W is industrial water consumption, α is fixed assets elasticity, β is labor force elasticity, λ is water use elasticity.
2. Water elasticity calculation
   The water elasticity is obtained by partial derivative of production function. In order to eliminate the influence of scale elasticity, introducing the converted water elasticity λ'.

$$\lambda = \frac{\partial Q}{Q} \bigg/ \frac{\partial W_1}{W_1} \tag{5}$$

$$\lambda' = \frac{\lambda}{\alpha + \beta + \lambda} \tag{6}$$

3. Benefit calculation
   The benefit per cubic meter is equal to partial derivative of industrial water. Hence, the benefit of industrial water supply can be written as

$$B_W = \frac{\partial Q}{\partial W_1} = \lambda' \frac{Q}{W_1} \tag{7}$$

$$E_1 = B_W \times W_1 \tag{8}$$

where, E_1 is the benefit of industrial water supply, B_W is the benefit per cubic meter of industrial water supply.

1.2.2 The Benefit of Agriculture Irrigation

In the economic evaluation of water conservancy construction project (SL 72-2013), it is recommended to use the benefit sharing coefficient method, the shadow price method and the water loss method to calculate the benefits of agricultural irrigation. However, the shadow price method has a certain geographical limitation, because it needs a large amount of data. The water loss method also requires a large number of long series of data, and involves optimal allocation problem. Therefore, the benefit sharing coefficient method is used to quantify the benefits of agricultural irrigation, which is the most widely used. The key of this method lies in the determination of the benefit sharing coefficient (Wang 2011). The sharing coefficient can be determined by statistical method and test method. The result of the test method is very limited, and the test cost is higher, so the statistical method is more convenient. To calculate benefit of agriculture irrigation in follow equation

$$E_2 = Vqk = (I/W_2)fqk \tag{9}$$

where, E_2 is the benefit of agricultural irrigation, V is the value of unilateral water, q is the volume of increased water supply, k is the utilization ratio of water resources, I is the industrial added value, W is irrigative water consumption, f is allocation coefficient.

1.2.3 The Benefit of Domestic Water Supply

At present, there is no specific specification for the calculation method of the benefit of domestic water supply (Fan and Nie 2014). The energy price method applied in this paper, and the price consists of water price of resource, water price of project and water price of environment. The energy input of domestic water consumption includes natural environmental resources and social economic resources investment. Through the conversion rate of the input indicators can be converted into a unified energy value, and it can get all the energy input on the living water supply. The total energy value can be expressed in the form of money according to the value of the currency ratio, which can get the water price under the energy calculation method. Thus, the prices and benefit can express as follows

$$P = P_z + P_g + P_h \tag{10}$$

$$\begin{cases} P_z = \dfrac{C_{yzr}}{Q_g \cdot EDR} \\ P_g = \dfrac{C_{gzr} + C_{gsj}}{Q_g \cdot EDR} \\ P_h = \dfrac{C_{pzr} + C_{psj}}{Q_p \cdot EDR} \end{cases} \tag{11}$$

$$E_3 = P \cdot W_3 \tag{12}$$

where, P is living water price of city town residents, P_Z is water price of resource, P_g is water price of project, P_h is water price of environment. $C_{yzr}, C_{gzr}, C_{gsj}$ are the energy of the material, the natural environmental resources, and the social economic resources that have been invested to form water resources. Q_g is the volume of water supply, Q_p is the volume of sewage discharge.

EDR is the ratio of energy to currency, E_3 is the benefit of domestic water supply, W is domestic water consumption.

1.2.4 The Benefit of Ecological Water Supply

The benefit of ecological water supply is an important part of the utilization of flood resources, but it is difficult to quantify because of its complexity and wide scope. Most quantitative methods are generally considered from an economic point of view, and only for a certain aspect of the ecological benefit. The advantage of emergy analysis method is that it can quantify multiple benefits (Buenfil 2001). The benefit of ecological water supply is calculated as follows:

$$E_4 = \frac{W_4}{W}\left(\tau_s NR + m_i\tau_i + E\tau_q + W_t\tau_t\right) \tag{13}$$

where, E_4 is the benefit of ecological water supply, W_4 is ecological water consumption, W is total water consumption. $\tau_s, \tau_i, \tau_q, \tau_t$ are the transformity of species, the i-th pollutant, steam and water. N is the number of species of aquatic organisms, R is the proportion of the active area of aquatic life in global area, m_i is the pollution receiving capacity of the i-th pollutant, E is water evaporation energy, W_t is annual water storage capacity of lakes or wetlands.

2 Case Study

2.1 General Situation of Lianyungang

Lianyungang city is located in the northeast of Jiangsu Province, and belongs to Huaihe valley. Xinyi river and Xinshu river which are the main flood channel of Huaihe valley flow through the Lianyungang into the sea. The channels undertake the flood discharge task nearly eighty thousand square kilometers from upstream. Three transit channels in the city are Xinyi river, Xinshu river and Longwang river. The average annual flood of these rivers reached about six billion cubic meters, which provides a guarantee for flood resources utilization. In addition, the city has many reservoirs, among them Shilianghe reservoir plays an important role in flood regulation and storage. The reservoir has about 15365 km^2 catchment area, and 5.31 × 108 m^3 total storage capacity. Design and check flood levels of the reservoir are 26.81 and 27.95 m respectively. The maximum discharge is 10131 m^3/s.

2.2 Risk and Benefit Estimation of Flood Resources Utilization

According to the statistical yearbook of Lianyungang and the data of water resources bulletin, the proportion of water resources in industry, agriculture, life and ecological water supply are 16.5%, 69%, 12.1%, and 2.4%. The increased water storage was allocated in accordance with the above ratio, and then multiplied with water benefit per cubic meter, in this way the benefit values of different risks were obtained. Furthermore, the original limited water level of Shilianghe reservoir was 23.5 m, and values were calculated when the water level was increased by 0.2 m. The results of different types of water benefits are shown in Tables 1 and 2. The relation curve between benefit value and water level is given at Fig. 1.

Table 1. Benefit per cubic meter of flood resources utilization

Type	Water for industry	Water for agriculture	Water for life	Water for ecology
Benefit per cubic meter (RMB/m^3)	2.4	0.227	6.05	39.77

Table 2. Risk and benefit of flood resources utilization

Limited water level (m)	Increased water storage ($10^4 m^3$)	Water for industry		Water for agriculture		Water for life		Water for ecology		Total Benefit (10^8 RMB)
		Water allocation ($10^4 m^3$)	Benefit (10^4 RMB)	Water allocation ($10^4 m^3$)	Benefit (10^4 RMB)	Water allocation ($10^4 m^3$)	Benefit (10^4 RMB)	Water allocation ($10^4 m^3$)	Benefit (10^4 RMB)	
23.5	0.00	0.00	0.00	0.00	0.00	0.00	0.00	0.00	0.00	0.000
23.7	1069	176.39	423.34	737.61	167.44	129.35	782.57	25.66	1020.5	0.239
23.9	2138	352.77	846.65	1475.22	334.87	258.70	1565.1	51.31	2040.6	0.479
24.1	3291	543.02	1303.3	2270.79	515.47	398.21	2409.2	78.98	3141.0	0.737
24.3	4526	746.79	1792.3	3122.94	708.91	547.65	3313.3	108.62	4319.8	1.013
24.5	5761	950.57	2281.4	3975.09	902.35	697.08	4217.3	138.26	5498.6	1.290
24.7	6996	1154.34	2770.4	4827.24	1095.8	846.52	5121.5	167.90	6677.4	1.567
24.9	8231	1358.12	3259.5	5679.39	1289.2	995.95	6025.5	197.5	7856.2	1.843
25.1	9548	1575.42	3781.0	6588.12	1495.5	1155.31	6989.6	229.2	9113.3	2.138
25.3	10947	1806.26	4335.0	7553.43	1714.6	1324.59	8013.8	262.7	10449	2.451

According to Table 2 and Fig. 1, reservoir storage capacity increases, which makes all kinds of water benefit values increase with the rising of the flood control level. Obviously, the allocation proportion and consumption of ecological water supply reaches minimum, but growth rate of benefit value is fastest, and the rate of benefit and water input reaches maximum. While the agricultural water consumption is on the contrary.

Different flood control level corresponds to different risk rate. Flood level meeting the design standards corresponding to reservoir flood control level is found on the basis

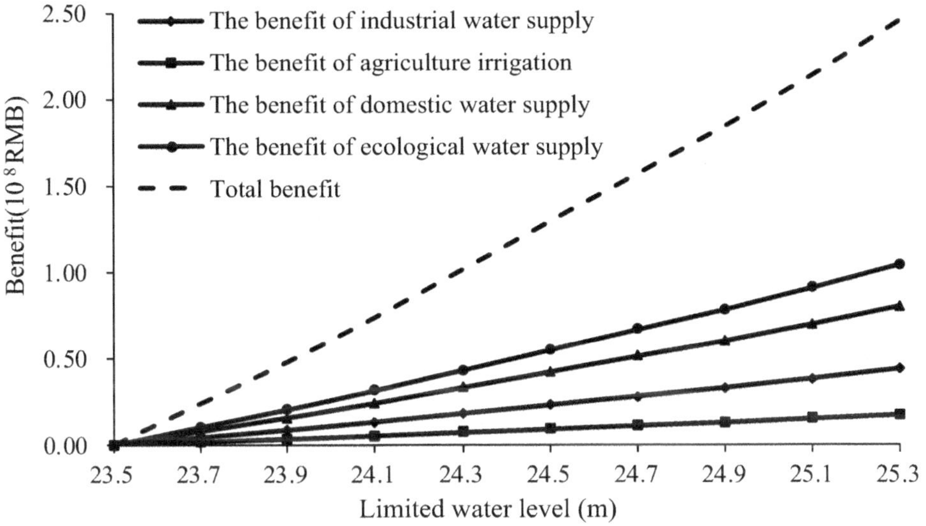

Fig. 1. Relation Curve of benefits and limited water level

of historical statistical datas of Shilianghe reservoir upstream, and flood probability is regarded as risk rate of dam safety corresponding to the flood control level; to calculate the risk rate when water discharge is more than downstream river safety discharge on the condition that reservoir water discharge is not more than safety discharge of Xinshu river. To insure risk rate of dam safety is different from the risk rate above, the difference between flood routing of maximum water level and standard value of reservoir design is less than 26.81 m. The risk rate of dam safety, the risk rate when water discharge is more than downstream river safety discharge and compositive risk rate are calculated according to formula 1, 2 and 3 respectively. The risk loss from dam safety risk is very heavy among total risk loss according to relevant documentation, therefore the weight of dam safety risk rate is 0.6, the weight of risk rate when water discharge is more than downstream river safety discharge is 0.4. The result of risk rate is given at Table 3, curve between risk rate and change of water level is given at Fig. 1. The relationship between compositive risk rate and total benefit is given at Fig. 3 (Fig. 2).

According to Table 3 and Fig. 1, we find that the risk rate of dam safety and excessive discharge are gradually increased with the rise of limited water level. The dam safety risk rate changes gently and the rate is less than 5% when limited water level below 24.9 m. Therefore, to take 24.9 m as the limited water level could guarantee the risk within control. By the calculation of flood routing, discharge volume of the reservoir have significant effects on the risk rate. The risk rate reduced by about 5% as the discharge volume increases 200 m^3/s when limited water level over 24.5 m. If we continue raising limited water level with keep the dam flood safety risk rate unchanged, it will increase the risk of downstream channel, and make the risk transfer

Table 3. Relationship between compositive risk rate and total benefit

Limited water level (m)	Total Benefit (10^8 RMB)	The risk rate of dam safety (%)	The risk rate of excessive discharge (%)	Comprehensive risk rate (%)
23.5	0.000	1.0	10.2	4.7
23.7	0.239	1.1	10.3	4.8
23.9	0.479	1.4	10.4	5.0
24.1	0.737	1.7	11.2	5.5
24.3	1.013	2.1	12.5	6.3
24.5	1.290	2.8	13.8	7.2
24.7	1.567	3.5	16.5	8.7
24.9	1.843	4.2	19.1	10.2
25.1	2.138	6.3	23.0	13.0
25.3	2.451	9.2	28.1	16.8

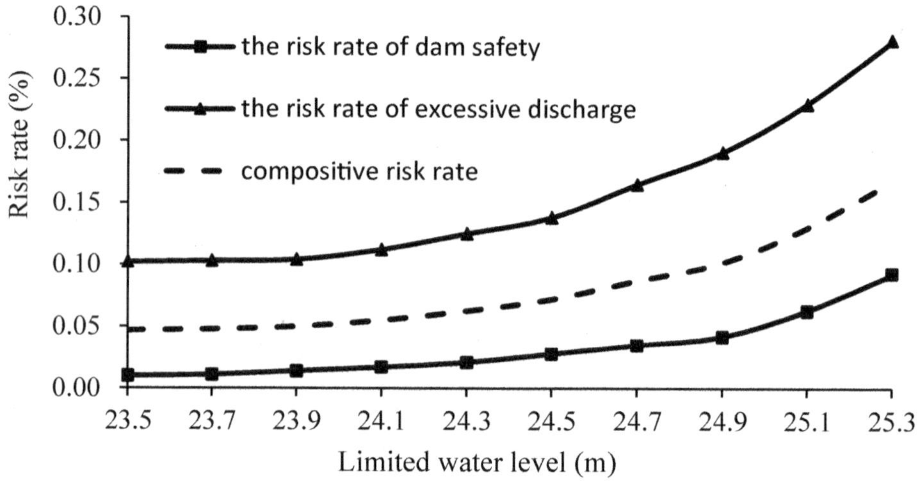

Fig. 2. Relation Curve of risk rates and limited water level

from the dam to downstream channel. Figure 3 shows water capacity is increased in reservoir with rise the limited water level, and the comprehensive risk rate is also gradually increased at the same time. After limited water level reaches 23.9 m, the total benefit value continues growth with the increase of comprehensive risk rate, but the trend gradually leveled off. Under the conditions of minimum risk and maximum benefits, the result demonstrated that the control range of the limit water level is 23.9 m–24.9 m at Shilianghe reservoir.

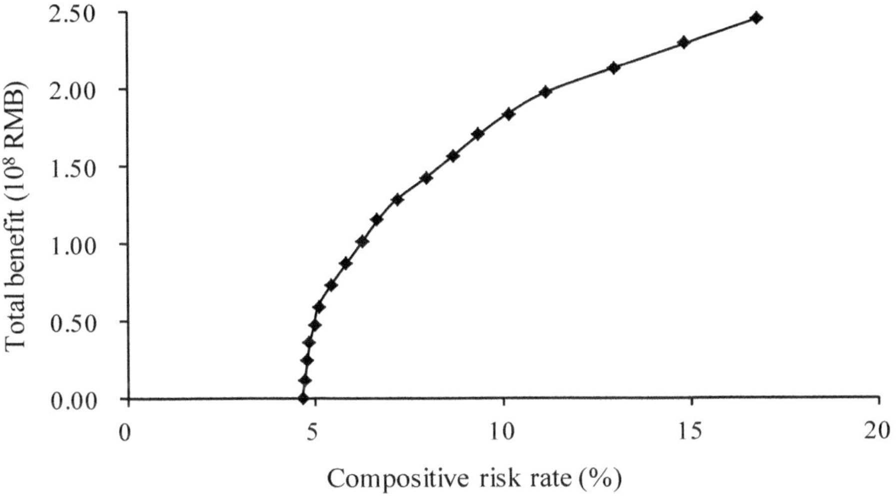

Fig. 3. Relation Curve of compositive risk rate and total benefit

3 Conclusions

On the premise of considering flood risk factors, this paper quantified the benefits of flood resources utilization. The influence factors include the risk rate of dam safety and the risk rate that discharge flow overtakes the safety standard, among them the former is the most important. This paper outlines some methods to quantify benefit. The benefit of industrial water supply was quantified by C-D production function method. The benefit of agricultural irrigation was quantified by sharing coefficient method. The benefit of domestic water supply and ecological water supply were quantified by emergy analysis method. To apply these methods to calculate flood benefit of Shilianghe reservoir. Based on relationship graph between water level and benefit and risk, we can find out that benefit and risk increase as the limited water level increases, and the growth rate of risk has gradually increased. The reasonable control range of the limit water level was obtained under minimum risk and maximum benefits of Shilianghe reservoir. In order to determine the exact value of limited water level, multi-objective models aimed at minimum risk and maximum benefit should be studied further.

Acknowledgements. This work was the National Key Research and Development Program of China (2016YFC0400909), and Water Conservancy Science and Technology Project of Jiangsu Province (2014069).

References

Wang, Z., Cheng, L., Liu, Y., et al.: Evaluation method for status and potential of flood resource utilization in a basin. J. Hydraul. Eng. **45**(4), 474–481 (2004)
Wang, Y., Hu, Q., Zhang, S., et al.: Study on rain and flood resources utilization assessment and utilization model. China Water Resour. **15**, 13–16 (2009)

Roy, B., van Remco, E.K.: Integrated ecological, economic and social impact assessment of alternative flood control policies in the Netherlands. Ecol. Econ. **50**, 1–21 (2004)

Xu, W., Li, K.: Economic value evaluation of utilization of urban rainwater resource. Environ. Sci. Technol. **07**, 190–194 (2009)

Feng, F., He, H., Ni, G.: Study on quantitative evaluation of social benefits of river flood resources utilization. J. Hydroelectr. Eng. **32**(4), 25–31 (2013)

Fang, H., Cheng, Y., Bian, Z.: Evaluation on flood water resources utilization and benefit for Gaotang-Ahu Reservoir Irrigation District. Water Resour. Hydropower Eng. **45**(8), 11–14 (2014)

Feng, P., Han, S.: Flood risk analysis of heightening limited water level of reservoir in flood season. J. Tianjin Univ. Sci. Technol. **40**(5), 525–529 (2007)

Huang, X., Lin, J., Fang, G., et al.: Research on limited water level by stage of Shilianghe Reservior. Water Resour. Power **33**(8), 42–45 (2015)

Wu, Z., Lv, C.: Calculation of benefit sharing coefficient of industrial water supply using energy method. Water Sav. Irrig. **6**, 2–15 (2009)

Wang, Z.: Estimating the economic benefits of Taihu Irrigation Project based on share coefficient method. J. Irrig. Drain. **30**(5), 139–142 (2011)

Fan, Y., Nie, H.: Analysis of degree of water contribution to industry and benefits of ecological water in Shanxi Province. Sci. Soil Water Conserv. **02**, 99–104 (2014)

Buenfil, A.A.: Emergy evaluation of water. Department of Environmental Engineering Sciences, University of Florida, Gainesville, US (2001)

Exploration of Compiling Principle Practice Courses Based on System Ability Development

Yanqiu Wang and Xiaofei Yan(✉)

College of Information Science and Engineering, Zaozhuang University, Zaozhuang 277160, China
Ww_yanqiu@163.com

Abstract. The training goal of students majoring in computer science in colleges and universities is embodied in "application". Therefore, it is necessary to increase the practical aspects of the compilation principle courses which focus on the theoretical knowledge of computer major. How to teach theoretical knowledge at the same time, vigorously develop students' programming skills? In view of this realistic problem, this paper puts forward a targeted curriculum practice plan based on the system ability cultivation through analyzing the acceptability of students and the characteristics of compiling principle courses, which is of great help to improve the students' ability of solving computer problems and autonomous learning, Improve the overall quality of students and professional ability has a positive effect.

Keywords: Compiling principle courses · System ability development
Compiling principle reform

1 Introduction

Please note that the first paragraph of a section or subsection is not indented. The first paragraphs that follows a table, figure, equation etc. does not have an indent, either. As the enrollment of undergraduates increases substantially, the employment competition for college graduates intensifies, and the employers' requirements for talents continue to increase. Many graduates in computer science and related majors expose themselves to poor hands-on ability during their employment, and their ability to analyze problems is weak [1]. Innovation is not strong and other issues. The emergence of these problems largely reflects the inadequacy of college teaching in disciplinary practice (especially curriculum practice). The training of computer science students from the popular software learning, computer common knowledge to teach, into the training of innovative capabilities, knowledge, ability, quality of the three. Compiling principle is the classic core course of computer science and technology specialty, and it is also a compulsory course for computer professional quality and ability training. However, due to the abstracted theory content, complicated algorithm process and difficult experiment of the course [2], the students' learning enthusiasm is not high ineffective. Therefore, how to improve the teaching effect, improve the quality of teaching and improve students' abilities are the teaching topics that need to be studied by the teachers of compiling principles [3].

J. Abawajy et al. (Eds.): ATCI 2018, AISC 842, pp. 1325–1329, 2019.
https://doi.org/10.1007/978-3-319-98776-7_164

2 "Principles of Compilation" Course Features

"Compilation principle" is the basic course of computer science and technology, belongs to the software technology series. The contents of the course contain not only the problem of abstraction but also the more mature theories. Besides the relevant theories, knowledge and techniques involved in compiling compiler, it also contains typical ideas and methods of solving basic problems and typical algorithms, "Program design", "data structure and algorithm", "operating system" and other courses, and then from the system level to recognize programs, algorithms [4], a course. Therefore, compiling principle is a basic principle that computer science and technology students must master. Compiling Principle is a very important professional course in computer science and technology. In addition, the principles and techniques of compiling a compiler are quite general, as Alfred V. Aho wrote in Compiled: Principles, Techniques, and Tools, a compilation of classic textbooks: "Compiler theory and techniques are also used in compiler design In many fields, these principles and techniques are often used many times in the career of a computer scientist." However, subject to the limitations of the subject itself, "Compilation Principle" in the experiment involves many other knowledge related to the course, comprehensive, coherent, a compiler to achieve a complete realization of a certain degree of computer professional basis; the same time the course is more difficult, The course duration is limited. Therefore, how to effectively arouse students 'enthusiasm, initiative and enthusiasm in learning Compilation Principle so that students can improve students' professional abilities through the learning of this course are always issues that need to be researched and attended by teachers of Compiling Principles [5]. In particular, The teaching and practice of curriculum aimed at ability cultivation is the main direction of teacher inquiry [6].

3 System Capacity Development

Systemic capabilities of computer science students include the system's cognitive, design, development and application capabilities. Through the first two years of college, basic knowledge of computers and program design learning, students have a certain degree of programming and implementation capabilities, through the "Computer Principle", "operating system" course after learning with hardware and system concepts, There is still no real full participation in the design, development, operation, and maintenance of a larger piece of software, and the Compilation Principles course is a systematic, quantitative, and rigorous method of applying to a complex, large-scale system The model is the best example of understanding and comprehending the abstraction, modularization and localization of system design, the consistency and certainty of system data, the completeness and verifiability of system functions. However, the goal of "Compilation Principle" is to let students grasp and understand the structure, work flow and compilation principles and implementation techniques of compilation system so as to master the ability to analyze, design, implement and maintain the compilation system. Based on the "compilation principle" of the curriculum objectives, the teaching methods are generally based on the compiler theory and the implementation of technology-based teaching, algorithm design to strengthen the understanding of the principle, the curriculum experiment is through the compiler decomposition programming, the final combination of a Write a simple compiler or augment a simple compiler.

In practice, the preparation of detailed questions is required. In the course of practice, in order to take account of the differences of students 'foundation and improve students' interest in learning and experimenting, the design of experimental project topics should not only be difficult, selective, enlightening and relevant. For example, in order to ensure that students can complete at least two analyzers, students and groups with higher abilities can use a variety of algorithms to complete and conduct comparative analysis. Regardless of the choice of program, to complete the preparation of the three closely linked procedures are a better training of students to improve their system capabilities, so that students feel exactly the compiler has a considerable size and complexity of the system to resolve the problem clearly Sexuality, and controllability of system size and complexity to enhance students' ability and confidence to control large, complex system designs.

4 All-Round Experimental Teaching Reform

In view of the different basic conditions of students and the common compilation principles in computer practice teaching aspects of the ubiquitous problem, we put forward a set of practical teaching reform model, the main contents include the following aspects.

4.1 Improve Experimental Teaching Methods and Teaching Methods

In compiling the principle and practice of teaching process, through continuous exploration, found that the best way to teach students according to their aptitude. For the experimental projects proposed in the above questions, students with strong learning ability can let them write small compilers by themselves. A complete compiler includes front-end parts such as lexical analysis, syntax analysis, semantic analysis, intermediate code generation, and code optimization, as well as executable code backends that fit into a specific computer architecture. The general principle of the compilation of practice most of the arrangements for the hours of about 32 h, so let a group of students in such a short period of time to complete the front and back two parts is unrealistic, allowing students to complete the front-end alone. For students with general learning ability, they can use LEX, YACC, Java CC and other tools to compile the compiler during the practice of compiling program. Because grammar analysis programs such as YACC are based on the bottom-up analysis method, we should give more time and energy to LL analysis methods such as compiler principles teaching so that students can use these tools to generate compiler programs, A clear understanding of how they work. For students with weaker learning ability, you can use the method of analyzing existing compiler programs. In other words, they do not ask themselves to design and develop a simple compiler, but understand how it works by reading existing compilers. Even allowing them to input and debug existing programs line by line to find out the typographical errors and unreasonable aspects of existing programs all play a role in improving their hands-on ability. Specific compiler can choose PL/0 grammar program code, or Pascal-S grammar program code. Of course, other similar compiler programs are available. For some students, but also allow them to use the knowledge learned to write class compiler tools, such as smart editors, formula editors. All attempts should be encouraged as long as they are able to apply the knowledge of compilation techniques to specific programs. However, such attempts must be conducted under the guidance and supervision of the teacher so as not to cause students to start out as well but encounter specific techniques Insurmountable problems can lead to frustration.

4.2 Construction of Practical Teaching Materials

Due to the emphasis on cross-integration of the contents of the professional courses, and for the mastery of other related courses to improve the overall system capacity of students, so in the preparation of practical teaching materials or practice syllabus to be amended, we must be specific to each specific The practice content clearly points out the related knowledge of the course fusion point, and if necessary, it is best to do some detailed introduction to the necessary knowledge of related points, such as using the method of recording video in advance and document, and some expansion contents You can provide students with reference materials, such as video, MOOC website, guide students in the face of problems learn to seek help solve the problem. We recommend writing detailed practice guides for professional courses, giving a summary of ideas for each practice project, and encouraging students to find alternative ways of solving problems while discovering points for follow-up courses so that they can be prompted, to some extent, But also to mobilize students' interest and enthusiasm.

4.3 To Build a New Experimental Course System

The ultimate goal of reasonable reform of existing practice teaching content is to strengthen students' understanding of basic knowledge and improve their comprehensive application ability so as to better stimulate the initiative and creativity of learning. In revising the existing course practice outline, we pay attention to the integration and optimization of the practical aspects of the related course design, and design a set of moderate difficulty, content-rich, maneuverable, multi-course knowledge integration, and practical work standards, Extensible and modularized professional course practice syllabus, these practice contents can be carried out in many courses' practice in stages, which can effectively solve the contradiction between limited study hours and teaching contents and greatly improve the design The proportion of practice. Take "embedded system", "operating system" and "compiling principle" as an example. In order to achieve the integration between three courses, a system based on a certain microprocessor platform must be implemented in practice. And application of deep guidance, making the boring theoretical teaching into a fun, challenging and practical work.

4.4 Practice the Reform of Curriculum Assessment Methods

We focus on the characteristics of compiler theory course, the use of MOOC online test mode and teaching mode assessment. The experiment mainly adopts the form of staged evaluation, combining the performance of the classroom Q & A, the achievement of the experimental practice and the completion of the work. The staged evaluation is based on the overall plan formulated at the beginning of the semester, and according to the selection of different groups, In the staged teaching mode to enhance students' knowledge of the situation, to encourage students to grasp and digest the knowledge on the peace process, put an end to pre-test assault situation. At the same time, we collect the suggestions and comments given by the students to the teachers in stages, make quantitative and qualitative evaluations, and improve and correct the teaching of the next phase so as to promote the improvement of teaching methods.

5 Conclusion

In this paper, based on the characteristics of "compilation principle" course, according to the training requirements of computer professionals, with the goal of computer system realization ability, the experiment teaching is explored and practiced, and the better teaching effect is achieved. This not only improves the students' compilation principle course's interest and enthusiasm, while enhancing students' abilities to comprehensively solve abstract problems, as well as through the relevant courses.

Acknowledgements. This research was supported by Zaozhuang University Excellent Course Project (NOS: 2015XJP024 and 2015XJP025). The authors would like to thank the entire research group members for their creative work and thank the anonymous reviewers for their valuable comments.

References

1. Jiang, M.: Teaching reform of constructivism and compiler principles. Contemp. Educ. Theory Pract. **6**(12), 89–93 (2013)
2. Wang, W.: Application-oriented teaching reform and exploration of college compiling principle. Off. Autom. **12**(21), 120–124 (2017)
3. Ruan, Z.: Principles of teaching reform in compiling principles. J. Hangzhou Dianzi Univ. (Soc. Sci.) **12**(04), 67–70 (2007)
4. Xu, J., Deng, H.B.: Comprehension principle teaching reform. Agric. Netw. Inf. **12**(04), 12–15 (2016)
5. Hu, X.: Compilation principles of curriculum adjustment and transformation. Comput. Educ. **12**(11), 45–50 (2006)
6. Lu, X.: Comprehension principles of curriculum compilation and transformation. Sc. Technol Inf. **12**(14), 20–23 (2010)

Research on the Application of Regional Cultural Symbols in Industrial Design

Zhao Yu(✉) and Xun Lin

Shanghai Jiao Tong University, Shanghai, Minhang 200240, China
Yuzhao888999@163.com

Abstract. With the development of globalization, the disintegration of culture has already presented in many places of China. Many cities began to pay attention to the regional culture in order to avoid the loss of their own characteristics. The regional culture should meet the requirements of contemporary design. They try to make a unity of globalization and localization and a unity of traditional culture and contemporary science. Through the analysis of the key points of regional culture development, this paper explores the application of regional culture in the local economy by taking the Guiguzi culture as an example. We elaborated the development content of regional culture in regional revitalization. The establishment of culture channel, extraction of cultural symbols, endowed with spiritual connotation and combined with the new technology and new media will make a unity of the region culture and contemporary design. Making the region culture integrate into the urban personality for the Regional revitalization.

Keywords: Cultural symbols · Guiguzi culture
Regional economic revitalization · Regional economic revitalization design

1 Introduction

Parkinson, President of the Royal Institute of Architects have argued: "It's pity that there is a great danger in the world: our city trends to be same one. We live in a fun life by the diversity and local characteristics". Some people believe that this is a manifestation of economic globalization, but economy and culture have their own characteristics. Western culture and oriental culture also have significant differences. The cultural integration will not happen, nor can western culture replace oriental culture. The urban characteristic is not based on the multiple needs of human beings. It is based on the objective difference between human material and social existence [1]. But only when it is national, it is world. Only it is preserves the local characteristics, it is belong to the requirement of the globalization. It is necessary for a city to excavate the differences between themselves and other city to highlight their own characteristics, attract tourists and bolster the regional economy in the background of the Belt and Road Initiative. The development of regional cultural symbols is an integral part of regional economic and social development. The various cultural symbols have become an indispensable driving force for the development of regional economy [2]. The case of one city is usually suitable for only this area. The region is the combination of its history, natural landscape, cultural style, the cultural connotation and the life style of a

J. Abawajy et al. (Eds.): ATCI 2018, AISC 842, pp. 1330–1336, 2019.
https://doi.org/10.1007/978-3-319-98776-7_165

place. We need to extract the distinctive regional cultural symbols and transform them. And then we must mix the regional culture symbols into urban characteristics to achieve regional economic revitalization.

2 Extracting and Transform the Regional Cultural Symbols

2.1 Extracting the Regional Cultural Symbols

Regional cultural symbols are the achievements of creation created by physical and mental labor during a long period historical development in a certain region. People tend to pay attention to the material culture, especially some large natural or cultural landscape, such as Jiuzhaigou, Shaolin Temple, the Imperial Palace and so on. Which make someone argue that there is no cultural symbol in their city. Actually the finding of cultural symbol and city character is necessary before carrying out the regional economic revitalization. Through the fieldwork to let people find out their own regional characteristics. These characteristics may be some small but distinctive material culture or some intangible cultural symbols. Many people don't know how to deal with the intangible cultural symbols. In fact, intangible culture is a key to attract tourists and exaggerate the cultural atmosphere. Nowadays, visitors are no longer focusing on the famous scenic spots and historical sites. More and more tourists want to have a unique and cultural experience. Tourism resources need to show the deep-level culture by the intangible culture which is touching the core of culture. The intangible tourism resources can improve the level of cultural tourism products, the establishment of regional culture symbol, exaggerate the cultural atmosphere, and enhance the cultural experience [3].

2.2 Transforming the Regional Cultural Symbols

After defining the regional cultural symbols, we need to develop and transform them. Regional cultural symbols are the inherent characteristics of a place, but not all of them can be used to attract tourists and enhance their image. We need to select the attractive and geographical cultural symbols for the widely open purpose. On the other hand, the development and transformation should be making sure suitable. We should make sure the international and national traditional culture form, the modern science and technology are combined to the local traditional culture. But we should pay attention that the transformation of regional cultural symbols is not only to meet the needs of tourists, but to enrich the history, nature and life culture of every place. Only when this introverted conception is consistent with the requirements of tourists can we achieve the regional economic revitalization of sustainable development and avoid the embarrassing situation of "thousands of cities but one mode" [4].

3 Case Analysis – Taking the Development of Cultural Symbols Guiguzi as an Example

To extract local characteristic culture, we can excavate the history and geography of the area firstly. When there are many cultures in a region, we need to select the key points which are attractive, unique and potential to develop. In this paper, we take the application of Guiguzi cultural symbols in Hebi regional economy as an example. Try to explain how to integrate the traditional culture and tourism product, how to extract the local intangible culture and carry out regional economic revitalization design.

3.1 Translating the Legends into a Cultural Channel

To show the charm of intangible culture, the first step is to make "stangers" understand and get interested in this culture. In the absence of cultural or natural landscapes, passing legends is a more effective way to convey the cultural connotations. Especially in the background of the Belt and Road Initiative, we have to face the tourists from all over the world. Establishing a culture channel to convey the designer's ideas by some well-known legends is the only way to make the foreigners come to experience the cultural charm rapidly.

Hebi, Henan Province, as an ancient city in Shang Dynasty is full of cultural legends. But compared with the Luoyang and Kaifeng, Hebi is lack of the giant ancient buildings. It's like that Hebi is lack of city characteristics. While there is a good intangible culture named the Guiguzi culture.

Guiguzi, whose real name is Wang Xu, also known as Wang Chan was born in Chu during the spring and autumn period. Who is the famous thinker and strategist, military strategist, educator, and the originator of political strategists of the Warring States period, Hebi choose Guiguzi as the cultural symbols to spread Hebi's unique city culture and realize the regional economic revitalization design.

Guiguzi has been added a lot of mysterious, legendary due to the rendering of Taoism and folk legends. Although his theory is not widely spread in the long history of Confucianism, the stories of him and his disciples have been widely circulated such as Zhang Yi, Sun Bin, Pangjuan. These stories have very strong appeal for all ages' tourists while Guiguzi's image was later carried out many mysteries. The Chinese astrology, divination, physiognomy and Shushu took Guiguzi as the founder. In addition, there are many unknown folk stories like his mother ate millet to give a birth to Guiguzi and white ape steals Guiguzi's mumbo-jumbo. These "orthodox" stories and folk stories both enhance the charm of Guiguzi cultural symbol but also its connotation and interest. Through these stories, the designers make the appropriate design for the landscape, architecture and product in Hebi City. Not only rendering the Guiguzi cultural atmosphere, but also enhance the unique aesthetic taste. As the yuan blue and white vase "Guiguzi goes downhill" which described the story that Guiguzi rescues his student bin is bought by the British businessman Giuseppe Eskenazito as 230 million yuan. It was the highest price for China art auction record at the time. Reflects the interest and cultural connotation of the Guiguzi is quite attractive to foreign tourists. In this way, the Hebi city spread the Guiguzi cultural symbol rapidly and established a culture channel effectively. Let the visitors more interested and more capable of understanding the connotation of cultural symbols on Guiguzi.

3.2 The Cultural Symbol of Guiguzi

Human beings transform the regional culture into symbols to convey information which is called regional cultural symbols. Cultural symbols can both come from the natural and cultural landscape, but also from the abstract culture and spirit. For the revitalization of regional economy, the cultural symbols are selected to show the excellent humanistic spirit, folklore and the local life. Guiguzi is a good cultural symbol for Hebi area. And for the Guiguzi cultural symbol, we need to select the best carrier to convey the connotation of Guiguzi culture. It is possible to clear the design direction and cultural atmosphere by extract the specific cultural symbols. Here is an example for the improper selection of cultural symbols. Jinshan district has one fishing village in the south of Shanghai and there is an ancient city named Cannes. The local tourism selected sailing and port as their cultural symbol without thinking that Jinshan is near to the Gulf of Hangzhou. The majority of tourists come from Ningbo which is the famous China port long time. Therefore, the cultural symbols of Jinshan cannot highlight the characteristics of their own areas, while reduce the attraction of culture itself.

In the past, the symbol extraction of Guiguzi is always the yin and Yang of Taoism. Although the relationship of Guiguzi and Taoism is close, the Taoism is unable to highlight the unique cultural connotation of Guiguzi. The extraction of cultural symbols should be analyzed from the three aspects: culture itself, the cultural atmosphere to be rendered and the design products. The Guiguzi theory is able to distinguish from the theory of yin and Yang, with the same period, rising to the same position with Confucianism or Taoism. It is precisely the Taoism on the basis to the high degree of philosophy. What's more it absorbs and evolves the Taoist philosophy, applies it to practice. It realizes the unity of theory and practice. Through this Guiguzi forms its own unique cultural connotation. And Guiguzi is not always been associated with Taoist temples. So the ancient Chinese military culture is more reasonable than the priest. According to the military culture to select the cultural atmosphere, design scene and tourism product, realize the culture fresh and interesting, the characteristics and cultural symbols. Finally, according to the customer demand, the regional economic revitalization needs to be challenged with the entire national culture by tourists from all over the world challenge. So we must fit our national culture with the aesthetic psychology and value orientation. Let people who don't understand Guiguzi field could understand the Guiguzi connotation and the ancient Chinese military culture. After determining the atmosphere and culture, Hebi city started from two aspects (the ancient Taoist philosophy and the military and cultural symbols) to support the products design (color, shape, texture and other aspects). The cultural symbols applied to regional economic revitalization improving the cultural charm by the design work to meet the user's cultural demand.

3.3 The Spiritual Connotation of Guiguzi Cultural Symbols

The regional cultural symbol has its unique charm because of the external performance and the spirit contained in them. Especially in an experiencing era, the design cannot really grasp the users if it can't convey its spiritual connotation. Taking the Guiguzi cultural symbols as an example, Hebi city should make use of the Guiguzi legends and cultural symbols. And more important is to show the cultural connotation of the theory of Guiguzi in tourism product. Must distinguish between the Guiguzi, Taoism and the hometown of celebrities. To analyze the uniqueness of Guiguzi cultural symbols mainly develop the "Guiguzi" book and other understanding of this theory.

Guiguzi is the Originator who emphasizing practical rather than academic education. His theory has been suppressed for thousands of years. First the core part of his theory has always been criticized for the political strategists as materialistic and treacherous. People feel they are lack of moral sense. In Han Fei Zi, Han judged the political strategists to be one of the five species of beetles who are bad for country. It is the inevitable choice mainly because the environment rather than their low moral character. "Guiguzi" described the theory of political strategists and their interest purpose in the law of the jungle era. After carefully reading, we can find out it's a philosophical book without moral limit. The book analyzes the relationship between people and people and social contradictions. The law of unity of opposites in it can guide us to deal with the problems in the modernization and development. It also plays a good role in balancing the development between urban and rural areas, and building a sustainable development mode with high efficiency and energy saving. At the same time, the diplomatic theory in Guiguzi like the "feiqian", "chuai", "mo" has great reference for our country's political, economic and military developments. "Guiguzi" contains a wealth of simple dialectics knowledge. These knowledge is proposed to deal with the practical problems. It not only provides literature for us to study the origin and development of ancient philosophy, but also widely applicable to our daily communication and real life. It provides us with the basic principles of doing things [5]. From a design point of view, the strategy of vertical and horizontal, great psychological tactics and theory of persuasion are mentioned in the book. All theories coincide with many popular theories of psychology and communication in today's society. Hebi designed for product and publicity to reflect the spiritual connotation of Guiguzi rather than a shell without cultural connotations. These designs enhance the level of design and cultural atmosphere. More importantly, it enhances user experience and cultural identity.

The theory of health knowledge in Guiguzi was recorded in "WBZD". It is said that one of his four subjects is "chushixue", which is "Take the lead, engage in self-cultivation, prolonging life, lift can be washed once." The theory of health care is more specific in the "Yin Fu Seven". Guiguzi emphasizes the monastic Yang de, mentality Qi, health guidance, and puts forward the method of bionic action fitting. These ideas contain a strong Taoist concept of health and it has a great influence on the thought of the health of the later Taoism [6]. These are the cultural connotation and the wisdom of Guiguzi as the core of Guiguzi culture. Health preservation and communication with the world are undoubtedly the most popular spiritual connotation of ordinary users, especially as the cultural symbols of Taoism. What is worth mentioning here is that there are a lot of dross Guiguzi cultural symbols in astrology, Bu Gua, Qigong. But for thousands of years folk artists handed down from age to age which shows the unique charm of its own Guiguzi. Local designers and decision-makers in Hebi should guide them to spread the positive world outlook and values recognized by the contemporary society. We cannot make them become the negative culture of feudalism and superstition. Similarly, for other traditional cultures, it is inevitably a theory that is not compatible with modern society because of the limitations of ancient science and technology. Designers should learn to identify what is excellent, intension culture, and what is low-grade and negative culture. We can't choose some vulgar tastes to cater to the audience. Doing this for regional development is tantamount to drinking poison. Only the culture with rich connotation and contemporary values can make the design full of humanistic concern, the economic sustainable development and realizing the revitalization of regional economy.

3.4 Guiguzi and New Technology

For users and tourists, an effective way to experience a culture means feeling intuitively. But cultural landscapes are often not touchable like Guiguzi culture. We need corresponding carriers to let tourists experience the connotation of culture. It is very necessary to seek the atmosphere or carrier that tourists can participate in. However, traditional tourist products are difficult to achieve the purpose of rendering atmosphere and retaining tourists except theme parks. The voice commentaries and artificial landscaping are costly, lack of novelty and monotonous. Therefore, it is particularly important to combine traditional design methods with the new technologies like VR and AR. VR technology can create virtual scenes to enhance visitors' sense of substitution. It enables visitors to feel the information and culture that designers want to show. AR technology can be combined with existing resources to restore stories. Like using the statue reproduce the learning scene of Sun Bin and Pang Juan in the Guigu cave. If we use AR technology to restore scenes, it will be more lifelike and excitement. In a word, the application of new technology can solve the following problems:

1. Prophase propaganda and marketing, and the display of tourist products.
2. Pre tourism experience in the form of Netscape.
3. Create a new entertainment project.
4. Scene reappearance and cultural experience [7].

It can be said that with the increase in the depth of tourists' demand, the tourism product design of new form and new technology will become an indispensable part of experience tourism. But we should take notice of that the combination of new technology and traditional culture cannot be abused. The subject and object of the application should be carefully chosen. Many cities have developed some inappropriate VR projects. These projects failed to enhance the cultural experience of tourists and broke the connection with traditional culture. They are incompatible with the whole scenic area, which affects the whole cultural atmosphere.

New media is also an important part of the development of regional culture symbols except VR and AR. Especially for the new generation tourists, the use of new media is often more important than tourism products themselves. The role of traditional TV plays and movies in promoting a local culture is self-evident. Such as "Joe courtyard" and Shanxi culture, "Shaolin Temple" and the Shaolin Temple in Songshan. We can see that with the development of new media and self-media, the effect of traditional media on people's actual communication has been reduced. Many new forms of media, such as net play, live broadcast and twitter, are highly concerned in the new generation of audiences. Many viewers will know what kind of food a restaurant has from live broadcast or which street has a shop that they are interested in. The publicity and exposure of culture and media cannot be ignored. Take the Guiguzi culture symbol as an example, the city of Hebi provides convenience for self - media podcasts. Let them create with local attractions or products. This kind of non-professional television production combined with regionalism and local cultural background has received an unexpected promotion effect.

4 Conclusion

Regional cultural symbols and contemporary design are like vertical and horizontal in Guiguzi. There is a clear goal when the two lines are handed over to a point. When the regional cultural symbols adapt to the aesthetic requirements of contemporary people, the regional revitalization design realize. Regional cultural symbols are inherited in the long historical process of development through continuous renovation and development. They must be the cultural essence of the rich connotation. It is the responsibility of designers to revive and develop our excellent regional culture. We should grasp the traditional culture and the pulse of the times. Identify the combination of traditional and contemporary, local and the world. Make regional cultural symbols alive to achieve a distinctive and sustainable regional economic revitalization.

References

1. He, X., Ruan, L.H.: On the creation of regional cultural symbols and urban characteristics. Chin. Foreign Archit. **32**(4), 15–18 (2004)
2. Zhang, F.: The concept and research path of "regional cultural symbols". Zhejiang Soc. Sci. **16**(4), 63–66 (2008)
3. Ouyang, Z.: Target on tourism development of intangible cultural heritage. Lanzhou University, Lanzhou (2012)
4. Guo, Q.S.: Traditional technology and regional economic revitalization. Observ. Reflect. (2011)
5. Liang, F.: Research of Guiguzi. Shandong Normal University, Ji'nan (2011)
6. Chu, W.: Guiguzi and Taoism Relations Research. Nanjing University, Nanjing (2014)
7. Du, W.: The development and innovation of humanistic tourism resources under the perspective of experience economy. Anyang Norm. Univ. Sch. **6**(2), 79–83 (2017)

Component Separation with Gaussian Moments

Yongjian Zhao(✉)

Shandong University, Weihai 264209, China
zhaoyj@sdu.edu.cn

Abstract. Blind signal processing (BSS) is a fairly new and generally applicable technique. A very intuitive and important principle for the BSS problem is to maximize/minimize non-Gaussianity. Gaussian moments are introduced here as a quantitative measure of non-Gaussianity for a random variable. A proper contrast function is presented correspondingly that has not asymptotic bias even in the noisy context. After maximization of the contrast function, an improved BSS algorithm is introduced correspondingly. Computer simulations demonstrate the efficiency of the proposed approach.

Keywords: Component · Moment · Separation · Kurtosis · Source Independence

1 Introduction

Over the past decades, blind signal separation (BSS) is increasing popularity in the field of biomedical signal processing, speech enhancement and telecommunication. In fact, BSS is a statistical and computational technique for revealing hidden factors that underlie a set of random variable, signals or measurements. The main task of BSS is to recover original source from its mixture and establish a system using only the recorded data after the sources pass through the system [1]. The term "blind" denotes that neither the original sources nor the system is known in advance. Indeed, the blindness is not complete since it is assumed that the original sources are independent and have some statistical properties. In contrast to correlation-based transformations such as principal component analysis and factor analysis, BSS not only decorrelates the signals but also reduces statistical independence, attempting to make the signals as independent as possible [2]. Therefore, BSS is a much more powerful technique capable of findings the underlying components or factors when these classic methods fail completely.

Recently, several methods have been developed for the BSS problem, which are generally based on the second or higher order statistics of the data [3–5]. Purely second-order statistical (SOS) methods do not limit the number of Gaussian sources that can be separated to one. Therefore, such methods have the advantage of require shorter data records due to their reduced small sample estimation errors. It must be mentioned that the SOS based methods are derived on certain assumptions about the original sources, such as time correlations, statistical non stationarity, linear predictability, etc. In other words, the SOS based methods are derived in a semi-blind

J. Abawajy et al. (Eds.): ATCI 2018, AISC 842, pp. 1337–1342, 2019.
https://doi.org/10.1007/978-3-319-98776-7_166

context. However, such assumptions are not always available in practice. Higher order statistics (HOS) method generally represents the promising technique for BSS problem, since it requires few assumptions apart from the statistical independence of the original sources [6–8]. That is to say, the HOS based methods address the BSS problem in a truly blind context. The Central Limit Theorem tells that the distribution of a sum of non-Gaussianity independent random variables tends toward a Gaussian distribution. Thus a very intuitive and important principle of signal separation is to maximize/minimize non-Gaussianity. To use non-Gaussianity for the BSS problem, one must introduce a quantitative measure of non-Gaussianity for a random variable [9]. Kurtosis is the name given to the forth-order cumulant of a random variable [2, 4]. Recently, kurtosis has become a practical measure of non-Gaussianity. Since kurtosis is immune to Gaussian noise, it is almost trivial to present one-unit approaches for the BSS problem. However, kurtosis based algorithms are very sensitive to outliers and have large mean square errors for super-Gaussian data [4, 5].

In practical applications, it is necessary to introduce a measure of non-Gaussianity that has better statistical properties than kurtosis. In this paper, we introduce the Gaussian moments as the expectation of the Gaussian function with different scale parameters. The Gaussian moments can be estimated from noisy measurements, which is important in practice. This enables us to utilize the Gaussian moments as one-unit contrast function that has not asymptotic bias even in the noisy context. After maximization of the contrast function, an improved BSS algorithm is introduced correspondingly. Computer simulations demonstrate the efficiency of the proposed approach.

2 Problem Formulation and the Proposed Algorithm

The BSS problem in noisy environments can be expressed as

$$x = As + n \tag{1}$$

where $x = (x_1, x_2, \cdots, x_n)$ is the vector of observed random variables, $s = (s_1, s_2, \cdots, s_n)$ is the mutually independent source vector, A is an unknown constant matrix, called the mixing matrix. The vector n is Gaussian noise, which is often omitted in traditional BSS methods for noise-free case. The main principle for the noisy BSS problem is to find a specific vector w and then apply a demixing operation by vector w to the mixtures x so that

$$y = w^T x = w^T As + w^T n \tag{2}$$

becomes the estimated source signal.

Preliminary whitening is a primary step for the BSS problem, which makes algorithm simpler and faster through decorrelating the observed mixtures [2, 3]. It must be mentioned that the effect of noise must be taken into account in noisy cases. Denoting by $M = E\{xx^T\}$ the covariance matrix of the observed noisy mixture, the ordinary preliminary whitening for noise-free case should be replaced by $\tilde{x} = (M - R_n)^{-\frac{1}{2}}x$, where R_n is the noise covariance matrix. After the whitening step, the quasi-whitened data $\tilde{x}$ satisfies the following noisy BSS model:

$$\tilde{x} = Bs + \tilde{n}, \tag{3}$$

where B is an orthogonal mixing matrix and $\tilde{n}$ is a linear transformation of the original noise in (1). The new noise covariance matrix for model (3) can be expressed as $\tilde{R}_n = E\{\tilde{n}\tilde{n}^T\} = (M - R_n)^{-\frac{1}{2}} R_n (M - R_n)^{-\frac{1}{2}}$. For convenience, in the following we assume that the observed sensor signals are quasi-whitened data.

To solve the BSS problem, one must introduce a proper contrast function that is greatly dependent on the parameters of the neural network model (1) or (3) [2, 4]. Non-Gaussianity is a simple and intuitive principle for constructing the contrast function. Recently, several BSS methods have been presented in which the higher-order statistics are taken into account through the contrast function:

$$J_G(w) = \left| E\{G(w^T x)\} - E\{G(v)\} \right|, \tag{4}$$

where G is practically any nonquadratic function, and v is a Gaussian variable of zero mean and unit variance. In reality, the contrast function in (4) can be interpreted as approximations of differential entropy. Find the maxima of this function under the constraint $\|w\| = 1$ will deduce only one source signal. The function does not need any prior information of the original sources. However, it is only suitable for noise-free context. In practice, noisy measurements are common and should be taken into account when developing BSS methods.

The most promising approach for the noisy BSS problem is bias removal techniques [1, 2]. In other words, the BSS methods for noise-free cases should be modified so that the bias due to noise can be removed, or at least reduced. The function in (4) can be suitable for noisy cases only when we can compute $J_G(w^T \tilde{x})$, that has been deduced for noise-free case, on the basis of the noisy observations.

Denote by

$$\varphi_c(x) = \frac{1}{c}\varphi\left(\frac{x}{c}\right) = \frac{1}{\sqrt{2\pi}c}\exp\left(-\frac{x^2}{2c^2}\right) \tag{5}$$

the Gaussian density function of variance c^2, and by $\varphi_c^{(m)}(x)$ the mth ($m > 0$) derivative of $\varphi_c(x)$. Denote further by $\varphi_c^{(-m)}(x)$ the mth integral of $\varphi_c(x)$. We can deduce $\varphi_c^{(-m)}(x) = \int_0^x \varphi_c^{(-m+1)}(\zeta)d\zeta$ where $\varphi_c^{(0)}(x)$ is expressed as $\varphi_c^{(x)}$.

Theorem 1. Let v be any non-Gaussian random variable and denote by n an independent Gaussian noise variable with variance δ^2. The function ϕ is formulated with (5). For any constant $c > \delta^2$, we can deduce

$$E\{\varphi_c(v)\} = E\{\varphi_d(v + n)\} \tag{6}$$

with $d = \sqrt{c^2 - \delta^2}$. Furthermore, (6) still satisfies when φ is substituted by $\varphi^{(m)}$ for any integer index m.

This theorem indicates that we can recover a specific source signal from noisy mixtures by maximizing a general contrast function with form of (4). The estimation of $E\{G(w^T x)\}$, deduced only for noise-free context before, is made possible by using $G(u) = \varphi_c^{(m)}(u)$. The statistics of $E\{\varphi_c^{(m)}(w^T x)\}$ are called the Gaussian moments of the data. The Gaussian moments have many useful properties for estimation of underlying variables from noisy observations [2, 5]. Here we introduce the following one-unit contrast function for the noisy BSS problem:

$$\begin{cases} \max\varphi(w) = \left| E\{\varphi_{d(w)}^{(m)}(w^T \tilde{x}) - E\{\varphi_c^{(m)}(v)\} \right| \\ s.t. \quad \|w\| = 1 \end{cases} \tag{7}$$

where $d(w) = \sqrt{c^2 - w^T \tilde{R}_n w}$, and $\tilde{x}$ are the quasi-whitened data. This gives an asymptotically unbiased method for estimating the noisy BSS model. Since $\varphi_c^{(m)}(\tilde{x}) = \varphi^{(m)}(\frac{\tilde{x}}{c})c^{(-m-1)}$ and $d(w) = \sqrt{c^2 - w^T \tilde{R}_n w}$, the gradient of $E\{\varphi_{d(w)}^{(m)}(w^T \tilde{x})\}$ with respect to w can be deduced as

$$\begin{aligned} \nabla_w \varphi_{d(w)}^{(m)}(w^T \tilde{x}) = {} & \tilde{x}\varphi_{d(w)}^{(m+1)}(w^T \tilde{x}) + \tilde{R}_n w (c^2 - w^T \tilde{R}_n w)^{-1} (w^T \tilde{x}) \varphi_{d(w)}^{(m+1)}(w^T \tilde{x}) \\ & + \tilde{R}_n w (m+1)(c^2 - w^T \tilde{R}_n w)^{-1} \varphi_{d(w)}^{(m)}(w^T \tilde{x}). \end{aligned} \tag{8}$$

For all integer m, we can deduce

$$(m+1)\varphi^{(m)}(\tilde{x}) + \tilde{x}\varphi^{(m+1)}(\tilde{x}) = -\varphi^{(m+2)}(\tilde{x}). \tag{9}$$

Therefore, the gradient in (8) can be further expressed as

$$\nabla_w \varphi_{d(w)}^{(m)}(w^T \tilde{x}) = \tilde{x}\varphi_{d(w)}^{(m+1)}(w^T \tilde{x}) - \tilde{R}_n w \varphi_{d(w)}^{(m+2)}(w^T \tilde{x}). \tag{10}$$

By Theorem 1, we can deduce

$$E\{G''(w^T \tilde{x})\} = E\{\varphi_{d(w)}^{(m+2)}(w^T \tilde{x})\}. \tag{11}$$

Then we can obtain the following one-unit algorithm with bias removal for quasi-whitened data:

$$\begin{cases} w^+(l+1) = E\{\tilde{x} g w^T \tilde{x})\} - (I + \tilde{R}_n) w(l) E\{g'(w^T \tilde{x})\} \\ w(l+1) = w^+(l+1)/\|w^+(l+1)\|. \end{cases} \tag{12}$$

where l denotes iteration index, and the function $g(\cdot)$ expresses the derivative of $G(\cdot)$ in (4). The typical choice of function $g(\cdot)$ is $g_1(u) = \tanh(u)$ or $g_2(u) = u\exp(-u^2/2)$. In fact, $g_1(u)$ is an approximation of $\varphi^{(-1)}$, which is the Gaussian cdf.

3 Simulations

To illustrate the algorithm presented here, we adopted four original sources as shown in [4]. The noise covariance was set to 0.25I. At each simulation, a 4×4 mixing matrix was randomly generated and normalized. That is to say, the total energy of the signals was equal to one and the signal-to-noise ratio (SNR) is 4. To make comparison, we ran the following signal separation algorithms in turn: fast ICA for noisy model (FastNM) [3], gradient algorithm for unifying model (GradUM) [4], and the algorithm presented here (NEWBSS) in the simulations. We normalized the original sources with $\|s_i\|_2 = 1$ and the separated sources with $\|\tilde{s}_i\|_2 = 1$ $(i = 1, 2, 3, 4)$. As the inferred sources have ambiguity of the sign, we further defined the separating errors as $\min\{\|\tilde{s}_i - s_i\|_2, \|\tilde{s}_i + s_i\|_2\}$. We ran the three algorithms in turn over 50 independent times. Table 1 shows the average errors of the four separating signals. One can see that the algorithm presented here is better than the existing algorithms.

Table 1. The average errors of the four separating sources by the three algorithms.

Algorithms	Source1	Source2	Source3	Source4
FastNM	0.1869	0.1879	0.1967	0.1879
GradUM	0.1837	0.1972	0.1856	0.1866
NEWBSS	0.2098	0.2188	0.2066	0.2077

4 Conclusions

In this paper, we consider the signal separation problem under blind environment when Gaussian noise is present. An improved quantitative measure of non-Gaussianity, called Gaussian moments, is introduced correspondingly. After analyzing the useful property of Gaussian moments, a contrast function is deduced which can be suitable in the noisy context. Then a one-unit algorithm is presented with bias removal for quasi-whitened data. As illustrated by the computer simulations, the method proposed here achieves better results than the previous methods. In real systems, noise always exists, so the proposed method seems more practical than the previous methods.

Acknowledgement. This paper is partially supported by Shandong Provincial Natural Science Foundation, China (ZR2017MA046) and the National Science Foundation of China (11473019). The authors gratefully acknowledge the helpful comments and suggestions of the reviewers, which have improved the presentation.

References

1. Cichocki, A., Amari, S.: Adaptive Blind Signal and Image Processing. Wiley, New York (2003)
2. Barros, A.K., Cichocki, A.: Extraction of specific signals with temporal structure. Neural Comput. **13**(9), 1995–2003 (2001)

3. Hyvärinen, A., Karhunen, J., Oja, E.: Independent Component Analysis. Wiley, New York (2001)
4. Hyvärinen, A.: A unifying model for blind separation of independent sources. Signal Process. **85**, 1419–1427 (2005)
5. Zhao, Y.J., Liu, B.Q., Wang, S.: A robust extraction algorithm for biomedical signals from noisy mixtures. Front. Comput. Sci. China **5**(4), 387–394 (2011)
6. Leong, W.Y., Mandic, D.P.: Noisy component extraction (NoiCE). IEEE Trans. Circuits Syst. **57**(3), 664–671 (2010)
7. Lu, W., Rajapakse, J.C.: ICA with reference. Neurocomputing **69**(16–18), 2244–2257 (2006)
8. James, C.J., Hesse, C.W.: Independent component analysis for biomedical signals. Physiol. Meas. **26**(1), 15–39 (2005)
9. Liu, W., Mandic, D.P.: A normalized kurtosis-based algorithm for blind source extraction from noisy measurements. Signal Process. **86**(7), 1580–1585 (2006)
10. Yu, D., Peng, L.: When does inferring reputation probability countervail temptation in cooperative behaviors for the prisoners' dilemma game? Chaos Solitons Fractals **78**, 238–244 (2015)

Study on Key Design Parameters of Calcinations System in Rotary Lime Kiln Based on Computer Simulation

Tingzhong Wang(✉) and Lingli Zhu

College of Information Technology, Luoyang Normal University, Luoyang 471934, China
wangtingzhongedu@163.com

Abstract. At present, the lime technology of rotary kiln in our country lacks basic research. There is no systematic study on the influence of key design conditions such as raw materials and fuel on the design parameters and it lacks the master of the core design parameters. Based on the results of experiment and computer aided engineering simulation, an intelligent design system for rotary kiln is developed to meet the requirements of raw materials and fuel. In this system, the key design parameters of rotary kiln are given quickly and optimally under the condition of the user using raw materials and fuel, and the stress condition of the key equipment of rotary kiln is checked.

Keywords: Computer simulation · Rotary Lime Kiln · Calcinations system Design parameters

1 Introduction

Lime is an important industrial raw material for steelmaking, electrolytic aluminum, paper making and calcium carbide smelting. According to statistics of the lime Association, the total lime production in China reached 1.1 million tons in 2013, accounting for 50% of the world's total output, which is expected to have 10% growth in 2014 to reach 1.2 billion tons.

Although China has been in the leading position in the world in terms of production capacity, the technology and equipment of lime production is still the leading position of foreign advanced technology [1]. According to statistics, the proportion of foreign advanced equipment accounted for nearly 47% in 2013, domestic advanced equipment accounted for 11.8%, domestic general and backward equipment accounted for 41.2%. With the demand of national energy saving and emission reduction, the quality requirements of lime are higher and higher in the downstream products of lime, such as iron and steel, alumina, calcium carbide, etc. It is expected that the proportion of foreign advanced technology and equipment will be further increased in the future.

Therefore, considering the stability, economy and calculation precision, the K-ε model can be used to analyze the flow field in the preheater. According to the flow of flue gas inside the preheater, the internal structure of the preheater is optimized, and the

J. Abawajy et al. (Eds.): ATCI 2018, AISC 842, pp. 1343–1348, 2019.
https://doi.org/10.1007/978-3-319-98776-7_167

structure of the preheater is re-simulated and analyzed until the structure of the preheater with the optimal structure is obtained.

Domestic rotary kiln lime production technology, mainly imported foreign technology and equipment. For different raw materials and fuels, they are still stereotyped and lack of innovation, resulting in different users of the same equipment using different effects, sometimes good and bad. Finally, they can only have headache and foot pain to heal their feet and feet, over and over again. However, the technical level of lime in China has not been raised in real sense. The reason is the lack of basic research on lime technology of rotary kiln in China. The influence of key design conditions, such as raw materials and fuel, on the design parameters is not systematically studied, and the core design parameters are not grasped.

2 Analysis of Raw Material and Fuel System for Lime Production in Rotary Kiln

The production of lime in rotary kiln has certain requirements for raw materials and fuel conditions, especially for the calorific value of gas fuel, the granularity of raw materials and the ratio of granularity, etc. Therefore, the adaptability of rotary kiln to raw material and fuel will be studied, and the premise of producing lime raw material and fuel in rotary kiln will be further standardized.

The calcination technology of rotary kiln in China is developed on the basis of imported equipment [2]. It is divided into two kinds of equipment systems, one is rotary kiln system with vertical preheater and vertical cooler, WISCO was imported from KMAG (West Germany) in 1976 and Baosteel from KVS (KVS) in 1995. The production capacity of WISCO is 600 tons per day.

The other is rotary kiln system with grate preheater and grate cooler (referred to as Ribor type), which is two production line equipment imported by Shanghai Baosteel from Mitsubishi heavy Industry in 1980. The production capacity of each production line is 600 tons per day.

$$H_{1,L}(f) = 2\sin^{L}(\pi f)\sum_{l=0}^{L/2-1}\binom{L/2-1+l}{l}\cos^{2l}(\pi f) \quad (1)$$

The imported active lime rotary kiln equipment has been put into production since 1979, which is the beginning of industrial production of active lime in China. After years of development, as is shown by Eq. (1), where H is the daily production of active lime in China has gradually increased from 600 tons to 1200 tons per day, and the development trend of large-scale lime industry has accelerated the development of China's lime industry.

Because the lime technology of our country is imported from abroad, the design of the equipment is developed on the basis of surveying and mapping, some equipments are completely copied and some large equipments are simply amplified. Therefore, the main problems are as follows: (1) for some key equipments, because they do not master the core technology, they can only be designed by experience, which makes the domestic calcinations of lime rotary kiln appear bulky and have high manufacturing cost.

(2) Since the theory of calcinations in lime rotary kilns has not been studied, the raw materials and fuels used in production have not formed a certain standard, resulting in market confusion and serious waste of energy and resources.) The key design parameters of some equipment are not well grasped [3]. Most of them are based on the calculation theory of cement industry, which obviously does not accord with the law of lime production. Because of the insufficient research on basic theory, the energy consumption level of the whole system of lime rotary kiln is still far behind that of foreign advanced level.

The vertical preheater is the most important equipment for producing lime in rotary kiln. Whether the lime can get the ideal predecomposition rate in the preheater is the key to the whole firing system. The key to determine the predecomposition rate of lime is the flow of hot air in the preheater, but the internal flow is invisible.

At present, the hot air flow can only be judged subjectively, the internal structure of the preheater is modified, and the preheating effect is improved. Fluent provides a visual analysis of the hot air flow line field, which can reflect the hot air flow situation in the whole preheater more intuitively [4]. By optimizing the key structure parameters of the preheater, the lime is heated uniformly and the exhaust gas outlet temperature is reduced.

3 Study on Key Design Parameters of Calcinations System in Rotary Lime Kiln Based on Computer Simulation

The residence time of lime in rotary kiln determines the firing quality. At present, there is no uniform formula for calculating the final residence time of lime in rotary kiln. Most of the formulas of cement rotary kiln are applied to lime system, but the composition and particle size of lime and cement are obviously different, so the calculation time is not accurate enough, which will bring inaccuracy to the design of key parameters of rotary kiln.

Therefore, EDEM software is adopted to simulate the residence time of lime in rotary kiln, and the residence time of lime in rotary kiln is analyzed according to different diameter of lime. At the same time, the residence time of lime was measured by experimental rotary kiln, and the current residence formula of lime rotary kiln was compared and revised. Experimental is study on residence time of Lime in Rotary Kiln.

The mechanical analysis of the supporting device of lime rotary kiln based on ANSYS shows that rotary kiln is one of the key equipment for claiming lime, and it is also the most important equipment in the firing system. At present, some large rotary kilns are used in mature cement rotary kilns, which do not consider the factors such as light lime weight and low filling rate, which result in the excessive weight of rotary kilns.

Therefore, the finite element analysis software ANSYS is used to analyze the force of the supporting device of the rotary kiln to optimize the weight of the supporting device of the rotary kiln to the maximum and reduce the weight of the rotary kiln equipment.

According to the results of test and simulation analysis, the original calculation formula is optimized [5]. The finite element software package of finite element analysis for lime rotary kiln supporting device based on ANSYS is a multi-purpose computer program for finite element method, which can be used to solve the structure and fluid. Power, as is shown by Eq. (2), where z is electromagnetic and collision problems [6].

$$\begin{aligned} z_1^k &= \{z(1), z(2), \ldots, z(k)\} \\ &= \{Z(1), Z(2), \ldots, Z(m-1);\ z((m-1)M+1), \ldots, z((m-1)M+s)\} \\ &= \left\{Z_1^m;\ z(m,1), z(m,2), \ldots, z(m,s)\right\} \end{aligned} \tag{2}$$

According to the results of experiment and computer aided engineering simulation, a rotary kiln intelligent design system is developed to meet the requirements of fuel and raw materials. In this system, the key design parameters of rotary kiln are given quickly and optimally under the condition of the user using raw materials and fuel, and the stress condition of the key equipment of rotary kiln is checked.

The research of this paper mainly realizes the following three objectives: The main contents are as follows: (1) to form a set of raw materials of rotary kiln calcinating lime system, the basic standard requirements of fuel and the basic conditions of calcining lime in rotary kiln are defined. (2) The key design parameters of lime shaft preheater and rotary kiln are studied, and the formula for calculating residence time of lime rotary kiln based on lime system is completed. (3) An intelligent design system of lime rotary kiln system is completed.

In this paper, the outlet temperature of preheater exhaust gas is reduced from 320 °C to 230 °C, and the whole weight of the optimized rotary kiln is reduced by 5%. An intelligent design system for calcinating lime system of rotary kiln is developed according to the raw material and fuel information provided by the user.

4 Experiments and Analysis

The raw material for lime production of rotary kiln is analyzed, limestone is the raw material for producing lime, its chemical characteristics determine how to calcine lime, i.e., it is not burnt, but can reach the lowest energy consumption and produce the highest quality lime.

Therefore, the calcination environment of limestone in rotary kiln system is simulated by the method of Maverick furnace calcinations test, and different heating curves are drawn to judge the calcination characteristics of limestone with different chemical composition and granularity, and the optimal calcination curve is obtained. The curve is applied to the actual rotary kiln calcination system.

Different fuels have different thermodynamic characteristics. By analyzing the characteristics of common fuels, the heat value and flame temperature of common fuels are calculated, and the radiation temperature is calculated according to the characteristics of rotary kiln [7]. According to the comparison of calcination curves obtained from limestone test, the lowest and highest requirements of limestone for flame were found out.

Form a set of different chemical composition, granularity of limestone, different calorific value and combustion flame temperature of rotary kiln calcinations specification and requirements, as is shown by Eq. (3), where f(s) is to determine the minimum and highest requirements for lime calcinations in rotary kilns [8].

$$f\left(s-1,\frac{k}{2}\right)=\frac{1}{\sqrt{2}}f(s,k)+\frac{1}{\sqrt{2}}f(s,k+1) \tag{3}$$

Based on the research of heat flow field and structure optimization of FLUENT vertical preheater, fluent software includes a separation solver based on pressure, a coupling solver based on pressure, an implicit solver based on density and an explicit solver based on density. Multi-solver technology enables FLUENT software to simulate complex flow fields from incompressible to hypersonic.

Including a variety of optimized physical models, such as computational fluid flows and heat conduction models (including natural convection, steady and unsteady flows, laminar flows, turbulence, turbulence, incompressible and compressible flows, periodic flows), Rotational flow and time dependent flow, radiation model, phase transition model, discrete phase transition model, multiphase flow model, chemical component transport and reaction flow model, etc.

The three-dimensional model of preheater is established. The boundary condition simulation analysis is given. The k-ε model is an improved scheme of the standard k-ε binary equation model, as is shown by Eq. (4), where W is the most widely used turbulence model at present, and can be used in various kinds of flow simulation effectively. It includes jet and mixing flow, pipeline flow, and boundary layer flow and so on.

$$W_{j,k}=2^{j/2}\tilde{W}_{j,2^j(n+1)-1}\ n=0,1,\ldots,N_j-1 \tag{4}$$

Experimental study of Lime residence time in Rotary Kiln EDEM is the first universal CAE software designed with modern discrete element model to simulate and analyze particle system processing and production operation. By EDEM's Particle Factory TM technique, it can not only simulate the collision between particles and vessel walls, but also simulate and analyze the interaction between particle groups.

In the simulation analysis, firstly, the shape and physical properties of limestone are defined, the mechanical geometry CAD model is introduced, the dynamic properties of geometry are defined, the particle generation factory is defined by Particle Factory tool, the solution is solved, and the time step is set. Post-processing is to generate 3 D video animation, section image.

The experimental results show that because the solid particle is discontinuous, it will diffuse along the axis of rotary kiln at the same time, and under the same working condition, Even the materials entering into the rotary kiln at the same time have different residence time in the rotary kiln, which is similar to normal distribution, so the average residence time of the material is adopted.

The three-dimensional model of the supporting device of rotary kiln is established. System optimization and simulation analysis verify, calculate the force and distribution of lime rotary kiln, as the boundary condition applied to the supporting device, through the simulation analysis of the deformation and displacement of the supporting key parts, determine whether the supporting device is suitable, The optimum parameters are obtained by modifying the structural parameters of the supporting device several times.

5 Summary

The intelligent design system of rotary kiln, according to the result of computer aided design, obtains a lot of knowledge that can be used in the intelligent design system of rotary kiln. The knowledge of simulation results is added to the intelligent design system by the combination of object oriented representation, and the function of intelligent design system of rotary kiln is completed by combining rule reasoning with case-based reasoning.

Acknowledgements. This paper is supported by Scientific and technological key projects in Henan Province (152102210123).

References

1. Qiao, B., Zhu, L., Lei, X.: Simulation of limestone residence time in rotary kiln based on EDEM software. Min. Mach. **10** (2010)
2. Serroukh, A., Walden, A.T.: Wavelet scale analysis of bivariate time series I: motivation and estimation. J. Nonparametr. Stat. **13**, 1–36 (2011)
3. Zhu, L., Zhang, K., Qiao, B.: Research on key parameters of active lime rotary kiln and development of intelligent design system. Min. Mach. (1) (2012)
4. Wang, Z., Daniel, W.C., Liu, X.: Variance-constrained control for uncertain stochastic systems with missing measurement. IEEE Trans. Syst. Man. Cybern. Part A Syst. Hum. **35** (5), 746–753 (2005)
5. Zhu, L., Zhi, X., Wang, G., Qiao, B.: Simulation analysis and structural optimization of flue gas flow field of active lime calciner preheater based on FLUENT. Min. Mach. **4** (2013)
6. Zhu, L., Zhi, X., Qiao, B.: Research and development of active lime precalciner. Manufacturing Automation, 4 April 2014
7. Magni, L., de Nicolao, G., Magnani, L., et al.: A stabilizing model-based predictive control algorithm for nonlinear systems. Automatica **37**(9), 1351–1362 (2011)
8. Zhang, K., Zhu, L., Qiao, B., Sun, S.: Experimental study on calcination system of high precalcinable lime rotary kiln. Min. Mach. **41**(6) (2013)
9. Peng, L., Fang, W.: Heterogeneity of inferring reputation of cooperative behaviors for the prisoners' dilemma game. Physica A **433**, 367–378 (2015)
10. Li, J., Ren, Q.: Study on supply air temperature forecast and changing machine dew point for variable air volume system. Build. Energy Environ. **27**(4), 29–32 (2008). (in Chinese)

Synchronization of Chay Neuron System Under Gap Junction Coupling

Di-cong Wang(✉) and Kai-jun Wu

School of Electronic and Information Engineering, Lanzhou Jiaotong University, Lanzhou 730070, China
923118494@qq.com, wkj@mail.lzjtu.cn

Abstract. Power generation pattern and discharge pattern of one pair of Chay neuron models under gap junction coupling as well as synchronization problem of the coupling system were studied based on Chay neuron model. The study finds that appropriate coupling strength can transform Chay neuron model from chaotic discharge pattern into ordered periodic discharge pattern, and moreover, it can induce synchronous behaviors of neuron system under gap junction coupling, which contributes to further follow-up study of complicated dynamical behaviors of biological neural network system and its mechanism to a certain degree.

Keywords: Chay neuron model · Gap junction · Coupling · Synchronization

1 Introduction

Basic structural units of nervous system are neurons, whose discharge activity is involved in complicated physical and chemical process with abundant nonlinear dynamical behaviors [1]. The enormous number of neurons constitute complicated nervous system and various units are closely combined through gap junction and chemical junction so as to form a complicated high-dimensional, multilayer, multi-time-scale and multifunctional information network structure, which gives rise to complicated network dynamical behaviors. Some new problems are generated in discharge activities of nervous system and study of information behaviors. In order to reveal nonlinear nature of neurocognitive activities more profoundly, scholars have taken full consideration of complexity and dynamical characteristics of nervous system during study process using the concept, theory and method of nonlinear dynamics. It's generally believed that subjects of nonlinear sciences include chaos, bifurcation, soliton, fractal, pattern and complexity, etc.

Most studies related to synchronization were based on coupling periodic system. Wang [2] expounded research progress of neural system in aspect of synchronous transition dynamics in recent years and revealed complicated synchronous transition dynamical behaviors presented by neuron system under the effects of different parameters like coupling, time lag and network topology as well as possible dynamical mechanism. Yang [3] studied that two electrical coupling pancreatic β cells had combined discharge of synchronous clusters using minimum phantom cluster discharge model. Deng [4] proposed an algorithm based on PI iterative learning control for

J. Abawajy et al. (Eds.): ATCI 2018, AISC 842, pp. 1349–1354, 2019.
https://doi.org/10.1007/978-3-319-98776-7_168

synchronous control problem of Hodgkin-Huxley (HH) neurons, simulated synchronous control of principal and subordinate neurons under four different circumstances and verified feasibility and effectiveness of the algorithm. Yuan [5] studied influence of cranial magnetosonic stimulation on neuron synchronization based on HR coupling system and his study results indicated that external magnetosonic stimulation could effectively inhibit neuron synchronization under appropriate parameter conditions. Since 1990 when two scholars Pecora and Carrol found synchronization of chaotic systems [6, 7], synchronization of chaotic systems has drawn much concern from many domestic and foreign scholars, showing powerful vitality. Hindmarsh-Rose neuron taken as the model, Wu [8] put forward three schemes—feedback coupling, layered feedback coupling and layered local mean-field feedback coupling—in order to straighten out influences of different coupling schemes on network synchronization. Taking hysteretic chaotic neuron and hysteretic chaotic neural network as study objects.

In this paper, numerical model constituted by two coupling Chay neurons was taken as the study object, and computational simulation of the model was conducted through C language programming and Matlab software simulation. Power generation and discharge patterns of one pair of gap junction Chay neuron model were discussed, and influence of certain coupling strength on synchronization of neuron systems was analyzed.

2 Synchronous Characteristics of Chay Neuron Systems Under Gap Junction Coupling

Gap junction commonly exists in nervous system. As the simplest form of information transfer between neurons, gap junction plays a significant role in information transfer process between coupling neurons. A coupling neuron system constituted by two-group gap junction coupling Chay neuron models was established in this paper, and its expression is shown as follow:

$$\frac{dV_1}{dt} = g_I m_\infty^3 h_\infty (V_I - V_1) + g_{K,V}(V_K - V_1)n_1^4 + g_{K,C}\frac{C_1}{1+C_1}(V_K - V_1) + g_L(V_L - V_1) + H_{syn}(V_2 - V_1) \quad (1)$$

$$\frac{dC_1}{dt} = \rho\left[m_\infty^3 h_\infty (V_C - V_1) - k_C C_1\right] \quad (2)$$

$$\frac{dn_1}{dt} = \frac{n_\infty - n_1}{\tau_n} \quad (3)$$

$$\frac{dV_2}{dt} = g_I m_\infty^3 h_\infty (V_I - V_2) + g_{K,V}(V_K - V_2)n_2^4 + g_{K,C}\frac{C_2}{1+C_2}(V_K - V_2) + g_L(V_L - V_2) + H_{syn}(V_1 - V_2) \quad (4)$$

$$\frac{dC_2}{dt} = \rho\left[m_\infty^3 h_\infty (V_C - V_2) - k_C C_2\right] \tag{5}$$

$$\frac{dn_2}{dt} = \frac{n_\infty - n_2}{\tau_n} \tag{6}$$

Power generation and discharge patterns of two Chay neuron models were discussed under the circumstance without coupling strength ($H_{syn} = 0$). Image of the solutions solved from equation set on (V_1, V_2) plane is shown in Fig. 1, which is random motion image, and then the system is under synchronous state. In time-history graph of time-dependent V_1 and V_2 in Fig. 2, it can be seen that values of V_1 and V_2 are different at the same moment and neurons are under chaotic discharge state, indicating that the system is under chaotic state. Moreover, it can also be found from synchronous difference graph of coupling neurons as shown in Fig. 2 that time-dependent change of synchronous difference values is always not zero, indicating that two neuron models are not under synchronous state. To sum up, it can be obtained that neuron system is under chaotic discharge state and nonsynchronous state when $H_{syn} = 0$.

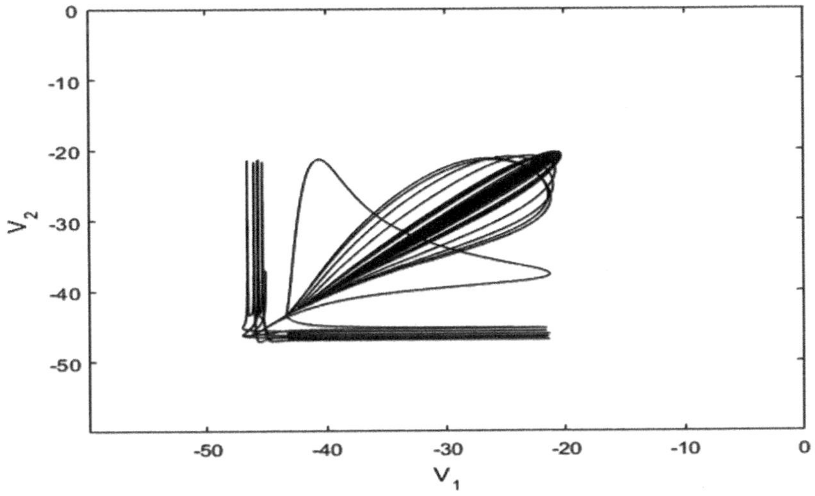

Fig. 1. Phase plane diagram of $H_{syn} = 0$

Secondly, change of synchronization of gap junction coupling neurons when $H_{syn} = 5$. Phase plane plot of neuron system related to V_1 and V_2 in Fig. 3 shows that system phase plane plot is a straight line with slope being 1, values of V_1 and V_2 are the same, and the system is under synchronous state at the time. Through the observation of time-dependent time-history graph of V_1 and V_2 as shown in Fig. 4, the system goes through random discharge activity, indicating that the two neuron models are under chaotic state. System membrane potentials V_1 and V_2 are the same at the same time,

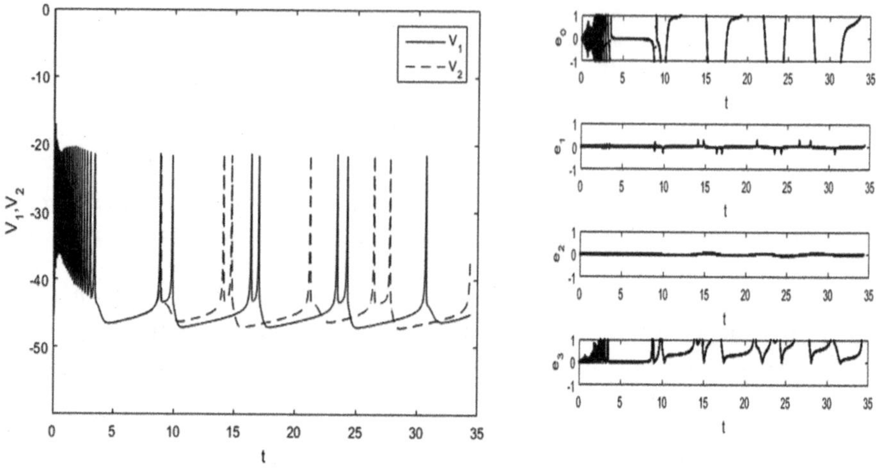

Fig. 2. Time history diagram and synchronization difference diagram for $H_{syn} = 0$

and the neuron system is under synchronous state. As shown in synchronous difference graph in Fig. 4, time-dependent change of synchronous difference is always zero, verifying that the neuron system is under synchronous state. To sum up, the system is under chaotic discharge state and synchronous state when $H_{syn} = 5$.

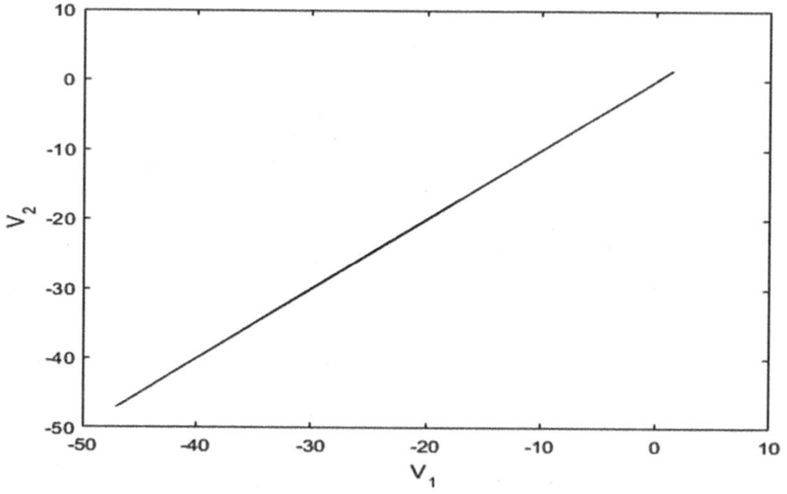

Fig. 3. Phase plane diagram of $H_{syn} = 5$

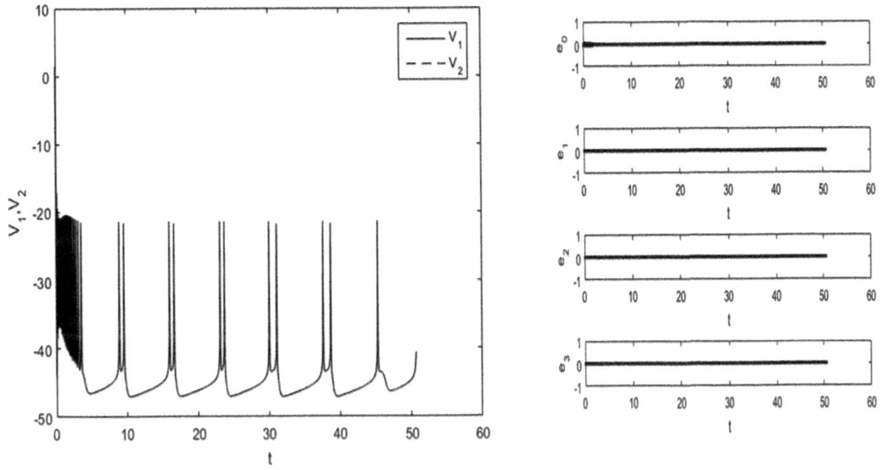

Fig. 4. Time history diagram and synchronization difference diagram for $H_{syn} = 5$

3 Conclusion

Influence of coupling strength on power generation and discharge patterns of neuron models and synchronization of neuron system were studied based on Chay neuron model in this paper. The study shows that different coupling strengths will generate different influences on power generation and discharge patterns of Chay neuron system and its synchronization. Chaotic discharge pattern of neuron models can be transformed into ordered periodic discharge pattern so as to induce synchronous state of coupling neuron system under nonsynchronous state by adjusting coupling strength of two neurons. This study is of certain reference value to studying influence of coupling strength on coupling neuron system and exploring into abundant dynamical behaviors of neurons.

Acknowledgements. This work was supported by the Natural Science Fund Project of Gansu Province (No. 17JR5RA085).

References

1. Wang, Q.Y., Shi, X., Lu, Q.S.: Synchronous Dynamics of the Neuron Coupling System. Science Press, Beijing (2008)
2. Wang, Q.Y., Zhang, H.H.: Advances of synchronization transition in neuronal networks. Adv. Mech. **43**(1), 149–162 (2013)
3. Yang, Z.Q., Zhang, X.: Compound bursting in an electrical coupling model with three different time scales. Acta Phys. Sin. **62**(17), 120–127 (2013)
4. Deng, B., Wang, X.J., Wang, J., et al.: Iterative learning control of synchronization of Hodgkin-Huxley neurons. Appl. Res. Comput. **31**(09), 2656–2660 (2014)

5. Yuan, Y., Sun, H.B., Chen, Y.D.: Study on desynchronization of coupled neurons with transcranial magneto-acoustical stimulation. Chin. J. Biomed. Eng. **36**(04), 502–506 (2017)
6. Pecora, L.M., Carroll, T.L.: Driving systems with chaotic signals. Phys. Rev. A At. Mol. Opt. Phys. **44**(4), 2374–2383 (1991)
7. Pecora, L.M., Carroll, T.L.: Synchronization in chaotic systems. Phys. Rev. Lett. **64**, 821–824 (1990)
8. Wu, W., Tang, G.: Synchronization of chaotic neuronal networks under different couplings. Acta Phys. Sin. **61**(07), 53–61 (2012)
9. Zhai, D.H.: The Study of Firing Activities in Coupled Neurons System. North China University of Technology, Beijing (2011)
10. Peng, L., Fang, W.: Heterogeneity of inferring reputation of cooperative behaviors for the prisoners' dilemma game. Physica A **433**, 367–378 (2015)

Low Carbon Economy Boosts New Development of Environmental Cost Accounting

Ran Wei(✉)

Shandong Women's University, Jinan 250300, China
varan@163.com

Abstract. With the increasing awareness of human environmental protection and sustainable development, low-carbon economic models have gradually become the mainstream mode of economic development, which has attracted worldwide attention. Under such a background, companies can adapt to the low-carbon economy only by carrying out environmental cost accounting and achieve the sustainable development with low pollution, low emissions and high performance. Low carbon is a corporate social responsibility, and it is also a kind of self-rescue in the environmental crisis. To comply with the requirements of a low-carbon economy and environmental protection, companies implementing environmental accounting can promote the realization of low-carbon economic goals. After more than 20 years of development in our country, environmental cost accounting has achieved remarkable results, but it pay more attention to a theoretical study. Nowadays, the rapid development of low-carbon economy has prompted companies to pay more attention to environmental cost accounting. Based on the perspective of low-carbon economy, this paper puts forward several measures to promote environmental cost accounting to practice on the basis of analysing the new changes in environmental cost accounting.

Keywords: Low-carbon economy · Environmental cost · Carbon emission cost

1 Introduction

With the rapid economic and social development, resources and the environment have paid a huge price. At present, countries around the world have problems of environmental pollution and ecological destruction to varying degrees. With the globalization of environmental issues, both theoretical and practical circles have studied and focused on issues related to environmental costs from different perspectives. Especially the proposal of the "low-carbon economy" theory has become a catalyst for the research and development of environmental cost accounting. This article is based on a low-carbon economic perspective and briefly discusses how environmental cost accounting can move toward practice.

J. Abawajy et al. (Eds.): ATCI 2018, AISC 842, pp. 1355–1361, 2019.
https://doi.org/10.1007/978-3-319-98776-7_169

2 The Impact of Low Carbon Economy on Enterprises

2.1 The Meaning of the Low-Carbon Economy

On February 24, 2003, British Prime Minister Tony Blair published a white paper entitled "our Future Energy-Establishing a Low-Carbon economy", which proposed the concept of low-carbon economy for the first time. After the Copenhagen World Climate Conference, our government and all walks of life began to pay attention to low-carbon economy from various angles. The essence of low-carbon economy is energy efficiency and clean energy structure, the core is energy technology innovation and institutional innovation, the goal is to mitigate climate change and promote human sustainable development. The low-carbon economy has abandoned the previous development model of pollution first. It is essentially a new revolution in energy consumption patterns, economic development patterns, and human life styles and is a great advance that the industrial civilization turns to ecological and economic civilization in the human society.

2.2 The Low-Carbon Path of Enterprise Development

The core concept of low-carbon economy meets the connotation of China's strategy of developing circular economy, energy saving and emission reduction, and sustainable development. It is the extension of implementing the scientific development view, building a resource-saving and environment-friendly society and changing the mode of economic development. In the 13th Five-Year Plan, Chinese government put forward a plan for the overall improvement of the quality of the ecological environment and the improvement of low carbon level. For enterprises, the development of low-carbon economy means the adjustment of energy structure, the application of technologies such as "carbon neutralization", and also means that enterprises will increase the cost of environmental protection investment in the short term. But under the guidance of low-carbon economy, driven by consumption, policy, responsibility, market, development and other factors, enterprises will have more initiative in energy saving and emission reduction. Companies will pay more attention to carbon emissions from material procurement to product production and marketing, and then achieve the goal of reducing carbon emissions from upstream to downstream.

3 The Role of Low Carbon Economy in the Development of Environmental Cost Accounting

Environmental cost accounting has been developing in China for more than 20 years. Theoretical research has made outstanding achievements, but it has been difficult to advance the practice. The continued promotion of a low-carbon economy will push environmental cost accounting to a new level.

3.1 The Importance of Environmental Cost Accounting has been Enhanced

The emphasis on environmental issues is the internal driving force for the development of environmental cost accounting. At present, environmental cost accounting is still in the stage of theoretical research in our country. It has not really been applied to practice. The development of low-carbon economy has grown rapidly, providing opportunities for the development of environmental cost accounting from theory to practice. In a low-carbon economic society, companies must constantly change the mode of development, adjust the energy structure, and study energy-saving and emission-reduction technologies. These all have a quiet impact on the development of environmental cost accounting. Enterprises attach importance to the development of low-carbon economy, not only should pay attention to the study of environmental cost accounting theory, but also pay more attention to the extensive application of environmental cost accounting in practice.

3.2 Environmental Cost Accounting has Become the Foothold of the Impact of Low-Carbon Economy on Corporate Accounting

The advocacy of a low-carbon economy has enabled environmental cost accounting to be based on the original, more theoretically rich, and more closely linked with practice. The impact of low-carbon economy on the economic business of enterprises needs to be reflected in accounting, and accounting methods need to be used to record this new change. Since the fundamental cause of the low-carbon economy is the emission of carbon dioxide and other greenhouse gases, the issue of carbon emissions and the impact of carbon emissions on the climate can all be attributed to environmental issues. The related costs arising from environmental issues are recorded by the environmental cost accounting. Therefore, in all the impacts of low-carbon economy on the enterprise, the issue of cost can be completely recorded by environmental cost accounting, and the process of record is also the process of change and improvement of environmental cost accounting itself.

3.3 Low-Carbon Economy is the Help of Environmental Cost Accounting to Practice

At present, the bottleneck of the development of environmental cost accounting in China is that environmental cost accounting cannot be widely used in practice because many companies believe that implementing environmental cost accounting means that companies have more environmental responsibilities and more cost. Therefore, many companies lack the enthusiasm they deserve, and they have chosen to ignore the environmental cost accounting. The current multi-investment of enterprises is a payment in advance for less investment of the future. When a low-carbon economy becomes inevitable, rather than a choice, the emission costs of carbon dioxide and other greenhouse gases is also a task of top priority. This means that the cost accounting for carbon emissions is the main content of environmental cost accounting, in the low-carbon economy. Under the impetus of low-carbon economy, we will soon enter the field of practice from a theoretical perspective.

4 The Correction of Low Carbon Economy to Current Environmental Cost Accounting

Under the low carbon economy, carbon emissions have become the focus of attention of enterprises, and the accounting of the related costs of carbon emissions is also concerned. In order to standardize the cost accounting of carbon emissions, we can establish a scientific accounting system of carbon emissions. However, due to the complexity of carbon classification and measurement, the study of carbon emission cost is still in its infancy, and has not formed an independent conceptual framework. In other words, it is very difficult to establish a separate carbon emission accounting system when environmental cost accounting has not been fully promoted. However, the research on the cost of carbon emissions has already existed. The feasible way is to revise the current environmental cost accounting system, and to incorporate some contents of carbon emission cost accounting into the framework of the current environmental cost accounting in order to ensure the authenticity and integrity of environmental cost accounting information in low-carbon economy.

4.1 The Accounting Content of Enterprise Environmental Cost Accounting will be Extended

Environmental cost is a dynamic expansion of the concept, it reflects the content which is closely related to its environment, when the external environment changes, environmental cost accounting should also make corresponding adjustments. At present, the environmental cost of enterprise accounting only has a few items, such as sewage charges, green fees, compensation, fines and so on. If low-carbon economic factors are considered, the content of environmental cost accounting will be extended to reduce the cost of carbon emissions, such as carbon capture and sequestration technology research and development costs, carbon abatement equipment depreciation costs, carbon emissions trading costs and so on.

4.2 The Accounting Method of Enterprise Environmental Cost Needs to be Changed

The environmental cost accounting in low carbon economy is a kind of development and innovation to the existing environmental cost accounting, and it is the enrichment and perfection of the existing environmental cost accounting theory. Under the influence of low-carbon economy, the content of environmental cost will be more extensive, and the connotation and extension of environmental cost accounting will be expanded accordingly. Therefore, it will also directly affect the innovation of environmental costs in the identification, measurement, collection, distribution, reporting and other links. The key to environmental cost accounting and control is to consider the characteristics of enterprise resources, energy consumption and its impact on the environment. When the above three aspects change, the accounting and control must change. Enterprise environmental cost accounting should also make corresponding changes.

4.3 The Contents and Methods of Environmental Cost Information Disclosure will also Change

Because in a low-carbon economy, stakeholders, such as the government and the public, will pay more attention to the carbon reduction obligations of enterprises and the control measures they take. Therefore, combined with the development of low-carbon economy, enterprises' environmental cost reporting and information disclosure format and content should be changed on the existing basis.

5 The Basic Way of Environmental Cost Accounting to Practice in Low Carbon Economy

With the rapid development of low-carbon economy, we should take low-carbon economy as an opportunity to find an effective way to enrich and improve the existing environmental cost accounting theory while achieving the goal of low-carbon development.

5.1 Establish a Carbon Emission Control System

In the measurement of carbon emission cost, the determination of carbon emissions is the key. The accuracy of carbon emissions directly affects the accuracy of carbon emission cost calculation. Carbon dioxide emission is essentially an energy consumption problem. The greater the energy consumption is, the greater the total amount of carbon dioxide emissions is; the higher the energy efficiency is, the lower the energy consumes per unit of output is, and the less carbon dioxide emissions is; the higher the proportion of clean energy in the same energy consumption is, the higher the amount of carbon dioxide emissions is, and the less the carbon dioxide will discharge. Therefore, there are two main ways to mitigate the impact of carbon dioxide emissions: the one is to reduce carbon dioxide emissions from the source; the other is to capture, store and use the carbon dioxide emissions. Therefore, enterprises should use these two channels to monitor carbon emissions and provide a data basis for correctly calculating the cost of carbon emissions.

5.2 Strengthening the Research on the Theory of Environmental Cost Accounting

Under the low carbon economy, it is necessary to reconstruct the environmental cost system of enterprises, which requires us to constantly strengthen the research on low carbon economy and environmental cost theory and methods, and improve the existing environmental cost accounting system. Facing the change of the situation, there are still many difficult problems in environmental cost accounting. This requires research institutions and researchers to focus on these new issues. And in the research process, we should pay attention to absorb foreign advanced research results, learn from the developed countries to develop low-carbon economy under the successful experience of environmental cost accounting.

5.3 Establish and Improve Relevant Laws and Regulations

A good legal environment is a necessary condition for the development of low-carbon economy and environmental cost accounting. According to the development of the situation, the national legislature should further establish and improve the environmental legal system to provide legal support for the development of environmental cost accounting. The standard system of environmental cost accounting, like financial accounting, should have a special standard, which can exist in the form of environmental cost accounting system and environmental cost accounting standards. On the one hand, it can be used to standardize the whole accounting process of environmental cost accounting from vouchers to statements, on the other hand, it can also be used to standardize the related contents of environmental cost accounting information disclosure. This special environmental cost accounting system and standards have more practical operability to prevent the short-term behavior of enterprises, to promote and standardize the healthy development of environmental cost accounting, therefore, it plays a vital role.

5.4 Advocate Comprehensive Low-Carbon Environmental Education

At present, our government, while vigorously developing low-carbon economy, emphasizes the harmony and friendship with the ecological environment. However, the relationship between low-carbon economy and the environment is not deeply understood by the public. Some enterprises still do not hesitate to harm the environmental interests to obtain immediate economic benefits, and do not set up the concept of low-carbon economy at all. Therefore, we need to spread the concept of low-carbon economy in the whole society, carry out environmental protection education, popularize environmental protection knowledge, and increase the intensity of environmental protection and low-carbon economy propaganda. In particular, enterprises should realize the importance of environmental protection and the necessity of environmental responsibility. Only in this way can environmental cost accounting really be accepted by enterprises.

5.5 Introduction of Reward and Punishment Mechanism

At present, due to the lack of corresponding rewards and penalties, the cost of some companies' inaction in environmental protection is too low, which objectively is a connivance for enterprises to ignore environmental responsibility and shirk social responsibility. The new "Environmental Protection Law" implemented on January 1, 2015 gave many new regulatory powers and means to the environmental protection department. The measures such as daily punishment, seizure, restriction of production and suspension of production have made the "Environmental Protection Law" a sharp sword. At the same time, to take measures to protect the environment, energy conservation and emission reduction should be given to low-carbon enterprises such as subsidies and other corresponding incentives to encourage enterprises to take the path of low-carbon energy saving and emission reduction.

References

1. Mohr, R.D., Shvawantee, S.: Distribution of environmental costs and benefits, additional distortions, and the porter hypothesis. Land Econ. **84**, 6–9 (2008)
2. Alewine, H.C., Dan, N.: How does environmental accounting information influence attention and investment. Int. J. Account. Inf. Manag. **21**(1), 22–52 (2013)
3. Menguc, B., Seigyoung, A., Lucie, O.: The interactive effect of internal and external factors on proactive environmental strategy and its influence on a firm's performance. J. Bus. Ethics **94**(2), 279–298 (2010)
4. Chao, W.: Application research of activity cost method in environmental cost: a case study of coal enterprises. Friends Account. (5), 110–113 (2016)
5. Guangda, Y.: Design of eco-environmental cost compensation standard for industrial industry in China-measurement method based on cost of environmental damage. Account. Res. (11), 88–95 (2014)
6. Tingru, Y.: Current Situation and Development of Environmental Accounting in China. Jilin University, Changchun (2015)
7. Xin, C.: Research on Environmental Accounting from the Perspective of Low Carbon Economy. Shenyang University, China, Shenyang (2017)
8. Li-yan, W.: Environmental cost and GDP effectiveness. Account. Res. (3), 3–11 (2015)
9. Liu J., Cao J.: Exploration of integrating environmental cost into enterprise cost accounting under the new normal condition. Friends Account. (13), 73–76 (2017)
10. Jin, S.: The growth of low-carbon economy. World Environ. (10), 30–32 (2007)
11. Xinghua, S., Weiping, W.: Discussion on the practice of green accounting in China. Account. Res. (6), 59–61 (2000)

Author Index

J. Abawajy et al. (Eds.): ATCI 2018, AISC 842, pp. 1363–1367, 2019.
https://doi.org/10.1007/978-3-319-98776-7

GPSR Compliance

The European Union's (EU) General Product Safety Regulation (GPSR) is a set of rules that requires consumer products to be safe and our obligations to ensure this.

If you have any concerns about our products, you can contact us on ProductSafety@springernature.com

In case Publisher is established outside the EU, the EU authorized representative is:

Springer Nature Customer Service Center GmbH
Europaplatz 3
69115 Heidelberg, Germany

Batch number: 10761593

Printed by Printforce, the Netherlands